INTEGRATED CIRCUITS
FOR WIRELESS COMMUNICATIONS

IEEE Press
445 Hoes Lane, P.O. Box 1331
Piscataway, NJ 08855-1331

IEEE Press Editorial Board
Roger F. Hoyt, *Editor-in-Chief*

J. B. Anderson	S. Furui	P. Laplante
P. M. Anderson	A. H. Haddad	M. Padgett
M. Eden	R. Herrick	W. D. Reeve
M. E. El-Hawary	S. Kartalopoulos	G. Zobrist
	D. Kirk	

Kenneth Moore, *Director of IEEE Press*
John Griffin, *Acquisition Editor*
Marilyn G. Catis, *Assistant Editor*
Denise Phillip, *Production Editor*

IEEE Solid-State Circuits Society, *Sponsor*
SSC-S Liaison to IEEE Press, Stuart K. Tewksbury

Cover design: William T. Donnelly, *WT Design*

Technical Reviewers
Dr. Jose F. Kukielka, *Alcatel Spain*
Josef Fenk, *Siemens AG Semiconductors High Frequency Products*
Dr. Paul Davis, *Lucent Technologies*

Books of Related Interest from IEEE Press...

CMOS: Circuit Design, Layout, and Simulation
R. Jacob Baker et al.
1998   Hardcover   944 pp   IEEE Order No. PC5689   ISBN 0-7803-3416-7

HIGH-TEMPERATURE ELECTRONICS
Edited by Randall K. Kirschman
1998   Hardcover   1392 pp   IEEE Order No. PC5735   ISBN 0-7803-3477-9

INTEGRATED CIRCUIT MANUFACTURABLILITY: The Art of Process and Design Integration
Edited by José Pineda de Gyvez and Dhiraj K. Pradhan
1999   Hardcover   336 pp   IEEE Order No. PC4481   ISBN 0-7803-3447-7

LOW-POWER CMOS DESIGN
Edited by Anantha Chandrakasan and Robert Broderson
1998   Hardcover   644 pp   IEEE Order No. PC5703   ISBN 0-7803-3429-9

INTEGRATED CIRCUITS FOR WIRELESS COMMUNICATIONS

Edited by

Asad A. Abidi
University of California, Los Angeles

Paul R. Gray
University of California, Berkeley

Robert G. Meyer
University of California, Berkeley

IEEE Solid-State Circuits Society, *Sponsor*

A Selected Reprint Volume

The Institute of Electrical and Electronics Engineers, Inc., New York

TK
7874
.I59432
1999

This book may be purchased at a discount from the publisher
when ordered in bulk quantities. Contact:

IEEE Press Marketing
Attn: Special Sales
445 Hoes Lane, P.O. Box 1331
Piscataway, NJ 08855-1331
Fax: 1-732-981-9334

For more information on the IEEE Press,
visit the IEEE home page: http://www.ieee.org/

©1999 by the Institute of Electrical and Electronics Engineers, Inc.,
3 Park Avenue, 17th Floor, New York, NY 10016-5997

*All rights reserved. No part of this book may be reproduced in any form,
nor may it be stored in a retrieval system or transmitted in any form,
without written permission from the publisher.*

Printed in the United States of America

10   9   8   7   6   5   4   3   2

ISBN 0-7803-3459-0

IEEE Order Number: PC5716

Library of Congress Cataloging-in-Publication Data

Integrated circuits for wireless communications / edited by Asad
　Abidi, Paul Gray, Robert Meyer.
　　　p.　cm.
　"A selected reprint volume."
　"IEEE Solid-State Circuits Society, sponsor."
　Includes bibliographical references and index.
　ISBN 0-7803-3459-0 (cloth)
　1. Integrated circuits.　2. Wireless communication systems—
Equipment and supplies.　3. Radio—Transmitter-receivers.
I. Abidi, Asad, 1956–　.　II. Gray, Paul R., 1942–　.　III. Meyer,
Robert G., 1942–　.　IV. IEEE Solid-State Circuits Society.
TK7874.I47145　1998
621.382—dc21　　　　　　　　　　　　　　　　98-6333
　　　　　　　　　　　　　　　　　　　　　　　　CIP

CONTENTS

Foreword   xi

Chapter 1   Introduction and Overview   1

Low Power Radio-Frequency ICs for Portable Communications   3
A. Abidi (*Proceedings of the IEEE,* April 1995).
Future Directions in Silicon ICs for RF Personal Communications   29
P. R. Gray and R. G. Meyer (*Proceedings of the IEEE,* May 1995).
The Challenges for Analogue Circuit Design in Mobile Radio VLSI Chips   37
J. Sevenhans and D. Rabaey (*Microwave Engineering Europe,* May 1993).

Chapter 2   Transmitter and Receiver Architectures   43

A Third Method of Generation and Detection of Single-Sideband Signals   45
D. K. Weaver, Jr. (*Proceedings of the IRE,* December 1956).
The Phasing Method for Sideband Selection in Broadcast Receivers   48
R. C. V. Macario and I. D. Mejallie (*EBU Review,* June 1980).
FM Receivers for Mono and Stereo on a Single Chip   55
W. G. Kasperkovitz (*Philips Technical Review,* 1983).
Advanced Low Voltage Single Chip Radio IC   69
T. Okanobu, H. Tomiyama, and H. Arimoto (*IEEE Transactions on Consumer Electronics,* August 1992).
Fully Integrated Radio Paging Receiver   80
I. A. W. Vance (*IEEE Proceedings,* February 1982).
A Single-Chip VHF and UHF Receiver for Radio Paging   85
J. F. Wilson, R. Youell, T. H. Richards, G. Luft, and R. Pilaski (*IEEE Journal of Solid-State Circuits,* December 1991).
Linear Transceiver Architectures   92
A. Bateman, D. M. Haines, and R. J. Wilkinson (*Vehicular Technology Conference Proceedings,* May 1988).
Direct Conversion Transceiver Design for Compact Low-Cost Portable Mobile Radio Terminals   99
A. Bateman and D. M. Haines (*Vehicular Technology Conference Proceedings,* May 1990).
A New Incoherent Direct Conversion Receiver   105
G. Schultes, A. L. Scholtz, E. Bonek, and P. Veith (*Vehicular Technology Conference Proceedings,* May 1990).
Personal Communications Transceiver Architectures for Monolithic Integration   112
E. Bonek, G. Schultes, P. Kreuzgruber, W. Simbürger, P. Weger, T. C. Leslie, J. Popp, H. Knapp and N. Rohringer
    (*International Symposium on Personal, Indoor, and Mobile Radio Communications,* 1984).
An All-CMOS Architecture for a Low-Power Frequency-Hopped 900 MHz Spread Spectrum Transceiver   118
J. Min, A. Rofougaran, H. Samueli, and A. A. Abidi (*IEEE 1994 Custom Integrated Circuits Conference,* 1994).
Spread-Spectrum Technology for Commercial Applications   122
D. T. Magill, F. D. Natali, and G. P. Edwards (*Proceedings of the IEEE,* April 1994).
Digital-Conversion Radio Transceivers for Digital Communications   135
A. A. Abidi (*IEEE Journal of Solid-State Circuits,* 1995).
A New Radio Receiver System for Personal Communications   147
R. Okanobu, D. Yamazaki, and C. Nishi (*IEEE Transactions on Consumer Electronics,* August 1995).

Chapter 3   Receivers   155

Receiver RF Design Considerations for Wireless Communications Systems   **157**
K. Hansen and A. Nogueras (*International Symposium on Circuits and Systems,* May 1996).

Design and Performance of Low-Current GaAs MMIC's for L-Band Front-End Applications   **161**
Y. Imai, M. Tokumitsu, and A. Minakawa (*IEEE Transactions on Microwave Theory and Techniques,* February 1991).

Ultra-Low DC Power Consumptions in Monolithic L-Band Components   **167**
K. Cioffi (*IEEE Transactions on Microwave Theory and Techniques,* December 1992).

GaAs Monolithic 1.5-2.5 GHz Image Rejection Receiver   **173**
A. Bóveda, G. L. Bonato, and O. Ripollés (*International Symposium on Circuits and Systems,* 1992).

Low Power GaAs ICS for Mobile Communication Equipment   **177**
H. Saki, A. Tezuka, Y. Mori, M. Sagawa, T. Katoh, J. Itoh, and K. Fujimoto (*International Symposium on GaAs and Related Compounds,* 1992).

Ultra Low Power Low Noise Amplifiers for Wireless Communications   **183**
E. Heaney, F. McGrath, P. O'Sullivan, and C. Kermarrec (*GaAs IC Symposium,* October 1993).

Low Current GaAs Integrated Down Converter for Portable Communication Applications   **186**
V. Nair, R. Vaitkus, D. Scheitlin, J. Kline, and H. Swanson (*GaAs IC Symposium,* October 1993).

X-Band Monolithic Series Feedback LNA   **190**
R. E. Lehmann and D. D. Heston (*IEEE Transactions on Microwave Theory and Techniques,* December 1985).

An RF Front-End for Digital Mobile Radio   **197**
J. Fenk, W. Birth, R. G. Irvine, P. Sehrig, and K. R. Schön (*IEEE 1990 Bipolar Circuits and Technology Meeting,* September 1990).

A Si Bipolar Monolithic RF Bandpass Amplifier   **201**
N. M. Nguyen and R. G. Meyer (*IEEE Journal of Solid-State Circuits,* January 1992).

A 1-GHz BiCMOS RF Front-End IC   **206**
R. G. Meyer and W. D. Mack (*IEEE Journal of Solid-State Circuits,* March 1994).

A One-Chip 2GHz Single Superhet Receiver for 2Mb/s FSK Radio Communication   **212**
V. Thomas, J. Fenk, and S. Beyer (*IEEE International Solid-State Circuit Conference,* February 1994).

A Low-Voltage Silicon Bipolar RF Front-End for PCN Receiver Applications   **215**
J. R. Long, M. A. Copeland, P. Schvan and R. A. Hadaway (*IEEE International Solid-State Circuits Conference,* February 1995).

A Highly Linear 1-GHz CMOS Downconversion Mixer   **218**
P. Y. Chan, A. Rofougaran, K. A. Ahmed, and A. A. Abidi (*European Solid-State Circuits Conference,* September 1993).

A 1.5 GHz Highly Linear CMOS Downconversion Mixer   **222**
J. Crols, and M. S. J. Steyaert (*IEEE Journal of Solid-State Circuits,* July 1995).

A Fully Integrated 900 MHz CMOS Double Quadrature Downconverter   **229**
J. Cros and M. Steyeart (*IEEE International Solid-State Circuits Conference,* February 1995).

Design Techniques for 1GHz Downconversion ICs Fabricated in a 1 μm 13 GHz BICMOS Process   **232**
W. D. Mack and R. G. Meyer (*Analog Circuit Design,* 1994).

A 1 GHz CMOS RF Front-End IC for a Direct-Conversion Wireless Receiver   **248**
A. Rofougaran, J. Y.-C. Chang, M. Roufougaran and A. A. Abidi (*IEEE Journal of Solid-State Circuits,* July 1996).

A 100 MHz IF Amplifier/Quadrature Demodulator for GSM Cellular Radio Mobile Terminals   **258**
I. A. Koullias, S. L. Forgues, and P. C. Davis (*IEEE 1990 Bipolar Circuits and Technology Meeting,* September 1990).

A Cellular Analog Front End with a 98dB IF Receiver   **262**
L. Longo, R. Halim, B-R. Homg, K. Hsu, and D. Shamlou (*IEEE International Solid-State Circuits Conference,* February 1994).

A 270-kb/s 35-m W Modulator IC for GSM Cellular Radio Hand-Held Terminals   **265**
J. J. J. Haspeslagh, D. Sallaerts, P. P. Reusens, A. Vanwelsenaers, R. Granek and D. Rabaey (*IEEE Journal of Solid-State Circuits,* December 1990).

A CMOS Limiting Amplifier and Signal-Strength Indicator   **273**
S. Khorram, A. Rofougaran and A. A. Abidi (*Symposium on VLSI Circuits,* June 1995).

Characterization of a Microwave Silicon single-Chip Direct Conversion RF Transceiver   **275**
W. Simburger, H. Knapp and P. Weger (*European Microwave Conference,* 1995).

A Single-Chip GaAs RF Transceiver for 1.9-GHz Digital Mobile Communication Systems **287**
K. Yamamoto, K. Maemura, N. Kasai, Y. Yoshii, Y. Miyazaki, M. Nakayama, N. Ogata, T. Takagi and M. Otsubo (*IEEE Journal of Solid-State Circuits,* December 1996).

A 1.9-GHz Single Chip IF Transceiver for Digital Cordless Phones **296**
H. Sato, K. Kashiwagi, K. Niwano, T. Iga, T. Ikeda, K. Mashiko, T. Sumi, and K. Tsuchihashi (*IEEE Journal of Solid-State Circuits,* December 1996).

A Direct-Conversion Receiver for 900 MHz (ISM Band) Spread-Spectrum Digital Cordless Telephone **302**
C. D. Hull, J. L. Tham, and R. R. Chu (*IEEE Journal of Solid-State Circuits,* December 1996).

A 2.7-V 900-MHz CMOS LNA and Mixer **311**
A. N. Karanicolas (*IEEE Journal of Solid-State Circuits,* December 1996).

A 12 mW Wide Dynamic Range CMOS Front-End for a Portable GPS Receiver **317**
A. R. Shahani, D. K. Shaeffer and T. H. Lee (*1997 IEEE International Solid-State Circuits Conference,* February 1997).

Chapter 4   Transmitters and Transceivers   321

An Integrated Si Bipolar RF Transceiver for a Zero IF 900 MHz GSM Digital Mobile Radio Frontend of a Hand Portable Phone **323**
J. Sevenhans, A. Vanwelsenaers, J. Wenin and J. Baro (*IEEE 1991 Custom Integrated Circuits Conference,* May 1991).

BBTRX: A Baseband Transceiver for a Zero IF G5M Hand Portable Station **327**
D. Haspeslagh, J. Ceuterick, L. Kiss, J. Wenin, A. Vanwelsenaers, and E. Enel-Rehel (*IEEE 1992 Custom Integrated Circuits Conference,* May 1992).

A 2.7V 800 MHz-2.1 GHz Transceiver Chipset for Mobile Radio Applications in 25 GHz ft Si-Bipolar **331**
W. Veit, J. Fenk, S. Ganser, K. Hadjizada, S. Heinen, H. Herrmann, and P. Sehrig (*1994 Bipolar/BiCMOS Circuits and Technology Meeting,* October 1994).

An Analog Radio Frontend Chip Set for a 1.9 GHz Mobile Radio Telephone Application **335**
J. Sevenhans, D. Haspeslagh, A. Delarbre, L. Kiss, Z. Chang, and J. F. Kukielka (*IEEE International Solid-State Circuits Conference,* February 1994).

A 900 MHz Transceiver Chip Set for Dual-Mode Cellular Mobil Terminals **338**
I. A. Koullias, J. H. Havens, I. G. Post, P. E. Bronner (*IEEE International Solid-State Circuits Conference,* February 1993).

A 2.7V to 4.5V Single-Chip GSM Transceiver RF Integrated Circuit **341**
T. Stetzler, I. Post, J. Havens, and M. Koyama (*IEEE International Solid-State Circuits Conference,* February 1995).

A 2.7V GSM Transceiver ICs with On-Chip Filtering **344**
C. Marshall, F. Behbahani, W. Birth, A. Fotowat, T. Fuchs, R. Gaethke, E. Heimerl, S. Lee, P. Moore, S. Navid, and E. Saur (*IEEE International Solid-State Circuits Conference,* February 1995).

A 2.4 GHz Single Chip Transceiver **347**
M. Devlin, B. J. Buck, J. C. Clifton, A. W. Dearn, and A. P. Long (*IEEE 1993 Microwave and Millimeter-Wave Monolithic Circuits Symposium,* 1993).

The Future of CMOS Wireless Transceivers **351**
A. A. Abidi, A. Rofougaran, G. Chang, J. Rael, J. Chang, M. Rofougaran, and P. Chang (*1997 IEEE International Solid-State Circuits Conference,* February 1997).

A 1.9 GHz Wide-Band IF Double Conversion CMOS Integrated Receiver for Cordless Telephone Applications **354**
J. C. Rudell, J-J. Ou, T. B. Cho, G. Chien, F. Brianti, J. A. Weldon, and P. R. Gray (*1997 IEEE International Solid-State Circuits Conference,* February 1997).

A 2.5 GHz BiCMOS Transceiver for Wireless LAN **357**
R. G. Meyer, W. D. Mack, and J. J. E. M. Hageraats (*1997 IEEE International Solid-State Circuits Conference,* February 1997).

A 2.7V DECT RF Transceiver with Integrated VCO **360**
G. C. Dawe, J-M. Mourant and A. P. Brokaw (*1997 IEEE International Solid-State Circuits Conference,* February 1997).

A 2.7V 2.5 GHz Bipolar Chipset for Digital Wireless Communications **363**
S. Heinen, K. Hadjizada, U. Matter, W. Geppert, V. Thomas, S. Weber, S. Beyer, J. Fenk and E. Matschke (*1997 IEEE International Solid-State Circuits Conference,* February 1997).

A 2.7V GSM RF Transceiver IC   **365**
  K. Irie, H. Matsui, T. Endo, K. Watanabe, T. Yamawaki, M. Kokubo and J. Hildersley (*IEEE International Solid-State Circuits Conference,* February 1997).

Chapter 5   Power Amplifiers and RF Switches   369

A Theoretical Analysis and Experimental Confirmation of the Optimally Loaded and Overdriven RF Power Amplifier   **371**
  D. M. Snider (*IEEE Transactions on Electronic Devices,* December 1967).
High Efficiency Transmitting Power Amplifiers for Portable Radio Units   **378**
  R. Nojima, S. Nishiki and K. Chiba (*IEICE Transactions,* June 1991).
A UHF Band 1.3W Monolithic Amplifier with Efficiency of 63%   **385**
  R. Takagi, Y. Ikeda, K. Seino, G. Toyoshima, A. Inoue, N. Kasai and M. Takada (*IEEE 1992 Microwave and Millimeter-Wave Monolithic Circuits Symposium,* 1992).
High Performance Integrated PA, T/R Swich for 1.9 GHz Personal Communications Handsets   **389**
  P. O. Sullivan, G. St. Onge, E. Heaney, F. McGrath and C. Kermarrec (*GaAsIC Symposium,* October 1993).
A 3.5V, 1.3 W GaAs Power Multi-Chip IC for Cellular Phones   **392**
  M. Maeda, M. Nishijima, H. Takehara, C. Adachi, H. Fujimoto and O. Ishikawa (*IEEE Journal of Solid-State Circuits,* October 1994).
A 900 MHz CMOS RF Power Amplifier with Programmable Output   **399**
  M. Rofougaran, A. Rofougaran, C. Ølgaard and A. A. Abidi (*1994 Symposium on VLSI Circuits Digest of Technical Papers,* 1994).
Highly Efficient UHF-Band Si Power MOSFET for RF Power Amplifiers   **401**
  I. Yoshida, M. Katsueda, S. Ohtaka, Y. Maruyama, and T. Okabe (*Electronics and Communications in Japan,* 1994).
Analysis of Phase Characteristics of a GaAs FET Power Amplifier for Digital Cellular Portable Telephones   **410**
  R. Ishizaki, H. Ikeda, Y. Yoshikawa and T. Uwano (*Electronics and Communications in Japan,* 1994).
Low Voltage, High Power T/R Switch MMIC Using LC Resonators   **419**
  T. Tokumitsu, I. Toyada and M. Aikawa (*IEEE 1993 Microwave and Millimeter-Wave Monolithic Circuits Symposium,* 1993).
A GaAs High-Power RF Single-Pole Double-Throw Switch IC for Digital Mobile Communications System   **423**
  K. Miyatsuji, S. Nagata, N. Yoshikawa, K. Miyanaga, Y. Ohishi and D. Ueda (*1994 IEEE International Solid-State Circuits Conference,* February 1994).

Chapter 6   Oscillators   427

A Simple Model of Feedback Oscillator Noise Spectrum   **429**
  D. B. Leeson (*Proceedings of the IEEE,* February 1966).
Low Current Oscillator Design for 900 MHZ GSM Applications   **431**
  S. Beyer and G. Lipperer (*Microwave Engineering Europe,* October 1991).
Design of a Low-Phase Noise VCO for an Analog Cellular Portable Radio Application   **434**
  T. Uwano, T. Ishizaki, Y. Nakagawa, and T. Nakamura (*Electronics and Communications in Japan,* 1994).
CMOS VCOs for PLL Frequency Synthesis in GHz Digital Mobile Radio Communications   **441**
  M. Thamsirianunt and T. Kwasniewski (*IEEE 1995 Custom Integrated Circuits Conference,* May 1995).
A CMOS 1.8 GHz Low-Phase-Noise Voltage-Controlled Oscillator with Prescaler   **445**
  J. Craninckx and M. Steyaert (*1995 IEEE International Solid-State Circuits Conference,* February 1995).
A Low-Power CMOS Digitally Synthesized 0-13 MHz Agile Sinewave Generator   **448**
  G. Chang, A. Rofougaran, M-K. Ku, A. A. Abidi and H. Samueli (*1994 IEEE International Solid-State Circuits Conference,* February 1994).
A 900 MHz CMOS LC-Oscillator with Quadrature Outputs   **451**
  A. Rofougaran, J. Rael, M. Rofougaran, and A. A. Abidi (*1996 IEEE International Solid-State Circuits Conference,* February 1996).
A Balanced 1.5 GHz Voltage Controlled Oscillator with an Integrated LC Resonator   **453**
  L. Dauphinee, M. Copeland and P. Schvan (*1997 IEEE International Solid-State Circuits Conference,* February 1997).

A 1.8 GHz CMOS Voltage-Controlled Oscillator   **456**
B. Razari (*1997 IEEE International Solid-State Circuits Conference,* February 1997).

Chapter 7   Components, Technology, and Modeling   **459**

Design of Planar Rectanular Microelectronic Inductors   **461**
H. M. Greenhouse (*IEEE Transactions on Parts, Hybrids, and Packaging,* June 1974).
Miniature Multilayer Spiral Inductors for GaAs MMICs   **470**
M. W. Geen, G. J. Green, R. G. Arnold, J. A. Jenkins and R. H. Jansen (*GaAs IC Symposium,* 1989).
Si IC-Compatible Inductors and *LC* Passive Filters   **473**
N. M. Nguyen and R. G. Meyer (*IEEE Journal of Solid-State Circuits,* August 1990).
Large Suspended Inductors on Silicon and Their Use in a 2-μm CMOS RF Amplifier   **477**
J. Y.-C. Chang, A. A. Abidi and M. Gaitan (*IEEE Electron Device Letters,* May 1993).
High Q Inductors for Wireless Applications in a Complementary Silicon Bipolar Process   **480**
K. B. Ashby, W. C. Finley, J. J. Bastek and S. Moinian (*Bipolar/BiCMOS Circuits and Technology Meeting,* 1994).
New Development Trends for Silicon RF Device Technologies   **484**
N. Camilleri, D. Lovelace, J. Costa and D. Ngo (*IEEE 1994 Microwave and Millimeter-Wave Monolithic Circuits Symposium,* 1994).
Miniaturized RF-Circuit Modules for Land Mobile Communication Equipment   **488**
Y. Mori, H. Yabuki, M. Ohba, M. Sagawa, M. Makimoto, and I. Shibazaki (*Vehicular Technology Conference,* 1990).
Miniaturized SAW Devices for Radio Communication Transceivers   **494**
M. Hikita, T. Tabuchi, and A. Sumioka (*IEEE Transactions on Vehicular Technology,* February 1989).
Recent Developments of Dielectric Resonator Materials and Filters in Japan   **501**
K. Wakino (*Ferroelectrics,* 1989).
RF Front End Circuit Components Miniaturized Using Dielectric Resonators for Cellular Portable Telephones   **519**
T. Nishikawa (*IEICE Transactions,* June 1991).
Bonding Pad Models for Silicon VLSI Technologies and Their Effects on the Noise Figure of RF NPNs   **526**
N. Camilleri, J. Kirchgessner, J. Costa, D. Ngo and D. Lovelace (*IEEE 1994 Microwave and Millimeter-Wave Monolithic Circuits Symposium,* 1994).
Wideband Characterization of Mutual Coupling Between High Density Bonding Wires   **530**
H-Y. Lee (*IEEE Microwave and Guided Wave Letters,* August 1994).
High-Speed Characteristics of Multilayer Ceramic Packages and Test Fixtures   **533**
D. H. Smith and R. M. Savara (*GaAs IC Symposium,* 1990).
An Equivalent Circuit for a Microwave Surface Mount Package   **537**
D. W. Hughes and D. M. Jackson (*Microwave Journal,* October 1991).
Electrical Characterization of Packages for Use with GaAs MMIC Amplifiers   **542**
S. R. Smith and M. T. Murphy (*IEEE MTT-S Digest,* 1993).
Frequency Limitation on an Assembled SO8 Package   **546**
F. Ndagijimana, J. Engdahl, A. Ahmadouche, and J. Chilo (*Electronic Components and Technology Conference,* June 1993).
Multifunction Silicon MMIC's for Frequency Conversion Applications   **552**
K. J. Negus and J. N. Wholey (*IEEE Transactions on Microwave Theory and Techniques,* September 1990).
Intermodulation in High-Frequency Bipolar Transistor Integrated-Circuit Mixers   **560**
R. G. Meyer (*IEEE Journal of Solid-State Circuits,* August, 1986).
Blocking and Desensitization in RF Amplifiers   **564**
R. G. Meyer and A. K. Wong (*IEEE Journal of Solid-State Circuits,* August 1995).
Modeling of the MOS Transistor for High Frequency Analog Design   **567**
P. J. V. Vandeloo and W. M. C. Sansen (*IEEE Transations on Computer-Aided Design,* July 1989).

Chapter 8   System Applications   **579**

A Double-Conversion Broad Band TV-Tuner with GaAs ICs   **581**
J-E. Muller, U. Ablassmeier, J. Schelle, W. Kellner and H. Kniepkamp (*GaAs IC Symposium,* 1984).

Hand-Held Portable Equipment for Cellular Mobile Telephone **585**
Y. Tamura, T. Maru, N. Hirasawa, H. Okuno, M. Komoda, M. Hotsumi, H. Matsumoto and K. Kimura (*NEC Research and Development,* October 1987).

Compact Size Numeric Display Pager with New Receiving System **594**
K. Yamasaki, S. Yoshizawa, Y. Minami, T. Asai, Y. Nakano, and M. Kuraoda (*NEC Research and Development,* January 1992).

Development of Advanced Mobile Telephone P3 (Personal Pocket Phone) **603**
Y. Tamura, A. Yonehata, S. Miyazaki, F. Kobayasi, Y. Fukuda and J. Kamishiro (*NEC Research and Development,* July 1990).

Radios for the Future: Designing for DECT **614**
B. Madsen and D. E. Fague (*RF Design,* April 1993).

Techniques for Open Loop Modulation of a Wideband VCO for DECT **620**
D. E. Fague, A. Dao, and C. R. Karmel (*RF Expo West,* 1994).

Performance Evaluation of a Single Chip Radio Transceiver **624**
K. M-D. Hess and D. E. Fague (*Vehicular Technology Conference,* April 1996).

Chip Set Addresses North American Digital Cellular Market **628**
M. M. Sera (*RF Design,* March 1994).

A Spread Spectrum Cordless Telephone **633**
K. Tanaka (*Electronics and Communications in Japan,* 1995).

A 1.9 GHz GaAs Chip Set for the Personal Handyphone System **643**
F. McGrath, K. Jackson, E. Heaney, A. Douglas, W. Fahey, R. G. Pratt, and T. Begnoche (*IEEE Transactions on Microwave Theory and Techniques,* July 1995).

Design Study on RF Stage for Miniature PHS Terminal **655**
H. Tsurumi, T. Maeda, H. Tanimoto, Y. Suzuki, M. Saito, K. Yoshihara, K. Ishida, and N. Uchitomi (*IEICE Transactions on Electronics,* May 1996).

Author Index   **661**

Subject Index   **665**

About the Editors   **673**

FOREWORD

FOR most of the more than three decades since the inception of integrated circuit (IC) technology, the design of silicon components for the radio frequency (RF) portions of communications transmitters and receivers has occupied a relatively small niche in the overall spectrum of mixed-signal IC applications. Traditionally, ICs at a modest level of integration are found in consumer radio and television circuits and in certain military communications systems. The expertise to carry out the successful design of these high-frequency circuits is concentrated in a small segment of the silicon IC design community.

This situation is changing rapidly, driven by the pervasiveness of RF communications as an important element of the overall communications infrastructure. An increasing percentage of voice, data, image, and video communications to individuals is taking place untethered, over wireless media, and this percentage clearly will continue to increase for the foreseeable future. Scaled IC technologies will make it possible to create high-frequency transceiver circuits at lower cost and higher integration levels than ever. This change will force a larger number of IC designers to involve themselves in RF circuits. The worldwide shortage of experienced engineers in the field of RF communications is a critical problem.

In this volume, we attempt to provide a comprehensive, up-to-date compilation of published papers in the field as a reference for design engineers new to the field. While RF design may be new to most mixed-signal design engineers, the few workers in the field in Europe, the United States, and Japan have produced, over the past few decades, a steady stream of innovations in basic technology, circuit design, and transceiver architecture; this work has had considerable impact on present-day communications transmitter and receiver implementations. These innovations have been described in disparate journals and conferences, many times not coming to the attention of IC designers. We have attempted to gather papers from as many of these sources as possible. Limitations on the physical size of the present compilation prevents us from including many interesting and innovative contributions. We hope that our choices will give an overall perspective to someone coming to the field for the first time, as well as provide a valuable compendium of references for more experienced RF IC designers.

The *Introduction and Overview* section consists of recent survey articles covering some aspects of RF-IC design from the viewpoint of silicon IC technologies. In the *Architectures of Transmitters and Receivers* section, two papers describe single-sideband architectures well suited to IC implementation, and several others that follow discuss either the actual implementation of highly integrated receivers and transmitters or the technical challenges involved in realizing them. The *Receivers* section is a chronological collection of in-depth papers on this topic spanning various IC technologies. The selected papers were usually the first, as far as we know, to describe a particular receiver circuit. The topic of receivers encompasses the RF front end, intermediate-frequency (IF) subsystems, and key baseband components which process the modulated received signal. Comparatively fewer papers comprise the *Transmitters and Transceivers* section, mainly because the subject has only become relevant with the advent of the two-way radio telephone. The lack of data though, is also an indication of the difficulty of integrating a transceiver on a chip. *Power Amplifiers and RF Switches* in the past have always constituted a separate topic, with its own terminology, special technology considerations to deliver high power, and unique circuit issues; they therefore warrant a section of their own.

Some radio engineers believe that a receiver is built around the frequency synthesizer and is ultimately limited by it—hence, the separate section in this volume on *Oscillators,* which includes some of the key publications on conventional oscillators and recent papers on alternative circuits well suited to integration. A discussion of radio circuits, whether highly or modestly integrated, would be incomplete without covering the essential role of certain passive components such as inductors and bandpass filters, and the detrimental role of others such as package and transistor parasitics. Papers on these topics, and others on how to model RF impairments, are included in the section *Components, Technology, and Modeling*. Finally, we must face the overarching question of how the various subcircuits have to work together to make a wireless system fulfill some difficult specifications. In the section on *Systems Applications,* we have selected hardware descriptions of recently deployed radio-communication standards or those currently in development.

After many years of relatively slow evolution, radio circuit design today is at a turning point. Foremost in the minds of many newcomers to the field are questions such as: Will the single-chip transceiver, with almost no RF or IF passive components, ever become a reality? How can one take advantage of smart base-stations and microcells to simplify handset design? Is CMOS a viable alternative to Si bipolar or GaAs technology for transceivers? It is difficult to answer these questions decisively, as much of the work of proving or disproving any particular approach remains to be done. We hope that this collection of publications will usefully inform and guide the investigator venturing into this exciting and rapidly moving new applications area.

Asad A. Abidi
Paul R. Gray
Robert G. Meyer

I

INTRODUCTION AND OVERVIEW

THE three papers in this first section present surveys of the present state of the art in analog IC design for radio transceivers and evaluations of new trends in the field, as well as projections of future developments emerging from current research activity.

Low-Power Radio-Frequency IC's for Portable Communications

ASAD A. ABIDI, MEMBER, IEEE

Invited Paper

The contributions of integrated circuits to the RF front-end of wireless receivers and transmitters operating in broadcast and personal communications bands are surveyed. It is seen from this that when IC's enable a rethinking of the RF architecture, the wireless device can sometimes become significantly smaller, and consume much less power. Examples are taken from FM broadcast receivers, pagers, and cellular telephone handsets.

Many semiconductor technologies are competing today to supply RF-IC's to cellular telephones. The various design styles and levels of integration are compared, with the conclusion that single-chip silicon transceivers, combined with architectures which substantially reduce off-chip passive components, will likely dominate digital cellular telephones in the near future.

The survey also projects future trends for IC's for miniature spread-spectrum transceivers offering robust operation in the crowded spectrum. With sophistication in baseband digital signal processing, its increasing interaction with the RF sections, and with increasing experience in simplified radio architectures, all-CMOS radios appear promising in the 900 MHz to 2 GHz bands. A specific CMOS spread-spectrum transceiver project underway at the author's institution is discussed by way of example.

I. INTRODUCTION

The portable revolution is upon us today. It promises to empower individuals throughout the world by giving them low-cost access to information wher ever they may be, thus allowing them to make informed decisions and to be more productive in business and at home, without necessarily being tied down to a physical location. It is expected that in the near future, individuals will be equipped with capabilities of local computing and of communications enabling them to perform almost all the tasks that today require the equipment on the office desktop: the telephone, the computer, its connection to a ubiquitous wired network, the fax machine, and so on [1].

Manuscript received November 21, 1994; revised January 13, 1995. This work was supported in part by the US Advanced Research Projects Agency, and in part by a consortium of semiconductor companies under the State of California MICRO Program.
The author is with the Electrical Engineering Department, University of California, Los Angeles, CA 90024-1594 USA.
IEEE Log Number 9409344.

The portable revolution has been many years in the making. The personal broadcast radio receiver and cassette player, as pioneered by Sony, was a runaway global success. The user could construct a private audio environment anytime and anywhere, using a device so small and light that its presence was easily forgotten. Portable computers brought about the next wave of change. The personal computer has had such a large impact on the broad working habits of individuals that without it they are lost. The luggable computer has evolved in a matter of one decade into the portable, the notebook, and the subnotebook. With the establishment of a wide-area network of radio paging transmitters, the personal radio pager has also became very popular since the mid-1980's. Today, the pager network spans the entire continental US and many other parts of the world, enabling the user to receive alphanumeric electronic mail messages. Two-way paging is actively under development. The cellular telephone became widely available shortly thereafter. The user could hook into the international switched-telephone network through the nearest cellular base station with a portable transceiver, which too has scaled down remarkably in size until its weight and volume is the smallest practical [2].

There are many competing visions of how these various portable devices and services will evolve and integrate in the next few years. These are covered extensively in the popular press, and in numerous keynote speeches in technical meetings. A common theme is that users will want a multimedia terminal, capable of wireless access to a global network which can transport communications, images, and databases to the user in an on-demand, interactive fashion [3], [4]. Such a terminal will have capabilities of computing, image acquisition and display, and obviously, of communications. It will likely derive as a hybrid of the various portable technologies available today.

What obstacles must be overcome to realize this vision? There include the very highly integrated electronics, effective displays, and a philosophy of design based on low power dissipation to prolong the battery life of the portable device. Sometimes single-battery operation will impose the additional constraint of operation at low voltage, as low as 1 V, which will require entirely new ways of doing electronic circuit design.

In the past few years, most designers of mass-market digital IC's have been preoccupied with low-power operation [5]. Principles such as operating CMOS logic at the lowest possible supply voltage have become widely known, and power-down modes, gear-shifting of operating clock frequencies, pipelining and parallelism, subthreshold operation, and other such methods which were once the province of specialized areas such as electronic wristwatches and implantable biomedical devices are becoming commonplace [6]. Studies into the fundamental thermodynamic limits to the energy required for computation are being initiated or revived. There is good reason to believe that all this activity will lead to significant improvements in the conventional circuit and system design styles for digital signal processing and computation.

How will this activity affect the communications aspects of the portable device? What similar principles to low-power digital design are there for energy-efficient wireless communications transceivers? These questions do not have simple answers. Low-power communication systems will result from use of the correct architectures, a sensible partition between analog and digital signal processing, low-power circuit techniques everywhere, and a judicious division between active and passive components. There is still not a widely known, integrated vision on this subject. Furthermore, there remains a gap between the IC design used in the portable applications described above, and the new designs that will be required over the next few years for advanced portable communicators. Consumers are demanding a great deal more functionality and performance, which is stressing present-day technology to its limits, and wireless communicator design itself is in transition from the classic analog modulation techniques used over the past 50–70 years to more sophisticated methods using digital signalling formats and signal-processing methods in transceivers.

This paper summarizes the key developments in the discipline so far, and from them forecasts wireless IC design trends in the near future.

II. Key Signal-Processing Issues in Wireless Transceivers

Were it not for the advent of the portable communications revolution, radio technique would almost certainly have become a lost art. The first edition of the last definitive textbooks on the subject dates to 1943 [7]. Whereas once radio engineering was synonymous with electronics [8], few university electronics curricula today offer a course on radio communications circuits. Only a few modern textbooks on radio design have been written in the past 25 years [9]–[13]. Radio communication methods, at least for nonmilitary applications, have remained relatively unchanged since World War II, and the evolutionary improvements in consumer equipment mainly owes to the use of high-frequency discrete transistors [14], smaller passive components, and building-block IC's which improve the long-term reliability and manufacturability of radio and TV receivers. The major impact of IC technology in these consumer items has probably been at baseband, in adding more user features. In contrast, the front-end radio architectures have evolved almost not at all in the past 40 or so years. For instance, IC's have contributed digital volume-control, digital frequency-tuning, features to alleviate manual effort on the part of the user, but the RF and IF sections still contain discrete and passive components in rather conventional architectures.

Why is this? It is partly because radio frequencies were too high for the low-cost IC technologies traditionally used in the consumer electronics industry. It is also because advances in component packaging alone have led to rapid downscaling in the size of consumer devices, often obviating the need to rethink the electronics. As a result, only a few individuals in a handful of institutions worldwide have concerned themselves with thinking about these problems. Today, as conventional solutions no longer suffice for the future wireless communications devices, there is a rekindling of interest in this subject, which has led to much rediscovery and some invention. Baseband IC designers are now attempting to apply familiar techniques to wireless, while microwave IC designers are exploring what to them are low-frequency commercial opportunities for their technologies.

To set the stage for further discussion, some of the unique problems of radio receivers and transmitters are first described. Unlike familiar wireline communications, the wireless environment accommodates essentially an unlimited number of users sharing different parts of the spectrum, and very strong signals coexist next to the very weak. The radio receiver must be able to select the signal of interest, while rejecting all others. It must do so using less than perfect active and passive components. There are two important problems in the receiver: *image-rejection* and *dynamic range*. Image-rejection relates to the receiver's ability to select the desired signal from the array of signals occupying the spectrum. Ideally, it might do so with a tunable bandpass filter, whose center frequency could be positioned at will in the RF, and whose passband was one channel wide. A filter with this small a fractional passband does not exist. Instead, a practical RF bandpass filter, which may or may not be tunable, will preselect an array of radio channels including the one of interest (Fig. 1). The other preselected channels are then removed at a lower intermediate-frequency (IF), by translating them in frequency with a downconversion mixer, and centering the desired channel within a bandpass filter at IF. The other mixer input is a frequency-tunable local oscillator (LO), offset by IF from the desired channel. As the preselected band after downconversion will very likely occupy an interval greater than (0, IF) on the frequency-axis, the IF bandpass filter will select both the desired channel, and another *image* channel the mixer has translated to –IF. The subsequent detector circuit will be unable to distinguish

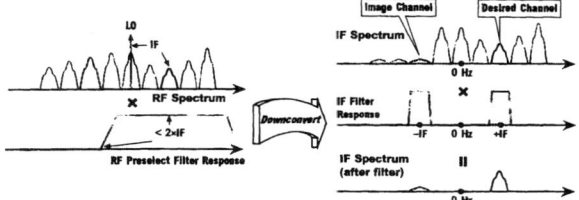

Fig. 1. The image-reject problem in radio receivers. The RF preselect filter passband must be determined with prior knowledge of the IF. The preselect filter is responsible for image-channel suppression before downconversion.

between the desired and the image channels, and therefore its output will be the result of the superposition of both. However, if the stopband of the preselect filter lies less than $2 \times$ IF away from the desired channel it will attenuate the image, so only the desired channel will contribute energy at IF. The receiver designer first studies the available filter technologies, and then chooses an appropriate IF which yields an acceptable image suppression. A high IF relaxes the prefilter passband specification, but it also means that the downconverted signal requires high-frequency amplifiers, which are usually power-inefficient. Further, the IF filter requires a smaller fractional passband. In such cases, following image rejection at this high IF, the channel may be selected after downconversion to a second, lower IF. In such a double superheterodyne, or dual-conversion receiver, the first IF may actually lie at a higher frequency than the incoming RF to make image rejection easier.

The noise-level and nonlinearity in the RF amplifier and first mixer usually set the receiver dynamic range. Consider reception of a weak channel surrounded by large undesired channels in the preselection band (Fig. 2(a)). First, the input-referred noise of the receiver directly adds to the sought signal, corrupting its signal-to-noise ratio (SNR). Second, the large adjacent channels will experience the nonlinearities in the RF amplifier and mixer, and some of the products of the ensuing intermodulation distortion may overlap the desired channel. The IF filter cannot reject these unwanted products, which, like noise, will degrade the received SNR. As the receiver frequency response is normally bandpass, its nonlinearity is measured by applying two tones of equal amplitude closely spaced in frequency (f_1 and f_2) at it input, and measuring the rise in the third-order intermodulation products (at $2f_1 - f_2$ and $2f_2 - f_1$) with input level (Fig. 2(b)). All other intermodulation tones usually lie outside the receiver passband. On a logarithmic plot, the third-order intermodulation level rises at a slope of 3 relative to the fundamental tone at the output. The two lines intersect at a point called the input-referred 3rd-*order intercept* (IP3). The intercept point is usually extrapolated from measurements at low levels, because the receiver front-end will saturate at large inputs. The 1-dB *compression point*, the input level which causes the receiver gain to drop by 1-dB relative to the small-signal gain, specifies the onset of saturation. The input-referred noise-level may be included in this plot to define a spurious-free dynamic range (SFDR), although this is rarely used in radio

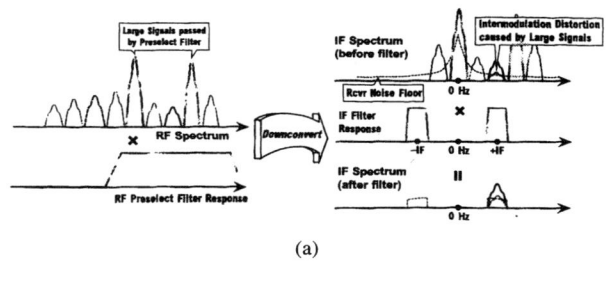

(a)

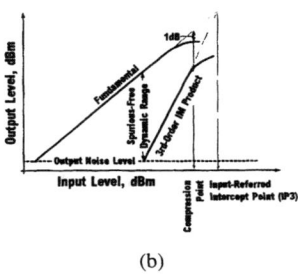

(b)

Fig. 2. The dynamic range problem in the radio receiver. (a) Adjacent large signals may create intermodulation products superimposed on the desired channel. Receiver noise floor is fundamental limit to sensitivity. Receiver dynamic range is specified (b) in terms of extrapolated 3rd-order intercept point, and noise level.

specification. Usually the input noise-figure (NF) and the input-referred intercept point (IP3) are separately specified.

Similar specifications apply to the transmitter, which operates at much larger signals. Suppose, as is almost always the case, that the transmitter is required to emit a single-sideband, suppressed-carrier output. Owing to circuit imperfections, it may also emit small amounts of the carrier and the unwanted sideband, which typically lie in the passband of the subsequent RF filter (Fig. 3). These unwanted emissions may become interferers for adjacent channels. Nonlinearities in the power amplifier may also produce emissions of intermodulation products at other frequencies. Phase-noise in the local oscillator responsible for upconverting to RF will convert to noise added to the signal, and the amplitude of this noise increases with the transmitted signal. This noise could possibly overwhelm nearby weak channels. Transmitter performance is usually specified in terms of the relative levels of unwanted signals to the desired signal, and in terms of absolute spectral density of output noise at maximum output power.

To understand the rationale underlying receiver architecture, let us use as an example the familiar broadcast FM receiver. The architecture to be described is the same that Armstrong, the inventor of FM and the superheterodyne, had originally proposed for FM reception. The desired channel consists of a carrier in the 88–108 MHz band, modulated by up to ± 75 kHz. Neighbouring channels are spaced apart by 200 kHz. The receiver must select the desired channel while rejecting nearby channels, and it must be sensitive to a signal of a few tens of microvolts induced on the antenna. A simple FM antenna is wideband, and will pick up signals well outside the broadcast FM band. The low-noise bandpass amplifier in the front-end may at best mildly attenuate the out-of-band signals—the

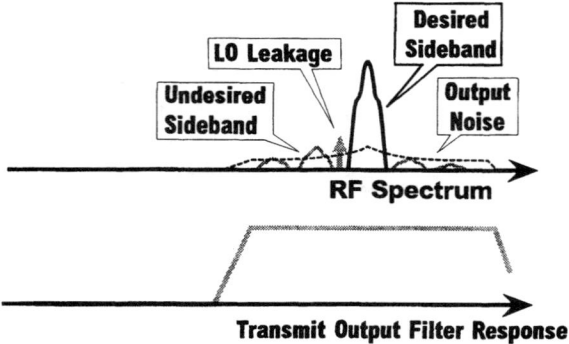

Fig. 3. Transmitter imperfections (such as mismatches in a quadrature upconverter) result in the appearance of spurii in the emitted spectrum (referred to as "spectral regrowth"). The spurii may not be removed by the output filter, and may superimpose on adjacent channels. Emitted noise might overwhelm weak adjacent channels.

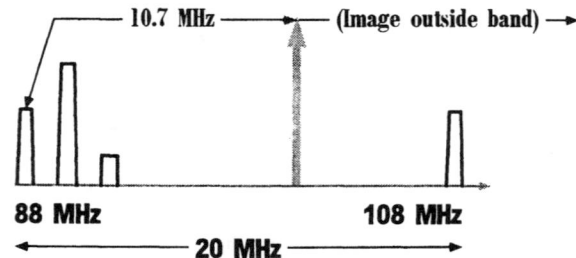

Fig. 4. The 10.7 MHz intermediate frequency conventionally used in FM receivers is the lowest frequency which will guarantee that the image lies outside the FM broadcast band. The FM demodulator will reject a (non-FM) image channel entering the receiver. 10.7 MHz passive bandpass filters in the IF strip select the desired channel.

actual channel selection must be done elsewhere. The RF amplifier uses an inductive load to resonate with the transistor and tuning capacitances, thereby transforming an inherently low pass characteristic to bandpass centered on the frequencies of interest. The transistor $f_{\max}$ limits the highest frequency at which such a tuned amplifier can still provide a gain greater than unity. This figure-of-merit is familiar to microwave circuit designers and to device designers, whereas baseband IC designers deal more often with transistor f_T, the capacitance-limited unity current-gain frequency. In bipolar IC processes not optimized for small-signal high-frequency use, $f_{\max}$ is comparable to f_T, whereas in the best RF processes it may be twice f_T [15], [16].

It is impossible to select the desired channel at RF, because no tunable filters exist with the required fractional bandwidth of 0.15% at 100 MHz. Therefore, following sufficient amplification at RF to overcome the noise-level of the following circuits, the signal is mixed down by a variable-frequency local oscillator to a lower IF. Furthermore, a filter to select the desired channel at IF will have a fixed center frequency, and the fractional bandwidth in the passband will be more reasonable. From this perspective, it is desirable to use as low an IF as possible. However, image-rejection poses yet another constraint on choice of IF. Conventional broadcast FM receivers use an IF of 10.7 MHz as a compromise. This IF guarantees that the image always lies outside the FM band (Fig. 4). It is unlikely, however, that the preselect filter can suppress this image, which will therefore either add noise or AM to the desired signal. However, the subsequent FM detector is inherently insensitive to both these forms of impairment. In this way, a medium-valued IF is made possible by exploiting properties of the detector, and thereby the fractional passband specification for the preselect filter is relaxed. The first mixer downconverts the entire FM band, with the desired channel centered at 10.7 MHz. A varactor-tuned Colpitts oscillator may be used as the first local oscillator. Prior to detection, a cascade of identical, fixed-frequency ceramic bandpass filters, each with a 200 kHz passband centred at 10.7 MHz, passes the desired channel while rejecting neighbouring channels [17]. In a high-quality receiver, the RF amplifier may be a discrete GaAs MESFET with a tuned load ganged to the LO tuning element, and another ganged tuned circuit may couple the antenna signal into the receiver [17].

To transplant this style of discrete radio-circuit design to IC's, one would have to implement LC tuned circuits and filters on silicon. One can indirectly surmise these concerns in the Motorola series of IC design textbooks from the 1960's, which discuss loss in spiral metal inductors fabricated on silicon substrates [18], as well as issues relating to simulated inductors for active filters at IF for radios, made with gyrators and capacitors [19]. However, on-chip spiral inductors of useful values were found to suffer excessive capacitance to the substrate, which lowered their self-resonant frequency to the point that they were not usable beyond the VHF band. It gradually became part of the collective consciousness of IC designers that on-chip tuned circuits are generally impractical.

When useful tuned amplifiers did appear on monolithic integrated circuits, it was not for the VHF to UHF range of relevance to consumer applications. Instead, it was military applications at much higher frequencies which drove the development of monolithic microwave integrated circuits (MMIC's) during the 1980's. MMIC's typically use MESFET's as the active device on semi-insulating GaAs substrates. This technology has enabled miniature radar, remote sensing, and communications at frequencies up to tens of GHz. The on-chip wavelengths are so small that monolithic distributed circuits may be built. MMIC's take advantage of the semi-insulating substrate in two important ways. Transistors on these substrates have lower parasitic capacitance, which means that they amplify to higher frequencies. It is also possible to build low-capacitance interconnect with airbridge structures, and high-frequency passive components such as spiral inductors required for narrowband tuned circuits. Thus on MMIC's, the board-level design styles used hitherto by radio- and microwave-engineers could be miniaturized. However, over its many years of existence, GaAs MMIC technology has not had the major impact on consumer electronics that its adherents had hoped for. Makers of consumer electronics favor silicon IC technology wherever feasible because of its low-cost, high yields, and the relative ease of mixing analog and digital circuits on a large scale.

It is anticipated that by the year 2000 about 300 million portable consumer wireless devices will be in use [20]. What IC technologies will enable the RF front-end of these devices? Do miniaturization and long battery-life call for architectural innovations in transceivers? What new circuit design styles will evolve in response? There is much curiosity and speculation on these matters, yet little is generally known about RF-IC design, or on the possible impact of large-scale integration and power-reduction strategies in the front-end of wireless transceivers. This paper presents a brief survey of the use of IC technology in wireless receivers and transmitters since the 1970's to date, and from this projects some future trends. RF-IC's are roughly defined as integrated circuits operating in the band of frequencies from 400 MHz to 2500 MHz, which covers most consumer wireless communication devices. As opposed to MMIC's, which were almost exclusively fabricated on III-V compound semiconductor substrates at small-scales of integration, mature silicon technologies will play a large, if not the dominant, role in RF-IC fabrication. It is the author's belief that in response to pressing demands for ubiquitous wireless access, both the underlying semiconductor technology and the design styles will rapidly evolve to realize the single-chip "VLSI radio" in the not too distant future.

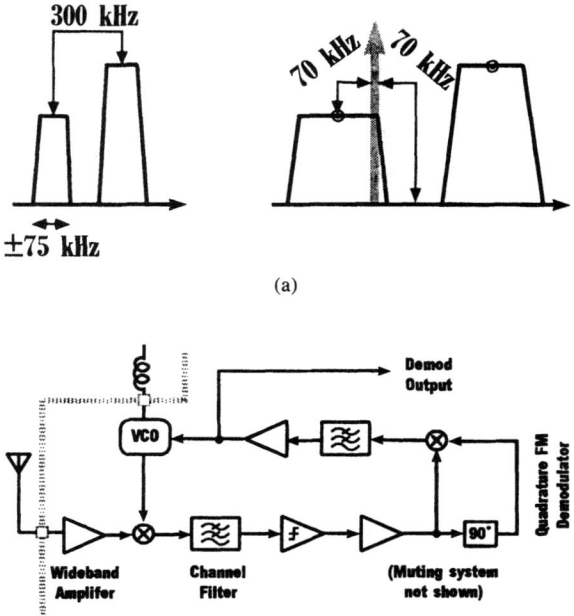

Fig. 5. (a) An alternative choice of IF in the FM band, which places the image in the gap between adjacent channels. The IF strip, including the channel filters, now operate at a 70 kHz frequency. (b) An FM receiver using a 75 kHz IF. The channel filter is an active-RC implementation on-chip. A frequency feedback loop compresses the incoming frequency swing. Except for a tuning inductor, no high-frequency off-chip components are required.

III. IC'S IN BROADCAST RADIO RECEIVERS

The two-way wrist radio has fascinated the popular imagination since its introduction in the popular American cartoon strip, *Dick Tracy*, in 1946 [21]. This is the ultimately unobtrusive piece of consumer electronics. Let us now see how feasible it is to build an FM receiver of this size with microelectronics technology. For the average user to accept such a radio, its selectivity and sensitivity must be comparable to that of tabletop models. If in the FM receiver described in the previous section all the transistors in the receiver electronics were to be integrated on to one silicon chip, the radio would still need a considerable number of off-chip tunable inductors and ceramic filters, and in spite of state-of-the-art miniature packaging, the components could not plausibly all fit into a wristwatch. Neither would the power dissipation be commensurate with the life of a wristwatch battery.

The first generation of silicon bipolar IC's developed in the late 1970's for the IF and baseband portions of broadcast receivers more or less contained the transistors of conventional receivers assembled on to one or more IC's [22]–[24]. However, integration did afford freedom to use transistor-rich circuits for higher performance. Circuit techniques such as double-balanced mixers using the Gilbert analog multiplier, phase-locked loops as FM demodulators, and balanced on-chip signal paths to attain greater immunity to pickup and common-mode noise became widely used as a result. By eliminating many of the coupling coils and other noncritical discrete components found in older radio circuits, IC's contributed to lowering the cost of assembling and aligning the final product. In the RF section, though, the receivers still used the conventional 10.7 MHz IF superheterodyne architecture implemented with shielded discrete-component circuits.

In the early 1980's, Kasperkovitz at Philips [25] made the first significant explorations into alternative architectures for highly integrated radio receivers. He realized that to reduce receiver size and power dissipation, it was very important to eliminate the many off-chip passive components. If certain passive inductors, capacitors, and resistors could not be eliminated, they could at least be packaged in surface-mount outlines for very small size. However, neither the volume of the IF ceramic filters could be readily scaled down, nor could their characteristic impedance be scaled much above 50Ω. Each filter requires an on-chip analog driver of comparable impedance. Alternatively, the filter may be realized on-chip as an active bandpass circuit of sufficient selectivity and dynamic range. Although a gyrator-capacitor based active filter is possible in principle, small phase-shifts in the gyrator transistors at the 10.7 MHz IF can seriously upset the filter passband shape. Active resonators are also known to suffer from a larger internal noise level than their passive counterparts, and this discrepancy worsens with increasing pole-Q and pole frequency [26]. Kasperkovitz solved the problem with an *architectural innovation*, by dramatically lowering the IF from 10.7 MHz to 70 kHz. This makes it a great deal easier to implement an IF active channel-select filter, which now need only be lowpass. Further, at a given dynamic range, the power dissipation in an active filter also scales down with the IF [26]. The low IF eliminates the off-chip channel-select filter and reduces power dissipation, both very desirable properties. But what of the image frequency, the principal reason for the choice of 10.7 MHz?

At a 70 kHz IF, the image frequency lies half-way to the adjacent FM channel (Fig. 5(a)). The image therefore is the inter-channel noise in the FM band. As an RF preselect filter cannot possibly reject an image this close to the desired signal, it will pass unattenuated to worsen the received signal-to-noise ratio (SNR) by 3-dB. Another consequence of this choice of IF is that after the first downconversion, an instantaneous frequency deviation in the received FM signal of more than 70 kHz will alias around dc to produce distortion. This is avoided by compressing the frequency deviation to $\pm$ 15 kHz with a negative feedback frequency-locked loop prior to downconversion (Fig. 5(b)). The FM mono/stereo radio [27]–[29] requires, in addition to the single-chip receiver, only 15 small capacitors and two inductors, and this collection of parts readily fits inside a wristwatch. A miniature earphone is plugged into the watch, and the earphone lead serves as the antenna. The radio when active drains 8 mA from a 4.5 V supply.

Sony, one of the world's leading makers of miniature radios, has also recently modified its integrated FM radio-receiver architecture from the conventional 10.7 MHz IF [30] to low-IF [31]. In the new architecture, a high first-IF of 30 MHz is used, so any out-of-FM band image falls in the stopband of a fixed 80 to 110 MHz bandpass preselect SAW filter after the antenna. A wideband IF amplifier boosts the received signal level of *all* FM channels *without* any filtering—amplification at 30 MHz is not a problem on this modern silicon bipolar IC process. Further, at a 30 MHz IF the entire broadcast FM band falls to one side of the LO frequency, which means that any channel in the FM band may be selected after the first downconversion. Following this, another mixer converts to a low second-IF of 150 kHz, and thereafter an on-chip 9th-order active-RC lowpass filter rejects adjacent channels. The low second-IF, however, will pass an image FM channel as well as the desired channel, and as there is no filtering at all at the first IF, an *image-reject mixer* is used for the second downconversion (Fig. 6). The variable-frequency first LO tunes the desired channel, while the second LO, which must produce quadrature outputs for the image-reject mixer, is at a fixed frequency.

The image-reject mixer provides a trigonometric solution to a difficult filtering problem [32]. The desired channel and its image are frequency-converted into two paths by mixers driven by quadrature phases of an LO. The mixer outputs are then phase-shifted 90° with respect to one another. The sum of these two signals will select the desired channel and suppress the image, while, *vice-versa*, the difference will select the image. The extent of image suppression depends on the gain matching of the two paths, and on the phase-accuracy of the LO quadrature outputs. For these reasons, this concept has only become practical with IC technology,

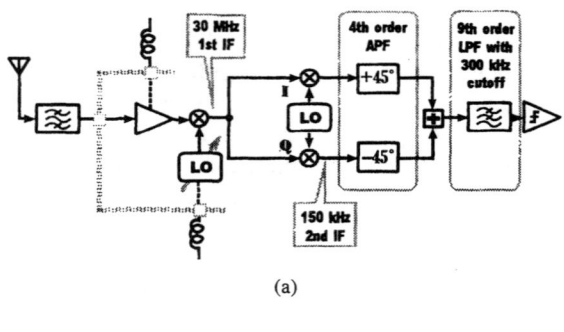

(a)

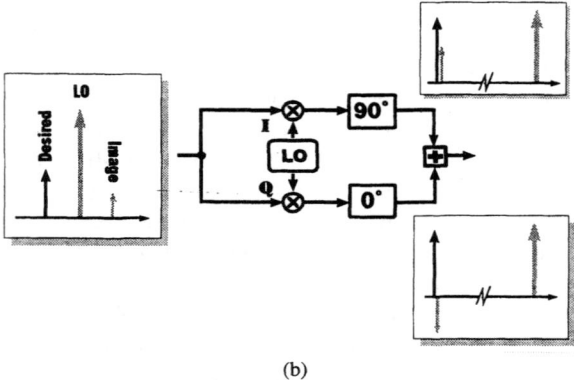

(b)

Fig. 6. (a) An alternative architecture for a single-chip FM receiver, with a 30 MHz IF chosen for strong image-rejection, and to translate the all channels in the FM band to IF on to one side of dc on the frequency-axis. The image-rejection mixer at 150 kHz selects the desired channel, while rejecting the undesired one 300 kHz away. Aside from a noncritical RF preselect filter, the remaining filters are active on-chip, including the phase-shifts in the two arms of the mixer. (b) The image-rejection downconversion mixer. The image and desired tones are at positive- and negative-frequency offsets from the local oscillator, and are discriminated in the mixer by a relative inversion of polarities in the two arms.

where the two paths are well matched on-chip and track each other over temperature. Image suppression on the Sony chip is limited to about 40–45 dB by residual gain mismatch in the two paths. An allpass active RC-CR filter produces 90° phase-shifted versions of the downconverted input. This receiver drains about 15 mA from voltages as low as 0.9 V in either FM or AM mode. All the necessary transistors are integrated on-chip—the RF amplifier portion, however, uses an off-chip load inductor and the local oscillator needs an off-chip LC tuned circuit.

The circuit techniques which enable sub-1 V operation are also interesting (Fig. 7). The antenna signal drives the emitter of a common-base NPN, which forms the tail of a differential pair. The input resistance of the common-base stage matches the antenna impedance. The signal develops at one inductively loaded collector of the differential pair, while the other collector dumps a fraction of the signal current into the supply in response to an AGC differential control voltage. Following amplification by a resistively loaded differential pair in cascade, the balanced RF signal is level-shifted into the first mixer, a simplified double-balanced Gilbert-cell with resistors instead of current sources in the tails.

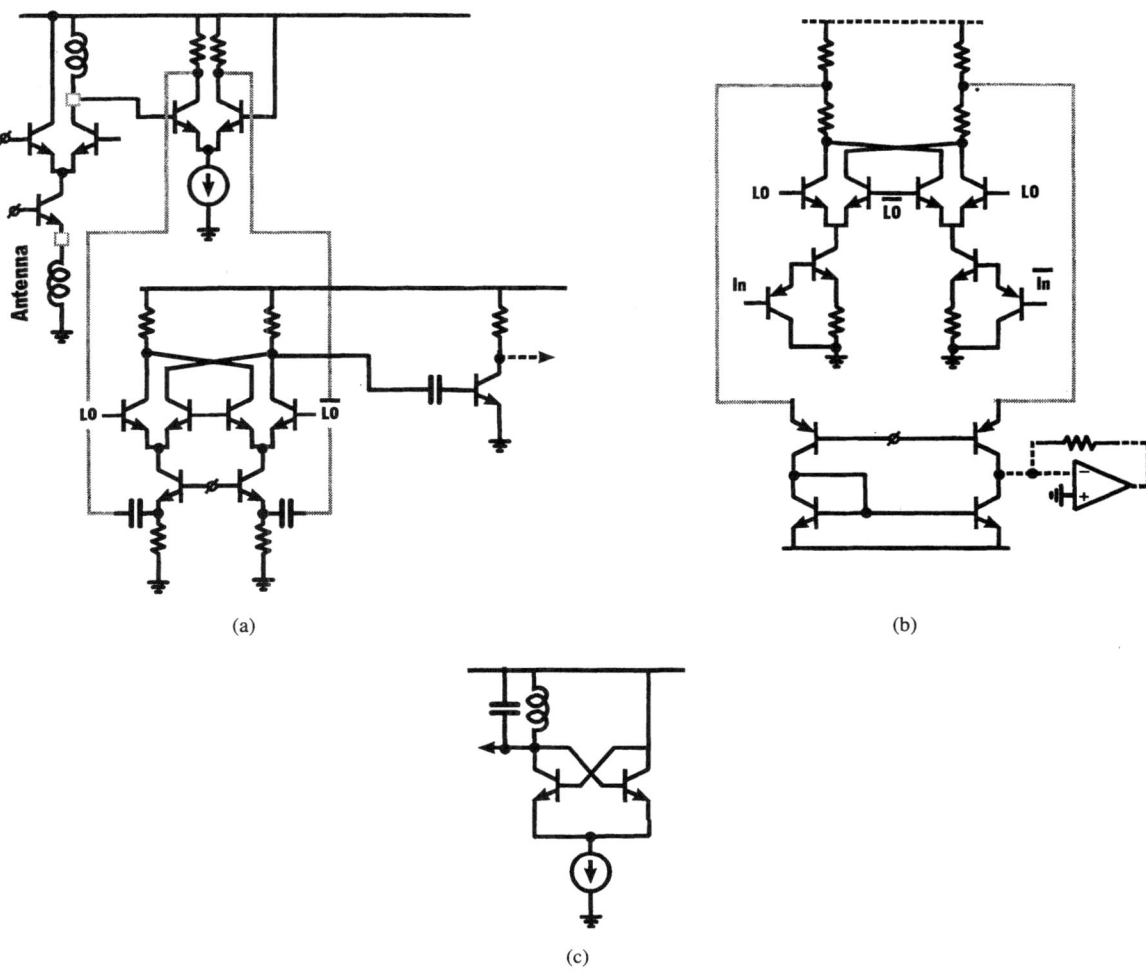

Fig. 7. Low-voltage circuits capable of operation with a 0.9-V supply. (a) RF amplifier. Note how the antenna carries the RF amplifier bias current, and how the on-chip capacitors level-shift the RF signal path. (b) IF amplifier. As the signal is downconverted in frequency, active level-shift and op amp circuits appear. (c) The cross coupled differential pair negative resistance is popular for LC oscillators.

As inductors do not drop a dc voltage, they make convenient loads for stacked transistor circuits operating at a low supply voltage. The signal on the free end of the inductor will swing above the supply. This is the case in the local oscillator, which implements a negative resistance of $-1/g_m$ with a cross coupled differential pair, causing an LC tuned circuit across it to oscillate. The oscillation amplitude is limited by the differential pair nonlinearity to a small multiple of kT/q.

As the signal propagating through the receiver downconverts in frequency, low-frequency circuit techniques and devices with a lower f_T are used. For instance, the 30 MHz IF signal couples into the second mixer through PNP emitter followers. At the 150 kHz IF, common-base PNP's and current mirrors level-shift down the signal, which is further amplified in an op amp.

This receiver is a significant example of how architectural rethinking combined with appropriate IC design styles has resulted in a very different solution to the well established broadcast FM receiver.

IV. IC's IN WIRELESS PAGING RECEIVERS

Miniature wireless communicators to page people on the move were first developed in the late 1950's. The Bell System's Bellboy™ paging receiver [33] anticipated many of the concepts underlying today's pagers. The system operated at a 150 MHz RF, addressing a receiver by frequency-modulating the carrier with a unique set of three tones, which the intended receiver recognized at baseband by the simultaneous response of three passive reed resonators. The superheterodyne receiver operated at 4 V using a total of only ten transistors [34], a notable early example of low-power and low-voltage circuit design. The low IF of 6 kHz meant that simple, capacitively coupled 10 kHz lowpass (rather than bandpass) filters could select the desired channel. Also, all low frequency amplifiers after the RF section used transistors biased at small currents. The two stages in the cascode RF amplifier were inductively coupled to share, or reuse, the same bias current, a power-saving method found even in todays MMIC's.

Paging receivers have been in continuous evolution since then. They are supported by a sophisticated nationwide wireless infrastructure. The modern pager uses digital signalling at rates anywhere from 500 to 1200 b/s, encoding binary data with a simple positive- or negative-offset of the carrier frequency—the binary frequency-shift keyed (FSK) modulation. In spite of market pressure to reduce the battery drain and miniaturize the unit, paging receivers until the early 1980's used conventional radio architectures without exploiting the powerful simplifications implied by this signaling scheme. Such a double-superheterodyne receiver [35] might consist of a *first upconversion* of the received signal to suppress the image channel with an RF crystal bandpass filter, and then a *downconversion*, followed by channel-selection with a ceramic bandpass filter. An analog frequency-discriminator demodulated the FSK. Much as in broadcast FM receivers, this architecture required tuned amplifiers and passive filters, which constrained further miniaturization of the paging receiver.

Vance at ITT Standard Telecommunication Laboratories first realized that by taking the idea of low-IF to its limit of *zero-IF*, an FM receiver could be scaled down to one chip with only one or two passive RF components. A quadrature downconversion mixer could discriminate positive- and negative-frequency modulation centered around dc [36]. There is now *no* image to be rejected, and a lowpass filter suppresses adjacent channels. If applied to analog FM as in broadcast signals, however, this scheme suffers from the dc offsets in the amplifiers and their flicker noise, which will seriously corrupt the SNR at mid-channel. Furthermore, the limiting amplifier which analog FM receivers use in place of AGC does not respond to dc inputs, because there are no zero-crossings. In paging receivers, however, the spectral energy clusters in two lobes on either side of dc owing to the relatively large modulation index, and zero-IF (or direct-conversion) is exactly the right solution [37], [38]. This remarkably simple receiver (Fig. 8) consists only of a quadrature demodulator, lowpass filters in each arm, limiters, and a D-type flip-flop detector. When integrated on an early bipolar chip, it drains 2.5 mA from 1.8 V when active, although in standby the current drain falls to a mere 50 μA. Large-value off-chip capacitors are used for ac coupling and for the lowpass filters. A second low-frequency digital CMOS IC performs all the user interface functions. Data is encoded by offseting the carrier frequency by ± 4.5 kHz, so capacitive coupling with a corner frequency below 1 kHz blocks out receiver dc offsets and lower frequency flicker noise from corrupting the signal-to-noise ratio. A 10 kHz lowpass filter suppresses high frequency out-of-band noise and adjacent channels. Vance also pointed out that data in a binary-FSK signal downconverted in quadrature to dc may be recovered by a simple flip-flop, when the limited output from one arm of the downconversion mixer is applied to the D-input, and from the other arm to the Clock-input. Although exceptionally simple, this flip-flop detector makes instantaneous decisions and therefore has poor immunity to a single noise spike. More sophisticated detectors must be used to get an acceptably low bit-error

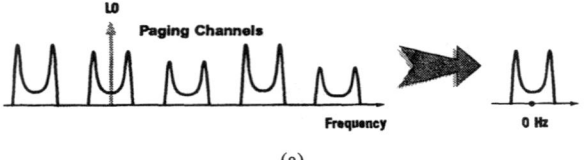

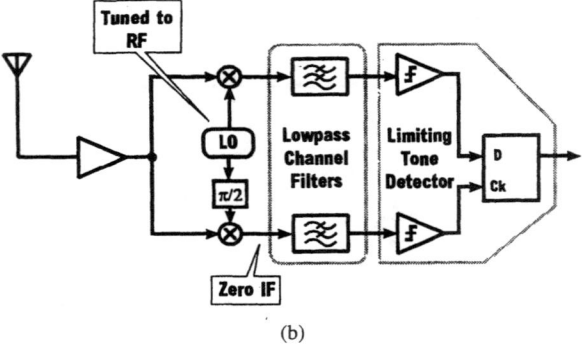

Fig. 8. A direct-conversion single-chip receiver for FSK demodulation. (a) There is no image channel here, and all channel-selection filtering is on-chip at baseband. (b) Note that the receiver requires no AGC, only limiters, as all information is contained in the zero crossings. The signal path must be capacitively coupled to suppress the undesirable dc offsets in the receiver electronics. Quadrature paths discriminate positive and negative frequency offsets.

rates at typical received SNR's. For instance, the inputs of two flip-flops may be cross coupled to the quadrature channel outputs, and the decision may be derived from the analog average of the two outputs [37]. The optimum binary-FSK detector in the presence of Gaussian noise correlates the downconverted signal with the two possible offset frequencies, integrates the output, and declares a valid bit when one of the integrator outputs crosses a threshold [39]. Even the most complex detector will dissipate a small power, because it operates at the low baseband data rate.

Pager IC's from Philips originally used a frequency-offset receiver principle [40], [41], whereby the local oscillator frequency is adaptively offset from the received carrier by 2 kHz, thus converting the FSK tones to 2.5 kHz and (aliased to) 6.5 kHz. This avoids a quadrature downconversion to differentiate between positive and negative frequency, but it requires a fairly sophisticated automatic frequency control. A frequency discriminator detects data. In addition to the local oscillator crystal (operated here at its fifth-overtone), this chip requires three off-chip tuned circuits. Philips later recognized the simplicity of a zero-IF receiver for this application [42]. In this single-chip receiver, the RF amplifier requires one off-chip inductor, the quadrature phase-shift circuits in the mixer another; both inductors are combined into one off-chip signal path. The channel-select filters are entirely on-chip. They consist of a third-order active *RC* lowpass filter, followed by a 7th-order gyrator-based lowpass filter with a 15 kHz cutoff. Most of the signal amplification occurs at these low frequencies.

NEC's paging receivers evolved from the superheterodyne [35], [43] to direct-conversion [44] in an effort to reduce receiver volume and parts count. Others have developed similar zero-IF bipolar integrated front-ends [45], [46]. A notable feature of the chips is that much of the die area is taken up by the capacitors for ac coupling the signal path and for the on-chip lowpass filters. Most pager IC's operate at supplies of 2 V to as low as 1V, and are implemented in silicon bipolar technology. In addition, full-featured pagers require 20 000 gate-equivalent digital IC's for the user interface [47], low-voltage EEPROMs for customization and software, and capability to drive a liquid-crystal display. The basic paging receivers can fit within a wristwatch [48]. The direct-conversion FSK digital paging receiver concept has also been successfully used at very low carrier frequencies (100's of kHz) with much lower data rates in implanted devices for biomedical applications [49].

V. IC'S IN CELLULAR TELEPHONE TRANSCEIVERS

Mobile and handheld cellular telephones are the first widespread two-way radios for consumer use. They were preceded by cordless telephones for local-area use. These wireless telephones must meet stringent demands for low weight and volume, long battery life, low cost, and reliable network access to be successful with consumers. In contrast, walkie-talkie transceivers were always aimed at specialized markets, and did not face these pressures for miniaturization. The average consumer, for instance, will not voluntarily accept a transceiver of the size and weight that policemen or soldiers carry as part of their outfit. Further, to support large numbers of users in a crowded radio spectrum, wireless telephones use more internal signal processing than other common transceivers, and must be capable of connecting to the public switched-telephone network [50]. Features such as digitally selected channels, direct-sequence spread spectrum, and diversity-selection are now becoming common. The transceivers perforce must use highly integrated, low-power electronics. Thus it may be said that with the advent of the modern cellular telephone the conventions of wireless design are being reexamined, and sometimes rewritten.

The first generation of cellular telephones carried voice signals by analog frequency modulation of a carrier. In the US AMPS system, for example, the handset receives at a carrier selected from the 869–894 MHz band, while it transmits on a carrier in the 824–849 MHz band. These 25 MHz wide bands are separated by 45 MHz between the uplink and downlink, enabling the user to talk and listen at the same time much as on the wired telephone (users are not too fond of the "over, over-and-out" protocol). An antenna duplexer suppresses coupling from the transmitter into the sensitive receiver, acting as the equivalent of a two-to-four wire hybrid transformer in a telephone. This duplexer is a passive three-port designed to pass energy in the receive frequency band from the antenna port to the receive port, while attenuating energy in the transmit band from the transmitter into the receiver. It is either made with high-dielectric ceramic resonators [51], or with a SAW filter and coaxial resonator in parallel [52]. The receive portion of a conventional handset resembles a broadcast

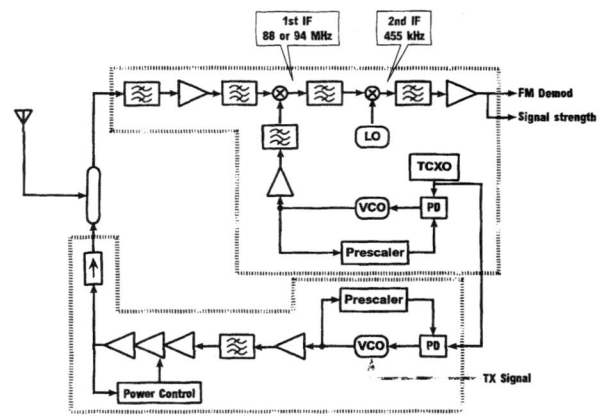

Fig. 9. The transmit and receive front-end of an analog cellular telephone. Receiver uses conventional double-superheterodyne architecture, while transmitter is one-step upconversion.

FM receiver (Fig. 9). If the local oscillator at the first mixer lies at a higher frequency than 894 MHz, the image is guaranteed to lie outside the AMPS band. Furthermore, an LO offset of more than 45 MHz ensures that the image is attenuated by the receive-band SAW filter, suffering at least 20 dB loss in each of the two filters in the handset [53]. A first IF of 90 MHz is therefore often used; this also avoids problems caused outside the handset by parasitic LO leakage through the antenna [54]. The desired channel is selected by locally synthesizing the first IF. This is followed by downconversion to a fixed second IF of 455 kHz, then demodulation by a frequency discriminator.

The first-generation of small-scale IC's for portable communication devices offered building-blocks for the intermediate-frequency chain, such as the mixer and local oscillator for conventional single or double-superheterodyne receivers [55], the IF amplifier chain and signal-strength indicator [55], or a standalone image-reject mixer [56]. Today, almost every major semiconductor company with an interest in the communications market offers building-block IC's at this scale of integration.

Although IC's entered the IF portions, the RF front-end circuits continued to be made from discrete components. An RF amplifier and a first mixer may be mounted with the associated filters on a dense miniature board [57], [58]. Discrete bipolar transistors responded to needs for portable RF applications, by offering, for instance, a low noise figure and f_T exceeding 5 GHz at less than 1 mA bias currents [59], [60]. Manufacturers of passive components, too, have steadily scaled-down their package sizes for high-density board mounting.

There is little argument, though, that the RF front-end components must also be integrated to reduce power dissipation. In competing with passive solutions, the RF-IC's must cross some important thresholds of low-price and high-performance [61]. However, they offer the prospect of an order-of-magnitude reduction in physical volume of the front-end electronics, and power savings will accrue by routing RF signals at a high impedance on-chip, while eliminating the low characteristic impedance interconnects

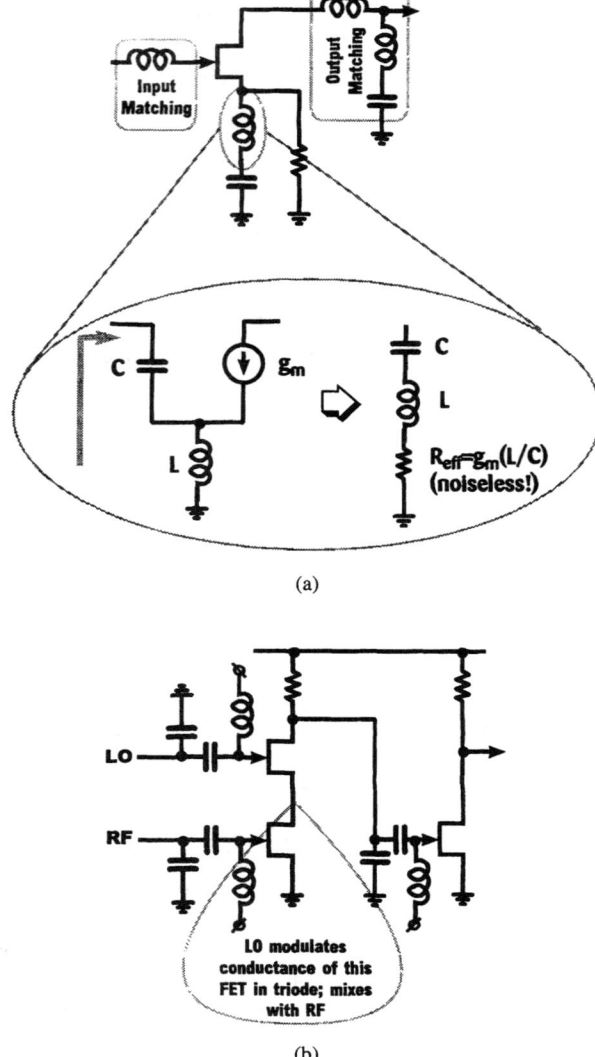

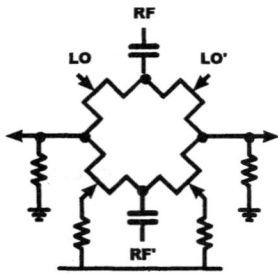

Fig. 11. A four-FET commutating switch mixer. The balanced LO signal switches the FET's, which chop the RF signal to produce the sum and difference frequencies.

Fig. 10. Typical standalone GaAs MMIC's implementing (a) Low noise amplifier and (b) downconversion mixer. Concept of series-feedback inductor for low-noise matching illustrated.

between discrete packages. Various reasons have been advanced for why GaAs MMIC technology is now the right choice for cellular telephones [62]–[64]. They are summarized as follows: first, that owing to the semi-insulating nature of the GaAs substrate, reasonable-size inductors may be integrated with transistors to make high-frequency monolithic tuned circuits, which allows for lower current operation at a given frequency than would be possible with RC broadbanding techniques; and, second, that MESFET's afford lower noise figures at a given bias current than a bipolar transistor in a comparable silicon bipolar technology. Most GaAs MMIC's integrate front-end components for cellular applications at a small scale. For instance, a chip may integrate a tuned RF low-noise amplifier in the 900 MHz band, or a mixer and a local oscillator [64]–[72].

The typical RF amplifier may consist of only one or two FET's, with LC matching circuits on the input and the output ports for standalone operation in a 50 Ω environment (Fig. 10(a)). A powerful and popular method to match the capacitive FET input is to insert a series feedback inductor, L, in the FET source, which at high frequencies contributes a *resistance* $g_m L/C_{GS}$ at the input port [73]. This method is preferred to resistive feedback found in wideband amplifiers for impedance matching, because unlike feedback resistors, the inductor does not degrade the noise figure. MMIC's from Matsushita favour the use of dual-gate MESFET's with RC matching instead of inductors [74], possibly because spiral inductors consume too large a chip area. GaAs IC designers must closely watch their chip real-estate to remain competitive in price. The various low-noise amplifiers operate in the 0.9–2 GHz bands, with gains of 15–20 dB, and noise figures of around 3 dB. These submicron MESFET IC's drain anywhere from 3–5 mA.

The received RF is typically downconverted by the local oscillator modulating the conductance of the mixer FET through a cascode FET (Fig. 10(b)). The mixers yield conversion gains greater than 10 dB, noise figures of 10–12 dB, and third-order input-referred intercept points of -5 to 0 dBm. Dual-gate MESFET's are naturally suited for mixer use, and offer a similar performance [74]. A recent 0.7 μm GaAs MMIC offers an LNA-mixer pair draining 3 mA from 3 V, with the LNA producing 13 dB gain, 3.6 dB noise figure, and an input-referred IP3 of -11 dBm [75].

Slicing across the system a little differently, another MMIC implements the downconversion and upconversion mixers for the receiver and transmitter, respectively, integrating their shared local oscillator on the same substrate [76]. The four-FET switch mixer (Fig. 11) is very linear, but it suffers from two disadvantages: it requires a large local oscillator drive to turn the switches on and off [77], and unlike the bipolar transistor Gilbert-cell analog multiplier, the switch mixer is lossy, requiring additional signal amplification from the following stages.

The current generation of analog cellular telephones transmits power levels of more than 1W (30 dBm). RF power amplifiers are usually packaged in a separate module with some form of integral heat-sink. A preamplifier, or in radio terminology, exciter, in the transmitter section boosts the modulated carrier level close to 0 dBm to drive the power amplifier input. The power amplifier module itself usually consists of a cascade of two or three FET's, tapering up to a single large-size FET which will deliver the required signal current into the antenna load. Furthermore, as the power amplifier is the largest single source of battery drain, it must have a *high conversion-efficiency*. Much as in baseband power amplifiers, an efficient RF power amplifier is biased close to cutoff to reduce the dc standing current, and then driven by the input signal in Class A-B or Class-B mode. Narrowband filters at the amplifier output remove harmonic distortion caused by nonlinear operation at RF [78]. These filters may be merged into the passive matching networks required for optimum power transfer from the amplifier to the load. As designers of RF power amplifiers have observed [10], [79], [80], much of their work consists of synthesis and iteration of the interstage, input, and output *matching networks*. It is easier to design efficient power amplifiers for constant-envelope modulations, such as analog FM or digital FSK, where distortion may be tolerated because the useful information is all contained in the zero crossings. Further, it is argued that owing to the lower parasitic capacitance of a GaAs MESFET relative to silicon devices, a GaAs power amplifier at 1W power levels affords a higher efficiency (~60%) compared to silicon bipolars or FET's (~ 45%) [47]. A monolithic power amplifier consisting of a four-stage cascade of MESFET's with on-chip lumped LCR input and inter-stage matching networks, delivers 1W at 900 MHz at 63% efficiency from a 5.5 V supply [80]. The distributed element output matching network at 900 MHz would be exhorbitantly large on an IC, and is therefore printed on an off-chip alumina substrate. The high efficiency is attributed to an improved method of suppressing the 2nd harmonic at the amplifier output. This multi-component module approach to mate IC's with matching networks is widely used in GaAs power amplifiers [81]. Another 900 MHz power module uses two discrete MESFET's, wirebonded to a hybrid IC containing a combination of distributed circuits and chip capacitors and resistors in the matching network, to attain 65% efficiency when delivering over 1.3 W from a 4.7 V supply [82]. The output FET is 12 mm wide, with a 1 μm channel length. A separate negative supply is required in many of these MESFET power amplifiers to bias the gate. The circuit was recently modified [83] to produce the same output power equally efficiently, now with a single 3.5 V supply and FET's of 0.6-μm channel length. The desired low-voltage operation, much more suitable for one battery, is the result of an improved device structure. The matching networks in the RF signal path are fabricated on separate passive-only GaAs IC's, which the authors claim are three times cheaper than GaAs IC's with FET's, while the bias networks reside on a miniature PC board. The various components are wirebonded to one another, and mounted on an AlN substrate with ten times higher thermal conductivity than alumina. Power amplifiers may also be built with silicon NMOSFET's: among the major semiconductor vendors, Hitachi has pursued this option. A MOSFET gracefully accepts large voltage swings on the gate, without possibility of Schottky conduction as in a MESFET gate. With an offset-gate FET structure and 0.8-μm channel length, a MOSFET power amp delivers 2 W

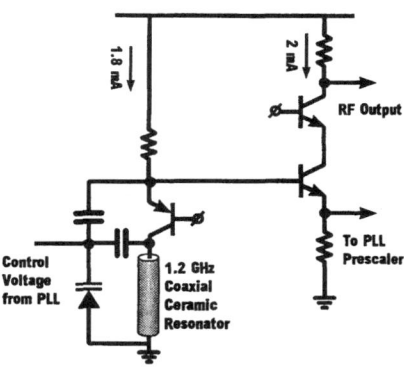

Fig. 12. Colpitts oscillator with coaxial resonator. Varactor controls frequency with control voltage. Series-connected buffer distributes RF oscillation to mixer and to PLL prescaler.

at 1.5 GHz with 55% efficiency from a single 6-V supply [84]. The power amplifier module embeds these FET's into matching networks [85].

The foregoing discussion is not meant to imply that cellular telephones mainly use GaAs MMIC's in the RF front-end. Silicon bipolar and BiCMOS technologies have made tremendous strides in improving f_T, and today silicon MMIC's offer comparable performance to what is available in GaAs. A recent 1 GHz BiCMOS LNA-mixer combination [86] uses familiar circuits to baseband designers, such as a Gilbert-multiplier type mixer, and the chip yields a comparable performance to the MMIC's described above. The RF signal path consists of bipolar circuits only, while the FET's are used as switches to select various power-down modes. There are no on-chip inductors to tune the low-noise amplifier, which is wideband; instead, the LNA output is routed off-chip into a passive bandpass filter, then returned to the chip for downconversion. This is part of a complete chipset from Philips for a digital cellular telephone handset [87].

The first LO in a cellular telephone receiver is programmed to the incoming RF channel with a phase-locked loop synthesizer, while the second LO at IF is fixed in frequency. The transmit LO oscillates in yet another frequency band. Both the receiver and the transmitter therefore require 900 MHz voltage-controlled oscillators. These are most often implemented with bipolar transistors, whose very small flicker noise means low phase-noise sidebands in the VCO. A stripline resonator sets the nominal frequency of a Colpitts oscillator, and this is voltage-controlled by a varactor diode in parallel [54], [57], [88]. Phase noise levels of -110 to -120 dBc/Hz are attained at a 50 kHz offset from the oscillation frequency. The VCO application has created a brisk demand for varactors with large voltage-coefficient and low-loss [60]. The VCO and its buffer may be connected in series to reuse the bias current between the two stages [57], [88] (Fig. 12). The nominal oscillation frequency may be slaved in a frequency multiplying PLL to a 12 or 15 MHz crystal oscillator.

The Motorola MicroTac, first introduced in 1989, set an industry standard for a miniature cellular hand set. Small

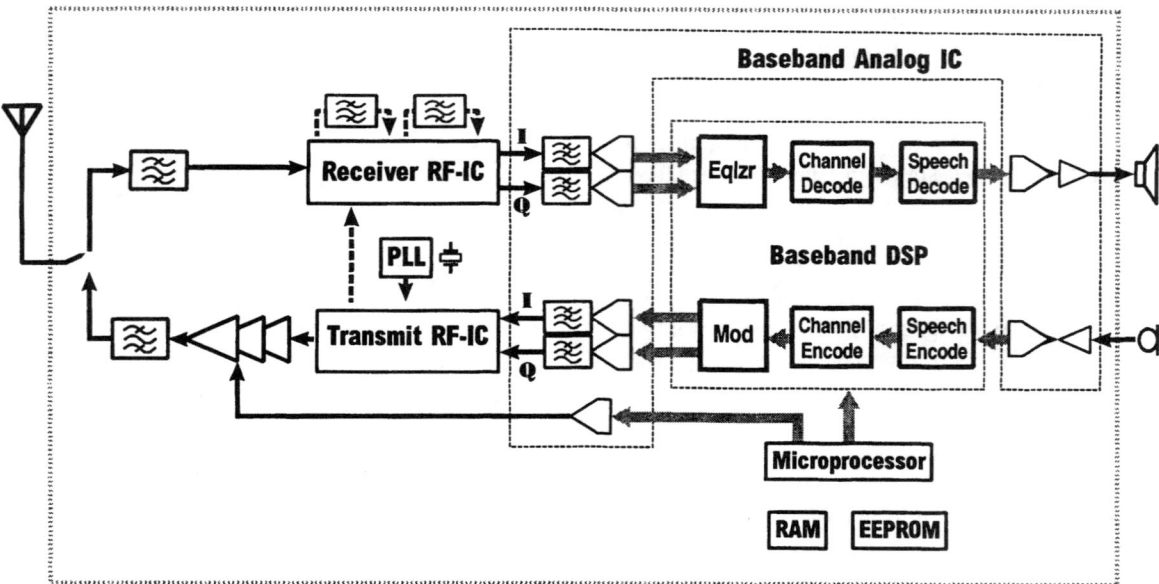

Fig. 13. Block diagram of digital cellular telephone. Chip partitions reflects current state-of-the-art technology.

handsets [89] continue to use conventional architecture, but attain a small size with discrete filters in miniature packages [52], [53], [58], [90], [91], and with lower power electronics which reduces battery weight. The electronics entail carefully designed standby modes, better software for the on-board microprocessors, and, among other items, lower power PLL-prescalers, digital counters which must operate all the time for frequency synthesis and which often posed a significant power drain [92], [93]. In the transmit mode, greater efficiency is sought in the power amplifier to prolong battery life, as well as smart control circuits to maintain maximum efficiency across the range of output power levels [89]. It was generally agreed that by 1992 the handset volume should scale down to less than 150 cc, and that its weight should be less than 230 gm [20], [94]. NTT demonstrated the prototype of such a telephone in 1991 [95], while noting that the volume should be no smaller than this target both for the sake of ergonomics, and to ensure adequate heat removal from the package so that the internal temperature rises no more than 15°C. These small telephones might be powered by a 6-V NiCd battery with 400 mA · h capacity.

With the emergence of the IS-54 standard, there is some convergence in the US on a digital cellular technology. Upward compatibility is sought with the existing analog cellular bands, which is why handsets conforming to IS-54 are referred to as dual-mode cellular. A recent 12 GHz f_T silicon bipolar transceiver IC from AT&T uses a conventional double-superheterodyne architecture with an 80 MHz first-IF, and after IF amplification the signal path bifurcates into a conventional 455 kHz second-IF FM demodulator, or a digital I-Q PSK detector [96]. An interesting alternative is to downconvert the second-IF, without preceding AGC, in a delta-sigma A/D converter for subsequent digital baseband signal processing [97].

VI. Digital Cellular and Cordless Telephones

Analog cellular telephones use the frequency spectrum inefficiently. The modulation schemes consume a large bandwidth, and every cellular telephone transmits at constant power all the time it is in use, thereby appearing as an interferer to other users at nearby frequencies. As a result, the spectrum allotted to cellular phones in large metropolitan areas nears exhaustion.

Digital cellular telephones are one solution to better utilize the scarce spectrum. They use more efficient modulation schemes, such as minimum frequency-shift keying or phase-shift keying, and multiple users may share timeslots on the same part of the spectrum. The complex modulation formats used in these telephones and the greater capabilities required to withstand nearby blocking signals are prompting large-scale integration of the RF and IF electronics. Examples of digital telephony standards are the European GSM, North American Digital Cellular, and the emerging Japanese personal handy phone (PHP) [98]. The salient chareteritics of these various systems, as well as key digital cordless standards, DECT and CT-2, are summarized in Table 1.

The typical handset involves an RF/IF front-end, followed by a baseband digital signal processor (Fig. 13). The first significant set of RF-IC's for GSM handsets appeared in 1990. Notable among them is a receiver and transmitter 7.5 GHz f_T silicon bipolar chip-set from Siemens (Fig. 14) [99], [100]. The double-superheterodyne receiver uses a selectable first-IF of anywhere from 45–90 MHz, which places the image channel in the stopband of the 25 MHz-wide RF preselect bandpass filter. A 71 MHz IF, for instance, guarantees that the image lies outside the GSM band. The desired 200 kHz channel is selected by a SAW filter at the first IF. The receiver IC provides an on-chip

Table 1

	GSM Europe	NADC North America	J-PHP Japan	CT-2 Europe, Asia	DECT Europe
Downlink Frequency Band	935–960 MHz	869–894 MHz	1.9 GHz	864–868 MHz	1.88–1.9 GHz
Uplink Frequency Band	890–915 MHz	824–849 MHz	1.9 GHz	864–868 MHz	1.88–1.9 GHz
Multiple Access Method	TDMA	TDMA	TDMA/TDD	FDMA/TDD	TDMA/TDD
Modulation	GMSK	$\pi/4$-DQPSK	$\pi/4$-QPSK	B-FSK	GMSK
Speech data rate	13 kb/s	8 kb/s	32 kb/s	32 kb/s	32 kb/s
Handset output power	3.7 mW $\to$ 1W	2.2 mW $\to$ 6 W	10 mW	1 mW $\to$ 10 mW	250 mW
Modulation rate	271 kb/s	49 kb/s	384 kb/s	72 kb/s	32 kb/s
Channel spacing	200 kHz	30 kHz	300 kHz	100 kHz	1.762 MHz
Burst length	156 b	324 b			424 b

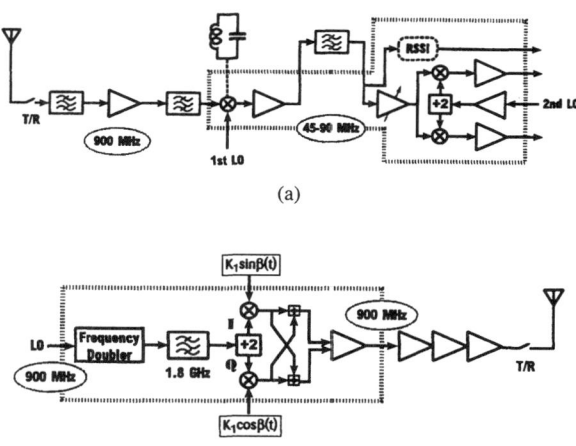

Fig. 14. Siemens receiver and transmitter IC's for GSM handsets. (a) Receiver is double-superheterodyne architecture. Requires one off-chip tuned circuit, SAW resonator at first IF, and low-noise amplifier. (b) Transmitter doubles LO frequency and then divides-by-2 to accurately obtain quadrature phases.

buffer to drive this filter. Following an RF amplifier with an off-chip LC tuned load, the input signal is downconverted by a Gilbert-cell mixer. There is AGC with 70 dB range and a signal-strength indicator at the first IF, and the selected channel is then downconverted to baseband in a quadrature demodulator to acquire both amplitude and phase for GMSK vector demodulation. The on-chip gain may be as large as 80 dB at the highest VGA setting, but as it is distributed in different frequency bands, there is little on-chip crosstalk and the monolithic receiver operates stably. The receiver uses a fully balanced signal path, and an input noise figure of 7 dB. It drains 27 mA from 5 V in active mode, and 10 μA in standby mode.

In the transmitter IC, a precision quadrature upconversion mixer produces a single-sideband, suppressed-carrier QPSK output from a baseband vector input. An RF differential input amplifier selects one sideband from the upconverted outputs in the two arms of the mixer. As the other sideband would occupy a nearby channel's spectrum, it must be adequately suppressed (relative to the level of the wanted sideband). Imperfect cancellation of the unwanted sideband arises from gain mismatch in the I and Q channels of the upconversion mixer, and in deviations from quadrature in the two LO outputs. The goal in this transmitter was quadrature phase-errors of less than 2°, which with adequate gain matching, implies unwanted sideband suppression of at least 40 dB relative to the wanted sideband. The stringent GSM requirements on low LO phase noise are only met with a 900 MHz off-chip coaxial-resonator based, varactor-tuned, Colpitts oscillator [101], which inherently provides a single-phase output. There are several methods to accurately derive quadrature phases. Siemens uses the double-frequency method. An on-chip nonlinear element doubles the oscillator frequency to 1.8 GHz, following which an on-chip bandpass filter removes harmonics. Then, the oscillation is divided by 2 by positive- and negative-edge triggered flip-flops, which will yield quadrature outputs with a phase-accuracy set by how close the duty-cycle of double frequency oscillation is to 50%. Combined with an output stage to drive an off-chip power amplifier module, the IC drains 40 mA. Using a similar superheterodyne architecture with 71 MHz IF, AT&T Microelectronics has recently combined the GSM receiver and transmitter blocks, including the frequency synthesizer, on to one chip [102].

In an effort towards even greater miniaturization at Alcatel, the transmitter and receiver sections are both integrated on to one 9 GHz f_T silicon bipolar chip (Fig. 15) [103], [104]. From the point of view of integration, the main advantage of the zero-IF, direct-conversion architecture is that it eliminates the IF passive high-frequency bandpass filters. The disadvantage, on the other hand, is that the downconverted signal has energy at dc, to which will add the receiver's offsets and low-frequency noise. An off-chip low-noise amplifier drives directly into quadrature (I-Q)mixers. The mixer dynamic range is wide enough so that a large blocking signal only 3 MHz away from the desired signal produces insignificant intermodulation distortion. The bipolar Gilbert-cell mixer is linearized with emitter degeneration resistors, which also degrade its noise figure. The receiver section requires large (650 pF) off-chip capacitors at its output for anti-alias lowpass filters, before the baseband signal is sent to a companion mixed-signal CMOS IC. This baseband CMOS IC contains a high-order switched-capacitor lowpass filter for channel selection, a digital GMSK detector, and various DSP functions [105]. An RC and CR network with off-chip trimming shifts the phase of an external local oscillator by precisely $\pm 45°$ to produce the quadrature drive to the mixers. Upconversion mixers in the transmitter drive a power amplifier module.

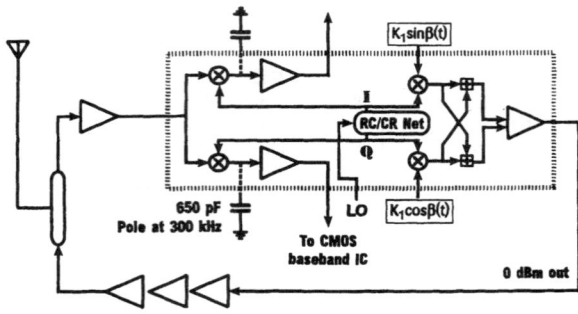

Fig. 15. Alcatel direct-conversion single-chip transceiver. Channel selection done by switched-capacitor lowpass filters in companion CMOS mixed-signal baseband IC. On-chip, trimmed RC-CR network generates quadrature phases for mixers.

The chip drains 25 mA from 5 V in the receive mode, and 45 mA in the transmit mode, rather comparable figures to the previously decribed Siemens chip set. The transmitter suppresses the unwanted sideband and carrier by about 40 dB.

There is increasing interest in the 1.9 GHz band for digital cellular telephony. All the current approaches (Table 1) use time-divison duplexed (TDD) receive and transmit frames, wherein users are assigned different time slots. Two-way communications take place over the same frequency band. A host of IC's is appearing to serve the European DECT standard. The majority cater to a superheterodyne receive-architecture with 110 MHz IF [106]–[109]; some of the literature [107], [109] also describes procedures for system design. The RF sections are usually integrated at small scales, embodying, say, the low-noise amplifier and first mixer on one chip, the quadrature modulator on another, and the exciter and power amplifier on a third and fourth chip. RF-IC's operating at 1.9–2.5 GHz exist in both GaAs [68], [110]–[112] and silicon [106], [113]–[115] technologies. Some of these standalone GaAs IC low-noise amplifiers achieve impressive gain and noise figure with 1–2 mA current drain from 3 V [69], [116]. At the system level, though, functionality and overall dynamic range in these short-haul wireless links takes precedence over raw component performance. Thus system-level input noise figures of 10–15 dB are acceptable [109], as are input-referred intercept points of -16 dBm, and a 20 dB power amplifier control range [112]. The less than stringent system specifications lead to simplifications in the transmit path, such as direct VCO-modulation by the baseband signal by opening the transmit PLL over the duration of the transmit-frame [107].

The power amplifier continues to be built almost exclusively as a separate GaAs IC, and at 1.9 GHz the on-chip wavelength is short enough that it is now possible to integrate distributed matching networks [68], [117], [118]. Efficiencies of 50% are attained when delivering almost 1 W to the load from 3 V. Either the transmitter or the receiver is active in TDD digital transceivers, so fast-switching power-down modes are designed into the various components. In particular, the switching trajectory must be shaped to suppress spurious emissions when the power amplifier is switched on and off [85]. Finally, the transmit and receive signals are directed to the antenna via a three-port passive circulator, or preferably through a low-loss, monolithic transmit/receive (T/R) RF switch [112], [119], [120]. A good FET switch must not appreciably distort the RF signal, or incur more than 1-dB insertion loss when ON, yet when OFF it should offer at least 30 dB isolation. Microcell applications, such as the Japanese handy phone, require an average output power of only 10 mW (20 dBm), although the $\pi/4$-QPSK modulation requires the power amplifier to handle larger peak powers. An exciter amplifier, power amplifier, and T/R switch have been integrated together [121]. The exciter and power amplifier attain a 44% efficiency at 23 dBm output from a 4.8-V supply, and the on-chip matching network uses LC lumped elements. Although matching networks will eliminate harmonic distortion, they are ineffective in suppressing near-carrier spurii produced by intermodulation distortion. Feedback linearization techniques have been proposed which predistort the digitally synthesized exciter input waveform to anticipate the power-amplifier nonlinearities [122]. Use of these techniques makes a highly nonlinear but efficient power amplifier appear linear. These remain at the experimental stage today, and with the move to microcells and low emitted power levels, they may soon not be necessary in handsets.

The building-block IC's described so far are important advances in realizing small, relatively low-power transceivers, but the ultimate goal remains to integrate the entire transceiver on to a single-chip. To this end, some early breadboard-level experiments show that direct-conversion receiver architectures seem well suited to the DECT application [123], [124]. A 16 GHz f_T silicon bipolar IC from Alcatel operating at 1.9 GHz contains a complete direct-conversion DECT transceiver [125]. The architecture resembles the previously described Alcatel direct-conversion GSM transceiver [103], requiring a separate low-noise amplifier in the receive path, a power amplifier module, and a baseband CMOS mixed-signal signal processor IC. From a 5-V supply, the transceiver drains 50 mA in receive mode and 80 mA in transmit mode. One of the challenges in realizing this system was a fast-switching PLL frequency synthesizer with low phase-noise and low spurious output levels, which was built here with an improved charge-pump and loop filter.

Siemens has extended its dual-conversion GSM transmitter/receiver chip set [99], [126] to a generalized front-end for digital cellular telephones operating in any part of the RF spectrum from 800 MHz to 2.1 GHz [127]. The 25 GHz f_T silicon bipolar chip set ope rates at a supply as low as 2.7-V, the transmitter chip draining 60 mA and the receiver 33 mA. Both have power-down modes. An external power amplifier module is required. A notable low-voltage circuit on this chip is a modified Gilbert-cell mixer, with resistors instead of current sources to the negative supply [31]. To accommodate a wide input dynamic range, the gain of the RF low-noise amplifier in the receiver may be switched from -5 dBm to $+15$ dBm, a powerful technique that has also been used in paging receivers [46]. This assumes that the recever does not instantaneously require a very wide dynamic range, but either receives mostly strong signals or mostly weak ones. The downconversion mixer, with a 13 dB noise figure and -3 dBm input IP3, is followed by an IF variable-gain amplifier with an 80 dB digitally programmable range [126]. The on-chip RF oscillator requires an off-chip resonator and varactor, and its phase noise at a 2 kHz offset is -88 dBc/Hz. A balanced signal path is used throughout. One good measure of the quality of on-chip gain matching in the I and Q paths in the direct upconverter, as well as of the quadrature accuracy of the upconversion clocks, is given by the relative levels of spurious emissions from the transmitter IC. The unwanted sideband is suppressed by 48 dB, while the RF carrier tone, emitted due to dc offsets in the two paths, is 37 dB down. A 3rd-order modulation tone at -46 dB appears due to mixer nonlinearity. The absolute output noise level at a 25 MHz offset is -141 dBm/Hz.

VII. Direct-Conversion Transceivers and Their Problems

A receiver with zero-IF is called a *direct-conversion* receiver. When the local oscillator is synchronized in phase with the incoming carrier frequency, this is also referred to as the *homodyne*. This architecture is sufficiently promising for single-chip transceivers to warrant a separate section to its study.

The desired channel is translated by the first mixer to a 0 Hz center frequency, and instead of adjacent channel rejection with a bandpass resonator, a more flexible and easier to implement lowpass filter is required—in effect, a bandpass filter centered at dc when the negative frequency axis is included. With zero-IF, there is *no* image frequency. As early as 1924, radio pioneers had considered homodyne architectures for crude receivers requiring only a single vacuum-tube, but it was in 1947 that a homodyne was first used to full effect, with a high-order lowpass filter for channel-selection, in a measuring instrument for carrier-based telephony [32].

A. Direct-Conversion Single-Sideband Synthesizers

For reasons of spectral efficiency, the transmitted signal in digital communications is always single-sideband with suppressed carrier. This is most often produced with the so-called *phasing method* [128]. The modulated signal is first synthesized in quadrature at baseband, *directly upconverted* into two paths by a quadrature LO centered at the carrier frequency, and added or subtracted to select either the upper or lower sideband (Fig. 16(a)).

The unwanted side band is suppressed to an extent limited by the gain mismatch in the two upconversion paths, and by departures from quadrature in the two LO outputs (Fig. 16(b)). Unequal dc offsets in the paths produce an output signal at the LO frequency. The unwanted sideband and LO leakage are spurious but unavoidable components of the transmitted spectrum. Although on the same IC the

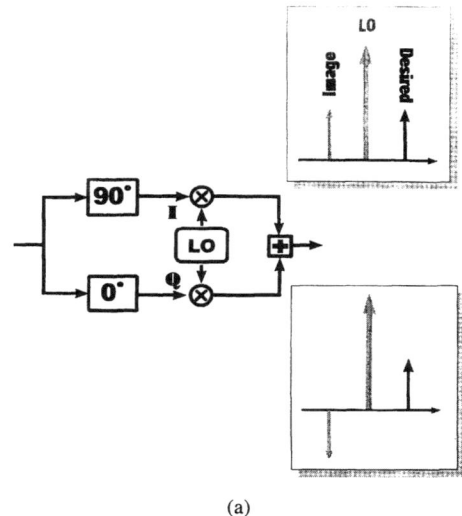

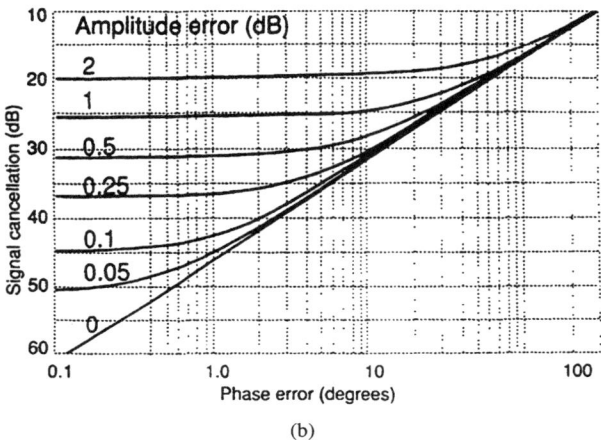

Fig. 16. (a) Single-sideband direct upconverter using the phasing method (also called the quadrature modulator). Extent of suppression of unwanted (image) sideband versus gain mismatch and phase-errors from quadrature in two arms of modulator.

two upconversion paths are well matched, a gain mismatch as small as 1% (0.1 dB) will limit suppression of the unwanted sideband to 45 dB. With this gain mismatch, a phase-error of up to 1° may be tolerated between the two LO outputs. These mismatches may be trimmed down at time of transceiver manufacture, or self-calibrated with loopback modes which are activated during idle times to sense and suppress the unwanted spurii.

An off-chip resonator, often connected to an on-chip unbalanced oscillator circuit, may also become a cause of spurious RF leakage if it couples energy into the power amplifier or the antenna. Frequency-offset upconversion schemes have been proposed [129] to combat this coupling problem. Other spurious output tones may arise from parasitic remixing of the modulated output with the baseband signal, and by intermodulation distortion in the output stage [96]. Balanced circuit topologies, on-chip LO's requiring no external resonators [130], and lowered transmit power levels in microcells, are all expected to lessen the magnitude of the spurious leakage problem.

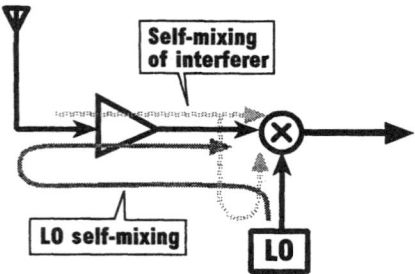

Fig. 17. Sources of leakage in a direct-conversion receiver, and how through self-mixing they create dc offsets.

Cellular wireless systems operate best with tightly regulated power control, the handsets transmitting only the minimum power required for reliable reception by the base station, and *vice-versa*. Whereas in the receiver the front-end is the main source of noise, in a transmitter the upconversion LO phase-noise appears as added noise on the emitted signal. Direct upconversion has the advantage over two-step schemes that only *one* LO contributes noise.

B. Direct-Conversion Receivers for GSM Digital Cellular Telephones

Among the various RF-IC suppliers for the European GSM digital cellular telephone handset, only Alcatel at present uses a direct-conversion receiver architecture [103], [125], [131]. This results in a relatively small silicon bipolar RF front-end chip, and the remainder of the signal processing, including lowpass channel-select filtering, is at baseband in mixed-mode CMOS. Why are the others reluctant to use direct-conversion in their receivers? Given the many decades of superheterodyne experience, the most likely reason is conservatism. But direct-conversion also suffers from some unique problems.

A well known problem is that spurious LO leakage from the receiver into the antenna becomes an in-band interferer to other nearby receivers tuned to the same band. Superheterodynes, with their frequency-offset LOs, do not suffer from this problem. However, experimental studies suggest that with standard shielding in the receiver, this problem is not so severe as to handicap the use of direct-conversion [132].

A more serious problem is dc *offset* in the receiver. Offset arises from three sources [131]: transistor mismatch in the signal path; the LO signal leaking to the antenna because of poor reverse isolation through the mixer and RF amplifier, then reflecting off the antenna and self-downconverting to dc through the mixer (Fig. 17); and a large near-channel interferer leaking into the LO port of the mixer, then self-downconverting to dc. Good circuit design may reduce these effects to a certain extent, but they cannot be eliminated.

The spectrum of the GMSK modulation used in GSM has a peak at dc. Offsets will directly add to the spectral peak of the downconverted signal. These offsets are usually much larger than the rms front-end noise, and may therefore significantly degrade the SNR at the detector.

To remove the offsets by ac-coupling the receiver will require impractically large capacitors, if the signal-bearing spectrum around dc is not to be sacrificed. However, they may be compensated with DSP-based self-calibration [105], [133]. The baseband signal processor makes long-term measurements on the dc level in the receiver, and subtracts off this level from the downconverted signal. The spectrum loss around dc is only a few hertz, and digital filtering does not distort the midchannel group delay in the receiver.

C. Local Oscillators with Quadrature Outputs

A *carrier-frequency* local oscillator with quadrature outputs is a key circuit component in direct-conversion transmitters and receivers. Usually this oscillator is tuned by an LC circuit or an off-chip resonator, and inherently produces a single-phase output. Quadrature phases are often derived by passing the oscillator output through a CR and an RC network, whose time constant is equal to the oscillation period. The two resultant outputs are then phase-shifted by $+45°$ and $-45°$, respectively. Inaccuracies in the actual values of R and C will lead to errors in quadrature, and are compensated by some form of on-chip trimming [103], [125].

An alternative method is to divide an oscillation at twice the carrier frequency with positive- and negative-edge triggered flip-flops [99]. The resulting two outputs at the desired frequency are in quadrature. Residual phase errors caused by unequal delays in the two flip-flops, and by departures from 50% duty cycle in the double-frequency oscillation, are found to be less than 1°. However, the double-frequency portions of the circuit may become a speed bottleneck.

Quadrature outputs may also be derived from a polyphase oscillator, such as a variable-frequency ring oscillator. In a four-delay stage ring oscillator, taps at diametrically opposite points will yield quadrature phases at all frequencies [134] (Fig. 18). In oscillators with an odd-number of unit delays, they may be synthesized from two taps by a voltage-controlled phase-shifter in feedback around a quadrature-sensing circuit [135]. Mismatches in transistor characteristics limit the attainable phase accuracy. This type of resonator-less oscillator is practical when the specifications on phase-noise are not too stringent (say up to -80 dBc/Hz at a 100 kHz offset from the carrier). The free-running oscillator must be slaved to a crystal reference in a frequency-locked loop, whose loop gain and bandwidth are specifically designed to suppress phase noise.

When circuit techniques cannot contribute further improvements to gain- and phase-matching in the two arms of a quadrature modulator or demodulator, digital adaptive algorithms may be used at baseband to sense and compensate for these errors [136].

VIII. IC'S FOR SPREAD-SPECTRUM WIRELESS TRANSCEIVERS

Spread-spectrum communications offer greater user capacity than narrowband techniques in a given piece of

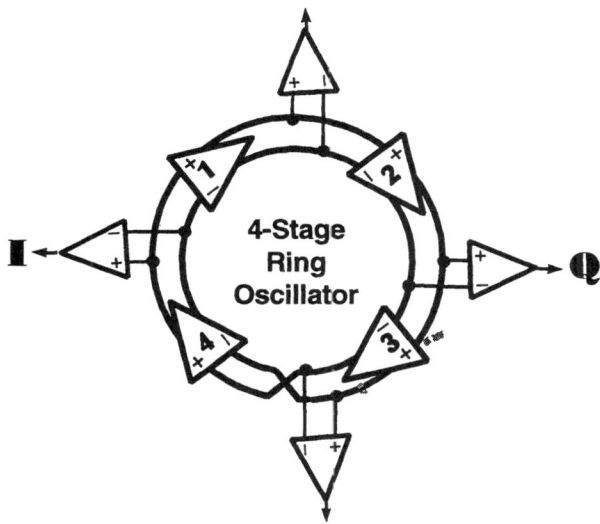

Fig. 18. A four-phase voltage-controlled ring oscillator. Diametrically opposite taps give quadrature phase at any frequency. Unequal loadings and device mismatch set phase errors. VCO must be embedded in PLL whose closed-loop bandwidth is tailored to suppress oscillator noise.

wireless spectrum. Much has been written on this subject [137]. Spread-spectrum techniques were developed during World War II as a form of secure communication with low probability-of-intercept and resilience to jamming [138]. Over the years, this technology has been further developed and refined for military communications. With the wireless revolution at hand and the IC technology now available to implement complex transceivers, spread-spectrum has awakened commercial interest [139]. Spread-spectrum communications make it easy for users to access the wireless channel. Whereas conventional narrowband wireless communications require a careful discipline, which the FCC or other government agencies enforce by issuing licenses so as to prevent use of the same frequency by nearby operators, for spread-spectrum use the FCC has allocated certain *unlicensed* bands in the US, referred to as the Instrumentation, Scientific, and Medical (ISM) bands, wherein the user is only required to spread spectrum by a minimum amount, and not to exceed an upper limit on transmitted power. The extent of spectrum spreading may be measured by the amount of *processing gain* in the receiver required to de-spread and detect the signal [140].

As analog cellular telephones begin to saturate the available radio frequency allocations, spread-spectrum techniques are being deployed in cordless telephones and wireless modems [141]. These devices usually operate in the two lower ISM bands: 902–928 MHz and 2.4–2.48 GHz. A ubiquitous wireless environment is envisioned, in which mobile users, wherever they may be, can access data and communications services through an intricate network of base stations [2]. The greatest hardware challenge in realizing this scenario is the development of a low-power, miniature handset.

There are two different methods to spread the spectrum of a signal: by *direct-sequence* modulation, or by *frequency*

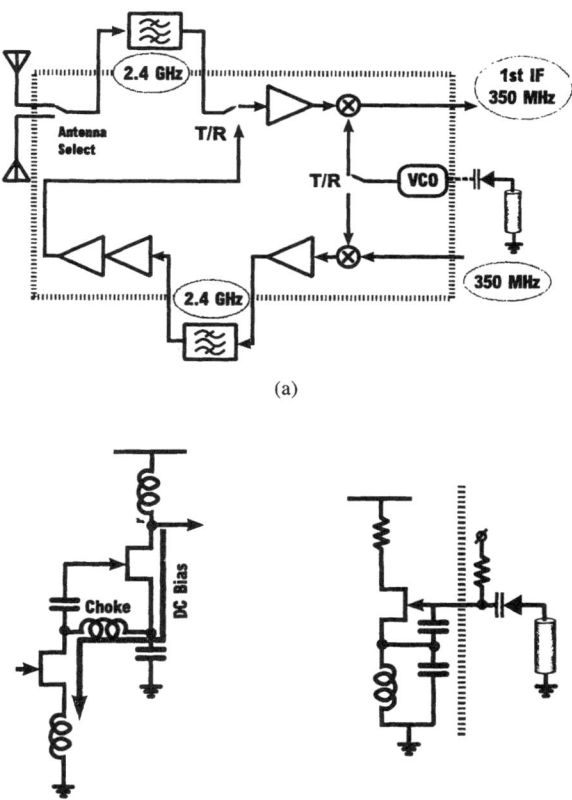

Fig. 19. Front-end GaAs MMIC transceiver for spread spectrum communications in 2.4 GHz band. Receiver is double-superhetorodyne. Note use of off-chip filters. (b) Low noise amplifier in transceiver IC (left) reuses bias current in two stages. Clapp voltage-controlled local oscillator (right) combines on-chip LC network with off-chip resonator and varactor.

hopping [140]. Direct sequence is conceptually the simpler, as well as the more straightforward to implement. Each data bit (either +1 or −1) at the transmitter multiplies a prescribed sequence of bits, or *chips* as they are called. The chip sequence is selected for very low autocorrelation, and is often referred to as a pseudo-noise, or PN, sequence. Each user is either assigned a unique PN sequence with very low cross correlation with other user sequences, or a time-shifted version of some long PN sequence. The receiver, after an initial acquisition search to align itself with the start of the PN sequence, correlates the incoming sequence with the pattern it knows to be its own. On detection of a correlation peak, the sign of the peak signals the source data bit. The longer the PN sequence for each bit, the greater the spreading factor, and the more reliable the detection. Here, the processing gain is the sequence length per bit. The spectrum of the transmitted data, composed from a concatenation of PN sequences, is noise-like; thus the term *spread-spectrum*. Several baseband and IF modem IC's have already appeared to support BPSK and QPSK direct-sequence spread-spectrum transceivers [142]–[144].

Whereas direct-sequence modulation spreads the spectrum by randomizing the waveform, frequency-hopping spreads spectrum by *hopping* the carrier according to a

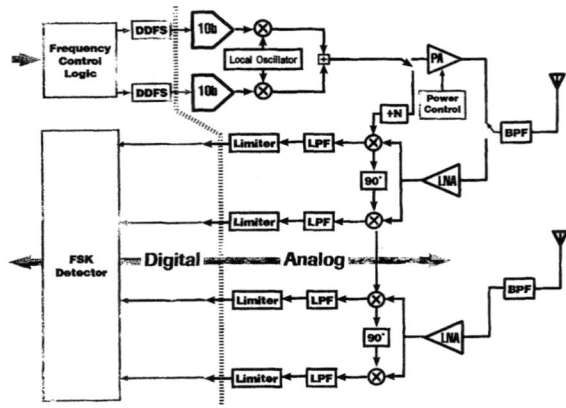

Fig. 20. The UCLA frequency-hopped spread-spectrum transceiver. This uses direct-conversion in both transmit and receive path, and binary-FSK modulation. In the transmitter, the spread-spectrum waveform i synthesized at baseband, and upconverted to RF. During receive mode, the DDFS under control of a synchronization loop hops with the incoming data, and demodulates it to baseband. The entire mixed analog-digital, baseband/RF system is to be integrated on a 1-μm CMOS IC.

prescribed sequence across the entire band. The effect in the frequency-domain is similar to direct-sequence, except for one fundamental difference: with direct-sequence, spreading across a wide bandwidth requires each data bit to be mapped into a long PN sequence. The resultant high "chip-rate" requires high-speed signal processing at the receiver front end. On the other hand, if the carrier frequency is hopped with a wideband synthesizer, the output spectrum may span an arbitrarily wide frequency range, even at low bit rates. Therefore, if a fast and agile frequency synthesizer is readily available, the receiver front-end operates at the actual data rate in frequency-hopped spread-spectrum, rather than at the considerably higher chip-rate in a direct-sequence receiver. The former will very likely lead to a low-power solution.

As a notable recent example of a miniature spread-spectrum transceiver, Plessey has introduced a 700 kb/s frequency-hopped device operating in the 2.4 GHz ISM band. The transceiver is entirely contained on a 2″× 3″ PCMCIA card for insertion into notebook computers [109], [145]. FSK data modulates the carrier frequency, and a variable-modulus PLL synthesizer slowly hops the carrier to spread the spectrum across the 80 MHz band. All the active devices in the transceiver are on three IC's, consisting of a GaAs RF front-end, a silicon bipolar IF receiver, and a CMOS IC for the hopping-frequency synthesis. As in any spread-spectrum two-way communication system, transmission and reception is time-division-duplexed on the same frequency band. The receiver architecture is a conventional double-superheterodyne. In addition, the transceiver requires 50 passive components, including six rather bulky filters, and when transmitting 100 mW RF power it dissipates more than 1 W.

By the norms of GaAs MMIC's, the single-chip IC front-end in this transceiver is highly integrated (Fig. 19(a)) [146]. It includes the power amplifier, and drivers for 2-GHz passive filters in the transmit and receive paths. Interesting features are the dc series connection of the single-ended two-stage low-noise amplifier, which shares the same bias current in both stages through a bypass inductor; the Clapp VCO; and the four-FET switched mixers in the transmit and receive paths (Fig. 19(b)). In its first version, the IC includes more than 20 on-chip spiral inductors, perhaps the largest number on any MMIC at the time of this design. A second bipolar IC processes the 350 MHz IF signal. In receive mode, the GaAs front-end IC drains 30 mA from a 5 V supply. The receiver selects one of two external antennas through an on-chip RF switch to attain spatial diversity. This work has spurred a flurry of similar GaAs IC's at the same level of integration [147]–[149], all operating at ±5 V supplies, with similar functionality and performance. One of them [149] uses a very high IF of 915 MHz, so that spurious products at the transmitter output lie well in the stopbands of the output filter, and where the IF signal is processed by a 915 MHz cellular-telephone type IC.

IX. IC'S TO ENABLE FUTURE TRANSCEIVERS

What will portable wireless communicators of the future look like? We may draw some conclusions from the foregoing summary of RF-IC developments. The crowded spectrum means that future wireless communicators will predominantly use spread-spectrum techniques. Coupled to this is a need for low-power dissipation, which will force *architectural innovations* and *higher levels of integration* in the electronics. As an example of such an advanced transceiver, the author with his colleagues and their graduate students at UCLA is investigating the architecture and circuit design of a frequency-hopped, binary frequency-shift keyed, zero-IF, all-CMOS two-chip transceiver capable of delivering up to 160 kb/s (the base ISDN rate) in the 900 MHz ISM band [39], [150]. The transceiver architecture is inspired by the modern paging receiver, a very low energy wireless device widely used today. The transceiver implementation (Fig. 20) will freely mix analog and digital circuits, which makes CMOS the IC technology of choice. Further, to avoid the routing of high-frequency signals off- and on-chip and thereby save the power the buffers would use to drive stray capacitance and off-chip low-impedance lines, it is preferred to integrate all the blocks, *including* the RF front-end. Finally, it is most desirable to use an unmodified, standard production CMOS process.

Low-power operation requires the entire system to operate on a 3 V supply. This supply voltage cannot be any lower because in many places the analog circuits contain stacked cascode transistors, and the FET threshold voltages are about 0.8 V. RF amplifiers are normally tuned with inductor loads, which on silicon IC's are almost always off-chip passive components [151]. However, this is itself wasteful of power, because in addition to the current supplied into the inductor, the circuit must also drive the various parasitic capacitances in off-chip routing the RF signal.

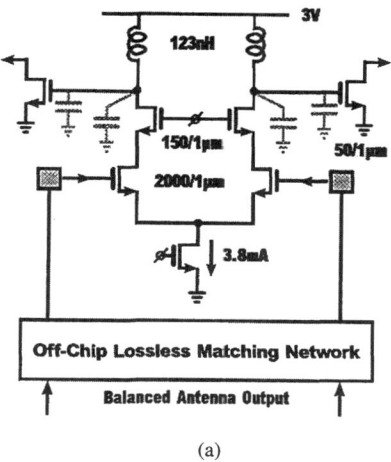
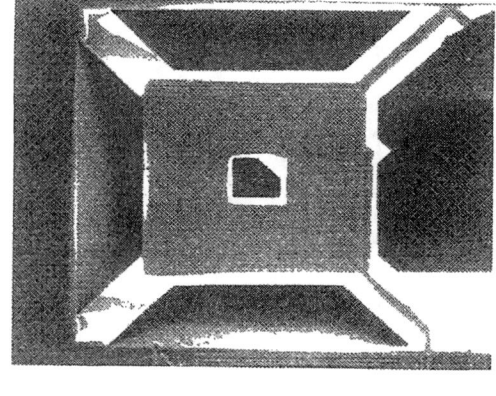

Fig. 21. (a) Balanced input CMOS RF amplifier. Cascode stage driving on-chip inductor substantially enhances voltage gain. Outputs common-source stages are for measurement only. LNA will directly drive mixer. (b) A microphotograph of a 110 nH spiral inductor suspended over a pit on substrate.

Spiral inductors on silicon substrates suffer from a large capacitance to the conducting silicon substrate. However, after many years of the belief that no useful inductors could be made on silicon, it was found that spiral inductors as large as 10 nH with self-resonance beyond 2 GHz, could be fabricated with the standard interconnect metallization. These inductors were used to build passive filters, a tuned amplifier, and an LC voltage-controlled oscillator on a silicon bipolar IC [152]–[154]. This work has rekindled interest in the design and modelling of small value spiral inductors on silicon (usually bipolar) IC's, with values in the range of 1–10 nH [155], [156]. In some instances, higher inductor Q's are obtained with thicker metallization, or higher resonant frequencies with thicker oxides. In search of larger value inductors for low-power, high-gain amplifiers, we have developed a fabrication method whereby the silicon substrate is selectively removed under the inductor, substantially reducing the capacitance to the substrate, and thus extending self-resonance to a higher frequency (Fig. 21(b)) [157]. The self-resonant frequency is now almost as high as it is on a semi-insulating GaAs substrate. Using two such inductors as loads in a balanced circuit, a 900 MHz RF amplifier has been built in 1-μm CMOS with a 30 dB gain draining only 3 mA (Fig. 21(a)). The amplifier IP3 is about 0 dBm, owing to the wide linear range of the MOS differential pair.

The front-end mixer in a direct-conversion receiver must be highly linear to suppress unwanted intermodulation produced by interferers. Also, as the local oscillator frequency is centered at RF in this direct-conversion receiver, inadequate reverse isolation or shielding may cause this frequency to leak through the antenna and to interfere with other nearby receivers tuned to the same RF. We use an unusual mixer—one that happens to be particularly well suited to MOS implementation—to circumvent both these problems. The desired signal is downconverted to baseband by sub-sampling the incoming RF [158]. The mixer circuit is a track-and-hold, with such a wide track-mode bandwidth that it can follow the RF waveform, and a short enough aperture to acquire the *instantaneous* value of the RF waveform on receipt of the sampling clock edge (Fig. 22). The received RF in our system is a 26 MHz wide spread-spectrum centered on a 915 MHz carrier, so by sub-sampling this waveform at a clock rate of 52 MHz or higher, the spectrum is translated to baseband without aliasing. However, the mixer also acquires wideband noise accompanying the signal from dc up to the gigahertz track bandwidth, and aliases the rms noise into a bandwidth at half the sample-rate, thereby raising the baseband noise spectral density by the ratio of the track-bandwidth to the sample-rate. Sub-sampling mixers therefore have a higher noise figure than conventional mixers. However, as such a mixer tends to be more linear than an analog multiplier, its dynamic range is also large, provided a high-gain, low-noise RF amplifier precedes it. The overall front-end spurious-free dynamic range is jointly set by the individual specifications of the RF amplifier and mixer. In this 1 μm CMOS prototype, the mixer circuit draws 4 mA from 3-V to acquire samples of a 900 MHz modulated waveform at a 50 MHz rate, which it then translates to baseband. The linearity, as measured by a +26 dBm third-order intercept, exceeds that of most continuous-time 900 MHz monolithic mixers, and for the fundamental reasons given above, the noise figure is 18 dB.

A low-power, agile frequency synthesizer with high spectral purity is required in a frequency-hopped transceiver. Conventional synthesizers based on a phase-locked loop with a variable-modulus divider in feedback suffer from limited agility, because the loop bandwidth may limit the speed of frequency switching. It is also difficult to design a VCO which both covers a wide range of frequency spreading, and produces a pure spectrum. The DDFS-DAC combination is an alternative solution to this problem, which so far has only been used by the military in spread-spectrum communication systems. A direct-digital frequency synthesizer (DDFS) [159] consists of an accumu-

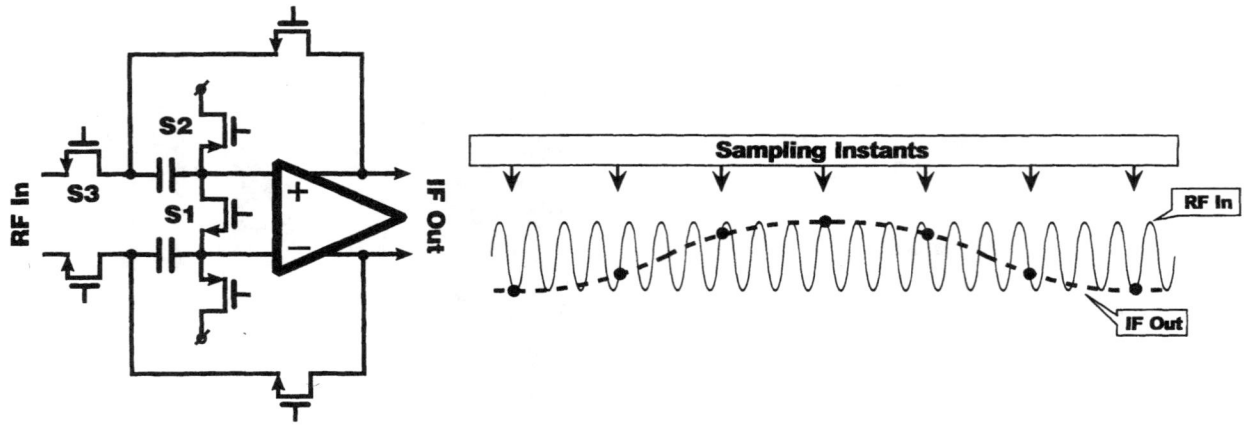

Fig. 22. A downconversion mixer based on direct sampling of RF. During track mode, switches are closed, and bandwidth of passive circuit is about 1 GHz. On receipt of sample clock, switches open within very short aperture time, capuring instantaneous value of RF waveform. Op amp removes switch charge injection, and buffers to subsequent circuits.

lator and ROM, which produces a sequence of digital words representing samples of a sinewave at a frequency set by an input control word. A D/A Converter converts the digital samples into a discrete-time analog sinewave, and a subsequent analog filter may smooth this into a continuous-time sinewave. Roundoff errors in the DDFS and nonlinearity in the DAC and filter will limit the attainable spectral purity of the synthesized sinewave [159]. In practice, DAC imperfections may contribute the largest nonlinearity in the system. A characteristic feature of digitally synthesized sinewaves is that the major imperfections do not necessarily produce harmonic tones, but instead spurious tones at rational multiples of the fundamental. These spurious tones often cluster around the fundamental and cannot be filtered. The DAC can add spurious tones of its own. We use a highly efficient DDFS combined with a high-speed, low-power charge-redistribution DAC to produce an FSK-modulated, frequency-hopped spread-spectrum at baseband, which a fixed-frequency local oscillator upconverts to RF. A CMOS implementation of a DDFS-DAC at 3 V dissipates only 40 mW when clocked at 50 MHz (of which the quadrature ROM-accumulator accounts for 35 mW), with worst-case spurious levels of −57 dBc or less across the entire spreading range (Fig. 23) [160].

The 915 MHz local oscillator is a four-stage MOS ring oscillator locked in a PLL to a lower frequency crystal reference. The quadrature outputs at 915 MHz for the image-reject mixers are tapped at diametrically opposite points [130], [134] to a phase accuracy of better than 1°. As the local-oscillator operates at a fixed frequency, it is locked to a low-frequency crystal reference in a PLL solely optimized to reduce phase noise. At a 100 kHz offset, a phase noise level of −75 dBc/Hz is measured on a prototype.

The power amplifier is a binary-weighted array of FET's biased near threshold, which is digitally selected to deliver power levels from −15 dBm to +15 dBm through a matching network to the antenna with a 40% conversion efficiency (Fig. 24) [161].

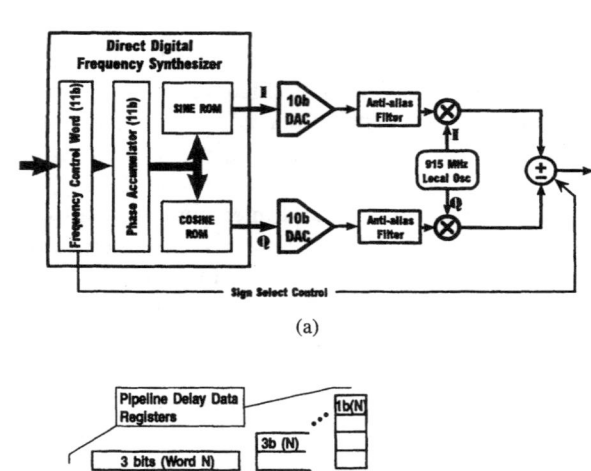

(a)

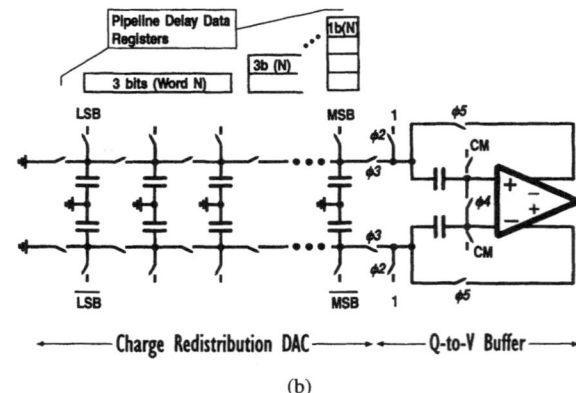

(b)

Fig. 23. (a) An agile frequency synthesizer. The DDFS produces digital samples of a sine and cosine, at a frequency set by an input word. DAC's convert these to analog domain, and after quadrature upconversion, either the upper or lower sideband is selected. Thus 0–13 MHz at baseband frequency produces output spanning 902–928 MHz. First-order anti-alias filter suffices if DDFS/DAC operates at 80 MHz. (b) DAC uses passive, pipelined architecture.

The transceiver achieves spatial diversity with two individual receive channels, which are powered-up continuously and connected to two different types of miniature antennas. Data decisions are made on an equally weighted sum of their baseband outputs. A high-order switched-capacitor elliptic lowpass filter selects the desired channel, which the on-board DDFS-based synthesizer has dehopped

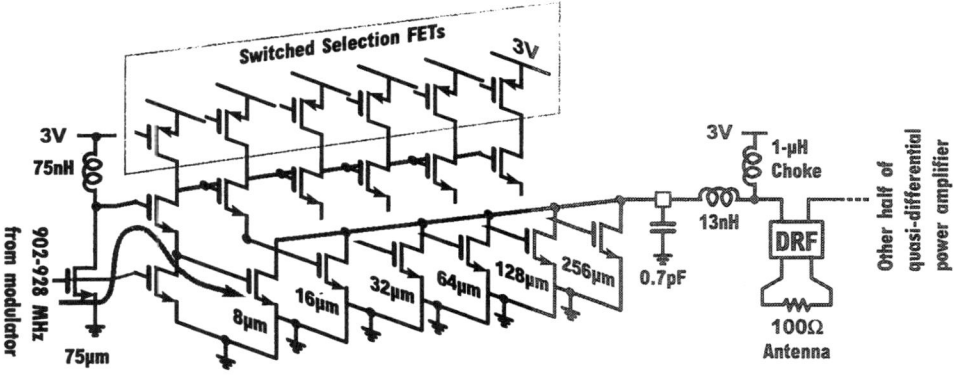

Fig. 24. A digitally controlled power amplifier. The binary-weighted array of output NMOSFET's drive the antenna load differentially, to deliver up to +15 dBm from a 3-V supply. Large on-chip inductor used as exciter load gives swings above power supply. PMOSFET's selectively activate source follower buffers, and thus total delivered power.

to dc in the course of downconversion. The capacitively coupled output is then amplitude-limited to produce a one-bit stream fed to an optimal FSK digital detector [39]. The detector quantizes the zero-crossings in time by oversampling the limited baseband data, and seeks correlations between one-bit representations of the two possible data values. A logarithmic amplifier connected to taps on the limiting amplifier chain measures the received signal-strength [162]. Operating from a 3-V supply, the 1-μm CMOS transceiver is expected to drain 70 mA in receive mode, and 100 mA in transmit mode (the power amplifier is on-chip). Considering that the transceiver communicates on a 26 MHz-wide spectrum, this is indeed a low power dissipation.

This research project is now exploring issues related to the single-chip integration of this transceiver. Such a highly integrated CMOS "VLSI radio" would represent a major step forward in the evolution of the radio integrated circuits described in this paper.

X. CONCLUSIONS

This survey has covered some of the key existing and emerging communication applications which have prompted advances in RF IC's. The emphasis was on widespread *portable consumer* applications. This has excluded coverage of integrated television tuners, for instance, which were historically the principal motivation for the development of UHF RF-IC's by the consumer electronics industry in the mid-1970's [163]. The fact that about 18 million tuners are produced every year means that there has been a steady stream of innovations in the building-block and architectures for this application [164]–[178]. Then there are emerging areas, such as personal GPS receivers, where RF integration and low-power will be the key for acceptance by consumers [179], [180]. Here, too, RF-IC designers are responding with integrated front ends [181]–[184], although the greater challenges to realize a system with an acceptable precision may lie in the baseband signal processing.

The lucrative global wireless market is attracting a great deal of attention from circuit designers across the industry. The RF range of interest spans 400 MHz to 2.5 GHz. There is a multi-pronged industry-wide assault to provide the right solutions. The solution must be low-cost, low-power, and it must give high functionality. MMIC designers bring to RF-IC's a microwave style derived from small-scale circuits which operated at much higher frequencies. On the other hand, as the frequency range lies within the capability of modern silicon IC technologies, there is a response from this community. Bipolar and BiCMOS RF-IC's are often extensions of baseband style circuits to high-frequency, often with more functionality than GaAs MMICs offer. A low transistor f_T is no longer a handicap, and experience rapidly accumulates on how to solve some of the unique problems of silicon substrates, such as high-frequency losses in the substrate [185] and on-chip coupling problems between circuit blocks. In the not too distant future, as microprocessors and memories drive the linewidths of silicon IC processes to deep submicron, RF-IC's might be designed as untuned wideband circuits, in the style of IC's for video applications today.

Small-scale GaAs MMIC's might sometimes perform better as standalone components, but embedded in a system such as a digital cellular telephone or a spread-spectrum transceiver, they require a large overhead in support circuits, and thereby lose an edge in certain specifications, such as power dissipation. Silicon RF-IC's offer higher levels of integration, but even within the silicon IC design community there are differing points of view. While integrated silicon bipolar transceivers are to be found in the next generation of products, CMOS RF solutions, albeit at an exploratory stage, now look very promising. They combine elements from well established techniques in voiceband and video IC's, and are unfettered in partitioning tasks between analog and digital signal processing, while using off-chip passive components when necessary. Digital control components and baseband signal processors are readily integrated with, indeed merged into, the RF and IF sections. This array of competing technologies offers communication system designers more creative opportunity, and the best wireless transceiver solutions may well emerge from system design evolving together with architecture, circuits, antennas, and power-allocation plans.

What are the prospects for this multitude of approaches to low-power wireless communications? If we project on the experience gained from baseband and video telecommunications circuits, the future wireless transceiver may be only one BiCMOS or CMOS large-scale, mixed analog-digital IC, requiring one passive filter and a crystal resonator. Baseband digital signal processing may make up for imperfections in the front-end RF sections. As it will likely be used in a microcell environment at low transmit power levels, the transceiver may be encapsulated in a plastic package. There is no fundamental reason why it should need more than a single 1.5 V battery as the power source. After decades of thinking of it only as science-fiction, the electronics world is prodded forth by the communications explosion into making the two-way wristwatch-size radio a reality. A large community researchers, not trained in the classical radio art, is busy reexamining the conventions, and at time challenging the received wisdom. We are poised at a turning point in the evolution of radio-circuit engineering.

ACKNOWLEDGMENT

My own involvement in RF-IC design is considerably enriched by collaborations with the many talented colleagues and graduate students participating in the Low-Power Wireless Transceiver Program in the UCLA Electrical Engineering Department. Professors Henry Samueli, Greg Pottie, and Yahya Rahmat-Samii have made an exemplary teamwork possible on this project.

The recent financial stringencies at UC have not so far eroded the marvellously rich Engineering Collection and online facilities at the UCLA Science and Engineering Library. It would be impossible to write this paper without ready access to these resources. Karen Andrews and Aggi Raeder, two of the expert librarians there, enthusiastically helped me in using the latest tools to peruse the literature, and amazed me with the ease with which they could track down obscure references from all corners of the world.

I thank Prof. Bob Meyer for a very thorough reading of the draft manuscript and for his thoughtful comments on it, and also Taiwa Okanobu and Mark McDonald who made useful suggestions. I am very grateful to Dr. Ran-Hong Yan, the Guest Editor of the *Proceedings of the IEEE* who solicited this manuscript, for his extraordinary ability to obtain speedy reviews and patiently absorb delays on my part in bringing this work to completion.

REFERENCES

[1] G. H. Heilmeier, "Personal communications: *Quo Vadis*," in *Int. Solid State Circ. Conf.*, San Francisco, 1992, pp. 24–26.
[2] D. C. Cox, "Universal digital portable radio communications," *Proc. IEEE*, vol. 75, pp. 436–476, Apr. 1987.
[3] S. Sheng, A. Chandrakasan, and R. W. Brodersen, "A portable multimedia terminal," *IEEE Comm. Magazine*, vol. 30, no. 12, pp. 64–75, Dec. 1992.
[4] B. A. Beatty, "The IBM personal communicator—Design considerations," *SOUTHEASTCON*, Miami, FL, 1994, pp. 432–435.
[5] A. Matsuzawa, "Low-voltage and low-power circuit design for mixed analog/digital systems in portable equipment," *IEEE J. Solid-State Circ.*, vol. 29, pp. 470–480, Apr. 1994.
[6] E. A. Vittoz, "Low-power design: Ways to approach the limits," in *Int. Solid State Circ. Conf.*, San Francisco, 1994, pp. 14–18.
[7] F. E. Terman, *Radio Engineers' Handbook*. New York: McGraw-Hill, 1943.
[8] W. L. Everitt, Ed., *Fundamentals of Radio and Electronics*, 2nd ed. Englewood Cliffs, NJ: Prentice-Hall, 1958.
[9] K. K. Clarke and D. T. Hess, *Communication Circuits: Analysis and Design*. Reading, MA: Addison-Wesley, 1971.
[10] H. L. Krauss, C. W. Bostian, and F. H. Raab, *Solid State Radio Engineering*. New York: Wiley, 1980.
[11] W. H. Hayward, *Introduction to Radio Frequency Design*. Englewood Cliffs, NJ: Prentice-Hall, 1982.
[12] U. L. Rohde and T. T. N. Bucher, *Communications Receivers: Principles & Design*. New York: McGraw-Hill, 1988.
[13] D. O. Pederson and K. Mayaram, *Analog Integrated Circuits for Communications: Principles, Simulation, and Design*. Boston: Kluwer, 1991.
[14] M. Ibuka, "How Sony developed electronics for the world market," *IEEE Trans. Eng. Manage.*, vol. EM-22, pp. 15–19, Feb. 1975.
[15] A. Laundrie and K. Negus, "GPS receiver chip targets low-power commercial markets," in *Microwaves & RF*, Aug. 1992, pp. 135–138.
[16] N. Camilleri, D. Lovelace, J. Costa, and D. Ngo, "New development trends for silicon RF device technologies," in *Microwave & Millimeter-Wave Monolithic Circ. Symp.*, San Diego, CA, 1994, pp. 5–8.
[17] W. Kanow and I. Siewert, "Integrated circuits for hi-fi radios and tuners," in *Electronic Components and Applications*, vol. 4, no. 1, pp. 11–27, Nov. 1981.
[18] R. M. Warner and J. N. Fordemwalt, Eds., *Integrated Circuits: Design Principles and Fabrication*. New York: McGraw-Hill, 1965.
[19] C. S. Meyer, D. K. Lynn, and D. J. Hamilton, Eds., *Analysis and Design of Integrated Circuits*. New York: McGraw-Hill, 1968.
[20] Y. Hishida and M. Kazama, "Mobile communication system: Its semiconductor technology," in *Hitachi Review*, vol. 42, no. 3, pp. 95–100, June 1993.
[21] B. Crouch, Ed., *Dick Tracy: America's Most Famous Detective*. Secaucus, NJ: Citadel, 1987.
[22] L. Blaser and T. Taira, "An am/fm radio subsystem IC," *IEEE Trans. Consumer Electron.*, vol. CE-23, pp. 129–136, Feb. 1977.
[23] W. Peil and R. J. McFayden, "A single chip am/fm integrated circuit radio," *IEEE Trans. Consumer Electron.*, vol. CE-23, pp. 424–431, Mar. 1977.
[24] O. L. Richards, "A complete am/fm signal processing system," *IEEE Trans. Consumer Electron.*, vol. CE-24, pp. 34–38, Jan. 1978.
[25] W. G. Kasperkovitz, "An integrated fm receiver," *Microelectronics Reliability*, vol. 21, no. 2, pp. 183–189, 1981.
[26] A. A. Abidi, "Noise in active resonators and the available dynamic range," *IEEE Trans. Circ. and Syst.*, vol. 39, pp. 296–299, Apr. 1992.
[27] W. G. Kasperkovitz, "Fm receivers for mono and stereo on a single chip," *Philips Tech. Rev.*, vol. 41, no. 6, pp. 169–182, 1983.
[28] W. H. A. van Dooremolen and M. Hufschmidt, "A complete fm radio on a chip," in *Electronic Components and Applications*, vol. 5, no. 3, pp. 159–170, 1983.
[29] J. Brilka and G. Sieboerger, "A single-chip fm radio," in *Int. Conf. on Consumer Electron.*, pp. 156–157, 1984.
[30] T. Okanobu, T. Tsuchiya, K. Abe, and Y. Ueki, "A complete single-chip am/fm radio integrated circuit," *IEEE Trans. Consumer Electron.*, vol. CE-28, pp. 393–408, Mar. 1982.
[31] T. Okanobu, H. Tomiyama, and H. Arimoto, "Advanced low-voltage single chip radio IC," *IEEE Trans. Consumer Electron.*, vol. 38, pp. 465–475, Mar. 1992.

[32] D. G. Tucker, "The history of the homodyne and synchrodyne," *J. Brit. Inst. of Radio Eng.*, vol. 14, no. 4, pp. 143–154, 1954.
[33] D. Mitchell and K. G. van Wynen, "A 150-mc personal radio signaling system," *Bell Syst. Tech. J.*, vol. XI, no. 5, pp. 1239–1257, 1961.
[34] A. E. Kerwien and L. H. Steiff, "Design of a 150-megacycle pocket receiver for the BELLBOY personal signaling system," *Bell Syst. Tech. J.*, vol. XLII, no. 3, pp. 527–565, 1963.
[35] K. Nagata, M. Akahori, T. Mori, and S. Umetsu, "Digital display radio paging system," in *NEC Res. & Devel.*, no. 68, pp. 16–23, Jan. 1983.
[36] I. A. W. Vance, "An integrated circuit VHF radio receiver," *The Radio and Electronic Engineer*, vol. 50, no. 4, pp. 158–164, 1980.
[37] ———, "Fully integrated radio paging receiver," *IEE Proc.*, vol. 129, pt. F, no. 1, pp. 2–6, 1982.
[38] R. C. French, "A high technology VHF radio paging receiver," in *Int. Conf. on Mobile Radio Syst. and Tech.*, York, UK, 1984, pp. 11–15.
[39] J. Min, A. Rofougaran, H. Samueli, and A. A. Abidi, "An all-CMOS architecture for a low-power frequency-hopped 900 MHz spread-spectrum transceiver," in *Custom IC Conf.*, San Diego, CA, 1994, pp. 379–382.
[40] K. Holtvoeth, "A fully integrated low-power FSK receiver for VHF and UHF paging," *IEEE Trans. Consumer Electron.*, vol. 35, pp. 707–713, Mar. 1989.
[41] S. Drude and T. Rudolph, "New chip-set reduces power consumption of radio pagers," *Electron. Components and Applications*, vol. 9, no. 2, pp. 112–124, 1989.
[42] J. F. Wilson, R. Youell, T. H. Richards, G. Luff, and R. Pilaski, "A single-chip VHF and UHF receiver for radio paging," *IEEE J. Solid State Circ.*, vol. 26, pp. 1944–1950, Dec. 1991.
[43] K. Nagata, D. Ishii, T. Mori, T. Oyagi, and T. Seo, "Slim digital pagers, in *NEC Res. & Develop.*, no. 74, pp. 55–64, July 1984.
[44] K. Yamasaki, S. Yoshizawa, Y. Minami, T. Asai, Y. Nakano, and M. Kuroda, "Compact size numeric display pager with new receiving system," in *NEC Res. and Develop.*, vol. 33, no. 1, pp. 73–81, Jan. 1992.
[45] A. Burt, "Direct conversion receivers come of age in the paging world," in *GEC Rev.*, vol. 7, no. 3, pp. 156–160, 1992.
[46] S. Tanaka, A. Nakajima, J. Nakagawa, and A. Nakagoshi, "High-frequency, low-voltage circuit technology for VHF paging receiver," *IEICE Trans. Fundamentals of Electron., Commun. and Comp. Sci.*, vol. E76-A, no. 2, pp. 156–163, 1993.
[47] B. F. Cambou, "MMIC consumer application and production," in *Microwave & Millimeter-Wave Monolithic Circ. Symp.*, Albuquerque, NM, 1992, pp. 1–3.
[48] H. Komatsu, R. Norose, and S. J. Symonds, "The wristwatch as a personal communications device," in *Pacific Telecommun. Conf.*, Honolulu, 1993, pp. 242–244.
[49] M. Pardoen, R. Pache, and E. Dijkstra, "Direct-conversion receiver provides CP-FSK operation," in *Microwaves & RF*, Mar. 1994, pp. 151–155.
[50] G. J. Lomer, "Telephoning on the move—Dick Tracy to Captain Kirk," *IEE Proc.*, vol. 134, pt. F, no. 1, pp. 1–8, 1987.
[51] K. Wakino, "Recent developments of dielectric resonator materials and filters in Japan," *Ferroelectrics*, vol. 91, pp. 69–86, 1989.
[52] M. Hikita, T. Tabuchi, and A. Sumioka, "Miniaturized SAW devices for radio communication transceivers," *IEEE Trans. Vehic. Tech.*, vol. 38, pp. 2–8, Jan. 1989.
[53] K. Eda, "Ultra-small SAW filter works in mobile communications," in *J. Electron. Eng.*, pp. 54–57, May 1992.
[54] Y. Tamura, T. Maru, N. Hirasawa, H. Okuno, M. Komoda, and M. Hotsumi, "Hand-held portable equipment for cellular mobile telephone, in *NEC Res. and Devel.*, no. 87, pp. 34–43, Oct. 1987.
[55] D. Anderson and R. J. Zavrel, "RF ICs for portable communications equipment," in *Electron. Components and Appl.*, vol. 7, no. 1 pp. 37–44, 1985.
[56] I. A. Koullias, S. L. Forgues, and P. C. Davis, "A 100 MHz IF amplifier/quadrature demodulator for GSM cellular radio mobile terminals," in *Bipolar Circ. and Tech. Mtg.*, Minneapolis, MN, 1990, pp. 248–251.
[57] Y. Mori, H. Yabuki, M. Ohba, M. Sagawa, M. Makimoto, and I. Shibazaki, "Miniaturized RF-circuit modules for land mobile communication equipment," in *IEEE Vehic. Tech. Conf.*, Orlando, FL, 1990, pp. 65–70.
[58] M. Hikita, A. Yuhara, and K. Oda, "New low-loss, miniature SAW filters," in *Hitachi Rev.*, vol. 42, no. 3, pp. 119–124, June 1993.
[59] L. Scharf, H. Frank, and T. Pollakowski, "High f_T and low noise," in *Siemens Components*, vol. 25, no. 4, pp. 119–122, 1990.
[60] S. Yamada and K. Oguri, "Low-power front-end oriented discrete semiconductor devices," in *Hitachi Rev.*, vol. 42, no. 3, pp. 101–106, June 1993.
[61] C. Huang, "Where do integrated circuits replace discretes?," in *Applied Microwaves & Wireless*, pp. 5–6, Summer 1994.
[62] S. P. MacCabe, "GaAs MMICs for cellular telephones, GPS and DBS markets," in *WESCON*, San Francisco, 1989, pp. 279–281.
[63] M. Rocchi, "GaAs & Si MMIC building blocks: A moot point revisited," *Microelectron. Eng.*, vol. 15, pp. 685–692, 1991.
[64] Y. Aoki, K. Kobayashi, and W. Kennan, "Monolithic microwave ICs (MMICs) for telecommunication," in *Fujitsu Sci. & Tech. J.*, vol. 30, no. 1, pp. 32–39, June 1994.
[65] Y. Imai, M. Tokumitsu, A. Minakawa, T. Sugeta, and M. Aikawa, "Very low-current and small size GaAs MMICs for L-band front-end applications," in *GaAs IC Symp.*, Monterey, CA, 1989, pp. 71–74.
[66] Y. Imai, M. Tokumitsu, and A. Minakawa, "Design and performance of low-current GaAs MMICs for L-band front-end applications," *IEEE Trans. Microwave Theory and Tech.*, vol. 39, pp. 209–215, Feb. 1991.
[67] K. R. Cioffi, "Ultra-low dc power consumptions in monolithic L-band components," *IEEE Trans. Microwave Theory & Techn.*, vol. 40, pp. 2467–2472, Dec. 1992.
[68] C. Kermarrec, M. Takano, F. McGrath, P. O'Sullivan, G. Dawe, E. Heaney, and G. St. Onge, "High performance, low cost GaAs MMICs for personal phone applications," in *Int. Symp. on GaAs and Related Compounds*, Karuizawa, Japan, 1992.
[69] E. Heaney, F. McGrath, P. O'Sullivan, and C. Kermarrec, "Ultra low power low noise amplifiers for wireless communications," in *GaAs IC Symp.*, San Jose, CA, 1993, pp. 49–51.
[70] S. Hara, K. Osato, A. Yamada, T. Tsukao, and T. Yoshimasu, "Miniaturized low noise variable MMIC amplifiers with low power consumption for L-band portable communication applications," in *Microwave and Millimeter-Wave Monolithic Circ. Symp.*, Atlanta, GA, 1993, pp. 67–70.
[71] T. Ohgihara, S. Kusunoki, M. Wada, and Y. Murakami, "GaAs JFET front-end MMICs for L-band personal communications," in *Microwave and Millimeter-Wave Monolithic Circ. Symp.*, Atlanta, GA, 1993, pp. 9–12.
[72] C. Kusano, E. Hase, and S. Sakamoto, "Low-power, low-noise GaAs MMIC," in *Hitachi Rev.*, vol. 42, no. 3, pp. 107–112, June 1993.
[73] R. E. Lehmann and D. D. Heston, "X-band monolithic series feedback LNA," *IEEE Trans. Microwave Theory and Tech.*, vol. MTT-33, pp. 1560–1566, Dec. 1985.
[74] H. Sakai, A. Tezuka, Y. Mori, M. Sagawa, T. Katoh, J. Itoh, and K. Fujimoto, "Low-power GaAs ICs for mobile communication equipment," in *Int. Symp. on GaAs and Related Compounds*, Karuizawa, Japan, 1992.
[75] V. Nair, R. Vaitkus, D. Scheitlin, J. Kline, and H. Swanson, "Low current GaAs integrated down converter for portable communication applications," in *GaAs IC Symp.*, San Jose, CA, 1993, pp. 41–44.
[76] C. Woo, A. Podell, R. Benton, D. Fisher, and J. Wachsman, "A fully integrated transceiver chip for the 900 MHz communication bands," in *GaAs IC Symp.*, Miami Beach, FL, 1992, pp. 143–146.
[77] P. E. Chadwick, "High performance integrated circuit mixers," in *Radio Receivers & Assoc. Syst.*, London, 1981, pp. 1–9.
[78] D. M. Snider, "A theoretical analysis and experimental confirmation of the optimally loaded and overdriven RF power amplifier," *IEEE Trans. Electron Devices*, vol. ED-14, pp. 851–857, Dec. 1967.
[79] J. J. Komiak, "Design and performance of an octave band 11 watt power amplifier MMIC," *IEEE Trans. Microwave Theory and Techn.*, vol. 38, pp. 2001–2006, Dec. 1990.
[80] T. Takagi, Y. Ikeda, K. Seino, G. Toyoshima, and A. Inoue, "A UHF band 1.3 W monolithic amplifier with efficiency of 63%," in *Microwave and Millimeter-Wave Monolithic Circ. Symp.*, Albuquerque, NM, 1992, pp. 35–38.

[81] M. Gat, D. S. Day, S. Chan, C. Hua, and J. R. Basset, "A 3.0 watt high efficiency C-band power MMIC," in *GaAs IC Symp.*, Monterey, CA, 1991, pp. 331–334.

[82] O. Ishikawa, M. Maeda, K. Nishii, M. Yanagihara, T. Yokoyama, Y. Ikeda, and Y. Ota, "Cellular telecommunication GaAs power modules," in *Applied Microwave and Wireless*, vol. 4, no. 3, pp. 83–88, Fall 1992.

[83] M. Maeda, M. Nishijima, H. Takehara, C. Adachi, H. Fujimoto, and O. Ishikawa, "A 3.5 V, 1.3 W GaAs power multi-chip IC for cellular phones," *IEEE J. Solid-State Circ.*, vol. 29, pp. 1250–1256, Oct. 1994.

[84] I. Yoshida, M. Katsueda, S. Ohtaka, Y. Maruyama, and T. Okabe, "Highly efficient 1.5 GHz Si power MOSFET for digital cellular front end," in *Int. Symp. on Power Semicond. Devices*, Tokyo, 1992, pp. 156–157.

[85] T. Okabe and K. Kobayashi, "Radio-frequency MOS power module," in *Hitachi Rev.*, vol. 42, no. 3, pp. 113–118, June 1993.

[86] R. G. Meyer and W. D. Mack, "A 1 GHz BiCMOS RF front-end IC," *IEEE J. Solid-State Circ.*, vol. 29, no. 3, pp. 350–355, 1994.

[87] M. N. Sera, "Chip set addresses North American digital cellular market," in *RF Design*, vol. 17, no. 3, pp. 54–62, Mar. 1994.

[88] Y. Funada, "VCO techniques enable compact, high-frequency mobilecom," in *J. Electron. Eng.*, Dec. 1993, pp. 40–43.

[89] Y. Tamura, A. Yonehata, S. Miyazaki, F. Kobayasi, Y. Fukuda, and J. Kamishiro, "Development of advanced mobile telephone P3 (personal pocket phone)," in *NEC Res. and Develop.*, no. 98, pp. 60-70, July 1990.

[90] H. Ohmori, "SAW filters hold key to creating small, portable equipment," in *J. Electron. Eng.*, pp. 63–67, May 1992.

[91] G. Kainz, "Microwave ceramics boosts miniaturization," in *Siemens Components*, vol. 27, no. 5, pp. 11–14, Sept./Oct. 1992.

[92] Y. Kado, M. Suzuki, K. Koike, and Y. Omura, "A 1 GHz/0.9 mW CMOS/SIMOX divide-by-128/129 dual-modulus prescaler using a divide-by-2/3 synchronous counter," *IEEE J. Solid-State Circ.*, vol. 28, pp. 513–517, Apr. 1993.

[93] T. Seneff, L. McKay, K. Sakamoto, and N. Tracht, "A Sub-1 mA 1.5 GHz silicon bipolar dual modulus prescaler," *IEEE J. Solid-State Circ.*, vol. 29, pp. 1206–1211, Oct. 1994.

[94] M. Morishima, "Present and future of mobile communications in Japan," in *Microwave & Millimeter-Wave Monolithic Circ. Conf.*, Atlanta, GA, 1993, pp. 3–6.

[95] I. Shimizu, S. Urabe, K. Hirade, K. Nagata, and S. Yuki, "A new pocket-size cellular telephone for NTT high-capacity land mobile communication system," in *Vehic. Tech. Conf.*, St. Louis, MO, 1991, pp. 114–119.

[96] I. A. Koullias, J. H. Havens, I. G. Post, and P. E. Bronner, "A 900 MHz transceiver chip set for dual-mode cellular radio terminals," in *Int. Solid State Circ. Conf.*, San Francisco, 1993, pp. 140–141.

[97] L. Longo, R. Halim, B.-R. Horng, K. Hsu, and D. Shamlou, "A cellular analog front end with a 98 dB IF receiver," in *Int. Solid State Circ. Conf.*, San Francisco, 1994, pp. 36–37.

[98] D. J. Goodman, "Second generation wireless information networks," *IEEE Trans. Vehic. Tech.*, vol. 40, pp. 366–374, Mar. 1991.

[99] J. Fenk, W. Birth, R. G. Irvine, P. Sehrig, and K. R. Schon, "An RF front-end for digital mobile radio," in *Bipolar Circ. and Tech. Mtg.*, Minneapolis, MN, 1990, pp. 244–247.

[100] E. Badura, "New integrated circuits for GSM mobile radio," in *Siemens Components*, vol. 26, no. 2, pp. 72–76, 1991.

[101] S. Beyer and G. Lipperer, "Low-current oscillator design for 900 MHz GSM applications," in *Microwave Eng. Europe*, Oct. 1991, pp. 35–41.

[102] R. E. Jenkins, "Transceiver chip simplifies GSM cellular design," in *RF Design*, June 1994, pp. 26–35.

[103] J. Sevenhans, A. Vanwelsenaers, J. Wenin, and J. Baro, "An integrated Si bipolar RF transceiver for a zero IF 900 MHz GSM digital radio front-end of a hand portable phone," in *Custom IC Conf.*, San Diego, CA, 1991, pp. 7.7/1–4.

[104] J. Sevenhans and A. Vanwelsenaers, "A silicon transceiver for a 900 MHz GSM hand set," in *Microwave Eng. Europe*, Dec./Jan. 1992/1993, pp. 59–63.

[105] D. Haspeslagh, J. Ceuterick, L. Kiss, and J. Wenin, "BBTRX: A baseband transceiver for a zero IF GSM hand portable station," in *Custom IC Conf.*, San Diego, CA, 1992, pp. 10.7.1–10.7.4.

[106] M. McDonald, "A DECT transceiver chip set," in *Custom IC Conf.*, San Diego, CA, 1992, pp. 10.6.1–10.6.4.

[107] B. Madsen and D. E. Fague, "Radios for the future: Designing for DECT," in *RF Design*, Apr. 1993, pp. 48–53.

[108] M. Eccles, "Gigahertz systems on a chip," in *Electron. World+Wireless World*, Aug. 1993, pp. 662–663.

[109] I. White, "UHF technology for the cordless revolution," in *Electron. World+Wireless World*, July/Aug. 1993, pp. 542–546 and 657–661.

[110] A. Boveda, G. L. Bonato, and O. Ripolles, "GaAs monolithic 1.5–2.5 GHz image rejection receiver," in *Int. Symp. on Circ. & Systems*, San Diego, CA, 1992, pp. 232–235.

[111] A. Boveda, F. Ortigoso, and J. I. Alonso, "A 0.7–3 GHz GaAs QPSK/QAM direct modulator," in *Int. Solid State Circ. Conf.*, San Francisco, 1993, pp. 142–143.

[112] M. Williams, F. Bonn, C. Gong, and T. Quach, "GaAs RF ICs target 2.4 GHz frequency band," in *Microwaves & RF*, July 1994, pp. 111–118.

[113] M. D. McDonald, "A 2.5 GHz BiCMOS image reject front-end," in *Int. Solid State Circ. Conf.*, San Francisco, 1993, pp. 144–145.

[114] J. L. Wang, T. Tsuchiya, H. Takeuchi, S. Shirotori, S. Miyazaki, and T. Nakata, "A miniature ultra low-power 2.5 GHz down-converter IC for wireless communications," in *NEC Res. and Develop.*, vol. 35, no. 1, pp. 46–50, Jan. 1994.

[115] T. Tsukahara, M. Ishikawa, and M. Muraguchi, "A 2 V 2GHz direct-conversion quadrature modulator," in *Int. Solid State Circ. Conf.*, San Francisco, 1994, pp. 40–41.

[116] M. Nakatsugawa, Y. Yamaguchi, and M. Muraguchi, "An L-band ultra low power consumption monolithic low noise amplifier," in *GaAs IC Symp.*, San Jose, CA, 1993, pp. 45–48.

[117] D. Ngo, B. Beckwith, P. O'Neil, and N. Camilleri, "Low voltage GaAs power amplifiers for personal communications at 1.9 GHz," in *MTT-S Int. Microwave Symp.*, Atlanta, GA, 1993, pp. 1461–1464.

[118] T. Kunihisa, T. Yokoyama, H. Fujimoto, K. Ishida, H. Takehara, and O. Ishikawa, "High efficiency, low adjacent channel leakage GaAs power MMIC for digital cordless telephone," in *Microwave and Millimeter-Wave Monolithic Circ. Symp.*, San Diego, CA, 1994, pp. 55–58.

[119] T. Tokumitsu, I. Toyoda, and M. Aikawa, "Low voltage, high power T/R switch MMIC using LC resonators," in *Microwave and Millimeter-Wave Monolithic Circ. Symp.*, Atlanta, GA, 1993, pp. 27–30.

[120] K. Miyatsuji, S. Nagata, N. Yoshikawa, K. Miyanaga, Y. Ohishi, and D. Ueda, "A GaAs high-power RF single-pole double-throw switch IC for digital mobile communication system," in *Int. Solid State Circ. Conf.*, San Francisco, 1994, pp. 34–35.

[121] P. O'Sullivan, G. St. Onge, E. Heaney, F. McGrath, and C. Kermarrec, "High performance integrated PA, T/R switch for 1.9 GHz personal communications handsets," in *GaAs IC Symp.*, San Jose, CA, 1993, pp. 33–35.

[122] A. Bateman, D. M. Haines, and R. J. Wilkinson, "Linear transceiver architectures," in *IEEE Vehic. Tech. Conf.*, Philadelphia, 1988, pp. 478–484.

[123] G. Schultes, A. L. Scholtz, E. Bonek, and P. Veith, "A new incoherent direct conversion receiver," in *IEEE Vehicle Tech. Conf.*, Orlando, FL, 1990, pp. 668–674.

[124] G. Schultes, E. Bonek, P. Weger, and W. Herzog, "Basic performance of a direct conversion DECT receiver," *Electron. Lett.*, vol. 26, no. 21, pp. 1746–1748, 1990.

[125] J. Sevenhans, D. Haspeslagh, A. Delarbre, L. Kiss, Z. Chang, and J. F. Kukielka, "An analog radio front-end chip set for a 1.9 GHz mobile radio telephone application," in *Int. Solid State Circ. Conf.*, San Francisco, 1994, pp. 44–45.

[126] V. Thomas, J. Fenk, and S. Beyer, "A one-chip 2 GHz single superhet receiver for 2 Mb/s FSK radio communication," in *Int. Solid State Circ. Conf.*, San Francisco, 1994, pp. 42–43.

[127] W. Veit, J. Fenk, S. Ganser, K. Hadjizada, S. Heinen, H. Herrmann, and P. Sehrig, "A 2.7 V 800 MHz-2.1 GHz transceiver chipset for mobile radio applications in 25 GHz f_T Si-bipolar," in *Bipolar Circ. and Tech. Mtg.*, Minneapolis, 1994, pp. 175–178.

[128] D. K. Weaver, "A third method of generation and detection of single-sideband signals," *Proc. IRE*, vol. 44, no. 12, pp. 1703–1705, 1956.

[129] K. Negus, B. Koupal, J. Wholey, K. Carter, D. Millicker, C. Snapp, and N. Marion, "Highly integrated transmitter RFIC with

monolithic narrowband tuning for digital cellular handsets," in *Int. Solid State Circ. Conf.*, San Francisco, 1994, pp. 38–39.
[130] A. A. Abidi, "Radio-frequency integrated circuits for portable communications," in *Custom IC Conf.*, San Diego, CA, 1994, pp. 151–158.
[131] J. Wenin, "ICs for digital cellular communication," in *Europe. Solid State Circ. Conf.*, Ulm, Germany, 1994, pp. 1–10.
[132] H. Tsurumi and T. Maeda, "Design study on a direct conversion receiver front-end for 280 MHz, 900 MHz, and 2.6 GHz band radio communication systems," in *IEEE Vehic. Tech. Conf.*, St. Louis, MO, 1991, pp. 457–462.
[133] A. Bateman and D. M. Haines, "Direct conversion transceiver design for compact low-cost portable mobile radio terminals," in *IEEE Vehic. Tech. Conf.*, San Francisco, 1989, pp. 57–62.
[134] A. W. Buchwald and K. W. Martin, "High-speed voltage-controlled oscillator with quadrature outputs," *Electron. Lett.*, vol. 27, no. 4, pp. 309–310, 1991.
[135] S. K. Enam and A. A. Abidi, "NMOS ICs for clock and data regeneration in Gb/s optical-fiber receivers," *IEEE J. Solid State Circ.*, vol. 27, pp. 1763–1774, Dec. 1992.
[136] J. K. Cavers and M. W. Liao, "Adaptive compensation for imbalance and offset losses in direct conversion transceivers," *IEEE Trans. Vehic. Tech.*, vol. 42, no. 4, pp. 581–588, 1993.
[137] J. T. Taylor and J. K. Omura, "Spread spectrum technology: A solution to the personal communications services frequency allocation dilemma," in *IEEE Commun. Magazine*, vol. 29, no. 2, pp. 48–51, Feb. 1991.
[138] R. A. Scholtz, "The origins of spread-spectrum communications," *IEEE Trans. Commun.*, vol. COM-30, pp. 822–854, Oct. 1982.
[139] D. T. Magill, F. D. Natali, and G. P. Edwards, "Spread-spectrum technology for commercial applications," *Proc. IEEE*, vol. 82, pp. 572–584, Apr. 1994.
[140] R. C. Dixon, *Spread Spectrum Systems*. New York: Wiley, 1976.
[141] M. Leonard, "Wireless data links broaden LAN options," in *Electronic Design*, pp. 51–58, Mar. 1992.
[142] D. T. Magill, "A fully-integrated, digital, direct-sequence, spread spectrum modem ASIC," in *Int. Symp. on Personal, Indoor and Mobile Radio Commun.*, Boston, 1992, pp. 42–46.
[143] R. Jain, H. Samueli, P. T. Yang, and C. Chien, "Computer-aided design of a BPSK spread-spectrum chip set," *IEEE J. Solid-State Circ.*, vol. 27, pp. 44–58, Jan. 1992.
[144] C. Chien, P. Yang, E. Cohen, R. Jain, and H. Samueli, "A 12.7 Mchip/s all-digital BPSK direct sequence spread-spectrum IF transceiver in 1.2 μm CMOS," in *Int. Solid-State Circ. Conf.*, San Francisco, 1994, pp. 30–31.
[145] M. Leonard, "PCMCIA-sized radio links portable WLAN terminals," in *Electron. Design*, pp 45–50, Aug. 1993.
[146] L. M. Devlin, B. J. Buck, J. C. Clifton, A. W. Dearn, and A. P. Long, "A 2.4 GHz single chip transceiver," in *Microwave & Millimeter-Wave Monolithic Circ. Symp.*, Atlanta, GA, 1993, pp. 23–26.
[147] T. Apel, E. Creviston, S. Ludvik, L. Quist, and B. Tuch, "A GaAs MMIC transceiver for 2.45 GHz wireless commercial products," in *Microwave & Millimeter-Wave Monolithic Circ. Symp.*, San Diego, CA, 1994, pp. 15–18.
[148] B. Khabbaz, A. Douglas, J. DeAngelis, L. Hongsmatip, V. Pellicia, W. Fahey, and G. Dawe, "A high performance 2.4 GHz transceiver chip set for high volume commercial applications," in *Microwave & Millimeter-Wave Monolithic Circ. Symp.*, San Diego, CA, 1994, pp. 11–14.
[149] M. S. Wang, M. Carriere, P. O'Sullivan, and B. Maoz, "A single-chip MMIC transceiver for 2.4 GHz spread spectrum communication," in *Microwave & Millimeter-Wave Monolithic Circuits Symp.*, San Diego, CA, 1994, pp. 19–22.
[150] J. Min, A. Rofougaran, V. Lin, M. Jensen, H. Samueli, A. A. Abidi, G. Pottie, and Y. Rahmat-Samii, "A low-power CMOS hand-held frequency-hopped spread spectrum transceiver hardware architecture," in *Virginia Tech's 3rd Symp. on Wireless Personal Commun.*, Blacksburg, VA, 1993, pp. 10/1-8.
[151] M. Sakakura and S. Skiest, "Ultra-miniature chip inductors serve at high frequency," in *J. Electron. Eng.*, pp. 48–51, Dec. 1993.
[152] N. M. Nguyen and R. G. Meyer, "Si IC-compatible inductors and LC passive filters," *IEEE J. Solid-State Circ.*, vol. 25, pp. 1028–1031, Apr. 1990.
[153] ———, "A silicon bipolar monolithic RF bandpass amplifier," *IEEE J. Solid-State Circ.*, vol. 27, pp. 123–127, Jan. 1992.
[154] N. M. Nguyen and R. G. Meyer, "A 1.8 GHz monolithic LC voltage-controlled oscillator," *IEEE J. Solid-State Circ.*, vol. 27, pp. 444–450, Mar. 1992.
[155] D. Lovelace, N. Camilleri, and G. Kannell, "Silicon MMIC inductor modeling for high volume, low cost applications," in *Microwave J.*, vol. 37, no. 8, pp. 60–71, Aug. 1994.
[156] K. B. Ashby, W. C. Finley, J. J. Bastek, S. Moinian, and I. A. Koullias, "High Q inductors for wireless applications in a complementary silicon bipolar process," in *Bipolar Circuits and Tech. Mtg.*, Minneapolis, 1994, pp. 179–182.
[157] J. Y.-C. Chang, A. A. Abidi, and M. Gaitan, "Large suspended inductors on silicon and their use in a 2 μm CMOS RF amplifier," *IEEE Electron Device Lett.*, vol. 14, pp. 246–248, May 1993.
[158] P. Y. Chan, A. Rofougaran, K. A. Ahmed, and A. A. Abidi, "A highly linear 1 GHz CMOS downconversion mixer," in *Europe. Solid-State Circ. Conf.*, Sevilla, Spain, 1993, pp. 210–213.
[159] H. T. Nicholas and H. Samueli, "A 150 MHz direct digital frequency synthesizer in 1.25 μm CMOS with -90 dBc spurious response," *IEEE J. Solid-State Circ.*, vol. 26, pp. 1959–1969, 1991.
[160] G. Chang, A. Rofougaran, M. K. Ku, A. A. Abidi, and H. Samueli, "A low-power CMOS digitally synthesized 0–13 MHz agile sinewave generator," in *Int. Solid-State Circ. Conf.*, San Francisco, 1994, pp. 32–33.
[161] M. Rofougaran, A. Rofougaran, C. Olgaard, and A. A. Abidi, "A 900 MHz CMOS RF power amplifier with programmable output," in *Symp. on VLSI Circ.*, Honolulu, 1994, pp. 133–134.
[162] K. Kimura, "A CMOS logarithmic amplifier with unbalanced source-coupled pairs," *IEEE J. Solid-State Circ.*, vol. 28, pp. 78–83, Jan. 1993.
[163] S. Watanabe, "Technology transfer for high frequency devices for consumer electronics," *IEEE Trans. Microwave Theory and Tech.*, vol. 40, pp. 2461–2466, Dec. 1992.
[164] U. Ablassmeier, W. Kellner, and H. Kniepkamp, "GaAs FET upconverter for TV tuner," *IEEE Trans. Electron Devices*, vol. ED-27, pp. 1156–1159, June 1980.
[165] K. Torii, S. Fujimori, S. Komatsu, S. Shimizu, K. Yoshihasa, J. Nagai, and Y. Yamagata, "Monolithic integrated VHF TV tuner," *IEEE Trans. Consum. Electron.*, vol. CE-26, pp. 180–187, 1980.
[166] J.-E. Muller, U. Ablassmeier, J. Schelle, W. Kellner, and H. Kniepkamp, "A double-conversion broad band TV-tuner with GaAs ICs," in *GaAs IC Symp.*, 1984, pp. 97–100.
[167] P. Dautriche, B. Y. Lao, C. Villalon, V. Pauker, M. Bostelmann, and M. Binet, "GaAs monolithic circuits for TV tuners," in *GaAs IC Symp.*, Monterey, CA, 1985, pp. 165–168.
[168] J. Fenk and R. Tauber, "TV VHF/hyperband tuner ICs," *IEEE Trans. Consum. Electron.*, vol. CE-32, pp. 723–733, Apr. 1986.
[169] H. van Glabbeek, N. Baars, H. Schreurs, and W. Zwijsen, "VHF, hyperband, and UHF mixer/oscillator sections of a TV tuner on one IC," *IEEE Trans. Consum. Electron.*, vol. CE-33, pp. 619–622, Apr. 1987.
[170] T. Nakatsuka, S. Sakashita, K. Goda, and S. Nambu, "A GaAs RF amplifier IC for UHF TV tuners," *IEEE Trans. Consum. Electron.*, vol. 34, pp. 366–371, Feb. 1988.
[171] H. Mizukami, K. Sakuta, H. Hatashita, T. Nagashima, and K. Shinkawa, "A high quality GaAs IC tuner for TV/VCR receivers," *IEEE Trans. Consum. Electron.*, vol. 34, pp. 649–658, Mar. 1988.
[172] T. Ducourant, P. Philippe, P. Dautriche, V. Pauker, C. Villalon, M. Pertus, and J.-P. Damour, "A 3-chip GaAs double-conversion TV tuner with 70 dB image rejection," in *Microwave & Millimeter-Wave Monolithic Circ. Symp.*, Long Beach, CA, 1989, pp. 87–90.
[173] H. Yagita, A. Terao, M. Tsuneoka, A. Watanabe, T. Tambo, and S. Nambu, "Low noise and low distortion GaAs mixer-oscillator IC for broadcasting satellite TV tuner," in *GaAs IC Symp.*, San Diego, CA, 1989, pp. 75–78.
[174] P. Philippe and M. Pertus, "A 2 GHz enhancement-mode GaAs down-converter IC for satellite TV tuner," in *Microwave Theory & Techniques Symp.*, Boston, MA, 1991, pp. 73–76.
[175] H. Mizukami, H. Ikedo, K. Ideno, T. Nagashima, and S. Yamada, "Low supply voltage GaAs ICs for a TV tuner," in

Int. Symp. on GaAs and Related Compounds, Karuizawa, Japan, 1992.
[176] K. Ideno, H. Mizukami, T. Nagashima, and K. Sakamoto, "A wide band GaAs IC for a DBS/TV receiver," in *Int. Symp. on GaAs and Related Compounds*, Karuizawa, Japan, 1992.
[177] K. Washio, T. Okabe, K. Norisue, and T. Nagashima, "An all-band TV tuner IC with 10 GHz 100 V mixed analog/digital Si bipolar technology," *IEEE J. Solid-State Circ.*, vol. 27, pp. 1264–1269, Sept. 1992.
[178] N. Scheinberg, R. Michels, V. Fedoroff, and D. Stoffman, "A GaAs upconverter integrated circuit for a double conversion cable TV set-top tuner," *IEEE J. Solid-State Circ.*, vol. 29, pp. 688–692, June 1994.
[179] J. Hurn, *GPS: A Guide to the Next Utility*. Sunnyvale, CA: Trimble Navigation, 1989.
[180] W. O. Dussell and J. Medina, "Using creative surface mount manufacturing and packaging as a competitive weapon," in *Surface Mount Int. Conf. & Expo.*, San Jose, CA, 1993, pp. 241–245.
[181] R. M. Herman, C. H. Mason, H. P. Warren, and R. A. Meier, "A GPS receiver with synthesized local oscillator," in *Int. Solid-State Circ. Conf.*, New York, 1989, pp. 194–195.
[182] R. M. Herman, A. Chao, C. H. Mason, and J. R. Pulver, "An integrated GPS receiver with synthesizer and downconversion functions," in *Int. Microwave Symp.*, Boston, 1991, pp. 883–886.
[183] R. Benton, M. Nijjar, C. Woo, A. Podell, G. Horvath, E. Wilson, and S. Mitchell, "GaAs MMICs for an integrated GPS frontend," in *GaAs IC Symp.*, Miami Beach, FL, 1992, pp. 123–126.
[184] K. J. Negus, R. A. Koupal, D. Millicker, and C. P. Snapp, "3.3 V GPS receiver MMIC implemented on a mixed-signal, silicon bipolar array," in *Microwave & Millimeter-Wave Monolithic Circ. Symp.*, Albuquerque, NM, 1992, pp. 209–212.
[185] N. Camilleri, J. Kirschgessner, J. Costa, D. Ngo, and D. Lovelace, "Bonding pad models for silicon VLSI technologies and their effects on the noise figure of RF NPNs," in *Microwave & Millimeter-Wave Monolithic Circ. Symp.*, San Diego, CA, 1994, pp. 225–228.

Future Directions in Silicon ICs for RF Personal Communications

P. R. Gray, R. G. Meyer
Electrical Engineering and Computer Sciences
University of California, Berkeley

Abstract

This paper presents an overview of technical challenges in achieving higher integration levels, lower power dissipation, smaller form factor, and lower cost in portable battery-powered RF transceivers for personal communications applications. Specific emphasis is placed on silicon integrated circuits for transceivers in the 800MHz-2.5GHz range of frequencies.

Introduction

Digital radio personal communications devices utilizing the bands between 800MHz and 2.5GHz will play an increasingly important role in the overall communications infrastructure in the next decade. In addition, the bands above 2.5GHz, where large, relatively untapped blocks of spectrum are available, will receive increasing use as transceiver costs in this frequency range are brought down. Compared to other types of integrated circuits, the level of integration in the RF sections of such transceivers is still relatively low. Considerations of power dissipation, form factor, and cost dictate that the RF/IF portions of these devices evolve to higher levels of integration than is true at present.

In this paper, we attempt to identify some of the key barriers to realizing these higher levels of integration, and discuss several of the avenues currently being pursued for achieving that objective for portable personal digital RF communications devices such as cellular telephones, cordless telephones, wireless PBXs and wireless LANs utilizing the bands of frequencies between 800MHz and 2.5GHz. The emergence of established standards and the rapid growth of deployment make these very attractive potential applications for high-integration dedicated integrated circuits. Other important applications include the services to be offered by a whole spectrum of providers using both unlicensed bands as well as the licensed part of the new PCS band now being allocated around 1.8 GHz. A good overview of these applications is given in [1].

Typical Present RF Transceiver Implementation

The vast majority of currently-manufactured transceivers for the applications mentioned above utilize single- or dual-conversion configuration for the receive path. Baseband channel bandwidths range from 10kHz to 1-2MHz. Good examples of systems at the current state of the art are surveyed or described partially in [2][3][4][5][6]. A typical example of such a transceiver as might be used in a frequency-hopped wireless LAN application is shown in block diagram form in Fig. 1.

The conventional architecture in Fig. 1 is not particularly amenable to higher levels of integration. Image rejection considerations usually dictate that the first intermediate frequency (IF) be on the order of 10% of the carrier frequency, with for example at least 70 MHz in 900 MHz receivers, and higher in higher frequency receivers. The use of complex signal representation at IF, as in image reject mixers, can improve image rejection by a considerable margin[47] and simplify the passive RF image reject filter, but the difficulty of phase and amplitude matching at IF usually limits the image rejection in such mixers to values on the order of 20dB, and as a result the IF must still be kept fairly high to preserve image rejection without requiring expensive and lossy ceramic RF filters. In many applications, two ceramic RF filters are required for adequate image rejection, one preceding and one following the low-noise amplifier (LNA). Depending on the modulation scheme used, the range of frequencies over which the receiver must be tuned, and the amplitude of the near-carrier interfering signals, a second frequency conversion may be performed, translating the signal down to a second IF on the order of 10-20% of the first IF.

Most current implementations also utilize external varactor-tuned LC resonators to provide the tuning element of the voltage-controlled oscillator (VCO) or VCOs which, in conjunction with a crystal reference, provides for frequency synthesis of the local oscillator (LO). The relatively high Q required for these resonators stems from phase noise considerations, discussed later.

These particular aspects of receiver architecture have fundamental implications for receiver integration level. Unfortunately, the required high-Q, low-noise, low-distortion bandpass IF filter (70-100 MHz for 900 MHz receivers, for example) is well beyond the capabilities of current low-power integrated filter technologies. As a result, external high-Q passive filters are generally used, usually implemented with SAW filters, ceramic filters, or in some cases LC filters. Because of the frequencies involved and the package parasitics usually present, considerable power dissipation is involved in taking IF signals off chip into these devices. Also, available on-chip spiral inductors in standard silicon technologies have Qs limited to 5-10 at the frequencies of interest, only adequate in some situations for implementation of the low-phase-noise VCO required in conventional synthesizer architectures.

In the transmit direction, the use of direct carrier modulation has become widespread[29], so that channel shaping filters can be implemented at baseband. In current practice the transmit power amplifier is usually implemented with GaAs discrete devices or simple ICs with a number of external inductors used for tuning and impedance matching. A narrowband external passive filter is usually required to limit transmit energy to the desired band.

The major challenge in RF transceiver design is to more effectively utilize scaled technologies to improve the integration level of RF transceivers, with resulting further improvements in power dissipation, form factor, and cost. Efforts are underway in industrial laboratories and universities around the world, taking various avenues toward this goal. The most promising approaches involve direct-conversion or low-IF receiver architectures that eliminate external IF filtering, new approaches to frequency synthesis that eliminate the need for external VCO resonators, and more effective utilization of on-chip spiral inductors available in near-standard IC technology to provide the tuning function essential to low-power realizations of RF functions. A hypothetical single-chip transceiver that might result from success in these areas is illustrated in block diagram form in Fig. 2.

Direct-Conversion, Quasi-direct Conversion, and Low-IF Receiver Architectures

A promising direction in architectures for higher integration in RF transceivers is the use of zero IF, low-IF, or quasi-IF configurations in the receiver, following the pager model, and the use of direct modulation in the transmit path. These configurations have been investigated intensively for many years (see for example [10][15]) but have made their way into practice in only a few specialized applications[9][11][12][13][14][15][16][17][18][19][20]. These configurations eliminate the external IF filtering function since the IF filter is replaced by two (I and Q) lowpass filters in the case of zero IF and quasi-IF receivers, or by a low-frequency, low-Q bandpass IF filter in the low-IF case.

The most severe problems in direct conversion receivers result from the fact that the baseband signal often contains low-frequency information that must be distinguished from DC and low-frequency errors that arise in the baseband signal path. One important error source is the device-mismatch-induced DC offset and 1/f noise of the signal path itself. For reasons of large-signal blocking performance, the gain of the LNA is usually restricted to the 20dB range, so that the wanted signal level reaching the mixer under weak signal conditions is on the order of 100 microvolts in amplitude. The accumulated DC offset referred to the mixer input can easily be 10mV, 100 times larger than the signal. Another important contributor is LO leakage, resulting from the fact that since the LO is at the carrier frequency, any energy from it reaching the RF path demodulates to a DC offset. Because the effects of LO leakage can be a function of the impedance seen at the antenna, these DC offsets can vary with time in an unpredictable manner. The problem is more severe in frequency-hopping receivers because the carrier leakage is different at each hop frequency, giving a time variation to the DC offset that is induced. The cumulative effect of carrier feedthrough and DC offsets is to superimpose large, possibly time-varying additive errors on the

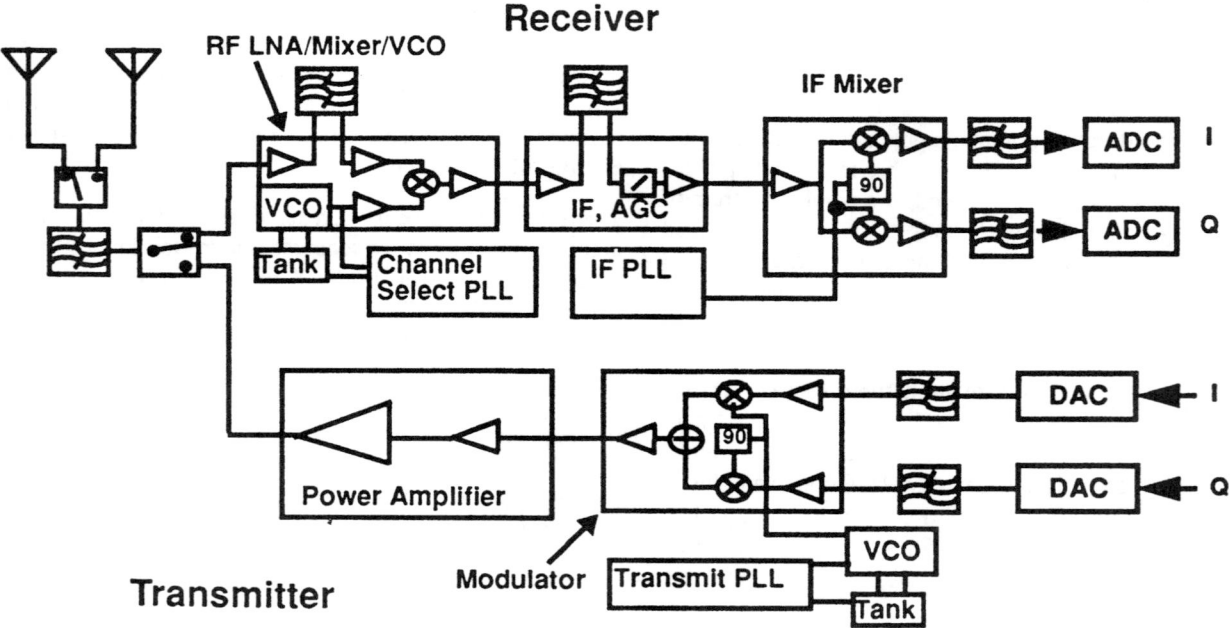

Figure 1. Block diagram of a typical multi-chip, multi-technology transceiver implementation

small wanted low-frequency AC signal in the baseband signal path.

Numerous approaches have been tried to attack these problems. Most have been attempted within the context of a more conventional, bipolar technology multi-chip receiver implementation using analog baseband filtering and demodulation. However, a closer coupling between the demodulation process and the RF and baseband analog signal path may well allow the separation of the DC offsets to be carried out using an adaptive approach that combines this function with carrier recovery, symbol timing recovery, automatic gain control (AGC), and data detection in a mixed analog-digital implementation. Most TDMA systems, for example, utilize a preamble in the frame structure which when demodulated to baseband has either zero or known DC content, allowing adaptive, frame-by-frame DC offset removal. The problem of 1/f noise can be attacked in a number of ways, one of which is to simply use correlated double sampling or chopper stabilization of the active elements in the baseband signal path. A/D conversion of the baseband signal at high resolution is a requirement for this approach.

Most of the benefits of homodyne receivers accrue if the IF is translated to a low but nonzero value instead of to DC as in homodyne receivers. The IF needs to be low enough that normal monolithic filtering techniques such as g_m/C continuous filters or switched-capacitor filters can be used. The advantage of this approach over homodyne receivers is that the problems of DC offset and 1/f noise are greatly reduced. However, a new problem of image rejection of the relatively close-in image frequency is introduced. This image energy must be eliminated through the use of an image-rejection mixer configuration following the LNA. Since this mixer will have to provide image rejection on the order of 60-70dB in some applications, phase shift accuracy and path matching accuracy within the mixer must be extremely precise. Progress has been made in this area in recent research [21].

Another important variation is the quasi-IF or "vestigial IF" approach in which the entire band of frequencies to be tuned by the receiver is translated down to IF in the first mixer, and then subsequently translated directly to baseband in the second mixer with little or no IF selective filtering. The channel-select filtering is done at baseband with a lowpass filter following translation to baseband. This technique has several important advantages. The first local oscillator can be implemented as a fixed frequency oscillator, making it easier to realize the required phase noise performance. The second LO, used to tune the desired channel, is at much lower frequency and its phase noise contributions, as well as the spurs associated with the narrow channel spacing and associated low comparison frequency, can be made much smaller. The carrier feedthrough problem is also eliminated. The technique eliminates the IF filter, but retains the image reject problem at RF and also many of the DC offset and drift problems of direct conversion receivers since adjacent channel blocking signals are carried to baseband and as a result most of the gain applied to the desired signal is done at baseband. The example receiver in Fig. 2 has this configuration.

Adaptation in Receiver Implementation

Conventional mixed-technology low-integration receivers have used RF/IF signal paths with relatively fixed functionality except for AGC and one or two other parameters. Higher integration implementations offer the possibility of much greater use of adaptive blocks. This capability may allow a signal transceiver to effect large savings in power, and to interface with more than one type of RF systems.

Power-Adaptive Transceivers

Most important power-dissipating elements in high-integration communications transceivers have a minimum dissipation requirement that is a function of the distance of the transceiver from the base station. In cellular phones, for example, the transmit power of the power amplifier is routinely varied adaptively depending on distance from the base station, both to save power and to reduce interference to other users.

Improvements in overall average power dissipation could result from a wider application of the power adaptation concept. For example, the LNA power dissipation is dictated by the requirement for low noise figure and good input matching to accommodate the weakest signal that will be encountered far from the base station. When the signal is stronger, the LNA could be adaptively powered back to a lower power setting. In some current receivers a variation on this approach is now employed in which the LNA is simply bypassed and powered down for very strong signals. Similar adaptive power reductions may be possible in the mixer and baseband signal path as well. Another large dissipater is the synthesizer VCO, which must have very low phase noise for many applications. This noise is less important when the received RF signal is stronger, since the mixer-aliased VCO phase noise from the adjacent channel can be larger in absolute terms. As a result, power savings may be available by reducing VCO power and allowing larger phase noise. Similarly, the dynamic range required in the baseband signal path is greatest when the desired signal is weakest. Optimum distribution of gain and dynamic range through the baseband path is a function of both the desired signal strength and the interfering signal strength, both of which are easily detected in a digital implementation. For strong desired signal conditions, smaller filtering capacitors could be adaptively used in the path, allowing operation of the active devices at lower bias current for the same bandwidth.

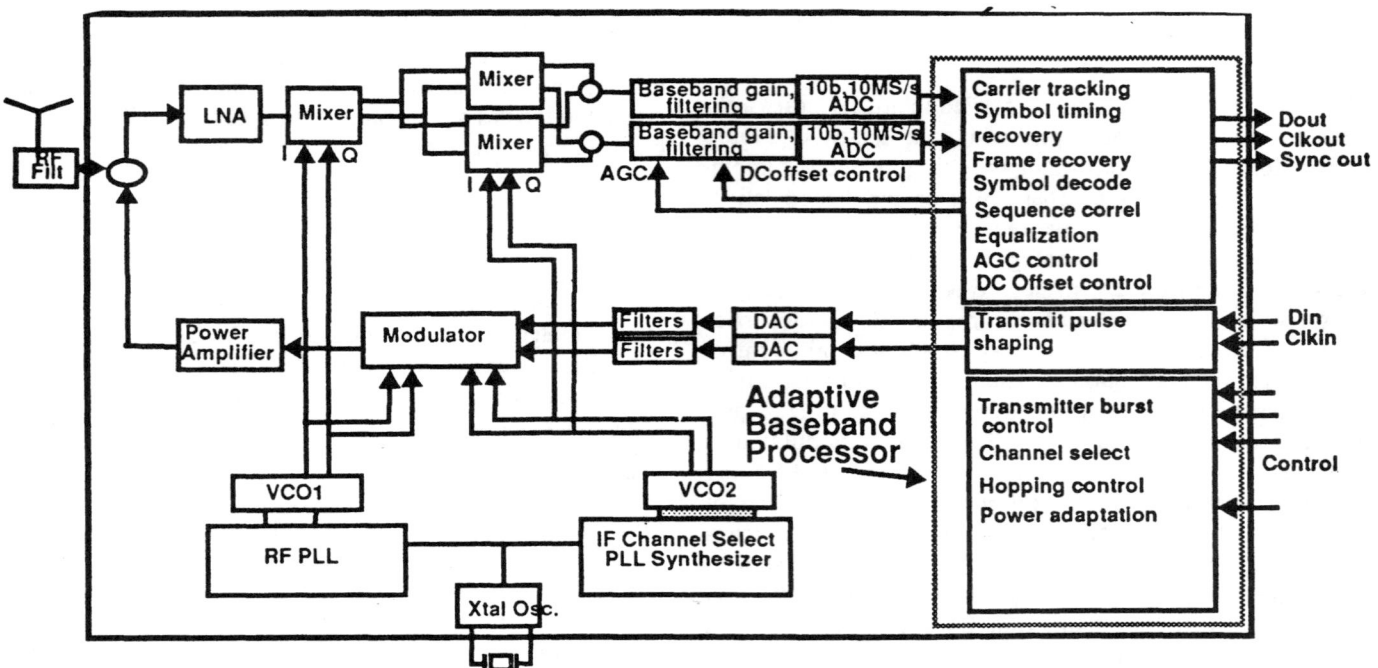

Figure 2. Block diagram of a possible future high-integration adaptive transceiver

Standard-Adaptive Transceivers

Great benefits could potentially accrue from a unification of approaches to data communications transceiver design. From an applications viewpoint, there is no fundamental reason that a single transceiver device could not provide the functionality of multiple communications standards at multiple frequencies. This might allow-for example, a single hand-held device to perform the functions of cellular phone and cordless phone compatibly with the varying standards for such service in Japan, North America, and Europe. Certainly multiple RF ports and antennas would be required for optimum performance at the different wavelengths, but following this a single transceiver should be able to adapt the different transmit power levels and modulation schemes. Central to achieving this would be an architecture in which a major portion of the IF (baseband in homodyne receivers) signal processing is performed in the digital domain, as well as carrier recovery, symbol timing recovery, equalization, DC offset control, power control, and so forth. A high-integration CMOS or BiCMOS implementation would be essential.

A secondary benefit would be a great reduction in overall engineering costs associated with development and productization of separate RF core functions for every standard and every type of device. The availability of a generic RF core function implemented in a VLSI digital compatible process such as scaled CMOS would be widely useful in implementing RF communications functions.

Circuit Approaches for Integrated Synthesizers

Most current transceiver implementations utilize external varactor-tuned resonators to implement the VCO function in one or two local oscillators that are slaved to a crystal reference through fixed or programmable divider chains in a phase-locked loop (PLL). Incorporation of these oscillators on-chip is an important goal in increasing receiver integration level and thereby reducing power. Often the divider is alternately switched between two adjacent divisor values at a high rate to achieve an effective interpolated value of division when very fine channel spacing is required (fractional-n synthesis). In the case of the US analog cellular standard, for example, the channel spacing is only 30kHz at 900 MHz, requiring the synthesizer to tune in very fine steps. This in turn requires the VCO frequency to be divided by a large integer, resulting in a low rate of comparison of current VCO phase with crystal oscillator phase. This in turn requires a low loop bandwidth.

Synthesizer PLL loop bandwidth is an important parameter in receiver design because of its influence on the local oscillator phase noise spectrum. The phase noise present in the local oscillator signal generated by the synthesizer contributes directly to phase noise on the IF or baseband signal after frequency conversion, and as a result directly degrades the effective signal-to-noise ratio (SNR) of phase-modulated signals. More importantly, LO phase noise mixes with adjacent channel signal energy, degrading overall receiver SNR and limiting receiver blocking performance. Finally, in the transmitter the phase noise of the LO contributes noise energy outside the band of the channel being transmitted. Spurious transmitted energy at adjacent channel frequencies must be closely controlled in most systems. With proper PLL loop design, the phase noise of the synthesizer is dominated by the phase noise of the crystal reference for frequencies far below the PLL loop bandwidth, and by the inherent phase noise of the VCO itself for frequencies far beyond the PLL loop bandwidth.

In conventional synthesizers realizing fine channel spacing, the phase comparison frequency is low and loop bandwidth is low. As a result, synthesizer phase noise in the regions of interest is dominated by inherent VCO phase noise. For an LC oscillator, the ratio of the internally generated VCO phase noise power to the carrier power can be shown to be directly related to the ratio of the amount of energy stored per cycle to the thermal energy kT. The energy stored is Q times the energy which must be supplied per cycle, and phase noise is directly related to the inverse of resonator Q, a fact predicted by a number of analyses of phase noise in oscillators.[31] Thus VCOs used to generate LO signals in phase-noise-critical applications almost always use some kind of external high-Q resonator.

A number of approaches show promise for realizing the VCO function on-chip. These include the use of on-chip spiral inductors[46], and the use of synthesizer configurations that allow wide PLL bandwidth so that the phase noise of the VCO is suppressed in the range of interest[27][28].

The use of on-chip inductors and varactors to implement the VCO has been demonstrated [14] using aluminum spiral inductors in standard bipolar technology. Because of the limited Q available, the phase noise achievable is not as low as required in some applications. Other alternatives with higher Q include the use of bond wire inductance [24][52] and the use of plated-up gold inductors over thick oxide [29]. Even with higher-Q inductors, the realization of a low-resistance, wide-range varactor tuning capacitance using standard IC technology is difficult. The simultaneous realization of high Q and tunability, together with either wide tuning range or high center frequency accuracy, is a very difficult task. The use of structures occurring in standard IC technology (such as bipolar base-emitter and base-collector capacitors) to perform the varactor function tends to introduce series resistance losses that reduce the Q to values below 10.

Ring-Oscillator-Based VCOs and Noise-Optimized Synthesizer Architectures

Ring oscillators are particularly attractive for LO generation because they inherently provide the quadrature clocks required for direct conversion and quasi-direct conversion receivers. Recent progress has been made in understanding the fundamental limits on noise in CMOS ring oscillator VCOs and the relationship between phase noise and power dissipation. In effect, for a given power dissipation a ring oscillator has a phase noise approximately equivalent to an LC oscillator with a Q of unity[27][28]. Each factor of 10 reduction in phase noise power with respect to the carrier requires a factor of 10 increase in power dissipation, all else being equal. The use of more advanced technology or lower supply voltages does not greatly alter this situation because of the fundamental processes involved. It appears that a phase noise level of about -106dBc per hertz should be achievable at 1MHz away from the carrier with about 50mW power dissipation in a 4-stage differential ring oscillator operating on a 3 volts supply at 1.8GHz. While this is adequate for some systems, it is inadequate for most digital radio communications applications.

The effects of close-in VCO phase noise can be minimized if PLL loop bandwidth, phase comparison frequency, and loop order can be kept high. A promising approach to achieving this is to use techniques of VCO phase interpolation in order to be able to make more frequent comparison of VCO phase with the instantaneous phase of the crystal reference. Discrete phase interpolation and noise-shaping M/N interpolation [49] are two examples of approaches to this goal. It appears likely that means can be found to increase the effective loop bandwidth by a large factor, thereby greatly reducing the effect of intrinsic VCO phase noise on the LO signal, in turn allowing the use of a ring oscillator VCO at least for some of the applications. In effect, this amounts to making the close-in VCO phase noise more dependent on the (very good) phase noise of the crystal reference and less dependent on the phase noise performance of the VCO. This approach has the additional advantage of preserving a wide loop bandwidth for frequency agility. Also, the short loop time constant potentially allows the powering down of the ring oscillator during inactive frame times in TDMA systems, resulting in power savings that offset the higher inherent active power of the ring oscillator.

The use of quasi-IF receiver architectures, in which the first VCO frequency is fixed, also relieves the phase noise problem because the comparison frequency and loop bandwidth can be kept much higher in a fixed-frequency first VCO. The second VCO/synthesizer performs the tuning function, but its impact on receiver phase noise is smaller since it operates at a much lower frequency.

Direct Digital Synthesis of the LO Signal.

Direct digital synthesis of the LO signal using a DAC, ROM, and phase-accumulation synthesizer is a very attractive alternative due to the excellent frequency agility achieved. Considerations of power dissipation and technology speed capability limit such waveform synthesis to about 100MHz and below for typical current technologies. However, the synthesized waveform can be used as a frequency offset added to a carrier generated with a fixed-frequency VCO and PLL[18][48]. Because the generated frequency is a fixed integer multiple of the crystal reference, the bandwidth of this loop can be kept high, allowing low phase noise in this fixed reference signal. This small difference frequency is used to translate the fixed carrier from the first PLL to the final carrier frequency using a quadrature modulator. It appears that this will be a very effective solution for at least the subset of applications where modest phase noise and modest spurious output component requirements can be tolerated. A similar approach using a secondary PLL to generate the offset frequency has recently been proposed[29].

Low-Power Baseband Signal Conversion and Processing

Depending on the type of transceiver, baseband operations of IF filtering, equalization, timing recovery, symbol constellation decoding, signal correlation, symbol generation, quadrature modulation, frequency synthesis and so forth are required. A major body of current research is aimed at performing more of these functions in the digital domain than is currently the case, with resultant improvement performance, adaptability, and manufacturability. The principal trade-offs are the incurred penalty in power dissipation and die area of the digital implementation, and the cost and die area of the required A/D converter. Rapid progress has been made in the implementation of these types of functions in VLSI CMOS[19][37][38][39].

Many benefits accrue in pushing baseband signal processing into the digital domain, particularly for multistandard adaptive transceivers. For direct conversion receivers, the composite baseband signal contains all the large adjacent-channel blocking signals, and as a result an all-digital implementation of the baseband signal processing would require two A/D converters of greater than 80dB dynamic range and 20MHz effective sampling rate. Some combination of analog and digital filtering will be optimum. For at least the higher-frequency portions of this set of applications, low-power, high-speed approaches such as pipelining will be required. Finding techniques for reducing the power dissipation of these A/D converters is a key goal. Current state of the art for this class of converters is about 1mW/MHz of sample rate at 10 bits [35].

Technologies for High-Integration RF Transceivers

Current transceiver implementations usually use a mix of technologies, with GaAs for the power amplifier and perhaps for the LNA, bipolar or BiCMOS for the mixer and IF functions, and CMOS for the baseband processing. High-integration implementations will require use of a single technology for most of the functions. GaAs will continue to play a very important role at the higher frequencies, but it appears likely that high-integration all-silicon solutions will evolve at the lower end of the spectrum.

Alternative technologies for a transceiver at the integration level of Fig. 2 are BiCMOS and CMOS. Bipolar and BiCMOS solutions are attractive because of the inherent capability of bipolar transistors to provide high g_m at low current, and because of the well-developed family of circuit techniques for RF design using bipolar technology[23]. There is also considerable interest in utilizing CMOS for high-integration transceivers, particularly within the university community. Because of the potentially lower cost of a CMOS implementation, efforts to overcome the poorer characteristics of CMOS for RF by utilizing alternative receiver architectures, taking advantage of the high f_{max} of the NMOS device, and using more adaptation in the receiver may pay large dividends. The continued scaling of CMOS technology, with 0.1 micron devices with f_ts of near 100GHz recently demonstrated,[42][43] should eventually allow this approach.

Silicon-germanium technologies now evolving have the potential to provide bipolar devices with substantially larger f_t and f_{max} than the best current bipolar technologies[44]. The impact of this development is likely to be felt mostly at the high end of the frequency spectrum under discussion here.

Compatible High-Q Inductors and Resonators in Silicon

For reasons explained earlier, on-chip inductors are essential to low-power RF design. Present practice makes extensive use of wirebond inductance and spiral inductors in silicon in products currently in production. Improvement of the implementation of and the modeling of integrated inductors is a key goal.

Bond Wire Inductors

Bond wires provide inductance on the order of 1-4nH, depending on length, at 2 GHz, with Q on the order of 50. Higher inductance is realizable with unusual pad placements.[24][52] While pad-pad jump bonds are possible, they are not compatible with most automated bonding equipment. Inductor tolerance is a function of die attach, bonder mechanical accuracy, and wire diameter, and is in the +/-20% range at present. Adjustments to mechanical assembly procedures could improve this significantly. Matching of bond wires on one side of the die to the same post is also better than 20%. Mutual coupling effects of adjacent wires must be accounted for through electromagnetic analysis of the configuration with a commercial package.

Creative use of bond wire inductance in SO and SSO packages with (perhaps) custom lead frames is a highly promising approach to implementing matching and tuning inductors in 1 and 2 GHz LNAs, VCOs, and power amplifiers. These elements are already widely used to perform part of the impedance matching function at LNA inputs and in power amplifiers. The tolerance is a major problem, since there is no good high-Q variable capacitive element available.

Spiral Inductors on Silicon

Extensive work has been carried out on spiral inductors on silicon[45][46]. These devices can provide implement inductors up to about 10nH range in reasonable area, with Q limited to about 5 at 1 GHz and 10 at 2 GHz by metal series resistance for standard technologies. Self-resonance due to the large parasitic capacitance to the substrate is a substantial problem, and drops to about 2GHz for a 10nH inductor in typical technology. Since the inductor is usually being used to match impedance or tune a gate or base diffusion capacitance, the parasitic capacitance can usually be incorporated in the design process as long as the self-resonant frequency is far above the frequency of interest.

Two approaches have been described in recent work for improving the performance of spiral inductors. In one approach, [25] a pit is etched under the spiral to remove the substrate capacitance and resulting self-resonance. The resulting structures have moderate Q but good self-resonance characteristics. Another approach is the deposition of thick oxide following normal IC fabrication, and the deposition of highly conductive interconnect metallization such as gold to form a high-Q spiral with reasonable self-resonance behavior. [29] This approach has high promise because the subsequent processing is relatively non-invasive to the underlying silicon and involves only low-temperature deposition and masking steps.

Package and Substrate Modeling

Perhaps the greatest single barrier to higher integration in RF transceivers is undesired interactions through substrate and package coupling. The problem can be addressed through a number of architectural and circuit approaches, but a critical missing link is a substrate and package modeling and simulation methodology to allow accurate prediction of these coupling effects prior to fabrication. Rapid progress is being made in this area [33][41][50].

Summary

Prospects for continued progress in high-integration, low-cost RF transceivers is excellent. A key requirement for progress is close collaboration between transceiver and system designers and architects, RF and digital circuit designers, and device and package modeling engineers, so that opportunities for innovation with new architectural approaches can be identified and exploited.

Acknowledgments

Research sponsored by NSF under grant MIP9101525, ARPA under contract J-FBI-92-150, and ARO under grant DAAHO4-93-G-0200. The help of numerous graduate students in the RF group at UC Berkeley is gratefully acknowledged.

References

1. J. Rapeli, "IC Solutions for Mobile Telephones," book chapter in *Design of VSLI Circuits for Telecommunications and Signal Processing*, Kluwer, June 1993.
2. D. Rabaey and J. Sevenhaus, "The challenges for analog circuits design in mobile radio VLSI chips," *Proceedings of the AACD*(Leuven), pp 225-236, March 1993.
3. L. Longo, et al, "A cellular analog front end with a 98dB IF receiver," *Digest of Technical Papers*, 1993 International Solid-State Circuits Conference, San Francisco, CA, Feb. 1994.
4. K.R. Lakshmikumar, D.W. Green, K. Nagaraj, K.-H. Lau, and others, "A baseband codec for digital cellular telephony," *IEEE Journal of Solid-State Circuits*, Dec. 1991.
5. M. Rahier, et al, "VLSI components for the GSM Pan-European digital cellular radio system," 4th Nordic Seminar on Digital Mobile Radio Communications DMR IV, 1990, Oslo, Norway.
6. I. A. Koullias, et al, "A 900 MHz transceiver chip set for dual-mode cellular radio mobile terminals," *Digest of Technical Papers*, 1993 International Solid-State Circuits Conference, San Francisco, CA, Feb. 1994.
7. S. Sheng, R. Allmon, L. Lynn, I. O'Donnell, K. Stone, R. W. Broderson, "A monolithic CMOS radio systems for wideband CDMA communications," *Wireless '94*, Calgary, Canada, July 1994.
8. V. Thomas, et al, "A one-chip 2 GHz single-superhet receiver for 2Mb/s FSK radio communications," *Digest of Technical Papers*, 1993 International Solid-State Circuits Conference, San Francisco, CA Feb. 1994.
9. J. Sevenhans, et al, "An analog radio front-end chip set for a 1.9 GHz mobile telephone application," *Digest of Technical Papers*, 1993 International Solid-State Circuits Conference, San Francisco, CA, Feb. 1994.
10. J.K. Cavers, M.W. Liao, "Adaptive compensation for imbalance and offset losses in direct conversion transceivers," *IEEE Transactions on Vehicular Technology*, Nov. 1993.
11. J. Sevenhans, et al, "An integrated Si bipolar RF transceiver for a zero IF 900 MHz GSM digital mobile radio single chip RF up and RF down converter of a hand portable phone," *Digest of Technical Papers*, 1991 Symposium on VLSI Circuits, Honolulu, HI, June 1991.
12. P. Weger, et al, "Completely integrated 1.5 GHz direct conversion receiver," *Digest of Technical Papers*, 1994 Symposium on VLSI Circuits, Honolulu, HI, June 1994.
13. K. Voudouris, J.M. Noras, "Direct conversion receiver for the TDMA mobile terminal," IEE Colloquium on 'Personal Communications: Circuits, Systems and Technology' London, UK, 22 Jan. 1993.
14. Plessey GP1010 GPS Receiver Preliminary Data Sheet, Oct. 1992.
15. A. Bateman, D.M. Haines, "Direct conversion transceiver design for compact low-cost portable mobile radio terminals," 39th IEEE Vehicular Technology Conference, San Francisco, CA, 1-3 May 1989.
16. J. Min, et al, "An all-CMOS architecture for a low-power frequency-hopped 900 MHz spread spectrum transceiver," *Digest of Technical Papers*, 1994 Custom Integrated Circuits Conference, San Diego, CA, June 1994.
17. A. Vanwelsenaers, D. Rabaey, E. Vanzieleghem, J. Sevenhans, and others, "Alcatel chip set for GSM handportable terminal," *Proceedings of 5th Nordic Seminar on Digital Mobile Radio Communications DMR V*, Helsinki, Finland, 1-3 Dec. 1992.
18. A. Abidi, "Radio frequency integrated circuits for portable communications", *Digest of Technical Papers*, 1994 Custom Integrated Circuits Conference, San Diego, CA, June, 1994.

19. A. Chandrakasan, S. Sheng, R. W. Broderson, "Design considerations for a future portable multi-media terminal," *Third-Generation Wireless Information Networks*, Kluwer Academic Publisher, 1992.

20. P. Baltus and A. Tombeur, "DECT Zero-IF receiver front-end," *Proceedings of the AACD* (Leuven), pp 295-318, March 1993.

21. M. Steyaert and J. Crols, "Analog integrated polyphase filters," *Proceedings of the AACD* (Eindhoven), March 1994.

22. M. Thiriamsut, et al, "A 1.2 micron CMOS implementation for a low-power 900 MHz mobile telephone radio-frequency synthesizer," *Digest of Technical Papers*, 1994 Custom Integrated Circuits Conference, San Diego, CA, June 1994.

23. R.G. Meyer, W.D. Mack, "A 1- GHz BiCMOS RF front-end IC," *IEEE Journal of Solid-State Circuits*, March 1994.

24. M. Steyaert, J. Craninckx, "A 1.1 GHz oscillator using bondwire inductance," *Electronics Letters*, 3 Feb. 1994.

25. J.Y.-C. Chang, A.A. Abidi, M. Gaitan, "Large suspended inductors on silicon and their use in a 2-μm CMOS RF amplifier," *IEEE Electron Device Letters*, May 1993.

26. S. Sampei, K. Feher, "Adaptive DC-offset compensation algorithm for burst mode operated direct conversion receivers," Vehicular Technology Society 42nd VTS Conference. Frontiers of Technology. From Pioneers to the 21st Century, Denver, CO, 10-13 May 1992.

27. T. Weigandt, et al, "Analysis of timing jitter in CMOS ring oscillators," *Digest of Technical Papers*, 1994 International Symposium on Circuits and Systems, London, UK, June 1994.

28. B. Kim, et al, "DLL/PLL system noise analysis for low jitter clock synthesizer design," *Digest of Technical Papers*, 1994 International Symposium on Circuits and Systems, London, UK, June 1994.

29. K. Negus, et al, "A highly integrated transmitter IC with monolithic narrowband tuning for digital cellular handsets," *Digest of Technical Papers*, 1993 International Solid-State Circuits Conference, San Francisco, CA Feb. 1994.

30. C. Nguyen, "Integrated filters using micro-mechanical resonators", Ph.D. Dissertation, University of California, Berkeley, CA, Nov. 1994.

31. J.K.A. Everhard, "Low-noise power-efficient oscillators, theory and design," *IEEE Proceedings*, Aug., 1986.

32. G. Chang, et al, "A low-power CMOS digitally synthesized 0-13 MHz agile sinewave generator," *Digest of Technical Papers*, 1994 International Solid-State Circuits Conference, San Francisco, CA, Feb. 1994.

33. R. Gharpury and R.G. Meyer, "Analysis and simulation of substrate coupling in integrated circuits," *International Journal of Circuit Theory and Applications*, to be published.

34. C.D. Hull, R.G. Meyer, "A systematic approach to the analysis of noise in mixers," *IEEE Transactions on Circuits and Systems I: Fundamental Theory and Applications*, Dec. 1993.

35. T. Cho, et al, "A 10-bit, 20MS/sec, 35mW pipeline A/D converter," *Digest of Technical Papers*, 1994 Custom Integrated Circuits Conference, San Diego, June 1994.

36. F. Lu, H. Samueli, J. Yuan, C. Svensson, "A 700- MHz 24-b pipelined accumulator in 1.2-μm CMOS for application as a numerically controlled oscillator," *IEEE Journal of Solid-State Circuits*, Aug. 1993.

37. R. Jain, H. Samueli, P.T. Yang, C. Chien, and others, "Computer-aided design of a BPSK spread-spectrum chip set," *IEEE Journal of Solid-State Circuits*, Jan. 1992.

38. B.-Y. Chung, C. Chien, H. Samueli, R. Jain, "Performance analysis of an all-digital BPSK direct-sequence spread-spectrum IF receiver architecture," *IEEE Journal on Selected Areas in Communications*, Sept. 1993, vol. 11, no.7:1096-107.

39. B.C. Wong, H. Samueli, "A 200- MHz all-digital QAM modulator and demodulator in 1.2-μm CMOS for digital radio applications," *IEEE Journal of Solid-State Circuits*, Dec. 1991.

40. M. Rofougaran, et al, "A 900 MHz CMOS power amplifier with programmable output," *Digest of Technical Papers*, 1994 Symposium on VLSI Circuits, Honolulu, HI, June, 1994.

41. M.J. Loinaz, D.K. Su, B.A. Wooley, "Experimental results and modeling techniques for switching noise in mixed-signal integrated circuits," *Digest of Technical Papers*, 1992 Symposium on VLSI Circuits. Seattle, WA, 4-6 June 1992.

42. R.H. Yan, K.F. Lee, D.Y. Jeon, Y.O. Kim, and others, "High performance 0.1- micron room temperature Si MOSFETs," *Digest of Technical Papers*, 1992 Symposium on VLSI Technology, Seattle, WA, 2-4 June 1992.

43. C. Jian, S. Parke, J. King, F. Assaderaghi, and others, "A high speed SOI technology with 12 ps/18 ps gate delay operating at 1.5V," *Proceedings of IEEE International Electron Devices Meeting*, San Francisco, CA, 13-16 Dec. 1992.

44. J. D. Cressler, et al, "Silicon-germanium heterojunction bipolar technology: the next leap for silicon?," *Digest of Technical Papers*, 1994 ISSCC, San Francisco, CA, Feb. 1994.

45. N.M Nguyen, R.G Meyer, "Si IC-compatible inductors and LC passive filters," *IEEE Journal of Solid-State Circuits*, Aug. 1990.

46. N.M. Nguyen, R.G. Meyer, "A 1.8- GHz monolithic LC voltage-controlled oscillator," *IEEE Journal of Solid-State Circuits*, March 1992.

47. M. McDonald, "A 2.5 GHz BiCMOS image-reject front-end," *Digest of Technical Papers*, 1993 International Solid-State Circuits Conference, San Francisco, CA, Feb. 1994.

48. G. Chang, et al, "A low-power CMOS digitally synthesized 0-13 MHz agile Sinewave generator," *Digest of Technical Papers*, 1993 International Solid-State Circuits Conference, San Francisco, CA, Feb. 1994.

49. T. Riley, et al, "Delta-Sigma modulation in fractional-N frequency synthesis," *IEEE Journal of Solid-State Circuits*, May 1993.

50. D.K. Su, M.J. Loinaz, S. Masui, B.A. Wooley, "Experimental results and modeling techniques for substrate noise in mixed-signal integrated circuits," *IEEE Journal of Solid-State Circuits*, April 1993.

51. A. Mansell, A. Bateman, "Practical implementation issues for adaptive predistortion transmitter linearization," *Digest of Technical Papers*, IEE Colloquium on 'Linear RF Amplifiers and Transmitters' London, UK, 11 April 1994.

52. J. Craninckx, M. Steyart, "A CMOS low-phase-noise voltage-controlled oscillator with prescaler," *Digest of Technical Papers*, 1995 International Solid-State Circuits Conference, San Francisco, CA, Feb. 1995.

The challenges for analogue circuit design in mobile radio VLSI chips

Jans Sevenhans and Dirk Rabaey from Alcatel Bell Telephone describe the present status of mobile phone ASIC's. Radio analogue design expertise is on the move from printed circuit board to large scale silicon monolithic.

Mobile radio telephony is becoming a driving application for analogue circuit design using silicon CMOS and RF bipolar technology.

Similar things are happening for several wireless personal communication systems. Basically the cellular radio telephone, the wireless PABX and the wireless SLIC bringing the same challenges to analogue circuit design: ie maximum integration of the basic radio functions into 1 or 2 silicon chips, CMOS, Bipolar or BICMOS or GaAs. The analogue circuit designer for radio telephone applications will need all the state of the art analogue design know how available today, from RF-mixers and GHz range low noise amplifiers and local oscillator synthesizers over base band 100kHz CMOS analogue to low frequency speech analogue to digital conversion. And for all these circuits the message is: minimum power consumption for battery autonomy, minimum silicon area for maximum functional integration per die to obtain a small, low cost pocket size radio telephone.

For the asic integration of a radio transceiver the "One asic per transceiver function" approach was regarded as the optimum architecture to obtain the maximum" per die integration level". In the next generation receiver and transmitter functions operating at the radio frequency are integrated in one RF-bipolar ASIC and the baseband circuitry in a CMOS mixed analogue/digital asic doing the analogue filtering, automatic gain control and A/D conversion. The synthesizer can be partitioned with the VCO and prescaler in the RF front-end IC and the two modulus counter in the CMOS analogue baseband IC but the integration of the VCO leads to some practical problems that will be discussed further on.

The low noise radio receiver

A high performance radio receiver basically consists of the following functionalities:

A low noise amplifier (LNA) to bring the femto Watt radio input signal up to the sufficient level to cope with the thermal noise of the mixers. The challenge here for analogue radio design is to build low cost silicon bipolar low noise amplifiers with very low power consumption, about 5mA to provide +/-20dB switchable gain and a 3dB or better noise figure and an input third order intercept point above 14dBm to handle the high intermodulation radio input signals and the blocking levels specified by GSM.

One or two cascaded mixer stages bring the radio signal down from RF to baseband with sufficient gain to further process the analogue receiver signal in noisy standard CMOS technology.

The challenge in the 900MHz or 1.8GHz mixers is first to integrate the set of I and Q mixers into one silicon die together with the quadrature phase shifter that provides the 90° local oscillator signals to drive the I and Q quadrature mixers.

In general we can say that to build a radio receiver with an acceptable Bit Error Rate, we need a low noise, low power (<25mA, 3 to 5V) analogue radio front end with 40dB total switchable gain in the low noise amplifier and the mixers before we can add 50nVHz$^{-1/2}$ CMOS thermal noise to the baseband radio receive signal.

For this low noise radio receiver the choice of technology is between advanced Silicon RF bipolar with 10 to 20GHz f_T or GaAs. The availability of those RF

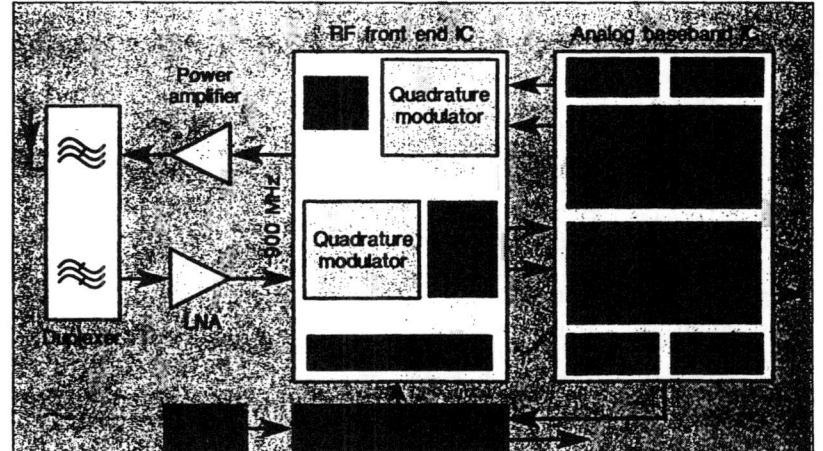

Figure 1: Block diagram of the radio analogue front end circuitry in a mobile radio telephone

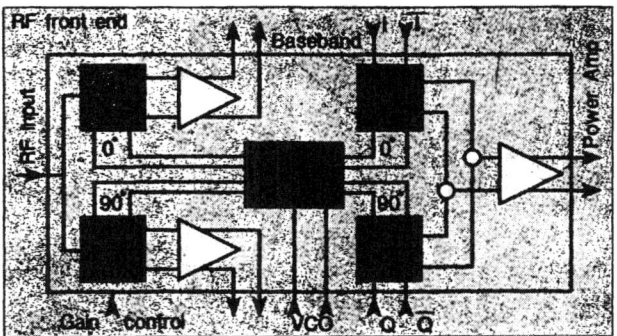

Figure 2: Monolithic silicon bipolar RF front end

Reprinted with permission from *Microwave Engineering Europe*, J. Sevenhans and D. Rabaey, "The Challenges for Analogue Circuit Design in Mobile Radio VLSI Chips," pp. 53-59, May 1993. © Miller Freeman Ltd.

bipolar technologies with Silicon foundries in Europe, USA as well as in Japan has lead the majority of integrated radio designs to the use of RF-bipolar for the transceiver radio frontend of the wireless personal communication terminals.

CMOS filtering, AGC and analogue to digital conversion technique cover all the basic needs for the base band signal processing of a high performance radio receiver. Several options are open to make a compromise between filtering, AGC and A/D dynamic range. In the limit one could build a receiver out of two 0.9GHz sigma delta a/d convertors in RF-Silicon bipolar or GaAs right after the LNA operating on quadrature phases of the local oscillator. In this case you need over 80dB linear range for the a/d convertor and less than 5nVHz$^{-1/2}$ equivalent input noise for the sigma delta modulator A/D convertors. In this extreme compromise another challenge is to build a 0.9GHz decimator first filter stage before the first downsampling. All of the filtering and AGC is then a job for a fast digital signal processor. Also undersampling of the radio-signal is a promising technique for futuristic radio receivers.

But today's solution is to use simple 8 bit a/d conversion and a classical rational AGC algorithm in combination with dedicated filtering between all the gain stages. The challenge of analogue radio design is to provide sufficient gain on the wanted signal to overcome the noise of the active filters that have to suppress the blocking signals and adjacent channels to protect the linear performance of the gain stage to follow, in other words a lot of delicate analogue CMOS design to provide a dB of filtering for each dB of gain until your receive signal reaches the A/D convertors input.

For accurate analogue filtering switched capacitor filters are the best as long as we have sufficient supply voltage to cover the signal

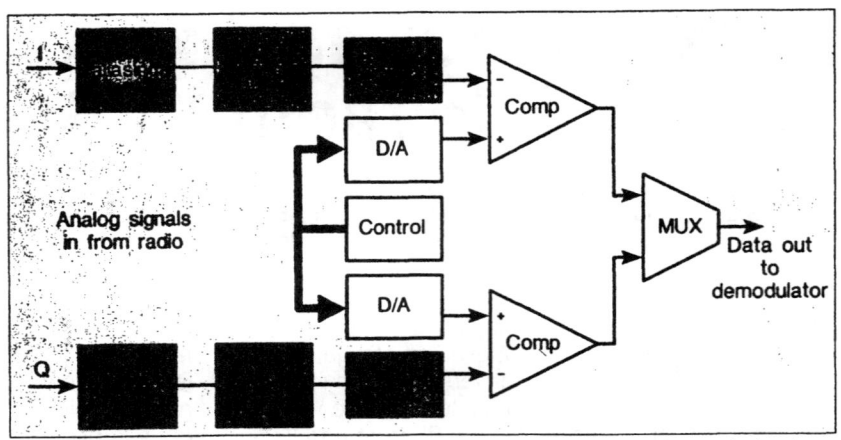

Figure 3: *GSM receive channel filter and successive approximation switched capacitor A/D*

amplitude and the thresholds of the switches. So with 5V supply and 1V threshold for the switches we still are 0.5V away from hard compression and signal clipping. With 3V supply voltage the trade off is a bit different. In MAD-design (Mixed Analogue and Digital) in 0.5 micron CMOS the 3V power supply is a technological maximum rating that analogue design will have to live with. The use of time continuous filters, OTA-C or MOSFET-C filters can become a solution as we go into 3V CMOS radio analogue. At least no switches threshold voltage is consuming any linear range between the supply rails. And CMOS transconductors with 60dB linear range for noise and distortion on 5V supply voltage will have to improve in the next few years to make them applicable for 3V CMOS radio analogue filter design.

For the receiver architecture the choice is between heterodyne and homodyne demodulation.

Heterodyne receivers have the advantage that the local oscillator frequency is different from the radio signal frequency on the antenna. Good design practice has lead to the choice of 71MHz as a standard for the IF (intermediate frequency) in GSM heterodyne receivers, because the 900MHz GSM band has a width of 70MHz: 890-915MHz for the base station receiver and 935 to 960MHz for the mobile terminal receiver. A second advantage of heterodyne radio receivers is the opportunity to filter the radio signal at the 71MHz IF. These IF filters can suppress the neighbouring channels and blocking signals to optimise the use of the available linear range in the rest of the receive chain. AGC amplifiers at IF have the inherent advantage that offset problems are easily solved by AC-coupling trough small (on chip) couple capacitors having transient time constants in the 10nsec range.

The only problem to integrate an IF receiver in a monolithic device is the integration of the IF filters. Passive filtering (SAW) is the solution for high Q filtering in a low noise receiver and the SAW filter drivers are very power hungry. Time continuous or sampled analogue filtering at 71MHz is very well feasibly in RF bipolar or GaAs biquads but the noise figure of a high Q active filter is going up with the Q to G ratio: quality factor to gain of one filter section.

The difficulty to integrate low noise IF band filtering is the main reason to work without an IF for a low noise radio receiver.

In a homodyne receiver, as we go directly to baseband in one mixer stage, all the filtering is well know base band filtering with low Q-factors, easy to integrate in CMOS time continuous or sampled data filters, switched capacitor or switched current circuits. Also the use of a homodyne receiver avoids the need for an image filter that prevents an IF receiver to create an unwanted response to a spurious signal at 2 times the intermediate frequency away from the wanted signal frequency.

The homodyne receiver, simple and compact as it is, with all it's advantages for low Q low pass filtering has one major draw back: OFFSET.

Demodulation of local oscillator leakage in the radio front end and self detection of high blocking levels are a major problem for radio communication systems like GSM that use a base band signal spectrum with important DC and low frequency content. In those systems the high pass action of an AC-coupling is cutting too much energy away from the signal spectrum. For this reason a homodyne receiver for GSM cellular radio mobile communication needs a sophisticated offset cancelling algorithm to cope with 3 types of offset: static offset resulting from transistor and resistor matching errors, dynamic offset resulting from the mixer demodulating the local oscillator leaking to the antenna signal and a second dynamic offset resulting from the high blocking signals self mixing as the antenna signal is leaking to the local oscillator input of the mixers.

Over the past few years elegant solutions have been developed to overcome this offset problem by monitoring the offset in the digital base band signal, low passing it and feeding the correct offset subtraction back to the input of the CMOS AGC.

The offset problem has been the dominant reason for a lot of radio telecom companies not to use the homodyne receiver for GSM and other radiocom receivers in terminals and base stations. THe implementation of sophisticated offset algorithms in the receiver DSP, coupled with D/A convertors to subtract the correct offset value have overcome this problem.

The voice codec

The other interface of the radiotelephone is the voice interface with the microphone and earpiece. This is a classical speech interface with 8kHz speech sampling with 13 bit accuracy.

The microphone signal is going to a second order switched capacitor sigma delta a/d convertor operating at 1MHz. The 1 bit of 1MHz PDM signal is then filtered and downsampled to 32kHz in the decimator and further lowpass filtered before downsampling to the 8kHz linear speech.

The earpiece signal coming from the GSM speech transcoder is passing through a symmetrical signal path. A bandpass filter operating at 32kHz is oversampling the speech signal 4 time and a linear interpolator increases the speech sample rate up to 256kHz before the digital sigma delta modulator converts the speech signal to the 1 bit 1MHz PDM. A switched capacitor a/d convertor then provides the analogue speech signal to the earpiece.

This voice interface is very similar to the analogue interface you find in every plain old telephone. This is the reason why in today's radiotelephones you find the voice codec as a separate component usable in any voice interface. Nothing but the silicon area per die prevents us from integrating the voice interface on a single chip with the baseband CMOS radio transceiver.

The low noise radio transmitter

The noise constraints for the transmitter are very high as well: 25dB above the $-174 dBmHz^{-1}$ thermal noise floor.

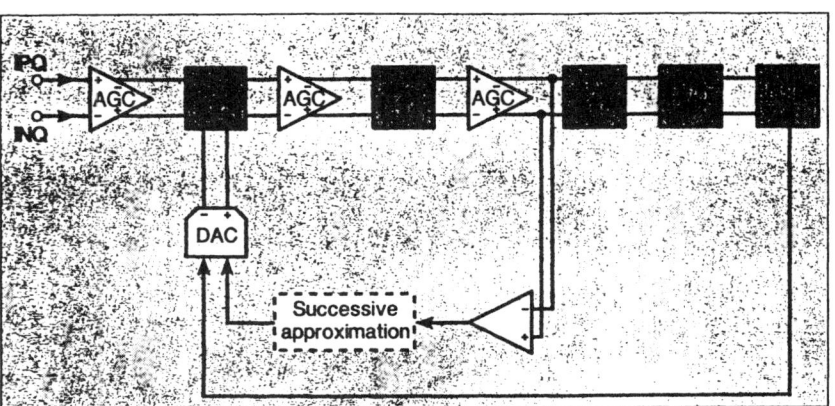

Figure 4: At RF the homodyne receiver by it's simplicity and compactness is certainly an advantageous solution for battery powered pocket radio telephones.

The noise constraints in the transmitter are set to prevent mobile transmitters to jam each other and the receiver of the base stations for distant users. From this point of view the homodyne transmitter was chosen: only one mixer state to contribute noise to the transmitter signal.

The quadrature phase shifter on the 900MHz local oscillator signal is a delicate aspect of the homodyne transmitter where a phase accuracy of .5° is required over the 70MHz range of the GSM receive and transmit band.

The power amplifier is still a big challenge for monolithic integration. For handportable phones the output power is between 0.8 and 2 Watt in a 1:8 burstmode duty cycle for GSM. Depending on the power efficiency of the class AB amplifier this corresponds to a power dissipation of less than half a Watt in one die, which is certainly not an obstacle for a plastic package. To provide 2W to an output transformer from a 3V supply it is clear that output currents will be in the range of ampères, forcing the designer to use 10 pins in parallel when using plastic flat packages. But the real challenge here for the analog radio design is in the efficiency of the class AB power amplifier. Today expensive hybrid modules are used for this application. But as the mobile radio telephone market will expand in the near future, RF MMIC design houses will spend the effort to develop monolithic solutions in Silicon or GaAs.

The control of the output power as a function of temperature and aging is another challenge for analogue design to measure the output power with an RF peak detector

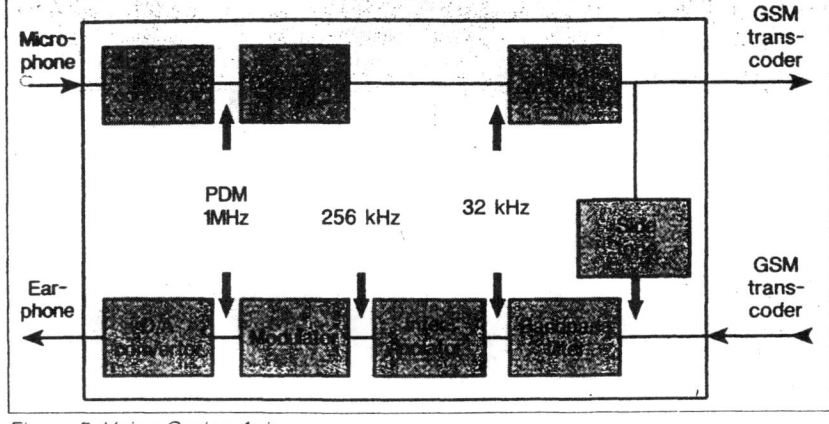

Figure 5: Voice Codec Asic

and adjust the gain of the power amplifier in a stable feedback loop. In addition to this the loop must have sufficient bandwidth to follow the burst ramp up of the time multiplexed GSM transmit signal.

The base band signal generator in a GSM transmitter basically consists of a D/A convertor and a ROM containing the Gaussian shaped quadrature I and Q signals. The challenge for analogue design here is the low power and low voltage constraint. The use of 3V supply voltage in the digital circuitry is driving the analogue design into the 3V range as well, and the RF bipolar transmit mixers need 1V signal amplitude driven to the base band input pins of the Gilbert cell mixers to keep up the signal to noise ratio of the transmitter. A signal amplitude of 1V is easy in a 5V design, but in a 3V design you need a rail to rail amplifier for this.

The local oscillator synthesizer

State of the art mobile radiotelephones use phase modulation on a carrier in the lower GHz spectrum: 0.9GHz for GSM, 1.8GHz for PCN and DECT up to 3.5GHz for other wireless telephone systems in the near future. The radiotelephone terminals have to rely on temperature controlled reference crystal oscillators (TCXO) to provide the local oscillator signal with a frequency accuracy of 5ppm. These reference oscillators, operating at 13 or 26MHz for GSM are not the first candidate for further integration. The challenge for radio analogue monolithic integration is in the VCO and the prescaler.

The VCO is very difficult to integrate in an RF bipolar or GaAs front end because of the resonator. Resonators integrated on silicon have been reported with Q factors not higher than 10, which is far too low to obtain a VCO with -100dBc phase noise at

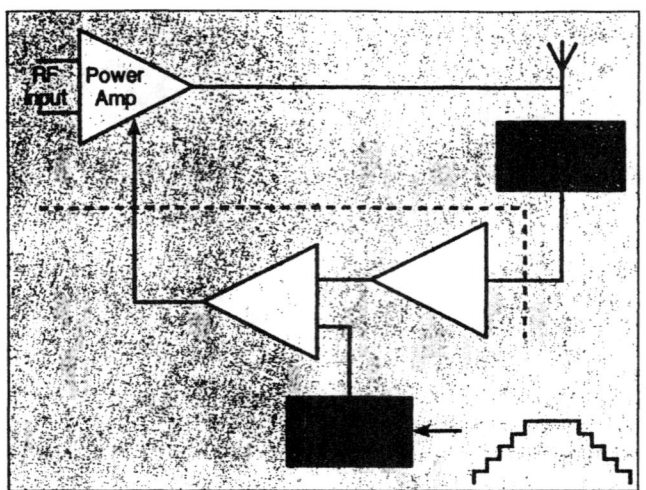
Figure 6: RF power amplifier with a feedback output level control loop.

10kHz away from the carrier. The Q-factor of the spiral inductors on silicon is physically limited to very low values because in the substrate the electromagnetic field of the spiral inductor is damped in substrate currents.

This damping of RF electromagnetic fields in the bulk is a good thing to provide isolation between functional blocks on a Silicon MMIC, but for an LC resonator it is killing the Q-factor. Probably clever designers will come up one day with high Q girator solutions or biquad resonators in 20GHz RF bipolar or GaAs technology, but until then we are stuck with ceramic resonators for high Q VCO's and monolithic BICMOS synthesizers can integrate only the RF divide by 64/65 two modulus prescaler in the bipolar part and the low frequency (15 to 50MHz) CMOS two modulus divider, the charge pump and the phase comparator. An other challenge for CMOS analogue radio design is the integration of the RF prescaler in submicron CMOS.

As the f_T of submicron CMOS is well beyond a GHz, a full CMOS 0.9GHz and 1.8GHz synthesizer is the next step for analogue radio design to further reduce the cost of a mobile radio telephone.

Technology aspects

Radio analogue design is in a turbulent evolution today because of the availability of RF-bipolar technology up to 20GHz f_T. The emphasis is moving from strip line and radio board design towards single chip full radio integration. Single chip radio is still an overstatement for mobile radio telephones today and the economic aspects are not proven yet, but the experts now know that in less than 100mm^2 submicron BICMOS it is possible to integrate the analogue radio for GSM including the synthesizer with prescaler, the RF-mixers, the base band filters and AGC as well as the receive and transmit A/D and D/A converters.

For the low noise amplifier, the VCO and the power amplifier there is still some reluctance to go for silicon monolithic integration. Micromodules are certainly a valid alternative and an intermediate stage on the route towards monolithic.

The trade off between GaAs and Silicon is also a moving compromise, 3 years ago monolithic microwave integrated circuits were a GaAs monopoly, today silicon has proven to be more cost effective, at least for applications below 5GHz. But recent publications show 73GHz f_T for SiGe polysilicon emitter heterojunction bipolar transistors.

Submicron CMOS developments are more important for the digital part of the radio system because in analogue the total silicon area will not benefit from the minimum dimensions to the same extent.

Matching and other requirements in accurate analogue circuits prevent us from using minimum gate-length transistors and

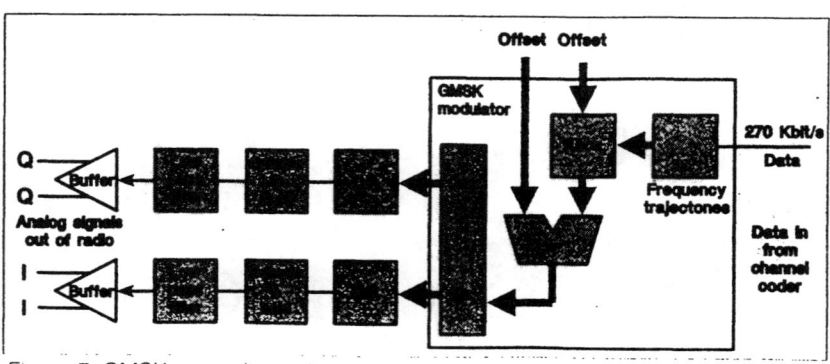
Figure 7: QMSK transmit quadrature pulls shaper.

minimum width poly for resistors and capacitors.

Conclusion

The challenge in analogue design for mobile telephony has a very broad spectrum: provide RF mixers for 3V supply in advanced bipolar technologies, come up with an active resonator with very low phase noise, design a monolithic 2W RF power amplifier, design CMOS 3V time continuous filters for the receive blocking and band filters. But as the silicon bipolar technology continues to catch up with GaAs for the RF building blocks and submicron CMOS trespasses the GHz border, some of the traditional compromise radio architectures will be questioned and new techniques will come up in analogue as well as digital radio design.

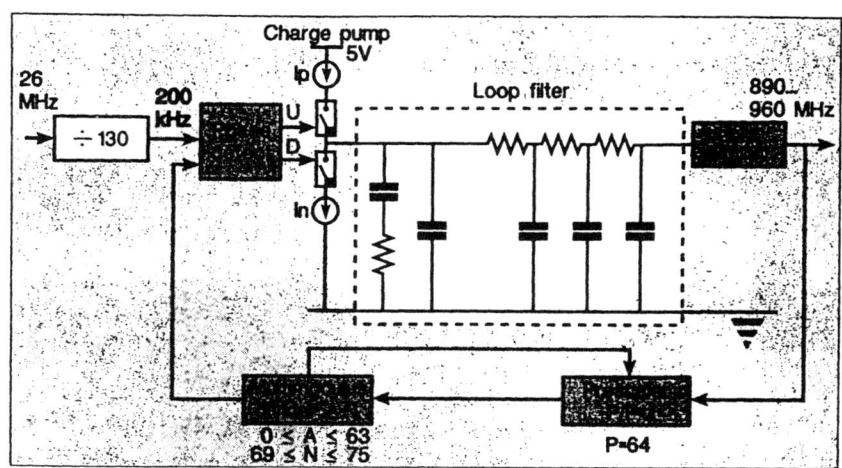

Figure 8: Dual modulus local oscillator synthesizer

References

[1] E F Crabbe, J H Comfort, Wai Lee, J D Cressler, B S Meyerson, J Y C Sun, J M C Stork: "73GHz Self Aligned SiGe-Base Bipolar Transistors with Phosphorus-doped Polysilicon Emitters". IEEE Electronic device letters, Vol 13, No 5, May 1992.

[2] D Haspeslagh, A Delarbre, E Moerman, I Girard: "A General Digital Signal Processor in 1.2um CMOS with onchip D/A and A/D conversion for use in a GSM handportable terminal". ESSCIRC, Milano, September 1991.

[3] H Busschaert, P Reusens, G Van Wauwe, M De Langhe, R Van Camp, C Gouwy, L Dartois: "A power efficient channel coder/decoder chip for GSM terminals". IEEE-CICC Conference, 12-15 May 1991, San Diego.

[4] E Vanzieleghem, L Dartois, J Wenin, A Vanwelsenaers, D Rabaey: "SPEECH: A Single Chip GSM Vocoder". IEEE-CICC 1992.

[5] A Vanwelsenaers, J Wenin, J Sevenhans, D Haspeslagh, J Haspeslagh, A Chateau, F Gerard, H BUsschaert, E Vanzieleghem, V Andreiu: " Alcatel Chipset for GSM Handportable Terminal" Digital Mobile Radio Conference (DMR 5), Helsinki, Finland, 1-3/12/92.

[6] Johan J J Haspeslagh D Sallaerts, P P Reusens, A Vanwelsenaers, R Granek D Rabaey,: "A 270Kbit/s 35mW Modulator IC for GSM Cellular Radio Handheld Terminals". IEEE journal of solid-state circuits, Vol 25, No 6, December 1990.

[7] J Sevenhans, A Vanwelsenaers, J Wenin, J Baro: "An integrated SI bipolar RF transceiver for a Zero-IF 900MHz GSM digital mobile radio front end of a handportable phone". IEEE-CICC Conference 12-15 May 1991, San Diego.

2

TRANSMITTERS AND RECEIVER ARCHITECTURES

THE selection of the basic system arrangement or architecture of a radio transceiver is a crucial first step in the design process. This choice is influenced by many factors, such as the projected cost of hardware production, power supply requirements, the physical size of the resulting electronics, and expected manufacturing yield. The decades-long reign of the classical superheterodyne architecture as the popular choice of receivers arises from the relative ease with which it meets the stringent requirements of noise control, spurious signal tolerance, and performance robustness. However, its need for a relatively large number of expensive and bulky passive filters has prompted an ongoing search for alternative choices. The continuing evolution of electronic technology from tubes to discrete transistors to ever-higher performance levels in integrated circuit realizations has opened the way for new architectures that were impractical in the past because of their need for large numbers of high-frequency devices with close matching. The capability of manufacturing large numbers of well-matched devices with very high frequency performance is a major feature of modern IC technologies. An excellent example of this trend is the current activity aimed at developing manufacturable receivers based on direct conversion, which eliminates the large and expensive IF filter external to the IC chip. The issues and tradeoffs involved in defining the architecture of transceivers are discussed in this section.

A Third Method of Generation and Detection of Single-Sideband Signals*

DONALD K. WEAVER, JR.†, ASSOCIATE MEMBER, IRE

Summary—This paper presents a third method of generation and detection of a single-sideband signal. The method is basically different from either the conventional filter or phasing method in that no sharp cutoff filters or wide-band 90° phase-difference networks are needed. This system is especially suited to keeping the signal energy confined to the desired bandwidth. Any unwanted sideband occupies the same band as the desired sideband, and the unwanted sideband in the usual sense is not present.

THE PURPOSE of this paper is to present a third basic method of generation and detection of single-sideband signals. Two methods are commonly used today. A block diagram of the first of these, the filter method, is shown in Fig. 1. The input signal (a speech waveform, for example) is applied to a balanced modulator along with the first translating or carrier frequency. The two normal sidebands appear in the output of the balanced modulation, but the carrier frequency is balanced out. The purpose of the filter is to select one sideband and reject the other. When the desired frequency location of the single-sideband signal is high compared with the original location of the input signal (*e.g.*, translating speech to the hf region), it becomes very difficult to obtain filters that will pass one sideband and reject the other. To avoid this, the translation is done in several steps so as to ease the filter requirement.

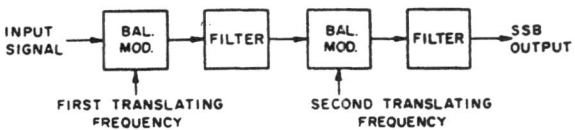

Fig. 1—Filter method of SSB generation.

Fig. 1 shows a system using two translational steps. In many radio transmission systems, three to five translational steps may be used. The detection problem is simply an inverse operation; that is, the arrows in Fig. 1 could be reversed. In detection, balanced modulators are not necessary, and ordinary converter circuits are satisfactory.

The second method, generally called the phasing method, is shown in Fig. 2. The input signal is applied to a wide-band 90° phase-difference network. This network passes all frequencies of the input signal uniformly in amplitude. However, the phase response is such that a sinusoidal input whose frequency falls anywhere within the input signal frequency band will result in two equal amplitude sinusoidal signals whose phases differ by 90°. These quadrature signals are applied to a pair of balanced modulators. The translating carrier frequency is also divided into two 90° components. When the output signals from these two balanced modulators are added, one set of sidebands will add in phase, generating the desired signal, while the other sideband will cancel itself out. By subtracting instead of adding, it is possible to change sidebands.

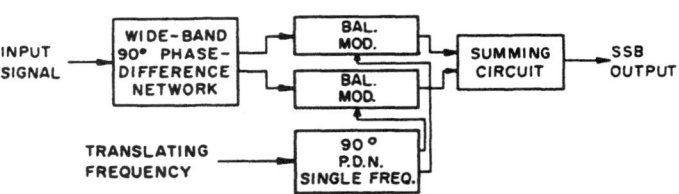

Fig. 2—Phasing method of SSB generation.

As this is a balancing method and does not require any sharp cutoff filters, it is possible to generate the desired sideband in a single translational step regardless of how high the final signal frequency may be. However, the degree to which the undesired sideband may be suppressed depends upon accurate balancing and requires very careful control of amplitudes and phases. As a practical matter it is quite easy to realize 20-db suppression, reasonable to expect 30 db, and quite difficult to go beyond 40 db. Suppression of 60 to 80 db or more can be realized using the filter method, but extreme care in maintaining low intermodulation in linear amplifiers is necessary if this degree of suppression is to exist in the final radiated signal.

The design and construction of a wide-band 90° phase-difference network is not a familiar art with most circuit designers, and this often acts as a roadblock to using the phasing method.

A block diagram showing the new method of single-sideband signal generation is shown in Fig. 3. The input signal e_s is confined to a bandwidth W with the lower band limit f_L as shown in Fig. 4. The band center is f_0.

$$f_0 = f_L + W/2. \quad (1)$$

For convenience let the input signal be expressed as a summation of sinusoidal terms.

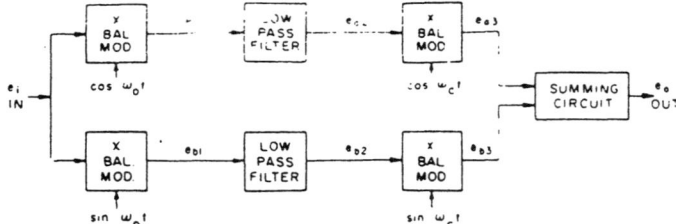

Fig. 3—Single-sideband generator

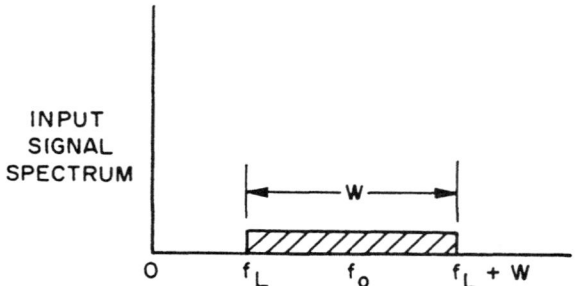

Fig. 4—Input signal spectrum.

Note that the modulating or carrier frequency of the first pair of balanced modulators is the center frequency of the input spectrum. The outputs of the first two balanced modulators are

$$e_{a1} = 2e_i(t) \cos \omega_0 t \qquad (3)$$

$$e_{b1} = 2e_i(t) \sin \omega_0 t, \qquad (4)$$

where

$$\omega_0 = 2\pi f_0. \qquad (5)$$

The coefficient 2 is used for convenience and can be considered a property of the balanced modulators. Substituting (2) into (3) and (4) and expanding gives

$$e_{a1} = \sum_{n=1}^{N} E_n \cos\left[(\omega_n - \omega_0)t + \phi_n\right]$$
$$+ E_n \cos\left[(\omega_n + \omega_0)t + \phi_n\right] \qquad (6)$$

$$e_{b1} = \sum_{n=1}^{N} -E_n \sin\left[(\omega_n - \omega_0)t + \phi_n\right]$$
$$+ E_n \sin\left[(\omega_0 + \omega_n)t + \phi_n\right]. \qquad (7)$$

The frequencies $f_n = \omega_n/2\pi$ are restricted to the original bandwidth W

$$f_L \leq f_n \leq f_L + W. \qquad (8)$$

Hence the spectrum of the signals e_{a1} and e_{b1} is as shown in Fig. 5. The low-pass filter passes the frequencies from zero to $W/2$. From $W/2$ to $2f_0 - W/2$ there should be no signal energy which provides a convenient transition region for the filter. Above $2f_0 - W/2$ the filter should

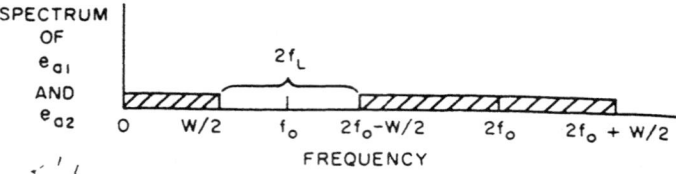

Fig. 5—Spectrum from first balanced modulators.

have adequate attenuation to eliminate the high-frequency components from the balanced modulators. Using such a filter the expressions for the filter output voltages are

$$e_{a2} = \sum_{n=1}^{N} E_n \cos\left[(\omega_n - \omega_0)t + \phi_n\right] \qquad (9)$$

$$e_{b2} = \sum_{n=1}^{N} E_n \sin\left[(\omega_n - \omega_0)t + \phi_n\right]. \qquad (10)$$

These two low-frequency functions are then applied to another pair of balanced modulators. However, in this case the translating frequency ω_c is the band center of the desired single-sideband signal. This is generally a high frequency compared with any of the frequencies of the original signal. The expressions for the outputs of this second pair of modulators are

$$e_{a3} = e_{a2} \cos \omega_c t \qquad (11)$$

$$e_{b3} = e_{b2} \sin \omega_c t. \qquad (12)$$

Substituting (9) and (10) into (11) and (12), and expanding gives

$$e_{a3} = \sum_{n=1}^{N} \frac{E_n}{2} \cos\left[(\omega_c + \omega_n - \omega_0)t + \phi_n\right]$$
$$+ \frac{E_n}{2} \cos\left[(\omega_c - \omega_n + \omega_0)t - \phi_n\right] \qquad (13)$$

$$e_{b3} = \sum_{n=1}^{N} \frac{E_n}{2} \cos\left[(\omega_c + \omega_n - \omega_0)t + \phi_n\right]$$
$$- \frac{E_n}{2} \cos\left[(\omega_c - \omega_n + \omega_0)t - \phi_n\right]. \qquad (14)$$

Finally, adding (13) and (14) gives the desired single-sideband output.

$$e_0 = e_{a3} + e_{b3} \qquad (15)$$

$$e_0 = \sum_{n=1}^{N} E_n \cos\left[(\omega_c + \omega_n - \omega_0)t + \phi_n\right]. \qquad (16)$$

Note that the frequency normally referred to as the carrier corresponds to $\omega_c - \omega_0$ and that the frequency ω_c is the center of the single sideband. Fig. 6 shows the spectrum of e_0.

This method of single-sideband generation does not need either sharp cutoff filters or wide-band 90° phase-

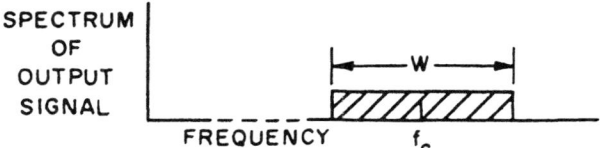

Fig. 6—Spectrum of output signal.

difference networks. Imperfections in the phasing or balancing do not result in the presence of the unwanted sideband in its usual location. Instead, the unwanted sideband occupies the same band of frequencies as the desired sideband, except that it is inverted. This is a very useful property of this system when channel conservation is an important reason for using single-sideband signals.

Fig. 7 shows the circuit of a single-sideband generator using this method. The input signal is a typical speech signal whose energy is confined to a band from 300 to 3300 cps. Care must be taken in the first pair of balanced modulators to keep the input signal component (linear term) from appearing in the output. The two low-pass filters pass all frequencies up to 1500 cps and provide adequate attenuation above 2100 cps. In the second pair of balanced modulators the rf oscillator signal must be accurately balanced out to keep it from appearing in the output.

Two tone tests indicated that undesired signal components were all more than 30 db below the desired signals. The input signal level was in the range 0.1 to 1.0 volt. Listening tests using speech and music indicated good quality. No difficulty was encountered in balancing the modulators or in phasing the translating signals. The balanced modulators, filters, and transformers can be packaged in a very small unit. As the circuit is bilateral, it can be used in demodulation as well as in generation of single-sideband signals. The lack of critical or expensive elements, combined with the ease of adjustment and the ruggedness and reliability of a passive circuit (such as the one shown in Fig. 7) makes this method attractive for application in future single-sideband systems.

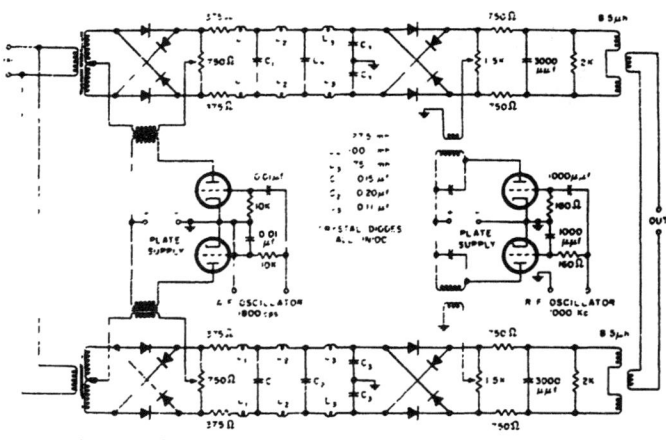

Fig. 7—Single-sideband generator.

Bibliography

[1] Polkinghorn, F. A., and Schlaack, N. F. "A Single Sideband Shortwave System for Transatlantic Radiotelephony." *Bell System Technical Journal*, Vol. 14 (July, 1935) and Proceedings of the IRE, Vol. 23 (July, 1935), pp. 701–718.
[2] Oswald, A. A. "A Short-Wave Single-Sideband Radiotelephone System." Proceedings of the IRE, Vol. 26 (December, 1938), pp. 1431–1454.
[3] Roetken, A. A. "A Single-Sideband Receiver for Short Wave Telephone Service," Proceedings of the IRE, Vol. 26 (December, 1938), pp. 1455–1465.
[4] Dome, R. B. "Wide-Band Phase Shift Network." *Electronics*, Vol. 19 (December, 1946), pp. 122–115.
[5] Bray, W. J., Lillicrap, H. G., and Lowry, W. R. H. "The Design of Transmitter Drives and Receivers for Single Sideband." *Journal of the Institution of Electrical Engineers*, Vol. 94, Part IIIA (December, 1947), pp. 298–312.
[6] Villard, O. G. "Simplified Single Sideband Reception." *Electronics*, Vol. 21 (May, 1948), pp. 82–85.
[7] Villard, O. G. "A High-Level Single Sideband Transmitter." Proceedings of the IRE, Vol. 36, (November 1948), pp. 1419–1425.
[8] Polkinghorn, F. A. "Commercial SSB Radiotelephone Systems." *Communications*, Vol. 28 (December, 1948), pp. 24–27.
[9] Kerwien, A. E. "Design of Modulation Equipment for Modern Single-Sideband Transmitters." Proceedings of the IRE, Vol. 40 (July, 1952), pp. 797–803.
[10] Kahn, L. R. "Single-Sideband Transmission by Envelope Elimination and Restoration." Proceedings of the IRE, Vol. 40 (July, 1952), pp. 803–806.
[11] Honey, J. F. "Performance of AM and Single Sideband Communications," *Tele-Tech.*, Vol. 12 (September, 1953), pp. 64–66, 147–149.
[12] Weaver, D. K. Jr. "Design of RC Wide-Band 90-Degree Phase-Difference Network." Proceedings of the IRE, Vol. 42 (April, 1954), pp. 671–676.

The phasing method for sideband selection in broadcast receivers

R.C.V. Macario and I.D. Mejallie*

1. Separation of sidebands

If single-sideband is ever introduced as a means of broadcast transmission modulation [1], the phasing method of sideband separation [2] is the one most likely to be adopted for receiver technology. There are four good reasons for this, the first being that digital tuning of receivers will be an established technology, and hence a reference carrier will automatically be available in the receiver. Secondly, the phasing method can give extremely good separation of the lowest audio frequencies, down to some 40 Hz, between the wanted and unwanted sidebands. Also, the overall audio-frequency response can be tailored to achieve the required separation and, finally, the response will not drift with age or other environmental factors as can happen with crystal filters, etc. Precisely tailored audio filters are being produced in the form of single integrated circuits** for the telecommunications market, and one may justifiably suppose that the designs could be easily adjusted to slightly different specifications.

The present article is intended to point out certain features of the phasing method, as applied to receivers, which do not appear to have been made too clear in the literature on the subject.

The phasing method of sideband separation is based on the following theory. Consider a carrier of frequency f_c which is amplitude-modulated by an audio tone of amplitude m and frequency f_a. The expression for the signal may be written as:

$$\cos \omega_c t + m_u \cos (\omega_c + \omega_u) t + m_l \cos (\omega_c + \omega_l) t \quad (1)$$

where $\omega_c = 2\pi f_c$ and $\omega_u = \omega_l = 2\pi f_a$, and where c denotes the carrier, u the upper sideband and l the lower sideband.

If this signal is now demodulated by a frequency-synchronous carrier of arbitrary phase represented by the expression $\cos(\omega_c t + \varphi_d)$ (where φ_d is the phase difference between the received carrier and the demodulating carrier), then the product components at the output, after removal of the RF and DC components, are:

$$\frac{m_u}{2} \cos [\omega_u t + \varphi_p + \varphi_d] \text{ and } \frac{m_l}{2} \cos [\omega_l t + \varphi_p - \varphi_d] \quad (2)$$

where φ_p is a phase-shift introduced in the audio path.

In a single path system these two components will add or cancel and separation of the sidebands is not possible.

---

* Messrs Macario and Mejallie are with the University College of Swansea, England.
** For example, Intel 2912, National Semiconductor AF100, Motorola MC 14414.

However, if a multipath system is employed, in which both φ_d and φ_p can be manipulated and then the paths are suitably combined, one sideband can be cancelled whilst the other is recovered.

2. The two-path method

This well-known method is shown in *Fig. 1*, and a practical circuit outline is shown in *Fig. 2* [3, 4]. Unless balanced modulators are used for the demodulators, low-pass filtering ahead of the phasing networks is essential in order to remove the predominant RF and IF components which otherwise will cause non-linearity in the operational amplifiers.

Referring to *Fig. 1*, it is arranged that, in the upper path

$$\varphi_d = 0 \quad \text{and} \quad \varphi_p = \varphi \qquad (3)$$

The two sideband components will be identical in the upper path and cannot be differentiated. In the lower path however, there will be a phase difference of 180° between the two sidebands because

$$\varphi_d = 90° \quad \text{and} \quad \varphi_p = \varphi + 90° \qquad (4)$$

Hence by either adding, or subtracting, the outputs from the two paths, one or other of the sidebands is recovered. (The result does not necessarily apply, however, if there is a phase error between the demodulating carrier and the received carrier.)

The method is very attractive but, as has been mentioned from time to time [5, 6], it is limited because of the difficulty of maintaining:

– the phase quadrature of the demodulating carrier;
– the 90° phase difference between the two paths;
– an exactly equal amplitude balance between the two paths.

Examination of a practical circuit *(Fig. 2)* indicates how phase and amplitude limitations can arise and that it may well be difficult to maintain adequate performance over a long period of consumer use.

Weaver [7] first showed that the best compromise to a 90° difference network, using a discrete network arrangement, consists of meeting the 90° condition periodically over the required phase cancellation band. As his data, shown in *Fig. 3*, demonstrate, to meet a specified sideband discrimination over a wide bandwidth, many sections are required.

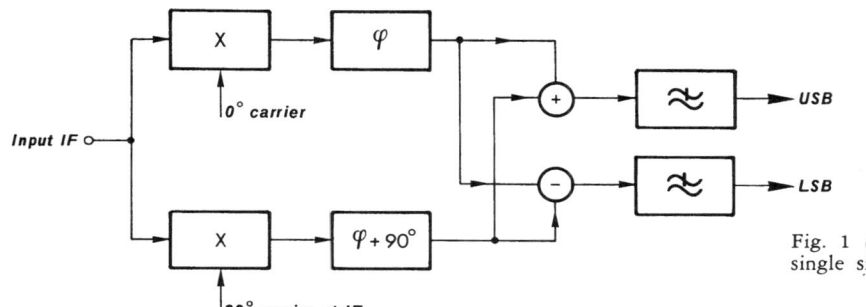

Fig. 1 *(left)* : The two-path phasing method of single sideband demodulation.

Fig. 2 *(below)* : A practical two-path phasing demodulator using operational amplifiers.

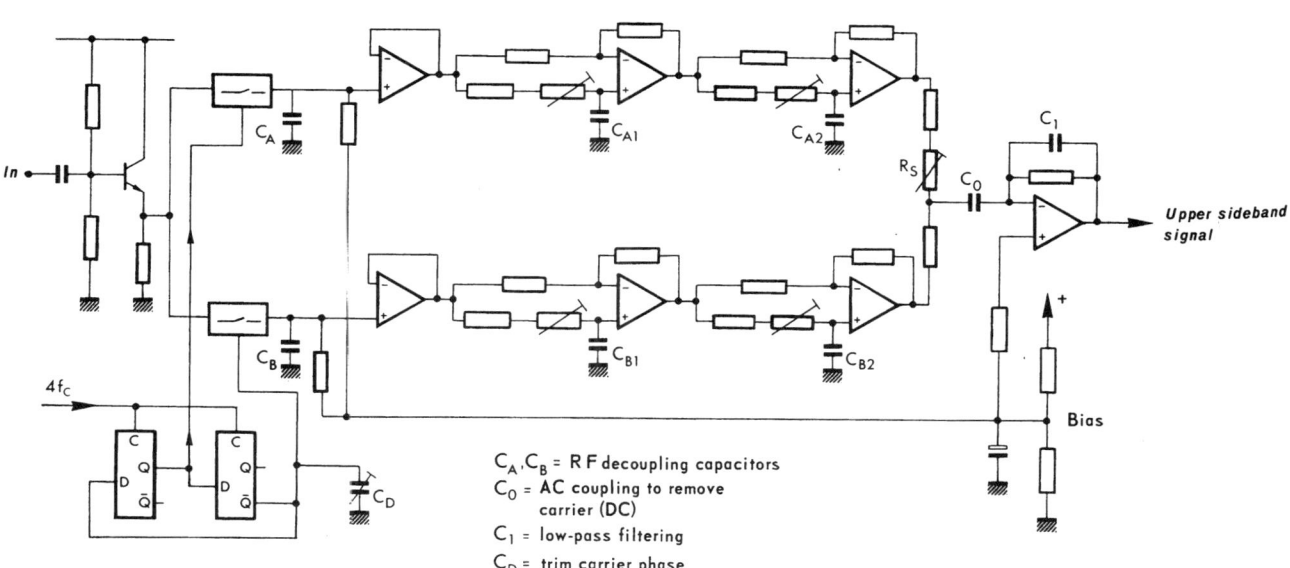

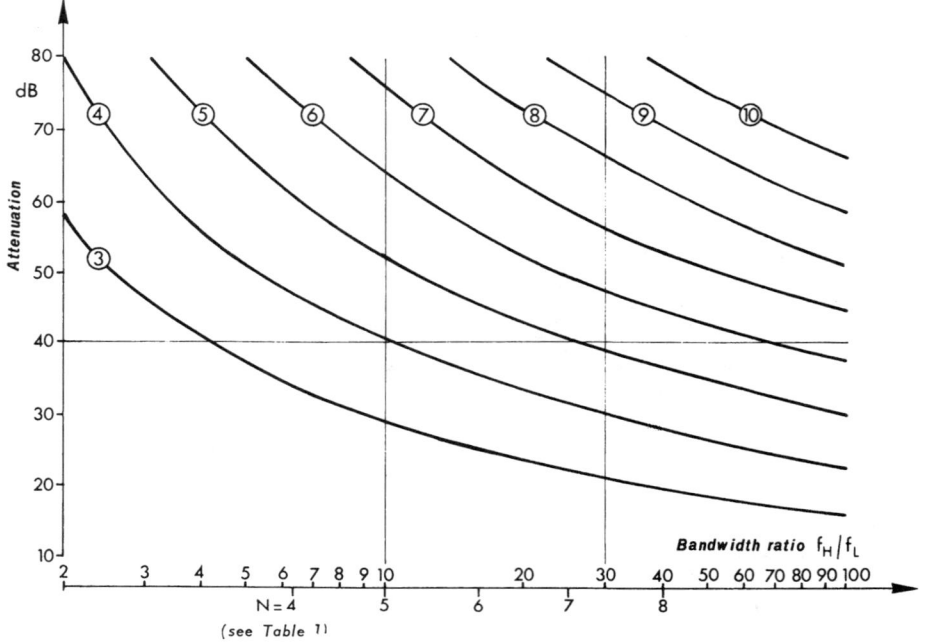

Fig. 3. — Maximum achievable discrimination as a function of the band-edge frequencies, f_H/f_L, for networks of various orders. The numbers in the circles indicate the sections.

Tables of values for R and C are given by Bedrosian [8]. Each value of RC places a pole in the stop band (*Fig. 4*) and if these are equally spaced, the overall discrimination equals that predicted by *Fig. 3*.

For example, choosing the required bandwidth to be 100 to 4500 Hz, at least six sections (three in each path) are required to achieve a stop band attenuation exceeding 40 dB. Because of component tolerances, this performance cannot be completely achieved and allowance for this fact must be made. This aspect is discussed in detail below.

3. Four-path method

Instead of attempting to maintain two identical paths 180° out-of-phase, in the four-path method a 90° phase difference is maintained between adjacent paths and the overall 360° condition is automatically maintained by cross coupling. The method is better known as the polyphase method [9]; it is shown in *Fig. 5*. Each four-pair RC section gives a null at $f_p = 1/2\pi RC$ in the response (*Fig. 4*).

The number of sections is equal to the number of nulls and odd numbers are just as easily implemented

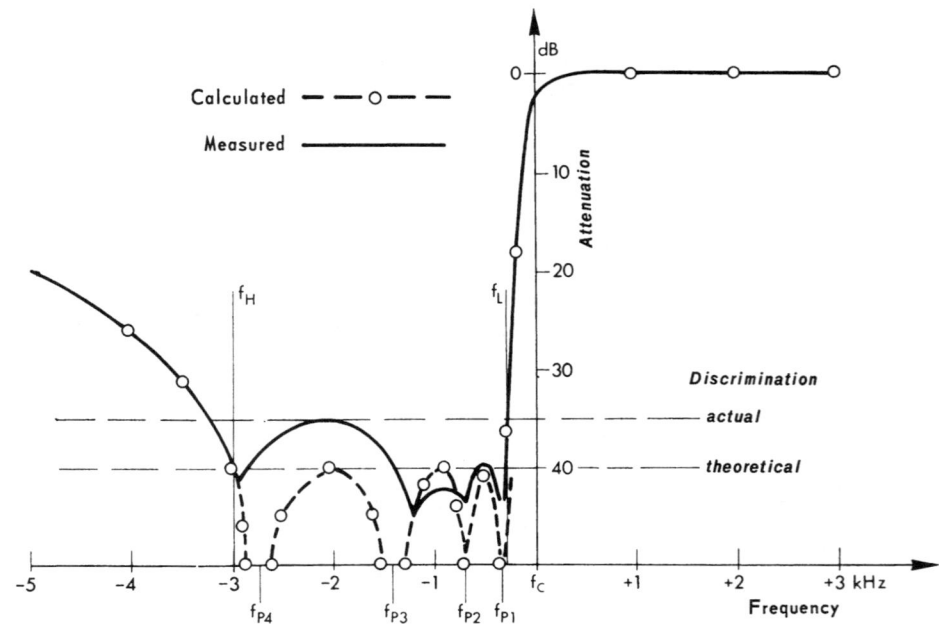

Fig. 4. — Width of the stop band of a phasing demodulator and the pole frequencies.

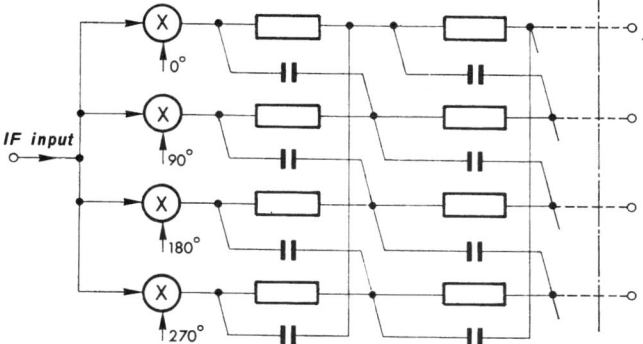

Fig. 5. — The polyphase demodulation circuit.

as even numbers, which is not so for the two-path method.

The polyphase network has to be driven by four quadrature mixers, but these may be implemented in exactly the same way as for the two-path method. An important difference, however, is that the unwanted sideband is now cancelled within the polyphase network and not outside it, as in the two-path method. This feature appears to have been overlooked in previous descriptions.

3.1. Method of cancellation

Provided each phase is voltage driven, the open circuit output voltage, for lines A and D, for example, is equal to:

$$V_{OA} = \tfrac{1}{2} V_{IA} \underline{/-45°} + \tfrac{1}{2} V_{ID} \underline{/+45°} \qquad (5)$$

as can be seen by referring to *Fig. 6* and assuming that the frequency of operation is such that $|X_c| = R$. Using this result, one can now construct the four outputs for the lower and upper sidebands, respectively, for the polyphase construction, as shown in *Fig. 7*.

For the upper sideband component, the phases of the four post-mixer audio signals are 0°, 90°, 180° and 270° respectively. For the lower sideband component, the phases become 0°, 270°, 180° and 90° respectively.

By repeatedly using equation (5) for each phase line output, we find that the USB component appears on all four lines, in phase quadrature, and that the LSB component never appears at all, having been cancelled within the polyphase network (at each pole frequency). To obtain the LSB, and cancel the USB, input connections B and D must simply be reversed.

It also follows that to obtain the LSB and the USB components simultaneously, two polyphase networks must be used, as shown in the practical circuit construction in *Fig. 8*.

Here, single-stage active low-pass filters using operational amplifiers remove the IF signal (which would normally pass straight through the polyphase network); they also provide the voltage source inputs. The output can be taken from any one or more phases as indicated.

A larger output and better discrimination can be obtained, if required, by using additional circuitry, as shown in *Fig. 9*. Here the four output phases are added to give twice the previous output.

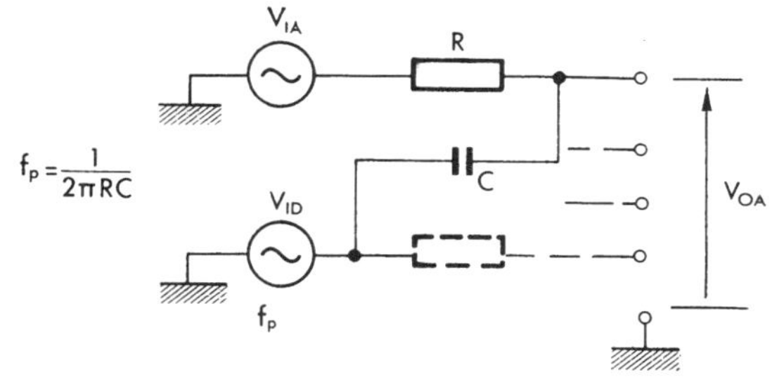

Fig. 6. *(left)* — One element pair within the polyphase circuit at a pole frequency.

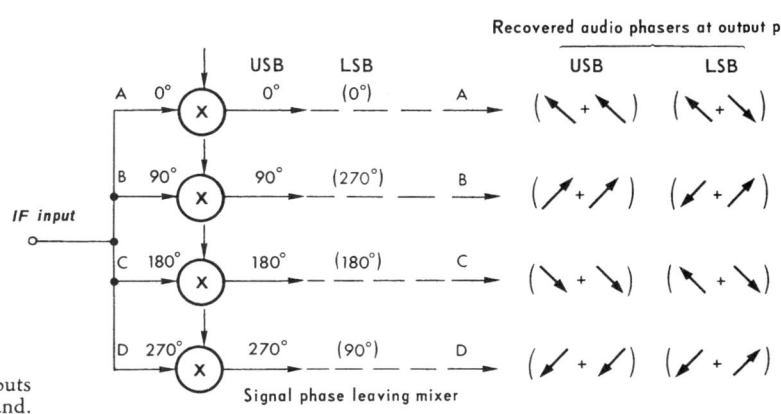

Fig. 7. *(right)* — The phase line outputs according to the applied sideband.

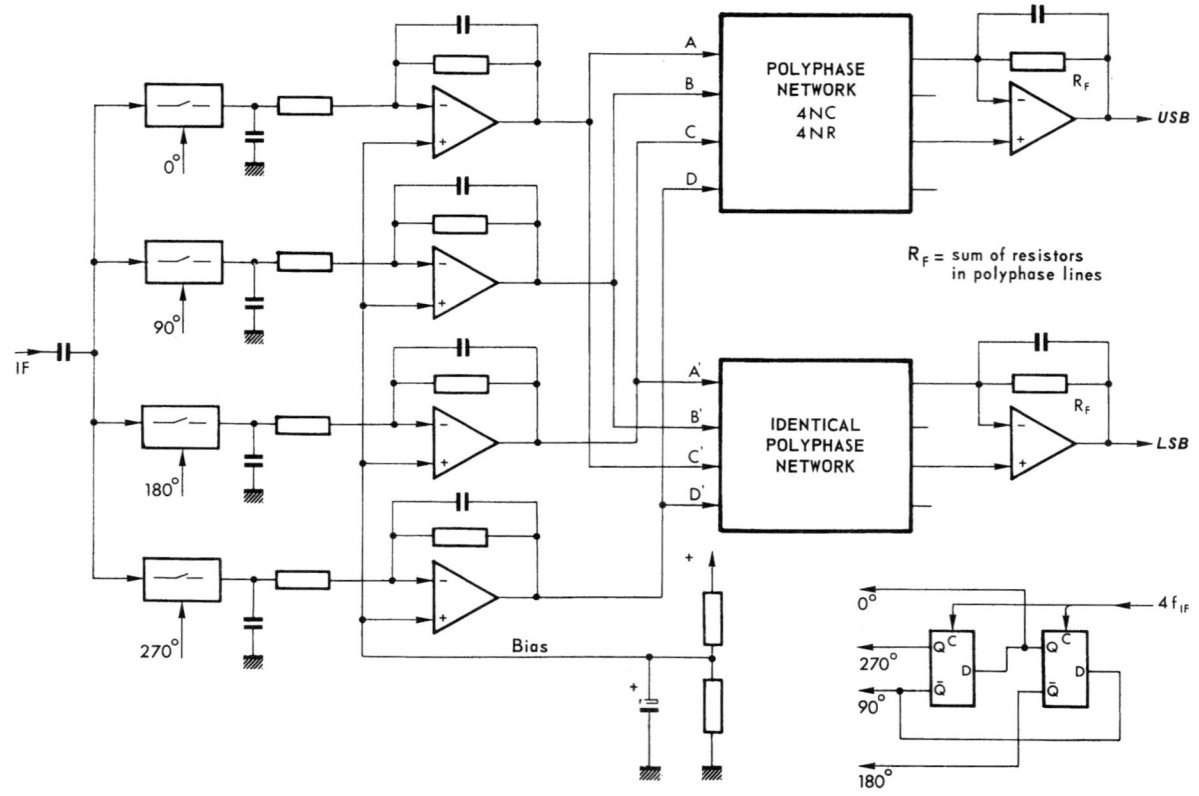

Fig. 8. — A practical four-path phasing demodulation using operational amplifiers.

3.2. Preferred component values

Each null in the rejected sideband response corresponds to an RC value. The nulls should be placed at equal frequency ratios and hence the values of RC will lie on a logarithmic scale, or on the scale of preferred values if the higher and lower frequencies f_H and f_L are suitable chosen.

The choice of frequencies and f_H/f_L ratios is rather restricted, but suitable values for broadcast receivers can be found. The necessary equations are derived in the *Appendix*, whilst the allowable values of f_L and f_H are given in *Table 1*.

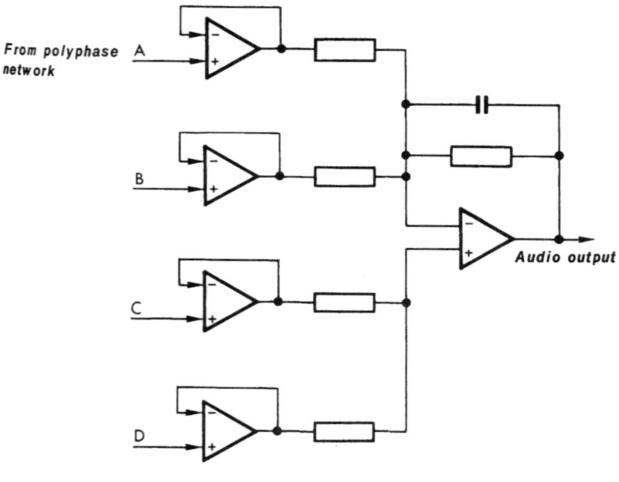

Fig 9. — Circuit for adding all phases to obtain maximum audio output.

Table 1. — Values of f_L for various values of f_H as a function of the order N of the polyphase network

Choice of f_H (Hz)	f_L (Hz)				
	N = 4	5	6	7	8
1000	160	100	63		
1250	200	125	80		
1600	250	160	100	63	
2000	320	200	125	80	
2500	400	250	160	100	64
3200	500	320	200	125	80
4000		400	250	160	100
5000			320	200	125
6300				250	160
8000					200
Discrimination (dB) from Fig. 3	47	52	56	59	66

If 5 % preferred values of R or C are accepted, more flexibility is possible in the choice of the higher and lower frequencies, although the phasing network discrimination will always depend on the network order N and the ratio f_H/f_L. If one aims at values of f_H and f_L which do not conform to those of *Table 1*, best values of R or C can be calculated using equation (8) in the Appendix, but they will not be preferred values. For example, if we choose $f_H = 4000$ Hz and $N = 6$, *Table 1* gives a value for f_L of 250 Hz. If a value of 10 nF is selected for C, then the corresponding values of R will be 50 $k\Omega$, 32 $k\Omega$, 20 $k\Omega$, 12.5 $k\Omega$, 8 $k\Omega$ and 5 $k\Omega$. On the other hand, choosing $f_H = 4000$ Hz, $N = 6$ and $f_L = 100$ Hz, the calculated values of R with $C = 10$ nF, are 117 $k\Omega$, 63 $k\Omega$, 34 $k\Omega$, 18.5 Ω, 10 $k\Omega$ and 5.4 $k\Omega$.

3.3. Choice of phasing network

From a consumer product viewpoint, comparison between the two-path and the four-path phasing circuits on a component count-basis certainly suggests that the two-path method could be produced at a very low cost, whereas adoption of the four-path method will always incur the penalty of requiring a large number of high-value capacitors.

However, the great merit of the polyphase method, as originally pointed out by Gingell [9], is the self-complementary error-cancelling mechanism within the polyphase network itself which reduces errors by " an order of magnitude ".

The effects of phase and amplitude errors were carefully calculated by Norgaard [6] and be gave the following relations :

$$\text{path amplitude error} \quad \alpha_{dB} = 20 \log \frac{A-B}{A+B} \quad (6)$$

$$\text{phase difference error} \quad \beta_{dB} = 20 \log \tan \varphi/2$$

These are plotted in *Fig. 10 (a)* and *(b)* respectively and appear more or less as straight lines. Both these results apply to the discrimination at a pole frequency, but since there are only a finite number of poles, the discrimination that is actually achieved will be limited to the data of *Fig. 3*.

For example, limits of 40 dB and 60 dB are drawn in *Fig. 10*. The effect of errors will be that the overall discrimination curves will follow the error curves as they approach each other : we have assumed a 3-dB approach difference.

With practical phasing circuits, it is difficult to reduce the phase errors below 1° and the amplitude imbalance to less that 1 % without very careful adjustment [4]. Clearly, it is not practical to expect discrimination in excess of 40 dB no matter how the two-path phasing circuit is arranged.

For the polyphase circuit, where we have the reported order-of-magnitude improvement to the errors, *Fig. 10* shows that the commonly reported discrimination of 60 dB is indeed practical.

A further important attribute of the four-path method is that the same sideband appears at the output regardless of any phase difference between the local carrier oscillator and the received carrier phase. This may easily be appreciated by just shifting all four paths (*Fig. 5*) by any angle θ. In the two-path method, however, the introduction of a phase shift of $\theta = 90°$, for example, interchanges the sidebands. Clearly, this would not be acceptable, and a very reliable carrier phase and frequency locking circuit is necessary in a receiver using such a demodulator.

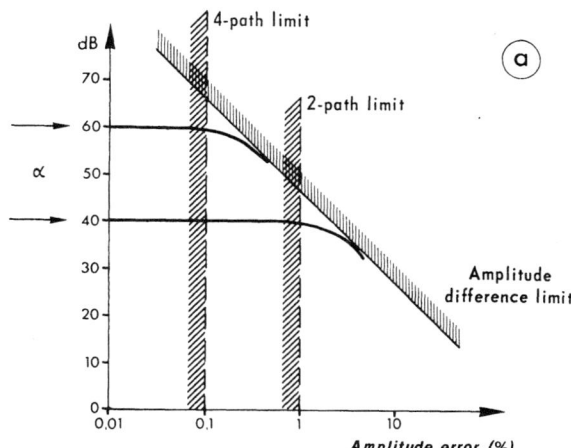

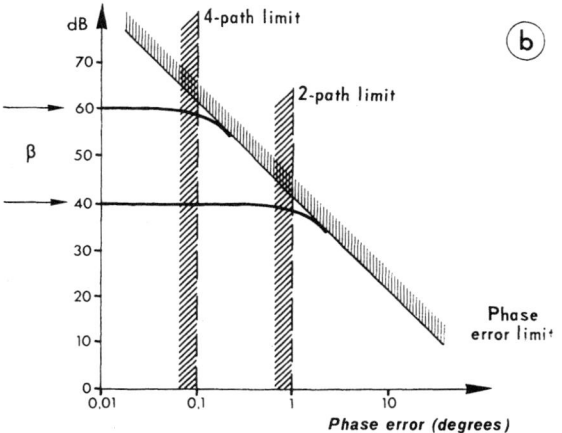

Fig. 10. — Discrimination degradation due to errors in the phasing networks :
a) due to path amplitude unbalance
b) due to path phase difference errors.

4. Conclusions

We have shown what can be achieved by the use of phasing demodulators and the extent of the circuitry required for their realisation. Given that the introduction of SSB in some of the broadcasting bands is widely advocated, and that any such development will require the mass production of receivers fitted with SSB demodulators, there are convincing reasons why, if a phasing method is chosen, preference should be given to the four-path method.

BIBLIOGRAPHICAL REFERENCES

[1] *Technical bases for the World Administrative Radio Conference 1979.*
SPM Report, Geneva 1978, §§ 4.1.5. onwards.

[2] Hartley, R.V.L.: United States Patent No. 1666, 206, 1925.

[3] Dickey, R.K.: *Outputs of op-amp networks having fixed phase difference.*
Electronics, 21st August 1975, p. 82.

[4] Macario, R.C.V.: *Meeting SSB mobile radio specifications with op-amp phasing networks.*
IERE Conference on land mobile radio, No. 33, November 1975.

[5] Saraga, W.: *The design of wide-band phase splitting networks.*
Proceedings of the IRE, July 1950, p. 754.

[6] Norgaard, D.E.: *The phase-shift method of single-sideband signal reception.*
Proceedings of the IRE, December 1956, p. 1735.

[7] Weaver, D.K.: *Design of RC wide-band 90-degree phase-difference networks.*
Proceedings of the IRE, April 1954, p. 671.

[8] Bedrosian, S.D.: *Normalized design of 90° phase-difference networks.*
IRE Transactions on Circuit Theory, June 1960, p. 128.

[9] Gingell, M.J.: *Single sideband modulation using sequence asymmetric polyphase networks.*
Electronic Communications, Vol. 48, 1973, p. 21.

[10] Hudson, R., Castle, R. and Krauss, H.: *An interference-rejecting AM receiver.*
IEEE Transactions CE-24, No. 3, August 1978, p. 235.

Appendix

The poles should be spaced equally on a logarithmic scale in the stop band, defined as the range of frequencies from f_L to f_H as indicated in *Fig. 11*. Let F = logarithm of frequency f.

If ΔF = pole spacing, the M-th pole will be at

$$F_{PM} = F_L + \frac{(2M-1)}{2} \Delta F.$$

Therefore $F_{P1} = F_L + B/2N$.

Taking antilogs, thus reverting to frequencies:

$$f_{P1} = f_L \cdot (b)^{1/2N} \quad (7)$$

where $\quad b = f_H / f_L \quad (8)$

Also $\quad f_{PM} = f_L \cdot (b)^{(2M-1)/2N} \quad (9)$

where $\quad f_{PM} = (2\pi RC)^{-1} \quad (10)$

We are interested in having R and C as preferred values, i.e. on the scale $10^{p/10}$ where n = integer.

This then puts a constraint on the choice of f_L and b, and hence f_H. Comparing equations (9) and (10), we see a need to satisfy the conditions that

$$(b)^{(2M-1)/2N} = (10)^n$$

where N, M and n are integers.

This is achieved if the bandwidth b is chosen as a preferred value $b = (10)^{p/10} \quad (11)$

where $\quad p = 2N \quad (12)$

N being the order of the phasing network. This condition restricts the f_H / f_L ratio as indicated in the table below:

Value of N	Value of p	f_H/f_L
4	8	6.3
5	10	10
6	12	16
7	14	25
8	16	40

These values are used to calculate f_L, from f_H, for different values of N (*Table 1*).

The corresponding pole frequencies will be found from equations (9), (11) and (12):

$$f_{PM} = f_L \cdot (10)^{(2M-1)/10} \quad (13)$$
$$M = 1, 2, \ldots, N.$$

which will also be on the preferred value scale (see *Table 1*, column 1) as will the values of R or C.

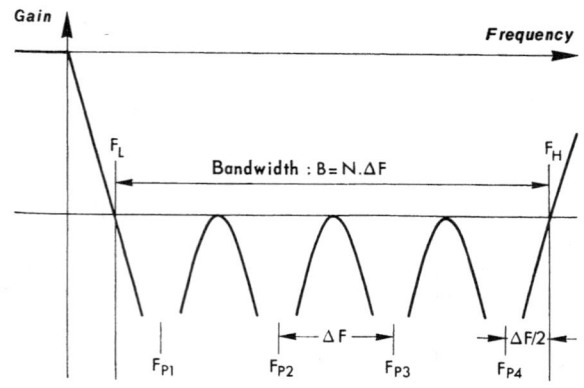

Fig. 11. — The position and definition of stop band limit frequencies f_L and f_H, and the pole frequencies.

PHILIPS TECHNICAL REVIEW

Vol. 41, 1983/84, No. 6

FM receivers for mono and stereo on a single chip

W. G. Kasperkovitz

Most households now possess a number of radio receivers. This wide availability of a once-expensive technical product has come about only through a phenomenal fall in the cost of manufacture. This in turn is the result of greatly increased production runs — now into millions — but more than anything it is due to new technological developments. Every now and then a fundamental advance is made, a step towards other, and cheaper, methods of achieving the required functions in the receiver. One such earlier step was the change from thermionic valves to transistors. In the new technology, costs are gradually being brought down by greater efficiency and economies of scale in production, but in the long run this downward trend in production costs must stop and another fundamental step forward becomes necessary. The article below describes such a new step: the integration of a complete FM mono receiver on a single chip. The number of trimming points — an important cost factor — has been reduced to one: the externally connected oscillator inductor. The other external components are a few capacitors and resistors, which do not have to meet any particularly arduous requirements. The new chip (TDA 7010T) is so small that extreme miniaturization is possible — and some wrist-watches on the market now contain a built-in FM receiver based on this chip. But even in this new technology further improvements are on the way; the successor, TDA 7020T, already in pilot production, will work on lower battery voltages and requires fewer external components. The article also presents a fundamental solution of the next step forward: a completely integrated FM stereo receiver.

Introduction

Although the first attempts to use integrated circuits in radio receivers to reduce the cost were made more than 15 years ago, most receivers built today still consist of a printed-circuit board with manually inserted discrete transistors, diodes, resistors, capacitors and inductors; the inductors generally have to be aligned. Integrated circuits are now frequently included in high-performance receivers to make tuning easier, through additional features such as noise muting, deviation muting (off-tune signal suppression), station preset, frequency display, frequency synthesis, touch control and remote control. But the basic circuit used in most of the less-expensive portables and clock radios has remained essentially unchanged for ten years or more.

This gives some idea of the difficulties encountered in designing a successful substitute for inductors that will combine high Q-factors with high signal-to-noise ratios at high operating frequencies. In principle, gyrators [1] can be used instead of inductors. However, with increasing operating frequencies these circuits become less attractive because of the limitations of dynamic range and Q-factor and because of their complexity and power dissipation.

Instead of designing an integrated radio receiver in which the inductors in the resonant circuits are replaced by active RC circuits, it may be possible to adopt other approaches that are less handicapped by the limitations of IC technology. Various systems have been proposed in which fewer inductors are necessary

Dr Ing. W. G. Kasperkovitz is with Philips Research Laboratories, Eindhoven.

[1] B. D. H. Tellegen, The gyrator, a new electric network element, Philips Res. Rep. **3**, 81-101, 1948.

Reprinted with permission from *Philips Technical Review*, W. G. Kasperkovitz, "FM Receivers for Mono and Stereo on a Single Chip," Vol. 41, No. 6, pp. 169-182, 1983. © Philips Research Laboratories.

for the receiver unit [2]–[7]. Compared with existing superheterodyne receivers the proposed receiver circuits use a significantly lower intermediate frequency, so that the intermediate-frequency bandpass filters may be replaced by *RC* filters that can be partly or completely integrated.

Depending on the choice of intermediate frequency, two classes of problems are encountered in FM receivers. With an intermediate frequency of zero [4]–[6] image-reception problems in superheterodyne receivers are eliminated. With these receiver systems, however, the amplitude of the i.f. signal cannot be limited before demodulation. This results in a 3-dB lower signal-to-noise ratio in the i.f. stage and in less suppression of unwanted amplitude modulation of the demodulated signal.

This problem is not encountered in a receiver system that has an intermediate frequency of 140 kHz [7], where a limiter/amplifier is used before the demodulator to suppress AM noise and peak distortions of the r.f. signal. With this system, however, there is an image response at a spacing of 280 kHz, almost coinciding with the centre frequency of a neighbouring channel at a spacing of 300 kHz. Because of this there are also two tuning positions for each station (quite apart from the double response on the outer slopes of the demodulator characteristic, which is dealt with below).

To be acceptable to the customer an integrated FM receiver should at least avoid the shortcomings of the proposed systems with low intermediate frequency. But if the shortcomings of today's superheterodyne receivers could also be eliminated, such an integrated receiver, in addition to being cheaper to produce, would have advantages for the user as well.

Compared with AM receivers, FM receivers have one fundamental weakness, which is that each station has at least three tuning positions. *Fig. 1* shows the amplitude of the demodulated audio signal of a typical superheterodyne portable receiver as a function of the tuning frequency, with the r.f. voltage V_{ant} at the antenna input as the parameter; the modulation is a constant 1000-Hz tone. Besides the correct response at the centre of fig. 1, there are two spurious responses, characterized by a reduced signal-to-noise ratio and increased harmonic distortion of the audio signal. The demodulation takes place on the sides of the demodulator curve, which are referred to here as 'noise slopes', because of the high noise level (see *fig. 2*). The position and relative intensity of the spurious responses depend on the antenna input voltage and the selectivity of the receiver. They are separated from the range for correct tuning by minima in the output voltage. It might seem as if these minima would make it easy to identify the spurious responses. However, this is not the case in the far more complex situation found in practice, where the modulation is not constant and sinusoidal as in fig. 1, but strongly time-dependent and often with significant overlap between the spurious responses of adjacent channels. These effects combine to give rather unsatisfactory tuning behaviour, especially for simple tuning by ear.

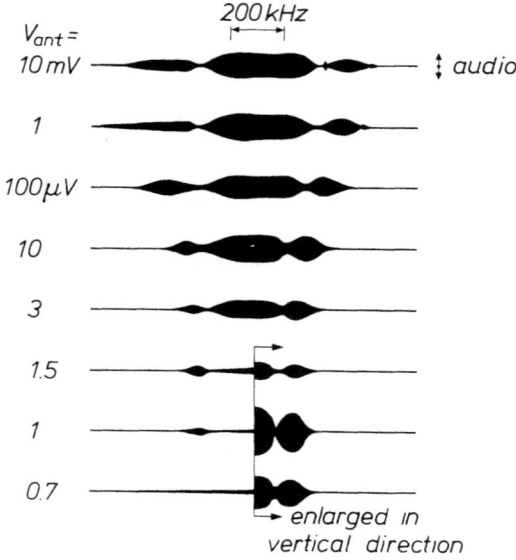

Fig. 1. The tuning behaviour of a conventional mono receiver at different antenna input voltages V_{ant}. The demodulated audio signal (vertical), here a 1000-Hz tone (*audio*), is shown as a function of the tuning frequency (horizontal). The audio signal has a maximum at the correct tuning position, and two secondary maxima to the left and right (on the outer slopes of the demodulator characteristic; see fig. 2). This makes it difficult to tune by ear. In the integrated FM receiver the two spurious responses are suppressed (see fig. 10).

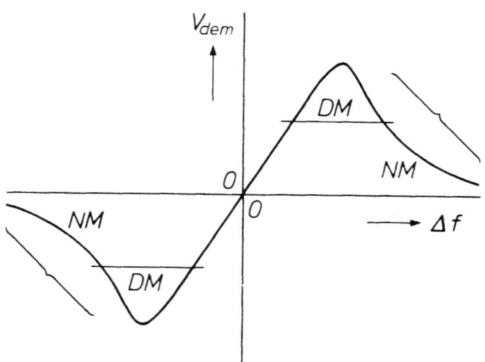

Fig. 2. Demodulator characteristic. This gives the relation between the demodulator output voltage V_{dem} and the frequency deviation Δf. The central part of the curve has the desired linear relation between the two. On detuning from the centre frequency, demodulation can also take place on the slopes *NS*; this is accompanied by distortion and noise. In our monolithic FM receiver these spurious responses are suppressed by a special circuit that reacts to frequency deviations (*DM*, 'deviation muting') or to excessive noise (*NM*, 'noise muting').

In the high-performance class of FM receivers squelch systems like noise muting [8] and deviation muting [9] are built in to suppress the spurious responses, and tuning meters are used to locate the correct tuning position. These features help to make tuning very much easier. Although cost reduction was the main objective in the integrated FM receiver presented here, it seemed a good idea to include these or similar features in the system. This had to be done, however, without unduly increasing the complexity, the power consumption or the number of external components and pins of the integrated circuit.

The integrated mono receiver circuit described in this article is now being marketed under the type designation TDA 7000 (or TDA 7010T) [10]. An improved design has recently been brought out under the type designation TDA 7020T. Some data are given in *Table I*. *Fig. 3* shows a wrist-watch, now commercially available, which contains an FM mono receiver built around this integrated circuit.

A question that arises, of course, is whether it would also be possible to build an integrated FM stereo receiver. This is a much more complicated exercise in view of the large bandwidth (53 kHz) of the stereo multiplex signal compared with the bandwidth of the mono signal (15 kHz). Nevertheless, it does seem feasible, and a complete integrated circuit for a stereo receiver is now being studied. A block diagram of the design is presented at the end of this article.

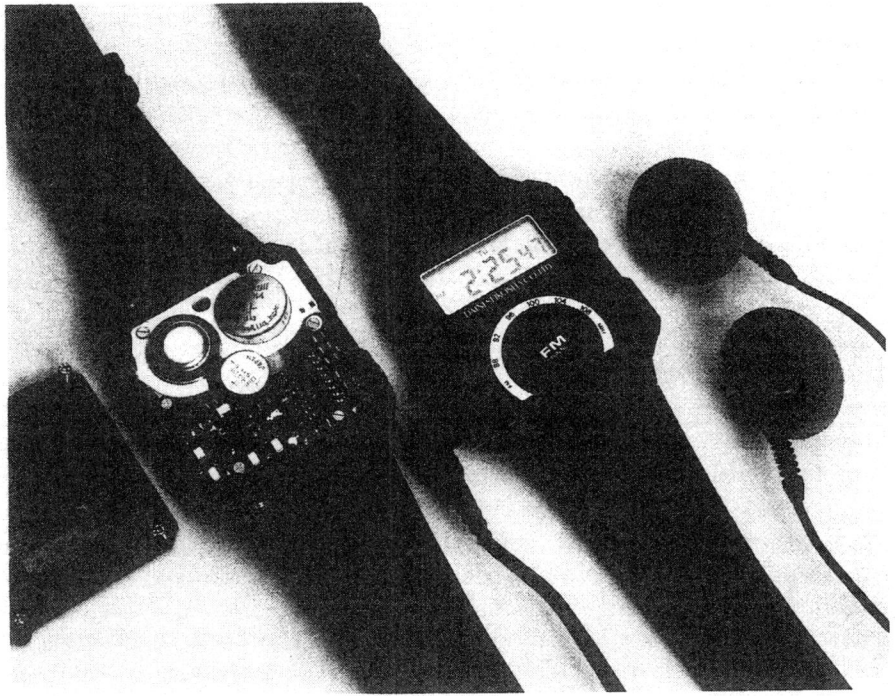

Fig. 3. A wrist-watch with a built-in FM receiver IC. The headphone lead also acts as the antenna.

Table I. Data for the integrated FM receivers TDA 7000/7010T and TDA 7020T.

	TDA 7000/7010T	TDA 7020T	
Supply voltage	4.5	3	V
Supply current	8	5.7	mA
Frequency range	1.5-110	0.5-110	MHz
Sensitivity (amplitude limiting −3 dB, input impedance 75 Ω, signal mute not operating)	1.5	1.5	μV
Max. input voltage (total harmonic distortion 10%, $\Delta f = \pm 75$ kHz, input impedance 75 Ω)	200	200	mV
Audio output voltage Minimum load	90 22000	100 35	mV Ω

[2] D. K. Weaver, Jr., A third method of generation and detection of single-sideband signals, Proc. IRE **44**, 1703-1705, 1956.
[3] J. P. Costas, Synchronous communications, Proc. IRE **44**, 1713-1718, 1956.
[4] J. G. Williford, U.S. patent No. 3 568 067.
[5] I. Vance, British patent No. 1 530 602.
[6] I. A. W. Vance, An integrated circuit VHF radio receiver, Proc. Int. Conf. on Land mobile radio, Bailrigg 1979, pp. 193-204.
[7] G. G. Gassmann, Ein neues Empfangsprinzip für FM-Empfänger mit integrierter Schaltung, Radio Mentor **32**, 512-518, 1966 (in German).
[8] J. Craft, U.S. patent No. 3 714 583.
[9] I. Fukushima et al., U.S. patent No. 3 851 263.
[10] W. H. A. van Dooremolen and M. Hufschmidt, A complete f.m. radio on a chip, Electron. Components & Appl. **5**, 159-170, 1983.

Basic features of the FM mono receiver

The integrated FM mono receiver described here operates with an intermediate frequency of 70 kHz. This value would lead to unacceptable harmonic distortion at a frequency swing of ± 75 kHz, which is the maximum value allowed in FM broadcasting. This is evident, since there is no room in the intermediate-frequency band for a deviation of −75 kHz, which would extend into the range of 'negative frequencies'. The frequency scale is then 'folded' around 0 Hz (see fig. 13), resulting in unacceptable harmonic distortion. For this reason 'frequency feedback' is used in our system, to reduce the maximum frequency swing of the intermediate frequency to ±15 kHz. This frequency feedback is obtained by using the demodulated audio signal to control the frequency of the local oscillator in our superheterodyne receiver. This is done in such a way that the oscillator frequency 'travels some of the way' with the frequency deviation of the transmitter. If the transmitter has a frequency deviation of ± 75 kHz, then the oscillator is given a deviation of ± 60 kHz, resulting in an i.f. signal with a frequency deviation of only ± 15 kHz.

With our choice of intermediate frequency, image reception occurs at a spacing of 140 kHz, i.e. at the edge of the received channel. Background noise in this part of the frequency band is equal in strength to the background noise of the received signal. Thus, with a given input signal and a given bandwidth, the signal-to-noise ratio at the output of the mixer stage is in principle 3 dB lower than in systems with complete suppression of the image response. This entails an increase in the noise figure, but the increase is partly compensated by the reduced i.f. bandwidth [11], and partly by a lossless coupling between the r.f. amplifier and the mixer stage.

An advantage of using an intermediate frequency not equal to zero is that amplitude modulation in the received signal can be effectively suppressed. A high-gain limiter/amplifier is included before the demodulator. This gives good AM suppression and automatic volume control (AVC), even for weak input signals.

The circuit also includes 'deviation muting', a system that suppresses the audio signal if the tuning is incorrect or if the input signals are comparable with the input noise. It is based on the correlation between the i.f. signal and a delayed and inverted version of the i.f. signal. A qualitative description of the system is given in *fig. 4*. The signal *IF'* is derived from the amplitude-limited intermediate-frequency signal *IF* by delaying this signal by one half of its period at correct tuning, and inverting the delayed signal. This means that for correct tuning the two signals are identical (fig. 4a), giving high correlation. In this situation the demodulated audio signal is applied to the audio output. If the tuning is incorrect (fig. 4b), one half of the period of the signal *IF* no longer corresponds to the delay between the two signals *IF* and *IF'*. In this situation the correlation between the two signals is small or negative, and the demodulated audio signal is not applied to the output. In this way the spurious responses for large input signals are suppressed.

If the input signal is comparable in level to the input noise, the two signals *IF* and *IF'* are as shown in fig. 4c. Because of the low Q of the i.f. filter (about 0.7) the intervals between the successive zero crossings of the i.f. signal fluctuate considerably, again giving a low correlation between the two signals and hence muting of the demodulated audio signal. In this way the skirts of the spurious responses at large input signals (top of fig. 1) and the entire spurious responses for small input signals (bottom of fig. 1) are suppressed. Fig. 2 illustrates what is meant here by 'skirts'.

This muting system that comes into operation when correlation is low combines the characteristics of signal muting operated by noise (in which the noise on the envelope of the i.f. signal is detected [8]), and signal muting operated by incorrect tuning (which depends on the automatic frequency control [9]). The correlation system can be used in our receiver because the compressed frequency swing is never so large as to disturb the correlation seriously. In addition, advantage is taken of the delay network that is part of the demodulation circuit; this gives a delay of a quarter of a period. At the low intermediate frequency it is not necessary to use a tuned circuit either for this delay or for the additional quarter-period delay also required

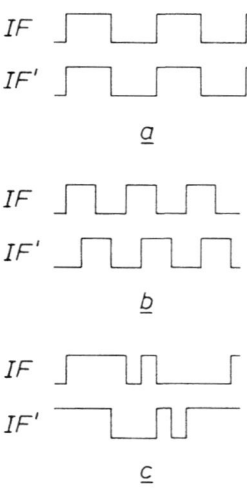

Fig. 4. Correlation measurement for signal muting. The i.f. signal is delayed by half a period, inverted (*IF'*) and compared with the original signal (*IF*). For correct tuning there is maximum correlation (*a*), for incorrect tuning less or no correlation (*b*), in the presence of noise alone there is no correlation (*c*). In cases *b* and *c* the signal is not supplied to the audio output.

for the muting, so that the entire signal-muting system requires few extra components and takes very little extra current. In our system the muting threshold is about five times as low as the muting thresholds of the most advanced systems based on detection of the noise on the envelope of the i.f. signal. It can therefore be high dynamic range at the output of the mixer, to a constant amplitude. To obtain the high gain of 90 dB at low current, LA_1 operates into a high impedance, and a buffer amplifier A_2 is added; this has a sufficiently low output impedance to reduce unwanted crosstalk between multipliers M_2 and M_3.

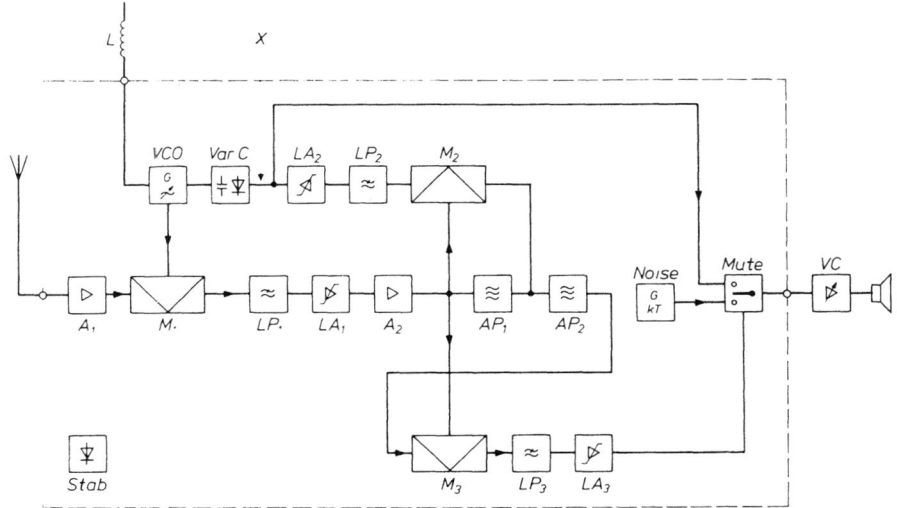

Fig. 5. Block diagram of integrated FM mono receiver. $A_{1,2}$ amplifier. $M_{1,2,3}$ multiplier circuit (mixer). $LP_{1,2,3}$ lowpass filter $LA_{1,2,3}$ limiter/amplifier $AP_{1,2}$ allpass network (gives 90° phase shift). *VarC* varactor. *VCO* voltage-controlled oscillator. *Noise* noise generator. *Mute* signal/noise switch. *Stab* stabilized direct voltage supply. *L* inductor (tunes VCO to the FM band). *X* point where the frequency-locked loop is regarded as open in the treatment of open-loop behaviour. *VC* volume control.

used in portable receivers, where many of the input signals are only slightly above the input noise level.

A noise generator is used in combination with the muting system to give an audible tuning indication in the absence of a tuning meter.

The integrated FM mono receiver

Fig. 5 shows a block diagram of the integrated circuit with its three essential connections to the outside world: the antenna input, the audio output and the connection to the circuit used for tuning to the desired station.

The antenna is connected to the input of the broadband amplifier A_1, which amplifies throughout the entire FM band. The output of A_1 is connected to the r.f. mixer M_1, which performs the conversion to the intermediate frequency. The signal gain from antenna input to mixer output is about 26 dB.

The i.f. signal is filtered by a fourth-order active lowpass filter (LP_1) to suppress signals outside the selected channel [12]. The response of this filter is shown in *fig. 6*. The limiter/amplifier LA_1, with a gain of more than 90 dB, limits the i.f. signal, which has a

Demodulation and limitation of the frequency swing are achieved by converting the frequency into a voltage, which is then used to correct the voltage-controlled oscillator *VCO*. Associated with this oscillator are an integrated varactor diode *VarC* (a variable-capacitance diode) and the external resonant circuit

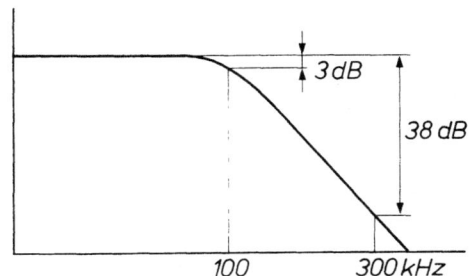

Fig. 6. Response of the lowpass filter LP_1. This fourth-order filter passes the band from 0 Hz to 100 kHz and acts as i.f. band filter in the integrated FM mono receiver, in which it determines the selectivity. A channel at a spacing of 300 kHz is attenuated by 38 dB.

---

[11] L. H. Enloe, Decreasing the threshold in FM by frequency feedback, Proc. IRE **50**, 18-30, 1962.
[12] A Sallen and Key configuration is used for this filter because it gives the best compromise between selectivity, current taken and signal-to-noise ratio. See R. P. Sallen and E. L. Key, IRE Trans. **CT-2**, 74-85, 1955.

mentioned above. The frequency is converted into a voltage by means of the multiplier M_2, in which the i.f. signal is multiplied by a version of the same signal shifted in phase by 90°. The phase shift takes place in the allpass filter AP_1. The combination of M_2 and AP_1 is called an FM quadrature detector (FMQD).

A mathematical description of quadrature demodulation is given on page 179, with a calculation of the signal delay in the quadrature demodulator.

The lowpass filter LP_2 is included in the frequency-locked loop thus formed. In addition to the audio frequency signal mentioned above, a d.c. signal also appears in the loop when the i.f. frequency does not have a value such that the delay in AP_1 is exactly a quarter-period of the carrier. This direct voltage provides automatic frequency control (AFC) by acting on the voltage-controlled oscillator VCO. The loop also contains a low-gain limiter/amplifier LA_2 to control the locking range of the AFC. Audio signals are extracted after this limiter/amplifier. The limiter does not distort the audio signals because as soon as it starts to operate the AFC is switched off; the receiver is then no longer tuned to the transmitter and the signal is suppressed.

The system that suppresses the signal in the absence of correlation (the correlation muting system) consists of two identical allpass filters AP_1 and AP_2, mixer M_3, lowpass filter LP_3 and limiter/amplifier LA_3. The correlation between the limited i.f. signal at the input of AP_1 and its delayed and inverted version at the output of AP_2 takes place in mixer M_3. The amplified and limited output signal of the correlator is used as a muting signal to suppress the audio signal for spurious responses.

Fig. 7 shows some of the control voltages in the integrated FM receiver as they would appear if the frequency-locked loop were opened at X in fig. 5; the responses shown are purely qualitative and relate to a particular level of the antenna signal. In fig. 7a and b the output voltages of demodulator M_2 and low-gain limiter/amplifier LA_2 are shown as a function of the difference between the antenna-signal frequency f_{ant} and the oscillator-signal frequency f_{osc}. With the loop closed the control makes this difference equal to the centre frequency f_c, giving correct tuning.

The output voltages of the correlator M_3 and of limiter/amplifier LA_3 are shown in fig. 7c and d as a function of $f_{ant} - f_{osc}$. There are two bands of $f_{ant} - f_{osc}$ where the signal is not suppressed but is applied to the output. One is centred around f_c, the point of correct tuning. The other band with no muting is defined by $-f_2 < f_{ant} - f_{osc} < -f_1$. This means that a

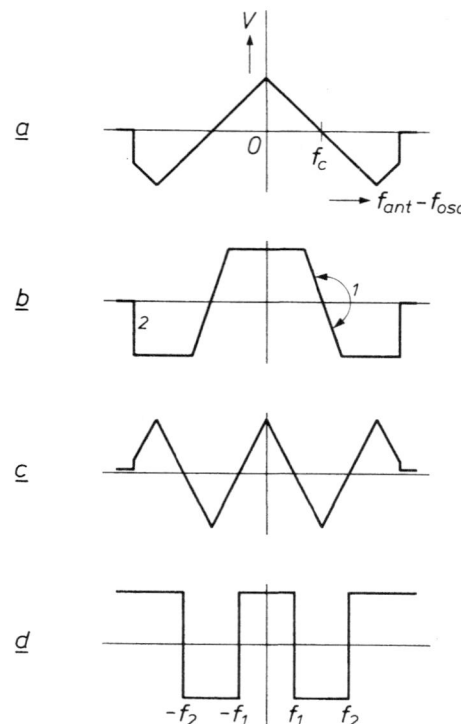

Fig. 7. Signal voltages V for frequency control and signal suppression (muting). Schematic representation of open-loop behaviour. *a)* Output voltage of mixer M_2 as a function of the difference $f_{ant} - f_{osc}$ between received frequency and oscillator frequency. f_c is the desired intermediate frequency to which $f_{ant} - f_{osc}$ is tuned in the closed loop. *b)* The same voltage after limiting and amplification in LA_2. Stable tuning positions are found on paths *1* and *2*, because if the value of $f_{ant} - f_{osc}$ is too high the control voltage is negative, so that $f_{ant} - f_{osc}$ is reduced again; the tuning position on path *2* (the slope) is one of the spurious responses. *c)* Output voltage of mixer M_3, used for deviation muting. *d)* The same voltage after limiting and amplification in LA_3. When the voltage is high, the output signal is suppressed (muted), thus eliminating the spurious respons on path *2*. The voltage is low in the desired frequency range between f_1 and f_2, and also at the image frequencies between $-f_2$ and $-f_1$; however, stable tuning is impossible at the image frequencies, because of the frequency control.

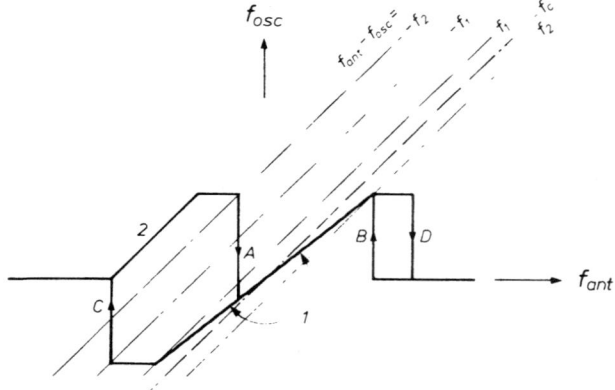

Fig. 8. Frequency control and signal muting with closed loop. (See fig. 7.) f_c desired intermediate frequency. *1* frequency trajectory at the correct tuning position; the frequency point moves to and fro along this path as a result of the frequency modulation. Because of the frequency feedback this line is not horizontal but sloping, and f_{osc} also varies. *Shaded area:* intermediate frequency bands where the audio signal is supplied to the receiver output; everywhere else the signal muting is operative. *2* second stable tuning position, but the signal is not supplied to the audio output. *A*, *B* frequency locking. *C*, *D* frequency release.

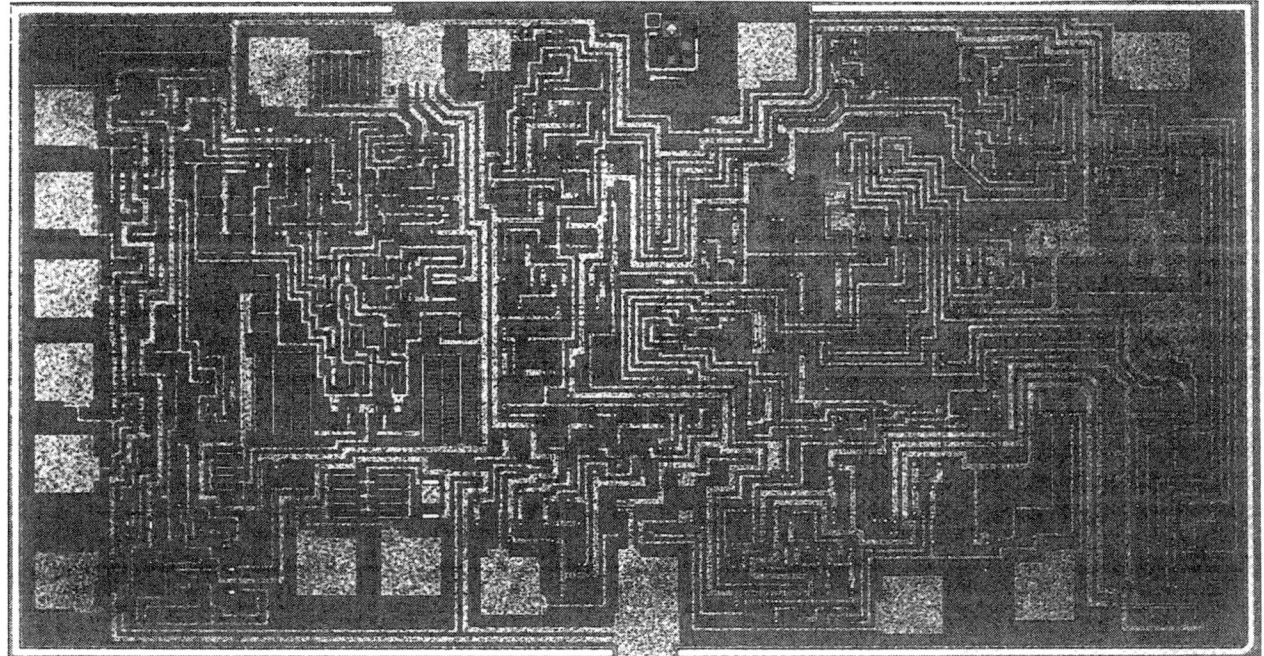

Fig. 9. Photomicrograph of the integrated FM mono receiver circuit TDA 7000/7010T. Chip area 3.5 mm².

double tuning response is possible with an open loop. The second (spurious) response is at the image frequency.

When it is closed the frequency-locked loop gives positive feedback at the image frequency; a deviation from the image frequency generates a control voltage whose polarity is such that the deviation increases. Double tuning is therefore impossible with the loop closed.

This is illustrated in detail in *fig. 8*, which shows the paths taken by the control point as it moves towards the point of correct tuning in the plane f_{ant}, f_{osc}. The shaded areas indicate that there is no muting. The figure relates to the same antenna signal as in fig. 7.

With correct tuning (path *1*) the demodulated audio signal — which is a linear function of f_{ant} and also of f_{osc} — is not muted but is applied to the audio output (see fig. 7*d*). On path *2*, the noise slope of the demodulator characteristic, the frequency feedback is negative, so that a stable tuning position is also possible here. This path, however, lies in a region where signal muting is operative.

If the deviation from the point of correct tuning is too great the frequency-locked loop loses its hold on the oscillator frequency (fig. 8 *C*, *D*). As the point of correct tuning is approached, the transmitter eventually becomes 'locked' in the loop (fig. 8 *A*, *B*). Both events are accompanied by a sudden jump in the loop voltage, which would be audible in the audio signal if no countermeasures were taken. These transients in the audio signal are suppressed in two different ways.

The lock operation *B* and the loss of lock *D* take place in a region where $f_{ant} - f_{osc}$ is greater than f_2. In this region the signal muting is permanently in operation. The situation is different with loss of lock indicated by *C*; here the signal muting is not operative during the complete transition. The transition starts in a region where the mute is operative, then passes the region $-f_2 < f_{ant} - f_{osc} < -f_1$, where the signal is not muted and ends up in a region where the signal is muted again. To make this transition inaudible the time constant of the lowpass filter LP_3 (see fig. 5) has been made such that the input voltage of LA_3 remains positive during the short time interval necessary for crossing the region $-f_2 < f_{ant} - f_{osc} < -f_1$.

In a similar way the time constant of LP_3 also determines what can be heard during the locking operation *A*, where a transition takes place from the noise generator to the demodulated audio signal. Here the control point passes through four different regions with alternating states of the muting signal. A proper choice of the time constant ensures a smooth acoustical transition from the region $f_{ant} - f_{osc} < -f_2$, where the signal from the noise generator is passed to the output, to the region $f_{ant} - f_{osc} > f_1$ where the desired transmitter is heard at the output.

Measurements on the FM mono receiver

Fig. 9 shows a photograph of the integrated FM mono receiver circuit on a single chip. All the functions indicated in the block diagram in fig. 5 have been optimized for chip area, power consumption and supply voltage variations. The current taken by the IC is 8 mA at 4.5 V, but it operates at any supply voltage between 3 V and 18 V. The chip area is 4.5 mm² and the number of bonding pads is 18.

Measurements have been made on an experimental receiver containing this IC. Apart from the integrated circuit, the receiver consists of a resonant circuit used for tuning to the desired station, a number of ceramic capacitors and a resistor. The resonant circuit is tuned by a varactor. No trimming is required apart from the adjustment of the direct-voltage range for tuning the resonant circuit through the FM band. Because of the low intermediate frequency i.f. trimming is unnecessary: the tolerance on the values of the most critical of the fixed capacitors is about 100 times greater than the tolerance on the LC products of the resonant circuits used in a conventional receiver.

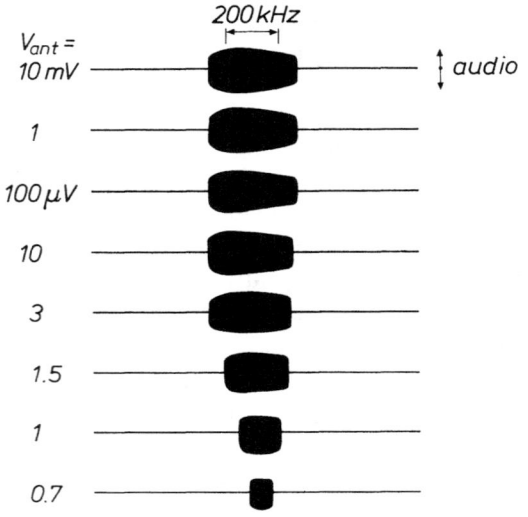

Fig. 10. Tuning behaviour of the integrated FM mono receiver (see also fig. 1).

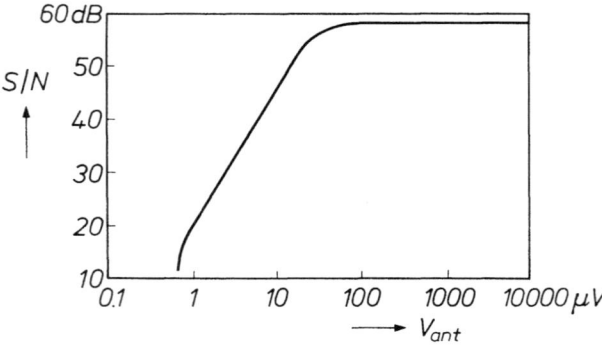

Fig. 11. Signal-to-noise ratio S/N of the demodulated signal (frequency 1 kHz, carrier frequency swing ± 22.5 kHz) as a function of the antenna voltage V_{ant} (across an input impedance of 40 Ω). At 1.5 µV the signal-to-noise ratio is 26 dB (this is the 'quieting sensitivity' as defined in the CCIR standard).

Fig. 10 shows the tuning behaviour of the experimental receiver. There are three improvements as compared with the tuning of a typical portable receiver, as shown in fig. 1:

- No spurious responses, at large or small input voltages. This is the result of the correlation muting system.
- Large range of correct tuning, even at small input voltages. This is achieved with the AFC. This function is illustrated by fig. 8, where a large variation in f_{ant} — which is equivalent to a large variation in the tuning frequency — is reduced to a small variation in the intermediate frequency $f_{ant} - f_{osc}$.
- No degradation of the audio signal for small antenna signals. This is due to the high gain of the limiter/amplifier LA_1 before the demodulator, which limits the i.f. signal even at input voltages lower than 1 µV. In conventional portable receivers, limiting at such low voltages results in stability problems because higher harmonics of the limited i.f. signal are radiated to the antenna. With the low intermediate frequency used in the design discussed here, it was possible to eliminate this radiation problem by careful layout of the integrated circuit.

In *fig. 11* the signal-to-noise ratio of the demodulated audio signal is shown as a function of the antenna input voltage; the frequency swing of the received signal is ± 22.5 kHz and the modulation frequency is 1 kHz. At 1.5 µV across an input impedance of 40 Ω the signal-to-noise ratio is 26 dB ('quieting sensitivity' as defined in the CCIR standard). The muting threshold is 0.7 µV, which practically coincides with the threshold of FM reception; below this level the signal-to-noise ratio decreases rapidly with decreasing input voltage. This means that with the correlation-muting system only audio signals of unacceptable reception quality are suppressed.

The fourth-order active RC filter LP_1 (figs. 5 and 6) has internal resistors and external capacitors. It provides 38 dB of attenuation for a neighbouring channel at a spacing of 300 kHz. The total harmonic distortion of the demodulated audio signal is defined by the characteristic curve of the integrated varactor diode. It has been measured as 1.8% at an input voltage of 1 mV and a frequency swing of ± 75 kHz. A better figure can be obtained by increasing the area of the integrated varactor.

Design of an integrated FM stereo receiver

Stability of frequency-locked loop

The bandwidth of the FM stereo signal that modulates the carrier is much larger than that of the FM mono signal (see *fig. 12*). As laid down in the international standards, the stereo signal, often called the stereo multiplex signal, consists of a baseband signal equal to the sum of the signals in the left and right channels (about 0-15 kHz), a pilot tone (19 kHz) and a subcarrier (38 kHz), modulated in amplitude by the difference between the left and right signals. This AM subchannel covers a band of 23 kHz to 53 kHz; the subcarrier is suppressed at the transmitter end and is reconstructed in the receiver from the pilot tone by frequency doubling [13].

The large bandwidth of the FM stereo multiplex signal (0-53 kHz) would lead to difficulties in a receiver like the one shown in fig. 5. The problem is connected with the stability of the frequency-locked loop, which

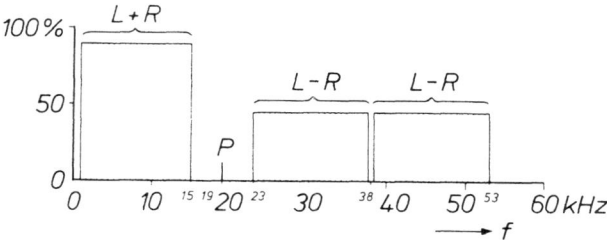

Fig. 12. Frequency spectrum of the stereo multiplex signal. The figure shows: from about 0 Hz to 15 kHz the stereo sum signal $L+R$, at 19 kHz the pilot tone P, from 23 kHz to 53 kHz the stereo difference signal $L-R$, amplitude-modulating a carrier at 38 kHz; this carrier is not transmitted. The vertical scale gives the contributions of each component to the frequency swing.

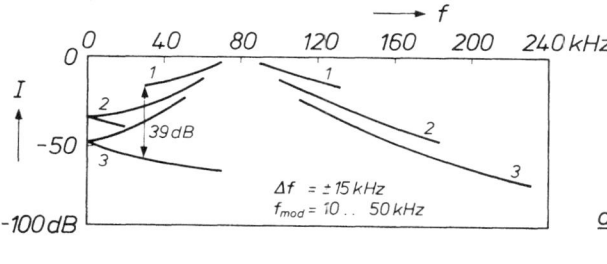

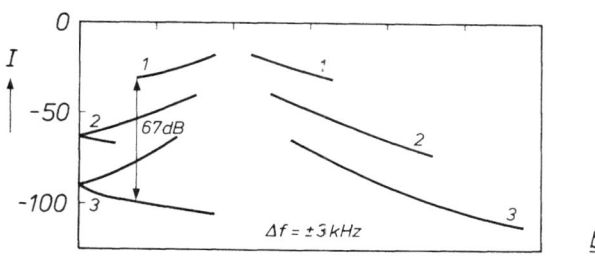

Fig. 13. Sideband 'folding'. The sideband level I for an FM signal is plotted against the frequency f (after conversion to an intermediate-frequency band centred on 80 kHz). The reference level of 0 dB corresponds to the unmodulated carrier. The diagram shows the sidebands of first, second and third order (1, 2, 3) for modulation by a sinusoidal tone at a frequency f_{mod} of 10 kHz to 50 kHz. The lower sidebands of second and third order extend further than the i.f. band permits; they are 'folded' around 0 Hz, causing harmonic distortion on demodulation. a) Weakly compressed or uncompressed frequency swing $\Delta f = \pm 15$ kHz. b) Highly compressed frequency swing $\Delta f = \pm 3$ kHz. The sidebands 2, 3 of second and third order are much weaker and the harmonic distortion is therefore less.

is responsible for compression of the frequency swing. For reception of the stereo multiplex signal, frequency-swing compression is even more necessary than for the mono signal, because of the larger bandwidth. The improvement resulting from frequency-swing compression is illustrated in fig. 13, where the level of the sidebands of an FM signal with a swing $\Delta f = \pm 15$ kHz (little or no compression) is compared with the level at $\Delta f = \pm 3$ kHz (high compression). A frequency swing of ± 15 kHz is the contribution from the encoded stereo difference signal if one of the channels is fully modulated and the other is quiescent. Because of the low intermediate frequency (80 kHz) assumed in fig. 13, the lower sidebands are 'folded' around the limit of 0 Hz, giving rise to harmonic distortion. Since the sidebands of higher order (2, 3) are disproportionately weaker when the swing is smaller, the resulting harmonic distortion is less.

The frequency-locked loop is a negative feedback system, which is only stable when the phase shift is less than 180° at the frequencies where the loop gain is greater than or equal to one. If the phase shift in this frequency range is 180°, the loop will oscillate spontaneously.

Now a phase shift is unavoidable in the filters used in this control loop: these are the i.f. filter (LP_1 in fig. 5), which determines the selectivity, the FM detector (AP_1 and M_2) and the loop filter (LP_2) that limits the bandwidth of the loop. To obtain a quantitative idea of the way in which the selectivity, the bandwidth of the closed frequency-locked loop and the transfer characteristic are affected by the stability condition, it is useful to have a simplified model of the frequency control system in the FM receiver. Let us first take another look at the mono receiver as shown in fig. 5.

Frequency control of mono receiver

Fig. 14 shows the model, including the transmitter. This is represented by the voltage-controlled oscillator VCO_1, to which the mono information V_i is supplied (there is no multiplex signal). VCO_1 delivers a frequency-modulated carrier. In the analysis that follows, the frequency of the unmodulated carrier is not relevant; we are more concerned with the frequency-deviation f_i. We shall therefore consider this alone in the following and merely state that VCO_1 delivers a signal f_i. This frequency deviation f_i is proportional to V_i.

All non-essential subfunctions in the receiver, such as the amplifiers and the signal muting, have been omitted. The oscillator VCO_2 is controlled by the

---

[13] N. van Hurck, F. L. H. M. Stumpers and M. Weeda, Stereophonic radio broadcasting, I. Systems and circuits, Philips Tech. Rev. **26**, 327-339, 1965.

voltage V_o, which is produced by the control system, which gives it a frequency deviation f_o. The mixer M_1 produces the difference signal $f_i - f_o$; this passes through the i.f. filter LP_1, which determines the selectivity, and goes to the frequency demodulator AP_1/M_2, whose output signal $(V_i - V_o)_{del}$ is delayed. This signal passes through the loop filter LP_2 (bandwidth B) and its output signal is the control voltage V_o mentioned above.

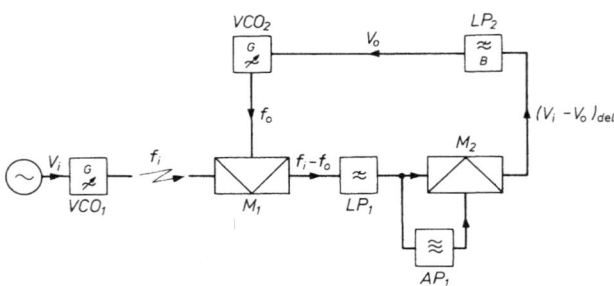

Fig. 14. Model of the frequency-locking system in the FM mono receiver. V_i mono signal. VCO_1 transmitter, frequency-modulated by V_i. f_i frequency deviation at receiver input. M_1 mixer. LP_1 i.f. filter. M_2 mixer for demodulation. AP_1 allpass filter (delay network) giving a 90° phase shift for the intermediate frequency. LP_2 loop filter. V_o control voltage. VCO_2 voltage-controlled oscillator. f_o deviation of the oscillator frequency. $(V_i - V_o)_{del}$ signal resulting from demodulation of $f_i - f_o$ and delayed by LP_1 and AP_1.

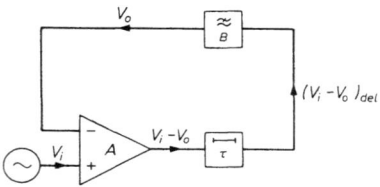

Fig. 15. Simplified version of fig. 14. The conversions from voltage to frequency deviation and frequency deviation to voltage have been omitted. The gain A is the product of the conversion factors of VCO_2 and AP_1/M_2.

The delay of $(V_i - V_o)_{del}$ is partly caused by the group delay of LP_1 and partly by the delay of the i.f. signal in AP_1. Both delays are added together to form a single delay τ.

When the voltage-frequency conversion in VCO_1 and VCO_2 is combined with the frequency-voltage conversion in demodulator AP_1/M_2, the result is a particularly simple model of an FM transmitter and FM receiver with a frequency-locked loop (fig. 15). The transmitter now consists only of the voltage source V_i, and the receiver consists of a differential amplifier of gain A, a delay element with delay τ and a loop filter of bandwidth B. The gain A is the product of the conversion gains of VCO_2 and AP_1/M_2.

The transfer function from transmitter to receiver is determined by these three quantities A, τ and B. The swing compression is $1/(1 + A)$. The part of the delay τ arising in LP_1 is connected with the selectivity, and the part arising in AP_1 is connected with the intermediate frequency. Finally, B is a quantity that determines the stability of the frequency-locked loop.

The contribution of the filter LP_1 to the delay τ can be determined quantitatively with the aid of fig. 16. The curves Bu represent the amplitude characteristic and the group delay of the filter as designed for the integrated FM mono receiver; it is a fourth-order Butterworth filter with a cut-off frequency of 100 kHz. It can be seen in fig. 16b that the group delay in the

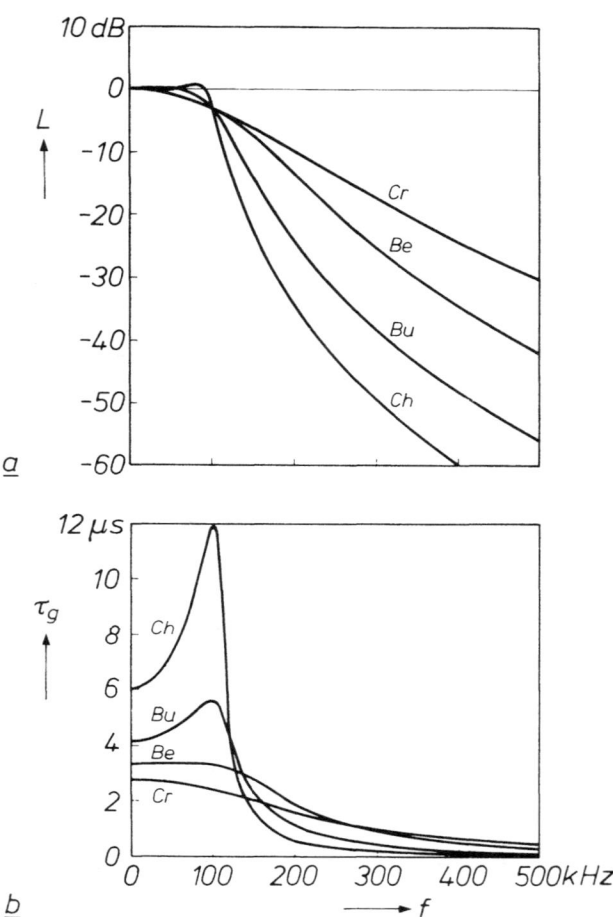

Fig. 16. Amplitude level L and group delay τ_g of the output signal as a function of frequency f for a fourth-order lowpass filter with a cut-off frequency of 100 kHz. Cr critically damped filter Be Bessel filter Bu Butterworth filter Ch Chebyshev filter. The selectivity of the filters increases in this order, but at the same time the group delay in the passband increases.

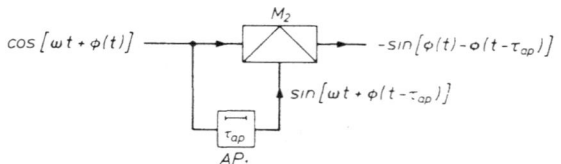

Fig. 17. Signal delay in the demodulator circuit AP_1/M_2. The delay τ_{ap} in the allpass filter AP_1 is a quarter of a period at the intermediate frequency. Because of the symmetry of the expression for the output signal a delay of $\frac{1}{2}\tau_{ap}$ can be assigned to it.

passband is reasonably constant and that LP_1 may be replaced by a frequency-independent delay element.

A further delay occurs in the frequency demodulator, which consists of the allpass filter AP_1 and the multiplier M_2 (*fig. 17*). The output signal of M_2 is the product of the i.f. signal and a delayed version of it obtained from AP_1; the delay in AP_1 corresponds to a phase shift of 90° at the nominal intermediate frequency. The symmetry of the contributions from the two input signals to the output signal suggests that the delay of the output signal is the average of the delays of the two input signals, i.e. half the group delay of AP_1.

This may be substantiated by a mathematical description of quadrature modulation. The product of the two input signals to M_2 (see fig. 17), taking τ_{ap} as the delay in AP_1, is:

$$\cos\{\omega t - \phi(t)\}\sin\{\omega t - \phi(t - \tau_{ap})\}$$
$$= \tfrac{1}{2}\sin\{2\omega t - \phi(t) - \phi(t - \tau_{ap})\}$$
$$- \tfrac{1}{2}\sin\{\phi(t) - \phi(t - \tau_{ap})\}$$

The first term is at twice the intermediate frequency and is suppressed by the loop filter. The second term represents the low-frequency output signal; the variation of the phase $\phi(t)$ with time contains the audio information, since $d\phi/dt$ is the instantaneous value of the frequency deviation and hence of the audio signal. We can expand $\phi(t)$ and $\phi(t - \tau_{ap})$ about the time $t - \tfrac{1}{2}\tau_{ap}$ in a Taylor series and then subtract:

$$\phi(t) = \phi + \frac{\tau_{ap}}{2}\frac{d\phi}{dt} + \frac{\tau_{ap}^2}{8}\frac{d^2\phi}{dt^2} + \frac{\tau_{ap}^3}{48}\frac{d^3\phi}{dt^3} + ..$$
$$\phi(t - \tau_{ap}) = \phi - \frac{\tau_{ap}}{2}\frac{d\phi}{dt} + \frac{\tau_{ap}^2}{8}\frac{d^2\phi}{dt^2} - \frac{\tau_{ap}^3}{48}\frac{d^3\phi}{dt^3} + ..$$
$$\phi(t) - \phi(t - \tau_{ap}) = \tau_{ap}\frac{d\phi}{dt} + \frac{\tau_{ap}^3}{24}\frac{d^3\phi}{dt^3} + ..$$

All the terms on the right-hand side have the argument $t - \tfrac{1}{2}\tau_{ap}$. The terms of third and higher order are small, so that, to a good approximation,

$$\phi(t) - \phi(t - \tau_{ap}) \approx \tau_{ap}\frac{d\phi(t - \tfrac{1}{2}\tau_{ap})}{dt}.$$

The output signal thus carries the information of the time $t - \tfrac{1}{2}\tau_{ap}$, so that the delay in the demodulator circuit amounts to half the delay in the allpass filter AP_1.

Now that we know the total delay τ in the frequency-locked loop in fig. 15, we can calculate the effect of the bandwidth B of the loop filter on the transfer function V_o/V_i. *Fig. 18* gives three examples for bandwidths of 2 kHz, 3 kHz and 5 kHz. As might be expected, the -3-dB bandwidth of the closed loop increases and at the same time the stability decreases with increasing bandwidth B of the loop filter. Although a -3-dB bandwidth of 53 kHz is obtained at $B = 5$ kHz, there is a peak of 8 dB because of the reduced stability, and this is unacceptable for stereo reception. The bandwidth of 53 kHz is necessary, however, if the entire stereo multiplex signal is to contribute to the frequency-swing compression. The peaking can of course be reduced by making τ smaller, but this necessitates a less selective i.f. filter (see fig. 16, curves Be or Cr); in many situations there would then be unacceptable interference from strong adjacent stations.

There is therefore a dilemma between undistorted stereo reception on the one hand and sufficient selectivity on the other. We have succeeded in finding a solution, which will be explained here with the aid of a simplified model of a stereo transmitter and receiver.

Frequency control of a stereo receiver

The essence of our solution to the stability problem caused by the large bandwidth of the stereo multiplex signal is that the stereo difference signal that modulates the suppressed 38-kHz subcarrier does not pass through the loop filter. Instead it is first demodulated and passed through a lowpass loop filter of its own. The output is then remodulated, again producing an AM signal with suppressed carrier at 38 kHz. This in turn is added to the stereo sum signal, to provide oscillator control.

A simplified model of the system of frequency control in the combination of stereo transmitter and stereo receiver is illustrated in *fig. 19*. Here the signal from the stereo transmitter consists only of a subcarrier at 38 kHz, modulated in amplitude by V_i, the difference between the right and the left channels. The stereo receiver consists for the most part of the same components as the greatly simplified mono receiver in fig. 14: VCO_2, M_1, LP_1 and AP_1/M_2. Here, however, the loop filter LP_2 is replaced by mixer M_3, lowpass filter LP_3 and mixer M_4. The large-signal input to mixer M_4 is a 38-kHz subcarrier in phase with the subcarrier of the transmitter. The large-signal input to mixer M_3 is a delayed version of the subcarrier; the delay τ is equal to the delay in LP_1 and AP_1/M_2.

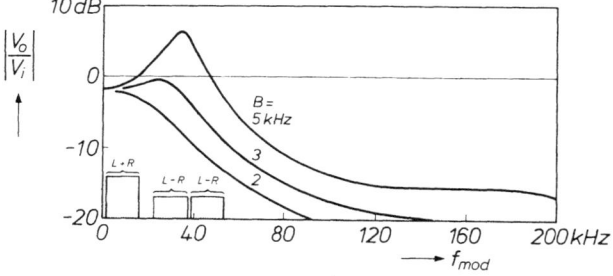

Fig. 18. Transfer function V_o/V_i in the closed frequency-locked loop; f_{mod} is the modulation frequency. The bandwidth B of the loop filter must be 5 kHz for all modulation frequencies up to 53 kHz of the stereo multiplex signal in fig. 12 to be passed, so that they can contribute to the swing compression. However, there is then an undesired peak of 8 dB, a sign of reduced stability.

The operation of the model is as follows. In M_0 the difference signal V_i modulates the subcarrier to produce V_i'; this then modulates VCO_1 to give the frequency deviation f_i. In a similar way the subcarrier is modulated in M_4 by the difference signal V_o obtained from the demodulation, resulting in a signal V_o', which

agreement between model and experiment up to modulation frequencies of about 30 kHz with an open loop, and to about 80 kHz with the loop closed. The highest frequency in the stereo difference signal is 15 kHz, which can be processed without difficulty by a receiver of this type.

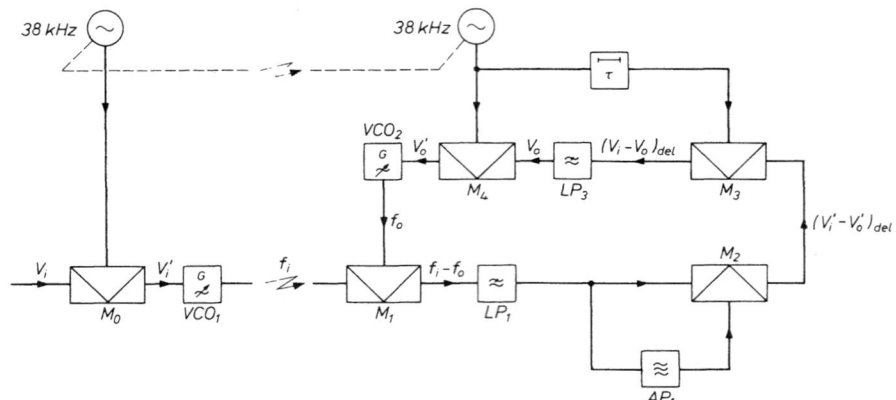

Fig. 19. Model of the frequency-control system in the FM stereo receiver. It is assumed that only a stereo difference signal V_i is transmitted; this is denoted as V_i' after modulation of the 38-kHz subcarrier. The model contains all the elements of the mono receiver in fig. 14. In the control loop now, however, demodulation of the 38-kHz subcarrier takes place in M_3 before the loop filter and remodulation to 38 kHz takes place in M_4 after the loop filter. In this model the 38-kHz generators in transmitter and receiver are assumed to be in phase; the signal delay τ in LP_1 and AP_1/M_2 makes it necessary to give the 38-kHz carrier the same delay to ensure correct phase in the demodulation in M_3.

then modulates VCO_2; the result is f_o. The output signal of mixer M_1 is the difference $f_i - f_o$. After filtering (in LP_1) and first demodulation (AP_1/M_2) this signal gives a delayed version $(V_i' - V_o')_{del}$ of the difference between the control signal V_i' of VCO_1 and V_o' of VCO_2. After a second demodulation in M_3, including compensation for the phase error introduced by LP_1 and AP_1/M_2, a delayed version $(V_i - V_o)_{del}$ of the difference between the difference signals V_i in the transmitter and V_o in the receiver is obtained. After filtering in loop filter LP_3 the output signal V_o is obtained in the same way as in the mono receiver.

Modulation by the 38-kHz subcarrier in M_0 and M_4 and corresponding inverse demodulation in M_3 take place without delay. The only delays in the receiver are the group delay of LP_1 and the delay in the demodulator AP_1/M_2. If the sum of these delays is equal to τ, and if the bandwidth of the loop filter LP_3 is equal to B, the transfer function V_o/V_i in the stereo receiver is the same as the transfer function in the highly simplified model of a mono receiver in fig. 15. The stability problem in the stereo receiver is therefore essentially the same as for the mono receiver and can therefore be solved with selectivity maintained.

Fig. 20 shows the calculated and measured transfer function V_o/V_i of a stereo receiver as in fig. 19 with open loop (*OL*) and closed loop (*CL*). There is good

A real stereo receiver must of course deal with the sum signal as well as the stereo difference signal. This means that the receiver must contain a loop as shown in fig. 14 in parallel with the loop shown in fig. 19.

In a practical receiver the subcarrier locking is obtained by means of a 19-kHz pilot tone, which has a delay τ with respect to the transmitter signal after demodulation. The regenerated 38-kHz subcarrier thus has the correct phase for the demodulation in M_3, but this phase must be advanced by an amount

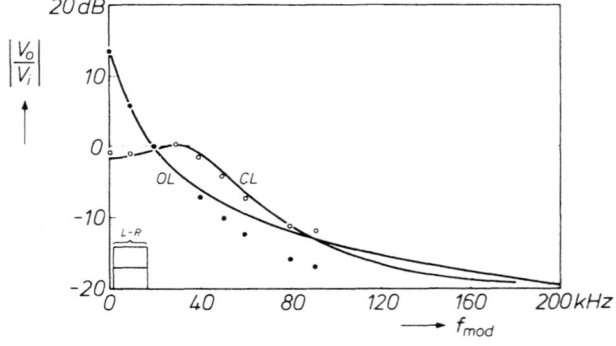

Fig. 20. Transfer function V_o/V_i in the open frequency-locked loop (*OL*) and in the closed loop (*CL*) in the stereo receiver; f_{mod} is the modulation frequency. The points indicate values calculated from a model like the one of fig. 19; the curves were measured for an experimental circuit. The stereo difference signal $L - R$ that the filter must pass has a bandwidth of 15 kHz.

$2\pi \times 38\,000 \times \tau$ for the remodulation in M_4. If, as assumed above, the i.f. filter LP_1 is a fourth-order Butterworth filter and the intermediate frequency is 70 kHz, the phase correction required at 38 kHz can even amount to about 90°. Without this phase correction the stereo difference signal modulated at 38 kHz would not contribute to the swing compression.

Block diagram of an integrated FM stereo receiver

The block diagram of the FM stereo receiver is given in *fig. 21*. The input stages are the same as in the FM mono receiver; the main additions are the more elaborate loop-filter network *LoopFi*, the subcarrier various paths in the loop-filter network. The lowpass filter LP_2 with a cut-off frequency of 5 kHz passes the baseband, but with an attenuation increasing by 6 dB/octave above 5 kHz (first-order filter). The multiplier circuit M_3 receives the 38-kHz subcarrier from the subcarrier regenerator and multiplies it by the multiplex signal. This results in demodulation of the difference signal, which is then limited in bandwidth in the first-order lowpass filter LP_3 (cut-off frequency also 5 kHz); after band limiting it is used for remodulating the amplitude of the subcarrier (multiplier M_4, amplitude modulation with suppressed carrier again). In this way a band filter centred on 38 kHz is obtained.

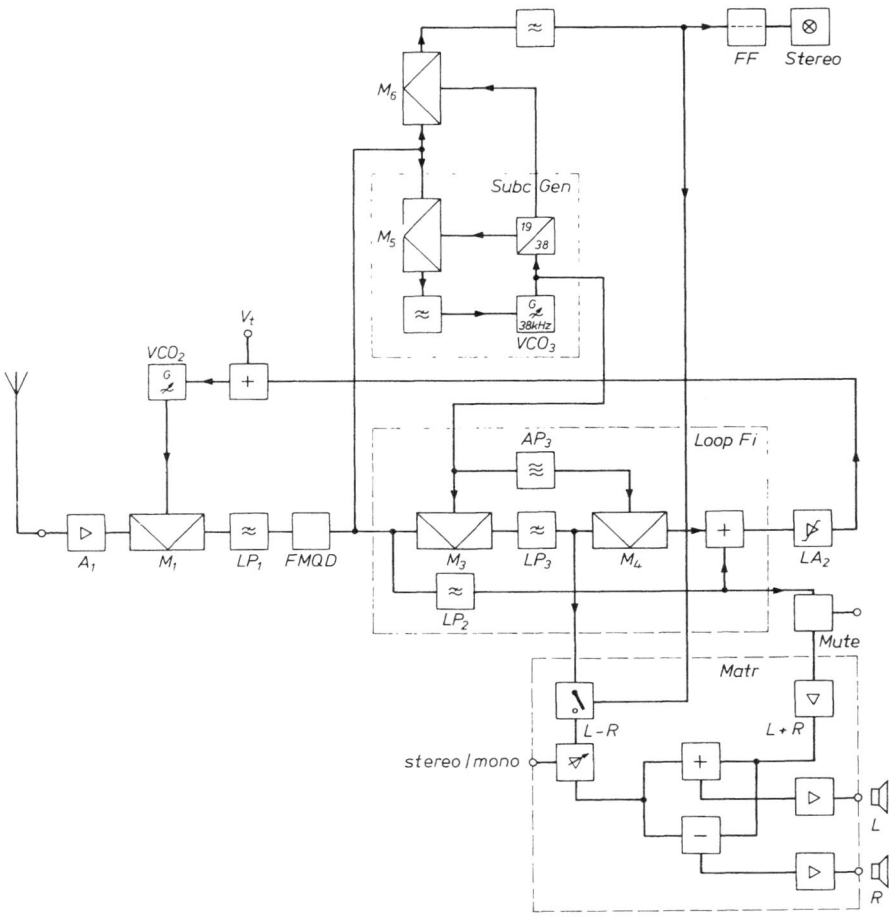

Fig. 21. Block diagram of an FM stereo receiver for integration. Many of the components are the same as in the mono receiver (see fig. 4). *FMQD* FM quadrature detector. *LoopFi* loop filter. *SubcGen* subcarrier generator. *Matr* matrix network (in which the left and right signals L and R are recovered from $L + R$ and $L - R$). *FF* bistable circuit. *Stereo* stereo indicator. *Mute* muting for off-tune signals.

regenerator *SubcGen* and the matrix network *Matr*, in which the signals for the left and right channels are recovered from the sum and difference signals.

The stereo multiplex signal that appears at the output of the FM quadrature detector *FMQD* takes In fact there is a temporary transformation from 38 kHz to 0 Hz, so that a simple lowpass filter is sufficient for limiting the passband.

A filtered multiplex signal is produced in an adder circuit and is returned via limiter/amplifier LA_2 to the voltage-controlled oscillator VCO_2 to give the required swing compression. The external tuning voltage V_t is added to the control signal.

The allpass filter AP_3 compensates for the delay of the 38-kHz carrier in the i.f. filter and the FM detector. As explained earlier, the phase of the subcarrier is advanced here by an amount τ, so that in M_4 the stereo difference signal modulates a carrier of the same phase as the subcarrier in the input signal of M_1.

The regeneration of the subcarrier for demodulating and remodulating the stereo difference signal in the loop-filter circuit *LoopFi* takes place in the stereo subcarrier regenerator *SubcGen*. This contains a phase-locked loop in which the 19-kHz pilot tone is multiplied by a 19-kHz signal, obtained by frequency division from a 38-kHz voltage-controlled oscillator (VCO_3). Any phase deviations cause a d.c. component in the product signal; this d.c. component is passed by a lowpass filter and is used to control the tuning of the oscillator.

The presence of the 19-kHz pilot tone is detected with the aid of multiplier M_6. Its output signal passes through a lowpass filter and controls a bistable circuit *FF*, which in turn switches on a stereo indicator; the same output signal operates the mono/stereo switch.

The demodulated stereo difference signal goes via this switch to a variable-gain amplifier and then to the matrix network, in which the signals for the left and right audio channels are derived from the difference and sum signals by addition and subtraction. The variable-gain amplifier makes it possible to vary the stereophonic effect (*stereo/mono*) in the reproduction. For a weak signal in noise it is better to attenuate the stereo difference signal; this reduces the stereo effect but improves the signal-to-noise ratio. On the other hand the stereo effect can be accentuated by amplifying the stereo difference signal more than proportionately, so that the sound sources seem to be further apart than the loudspeakers.

Summary. The article describes the integration of a complete FM mono receiver on a single chip. The external components are a single inductor (for the oscillator) and some capacitors and resistors. The use of a very low intermediate frequency (70 kHz) enables the i.f. bandpass filters to be replaced by lowpass filters. Harmonic distortion in such a limited i.f. band is avoided by compressing the frequency swing to ± 15 kHz. The IC is currently being marketed as type TDA 7000/7010T; an improved version, TDA 7020T, which operates with a lower battery voltage, is in pilot production. A monolithic FM stereo receiver is now being studied. In this receiver two frequency bands are selected from the stereo multiplex signal for oscillator control, since a frequency-locked loop with the full stereo multiplex bandwidth is not stable because of the group delay in a sufficiently selective i.f. filter.

ADVANCED LOW VOLTAGE SINGLE CHIP RADIO IC

Taiwa Okanobu, Hitoshi Tomiyama, and Hiroshi Arimoto*
Personal Telecommunications Group, Sony Corporation
1-7-4 Kohnan, Minato-ku, Tokyo, 108 Japan
* Semiconductor Group, Sony Corporation, Atsugi, Japan

ABSTRACT

We have developed the technology including intermediate frequency (IF) filters in a radio IC by getting IF low. Degrading of image interference characteristic can be avoided by image cancelling circuits by Inphase/Quadrature mixers and Phase Shift Network (P.S.N). We have developed new two ICs for FMstereo/AM and AM only radio using this technology. FM stereo/AM radio IC can work at a supply voltage of 0.95 volt.

INTRODUCTION

We made a complete single chip radio IC for reducing external parts and adjustment points. But Intermediate Frequency(IF) filter was not included. At conventional radio system, IF filter needs sharp selectivity, we use IFTs or ceramic filters. As IF is high (455kHz or 10.7MHz), so it's very difficult to include filters in IC. Because high frequency filter circuits is too large scale and unstable. But at low IF system, image frequency is very close to desired frequency, so image interference characteristic is degraded.

New IC system includes image cancelling circuits. The circuits have two mixers inquadrature and Phase Shift Network (PSN). So, IF can be very low without image interference characteristic getting worse. Therefore, Selectivity is sufficient in spite of active filter circuits having a 10~20 Q factor which can be fabricated in IC. And we have achieved same performance IC as an usual radio using ceramic filters.

At FMstereo/AM radio IC, FM IF is 150kHz, AM IF is 55kHz. Both frequencies are very close, therefore PSN and mixers as image cancelling circuits are used both for FM and AM. FM IF is very low , so pulse count detector circuits can be realized as FM detector, although supply voltage is 0.95 volt. The adjustment point and external parts of FM detector is eliminated. FM signal to noise ratio is realized on 60dB.

The drawback of low IF system is the beat which occurs by double IF and FM MPX sub carrier. To avoid the beat, linear multiplier is used for FM MPX decoder.

At AM only radio IC, pins of this IC decrease by the technology including IF filter. So 8 pins AM only radio IC can be realized, which has an audio power amplifier. The audio power output is more than 100mW. This IC is suitable for a portable AM radio.

IMAGE CANCELLING CIRCUITS

At a conventional radio system, IFT or ceramic filter was used for IF filter, Active C R filter which can be fabricated in IC, has low Q factor, about under 20. To achieve sufficient selectivity using such low Q factor filter, IF must be low. At low IF system, image signal can not be decreased by the filters before mixer. Therefore, image cancelling circuit is needed to make IF low.

Fig.1 is a block diagram of image cancelling circuit. Two 90 degrees phase-shifted signals LO1 and LO2 are fed to Mixer Q and Mixer I, respectivery. Received signal (DSR or IMG) is converted to IF1 and IF2 by the mixers. And next, phase of IF2 increases 90 degrees by PSN. It is IF2'. After these signals (IF1 and IF2') are added, desired signal appears and image signal is cancelled. These mixers are used both for AM and FM. LO1 and LO2 varied from 500kHz to 60MHz, so we use 1/2counter in order to make two 90 degrees phase-shifted signals LO1 and LO2.

Fig.2 and Fig.3 describe magnitude and phase of these signals. Fig.2 shows desired received signal and Fig.3 shows image signal.

Image signal IMG is higher than LO2, so signal IF2, mixed by LO2 and IMG is 90 degrees lead to signal IF1. IF2' is 90 degrees lead to signal IF2. Therefore, IF1 and IF2' are contrary signals. Therefore, image signal is cancelled.

At real IC system, image rejection ratio is decided by these 3 reasons.

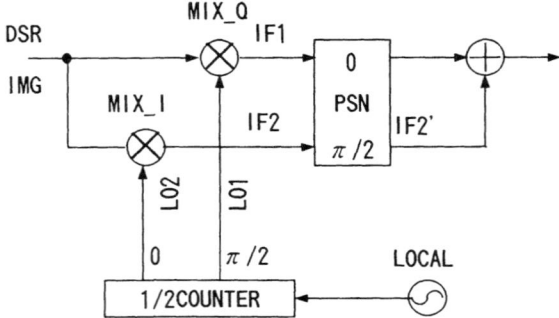

Fig. 1 IMAGE CANCELLING CIRCUIT BLOCK DIAGRAM

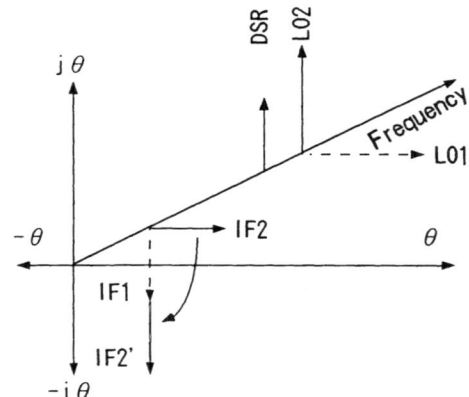

Fig. 2 MAGNITUDE AND PHASE OF DESIRED SIGNALS

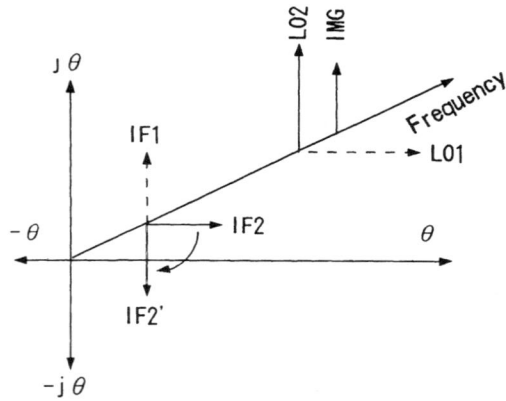

Fig. 3 MAGNITUDE AND PHASE OF IMAGE SIGNAL

1) Gain difference between two mixers
2) Phase error of PSN
3) Phase error of counter

1) Gain difference between two mixers

Gain difference between two mixers are decided by relative error of resistors or transistors in the IC. At FMstereo/AM radio IC, mixer gain is controlled by external terminal. At AM only radio IC, there's no such terminal to decrease a pin. 30dB image rejection ratio is realized by image cancelling circuits in spite of no gain adjustment, in addition to about 20dB by MW ferrite bar antenna. So sufficient rejection over 50dB is accomplished at AM only radio IC.

2) Phase error of PSN

At FMstereo/AM radio IC, PSN is used for both AM and FM, so wide band PSN is needed. PSN circuits is 4th order PSN which consists of two 2nd order all pass network shown at Fig. 4. The characteristic is shown at Fig. 5. If phase of EA is same as phase of EB, the phase difference is 90 degrees with 1 degree phase error from 45kHz to 300kHz. The error corresponds to 40 dB image rejection ratio. At AM only radio IC, PSN is 3rd order all pass network from 45kHz to 65kHz.

3) Phase error of counter

Two 90 degrees-phase-shifted signals are obtained by 1/2counter. So the counter is triggered by both rising up and falling down of oscillator signal. The duty factor of the oscillator signal is important to get two 90 degrees-phase-shifted signals. To get constant duty factor, oscillator signal level is controlled to oscillate as constant level.

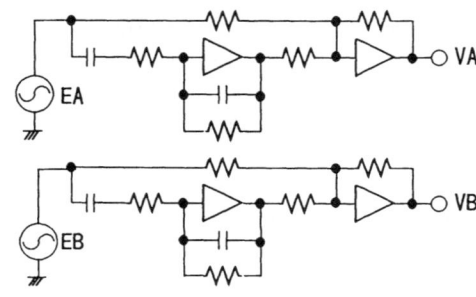

Fig. 4 4TH ORDER PHASE SHIT NETWORK

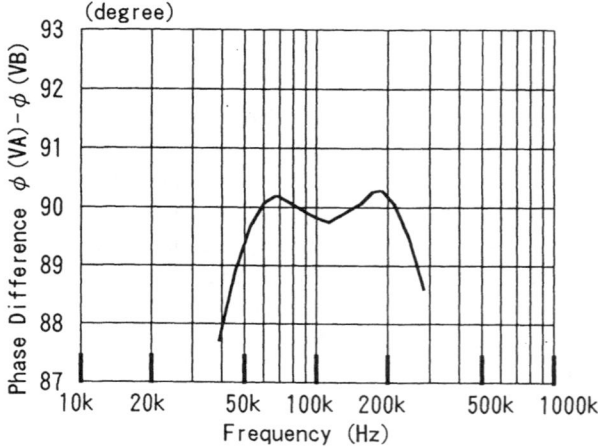

Fig. 5 PHASE SHIFT NETWORK CHARACTERISTIC

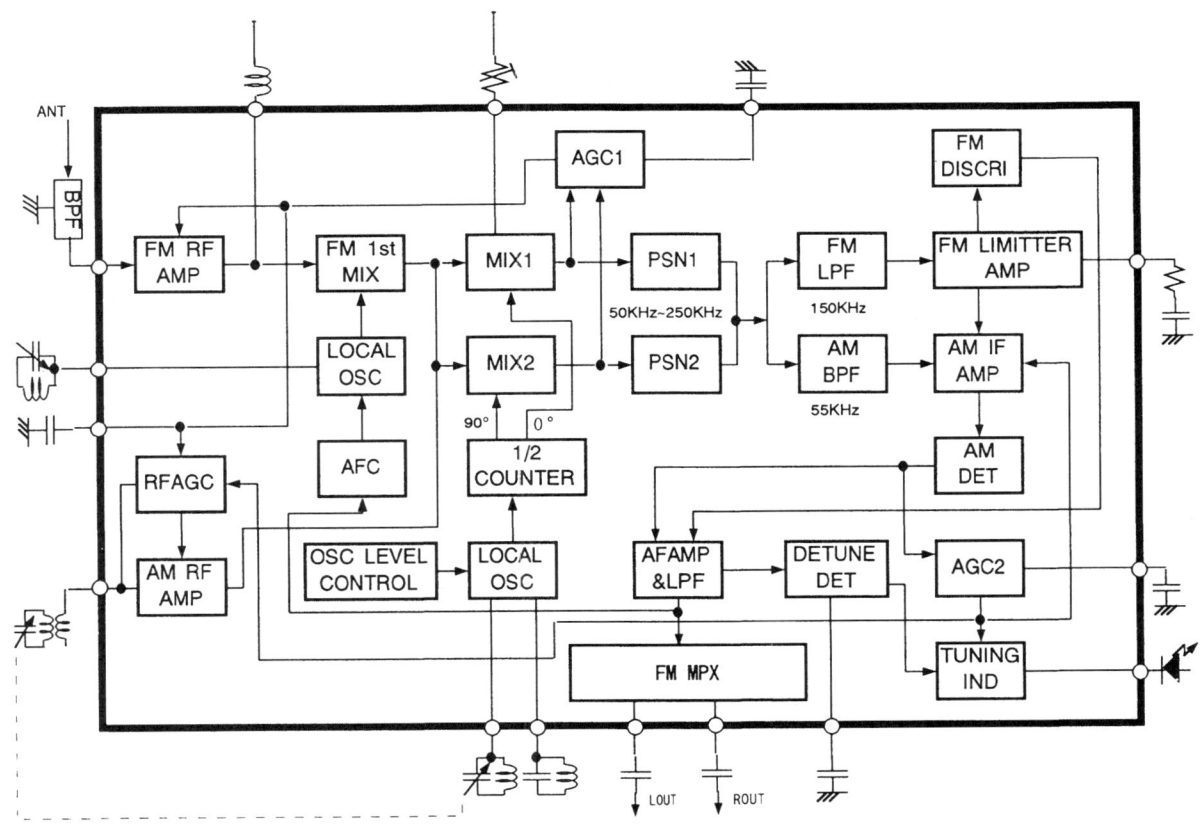

Fig. 6 BLOCK DIAGRAM OF LOW VOLTAGE FMSTEREO/AM RADIO IC

LOW VOLTAGE FMSTEREO/AM RADIO IC

At first, I will describe about low voltage FM stereo / AM radio IC.

The IC has all functions for FM and AM radio and FM stereo decoder. It operates only at a supply voltage of 0.95 volt, and intermediate frequency filter is included. Image cancelling circuit consists of in-phase and quadrature mixers.

The figure 6 shows a block diagram of low voltage FM stereo / AM radio IC. FM radio receiver is dual conversion super heterodyne, 1st IF is 30 MHz and 2nd IF is 150 kHz. To make two 90 degrees phase shift signals, local oscillator signal is divided by 1/2 counter.

FM received signal passes through band pass filter to FM RF amplifier. The signal is converted to 1st IF signal about 30 MHz by FM 1st mixer. So image rejection ratio is better than a conventional FM radio in which 1st IF is 10.7 MHz.

1stIF signal is converted to 2nd IF signal 150 kHz by in-phase/quadrature mixer, and 2nd IF signal is selected by FM low pass filter. 2nd IF signal is amplified by FM limiter amplifier, and detected by pulse count circuit. The detected signal is decoded to FM stereo signal by FM MPX decoder.

AM received signal is converted to IF signal 55 kHz by in-phase/quadrature mixers, and selected by band pass filter. AM IF signal is amplified by AMIF, and detected by AMDET. Gain of AMIF and AMRF is controlled by AGC2 to keep audio output signal level constant. Phase shift network and band pass filter are realized by active circuits, so input signal level is limited under dynamic range by AGC1.

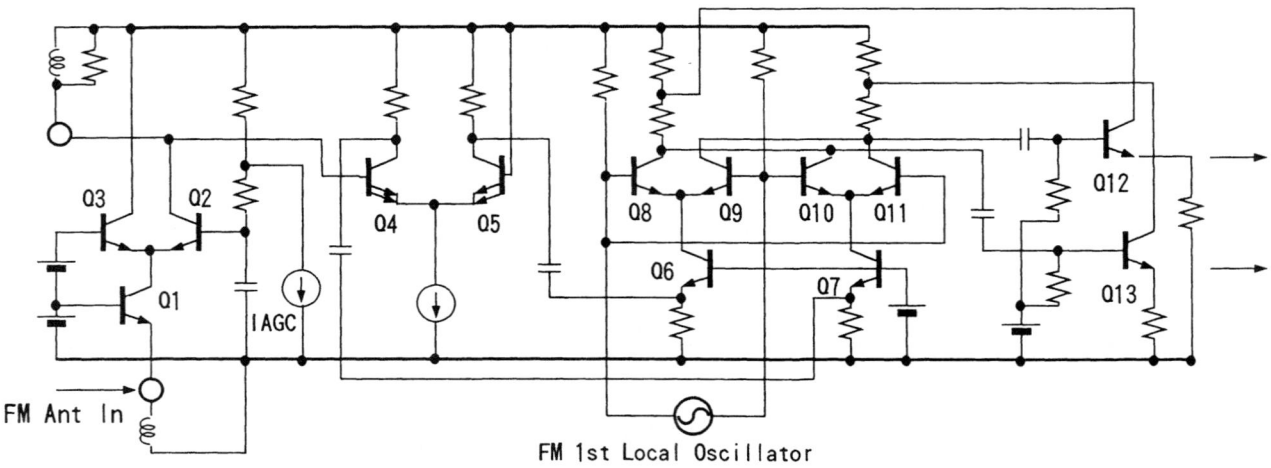

Fig. 7 FM RF AMPLIFIER AND 1ST MIXER

FM RF AMP

The FM RF AMP consists of common-base amplifier and cascade configuration by Q1 and Q2. And its gain is 10 dB. The gain of FM RF AMP is controlled by base voltage of Q2 varied with AGC voltage.

The mixer is a double-balanced mixer consisting of Q4~Q11. The collector output of emitter follower Q12 and Q13 are fed back positively to compensate a gain loss of the emitter follower.

At conventional FM radio IC, The output stage of RF AMP is a tunable bandpass filter to get good sensitivity and good interference characteristic. In this IC, by making circuits low noise and improving the circuits linearity, same characteristic as conventional one is achieved in spite of untunable bandpass filter. As for image interference characteristic, sufficient image signal rejection can be realized because FM 1st IF is 30MHz

The local oscillator is a positive feedback type. Its frequency is controlled by AFC (automatic frequency control) circuit to avoid frequency drift.

I, Q MIXER and PSN

Two mixers inquadrature are used as FM 2nd mixer and as AM mixer. These mixers are double-balanced type. Fig. 8 shows one mixer circuit. Mixer output flows into buffer amplifier. The gain of mixer is controlled by external variable resistor R1. So image rejection ratio is adjusted by the resistor.

PSN is 4th order PSN as described before.

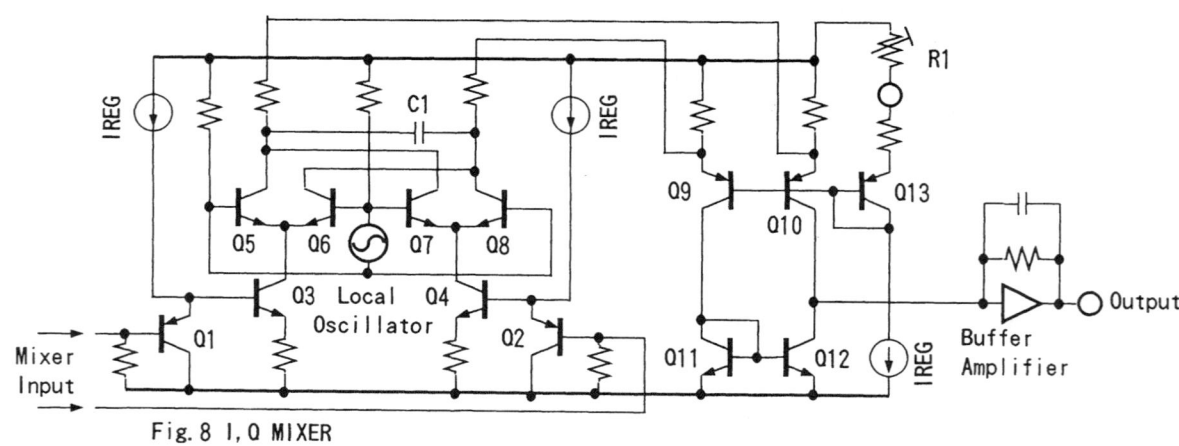

Fig. 8 I, Q MIXER

FM LPF

IF filters consists of active filters with capacitors and resistors.

FM LPF is 9th order low pass filter with 300kHz cutoff frequency shown in Fig. 9. The filters characteristic is shown at Fig. 10.

As a result, this FM effective selectivity shown in Fig. 11 is realized. Adjacent channel suppression is over than 40dB, corresponds to using 3 elements ceramic filter shown as an interrupted line in Fig. 11. Image signal -300kHz away from center frequency is 40 dB attenuated.

AM BPF

Fig. 13 illustrates AM band pass filter characteristic. The filter consists of three 2nd order biquad 55kHz band pass filters shown at Fig. 12, and adjacent channel suppression 10kHz away from center frequency is 35 dB.

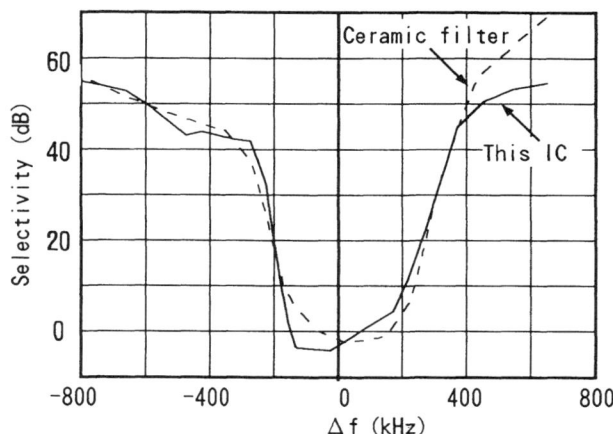

Fig. 11 FM EFFECTIVE SELECTIVITY

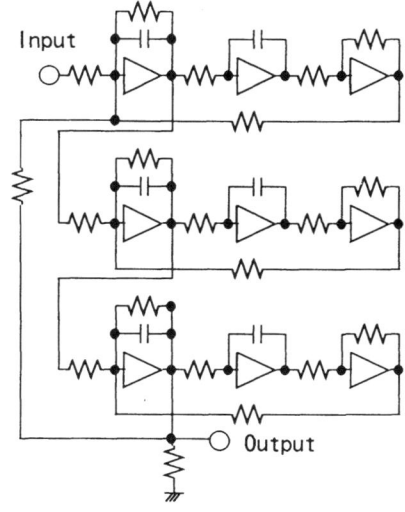

Fig. 12 AM BPF

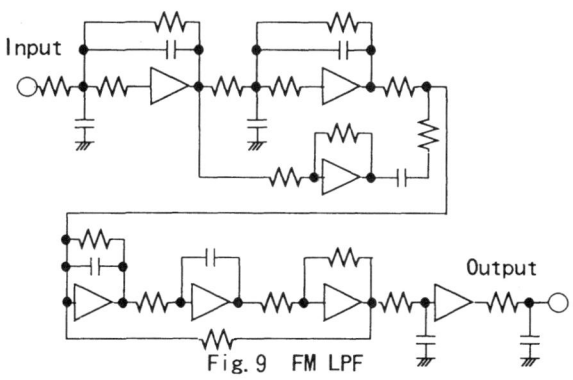

Fig. 9 FM LPF

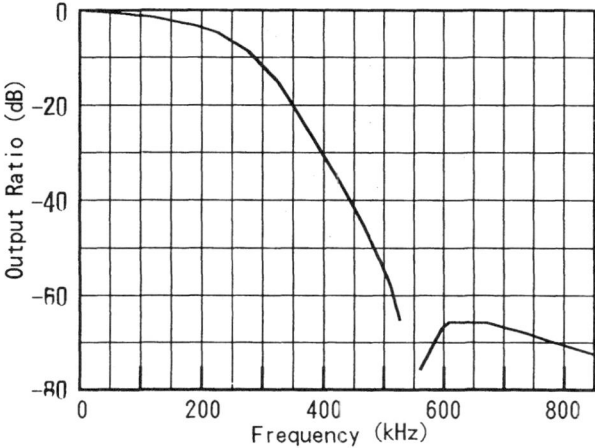

Fig. 10 FM LPF CHARACTERISTIC

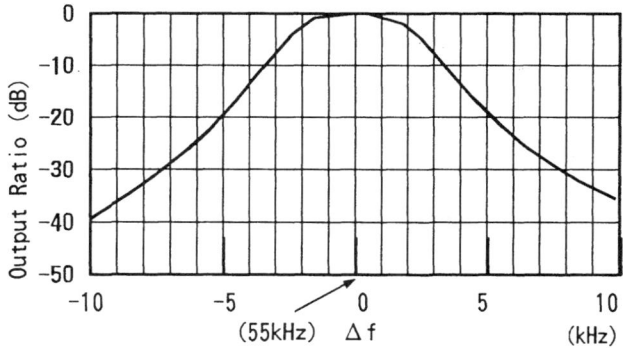

FIG. 13 AM BPF CHARACTERISTIC

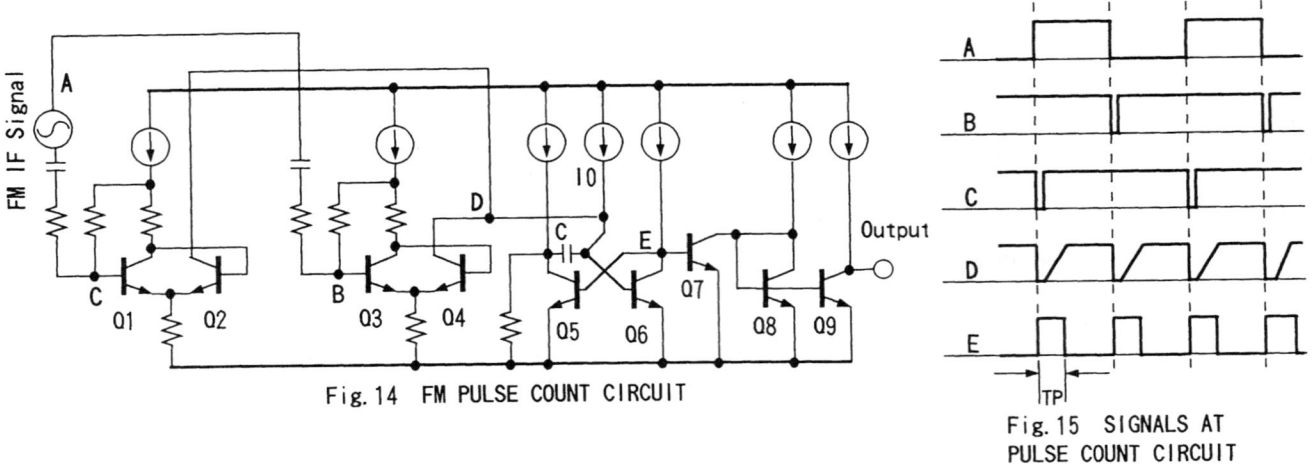

Fig. 14 FM PULSE COUNT CIRCUIT

Fig. 15 SIGNALS AT PULSE COUNT CIRCUIT

Pulse Count Circuit

The fig. 14 is a pulse count circuit of this IC. The differential outputs of FM IF amplifier are added to two trigger circuits consisting of Q1~Q4 respectively, so mono-stable multivibrator consisting of Q5 and Q6 is triggered by both rising up and falling down of the FM IF signal.

Therefore, sampling frequency is the double of IF, and Nyquist frequency is IF. So sufficient band width for FM stereo MPX signal is achieved. Pulse width TP decided by I0 and C is 875 nano second, so peak separation is 500 kHz.

Fig. 15 shows signals of each stages.

FM MPX Decoder

At FM stereo decoder, mixed by double intermediate frequency about 300 kHz and harmonics of FM MPX sub carrier 38 kHz, the beat will occur. To prevent the beat, in this IC 4th order low pass filter is included at detector output and FM detected signal is decoded to left and right output by 38 kHz sin wave.

Figure 16 is FM MPX decoder circuit. Input signal Vin is converted to Iin by negative feedback circuit consisting of Q1~Q3. And Iin is decoded to left signal and right signal by stereo MPX decoder consisting of Q4~Q9. Sub carrier signal which inputs to base of stereo MPX decoder is logarithmically transferred signal. So collector current of stereo MPX decoder is proportional to the stereo sub carrier Isub. Therefore the decoder is linear multiplier.

Logarithmic amplifier consists of Q10~Q13. By feedback circuit (Q10~Q12) current Isub flows into transistor Q10, so output signal is logarithmically transferred.

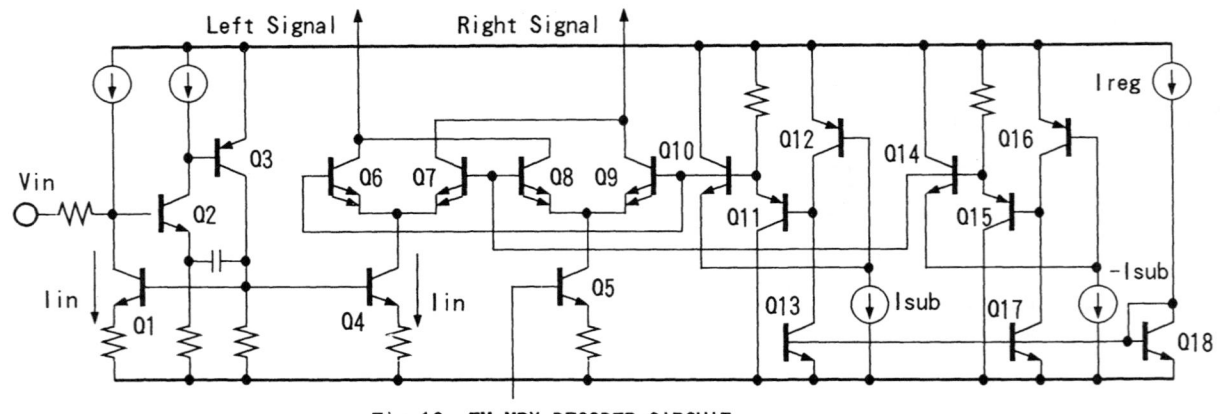

Fig. 16 FM MPX DECODER CIRCUIT

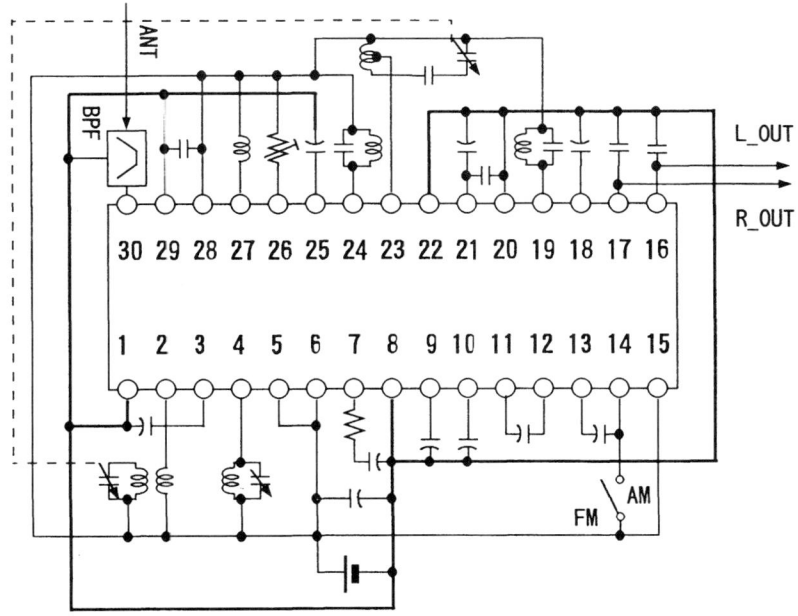

Fig.17 PERIPHERAL COMPONENTS OF LOW VOLTAGE FMSTEREO/AM RADIO IC

Fig.18 MICROPHOTOGRAPH OF LOW VOLTAGE FMSTEREO/AM RADIO IC

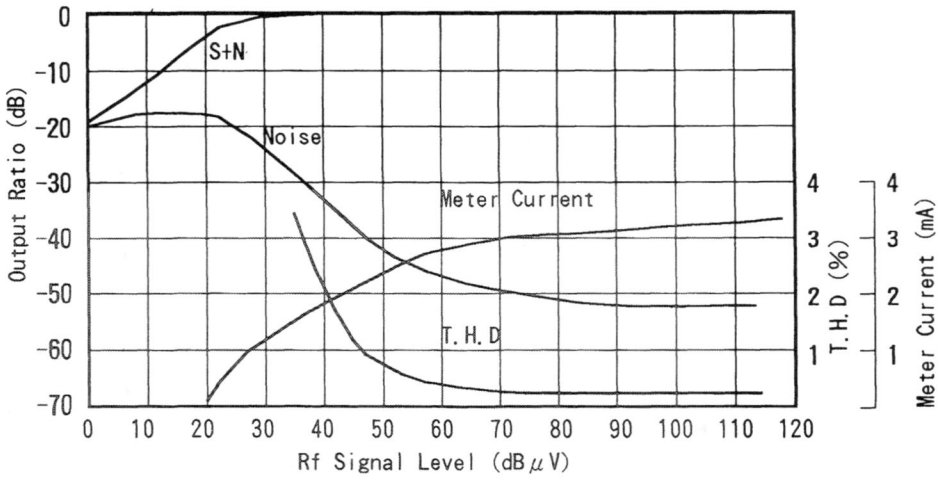

VCC=1.5V
Frf=1000kHz
Fmod=400Hz 30%

Fig. 19 AM S/N PERFORMANCE

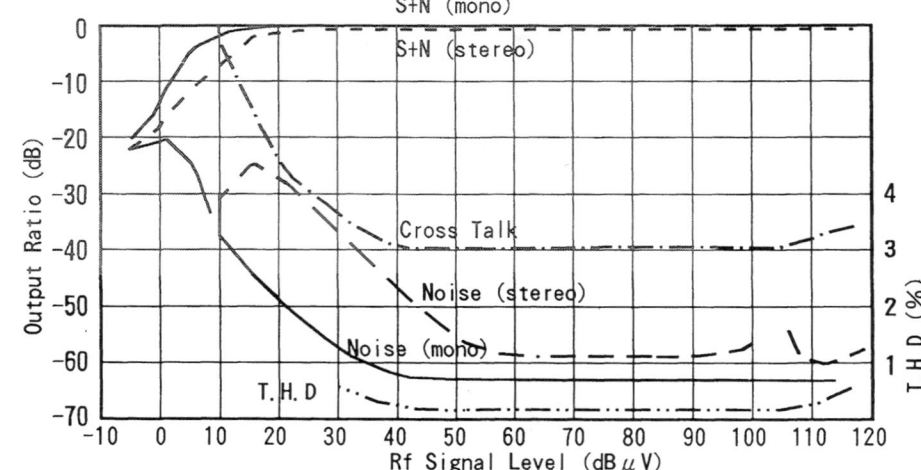

VCC=1.5V
Frf=100MHz
Fmod=400Hz 22.5kHz dev

Fig. 20 FM S/N PERFORMANCE

Table.1 Static Characteristic

Operating Voltage	0.9v~4.5v
Current Consumption (FM)	16mA
Current Consumption (AM)	14mA

Table.2 FM and AM Characteristic

FM Selectivity(Δf=400kHz)	40dB
FM Image Rejection	39dB
AM Selectivity(Δf=10kHz)	35dB
AM Image Rejection	41dB

Overall Characteristic

Fig. 17 shows peripheral components of low voltage FMstereo/AM radio IC. The IC is 30 pins small outline package.

Fig. 18 shows a microphotograph of the IC. The chip size is 4.37mm×2.84mm, and the number of devices is 1360. There are some rectangles on the chip. These are MIS capacitors. Capacitors in the PSN and BPF as AM IF filter are from 50 pF to 100 pF. So capacitors occupy such large space.

Fig. 19 shows AM S/N performance of the IC. S/N6dB sense is 10dBμV, S/N is 50dB, T.H.D is 0.3%.

Fig. 20 shows FM S/N performance of the IC. S/N30dB quieting sense is 10dBμV, S/N is 60dB, T.H.D is 0.1%.

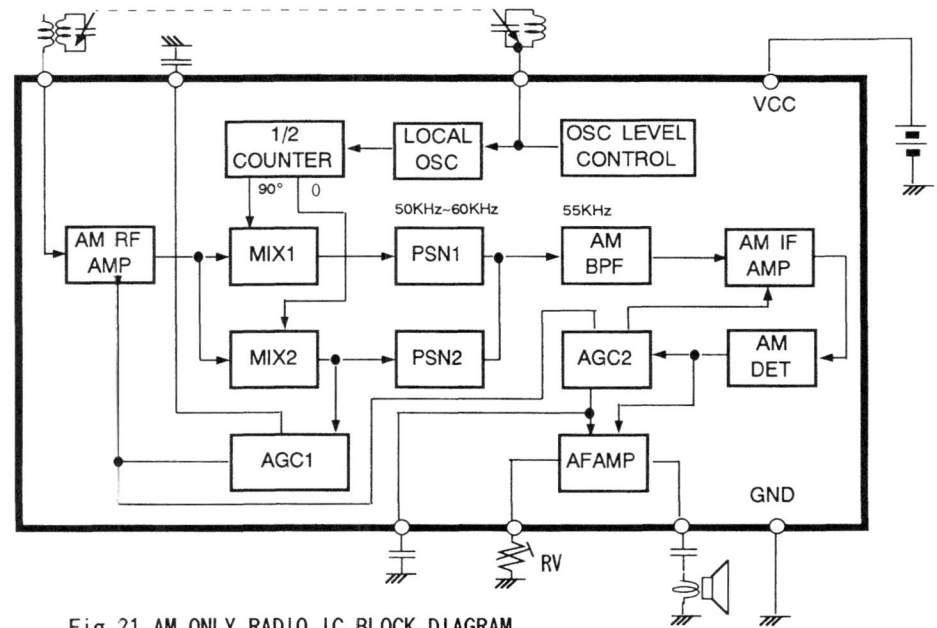

Fig. 21 AM ONLY RADIO IC BLOCK DIAGRAM

AM ONLY RADIO IC

This figure 21 shows a block diagram of 8 pins AM only radio IC. IF filter is included in the IC using image cancelling technology. Audio power amplifier is included, which can sound speaker.

AM received signal is amplified by AM RF, and converted to IF signal by Mixer 1 and Mixer 2. 1/2counter is used to make two 90 degrees phase shifted local oscillator signals. In order to receive AM broadcast from 150kHz to 30MHz, the counter operates from 150kHz to 60MHz. Image cancelling circuits consists of Mixer 1, Mixer 2, PSN1 and PSN2. IF signal is selected by AM BPF which is two 2nd order 55kHz biquad BPF. The selected signal is amplified by AM IF and detected by AM DET. The audio signal is amplified by AF AMP which sounds a speaker.

In this IC, we make much effort to decrease pins of an AM radio IC.

At an AM radio IC which has an audio power amplifier, most difficult problem is the feedback of audio signal and its harmonics to RF AMP or MIXER. To solve the problem, usually ripple rejection circuits or regulator are used. At this IC, the amplifiers are balanced type. So the ripple of supply voltage doesn't influence received signal. Therefore, The pins for ripple filter and for regulator are eliminated.

The input signal level of PSN and BPF is limited by AGC1 to prevent too large signal input. By minimizing the variety of the voltage of AGC1, the voltage is used as reference voltage of amplifiers in the IC. These amplifiers are used in PSN and BPF.

AM detected signal consists of DC voltage (AGC voltage) corresponding to carrier signal and AC voltage (audio signal). To connect AM detector output and audio amplifier input, capacitor is needed. The capacitor is too big (several μF) to include in IC. At this AM only radio IC, audio amplifier input is differential type. AM detected signal is fed to one side input and AGC voltage is fed to the other side input, so only audio signal is amplified by the audio amplifier. As the result, AM detector output and audio power amplifier input is connected in the IC.

Fig. 22 shows characteristic of band pass filter as AM IF filter. Adjacent channel suppression 10kHz away from center frequency is about 20dB.

Fig. 23 shows input characteristic. S/N6dB sensitivity is 10dBμV. S/N is 50 dB and total harmonic distortion(T.H.D) is 0.5%.

Fig. 24 shows a photograph of an AM radio using the AM only radio IC. Chip size of the IC is 1.9mm×2.4mm.

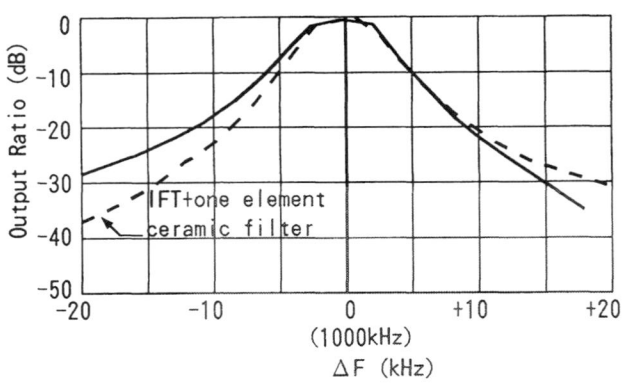

Fig. 22 AM SELECTIVITY

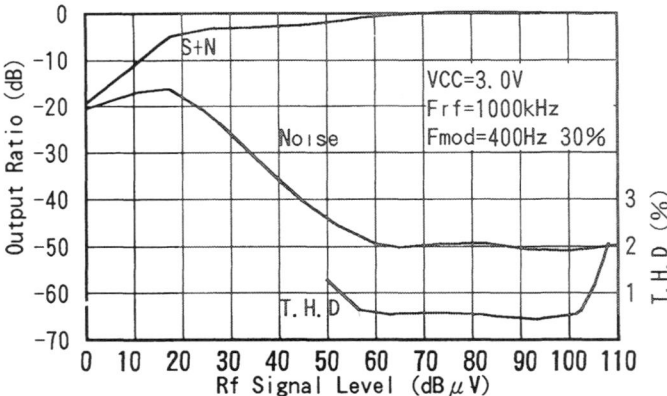

Fig. 23 AM S/N PERFORMANCE

CONCLUSION

The technology including IF filters in bipolar monolithic IC has been discussed. These filters are fabricated by existing process with diffused resistors and MIS (Metal Insulator Semiconductor) capacitors. So the cost is not so high to make a radio IC.

This technology including IF filters in IC can be used not only for a radio, but also for communication receiver.

ACKNOWLEDGEMENTS

The authors would like to express their appreciation to S. Horigome (Personal Telecommunications Group, Sony Corp.), R. Mizoguchi, C. Nishi (Bipolar division, Semiconductor Group, Sony Corp.) for their kind support and valuable advice in developing these ICs.

We also wish to express our thanks to engineers who were engaged in the development of radio for their co-operations.

Fig. 24 Photograph of AM radio using the AM only radio IC

REFERENCES

1) T. Okanobu, T. Tsuchiya, K. Abe and Y. Ueki, "A complete single chip AM/FM radio integrated circuit," IEEE Trans. Consumer Electronics, vol. CE-28, pp. 652-664, Dec. 1983.

2) H. Kuwahara, et al. "Very low voltage operation IC's for portable radio receiver use" IEEE Trans. Consumer Electronics, vol. CE-32, No. 1, February 1986

3) Bang-sup Song, "A Narrow-Band CMOS FM Receiver Based on Single-Sideband Modulation IF Filtering" IEEE Journal of Solid-State Circuits, vol. sc-22, No. 6, December 1987.

Fully integrated radio paging receiver

I.A.W. Vance, B.Eng., M.Sc

Indexing terms: *Radiocommunication, Integrated circuits*

Abstract: A review is made of radio paging receiver requirements and a new receiver architecture is described which uses a direct conversion (zero-IF) principle. This circuit is shown to have advantages over the conventional superheterodyne with high-frequency IF and in particular is shown to be suitable for monolithic realisation. A single chip which performs as a complete high-sensitivity VHF receiver is presented, and the resulting miniature pager is shown.

1 Introduction

Radio paging is an expanding sector of the telecommunications market, and in particular a large increase in the use of wide-area paging systems is taking place at the present time. Receivers for wide-area applications place severe demands on the circuits and technology used owing to the necessity of achieving high levels of performance within constraints of size and power consumption. Table 1 gives a set of parameters as typically specified. Some of these have direct relationships to the physical size of the receiver, for example sensitivity via antenna size, spurious responses via constraints on realisable 'Q' of tuned circuits and, more obviously, power consumption via battery size.

Table 1: Typical specifications for wide-area radiopaging receivers

Frequency bands, MHz	87, 150, 450, 900
Channel spacing, kHz	25
Modulation format, kHz, NRZ FSK	± 4.5
Data rate, bit/s	500
Sensitivity, μV/m	10*
Next channel rejection, dB	65
Next channel blocking level, dB	60
± 20 MHz blocking level, dB	100
Spurious response rejection, dB	70
Battery life, week	13
Temperature range, deg C	−5 to + 50

*8 position test on body — see Reference 15

However, there is a trend in the industry towards smaller pagers, despite an increase in functions being offered and a requirement to mass produce the ever increasing quantities. Furthermore, paging receivers exist in a rather unprotected environment, where they are liable to be dropped, thrown and generally roughly treated.

This paper describes the implementation of a radio paging receiver which attempts simultaneously to ease the solution of all the above mentioned problems by the use of the integrated circuit medium.

2 Receiver fundamentals

The requirements for small size, high reliability and low assembly cost in the radio paging receiver lead directly to the idea of a one-chip integrated-circuit realisation if this is possible. The digital decoding circuitry which follows the receiver portion is already highly integrated in modern pagers, and so a complementary radio chip is a very attractive proposition.

Conventional pagers have for many years relied on the single- or double-conversion superheterodyne receiver with a variety of intermediate frequencies being used, depending on the input signal band. Full integration of this type of receiver has not been attractive for a number of reasons. The most important of these is that such a chip fails to replace most of the components and becomes little more than the transistors of the discrete circuit made on one IC [1].

Existing circuits (e.g. Reference 2) integrate the intermediate frequency and demodulator functions of a conventional superheterodyne receiver, but further integration of this to include the front end is not attractive owing to the small impact on size and power consumption achieved relative to a discrete component version.

It has been proposed previously [3–5] that the direct conversion (zero-IF) type of receiver is optimum for monolithic construction, and this proves to be especially true for the modulation format usually adopted for digital paging systems, i.e. nonreturn to zero direct frequency-shift keying (FSK).

The zero-IF receiver has the advantage that it has no 1st-order spurious responses (i.e. image), and thus the problem of obtaining VHF selectivity in a small physical volume is greatly eased. Image and IF related responses other than first order are also eliminated, for example IF difference, IF breakthrough and images of harmonics. Furthermore, most of the receiver's gain is obtained at the lowest possible frequency, resulting in minimal power consumption and greater tolerance over printed circuit board layout as far as stability is concerned. The major benefit, however, is that the circuit blocks which need to be realised for this type of receiver are more compatible with integrated-circuit techniques.

It is interesting to note that the first reported solid-state paging receiver, while completely discrete in construction [6], also used a quasidirect conversion technique. The IF used was low frequency, 5 kHz, such that the image was within the channel bandwidth but no sideband folding occurred.

A similar system has also been proposed [7] with an in-channel IF and the use of frequency feedback to avoid spectrum fold-over.

However, both these methods incur a noise figure penalty, up to 3 dB, unless an image rejection mixer is used if the noise figure is determined by the premixer gain. (They also clearly have assymetrical out-of-band characteristics.)

In pagers, constraints of antenna size and current consumption do not usually allow such a relaxation of inherent sensitivity.

Within the class of true zero-IF systems, the phase-locked loop demodulator operating at RF should be considered [8]. This is conceptually simple but, owing to closed-loop stability requirements, it is not possible to obtain the required next-channel rejection, and miniature RF filters with adequate selectivity to separate a 25 kHz channel at 150 MHz are only just becoming state of the art via surface-acoustic-wave devices [9].

The system described below follows many of the principles of previous work but represents a new [10] class of direct conversion receivers, i.e. a true zero IF using in-phase and quadrature channels but with limiting amplifiers in each arm, and a digital demodulator.

Paper 1729F, received 10th September 1981
The author is with Standard Telecommunication Laboratories Limited, London Road, Harlow CM17 9NA, England

3 Zero-IF pocket paging equipment receiver

3.1 Principle of operation

Fig. 1 shows a block diagram of the principle of the new paging receiver. The incoming RF signals are mixed in two mixers with a local oscillator nominally on the centre frequency of the signal to be received and split into quadrature paths as shown. At the output of the mixers, two lowpass filters form the channel selectivity. These filters have a passbandwidth of one half the bandwidth of the RF signal spectrum, since information from both the high side and low side of the local oscillator is folded over into this frequency band. Thus, in the typical case of FSK with a deviation of ± 4.5 kHz, the filters have a corner frequency of between 6 and 8 kHz to allow for oscillator drift etc. The low-frequency outputs from the mixers, after selection by these filters, are fully limited in two high-gain limiting amplifiers and then treated as digital signals and processed by a digital demodulator which consists in its most simple form of a D-type flip-flop, as shown.

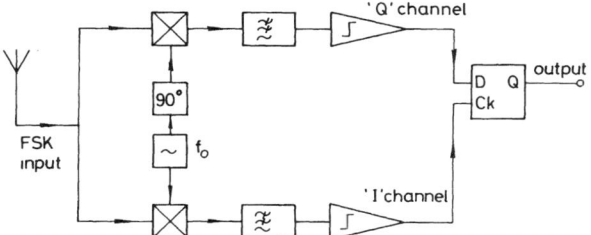

Fig. 1 *Principle of FSK receiver, block diagram*

The operation of this circuit may be easily seen by the following consideration. In the steady-state conditions with the input signal in, say, the 'mark' condition, i.e. nominally 4.5 kHz below the local oscillator, the outputs from the two mixers will consist of two sine waves at 4.5 kHz in quadrature. After the limiting amplifiers, the data at the 'clock' and 'D' inputs of the flip-flop are thus 4.5 kHz square waves in quadrature, as illustrated in Fig. 2a. If the D-type is positive-edge clocked and the relative quadrature condition is as shown in

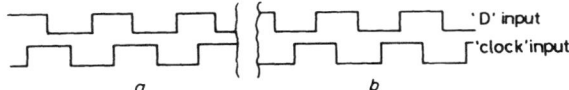

Fig. 2 *Waveforms at digital demodulator*

a Input below f_0, $Q = 1$
b Input above f_0, $Q = \phi$

the Figure, then the 'Q' output will be a constant logical 1 condition.

If the input signal is now considered in the 'space' condition, i.e. 4.5 kHz *above* the local oscillator, the same conditions as above apply except that the relative phase of the data at the flip-flop inputs has shifted by 180°, i.e. 90° lag instead of 90° lead (Fig. 2b). This is due to mixing in the RF mixers from the high side instead of the low side as previously. The 'Q' output is now clearly a continuous logical 0.

Thus the ouput of the flip-flop changes state according to whether the frequency of the input signal is above or below the local oscillator. This is as required for an FSK demodulator.

4 Detailed performance of the system

The method just described of receiving and demodulating signals will be expected to have performance differences from the conventional superheterodyne. These will be discussed in order to highlight the improved areas and to show that the nonoptimal areas have a negligible impact on the application in hand.

4.1 Sensitivity

For wide-area pagers, high sensitivity is a major requirement. One of the objectives of the present work was to effect a dramatic decrease in size and weight of the complete pocket equipment, so as to increase its acceptability to the user. This implies a lower fundamental limit on the power received by the antenna, however engineered, thus placing greater demands on the actual receiver's performance.

Five factors contribute to the sensitivity obtainable: namely the RF noise figure, the predetection noise bandwidth, the demodulation process, the postdetection filtering and the error tolerance of the digital decoder circuitry. These last two items will be assumed to be optimised independently of the receiver, and similarly the RF noise figure obtainable is not particular to this method. Hence we shall consider the noise bandwidth and the demodulator characteristics.

4.2 Predetection bandwidth

The predetection noise bandwidth will be determined by that of the lowpass filters, provided that the front-end gain is sufficient to overcome the wideband noise of the limiting amplifiers. This condition is easily met in a minimum power consumption design, as the amplifier bandwidth will be as small as is practicable within IC tolerance considerations. It is also a necessary condition to ensure maximum sensitivity.

The effective noise bandwidth is thus twice the physical bandwidth of the filters, since noise is mixed from both above and below the local oscillator frequency. The split into two channels can be ignored, given the proviso as above that all the noise originates from the front end.

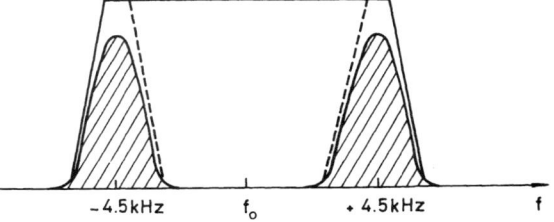

Fig. 3 *Spectrum of slow FSK signal with filtering*

---- optimum filter
—— conventional filter

However, for an FSK signal with a large modulation index, the power spectrum is entirely concentrated in two peaks at plus and minus the deviation frequency. The predetection filter would thus optimally have a shape as shown in Fig. 3 matched to the signal spectrum and a receiver so configured would achieve the same increase in low signal-level sensitivity over a simple flat filter as would a fully optimised threshold extension frequency feedback system. This is important in paging systems, as the sensitivity specifications result in a measurement somewhere near the bottom of the FM threshold. The exact point depends on the coding, error tolerance and the performance of the data synchronisation circuits (see, for example, Reference 11). However, with the POCSAG code [12] used in 25 kHz channel spacing format at 512 bit/s, the point at which there is a 0.5 probability of receiving a paging call with an optimum decoder occurs when the signal/noise ratio in the IF is about -1 dB. In the exemplary case used above, some 9 dB of improvement is theoretically available at this point over a flat IF filter.

With the zero-IF system, a good approximation to this ideal filter can be made by introducing a highpass corner into the two channels; this, when reflected around the oscillator frequency, gives the desired response. In practice, with receiver and transmitter frequency offsets taken into account, a more modest but still useful improvement in sensitivity can be obtained over a conventional superhet design, and a DC block is necessary at this point in any case.

Of particular interest here is the effect of deliberate transmitter offsets employed in a simultaneous multitransmitter system [13, 14]. As an example, with ± 1.5 kHz receiver oscillator allowance and no offset, a flat filter bandwidth of ± 6 kHz is needed, or a two-peak filter bandwidth for this system of ± 3 kHz.

When transmitter offset is used, not only does it degrade sensitivity at the postdetection slicing stage (see below), but clearly it also requires an increased noise bandwidth. While both flat and two-peak filters become wider, the latter does so at twice the rate of the former. Hence, with ± 1 kHz maximum offset, the bandwidth ratio becomes ± 7 kHz to ± 5 kHz. The analysis by Hatton et al. [14] assumes a receiver with noise characteristics of the type described in this paper.

These results are summarised in Table 2.

Table 2: Predetection bandwidth advantage of zero-IF system

Receiver allowance ppm	Transmitter allowance kHz	Bandwidth advantage dB
0	0	9
5	0	6
10	0	3
10	± 1	1.5

conditions: ± 4.5 kHz deviation, 512 bit/s data, 150 MHz

4.3 Demodulator characteristics

The demodulator is fundamentally digital in nature, and so its output has only two states. Thus the output voltage is independent of the frequency deviation and also of any offset from the on-tune condition. The primary use of the circuit is with the POCSAG [12] code, which is direct non-return-to-zero FSK. With such a DC coupled code, a fixed slicing level has to be adopted if no coding restrictions are to be imposed; this implies a reduced noise margin for data in one direction with a conventional demodulator when off tune. In the present case, no such reduction occurs for this reason.

However, set against this is the fact that, if an error occurs, it is a full-height voltage step which will not be 'corrected' until the next correctly phased clock edge on the D-type. With the circuit as shown, this error pulse length would be on average half a cycle of the deviation frequency.

This pulse length can be reduced by utilising a more complex circuit than that of Fig. 1, in which information is extracted from both the positive- and negative-going edges in both channels. Many circuits are possible, but of particular

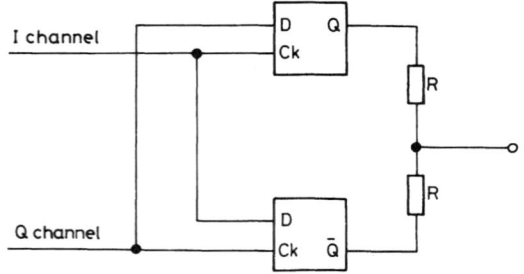

Fig. 4 *Improved digital demodulator circuit*

interest is that of Fig. 4, where two flip-flops are clocked off opposite channels in a symmetrical arrangement. The outputs are added resistively. This has the advantage that a noise error occurring in one channel will cause only a half-height output error. (Note that even front-end originated noise appears independently in the two channels.)

A further characteristic of the system is that offsets and changes of deviation do manifest themselves in the form of edge jitter on the output pulses. This is considered more fully in the next Section.

The result is that this demodulator has an output waveform and noise spectrum quite unlike a high-frequency discriminator and needs different filtering and decision-making circuitry for optimum sensitivity. It is difficult to prove that analysis is accurate for the very low signal/noise case needed here, but practical measurements of bit error rates corrected for IF bandwidth (Section 4.2) made with two systems using a common noise dominant RF stage gave sensitivities within 1 dB. Absolute sensitivity at 512 bit/s, ± 4.5 kHz deviation, with ± 2.5 kHz receiver and transmitter offset allowance, is − 132 dBm for 0.5 probability of POCSAG paging success with an optimum decoder circuit.

4.4 Data-rate limitations and jitter

The zero-IF system shown is suboptimal in that it uses limiting amplifiers. The effect of this may be seen by viewing the whole circuit as follows.

The operation of one limiter if such that the state of the output of the amplifier changes for each 180° of phase change of the RF input signal. The other limiter behaves similarly but interleaved in quadrature. This is illustrated in Fig. 5, which

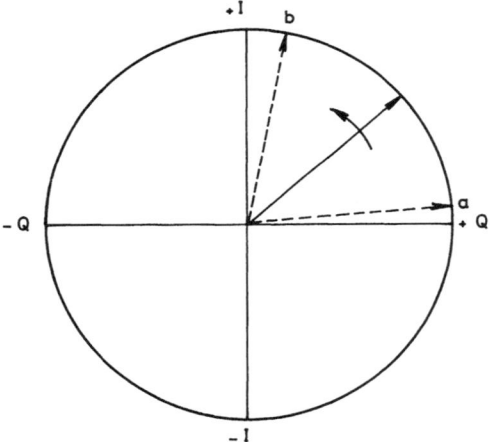

Fig. 5 *Phasor diagram for in-phase and quadrature sampling*

is a phasor diagram; '*I*' represents the in-phase channel and '*Q*' the quadrature channel. Each time the input signal vector moves from one quadrant to the next, an edge results in one channel; continuous movement then causes the sequences as illustrated in Fig. 2 previously. When the input signal passes through the local oscillator frequency, the phasor stops and then starts to rotate in the opposite direction. Since the signal frequency and the time of occurrence of this transition will be in general asynchronously related, the position of the phasor at the zero-frequency crossing point will be randomly distributed around the phase diagram. Now, if the zero frequency point occurs, for example, at position 'a' in Fig. 5 for a phasor that was initially rotating counterclockwise, then an edge will occur almost immediately after the true zero crossing as the phasor repasses the + Q axis. If, however, point 'b' had been reached, then a delay will occur before the + Q axis is recrossed.

Information is only updated when an edge occurs in one limiting amplifier output, and thus a variable edge jitter will result on the demodulated output pulses.

For an instantaneous phase-continuous transition, this delay τ will lie within the range

$$0 \leq \tau \leq \frac{4}{\Delta f_1}$$

where Δf_1 is the frequency deviation in the direction relevant to the edge being considered, and τ will be uniformly distributed within this range. Positive- and negative-going edges will only have equal peak jitter for the on-tune condition.

In the case of FSK signals with shaped transitions, the maximum delay is the time at which $90°$ of phase shift has been accumulated. This may easily be found by integrating the rate of phase change to give the time, providing, of course, that the risetime shaping is well defined.

Example

For a signal with ± 4.5 kHz deviation in the on-tune condition with instantaneous transitions the jitter produced is from 0 to $56\,\mu s$. For a bit rate of 500 bit/s (i.e. a bit of 2 ms), this represents an edge jitter of 2.8%.

A second limitation may be due to the particular realisation of the demodulator process. The simple, one D-type, circuit considered in Section 2 only clocks on positive edges of one channel and hence has a maximum pulse edge delay of four times that calculated above for instantaneous transitions. (For shaped transitions, a smaller ratio will apply since the first $90°$ takes longer to accumulate than the subsequent phase quadrants.) Between the simplest circuit and the fundamental limit of the phase sampling in the limiting amplifiers, a trade-off between complexity and operating speed may be made.

As related to the typical paging example, it may be seen that the pulse distortion due to this cause can readily be kept below 10% and thus have a negligible effect on performance.

However, this nonideality implies that there is a lower usable limit to the modulation index which may be used with the system. The value of the limit will be determined by the performance and design of the postdetection stages.

4.5 Selectivity and spurious responses

It will readily be apparent from the discussion of Section 4.3 that a correctly modulated signal on the adjacent channel will not be demodulated since it fails to cross the local-oscillator frequency. That is, the next channel rejection of this receiver is infinite, even with no channel filters. The lowpass filters are needed only to give the necessary blocking performance and to establish the noise bandwidth.

This behaviour is particularly useful in paging systems where adjacent channel responses constitute false calls, whereas blocking results in lost calls, the former representing a greater nuisance to the user.

Image rejection is similarly infinite, and input filtering is necessary only to ensure adequately low spurious responses to oscillator harmonics, and for protection against front-end blocking. It is possible to control both of these effects other than with an input filter; the blocking by designing for large-signal handling and the spurious by control at the oscillator.

4.6 Spurious radiation

In the system described, the local oscillator is tuned to the input channel; thus not only is there no possibility of isolating it from the input by filtering, but also the antenna is peaked at this frequency. Suppression can be effected by three contributing factors:

(a) low-level oscillator injection
(b) high degree of balance in mixers
(c) reverse isolation of RF stages.

In practice, careful attention to detail has enabled this problem to be solved with margin in hand.

The integrated-circuit form provides further advantages in this respect. The miniaturised parts and connections mean that unintentional 'antennas' are greatly reduced in radiating efficiency, affecting both transmission and reception from and to the circuitry.

It should also be noted that, should there be any remaining local-oscillator radiation, it will be on the allocated channel only.

5 Monolithic realisation

The receiver using the above architecture has been integrated on a single silicon chip some 2 mm by 2.5 mm which contains all the functions including local oscillator, RF stage, main gain blocks, demodulator, and data slicing circuit. This is then a true one-chip receiver having the antenna connected to one pin and fully filtered sliced data presented at another. Fig. 6 shows the chip, which has been designed to conservative rules to enable multisourcing in high-frequency bipolar processes.

The primary advantage of the direct conversion receiver is that the off-chip component count is minimised and consists mainly of low-frequency decoupling capacitors and the two lowpass filters. About 20 components are needed to make a complete receiver.

The use of integrated-circuit design techniques allows performance improvements to be obtained over simple discrete designs. For example, it allows the provision of such details as constant current operation over varying supply voltage to be included and stabilisation of current, voltage and temperature performance in the oscillator sections. Low power consumption is achieved by obtaining most of the circuit gain in the LF amplifiers, but even at the bandwidths used careful design is necessary to ensure stability. Note, for example, the 'ringmain' earth around the periphery of the chip. The circuit operates down to 1.8 V supply and draws some 2.5 mA when fully operational. In stand-by (battery save) mode, this current drain is reduced to $120\,\mu A$.

6 Conclusions

A novel radio receiver architecture has been described which is suitable for large modulation-index frequency-shift keyed signals. It is also readily adaptable to an integrated-circuit design, even more so than previously published zero IF configurations.

The combination of this new system with advanced inte-

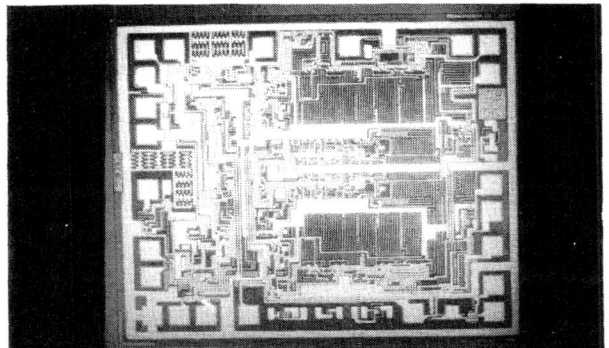

Fig. 6 *VHF radio receiver chip*

Fig. 7 *Miniature pocket pager*

grated design techniques has allowed the realisation of a true one-chip paging receiver capable of meeting stringent wide-area specifications, and operating throughout the VHF bands. In particular, the new equipment has exceptional spurious response rejection and high sensitivity.

The resulting receiver (Fig. 7) is substantially smaller (only 55 cm^3) than would have been possible using conventional techniques and has a complete lack of screens, multiboard or hybrid-type construction. The production process has been deskilled by removing the radio-frequency critical components, and it is predicted that better reliability will be seen in service owing to the more compact form and the lower component count.

It is clear that many other applications exist for this circuit and that significant advantages can be obtained in systems design given the availability of a truly mass-producible receiver.

It seems reasonable to look forward to similar receivers for other modulation formats and specifications.

7 Acknowledgments

I should like to thank my many team colleagues at STL and STC who have contributed a wide range of skills on this programme and made it possible.

8 References

1 TANABE, K., KANNO, M., SUZUKI, J., OKAMOTO, M., MURAKAMI, D., OKUBO, T., and IDESAWA, M.: 'Development of VHF/FM radio front-end IC', *IEEE Trans.*, 1980, **CE-26**, pp. 677–692
2 Plessey Semiconductors, data sheet for SL6690, 1980
3 VANCE, I.A.W.: 'An integrated circuit v.h.f. radio receiver', *Radio & Electron. Eng.*, 1980, **50**, pp. 158–164
4 VANCE, I.A.W., and NEALE, M.W.: 'Large scale integrated circuits for v.h.f. receivers'. IEE Digest 1980/41 conference on the impact of new l.s.i. techniques on communication systems, pp. 28–34
5 GOATCHER, J.K., NEALE, M.W., and VANCE, I.A.W.: 'Noise considerations in an integrated circuit v.h.f. radio receiver'. Proceedings of conference on radio receivers and associated systems, 1981, IERE no. 50., pp. 49–59
6 KERWEIN, A.E., and STEIFF, L.H.: 'Design of a 150 Megacycle pocket receiver for the Bellboy personal signalling system,' *Bell Syst. Tech. J.*, 1963, **42**, pp. 527–565
7 KASPERKOVITZ, D.: 'An integrated f.m. receiver', in KAISER, W.A., and PROEBSTER, W.E. (eds.): 'Electronics to microelectronics' (North Holland Publishing Co., 1980.), pp. 623–625 (preprints of EUROCON '80)
8 GREBENE, A.B.: 'An integrated frequency-selective AM/FM demodulator', *IEEE Trans.*, 1971, **17**, pp. 71–80
9 MOORE, P.A., MURRAY, R.J., WHITE, P.D., and GARTER, J.: Surface acoustic wave filters for use in mobile radio'. Proceedings of conference on radio receivers and associated systems, 1981, IERE no. 50, pp. 19–28
10 British Patent Specification no. 1 517 121
11 MABEY, P.J.: 'Digital signalling for radio paging', *IEEE Trans.*, 1981, **VT-30**, pp. 85–94
12 TRIDGELL, R.H.: 'The application of coding techniques to radio-paging'. Proceedings of 1980 international Zurich seminar on digital communication, pp. C9.1–9.5
13 MIZUKAMI, T., MASAKI, M., NAGATA, K., and HIKOSAKA, K.: 'NEC digital paging system'. NEC Research and Development no. 50. 1978, pp. 1–10
14 HATTON, T., HIRADE, K., and ADACHI, F.: 'Theoretical studies of a simulcast digital radio paging system using a carrier frequency offset strategy', *IEEE Trans.*, 1980, **VT-29**, pp. 87–95
15 MITCHELL, D., and VAN WYNEN, K.G.: 'A 150 Mc personal radio signalling system', *Bell Syst. Tech. J.*, 1961, **40**, pp. 1239–1257

A Single-Chip VHF and UHF Receiver for Radio Paging

John F. Wilson, Richard Youell, Tony H. Richards, Gwilym Luff, and Ralf Pilaski

Abstract —We describe a single-chip radio receiver for VHF and UHF digital wide-area paging transmissions up to 500 MHz with FSK data rates up to 1200 Bd. All channel filtering is on-chip and the IC requires only 28 surface mounted external components and a quartz crystal to make a complete receiver. With -126-dBm sensitivity, 70-dB adjacent channel rejection, and 60-dB intermodulation immunity, it satisfies all known pager specifications, worldwide, using the POCSAG paging code. High-dynamic-range mixers, integrated gyrator filters, small-area high-pass filters, and an efficient FSK demodulator combine to give good performance with a current consumption of only 2.7 mA from a 2-V supply. High-density on-chip capacitors in a bipolar process designed for analog RF applications give a chip size of only 4.6×3.8 mm.

I. Introduction

THIS PAPER describes a single-chip radio receiver for VHF and UHF wide-area paging transmissions up to 500 MHz with FSK data rates up to 1200 Bd. Radio pagers are a demanding integrated circuit application on the borderline between professional and consumer products. Pagers are sold directly to end users in a fiercely competitive market, competing on cost, size, and user features. Physical space is tight, and the pagers must operate for months from small batteries. The digital parts of the pager are already subject to considerable integration, and one way of reducing the cost and physical size of a pager is to integrate the receiver function. We need to considerably reduce the total number of components including bulky and expensive parts such as crystal and ceramic filters to justify the effort of developing a custom IC. The chip described here achieves this by integrating all the channel filtering and active elements of a radio receiver onto a single IC that requires only 28 surface mounted external passive components and a quartz crystal to act as a radio receiver suitable for FSK data formats up to 1200 Bd.

Previous single-chip receivers [1]–[3] have all had various deficiencies, such as operation only up to 200 MHz, off-chip channel filtering components, operation only at low baud rates, and insufficient dynamic range, which have not allowed them to fully meet the international wide-area paging specifications.

II. Zero IF

Most pager receivers are of a conventional superheterodyne design, gaining selectivity from quartz crystal and ceramic filters. These are bulky and expensive components which we wish to replace with on-chip filters. These integrated filters must inevitably be continuous-time filters such as RC or transconductance-C filters with limited total capacitance. Switched-capacitor techniques are not suitable due to their high noise contribution and use of clocks, neither of which is compatible with the handling of microvolt signals in receiver IF's. The dynamic range of such filters is inversely proportional to their Q, so they must operate at a very low IF frequency to have a large dynamic range. However, a low IF frequency requires very selective filtering before the mixer to eliminate unwanted image responses. These opposing requirements are reconciled in the zero IF or direct conversion receiver [4]. Here the IF center frequency is the lowest possible (zero) and the channel filters are low-pass filters with modest Q values. The local oscillator is therefore at the channel frequency and no image filtering is required before the mixers, thus minimizing off-chip RF components. The penalty is that two identical IF channels carrying signals in phase quadrature are now needed in order to be able to distinguish between signals above and below the local oscillator frequency. With on-chip filtering this duplication is of on-chip circuit blocks and not of external components. Another way of viewing the zero IF receiver is as processing the received signal as a complex low-pass (I and Q) signal between the mixer output and the demodulator.

An unavoidable problem with this architecture is that the receiver will receive interference from its own local oscillator producing a dc component in the IF; this and circuit dc offsets must be removed by high-pass filters in the IF circuitry. However, FSK paging signals use low data rates, 512 and 1200 Bd, with frequency shifts of ± 4.5 kHz, and thus have a spectrum where most of the signal energy is concentrated about the FSK tone frequencies and little at the center (Fig. 1). The high-pass filters can therefore have relatively high cutoff frequencies without affecting receiver operation. Fig. 2 shows the resulting block diagram of the paging receiver.

Manuscript received May 1, 1991; revised August 27, 1991. This work was performed at Philips Radio Communications Systems Ltd.
J. F. Wilson, R. Youell, and A. H. Richards are with Philips Radio Communications Systems Ltd., Cambridge CB4 1DP, England.
G. Luff was with Philips Radio Communications Systems Ltd., Cambridge CB4 1DP, England. He is now with the European Research Laboratory, Communications Division of Motorola Ltd., Basingstoke, Hampshire RG22 4PD, England.
R. Pilaski is with Philips Semiconductors RHW, D-2104 Hamburg 92, Germany.
IEEE Log Number 9103701.

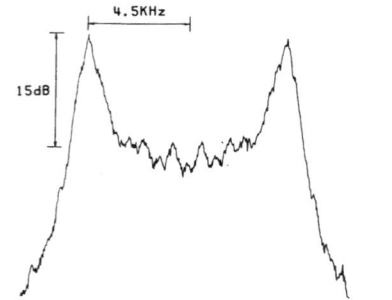

Fig. 1. Frequency spectrum of POCSAG coded FSK signal.

In a complete pager the demodulated digital output is connected to a CMOS decoder IC [5] which performs digital filtering, clock recovery, data decoding, and all the other functions required for a "bleep only" pager.

III. RF STAGES

The RF amplifier (Fig. 3) is a fully differential cascode arrangement. The tuned loop antenna, commonly used in paging, is designed to be a good match for direct connection to the bases of the input transistors. A balanced tuned circuit is used at the output with a matching network to feed the phase splitter. The 3-dB noise figure at frequencies up to 500 MHz is achieved by careful optimization of device size in a process designed for good RF analog performance.

A low-Q lumped component phase splitter generates the quadrature phase shifted signals needed for the complex baseband processing. The following common-base input tree mixers present a low input impedance to the phase splitter over a wide frequency range. The common-base input also gives a better total dynamic range performance than the usual emitter-coupled pair input for a given current drain. Fig. 4 shows the phase splitter and mixer stages.

The mixer local oscillator input comes from a conventional Colpitts crystal oscillator and frequency multiplier chain. The current through all the RF stages can be set by external resistors to suit the operating frequency.

IV. IF FILTERING AND AMPLIFICATION

These stages operate at audio frequencies and provide all the adjacent channel selectivity of the receiver. They provide a 9-kHz passband to accommodate the nominal 4.5-kHz IF frequency (produced by the ±4.5-kHz frequency deviation of the POCSAG coded FSK signal) and the frequency drift of the local oscillator. The 85-dB stopband attenuation starts at 15 kHz to allow use in European 20-kHz channels.

A cascade of on-chip filters of increasing complexity is used, each reducing the level of the interfering signals to within the dynamic range of the following filter. The first filter is a passive RC network at the mixer output which protects the low-noise preamplifier. Together with the following fully differential Sallen and Key stage, this forms a third-order elliptic filter with 18-dB stopband attenuation. These filters have a good dynamic range but their cutoff frequency is fixed and depends upon both the capacitance per unit area of the on-chip capacitors and the sheet resistance of the base diffusion. Between this filter and the gyrator channel filter is a high-pass filter of a similar type to those in the limiter but with a current mirror load to give a single-ended current output.

The third filter is a seventh-order gyrator-capacitor [6] elliptic filter. It implements a doubly terminated LC filter taken from standard tables [7] so as to be relatively insensitive to component value variations. Constructing each gyrator from linearized transconductance amplifiers (transconductors) with a differential input and single-ended output and then merging transconductors leads to the leapfrog-like structure of Fig. 5. This is still a gyrator filter in the sense that the capacitor configuration of the original LC filter can be seen in the lower part of the circuit, and the network of transconductors and capacitors in the upper part simulates a network of inductors. Using a gyrator filter rather than a leapfrog filter with active integrators gives many fewer components in the signal path, which simultaneously reduces noise and power consumption. Also each "top" zero forming capacitor in the prototype LC filter is represented by one capacitor in the gyrator filter. This saves chip area compared to a true leapfrog filter in which four capacitors of that value are required [8].

The gyration resistance of the filter is equal to the termination resistance of 150 kΩ, and is set by an external resistor that defines PTAT bias current to the transconductors. Changing this resistor alters the cutoff frequency of the gyrator filters, and with a 100-kΩ external resistor the chip is suitable for 12.5-kHz channel spacing.

Base current compensation is applied to every transconductor. A voltage limiting circuit restricts the voltage swing at the third pole of the filter to ±45-mV peak to speed up recovery from overloading by large low-frequency signals. Each gyrator filter takes 3 μA from the 2-V supply.

V. LIMITING AMPLIFIER

Hard limiting can be used in each channel of the zero IF receiver because of the high modulation index of the POCSAG coded FSK transmissions. The limiting amplifiers provide 75 dB of small-signal gain between the filter output and the demodulator and contain high-pass filtering to remove accumulated dc offsets.

The high-pass filters use a new circuit [9] to provide low cutoff frequencies (150 Hz) with small on-chip capacitors (330 pF). The circuit is derived from the single-ended high-pass filter circuit in Fig. 6(a) which is used to make one-pin high-pass filters. Fig. 6(b) shows the differential version, where two of the circuits of Fig. 6(a) are connected in parallel and the two capacitors to ground are replaced by a single differentially connected capacitor of half the value needed in the single-ended circuit. There is

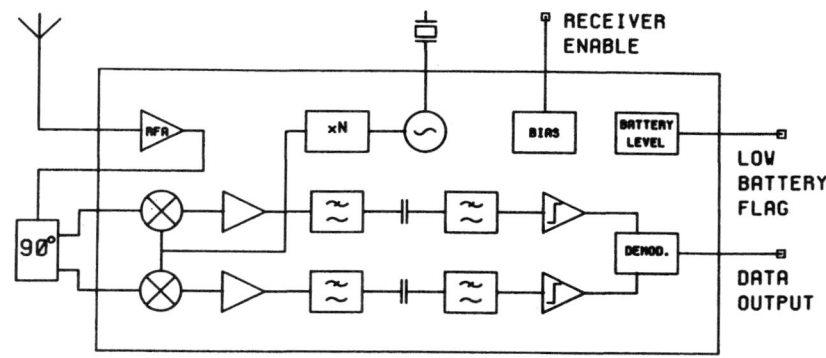

Fig. 2. Receiver block diagram.

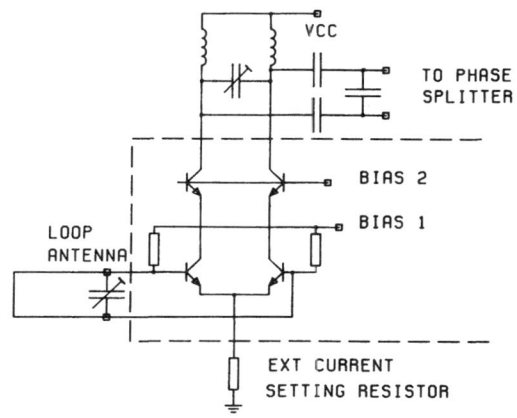

Fig. 3. RF amplifier.

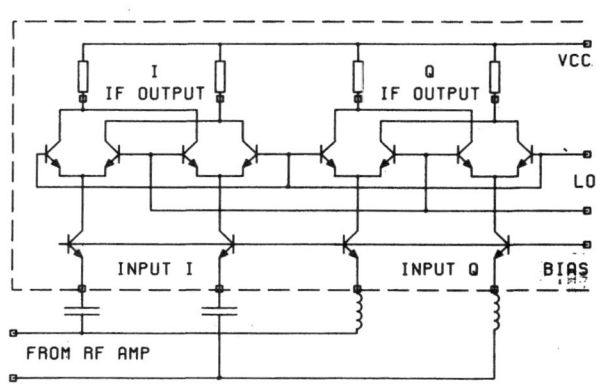

Fig. 4. Phase splitter and mixers.

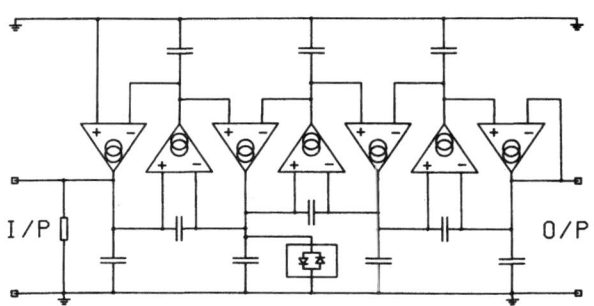

Fig. 5. Gyrator filter structure.

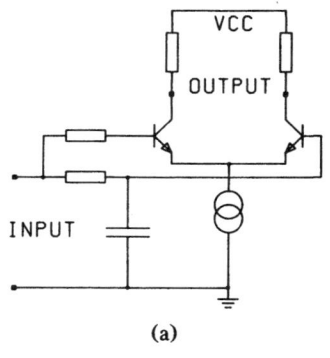

(a)

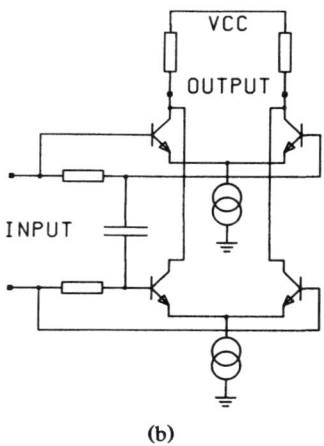

(b)

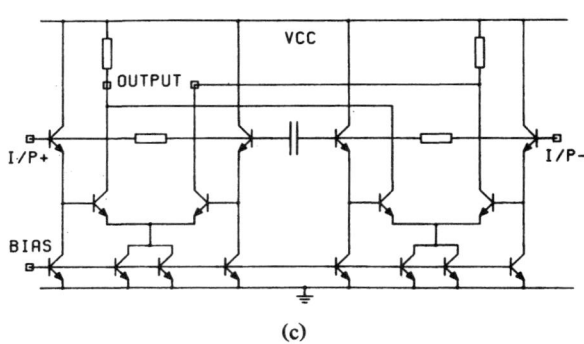

(c)

Fig. 6. (a) Single-ended dc block. (b) Differential dc block. (c) Actual differential dc block.

now no dc bias voltage across the capacitor, so when the circuit is turned on the capacitor can charge to its final voltage very quickly. The base-current-induced voltage drops across the resistors cancel at the filter output. This allows the use of high-value pinch resistors and small capacitors, which leads to the complete limiter circuit, using the practical dc block circuits of Fig. 6(c), occupying an area of only 1.6 mm². The dc blocks interface easily to the differential pairs, which provide the gain and progressive limiting required. The outputs of this amplifier are fully limiting on the receiver's own noise so a digital demodulator can be used.

The low frequency of operation allows microamp bias currents and pinch resistors to be used to give a total supply current for the limiter of only 40 μA.

VI. Demodulator and Data Output

The information in the limiter outputs requires demodulation to recover the binary data from the FSK signal in the quadrature baseband I and Q channels. The rate of change of the phase of the received signal is represented in the limiter outputs by a sequence of zero crossings. If the I-channel signal leads the Q-channel signal in phase, the FSK tone frequency lies above the local oscillator frequency (POCSAG data "0"). If the I channel lags the Q channel, the FSK tone frequency lies below the local oscillator frequency (POCSAG data "1").

The demodulator is a micropower (2 μA/gate) current-mode logic circuit which implements the modified differentiate and multiply algorithm shown in Fig. 7(a) [10]. At each zero crossing in either channel a pulse is produced by a differentiator which is multiplied by the signal in the other channel. The combination of the two multiplier outputs is a series of short pulses (Fig. 7(b)) whose polarity indicates whether the received signal frequency is above or below the receiver center frequency. An hysteresis circuit holds the polarity of the last pulse until the next pulse occurs. This demodulator uses all the information contained in the hard-limited IF signals to give a bit error rate of 3×10^{-2} for an input signal-to-noise ratio of 0.8 dB (in the 18-kHz RF bandwidth) with a single-bit digital filter in the companion decoder IC providing post-detection filtering.

The CMOS-compatible data output buffer acts as a constant current source during data transitions to give controlled output slew rates. This minimizes interference to the microvolt level signals in the IF, which are in a similar frequency range to that of the demodulated data. The data output block also contains a battery endpoint detector that monitors the chip's supply voltage and gives a logic signal to the decoder IC if the supply voltage falls below a nominal 2.05 V. This voltage comparator uses an identical controlled slew-rate buffer to the data output.

VII. Housekeeping

A central bandgap reference and housekeeping circuit provides bias currents to all other circuit blocks. These are distributed as scaled currents to avoid errors caused

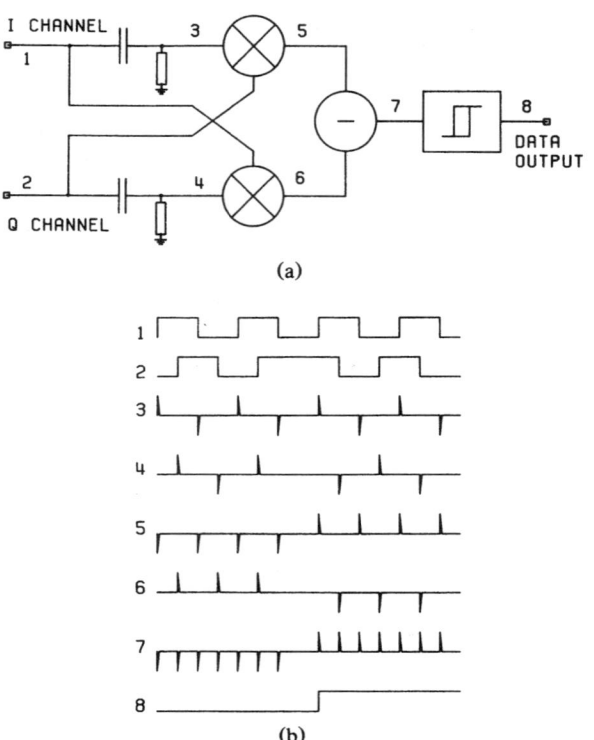

Fig. 7. (a) FSK demodulator. (b) Demodulator waveforms.

by voltage drops in the V_{EE} wiring. A V_{be}-related bias voltage is generated for the RF circuits, which allows their bias currents to be set by external resistors to ground. The comparatively large current of 40 μA is used in these bias circuits to reduce the noise in the bias supplies. The measured noise on the bandgap reference output was 168 nV/√Hz.

The POCSAG paging code allows the page receiver to be turned off for much of the time to increase the life of the pager's battery. To support this the entire IC can be powered up or down in 5 ms by enabling or disabling the main bandgap reference with a logic input.

VIII. Process and Layout

The chip was processed in a 2-μm junction-isolated bipolar process with high-density (1 nF/mm²) dielectric capacitors. RF devices with 6-GHz F_T and low noise figures at UHF are available as well as lateral p-n-p transistors with high beta and 40-MHz F_T. The well-controlled high-density capacitors are required for the on-chip filtering and the good lateral p-n-p's are needed in the gyrator filter transconductors.

Two-level aluminum interconnect and a grid-based layout style made a rapid and compact layout possible. The chip microphotograph in Fig. 8 shows the duplicated I and Q channels as well as the block floor plan. The pin-out was chosen to give a good PCB layout for the external RF components with the RF and power connections made via the short central leads of an SO28 pack

Fig. 8. Chip microphotograph.

age. The large unbroken areas are capacitors which occupy over half of the 18-mm² chip area. The channel filter, limiter, demodulator, and housekeeping blocks occupy the chip core, with the RF circuits, IF preamplifiers, and logic output drivers between the pads. Each circuit block has independent power connections to the V_{CC} and V_{EE} pads, and is encircled by a substrate connection which is connected to its own "quiet" V_{EE} pin. Signal interconnections between circuit blocks are either differential or made by current transfer. There are no stability or signal breakthrough problems on the chip despite the presence of 120 dB of small-signal gain and 85 dB of stopband attenuation.

IX. TEST

This chip is difficult to test as it is an analog "system on a chip": the signal enters the chip as low-level RF at the mixer inputs and next emerges as demodulated digital data at the data output pin. Buffered test points are provided at the gyrator filter outputs for chip test and receiver alignment, and test multiplexers can route the limiter outputs to the data output and battery level pins. The RF circuits are quite accessible and can be dc tested, but the IF circuits have to be functionally tested by injecting an audio signal at the mixer inputs. The filter responses are observed at the IF test points and the limiter performance observed at the digital outputs via the test multiplexer. This increases the chip test time as measurements have to be made of the filter attenuation at several frequencies, some of them low. Extra bond pads give access to the IF preamplifier outputs and housekeeping bias voltages at wafer probe.

The first mask set also included a low-frequency test chip in which all the RF circuitry was omitted and individual IF functions were brought out to package pins and an RF test chip containing only the RF circuits. This made possible the verification of subcircuit performance against circuit-level simulations. The IF test circuits proved very difficult to measure because of their high gain and high selectivity at high impedance, which made them very susceptible to capacitive coupling between package pins.

X. DESIGN METHODOLOGY

This IC is specified as a paging receiver by its sensitivity and its immunity to intermodulation, blocking, and adjacent channel signals. Individual stages, however, are specified by performance measures such as power gain, noise figure, third-order intermodulation intercept point, and adjacent channel attenuation. To predict the performance of the complete receiver from these figures for individual stages, a radio receiver architecture analysis program was developed. This uses equations for the noise figure and third-order intermodulation intercept point of cascaded stages to calculate the overall performance.

This type of system simulation is particularly necessary with receivers using active filters as, unlike LC, quartz, and ceramic filters, their contribution to the receiver IP3 figure is considerable. This means that it is no longer possible to provide enough gain in the RF amplifier stage to reduce the noise figure contribution of succeeding stages to negligible proportions as this would result in unacceptable IP3 performance. The program gives a complete analysis of the signal levels, noise figure, and IP3 at the input of each stage of the receiver together with the percentage contribution of each stage to the overall noise figure and IP3 performance. This allows decisions to be made about the initial specification of each stage, and design trade-offs to be made easily. Finally, the measured performance of the complete receiver can be checked against that of the individual circuit blocks on the test chips and against the simulation results.

Circuit blocks were simulated in ESPICE and PHILPAC, a Philips proprietary circuit simulator capable of Monte-Carlo statistical analysis and optimization. Fourier post processing of time-domain analyses of the RF, mixer, and filter stages was used to calculate intermodulation products. A functional breadboard was constructed which verified the receiver principles and was used to evaluate different demodulator designs.

Conventional full-custom analog layout methods were used. The complete layout was connectivity checked against a schematic capture derived netlist which had been used to simulate the power up sequence of the complete receiver. Principal node voltages at the end of the power-up sequence simulation were hand compared with those from the block-level simulations. These analyses in turn provided the input data to the architecture analysis program.

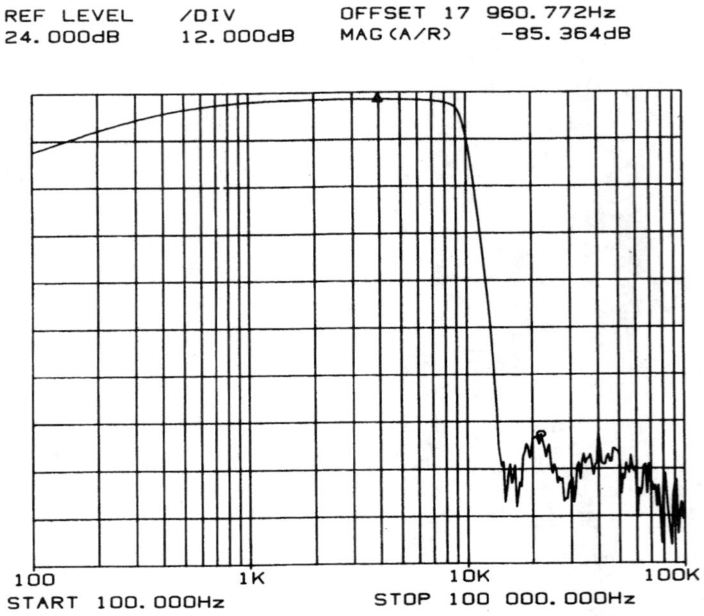

```
REF LEVEL      /DIV         OFFSET 17 960.772Hz
24.000dB    12.000dB         MAG(A/R)   -85.364dB
```

Fig. 9. IF filter measured frequency response.

TABLE I
TYPICAL PERFORMANCE AT 470 MHz (FROM A 50-Ω SOURCE)
MEASURED WITH 4-kHz DEVIATION 1200-Bd PRBS DATA

Operating Frequency	20–500 MHz
Data Rate	1200 Bd
Channel Spacing	20–30 kHz
Sensitivity	−126 dBm
Adjacent Channel	70 dB
Intermodulation	60 dB
Frequency Offset	±2.5 kHz
Blocking (1 MHz off)	82 dB
Deviation Acceptance	1.5 to 8 kHz
Power Consumption	2.7 mA at 2 V
Power Down Current	<1 μA
Chip Size	4.6×3.8 mm

XI. RESULTS

The first diffusion of the IC was fully functional and met all specifications except 1-MHz blocking. All the subblocks on the test chips performed as simulated. A new mask set was required for production because the original set was a multichip reticle including a complete receiver and the two test chips. In this second diffusion extra decoupling of the front-end bias was added to improve 1-MHz blocking and diodes were added to the RF amplifier input to protect it against ESD and large RF input signals. Fig. 9 shows the frequency response of the complete IF filtering as measured between the *I*-channel mixer input pins (with a 200-mV dc offset between the mixer LO inputs) and the *I*-channel IF test point. Table I shows the performance of the chip as measured on a 470-MHz test board which uses external components representative of those used in current pagers. This performance is achieved with a current drain of 2.7 mA from a 2.0- to 3.5-V supply. This voltage range gives a typical pager over three months operation from two AAA size alkaline cells. The chips operate to full specification in production 1200-Bd 470-MHz POCSAG pagers. At 930 MHz the mixer conversion gain is significantly reduced by the stray impedances of the SO28 package. Either an external RF amplifier can be added or the IC could be packaged differently for 930-MHz use.

XII. CONCLUSION

A highly integrated receiver for VHF and UHF FSK paging transmissions has been realized. It is engineered to consistently meet paging receiver specifications worldwide with a supply voltage of 2 V and minimal current drain. It represents a high state of analog system integration, with much functionality entirely on-chip.

The systematic design, large-scale analog simulation, and design verification techniques now available mean that analog LSI can have predictable and reliable performance when first fabricated. This reduces the large time risk that IC redesign cycles traditionally present to product developments dependent upon custom analog integrated circuits.

REFERENCES

[1] R. C. French, "A high technology VHF radio paging receiver", in *Proc. IEE Mobile Radio Syst. Tech. Conf.* (York, England), 1984 (IEE Conf. Publ. 238).
[2] SL6639 data sheet, Plessey Semiconductors, Swindon, U.K., 1990.
[3] UAA2033 data sheet, Philips Semiconductors, Eindhoven, The Netherlands, 1987.
[4] I. A. W. Vance, "Fully integrated radio paging receiver," *Proc. Inst. Elec. Eng. F*, vol. 129, no. 1, pp. 2–6, Feb. 1982.
[5] "PCF5001 POCSAG decoder IC," data sheet, Philips Semiconductors, Eindhoven, The Netherlands, 1991.
[6] J. O. Voorman, "The gyrator as a monolithic circuit in electronic systems," Ph.D. dissertation, Univ. Nijmegen, Nijmegen, The Netherlands, 1977.

[7] A. I. Zverev, *Handbook of Filter Synthesis*. New York: Wiley, 1967.
[8] G. M. Jacobs, D. J. Allstot, R. W. Brodersen, and P. R. Gray, "Design techniques for MOS switched capacitor ladder filters," *IEEE Trans. Circuits Syst.*, vol. CAS-25, no. 12, pp. 1014–1021, Dec. 1978.
[9] A. H. Richards, "D.C. blocking amplifier," European Patent Spec. 0397250A2, filed May 4, 1990.
[10] G. F. Luff, J. F. Wilson, and R. J. Youell, "Radio receiver," European Patent Spec. 0405676A2, filed June 25, 1990.

Linear Transceiver Architectures

A. Bateman, D.M. Haines, and R.J. Wilkinson

(Communications Research Group)
(University of Bristol, England)

Abstract

This paper describes the work in progress at the University of Bristol developing advanced mobile transceivers. The eventual goal is to produce efficient, compact, data compatible systems. Several methods under investigation are described, including the extensive use of digital signal processing.

1. Introduction

The future portable communications device will need to be low cost, low power, small, lightweight, resilient to multipath propagation and spectrally efficient, - a very stringent specification. Thankfully, some of these criteria are already met by existing systems but others are not. The present FM schemes for example, using superheterodyne transceiver architectures, are reasonably compact, exhibit low out-of-band emissions and possess good co- and adjacent-channel performance. They also perform acceptably well in the multipath fading environment with speech and low bit-rate constant-envelope data formats (FSK, MSK, etc.). Yet they are neither power or spectrally efficient, and are handicapped by their incompatibility with high capacity non-constant envelope data modulation such as 16-QAM or 32-QAM, now commonplace for fixed-link voice-band communications.

The present FM technology has attained its dominance due largely to its relative cheapness through mass production and the relative immunity of FM to multipath distortions. Yet to produce a smaller, cheaper, more flexible and most importantly, a more power and spectrally efficient radio system will require a radical departure in the areas of modulation format and transceiver architecture. A way forward in terms of architecture may be seen from the example of radio pagers. The latest radiopagers perform useful communication functions (100+ character messages) using a very compact, power efficient design. It should, of course, be borne in mind that most radiopagers are only receivers, but the key to a reduction in size and current consumption has been to move away from the superhet configuration to a direct conversion architecture. A single VLSI chip {1} can form the entire receiver in a radiopager, and although present devices use low data rates with FSK modulation, the savings in space and power are dramatic.

If direct conversion architectures are the key to space saving, then the modulation technique is the key to spectral efficiency. The most spectrally efficient technique will be one that employs *both* amplitude and phase for information transfer, namely a *linear modulation* (LM) system. When employed solely for analogue speech communications, this approach is commonly referred to as single sideband (SSB). The linear frequency translation process inherent in a linear modulation system has conventionally been achieved by double-balanced mixing and sideband filtering - a process necessitating a superhet transceiver configuration and costly, bulky, IF crystal filters. In future, it is likely that the frequency translation will be performed in only one stage (direct conversion), using either the *phasing* {2} or *Weaver* method {3} architectures. The most promising technique is the Weaver method with all mixing (image) products falling in the user's own channel. This greatly reduces the tolerance constraints placed on gain and phase matching in the conversion process. Further, no wideband phase-shift networks or sideband filtering are required.

The main problem with all linear modulation formats, however, is not the frequency translation, but the RF power amplification. In general, linear power amplifiers are grossly power inefficient if they are at all linear, or far from linear if they are at all power efficient. Neither category satisfies the system power efficiency or spectral efficiency criteria.

In the next two sections, methods of overcoming these disadvantages are described. The result is a proposed transceiver architecture which can make full use of the benefits of linear modulation and direct conversion techniques.

2. Transmitter Section

If the benefits of the close-spaced narrowband channels are to be realised, it is vital that the generation of spurious adjacent-channel signals in the transmitter be minimised. These unwanted signals may arise from the frequency translation process, or from inter-modulation distortion in the RF amplifier chain. The former may be avoided by using a Weaver architecture (Fig.1), while the latter may be reduced by the use of a linearised amplifier.

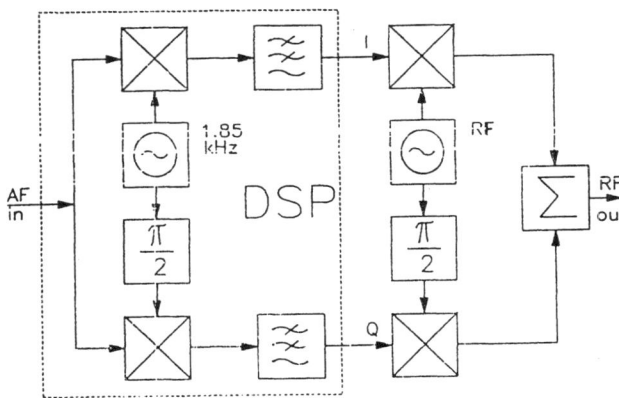

Fig.1 Weaver Method LM generator with digital filtering of unwanted components

In the modulation process, any image components generated through mismatches in the gain and phase of the quadrature paths will appear in the same channel as the desired signal, rather than in the adjacent channel. The unwanted products of the baseband mixing process can be attenuated by over 80dB by standard digital filtering and hence present no problem. A minimum bandwidth RF signal is thus produced.

To generate this signal at a suitable power level for mobile communications (25W) requires linear RF amplification of around 70dB. Any non-linearity in the amplifier will cause RF intermodulation products to appear in adjacent channels - hence the need for highly linear amplifiers.

Conventional linear amplifiers, in order to minimise distortion, have used devices biased to operate over only a small highly linear portion of their transfer characteristic. Unfortunately this is grossly inefficient both in terms of power conversion and cost - a technique hardly suitable for mobile and portable systems.

A major research topic at the University of Bristol is concerned with techniques for linearising the characteristics of *power efficient* amplifiers. There are three techniques which show considerable promise, Cartesian feedback, adaptive pre-distortion, and the LINC system. All three encompass the linear frequency translation process required for LM operation, and thus form the complete transmitter section.

2.1 Cartesian Feedback Loop (Fig.2)

For the Cartesian feedback configuration, the transmitter output is sampled just after the final RF amplifier, and synchronously demodulated, to recover quadrature cartesian components of the modulation. These signals are used to provide negative feedback, subtracting from the modulating signals (derived from the DSP) to generate a loop error signal,

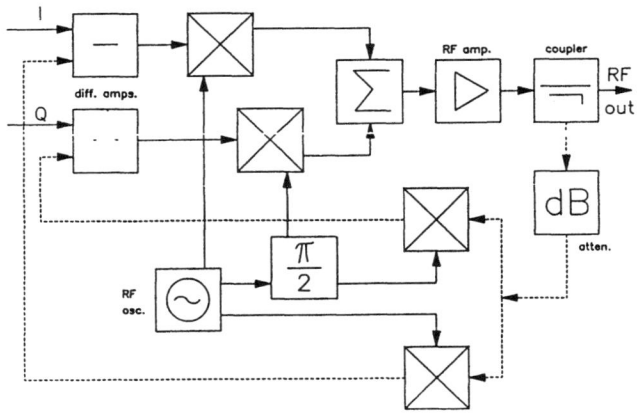

Fig.2 Cartesian Loop Transmitter

which drives the modulators. Provided that the loop gain is of sufficient magnitude, the feedback loop will, in theory, continuously correct for any non-linearity in the upconversion RF amplification stages (the amplifier must not contain a discontinuous transfer function). A major obstacle in the design of such a transmitter is maintaining the stability of the feedback loop: the RF amplifiers create a significant phase shift of the feedback signals, which varies with frequency and output level. If this phase shift becomes excessive, then oscillation will occur in the feedback loop. Two techniques have been developed to minimise this problem. The first is a voltage-controlled RF phase shift network which delays both quadrature local oscillators driving the down-conversion mixers. This (phase shift) may be adjusted, under software control, to minimise the phase error in the fed-back signals. The alternative is to digitise the fed-back signals, calculate the phase shift and then correct for it in software at baseband. The added attraction of digitising the fed-back baseband signals is that the system loop filter is software-configurable and can be optimised for the bandwidth of modulation in use. The major drawback of using a closed loop feedback control system is that which plagues all feedback control processes, namely a restricted loop bandwidth governed by loop time delay. For the transmitter, this defines the bandwidth of operation, typically 10% of the operating frequency.

Results achieved so far using the Cartesian feedback technique are shown in Fig.3. The considerable improvement

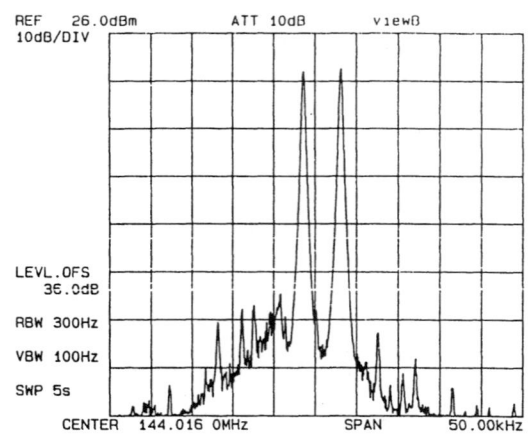

Fig. 3(b) Two-tone test result for the Cartesian Loop transmitter with feedback applied.

obtained for a two-tone test signal is clearly evident. Still further improvement is anticipated, and work by Petrovic {4} has demonstrated that two-tone performance with better than -70dB intermodulation products is possible.

2.2 Adaptive Predistortion

The configuration of the transmitter for adaptive predistortion is very similar to that of the Cartesian Loop, but this time, no continuous feedback is used. Instead, a calibration signal, generated by baseband digital signal processing, is fed through the transmitter, amplified, demodulated and fed back to the baseband processor (Fig.4). From the fedback information, the (gain and phase) distortion of the

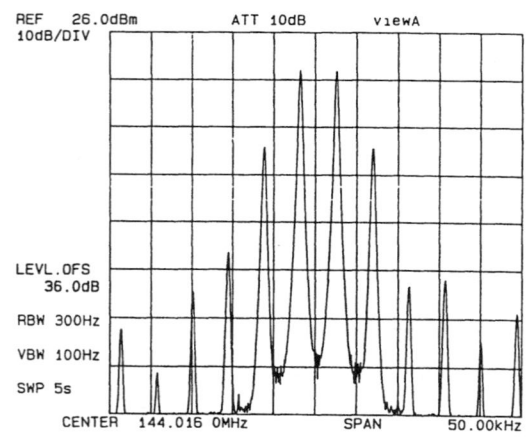

Fig.3(a) Two-tone test result for the Cartesian Loop transmitter with no feedback

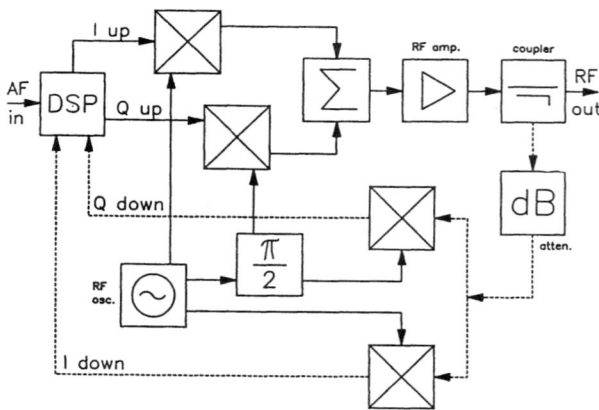

Fig.4 Adaptive Predistortion transmitter system.

amplifier is fully characterised and stored in memory in the form of a table of values. The digital processing then applies the measured data, together with a curve-fit routine, to predistort the modulating signals in a complementary manner to the distortion produced in the amplifier. In this way, the overall transmitter characteristic is linearised.

The calibration routine is very rapid (<100ms) and can thus be performed regularly to adapt for thermal effects on the amplifier. Since orthogonal (cartesian) components of the baseband signal are used, both gain and phase distortion are measured and corrected for. Being open-loop in operation, the channel bandwidth can be much wider than for the Cartesian feedback system. The major drawback with this adaptive approach however, is the interruption in transmission when system updating occurs. With keyed transmitter working, as prevails in private mobile radio networks, this presents little problem as updating can occur each time that keying occurs.

2.3 LINC

The LINC transmitter {5} (LInear amplification with Nonlinear Components) is fundamentally different from the other two systems in that there is no feedback of information from the amplifier output, and the amplifier itself can be grossly non-linear. The principle of operation is that the baseband processing accepts a *gain and phase modulated* input waveform, and generates two wideband *constant envelope phase-modulated* signals. These signals are up-converted by a pair of Weaver modulators, amplified through two well-matched non-linear amplifier chains and then summed (Fig.5). The complex signals are so generated that all undesired out-of-band components are in exact antiphase in the two amplifier chains and cancel at the output, while the wanted components are in phase and reinforce (Fig 6). The essence of the LINC approach is that the RF amplification is performed by highly efficient non-linear amplifiers (Class C, D or E), which operate on constant envelope signals. The difficulties with this method are the production of the complex phase-modulated signals, the design of two well-matched amplifier chains and a method of combining the two high-power signals from the amplifiers.

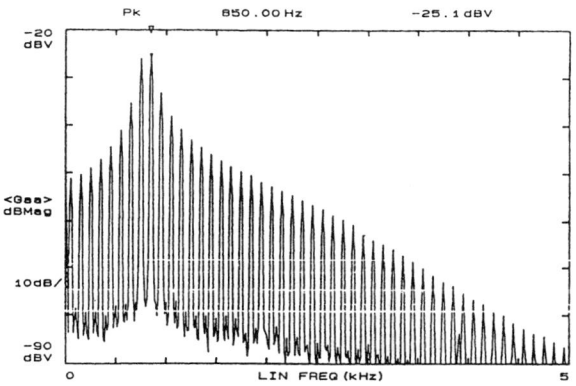

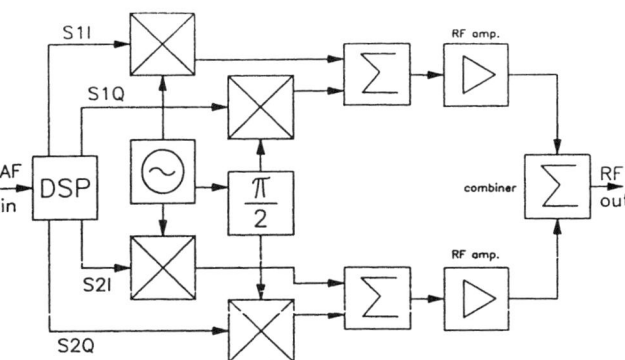

Fig.5 LINC transmitter system.

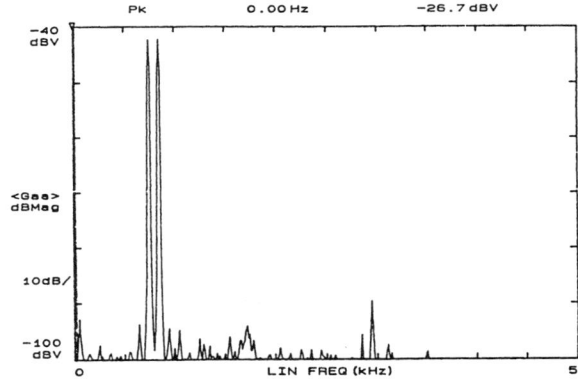

Fig.6 Simulated output from non-linear amplifier with two-tone input (above) and with antiphase cancellation (below).

Digital signal processing, with the capability of evaluation of non-linear functions in real time, has been used to solve the problem of producing the phase-modulated signals. The matching of the amplifiers, however, is not so readily achieved, and requires very careful design and alignment if the unwanted components are to be sufficiently well suppressed. One solution is to employ corrective feedback around the amplifier chain in a manner similar to the predistortion concept.

To date, the vector summation of the two amplifiers' outputs has been performed using a hybrid combiner, but this results in wasted power being dissipated in a resistive load. A better solution would be a 'chirex' network {6} optimised for the type of modulation in use. However for optimal performance, a purpose built dual amplifier with a combined output stage (voltage summation) is envisaged, which would match straight to the aerial. Work is in progress in this area.

3. Receiver Section

As for the transmitter, the most attractive architecture for receiver implementation is the direct conversion Weaver method with adjacent channel filtering performed by efficient baseband digital technology. Compared with the conventional technology (employing a crystal IF filter in a superhet architecture), the savings in space, cost and receiver complexity are enormous. Perhaps more importantly, however, is the new degree of flexibility available with a system that can define its channel bandwidth by means of a reconfigurable digital filter. One immediate benefit is that the channel selectivity can be greatly improved allowing either additional signalling bandwidth, or increased adjacent-channel interference immunity, for a given channel spacing.

The Weaver architecture applied to receiver design is, however, plagued by three main problems - carrier leakage/d.c. feedthrough, restricted dynamic range, and the need for accurate gain and phase matching between the I and Q channels. Fortunately, all of these problems can be overcome by the application of digital signal processing technology.

3.1 Carrier Leakage/DC Feedthrough

The most exacting problem with the Weaver receiver is d.c. offsets in the downconverted I and Q signal paths. For a Weaver system, a zero hertz component in the I and Q paths corresponds to a component in the centre of the audio band at the receiver output. Any d.c. error in the receiver, whether it be caused by RF local oscillator feedthrough of simple component d.c. drift, is thus manifest as an audio tone, (Fig.7). For weak received signals, this d.c. error can more than swamp the wanted signal, with consequent loss of communications.

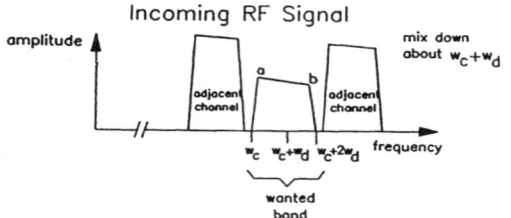

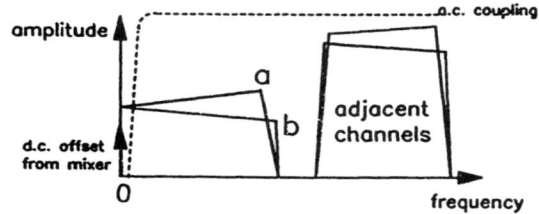

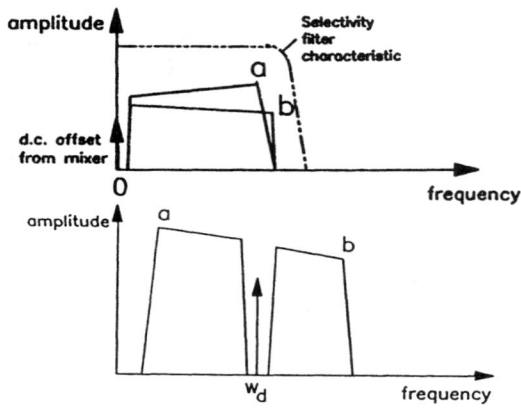

Fig.7 Diagrams showing the production of a notch in the reconstructed spectrum of a Weaver method receiver due to a.c. coupling or a tone due to d.c. offsets.

There is no way of overcoming this problem other than to remove the d.c. term by effectively a.c. coupling the receiver chain. Unfortunately this a.c. coupling appears as a notch in the audio output spectrum (Fig.7), which is clearly undesirable. There are three methods to counter this problem, whilst still retaining the essential a.c. coupling.

The first method is simply to make the resulting notch infintesimally narrow by using a very long averaging process for the d.c. suppression circuitry. With averaging times of several seconds, or even minutes, the amount of energy lost through a.c. coupling is negligible. The rate of averaging will be dictated by the rate of change of d.c. offsets in the receiver system - with careful design it should be possible to keep this within manageable levels.

The second solution is to 'wobble' the I and Q downconverted signals so that no one set of input components fall within the notch bandwidth. The effect of the notch is now spread across a wide range of frequencies with a consequent lessening of its effect. The wobble is easily introduced by modulating the RF local oscillator, and can be equally simply removed by a reverse modulation process in the digitised baseband signal - at which point the d.c. error has been taken out. For the user to experience no degradation from the wobbling/de-wobbling process, the I and Q signals paths must be carefully gain matched and exhibit a linear phase response. This requirement is dealt with in section 3.3.

The third, and most elegant, solution was first proposed by Stevenson {7}, and is based on the Transparent Tone-In-Band technique (TTIB) {8-10}. TTIB is a frequency manipulation process which allows a signal to be split into two or more frequency segments which can then be individually frequency shifted to form artificial notches in the waveform spectrum. A complementary process in the receiver can restore the frequency segments to their original positions, maintaining linear gain and phase across the entire signal band. Using a two segment TTIB system to create a central band gap in the Weaver transmitted spectrum permits a.c.-coupling in the receiver, without any loss of signal components. The reverse TTIB process is then performed on the receiver output.

3.2 Dynamic Range

The large dynamic range over which a mobile receiver is required to work is a rigorous specification. A conventional PMR radio set may be expected to operate over a range in excess of 100dB. In superhet designs, IF automatic gain control (AGC) is used to maintain a useful signal level at the receiver output, and it is reasonable to suppose that similar techniques can be used on the baseband I and Q outputs of a Weaver receiver. This in fact is the case providing that channel filtering can be performed prior to the AGC circuitry. As the channel filtering is best performed digitally, this implies that the I and Q signals must be digitised prior to the AGC system, which in turn imposes a stringent dynamic range requirement on the A/D converter. To achieve a usable dynamic range of 100dB would require 20 bit linear converters, and consequently 20-22 bit signal processing for the filtering tasks. This specification can be alleviated somewhat by allocating 10-20dB of the gain control to the front-end RF stage, with the added benefit that the dynamic range of the RF mixer circuitry can be relaxed. 16-bit processing then becomes a realistic objective. If this is still too complicated or costly a solution, then a stage of analogue AGC can be applied prior to the channel filtering, but at the expense of a reduced adjacent- channel rejection capability for the receiver. Both techniques are currently being investigated at Bristol.

3.3 Gain and Phase matching

To ensure adequate suppression of the image sideband in the receiver, (>40dB), gain and phase matching in the I and Q paths must be maintained to within 0.12dB and 0.8° respectively. This specification is considerably less severe than for the phasing receiver configuration, however, which must achieve image rejection of the order of 70- 80dB (0.002dB and 0.01° gain and phase error), as the image is now generated by an adjacent channel signal rather than originating from the user's own channel.

With the use of phase controlled hybrid power splitters for quadrature RF local oscillator generation, a phase error of less than 0.3° can be readily maintained over at least a 33% bandwidth and thus presents no problem for the Weaver system. Gain matching of the I and Q signal paths is also simply achieved in the baseband digital processing section of the radio by Hilbert transforming (say) the Q path signal and making a direct level comparison with the I signal. A gain correction signal can then be generated. The fact that these relatively simple measures can be taken is a direct result of using the Weaver architecture and not the phasing method.

4. Conclusions

The transmitter and receiver described here have the potential to become the basis for a mobile, hand-held "communicator" of the future. The development of the present system is limited by the fundamental characteristics of FM and those of superhet transceiver architectures. Direct conversion transceivers offer realistic potential for reductions in both size and cost of terminal equipment, and the parallelled introduction of linear modulation techniques provides the means of increasing power and spectral efficiency. Of particular importance in the system development is the Transparent Tone-In-Band processing, providing not only the means of overcoming d.c. errors in the Weaver receiver, but also forming the basis of a multipath fading correction system {8}, essential for high quality mobile communications. Developments in digital signal processing chips already in progress will allow all of the transceiver functions to be implemented on a single chip. The integrating of a modem and voice coding will also be possible in the foreseeable future.

5. References

{1} SL6637 Direct Conversion FSK Receiver Advance Information : Plessey Semiconductors Ltd 1987.

{2} Norgaard, Donald E. "The Phase Shift Method of Single Sideband Signal Generation" : Proc. IRE December 1956.

{3} Weaver, Donald K. "A Third Method of Generation and Detection of Single Sideband Signals" : Proc. IRE December 1956.

{4} Petrovic, V. "Reduction of Spurious Emission from Radio Transmission by means of Modulation Feedback" : IEE Conf. Radio Spectrum Conservation Techniques, Sept. 1983.

{5} Cox, D.C. "Linear Amplification with Non-Linear Components" : IEEE Trans. Communications, COM-22, 1974, pp1942-1945.

{6} Raab, F.H. "Efficiency of Outphasing RF Power Amplifier Systems": IEEE Trans. COM-33, 1985, pp1094-1099.

{7} Stevenson, Carl R. "DPMM Modulation Is 'Backward Compatible' with FM" : Mobile Radio Technology, vol.2 no.11, November 1984.

{8} McGeehan, J.P. and Bateman, A. "Theoretical and Experimental Investigation of Feedforward Signal Regeneration (FFSR) as a Means of Combatting Multipath Propagation Effects in Pilot-Based SSB Mobile Radio Systems": IEEE Trans. Vehicular Tech., 1983, VT-32, pp.106-111.

{9} Bateman, A. and McGeehan, J.P. "A Survey of Potential Narrowband Techniques for use in Satellite and Cellular Systems" : Second Nordic Seminar on Digital Land Mobile Radio Communications, 1986.

{10} McGeehan, J.P. and Bateman, A. "Phase-Locked TTIB: a new spectrum configuration particularly suited to transmission of data over SSB mobile radio networks" : IEEE Trans. Communications, COM-32 no.1, January 1984.

Direct Conversion Transceiver Design for Compact Low-Cost Portable Mobile Radio Terminals

A. Bateman & D.M. Haines
Centre for Communications Research, Queens Building
University of Bristol, University Walk
Bristol. BS8 1TR, U.K.

Abstract

The full benefit of using linear modulation methods for the transmission and reception of data in the mobile environment will only be realised when compact and power efficient linear transceiver equipment becomes a commercial reality. Moving from superheterodyne to direct conversion transceiver architectures is seen as a major step towards achieving this goal. Several characteristic problems with direct conversion technology have however hindered progress in this area.

This paper describes methods for overcoming many of the difficulties associated with direct conversion architectures and presents results from a working prototype developed by the Centre for Communications Research at Bristol University.

Introduction

The availability of high quality mobile telephone systems has proven to be an invaluable resource in many facets of business and social life. Yet, despite the economic benefits of a mobile phone, widespread acceptance has been hampered from the user perspective by the short battery life, terminal size and cost, and from the operator perspective by the shortage of spectrum, hindering service expansion. The spectral efficiency of radio communication systems is central to the future of mobile communications industry. It is predicted that in the UK alone, the present cellular telephone population of 1.5 million users will rise to 10 million by the year 2000 [1].

Relatively inefficient use of spectrum by the present FM system has manifested itself as crowding on the available channels. This is particularly prevalent in cities and on sections of motorways at times of peak demand. Overcrowding in the present UK TACS system results in long delays for connection and (occasionally) call termination. Battery life, whilst of less importance in vehicle-based equipment, has proved a decisive factor in the serviceability of hand held units. Present hand-held designs are a compromise between transmission range and battery life, and their performance falls short of similar vehicle-based transceivers. Also, even the smallest hand held units require a larger than average pocket.

Future public land mobile telecommunication systems (FPLMTS) will be required in both vehicle-based and hand-held form. The challenge of meeting the needs of power and spectrum efficiency in a compact package is a hard one. However, if any FPLMTS is to be truly universal and widely available, the eventual goal of a pocket-sized "personal communicator" must be realised. This has been recognised by several working groups [2,3], and a target of 200g weight and $200 cm^3$ volume has been suggested [4]. There is also some consensus that a FPLMTS should be based on low bit-rate (16 kbits/s or less) voice technology. The transceiver systems employed in any FPLMTS must be power efficient and suitable for use with data modulation. In addition, the modulation scheme adopted must be as efficient in its use of spectrum as possible.

Previous work at the Centre for Communications Research at Bristol [5,6,7] has shown the arguments for a linear modulation (LM) scheme being one of the most spectrally efficient. Other work [8,9] has supported these arguments. Yet existing LM equipment of sufficient quality is bulkier at present than comparable FM equipment, and more power hungry. To utilise the obvious advantages of LM, a departure has to be made from conventional transceiver designs. A *direct conversion* transceiver system eliminates many of the space critical components found in superheterodyne designs, and allows the engineer much scope for integration. There are, however, severe problems with this approach [10,11]. This paper shows how one type of direct conversion transceiver has been developed at Bristol to the point where a prototype is under evaluation. Test results are provided to show the performance of this transceiver to be equivalent to (and in some cases better than) existing systems.

The Transmitter

The requirements for a transmitter in a LM system are that it should have adequate output power, low out-of-band emissions and (if possible) be power efficient. The last two attributes are incompatible in conventional transmitter power amplifier designs. Linear amplifiers generally have large quiescent currents, and efficient Class C amplifiers are highly non-linear.

One method of meeting all the above requirements is the

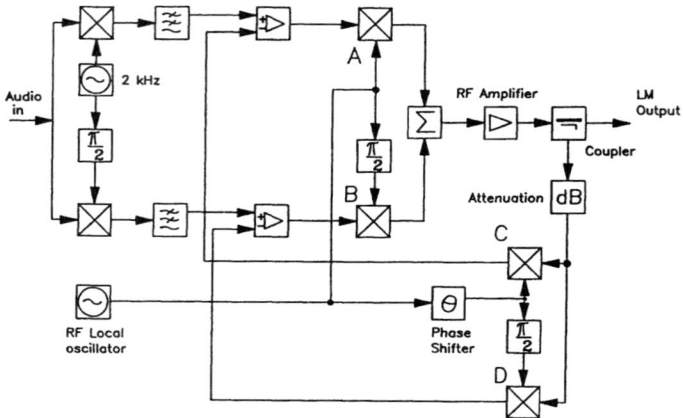

Figure 1: Cartesian Loop Transmitter with Phase Shifter

Cartesian Loop transmitter [12,13]. Figure 1 shows the entire transmitter in a Weaver method [14] configuration. This method was adopted as it allows image suppression which is superior to other techniques. It is far from being a new idea, but its development has been limited in the past because of three problems; loop stability, generation of accurate I and Q signals and maintenance of quadrature on the local oscillators.

These problems have been overcome by several developments. Firstly, loop stability is governed primarily by the phase shift needed between mixer pair AB and mixer pair CD. This phase shift corresponds to the delay through the power amplifier. Maintenance of the correct phase shift is therefore essential for spurious-free operation. A voltage-variable r.f. phase shifter θ has been incorporated into the design. This enables the relative phase of the two sets of local oscillators to be varied over a wide range, and also over a wide range of frequencies.

A related network has been developed to provide accurate quadrature for each set of local oscillators. Fine adjustment is possible by means of a voltage level, and quadrature can be maintained over a large bandwidth with minimal amplitude fluctuation.

Finally, a digital signal processing (DSP) system enables the production of very accurate I and Q baseband signals. This also allows control and calibration of the transmitter, during an initial calibration period and subsequently during operation.

The results of these developments are dramatic (Figure 2). An existing LM amplifier, biased in Class A or AB mode is palpably worse than the linearised Class C amplifier. Output powers of 20W PEP have been obtained at VHF with excellent suppression of spurii up to the bandwidth of the differential amplifiers. This may be termed the *linearising bandwidth*. This is different from the transmitter's *bandwidth of operation*, which is dependent upon the quadrature local oscillator networks, the phase shift network and the amplifiers. This can be very wide (several tens of MHz). There are, however, limits on the Cartesian Loop Transmitter's linearising bandwidth, governed primarily by the phase delay within the feedback control loop. With too wide a loop bandwidth or too high a loop gain, instability will occur. Linearised bandwidths of several hundred kHz have so far been achieved with this technique.

The above results have been reproduced at UHF (900 MHz). The nature of components at these frequencies means

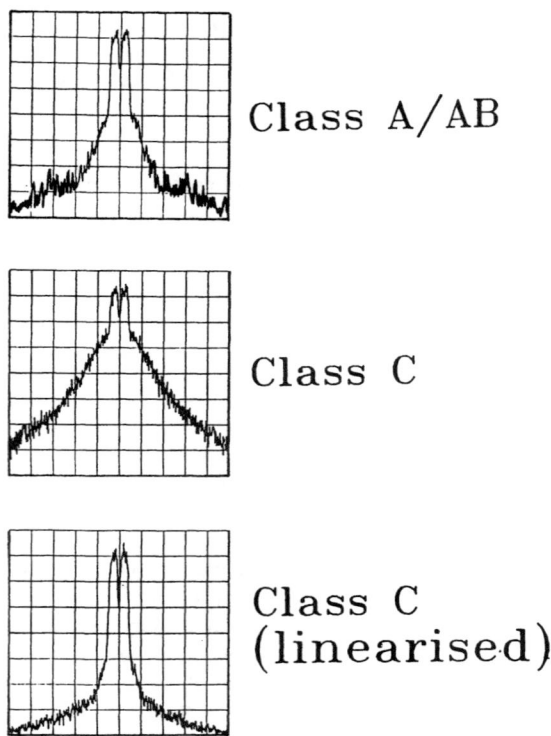

Figure 2: Outputs from different VHF amplifier types with a 4 kHz notch-filtered noise modulating signal

that the resulting transmitter may be even smaller in size than its VHF counterpart.

The gains in power efficiency are equally remarkable. Whereas a conventional LM amplifier (for 20W output) may have a quiescent current of about 100mA at 12V, the corresponding current in a linearised Class C amplifier is zero. LM is also generally more power efficient than FM, as Figure 3 shows. Here an FM system with a Class C power amplifier is compared with an LM system with a linearised Class C amplifier. Where the modulating signal has a high peak-to-mean ratio (such as speech) the overall gain in power efficiency is impressive.

A final advantage is that the Cartesian Loop Weaver method transmitter uses components which may all be reduced in size to chip or hybrid level integration.

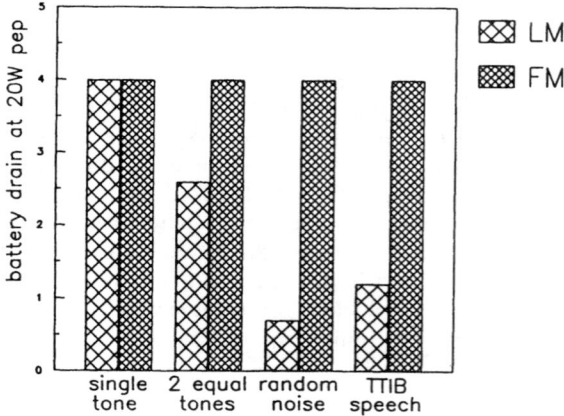

Figure 3: Comparison of battery drain (in amps) for an FM Class C amplifier and for an LM linearised Class C amplifier with different modulation types

The Receiver

A Weaver architecture has also been employed in the receiver section. There are severe difficulties inherent in this type of receiver [15], but they have largely been solved by careful design and advanced DSP techniques. These difficulties will be considered together with the requirements for sensitivity, dynamic range and carrier feedthrough.

Sensitivity

Determining the sensitivity requirement for a receiver is not straightforward. It is desirable to have system sensitivity as high as possible, but not to the point where other aspects of receiver performance are degraded, particularly strong signal handling. A guide to necessary sensitivity may be obtained by looking at two comparable radio sets; an Aerotron ACSB Pioneer 1000 system (Set 1), and a Securicor T530 FM Mobile (Set 2). Their relative sensitivities (for 12 dB SINAD) are given in Table 1. From these values, it is clear that a sensitivity of around −120 dBm is necessary [18].

The front-end of the direct conversion receiver is shown in Figure 4. This compares with the front-end of a superheterodyne (superhet) receiver shown in Figure 5. A noise figure summary for a typical superhet receiver is given in Figure 6, for a sensitivity of −120 dBm (noise figure of 7.2dB). The top portion of the figure shows noise temperature and gain data for the individual stages. The bottom portion shows how the total noise figure of the receiver is built up from the individual stage data. Equations for deriving noise figures in cascaded stages are well-documented elsewhere [19]. The output of the IF strip is used as a reference — it is assumed that after this stage little noise will be added. A corresponding noise figure analysis for the direct conversion receiver is shown in Figure 7. Here the input to the A/D converter is used as a reference.

This analysis shows that the direct conversion receiver relies heavily upon the preamplifier for even this modest sensitivity requirement. The presence of the signal splitter and passive mixers at such an early point in the signal path is undesirable. However, their inclusion (instead of active mixers) was necessary for improved intermodulation performance. The analysis also demonstrates that the audio amplifiers are pushed to the limits of their noise capabilities.

The required sensitivity can just be achieved. Future work will include the evaluation of higher quality active mixers and a very low noise audio preamplifier to enhance sensitivity.

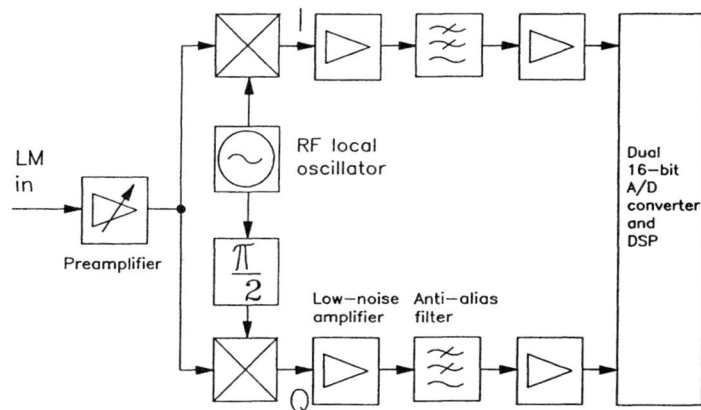

Figure 4: Digitally-implemented Weaver Direct Conversion receiver

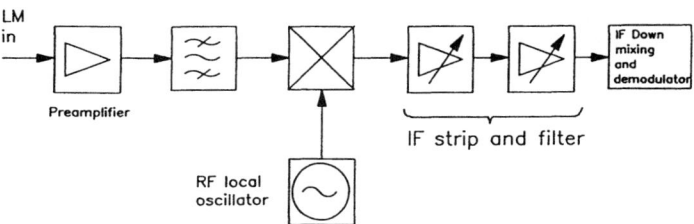

Figure 5: Single-conversion Superhet Receiver front-end components

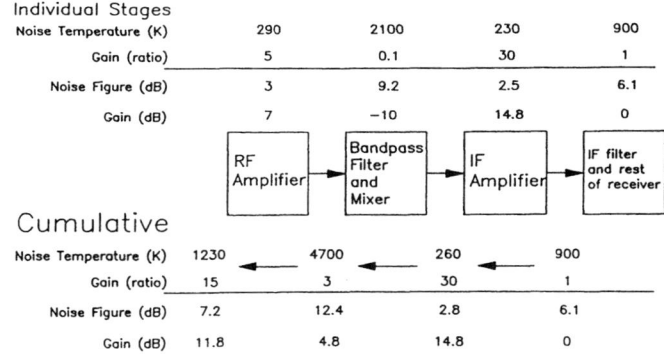

Figure 6: Superhet receiver noise figure analysis

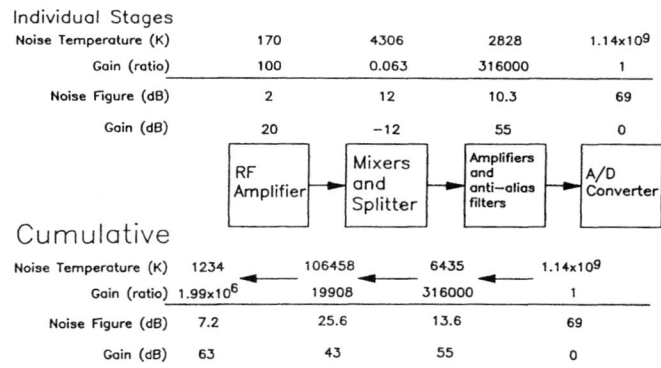

Figure 7: Direct Conversion receiver noise figure analysis

Transceiver		Sensitivity (dBm)
Aerotron 1000	with preamplifier	−125
	without preamplifier	−119
Securicor T530		−120

Table 1: Sensitivities of two commercial PMR transceivers

	Component			
	Preamplifier	Mixers	Amplifier/Filter	A/D Converter
Maximum Signal (dBm)	−20	−20	+23	+23
Minimum Signal (dBm)	−120	−130	−115	−60

Table 2: Dynamic ranges for Direct Conversion Receiver components

Dynamic Range

There are two aspects to the dynamic range of a receiver. Firstly, there is the range of signals in the wanted band that the receiver is required to tolerate. This may be as high as 120 dB. Secondly, there is the largest ratio of adjacent signal to wanted signal that can be processed before third-order distortion products swamp the wanted signal. This might reasonably be 70 to 80 dB [16] and is termed the spurious-free dynamic range (SFDR). Into this area also comes the question of adjacent channel rejection and selectivity.

SFDR will be considered first. It is affected by every component in the signal path. Referring to the direct conversion receiver front-end (Figure 4), each component can be assigned a dynamic range. That is, it can be given a range of signals which can be amplified without distortion (at the high end) or without disappearing into noise (at the low end). These are summarised in Table 2. Figure 8 shows how an incoming signal range from −120 dBm to −40 dBm can be translated to −67 dBm to 23 dBm for a 16-bit analogue to digital (A/D) converter. SFDR is directly applicable to A/D converters as well. The signal-to-noise (S/N) ratio of an n-bit A/D converter is given by [17]:

$$\left[\frac{S}{N}\right]_{dB} = 4.8 + 6n$$

For a 16-bit converter, the S/N ratio is approximately 96 dB. This does not take account of thermal noise. It also does not correspond to dynamic range. The smallest signal that may be accomodated 12 dB above the noise floor is $96 - 12 = 84$db below the full-scale signal. This assumes a perfectly ideal converter. Until recently, 16-bit converters at reasonable cost achieved, in practise, only 14 or 15-bit resolution. The advent of self-calibrating converters which provide true 16-bit resolution with very low differential non-linearity has made the construction of this receiver possible. Future oversampling converters which require minimal anti-alias filtering should enable the receiver to be simplified further. The point has now been reached where the linearity of the receiver is determined by other receiver components, in particular the audio amplifiers.

The direct conversion receiver developed uses digital signal processing (DSP) to perform several tasks usually undertaken by analogue circuitry. Adjacent channel filtering, automatic gain control (AGC), squelch control and demodulation are all carried out on one DSP chip. In particular, digital adjacent channel filtering is a significant advance as it allows linear phase low-pass filters to determine the channel characteristic. Existing LM (and FM) equipment suffers from the inclusion of a crystal IF filter. The gain and group delay characteristics of a typical 10.7 MHz IF filter, compared with similar characteristics for a digital filter, are shown in Figure 9. It is clear from these graphs that the digital filtering provides a much better behaved channel characteristic, particularly suited to data transmission.

In addition, digital channel filtering provides well-defined

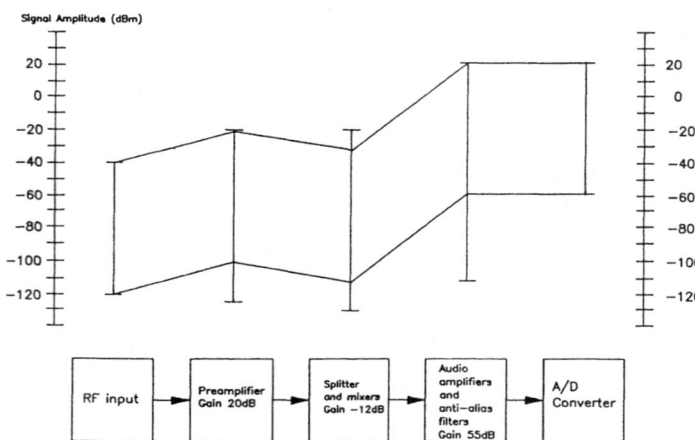

Figure 8: Translation of dynamic range from input to output in the Direct Conversion Receiver

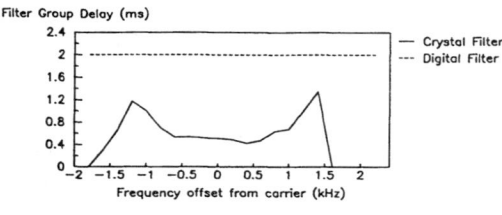

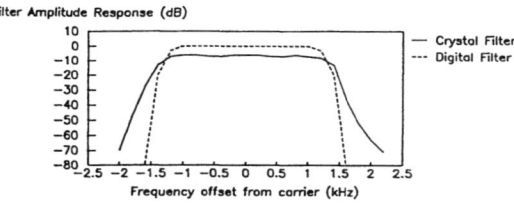

Figure 9: Comparison of Gain and Group Delay Characterisitcs for Digital and Crystal Channel Filters

adjacent channel performance. Figure 10 shows the selectivity[1] curves for the direct conversion receiver, set 1 and set 2. From these curves, it is clear that the direct conversion receiver has both a tighter and better defined selectivity performance than either of the other radios.

The direct conversion receiver has, as required, an SFDR of about 80 dB. This SFDR may be located between 0 and −120 dBm by varying the gain of the preamplifier. Under software control, the gain of the preamplifier may be adjusted from −20 to 20 dB, and thus the overall dynamic range of 120 dB is achieved.

---
[1]The input signal level is set to give 12dB SINAD. An interfering signal is then introduced at a variable frequency offset, and its level is adjusted to degrade the wanted signal to 6dB SINAD

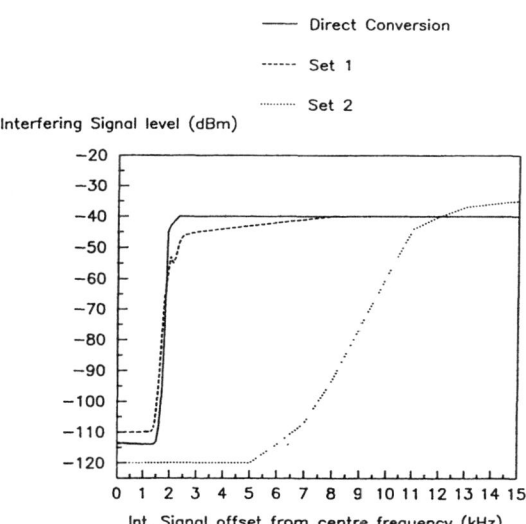

Figure 10: Adjacent Channel Rejection responses for the receivers under test

A related problem is that of gain and phase matching in the I and Q signal paths. Accurate quadrature of the local oscillators is maintained by a network similar to that outlined in the transmitter section. The amplitudes of the I and Q signals are compared in software and any required adjustments made on a continuous basis. These techniques have the effect that image suppression of 40dB or better is maintained at all times.

Carrier Leakage

Carrier leakage and feedthrough onto the I and Q signal paths results in a d.c. level being present. Also, the audio amplifiers and anti-alias filters may contribute some d.c. offsets. If fed through to the second mixing stage, this d.c. level appears as an undesirable tone in the centre of the band.

Three methods of addressing the carrier leakage/d.c. offset problem have been identified [15]. The most promising of these is d.c. correction — a software method whereby the incoming signal is averaged over a relatively long period, and the result subtracted from the signal. This method has been implemented, and works to a large extent. It is a substantial improvement over a.c. coupling for two reasons. If the I and Q paths are a.c. coupled at (say) a cut-off of 50 Hz, then a significant amount of information is lost in the notch created [10]. If the coupling is reduced to 5 Hz, the notch is narrower, but the group delay characteristic of the a.c. coupling filter adversely affects the channel characteristic. The DC correction technique differs from the above in that it is essentially a discrete system, correcting for d.c. error only at specific instants in time. The result is that the group delay response of the receiver is thus left largely unaffected. A notch is still created, but it can be made very narrow without introducing significant delay distortion. In a pilot-based system, such as TTIB, where the pilot is nominally in the centre of the frequency band (and thus appears at d.c. in a Weaver demodulator), the effects of the d.c. nulling notch can be overcome by introducing a small frequency offset (10 Hz or so) into the downconverted signal.

The d.c offsets on audio amplifiers and anti-alias filters are unfortunately not constant, and even very small variations have a significant effect if the incoming signal is also very small. Carrier leakage too is not constant, depending on environmental and circuit effects. Correction for d.c. offsets therefore has to occur at a rate sufficient to counteract the change. At present, correction is used in the direct conversion receiver at approximately 2–3 times per second. Careful front-end redesign and the use of low-drift amplifiers should allow this rate to be reduced at least ten times, with the result that an effective notch width of < 1 Hz is experienced. Other techniques, including local oscillator dither to reduce the d.c. content of the down converted input are under investigation. A combination of d.c. correction and dither may well prove to be the most practical solution.

Receiver Fading Performance

The test arrangement for assessing the direct conversion receiver's performance in a fading environment is shown in Figure 11. The test signal used was a sine wave, for simplicity. The SINAD results are shown in Table 3. Set 1 was employed, with its preamplifier connected. This demonstrates the superior sensitivity of Set 1, and the superior strong-signal handling of the direct conversion receiver. Preliminary tests with speech indicate similar conclusions. It is expected that future data trials will demonstrate the advantage of the linear phase channel characteristic.

Receiver		SINAD measurements		
	Signal Strength (dBm)	No fading	Moderate fading	Severe fading
Set 1	-27	18	19	9
	-47	23	21	12
	-77	23	21	12
	-107	18	16	10
Direct conversion receivers	-27	22	21	10
	-47	22	21	10
	-77	22	20	10
	-107	17	14	8

Table 3: Comparative SINAD measurements for different levels of fading

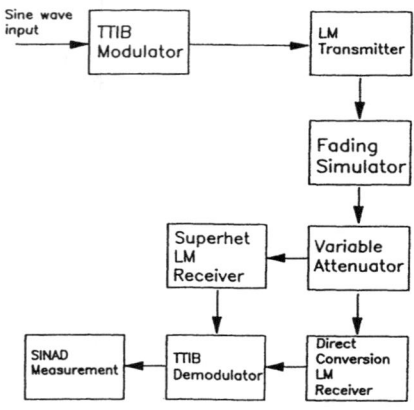

Figure 11: Test arrangement for fading comparison tests

Conclusions

The results presented in this paper show that a direct conversion transceiver system can be made to work at both VHF and UHF. The improved channel characteristics facilitated by digital channel filtering make this configuration particularly suited to data transmission. In addition, the integration potential of all the system components is the best available route to the concept of the 'personal communicator'.

Future work in this area will concentrate on techniques for automatic calibration and testing of the transmitter and receiver, the research and development of improved linear RF component technology, together with further evaluation of the prototype operation and reliability.

References

[1] P. Carpenter *From Mobile to Personal Communications* : Copenhagen Oct 1988 p1.7

[2] R. MacNamee, S. Vadgama, R. W. Gibson *Universal Mobile Telecommunications - A Concept* : 4th IERE conf 'Land Mobile Radio' Warwick 15 Dec 1987 p19

[3] R. Gibson, G. MacNamee, S. Vadgama *Universal Mobile Telecommunications System A Concept* : 'Telecommunications' USA vol 21 no 11 p23-6 Nov 87

[4] Draft Report M/8 Future Public Land Mobile Telecommunications Systems

[5] A. Bateman, J. P. McGeehan *Phase-Locked Transparent Tone-in-Band (TTIB): a New Spectrum Configuration Particularly Suited to the Transmission of Data over SSB Mobile Radio Networks* : IEEE Trans. COM-32, 1984 pp81-87.

[6] J. P. McGeehan, A. Bateman *Data Transmission over UHF Fading Mobile Radio Channels* : IEE Proceedings Pt F Vol 131, 1984 pp 364-374.

[7] H. Hammuda, J. P. McGeehan *Spectral Efficiency of Cellular Land Mobile Radio Systems* : IEEE Conf VT-88 Philadelphia June 1988 pp616-622.

[8] Y. Akaiwa, Y. Nagata *A Linear Modulation Scheme for Spectrum Efficient Digital Mobile Telephone Systems* : International Conference on Digital Land Mobile Radio Communications, Venice, July 1987.

[9] *AT& T Digital Cellular System Proposal* : Submission to EIA Technical Subcommittee TR 45.3 June 1988.

[10] C. J. Collier, C. R. Poole *Digital Correction of Channel Mismatch for a Digitally Implemented Direct Conversion Radio* : 4th IERE conf 'Land Mobile Radio' Warwick 15 Dec 1987 p19

[11] R. Zavrel *State-of-the-art IC's simplify SSB Receiver Design* : 'Electronic Components and Applications' Vol 7 No 4, pp223-228

[12] V. Petrovic *Reduction of Spurious Emission from Radio Transmitters by Means of Modulation Feedback* : IEE Conf on Radio Spectrum Conservation Techniques, September 1983, pp44-49.

[13] V. Petrovic *Application of Cartesian Feedback to HF SSB Transmitters* : IEE Conf on HF Communications Systems and Techniques, 1985, pp81-85.

[14] D. K. Weaver *A Third Method of Generation and Detection of Single Sideband Signals* : Proc IRE December 1956, pp1703-1705.

[15] A. Bateman, D. M. Haines, R. J. Wilkinson *Linear Transceiver Architectures* : IEEE Conf VT-88, Philadelphia June 1988.

[16] I. White *Modern VHF/UHF Front-End Design* : 'Radio Communication' April 1985, pp264-268.

[17] F. G. Stremmler *Introduction to Communication Systems* : Addison-Wesley, USA 1982 p511.

[18] J. N. Gannaway *The Effects of Preamplifiers on Receiver Performance and a Review of some Currently Available 144 MHz Preamplifiers* : 'Radio Communication' November 1981, pp1026-1031.

[19] H. Taub, D. L. Schilling *Principles of Communications Systems* : McGraw-Hill, USA 1987 p624.

A NEW INCOHERENT DIRECT CONVERSION RECEIVER

Gerhard Schultes[1], Arpad L. Scholtz[1], Senior Member, IEEE
Ernst Bonek[1], Senior Member, IEEE, Peter Veith[2], Member, IEEE

[1] Institut für Nachrichtentechnik und Hochfrequenztechnik
Technische Universität Wien, Gusshausstrasse 25 A-1040 Wien, Austria

[2] Siemens AG Österreich, WGS EKP2
Hainburgerstrasse 33, A-1030 Wien, Austria

I ABSTRACT

This paper presents a new type of incoherent homodyne receiver. It is designed for low cost mobile communication applications in the upper UHF area. The receiver operates on digital angle modulated signals at high data rates like GMSK of the proposed European DECT-standard as well as on more frequency efficient linear modulation techniques. We present the design of the RF and mixing circuits as well as the baseband filtering and the vector operation based demodulation circuit. At the end of this paper we show some computer simulations of the operating performance of this new receiver system. Results of actual measurements on the hardware will be presented at the conference.

II INTRODUCTION

Today's heterodyne receivers have reached a high technical standard. For applications in low cost high performance mobile personal communication systems they suffer from high production costs because of many expensive RF and IF components and a lot of adjustment procedures. The practicabilities of mechanical integration will also reach their limits in the near future. The alternative lies in the development of other receiver concepts like direct conversion (homodyne) receivers.

The receiver we describe in this paper, patent is applied for, is based on the incoherent direct conversion principle with differential data detection. Every incoming signal can be seen as a time variant vector in the complex plane. Our receiver converts this vector by incoherent mixing to the baseband. After analog to digital conversion the vector is delivered as a pair of binary numbers. The transmitted data is reconstructed from the relative change within one bit perion by straightforeward vector operations. This principle has the following advantages:

Direct conversion
+ reduces the number of RF circuits,
+ transfers signal processing into the baseband,
+ avoids mirror frequency problems

Incoherent reception
+ avoids expenses for carrier or modulation tracking,
+ allows AC-coupled baseband signal processing.

Differential detection
+ is insensitive against carrier to LO frequency offset and drift.

Further benefits:
+ Applicable for nearly all two or four level modulation techniques,
+ No adjustments necessary.
+ Full integrability in Si-bipolar or MOS technology.

A disadvantage is the loss of about 3 dB in sensitivity against synchronous demodulation.

Most of todays narrow band homodyne receiver prototypes suffer from problems caused by the 1/f- noise in baseband. Because of the high baseband bandwidth in our receiver and the AC- coupled signal pathes the 1/f-noise can be neglected.

The parameters of the communication link the receiver is designed for are given in Table 1.

III PRINCIPLE OF THE VECTORIAL RECEIVER

A functional block diagram of the vectorial receiver considered here is shown in Figure 1. The operation method is to convert the incoming signal (S_{RF}) with the center frequency f_E and the bandwidth B_E by a quadrature demodulator in one step down to the baseband.

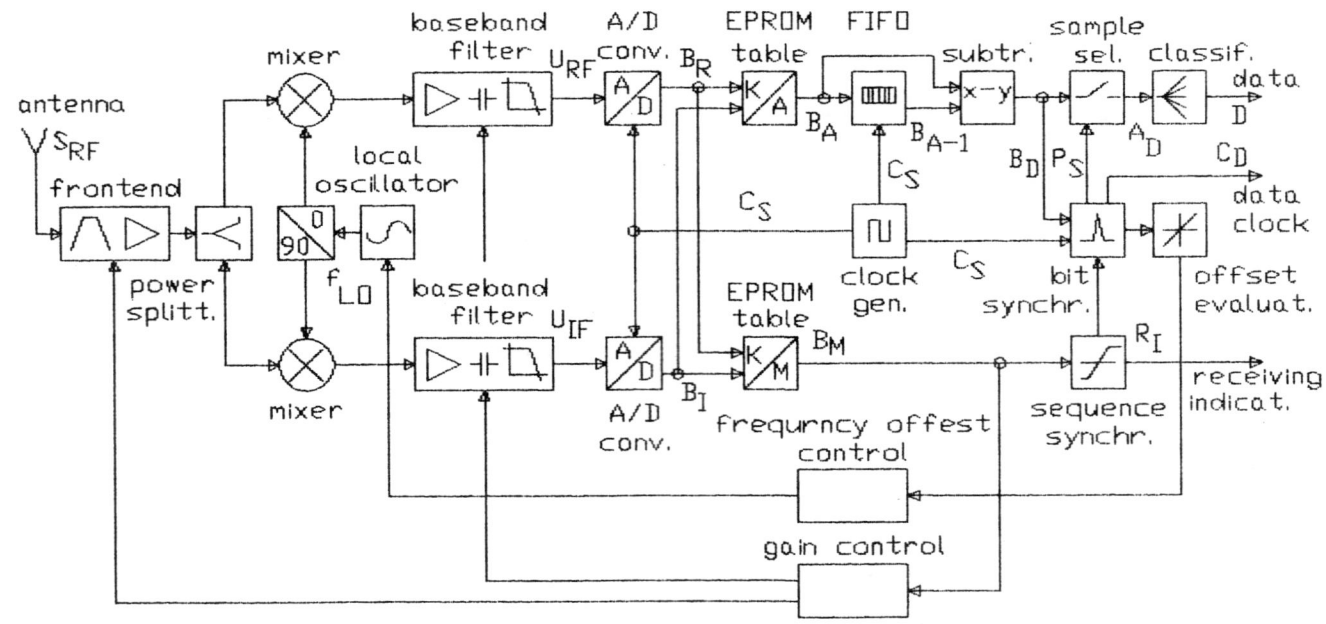

Fig. 1: Block diagram of the vectorial receiver

Transmission method:	TDM-FDMA
Modulation method:	GMSK, BT = r = 0.5
Transmission band:	1680+/- 15 MHz
Data rate R:	1.0 Mbit/s
Transmission sequence length T_S:	500 µs
Channel spacing:	1.5 MHz
RF channel bandwidth B_E:	1.2 MHz
Transmited power:	24 dBm
Maximum received power:	7 dBm
Maximum bit error rate for nearly undisturbed reception:	10^{-3}

Tab. 1: Communication link parameters

The local oscillator supplies the direct converter with two orthogonal signals of the frequency f_{LO} close to the incoming frequency. <u>Coherence is not necessary.</u> After conversion the received signal is delivered splitted in its real (U_R) and imaginary (U_I) part, superimposed with noise and adjacent channel signals in the baseband. Now the signals are filtered in AC-coupled lowpasses with a flat group delay. After sampling and analog to digital conversion the signal emerges as two binary numbers (B_R, B_I), ready for high speed digital processing. In GMSK and in most of the other linear and nonlinear (= constant envelope) two and four state modulation- schemes the information is or can be coded in the angle difference between two consecutive bits.

Because of this fact thr real and the imaginary part of the signal are converted to magnitude (B_M) and angle (B_A) by the means of two EPROM-tables. The magnitude is used for gain control and synchronisation purposes. The angle of the current bit is subtracted from the angle of the previous bit. This difference (B_D) is classified by an EPROM-table to reconstruct the transmitted data (D). Additionally the difference is useed for extracting the data clock (C_D) at the beginning of a transmission sequence.

IV DESCRIPTION OF THE RECEIVER BLOCKS

The Frontend

The purpose of the frontend (see Figure 2) is the preselection of the communication band and the low noise and low intermodulation amplification or attenuation of the incoming antenna signal for postprocessing in the following low dynamic direct converter stage . The frontend consists of a two stage GaAs-Mes-Fet amplifier with digitally selectable gain and two fix bandfilters of second and fourth order made of ceramic $\lambda/4$ resonators of high dielectric constant /1/. A summary of the technical data of the frontend is given in Table 2.

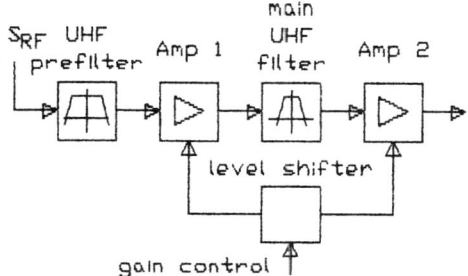

Fig. 2: Block diagram of the frontend

```
Center freqency              1680 MHz
Bandwidth (-1 dB):             30 MHz
Gain:                    -8,+7,+23 dB
1 dB input compression point:  +3 dBm
Noise figure:                   3 dB
Input third order
  intercept point:            +10 dBm
```

Tab. 2: Technical data of the frontend

The Direct Converter

The direct converter is the key component of every homodyne receiver. From todays point of view it is the most sensitive part related to intermodulation problems and noise figure in the receiver architecture. Passive mixers like Estabrook /2/ and Bateman /3/ used in her works have good large signal performance but they suffer from enormous LO power consumption and a conversion loss of at least 6 dB. The last causes noise problems at the design of the following low level baseband filtering circuits. So we decidet to use active, high conversion gain, four quadrant multipliers in 2 μm Si-bipolar technology from SIEMENS Munich AG /4/, /5/. Their conversion gain is freely selectable by the means of the collector resistance of the mixer output. A resistor of 50 Ω allows a conversion gain of 16 dB at 2 GHz. These multipliers have differential in- and outputs and can operate on all ports with frequencies from DC to 5 GHz. The LO power consumption is below -10 dBm. Unfortunatly their noise figure is about 10 dB. At the output of the mixer a magnitude of the unfiltered baseband signal consisting of the desired and adjacent channels is produced for gain control purposes. A detailed circuit diagram of one mixer is shown in Figure 3.

The direct converter, shown in Figure 4, uses two of these mixers. The RF inputs are coupled by a 0° Wilkinson power divider. The LO-signal is supplied via the same kind of divider but the signal to one mixer is delayed by 90° to generate orthogonal LO signals. A summary of the technical data of the direct converter is given in Table 3

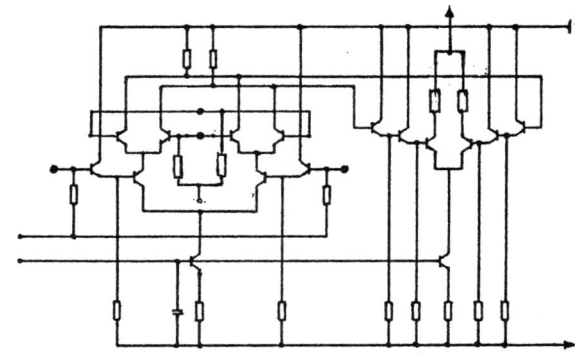

Fig. 3: Circuit of the active mixer

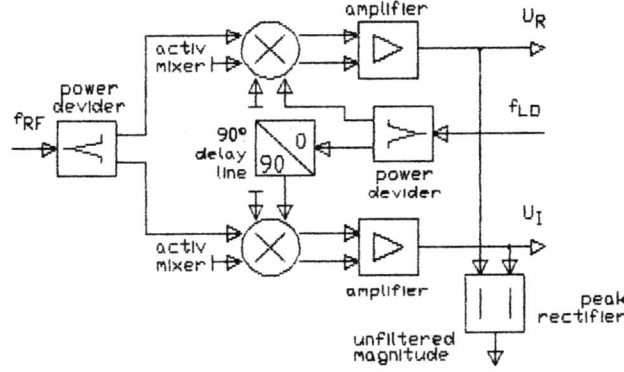

Fig. 4: Block diagram of the direct converter

```
Input frequency range:    1680+/-15 MHz
Conversion gain:                 30 dB
Noise figure:                    13 dB
1 dB input compression point:   -17 dBm
Input third order
  intercept point:               -9 dBm
Output frequency range:  500 Hz to 10 MHz
```

Tab. 3: Technical data of the direct converter

The Local Oscillator

The local oscillator is a synthesizer programable within the transmission band in steps of the channel spacing. The necessary frequency accuracy of the LO was evaluated by computer simulation of the complete receiver operation. Figure 10 shows the decrease in sensivity of the receiver as a function of the offset between RF and LO frequency related to the half data rate. For a sensivity decrease below 1 dB a LO frequency accuracy of 15 ppm is needed.

The Baseband Filter

The task of the baseband filter is the selection of the desired reception channel and the reduction of the enormous dynamic range of about 60 dB at the output of the direct converter to 6 dB at the inputs of the A/D-converters. A voltage gain of +48 dB to -18 dB in 6 dB steps is required. The lower cutoff frequency f_L of the filter results from the length of one transmission sequence T_S. Frequencies below $1/2 T_S$ are not necessary for transmitting information. The upper cutoff frequency is determined by two physical effects:

* Increasing upper cutoff frequency causes reception of more signal power.
  -> BER decreases when increasing bandwidth.

* Increasing upper cutoff frequency causes more noise and adjacent channel interferences.
  -> BER increases when increasing bandwidth.

The first effect certainly dominates for a bandwidth lower $R/2$, the second effect for a bandwidth higher than R. The optimum cutoff frequency is located between both effects. We used a computer simulation to determinate the optimum filter bandwidth. Figure 11 depicts the required signal to noise ratio at the input of the direct converter for a BER = 10^{-3} as a function of the upper cutoff frequency f_U of the baseband filter with rectangular transfer function. For this simulation we assumed a noisefree frontend and direct converter. The optimum frequency was found at $1.2 R/2$ or 600 kHz.

For our simulation and hardware experiments we used a tenth order Butterworth approximation with group delay equalizer for forming the filter described above. It was realized as an active filter in salln & key technology with low noise operational amplifiers. For dynamic equalisation amplifier stages with selectable gain are connected between the filter stages. A block diagram of the baseband filter is shown in Figure 5. Table 4 comprises its technical data.

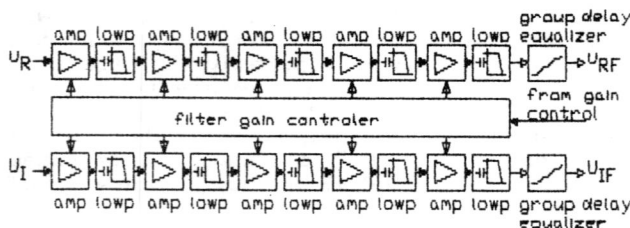

Fig. 5: Block diagram of the baseband filter

It would be possible to use a digital FIR-filter for channel selection /6/. Such a filter has the benefit of good integrability and constant group delay. Further improvement is expected from the poor resolution of available low cost high speed A/D-converters, gain switched amplifiers and anti-aliasing filters for dynamic range reduction prior to the converters.

Approximation: Butterworth with group delay equalizer
Freq. range (-3 dB): 1 kHz to 600 kHz
Gain: +48 to -18 dB
Gain step: 6 dB
Phase lnearity: < 15 °
Equivalent inp. noise voltage: 5 μV_{eff}

Tab. 4: Specifications of the baseband filter

Digitisation

After filtering, the real and imaginary part (U_{RF}, U_{IF}) of the received signal are sampled and converted into binary numbers (B_R, B_I) for high speed digital signal processing. If the optimum sampling time within the bit duration were known we would only need one sample per bit for data recovery. The correct sampling time, however, is at first unknown in our receiver. So the digitisation unit has to supply the demodulation and synchronisation circuits with a higher number of samples (oversampling). The sample related to the correct sampling time is selected in the decoding unit for data recovery. If we tolerate a deterioriation of the sensivity of less than 1 dB as consequence of the quantisation of the sampling time, the necessary oversampling factor can be evaluated to be 7. Therefore, in the following stages a resolution of 6 bits is necessary for digital demodulation.

For sampling, hold, and analog to digital conversion functions we used a pair of general purpose 8 bit flashconverters with output latch and an oversampling factor of 8. The receiver system clock (C_S) is chosen to be 8 times the data rate.

Signal processing

In GMSK and in most other two and four state modulation schemes the signal information is or can be coded in the angle difference between consecutive bits. To evaluate this angle difference (B_D) we convert the cartesian components, real and imaginary parts (B_R, B_I) of the signal, with EPROM- tables into magnitude and angle (B_M, B_A). Because of the symmetry of the cartesian system we only calculate one quadrant in tables to reduce the memory requirements. The magnitude is delivered to the synchronisation and gain control unit. The angle is coded in N Bit so that

2^N angle steps equal 360°. This coding allows arithmetic operations around the full angle without discontinuity. The angle (B_A) is supplied to the data decoder.

Decoding

The decoder has to calculate the angle difference (B_D) between the currently sampled angle (B_A) and the one stored during the previous bit period (B_{A-1}). The storage is performed by a programmable length dual port FIFO, the subtraction is made by a standard 8 Bit ALU. The discrete angle difference function (B_D) evaluated here is delivered to the synchronisation unit for sampling time determination. The optimum angle difference for data recovery (A_D) is selected in a 8 bit data latch by a sampling pulse (P_S) from the synchronisation unit and gets classified in a modulation technique dependent programmed EPROM. Table 5 shows the decision areas of the angle difference (A_D) for popular modulation schemes.

MSK, GMSK: $0° \leq A_D < +180°$, $-180° \leq A_D < 0°$

BAM, BPSK: $|A_D| \leq 90°$, $|A_D| > 90°$

QAM, QPSK: $-45° \leq A_D < +45°$, $+45° \leq A_D < +135°$
                         $+135° \leq A_D < +225°$, $-135° \leq A_D < -45°$

Tab. 5:   Decision areas of popular modulation formats by differential detection.

Synchronisation

The synchronisation unit performs the timing functions of the receiver. Figure 6 shows a simplified block diagram of the synchronisation circuits. By means of the magnitude (B_M) the transmission power up indicating the beginning of a transmission sequence is detected. At this point of time counters are started for measuring the synchronisation and data reception times. The data clock recovery circuit numerically differentiates the discrete angle difference function (B_D) supplied from the decoder. The sign function of the result is fed into a digital phase discriminator calculating the phase between this function and the receiver internal data clock. The result is an duty cycled logic signal (D_C) quantizised in 8 time steps. The duty cycle is proportional to the phase difference between the receiver internal data clock and the incoming bipolar synchronisation pattern. The phase value is integrated in an up/down counter over the complete synchronisation time. A following pulse positioning circuit uses the evaluated phase information for generating the correctly timed sampling pulses (P_S) for sample selection in the decoding unit. In this way the synchronisation can be performed very exactly even if the receiving conditions are poor.

The synchronisation is performed only once per reception sequence. The possible length of the reception sequence for valid synchronisation is mainly dependent on the frequency accuracy of the transmitter and receiver data clock and can be evaluated from

$$N = \frac{R}{2 \, O_S \, dF_{DATACLOCK}} . \qquad (1)$$

In (1) N is the sequence length in bit, R the data rate, $dF_{DATACLOCK}$ the clock frequency difference and O_S the over-sampling factor. For a clock accuracy of 15 ppm the sequence length becomes about 4000 bit.

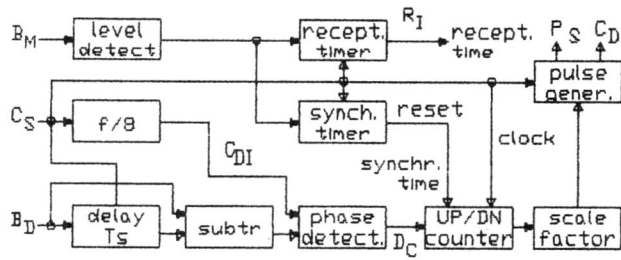

Fig. 6:   Block diagram of the synchronisation unit.

Gain and Frequency Control

Gain control is performed by an 8 bit single chip microprocessor which calculates the optimum partitioning of the gain to frontend and baseband filter as a function of the filtered and unfiltered magnitudes delivered to the CPU. The frequency offset information is also processed here and converted into a local oscillator frequency correction voltage.

V COMPUTER SIMULATIONS

To predict the operating performance of the receiver under the conditions of the communication link given in Table 1 we made a computer simulation of the complete receiver system under the assumption of noisefree frontend and direct converter.

As basic operating performances we evaluated:

* BER as a function of SNR comparend to ideal bipolar transmission.

* BER as a function of adjacent channel interference.
* BER as a function of cochannel interference.

The results are shown in Figures 7 to 9. The vectorial receiver needs about 4.5 dB more SNR to bring the same bit error rate of 10^{-3} than an ideal bipolar (BPSK-) transmission. If there is an interfering adjacent channel the interfering channel power must not exceed more than 25 dB the selected channel for undisturbed transmission. This low value of adjacent channel suppression is no poor performance of this receiver system but it is caused by the wide spectral sidelobes of the GMSK r = 0.5 power density spectrum /7/. A cochannel with about 6 dB lower power can easily be suppressed.

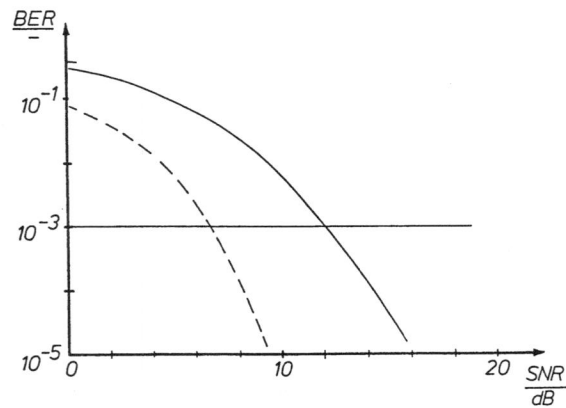

Fig. 7: Bit error rate as a function of of the signal to noise ratio (ideal BPSK transm. dotted)

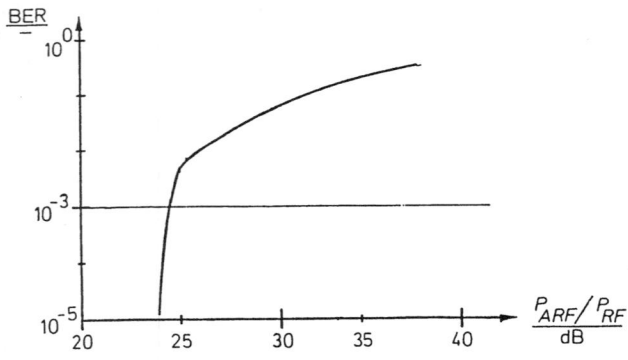

Fig. 8: Bit error rate as a function of adjacent channel interference

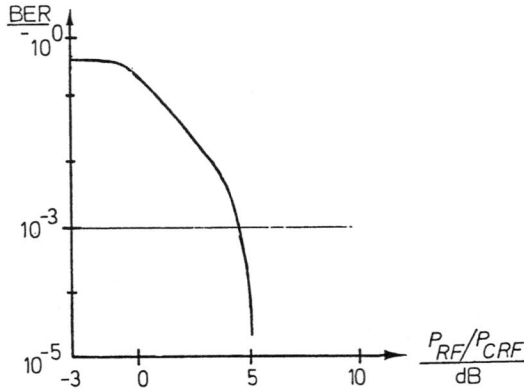

Fig. 9: Bit error rate as a function of cochannel interference

Further we evaluated the decrease of sensivity as a function of RF and LO frequency offset and the behaviour of the sensivity by variation of the upper cutoff frequency of the baseband filter. The resulting curves are shown in Figures 10 and 11.

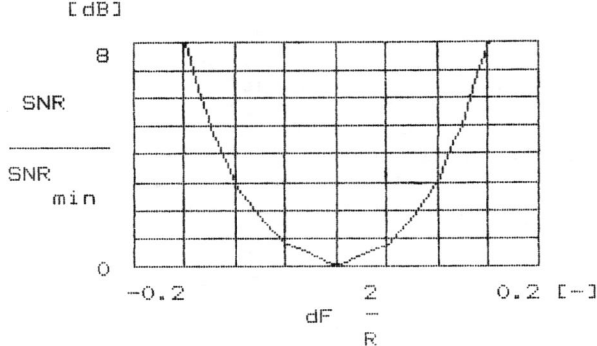

Fig. 10: Decrease of the sensivity as a function of LO frequency offset

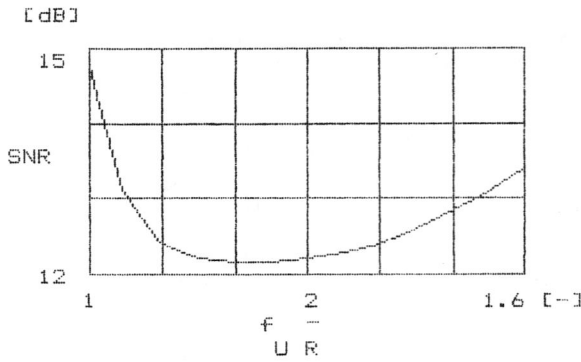

Fig. 11: Required SNR for BER = 10^{-3} as a function of the baseband filter bandwidh

VI CONCLUSION

We have shown the possibility to build a low cost incoherent homodyne receiver for mobile communication systems. The challenge remains to minimize such a system to the size of a few postage stamps.

VII REFERENCES

/1/ Siemens Comonents Jahrgang 27 Heft 6/89 (november/december) pp 223

/2/ P. Estabrook, B. Luignan "The Design of a Mobile Radio Receiver using Direct Converion Architecture", 39th IEEE Vehicular Technology Conference 1989, VOL I, pp. 63

/3/ A. Bateman, D. Haines, "Direct Concersion Transceiver Design for Compact Low- Cost Portable Mobile Radio Terminals" 39th IEEE Vehicular Technology Conference 1989, VOL I, pp. 57

/4/ P. Weger, L. Treitinger, E. Bertagnolli, E. Bonek, G. Schultes, "Silicon Bipolar Circuits for GHz Communication Systems, Mixers, Multipliers, Modulators" International 1990 Zürich Seminar on Digital Communication, March 1990

/5/ P. Weger L. Treitinger, "A Si bipolar 15 GHz static frequency devider and 10 Gbit/s multiplexer", ISSCC Dig. Tech. Papers 1989 pp. 222.

/6/ R. Maurer, J. Burgard, "A Synchronous Homodyn Receiver with Digitl Signal Processing in the Baseband for Receiving Vestigal Sideband Amplitude Modulated TV Signals in the UHF Band", Frequenz 43 1989 (9), pp. 234.

/7/ K. Murota, Kenkichi Hirade, "GMSK Modulation for Digital Mobile Radio Telephony", IEEE Trans on Comm. VOL. COM.-29, 7 July 1981 pp. 1044.

Personal Communications Transceiver Architectures for Monolithic Integration

E. Bonek, G. Schultes, P. Kreuzgruber, W. Simbürger, P. Weger*, T. C. Leslie**,
J. Popp*, H. Knapp, N. Rohringer

Institut für Nachrichtentechnik und Hochfrequenztechnik
Technische Universität Wien, Vienna
A-1040 Wien, Austria

*Siemens AG, Corporate Research and Development, Microelectronics,
D-81730 München, Germany

**Phoenix VLSI Consultants Ltd., Water Lane, Towcester, England

Abstract: The proper combination of architecture and integration concept will decide about the success of a transceiver realization. We discuss three realization concepts of monolithic integrated transceivers: i) a modular multichip solution, ii) a dedicated few chip solution and iii) a single chip solution of full custom library modules. We compare them with respect to size, power consumption, and production cost. Applied to heterodyne and direct conversion transceiver architectures, we find that direct conversion is more complex but superior for integration and cost.

We present first-time experimental results of a prototype 1.5GHz fully integrated direct conversion transceiver chip that uses a VCO at twice the required LO frequency for reduction of DC offset. A novel, mainly digital high-dynamic-range limiting baseband detector complements a direct conversion solution.

I. INTRODUCTION

The ultimate success of personal mobile radio will rely on the intelligent combination of knowledge about the radio channel on the one hand and VLSI design and production practices on the other. Prime considerations for the personal handheld will be low power consumption, translating into long intervals between battery recharge, weight and size, and cost. Evidently, most of these goals can only be achieved by integration of components as high as possible. What functions of a handheld can and should be integrated is an intricate question of the choice of semiconductor technology and of architecture. Architecture is how to perform the necessary functions. An intelligent architecture will group the necessary functions such that the constraints of die size, power consumption, price, reliability are met. One means is to keep the number of RF and analog functions to an absolute minimum and perform most signal processing as hardwired DSP.

Having in mind that alternative technology choices have been implemented (mixed Si/GaAs [1]), in this article we want to champion the bipolar/CMOS solution. That is, integrate the RF and analog functions in a high-speed bipolar chip and combine any other functionality in one or more dedicated CMOS chips. Of course, a true BiCMOS solution would be an interesting alternative. In principle, it offers even higher level of integration and more degrees of freedom for optimization, but is probably more difficult to implement.

In this paper, we will identify areas where improvements can be expected or where we think them particularly rewarding. We will concentrate on FDMA/TDMA transceivers (although some lines of thought apply to CDMA as well), and, further, on the RF frontend and on detection. Problems of architecture will be discussed by highly advanced realization examples.

We will highlight two very recent developments:

- the first completely integrated direct conversion 1.5 GHz transceiver RF chip [2]
- a digital incoherent quadrature detector for direct conversion with dynamic range as high as limiter/discriminators [3].

II. OVERALL ARCHITECTURE

Figure 1 shows a very general block diagram of a mobile radio transceiver. The RF frontend contains a receive (RX)-transmit (TX) switch, an RX filter, preamplifier, a down converter (mixer), a local oscillator (synthesizer

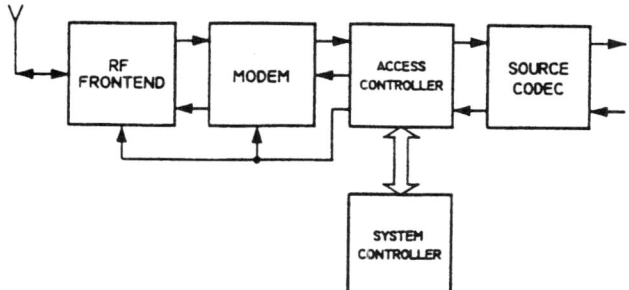

Figure 1. Radio transceiver general block diagram

consisting of VCO, prescaler and probably low-frequency synthesizer components), and, in the TX train, an up converter, power amplifier(s) and a TX filter. Usually the same frequency synthesizer is used as in the RX train.

The block <u>modem</u> combines the functions of channel selection, A/D conversion, detection in the RX train, and symbol shaping, D/A conversion in the TX train. Channel equalization and channel decoding/coding may be contained in the modem or, partly, in the next blocks, <u>access controller</u> and/or <u>source codec</u>. The <u>controllers</u> provide access timing and symbol synchronization, and control selection of the RF channel (MAC layer) and oversee logical functions of the transceiver.

Main issues are: Where to make the partition between analog and RF on one hand and digital signal processing on the other? How to meet radio-related specification requirements, such as for adjacent-channel suppression, spurious emissions and intermodulation, i. e. "system selectivity", and dynamic range. Where to place selectivity functions actually? In the next section we will show our preferred solution, which is direct conversion, as it provides the quickest way to come into baseband and perform signal processing there digitally.

In choosing a realisation, several decisions have to be taken. In principle, three different approaches are conceivable:

i) An n chip modular concept (n>4) that is fast available, but sub-optimum with respect to size, pin count and power consumption. Several chips, individually optimized for a single task [4] can offer second-sourcing; if interfaces can be agreed on and even use of different substrate materials (Si, GaAs [1]).

ii) A three to four chip full custom or semi custom solution. The full custom solution has high performance at low power and small size but low compatibility and flexibility. The semi custom solution is flexible but sub-optimum in chip-size, performance, and power consumption. Both ways are state of the art e.g. [5, 6, 7, 8, 2]. The approach of several dedicated function chips has

TABLE I
Comparison of building blocks: Direct Conversion - Heterodyne

Direct Conversion Transceiver	Heterodyne Transceiver
Active/Passive Discretes:	
Reference crystal	Reference crystal
RF duplex switch	RF duplex switch
RF-resonator (e.g. ceramic)	RF-resonator (e.g. ceramic)
RX prefilter	RX image rejection filter
	TX image rejection filter
TX-RF harmonics lowpass	TX-RF harmonics lowpass
Power amplifier	Power amplifier
	RX-IF-filter
	TX-IF harmonics lowpass
	Discriminator resonant circuit
Integrated Circuits:	
1 RF-BB-IC	1 RF-IF-IC
1 Access controller IC	1 Access controller IC
1 System controller IC	1 System controller IC
1 Source codec IC	1 Source codec IC

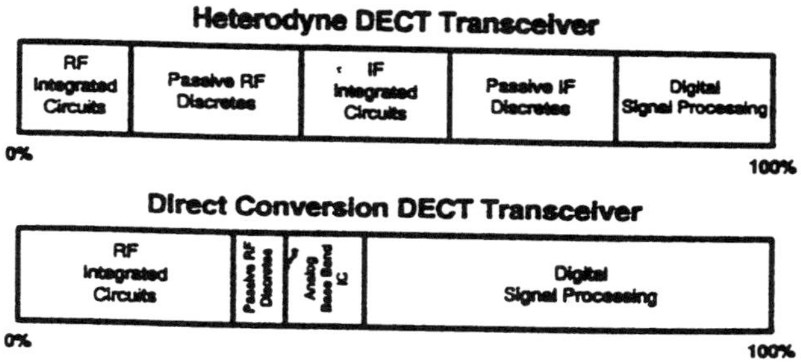

Figure 2. Normalized production expenses for heterodyne and direct conversion

been pursued with a few chips with a high level of integration.

iii) A versatile highly integrated one chip solution combining optimized full-custom functions e.g. RF and baseband signal processing units in a modular way to serve a particular system (e.g. GSM, DECT, DCS 1800, ADC, PDC, PHP, satellite, ...) best. In this approach the challenge is to minimize power consumption, number of pins, and chip area. The long sought-for subcircuit library for RF functions (mixer, VCO, preamp, ...) would be very helpful indeed for such an architecture. Strong support can be expected from hardwired or programmable DSPs. The greatest benefit is found in production cost and reliability. This seems to be the way to the future.

The RF transistor array described in a later section could be viewed as a first design step in the second direction.

III. DIRECT CONVERSION VS HETERODYNING

Direct conversion (or zero IF) can be implemented in the TX and in the RX trains. Direct up-conversion (FM or I/Q) has been accepted as a state-of-the-art solution. For a high-data rate TDMA system, we have demonstrated the viability of direct down-conversion for the first time in [9]. Recent comparisons between direct conversion and heterodyning have been published in [10].

Major advantages of direct conversion are:

- signal processing can be done in baseband, employing low power CMOS
- high level of integration possible
- no RF tuning
- hence low production cost

Heterodyning goes easily with e.g. the well-known, low power limiter/discriminator. Major problems are image rejection and a large number of IF discretes. Table 1 gives a comparison of both architectures by building blocks at realization level ii. Figure 2 shows the potential benefit of direct conversion transceivers in terms of normalized production expense. Why, then, has this principle not been used until very recently in actual circuits [6, 7, 2]?

Problems to be treated in depth with zero IF architecture for TDMA transceivers are:

- DC offset in baseband
- TDMA dynamics management
- synchronization speed
- (in)coherence of LO

The most common general prejudice about direct conversion concerns DC offset. Several mechanisms exist that produce DC offset in the quadrature down converter: LO self-mixing, RF self-mixing, and feedthrough of DC levels via LO and RF ports to the BB port. Unfortunately, these DC levels are RF-level dependent, frequency dependent, and, due to TDMA, functions of time. However, the problem can and has been overcome by several methods: e.g. differentiation and re-integration [11], or AC coupling and periodic discharge. In contrast to heterodyne solutions, direct conversion requires linear signal processing between mixer and detector in RX, and between symbol generator and up-converter in TX, even for constant-envelope formats. Of course, intermodulation has to be tightly controlled in heterodyning as well.

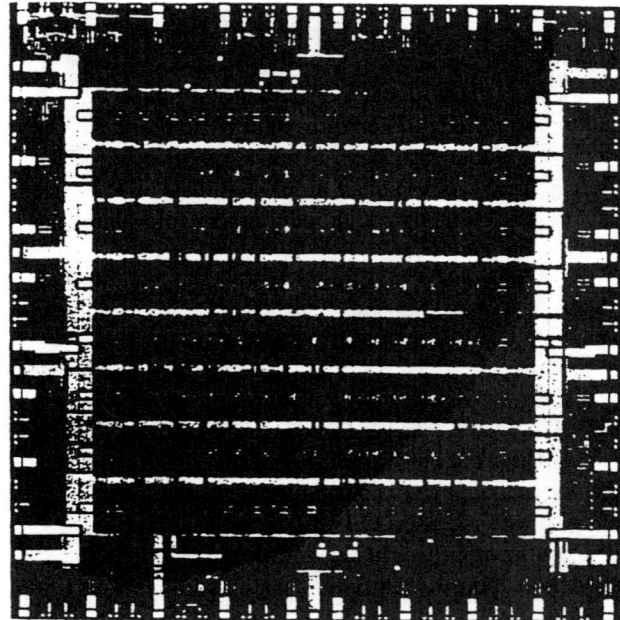

Figure 3. Microphotograph of the transceiver chip

Given the high dynamic range (80 to 100 dB) and Mbit/s data rates to be coped with, the baseband signal processing in the RF part has to be carefully engineered. Therefore, if you have a direct conversion RF frontend, you do not have the complete transceiver yet. Successful direct conversion transceiver design requires a complete direct conversion solution, from the antenna to modem output. A direct conversion architecture is more complicated than heterodyning, but, in last consequence, an extremely low-cost solution indeed.

IV. INTEGRATED DIRECT CONVERSION TRANSCEIVER

A transceiver was implemented on one 5k subarray of a 50k analog/digital transistor array built from different types of cells for realization of basic RF functions, and fabricated in a 0.8μm / 25GHz silicon bipolar technology (Fig. 3 [2]). The fully functional single-chip direct conversion transceiver is shown schematically in Fig. 4. For the first time, it contains *all* RF functions except the antenna, pre-selection filter, power amplifier, and VCO resonator, but including VCO, preamplifiers, and medium-power amplifier, which puts it into contrast to prior realizations.

The quadrature down and up converters in the RX/TX trains share the VCO and the phase-shift circuitry. The phase shifter is realized by a static 2:1 frequency divider, halving both frequency and phase of clock signals with 180 deg phase shifts at twice the LO frequency.

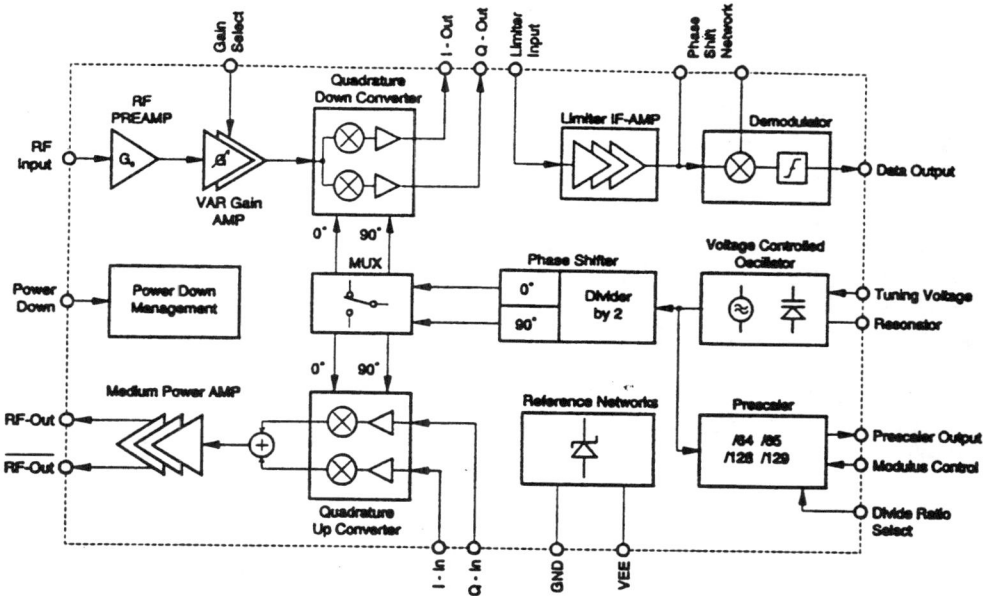

Figure 4. Fully functional single-chip direct conversion transceiver - simplified block diagram

Two major benefits follow from this: The transceiver can operate at any frequency, set by an external resonator to the VCO, from DC to the upper frequency limit of the static divider; and coupling from the VCO to the RX path is minimized, which diminishes the DC offset. The integrated dual modulus prescaler and an external low frequency PLL manage VCO fine tuning. An additional limiter/IF amplifier has been included for purpose of comparison.

The frequency response of the transceiver is flat up to 1.5GHz. Figure 5 demonstrates the dynamic range (i.e. -100 dBm to -20dBm) that could be coped with by the down converter, at two different gain settings. A low gain setting of the RF front-end is advisable. Management of dynamic range should be primarily transferred to baseband. From this, an increased dynamic range of the receiver, a simpler RF pre-amplifier, and reduced DC offset will result. Also, digital channel selection is possible. A convenient interface between RF and baseband is the complex baseband signal (I and Q output ports). Both, linear and constant envelope modulation formats can be handled by this architecture. As an example, the constellation of QPSK is shown in Fig. 6. Figure 7 shows the excellent intermodulation characteristics in the TX train, as a consequence of the first cross-quad design of the mixers at such high frequencies.

We want to point out that "single chip" pertains to RF functions only. Digital baseband processing, control functions, and the frequency selection circuit are left to a dedicated CMOS chip. Though the technology used is

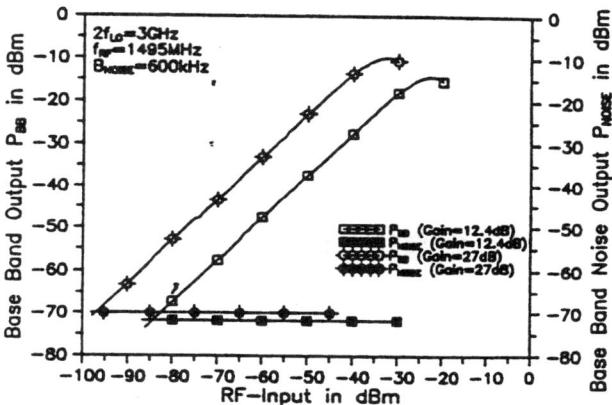

Figure 5. Measured single-tone compression- and noise behaviour of the receiver line

highly advanced, we did not utilize its full potential in frequency range nor in power saving, but chose a transistor array for test purposes. For instance, the highest operating frequency achieved was limited by packaging effects to an external 3 GHz resonator.

V. DETECTION

In this section we concentrate on incoherent detection of (G)FSK type TDMA signals. The issues are integratability (=price, size, and weight), tolerance to frequency offset between incoming and LO signals, and to intersymbol interference, and high dynamic range. For heterodyne reception there exists the well-proven limiter/discriminator

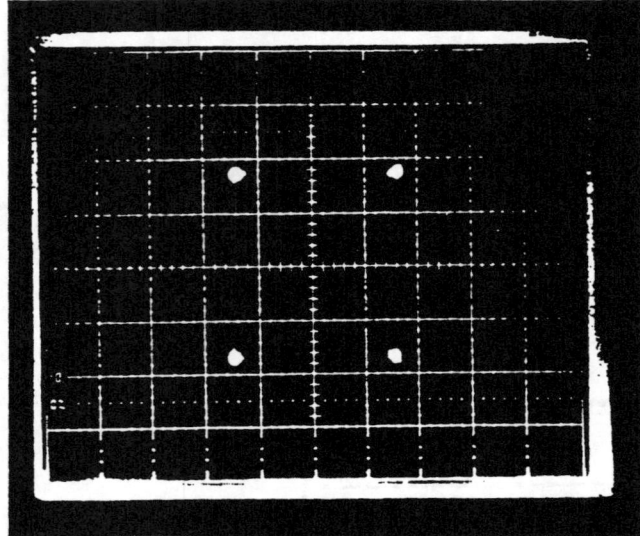

Figure 6. Measured QPSK-constellation (f_{RF}=1.5GHz/-50dBm, 20mV/div)

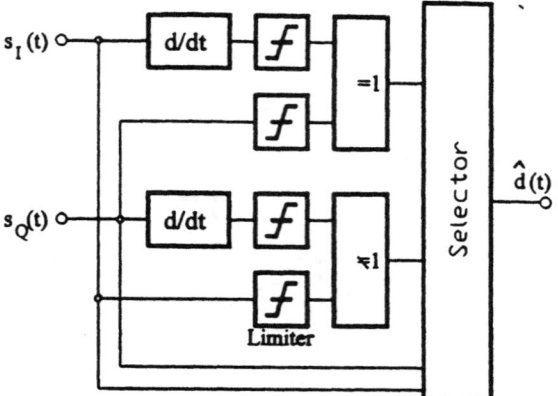

Figure 8. Balanced quotient detector

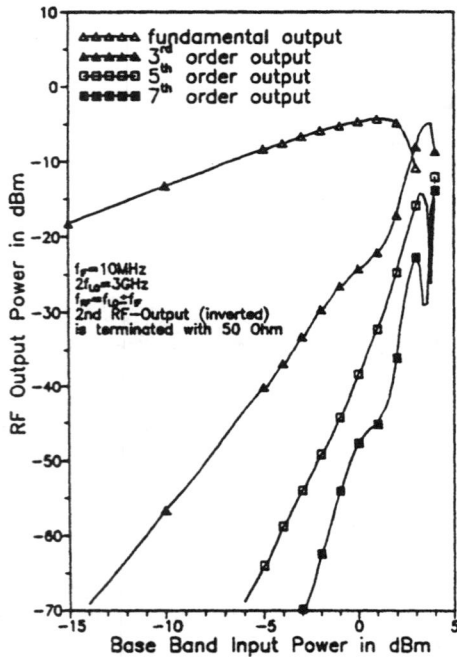

Figure 7. Measured two-tone intermodulation characteristic of the transmitter

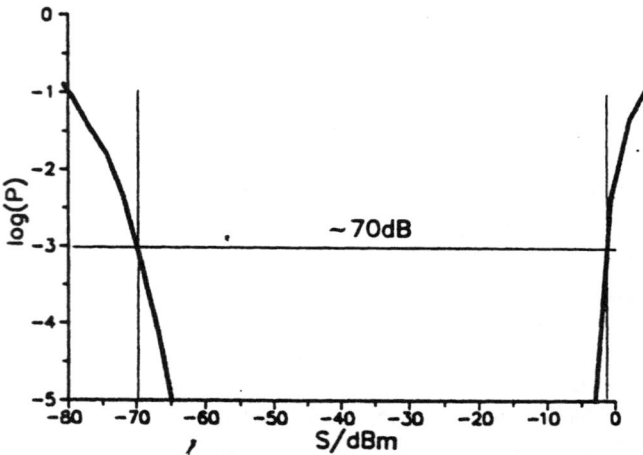

Figure 9. Dynamic range of the balanced quotient detector (S signal power, P bit error probability)

as a simple low-cost solution. For the direct conversion architecture, requiring complex baseband signal processing, the differential detector has provided good results [12].

From a new class of binary FSK detectors [3], the digitally implemented differentiating quotient detector, related in some way to the quadricorrelator, is shown in Fig. 8. Performing division of one differentiated signal by its quadrature signal using hard decision and an EXOR gate division gives the inherent high dynamic range of a limiter circuit. So far, we have achieved 70dB dynamic range at a BER of 10^{-3} in a test receiver at 1.8GHz for a GMSK signal with 1.152Mbit/s data rate and time-bandwidth product 0.5 (Fig. 9). This covers the full range required by the DECT (Digital European Cordless Telecommunications) standard, without the need for *any* gain control at RF. At an SNR = 15dB, a sensitivity degradation of 3dB versus differential detection is incurred. Altogether, the low complexity makes it a promising candidate for digitally processing complex baseband signals, i.e. it fits perfectly into the low cost architecture of direct conversion.

VI. CONCLUSION

In search for handheld mobile communication transceiver designs that consume little power and are cheap and reliable to produce, we demonstrated the various benefits of direct conversion, or zero IF, architectures by example of a highly advanced bipolar single-chip RF frontend and a GFSK detector. Direct conversion is suitable for linear and constant envelope modulation formats and for CDMA and TDMA signals. In the long run, the lowest possible count of discrete components will make direct conversion the preferred choice, despite expenses in design work and residual risk. We showed how use of a VCO at twice the required LO frequency reduced the DC offset problem, though it mandated high-speed processes. For FSK-like signals, we turned the sometimes considered awkward requirement for I/Q detection into a benefit for an elegant digital solution with high inherent dynamic-range.

In general architectural terms, several solutions with different technologies have come up recently. They struck the compromise between modularity and integration level in quite diverse ways. Still, a mobile transceiver having all functions integrated into a single chip with pins for only RF, data in/out, and battery, has come into reach by recent advances from all over the world.

VII. ACKNOWLEDGEMENT

Support of the Forschungsförderungsfonds für die gewerbliche Wirtschaft, FFF, Wien, Austria, is gratefully acknowledged.

VIII. REFERENCES

[1] M. Muraguchi, T. Tsukahara, M. Nakatsugawa, Y. Yamaguchi, T. Tokumitsu, "1.9 GHz-band low voltage and low power consumption RF IC chip-set for personal communications", in *Proceedings of the 44th IEEE Vehicular Technology Conference*, June 8-10, 1994, pp. 504-507.

[2] P. Weger, W. Simbürger, H. Knapp, T.C. Leslie, N. Rohringer, J. Popp, G. Schultes, A.L. Scholtz, L. Treitinger, "Completely Integrated 1.5 GHz Direct Conversion Transceiver", in *Proceedings of the 1994 Symposium on VLSI Circuits*, Hawaii, USA, June 09-11, 1994, pp. 135-136.

[3] P. Kreuzgruber, "A class of binary FSK direct conversion receivers", in *Proceedings of the 44th IEEE Vehicular Technology Conference*, June 8-10, 1994, pp. 457-461.

[4] G. Geiger, et al , *Siemens Components 5* , 1992, pp. 181 ff.

[5] V. Thomas, J. Fenk, S. Beyer, "A one-chip 2GHz single superhet receiver for 2Mb/s FSK radio communications", in *Proceedings of the IEEE International Solid-State Circuits Conference*, 1994, WP2.7, pp 42 ff.

[6] J. Sevenhans, Arnoul Vanwelsenaers, J. Wenin, J. Baro, "An integrated Si bipolar RF transceiver for a zero IF 900 MHz GSM digital mobile radio single chip RF up and RF down converter of a hand portable phone", *IEEE Custom Integrated Circuits Conference*, 1991, pp 7.7.1 ff.

[7] J. Sevenhans, D. Haspeslagh, A. Delarbre, L. Kiss, Z. Chang, J.F. Kukielka, "An integrated Si bipolar RF transceiver for a zero IF 900 MHz GSM digital mobile radio single chip RF up and RF down converter of a hand portable phone", in *Proceedings of the IEEE International Solid-State Circuits Conference*, 1994, WP2.8, pp 44 ff.

[8] D. E. Fague, Benny Madsen, Clay Karmel, and Andy Dao, "Performance evaluation of a low cost, solid state radio front end for DECT", in *Proceedings of the 44th IEEE Vehicular Technology Conference*, June 8-10, 1994, pp. 512-515.

[9] G. Schultes, A. L. Scholtz, E. Bonek, P. Veith., "A New Incoherent Direct Conversion Receiver", in *Proceedings of the 40th IEEE Vehicular Technology Conference*, Orlando, USA, May 6-9, 1990, pp. 668-674.

[10] G. Schultes, P. Kreuzgruber, and A.L. Scholtz, "DECT transceiver architectures: super-heterodyne or direct conversion?", in *Proceedings of the 43rd IEEE Vehicular Technology Conference*, May 1993, pp. 653-656.

[11] B. Lindquist, M. Isberg, and P. W. Dent, "A direct conversion receiver for TDMA systems", in *Proceedings of the Radio Vetenskaplig Konferens*, April 1993, pp. 189-192.

[12] G. Schultes, E. Bonek, A. L. Scholtz, W. Herzog, "Performance of a Self Synchronizing Direct Conversion DECT Receiver", *Electronics Letters*, Vol.27, No.19, 12th Sept 1991, pp 1715-1717.

An All-CMOS Architecture for a Low-Power Frequency-Hopped 900 MHz Spread Spectrum Transceiver

Jonathan Min, Ahmadreza Rofougaran, Henry Samueli, and Asad A. Abidi

Integrated Circuits and Systems Laboratory
Electrical Engineering Department
University of California, Los Angeles

Abstract

An all-CMOS single chip architecture of a frequency-hopped spread spectrum (FH/SS) transceiver is presented. The transceiver implements complete modem functions spanning a tuned RF front-end to a digital binary frequency-shift keying (FSK) modulator/demodulator, all operating from a 3-V supply. This FH/SS transceiver is intended for use in a wide variety of indoor/outdoor portable wireless applications in the 902-928 MHz ISM band [1]. System features such as dual antenna/branch diversity, fast frequency hopping, and adaptive power control are incorporated into the transceiver architecture to achieve robust wireless digital data transmission over multipath fading channels. This paper describes the architectural features of the transceiver and surveys several of the key building blocks which have been completed to date.

I. Introduction

A FH/SS code division multiple access (CDMA) technique has advantages in that a) it provides an inherent immunity to multipath fading; and b) the signal processing is performed at the hopping rate, which is much slower than the chip rate encountered either in a direct sequence (DS) CDMA or time division multiple access (TDMA) system, thereby potentially resulting in much lower receiver power consumption. Furthermore, FSK modulation with noncoherent detection in a frequency-hopped system results in a much simpler transceiver architecture as compared with coherent amplitude and phase modulation methods commonly used in DS and TDMA systems. The proposed transceiver has the objective of passing data at rates up to 160 kb/s, hopping over the entire 26 MHz bandwidth. The FH/SS transceiver specifications are summarized in Table I.

The principal channel impairments in multi-user radio systems are attenuation due to shadowing, multipath fading, and interference from other radios. To overcome these impairments without resorting to high transmitter power, the proposed architecture incorporates many advanced techniques. Two antenna/receiver branches are used to provide equal-gain spatial diversity. Frequency hopping using a quadrature direct digital frequency synthesizer is implemented to provide frequency diversity. Adaptive power control is also employed with a programmable gain power amplifier so that the minimum transmitted power required for reliable communication is used, and channel coding using a convolutional code will be employed to improve the bit error rate performance.

TABLE I. FH/SS TRANSCEIVER SPECIFICATIONS

Multiple Access Scheme	Frequency-Hopped CDMA
Modulation Scheme	Noncoherent Binary FSK
Duplexing Mode	Time Division Duplex
Channel Coding	Rate 1/2, K=6 Convolutional Code
Hopping Bandwidth	26 MHz (902-928 MHz)
Data Rates	2 - 160 kbps
Maximum Hop Rate	160 khop/s
Antenna/Branch Diversity	2
Maximum Range	500 meters
Transmission Power	1 µW - 20 mW
Active Power	300 mW (Tx), 225 mW (Rx)
Operating Voltage	3 V
IC Technology	0.8 µm CMOS

Conventional RF front-end building blocks in the 1 GHz frequency band are typically implemented in GaAs or silicon bipolar technologies due to the required gain and low noise requirements. With rapid advances in CMOS IC technology, however, we have demonstrated a CMOS monolithic RF amplifier [2]. By using CMOS for the RF front-end, the entire transceiver is integrated in a single technology, with the exception of the antennas and 900 MHz passive bandpass filters. This full integration of RF, IF, and baseband components into one monolithic CMOS chip will result in substantial power savings. Further power reduction comes from architectural optimizations which will be described in the following section.

II. Transceiver Architecture

A. RF/IF Architecture

An RF/IF architecture employing high-performance low-power circuit techniques is proposed for implementing the FH/SS transceiver. The combination of a direct digital frequency synthesizer (DDFS) and a digital-to-analog converter (DAC) is used to generate a single-sideband 26 MHz spread-spectrum waveform in the 902-928 MHz band. The two DDFS/DAC channels produce sinewaves in quadrature with frequency selectable from 0 to 13 MHz (Fig. 1). After anti-alias filtering, these sinewaves are respectively upconverted by quadrature outputs from a 915 MHz local oscillator. If the two upconverted outputs are added, the output frequency ranges from 915 to 928 MHz; if they are subtracted, it ranges from 902 to 915 MHz. This signed I-Q frequency synthesis architecture reduces the highest frequency required from the DDFS/DAC to 13 MHz, yet covers the desired 26 MHz hopping bandwidth. In addition, this single-sideband (SSB) modulation technique eliminates the need for image rejection filters with sharp roll-offs, and therefore lowers the overall power consumption. By using DDFS techniques coupled with the inherent matching of monolithic CMOS analog circuits, we expect to achieve a SSB image rejection of 50 dB. Since we are using an FSK modulation scheme, a simple highly efficient class-C differential power amplifier is used for transmission. Finally, a low-cost off-the-shelf 902-928 MHz dielectric resonator bandpass filter is used between the power amplifier and the antenna to reject the out-of-band harmonics and meet FCC transmission mask requirements.

Before down conversion, the received signal needs to be amplified using a low noise amplifier (LNA). The gain and noise figure of the LNA are critical factors in determining the overall transceiver performance. We have successfully developed a new technique to achieve a monolithic high-Q inductor in CMOS without altering the fabrication process [2]. This technique is applied to design an LNA with high gain at 915 MHz. The high-Q inductor amplifier also provides extra filtering for rejecting the out-of-band signals and noise. To minimize the number of high frequency components and thereby save power, the dehopping and down conversion is performed in a single "direct-conversion" stage. Direct conversion requires accurate I-Q dehopping carriers at 915 MHz. Conventionally, I-Q carrier generation has been implemented by RC passive elements [3]. However, the accuracy of the I-Q signals is not adequate enough for high performance receivers. To overcome this short coming, a new technique is proposed which uses the I-Q outputs of the DDFS/DAC and of the LO. Accurate I-Q dehopping carriers are generated by taking the real and imaginary parts of the multiplications of these two complex signals, as shown in Fig. 1. Addition or subtraction will again result in selection of the upper side-band or lower side-band of the I-Q dehopping carriers. These carriers are then divided into lower frequencies to be used in the sub-sampling down conversion mixers. One drawback with this scheme, however, is that the circuitry which generates the I-Q dehopping carriers must be balanced. The input stage of the power amplifier is modified to add both of the I-Q signals.

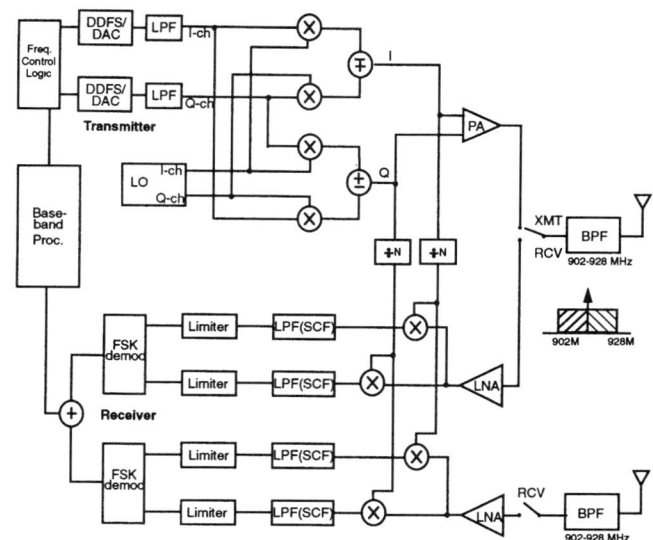

Fig. 1. FH/SS transceiver architecture.

This addition will result in another signal of the same frequency with a 45 degree phase shift only. Therefore, the transmitted signal carries the same information. The down conversion mixer must achieve a low intermodulation distortion (IMD). One way to reduce power and lower the IMD products is to use a sub-harmonic sampling circuit which is switched at an integer (small) fraction of the RF frequency (915 MHz). This method can be easily implemented using conventional switched-capacitor techniques which are well developed for CMOS technology. These techniques not only result in low IMD products but also result in lower overall power dissipation [4].

B. Baseband Architecture

Low power consumption requires that complex signal processing, such as adaptive equalization, be avoided if possible. At a given hop time, the FH system resembles a narrow band system unlike its DS counterpart. High rate frequency hoping with FSK modulation can be easily implemented by appropriately updating the frequency control word of the DDFS. The DDFS approach has several key advantages over conventional analog PLL-based synthesizers, the most important of which is being able to achieve rapid frequency changes with continuous phase transitions. The hop rate is no longer limited by the RF synthesizer settling time but by the system requirements.

Quadrature demodulation is required to efficiently use the entire 26 MHz transmission bandwidth. The automatic gain control (AGC) function is implemented using a combination of a lowpass filter (LPF), a hard limiter, and an oversampled 1-bit demodulator. The limiter works as a 1-bit quantizer. Due to odd harmonics produced by hard-limiting the sinewave, this approach limits the modulation scheme to binary FSK rather than M-ary FSK. However, the merits from power savings and hardware simplicity obtained from the absence of an expensive (in both power and complexity) linear variable gain amplifier (VGA) and a complex analog-to-digital converter (ADC) justifies this architecture for a portable transceiver design. The quadrature binary FSK modulation scheme also relieves the problem of group delay variations at different frequencies in the LPF. The LPF is implemented using a switched-capacitor technique for its precise tuning capability. When programmable bandwidths and data rates are required, the SCF also has a distinct advantage over the continuous-time filter since its bandwidths scale with the clock rate. An all-digital quadrature demodulation architecture using an oversampled 1-bit correlation detector has been chosen over a fast fourier transform (FFT) or matched filter method. This choice is based on the fact that the correlator provides a flexible design which can easily accommodate programmable data rates, and the same hardware can be used to demodulate both data and sync hops. Following binary FSK demodulation, the rest of signal processing is all performed digitally. This includes frame synchronization, PN acquisition, clock recovery, and frequency tracking.

III. Analog and Digital Circuits

Various low-power analog and digital transceiver building blocks, covering RF, IF, and baseband circuits all developed in CMOS technology, are overviewed in this section. These prototype circuits will form the basis of the complete monolithic transceiver which is currently under development.

A. Low Noise RF Amplifier - A low-power, low-noise RF amplifier in 2–μm N-well CMOS has been implemented and tested [2]. The amplifier has been measured to provide 14 dB of gain while dissipating only 6.9 mW (excluding output buffers) with a single 3-V supply. The amplifier has been designed to be a tuned amplifier in order to limit the noise contribution from the amplifier itself. The tuning of the amplifier was achieved by a large on-chip spiral inductor resonating with the parasitic capacitances. A novel technique has been applied to suspend spiral inductors over air-filled pits in order to attain over 100 nH of inductance on chip while providing a high self-resonant frequency. The noise figure of this amplifier was measured to be 6.1 dB (excluding input terminations), and the center frequency was 770 MHz. A new version of this amplifier also has been developed in 1-μm CMOS which is centered at 915 MHz. This amplifier has a gain of 30 dB with a noise figure of 4 dB.

B. Power Amplifier - A low-power and highly efficient class-C power amplifier has been designed in a 1-μm N-well CMOS process for the 902-928 MHz application. The power amplifier transmits up to 20 mW with an efficiency of better than 35%. The maximum power consumption is about 60 mW. An on-chip inductor has been used to improve the gain and efficiency of the amplifier. A differential scheme is used to improve the performance of the power amplifier and minimize the out-of-band harmonics. This power amplifier uses a 6-bit binary word to control transmission power level. This technique provides more than 36 dB dynamic power control capability. To extend further, the input swing is also controlled with an additional 15 dB range, thus providing a total of 50 dB programmable dynamic range.

C. Sub-Sampling Mixer - A low-distortion sub-sampling mixer using 1-μm CMOS technology has been designed and fabricated [4]. This sub-sampling mixer is to be used in the receiver after the low noise RF amplifier stage to perform direct down conversion. The measured third-order intercept is about 25 dBm, and the fundamental output is compressed by 1 dB at an input power of 12 dBm. The overall power consumption is less than 11 mW. The clock of the sub-sampling mixer needs to run at a fraction the carrier frequency. In our measurements, a 50-MHz clock has been used to downconvert the 900 MHz signals.

D. Switched-Capacitor Lowpass Filter - A low-sensitivity 5th order elliptic switched-capacitor LPF has been designed. This filter rejects undesired high frequency images from the sub-sampler output by -60 dB. A continuous-time prototype based on the doubly-terminated LC ladder filter structure was derived first, and then by applying the bilinear transformation, a parasitic-insensitive switched-capacitor ladder filter was synthesized. The resulting circuit has very low sensitivities to both element-value variations and parasitic capacitances. High dynamic range and minimum capacitance spread were obtained through capacitance scaling. A 3-V prototype circuit fabricated in 1.0-μm CMOS technology dissipates about 15 mW. Further circuit optimization will be performed in the next version to reduce the power dissipation.

E. Local Oscillator (PLL/VCO) - A low phase noise local oscillator (LO) at 915 MHz is being developed. The LO must be designed to generate these high clock frequencies by scaling a stable reference source. The LO must not only have low phase/spurious noise but it must also have a low power dissipation. These specifications are further complicated by the large operating temperature range and low-power supply (3 V). A phase-locked loop (PLL) was chosen to keep the phase noise to a minimum. The individual circuit blocks will be optimized for low power dissipation. The PLL consists of an external crystal used to generate the reference frequency, a phase/frequency detector, a charge pump, a voltage controlled oscillator (VCO) and a digital divider. Simulated power dissipation is about 16 mW.

F. Digital-to-Analog Converter - A monolithic low-power 40 Msample/s 10-bit digital-to-analog converter (DAC) with a 3-V supply has been designed and tested [5]. Conventionally, high-speed and high-precision DACs are implemented with current mode architectures which typically dissipate a large power. The proposed DAC employs a pipelined charge redistribution algorithm implemented differentially, in which the pipelined digital code controls MOS switches that charge the capacitors to some known reference voltage. Due to the inherent nature of a switched-capacitor circuit, there is no DC standing current running through the circuit itself. Thus, it is ideally suited for the low-power application. The core of the DAC circuit, excluding the digital clock generator and the output buffer, dissipates only 1 mW.

G. Direct Digital Frequency Synthesizer - A 3-V CMOS quadrature direct digital frequency synthesizer (QDDFS) has been fabricated [5]. The QDDFS synthesizes 10-bit output sine and cosine waves simultaneously at 40 Msample/s. Frequency is selected with an 11-bit word to obtain a 19.53 kHz resolution. The synthesizer can cover a bandwidth from DC to 20 MHz with a switching speed of 25 ns and a tuning latency of 2 clock cycles. Several techniques are employed to reduce the ROM storage. The worst-case spurious response for the proposed DDFS is -72.63 dB. The measured power dissipation of a 1.0-μm CMOS prototype at 40 MHz is about 25 mW.

H. FSK Demodulator - A low-power quadrature binary FSK correlation demodulator (Fig. 2) is in fabrication. Many architectural optimization techniques have been incorporated into the design. First, the front-end correlator uses only a 1-bit multiplier due to the fact that the incoming signal is hard limited to +/- 1's. Note that FSK demodulation requires only zero crossing information, not the amplitude. The input signal is oversampled according to the clock rate of a SCF/limiter combination (roughly 10 MHz). However, this is immediately decimated down to twice the baud rate following the integrate-and-dump block. The integration duration is half the baud time in order to generate the early and late signals for the clock recovery loop. A noncoherent (NC) demodulator requires a magnitude calculation unit, which is conventionally implemented with a pair of squaring multipliers followed by a summing node. In our architecture, an absolute value addition block replaces this squaring block. Thus, a truly multiplierless NC FSK demodulator has been implemented with little performance degradation. Simulated power dissipation in 1.0-μm CMOS technology is only 2 mW.

IV. Conclusions

In this paper, an efficient architecture for an all-CMOS FH/SS transceiver for use in low-power handheld communications applications has been presented, along with a description of the major analog and digital circuit components. Power savings of the transceiver comes from all three levels of optimization: system-level (FH over DS), architecture-level, and circuit-level (voltage scaling with parallelism and use of minimum geometry transistors when applicable). A highly integrated architecture of low voltage custom analog and digital CMOS components into a single chip, thereby minimizing the

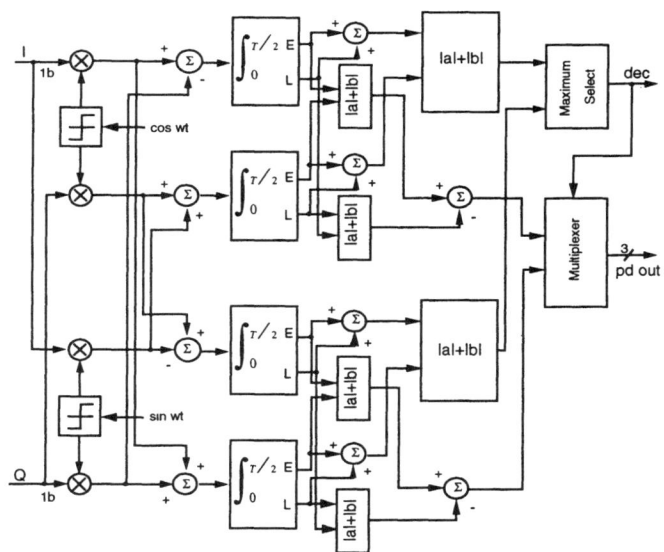

Fig. 2. Quadrature NC binary FSK demodulator.

use of discrete components, is also a key to achieving the minimum power consumption.

Acknowledgment

The authors wish to thank all members of the FH/SS project team at UCLA for their technical contributions. This work was supported by ARPA/ESTO under Contract No. DAA B07-93C-C501 and the State of California MICRO Program.

References

[1] J. Min, A. Rofougaran, V. Lin, M. Jensen, H. Samueli, A. A. Abidi, G. Pottie, and Y. Rahmat-Samii, "A Low-Power Hanheld Frequency-Hopped Spread Spectrum Transceiver Hardware Architecture," *IEEE Workshop on VLSI in Communications*, Lake Tahoe, CA, Sep., 1993.

[2] J. Y-C. Chang, A. A. Abidi, and M. Gaitan, "Large Suspended Inductors on Silicon and Their Use in a 2-μm MOS RF Amplifier," *IEEE Electron Device Letters*, vol. 14, no. 5, pp. 246-248, May 1993.

[3] M. D. McDonald, "A 2.5 GHz BiCMOS Image-Reject Front-End," *Technical Digest of IEEE 1993 International Solid-State Circuits Conference*, San Francisco, CA, pp. 144-145.

[4] P. Y. Chan, A. Rofougaran, K. A. Ahmed, and A. A. Abidi, "A Highly Linear 1-GHz CMOS Downconversion Mixer," *Proceedings of 1993 European Solid-State Circuits Conference*, Sevilla, Spain, pp. 210-213.

[5] G. Chang, A. Rofougaran, M-K. Ku, A. A. Abidi, and H. Samueli, "A Low-Power CMOS Digitally Synthesized 0-13 MHz Sinewave Generator," to appear in *Technical Digest of IEEE 1994 International Solid-State Circuits Conference*, San Francisco, CA, Feb. 1994.

Spread-Spectrum Technology for Commercial Applications

D. THOMAS MAGILL, MEMBER, IEEE, FRANCIS D. NATALI, SENIOR MEMBER, IEEE, AND GWYN P. EDWARDS, MEMBER, IEEE

Invited Paper

Only recently has our technology advanced to the point that commercial application of spread-spectrum signaling is economically feasible. This has motivated a number of companies and individuals to seek new ways to benefit from spread-spectrum techniques in commercial systems and products.

In this paper, we give a very brief overview of spread-spectrum signaling. We then consider applications to satellite mobile applications as well as indoor wireless applications. An overview is presented of several proposals to provide worldwide personal communications through satellite-based spread-spectrum systems, the tradeoffs to be considered, and the controversies involved. A description of a high-capacity wireless office telephone system serves to illustrate how spread-spectrum signaling may be useful in this environment. Finally, we describe a number of digital processing algorithms and devices that implement spread-spectrum signaling in a cost-effective manner.

I. INTRODUCTION

As our society has become increasingly interdependent, the transfer of information has become more and more important. Any product that can provide better, quicker, more mobile, or more convenient communications is destined for success. The general, personal desire to compute or communicate at any time, and from any place, has supported the growth of the cellular mobile telephone network, the emergence of the wireless PBX, and many other Personal Communications Service (PCS) experiments and proposals. The urge for untethered, mobile computing for professionals has led to the development of wireless Local-Area Networks (LAN's). Cordless telephones have become commonplace in the home as well as the work place.

The development of the spread-spectrum Global Positioning System (GPS) has spawned an increasing demand for mobile/portable position location products both for stand-alone use and in conjunction with a communications link. New products incorporating GPS are proliferating rapidly.

Manuscript received June 30, 1993; revised August 30, 1993.
The authors are with Stanford Telecom, Sunnyvale, CA 94089-1117.
IEEE Log Number 9215558.

As we attempt to make wireless technology available for a broader range of applications, we are faced with the challenges of spectrum overcrowding, privacy considerations, fading mobile channels, as well as the complexities of the propagation environment inside buildings, to name a few.

Many of these problems have been tackled for military applications where complex and costly solutions have often been considered acceptable, if the desired performance was achieved. Spread-spectrum (SS) communications has been one of the most intriguing and exciting technologies to emerge from these efforts. References [1] and [2] provide a brief introduction to SS signaling while [3] is an excellent three volume text.

SS may be defined as "a technique in which an auxiliary modulation waveform, independent of the information data, is employed to spread the signal energy over a bandwidth much greater than the signal information bandwidth." The signal is "despread" at the receiver using a synchronized replica of the auxiliary waveform.

SS signaling has properties which make it useful for

- signal hiding and noninterference with conventional systems
- anti-jam and interference rejection
- privacy
- accurate ranging
- multiple access
- multipath mitigation.

SS does not improve performance in the presence of Gaussian noise, and the signal does require increased bandwidth.

While some of the basic SS concepts were formulated earlier, World War II spurred researchers to find new ways of communicating that would be secure and work in the presence of jamming. Practical application of these techniques was limited by the technology available at the time (an excellent historical overview of the development

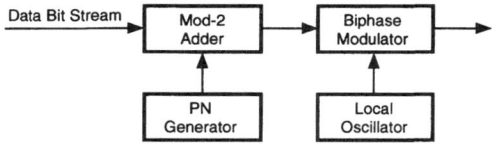

Fig. 1. Simple direct-sequence modulator.

of SS signaling is given in [3]). A general understanding of SS systems was slow to emerge after the war due to the strict classification that was imposed for many years, and to the inherent complexity of existing equipment, which served to limit its utility.

For many years SS was considered solely for military applications. However, rapid advances in LSI technology have made it possible to implement the complex functions required for SS within size and cost constraints that make it attractive for consumer products. In addition, the characteristics of SS signaling appear to be well-suited to mitigating the problems of mobile/portable personal communications.

This paper gives an overview of SS signaling, its advantages (and shortcomings), existing and proposed applications, and the technology that is making it all possible.

II. Direct-Sequence (DS) Spread Spectrum

There are two principal types of SS systems—frequency hopping (FH) and direct sequence (DS).

Direct-sequence signaling is accomplished by phase modulating the data signal with a pseudo-noise (PN), i.e., pseudo-random, sequence of zeros and ones, which are called chips. The chip modulation is most commonly binary phase-shift keying (BPSK) for simplicity and is achieved by mod-2 adding the PN chip sequence with the data as shown in Fig. 1. The number of PN chips per bit is a measure of the processing gain. Quaternary PSK (QPSK) chip modulation, while more complex, is sometimes used to prevent signal capture when a strong interferer drives the receiver into saturation.

Typically DS systems use BPSK data modulation since it is simple, and higher order PSK modulation formats offer no increase in processing gain. As we will see later QPSK data modulation can be useful for some specialized DS systems.

DS systems may be categorized as either long- or short-code systems. Long-code systems have a code length that is much longer than a data symbol so that a different chip pattern is associated with each symbol. The chip sequence is essentially random for a sufficiently long code. In code-division multiple access (CDMA) situations, the users will experience varying degrees of cross correlation which has an effect similar to random noise. In this case, it is generally necessary to implement accurate power control such that all user signals are received at nearly the same power level, otherwise, one signal can cause substantial interference to the others and reduce the capacity of the system. If the users have modest timing accuracy with respect to each other, it is possible for all users to share the same code with different time displacements for each user.

By contrast, short-code systems repetitively use the same sequence for each data symbol. Short-code systems normally use matched-filter detection while long-code systems use correlation detection. Thus short-code systems are capable of offering much more rapid acquisition and are well-suited for burst traffic, such as transaction systems.

Particular care must be taken in choosing short codes for the CDMA situation. Not only must the codes have good autocorrelation functions to permit reliable symbol synchronization, the codes must also have low cross correlation for all shifts. For short code lengths there are not many sequences which meet these properties. As a result, multiple access is often achieved by ALOHA or a carrier sense protocol in these systems.. The role of spread spectrum in these cases is usually to provide interference rejection and spectrum compatibility with other systems.

It is possible to achieve orthogonal CDMA operation with short codes if sufficiently precise timing is maintained. That is, the cross correlation between users is zero when the users are received in time synchronism. Typically this requires a star network configuration using orthogonal CDMA on the inbound (to the hub) links and orthogonal code-division multiplexing on the outbound links (the outbound signals are multiplexed on a single carrier in time synchronism). The orthogonal codes are frequently based on the Rademacher–Walsh functions but any orthogonal code set is acceptable. To maintain orthogonality on the inbound links it is necessary that all signals arrive at the hub timed to within a very small fraction of a code chip.

Forward error correction (FEC) encoding is usually employed in CDMA systems in order to increase the multiple-access capacity. Since the processing gain is not affected by the coding rate, performance is improved by the coding gain. Interleaving is also generally employed to combat burst errors due to interference or fading.

The CDMA system discriminates against multipath components that are delayed with respect to the line-of-sight path by more than one chip duration, thereby reducing fading due to multipath. Note that this capability is achieved without requiring the use of forward error correction coding. Obviously, a high chipping rate is required to discriminate against short delays.

One disadvantage associated with DS spread spectrum is its inability to deal with very-large interfering signals such as occur when an interferer is located very close to the receiver, i.e., the near/far problem. The interference is reduced by the available processing gain which is inadequate in many cases. One solution to this problem is to use notch filters, especially if the interference is narrowband. This can lead to excessive hardware complexity if there are a large number of interferers. Another approach is to employ a reduced chipping rate and multiple channels. Channel selection and filtering can help provide the desired rejection of the interference. If heavy, fixed interference is expected to be a problem, one may be better off with narrowband DS spread spectrum/FDMA than wideband DS spread spectrum. In this case, one may pay the price of increased susceptibility to multipath fading.

III. FH Spread Spectrum

FH spread spectrum divides the available bandwidth into N channels and hops between these channels according to a PN code known to both the modulator and demodulator. Since the hops generally result in a phase discontinuity (depending on the particular implementation), a noncoherent modulation format such as MFSK or differentially coherent PSK modulation is often employed. The increased energy efficiency associated with higher order MFSK, such as M equal to eight or sixteen, is gained at no increase in spread bandwidth. FH spread spectrum systems often employ forward error correction (FEC) coding to provide protection for those data bits transmitted at a hop frequency with interference or with a fade.

FH systems may be categorized as either slow- or fast-hopping (relative to the data symbol rate). With slow hopping there are multiple data symbols per hop and with fast hopping there are multiple hops per data symbol. Systems employing MFSK are generally fast hopping, while binary DPSK modulation is often used with slow hopping. In this case, the first symbol in a hop serves as a phase reference.

Fast hopping provides frequency diversity within a data symbol and, depending on the number of hops per symbol, coding may not be required. Note, however, that since MFSK detection is done on a per-hop basis, one encounters the loss associated with post-detection integration. One may wish to use FEC coding to compensate for this loss. By contrast, the coding system for slow hopping must provide frequency diversity through interleaving to deal with the error bursts.

IV. Comparison of DS- and FH-SS

The potential benefits of SS do not depend on how the spreading is achieved; however, there are many practical differences in both performance and implementation for different spreading techniques.

The bandwidth of the DS-SS signal is approximately twice the PN sequence clock rate. Wide spreading bandwidths require high clock rates which can present synchronization difficulties, as well as increased equipment cost and power consumption. The bandwidth of an FH system depends only on the tuning range and so can be hopped over a wide bandwidth with little difficulty.

Timing is generally much less critical in an FH system since hop rates usually range from a few hops per second to several thousand hops per second, compared to megahertz chipping rates in DS systems. In order to synchronize the transmitter/receiver pair, the receiver must search its initial time uncertainty until it is within a fraction of a hop or chip duration, at which time correlation will occur, and the signal can be detected. Since many fewer hops than chips occur in a given time interval, the FH code-acquisition times can be relatively short in comparison to some DS systems.

Another distinction between SS systems is that the spectrum of a DS signal looks relatively uniform (unless a very short code is employed) for observation bandwidths on the order of the symbol rate. The frequency hopper, on the other hand, is essentially a narrowband signal whose center frequency is changed frequently. This distinction may or may not be important, depending on the application.

Both DS- and FH-SS can be useful for multipath mitigation when properly designed. In the case of DS spreading, multipath returns that are delayed by a chip period or longer relative to the desired return are essentially uncorrelated and do not contribute to multipath fading. In the FH case, one transmits over a wide enough bandwidth to ensure that the channel is frequency-selective, i.e., while some frequencies are "faded," others are not. As the signal hops around some hops are lost while others get through. The challenge is to design the signal with enough redundancy, and in some cases interleaving, to maintain an acceptable average error rate.

V. SS for Commercial Applications

The proliferation of SS products for commercial applications began with

- the emergence of the GPS System
- the successful demonstration of SS for cellular telephones.
- the allocation of the Instrumentation, Scientific, and Medical (ISM) bands in the USA by the FCC.

In 1989, the FCC mandated special use of the ISM bands so that an SS system is permitted to operate without license so long as the system does not interfere with an existing system in these frequency bands. A very concise summary of the frequency allocations and use restrictions for the ISM bands is given in Tables 1 and 2.

VI. Satellite Mobile Applications

Satellites have the capability to provide line-of-sight coverage of large geographical areas. As a result, they represent an attractive means of delivering mobile and portable services to large numbers of users who could not be economically served with terrestrial based stations.

The Global Positioning System (GPS) is a case in point. GPS receivers, which have the capability to calculate position to within about 100 ft anywhere in the world, now sell for under $1000. The burgeoning navigation market indicates that the technology involved to provide satellite-based mobile services has come of age. A direct sequence CDMA spread-spectrum signal was chosen for GPS because it provided a means of incorporating accurate ranging, data transmission, multipath mitigation, multiple access, interference rejection, and access security (for the

Table 1 FCC Part 15, ISM Frequency Bands

Band	Bandwidth
902–928 MHz	26.0 MHz
2.4–2.4835 GHz	83.5 MHz
5.725–5.850 GHz	125.0 MHz

Table 2 FCC Part 15.247, Spread Spectrum Use of ISM Frequency Bands

	Direct Sequence	Frequency Hopped
Frequency Constraints	• At least 500 kHz must be occupied, up to a maximum of the allocated band.	• All subchannels must be occupied at some time. • Minimum number of channels is 50 in 902-MHz band and 75 in the other two bands. • Average dwell time on any band must not exceed. 4 s in any 20-s period for 902-MHz band and 0.4 s over a 30-s interval in the other bands. • Maximum bandwidth of a hopping channel is 500 MHz
Power	• 1 W maximum • Processing Gain (PG) of at least 10 dB (PG is the ratio of signal-to-noise ratio without spreading)	• 1 W maximum

high-accuracy P signal) in a convenient manner. This L-band system has proved to be both performance- and cost-effective.

The Qualcomm OmniTRACS system (Ku-band) provides a logical extension of this service with two-way data messaging as well as vehicle position reporting. The return link signal for this system employs a combination of direct-sequence CDMA and frequency hopping in order to ensure that users will not interfere with adjacent satellite systems.

The success of terrestrial mobile/portable cellular telephony has lead some industry experts to believe that a large market is waiting to be tapped by a similar satellite-based system that can provide service to areas that could not be economically served by conventional means. They expect personal communication and location services to grow to a multi-billion dollar industry within the decade. At the 1992 World Administrative Radio Conference (WARC-92), spectrum was allocated internationally for Mobile Satellite Service (MSS) in L-band 1610–1626.5 MHz (earth-to-space) and S-band 2483.5–2500 MHz (space-to-earth) on a primary basis. The band 1613.8–1626.5 MHz was also allocated for MSS downlinks on a secondary basis.

VII. Mobile Satellite Propagation

Traditionally, satellite systems have been used for point-to-point communications with directive antennas. This results in generally benign links with no significant fading (except for that due to rainfall attenuation). Until recently, the emphasis has been on power efficiency with bandwidth efficiency being of secondary importance. Initial shipboard mobile applications have also employed directional antennas. The omni-directional satellite mobile channel is radically different and presents some interesting design challenges.

L-band directional antennas are impractical at present for hand-held (and most mobile) units due to size and cost constraints. Omni-directional antennas do not reject scattered signals arriving from directions other than the direct path. It is these multipath signals which cause fading. Further, the user/satellite motion can result in the LOS path being obstructed by vegetation, terrain, or buildings which will attenuate the shadowed signal. Loo [4] has proposed that the channel be modeled for a rural environment as a log-normally-distributed direct path (due to foliage attenuation) and a Rayleigh-distributed multipath component.

Barts and Stutzman [5] suggest weighting the shadowed and unshadowed distributions by the appropriate fraction of shadowing and unshadowing to describe typical mixed path statistics, i.e.

$$G(R) = G_s(R)S + G_u(R)(1 - S)$$

where

$$G_s(R) = \int_R^\infty p_s(r)\, dr$$

is the cumulative fade distribution for the foliage-shadowed case and S is the fraction of shadowing. This model is generally agreed to be useful and provides a good fit to experimental data if the appropriate parameters are chosen. In fact, experimental propagation data are often described, for convenience, by giving the parameters for the above equations such that a best fit to the data is obtained [6], [7]. One of the advantages of the above modeling process is that it leads to convenient computer simulation algorithms [5], [8], [9].

Other researchers [10], [11] have chosen to model the shadowed signal as a Rayleigh distribution with a time-varying mean which is assumed to be log-normally distributed. Once again, this model is convenient for simulation purposes.

Goldhirsh and Vogel [12] have developed an Empirical Fade Distribution Equation (EFDE) for attenuation due to roadside trees in the mobile situation by curve fitting to various data [13]–[15]. The fade distribution curves generated using the EFDE are shown in Fig. 2 with elevation angle as a parameter. Similar data were presented by the US to a CCIR Study Group [13] for the heavily foliage-shadowed hand-held handset.

Terrestrial mobile systems in an urban environment often operate on multipath reflections that are stronger than the direct-path component. The satellite system is very different in this regard. Typically, the multipath delay spread is relatively small for appreciable received power. This is illustrated by data taken in downtown Chicago and the surrounding environment [16]. For example, it was found that only 1% of the time were multipaths with amplitude greater than -12 dB delayed by more than 100 ns relative to the direct path.

Table 3 Summary of Proposed System Parameters [17]

Company/System	Number of Satellites	Orbit Altitude (km)	Satellite Beams	Service Region
Constellation/Aries	48	1020	7	worldwide
Ellipsat /Ellipso	6, later 24	580 × 7800	8	worldwide
LQSS/Globalstar	48	1414	6	worldwide
Motorola/Iridium	66	780	48	worldwide
TRW/Odyssey	12	10 370	19	worldwide
AMSC	2	geostationary 62 W/139 W	4	N. America
Celsat/Celstar	2	geostationary 76 W/116 W	149	N. America

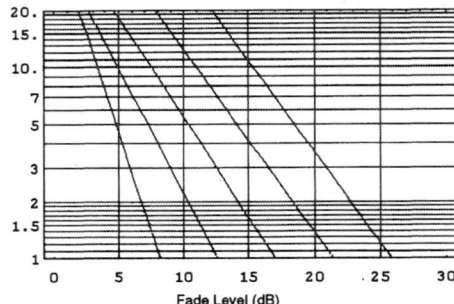

Fig. 2. Probability of exceeding a given fade level with elevation angle as a parameter based on the EFDE [12] for a heavily foliage-shadowed highway.

VIII. PERSONAL COMMUNICATION BY SATELLITE

Several companies have filed applications with the FCC to provide satellite-based personal communications networks with worldwide coverage. The applicants have proposed to provide a variety of services including voice, data, paging, facsimile, and position determination to mobile or hand-held subscriber users. The design of these systems is driven by the perceived marketing requirement to support a hand-held subscriber unit which is similar in appearance and performance to a cellular telephone.

The proposed systems include, among others, low earth orbit (LEO) systems such as IRIDIUM (Motorola) and Globalstar (Loral/Qualcomm), the medium earth (MEO) orbit Odyssey (TRW) system, and geostationary earth orbit (GEO) as proposed by AMSC and Celsat for coverage of North America.

A summary of proposed system constellations is given in Table 3 [17]. The required number of LEO and MEO satellites for worldwide coverage ranges from 66 for Motorola's IRIDIUM to 12 for TRW's Odyssey. These systems represent a major escalation in the size of commercial systems. Present proposals estimate systems with worldwide coverage to cost anywhere from one to over three billion dollars, depending on the complexity and sophistication of the system, a very sizable sum of private investment dollars.

The currently envisioned frequency plans, modulation and channelization schemes are summarized in Table 4 [17] for some representative systems. Direct sequence spread-spectrum signaling is being proposed by nearly all of the potential system developers, the notable exception being Motorola's IRIDIUM. Potential advantages include multipath mitigation, increased system capacity, and band sharing between systems with a minimum of coordination. However, the actual gain to be realized by using spread spectrum in the proposed systems is a matter of some controversy and the issues are being hotly debated by the interested parties.

The ability of spread-spectrum signaling to combat multipath fading depends on the amplitude and delay spread of the multipath returns, as well as the spread signal chipping rate. Goldhirsh and Vogel [14] note that the severe fading experienced by the mobile satellite user at lower elevation angles in some environments is primarily due to shadowing as opposed to multipath. They noted that due to shadowing the probability of exceeding 20- and 10-dB fade levels was 1% and 10%, respectively. For the multipath fading channel at an elevation angle of 45°, the comparable fade levels were 6 and 3 dB, respectively. Rubow [18] made similar observations for an elevation angle of 15°. SS signaling has the potential for reducing fading due to multipath, but can do nothing, of course, to mitigate the attenuation due to shadowing. This leads to the conclusion that only modest gains in combating fading can be achieved by spreading.

Multipath mitigation is achieved only when the multipath delay is equal to (or greater than) a significant portion of a chip period. The multipath delay data of [16] suggest that a chipping rate of at least 10 MHz will be required to be effective for reducing fading. None of the systems summarized in Table 4 currently propose a spread bandwidth greater than 5.5 MHz, implying a chipping rate of about half of that value (TRW is investigating the advantages of full band spreading). It should be noted that the multipath delay data of Fig. 4, taken in downtown Chicago, are limited in scope and more data of this type are required before a full assessment can be made. In summary, it appears that multipath mitigation through spectrum spreading is not

Table 4 Summary of MSS System Parameters

Company/System	Modulation	Multiple-Access Method (Forward Link)	Multiple-Access Method (Return Link)	Channelization (MHz)	Frequency Band (MHz)
Constellation	QPSK	Spread TDM	Channelized CDMA	16.5 forward 1 to 5 return	1610.1626.5 2483.5-2500
Ellipsat	OQPSK	Channelized CDMA	Channelized CDMA	1.1	1610.1626.5 2483.5-2500
LQSS	QPSK	Channelized CDMA	Channelized CDMA	1.25	1610.1626.5 2483.5-2500
Motorola	DE-QPSK	FDMA/TDMA	FDMA/TDMA	41.67 kHz	1616-1626.5 2483.5-2500
TRW	BPSK	Channelized CDMA	Channelized CDMA	5.5	1610.1626.5 2483.5-2500
AMSC	QPSK	CDMA (or FDMA/TDMA)	Channelized CDMA	5.5	1610.1626.5 2483.5-2500
Celsat	QPSK	Channelized CDMA	Channelized CDMA	1.25	1610.1626.5 2483.5-2500

of primary concern to any of the personal communication satellite systems presently proposed.

Another potential advantage of spectrum spreading is a possible increase in system capacity. Viterbi compared the spectral efficiency of CDMA with FDMA in [19] and found FDMA to be considerably superior to CDMA. He concluded that "When C/N_0 is at premium do not contribute further to the noise by having the users jam one another." Gilhousen et al. [9] claim that this result is reversed in the mobile satellite environment due to additional system considerations and the fact that, in general, the satellite system is more bandlimited than power-limited. Some of the factors that contribute to the CDMA capacity gain include: the voice activity factor, increased frequency reuse in conjunction with satellite multibeam antenna spatial discrimination, and the discrimination between multiple satellites with overlapping coverage. The actual system capacity is parameter-dependent and is very sensitive to received C/N_0 as well as the power control accuracy.

Some investigators, the authors among them, believe that the assumptions of [9] for voice activity factor, and gain due to polarization reuse are overly optimistic. Further, the importance of the link power limitation and percentage of users shadowed may have been underestimated. In reality, the capacity of the mobile satellite CDMA system is probably comparable to (or slightly less than) that of an FDMA or TDMA system. For example, satellite-user full duplex link capacities were estimated to be 2800 and 2300 for the CDMA Globalstar [20] and Odyssey [21] systems, respectively, as compared to 4070 links for the FDMA/TDMA IRIDIUM [22] system.

The "real" advantage of spread-spectrum signaling for mobile satellite systems may well be the ability to coexist with other users and systems with relatively little coordination. For example, CDMA users are relatively tolerant to unintentional interference by virtue of the inherent processing gain (assuming that the system is operating below full capacity). The use of "notches" against strong narrowband interferers can provide additional tolerance. Likewise, interference to other narrowband systems is minimized by the spectrum spreading. Further, CDMA systems can share the same band without coordination [17] (although individual system capacity is reduced). Once again, the effectiveness of system bandsharing is a matter of some controversy.

IX. INDOOR WIRELESS APPLICATIONS

A very large market is expected for indoor wireless communications. The economic advantages and convenience of untethered personal and computer communications are obvious and a sizable market already exists. However, the problems associated with the difficult propagation environment, interference, and the high user density in some buildings raise some formidable challenges for successful system implementation. Some of the channel considerations are discussed below and then an example spread-spectrum wireless telephone system is presented.

X. THE INDOOR CHANNEL

A wide range of parameters exist for the indoor channels in which a wireless system must operate. The average path loss exponential with distance can vary from somewhat less than the nominal LOS value of two to greater than five, depending on the environment and on the building. The SS system must be designed to operate with a wide range of received signal levels (RSL). One can experience radical changes in the RSL with a motion of a few inches—something that a portable handset might encounter even in a "static" environment with relatively small displacements such as changing ears.

An excellent UHF indoor radio channel modeling tool called SIRCIM has been developed at the Virginia Polytechnic Institute by Seidel and Rappaport [23]. This model generates time-varying CW fading signal levels, fading distributions, path-loss values, and wide-band impulse responses for a variety of indoor environments.

Signals in an indoor environment encounter fading due to both attenuation and multipath. Saleh and Valenzuela [24] report experiments in a medium-size office building that show a delay spread range of 200 ns with an rms value of no more than 50 ns. This would indicate that an SS bandwidth on the order of 10 MHz or greater is required for significant multipath mitigation. It is often desirable to use bandwidths lower than this for reasons of interference to/from other systems, as well as timing and power dissipation considerations. In this case, antenna diversity is critical to achieving the desired performance.

We conducted some informal measurements of RSL's on the second floor of our headquarters office building at a frequency of approximately 1.6 GHz. For the power level and the ranges we used, we found that about 90% of the locations yielded acceptable performance. However, if at the same location we used the better of horizontal or vertical polarization, the number of acceptable locations was increased to approximately 99%. (Better coverage could have been obtained by increasing the transmitter power.) Our results indicate that the two polarizations fade independently and that the use of polarization (or spatial) diversity is a very effective means of improving performance. Fortunately, a handset is large enough to support two antennas. One configuration uses a whip near the speaker and an Alford loop near the microphone providing a combination of spatial and polarization diversity in a compact form.

Little information is present in the open literature on the interference environment indoors for the ISM bands. However, some measurements have been taken outdoors in several cities at a variety of locations [25]. There can be substantial narrowband interference at quite high power levels. Based on measurements at 850 MHz one can expect the outside interference to be reduced by about 14 dB [26] for the lowest ISM band. There are interior sources of interference sources such as microwave ovens and X-ray machines, which will not be reduced by the exterior wall attenuation. The interference level may be sufficiently high that there is not sufficient processing gain to negate the interference even if the entire available bandwidth is utilized. It is desirable to occupy only a limited portion of the ISM band such that filters can be used to avoid interference to/from other systems. Use of spread spectrum in the ISM bands is only permitted by the FCC under the condition of noninterference with existing services.

XI. Wireless PBX System

In this section we describe a wireless PBX system based on DS spread spectrum and digital signal processing technology. A wireless PBX consists of a base station which is connected to the PBX and a collection of handsets or subscriber terminals. The wireless PBX system has a star topology which is very desirable in that it permits centralized control of power, time, and frequency as well as centralized net management. All outbound (from the PBX or hub) signals can be code-division-multiplexed on a common carrier while the inbound signals use code-division multiple access. Maintaining relative power, time, and frequency on the outbound channels is easy since these channels are collocated. However, achieving these objectives for the inbound traffic is more difficult owing to their remote location.

The 902- to 928-MHz ISM band was selected and a frequency channelized design was adopted to minimize interference effects. For reasons of voice quality and cost, 32-kb/s ADPCM encoding was selected as the voice digitization technique. Omni-directional (in azimuth) antennas were assumed necessary to provide the desired coverage for mobile users.

A high priority is to permit as many simultaneous channels as possible in the available bandwidth. The system must operate under the propagation situations described in detail above, i.e., serious fading due to multipath and a large dynamic range. Fading is certainly possible since the user may be walking (typically less than 1 m/s) or moving in his chair. A few inches of motion can cause a 30-dB fade. The wireless PBX system must solve these serious propagation problems, support the desired number of users, and comply with the FCC requirements for the ISM bands (§ 15.247).

XII. Multiple-Access Technique and Channel Crosstalk

For spectral efficiency, an orthogonal CDMA system based on the orthogonal Rademacher–Walsh (RW) functions was selected. The approach assigned a RW function to each channel but also mod-2 added a common PN sequence to each channel on the basis of one PN chip per one RW chip. This step is necessary to meet the FCC spectral density requirements and has the beneficial effect of randomizing such that all channels have similar spectra and encounter similar degradations.

The RW functions provide, in theory, zero cross correlation between the various functions. In practice, finite-bandwidth effects and time-base errors will create finite correlation between the signals. Simulation runs were performed with several filter types and bandwidths to assess the sensitivity to filtering effects. Based on these results, a fourth-order, 0.1-dB-ripple, Chebychev filter was selected for transmitter and receiver. With the ratio of 3-dB bandwidth-to-chipping rate of 1, 0.8, and 0.6, the rms cross correlation between "orthogonal" channels was -34.5, -32.2, and -31.4 dB, respectively.

Simulations were performed for different RW set sizes, e.g., 16, 32, 64, 128. We found that for a given filter the average cross-correlation value was essentially independent of the set size. However, the number of potential interferors is directly proportional to the set size. Thus for the case of significant filtering, it is wise to use relatively small RW sets, e.g., 16 or 32, so as to minimize the impact of filtering.

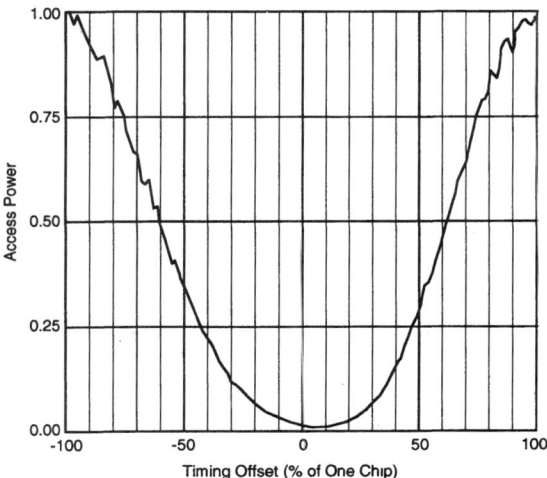

Fig. 3. Access noise power (normalized by processing gain) as a function of timing offset for the case of 16 simultaneous, equal power accesses.

The other major contributor to nonorthogonality is timing error. There are two types of timing error. First, there is a fixed timing error offset such as occurs on the inbound links to the station. The outbound links from the station are multiplexed and have no fixed time base error. Second, both inbound and outbound links encounter multipath spread. The delay spread sets a limit on the minimum crosstalk level which can be achieved no matter how accurate the network timing system is. We set up a simulation to evaluate the signal-to-interference ratio for the case of 300-ns delay spread with a uniform distribution and a mean offset of 2% of a chip duration. The simulation was performed for the following cases: 1) 16 chips/symbol at a 320-kchips/s rate and 15 other equal power users, 2) 64 chips/symbol at a 1.28-Mchips/s rate and 127 other users, and 3) 128 chips/symbol at a 2.56-Mchips/s rate and 63 other users. We found that the signal-to-interference ratios (in decibels) to be : 1) 23 dB, 2) 13 dB. and 3) 7 dB. respectively. From these results, which are based on reasonable assumptions, we conclude that the upper limit on the RW set size should be 32.

Figure 3 demonstrates the sensitivity to timing offset for the case of 16 chips/symbol when there are 15 other signals of equal power. For an offset of one chip there is no effective processing gain since the nominal value of 16 is reduced by a factor of 15 due to the other accesses. In order to provide adequate margin it is desirable to maintain a timing error better than one-tenth of a chip duration. The previous discussion on effects which contribute to channel crosstalk indicate that it is desirable to use the minimum chipping rate which will satisfy the FCC spectral density and bandwidth requirements.

XIII. Frame Structure

The previous section demonstrated that truly orthogonal CDMA cannot be achieved in a realistic environment. Further, the potential large variation in RSL's on the inbound links can make the system operation quite sensitive to channel crosstalk. Time-division duplexing (TDD) permits the same frequency to be used for both transmission and reception in both directions. The first half of the frame is used for outbound CDM transmissions and the second half carries the inbound CDMA transmissions. A TDD frame period of 10 ms was selected so that a subscriber handset can use the RSL from the base station as an accurate measure of the path loss and adjust the transmit power to achieve the desired signal level at the base station.

An advantage of the TDD format is that the antenna diversity system is simplified. The antenna diversity system uses a single omnidirectional at the base station and dual antenna diversity at the handset. Midway during the TDD frame the base station sends two identical pilot signals on the order wire (OW) channel. The handset sequentially receives the two signals—one-on-one antenna and the other on the other antenna. The handset measures the RSL and then selects the preferred antenna for one TDD frame.

XIV. Modulation Format

QPSK data modulation is employed since this doubles the number of frequency channels that could be obtained with BPSK data modulation. The QPSK signal is detected coherently using a block phase estimator [25] rather than the more familiar phase-lock loop techniques which perform poorly in fading environments [26].

For simplicity, BPSK chip modulation is used and spectral shaping achieved by post-modulation Chebychev filtering. Since there are 16 chips per data symbol the chipping rate is 640 kchips/s. With a normalized bandwidth of 0.78, the channel bandwidth is 1 MHz allowing 26 frequency channels in the lowest ISM band. This is more than enough to avoid interference problems and to support multiple adjacent cells with a frequency reuse plan. If a base station must support more than 15 simultaneous channels, then additional frequency channels must be employed.

XV. Network Control System

The base station serves as the point for centralized control of the handset frequency, transmit time base, and transmit power level. The base station monitors the power levels and the receive time bases of each of the RW channels and transmits time and power corrections on the outbound OW. The handset transmit frequencies are slaved to the base station carrier, there is no need for the base station to provide frequency corrections.

It is crucial for the base station to monitor the RSL of each access and to provide corrective information. While it is true that the handsets perform this function on an instantaneous feedforward basis, they rely on a constant receiver gain over a large-signal dynamic range. Temperature and aging effects will cause a slow variation in the gain and it is necessary for the base station to provide corrections if the RSL's are to be maintained to the desired accuracy of 1 or 2 dB.

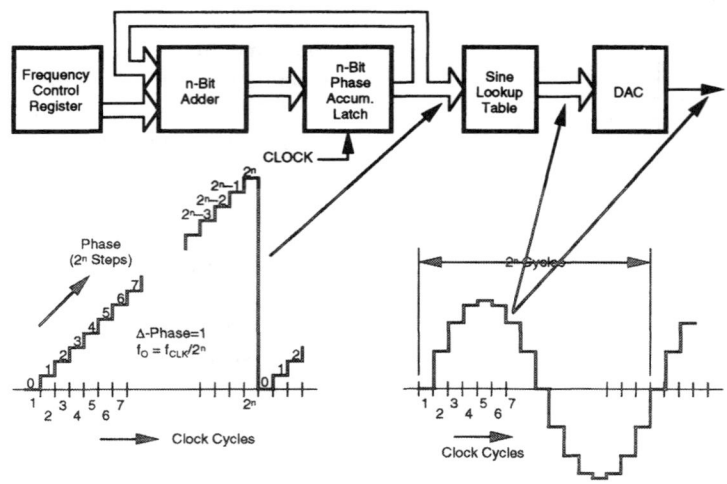

Fig. 4. Principle of Direct Digital Synthesis.

XVI. Technology for Spread-Spectrum Applications

A. Generating FH and DS Signals Using Direct Digital Synthesis

Direct digital synthesis (DDS) is an ideal technology for the generation of signals for both DS- and FH-SS systems due to the ease of modulating both the phase (PSK) and frequency (FSK and FH) of a DDS. A functional block diagram of a DDS is shown in Fig. 4.

A phase accumulator is used to address a sine look-up table (LUT), generating a digitized sine-wave output. This signal is usually fed into a digital-to-analog converter (DAC) to produce an analog output. The analog signal is then low-pass-filtered, resulting in a smooth continuous signal. The digital part of the DDS, i.e., the phase accumulator and LUT, is usually called a numerically controlled oscillator (NCO). The frequency of the output signal is determined by the number stored in the frequency control register, or phase step register. For example, in the n-bit system shown, if the phase step equals one the accumulator will count by ones, taking 2^n clock cycles to address the entire LUT and to generate one cycle of the output sine wave. This is the lowest frequency that the system can generate, and it is also the frequency resolution. Setting the phase step register equal to two results in the accumulator counting by twos, taking 2^n clock cycles to complete one cycle of the output sine wave. It can easily be shown that for any integer m, where $m < 2^{n-1}$, the number of clock cycles taken to generate one cycle of the output sine wave will be $2^n/m$, so that the output frequency will be

$$f_o = \frac{m \times f_{\text{clk}}}{2^n}.$$

A typical value for n is 32, although NCO's with 48-b resolution are also available, so that a DDS can have extremely high frequency resolution. For example, with $n = 48$ and $f_{\text{clk}} = 50$ MHz, the frequency resolution is 0.2 μHz! Note that when the value of the phase step register is changed, the output frequency changes instantaneously, i.e., from one clock cycle to the next, with no phase discontinuity in the output signal. In practice, all NCO's have a pipelined architecture resulting in a throughput delay such that the frequency change command does not occur immediately after the command. Frequency modulation for both hopping and FSK can easily be implemented by changing the phase step register value, and many NCO's incorporate features to simplify doing this, such as dual phase registers. These can be used to generate an FSK signal by loading the mark frequency into one register and the space frequency into the other; the FSK signal is then generated by selecting the appropriate register.

Phase modulation (PM) for PSK modulation is implemented by introducing an adder between the phase accumulator and the LUT. This modifies the absolute value of the phase without affecting the slope (which determines the frequency of the output signal). Modifying the MSB of the phase value changes the phase by 180°, modifying the next bit changes the phase by 90°, and so on, allowing simple BPSK or higher level PSK signals to be generated with the same ease.

Alternatively, by using a large number of bits for the PM function, linear PM can be implemented; 12 b is a typical value used. This capability can be used in conjunction with signal shaping for spectrum control. By filtering the data signal before modulation (using a raised cosine digital interpolating FIR filter, for example) the sidebands of the modulated carrier can be reduced to low levels without filtering at the output. This can be very advantageous in FH systems, where a narrowband filter cannot be used because of the need to hop the signal over a wide band. A similar technique can also be used when frequency modulating an NCO, since the modulation update rate capability of these devices is typically much greater than the data rate demands.

XVII. Using a DDS for Modulation and Carrier Generation in FH-SS.

As described in an earlier section, noncoherent FSK is usually used in FH-SS systems, and a DDS is ideal

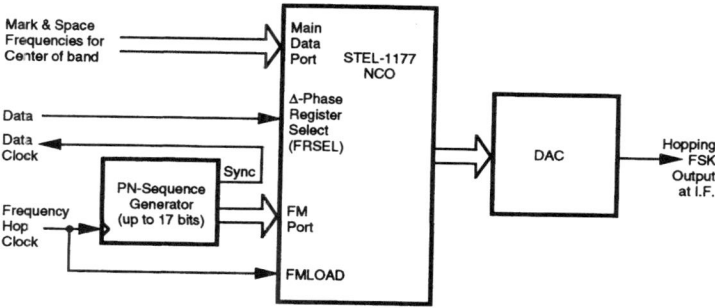

Fig. 5. Using an NCO to generate a frequency hopping FSK signal.

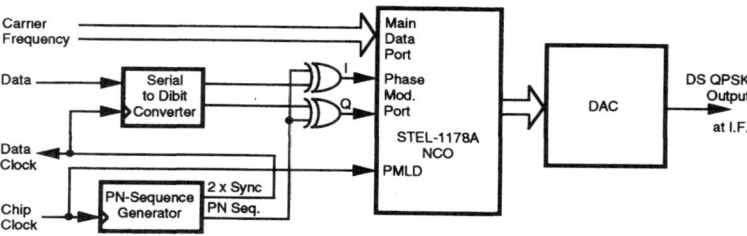

Fig. 6. Using a DDS to generate a QPSK direct sequence spread spectrum signal.

for implementing both the frequency hopping and the FSK functions. For example, at least one commercially available NCO incorporates dual-phase registers for easy FSK modulation as well as a separate FH port, as shown in Fig. 5.

With this NCO there are 2^{17} potential hop frequencies which can be changed at up to 15 MHz (the maximum hopping bandwidth is 7.5 MHz). Alternatively, the FH port, which has 16-b linearity, can be used in conjunction with an interpolating filter to shape the data themselves for spectrum control prior to modulation again taking advantage of the 15-MHz update rate capability. This device can also be phase-modulated with up to 12-b linearity, allowing it to be used for PSK FH-SS if required.

XVIII. Using a DDS for Modulation and Carrier Generation in DS-SS.

As described in an earlier section, PSK is commonly used in DS-SS systems. This is easily implemented with a DDS, as shown in Fig. 6 for BPSK chipping with either BPSK or QPSK data modulation.

XIX. Despreading and Demodulating Spread-Spectrum Signals.

Before a SS signal can be processed by a data demodulator it has to be despread, i.e., the spreading information has be removed. The way in which this is done depends on the spreading method, of course. Let us examine the methods for FH and DS SS in turn.

XX. Despreading FH SS Signals

The basic method of despreading an FH-SS signal involves generating a local-oscillator (LO) signal which hops in synchronism with the received signal carrier, resulting

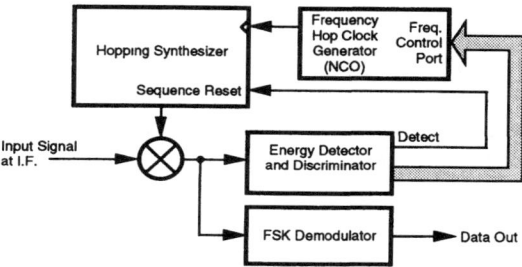

Fig. 7. Despreader for FH-SS.

in the signal being continuously converted to baseband as the frequency of the signal hops. Generating the hopping LO is readily accomplished using a DDS, in the same way as for generating the transmit signals described above. The difficult part, as with most SS signals, is synchronizing the hopping of the LO to the hopping IF signal.

There are two parts to the synchronization problem—initial acquisition and tracking. Details of the circuit implementation will depend on the FH and modulation formats but these functions are usually performed by the general approaches described below. Initial acquisition is generally achieved by a serial search procedure involving a single despreader as described above. However, rather than demodulating the data, the received energy over multiple symbols is compared with a threshold level. If the hopping patterns are aligned within ±1/2 hop duration, the detected power will exceed the threshold and synchronization acquisition will be declared and the FH receiver will enter the tracking mode.

In the tracking mode it is necessary for the FH receiver to develop a time discriminator characteristic. This entails generating early and late (e.g., by one-half hop duration) hopping patterns as well as the punctual hopping pattern.

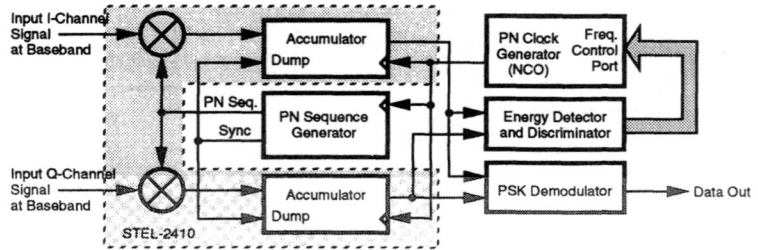

Fig. 8. Serial correlator despreader for direct sequence SS.

Time base tracking is obtained by taking the difference in detected energy levels in the early and late channels. When the received signal and receiver time bases are aligned the difference will be zero and NCO driving the PN generator(s) will maintain the correct frequency.

Before the advent of NCO's and DSP, the complexity of FH despreaders was sufficiently great that a single FH despreader was time-shared between the early and late positions. Since the FH despreader was never in the punctual position it was necessary to use quite small early/late time offsets to minimize the data detector loss. This meant that the time discriminator was quite narrow.

At present, through the use of DSP and ASIC technology, it is possible to implement parallel early, punctual, and late despreaders in a compact package. Thus it is possible to obtain optimal detection performance while maintaining a robust tracking loop. Figure 7 is a block diagram of a typical FH demodulator. Noncoherent or differentially coherent modulation techniques are generally used with FH.

XXI. Despreading DS-SS Signals

As with FH-SS signals, despreading DS-SS signals involves generating a replica of the spreading sequence at the receiver in synchronism with the sequence modulated on the incoming signal. The major difference is that with a digital implementation of DS-SS, the signals are generally despread at baseband, whereas FH signals are despread in the downconversion to baseband. Assuming that BPSK chipping is used, the spreader multiplies the transmitted signal by a given sequence of $\pm$ones. When the despreader is correctly synchronized it will multiply the signal by the same sequence of $\pm$ones, and since both $+1$ and -1 squared equal $+1$, the result is that the spreading sequence is removed from the signal. If the despreading sequence is not synchronized, or uses a different sequence, the signal will effectively be multiplied by a random sequence of $\pm$oness so that the signal will not be despread.

There are two primary methods of implementing a despreader for DS-SS signals. The first uses a serial correlator–accumulator technique, and is shown in Fig. 8.

The complex correlator–accumulator is shown in the shaded region. The incoming complex baseband signal is multiplied by the despreading sequence and the samples are integrated over a symbol period. This generates the cross-correlation factor of the two signals over that period, so that when they are in phase the result will be a large sum, and

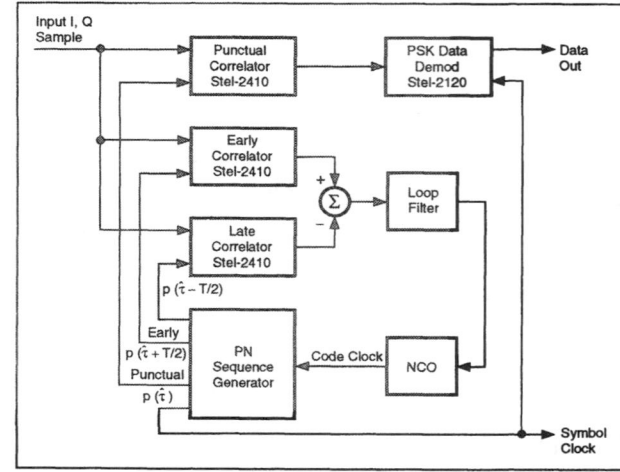

Fig. 9. Correlator despreader for direct-sequence SS.

when they are out of phase the sum will be much smaller, assuming that spreading sequences will good correlation properties have been selected. This technique has a lot in common with the method of despreading fast-hopping FH-SS signals described previously, since the major problem is synchronizing the locally generated reference sequence. As with the FH-SS system, the clock for the local sequence generator is first set to the nominal chipping rate of the incoming signal and the chipping sequences are made to precess with respect to each other until correlation occurs. When this occurs, the sum in the accumulator at the end of each symbol period will increase significantly. At this time the phase of the chipping sequence is adjusted to maximize the signal energy by adjusting the frequency of the clock for the local sequence generator under the control of the energy detector.

One method of controlling the phase of the chipping sequence is the "delay-lock loop" technique [29]. Figure 9 illustrates how three commercially available correlator ASICs can be used to create a delay-lock discriminator and a correlation despreader.

One correlator is used as the "on-time" correlator, and the other two are used as "early" and "late" correlators, with the despreading sequences being fed into them half a chip early and late, respectively. The difference between the outputs of these two correlators forms a timing discriminator function going to zero when the timing is optimal. This method is used very frequently because of its excellent performance.

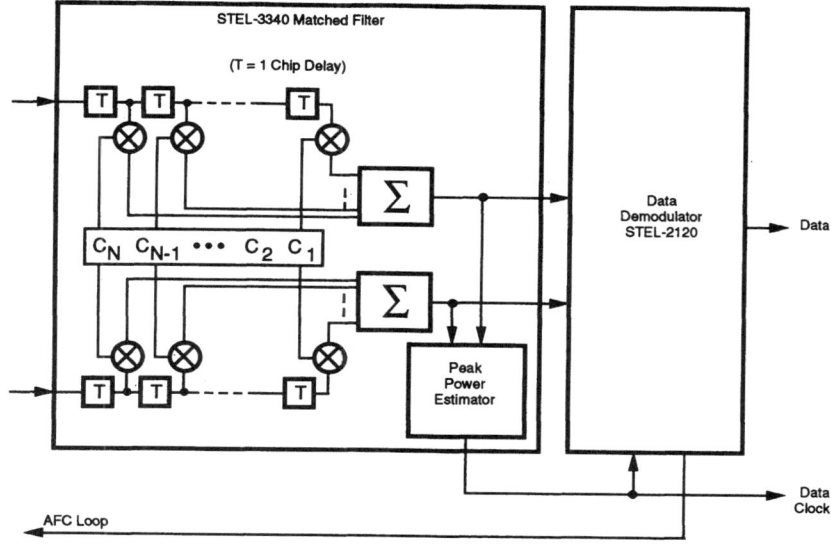

Fig. 10. Matched-filter despreader for direct-sequence SS.

The second method of despreading a DS-SS signal, which is most suitable for short-code systems, is with a matched filter. This is a parallel approach to computing the cross-correlation function between the reference and received sequences, with the consequence that it offers faster acquisition, at the expense of power consumption. The matched-filter based despreader is shown in Fig. 10.

In this system, the entire reference sequence is stored in a parallel register, instead of being generated serially. The incoming signal samples are also stored in another register of equal length, so that the $n-1$ previous samples of the signal are available, as well as the current sample, where n is the spreading code sequence length. The cross correlation between these two sequences (sum of products) is computed once per chip (as opposed to once per symbol in the correlator despreader). As the incoming signal moves down the signal register, chip by chip, its sequence will match the reference once per symbol, giving high cross correlation. At this time the symbol timing is automatically derived from the energy output of the correlation. Note that, depending on circumstances, no sequence acquisition process may be required; the signal is acquired during the first complete symbol received, i.e., when the threshold is crossed.

Although the code-matched filter technique is conceptually simpler than the correlator technique, its simplicity belies the complexity of its implementation, with two multipliers (typically 1 by 3 b) per chip of the code length. Thus this technique is limited in practice to relatively short spreading sequences, typically 64 chips or less, whereas the correlator despreader does not increase significantly in complexity as the length of the code increases, making it much more suitable for long-code system.

XXII. CONCLUSION

Spread-spectrum modulation offers several advantages such as precise timing, tolerance to interference, multipath amelioration, and the ability to share a common bandwidth with other signals. Until recently, the complexity of many spread-spectrum modems and navigation equipment had restricted their application to military use. However, due to the recent advances in the state of the art in digital signal processing and application-specific integrated circuits it is feasible now to use spread spectrum in low-cost commercial applications such as GPS receivers, satellite-based and terrestrial cellular telephone, and wireless PBX systems.

REFERENCES

[1] Cook and Marsh, "An introduction to spread spectrum," *IEEE Commun. Mag.*, pp. 8–16, Mar. 1983.
[2] Pickholtz, Schilling, and Milstein, "Theory of spread-spectrum communications—A tutorial," *IEEE Trans. Commun.*, pp. 855–883, May 1982.
[3] Simon, *et al.*, *Spread Spectrum Communications*. Rockville, MD: Comput. Sci. Press, 1985.
[4] C. Loo, "A statistical model for a land mobile satellite link," *IEEE Trans. Vehic. Technol.*, pp. 122–127, Aug. 1985.
[5] Barts and Stutzman, "Modeling and simulation of mobile satellite propagation," *IEEE Trans. Antennas Propagat.*, pp. 375–382, Apr. 1992.
[6] C. Loo, "Measurements and models of a land mobile satellite channel and their applications to MSK signals," *IEEE Trans. Vehic. Technol.*, pp. 114–121, Aug. 1987.
[7] J. S. Butterworth, "Propagation measurements for land mobile satellite system at 1542 MHz," Commun. Res. Cent., Dept. Commun., CRC Tech. Note 724, Aug. 1984.
[8] Irvine and McLane, "Symbol-aided plus decision-directed reception for PSK/TCM modulation on shadowed mobile satellite fading channels," *IEEE J. Selected Areas Commun.*, pp. 1289–1299, Oct. 1992.
[9] Gilhousen *et al.*, "Increased capacity using CDMA for mobile satellite communication," *IEEE J. Selected Areas Commun.*, pp. 503–514, May 1990.
[10] Hansen and Meno, "Mobile fading—Rayleigh and lognormal superimposed," *IEEE Trans. Vehic. Technol.*, pp. 332–335, Nov. 1977.
[11] Lutz *et al.*, "The land mobile satellite communication channel—Recording, statistics, and channel model," *IEEE Trans. Vehic. Technol.*, pp. 375–386, May 1991.
[12] Goldhirsh and Vogel, "An overview of results derived from mobile-satellite propagation experiments," in *Proc. Int. Mobile Satellite Conf.*, (Ottawa, Ont., Canada, 1990), pp. 219–224.

[13] ——, "Roadside tree attenuation measurements at UHF for land-mobile satellite systems," *IEEE Trans. Antennas Propagat.*, vol. AP-35, pp. 589–596, 1987.
[14] ——, "Mobile satellite system fade statistics for shadowing and multipath from roadside trees at UHF and *L*-band," *IEEE Trans. Antennas Propagat.*, pp. 489–498, Apr. 1989.
[15] ——, "Mobile satellite system propagation measurements at *L*-band using MARECS-B2," *IEEE Trans. Antennas Propagat.*, pp. 259–264, Feb. 1990.
[16] USA delegates at the ICU, "Impact of propagation impairments on the design of LEO mobile satellite systems providing personal communication services," CCIR Study Groups, US WP-8D-14, Oct. 1992.
[17] "Final Report of the Majority of the Active Participants of Informal Working Group 1 to Above 1 GHz Negotiated Rulemaking Committee," MSSAC-41.6 (Final), IWG1-81 (Final), FCC, Apr. 1993.
[18] W. Rubow, "MOBILESTAR field test program," in *Proc. Mobile Satellite Conf.* (Pasadena, CA, May 1988), pp. 189–194.
[19] A. J. Viterbi, "When not to spread spectrum—A sequel," *IEEE Commun. Mag.*, vol. 23, pp. 12–17, Apr. 1985.
[20] "GLOBALSTAR System Application," presented by Loral Cellular Systems Corp. before the Federal Communications Commission, Washington, DC, June 1991.
[21] "Application of TRW Inc. For Authority to Construct a New Communications Satellite System Odyssey," presented by TRW Inc., before the Federal Communications Commission, Washington, DC, May 1991.
[22] "Application of Motorola Satellite Communications, Inc. for IRIDIUM," presented by Motorola Inc. before the Federal Communications Commission, Washington, DC, Dec. 1990.
[23] T. S. Rappaport, S. Y. Seidel, and K. Takamizawa, "Statistical channel impulse response models for factory and open plan building radio communication system design," *IEEE Trans. Commun.*, vol. 39, pp. 794–807, May 1991.
[24] A. A. M. Saleh and R. A. Valenzuela, "A statistical model for indoor multipath propagation," *IEEE J. Selected Areas Commun.*, vol. SAC-5, no. 2, pp. 128–137, Feb. 1987.
[25] Wepman *et al.*, "Spectrum usage measurements in potential PCS frequency bands," NTIA Rep. 91-279, Sept. 1991.
[26] E. H. Walker, "Penetration of radio signals into buildings in the cellular radio environment," *Bell Syst. Tech. J.*, vol. 62, no. 9, pp. 2719–2734, Nov. 1983.
[27] D. Richer, "A block estimator for offset QPSK signaling," in *Nat. Telecommunications Conf. Rec.*, vol. 2, pp. 30-6–30-11, 1975.
[28] F. M. Gardner, "Hang-up in phase-lock loops," *IEEE Trans. Commun.*, vol. COM-25, pp. 1210–1214, Oct. 1977.
[29] J. J. Jr. Spilker and D. T. Magill "The delay-lock discriminator–An optimum tracking device," *Proc. IRE*, pp. 1403–1416, Sept. 1961.

Direct-Conversion Radio Transceivers for Digital Communications

Asad A. Abidi, *Senior Member, IEEE*

Abstract—Direct-conversion is an alternative wireless receiver architecture to the well-established superheterodyne, particularly for highly integrated, low-power terminals. Its fundamental advantage is that the received signal is amplified and filtered at baseband rather than at some high intermediate frequency. This means lower current drain in the amplifiers and active filters and a simpler task of image-rejection. There is considerable interest to use it in digital cellular telephones and miniature radio messaging systems. This paper briefly covers case studies in the use of direct-conversion receivers and transmitters and summarizes some of the key problems in their implementations. Solutions to these problems arise not only from more appropriate circuit design but also from exploiting system characteristics, such as the modulation format in the system. Baseband digital signal processing must be coupled to the analog front-end to make direct-conversion transceivers a practical reality.

I. INTRODUCTION

THE CURRENT interest in portable wireless communications devices is prompting research into new IC technologies, circuit configurations and transceiver architectures. Low-power miniature radio transceivers are sought to communicate digital data in cellular telephones, wireless networks, and radio messaging systems. While transistor *technology scaling* and *improved circuit techniques* will contribute evolutionary advances towards this goal, *architectural innovations* in the transceiver may lead to revolutionary improvements [1]. It is in this context that there is a resurgence of interest in direct-conversion.

The superheterodyne receiver, which Armstrong introduced in 1918 [2], is generally thought to be the receiver of choice owing to its high selectivity and sensitivity. Something like 98% of radio receivers use this architecture. In a superheterodyne receiver, the input signal is first amplified at RF in a tuned stage, then converted by an offset-frequency local oscillator to a lower intermediate frequency (IF), and substantially amplified in a tuned IF "strip" containing highly-selective passive bandpass filters. The role of the various filters is illustrated by the typical frequency plan of a superheterodyne receiver (Fig. 1). The IF must be sufficiently high so that the *image channel* lies in the stopband of the RF preselection filter or the antenna, otherwise the IF filter will pass this channel unattenuated in its own image passband. These considerations determine the familiar intermediate frequencies used in radio and TV receivers.

Manuscript received May 22, 1995; revised August 29, 1995.
The author is with the Electrical Engineering Department, University of California, Los Angeles, CA 90095 USA.
IEEE Log Number 9415818.

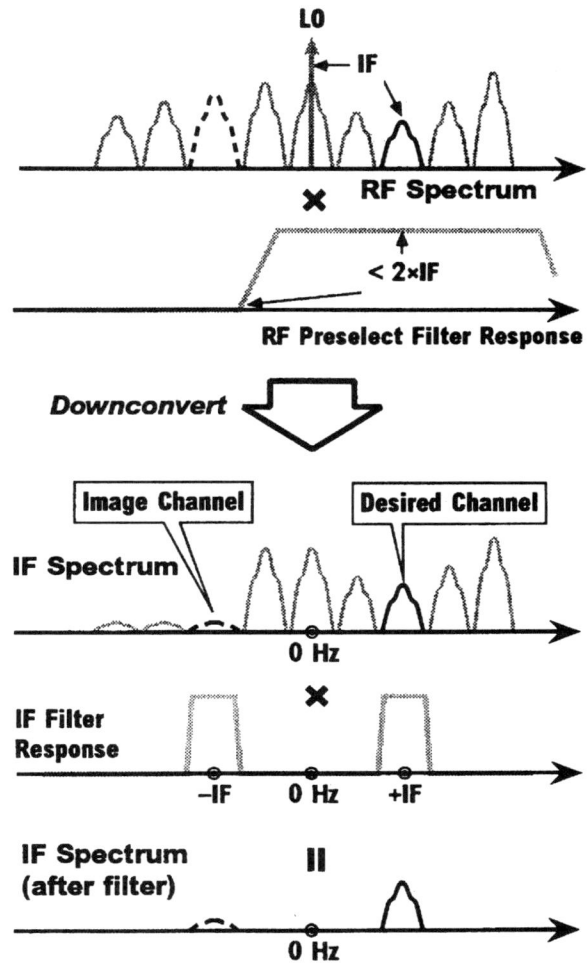

Fig. 1. Frequency plan of a superheterodyne receiver. Choice of IF is governed by width of preselect filter passband. The RF preselect and IF filters work together to select the desired channel.

In a standard broadcast FM receiver, for instance, the 10.7 MHz IF guarantees that the image channel lies outside the 20 MHz wide FM band. Therefore, even if the preselect filter inadequately suppresses the image, which is assumed not to be an FM signal, the subsequent frequency-discriminating detector will inherently tend to reject it. This relaxes the selectivity of the 100 MHz preselect filter, which may be constructed with either a single or ganged collection of *LC*-tuned circuits. Ceramic filters are an enabling technology for the 10.7 MHz FM IF. These small and cheap filters offer a narrow passband and good stopband attenuation, and they are widely used. The traditional 43.5 MHz IF in a TV receiver cannot suppress the image across the entire VHF, hyperband, and UHF bands, so as the receiver is tuned across various sub-bands, one of an array of RF narrowband filters is switched into the RF front-end [3] to suppress the image channel. Many analog cellular telephones use a 90 MHz IF. Amplification and filtering at these high intermediate frequencies between 10–100 MHz comes at the price of *power dissipation* because transistors must be biased at large currents to drive the parasitics and the low characteristic impedance of the passive IF filters. Further, the IF strip may require a large number of *off-chip passive* components, which add to receiver size. Although these are not serious problems for tabletop receivers—the easy alignment of a superheterodyne, resulting in a high selectivity, was always one of its strengths—they may become limitations in miniature, low power transceivers.

Wireless receivers must often handle very weak channels existing side-by-side with very strong channels in the same band. Thus, in addition to a minimum *stopband attenuation* to suppress the image channel, the filter must also have a wide *dynamic range*, that is, the ability to handle strong signals without distortion while remaining sensitive to weak signals above the intrinsic noise level in the passband. In this respect, passive filters are almost always superior, as the small-signal handling of active bandpass filters is limited by a fundamentally higher noise level [4], and nonlinearity in the active device tends to distort large signals. The dynamic range of active filters may only be increased at the expense of capacitor size and power dissipation.

Although most often a passive RF preselect filter attenuates the image channel, it may also be suppressed by selective *signal cancellation*. Here, the entire RF spectrum is downconverted to an IF in two identical mixers driven by quadrature phases of a local oscillator (LO). The downconverted spectra in the two branches are subjected to a 90° phase-shift relative to one another and then added (Fig. 2(a)). With appropriate design, this arrangement downconverts the desired channel to IF with the same phase in the two branches and the image channel to -IF but antiphase in the two branches. After addition, the desired channel appears at the output with double strength, while the image channel subtracts and disappears. This *image-reject downconverter* is the dual of Weaver's celebrated phasing method of sideband selection [5], which is discussed later in the section on transmitters. In practice, departures from quadrature in the two LO signals and gain mismatch in the two branches will limit the extent of signal cancellation (Fig. 2(b)). When used in a receiver, the effectiveness of this image-suppression method is further limited by the wide dynamic range of radio signals. If the unwanted image channel is much stronger than the desired one, then after imperfect signal cancellation, it may only be suppressed to a comparable level to the desired channel, resulting in an intolerably large interference. Nevertheless, this type of downconverter is used, for instance, in a single-chip broadcast FM receiver with a low IF of 150 kHz [6].

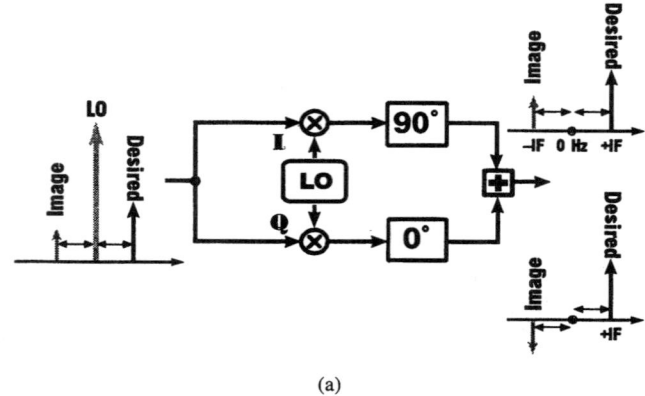

(a)

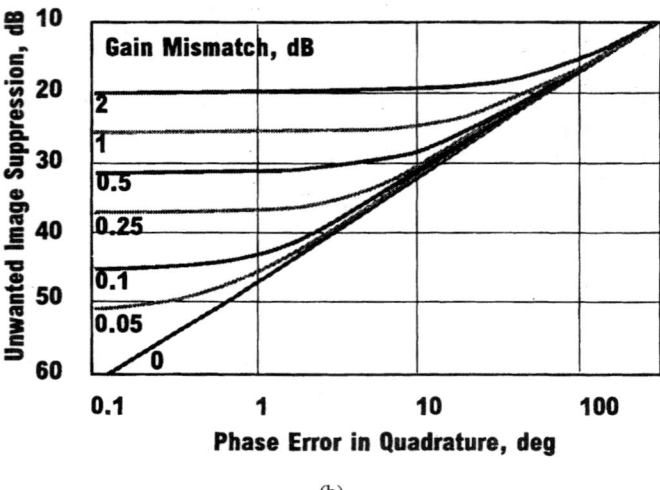

(b)

Fig. 2. (a) The image-canceling downconversion mixer. The desired signal appears in-phase in the two branches and the undesired signal anti-phase. Allpass filters may be used to synthesize 90° phase shift. (b) Unwanted signal suppression as a function of errors in phase from ideal quadrature, with gain mismatch in the two branches as a parameter.

II. THE DIRECT-CONVERSION ARCHITECTURE

Suppose that the IF in a superheterodyne is reduced to zero. The LO will then translate the center of the desired channel to 0 Hz, and the portion of the channel translated to the negative frequency half-axis becomes the image to the other half of the *same* channel translated to the positive frequency half-axis (Fig. 3). The downconverted signal must be reconstituted by a phasing method of the type described above, otherwise the negative-frequency half-channel will fold over and superpose on to the positive-frequency half-channel. Zero-IF, therefore, mandates quadrature downconversion into two arms and a vector-detection scheme. However, this scheme does not suffer from the strong-image problem when the image-reject downconverter is used in a nonzero IF heterodyne receiver, and the typical gain mismatches and phase errors in the two branches cause only a small loss in detected SNR. A lowpass filter, which is in effect a bandpass centered at dc when the negative frequency axis is included, may be used to select the desired channel and to reject all adjacent channels. Therefore, RF preselection may in principle be

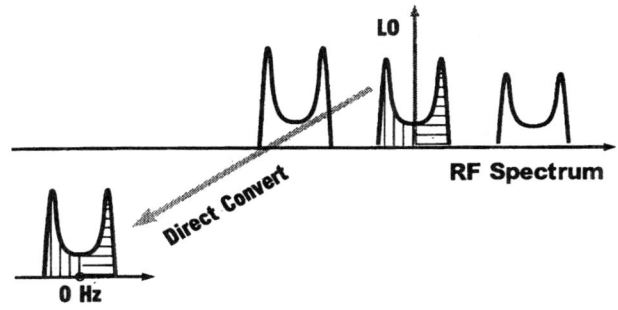

Fig. 3. Spectrum before and after direct-conversion.

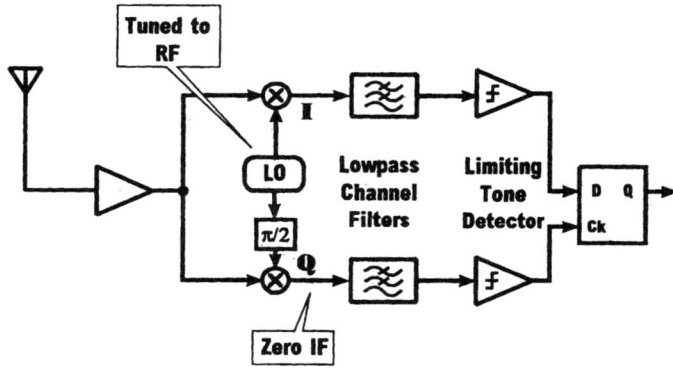

Fig. 4. Block diagram of a direct-conversion FSK receiver as may be used for radiopaging signals.

eliminated because there is no image channel. In practice, it is still required to suppress strong out-of-band signals that may create large intermodulation distortion in the front-end prior to baseband channel selection and to avoid harmonic downconversion. There is also the advantage that if a high-order active filter is used for channel selection, it will dissipate lower power and occupy a smaller chip area at a given dynamic range than an active bandpass filter with the same selectivity centered at a high IF [4]. All amplification past the front-end is also at baseband, and therefore consumes a small power. This zero-IF scheme is also called *direct-conversion*. When the local oscillator is synchronized in phase with the incoming carrier frequency, the receiver is called a *homodyne*.

As early as 1924, radio pioneers had considered use of homodyne architectures for single vacuum-tube receivers, but it was a homodyne measuring instrument for carrier-based telephony built in 1947 that first employed a high-order lowpass filter for channel-selection [7]. Thereafter, the concept lay dormant, until it was revived in 1980 in the radio-paging receiver, the first miniature digital wireless device for personal communication to attain widespread consumer use.

III. DIRECT-CONVERSION FSK RECEIVERS

Digital data in broadcast paging modulates the carrier by frequency-shift keying (FSK). The carrier frequencies may lie in the 400 MHz or the 900 MHz bands and binary data at 512 b/s or 1.2 kb/s rates shifts the carrier frequency by ± 4.5 kHz. This large modulation index results in a spectrum with two lobes symmetrically offset around the carrier (Fig. 3). Vance at ITT Standard Telecommunications Labs was the first to apply direct-conversion to this signal spectrum with a *single-chip* paging receiver [8], thus establishing a key concept for small, light paging receivers. Not all pagers today, though, use direct-conversion; some continue to use the superheterodyne implementation for higher performance [9].

Following a single-stage of RF amplification in this simple FSK receiver (Fig. 4), a local oscillator (LO) tuned to the incoming carrier downconverts the center of the desired paging channel to dc. In fact, quadrature phases of the LO downconvert the signal into two branches, labeled I and Q, enabling the detector to discriminate the signal at positive and negative frequencies (i.e., data 1's and 0's). A high-order *lowpass filter* in each branch with a cut-off at about 10 kHz selects the desired channel, while all other channels fall into the filter stopband. This may be integrated as a low-power active filter. A single-chip paging receiver from Philips uses a tenth-order continuous-time lowpass filter [10]. The data is encoded in the zero-crossings of the downconverted and filtered signal, which a limiter then amplifies to logic levels, eliminating the need for AGC. However, dc offsets directly add to the downconverted signal and may be so large as to disable zero-crossings in the limiter output, causing the receiver to fail to detect data. This problem is overcome by capacitively coupling the baseband signal path into the limiter to null these offsets. Some of the consequences of capacitive coupling are discussed in Section VI.

The detector is, in principle, only a flip-flop, driven at the D input by the I branch limiter and at the CK input by the Q branch limiter. The flip-flop output attains one steady-state if transitions at the CK input lead the transitions at the D-input and the other state if they lag. This corresponds exactly to a positive or negative frequency shift of the carrier, and thus, the data. Although this simple detector is found in many FSK receivers, it is susceptible to upset from a single noise impulse at the input. A more sophisticated detector oversamples the limiter output at a multiple of the data rate, thereby more finely quantizing the zero-crossing instant, and correlates this with quadrature phases of the expected frequency shift (4.5 kHz in the paging channel). The correlated output from the I and Q channels is integrated over a bit period, and the bit decision is made depending on which of the integrators first crosses a preset threshold. Correlation reduces the noise bandwidth. A one-bit implementation of this correlation detector in a spread-spectrum FSK receiver shows that it is very compact and dissipates a small power [11].

There are now many low-power, single-chip bipolar IC's implementing direct-conversion paging receivers [10], some operating at a supply as low as 1.1 V [12]. The on-chip capacitors required by the two active lowpass filters and ac coupling after downconversion occupy a large portion of the total die area. Aside from a reference crystal, the circuits only need some miniature off-chip inductors for the tuned RF amplifier loads and sometimes for the quadrature phase-shift network. The complete pager, including the microprocessor

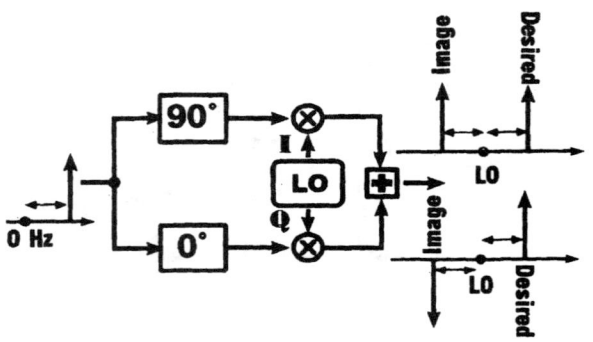

Fig. 5. A direct-upconversion mixer using the phasing method for selecting one sideband. Either one of the sidebands may be selected by combining the branches with the appropriate sign.

and display driver, may be a two-chip device with battery life in excess of six months.

One may appreciate the simplicity of a direct-conversion state-of-the-art paging receiver by comparing its inventory of parts with a superheterodyne implementation of the same built in a comparable technology [13]. The superheterodyne requires one more crystal, two trim capacitors, and a SAW filter, which together add a significant fraction to the total parts count, thus increasing the physical volume of the receiver and its power dissipation.

IV. DIRECT-CONVERSION SINGLE-SIDEBAND SYNTHESIZERS

For reasons of spectral efficiency, the transmitted signal in digital communications is usually single-sideband with suppressed-carrier. It would require an RF filter with a very sharp transition band to suppress one sideband on the modulated carrier while passing the other. The much more practical solution is the phasing method [5], where the modulated signal is first synthesized in quadrature at baseband, *directly-upconverted* into two branches by a quadrature LO centered at the carrier frequency, and added or subtracted to select either the upper or lower sideband (Fig. 5). The phasing method of sideband selection has been used for many years in single-sideband communication transceivers.

The unwanted sideband is suppressed to an extent limited by the gain mismatch in the two upconversion branches and by departures from quadrature in the two LO outputs (Fig. 2(b)). The dc offsets in the branches produce an output tone at the LO frequency. The unwanted sideband and LO leakage are unavoidable spurious emissions in the transmitted spectrum. Although the two upconversion branches will match well on the same IC, a gain mismatch as small as 1% (0.1 dB) limits unwanted sideband suppression to about 45 dB. With this gain mismatch, a phase-error of up to 1° is tolerable between the two LO outputs before the unwanted sideband grows further in relative amplitude. These mismatches may be trimmed at time of transceiver manufacture or self-calibrated with loopback modes that are activated during idle times to sense and suppress the unwanted sideband. Some trimming and adaptive methods are discussed in Section VII.

As the LO frequency in a direct-upconverter is centered in the transmit band, energy at this frequency may be spuriously radiated through parasitic unbalanced coupling into the power amplifier or antenna. For instance, the single-ended signal produced by an on-chip oscillator circuit tuned with an off-chip resonator may couple to the power amplifier input across pins of the RF package. Frequency-offset multi-step upconversion schemes, which are the dual of a heterodyne downconverter, have been proposed [14] to combat this coupling problem. However, as LO phase-noise in a transmitter appears as noise added to the emitted signal, a process called *reciprocal mixing* in the radio literature, direct upconversion has the advantage over a frequency-offset scheme that only *one* LO contributes noise. Other spurious output tones in a single-sideband transmitter may arise from parasitic remixing of the modulated output with the baseband signal and by intermodulation distortion in the output stage [15]. Balanced circuit topologies, on-chip LO's that require no external resonators [1], [16], and the lowered transmit power levels required in microcells, are all expected to lessen the magnitude of these problems.

V. DIRECT-CONVERSION RECEIVERS FOR DIGITAL CELLULAR TELEPHONES

Designers of portable digital cellular telephones are very interested in low-power radio architectures. Several integrated receiver and transmitter IC's conforming to established standards such as GSM and DECT have been developed in the past few years. This section summarizes some of their main features.

All transmitters in these portable phones use direct upconversion to produce a single-sideband output. In receivers, however, the superheterodyne architecture is more common. For instance, a 900 MHz bipolar IC GSM receiver from Siemens [17] downconverts the amplified RF signal from an off-chip low-noise amplifier to an IF of 45–90 MHz (Fig. 6). At this IF, the image lies in the stopband of the fixed RF preselect SAW filter. The amplified IF signal is sent to another off-chip SAW filter to reject adjacent channels. A quadrature mixer then downconverts the signal to baseband, and the vector baseband signal is finally detected. This architecture is preserved in later generations of this transceiver operating up to 2 GHz for DECT use [18], [19]. Other recent GSM transceivers build on a similar single superheterodyne architecture [20], in one case with a very high IF of 400 MHz [21]. Alcatel has publicized its use of direct-conversion in GSM and DECT receivers [22]–[24], although others [25] are exploring its possible use, and not all companies using direct conversion have published their experience. Alcatel's RF front-end is a relatively small silicon bipolar IC (Fig. 7), and the remainder of the signal processing, including lowpass channel-select filtering, takes place at baseband in a mixed analog-digital CMOS IC [26].

Given the many decades of familiarity with the superheterodyne, there will likely remain some reluctance towards adopting a new architecture until there is widespread experience in its effectiveness. However, direct-conversion also suffers from some unique problems to which the superheterodyne is immune. These are discussed in the following section.

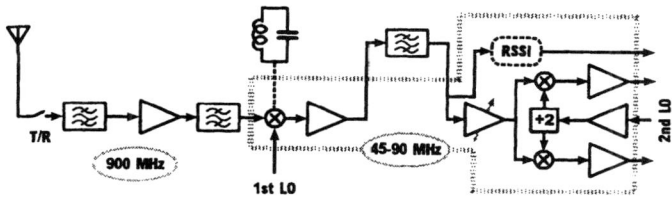

Fig. 6. A superheterodyne receiver for a digital cellular telephone.

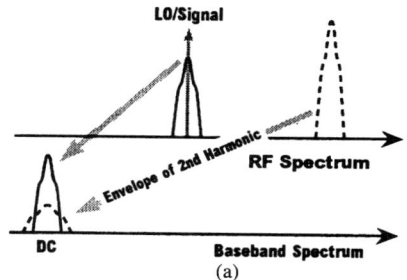

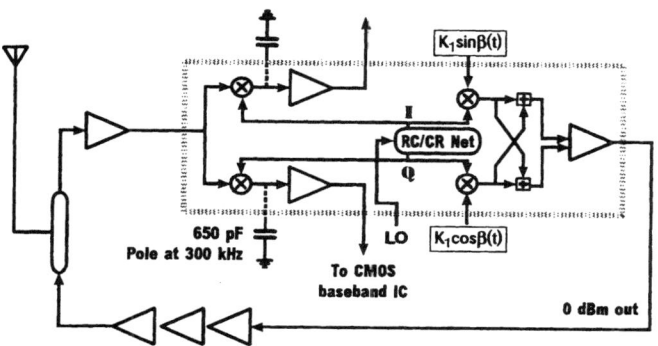

Fig. 7. A direct-conversion receiver and transmitter for a digital cellular telephone.

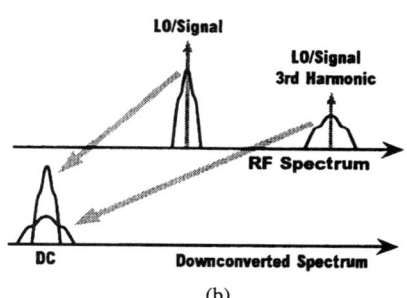

VI. PROBLEMS IN DIRECT-CONVERSION RECEIVERS

Among the problems in direct-conversion receivers, *spurious LO leakage* is probably best known. This arises because the LO in a direct-conversion receiver is tuned exactly to the center of the LNA and antenna passbands. In receive mode, a small fraction of the LO energy may make its way back to the antenna through the mixer and LNA, owing to their finite reverse isolation, or couple into the antenna through external leads, and then radiate out [27]. This becomes an in-band interferer to other nearby receivers tuned to the same band, and for some of them it may even be stronger than the desired signal. Regulatory bodies such as the FCC strictly limit the magnitude of this type of spurious LO emission. The problem is much less severe in a superheterodyne whose LO frequency usually lies outside the antenna passband. However, experimental studies [28] suggest that standard shielding in the receiver may control LO leakage to the point that it does not seriously handicap the use of direct-conversion.

Distortion produced by strong signals in the downconversion mixer will cause the sensitivity of a direct-conversion receiver to degrade more rapidly than of the superheterodyne. Second-order distortion in a single-ended mixer will rectify the envelope of an amplitude-modulated RF input such as QPSK data to produce spurious baseband spectral energy centered at dc, which then adds to the desired downconverted signal [25], [27] (Fig. 8(a)). This is particularly serious if the envelope is that of a large unwanted signal lying in the preselect filter passband, which has not yet been rejected by the baseband channel-select filter. The most effective solution is to use

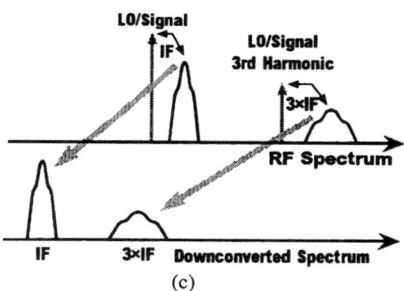

Fig. 8. Spurious downconversion caused by nonlinearity in a direct-conversion receiver. (a) Second-order distortion detects envelope of near-band interferer at baseband, overlaying desired signal. (b) LO harmonics downconvert signal harmonics to baseband, resulting in interference, (c) whereas in a superheterodyne downconverted harmonic products lie in IF stopband.

balanced circuits in the RF front-end, particularly the mixer, which will only create odd-order distortion.[1]

However, even in a balanced circuit, the third harmonic of the desired signal may downconvert the third LO overtone to create spurious dc energy competing with the fundamental downconverted signal (Fig. 8(b)), whereas in a superheterodyne this downconverted component lies in the stopband of the IF filter (Fig. 8(c)). This is a small effect to the extent

---

[1] There is no fundamental loss of noise figure in a balanced front-end. When the antenna signal drives a balanced low-noise amplifier through a power-splitting balun, the noise figure is exactly the same as directly driving the signal into a single-ended half circuit. However, the balanced circuit drains twice the current of the single-ended half circuit.

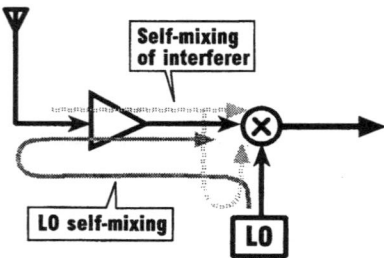

Fig. 9. Some sources of dynamic offset in a direct-conversion receiver. The LO may leak into the antenna, reflect off external objects, and downconvert to dc. A strong interferer may leak into the LO port and downconvert itself to dc. These dc offsets vary with physical location.

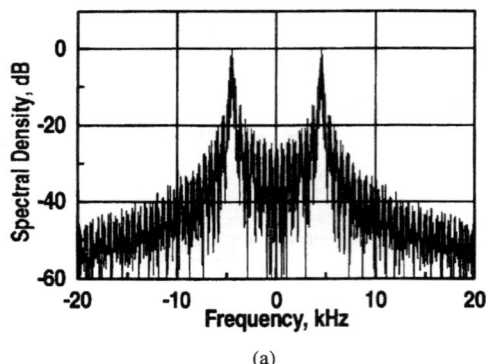

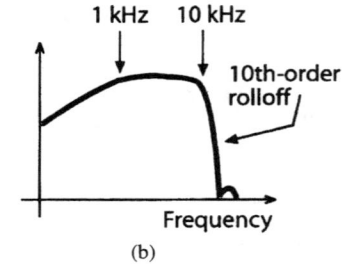

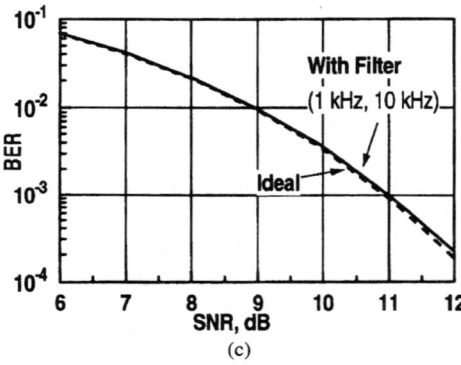

Fig. 10. (a) Downconverted spectrum of a paging channel carrying 500 b/s data with 4.5 kHz carrier frequency keying. (b) A typical channel-select filter in the receiver, with a 1 kHz lower cutoff frequency to null dc offsets. Channels are spaced apart by 25 kHz. (c) Receiver sensitivity simulation, comparing an ideal offset-free receiver containing 10 kHz brickwall filter, with receiver suffering from offset, but using the filter in (b).

that both are third-order terms. To suppress this harmonic downconversion, mixers with embedded overtone-rejection bandpass filters have been proposed [29]. If exactly the same immunity to spurious downconversion is sought as in a superheterodyne, the direct-conversion receiver must use a balanced mixer with inherently greater linearity. Again, this is not an insurmountable obstacle; mixer linearity and related design issues are discussed in Section VII.

Perhaps the most serious problem is that of dc *offset* in the baseband section of the receiver following the mixer. This offset appears in the middle of the downconverted signal spectrum, and may be larger than the signal itself, and much larger than thermal and flicker noise. For instance, a downconverted signal of a few hundreds of microvolts rms may compete with an offset of a few millivolts. Unless the offset is removed, the SNR at the detector input will be very low. Offset arises, first, from transistor mismatch in the signal path between the mixer and the I and Q inputs to the detector. To this will add other offsets peculiar to the wireless environment [24], [30]. As described above, the LO signal leaking from the antenna during receive mode may reflect off an external object and self-downconvert to dc in the mixer (Fig. 9). Similarly, a large undesired near-channel interferer in the preselect filter passband may leak into the LO port of the mixer and self-downconvert to dc. These offsets are insidious because their magnitude changes with receiver location and orientation, and in a frequency-hopping receiver, with the instantaneous LO frequency. It is unlikely whether through better circuit design alone, the reverse isolation of mixers and LNA's can be improved to the point that these offsets become on the order of thermal noise. Appropriate circuits must be built into the receiver to remove them.

In the direct-conversion paging receiver, dc offset is removed by *capacitively coupling* the baseband circuits up to, and including, the limiter. Owing to the high impedance levels of active filters, the coupling capacitors are small enough to be integrated on to the receiver IC [10], [31]. However, this simple solution only works because of the specific spectral features of wideband-FSK modulation [10]. The paging spectrum peaks at ±4.5 kHz and dips at dc by at least 25 dB relative to its level at the peaks (Fig. 10(a)). Simulations show that a first-order capacitive coupling, which produces a lower cutoff at 1 kHz in the channel-select filter preceding the limiting amplifier (Fig. 10(b)), causes no loss in receiver sensitivity (Fig. 10(c)) when the channel filter captures the valuable signal spectrum between 1–10 kHz. These are precisely the characteristics of the channel-select filter in a Philips single-chip paging receiver [10].

A small *frequency-error* between the transmitter and receiver LO's will cause the signal to downconvert asymmetrically around dc, and the capacitively-coupled filter may now place its notch in one of the two signal-bearing spectral lobes. For instance, a 5 ppm relative frequency-error at 900 MHz places the capacitive notch 4.5 kHz away from the center of the spectrum in the middle of one of the lobes. This limits the acceptable frequency tolerance on the crystal

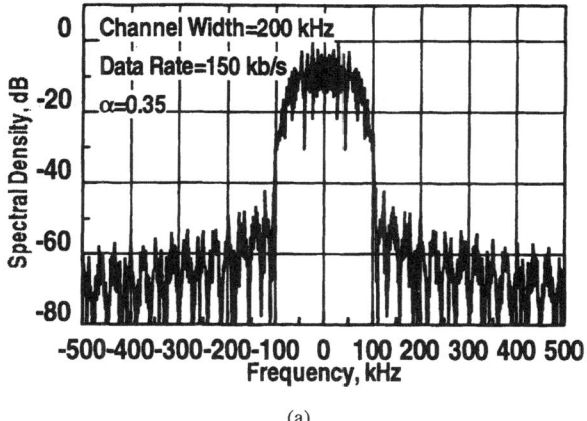

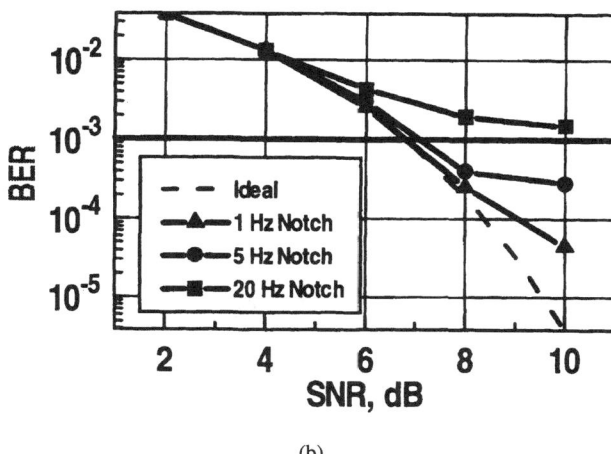

Fig. 11. (a) Typical direct downconverted spectrum of an efficiently modulated carrier, here a cosine-filtered BPSK. (b) The impact of a notch of varying widths at dc on receiver performance. At speech-quality BER of 10^{-3}, the receiver ceases to function if the notch width is 20 Hz.

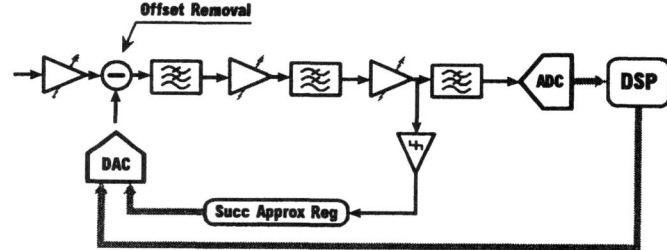

Fig. 12. Digital offset removal in a direct-conversion receiver. Offset is estimated from long-term average of digitized signal and subtracted from analog signal.

oscillator regulating the LO. In the absence of a sufficiently stable frequency reference, some form of automatic frequency-control must be used. The LO frequency error may be deduced from the long-term average of the frequency shifts keyed by pseudorandom data, which should ideally be zero after downconversion. This may be used as a correction signal on a varactor diode, or equivalent means of fine-tuning the LO. In a mostly digital implementation of a frequency-hopped FSK receiver, a numerically-controlled oscillator with resolution of a few hertz compensates frequency errors in the receiver clock as part of a frequency-acquisition loop [32].

Wideband FSK is spectrally inefficient, but it is used in paging because of the very low data rates and small duty cycle of use. However, digital cellular systems must fit continuous voiceband traffic at 200–1100 kb/s into 50–100 kHz wide channels, and to do so they use other spectrally-efficient modulation schemes. The GSM system, for instance, employs Gaussian-Filtered Minimum-Shift Keying (GMSK), a sort of optimized FSK. Other systems may use single-ended or differential Quadrature Phase-Shift Keying (QPSK). The spectra of all these modulation schemes peak at dc (Fig. 11(a)).

Following downconversion of the received signal to zero-IF, offsets will now directly add to the *peak* of the spectrum. It is no longer practical to null the offsets by capacitive coupling of the baseband signal path because signal energy will almost surely be lost from the spectral peak. Simulations on a representative 200 kHz wide spectrum (Fig. 11(b)) suggest that at a target bit-error rate of 10^{-3}, a 5 Hz notch at dc causes about 0.2 dB loss in receiver sensitivity, yet this notch need only widen to 20 Hz when the receiver will cease to function. It would require impractically large capacitors to produce this narrow notch, and the phase-distortion due to the CR coupling, which these simulations do not model, would cause the receiver performance to further deteriorate. Furthermore, during burst-mode communications, the capacitor would produce intolerably long transients.

DC offsets may be estimated and removed digitally [33], an approach Alcatel takes in their GSM receiver [26] (Fig. 12). The baseband signal is digitized and averaged in an 8-b DSP over a programmable time window. In addition, a 10-b successive-approximation A/D converter measures the analog signal polarity. These measurements are weighted together in a DAC, which subtracts off an estimate of the offset from the baseband signal. The spectrum loss around dc is only a few hertz, and digital filtering does not distort the group delay. The bandwidth of the feedback loop must be wide enough to accommodate the dynamic offsets described above. The settling time of the offset subtraction circuit may still cause loss of the first few symbols in a TDMA [30] or a frequency-hopping CDMA receiver.

Finally, when offsets have been satisfactorily nulled, flicker noise at the mixer output adds to the baseband signal. The SNR is lower than in a high-IF superheterodyne where only thermal noise is present. A bipolar transistor front-end may be superior in this respect to an FET circuit, but it is also possible to use autozero or double-correlated sampling to suppress flicker noise in MOS opamp-based circuits.

VII. COMPONENTS OF DIRECT-CONVERSION TRANSCEIVERS

A. Carrier-Frequency Local Oscillators with Quadrature Outputs

Direct-conversion transmitters and receivers need a local oscillator with quadrature outputs for vector modulation and demodulation, respectively. Whereas in a superheterodyne receiver, the signal is quadrature-downconverted by a low frequency second LO, the greater challenge in direct-conversion is to produce accurate quadrature phases with good amplitude match at the much higher *carrier-frequency*. An error in quadrature of less than 1° is usually desired. A conventional 900 MHz or 2 GHz oscillator tuned with a single LC circuit or an off-chip resonator only produces a single-phase output. Quadrature phases may then be derived by passing this through a phase-shift network, composed of an RC and a CR (Fig. 13(a)), which phase shifts by $-45°$ and $+45°$, respectively. When the time constant is equal to the oscillation period, the amplitudes of the two phases are also the same. Inaccuracies in the actual values of R and C will lead to errors in quadrature and are compensated for by some form of on-chip trimming such as the addition of voltage-controlled correction vectors (Fig. 13(b)) [22], [23]. Other circuits to successively improve the quality of quadrature have been proposed [15], but the simplest of them all is the cascaded four-branch RC polyphase network [34] (Fig. 13(c)). This may be thought of as a generalized RC-CR network with a four-phase input and output. When driven at the input by four unequally spaced phasors, for instance only a single phase sinewave and its inverse, with the other two inputs grounded, the circuit reinforces one sequence of the four phases of oscillation, say clockwise, and attenuates the other counterclockwise sequence [35]. With quite mild tolerances on its components, it is possible to derive quadrature phases from this network with order-of-magnitude improvements in phase accuracy over the RC-CR network. The polyphase RC network has been recently used in this way in a monolithic downconversion mixer [36] to attain effectively a 0.5° phase error and 0.5 dB amplitude error. The network is at its most powerful in suppressing the undesired sideband in a single-sideband upconverter, where it takes as its four inputs the balanced signals on the two mixer output branches (Fig. 5). Unwanted sideband suppression of as large as 60 dB has been reported [35].

Alternatively, quadrature phases are obtained at the outputs of positive- and negative-edge triggered flip-flops dividing a clock at *twice the carrier frequency* by two. In a direct upconversion IC for GSM where this scheme is used, output phase errors caused by unequal delays in the two flip-flops, and by departures from 50% duty cycle in the double-frequency oscillation, are found to be less than 1° [17] (Fig. 14). An on-chip bandpass filter improves duty cycle accuracy by suppressing harmonics of the double-frequency oscillation. However, the portions of the circuit operating at the double-frequency may become a speed bottleneck.

Quadrature outputs may also be tapped from a *polyphase oscillator*, such as a variable-frequency ring oscillator. In a four-delay stage ring oscillator, taps at diametrically-opposite points yield quadrature phases at every oscillation frequency [37] (Fig. 15). When the oscillator is composed of an odd-number of unit delays, the desired phases may be synthesized by interpolating two taps with a voltage-controlled phase-shifter in feedback around a quadrature-sensing circuit [38]. Mismatches in the delays of the unit cells limit the attainable phase accuracy. It is possible to keep phase errors to less than

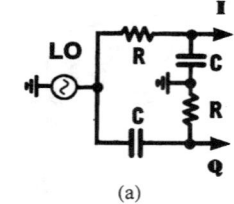

(a)

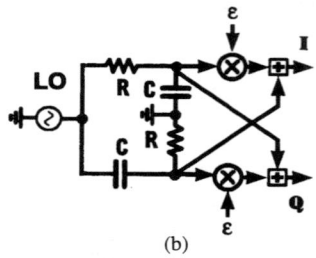

(b)

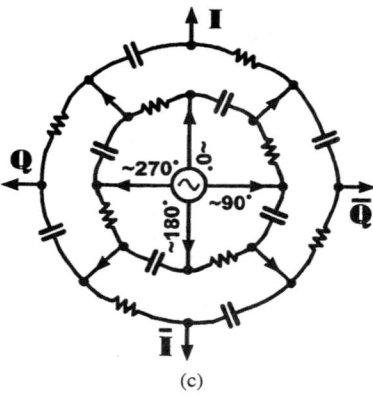

(c)

Fig. 13. Methods to derive quadrature phases from a single-phase oscillation. (a) The RC-CR phase-shift network, whose accuracy depends on matching of R's and C's. (b) An extension to add programmable vector correction under voltage control. (c) Two stages in cascade of an RC polyphase network, which may be driven at the input with only a single phase and its inverse, and which precisely produces four phases at the output even with lax component matching.

1° with careful balance of the capacitive loading on all the oscillator taps.

This form of resonator-less oscillator is attractive because it is fully integrable. It may be used when the specifications on phase-noise close to the carrier are not as stringent as they are, for instance, in analog cellular telephones [39], where the oscillator must use a passive, off-chip high-Q resonator. The free-running oscillator should obviously be designed for inherently low phase-noise, and for frequency stability it must also be slaved to a crystal reference in a frequency-locked loop. The loop gain and bandwidth should be chosen specifically to suppress phase-noise. The loop

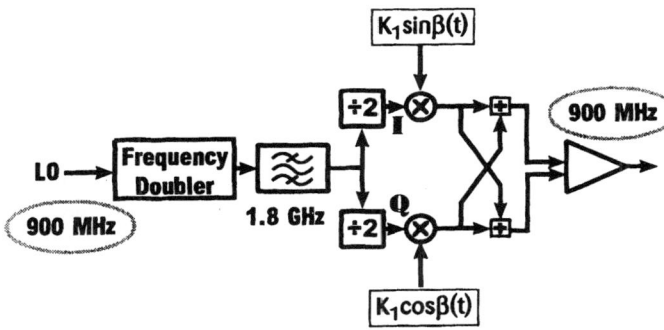

Fig. 14. A single-sideband upconversion circuit, where quadrature LO phases are derived by doubling LO frequency, and dividing down in ÷2 flip-flops triggered on opposite edges. Quadrature baseband modulated signal is digitally synthesized. Sideband selection in final differential RF amplifier.

Fig. 15. A polyphase, voltage-controlled ring oscillator. The small number of delay stages lead to a high frequency of oscillation. Quadrature phases at diametrically opposite taps. This oscillator is readily integrable.

gain, which is infinite at dc because of the phase-integrating VCO, typically starts with a second-order frequency roll-off, and slows down to a first-order frequency roll-off prior to crossing unity gain [40]. Over the region of second-order roll-off close to the oscillation frequency, the loop is able to very effectively suppress phase noise caused by low-frequency $(1/f)$ fluctuations in a FET oscillator. The unity-gain frequency of the loop is normally designed to be an order-of-magnitude lower than the reference frequency, which is derived, say, from a 20 MHz crystal oscillator. Thus, at a 100 kHz offset from the LO, the loop suppresses noise by 20 dB or so. Measurements on a three-stage 900 MHz CMOS voltage-controlled ring oscillator dissipating 20 mW show open-loop phase noise levels of −90 dBc/Hz at a 100 kHz offset from the carrier [41]. When the loop suppresses this further by 20 dB, the phase noise becomes comparable to the measured results at a 100 kHz offset in resonator-based oscillators for digital cellular use [20], [42].

Resonator-based oscillators, in which the off-resonance feedback loop gain to noise collapses at a rate proportional to the square of resonator Q, tend to convert less of a given voltage- or current-noise in the circuit into phase noise [49]. By combining a recently developed technology to fabricate large-value spiral inductors on CMOS substrates [50] with a special cross-coupling technique between two identical LC oscillators [51], it is possible to automatically derive quadrature outputs from a fully integrated LC resonator-based oscillator with lower phase noise than in a ring oscillator. However, owing to its usually large series winding resistance the on-chip inductor is relatively low Q, and it is therefore difficult to obtain dramatic improvements over the phase noise in a ring oscillator. The same approach may be more useful with high-Q off-chip inductors. Unlike the ring oscillator, strong voltage control of an LC oscillator's frequency requires varactor diodes.

To compare phase-noise specifications on the LO in a direct-conversion receiver with that in a superheterodyne, note that with a modest IF, the frequencies of the two LO's will be about the same. Therefore, for the same amount of allowed reciprocal mixing, the phase-noise specifications on the direct-conversion receiver LO are not significantly more stringent than on the first LO in the superheterodyne.

As an alternative to the various analog circuit techniques described so far, digital LMS adaptive algorithms at baseband may sense and compensate for phase- and gain-errors in quadrature upconverters and downconverters [43]. These algorithms may be used in receivers where a modest- to high-resolution A/D converter precedes the baseband signal conditioning, as is the trend in many modern wireless devices. Other methods to sense and correct unknown gain- and phase-errors have also been proposed, such as downconversion with a three-phase LO [44].

B. Mixers for Direct-Conversion

As was described above, direct-conversion receivers need a more linear mixer to attain the same performance as a superheterodyne. Mixers commutate the amplified RF signal with the LO. In the often-used bipolar mixer based on the Gilbert analog multiplier, this is a current-mode commutation. The subcircuit responsible for RF voltage-to-current conversion prior to the commutator usually determines the overall mixer linearity. Resistive degeneration of the differential pair V-to-I converter [22] improves mixer linearity, and indeed, the available dynamic range[2], at the expense of a higher noise figure. MOSFETs, on the other hand, make more linear open-loop V-to-I converters at high frequencies, due to the cancellation of the dominant quadratic nonlinearity in a balanced configuration. They are also excellent switches. Recent MOSFET implementations [36], [45] of 1 GHz downconversion mixers prove that they have a very wide dynamic range. However, a continuous-time MOSFET mixer will add a considerable flicker noise to an input it has downconverted to zero IF, thereby degrading receiver noise figure. Any mixer commutating at the carrier frequency in a direct-conversion receiver is also likely to leak LO energy into the antenna because of imperfect reverse isolation.

---

[2]Dynamic range of a differential pair increases with degeneration because the voltage linearity improves roughly proportionally to the degeneration, while noise, voltage increases as the square-root.

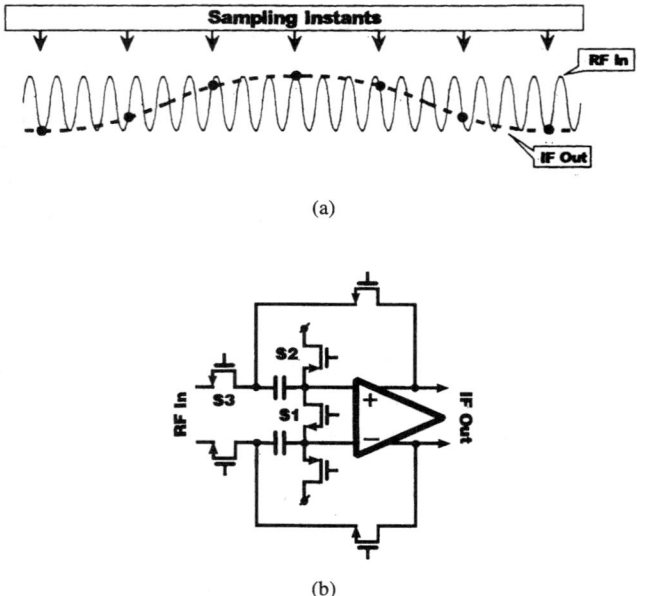

Fig. 16. (a) Downconversion by sampling a modulated RF carrier. A discrete-time (zero) IF signal is obtained when the sample rate is at least twice the modulation bandwidth and an integer divisor of RF carrier. (b) An implementation of the mixer. Switches $S1$ and $S3$, with the associated capacitors, form a wideband passive RF sampling network. Opamp is clocked at sample rate.

One way around these problems is to downconvert by *subsampling* the modulated RF [46] (Fig. 16(a)). The track-mode bandwidth of the sampling circuit must be greater than the input *carrier frequency*, whereas the (much lower) sampling rate must be at least twice the larger of the *modulation* or *spreading bandwidth* to downconvert without aliasing. The modulation appears at the output as a discrete-time beat between the RF input and sampling clock. Therefore, a sampling clock at an exact subharmonic of the input RF carrier accomplishes direct downconversion. The sampling clock is now the mixer's LO. Although one or more of the clock's high-order harmonics may leak into the LNA and antenna passband, any spurious radiation they produce is usually much weaker than due to an LO tuned to the carrier frequency.

The subsampling mixer tracks the input with a passive circuit consisting of FET switches and capacitors (Fig. 16(b)). The quadratic *I–V* characteristic of MOSFET switches may contribute some distortion, but this is alleviated by a balanced implementation. The sampling clock is nominally a square wave with 50% duty cycle, whose falltime sets the switch sampling aperture. In a 1-μm CMOS implementation, the aperture may be as short as 200 ps [46]. The *nonzero* aperture in FET turn-off limits only the acquisition bandwidth but causes no distortion. However, *signal-dependent charge* injected from the nonlinear FET capacitances onto the sampling capacitor will distort the held signal. With the proper switching sequence, the injected charge is reduced by the opamp gain in hold mode [47]. The opamp is designed to clock at the sampling rate and to fully settle in hold mode.

An undesirable consequence of subsampling is that the circuit also tracks noise and possible interferers lying outside the Nyquist bandwidth (half the sample rate) and aliases these into the Nyquist bandwidth. Although adjacent channel interferers that are downconverted without aliasing will lie in the stopband of the subsequent channel-select filter, far-off interferers may alias into its passband and corrupt the desired signal. An RF preselect filter with a passband no wider than the Nyquist bandwidth of the mixer must be used at the antenna to attenuate these far-away interferers. Now, as sample rates as high as 100 MHz have been demonstrated in 1-μm CMOS op-amp circuits, a preselect filter centered at, say, 900 MHz may have up to a 50 MHz wide passband. Although this is not a highly selective filter, it must have a large stopband attenuation. The mixer still aliases broadband LNA output noise, which is bandlimited only by the low-Q tuned LNA load. A second preselect filter, identical to the first, when inserted between the LNA and mixer eliminates this aliasing, but this is not a desirable solution in a highly integrated receiver. In the absence of this second filter, then, the noise spectral density multiplied by aliasing competes with the downconverted signal and raises the receiver noise figure [46] to typically 6 dB or higher. The signal processing now required at the high clock frequency mixer input is not dissimilar to that in a high-IF superheterodyne. However, direct conversion still retains an advantage, in that after simple prefiltering, the mixer output may be decimated down to a lower clock at the channel-select filter. With some form of autozero during track mode, the input flicker noise in the opamp may be also suppressed. Owing to its wide dynamic range and low LO leakage, this may prove to be a useful mixer in direct-conversion receivers. As a secondary advantage, the discrete-time mixer output readily interfaces to a wide dynamic range, auto-zeroed switched-capacitor CMOS channel-select filter operating at a reduced clock rate.

VIII. CONCLUSIONS

The direct-conversion receiver eliminates many off-chip components and may offer significant power savings by amplifying a received signal mostly at dc rather than at an IF of tens or even hundreds of MHz. Direct-conversion is already widely used in single-sideband transmitters. Several problems in direct-conversion receivers have been identified, of which static and dynamic dc offset are probably the most important. In a wideband FSK receiver, this offset is removed very simply. Otherwise, DSP-based offset-removal must supplement good RF and baseband analog design in the receiver. Inspired by the simplicity of the paging receiver, our research group at UCLA is developing a single-chip spread-spectrum transceiver that uses binary-FSK modulation on a frequency-hopped carrier [48]. This transceiver uses direct-conversion in the transmit and receive paths.

With more practical experience, direct-conversion is expected to be used widely in certain wireless applications. This paper has summarized some of the key problems unique to direct-conversion and has presented various solutions. It is unlikely that a direct-conversion receiver that does not embody some or all these solutions will ever be able to perform equally well with the superheterodyne. However, even with the added complexity of these solutions, the aggregate of direct conversion's advantages for a miniature, low-power radio transceiver is enough to warrant continued research and development.

ACKNOWLEDGMENT

The author is grateful to J. Sevenhans of Alcatel Bell Telephone for sharing his experience in direct-conversion receiver design and to J. Min of UCLA for the simulations underlying Figs. 10 and 11. Many graduate students at UCLA have contributed to some of the key RF-CMOS circuits described in this paper.

REFERENCES

[1] A. A. Abidi, "Low-power radio-frequency IC's for portable communications," *Proc. IEEE*, vol. 83, no. 4, pp. 544–569, 1995.
[2] L. Lessing, *Man of High Fidelity: Edwin Howard Armstrong, A Biography*. New York: Bantam Books, 1969.
[3] S. Watanabe, *Semiconductor Devices for Electronic Tuners*, vol. 13. New York: Gordon and Breach, 1991.
[4] A. A. Abidi, "Noise in active resonators and the available dynamic range," *IEEE Trans. Circuits Syst.*, vol. 39, no. 4, pp. 296–299, 1992.
[5] D. K. Weaver, "A third method of generation and detection of single-sideband signals," *Proc. IRE*, vol. 44, no. 12, pp. 1703–1705, 1956.
[6] T. Okanobu, H. Tomiyama, and H. Arimoto, "Advanced low-voltage single chip radio IC," *IEEE Trans. Consumer Electron.*, vol. 38, no. 3, pp. 465–475, 1992.
[7] D. G. Tucker, "The history of the homodyne and synchrodyne," *J. British Inst. Radio Engineers*, vol. 14, no. 4, pp. 143–154, 1954.
[8] I. A. W. Vance, "Fully integrated radio paging receiver," *IEE Proc.*, vol. 129, pt. F, no. 1, pp. 2–6, 1982.
[9] K. Yamasaki, M. Matai, M. Miyashita, K. Yonekura, M. Inagaki, and Y. Morita, "Credit card size numeric display pager with microstrip antenna for 900 MHz band," in *NEC Research & Development*, vol. 34, no. 1, pp. 84–95, Jan. 1993.
[10] J. F. Wilson, R. Youell, T. H. Richards, G. Luff, and R. Pilaski, "A single-chip VHF and UHF receiver for radio paging," *IEEE J. Solid-State Circuits*, vol. 26, no. 12, pp. 1944–1950, 1991.
[11] J. Min, H.-C. Liu, A. Rofougaran, S. Khorram, H. Samueli, and A. A. Abidi, "Low power correlation detector for binary FSK direct-conversion receivers," *Electron. Lett.*, vol. 31, no. 13, pp. 1030–1032, 1995.
[12] S. Tanaka, A. Nakajima, J. Nakagawa, A. Nakagoshi, and Y. Kominami, "High-frequency, low-voltage circuit technology for VHF paging receiver," *IEICE Trans. Fundamentals of Electron., Commun., Comput. Sci.*, vol. E76-A, no. 2, pp. 156–163, 1993.
[13] K. Yamasaki, K. Yoshizawa, Y. Minami, T. Asai, Y. Nakano, and M. Kuroda, "Compact size numeric display pager with new receiving system," in *NEC Research & Development*, vol. 33, no. 1, pp. 73–81, Jan. 1992.
[14] K. Negus, B. Koupal, J. Wholey, K. Carter, D. Millicker, C. Snapp, and N. Marion, "Highly integrated transmitter RFIC with monolithic narrowband tuning for digital cellular handsets," presented at *Int. Solid-State Circuits Conf.*, San Francisco, CA, 1994, pp. 38–39.
[15] I. A. Koullias, J. H. Havens, I. G. Post, and P. E. Bronner, "A 900 MHz transceiver chip set for dual-mode cellular radio terminals," presented at *Int. Solid-State Circuits Conf.*, San Francisco, CA, 1993, pp. 140–141.
[16] A. A. Abidi, "Radio-frequency integrated circuits for portable communications," presented at *Custom IC Conf.*, San Diego, CA, 1994, pp. 151–158.
[17] J. Fenk, W. Birth, R. G. Irvine, P. Sehrig, and K. R. Schon, "An RF front-end for digital mobile radio," presented at *Bipolar Circuits and Technol. Meet.*, Minneapolis, MN, 1990, pp. 244–247.
[18] V. Thomas, J. Fenk, and S. Beyer, "A one-chip 2 GHz single superhet receiver for 2 Mb/s FSK radio communication," presented at *Int. Solid-State Circuits Conf.*, San Francisco, CA, 1994, pp. 42–43.
[19] W. Veit, J. Fenk, S. Ganser, K. Hadjizada, S. Heinen, H. Herrmann, and P. Sehrig, "A 2.7 V 800 MHz–2.1 GHz transceiver chipset for mobile radio applications in 25 GHz f_T si-bipolar," presented at *Bipolar Circuits & Technol. Meet.*, Minneapolis, MN, 1994, pp. 175–178.
[20] T. Stetzler, I. Post, J. Havens, and M. Koyama, "A 2.7–4.5 V single-chip GSM transceiver RF integrated circuit," presented at *Int. Solid-State Circuits Conf.*, San Francisco, CA, 1995, pp. 150–151.
[21] C. Marshall, F. Behbahani, W. Birth, A. Fotowat, T. Fuchs, R. Gaethke, E. Heimerl, S. Lee, P. Moore, S. Navid, and E. Saur, "A 2.7 V GSM transceiver IC's with on-chip filtering," presented at *Int. Solid-State Circuits Conf.*, San Francisco, CA, 1995, pp. 148–149.
[22] J. Sevenhans, A. Vanwelsenaers, J. Wenin, and J. Baro, "An integrated Si bipolar RF transceiver for a zero IF 900 MHz GSM digital radio front-end of a hand portable phone," presented at *Custom IC Conf.*, San Diego, CA, 1991, pp. 7.7/1–4.
[23] J. Sevenhans, D. Haspeslagh, A. Delarbre, L. Kiss, Z. Chang, and J. F. Kukielka, "An analog radio front-end chip set for a 1.9 GHz mobile radio telephone application," presented at *Int. Solid-State Circuits Conf.*, San Francisco, CA, 1994, pp. 44–45.
[24] J. Wenin, "IC's for digital cellular communication," presented at *European Solid-State Circuits Conf.*, Ulm, Germany, 1994, pp. 1–10.
[25] C. Takahashi, R. Fujimoto, S. Arai, T. Itakura, T. Ueno, H. Tsurumi, H. Tanimoto, S. Watanabe, and K. Hirakawa, "A 1.9 GHz Si direct conversion receiver IC for QPSK modulation systems," presented at *Int. Solid-State Circuits Conf.*, San Francisco, CA, 1995, pp. 138–139.
[26] D. Haspeslagh, J. Ceuterick, L. Kiss, and J. Wenin, "BBTRX: A baseband transceiver for a zero IF GSM hand portable station," presented at *Custom IC Conf.*, San Diego, CA, 1992, pp. 10.7.1–10.7.4.
[27] N. C. Hamilton, "Aspects of direct conversion receiver design," presented at *Fifth Int. Conf. HF Radio Syst. & Technol.*, Edinburgh, Scotland, 1991, pp. 299–303.
[28] H. Tsurumi and T. Maeda, "Design study on a direct conversion receiver front-end for 280 MHz, 900 MHz, and 2.6 GHz band radio communication systems," presented at *IEEE Veh. Technol. Conf.*, St. Louis, MO, 1991, pp. 457–462.
[29] J. M. Moniz and B. Maoz, "Improving the dynamic range of Si MMIC Gilbert cell mixers for homodyne receivers," presented at *Microwave & Millimeter-Wave Monolithic Circuits Symp.*, San Diego, CA, 1994, pp. 103–106.
[30] G. Schultes, E. Bonek, A. L. Scholtz, and P. Kreuzgruber, "Low-cost direct conversion receiver structures for TDMA mobile communications," presented at *Sixth Int Conf. Mobile Radio and Personal Commun.*, Coventry, UK, 1991, pp. 143–150.
[31] A. Burt, "Direct conversion receivers come of age in the paging world," in *GEC Rev.*, vol. 7, no. 3, pp. 156–160, 1992.
[32] H.-C. Liu, J. Min, and H. Samueli, "A low-power baseband receiver IC for frequency-hopped spread spectrum applications," presented at *Custom IC Conf.*, Santa Clara, CA, 1995, pp. 311–314.
[33] A. Bateman and D. M. Haines, "Direct conversion transceiver design for compact low-cost portable mobile radio terminals," presented at *IEEE Veh. Technol. Conf.*, San Francisco, CA, 1989, pp. 57–62.
[34] M. J. Gingell, "Single sideband modulation using sequence asymmetric polyphase networks," *Electrical Commun.*, vol. 48, no. 1–2, pp. 21–25, 1973.
[35] R. C. V. Macario and I. D. Mejallie, "The phasing method for sideband selection in broadcast receivers," *EBU Rev. (Tech. Pt.)*, no. 181, pp. 119–125, 1980.
[36] J. Crols and M. Steyaert, "A fully integrated 900 MHz CMOS double quadrature downconverter," presented at *Int. Solid-State Circuits Conf.*, San Francisco, CA, 1995, pp. 136–137.
[37] A. W. Buchwald and K. W. Martin, "High-speed voltage-controlled oscillator with quadrature outputs," *Electron. Lett.*, vol. 27, no. 4, pp. 309–310, 1991.
[38] S. K. Enam and A. A. Abidi, "NMOS IC's for clock and data regeneration in Gb/s optical-fiber receivers," *IEEE J. Solid-State Circuits*, vol. 27, no. 12, pp. 1763–1774, 1992.
[39] T. Uwano, T. Ishizaki, Y. Nakagawa, and T. Nakamura, "Design of a low-phase noise VCO for an analog cellular portable radio application," *Electron. & Commun. in Japan*, vol. 77, pt. 2, no. 3, pp. 58–65, 1994.
[40] D. H. Wolaver, *Phase-Locked Loop Circuit Design*. Englewood Cliffs, NJ: Prentice-Hall, 1991.
[41] M. Thamsirianunt and T. A. Kwasniewski, "CMOS VCO's for PLL frequency synthesis in GHz digital mobile radio communications," presented at *Custom IC Conf.*, Santa Clara, CA, 1995, pp. 331–334.
[42] S. Beyer and G. Lipperer, "Low-current oscillator design for 900 MHz GSM applications," in *Microwave Eng. Europe*, Oct. 1991, pp. 35–41.
[43] J. K. Cavers and M. W. Liao, "Adaptive compensation for imbalance and offset losses in direct conversion transceivers," *IEEE Trans. Veh. Technol.*, vol. 42, no. 4, pp. 581–588, 1993.
[44] R. K. Loper, "A tri-phase direct conversion receiver," presented at *Military Commun. Conf.*, Monterey, CA, 1990, pp. 1228–1232.

[45] A. Rofougaran, J. Y.-C. Chang, M. Rofougaran, S. Khorram, and A. A. Abidi, "A 1 GHz CMOS RF front-end IC with wide dynamic range," in *European Solid-State Circuits Conf.,* Lille, France, 1995, pp. 250–253.

[46] P. Y. Chan, A. Rofougaran, K. A. Ahmed, and A. A. Abidi, "A highly linear 1-GHz CMOS downconversion mixer," presented at *European Solid-State Circuits Conf.,* Sevilla, Spain, 1993, pp. 210–213.

[47] Y.-M. Lin, B. Kim, and P. R. Gray, "A 13 b, 2.5 MHz self-calibrated pipelined A/D converter in 3-μm CMOS," *IEEE J. Solid-State Circuits,* vol. 26, no. 4, pp. 628–636, 1991.

[48] J. Min, A. Rofougaran, H. Samueli, and A. A. Abidi, "An all-CMOS architecture for a low-power frequency-hopped 900 MHz spread-spectrum transceiver," presented at *Custom IC Conf.,* San Diego, CA, 1994, pp. 379–382.

[49] D. B. Leeson, "A simple model of feedback oscillator noise spectrum," *Proc. IEEE,* vol. 54, no. 2, pp. 329–330, 1966.

[50] J. Y.-C. Chang, A. A. Abidi, and M. Gaitan, "Large suspended inductors on silicon and their use in a 2-μm CMOS RF amplifier," *IEEE Electron Device Lett.,* vol. 14, no. 5, pp. 246–248, 1993.

[51] A. Rofougaran, J. J. Rael, M. Rofougaran, and A. A. Abidi, "A 900 MHz CMOS *LC* oscillator with quadrature outputs," to be presented at *Int. Solid-State Circuits Conf.,* San Francisco, CA, 1996.

A NEW RADIO RECEIVER SYSTEM FOR PERSONAL COMMUNICATIONS

Taiwa Okanobu, Daisuke Yamazaki,* and Chikara Nishi*
Advanced Technology Development Div. 4
InfoCom Products Company, Sony Corporation
1-7-4 Kohnan, Minato-ku, Tokyo, 108 Japan
*Semiconductor Company, Sony Corporation

Abstract

A new radio receiver system for personal communications is described. This technology incorporates direct conversion into an integrated circuit in such a way that the need for discrete IF filters is avoided. It also employs a second IF functional circuit that contains a carrier leak detector and an adjustment-free system. The carrier leak detector suppresses the level of unwanted carrier leak and the adjustment-free system automatically compensates for any variation in circuit response that corrupts the performance of the receiver. The resulting design is particularly suitable for use in wireless devices such as pagers and cordless telephones.

Introduction

Wireless personal communication devices like pagers and cellular and cordless telephones are getting smaller, more compact, and more lightweight in design. These design features are realized due to improvements in device-mount technology and developments of various kinds of devices. The devices are generally integrated

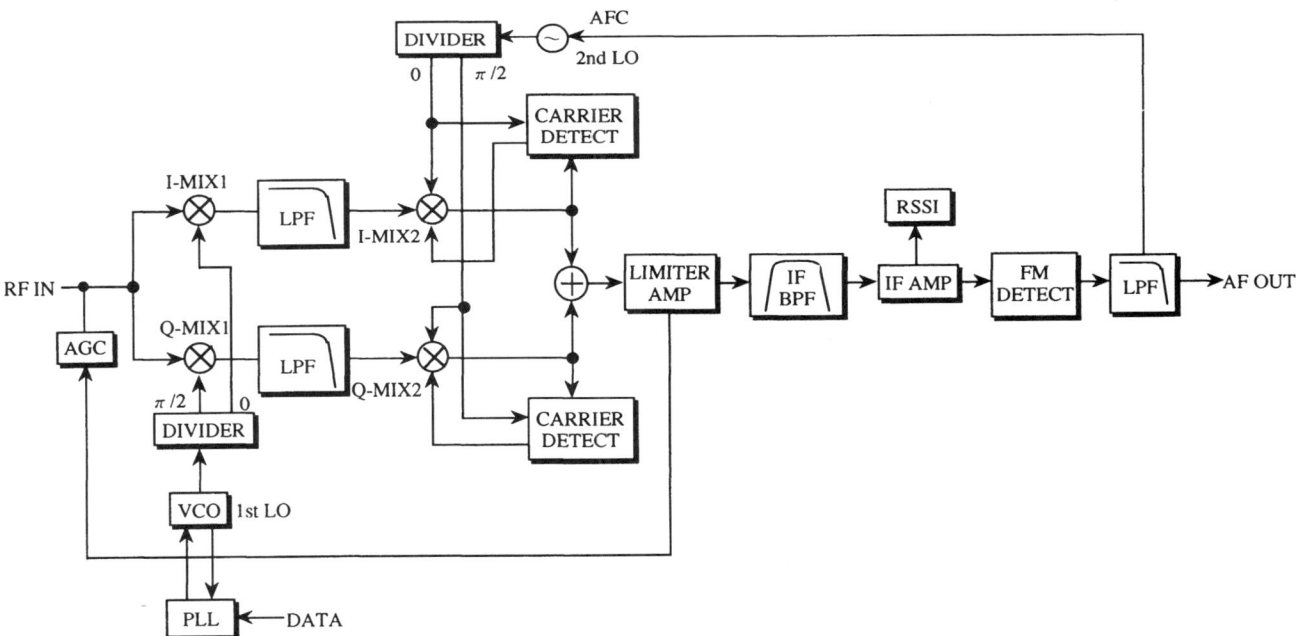

Fig. 1 A New Radio Receiver System for Personal Communications

circuits or ICs that pack more of the necessary system functions, thereby eliminating the need for more discrete components, or discrete devices where size is reduced as much as possible. In other words, it is the same system technology with just less device compositions. It is merely an optimization of available physical space and by no means a revolutionary technological change.

This paper describes a new radio receiver system that promises to reduce the size and weight of personal communication devices. The new system combines and incorporates into an IC the concept of direct conversion and a second IF (Intermediate Frequency) functional circuit. It allows the inclusion into the design approach of a circuit that functions as a discrete IF filter and know-how in traditional IC-discrete filter interface. No more discrete IF filters are needed.

The second IF functional circuit contains a carrier leak detector and an adjustment-free system. The carrier leak detector suppresses the level of unwanted carrier leak and the adjustment-free system guarantees that the receiver works properly under any variation in circuit response.

The new system also has the same image frequency cancel function, selectivity, and AMRR (Amplitude Modulation Rejection Ratio) characteristic with that of the conventional double-conversion radio receiver system.

The Configuration of the New Radio Receiver System

Fig. 1 shows the block diagram of the new radio receiver system. Compared with a conventional double-conversion radio receiver system shown in Fig. 2, there are no discrete IF filters. There is no discrete discriminator, either, because the FM (Frequency Modulation) detector implements the double-pulse count method. In the double-conversion receiver system, the frequency of the second local oscillator is high (over 10 MHz) because the first IF is high. The second

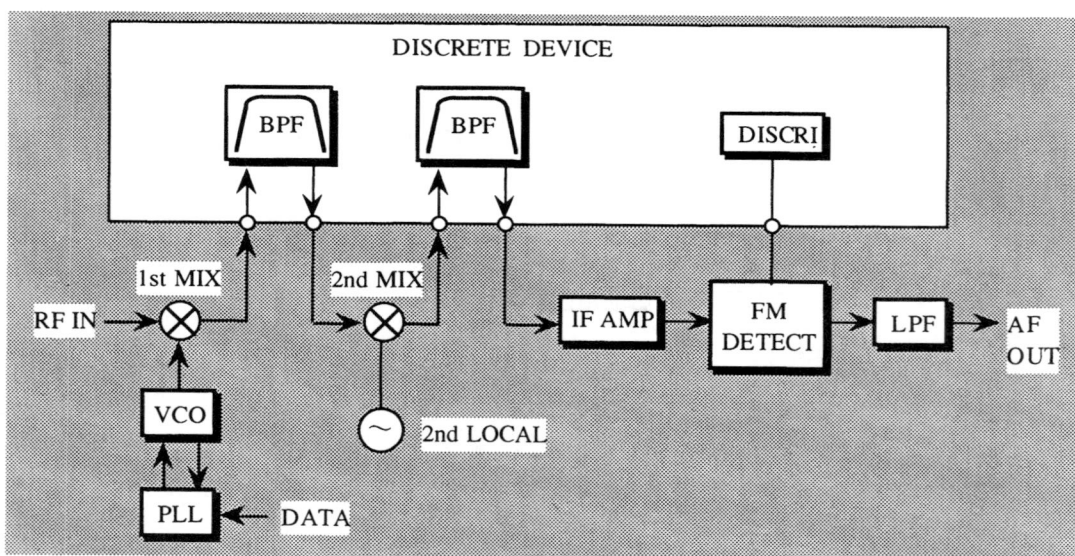

Fig. 2 Conventional Double-Conversion System

local oscillator requires stability and necessitates a crystal oscillator which adds to the number of discrete components. On the other hand, in the new system, since the frequency of the second local oscillator is just tens of kilohertz, it does not require much stability and can be built inside the IC.

This is how the new system works. The received RF (Radio Frequency) signal is fed into two subsystems we call I (for in-phase) and Q (for quadrature phase). The I subsystem consists of mixers I-MIX1 and I-MIX2 and a LPF (Low-Pass Filter), while the Q subsystem consists of mixers Q-MIX1 and Q-MIX2 and another LPF. In the I subsystem, the RF signal is mixed with a signal of the same frequency and phase from the first local oscillator in mixer I-MIX1. I-MIX1 in effect converts the received RF signal into a baseband signal. The output of I-MIX1 then passes through an active LPF which eliminates the image frequency leaving only the baseband signal. (Image frequencies will be explained in detail in the next section.) The baseband signal is then mixed in another mixer I-MIX2 with a signal from the second local oscillator which is in-phase but with a frequency of about 65 kHz $\pm$ 30%. The output of I-MIX2 is the output of the I subsystem.

The Q subsystem functions exactly like the I subsystem. The only difference is that the signal from the first and second local oscillator is in quadrature phase or $90°$ out-of-phase. In the first local oscillator, the VCO (Voltage Control Oscillator) and PLL (Phase-Locked Loop) which is controlled by a set of data from a microcomputer generates a signal twice the frequency of the RF signal. The divider then divides the frequency of the signal into two and produce two signals that are in quadrature phase. On the other hand, in the second local oscillator, the divider divides the frequency of the signal from the oscillator into four and produces the two in-quadrature-phase signals.

The IF outputs from the I and Q subsystem are added, cancelling another set of image frequencies as will be explained in the next section. The resulting sum of the signals is then fed to the limiter amplifier or limiter amp. The limiter amp suppresses any AM signal component that are due to effects of fading and multipath. It guarantees an AMRR of 40 dB for the system. The limiter amp is also coupled with an overload AGC (Automatic Gain Control) to form a loop. The AGC makes sure that the received RF signal does not exceed the dynamic range of the mixers and the active LPFs in the I and Q subsystem. With the AGC, nonlinearity does not occur in the operation of the active LPFs.

The output of the limiter amp is fed to the IF BPF (Band-Pass Filter). The IF BPF is a biquad-type filter that uses an operational amplifier or op amp. It determines the selectivity of the system. After passing the IF BPF, the IF signal is amplified in the IF amplifier and then fed to the RSSI (Radio Signal Strength Indicator) amplifier and FM detector. The FM detector employs a double-pulse count method. Unlike the quadrature detector circuit, it does not need a discrete discriminator. In the case of a double-pulse count detector, the distortion of the signal is small because the f-v characteristic is a straight line. Also, the detection efficiency is high because of the low frequency (65 kHz $\pm$ 30%) of

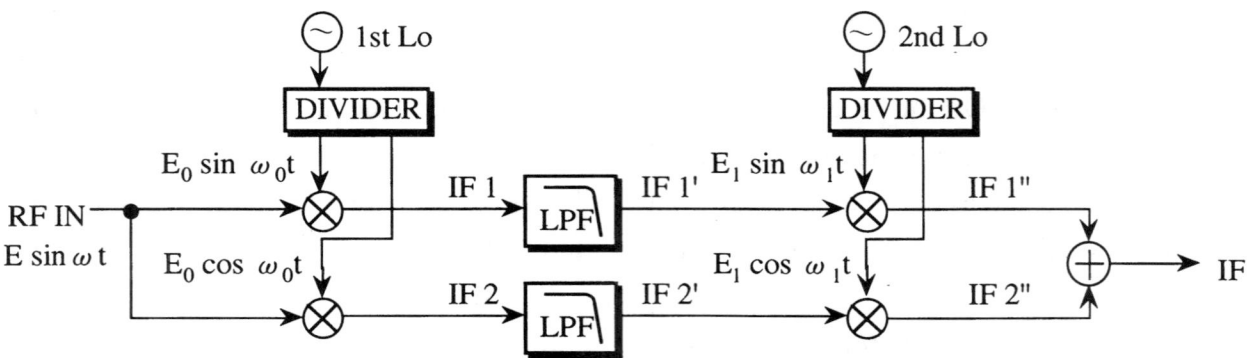

Fig. 3 New Image Frequency Signal Canceling System

the IF. After detection, the signal is converted to a baseband signal and passes through a LPF. The output is then fed to the baseband processing block of a personal communication device. The LPF is also coupled with the second local oscillator and acts as an AFC (Automatic Frequency Controller) and stabilizes the generated frequency.

The Direct Conversion for the Off-Chip-Filter-Less System

As mentioned in the previous section, in both the I and Q subsystem, the received RF signal is mixed with a signal of equal frequency generated by the PLL. The output is a so-called zero IF and the received signal is converted to a baseband signal. After passing through the LPF, it will be remodulated with a carrier from the second local oscillator (65 kHz ± 30%). Adding together the output of the two subsystems, an image frequency signal suppression of more than 40 dB can be achieved without any adjustment.

Let us suppose that an RF signal $E \sin \omega t$ enters the system as shown in Fig. 3. This is mixed with the output from the first local oscillator $E_0 \sin \omega_0 t$ and $E_0 \cos \omega_0 t$ which are in quadrature phase. The outputs IF1 and IF2 become

$$\text{IF 1} = \frac{E \cdot E_0}{2} \{ -\cos(\omega + \omega_0)t + \cos(\omega - \omega_0)t \}$$

$$\text{IF 2} = \frac{E \cdot E_0}{2} \{ \sin(\omega + \omega_0)t + \sin(\omega - \omega_0)t \}$$

IF1 and IF2 pass through LPFs and become IF1' and IF2'.

$$\text{IF 1'} = \frac{E \cdot E_0}{2} \cos(\omega - \omega_0)t$$

$$\text{IF 2'} = \frac{E \cdot E_0}{2} \sin(\omega - \omega_0)t$$

The higher frequency component $\omega + \omega_0$ in IF1 and IF2 is eliminated. This is the first set of image frequency.

Again mixing IF1' and IF2' with the signals from the second local oscillator $E_1 \sin \omega_1 t$ and $E_1 \cos \omega_1 t$ results to IF1" and IF2".

$$\text{IF 1"} = \frac{E \cdot E_0 \cdot E_1}{4} \{ \sin(\omega - \omega_0 + \omega_1)t + \sin(-\omega + \omega_0 + \omega_1)t \}$$

$$\text{IF 2"} = \frac{E \cdot E_0 \cdot E_1}{4} \{ \sin(\omega - \omega_0 + \omega_1)t \\ \sin(-\omega + \omega_0 + \omega_1)t \}$$

Adding IF1" and IF2" gives us IF

$$\text{IF} = \frac{E \cdot E_0 \cdot E_1}{2} \{ \sin(\omega - \omega_0 + \omega_1)t \}$$

The lower frequency component $\omega_1-(\omega-\omega_0)$ is cancelled in the addition process. This is the second set of the image frequency.

When $\omega = \omega_0$, IF becomes

$$IF = \frac{E \cdot E_0 \cdot E_1}{2} (\sin \omega_1 t)$$

If the RF signal is modulated, IF will be the carrier for the intelligence (data or audio) in the modulated signal.

The Carrier Leak Detector and the Adjustment-Free System

As mentioned previously, the new system incorporates a second IF functional circuit. It contains a carrier leak detector and an adjustment-free system.

(1) Carrier Leak Detector

I-MIX2 and Q-MIX2 in the I and Q subsystem are coupled with a carrier leak detector. The carrier leak detector suppresses the level of the carrier leak from the second local oscillator at the output of the mixer. It is the most important circuit in this new system. And without it, sensitivity is lost and the receiver can not pick up signals at very low level. How much the carrier leak is suppressed determines the sensitivity of the receiver. The carrier leak also degrades the circuit response of the system from the IF down to the end of the system. If there is a carrier leak, even though there is no RF signal received, the system will function as if there is one and will cause a system error.

Fig. 4 shows the carrier leak detector. Carrier leak occurs when there is a DC offset in MIX2 caused by unbalance between transistors that receives input signal and reference (input bias) in differential amplifier. This DC offset is also a result when the input signal has already received DC offset from previous circuit stage. When there is a DC offset, carrier from the second local oscillator leaks at the output of MIX2. To cancel this DC offset, the carrier leak is detected and amplified in the synchronous detector inside the carrier leak detector with the carrier as reference. The output of the synchronous detector passes through a LPF leaving only a DC signal.

In the figure, let us suppose that

$$\text{carrier} : E_1 \sin \omega_1 t$$
$$\text{carrier leak} : K \sin \omega_1 t$$

$E_1 \sin \omega_1 t$ and $K \sin \omega_1 t$ are mixed in the synchronous detector. Thus

$$E_1 \sin \omega_1 t \cdot K \sin \omega_1 t$$
$$= \frac{E_1 K}{2} \{-\cos(\omega_1+\omega_1)t + \cos(\omega_1-\omega_1)t\}$$
$$= -\frac{E_1 K}{2} \cos 2\omega_1 t + \frac{E_1 K}{2}$$

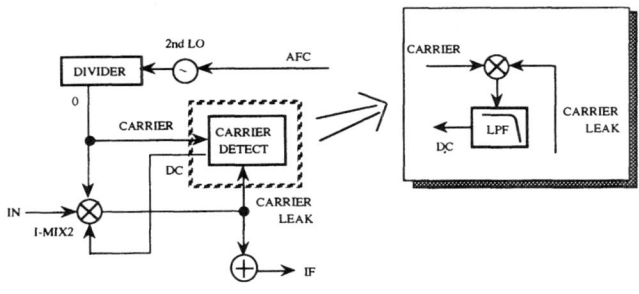

Fig. 4 Carrier Leak Suppressing System

When this passes through a LPF, the frequency component 2ω1 disappears. Therefore,

$$E_1 \sin\omega_1 t \cdot K\sin\omega_1 t = \frac{E_1 K}{2}$$

and becomes a DC signal that is directly proportional to the level of the carrier leak. This DC signal is used as feedback to the reference of MIX2 and cancels the DC offset caused by the unbalance. The carrier leak can be suppressed up to more than 60 dB and as a result, a sensitivity of -118 dBm can be guaranteed.

(2) Adjustment-Free System

The IF signal is demodulated in the FM detector after passing through the IF BPF. Because variations in the absolute value of resistors and capacitors due to the wafer manufacturing process and temperature changes affect circuit response, the FM detector and IF BPF forms a closed loop with the second local oscillator. The resistors and capacitors that determine the circuit response of the FM detector, IF BPF, and second local oscillator are designed with the same material, layout and geometry so that the closed loop will perform accordingly to changes in circuit response without compromising the characteristics of the whole system. Moreover, after demodulation, AFC is applied to the second local oscillator to eliminate the relative variation between the devices of the different circuits in order to assure a more stable system characteristic. All the adjustments to the system characteristics is done automatically by the system itself. No manual adjustment in the external discrete device is necessary to change and optimize the system characteristics. Thus, it is an adjustment-free system.

Selectivity

Since the new system is an off-chip-filter-less system, the selectivity is determined by the characteristics of the on-chip filters. The LPF of the I and Q subsystem is a fourth-order Butterworth LPF that uses an op amp. It is shown in Fig. 5 with its frequency characteristics in Fig. 6. There is a 30 dB attenuation at the point where the adjacent channel exists (at 15kHz).

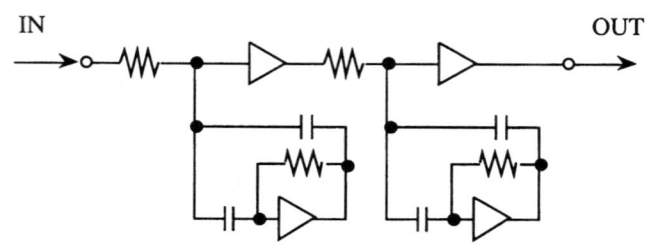

Fig. 5 LPF_I/LPF_Q
Fourth-Order Butterworth

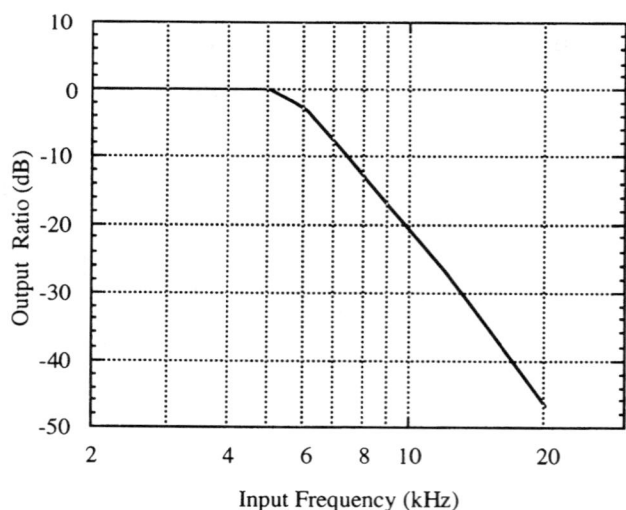

Fig. 6 LPF Frequency Characteristic

The IF BPF shown in Fig. 7 is also a biquad-type filter that uses an op amp. By combining low-pass, high-pass, and band-pass filter in its construction, there is a 50 dB attenuation at the adjacent channel point. This frequency characteristic is shown in Fig. 8.

Therefore, the whole system guarantees a selectivity of 80 dB (at 15 kHz).

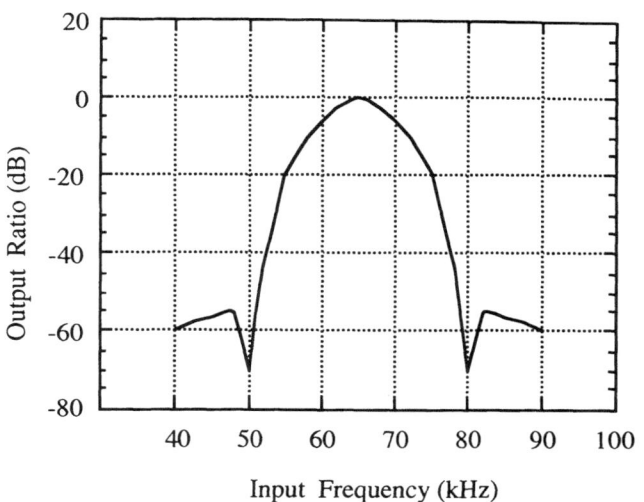

Fig. 8 IF BPF Frequency Characteristic

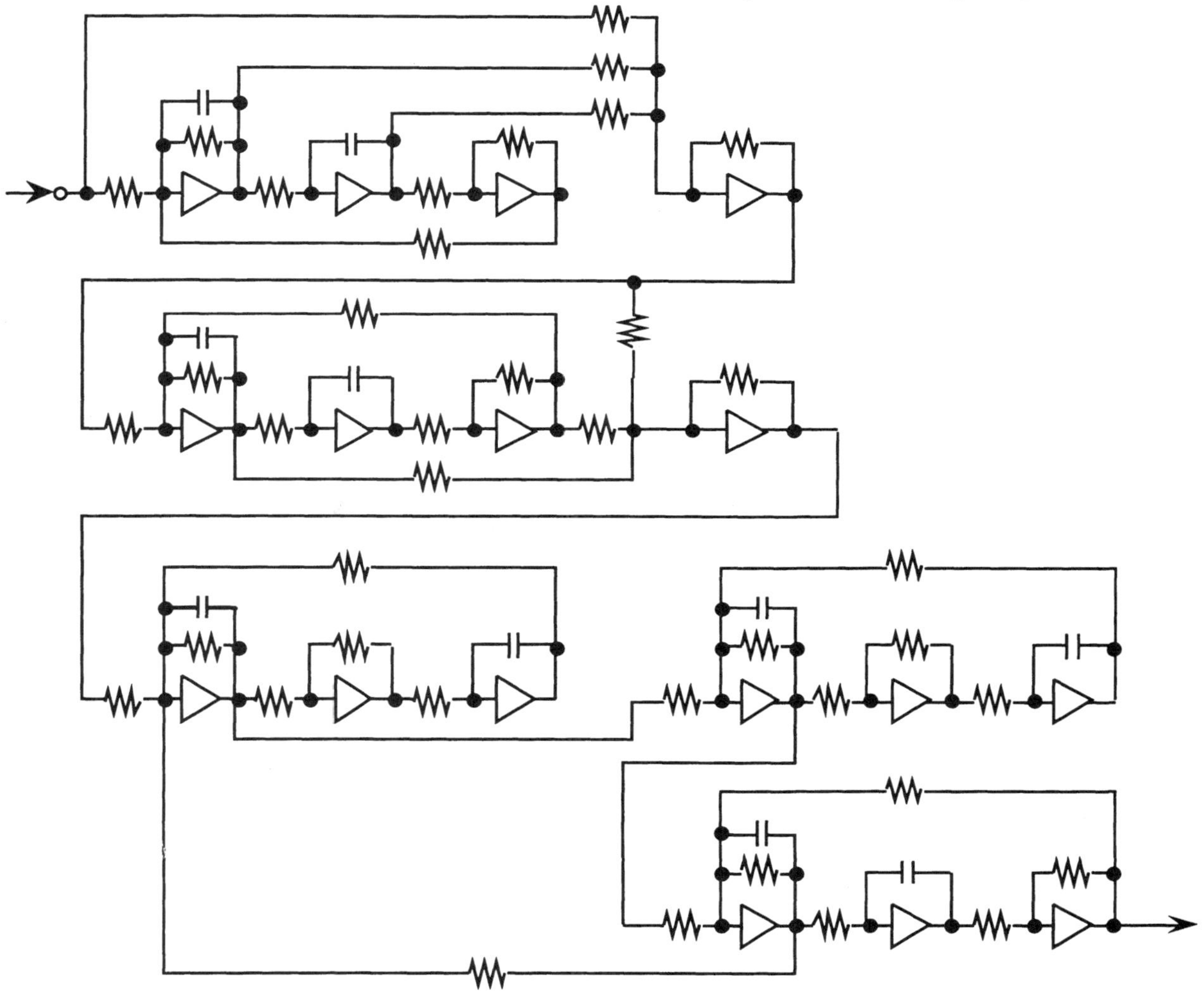

Fig. 7 Biquad-Type IF BPF

Characteristics of the New System

This section describes characteristics of the new radio receiver system as applied to analog cordless telephone operating at 46/49 MHz band in the U.S.A. The same characteristics hold for other types of personal communication devices.

Input-output characteristics is shown in Fig. 9. It also shows the THD (Total Harmonic Distortion) and RSSI voltage output. SINAD is -118 dBm, maximum S/N ratio is 60 dB, and THD is below 1%. The RSSI has an operating range of about 60 dB.

Fig. 10 shows the C/N ratio of the first local oscillator. C/N ratio is about 80 dB.

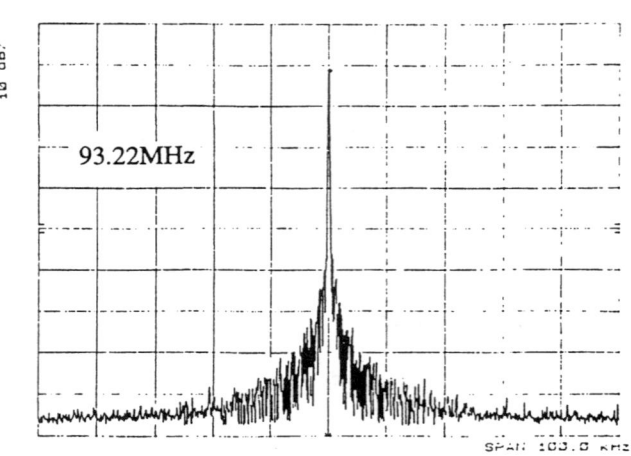

Fig. 10 First Local Oscillator VCO Spectrum

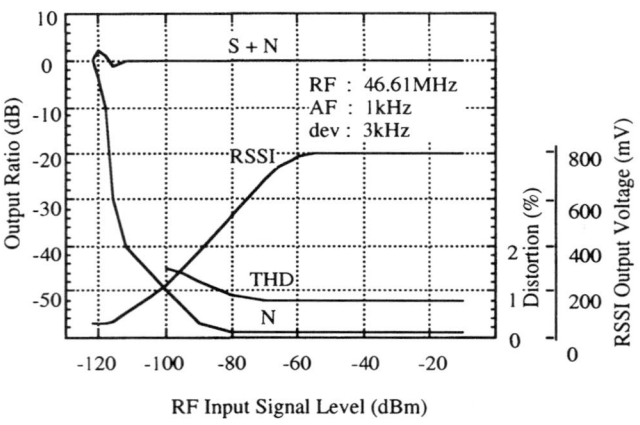

Fig. 9 I/O Characteristic

Conclusion

In conclusion, a new revolutionary technology in radio receiver system for personal communications is developed. This technology eliminates the need for external discrete IF filters by effectively implementing the concept of direct conversion and introducing new circuits that take into account the characteristics of semiconductor devices.

Acknowledgements

The authors would like to express their sincerest thanks to Mr. S. Horigome of InfoCom Products Company, Sony Corp., and Mr. Y. Yoshii and Miss Morizono of Semiconductor Company, Sony Corp. for their kind support in the project.

3

RECEIVERS

RECEIVER designs based on a wide variety of architectures, in fact, have many functions in common. A low-noise amplifier (LNA) is presently a universal requirement for the RF front end, as is the presence of one or more RF mixers to convert the high-frequency incoming signal to a much lower IF for further amplification and processing. Important specifications for these circuits are noise performance, interfering signal immunity, achievable dc power requirements, and overall performance stability in the presence of supply voltage and ambient temperature changes. The use of on-chip inductors has emerged as an important technique allowing increased gain, voltage headroom, and frequency selectively in the RF front end. Practical issues, such as package and substrate modeling, are crucial to the realization of successful designs. The actualization of the basic receiver functions over a range of technologies (GaAs, BiCMOS, and CMOS) is the focus of this section.

Receiver RF Design Considerations for Wireless Communications Systems

Ken Hansen (ekh002@email.mot.com) and Alexis Nogueras (ean002@email.mot.com)

Motorola, Inc.
Land Mobile Products Sector
8000 W. Sunrise Blvd.
Plantation, FL 33322

ABSTRACT

In order to design integrated circuit solutions for wireless communications systems, receiver RF design trade-offs must be understood. This paper will present the RF environment and RF design issues for receivers.

1. INTRODUCTION

Over the past 5-10 years, the growth in wireless communications has increased at a phenomenal rate. The growth has been primarily stimulated by the use of cellular telephone. However, there has also been substantial growth in the paging, two way land mobile, and cordless telephone communication businesses as well. Furthermore, with the FCC allocation of 120 Mhz in the 2Ghz range for PCS (Personal Communication Services), there will be a continued growth of wireless communications products in the future. This rapid growth has generated competition within the semiconductor industry for wireless communications solutions. Yet the RF environment and the RF design constraints that the silicon products must operate in is not well understood. In this paper, the RF environment and the key RF design issues will be described.

2. RADIO WAVE PROPAGATION

The wire-lined communications channel can be characterized as being stationary and predictable. However, in a wireless communications system the channel is constantly changing over time due largely to the movement of the mobile communications device and user. The reason for the change in the channel characteristics as a function of time is due to propagation effects. The RF power at a receiver will be a function of the distance and the geography (buildings, trees, hills) between the transmitter and the receiver. There are 3 types of propagation losses: 1) power law, 2) log normal shadowing, and 3) Rayleigh fading.

The power law effect is due to the power loss of an electromagnetic wave travelling through free space in a line of sight path from transmitter to receiver. Under the assumption that the transmitting antenna is in the far field of the receiving antenna, it is given by [1]

$$P_r = \frac{P_t G_t G_r}{[4\pi(r/\lambda)]^2} \quad (1)$$

where P_r = power at the receiver, P_t = power at the transmitter, G_t = gain of the transmit antenna, G_r = gain of the receiver antenna, r = distance the receiver is from the transmitter, and λ = wavelength of the RF. A more general expression can be derived to include reflections from the earth as shown in Figure 1. In this case it can be shown that [1]

$$P_r \approx P_t G_t G_r \left(\frac{h_b h_m}{d^2}\right)^2 \quad (2)$$

where h_b = height of the base antenna, h_m = height of the mobile

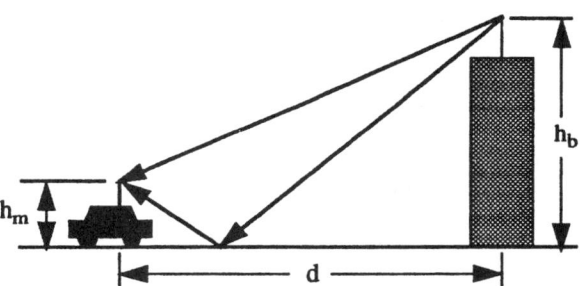

Fig.1 Propagation model for direct and reflected waves.

antenna, and d = the horizontal distance between the base and mobile. This result indicates that the relationship between the power at the receiver and the power at the transmitter is inversely proportional to the fourth power of distance. Therefore, for each doubling of distance from the transmitter, there is a 12 db loss in power at the receiver. This fairly simple model predicts reasonably accurately what is observed empirically in the field. Okumura [2] has collected field strength data in urban areas. For base station antennas less than 20 m over a range of 1 to 15 km, the power law exponent is 3.6 compared to 4 which the simple model would predict.

To a first order effect the power law expressions describe the power at the receiver. However, there is substantial variation in the receive power for small changes in distance. It has been found empirically that the distribution of received power measured at a constant distance from the base (circular pattern) is log normal with a typical standard deviation of 6.5 db. However, if distances over only a few wavelengths about a fixed distance are observed experimentally, the distribution of the envelope of the received power takes on a Rayleigh distribution and hence the name Rayleigh fading. The cause of Rayleigh fading is due to multi-path signals arriving at the antenna with different time delays (phase shift) and amplitudes. The vector sum of the incident waves produces the resultant power. By moving only a fraction of a wavelength, it is possible for the phases of the incident waves to substantially cancel the signal resulting in a significant attenuation (40db) of the received signal or a deep fade. The rate at which a fade occurs is a function of vehicle speed and is given by

$$t = \frac{0.5\lambda}{v} \quad (3)$$

where t = time between fades, λ = wavelength of RF signal, and v = velocity of the vehicle. For a vehicle travelling at 60mph receiving an 800 Mhz signal, a fade occurs every 7 msec or at a 143 Hz rate. It is possible for a portable user to locate itself in a fade. Communications can be interrupted if either the transmit or receive voice channel frequencies, or if the transmit or receive control channel frequencies is being affected by a fade.

In order to combat the effects of fading, the system must be designed with sufficient transmit power for the desired coverage area to prevent loss of communication during a fade. For the

International Symposium on Circuits and Systems, pp. 93-96, May 1996.

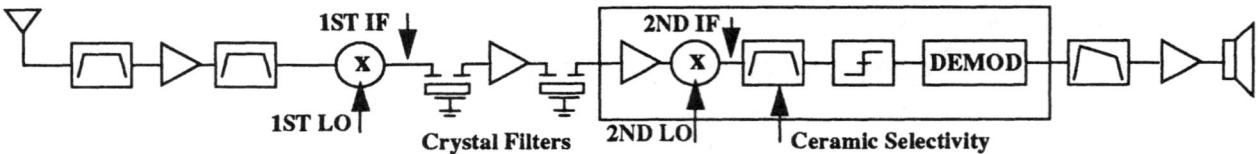
Figure 2: FM Receiver Block Diagram

---|F_1,G_1,IP_{1i}|---|F_2,G_2,IP_{2i}|--- ... ---|F_n,G_n,IP_{ni}|---

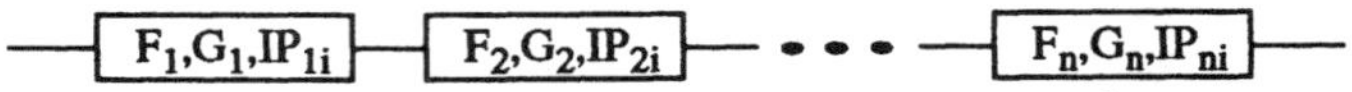

Figure 3: Cascaded Gain Stages

radio circuit designer, he must design circuits whose performance is insensitive to the rapid change of RF power or at least mitigates their effects.

3. RADIO INTERFERENCE

The goal of a receiver design is to optimize its performance in the presence of interferers. The primary sources of interferers are undesired transmitters at the same frequency (co-channel interferer) or at a frequency one channel away (adjacent channel interferer). However, there are other interference sources such as man-made noise from automotive ignition systems, power distribution or transmission lines, industrial equipment (motors, welders), and consumer products (fluorescent lights, TV local oscillators, garage door openers).

Co-channel interference, since it is at the receivers desired frequency, can only be minimized by geographic separation and by reducing the number of potential interferers. From the power law equation (2), a first order estimate of a carrier and an interferer can be made. For FM systems a minimum carrier to interferer (C/I) level of 17-18 db is required for acceptable voice quality. In the case of cellular systems, high capacities are achieved through frequency reuse and cell splitting which creates a co-channel interferer. A cellular system must be planned to achieve at least 17 db C/I to operate effectively.

A typical FM receiver block diagram is shown in Figure 2. The adjacent channel interferer is attenuated by the crystal and ceramic filtering. Crystal filtering typically provides 15-20 db of protection and the remainder (determined by system spec) is provided by the ceramic filtering.

4. RECEIVER BLOCK DIAGRAM

The receiver block diagram shown in Figure 2 consists of an RF front end, a 1st IF (Intermediate Frequency), a second IF, and an audio processing block. The RF selectivity is typically 3-8 Mhz wide depending on the application, consisting of 2 poles per stage with an insertion loss of 2-3 db per stage. The 1st mixer frequency translates the RF carrier to an IF frequency. All further signal processing can be done at single tuned frequencies. Each crystal filter typically consists of a 2 pole stage with 3 db of insertion loss. The 2nd IF circuitry is referred to as the receiver backend and is typically integrated including the selectivity that would be provided by the ceramic filters. These filters typically consist of 8 poles with a 12 db insertion loss for ceramics. The audio processing circuitry conditions the audio per the system specifications and is typically integrated. Although the block diagram is for an FM system, all of the RF/IF blocks are essentially identical for all of the digital modulation formats. Many of the digital modulation schemes carry amplitude (not necessary in FM) as well as phase information which requires a different solution for the demodulation function as compared to FM. This is typically handled with a DSP whose input is the output of the 2nd mixer. The purpose of the amplifiers is to provide isolation between stages and gain to improve the receiver sensitivity. The placement and design of the selectivity and amplifiers balance the requirements of receiver sensitivity and intermodulation performance which will be discussed next.

5. RECEIVER SENSITIVITY

A receiver's ability to receive a signal is defined in terms of SINAD. SINAD is measured at the input to the speaker and is defined to be

$$SINAD = \frac{S+N}{S+N+D} \qquad (4)$$

where S = signal, N = noise, and D = distortion. The measure of receiver sensitivity is the RF signal level required to produce 12 db SINAD. 12 db SINAD is an arbitrary level but is considered an accepted standard. The distortion is caused in part by the FM process of receiving only limited sidebands. This degradation is primarily fixed by the system specifications for channel spacing and frequency deviation. Noise degradations are caused by the inherent noise of electronic circuits. For RF circuits, noise performance is characterized by noise figure. Noise figure is loosely defined as

$$F = \frac{(S/N)_I}{(S/N)_O} \qquad (5)$$

where F = noise figure, $(S/N)_I$ = ratio of signal to noise at the input, and $(S/N)_O$ = ratio of signal to noise at the output. If several blocks are cascaded as shown in Figure 3 with a noise figure F and power gain G, the overall noise figure of the system is given by Friis' formula[3]

$$F = F_1 + \frac{F_2-1}{G_1} + \frac{F_3-1}{G_1G_2} + ... + \frac{F_n-1}{G_1G_2G_3...G_{(n-1)}} \qquad (6)$$

Therefore, the overall noise figure is highly dependent on the first stage. It should have a low noise figure and a high gain. Noise figure can be related to receiver sensitivity through a receiver rise sensitivity measurement. Rise sensitivity (R) is defined to be

$$R = \frac{S+N}{N} \qquad (7)$$

The rise sensitivity is typically measured at the lowest IF frequency. It has been experimentally found that a 6 db rise correlates to 12 db SINAD. For a signal generator driving a matched load it can be shown that the required voltage level to achieve a given rise sensitivity is [4]

$$v = \sqrt{FkTBRg(R-1)} \qquad (8)$$

where k = Boltzman's constant, T = temperature in ^0K, B = bandwidth, and Rg = matched load impedance. For the typical case where Rg = 50 Ω and R = 6db, a family of noise figure curves is plotted in Figure 4 as a function of IF bandwidth.

Equation (5) can be rewritten as

$$F = \frac{1}{G}(N_I/N_O) \qquad (9)$$

Considering the RF and IF selectivity elements as noiseless, it can be seen from equation (9) that the insertion loss of each element then equals the noise figure of the element. For example, a crystal filter with a 3 db insertion loss has a noise figure of 3db. From equation (6) for two cascaded blocks, if the first block has an insertion loss of 3 db, the overall noise figure is F_2 + 3 db. Therefore, a block with an insertion loss provides a db for db degradation in the overall noise figure at that point.

6. INTERMODULATION

Intermodulation occurs when two or more sinusoidal signals are applied to a non-linear circuit. Under these conditions, the output will consist of the fundamentals, harmonics, and other spurious frequencies. A classical way of analyzing the problem is to represent a non-linear transfer function as a power series [5]

$$v_0 = a_0 + a_1 v_i + a_2 v_i^2 + a_3 v_i^3 + \ldots + a_n v_i^n \quad (10)$$

where V_0 = output voltage, V_i = input voltage, and a_i = power series coefficient. If the input consists of two sinusoids of equal amplitude (replicates standard test conditions),

$$v_i = v_1 \cos\omega_1 t + v_1 \cos\omega_2 t \quad (11)$$

then

$$\begin{aligned}
v_0 =\ & a_0 + a_1 v_1 (\cos\omega_1 t + \cos\omega_2 t) \\
& + a_2 v_1^2 \left[1 + \tfrac{1}{2}\cos 2\omega_1 t + \tfrac{1}{2}\cos 2\omega_2 t \right. \\
& \left. + \cos(\omega_1+\omega_2)t + \cos(\omega_2-\omega_1)t\right] \\
& + a_3 v_1^3 \left[\tfrac{9}{4}\cos\omega_1 t + \tfrac{9}{4}\cos\omega_2 t \right. \\
& + \tfrac{1}{4}\cos 3\omega_1 t + \tfrac{1}{4}\cos 3\omega_2 t + \tfrac{3}{4}\cos(2\omega_1+\omega_2)t \\
& + \tfrac{3}{4}\cos(2\omega_2+\omega_1)t + \tfrac{3}{4}\cos(2\omega_1-\omega_2)t \\
& \left. + \tfrac{3}{4}\cos(2\omega_2-\omega_1)t\right] + \ldots
\end{aligned} \quad (12)$$

For this analysis to be valid, it is assumed that the circuit is weakly non-linear. In order for this condition to be met, the higher order terms must be small compared to the fundamental amplitude $a_1 v_i$. For most communications circuits this is a good assumption.

A frequency domain plot of equation (12) is shown in Figure 5. Second order terms in the frequency domain consist of a dc component, the second harmonics of each input source, and the sum and difference frequencies of the two input sources. Third order terms are located at the fundamental, the third harmonics of the input sources, and the sum and differences of twice one input source with the other input source. The intermodulation ratio (IM) is defined as the ratio of the desired amplitude to the undesired amplitude. The particular frequency components that are of interest are those created by the sum and difference frequencies of the two input sources. Thus, second order IM is $(a_1/a_2)(1/v_1)$ and third order IM is $(4/3)(a_1/a_3)(1/v_1^2)$. In a wireless communications application, third order IM is the primary concern. Consider the typical measurement for IM, that is to place two interfering signals at one and two channels away from the desired signal such that ω_d = 150 Mhz, ω_1 = 150 Mhz + 25 Khz, and ω_2 = 150 Mhz + 50 Khz where ω_d = the desired frequency and ω_1 and ω_2 are interferers. From equation (12), there is an IM product at $2\omega_1 - \omega_2$ = 150 Mhz at the desired frequency. There is little protection to this IM product. The RF passband will pass each interferer unattenuated. However, the crystal filter can provide some protection. It has been shown that the IM ratio is improved 2 dB for every dB attenuation one channel away from the desired and is improved dB for dB for attenuation two channels away from the desired. [6] The remainder of the IM performance can only be achieved by

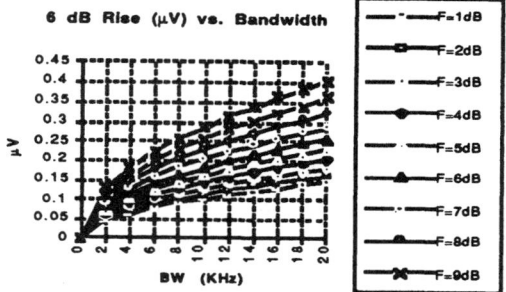

Figure 4: Receiver Rise Sensitivity vs. IF Bandwidth

designing circuits that are highly linear. For dual conversion receivers, second order IM is not important as the IM products fall well outside the RF passband. However, for direct conversion receivers, the second order IM product ($\omega_2-\omega_1$) will fall at the desired frequency.

In order to specify the IM characteristics of a circuit, two of the following must be specified; 1) IM ratio, 2) level of the input interferer (v_1), or 3) the level of IM generated (v_c). A convenient way to describe these relationships is a mathematical concept called the intercept point. Figure 6 shows the intercept point concept plotted on log-log paper. Each line is a plot of output power versus input power. The slope of the fundamental is one and the slope of the mth order IM product is m. The intersection of the 2 lines defines the intercept point. Both an mth order input intercept point (IP_{mi}) and an mth order output intercept point (IP_{mo}) can be defined as shown in Figure 6. In practice, the circuit will gain compress prior to reaching the intercept point levels.

From Figure 6 using simple geometric techniques, the relationship between IM and intercept point can be determined. Several expressions are listed below where all units are in dB:

$$IP_{2i} = IM_2 + v_1 \quad (13)$$
$$IP_{2i} = 2(IM_2) + v_c \quad (14)$$
$$IP_{3i} = (IM_3)/2 + v_1 \quad (15)$$
$$IP_{3i} = (3/2)(IM_3) + v_c \quad (16)$$

where IP_{2i} = second order input intercept point, IM_2 = second order IM ratio, IP_{3i} = third order input intercept point, and IM_3 = third order IM ratio.

Referring to Figure 4, the overall intercept point for cascaded stages can be calculated to be [7]

$$\frac{1}{IP_{ti}} = \left[\left(\frac{1}{IP_{1i}}\right)^q + \left(\frac{G_1}{IP_{2i}}\right)^q + \left(\frac{G_1 G_2}{IP_{3i}}\right)^q + \ldots + \left(\frac{G_1 G_2 \ldots G_{(n-1)}}{IP_{ni}}\right)^q\right]^{\frac{1}{q}} \quad (17)$$

where q=(m-1)/2, m=order of the intercept, and IP_{ni}=input intercept of nth stage. Therefore, from equation (17), it is undesirable to add gain in the receiver path. This is in direct conflict with the requirement for best receiver sensitivity discussed in Section 5. Trade-offs in the gain and selectivity of each stage must be made to optimize the system.

7. SPURIOUS RESPONSES

Receiver spurious responses are defined to be an apparent on-channel response to an undesired signal or group of signals. Spurious responses are caused by non-linearities in the receiver circuits. Intermodulation is a type of spurious response. Other

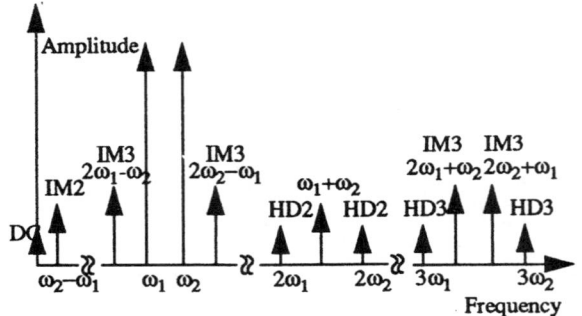

Figure 5: Spectral Relationship of IM components

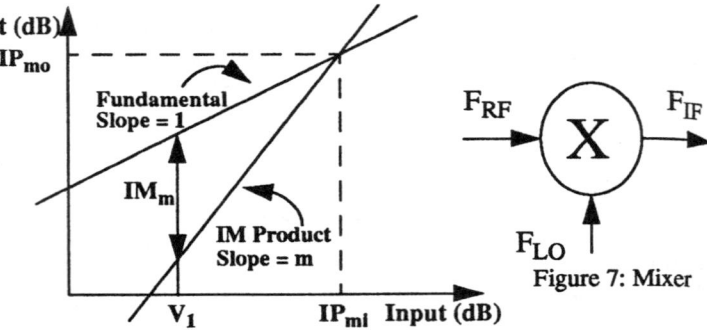

Figure 6: Intercept point

common spurious responses are the image, half-IF, and Able-Baker spurs. These spurs are generated at the mixer. A mixer function provides a frequency translation from the input frequency to an IF frequency. For the mixer shown in Figure 7

$$f_{IF} = f_{RF} - f_{LO} \quad (18) \text{ for low side injection or}$$
$$f_{iF} = f_{LO} - f_{RF} \quad (19) \text{ for high side injection}$$

where f_{RF} = RF frequency, f_{LO} = Local Oscillator (LO) frequency and f_{IF} = IF frequency.

The image spur (f_I) occurs at an IF frequency away from the LO frequency in the opposite direction from the RF frequency and is applied at the RF input of the mixer as an undesired signal. Therefore, the difference between the desired RF frequency and the image frequency is twice the IF frequency. Protection from the image can only be provided by selectivity ahead of a mixer. At low IF's, for example 455 KHz, the image is only 910 KHz from the desired input signal. For this reason 4 poles of crystal selectivity are typically required ahead of the second mixer.

The half IF spur (f_{HIF}) occurs at a frequency of 1/2 the IF frequency from the RF frequency towards the injection frequency. For low side injection

$$f_{HIF} = f_{LO} + f_{IF}/2 \quad (20)$$

Spurs are characterized by their order and are typically written as a spur of order (m, n) where m is the harmonic of the incoming spurious signal and n is the harmonic of the local oscillator required to produce a signal at the IF frequency. In the case of the half IF spur, it is of the order (2, 2).

Therefore,

$$2 \times f_{HIF} - 2 \times f_{LO} = f_{IF} \quad (21)$$

By rewriting this equation

$$f_{HIF} = f_{LO} + f_{IF}/2 \quad (22)$$

which is identical to equation (20).

The half IF spur is even more difficult to protect against than the image spur because it is located only half an IF from the desired RF frequency. In order to minimize its impact the second order non-linearities of the circuit have to be minimized.

Able-Baker spurs are defined to occur at frequencies where n and m are separated by one. Spurs of this type are likely to fall within the RF passband but tend to be higher order. It has been shown that the distance the spur is from desired is [8]

$$\Delta f = \frac{fd(n-m) - f_{IF}(n \pm 1)}{m} \quad (23)$$

where Δf = fs-fd, fs = spurious frequency, and fd = desired frequency. As Δf approaches 0 the spurious interferer falls inside the RF and IF passbands. When $\Delta f = 0$, from (23) it can be seen that the lower the IF frequency, the higher the order the spur will be. This is desirable as the magnitude of higher order harmonics will be substantially reduced.

The choice of an IF frequency requires trade-offs to be made. The first criteria is to choose an IF frequency that is not identical to a powerful transmitter (such as a commercial FM station) due to direct pickup. To minimize the impact of the image and half IF spurs, it has been shown that the IF frequency should be made as high as possible. However, to minimize the impact of Able-Baker spurs the IF should be chosen as low as possible. The designer must balance these trade-offs.

8. CONCLUSION

Due to the rapid growth rate of the wireless communications industry, the semiconductor industry has begun to generate silicon solutions for these applications. Most of these solutions have been focused at the baseband processing. As semiconductor manufacturers begin to increase the level of integration, the factors discussed in this paper will have to be considered. Gain distribution and the impact it has on receiver sensitivity and intermodulation have been described. The origin of spurious responses and the impact of the IF frequency selection on these spurs has been presented. The need for highly linear and robust circuits that can follow RF fades has been identified. It will be necessary to follow these guidelines to develop highly integrated wireless communications products.

REFERENCES

[1] G. C. Hess, Land Mobile Radio System Engineering, Norwood, MA, 1993

[2] Y. Okumura, et al., "Field Strength and It's Variability in VHF and UHF Land-Mobile Radio Service", Rev. Elec. Communication Lab, vol. 16, Nos. 9-10, Sept.-Oct., 1968, p. 841

[3] Taub, H. and D. L. Schilling, Principles of Communications Systems, New York, NY, McGraw-Hill Book Co., 1971

[4] J. Cramer, "Analytical Basis for Rise and Takeover Measurements", Motorola Internal Technical Report, April 1969

[5] C. Lynk and J. Ganzel, "Receiver Intermodulation Analysis and Calculations", Motorola Internal Technical Report, 1962

[6] J. Heck, "Intermodulation Distortion Analysis", Motorola Internal Technical Report, 1993

[7] R.C. Sagers, "Intercept Point and Undesired Responses", Correlations: An Engineering Bulletin from Motorola Inc., Vol. V, No. 1, pp. 35-55, 1985

[8] J. Glasser, "Receiver Spurious Responses and Other Considerations in the Selection of an Intermediate Frequency", Motorola Internal Technical Report, 1968

Design and Performance of Low-Current GaAs MMIC's for L-Band Front-End Applications

Yuhki Imai, Masami Tokumitsu, and Akira Minakawa

Abstract —GaAs MMIC's with very low current and of very small size have been developed for L-band front-end applications. The MMIC's fully employ lumped LC elements with uniplanar configurations. There are two kinds of MMIC's: a low-noise amplifier and a mixer. The low-noise amplifier has a noise figure of 2.5 dB and a gain of 11.5 dB. The mixer has a conversion gain of 12.5 dB with small LO power of −3 dBm. Total current dissipation of the two MMIC's is less than 8 mA with 3 V drain bias voltages.

I. Introduction

THE recent development of GaAs monolithic microwave integrated circuits (MMIC's) promises highly integrated, low-cost modules for system applications [1]. So far, however, power consumption of MMIC's has not been of great concern in designs, although it is very important in battery-operated instruments such as portable radio units. In portable radio communications, the need for GaAs MMIC's is obvious because MMIC's allow a substantial reduction in size and weight of the unit. For this application, MMIC's have to be designed to operate at very low current levels to conserve the battery drain in portable units [2], [3].

This paper reports the development of GaAs MMIC's with very low dissipation for L-band front-end applications. There are two kinds of MMIC's: a low-noise amplifier and a mixer. A key feature of the MMIC design is the use of lumped *LC* matching circuits with a uniplanar configuration to realize low current dissipation and small size. Here, we present the design, fabrication, and test results for the MMIC's developed.

II. Circuit Design

Fig. 1 shows the front-end module design, consisting of a low-noise amplifier and an FET mixer. The module was designed to use an external filter to reject out-of-band noise. Circuits were designed using the circuit simulator SPICE with a modified FET model and Touchstone.

A. Low-Noise Amplifier

The key issue in minimizing current dissipation is achieving a low noise figure and a high gain using FET's with the

Manuscript received May 15, 1990; revised September 17, 1990.
Y. Imai is with the NTT LSI Laboratories, 3-1, Morinosato Wakamiya, Atsugi-shi, Kanagawa 243-01, Japan.
M. Tokumitsu and A. Minakawa are with the NTT Radio Communications Systems Laboratories, 1-2356 Take, Yokosuka-shi, Kanagawa 238-03, Japan.
IEEE Log Number 9041093.

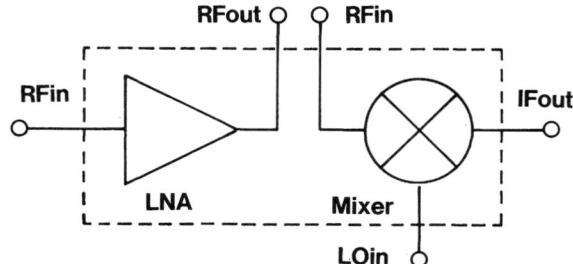

Fig. 1. Block diagram of L-band front-end module.

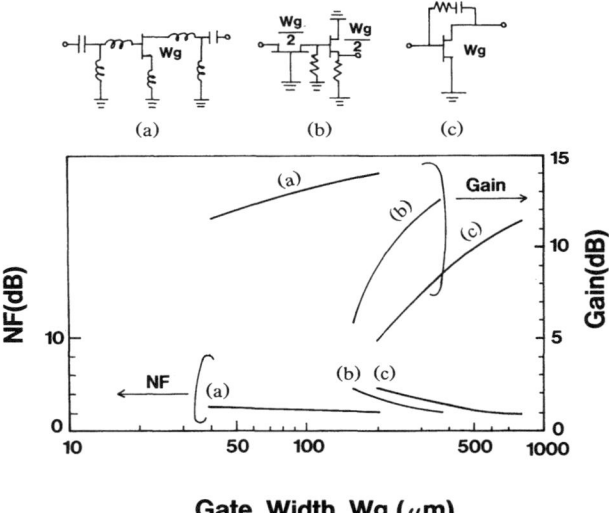

Fig. 2. Simulated gate-width dependencies of RF performance for low-noise amplifiers: (a) Lumped *LC* matching amplifier; (b) active matching amplifier; (c) *RC* feedback amplifier.

narrowest possible gate width. It was from this point of view that we approached our study of an optimum amplifier circuit. Fig. 2 shows three kinds of amplifier circuits and their simulated gate-width dependencies on RF performance. Circuit A fully employs lumped *LC* elements for impedance matching. Circuit B is an active matching amplifier that cascades common-gate and source-follower FET's [4]. Circuit C uses an *RC* feedback configuration for impedance matching [5]. Circuits B and C were generally adopted for UHF and L-band amplifier IC's for the broad bandwidths they allow. The performance of each circuit was

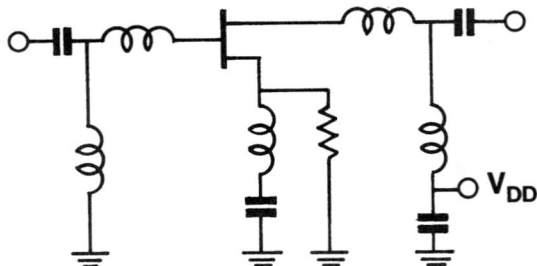

Fig. 3. Circuit schematic of low-noise amplifier.

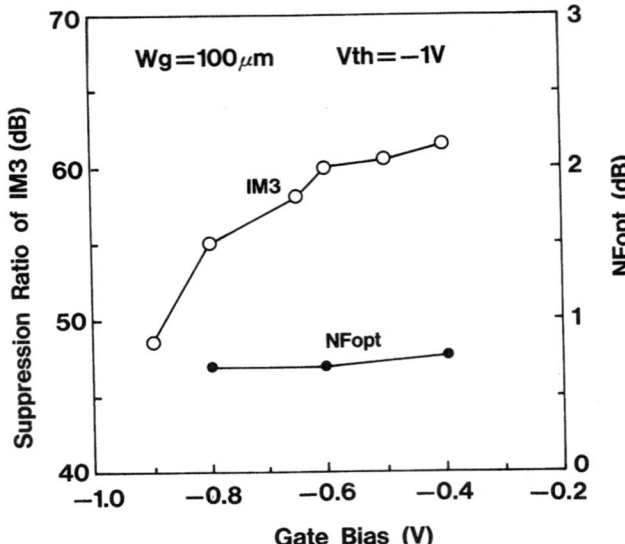

Fig. 4. Gate-bias dependencies of NF_{opt} and a suppression ratio of IM3.

simulated using an FET with 200 mS/mm transconductance and 50 mA/mm drain current. The noise figure and the gain were simulated at a frequency of 1.6 GHz with the matching condition that input and output reflection coefficients S_{11} and S_{22} were better than -10 dB. An active matching amplifier (circuit B) required a total gate width of greater than 350 μm to obtain better than 3 dB noise figure and 10 dB gain. For this circuit, a gate width of less than 150 μm was difficult to use because it severely degraded the matching characteristics. An RC feedback amplifier (circuit C) also required a large gate width of 500 μm to achieve the same performance because it needs a high transconductance to improve the noise figure and gain. On the other hand, a lumped LC matching amplifier (circuit A) was expected to have less than one fifth the gate width of these amplifiers to obtain a low noise figure and a high gain. Therefore, the lumped LC matching configuration allowed very low current operation of a low-noise amplifier using the narrow-gate-width FET.

Fig. 3 shows a schematic circuit of the low-noise amplifier. The amplifier used a 100-μm-gate-width FET with self-bias circuitry. Inductive series feedback at the source was employed to improve amplifier input VSWR and stability. Applying self-bias circuitry, the gate bias was optimized for low-current operation as well as high RF performance.

Fig. 4 shows the measured gate-bias dependencies of the FET optimum noise figure, NF_{opt}, and the suppression ratio of a third-order intermodulation product, IM3. IM3 was measured using two-tone equal-level input signals with -30 dBm power and 1.6/1.605 GHz frequencies. The gate width and threshold voltage of the FET were 100 μm and -1 V. Based on the results, a gate bias of -0.6 V was selected to provide a low-current operation with low NF_{opt} and IM3.

To use lumped LC matching elements in an L-band MMIC, a monolithic spiral inductor with a large inductance is necessary. Since a spiral inductor needs a high turn with a long line length to increase the inductance value, the parasitic elements are large and adversely affect the MMIC performance. Fig. 5 shows measured parasitic resistance and resonance frequency versus an inductance for monolithic spiral inductors. The line width and space of the spiral inductor were 5 μm to 10 μm to reduce the size. For an accurate MMIC design within 1–2 GHz frequency band, a spiral inductor with a resonance frequency greater than 4 GHz was used. On this condition, the spiral inductors employed in the MMIC's were held below 12 nH. The parasitic resistance also has a large influence on the MMIC performance, especially the amplifier noise figure owing to the loss of the input matching circuit. Therefore, the spiral inductor used three-level metal interconnect layers and a via metal to reduce the parasitic resistance. The total thickness was about 4 μm. The resistance was less than 5 Ω for the spiral inductor used.

RF performance of a lumped LC amplifier is mainly affected by the variations of the matching circuit elements. Fig. 6 shows the sensitivity of the circuit to variations of the inductor value. In the simulation, the values of all spiral inductors were changed uniformly. The variations of NF and gain were within 0.2 dB and 3 dB for each 10% increase and decrease of the inductor value. The predicted noise figure and gain for center values were 2.4 dB and 13 dB.

B. Mixer

A high conversion gain with a low current dissipation is required for a mixer. Built-in LO/RF isolation is also desirable to eliminate filter circuits. For this purpose, two kinds of mixers were studied, as shown in Fig. 7. Circuit A is a series-connected FET mixer with lumped LC matching circuits for RF and LO ports [6]. Circuit B is an analog multiplier based on Gilbert's cell [7] with RF and IF buffer amplifiers. Both circuits have good built-in LO/RF isolations. Fig. 7 shows the simulated current dissipation dependencies of the conversion gains for these mixers. The performance was simulated using an FET of 200 mS/mm transconductance and 50 mA/mm drain current. RF and IF frequencies were 1.6 GHz and 50 MHz with an LO power of -3 dBm. Circuit B required a current dissipation of greater than 17 mA to obtain a conversion gain. This is because large-gate-width FET's were required to increase the conversion gain. Circuit A, on the other hand, had a high conversion gain with a current dissipation of less than 5 mA. Therefore, circuit A is well suited to low-current operation with a high conversion gain.

Fig. 8 shows a schematic circuit of a mixer. RF and LO matching were provided using pi-type matching circuits to reduce the number of spiral inductors. A buffer amplifier was used to obtain high gain with small LO power and IF impedance matching. The gate widths of the FET's were 100 μm and 80 μm for the mixer and buffer amplifier, respec-

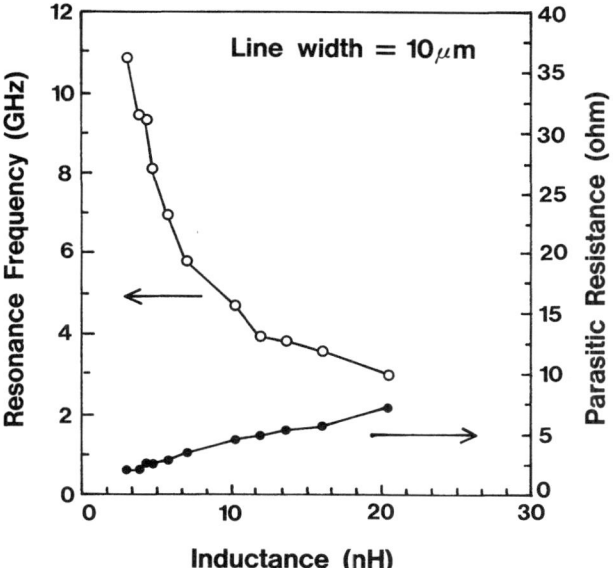

Fig. 5. Resonance frequency and parasitic resistance for monolithic spiral inductors.

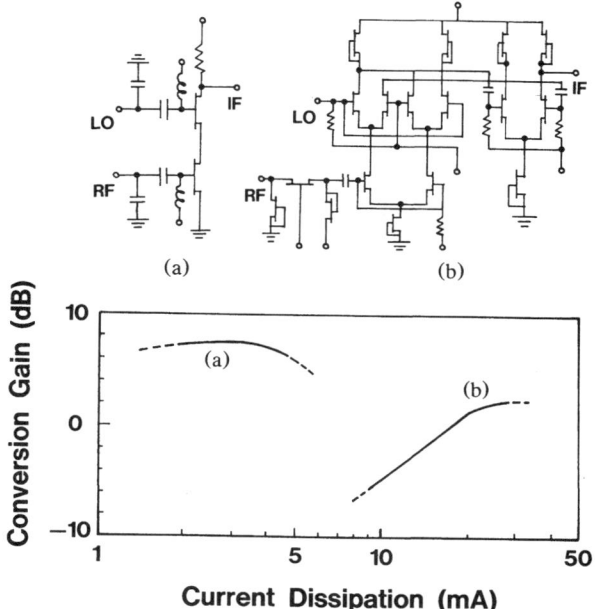

Fig. 7. Simulated current dissipation dependencies of conversion gain for mixers: (a) Series-connected FET mixer; (b) Analog multiplier based on Gilbert's cell.

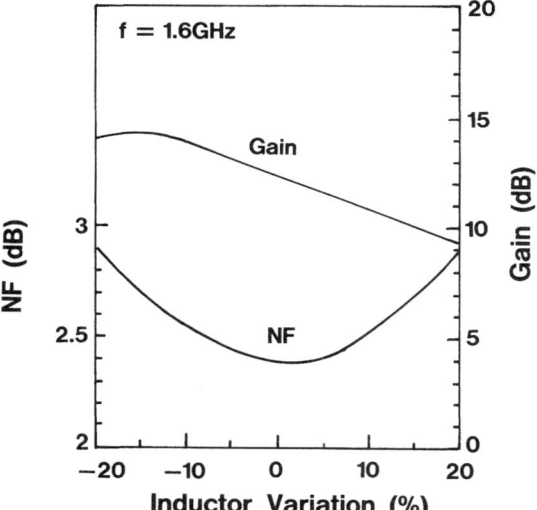

Fig. 6. Performance sensitivity to inductor values for a low-noise amplifier.

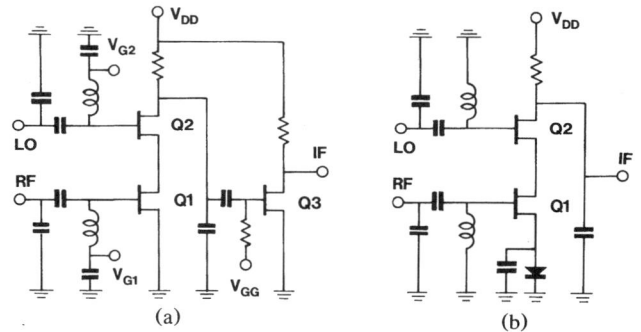

Fig. 8. Circuit schematics of mixers: (a) mixer without self-bias circuitry; (b) mixer with self-bias circuitry.

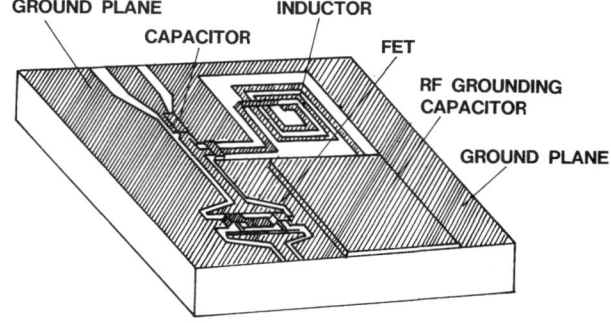

Fig. 9. Uniplanar MMIC configuration with lumped LC elements.

tively. A mixer with self-bias circuitry was also tested to simplify the power supply requirement. A schematic circuit is shown in Fig. 8. Because of low IF frequency, self-bias circuitry using a resistor with a bypass capacitor at an FET source degraded the conversion gain. Therefore, a level-shifting diode was used instead of a biasing resistor for the self-bias mixer. The mixer used a 200-μm-gate-width FET without a buffer amplifier. The predicted conversion gains for the mixers were 13 dB with a buffer amplifier and 5 dB without a buffer amplifier, respectively.

III. MMIC Fabrication

MMIC's were fabricated using advanced self-aligned implantation for n$^+$ layer technology (ASAINT) [8]. The gate length and the threshold voltage were 0.3 μm and -1 V. A uniplanar configuration was employed for the MMIC's [9]. A schematic of a uniplanar configuration with lumped LC elements is shown in Fig. 9. This configuration has no via holes and includes a coplanar ground plane close to each circuit element for easy access to an RF ground. It permits a compact MMIC layout because the large RF grounding

163

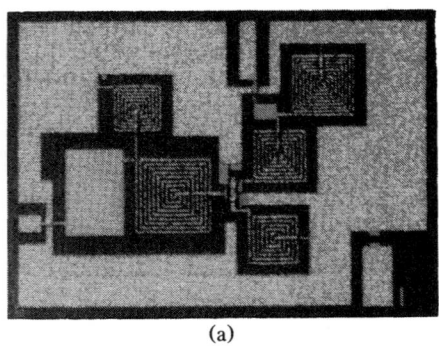
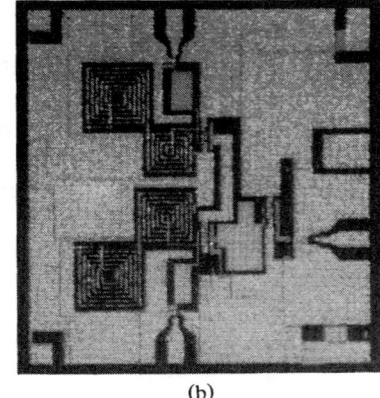

Fig. 10. Microphotographs of MMIC's: (a) low-noise amplifier; (b) mixer.

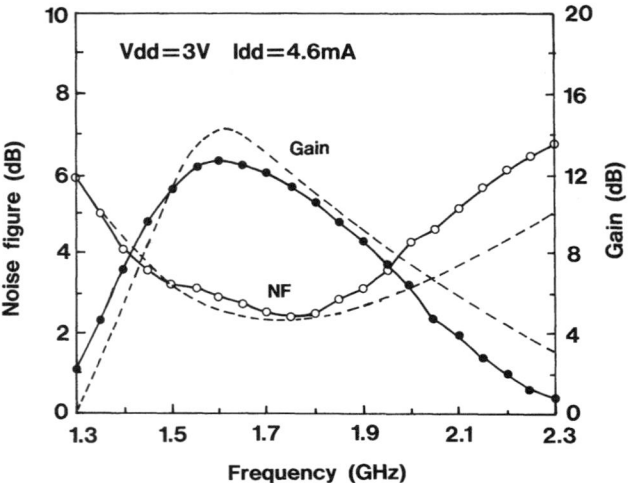

Fig. 11. Frequency dependencies of NF and gain for a low-noise amplifier. Solid lines: measured results. Dashed lines: simulated results.

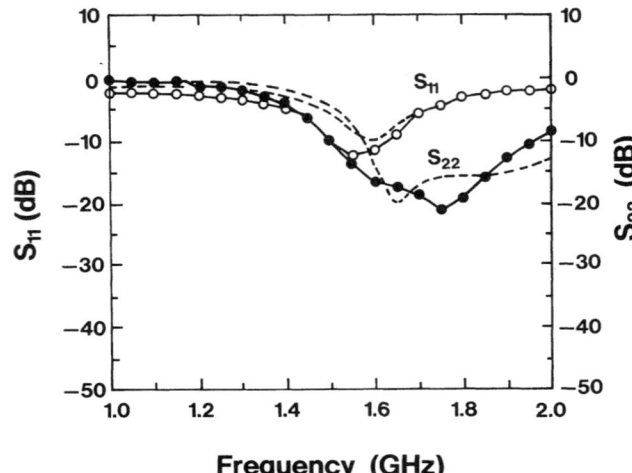

Fig. 12. Frequency dependencies of S_{11} and S_{22} for a low-noise amplifier. Solid lines: measured results. Dashed lines: simulated results.

capacitors can be easily integrated anywhere on the substrate. This feature is especially effective for L-band applications in which very large RF grounding capacitors are needed. Fig. 10 shows microphotographs of the low-noise amplifier and mixer using a lumped LC uniplanar configuration. Chip sizes were 1.5×2 mm^2 for the low-noise amplifier and 2×2 mm^2 for the mixer. A grounding capacitor with a capacitance exceeding 10 pF was used for each biasing line to eliminate an off-chip capacitor.

IV. Performance

A. Low-Noise Amplifier

Typical frequency dependencies of the noise figure and gain are shown in Fig. 11. Frequency dependencies of S_{11} and S_{22} are also shown in Fig. 12. The amplifier had a current dissipation of 4.6 mA and a single bias voltage of 3 V. The minimum noise figure was 2.5 dB with a gain of 11.5 dB. The amplifier had a gain of more than 11 dB and a noise figure of less than 3 dB in the 1.5–1.7 GHz frequency band. Simulated results are also shown in Figs. 11 and 12; they agree well with the measured results.

Amplifiers with different threshold voltages, ranging from -0.7 to -1.3 V, were measured to study current dissipation dependences of the RF performance. The results are shown in Fig. 13. The suppression ratio of IM3 was measured at -30 dBm input power with 1.6/1.605 GHz input signals. The gain and the noise figure were not degraded with current dissipation down to 3 mA. On the other hand, the suppression ratio of IM3 decreased as the current dissipation was reduced. The amplifier with 3.5 mA current dissipation provided a suppression ratio of more than 60 dB, which is acceptable for most front-end applications. A plot of the output power versus input power for this amplifier is shown in Fig. 14. The intercept point was 9 dBm. We compared the current dissipation with previously reported low-noise amplifier MMIC's operating in the L band and UHF bands and the results are shown in Fig. 15. The developed amplifier dissipated only one fourth, the current of previously reported MMIC's in the L and UHF bands [4], [5] [10]–[13].

B. Mixer

Typical frequency dependencies of the SSB noise figure and conversion gain for the mixer with buffer amplifier are shown in Fig. 16. LO power and IF frequency were -3 dBm

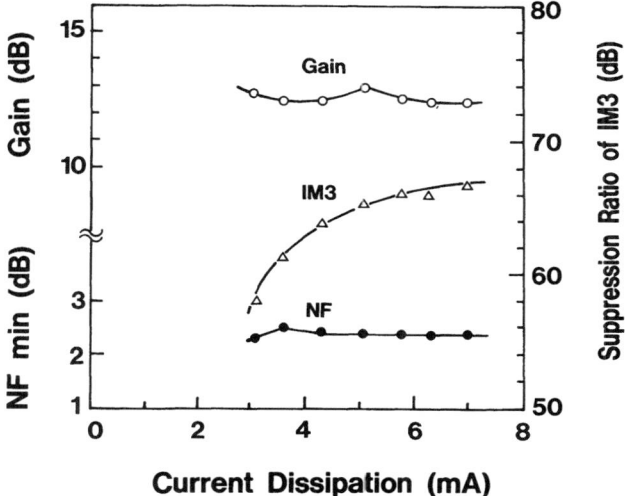

Fig. 13. *NF*, gain, and suppression ratio of IM3 versus current dissipation for amplifiers with different threshold voltages.

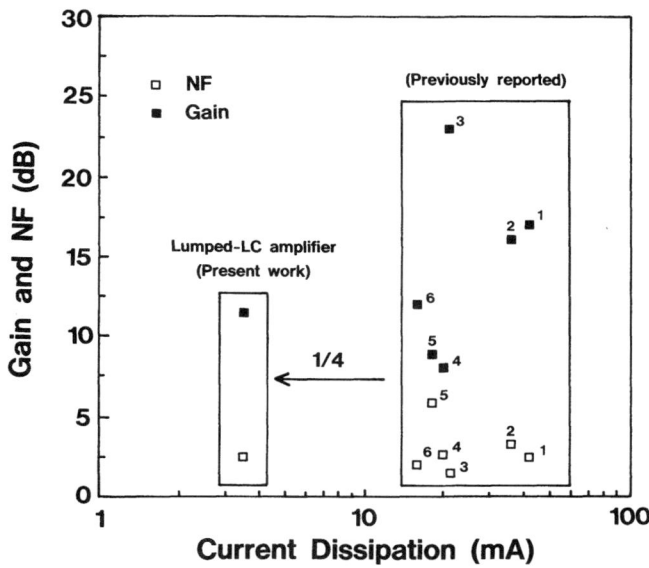

Fig. 15. Low-noise amplifier MMIC's operating in L bands and UHF bands. 1. $V_{dd} = 4$ V [13]. 2: $V_{dd} = 5$ V [12]. 3: $V_{dd} = 8$ V [5]. 4: $V_{dd} = 1$ V [11]. 5: $V_{dd} = 6$ V [4]. 6: $V_{dd} = 5$ V [10]. Present work: $V_{dd} = 3$ V.

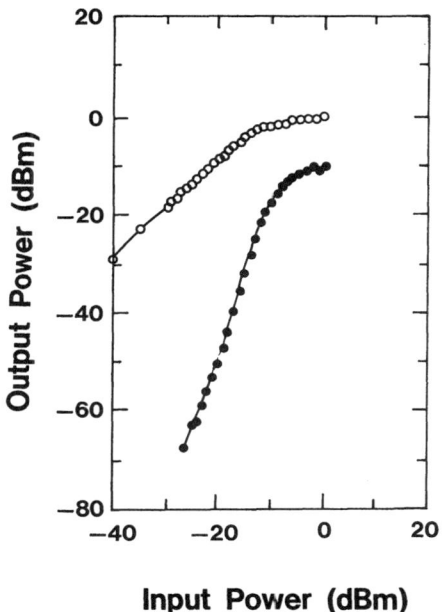

Fig. 14. Output power versus input power for amplifier with 3.6 mA current dissipation: ○ fundamental frequency; ● third-order intermodulation product.

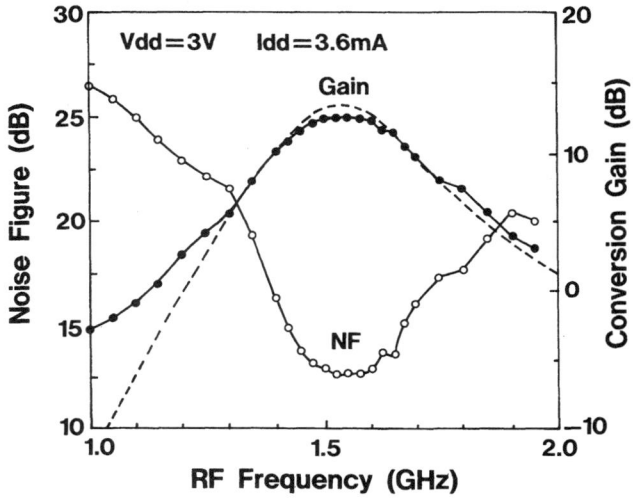

Fig. 16. Frequency dependencies of SSB noise figure and conversion gain for a mixer without self-bias circuitry. Solid lines: measured results. Dashed line: simulated result.

and 50 MHz. The current dissipation was 3.6 mA with 3 V drain bias and additional gate biases. The mixer had a maximum conversion gain of 12.5 dB with a noise figure of 12.6 dB. The simulated conversion gain using SPICE agreed well with the measured results. The suppression ratio of IM3 was measured under the same condition as the low-noise amplifier. The suppression ratio of IM3 and the gain were 46 dB and 12.5 dB for high-gain gate biases. For the medium-gain gate biases, the values were 63 dB and 8.3 dB. LO/RF isolation and LO/IF isolation were better than 20 dB and 30 dB, respectively.

Fig. 17 shows gate-bias dependencies of the conversion gain for a mixer with self-bias circuitry. RF and IF frequencies were 1.55 GHz and 50 MHz. The drain bias was 5 V. A conversion gain greater than 0 dB was obtained in the gate-bias ranges from 0 to 0.5 V for Q1FET and from −0.1 to 0.7 V for Q2FET, respectively. A self-bias mixer thus operated with only positive biases, which greatly simplifies the power supply. The current dissipation and conversion gain were 2.2 mA and 0 dB with 5 V drain and 0 V gate biases.

V. CONCLUSION

GaAs MMIC's with very low current and small size were developed for L-band front-end applications. Total current dissipation of the low-noise amplifier and mixer was 7.1 mA with a conversion gain of 25 dB using −3 dBm LO power. With these MMIC's, front-end-module size is reduced to about one fifth that of conventional hybrid circuit technolo-

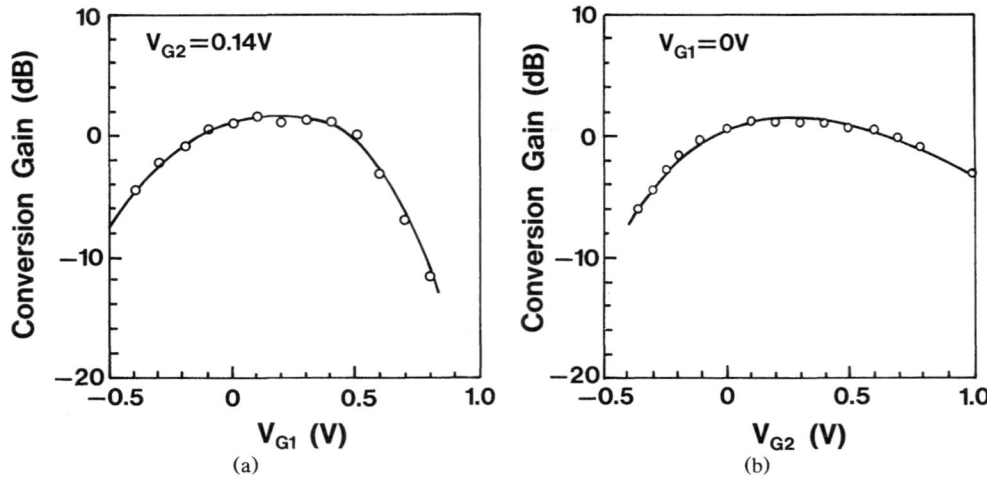

Fig. 17. Gate-bias dependencies of conversion gain for a mixer with self-bias circuitry. $VG1$: gate bias of Q1FET. $VG2$: gate bias of Q2FET.

gies. The developed MMIC's will have a great effect on telecommunication system applications, such as mobile radio units, thanks to their low current dissipation and small size and the elimination of circuit adjustments.

Acknowledgment

The authors wish to thank Y. Yamao for useful discussions on the system applications. They also thank T. Sugeta and M. Aikawa for their advice and encouragement.

References

[1] T. Hirota, M. Muraguchi, A. Minakawa, and K. Osafune, "A uni-planar MMIC 26-GHz-band receiver," in *IEEE GaAs IC Symp. Tech. Dig.*, 1988, pp. 185–188.
[2] V. Nair, "Low current enhancement mode MMICs for portable communication applications," *IEEE GaAs IC Symp. Tech. Dig.*, 1989, pp. 67–70.
[3] Y. Imai, M. Tokumitsu, A. Minakawa, T. Sugeta, and M. Aikawa, "Very low-current small-size GaAs MMICs for L-band front-end applications," *IEEE GaAs IC Symp. Tech. Dig.*, 1989, pp. 71–74.
[4] E. Kemppinen, E. Järvinen, and T. Närhi, "Design of an L-band monolithic GaAs receiver front-end with low power consumption," in *Proc. IEEE Int. Symp. Circuits Syst.*, 1988, pp. 2535–2538.
[5] M. Nishima et al., "A UHF GaAs multi-stage wideband amplifier with dual feedback circuits," in *IEEE GaAs IC Symp. Tech. Dig.*, 1987, pp. 223–226.
[6] C. Tsironis, R. Meierer, and R. Stahlmann, "Dual-gate MESFET mixers," *IEEE Trans. Microwave Theory Tech.*, vol. MTT-32, pp. 248–255, Mar. 1984.
[7] B. Gilbert, "A high-performance monolithic multiplier using active feedback," *IEEE J. Solid-State Circuits*, vol. SC-9, pp. 364–373, Dec. 1974.
[8] T. Enoki, K. Yamasaki, K. Osafune, and K. Ohwada, "0.3-μm advanced SAINT FET's having asymmetric N^+-layers for ultra high frequency GaAs MMIC's," *IEEE Trans. Electron Devices*, vol. 35, pp. 18–24, Jan. 1988.
[9] M. Muraguchi, T. Hirota, A. Minakawa, K. Ohwada, and T. Sugeta, "Uniplanar MMIC's and their applications," *IEEE Trans. Microwave Theory Tech.*, vol. 36, pp. 1896–1901, Dec. 1988.
[10] V. Pauker, P. Dautriche, A. Giakoumis, and A. Kazeminejad, "Normally-off MESFET analogue circuits," in *Proc. European Microwave Conf.*, 1984, pp. 631–633.
[11] J. Tajima, Y. Yamao, T. Sugeta, and M. Hirayama, "GaAs monolithic low-power amplifiers with *RC* parallel feedback," *IEEE Trans. Microwave Theory Tech.*, vol. MTT-32, pp. 542–544, May 1984.
[12] K. Osafune, N. Kato, T. Sugeta, and Y. Yamao, "A low-noise GaAs monolithic broad-band amplifier using a drain current saving technique," *IEEE Trans. Microwave Theory Tech.*, vol. MTT-33, pp. 543–545, June 1985.
[13] K. Honjo, T. Sugiura, T. Tsuji, and T. Ozawa, "Low-noise, low power dissipation GaAs monolithic broad-band amplifiers," *IEEE Trans. Microwave Theory Tech.*, vol. MTT-31, pp. 412–417, May 1983.

Ultra-Low DC Power Consumptions in Monolithic *L*-Band Components

Kenneth R. Cioffi, *Member, IEEE*

Abstract — A set of monolithic *L*-band components operating at milliwatt and sub-milliwatt dc power consumptions have been designed and fabricated. A maximum gain/power quotient of 19.1 dB/mW was recorded for a monolithic amplifier at a frequency of 1.25 GHz with a cascade of 2 MMIC amplifiers yielding a total gain of 15.3 dB on a total power consumption of just 800 μW. This is believed to be the highest gain/power quotient ever reported for a monolithic circuit at *L*-band. A four pole voltage controlled filter with low power amplifier gain stages showed a loss of 1.6 dB with 15% 3dB bandwidth on a power consumption of 6.75 mW at 1.575 GHz. A subsystem containing the chips was assembled and tested. The ultra-low power consumptions were obtained with a standard foundry process using an enhancement mode MESFET with a variety of design techniques. Yields obtained on two 4″ GaAs wafers were 96–100%.

I. Introduction

MINIMIZATION of dc power consumption is critical for prolonged battery life in portable RF applications. Presently, commercially available MMIC amplifier chips runs at power consumptions which are an order of magnitude or more higher than what is desired for many battery operated systems. Also, many of these chips are wastefully broadband which can increase the demands on front end desensitization and image reject filters. Interest in this area has been steadily increasing [1], [2], [5] because of an expanding commercial need in personal communications products such as pagers, wireless modems, and wireless local area networks, as well as defense related applications in portable communications. Portable Global Positioning System receivers and RF tags are of interest in both the defense and commercial sectors.

II. Design

Both device/component and circuit design optimization are important in obtaining minimum dc power consumption in MMICs. All of the design methodologies in this section are oriented toward reactive (L-C) matching networks rather than lossy (R-C) matching networks due to the improved gain and noise figure performance of this type of circuit. Specific amplifier designs are not discussed in favor of a general design methodology which applies to all of the amplifiers in this study. A voltage controlled filter which incorporated alternating bandpass and low power gain stages is discussed in the last section.

A. Device/Component Optimization

A 1 μM gate length MESFET process was chosen for this work because of its versatility and high reproducibility at the frequencies of interest. Since device transconductance is inversely proportional to device channel depth in the MESFET, shallower channel devices will yield improved performance at low power consumptions. For this reason, the gain potential of an enhancement-mode FET is superior to a depletion-mode FET at low biases given similar doping profiles. This type of device has been chosen for the circuits fabricated in this work. Other authors have previously incorporated E-mode MESFETs to achieve low dc power consumptions with excellent results [2], [3].

The device width must be chosen carefully in order to ensure the best circuit performance along with circuit yield. Smaller width devices theoretically appear to offer higher gain for the same power consumption because of their larger maximum stable gain at a particular power level; however, several factors degrade the performance as device width is decreased. At low power consumptions, as shown in Fig. 1, the device gain circles shrink as the power is reduced for a constant device width. Highly controlled element values are thus required to realize maximum device gain. Instabilities can easily result if the element values arre perturbed slightly from their intended values. In order to counteract this effect, the device width must be increased. Fig. 2 illustrates the effect of increasing the device width at a small constant power consumption of 1.50 mW. It can be seen that the low power matching problem is alleviated at larger device widths. Another problem with small width devices is that the higher impedance levels of the device require higher values of matching inductance which may translate into lower inductor Qs, higher noise figures, and lower gain. In contrast, a device which is too large will produce unacceptably small gain.

In order to determine the optimum device width, the effects of device width on circuit gain and yield are studied for an enhancement mode MESFET biased at 1 V and 1.5 mA. The *S*-parameter data was obtained by scaling data from a 300 μM device biased at $V_{ds} = 1.0$ V and $I_{ds} = 1.0$ to 3.0 mA in 0.5 mA increments. Data for 150, 180, 225, 300 and 450 μM devices at a power consump-

Manuscript received March 31, 1992; revised August 7, 1992.
The author was with Rockwell International, Inc., Anaheim, CA 92803. He is now with Wireless Access, Inc., 210A Twin Dolphin Drive, Redwood City, CA 94065.
IEEE Log Number 9203962.

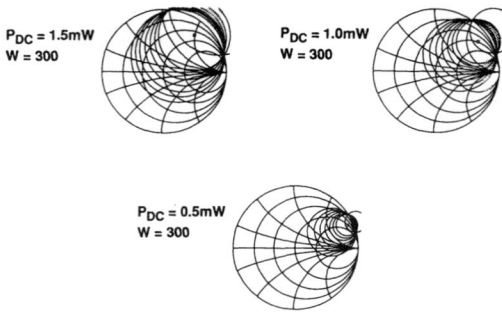

Fig. 1. Gain circles for three different device power consumptions at a constant width of 300 μM.

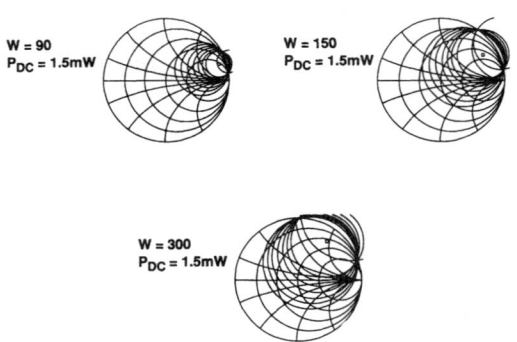

Fig. 2. Gain circles for three different device widths at a constant power consumption of 1.50 mW.

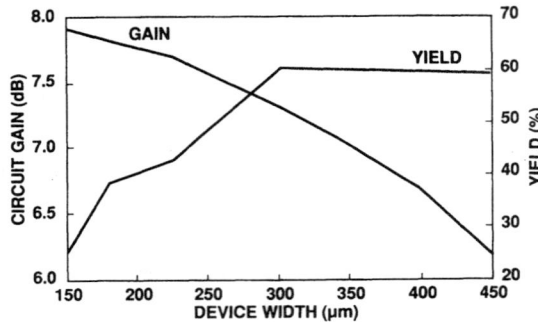

Fig. 3. Optimized circuit gain and corresponding yield vs. device width for a 1 μM enhancement mode MESFET at 1.5 mW total power consumption.

tion of 1 V and 1.5 mA was then derived from this data. Element values for a circuit with the narrowband topology discussed below were then optimized subject to the same optimization criteria for each of the device widths. Nonideal elements were used. Monte Carlo yield analysis was then performed assuming an angular S-parameter tolerance of 10° and capacitance variation of ±15%. A circuit was considered a failure if its gain varied by more than ±0.5 dB or its input or output VSWR exceeded 2:1 at a frequency of 1 GHz. The results of these simulations are shown in Fig. 3. It can be seen that very little improvement in gain was obtained at device widths below 200 μM at the cost of a drastic reduction in circuit yield. At device widths greater than 300 μM, no improvement in yield was obtained at the cost of a large reduction in circuit gain. A device width of 300 μM was chosen since only a small improvement in gain (≈ 0.5 dB) was obtained by reducing the device width to 200 μM at the cost of approximately 1/3 of the yield.

Choice of device bias point in generally a trade-off between power consumption, size, and system dynamic range and linearity requirements. In systems that do not have stringent linearity requirements (the majority of commercial portable and satellite communications systems), the device can be operated at a point where it will yield the highest gain/power consumption. Lower gain/stage at lower power consumptions will increase the number of required stages and, therefore, the amplifier size, however, the resulting reduction in the battery size of a battery operated system may be more significant in terms of overall size. Low power bias points may also be used in the first stages of a front end amplifier chain in a system that has higher linearity requirements so long as the overall linearity of the amplifier chain is preserved [7].

MESFETs obtain their highest gain/power efficiencies at low voltages around the "knee" (generally less than 1 V). This does not present any limitation on the use of MESFETs in systems with higher voltage supplies (typically 3 V to 5 V in battery powered systems) since device biasing can be easily achieved through voltage stacking (as demonstrated below) rather than current stacking as in conventional systems. The highest gain/power efficiencies for the 1.0 μM gate length enhancement mode MESFETs used in this work were obtained at voltages of 0.5 V and current consumptions of about 0.4 mA (total device power consumption of 200 μW).

Inductor Qs were optimized in order to realize the full potential of the minimal device gain at low power consumptions and to yield optimum noise figures. Thicker metal layers are desirable, however, when this is not an option, as with a fixed foundry process, wider line widths with narrow spacings yield higher Qs at the expense of circuit area. Fig. 4 shows a comparison of two inductors having equivalent low frequency inductance but much different geometries. The inductor Q is improved by a factor of 4 over the frequency of interest on the larger inductor. This area trade-off for performance may be acceptable given the application. The measured Q's for 2 μM metal thickness lines were between 13 and 19 for the inductors designed in this work. Typical inductor widths and spacing were 20–40 μM and 5–10 μM, respectively.

B. Amplifier Design

Two topologies, shown in Fig. 5, were used to obtain the results presented in this work. The two element matching network shown in Fig. 5(a) yields narrow band results, while the 3-element matching network gives a broader frequency response. The use of a source inductor is a standard low noise technique and is effective in shifting the noise figure circles closer to the gain circles resulting in low noise circuits with excellent gain and input match.

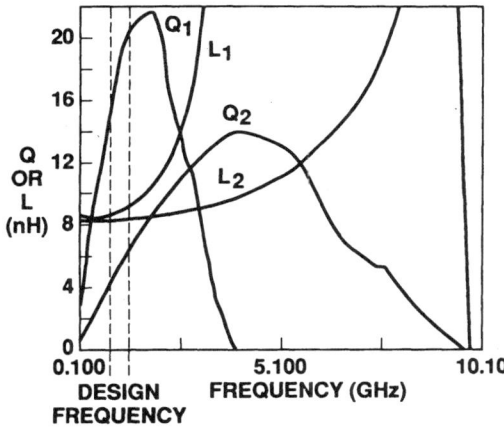

Fig. 4. Inductor values and Qs for two different size inductors; (Q1, L1) is 900 μM × 480 μM with a line width of 40 μM and spacing of 5 μM placed directly on a 25 mil GaAs substrate and (Q2, L2) is 190 μM × 190 μM with a line width and spacing of 5 μM suspended with air bridges over a 25 mil GaAs substrate.

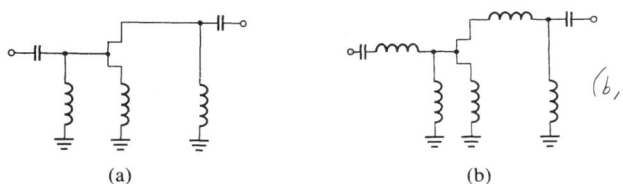

Fig. 5. (a) Narrowband and (b) broadband circuit topologies of the single-stage amplifiers fabricated in this work.

Bias networks were chosen to be efficient in terms of power consumption and at the same time allow sufficient temperature compensation and insensitivity to supply voltage variations. Active bias networks, although efficient in terms of area since they can reduce the number of required bypass capcitors, are not generally efficient in terms of power consumption. Simple resistive bias networks utilizing a source resistor were incorporated into the design in the self-biased circuits fabricated in this work. This type of network generally provides sufficient temperature compensation without the need for any additional circuitry. On board variable resistances supported operation on a variety of voltage supplies and at currents correlating to the noise figure and dynamic range requirements.

C. Voltage Controlled Filter Design

The block diagram and circuit diagram of a voltage controlled filter intended for receiver frequency hopping is shown in Fig. 6. Two single stage low power amplifiers having the topology shown in Fig. 5(a) were interspersed with two 2-pole filters. Each filter stage is a two resonator bandpass with Schottky diodes used as the voltage tuned capcacitive elements. A minimum bandwidth of 15% was determined to be a reasonable goal given the realizable Q's of the monolithic inductors and Schottky diode capacitance. The resonator inductance and capacitance values were 2.2 nH and a tunable 2.0–4.5 pF respectively. A depletion FET implant was used to form the diodes. A

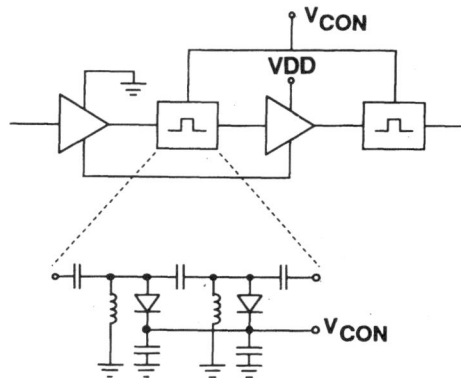

Fig. 6. (a) Block diagram of the 4-pole voltage controlled filter developed in this work and (b) circuit topology of one of the 2-pole filter blocks.

total diode area of 1200 μm × 1 μm was required to form the necessary capacitance values. Serial biasing was used for the two amplifiers stages to conserve on power consumption.

III. RESULTS

A. Amplifier Results

Amplifiers were designed for applications in the frequency range between 900 and 1575 MHz. The lumped element designs were fabricated on 25 mil 4" GaAs wafers without backmetallization. The results are summarized in Table I and plotted in Fig. 7 where they are compared with previously published data. The vertical scale uses a criterion which has been defined for comparison purposes as the circuit gain divided by its power consumption in mW. Input and output VSWRs of all of the amplifiers shown were better than 3:1 at the frequencies listed. Yields on the circuits were generally 100% with gain variations approximately ±0.3 dB/stage across two four inch GaAs wafers. A more complete set of statistical variations was given in [8].

The highest gain/power quotient was obtained on a two stage amplifier operating at 1.25 GHz on a total power consumption of 400 μW. Each of the two transistors in this amplifier were biased at 0.5 V with a current consumption of 400 μA. The gain and return loss of this amplifier are shown in Fig. 8(a). A self-biased version of this amplifier operated on a single 1.5 V supply at a current consumption of 0.5 mA with a gain of about 10 dB. Both versions of the amplifier were unconditionally stable at all frequencies. The non-self-biased version was designed to be easily cascaded using voltage stacking, i.e., the MESFET source was dc floated. In order to demonstrate the stacking feature, a two chip cascade was assembled, the results of which are shown in Fig. 8(b). In this circuit four stages are biased serially from a single 2 V supply to obtain a gain of better than 15 dB on a current consumption of only 400 μA.

The lowest noise figure of 1.6 dB was obtained on a single stage amplifier which operated at 1 GHz on a power consumption of just 5 mW. Input and output VSWRs were

TABLE I
TABULATED RESULTS FOR THE AMPLIFIERS DEVELOPED IN THIS STUDY

Circuit Description	Freq (GHz)	Bias Condition	P_{DC} (mW)	G (dB)	NF (dB)	Size (in^2)
1 GHz LNA	1.0	2.5 V @ 2 mA	5.0	10.3	1.6	0.52 × .112
	1.0	1 V @ 0.5 mA	0.5	6.0	2.5	
1 GHz Self-Biased Narrow-Band 2-Stage LNA	1.0	5 V @ 2 mA	10.0	19.6	2.2	.112 × .112
	0.9	2.5 V @ 0.5 mA	1.25	9.2	4.8	
Self-Biased GPS LNA	1.575	5 V @ 2 mA	10.0	8.5	1.7	.052 × .112
	1.575	2.5 V @ 1 mA	2.5	6.0	2.0	
2-Stage Amplifier	1.25	1 V @ .4 mA	0.41	7.2	6.0	.080 × .112
2-Stage Self-Biased Amplifier	1.25	1.5 V @ 0.5 mA	0.75	10.0		.080 × .112
Self-Biased GPS 2-Stage LNA	1.575	5 V @ 2 mA	10.0	17.4	2.2	.080 × .112
	1.575	2.5 V @ 0.5 mA	1.25	9.0	3.8	
Narrow-Band GPS LNA	1.575	2.5 V @ 2 mA	5.0	7.5	2.2	.052 × .052

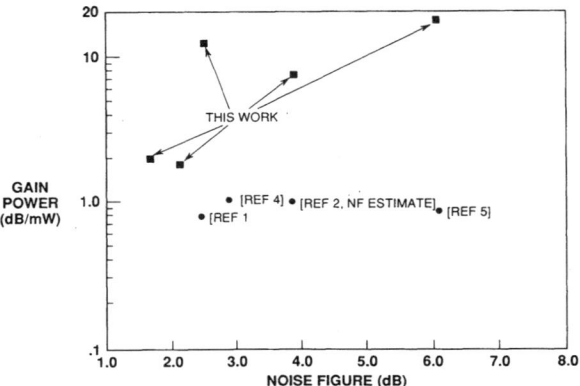

Fig. 7. Comparison of the results of the LNAs designed and fabricated in this work with previously published results.

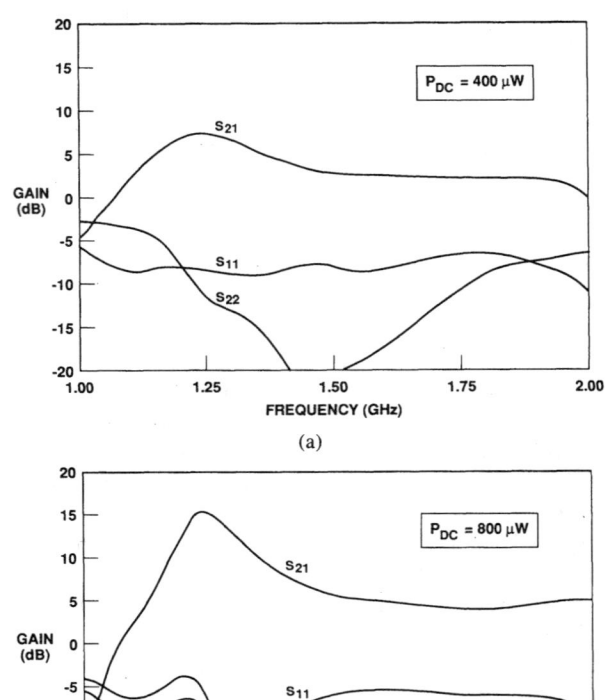

Fig. 8. (a) Gain and return loss of a two-stage low power MMIC amplifier biased at 1 V and 0.4 mA and (b) results for a two chip cascade of the circuit in (a) with 4 stage voltage stacking resulting in a total bias of 2 V at 0.4 mA.

better than 2:1. This amplifier used the broadband topology shown in Fig. 5(b) and had a usable bandwidth between 0.7 and 1.3 GHz. The transistor used in this circuit was a 450 μM width enhancement mode MESFET. This same amplifier operated at 500 μW with a noise figure of 2.5 dB and a gain of 6 dB. In order to demonstrate the feasibility of a low power transmitter for short range applications, this amplifier circuit was also characterized for large signal operation. Figure 9 shows measured power added efficiency and power output for the circuit. The nominal small signal bias condition for the amplifier was 2.5 V @ 2 mA. Gain is shown in Fig. 10. A maximum power added efficiency at 1 dB compression of 34% was obtained when the circuit was operated in class AB mode. Although the circuit was not specifically designed for large signal operation, its results demonstrate that reasonable efficiencies and gain can be achieved at very low power bias conditions.

Gain and noise figure for the self-biased two stage amplifier targeted for GPS applications are shown in Fig. 11. This amplifier had the topology illustrated in Fig. 5(b). The FETs were biased in series and an on-board variable bias network was available to adjust the amplifier to operate on different supply voltages and currents. The data for two bias conditions are illustrated. Input and output VSWRs were better than 3:1. The measured 1 dB compression point at 1.575 GHz was −2 dBm at 10 mW power consumption and −16 dBm at 1.25 mW power consumption.

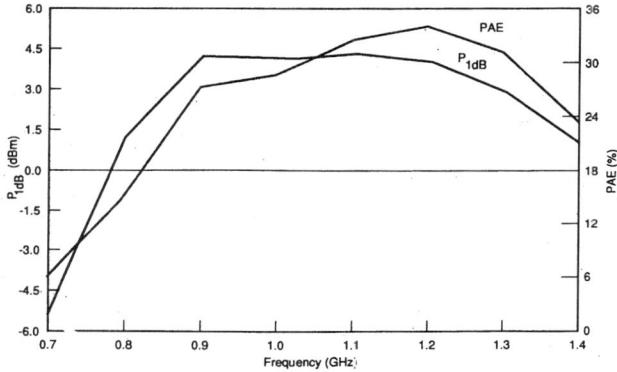

Fig. 9. Power added efficiency and output power at 1 dB gain compression for the low power broadband LNA.

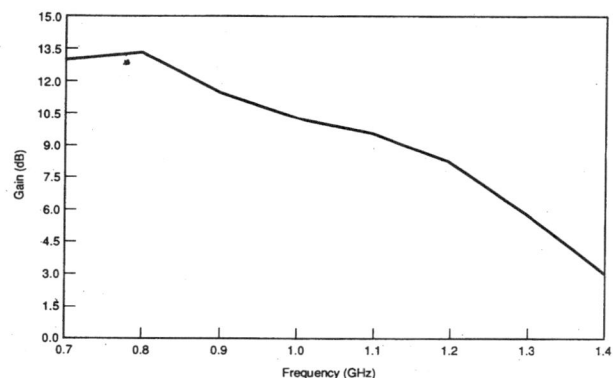

Fig. 10. Gain at 1 dB compression for the low power broadband LNA.

B. Filter Results

The 4-pole voltage controlled filter was mounted in a test fixture and characterized. A single supply voltage of 5 V was used to bias the amplifier stages. The total current from this supply was 1.35 mA for a total filter power consumption of 6.75 mW.

Frequency response and return loss as a function of the tuning voltage are shown in Fig. 12(a) and (b), respectively. The center frequency was tunable from 1.575 to 1.9 GHz with a tuning voltage ranging from -0.5 to 1.0 V. The bandwidth increased from 15% to 20% and the loss of the individual filter stages decreased as the center frequency increased. The overall filter response showed an initial decrease in loss as the center frequency was increased followed by an increase in loss at higher center frequencies. The latter increase in loss is due to the dominance of the amplifier response which exhibits a gain rolloff at frequencies beyond 1.5 GHz. The total loss varied from 0 to 3 dB over the frequency range.

The measured Q's for the 2.2 nH inductors used in the filter stages was 12 at 1.7 GHz. The low Q's resulted in maximum filter loss of about 7 dB for each of the 2-pole filter blocks. This loss was compensated by the amplifier stages. The diode capacitance varied from 2.0 pF to 4.8 pF with a Q of 54.2 to 16.8 as the voltage varied from -1.0 to 0.5 V.

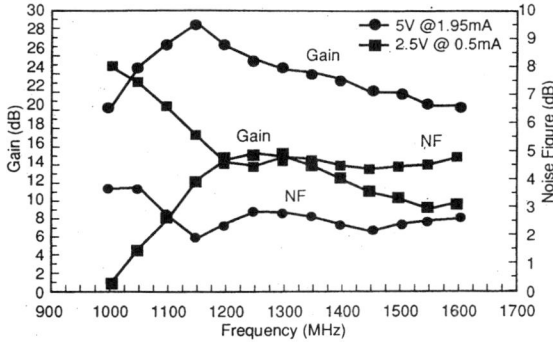

Fig. 11. Gain and noise figure for the 2-stage GPS amplifier.

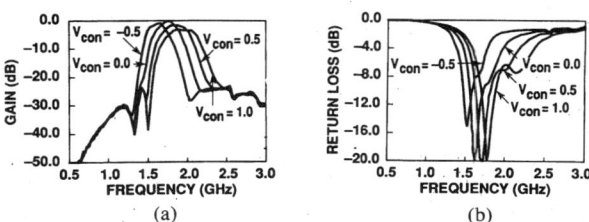

Fig. 12. (a) Frequency response and (b) return loss for the 4-pole voltage controlled filter tuned to center frequencies from 1.575 GHz to 1.9 GHz.

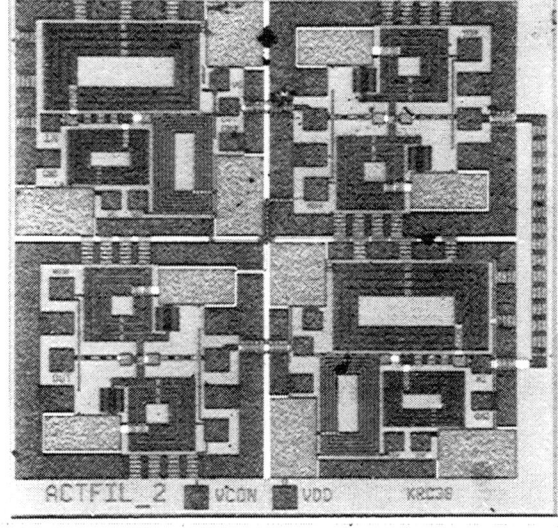

Fig. 13. Photograph of the 4-pole voltage controlled low power MMIC filter.

The filter was unidirectional due to the amplifier stages and showed a reverse transmission loss of greater than 25 dB. The size of each of the amplifier stages and each of the 2-pole filters in the layout is 1.3 mm × 1.3 mm. The fully integrated chip measured 2.8 mm × 2.8 mm and is shown in Fig. 13.

C. Module Results

A subsystem module was assembled using amplifier chips developed in this work. Two of the two-stage narrowband LNAs whose results were described in [8] were

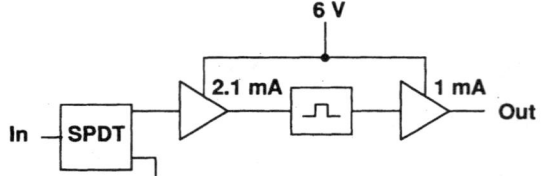

Fig. 14. Block diagram of the subsystem module formed with low power MMIC amplifiers.

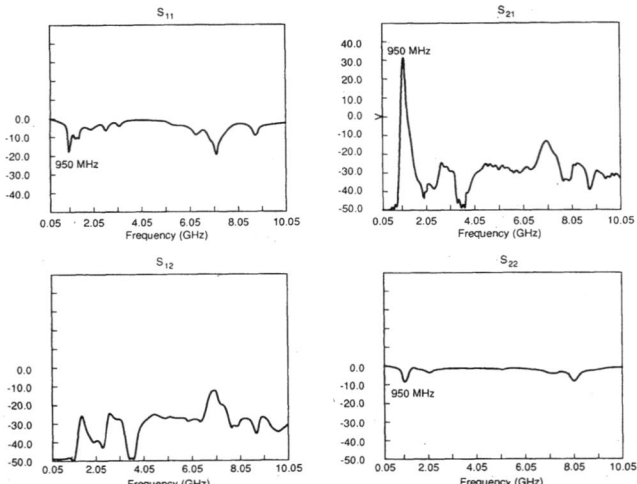

Fig. 15. Gain and return loss for the MMIC module.

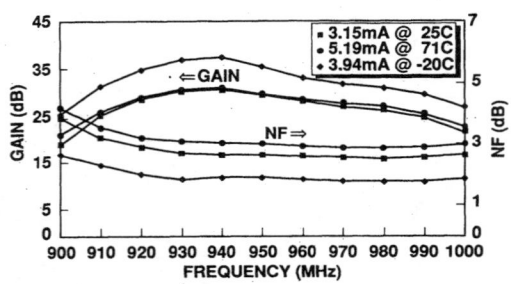

Fig. 16. Gain and noise figure for the MMIC module as a function of temperature.

combined with an interstage bandpass filter and commercial MMIC SPDT switch to form a subsystem module as illustrated in the block diagram of Fig. 14. The 25 mil thick GaAs chips were mounted on a 10 mil alumina substrate which contained via holes to ground. The interstage filter was also mounted to the alumina substrate in between the two amplifier chips and consisted of discrete inductors and capacitors. The alumina substrate containing the filter and LNAs was mounted on a test fixture with SMA connectors.

The gain and return loss from 50 MHz to 10.05 GHz is shown in Fig. 15. The module operated on a single 6 V supply with a total power consumption of 18.3 mW. The maximum gain for the module was 33 dB at 940 MHz which included about 5 dB of filter loss and 0.5 dB of switch loss. The 3 dB bandwidth for the subsystem was 40 MHz. Gain and noise figure variations as a function of temperature are shown in Fig. 16. Due to the temperature compensation of the mixed thin film and implanted resistor bias networks of the monolithic amplifiers, very little degradation in performance was observed as the temperature was increased. The size of the integrated module was $.375'' \times .450''$.

IV. Conclusion

A set of monolithic amplifiers operating on dc power consumptions of a few milliwatts down to less than a milliwatt have been demonstrated. In addition, a monolithic low power filter was demonstrated. The components were fabricated using a standard production ready foundry process and have achieved yields approaching 100%. Power consumption reduction in MMIC components is critical for the reduction of the overall size and cost of portable RF equipment.

References

[1] Y. Imai, M. Tokumitsu, and A. Ninakawa, "Design and performance of low-current GaAs MMICs for L-band front-end applications," *IEEE Trans. Microwave Theory Tech.*, vol. 39, pp. 209–215, 1991.

[2] V. Nair, "Low current enhancement mode MMICs for portable communication applications," in *IEEE GaAs MMIC Symp. Tech. Dig.* San Diego, Oct. 1989, pp. 67–69.

[3] P. Phillippe, and M. Pertus, "A 2 GHz enhancement mode GaAs down converter IC for satellite TV tuner," in *IEEE Microwave and Millimeter-Wave Monolithic Circuits Symp. Dig.*, June 1991, pp. 61–64.

[4] *Avantek Microwave Semiconductors GaAs and Silicon Products Data Book*, "Si bipolar amplifier," part no. INA-03170, 1991, pp. 4–18.

[5] H. Takeuchi et al., "A Si wide-band MMIC amplifier family for L-S band consumer product applications," in *IEEE MTT-S Int. Microwave Symp. Dig.* June 1991, pp. 1283–1284.

[6] K. W. Kobayashi, R. Esfandiari, M. E. Hafazi, D. C. Streit, A. Oki, and M. E. Kim, "GaAs HBT wideband and low power consumption amplifiers to 24 GHz," in *IEEE Microwave and Millimeter-Wave Monolithic Circuits Symp. Dig.*, June 1991, pp. 85–88.

[7] S. A. Maas, *Nonlinear Microwave Circuits*. Norwood, MA: Artech House, 1988, p. 172.

[8] K. R. Cioffi, "Monolithic L-band amplifiers operating at milliwatt and sub-milliwatt dc power consumptions," in *IEEE Microwave and Millimeter-Wave Monolithic Circuits Symp. Dig.*, June 1992, pp. 9–12.

GaAs MONOLITHIC 1.5 - 2.5 GHz IMAGE REJECTION RECEIVER

Angel Bóveda, G. Luca Bonato and Olga Ripollés

Telettra España S.A.
Av. de Cantabria 51, 28042 Madrid - Spain

ABSTRACT

Design and performance of a microwave receiver designed using GaAs monolithic technology is described. The receiver has been designed for use in digital radio links at 1.5-2.5 GHz. Its main features are low noise figure and image frequency rejection. The receiver consist of two GaAs chips: a low noise amplifier (LNA) and an image rejection mixer. The LNA achieves 19 dB gain with an associated noise figure of 1.7 dB. The image rejection mixer allows 20 dB of image suppression. The whole receiver exhibited a total gain greater than 31 dB and an associated noise figure lower than 2.5 dB. This performance is achieved over 1 GHz bandwidth. These results were obtained after the first processing iteration in the foundry.

INTRODUCTION

GaAs Microwave Monolithic Integrated circuits (MMICs) have promised high potential for reducing system cost. The objective of the receiver design is to develop a low cost front-end that results in improved system performance compared to more traditional microwave technologies using circuits for wide-band operation covering L-band and S-band frequencies. To achieve this goal, we use a well established GaAs technology with MESFETs of $0.8\mu m$ gate length. The front-end we have developed consists of two single chips: a low noise amplifier and an image rejection mixer that employes phase cancellation techniques to separate the downconverted products resulting from the undesired image and desired RF inputs. The low noise amplifier takes advantage of the low noise figure of GaAs transistors and improves the sensitivity of the full-receiver. Moreover, circuit symmetry obtained by mixer integration on a single chip improves the image rejection performance. An in-phase signal splitter, a quadrature signal-splittrer and two identical mixers are included in the image rejection mixer chip. This circuit generates two IF outputs that are combined in a external IF hybrid, which separates the desired signal and its image. The IF frequency can be chosen between 0 and 100 MHz selecting the components of the hybrid.

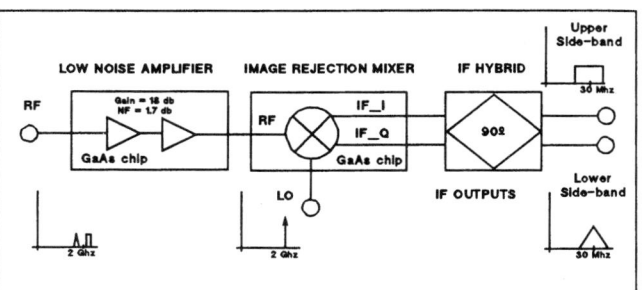

fig.1. Block diagram of the receiver

fig.2. Photograph of the Low Noise Amplifier chip.

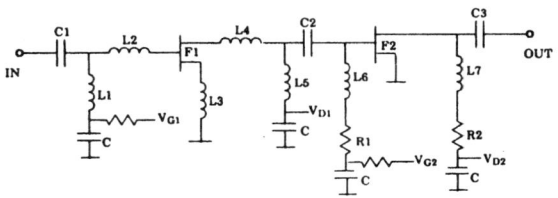

fig.3. Schematic of the Low Noise Amplifier.

International Symposium on Circuits and Systems, pp. 232-235, 1992.

LOW NOISE AMPLIFIER

Different topologies can be selected to design GaAs monolithic amplifiers. However, since the minimun noise figure is an essential requirement in our design, a reactive matching technique has been used. In this configuration a purely reactive network at the input of the amplifier matches the input line impedance (50 Ω) to the gate impedance of a common-source MESFET. The use of resistors in this network has to be avoided in order not to introduce additional termal noise. For the same reason, a parallel feedback configuration is not suitable for such application.

Two problems arise when designing reactive matching amplifiers; 1) there is a physical limit imposing a compromise between matching and bandwidth [1] and 2) in real MESFET, the optimum noise impedance, (that is the impedance to be presented to the gate for FET operating with minimum noise) is different from optimum input impedance [2].

These observations show the difficulty in achieving operation at minimum noise and input impedance matching, simultaneously.

Reactive Serial Feedback

These difficulties can be overcome including a small inductance between the MESFET source and ground. This inductance produces a change in the MESFET input and optimum noise impedances, so it is possible to make both impedances coincide (fig. 4)[2,3]. Moreover, the change in the value of the FET input impedance makes it easier to match. This configuration is rarely used in the design of amplifiers except in this case, because the introduced feedback limits the high frequency performance of the transistor, and brings about a response strongly decreasing with frequency.

Amplifier Design

In the chosen structure for the amplifier shown in fig. 3, reactive serial feedback is used in its first stage to achieve simultaneously good input matching and minimum noise operation of the FET. The input matching network is purely reactive. The second stage, designed to have gain increasing with frequency, compensates the opposite behavior of the first one to obtain an equalized amplifier gain. The use of resistors in the matching networks of the second stage is not a problem, because here the signal has been amplified already, and their contribution to the global noise figure is negligible.

Fig.2 shows the photograph of the low Noise Amplifier. The chip size is 1.6 x 1.8 mm.

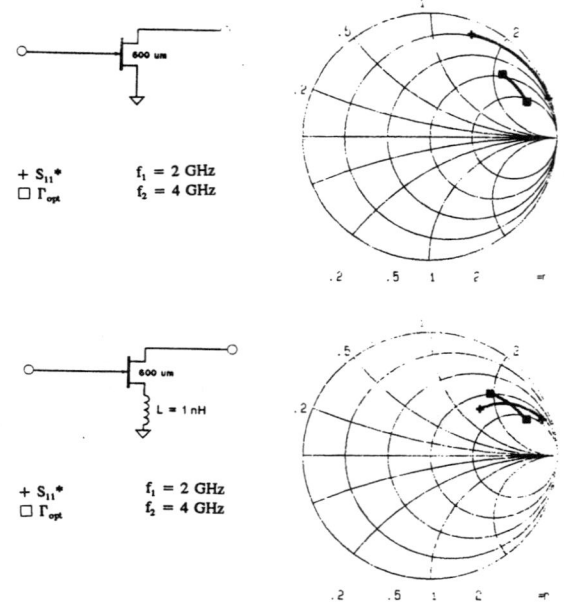

fig.4. Effect of the inductive serial feedback in the Input Impedance (S_{11}) and in the optimun noise impedance (Γ_{opt}) of a MESFET.

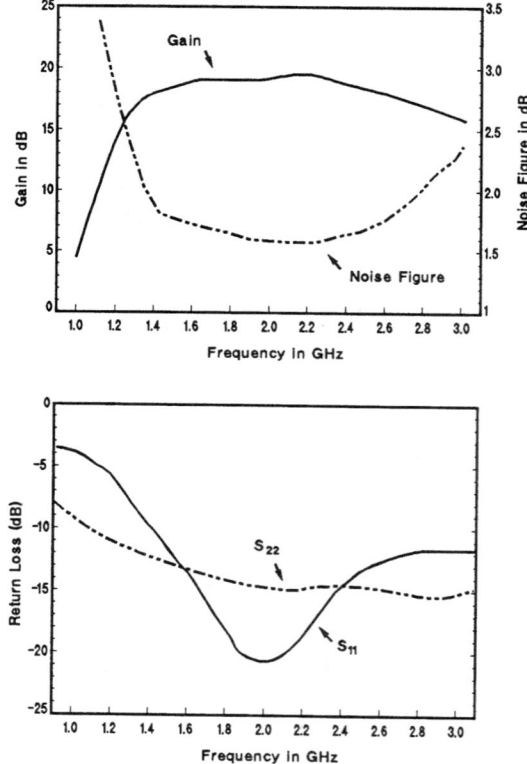

fig.5. Performance of the Low Noise Amplifier.

Amplifier Performance

A gain of 19 dB, a noise figure below 1.7 dB and good input matching (about -15 dB) is achieved over the full bandwidth (1.5-2.5 GHz). This performance is obtained using a low-cost 0.8 μm MESFET process. Fig. 5 shows the measured performance of the LNA.

SMALL SIZE LNA

In the LNA picture (fig.2) we can see that the most of the area is occupied by the inductors of the matching and bias networks. The chip size can be reduced if the bias inductors are replaced by active loads and the matching networks are simplified. This is made in the amplifier shown in fig. 6, designed for use in systems with less severe noise specifications.

In this version, the noise is not as good as in the first because the poorer matching networks and the noise introduced by the active loads, but the chip size is about one half the previous version (1.8 x 0.8 mm).

The performance is similar to the previous version, except the noise figure, that is slightly worse. Fig. 7 shows the measured Gain and Noise Figure of this Amplifier.

IMAGE REJECTION MIXER

In a heterodyne receiver, a radiofrequency signal f_{RF}, is mixed with a local oscillator tone f_{LO} to produce an Intermediate Frequency f_{IF}, so $f_{IF} = |f_{RF} - f_{LO}|$. There are two RF frequencies that produce the same IF, the Upper-sideband (USB) and the Lower-sideband (LSB). One of these bands is the desired signal band, the other is called the image-band, and if not eliminated will produce interference with the signal band.

Two procedures are used to avoid the interference of the image-band. The most simple is using filters. However as the signal and image bands are usually very close, these filters should be very selective. Because of this, they are expensive and difficult to manufacture, they operate only for a fixed local oscillator frequency, and it is not possible to integrate them in a monolithic circuit.

The best way to reject the image-frequency is by phase cancellation, by using two mixers in quadrature [4,5]. This configuration does not constrain the operation to only one LO frequency, and it is also suitable for monolithic implementation [3].

The block diagram of the image rejection mixer is shown in fig. 9. The RF signal is multiplied by two components delayed by 90° of the local-oscillator tone in two separated mixers, generating two IF signals. There is a different phase delay between the two IF components depending on whether they come from the signal or the image band, so they can be separated in a 90° IF hybrid.

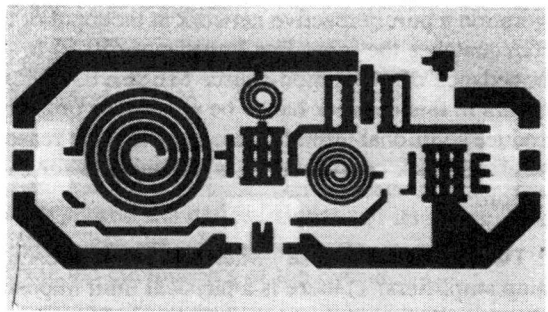

fig.6. Small size version of the LNA.

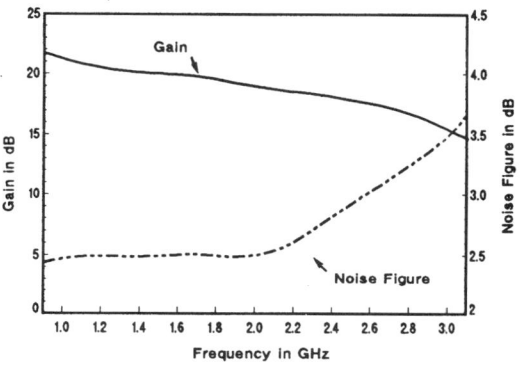

fig.7. Performance of the small size LNA.

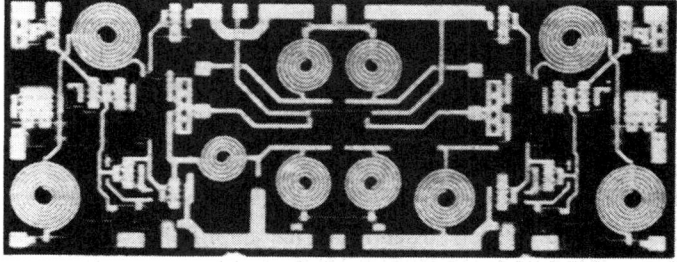

fig.8. Photograph of the Image Rejection Mixer.

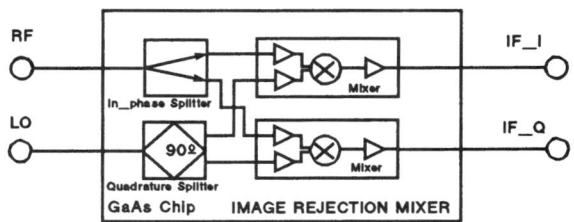

fig.9. Block diagram of the Image Rejection Mixer chip.

Figure 8 shows a photograph of the image rejection mixer chip. Its size is 1.2 x 3 mm.

IF Hybrid

The two outputs of the Image Rejection Mixer chip are combined in a 90° hybrid which separates the IF signals coming from the two sidebands. In our measurements a simple, lumped-type hybrid centered at 30 MHz has been used. The IF hybrid is made with standard low-cost, low-frequency components, which however would occupy a large area in the GaAs chip if they were integrated, so this aproach is unacceptable for reasons of cost. Another advantage of the non-integrated hybrid is that the IF frequency can be changed easily selecting the appropiate components for it. It is also possible to select the upper-sideband or the lower-sideband by changing the conection of the hybrid.

RECEIVER PERFORMANCE

The receiver was tested in all the RF bandwidth (1.5-2.5 GHz). For a selected IF frequency of 30 MHz the measured gain conversion was 30 dB with a LO power of 0 dBm. The image rejection was better than 17 dB all over the band. This performance can be improved up to 30 dB by a fine tune of the bias of the mixing FETs, and by using precision components for the IF hybrid.

The Noise Figure of the full-receiver was 2.5 dB. Fig. 10 shows the measured Conversion Gain and Noise Figure of the receiver. Fig. 11 shows the Image Rejection.

MMIC FABRICATION

The circuits, developed within a collaborative ESPRIT program, have been manufactured by Siemens, using its established MESFET production process, which includes 3" wafer, 0,8 μm self-aligned gate, and air bridges. Via-holes were avoided to ensure high-yield and low production cost.

CONCLUSION

A Monolithic Image-Rejection Receiver designed for use in 1.5-2.5 GHz radio links is presented. It is composed of two GaAs chips: the low noise amplifier and the image rejection mixer. The experimental results demonstrate a total conversion gain of about 31 dB with an associated noise figure of 2.5 dB, and more than 20 dB of image rejection.

ACKNOWLEDGEMENTS

The authors wish to acknowledge the efforts of the entire staff at Siemens Semiconductor Division during design and fabrication of the MMICs, and the Packaging Laboratory of Telettra SpA. for the assembling of the testing jigs.

This work has been supported by the European Community under ESPRIT 5018 Research Project.

REFERENCES

[1] R.M. Fano. "Theoretical limitations on the Broadband Matching of Arbitrary Impedances". Journal of the Franklin Institute, vol 249, pp 57-84, 139-154. Jan-Feb 1950.
[2] J.Enberg. "Symultaneous Input Power Match and Noise Optimization Using Feedback". European Microwave Conference Digest 1974, pp 385-389.
[3] R.Lehmann. "X-Band Monolithic Series Feedback LNA". IEEE Trans. Microwave Theory and Tech. vol MTT-33, Dec. 1985 pp 1560-1566.
[4] J. Van Heghe and R. Verbiest. "MMIC Receiver Front-End for 15 GHz Urban Link". 1990 IEEE GaAs IC Symposium, pp 267-270
[5] B.R. Halford. "Single-Sideband Mixers for Communication Systems". 1982 IEEE MTT Symposium Digest, pp 30-32.
[6] B.C. Henderson and J.A. Cook. "Image-Reject and Single-Sideband Mixers". MSN & CT, Aug. 1987. pp 75-83.

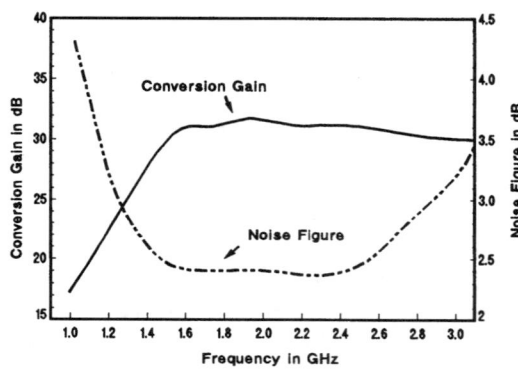

fig 10. Conversion Gain and Noise Figure of the receiver.

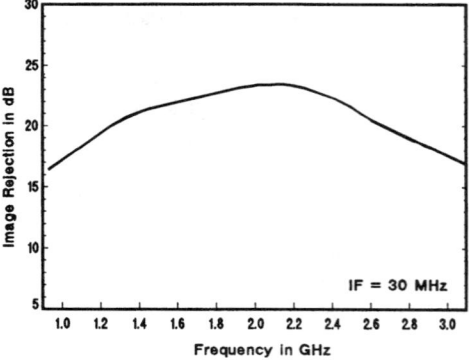

fig 11. Image Rejection of the receiver.

Low power GaAs ICs for mobile communication equipment

Hiroyuki Sakai, Akitoshi Tezuka, Yoshikazu Mori*, Morikazu Sagawa*,
Toshiaki Katoh, Junji Itoh and Kazuhisa Fujimoto

Semiconductor Research Center,
*Information and Communications Research Center,
Matsushita Electric Industrial Co., Ltd.
3-15 Yagumo-nakamachi, Moriguchi, Osaka 570, JAPAN
*3-10-1 Higashimita, Tama-ku, Kawasaki 214, JAPAN

ABSTRACT

A family of low power GaAs IC has been developed for RF section of mobile communication equipment by using asymmetric self-aligned LDD MESFET process. Circuit design based on dual-gate FET realized both excellent RF characteristics and low power dissipation (1/2 of Silicon IC). Fabricated receiver front-end IC showed conversion gain of 21dB for -15dBm local power, mixer/amplifier transmitter IC showed saturated output power of 10dBm, power splitter IC showed saturated output power of over 0dBm and inter-port isolation greater than 25dB, under 3V supply voltage. Dissipation currents were 4.5mA, 27mA, 3.5mA, respectively.

1. INTRODUCTION

With recent remarkable growth of land mobile communications, very compact and light weight equipments are strongly demanded. Integration technology[1-2] has contributed to the downsizing of mobile communication equipment. However, in order to miniaturize battery operated equipment, low power dissipation is very important as well. GaAs IC is a promising candidate both for integration and to decrease power consumption of the RF circuits operating up to GHz band.

In this paper, a family of low power GaAs IC's which has been developed for RF section of mobile communication equipment is reported.

2. DESIGN

Figure 1 shows the block diagram of RF section for mobile communication equipment. RF circuits of (1)receiver front-end, (2)up-mixer and PA driver amplifier, and (3)power splitter for local signal, shown in Fig. 1, are integrated in GaAs monolithic IC, respectively. In order to realize low voltage and low dissipation current operation, these IC's are designed by using simple

© 1993 IOP Publishing Ltd and individual contributors

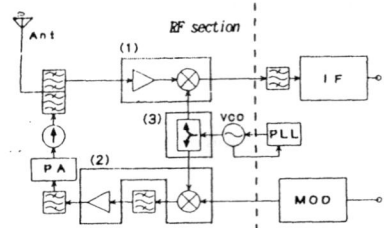

Fig. 1 Block diagram of mobile communication equipment (RF section)

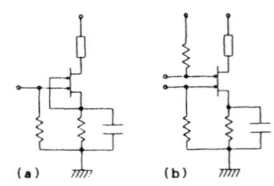

Fig. 2 Two basic circuits used in the IC's
(a) Basic circuit for an amplifier
(b) Basic circuit for a mixer

dual gate MESFET's circuits. The two basic circuits (amplifier and mixer[3]) used in these IC's are shown in Fig. 2. These circuits have the following features:
(1) Simple structure (suited for integration),
(2) Low voltage operation (FET's are not stacked),
(3) Superior characteristics in gain and noise,
(4) High stability and single power supply operation,
(5) Good isolation.

2.1 Receiver front-end IC

Figure 3(a) shows the block diagram of receiver front-end IC. Low noise amplifier, local amplifier and down-mixer are integrated in one GaAs chip. The mixer is connected to each amplifier by MIM (metal/insulator/metal) coupling capacitors on chip. In order to make this IC applicable to various frequency range, input and output matching circuits and load inductors are formed outside of the IC. In order to secure the isolation from output to input ports, both amplifiers are composed of dual gate MESFET and their second gates are effectively grounded as shown in Fig. 2(a). The mixer also has good isolation between RF and LO ports using the dual gate MESFET as shown in Fig. 2(b). There is no need to insert band pass filters, because the impedance matching circuit is designed to have narrow band width using high input impedance of GaAs MESFET's.

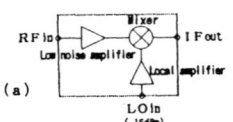

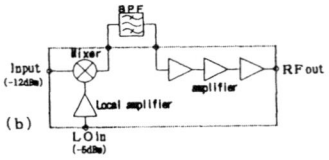

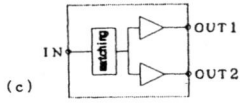

Fig. 3 Block diagram of the GaAs RF IC's
(a) Receiver front-end IC
(b) Mixer/amplifier transmitter IC
(c) Power splitter IC

2.2 Mixer/amplifier transmitter IC

Figure 3(b) shows the block diagram of mixer/amplifier transmitter IC, consisting of local amplifier, up-mixer and PA driver amplifier. This IC is designed to use an external filter to reject spurious frequencies. The circuit configuration of up-mixer is similar to the down-mixer of receiver front-end IC, but three stage single gate MESFET with inductive load is used for driver amplifier to obtain the enough output level for driving PA. The local amplifier consists of common gate FET circuit both for increasing conversion gain and impedance matching with local signal source. These circuits are connected to each other by MIM capacitors on chip, but load inductors are formed outside of the IC similar to that for receiver front-end IC.

2.3 Power splitter IC

When using high-density assembly technique to reduce the size of the equipment, it is difficult to secure enough isolation between each RF circuit. Power splitter IC is designed for use in dividing local signal from VCO to each RF block while keeping good isolation characteristics. This IC is used not only for dividing to receiver and transmitter described in Fig. 1, but also for dividing to PLL circuit, and/or to more than one receiver mixer for diversity receiving. Figure 3(c) shows the block diagram of this IC. Parallel dual gate FETs such as described in Fig. 2(a) are used to realize good isolation from output to input, and are directly connected by taking advantage of their high input impedance. Input matching circuit is configured by simple resistor on chip, because noise characteristics is not a severe problem for local signal. Load inductors are formed outside of the IC similar to that for receiver front-end IC.

3. FABRICATION

Dual gate FET's with high g_m characteristics at low drain current are necessary to realize the GaAs RF IC's which shows both excellent RF characteristics and low power dissipation. The dual gate MESFET's used in these IC are fabricated by asymmetric self-aligned LDD MESFET process[4]. Figure 4 shows cross sectional view of the dual gate MESFET. Gold metal has been deposited on the WSi refractory metal gate to reduce the gate resistance and improve the gain and noise characteristics. Gate length (L_g) is $1\mu m$. Figure 5 shows the typical I_d and g_m characteristics of this dual gate FET. The value of 22mS for g_m is obtained with gate width of $400\mu m$ at $V_{ds}=3V$, $V_{gs2}=0V$ and $I_{ds}=2mA$. Figure 6 shows photograph of fabricated IC chips. NPN type ESD(electrostatic discharge) protection diodes are integrated at each port of receiver front-end IC and mixer/amplifier transmitter IC. These IC's are sealed in compact plastic flat package shown in Fig. 7.

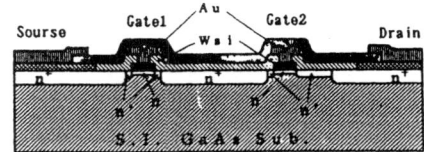

Fig. 4 Cross sectional view of dual gate MESFET

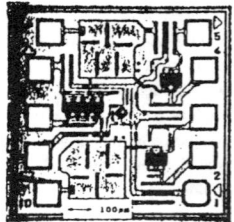

(a) Receiver front-end IC
(chip size: 1.05mm × 1.05mm)

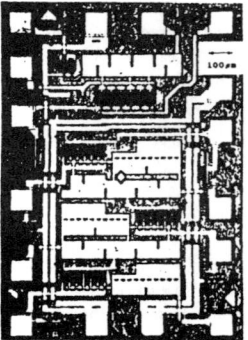

(b) Mixer/amplifier transmitter IC
(chip size: 1.20mm × 1.70mm)

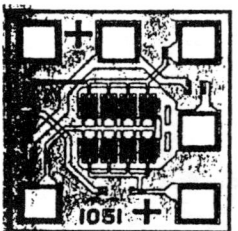

(c) Power splitter IC
(chip size: 0.58mm × 0.61mm)

Fig. 6 Photograph of fabricated IC chips

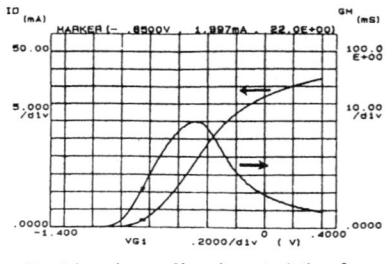

Fig. 5 I_d and gm vs V_{gs1} characteristics of dual gate MESFET
($V_{ds} = 3V$, $V_{gs2} = 0V$)

Fig. 7 Photograph of the GaAs IC package
(from left, receiver front-end IC,
mixer/amplifier transmitter IC,
power splitter IC)

4. PERFORMANCE of the GaAs IC's

RF characteristics of these GaAs IC's measured at 900MHz band are described below.

4.1 Receiver front-end IC

Figure 8 shows conversion gain and noise characteristics of receiver front-end IC. High conversion gain (more than 21dB) and low noise figure (less than 3.5dB) are obtained under low supply voltage down to 2V, when the local input level is only -15dBm. Isolation between the local input and the RF input are shown in Fig. 9. By using dual gate FET amplifier, good isolation of

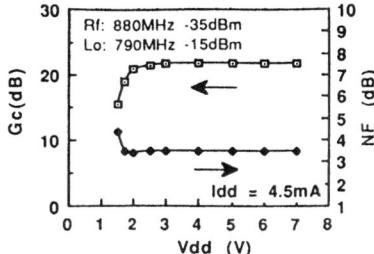

Fig. 8 Conversion gain and noise vs Vdd characteristics of receiver front-end IC

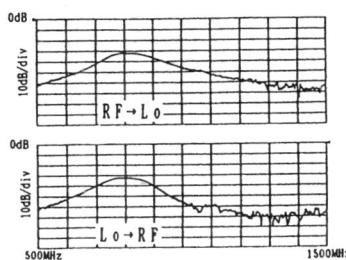

Fig. 9 Isolation characteristics of receiver front-end IC

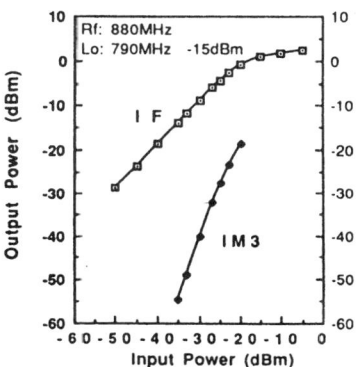

Fig. 10 Input-output characteristics of receiver front-end IC

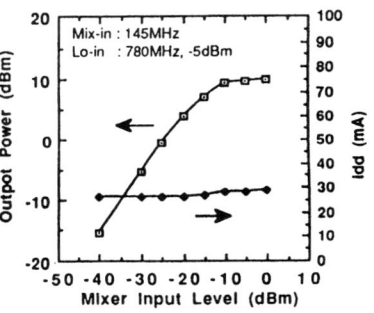

Fig. 11 Total Input-output characteristics of mixer/amplifier transmitter IC

more than 30dB is obtained. Input-output characteristics are shown in Fig. 10. Saturate IF output power of 2dBm and 1dB gain compression level of -4dBm are obtained. Third order intermodulation intercept point is 7dB a output level. Dissipation current is 4.5mA.

4.2 Mixer/amplifier transmitter IC

The total input-output characteristics c mixer/amplifier IC including an outside L(band pass filter is shown in Fig. 11. Tota output power of 8dB for mixer input leve -12dBm and local power -5dbm is obtaine under 3V supply voltage. Dissipation curren is 27mA. Even though the mixer and th amplifier are integrated into same chip spurious level is suppressed less than -80dBc.

4.3 Power splitter IC

Figure 12 shows isolation characteristic (1)from output to input and (2)between tw output ports of power splitter IC. Inter-por isolation is greater than 25dB. Output powe and dissipation current versus input level ar also showed in Fig. 13. Saturated output powe of over 0dBm and 1dB gain-compression leve of -6dBm are obtained. Supply voltage is 3 and dissipation current is less than 3.5mA.

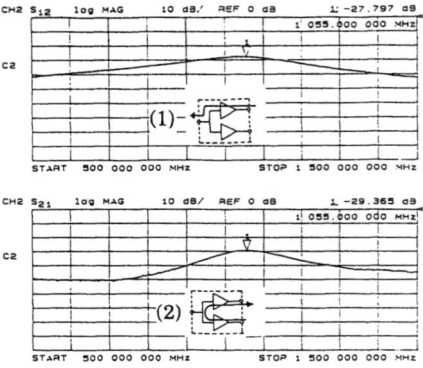

Fig 12 Isolation characteristics of power splitter IC
(1) Output to input isolation characteristics
(2) Isolation characteristics between output ports

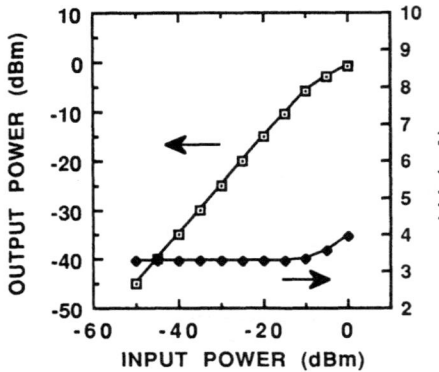

Fig 13 Input-output characteristics of power splitter IC

5. RELIABILITY

These IC's, sealed in compact plastic package, have been thoroughly tested for their reliability and have proven to be reliable in practical use ESD level of each port has been tested up to 100V under test condition of C=200pF and R=0Ω, thus confirming the effects of the protection diodes

6. CONCLUSIONS

A family of low power GaAs RF IC's consisting of receiver front-end IC, mixer/amplifier transmitter IC and power splitter IC have been developed by using asymmetric self-aligned LDD MESFET process Circuit design based on dual gate FET realized both excellent RF characteristics and low power dissipation under 3V supply voltage These IC's, sealed in compact plastic package, have been confirmed to have excellent reliability for practical use This IC family should make considerable contribution to miniaturize battery operated mobile communication equipment

ACKNOWLEDGMENT

The authors wish to thank T Takemoto, T Onuma and M Makimoto for their encouragement They also grateful to many colleagues at Matsushita Electronics Corporation and Matsushita Communication Industry Co , Ltd for their various support

REFERENCES

[1] Y Tamura et al ; "Silicon bipolar integration of RF section in cellular handheld telephones", MWE'91 Workshop Dig , pp 87-93, 1991

[2] H Suzuki; "A modern technologies for frequency synthesizers", MWE'91 Workshop Dig , pp 87-93, 1991

[3] S A Maas; MICROWAVE MIXERS, ARTECH HOUSE Inc

[4] H Yagita et al ; "Low noise and low distortion GaAs mixer-oscillator IC for broadcasting satellite TV tuner", IEEE GaAs IC Symp Tech Dig , pp 75-78, 1989

Ultra Low Power Low Noise Amplifiers for Wireless Communications

E. Heaney, F McGrath, P.O'Sullivan, C Kermarrec*

M/A-COM - IC Design Center
100 Chelmsford Street
LOWELL, MA 01851-2694
*Analog Devices

Abstract

This paper presents the design and performance results of surface mount plastic packaged, ultra low power, GaAs monolithic low noise amplifiers, for cordless phone receivers at 1.9GHz and spread spectrum receivers at 2.4GHz.
The results for the 1.9GHz amplifier show a small-signal gain of 14.5dB and Noise Figure of 1.5dB with a 4V supply while drawing only 3mA. The LNA was also tested with a 3V supply and with a current of 2mA the measured Gain and Noise Figure were 13.5dB and 1.75dB respectively. Thus this LNA is compatible with the voltage requirements of the latest generation of 3V CMOS logic circuits.
The results for the 2.4GHz LNA show a small-signal gain of 15dB and a Noise Figure 1.8dB drawing less than 5mA from a single +5 volt supply.
The LNA's are now being produced in volume in surface mount plastic packages for custom wireless communications applications.

Introduction

Present day wireless communications receivers demand low power consumption for portability and long battery life, while simultaneously requiring low noise figure and high dynamic range operation. This paper presents the design and results of the state-of-the art, very low power consumption GaAs MMIC Low noise amplifiers.
The designs are based on a cascode configuration including inductive feedback to the common source amplifier for simultaneous noise and input impedance matching, and capacitive feedback to the common-gate transistor to simultaneously optimize for linearity, gain and stability performance.

LNA Design Approach

Figure 1 shows the LNA circuit topology. The cascode configuration alllows both FET's to use the same current and eliminating the need for coupling capacitors. Active biasing is used to provide increased bias stability and improve yield. The biasing scheme allows the amplifier to be switched to the standby mode, necessary in digital TDMA receivers, which reduces current draw to 160uA. The design approach is now described.

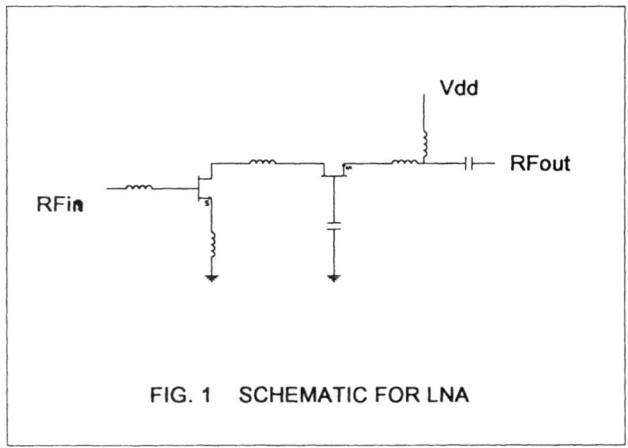

FIG. 1   SCHEMATIC FOR LNA

Figure 2 represents a simplified FET electrical equivalent circuit used for the first stage. It includes a feedback inductance L together with its equivalent input impedance.
The inductance L modifies the FET input impedance by adding a noiseless resistance $R = (gm/Cgs)*L$ in series with the input capacitance Cgs and inductance L. The input impedance Zin is then given by

$$Zin = LGm/Cgs + j(\omega L - 1/Cgs\omega)$$

It is worthwhile to notice that the ratio $gm/Cgs = 2\pi Ft$, (where Ft is the transistors unity current gain bandwidth). Since Ft is independent of FET width, in a 50 ohm system L can be calculated from

$$L = 50/2\pi Ft$$

and is independent of FET periphery.

This is a first order approximation and when Cgd is included in the FET equivalent circuit and the miller effect is taken into account, the required inductance increases. The degenerated transconductance should also remain high enough to guarantee the required transducer gain.

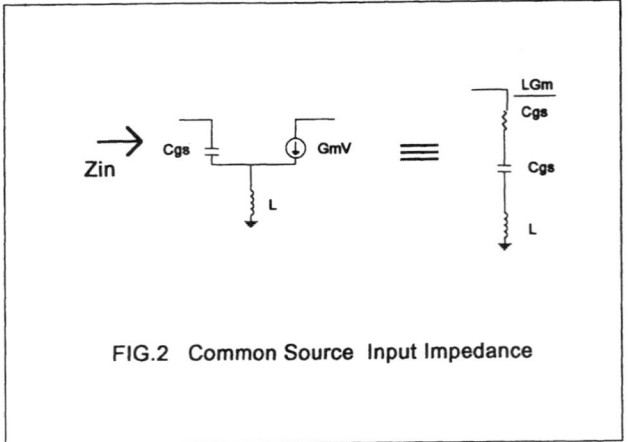

FIG.2 Common Source Input Impedance

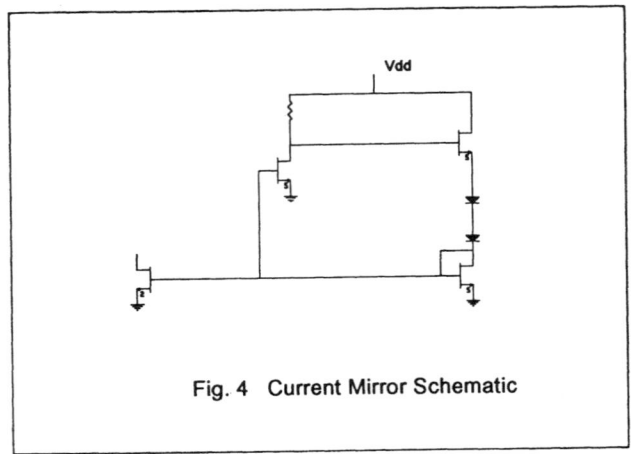

Fig. 4 Current Mirror Schematic

Figure 3 shows a simplified FET electrical equivalent circuit used for the second stage, including a feedback capacitor C with its equivalent input impedance. The capacitor C increases the input impedance by a factor $(1 + C_{gs}/C)$ and decreases the FET transconductance by the same factor. This design technique allows the designer to depart from the traditional cascode configuration (where the common source FET presents a unity voltage gain) and to split the RF voltage between the two cascode FETs for linearity optimization. The feedback capacitor also improves circuit stability.

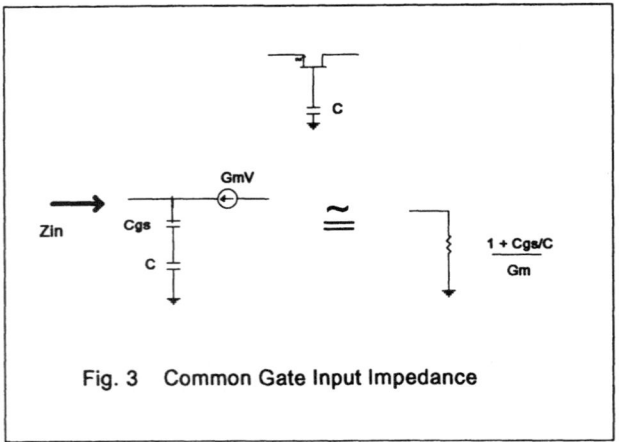

Fig. 3 Common Gate Input Impedance

The output matching has a dramatic impact on circuit linearity and power consumption required to achieve LNA third order intercept point. Ideally, the LNA output port should be matched for power and output match considerations. These conditions would require very high impedance transformation ratios out of reach of GaAs MMIC circuits due to size and inductor quality factor limitations. The impedance transformation ratio achieved is about 7, thus providing the FET with a 350 ohm load and 50 ohm output impedance. The layout of the output matching circuit is also critical because the drain of the common gate FET is at the highest impedance in the circuit. This makes it very susceptible to parasitic capacitances experienced in plastic packages and can cause a frequency shift in the output response.

Finally, an active bias circuit is used to apply the required gate voltage to the common source transistor to guarantee a constant drain current regardless of manufacturing process variations. Figure 4 shows the schematic for the active biasing. An increase in current in the common-source FET (due to process or temperature) is mirrored by an increase of current in the active bias circuit. This increase in current is amplified, inverted and fed back to reduce the voltage on the gate of the common-source FET.

Results

The LNAs were fabricated on an ion implanted 1 micron process featuring low noise figure and low maximum channel current.

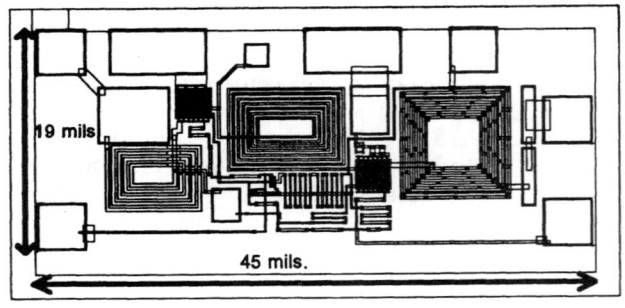

Fig. 5 Layout of 2.4GHz LNA

Figure 6 shows the measured Gain and Noise performance of the 1.9GHz amplifier. The gain is 14.5dB with a Noise Figure of 1.5db with an applied bias of 4V and drawing 3mA. The 1dB compression is -1.1dBm and the OIP3 is +11.2dBm when driven by two -35dBm tones.

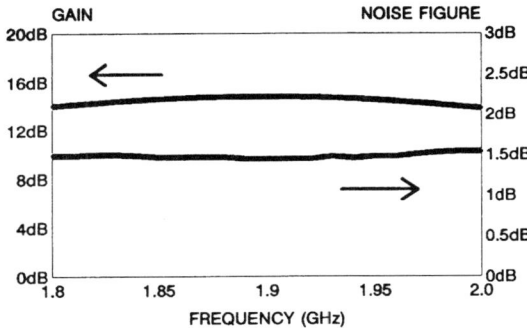

Fig.6 1.9GHz LNA Vdd=4V Id=3.2mA

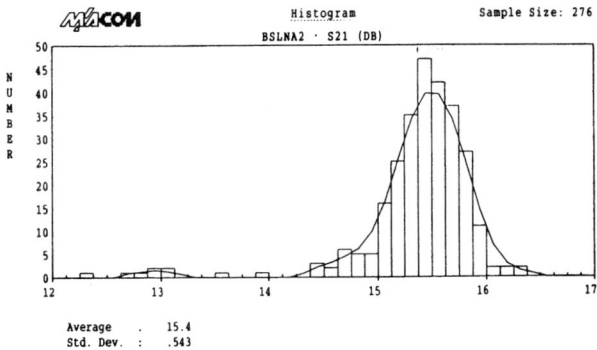

Fig 8 Manufacturing Distribution in Gain
for 4V 3mA LNA

Figure 7 shows the same device with a bias voltage of 3V and drawing only 2mA. The measured Gain and Noise Figure of 13.5dB and 1.7dB respectively, for a DC power requirement of 6mW, represent to the authors knowledge, the lowest DC power requirements ever published.

Conclusion

An ultra low-power low-noise amplifier for wireless communications has been designed, manufactured and tested. The LNAs exhibit good linearity and Noise Figure with exceptional DC power requirements. This is a key component for portable wireless receivers where sensitivity and dynamic range are important.

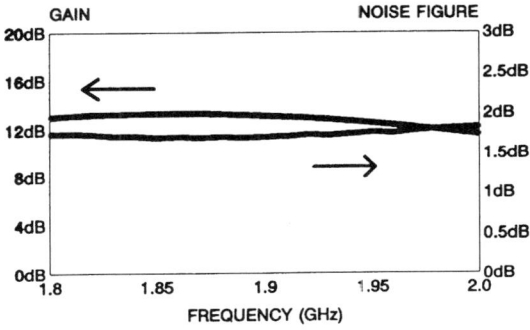

Fig.7 1.9GHz LNA Vdd=3V Id=1.9mA

The results for the 2.4GHz LNA show a small-signal gain of 15dB and a Noise Figure 1.8dB drawing less than 5mA from a single +5 volt supply. The 1dB compression point was measured at +4dBm with an OIP3 for two -20dBm tones at +13dBm.

The devices were also measured over temperature. The typical variation over temperature for Current, NF and Gain are 4.4uA/C, .004dB/C and -.009dB/C respectively.

Low Current GaAs Integrated Down Converter for Portable Communication Applications

V. Nair, R. Vaitkus, D. Scheitlin, J. Kline, and H. Swanson*

Motorola Inc., Phoenix Corporate Research Laboratories
*Linear Analog Technology Group
2100 E. Elliot Road, MD EL508, Tempe, Arizona, 85284

Abstract

A low current GaAs MMIC amplifier/mixer was designed and characterized for portable communication applications in the 900 MHz band. This single chip integrated front-end IC (90 mil X 110 mil) achieved -118 dBM sensitivity at the cellular band. The extremely low power dissipation, high level of integration, and very good RF performance of this monolithic IC make it an ideal candidate for portable communication applications.

Introduction

Receiver front-end ICs for portable communication systems such as cellular phones and portable radios are potential areas for utilization of GaAs technology. In portable communication units GaAs MMIC receivers lead to a reduction in the number of parts and interconnects and, hence, their size and weight. However, to conserve the battery drain in portable units, devices and circuits have to be designed to operate at very low current levels. Results for enhancement mode and depletion mode MESFET MMIC amplifiers with excellent RF performance at low current levels have been published by several authors(1-4).

This paper reports on the development of an integrated GaAs MMIC down converter with low power dissipation for portable communication applications. A single chip down converter consisting of a low noise amplifier and dual gate FET mixer was designed, fabricated and characterized. A key feature of the circuit is the utilization of on-chip lumped LC matching circuits to realize small size and to lower power dissipation.

Circuit Design

The goal of this project was to design and characterize an integrated amplifier/mixer chip to meet the front-end receiver needs in 800 - 900 MHz portable communication equipment. The down converter chip should operate at a total current of 3 mA at 3.0 V. In order to meet the system requirements, LNA should achieve ≥12 dB gain and ≤3.5 dB noise figure. The mixer should have on-chip matching except for the IF at 45 MHz. The down converter IC should also meet the receiver sensitivity requirement of portable applications (-112 dBm to -121 dBm)

Several GaAs MESFET lots were characterized and noise figure parameters were measured in a system specially designed for low current measurement (5). In order to minimize the current drain and to achieve the required RF performance, devices with gate widths of 400 µm, 600 µm, 800 µm and 1000 µm were modeled. Based on this study, a two stage LNA utilizing 600 µm MESFET and a dual gate FET mixer were designed to meet the design objectives. Dual gate MESFET was modeled as a cascode combination of two single gate MESFETs. Several matching techniques were also explored. Even though active matching would have decreased the chip size, it was not pursued due to higher power consumption. Off-chip matching approach was also not utilized since it resulted in larger chip size. An on-chip lumped element LC matching network was found to be the best approach to minimize the size and power dissipation.

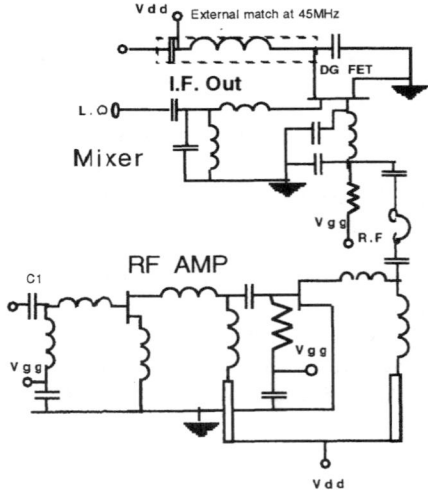

Figure 1. Down Converter Schematic

GaAs IC Symposium, pp. 41-44, October 1993.

The circuit was optimized to give the frequency response and return loss as close to the design goal as possible, while maintaining unconditional stability inside and outside the frequency pass band. The only external matching used in the circuit was for the mixer IF port at 45 MHz. Monte Carlo analysis was employed to study the sensitivity of the circuit to variations in component values due to fabrication tolerances. The circuit was laid out to facilitate on-wafer characterization of the amplifier and the integration of a single amplifier/mixer circuit. Fig.1 shows the schematic diagram of the GaAs down converter MMIC.

Fabrication

The wafers were fabricated using a 0.7 μm GaAs MESFET MMIC process. This process employs ion implanted MESFET with Ni/Ge/Au ohmics, Ti/PT/Au Schottky gate and two level metal interconnects with no airbridge, and no via holes. The Si_3N_4 MIM capacitors and 3μm thick Au inductors were fabricated for on-chip matching of the circuits. The fabricated die (chip size 110 mil X 90 mil) is shown in Fig. 2.

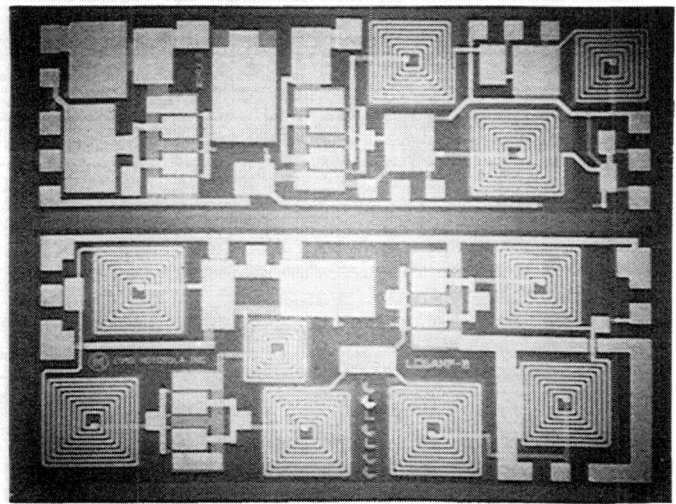

Figure 2. GaAs MMIC Down Converter

Results and Discussion

The gain, return loss, noise figure and intermodulation characteristics of the LNA were measured on-wafer using an integrated test system developed in-house. The measured gain and return loss of the amplifier are plotted in Fig. 3. Amplifier achieved a peak gain of 16 dB and better than 10 dB return loss at a drain current of 1.5 mA @ 3.0 V. The output third order intercept point for the amplifier was calculated from the two-tone measurement. The RF characteristics of the amplifier were measured at 3V and 1V for different current levels. Fig 4 shows gain, noise figure and the third order intercept point of the amplifier at different bias levels. At 880 MHz, LNA achieved 13 dB gain and 3.6 dB noise figure at Idd =1.5 mA and Vdd = 3.0V. At Idd = 2mA and Vdd = 1.0 V a gain of 12 dB and noise figure of 3.5 dB was achieved.

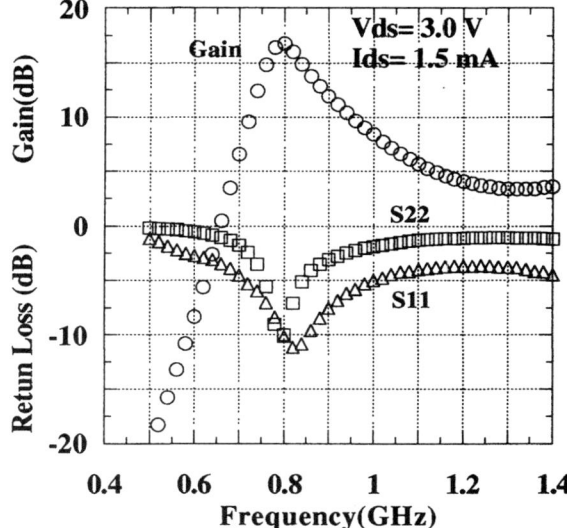

Figure 3. Measured Frequency Response of the LNA. Vds= 3.V, Ids = 1.5 mA

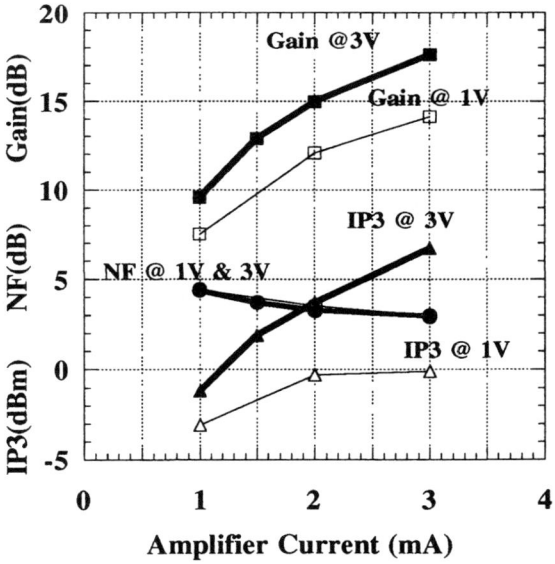

Figure 4. Gain, NF and output IP3 of the LNA for different bias levels (Frequency = 880 MHz)

The intercept point of the amplifier exhibited a strong dependence on the supply voltage. At 2mA of bias current, IP3 decreased about 4 dBm when the drain voltage was decreased from 3V to 1V.

In Fig. 5, the performance of this LNA is compared with other published results. In this technology trend chart, power consumption of MMIC amplifiers operating at UHF or L- band range and having a gain ≥ 10 dB are included. As shown in the figure, the GaAs MMIC amplifier reported in this paper has the lowest power consumption compared to other published results.

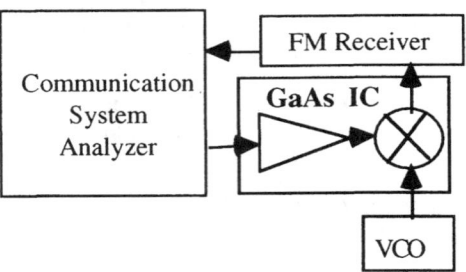

Figure 6. Receiver Sensitivity Measurement System Configuration

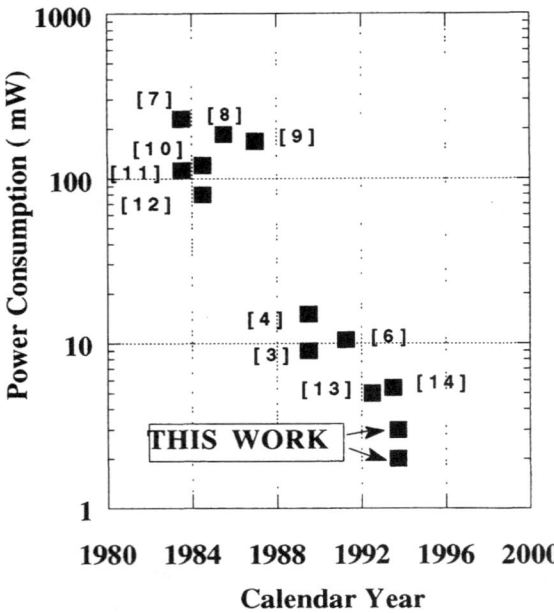

Figure 5. Monolithic LNA power consumption. (ICs Operating UHF or L bands). [The numbers shown in the parenthesis are the references listed at the end of the paper]

Down Converter Evaluation

In order to evaluate the performance of the mixer and the down converter ICs, the wafers were thinned, scribed and packaged. The circuits were tested in 14 pin SOIC and ceramic packages. A system level test was performed to evaluate the down converter IC. A SIgnal, Noise And Distortion (SINAD) measurement system (shown in Fig 6) consisting of a communication system analyzer, VCO, narrow band FM receiver and GaAs down converter IC was utilized to characterize the circuit.

The measured performance of the down converter is shown in Fig 7a and 7b. The measurement conditions were RF = 900 MHz, LO = 945 MHz and a modulation frequency of 1KHz with a frequency deviation of ± 8 KHz. A 12 dB SINAD corresponds to 25% distortion in the recovered audio

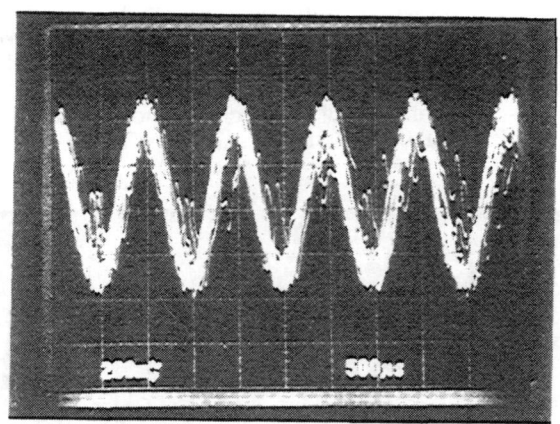

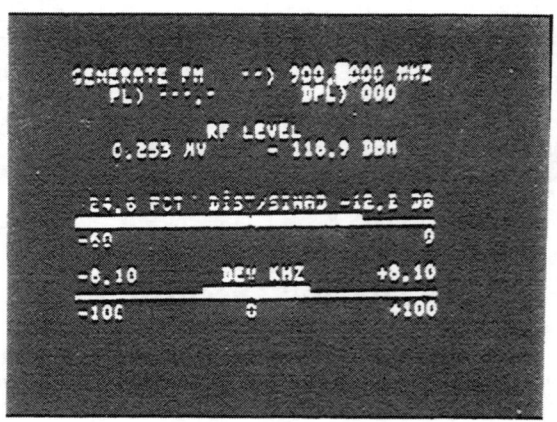

Figure 7. Down Converter Sensitivity - Measured Result. (a) Recovered Audio Signal at 12dB SINAD (Corresponds to 25% distortion). (b) Communication Sytem Analyzer Plot. RF = 900 MHz

signal (shown in fig. 7a). It is the specified measurement standard for sensitivity test of professional radio receivers such as land mobile and cellular radios. The GaAs down converter achieved a 12 dB SINAD sensitivity of -118.9 dBM (shown in Fig. 7b) at a total drain current of 2.7 mA @ 3.6 V

Conclusion

A fully monolithic single chip GaAs MMIC amplifier and mixer was designed, fabricated and tested for potential application in a portable communication system. This single chip integrated front-end IC (90 mil X 110 mil) achieved -118 dBm sensitivity in the 900 MHz cellular band. The extremely low power dissipation, high level of integration and very good RF performance of this monolithic IC make it an ideal candidate for portable communication applications.

Acknowledgments

We are thankful to J. Escher, C. Monroe, C. Weitzel and R.Potyka for their strong support and encouragement of this work. We also thank S. Dorn and D. Halchin for their characterization support. Help of C.E. Wu, K. Hilgers and N. Mellen is also appreciated.

References

(1) U. Ablassmeir, et al. "GaAs FET up converter TV tuner," IEEE Trans. Electron Devices, vol. ED-28, pp.1156-1159, Feb. 1980.

(2) K. Kanazawa, et al, " A Double- Balanced Dual Gate FET mixer IC for UHF Receiver Front-End Applications," IEEE Trans. Microwave Theory and Techniques, Vol. TM-33, pp. 1548-1554, Dec. 1985.

(3) V. Nair, "Low Current Enhancement Mode MMICs for portable Communication Applications," IEEE gas IC Symposium. Tech. Digest., 1989, pp. 67-70

(4) Y. Imai, et al, " Very Low Current Small Size GaAs MMICS for L-band Front-end Applications," IEEE GaAs IC Symposium. Tech. Digest., 1989, pp. 71-74.

(5) R. Lucero, et al. Noise Characterization Uncertainty of Microwave Devices Under Low Current Operation," 1989 MTT-S Symposium digest, pp., 893-896.

(6) Y. Imai, et al." Design and performance of Low-Current GaAs MMICs for L- band Front-End Applications," IEEE Trans. Microwave Theory and Tech., Vol. MTT-39, pp. 209-215, Feb. 1991

(7) K. Honjo et al. "Low Noise Low Power Dissipation GaAs monolithic Broad Band Amplifiers," IEEE Trans. Microwave Theory and Tech., Vol. MTT-31, pp. 412-417, May 1983

(8) K. Osafune et al. "Low Noise GaAs Monolithic Broadband Amplifier using a Drain Current Saving Technique," IEEE Trans. Microwave Theory and Tech., Vol. MTT-33, pp.543-545, June 1985

(9) M. Nishima et. al. " A UHF GaAs Multi-Stage Wide Band Amplifiers With Dual Feedback Circuits," IEEE GaAs IC Symposium. Tech. Digest., 1987, pp. 223-226.

(10) J. Tajima Et al. " GaAs Monolithic Low Power Amplifier with RC Parallel Feedbacks," IEEE Trans. Microwave Theory and Tech., Vol. MTT-32, pp. 542-544, May 1984.

(11) V. Paulker et al. " Normally Off MESFET Analogue Circuits," in Proc. European Microwave Conf., pp. 631-633, 1984

(12) K. R. Cioffi, " Monolithic L-Band Amplifiers Operating at Milliwatt and Submilliwatt DC power Consumptions" IEEE Microwave and Millimeter-Wave Monolithic Circuits Symposium Digest, pp. 9-12, June 1992

(13) S. Hira et al. " Miniaturized Low Noise Variable MMIC Amplifiers With Low Power Consumption For L-Band Portable Communication Applications," IEEE Microwave and Millimeter-Wave Monolithic Circuits Symposium Digest, pp. 67-70, June 1993

X-Band Monolithic Series Feedback LNA

RANDALL E. LEHMANN, SENIOR MEMBER, IEEE, AND DAVID D. HESTON, MEMBER, IEEE

Abstract — An X-band monolithic three-stage low-noise amplifier (LNA) employing series feedback has demonstrated 1.8-dB noise figure with 30.0-dB gain and an input VSWR less than 1.2:1 at 10 GHz. The key to this design is using monolithic technology to obtain an exactly repeatable series feedback inductance to achieve a simultaneous noise match and input VSWR match.

I. INTRODUCTION

IN CONVENTIONAL LNA's, the common-source FET input stage is presented with an optimum noise match (Z_{opt}) to achieve minimum noise figure at the expense of exhibiting high input VSWR. However, to achieve optimum noise figure and low-input VSWR simultaneously for a single-ended amplifier, series feedback provides the solution. This is the first reported demonstration of the use of series feedback in a monolithic microwave integrated circuit (MMIC) to achieve state-of-the-art X-band performance.

II. HISTORY

Strutt and Van Der Ziel in their 1942 article, "Suppression of spontaneous fluctuations in amplifiers and receivers for electrical communication and measuring devices," reported that a feedback inductor inserted into the cathode lead of a common-cathode high-vacuum triode circuit might enhance the signal-to-noise ratio at high frequencies [1].

In 1974, Jakob Engberg presented equations as well as optimization procedures for the design of two-port low-noise amplifiers [2]. Engberg describes how a combination of shunt and series feedback and proper output loading can be used to achieve $Z_{opt} = S_{11}^*$. As only lossless feedback elements are used, Engberg states that the minimum noise measure M_{min} remains constant. He reports that this theory has been verified at UHF frequencies using hybrid circuits. Other researchers [3]–[5] have demonstrated hybrid amplifiers that employ reactive feedback for improved performance.

Monolithic technology provides the key in obtaining an exactly repeatable series feedback inductance. A high-impedance microstrip transmission line can be accurately modeled and reproduced in large quantity. Optimization of bond wire lengths to achieve the correct feedback impedance, as in a hybrid amp, is eliminated.

Manuscript received May 1, 1985; revised July 1, 1985. This work was supported in part by Air Force Wright Aeronautical Laboratories/Avionics Laboratory under Contract F33615-82-C-1766.
The authors are with Texas Instruments, 13500 N. Central Expressway, Dallas, TX 75266.

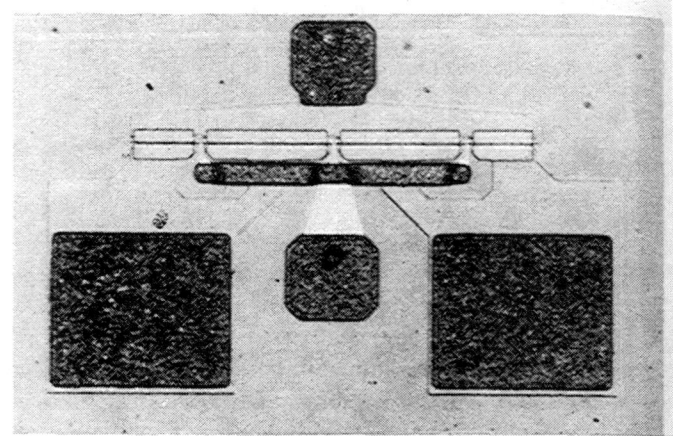

Fig. 1. "Monolithic-discrete" FET.

TABLE I
NOISE FIGURE AND GAIN DATA FOR "MONOLITHIC-DISCRETE" FET's AT 10 GHz

TOTAL NUMBER FETs TESTED	NFmin (dB)			Associated Gain (dB)		
	LO	HI	AVG.	LO	HI	AVG.
75	1.5	1.9	1.7	9.5	11.5	11.0

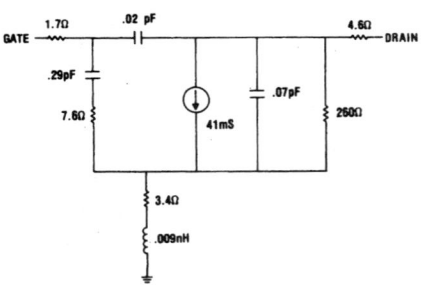

Fig. 2. "Monolithic-discrete" device model.

III. DEVICE CHARACTERIZATION

A 0.5-μm gate length, 300-μm gate width FET, shown in Fig. 1, is used in each of the three stages. The active layer is formed by ion implantation. The device incorporates reactively ion-etched vias through 0.15-mm-thick GaAs to provide source grounding. Gate and drain terminals are brought to single bond pads to facilitate implementation into a monolithic circuit. The "monolithic-discrete" device is processed identically to the final MMIC, including deposition of the correct silicon nitride thickness to be used for the metal–insulator–metal (MIM) capacitors. This en-

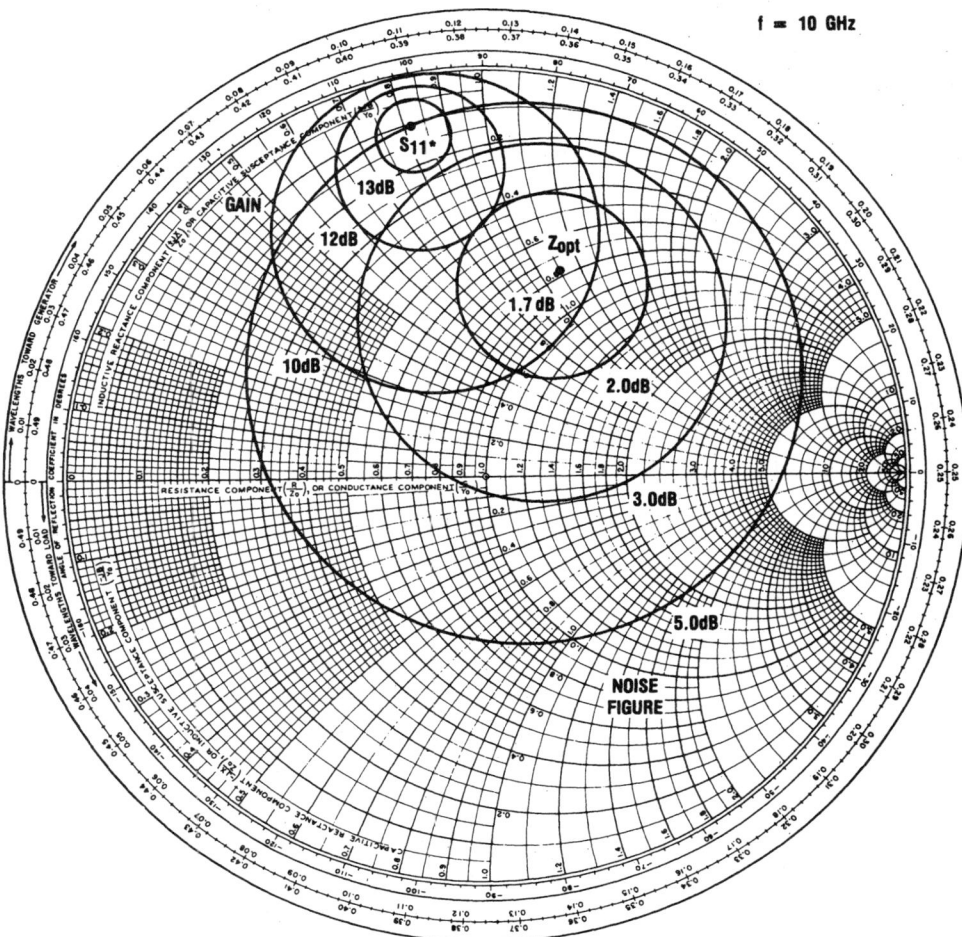

Fig. 3. 300-μm device noise figure and gain circles at 10 GHz.

sures that any change or increase in gate-drain and gate-source capacitances resulting from the increased capacitor dielectric thickness, which is greater than the standard discrete FET passivation thickness, is accounted for in the device characterization. The discrete FET is also fabricated on the same thickness GaAs as the MMIC so that the inductance of the vias can be properly modeled.

Several slices of these "monolithic-discrete" devices were evaluated for minimum noise figure (NF) and associated gain at 10 GHz. Table I shows a summary of the measured results from the 75 FET's tested. The minimum NF and associated gain are listed for both the best and worst FET measured, as well as the numerical average of all 75 FET's.

As the noise figure data indicates, the 75 monolithic discrete devices tested from four different slices exhibit similar RF performance. To examine the differences that exist from device to device, 16 FET's were selected for modeling. The equivalent circuit shown in Fig. 2 represents an average value model of the 16 FET's.

To aid in LNA designs and to better understand the device performance tradeoffs, the NF and gain circles of a typical device are plotted in Fig. 3. This device has a 1.7-dB minimum NF with 10.6 dB of associated gain at Z_{opt}. It is observed from Fig. 3 that a simultaneous conjugate match would result in a 5-dB NF. On the other hand, an optimum noise match would cause a 2.5-dB mismatch loss at the input, resulting in a 5:1 input VSWR.

IV. Circuit Design

Implementation of series feedback provides several advantages for low-noise amplifier design. Inductive reactance in the source lead of a common-source FET increases the real part of the input impedance. With proper impedance loading at the output of the FET, the conjugate of the FET input impedance and the optimum noise match impedance become coincident.

Impedance mapping on a Smith Chart is a useful design tool. Figs. 4 and 5 show the impedance maps of S_{11}^* and Z_{opt} as a function of inductive series feedback and output loading. In both plots, the inductive series feedback is incremented from 0.01 pH to 0.4 nH in 0.1 nH steps. In Fig. 4, the real part of the output load is varied from 25 to 50 to 100 Ω. In Fig. 5, the real part of the output load is fixed at 50 Ω and the imaginary part of the output load is

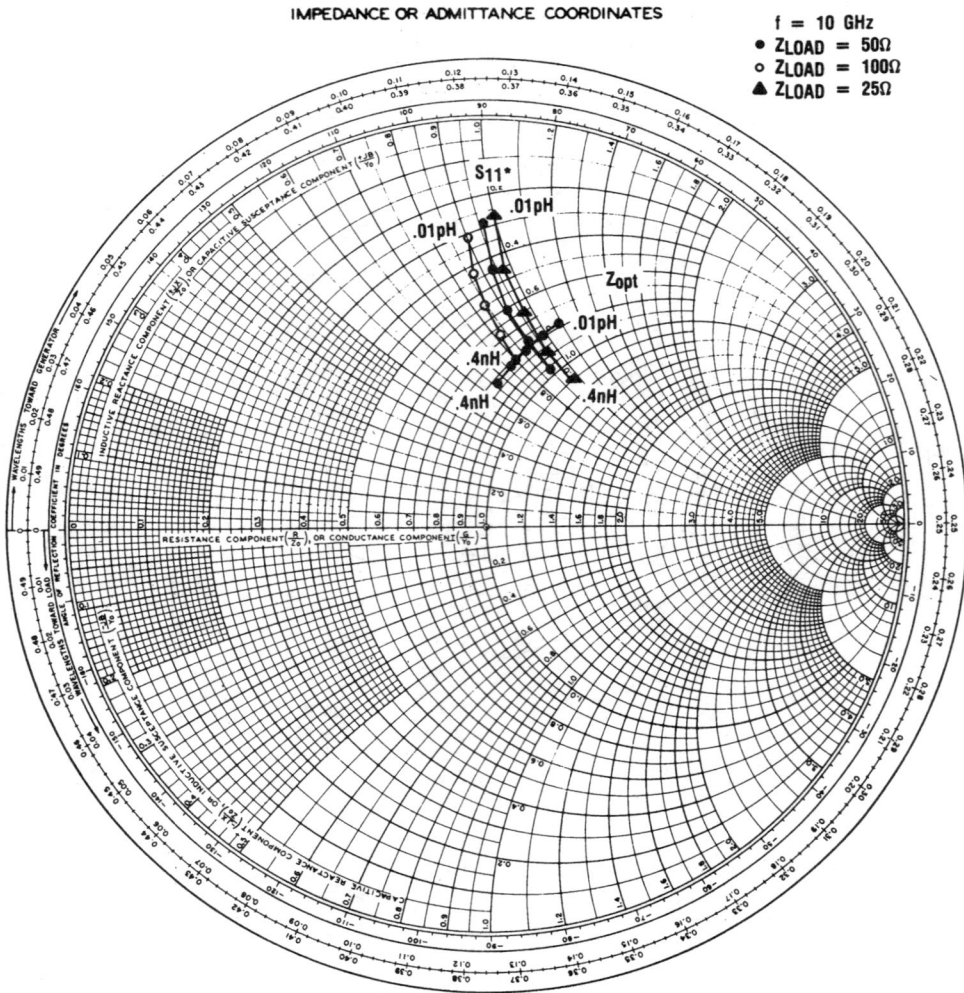

Fig. 4. Impedance mapping of S_{11}^* and Z_{opt} versus real part of output load and series feedback at 10 GHz.

varied ± 50 Ω. It is observed that S_{11}^* of an FET is altered by both the feedback and the output loading, whereas Z_{opt} is unaffected by the output load and only varies with feedback. Using mapping techniques, it is possible to find a combination of inductive series feedback and output loading that results in S_{11}^* being equal to Z_{opt} at 10 GHz. Fig. 6 shows the NF circles and S_{11}^* for 0.28 nH of series feedback and an output load of 50-j25 Ω. Examination of Fig. 6 reveals several distinct advantages of using inductive series feedback, rather than another form of feedback (i.e., shunt or series resistive feedback). First, it is apparent that a simultaneous noise and gain match can be obtained with a proper choice of inductive feedback and output load. The other advantages become apparent when Fig. 6 is compared with Fig. 3. The minimum NF at Z_{opt} decreases from 1.7 to 1.6 dB. This reduction in NF_{min} with increasing feedback is consistent with the reduction in gain due to the addition of feedback. Lossless inductive feedback adds no noise to the circuit and therefore the minimum noise measure (M_{min}) of the device plus lossless feedback should remain constant [2]. Minimum noise measure is computed as

$$M_{min} = \frac{F_{min} - 1}{1 - \frac{1}{G_{av}}}$$

where G_{av} is the available gain of the FET with the input noise matched. For the example above, a sample calculation of M_{min} is shown in Table II. This drop in NF_{min} with a constant M_{min} has been experimentally verified with discrete FET's at 10 GHz.

Inductive series feedback decreases the equivalent noise resistance (r_n) of the two-port (device plus feedback). Series feedback also decreases the sensitivity to changes in the intrinsic device properties giving the final circuit greater tolerance to process variations. Inductive series feedback requires no dc blocking capacitor as in the case of shunt feedback. The feedback inductor (high-impedance line) can be realized monolithically in a very repeatable, high-yield

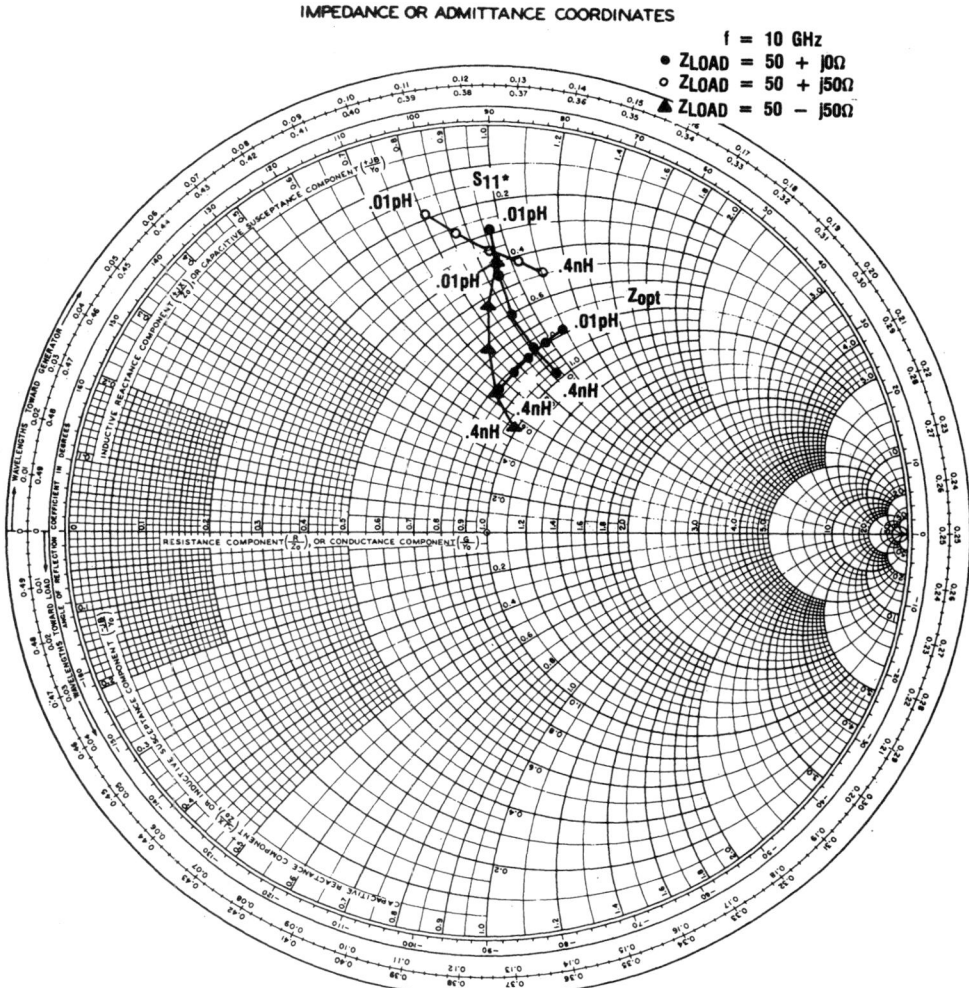

Fig. 5. Impedance mapping of S_{11}^* and Z_{opt} versus imaginary part of output load and series feedback at 10 GHz.

manner. The monolithic feedback element is fabricated on the semi-insulating GaAs substrate at the same process step as the RF transmission lines and the bottom plates of the MIM capacitors. No additional process steps or increased complexity are required.

The three-stage LNA circuit design is shown in Fig. 7. All RF matching and dc bias circuitry is included on the monolithic chip, shown in Fig. 8. The chip size is $3.0 \times 2.3 \times 0.15$ mm. Via holes are etched through the substrate and plated with gold to form low-resistance, low-inductance ground connections for the sources of the FET and the RF bypass capacitors. The gate and drain bias voltages are brought to common points at opposite corners of the chip. Gold–germanium–nickel–gold resistors are employed in the gate bias line to improve low-frequency stability.

V. RF Performance

The monolithic three-stage LNA with series feedback has demonstrated a 1.8-dB noise figure with 30.0-dB gain and an input VSWR less than 1.2:1 at 10 GHz. The X-band gain and noise figure response is shown in Fig. 9. Maximum noise figure is 2.0 dB from 8.5 to 11.5 GHz. From 9.0 to 11 GHz, input VSWR is less than 1.8:1. The input and output VSWR response is illustrated in Fig. 10. The amplifier, which is unconditionally stable, is operated at a drain bias of 3 V and a total drain current of 30 mA. Output power at 1-dB gain compression is 10 dBm.

Thirty-five LNA's from nine different slices have been evaluated for NF, gain, and VSWR. TAble III shows a summary of the results obtained from each slice at 10 GHz. LNA's were evaluated at bias conditions for minimum NF. The best and worst LNA measured as well as the average of all LNA's from that slice are recorded.

VI. Conclusions

An X-band monolithic three-stage LNA using series feedback has demonstrated excellent gain, noise figure, and input VSWR performance. Results from thirty-five LNA's

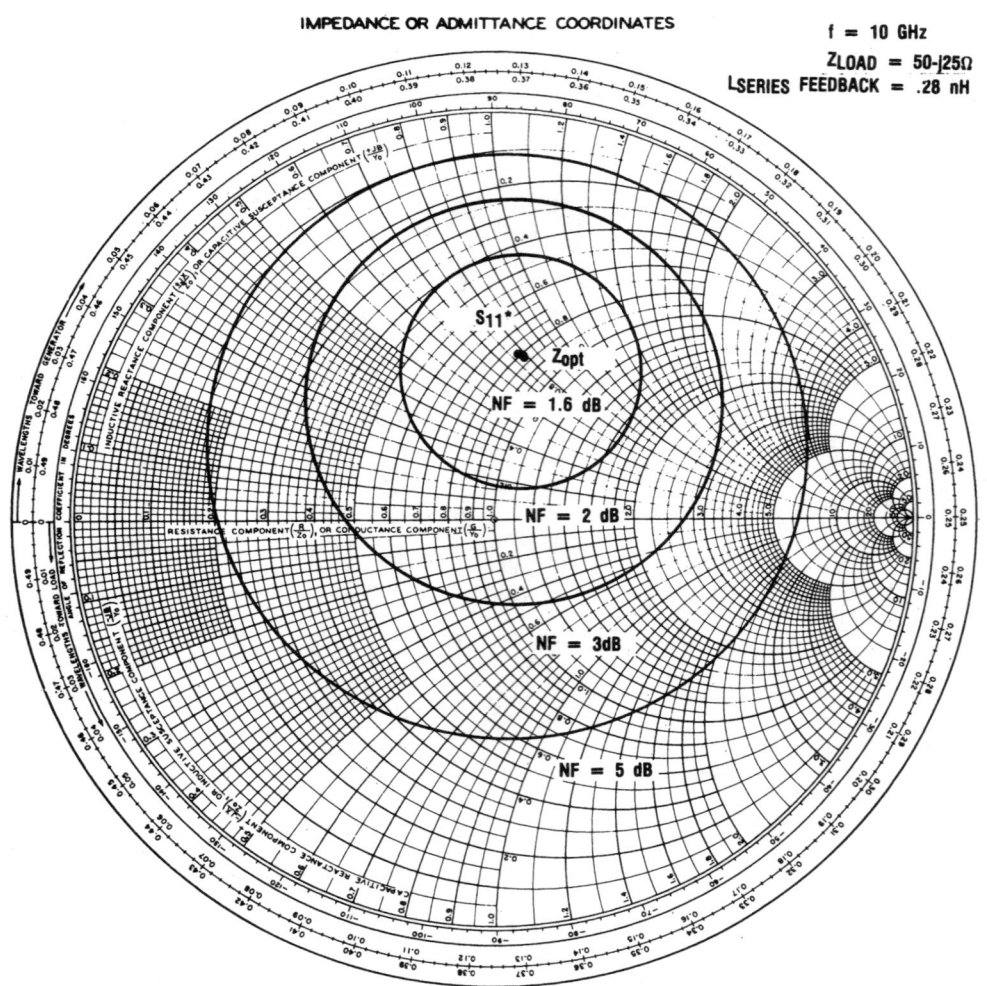

Fig. 6. NF circles and S_{11}^* for 0.28 nH of series feedback at 10 GHz.

TABLE II
CALCULATION OF MINIMUM NOISE MEASURE AT 10 GHz

SERIES FEEDBACK INDUCTANCE	NFmin (dB)	Gain (dB)	Mmin
.01pH	1.7	10.6	.52
.28nH	1.6	8.4	.52

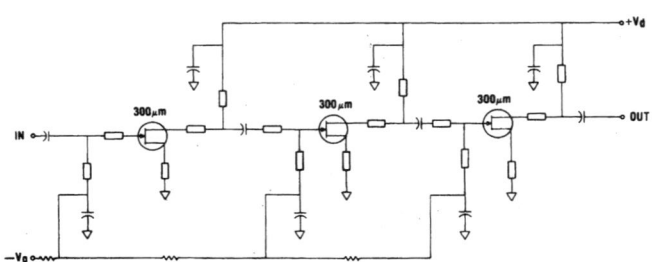

Fig. 7. Monolithic three-stage LNA circuit schematic.

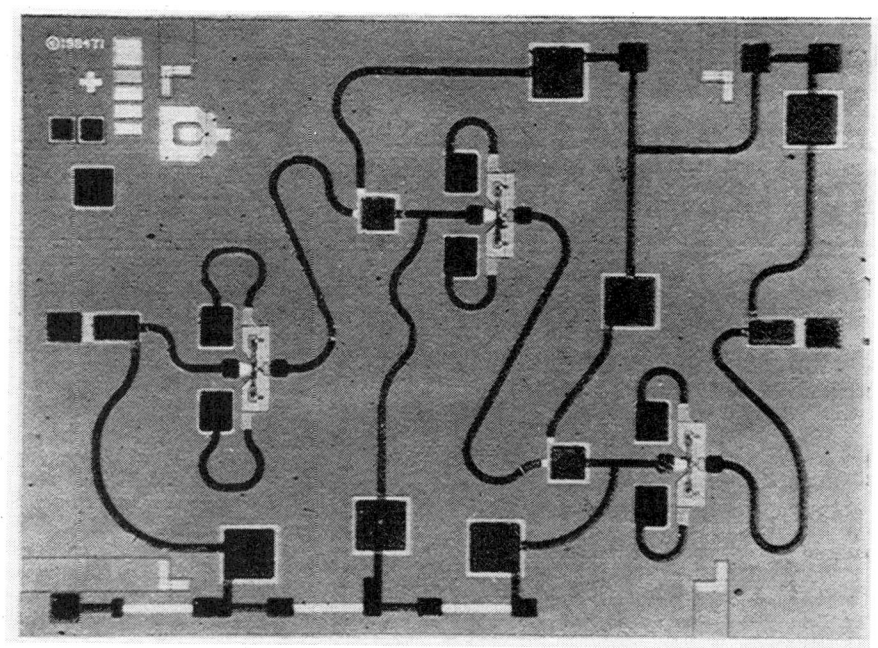

Fig. 8. Monolithic three-stage LNA.

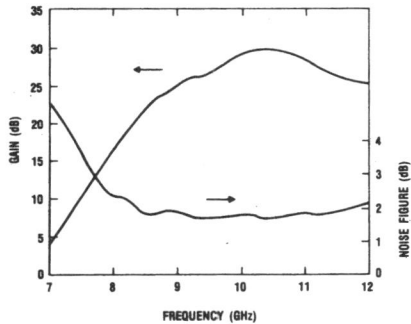

Fig. 9. LNA gain and noise figure performance.

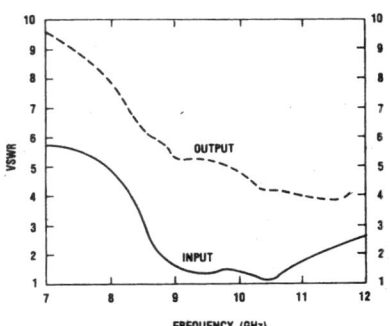

Fig. 10. LNA VSWR performance.

TABLE III
SLICE SUMMARY OF LNA MEASUREMENTS AT 10 GHz

SLICE #	# LNAs TESTED	NF (dB)			GAIN (dB)			INPUT VSWR		
		LO	HI	AVG.	LO	HI	AVG.	LO	HI	AVG.
1	10	1.9	2.2	2.0	28.2	32.4	30.0	1.1	1.4	1.2
2	4	1.8	2.0	1.9	29.5	32.0	30.9	1.2	1.4	1.3
3	4	1.9	2.0	1.9	28.3	32.1	30.4	1.1	1.4	1.2
4	4	1.9	2.2	2.0	26.5	31.6	28.7	1.1	1.2	1.2
5	2	2.3	2.4	2.3	27.2	27.6	27.4	1.3	1.3	1.3
6	4	2.0	2.1	2.1	28.7	30.8	30.0	1.1	1.3	1.2
7	5	2.1	2.4	2.2	25.1	29.3	26.9	1.1	1.4	1.2
8	1	2.2	2.2	2.2	30.0	30.0	30.0	1.4	1.4	1.4
9	1	2.0	2.0	2.0	30.1	30.1	30.1	1.4	1.4	1.4
TOTAL	35	1.8	2.4	2.0	25.1	32.4	29.4	1.1	1.4	1.2

from nine slices highlight the advantages of a series feedback design to achieve very repeatable performance.

ACKNOWLEDGMENT

The authors wish to thank R. E. Williams for processing support and S. F. Goodman and J. Wright for technical assistance.

REFERENCES

[1] M. J. O. Strutt and A. Van Der Ziel, "Suppression of spontaneous fluctuations in amplifiers and receivers for electrical communication and for measuring devices," *Physica*, vol. IX, no. 6, pp. 513–538, June 1942.
[2] J. Engberg, "Simultaneous input power match and noise optimization using feedback," in *Dig. Tech. Pap. Fourth Eur. Microwave Conf.*, Sept. 1974, pp. 385–389.
[3] L. Besser, "Stability considerations of low-noise transistor amplifiers with simultaneous noise and power match," in *IEEE MTT-S Int. Microwave Symp. Dig.*, 1975, pp. 327–329.
[4] K. Niclas, "Noise in broad-band GaAs MESFET amplifiers with parallel feedback," *IEEE Trans. Microwave Theory Tech.*, vol. MTT-30, pp. 63–70, Jan. 1982.
[5] R. W. Thill, W. Kennan, and N. K. Osbrink, "A low-noise GaAs FET preamplifier for 21 GHz satellite earth terminals," *Microwave J.*, pp. 75–84, Mar. 1983.

AN RF FRONT-END FOR DIGITAL MOBILE RADIO

J. Fenk, W. Birth, R.G. Irvine, P. Sehrig, K.R. Schön
Siemens AG, Semiconductor Group
Balanstr. 73
München, D-8000

Abstract

This paper presents a bipolar transceiver chipset for application in new generation digital mobile radio equipment.

Introduction

For the next generation of mobile radio equipment, which will be a digital system working around 900 MHz, a transmitter and a receiver circuit for the RF frontend have been developed. Fig. 1 shows a block diagram of the RF part and its connection to the digital (baseband) circuitry. The signal transmitted is a Gaussian-filtered minimum shift keying (GMSK) signal.

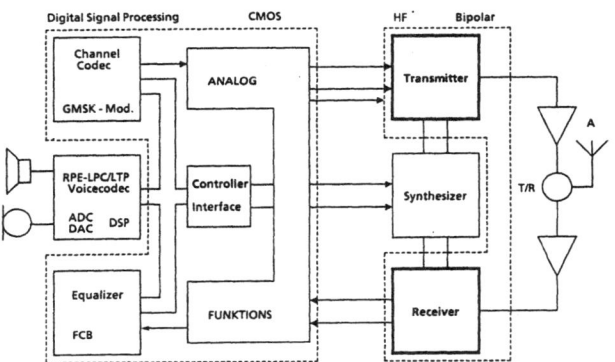

Fig. 1 Block diagram of the RF part and its connection to the digital circuitry

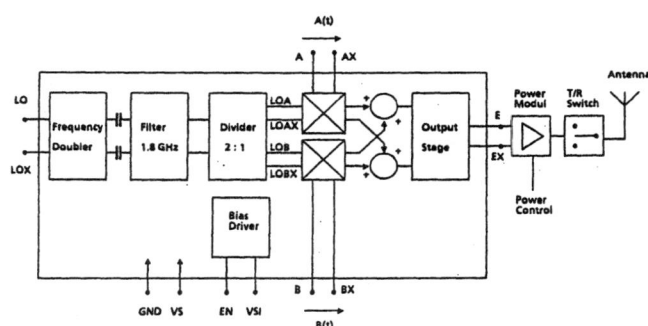

Fig. 2 Block diagram transmitter

Both circuits are realized in the bipolar technology OXIS-3 with a minimum feature size of 1.5 μm and a transit frequency of about 7.5 GHz.
The main features of the chip set are:
- a variety of applications
  - standards: GSM, NMT, TACS, AMPS
  - Modulation/Demodulation: AM, DSB, SSB, PM, PSK, FM, FSK, QAM, QPSK, GMSK
- simple switch over from one standard to another
- simple and reduced rf-design for the customer
- reduced peripheral components
- no alignment necessary
- reduced board design area
- reduced power consumption
- high modulation phase stability versus temperatur and product life

Transmitter Circuit

The main part of the transmitter circuit (Fig. 2) is a quadrature modulator followed by a linear amplifier with an output power of 0 dBm. For sufficient carrier and sideband suppression the modulator structures have to be as symmetrical as possible. For a low bit-error rate, the phase error of the 0° and 90° carriers must be less than 3°. External circuitry for phase shifting such as directional couplers, LC resonant circuits or delay lines is expensive and needs to be trimmed. To avoid this, a phase-shift circuit has been implemented.
A power-down switch reduces the current consumption from about 40 mA in the active mode to less than 10 μA in the standby mode.

Theoretical details

In the GSM digital mobile telephone system [1] an RF carrier with a constant magnitude is modulated in the phase $ß(t)$. This continuous phase modulation can be represented by a pointer rotating with phase as a result of modulation. The cartesian components

$$A(t) = K_1 \sin ß(t) \quad \text{and} \quad B(t) = K_1 \cos ß(t)$$

of this pointer are the baseband modulation input signals of the circuit. For realising eq.(8) in GSM recommendation 05.04. these terms are to be multiplied with the orthogonal LO carriers as follows

$$K_1 \cos ß(t) \times K_2 \cos \omega_0 t - K_1 \sin ß(t) \times K_2 \sin \omega_0 t = K_1 K_2 \cos(\omega_0 t + ß(t))$$

for values of $K_2 : 0 < K_2 \leq 1$.

Integrated solution

Fig.2 shows the block diagram for this function. Two Gilbert multipliers are driven by A(t) and B(t). The actual internally generated LO carriers LOA and LOB work in a switching mode. The outputs of both multipliers are combined at the summing points. The sum drives a linear output stage followed by the power module of the transmitter after filtering.
Output power modules which compress the top of the amplitude may be used. In order to avoid unwanted spectrum near the carrier caused by this compression, the transmitter circuits should not produce large amplitude ripples. Furthermore to avoid ripples high quality 0° and 90° carriers are required.
In the presented RF modulator (Fig.2) these requirements are fulfilled as follows: The transmission frequency f_0 at the differential inputs LO, LOX is first doubled and then filtered in a way that a sine wave is generated at $2f_0$ about 1.6 to 2.0 GHz.
This frequency is the clock for a 2:1 divider. At the outputs of the two latches of this divider, orthogonal carriers LOA and LOB for the modulator are provided with exactly the same power levels and switching slew rates. Delays in the layout are matched.

IEEE 1990 Bipolar Circuits and Technology Meeting, pp. 244-247, September 1990.

The signal paths from the doubled LO frequency to the mixer outputs are designed with high symmetry. Any differences in the metallisation are very small. In the chip photo Fig.3 the high symmetry of the layout and the shielding by supply lines and by substrate contacts can be seen.

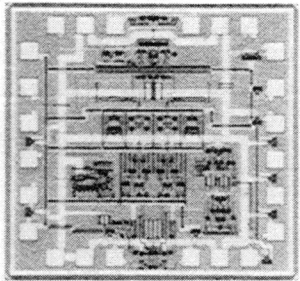

Fig. 3 Photomicrograph of the transmitter

Measurement results

Phase errors of less than +/-2° in the 0° and 90° carriers can be measured far away from the used GSM frequency band, e.g. at 800MHz and at 1GHz.

The spectrum for SSB sine modulation in Fig.4 contains a lot of information. The used carrier frequency f_0 is 900 MHz. This carrier has 43dB suppression. The baseband modulation signals for this test are

$A(t) = 0.5V \sin 2\pi f_m t$ and $B(t) = 0.5V \cos 2\pi f_m t$

with f_m = 67.8 kHz. Thus the lower side band LSB at f_0-f_m is suppressed by about 43dB while the upper side band USB at f_0+f_m has the full output power of 0 dBm. A third order product at f_0-3f_m is damped with a_{K3} = 43dB. The fifth order poduct at f_0+5f_m shows a_{K5} = 64dB. All even order distortion products are negligible.

The special situation of f_0+2f_m in Fig.4 is caused by SSB remodulation via cross talk. Spurious parts of the main power signal at f_0+f_m get back to the LO input. Those signals are SSB remodulated with f_m. The result can be seen at f_0+2f_m with the so called remodulation damping a_{rem} = 57dB. This value is identical to the crosstalk damping in the internal and external paths from the RF output to the LO input.

It is shown that high crosstalk damping at 900 MHz is possible on the chip. For measurements of spurious crosstalk effects a good shielding of the external circuitry is necessary.

Fig.5 and 6 show that there is only a very small influence of the LO input power on the carrier rejection and to the possible single sideband rejection.

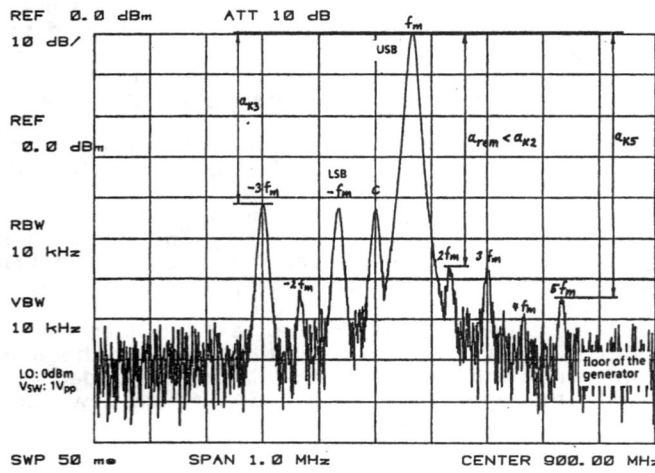

Fig. 4 Test spectra for SSB sine modulation

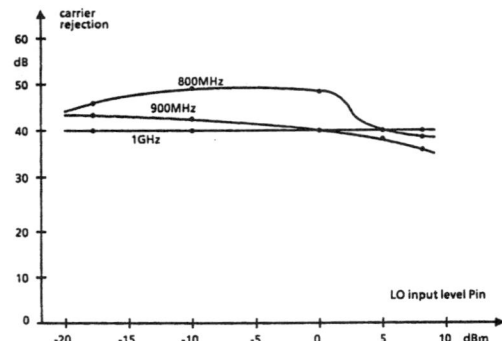

Fig. 5 Carrier rejection at modulation with 1Vpp versus LO input level with asymmetric LO input supply.

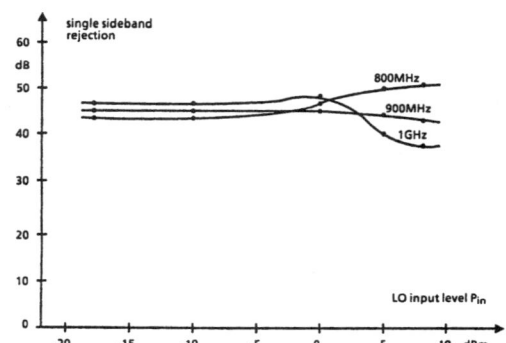

Fig. 6 Single sideband rejection versus LO input level. Test modulation signals: V(A,AX) = $U_m \sin 2 f_m t$; V(B,BX) = $U_m \cos 2 f_m t$; f_m = 67.7kHz; U_m = 500mV

Receiver Circuit

The receiver circuit consists of two electrically independent parts (Fig. 7, 8).

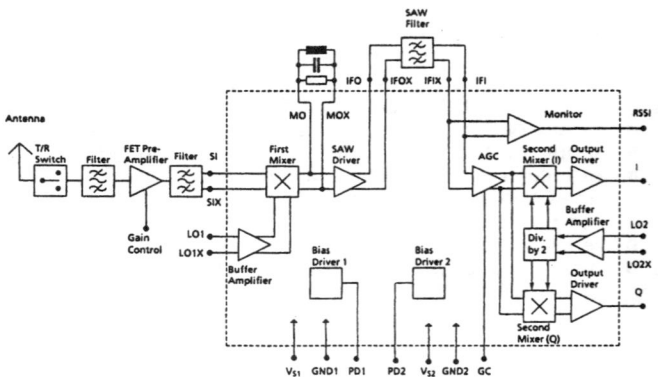

Fig. 7 Block diagram receiver

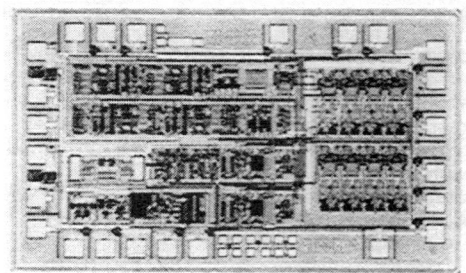

Fig. 8 Photomicrograph of the receiver

RF-front end mixer and SAW-driver circuit section

In the first part the RF signal is mixed down to an IF frequency between 45 and 90 MHz. A LO buffer amplifier with short rise and fall times minimizes mixer noise. After the SAW driver the signal is fed to an external SAW filter to filter signals outside the band. The overall gain from RF input to IF output is about 15 dB. The current of this section is reduced in the standby mode from about 13 mA to less than 10 µA by one power-down switch.

IF-circuit section

The second part contains an AGC amplifier with 10 to 80 dB gain and a quadrature demodulator to generate the I and Q components in the baseband. The double value of the second LO frequency is fed to a divider by two to provide the demodulator with 0° and 90° signals. A field strength monitor generates a dc level proportional to the logarithmic IF field strength.

For extremely high demands of cross-talk attenuation between IF output and IF input signals, IFO and IFI, respectively, two separate chips may be used for each function without higher power consumption.

As in the modulator, the current is reduced in the standby mode from about 11 mA to less than 10 µA by an additional power-down switch.

<u>Variable gain stage AGC</u>: To get a gain control range of about 70dB a 4 stage amplifier with variable gain was used.

To prevent DC-offsets capacitive coupling after every two stages was used.

Fig. 9 shows the principle of such a gain controlled amplifier. The gain controlled amplifier is made up of a differential pair with additional emitter resistors. Minimum gain is controlled by the resistors 2 x RE and maximum gain results when the diode bridge has maximum DC-current I. The diode bridge has the advantage that the differential pair is not influenced by the gain control current I.

The gain control voltage is transferred to a gain control current. For minimum current I the diode bridge with the additional resistors RD has a very high impedance, about 10 times more than 2 x RE therefore the amplifier has minimum gain. With maximum current I the diode bridge has minimum impedance about 4 times smaller than 2 x RE therefore the amplifier has maximum gain.

Fig. 10 shows the simulated and measured results of the gain versus gain control voltage.

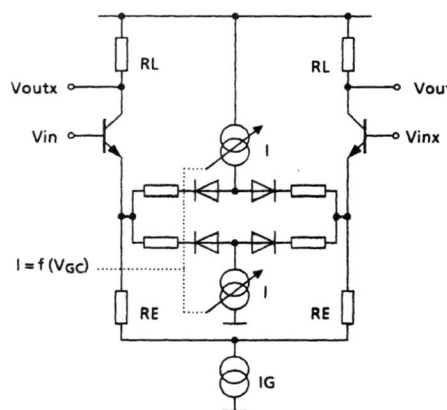

Fig. 9 The principle of such a gain controlled amplifiers

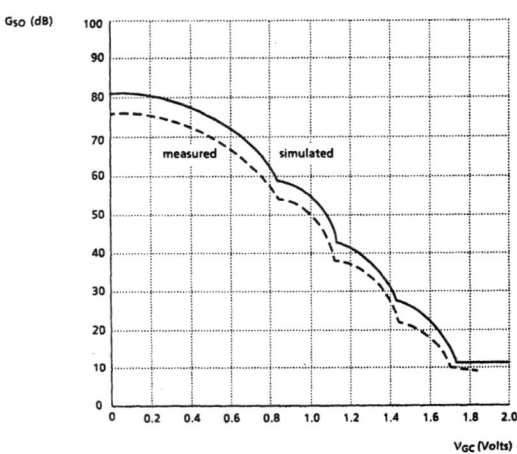

Fig. 10 The results of the gain versus gain control voltage

<u>RSSI - Fieldstrength Monitor</u>: The RSSI-block is designed for a dynamic range of 70dB resulting in an equivalent DC-voltage with a logarithmic indication. To realize this a 7 stage limiting amplifier is used with a gain of 10dB per stage and a field strength detector circuit. To prevent DC-offset problems all stages are AC-coupled.

One standard limiting amplifier stage and field strength detector circuit is shown in Fig. 11.

An input voltage V_{in} greater than $V_{inpp} > 4V_T$ results in a DC-voltage offset V_D compared to V_{ref}. Below this value $V_D = V_{ref}$. Due to the fact that $I_1 = I_2$, the detector output current is $I_Q = I_2 - I_1 = 0$.

For values of $V_{inpp} > 4V_T$ limiting occurs resulting in a slight increase of V_D. With a value of $V_{inpp} > 8V_T$ $I_1 = 0$ and $I_Q = I_2$.

All detector output currents flow through one load resistor. Figure 12 shows the RSSI-field strength voltage versus the input signal.

For small input signals (≤ -85 dBm) the measured results show a minimum voltage of about 0.8V due to the broad band noise of the input stage. By optimizing the noise figure of the input amplifier and reducing the overall bandwith of the amplifiers this value can be reduced to get the full dynamic range.

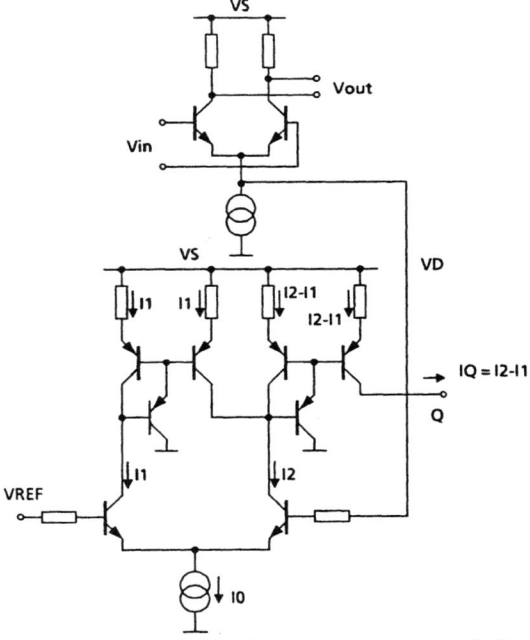

Fig. 11 One standard limiting amplifier stage and field strength detector circuit

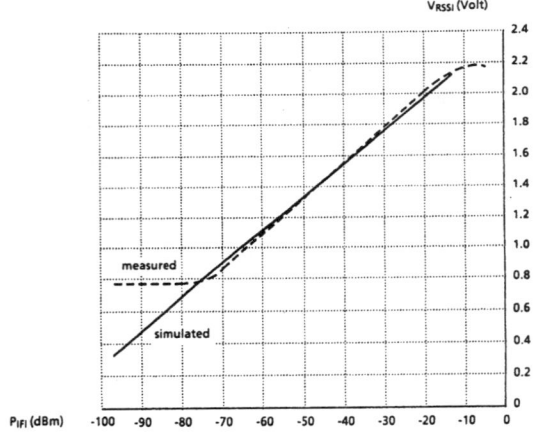

Fig. 12 The RSSI- Field strength voltage versus the input signal

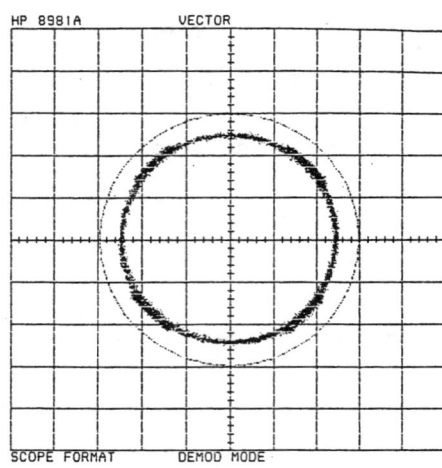

Fig. 13 The demodulated GMSK signal with the I- and Q-component

Measurements results with both circuits

First samples of both circuits have been examined. To test the resulting phase error of transmitter and receiver an exact GMSK quadrature signal with random pattern is fed to the transmitter. The output of the transmitter is connected directly with the receiver via an attenuator. Fig. 13 shows the demodulated GMSK signal with the I-component on the x-axis and the Q-component on the y-axis. The outer circle is a reference circle. The test result shows the almost negligible phase error.
In Fig. 14 an eye pattern of the demodulated I- and Q-signal is shown.

Conclusion

This test results demonstrates the high performance of both circuits for the use in the RF part of the digital mobile radio. With these transceiver chips, now available in SMD-packages, a very compact, size reduced RF-front-end module can be developed by the customers for mobile radio and hand portables with a lot of advantages compared to discrete solutions.

Acknowledgements

The authors wish to express their gratitude to Mr. Beyrich from SIEMENS Mobile Radio Systems Division for very helpful system design discussions and Mr. Ganser and Mr. Samson for their layout work.

References

[1] GSM recommendation 05.04.

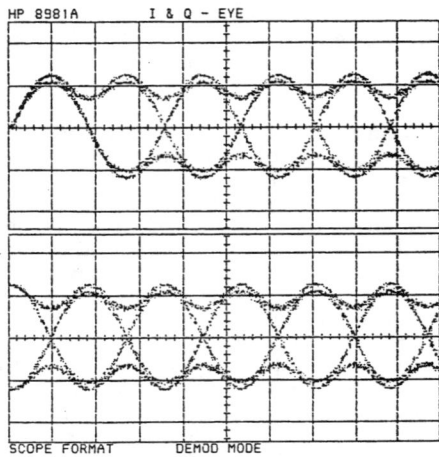

Fig. 14 An eye pattern of the demodulated I- and Q- signal

A Si Bipolar Monolithic RF Bandpass Amplifier

Nhat M. Nguyen and Robert G. Meyer

Abstract—The application of monolithic inductors to the realization of Si bipolar monolithic RF amplifiers is investigated. As a test vehicle, a bipolar monolithic bandpass amplifier was fabricated and characterized. A 4-nH silicon integrated inductor was used to achieve a peak S_{21} gain of 8 dB, a simulated noise figure of 6.4 dB, and a matched input impedance of 50 Ω in the frequency range of 1–2 GHz.

I. Introduction

RF AMPLIFIERS are widely used in many RF communication systems. Important characteristics include gain per stage, frequency response, and noise performance. In discrete circuits and also in microwave GaAs circuits [1], inductors are often used to boost the gain of RF amplifiers and to improve matching. In this paper we discuss the application of monolithic inductors in the design and fabrication of Si monolithic RF amplifiers in the L band (1–2 GHz).

II. Design Approach

A simplified ac circuit schematic of a widely used resistive feedback amplifier is shown in Fig. 1(a). The circuit incorporates a shunt feedback resistor R_F and a series feedback resistor R_E to achieve stabilized circuit gain, low output distortion, and simultaneous input and output impedance matching. In order to predict the frequency response of this circuit, we consider the simplified equivalent circuit shown in Fig. 1(b) which neglects the effects of C_{bc} and r_b of the bipolar transistor. It can be shown that the transimpedance gain is

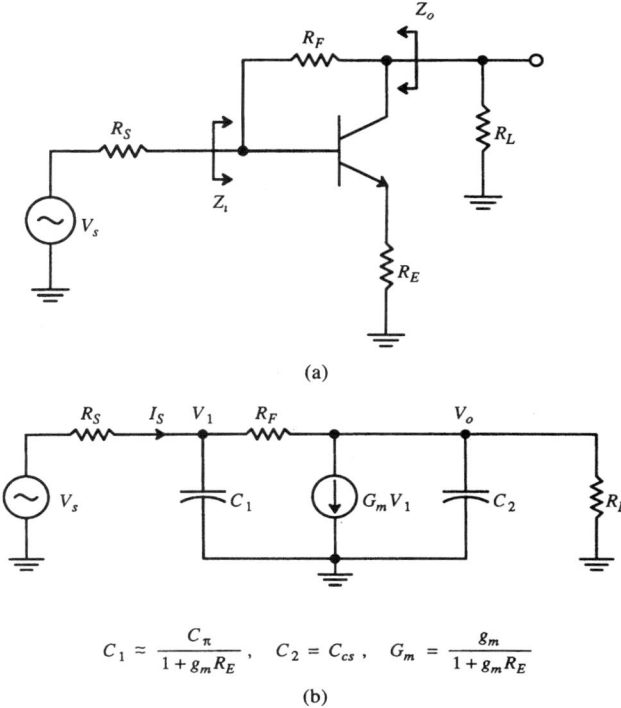

Fig. 1. (a) Series and shunt feedback amplifier and (b) simplified equivalent circuit.

$$\frac{V_o(s)}{I_S(s)} = \frac{(1 - G_m R_F) R_L}{(1 + G_m R_L) + s[C_1 R_F + (C_1 + C_2) R_L] + s^2 C_1 R_F C_2 R_L}. \quad (1)$$

The denominator of (1) can be written as

$$D(s) = \left(1 - \frac{s}{p_1}\right)\left(1 - \frac{s}{p_2}\right)$$
$$= 1 - s\left(\frac{1}{p_1} + \frac{1}{p_2}\right) + \frac{s^2}{p_1 p_2} = 1 + a_1 s + a_2 s^2. \quad (2)$$

If p_1 is a dominant pole, it can be approximated by

$$p_1 \approx -\frac{1}{a_1} = -\frac{1 + G_m R_L}{C_1 R_F + (C_1 + C_2) R_L} = f_{-3\text{-dB}}. \quad (3)$$

The midband gain of the circuit from (1) is

$$A_o = \frac{(1 - G_m R_F) R_L}{1 + G_m R_L}. \quad (4)$$

Equations (3) and (4) can be combined to yield the gain–bandwidth product. It is approximately equal to

$$GB \equiv A_o f_{-3\text{-dB}} \approx \omega_T R_L \quad (5)$$

where $\omega_T \approx g_m / C_\pi$ denotes the transition frequency of the bipolar transistor. It is seen from (5) that the gain-bandwidth product is a constant quantity, implying that if the −3-dB frequency is optimized well into the micro-

---

Manuscript received March 20, 1991; revised July 24, 1991. This work was supported by the U.S. Army Research Office under Grant DAAL03-87-K0079 and by a grant from IBM.

The authors are with the Department of Electrical Engineering and Computer Sciences and the Electronics Research Laboratory, University of California, Berkeley, CA 94720.

IEEE Log Number 9103864.

wave region, a considerable amount of circuit gain is sacrificed.

For the low-pass amplifier of Fig. 1(a) the -3-dB frequency is observed to be limited mainly by capacitance C_1. Beyond the -3-dB frequency, the terminal impedances are severely mismatched while the circuit gain is greatly reduced. The performance of such circuits can be optimized for use with band-limited RF signals by focusing on the passband of interest. For the circuit in Fig. 1(b), this can be accomplished with the addition of a shunt inductor across capacitor C_1. Effectively, this inductor transforms the "untuned" low-pass amplifier into a bandpass amplifier.

In order to understand how the shunt inductor used in this topology modifies the circuit bandwidth, we first consider the general representation of a low-pass amplifier shown in Fig. 2(a). The gain function can be derived to be

$$\frac{V_o(s)}{V_s(s)} = -G_m(R_L \| R_o) \frac{R_i}{R_i + R_S} \left(\frac{1}{1 - s/p_1}\right) \left(\frac{1}{1 - s/p_2}\right) \quad (6)$$

where

$$p_1 = -\frac{1}{(R_S \| R_i) C_i}$$

and

$$p_2 = -\frac{1}{(R_L \| R_o) C_o}. \quad (7)$$

A typical plot of (6) is shown in Fig. 2(b) where the assumed dominant pole p_1 determines the bandwidth of the circuit. Since the input resistance R_i must be matched to the source resistance R_S in order to minimize reflected signal loss, quantity C_i dictates the frequency p_1. We next consider the modified circuit shown in Fig. 3(a). It can be shown that

$$\frac{V_o(s)}{V_s(s)} = -G_m(R_L \| R_o) \left(\frac{1}{1 - s/p_2}\right)$$
$$\cdot \left[\frac{s(L_1/R_S)}{s^2 L_1 C_i + sL_1(1/R_i + 1/R_s) + 1}\right]. \quad (8)$$

A typical plot of (8) is shown in Fig. 3(b). At frequencies well below or above the center frequency ω_c ($= 1/\sqrt{L_1 C_i}$), the parallel $R_i L_1 C_i$ tank circuit behaves like a short circuit, and consequently, little signal power can be transmitted to the output. Denote the frequencies at which the peak gain drops by 3 dB by ω_- and ω_+, and the bandwidth of the circuit by $\omega_{-3\text{-dB}}$. If $p_2 \gg \omega_+$, it can be shown that [2]

$$\omega_{-3\text{-dB}} \equiv \omega_+ - \omega_- = \frac{\omega_c}{Q} \quad (9)$$

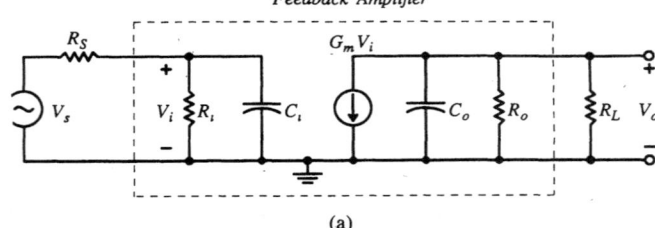

(a)

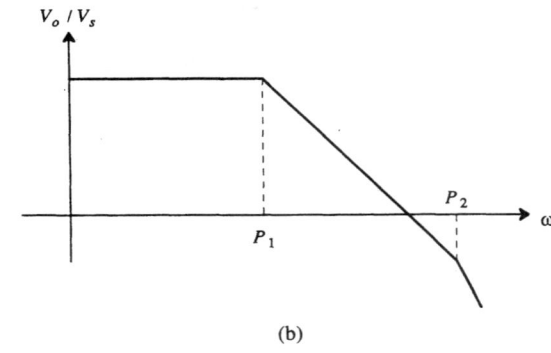

(b)

Fig. 2. (a) Low-pass amplifier and (b) frequency response.

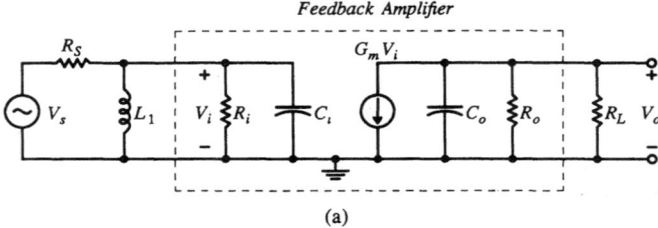

(a)

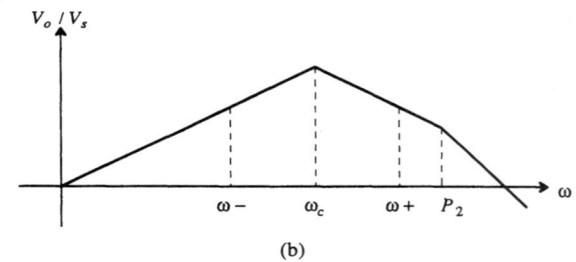

(b)

Fig. 3. (a) Bandpass amplifier and (b) frequency response.

where

$$Q = (R_S \| R_i) \sqrt{\frac{C_i}{L_1}}. \quad (10)$$

By substituting (10) into (9), we obtain

$$\omega_{-3\text{-dB}} = \frac{1}{(R_S \| R_i) C_i}. \quad (11)$$

It is interesting to note that the bandwidth of the modified circuit is equal to that of the original circuit.

The new circuit technique effectively transforms a low-pass frequency response to a bandpass response, centered around ω_c. This method is attractive because it achieves the required bandwidth without reducing circuit gain. In

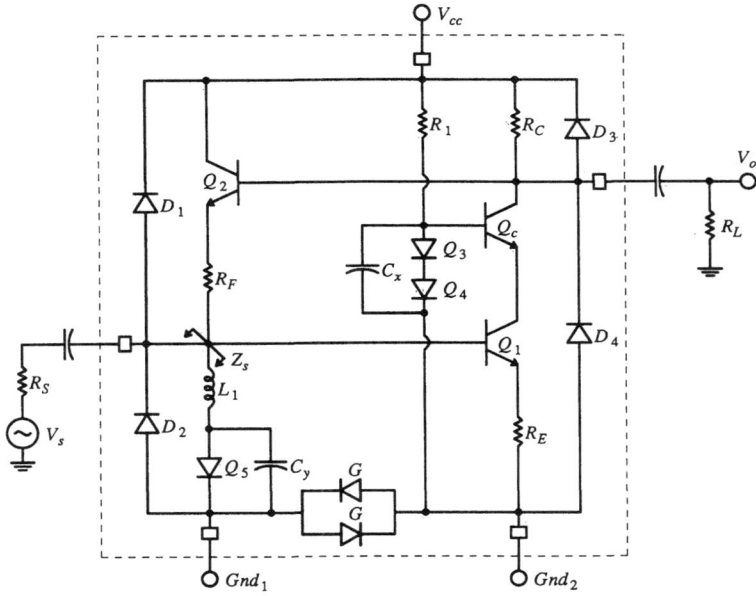

Fig. 4. Complete schematic of the L-band RF amplifier.

addition, it can also achieve a matched input impedance. By tuning out the device capacitance near the center frequency ω_c, the shunt inductor also improves the noise performance at high frequencies.

III. Circuit Configuration and Implementation

As an application of the above circuit technique, an L-band (1–2 GHz) bipolar monolithic amplifier has been realized. The complete circuit is shown in Fig. 4. Cascode transistor Q_c eliminates the Miller effect resulting from capacitance C_{bc} of transistor Q_1. Transistor Q_2 functions as a voltage buffer that minimizes forward transmission through the feedback resistor R_F. The only drawback to having the buffer Q_2 is that simultaneous impedance matching at both the input and output cannot be achieved. For optimum noise performance, R_E and the base resistance of Q_1 must be minimized, and R_F maximized. Diodes D_1–D_4 function as protection circuit against electrostatic discharges. The amplifier utilizes a 4-nH silicon integrated inductor [3] to achieve a peak S_{21} gain of 8 dB and a bandwidth of 1.2 GHz that extends from 700 MHz to 1.9 GHz. In order to reduce base resistance, the input transistor Q_1 is a large device and is fabricated with six 30×2-μm base strips, yielding a 17-Ω base resistance at 6 mA of collector bias current. The measured quiescent power dissipation is 130 mW from a single supply of 10 V. The circuit was fabricated in an oxide-isolated Si bipolar IC process with peak $f_T = 9$ GHz.

The amplifier's input resistance is matched to a system impedance Z_T of 50 Ω with careful selection of R_F and R_E according to the derived relationship

$$R_F = A_V Z_T \tag{12}$$

where $A_V \equiv G_m Z_T$, $G_m = (g_{m1})/(1 + g_{m1} R_E)$, and $R_S = R_L = Z_T$ (=50 Ω). The circuit bandwidth can be determined from the untuned amplifier (without L_1) shown in Fig. 5. By neglecting the base resistances of the transistors and by treating the cascode transistor Q_c as a unity current buffer, we can show that the admittance matrix of the two-port amplifier is

$$\begin{bmatrix} Y_{11} & Y_{12} \\ Y_{21} & Y_{22} \end{bmatrix} \approx \begin{bmatrix} \dfrac{1}{R_F} + sC_1 & -\dfrac{1}{R_F} \\ -\dfrac{1}{R_F}\dfrac{s}{\omega_t} + G_m & \dfrac{1}{R_F}\dfrac{s}{\omega_t} + sC_2 \end{bmatrix} \tag{13}$$

where

$$\begin{bmatrix} I_1 \\ I_2 \end{bmatrix} = \begin{bmatrix} Y_{11} & Y_{12} \\ Y_{21} & Y_{22} \end{bmatrix} \begin{bmatrix} V_1 \\ V_2 \end{bmatrix}$$

$$C_1 = \frac{C_{\pi 1}}{1 + g_{m1} R_E}$$

$$C_2 = C_{csC} + C_{bcC} + C_{bc2}.$$

The gain function can be derived to be

$$\frac{V_o(s)}{V_s(s)} = \frac{Y_S Y_{21}}{Y_{12} Y_{21} - (Y_{11} + Y_S)(Y_{22} + Y_L)}$$

$$\approx -\frac{A_V}{s^2 C_1 C_2 Z_T^2 + s Z_T \left[C_1 + C_2 \left(1 + \dfrac{1}{A_V} \right) \right] + \left(2 + \dfrac{1}{A_V} \right)} \tag{14}$$

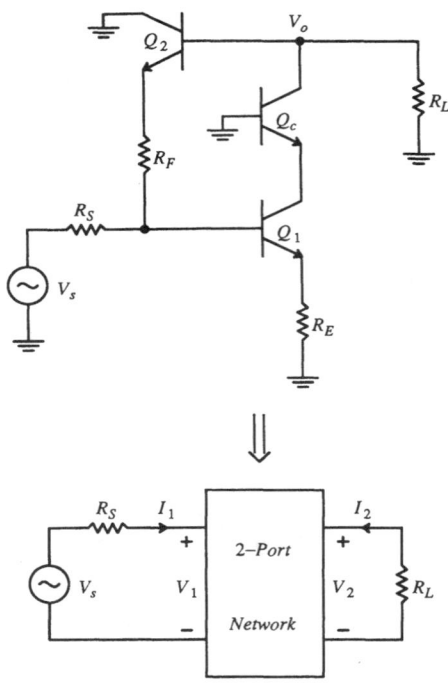

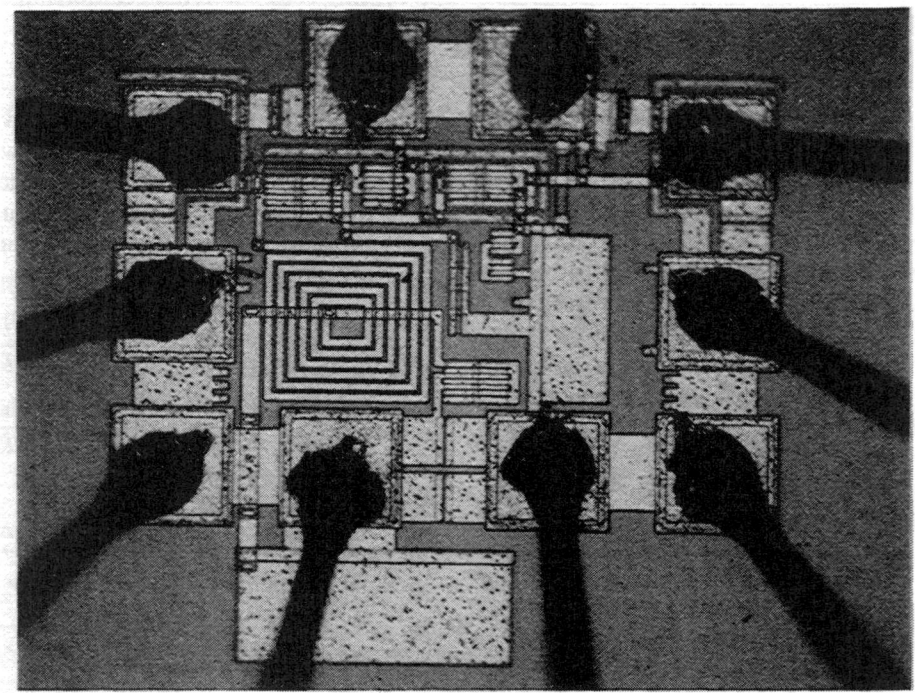

Fig. 5. Simplified circuit for frequency analysis.

Fig. 6. Die photograph of the bandpass amplifier.

where $Y_S = 1/R_S$ and $Y_L = 1/R_L$. From (14), the -3-dB frequency can be predicted assuming that the transfer function has a dominant pole. It is equal to

$$f_{-3\text{-dB}} \approx -\frac{2}{Z_T(C_1 + C_2)} \approx -2\frac{\omega_t}{A_V}. \quad (15)$$

A die photograph of the amplifier is shown in Fig. 6. In order to minimize electrical coupling through the substrate, a buried p-type layer was placed around the I/O pads and around the periphery of critical devices. Measured $|S_{21}|$ and $|S_{11}|$ are shown in Figs. 7 and 8, respectively, and are close to computer simulations and theoretical analysis. The amplifier was simulated assuming 3σ limits of $\pm 10\%$ on the capacitance values. When subject to these process variations, the circuit bandwidth is observed to shift at less than $\pm 1.5\%$. Simulations on the

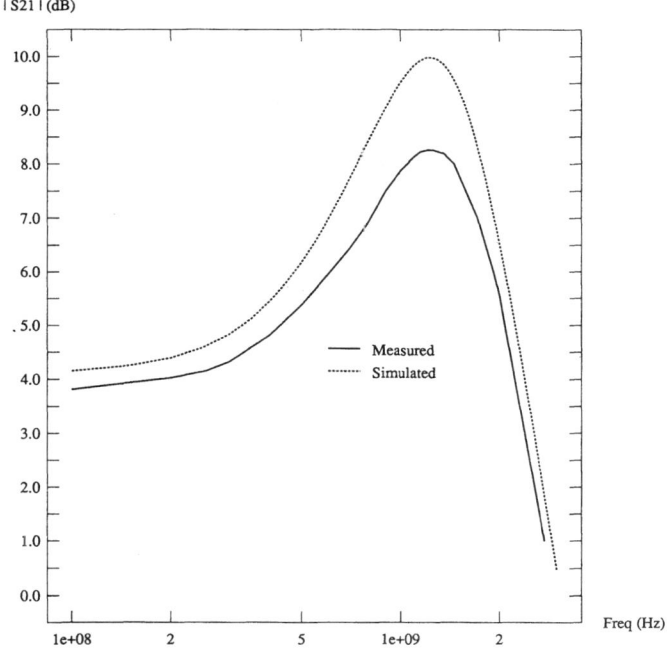

Fig. 7. Measured and simulated $|S_{21}|$ of the amplifier.

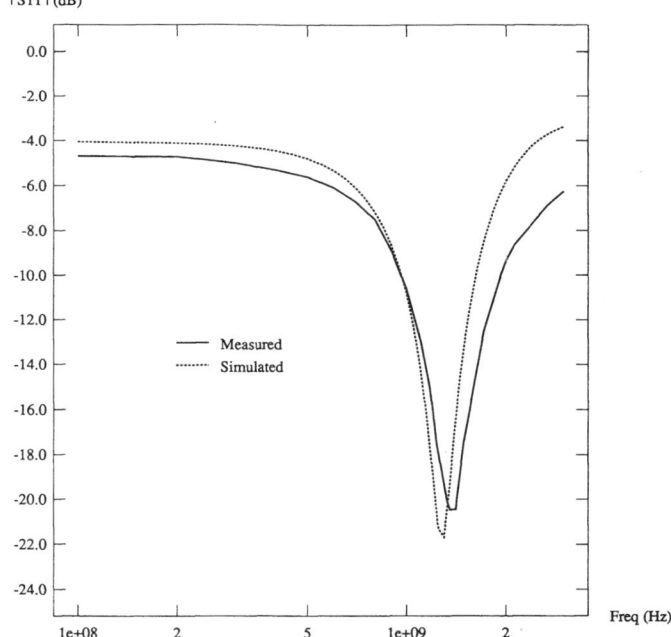

Fig. 8. Measured and simulated $|S_{11}|$ of the amplifier.

large-signal characteristic of the amplifier in the passband give a 1-dB gain compression point of +0.3 dBm (at source signal level of −4.0 dBm) and a third-order intercept of +17.2 dBm.

While frequency response is an important requirement in an RF amplifier, a good noise performance is equally significant since it places a limit on the smallest RF signal that can be detected. The noise figure is commonly used as a figure of merit in low-noise amplifiers. The bandpass amplifier has better noise performance than that of the corresponding low-pass amplifier since the source impedance Z_s as defined in Fig. 4 has an inductive reactance component which is closer to the optimum for lowest noise figure. The simulated noise figure has a minimum of 6.4 dB at the frequency 1.5 GHz, of which 0.5 dB is contributed by transistor Q_5 and resistive loss in L_1. For purposes of comparison, the untuned circuit has a −3-dB frequency of 1.2 GHz and a noise figure of greater than 7 dB at the frequency 1.5 GHz.

IV. Conclusion

Using a shunt integrated inductor, a Si bipolar monolithic L-band amplifier has been realized which achieves 8-dB gain, matched input impedance, and 6.4-dB simulated noise figure in the 1–2-GHz frequency band.

Acknowledgment

The authors wish to acknowledge the contributions of W. Mack of Signetics Corporation for layout assistance and characterization, and of Signetics for HS3 fabrication.

References

[1] R. Bayruns, T. Laverick, N. Scheinberg, and K. Li, "A 5-KΩ 2-GHz GaAs transimpedance amplifier with a low-noise active load," in *1991 ISSCC Dig. Tech. Papers*, pp. 272–273.

[2] L. O. Chua, C. A. Desoer, and E. S. Kuh, *Linear and Nonlinear Circuits*. New York: McGraw-Hill, 1987, ch. 9.

[3] N. M. Nguyen and R. G. Meyer, "Si IC-compatible inductors and *LC* passive filters," *IEEE J. Solid-State Circuits*, vol. 25, no. 4, pp. 1028–1031, Aug. 1990.

A 1-GHz BiCMOS RF Front-End IC

Robert G. Meyer, *Fellow, IEEE*, and William D. Mack, *Member, IEEE*

Abstract—An RF front-end IC containing a low-noise amplifier and mixer is described. On-chip temperature and supply-voltage compensation is used to stabilize circuit performance. Realized in a BiCMOS process, the circuit consumes 13.0–mA total current from a 5–V supply. The amplifier gain at 900 MHz is 16 dB, the noise figure is 2.2 dB, and the input third-order intermodulation intercept is -10 dBm. The mixer input third-order intermodulation intercept is +6 dBm with 15.8 dB noise figure.

I. INTRODUCTION

THE rapid growth of portable RF communication systems has led to a demand for low-cost high-performance front-end electronics (RF amplifiers, mixers, and local oscillators) for communication receivers in the gigahertz frequency range. Minimization of board and substrate area is also an important factor, leading to a trend to replace discrete solutions with integrated circuits of increasing integration levels. A further advantage of monolithic realizations of the front end is that sophisticated on-chip compensation techniques can be included at little extra cost to achieve much improved independence from supply voltage and temperature variations, thus improving worst-case system performance margins.

In this paper, a front-end IC for the 1-GHz band is described. The IC contains a temperature-compensated low-noise amplifier (LNA) and wide-dynamic-range mixer. The design is implemented in a silicon BiCMOS process.

II. SYSTEM APPLICATION

A typical application of the IC is shown in Fig. 1. A signal received from the antenna is passed through a first RF filter before being applied to the IC and amplified by the LNA. The signal is then passed to a second (external RF) filter before reentering its IC and being applied to the mixer, which is driven by an externally supplied local oscillator (LO). The mixer has a high impedance (open collector) output which is loaded externally and then passed to the IF filter and amplifier.

The IC is housed in a 14 pin small outline plastic package (SO 14) with pinout as shown in Fig. 2. The power supply is 5 V. Pin 8 is an enable pin which can be used to power down the LNA in the presence of strong signals (2-state gain control) to improve large-signal handling capability or during periods when the receiver is inactive in order to conserve battery power.

Manuscript received June 21, 1993; revised November 8, 1993. This work was supported by the U.S. Army Research Office under Grant DAAH04-93-G-0200.
R.G. Meyer is with the Department of Electrical Engineering and Computer Sciences and the Electronics Research Laboratory, University of California, Berkeley, CA 94720.
William D. Mack is with Philips Semiconductors, Sunnyvale, CA 94086.
IEEE Log Number 9216476.

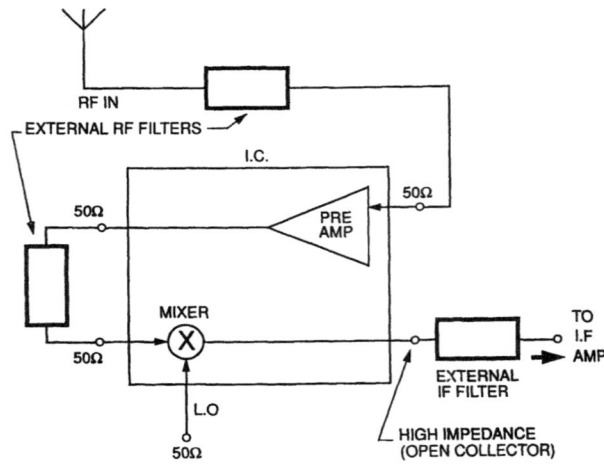

Fig. 1. Typical application of the front-end IC.

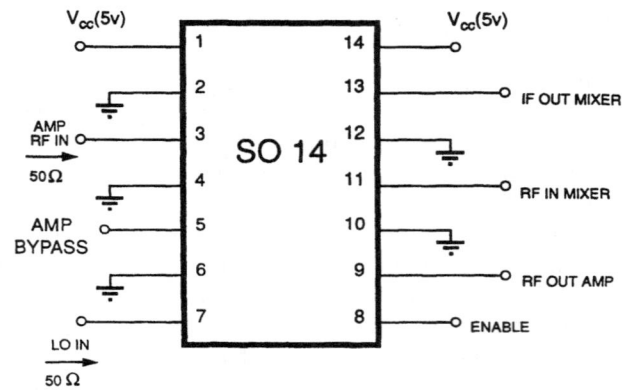

Fig. 2. Pinout of the IC.

III. FABRICATION PROCESS

The process used for fabrication was Philips Semiconductors QUBiC BiCMOS [1] process with peak n-p-n f_T of 13 GHz, minimum $C_{\mu 0}$ of 6 fF, and CMOS L_{eff} of 0.8 μm.

IV. PREAMPLIFIER (LNA) DESIGN

The design of a receiver preamplifier involves numerous tradeoffs. Important RF parameters include noise performance, intermodulation performance, and overload level. Gain must be sufficient to minimize mixer noise, but not so much as to cause premature mixer overload. Input and output matches should be adequate to eliminate the need for external matching elements. Gain stability over temperature and supply voltage variation is important as this affects worst case system

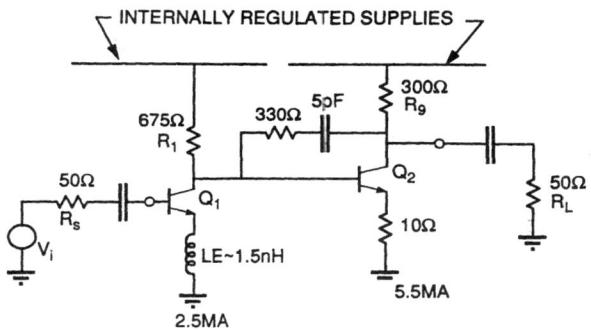

Fig. 3. Basic topology of the LNA.

specifications. Finally, supply current is critical in portable applications and must be minimized.

The basic topology of the LNA is shown in Fig. 3. In order to realize a nominal gain of 16 dB at 1 GHz, a two-stage amplifier was needed. The n-p-n bipolar transistors in the process were used because of their higher f_T (about twice) compared with the n-channel MOSFET's. Terminal matches to 50 Ω can be achieved in various ways, with differing effects on LNA noise figure. A common technique [2], applied to the second stage here, is to use combined shunt and series feedback. This tends to stabilize gain as well as the terminal impedances. The disadvantage of this technique is the noise figure degradation caused by the feedback resistors. This disadvantage is not critical in the second stage as its contribution to overall noise figure is reduced by the power gain of the first stage.

In order to minimize LNA noise figure, the input stage does not use resistive feedback at all, but relies on the presence of emitter bond-wire inductance to achieve an input match. The small-signal input impedance of a bipolar transistor becomes small at high frequencies due to the shunting effects of the input capcitance C_π and the Miller effect due to C_μ. Ultimately, the input impedance approaches the base resistance r_b, which is about 11 Ω in this case. (A large area input device of 36x the minimum size is used to minimize r_b, and thus the LNA noise figure.) However, the input impedance can be boosted to near 50 Ω by inclusion of finite emitter inductance L_E (which is inherently present in most packaging schemes) without degrading the noise figure (NF). This effect can be appreciated by noting that L_E raises the high-frequency input impedance of a bipolar transistor by the amount

$$Z_i \simeq \beta(j\omega)Z_E$$
$$= \frac{\omega_T}{j\omega}j\omega L_E$$
$$= \omega_T L_E. \quad (1)$$

Substituting $f_T \simeq 5$ GHz at the operating point and $L_E = 1.5$ nH gives $Z_i = 47\Omega$. The measured value of s_{11} so realized was about -10 dB at 900 MHz. Sensitivity to the expected maximum production variation of about ±10% in the L_E was examined by SPICE simulation. This showed expected variations in s_{21} of ±0.3 and ±0.6 dB in s_{11}.

The contributions to NF can be seen by using the approximate formula [3]

$$NF = 1 + \frac{r_b}{R_s} + \frac{1}{2g_m R_s} + \frac{2qI_c}{|\beta|^2}\frac{1}{4kT\frac{1}{R_s}} + \frac{2qI_B}{4kT\frac{1}{R_s}} \quad (2)$$

$$= 1 + \frac{r_b}{R_s} + \frac{1}{2g_m R_s} + \frac{g_m R_s}{2|\beta|^2} + \frac{g_m R_s}{2\beta_0}. \quad (3)$$

Substitution of data gives

$$NF \simeq 1 + \frac{11}{50} + \frac{5}{50} + \frac{5}{50} + \frac{5}{160}$$
$$= 1.62 dB$$

SPICE predicts 1.95 dB for the whole amplifier with the second stage making some contribution. The biggest single contribution is the r_b of Q_1.

The overload characteristics of the LNA are also very important. This aspect of performance is usually measured by the -1 dB gain compression point $P_{-1 dB}$, the third-order intermodulation intercept point (IM_{3INT}), and the -3 dB outband blocking level P_{OB}. These measures of distortion and overload are nominally related by fixed ratios [4] IM_{3INT}; being about 10 dB higher than $P_{-1 dB}$; and P_{OB} being about 1 dB higher than $P_{-1 dB}$. Due to the gain of the first stage, overload in the second stage is a major source of nonlinearity and is minimized by the feedback used to match the second stage, and also by using higher bias levels there (bias current of 5.5 mA versus 2.5 mA for the first stage). As a consequence, overload occurs essentially simultaneously in both stages. The measured and computed values of $P_{-1 dB}$ were both about -20 dBm at the input, with IM_{3INT} being about -10 dBm at the input. The value of P_{OB} was approximately equal to $P_{-1 dB}$. The value of $P_{-1 dB}$ corresponds to about 4 mA zero-to-peak signal swing at the output. This is close to the bias current in the output stage, and Q_2 is significantly compressed at this level. The measured value of $P_{-1 dB}$ corresponds to about 32 mV zero-to-peak signal at the input, which is somewhat above the theoretical value for a voltage-driven bipolar transistor [5] of $\sqrt{0.88}\frac{kT}{q} = $ 24 mV. The input stage is linearized to some degree by the finite driving source impedance and also by the effective high-frequency emitter degeneration contributed by L_E.

One disadvantage of using a nonfeedback stage (apart from the effect of L_E) for the input stage is the resulting temperature dependence of the gain. For a bipolar transistor, $g_m = \frac{qI_C}{KT}$, and this causes a 3–dB gain variation from -30 to +10°C. This variation is compensated as shown in Fig. 4 by on-chip generation of regulated power supplies with temperature dependence designed to first-order cancel the temperature variation of the gain. If I_{c1} is made proportional to absolute temperature (PTAT), then both theory and SPICE simulations predict the first stage gain to be approximately constant with temperature. This is achieved by applying a regulated voltage of $(2V_{BE} + V_{PTAT})$ to the base of Q_4 via a unity-gain buffer amplifier and a high-frequency noise filter composed of 2 kΩ series resistance and 10–pF shunt capacitance. An overall dc feedback loop forces $V_{CE1} = V_{BE2} + 10I_{C2} \simeq V_{BE2}$. Thus,

$$I_{C1} = \frac{V_{PTAT} + 2V_{BE} - V_{BE4} - V_{BE2}}{R_1}$$
$$\simeq \frac{V_{PTAT}}{R_1}. \quad (4)$$

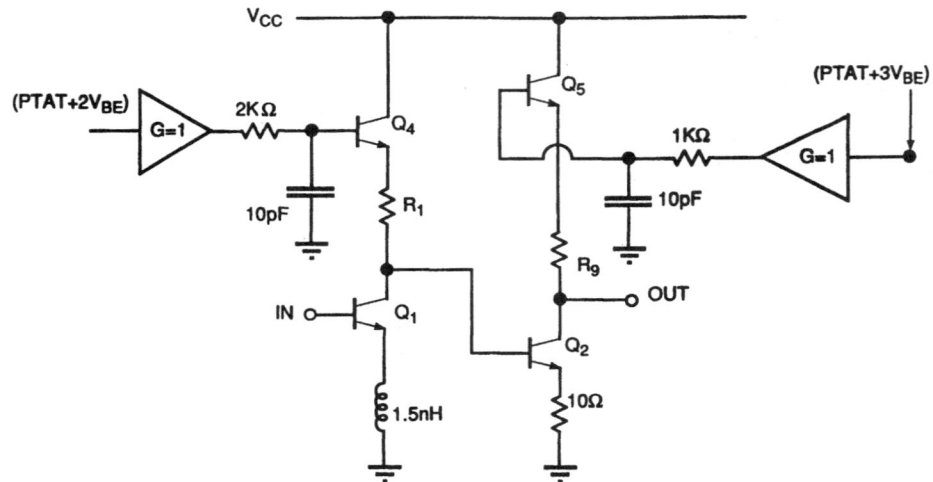

Fig. 4. LNA temperature compensation.

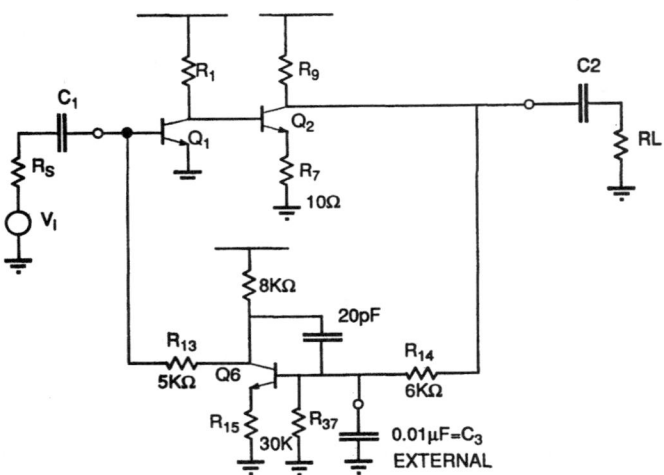

Fig. 5. Overall feedback bias loop for the LNA.

A similar technique compensates the gain of the second stage. The measured temperature dependence of the gain of the LNA was thus reduced to -0.008 dB/°C, which was very close to SPICE simulations. This is a factor of three improvement over an uncompensated amplifier.

The overall dc feedback loop which stabilizes the bias of the LNA is shown in Fig. 5. Transistor Q_1 acts as an inverter which forces

$$V_{CE2} = \frac{R_{14} + R_{37}}{R_{37}}(V_{BE6} + I_{C6}R_{15}) \simeq 1.2\,\text{V}. \quad (5)$$

Resistors R_{13} and R_{14} decouple the bias from the signal path while causing minimal loading. External capacitor C_3 decouples and stablizes the overall dc feedback loop which has a large low-frequency loop gain ($\simeq 400$) and three poles set by $C_1, C_2,$ and C_3. Capacitors C_1 and C_2 are nominally 100 pF for 900-MHz operation.

The bias generator for the circuit is shown in Fig. 6. A self-biased PTAT current is mirrored via Q_{14} and Q_{20} to generate the bias voltages for the LNA. A bias current equal to V_{BE}/R is also generated and fed to node in $\otimes$ in Fig. 7, which is the schematic of the two unity-gain op amp buffers used to bias Q_4 and Q_5. Since V_{BE} decreases as temperature rises, the bias currents in the buffers also decrease with temperature. This saves overall bias current as the output current capability of the buffers need only track the base currents of Q_4 and Q_5, which decrease as temperature rises.

A further current-saving feature is the shut-down capability of the LNA, which also acts as a 2-state gain control. This is achieved by turning on the normally off MOSFET's M_{10}–M_{15}. Devices M_{11}–M_{15} pull down various bias voltages in the LNA and turn off all devices in the signal path, where most bias current is consumed. Device M_{10} connected between input and output of the LNA is then switched on, and its resistance of about 50 Ω gives a through loss of about -6 dB with an input IM_{3INT} of +26 dBm. The MOS devices in the BiCMOS process proved extremely useful in realizing this function while causing little degradation to the RF characteristics of the front end. The RC cutoff frequency of the NMOS devices used as resistors is about 15 GHz.

V. MIXER

The mixer schematic is shown in Fig. 9. It comprises a differential pair Q_{56}–Q_{57} which is driven by an external local oscillator (LO) signal. This chops the signal current in Q_{58}, which is a common-base input buffer. The input impedance at base of $\frac{1}{g_m} \simeq 10\,\Omega$ is padded with $R_{45} = 30\,\Omega$ which along with bond wire and package inductance gives a high-frequency input impedance of approximately 50 Ω. This configuration also has excellent linearity and overload characteristics since the 50 Ω source resistance together with R_{45} yield effective emitter degeneration on Q_{58} of 80 Ω. This results in a high value of input P_{-1dB} (-3 dBm predicted and -4 dBm measured) and a high input IM_{3INT} (+6 dBm predicted and measured).

In order to stabilize the mixer parameters against temperature and supply-voltage variations, on-chip generated bias compensation is again used. This is shown in Fig. 10 where 100 μA of PTAT current is mirrored to produce a voltage

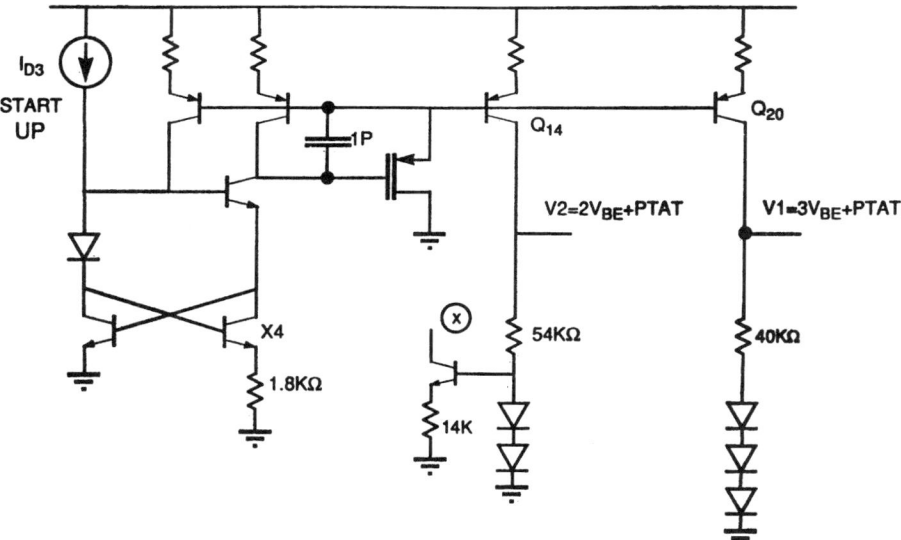

Fig. 6. Bias generation.

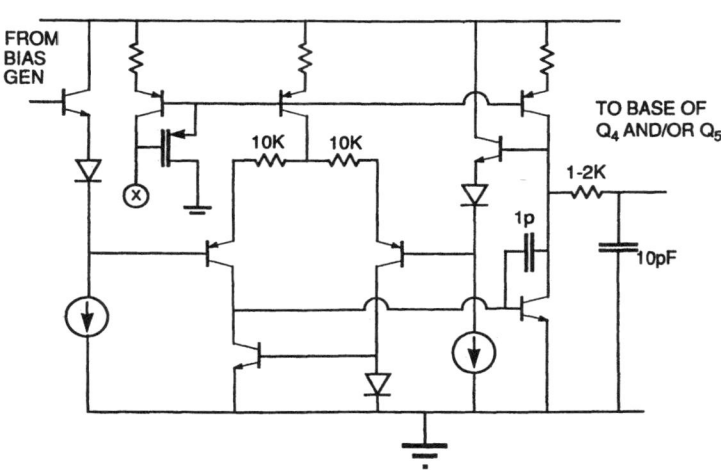

Fig. 7. On-chip unity-gain op amp buffer.

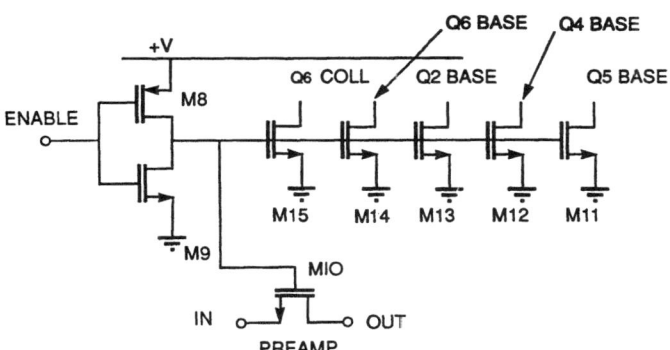

Fig. 8. LNA shut-down and bypass circuit.

$V_X = (V_{BE} + V_{PTAT})$ where $V_{PTAT} = 800 mV$. This voltage is then applied across R_{44} to give a 3.2 mA stable bias current in the mixer. On-chip noise filters again prevent bias circuit noise from entering the high-frequency signal path.

The mixer conversion gain can be calculated by noting that the input drive voltage V_i in Fig. 9 is converted to a current of $V_1/100\,\Omega$ and passed through the common base stage Q_{58} to the differential pair Q_{56}–Q_{57}. For large LO drive, this multiplies the input current by the switching function S(t) which ideally alternates between 0 and 1 at the LO frequency ω_0 and has a Fourier expansion

$$S(t) = 1/2 + \frac{2}{\pi}\cos\omega_0 t + \frac{2}{3\pi}\cos 3\omega_0 t + \cdots \quad (6)$$

The output current in the collector of Q_{56} is thus

$$I_{C56} = (I_{Q58} + \frac{V_s}{100}\cos\omega_s t)S(t) \quad (7)$$

where $I_{Q58} = 3.2mA$ and $V_i = V_s \cos\omega_s t$. Equation (7) shows that the output contains large components at the signal

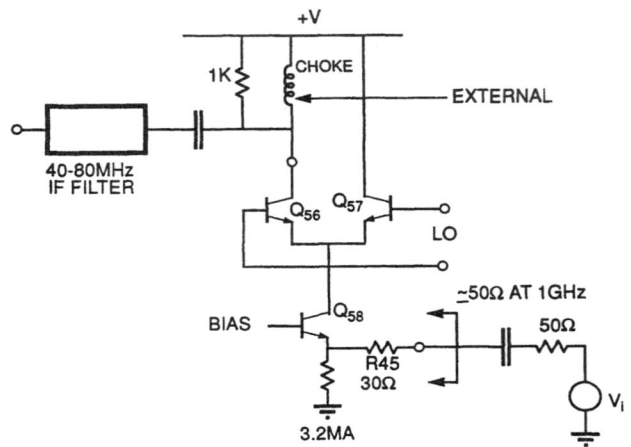

Fig. 9. Mixer schematic.

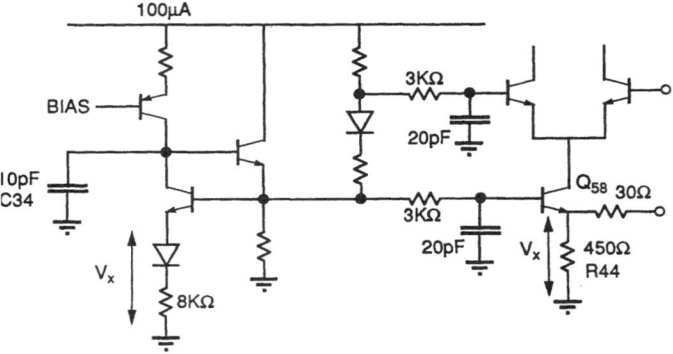

Fig. 10. Mixer bias.

and LO frequencies which must be filtered out by the IF circuit. (A double-balanced mixer greatly attenuates these unwanted output components, at the expense of higher total bias current.) The IF output current component is

$$I_{IF} = \frac{V_s}{100} \frac{1}{\pi} \cos(\omega_0 - \omega_s)t. \qquad (8)$$

The above analysis assumes ideal instantaneous switching between Q_{56} and Q_{57}. For a typical LO drive level of 0 dBm (about 600 mV peak–peak), this is a good assumption for the gain calculation and (8) is quite accurate. However, there is a small but finite time per cycle when both Q_{56} and Q_{57} are on together during the transition, and this is important in determining the mixer noise figure, as the differential pair then has a large gain to the output and injects its own amplified noise into the circuit. This contribution, together with that of Q_{58}, R_{45}, and the r_b of Q_{58} gives a measured mixer noise figure of 15.8 dB at 0 dBm LO drive. This can be reduced by reducing R_{45}, at the expense of input match, and/or by increasing the LO drive level to reduce the time spent in the active region by the differential pair Q_{56}–Q_{57}.

VI. LAYOUT CONSIDERATIONS

A die photograph of the IC is shown in Fig. 11. The LNA first and second stages are separated on the die by over

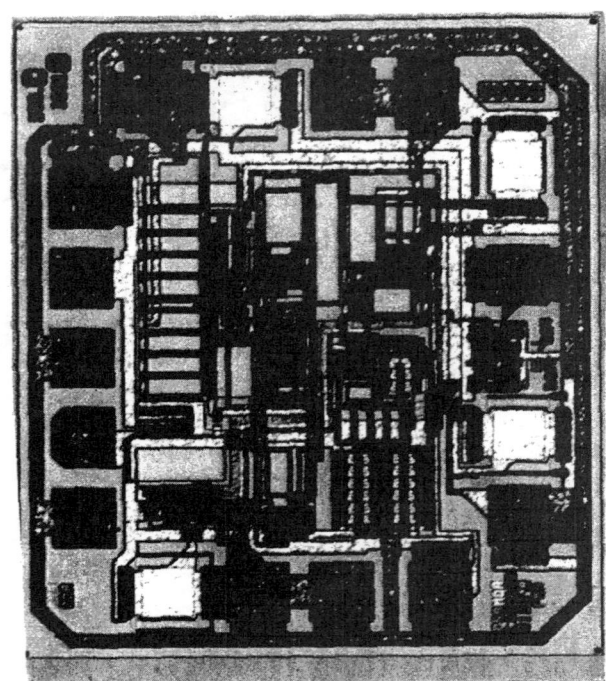

Fig. 11. Die photograph.

1 mm, and use grounds on opposite sides of the package. This minimizes feedback due to inductive coupling between bond wires and package pins that can cause gain peaking or instability in the LNA. The interstage metal interconnects are routed on top of a ground shield formed by a low-resistance n^+ layer beneath 2.2 μm of SiO_2 and the metal lines. The shield is contacted every 50 μm to maintain a low impedance to ground. All other high-frequency ports use a similar shield to minimize signal coupling. This is particularly important for the LO lines which carry large high-frequency signals which must be well shielded from sensitive points such as the LNA and mixer inputs. Pin selections must also be made carefully to minimize LO coupling to these points via mutual inductance and capacitance between bond wires and package pins. Extensive use is made of dense on-chip capacitors formed from the 220 Å MOS gate oxide to filter and bypass supply and internal bias voltages.

VII. MEASUREMENTS

At 27°C, $V_{cc} = 5V$ and at 900 MHz, the measured amplifier gain was 16.0 dB, noise figure 2.2 dB ($R_s = 50 \Omega$), and input IM_3 intercept -10 dBm. Input match was $s_{11} = -10$ dB and output match was $s_{22} = -15$ dB. These values showed no measureable variation for $V_{cc} = 4.5$–5.5 V. Addition of a 15 nH shunt inductor at the input of the LNA for noise matching improved the noise figure to 2.0 dB and s_{11} to -15 dB. Overall chip supply current was 13.0 mA at room temperature with a variation from 12 to 14 mA over -40° to 100°C. The mixer voltage gain into a 1 kΩ load was 10 dB. LO feedthrough to the RF input was -46 dB, and to the mixer input was -33 dB. All these measurements were taken without external filters. Measured amplifier noise figure versus frequency is

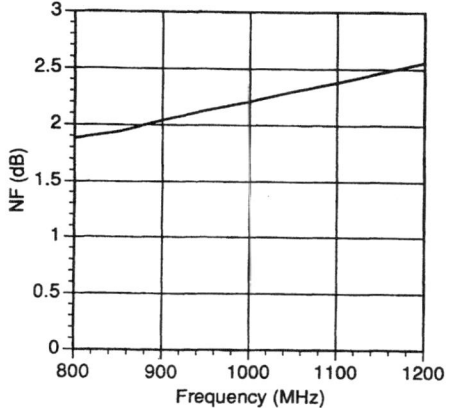

Fig. 12. Measured amplifier noise figure versus frequency with $R_s = 50~\Omega$.

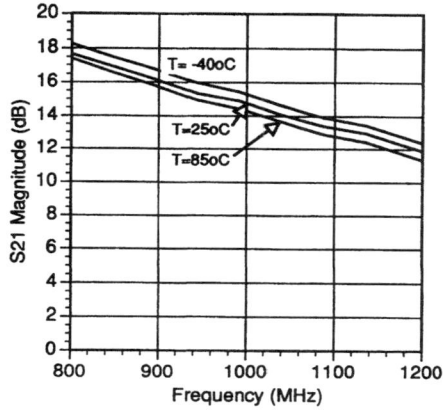

Fig. 13. Measured amplifier gain versus frequency and temperature.

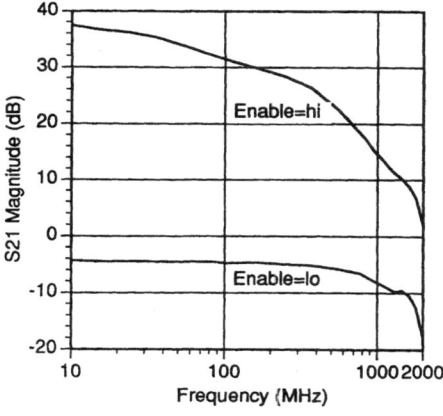

Fig. 14. Measured amplifier gain versus frequency in the enabled and disabled states.

shown in Fig. 12. Measured amplifier s_{21} versus frequency and temperature is shown in Fig. 13. Measured amplifier s_{21} in the enabled and disabled states is shown versus frequency in Fig. 14.

VIII. Conclusions

An RF front-end IC for the 900 MHz–1 GHz band has been described. The circuit comprises a temperature-compensated and supply-regulated LNA and mixer, and is realized in a BiCMOS process. The LNA gain and noise figure are 16.0 and 2.2 dB, respectively, and total chip supply current is 13.0 mA from a 5-V supply.

References

[1] J. L. de Jong, R. Lane, B. van Schravendijk, and G. Conner, "Single polysilicon layer advanced super high-speed BiCMOS technology," in *Proc. IEEE Bipolar Circuits Technol. Meeting*, Sept. 1989, pp. 182–185.
[2] C. D. Hull and R. G. Meyer, "Principles of wideband feedback amplifier design," *Int. J. High Speed Electron.*, vol. 3, no. 1, pp. 53–93, Mar. 1992.
[3] P. R. Gray and R. G. Meyer, *Analysis and Design of Analog Integrated Circuits*, third ed. New York: Wiley, 1993, ch. 11.
[4] K. A. Simons, "The decibel relationship between amplifier distortion products," *Proc. IEEE*, vol. 58, pp. 1071–1086, July 1970.
[5] D. O. Pederson and K. Mayaram, *Analog Integrated Circuits for Communication*. Kluwer, 1991.
[6] M. Steyaert and R. Roovers, "A single chip quadrature modulator," *IEEE J. Solid-State Circuits*, vol. 27, pp. 1194–1197, Aug. 1992.

WP 2.7: A One-Chip 2GHz Single Superhet Receiver for 2Mb/s FSK Radio Communication

Volker Thomas, Josef Fenk, Stefan Beyer

Siemens AG, Semiconductor Group, Munich, Germany

Mass production of mobile communication radios requires an electronics board with a minimum number of components, optimized cost and reliable performance within production tolerances. Reduction of the number of battery cells to 3 or 4 requires ICs with 3V supply and low power consumption.

A one-chip 2GHz single superhet receiver for digital data transmission in a 7.5 GHz bipolar process is optimized for low power consumption and a dynamic range of at least 70dB at the RF signal input. The IC is suitable for both radios and base stations of digital mobile radio, especially cordless phone systems up to 2GHz including the Digital European Cordless Telephone (DECT), wireless LAN systems and any other systems using incoherent FSK modulation [2].

Previously-published receiver circuits are not characterized for 2GHz applications, can not handle 110MHz IF or 1Mb/s data rate, or do not have a comparable level of integration [1]. The IC presented here consists of a 2GHz RF mixer, a 40 to 115MHz limiter with a radio signal strength indicator (RSSI), an incoherent FM demodulator, a sample-and-hold offset-compensation circuit and a baseband filter (block diagram, Figure 1). It is part of a 3-IC transceiver/synthesizer chip set.

The mixer has a single-balanced Gilbert cell architecture with a balanced IF open-collector output (Figure 2). The RF input impedance of 15Ω and the balanced LO input are matched to 50Ωs by SMD capacitances and printed stripline elements. The gain and the 3rd-order intercept point behave vice versa depending on the real collector load resistance between 200Ω and 2 kΩ. The resulting output resistance is matched to a 50Ω surface acoustic wave (SAW) filter via a transformer, that may be replaced by simpler components when low-cost SAW filters with a balanced high-impedance input become available. 3mA is sufficient for 6dB gain and a 3rd-order intercept point at input of -4dBm, measured at f_{RF}=1.89GHz and f_{LO}=1.78GHz.

The required limiter dynamic range is achieved by voltage amplification of 92dB from the limiter input to the (internal) output of the last stage. The parasitic coupling between the 50Ω input and the balanced mixer output and between the input and the demodulator tank interface are low enough to avoid stopband crosstalk and instability, respectively. The capacitive impedances of the on-chip C_{k1}, C_{k2} are 5 times the resulting real resistance of the tank circuit, attenuating the limiter output to the tank circuit pins by 20dB (Figure 1). The limiter input sensitivity is improved by adding an external limiter tank circuit that limits the bandwidth to 10MHz (dependent on the values of L and C). The IC measures the radio signal strength over a range of 80dB (Figure 3). Both a time-continuous and a peak-hold output (according to Reference 2) are implemented.

The coincidence demodulator, connected to an external tank circuit with Q=20, has a demodulation constant of 2V/MHz for f_{IF}=110.6MHz within a linear range of ±1.4MHz. A received DECT signal (frequency deviation of 288kHz) yields a 1.2V peak to peak voltage in the baseband. The demodulated signal is filtered by a 4th-order low-pass consisting of two on-chip operational amplifiers and external passive components. For a Gaussian frequency shift keying (GFSK) modulated DECT signal with BT=0.5, a rolloff of 1.4 is optimal for minimum bit error rate (BER).

Frequency deviations of the received carrier, the local oscillator, and the tank resonance frequency result in baseband offset, compensated by the sample and hold circuit. The decision threshold level is available as an extra signal. The change between sample and hold periods is controlled via the gating input by the processor. During reception of the synchronization field (bit sequence 0101) at the beginning of every burst, a control loop is closed (sample mode with a time constant of 5µs) [2]. Figure 4 shows the response of baseband output level versus IF input carrier frequency. During data reception, the integrator capacitor serves as a hold capacitor (hold mode). The maximum relative offset at decider input in relation to the peak-to-peak signal amplitude is 4% including a finite capacitor discharge in hold mode, temperature and voltage supply variations and a decider offset value of 20mV. This offset is equivalent to a carrier frequency offset of 23kHz for the uncompensated case with the demodulator parameters above.

The BER measured for different frequency offset values is shown in Figure 5. 0.001 BER is achieved for -90dBm IF input level. The rate corresponds to 13dB baseband signal-to-noise ratio. The diagram reveals maximum effective frequency offset caused by sample-and-hold inaccuracies does not increase BER by more than 20% at the sensitivity limit.

Table 1 summarizes the IC characteristics, including system performance data. The latter presume a SAW IF-filter with an insertion loss of 4dB, a front end gain of 11dB and a noise figure of 5dB between the antenna and the RF mixer input. The single superhet solution is the best compromise between double frequency conversion and the zero-IF (or homodyne reception). The results above prove that the main challenge of the single conversion architecture is an adequate IC design. Available SAW-filters achieve the selectivity required for digital systems. The advantages of this concept are less power-consuming active circuits (no 2nd-LO-synthesizer or 2nd mixer), fewer discrete components, no crosstalk between digital circuits and 2nd IF, and no on-chip VCO that may be sensitive to strong incoming interferers (injection locking). On the other hand, a zero-IF receiver would avoid passive IF filters entirely, but the problem of sufficient dynamic range and offset compensation suitable for low-cost mass production is not solved for 2GHz applications. DECT requirements for sensitivity of -83dB and intermodulation performance corresponding to an input intercept point of -27dBm are easily met, and enough margin is left to allow production tolerances, to improve the reception sensitivity further or to use an inexpensive bipolar LNA. Figure 6 is a chip micrograph.

References

[1] Fenk, J., et. al. , "An RF front-end for digital mobile radio", IEEE 1990 Bipolar Circuits and Technology Meeting 11.2, pp. 244-247.

[2] ETSI, "Radio equipment and systems. Digital European Cordless Telecommunications Common Interface. Part 2: Physical Layer", DE/RES 3001-2, 1992.

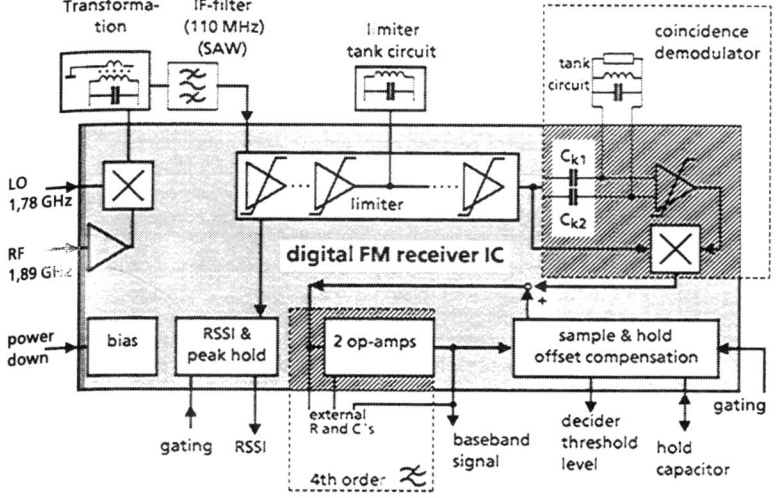

Figure 1: Block diagram.

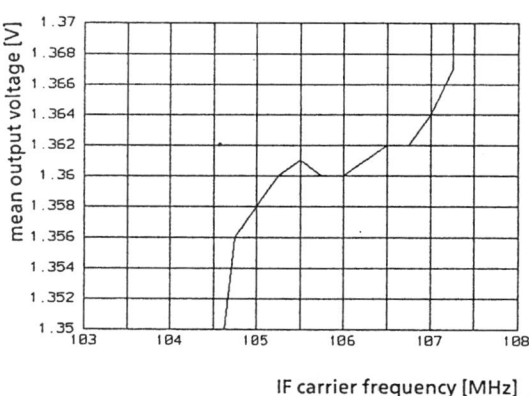

Figure 4: Baseband output level vs. IF input frequency in sample mode.

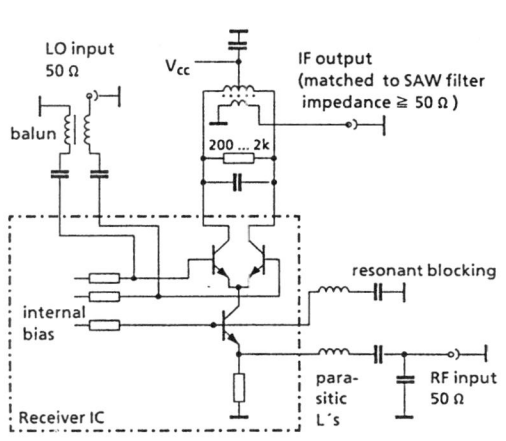

Figure 2: RF mixer including matching circuits.

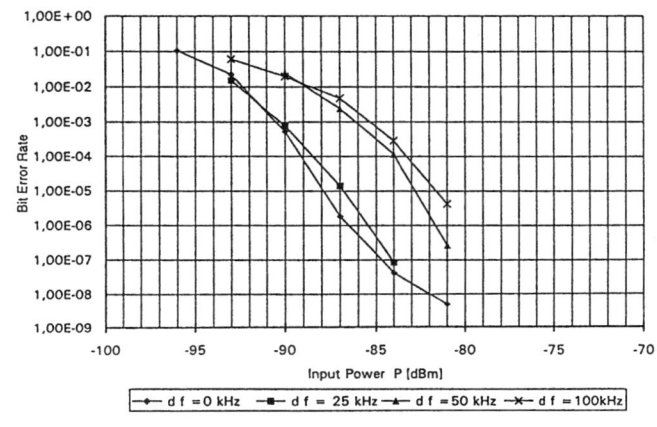

Figure 5: BER performance of limiter with and without frequency offset.

Figure 6: See page 307.

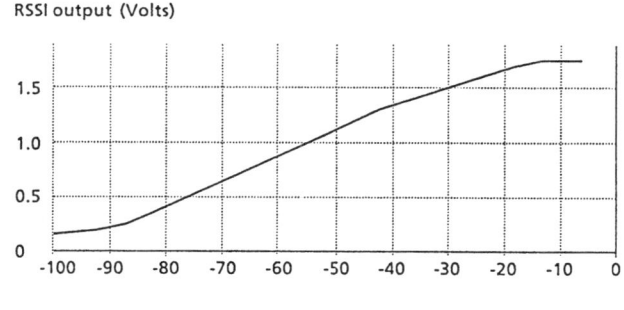

Figure 3: RSSI vs. IF input level.

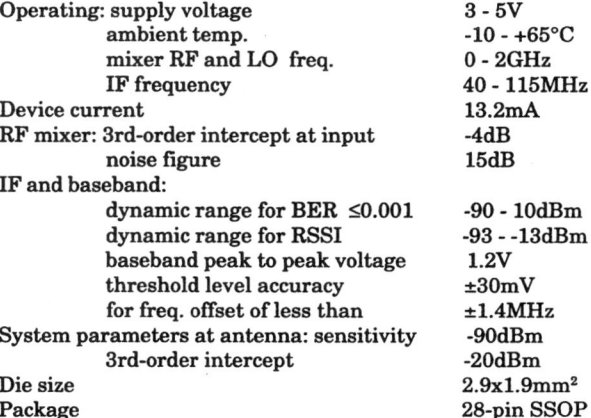

Operating: supply voltage	3 - 5V
ambient temp.	-10 - +65°C
mixer RF and LO freq.	0 - 2GHz
IF frequency	40 - 115MHz
Device current	13.2mA
RF mixer: 3rd-order intercept at input	-4dB
noise figure	15dB
IF and baseband:	
dynamic range for BER ≤0.001	-90 - 10dBm
dynamic range for RSSI	-93 - -13dBm
baseband peak to peak voltage	1.2V
threshold level accuracy	±30mV
for freq. offset of less than	±1.4MHz
System parameters at antenna: sensitivity	-90dBm
3rd-order intercept	-20dBm
Die size	2.9x1.9mm²
Package	28-pin SSOP

Table 1: Experimental results.

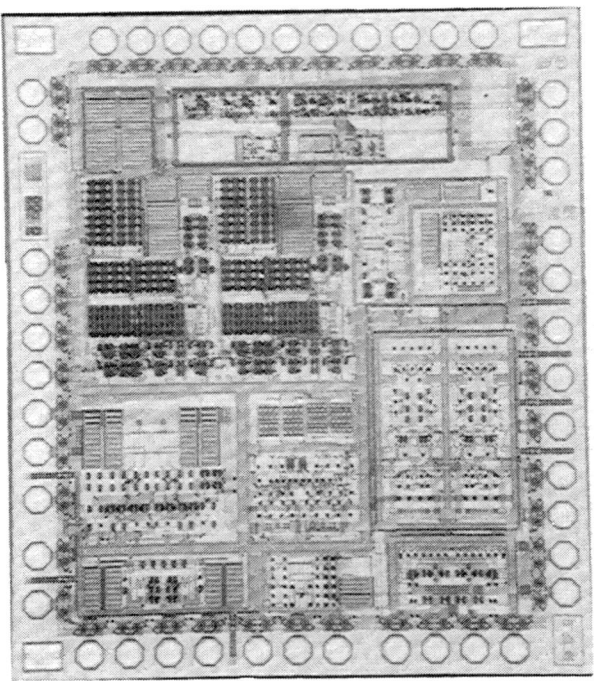

Figure 6: Radio bipolar front-end chip.

TA 8.3: A Low-Voltage Silicon Bipolar RF Front-End for PCN Receiver Applications*

John R. Long, Miles A. Copeland, Peter Schvan[1], Robert A. Hadaway[1]

Department of Electronics, Carleton University, Ottawa, Canada
[1]Telecom Microelectronics Centre, Northern Telecom Ltd., Ottawa, Canada

Monolithic microstrip transformers are used to perform the coupling and phase-splitting functions in a bipolar low-noise amplifier and mixer designed for 1.9GHz wireless receiver applications. These circuits are fabricated in a production 0.8µm BiCMOS process with a transistor transit frequency (f_T) of 11GHz [1]. Using reactive feedback and coupling elements in place of resistors significantly improves the noise figure through the reduction of resistor thermal noise, and also allows both the low-noise amplifier and the mixer to operate at supply voltages below 2V.

The schematic diagram of the doubly-balanced mixer is shown in Figure 1. The radio frequency (RF) input signal is split into antiphase and in-phase components by balun T1 and is then fed to the cross-coupled switching quad of transistors, Q1 to Q4. Bias current is fed from current source Q5 to the switching quad through the center-tap in the balun secondary. The signal current is chopped by the transistor quad at the local oscillator (LO) rate to down-convert the input signal from RF to the desired intermediate frequency (IF). The down-converted RF input signal is buffered to a 50Ω load using the on-chip Darlington buffer (Q6 and Q7), or alternately, the IF output can be impedance matched using an external matching network (e.g., transformer T2 shown in Figure 1). The input from the local oscillator is terminated on-chip by resistor R_{LO}.

The monolithic balun consists of 9 turns of 1µm-thick aluminum with a line width of 15µm and a line spacing of 1.8µm, and measures 425µm on a side. The balun is symmetrically wound so that the center-tap can be precisely located and the losses in both halves of the secondary winding equalized. The design of the balun can be seen in the micrograph of the complete mixer shown in Figure 4.

A computer program developed to model microstrip components on silicon is used for the design of the balun. The parameters for a physically-based, lumped-element model are determined from the layout geometry and process parameters, and this model can then be used in either a time-domain or a frequency-domain circuit simulation. The measured and simulated transmission coefficient (S21) from the balun primary input to the inverting secondary output is shown in Figure 2. Excellent agreement between the measured and simulated results has also been obtained for the primary to noninverting secondary output response.

The measured performance of the doubly-balanced mixer is summarized in Table 1. The packaged mixer, which has an active area of 0.85x0.65mm², is operated at a supply voltage of 1.9V and bias current of 2.5mA. Recently reported results for a low-power 1.9GHz monolithic mixer fabricated in a 0.5µm GaAs technology are also listed in Table 1 for comparison. A 3V power supply is required by the cascoded transistors of the GaAs mixer and hence that circuit dissipates more power. Noise figure and linearity (indicated by the high third-order intercept point, or IP3 at the input) of the silicon bipolar mixer is comparable to results demonstrated for the GaAs design. Quality of the input port match obtained using the monolithic balun is indicated by low measured input VSWR. A doubly-balanced configuration is used for the silicon mixer as compared to the singly-balanced GaAs design, resulting in a clearly superior port-to-port isolation. This is desirable to suppress spurious frequency components at the mixer IF output in a monolithic receiver implementation. The LO-RF isolation is degraded by parasitic signal paths on the chip, within the IC package and on the test fixture, and these signal paths are not accounted for in the simulation. Good agreement between measurement and simulation for all other specifications is obtained from both harmonic balance and time-domain circuit simulators.

The schematic diagram of the companion low-noise amplifier (LNA) is shown in Figure 3. Mutual coupling between the collector and emitter of bipolar transistor Q1, through the 1:4 step-up transformer T1, sets the voltage gain of the amplifier. Since the current flowing in the transformer secondary also flows through the primary windings, the power gain of the LNA is essentially the same as the voltage gain.

The inverting step-up transformer consists of 8 turns of 1µm thick aluminum with a line width of 15µm and a line spacing of 1.8µm. The transformer measures 400µm on a side and its layout is shown on the micrograph of Figure 4. The transformer turns ratio is photolithographically defined to high precision by the metal line width and line spacing used, which makes the power gain of the LNA stable, and independent of temperature and process variations. The active area required by the amplifier is 0.65x0.55mm².

The performance of the bipolar low-noise amplifier is summarized in Table 2, where the specifications for two 1.9GHz GaAs amplifiers are also shown for comparison. Measured data for a packaged LNA was not available at the time that this manuscript was prepared, so the experimental results that are listed in Table 2 are obtained from RF measurements of wafer-probed devices only. Good agreement between these on-wafer measurements and the computer simulations is achieved. Also, the measured noise figure of the bipolar transformer-coupled LNA (2.7dB in a 50Ω system) is comparable to that achieved by the other designs listed in Table 2 at 1.9GHz. The correspondence between the measured and simulated results for both the LNA and the mixer verifies the accuracy of the computer simulations, and indicates that performance close to the simulated values for linearity and port VSWR can be realized in practice for a properly matched and packaged device.

Performance comparable to some recently reported GaAs circuits was demonstrated for a silicon front-end fabricated in a production VLSI process. These performance gains have been achieved through the use of monolithic microstrip transformers fabricated in a silicon BiCMOS technology with standard metallization.

References

[1] Hadaway, R., et al., "A Sub-Micron BiCMOS Technology for Telecommunications", J. of Microelectronic Engineering, vol.15, pp. 513-516, 1991.

[2] Kusunoki, S., et al., "GaAs Front-End MMICs for L-Band Personal Communications", IEEE Microwave and Millimeter-Wave Monolithic Circuits Symposium Technical Digest, pp. 9-12, June, 1993.

[3] Heaney, E., et al., "Ultra Low Power Low Noise Amplifiers for Wireless Communications," Proceedings of the GaAs IC Symposium, pp. 49-51, Oct., 1993.

---

*This work was supported by grants from the Natural Sciences and Engineering Research Council of Canada (NSERC), Micronet and the Telecommunications Research Institute of Ontario (TRIO).

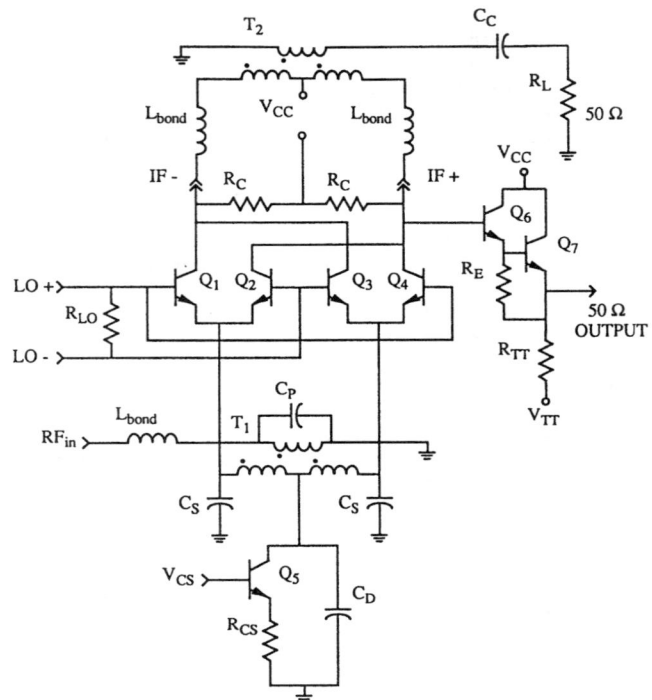

Figure 1: Schematic diagram of the doubly-balanced mixer.

Mixer	0.5 μm GaAs-JFET [2]	Measured 0.8μm Si-Bipolar (NT)	Simulated 0.8μm Si-Bipolar (NT)
Supply Voltage	3.0 V	1.9 V	1.9 V
Supply Current	4.0 mA	2.5 mA	2.5 mA
RF Frequency	1.9 GHz	1.9 GHz	1.9 GHz
LO Frequency	1.66 GHz	1.8 GHz	1.8 GHz
SSB NF (50Ω)	10.8 dB	10.9 dB	10.5 dB
Conversion Gain	5.7 dB	6.1 dB	6.3 dB
IP3 (input)	2.3 dBm	2.3 dBm	1.1 dBm
RF Port VSWR	1.2	1.17	1.15
LO-RF Isolation	13 dB	32 dB	80 dB
LO-IF Isolation	5 dB	47 dB	52 dB

Table 1: Comparison of mixer performance.

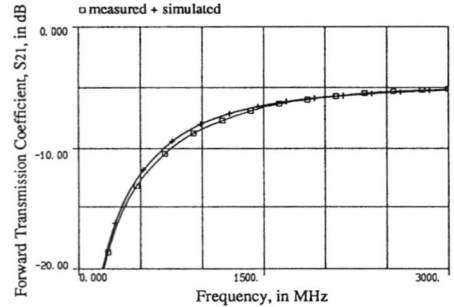

Figure 2: Measured and simulated S21 (input to inverting port) for the monolithic balun.

Low-Noise Amplifier	1μm GaAs-MESFET [3]	0.5μm GaAs-JFET [2]	Measured 0.8μm Si-Bipolar	Simulated 0.8μm Si-Bipolar
Supply Voltage	3.0 V	3.0 V	1.9 V	1.9 V
Supply Current	2.0 mA	4.0 mA	2.0 mA	2.0 mA
Frequency	1.9 GHz	1.9 GHz	1.9 GHz	1.9 GHz
Noise Fig. (50Ω)	-	2.8 dB	2.7 dB	2.9 dB
Min. Noise Fig.	1.75 dB	-	2.1 dB	1.9 dB
Gain	13.5 dB	18.1 dB	10.0 dB	10.5 dB
IP3 (input)	-	-11.1 dBm	-	-10 dBm
Input VSWR	-	1.5	-	1.15
Output VSWR	-	3.1	-	1.03
Isolation	-	21 dB	26 dB	30 dB

Table 2: Comparison of low-noise amplifier performance.

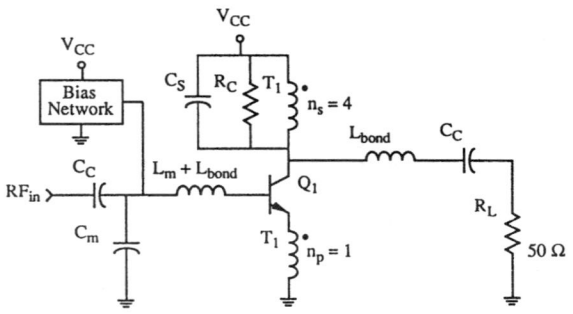

Figure 3: Schematic diagram of the low-noise amplifier.

Figure 4: See page 353.

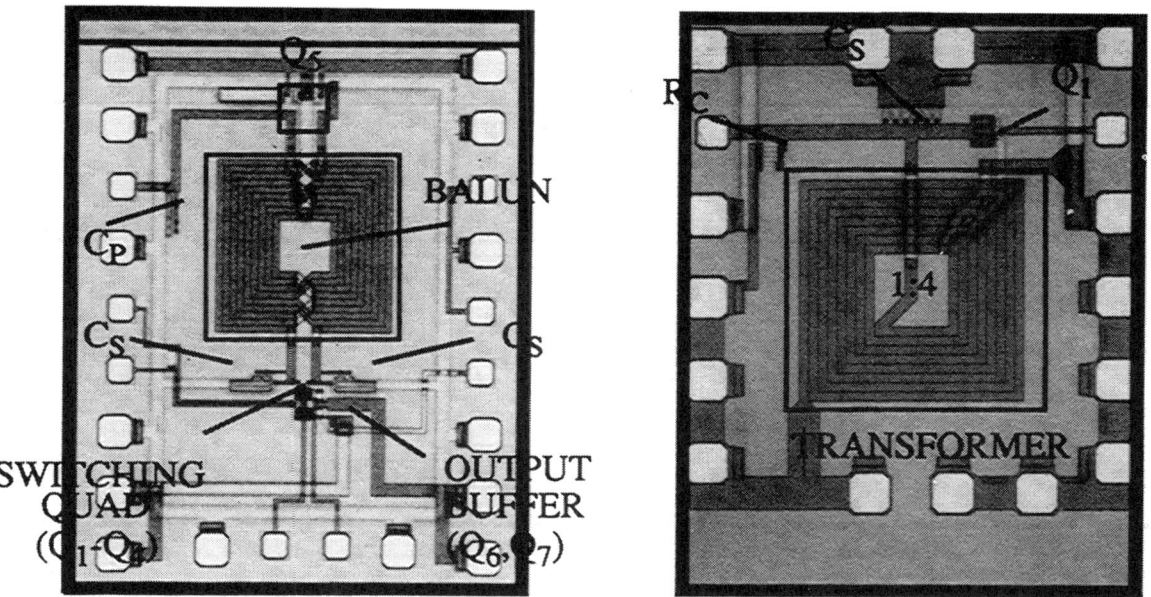

Figure 4: Front end test chip micrographs.

A Highly Linear 1-GHz CMOS Downconversion Mixer

P. Y. Chan, A. Rofougaran, K. A. Ahmed, and A. A. Abidi

Integrated Circuits & Systems Laboratory
University of California
Los Angeles, CA 90024-1594

ABSTRACT

A highly linear, closed loop CMOS architecture is demonstrated for the mixing of RF waveforms to baseband as sampled-data signals. In converting a 915 MHz RF waveform to a 20 MHz spread spectrum baseband signal, the measured third-order intercept lies at +27 dBm of input power. As the 1-μm CMOS prototype dissipates only 12 mW from a 5V supply, and is capable of operation at 3V, it is expected to be of use in advanced handheld wireless receivers.

Introduction

IC design faces a new challenge posed by the emerging requirements of advanced transceivers for wireless communications. First radio pagers, and then cellular telephones, have prompted highly integrated solutions for the required RF and IF components. It is widely felt today that ICs in advanced bipolar and GaAs technologies will provide the best solutions in terms of the sought performance at the lowest power dissipation. Several chipsets have been developed in these technologies [1-3]. We believe that for operation at RF in the 1 GHz band, even CMOS is a viable alternative. A highly integrated CMOS RF and IF chip eliminates the need to route signals on low impedance lines between chips, thus saving the power wasted in buffers. Our circuit designs are based on a total system solution for a frequency-hopped spread-spectrum transceiver, which optimally partitions signal processing between the analog and digital domains.

For use in the receiver, we have already demonstrated a low power 750 MHz CMOS RF amplifier with on-chip tuning elements [4]. The first mixer in the receiver poses an even a greater design challenge, because intermodulation distortion between an interference signal passing through the preselection filter and the desired signal may produce in-band spurious signals, and irrecoverably degrade the received signal to noise ratio. Thus, a mixer with a very wide dynamic range is sought. Mixer design is today a well acknowledged art unto itself.

Mixers: Bipolar, Switching, and CMOS

Conventional mixers for continuous-time waveforms downconvert by producing an output at the difference frequency between the RF and local oscillator (LO) inputs. For direct conversion to baseband, the LO is close in frequency to the RF, and thus itself at a high frequency. A mixer is more linear, and more effectively suppresses intermodulation distortion, either by how closely its input-output relationship approaches an analog multiplication, or if it is a commutating switch, by the strength of LO driving the switch and its turnoff aperture time [5]. For these reasons, commutating mixers do not often find use in low power, battery operated receivers, although they are quite common as standalone microwave components.

Mixers based on Gilbert's analog multiplier are ubiquitous in bipolar ICs [1,5]. Intermodulation distortion arises in this circuit because at RF it is difficult to predistort the input signal to cancel the exponential bipolar transistor characteristic, and also from imperfections such as the large base spreading resistance found in transistors with a high cutoff frequency. Nonlinearity in a balanced mixer is specified by the third-order intercept (IP3), the RF input power at which the third harmonic strength at the output is equal to the fundamental frequency. The larger the IP3, the lower the intermodulation distortion. The IP3 of mixers based on the Gilbert multiplier is usually somewhere below 0 dBm, and if simple circuits compensate some of the imperfections, it may be raised to 5 or perhaps 10 dBm. This is only marginally acceptable in spread spectrum receivers, which may be susceptible to very large narrowband interfering signals.

When the Gilbert cell topology is used in a GaAs FET mixer, cross-modulation terms appearing in the circuit input-output characteristic result in a lower IP3 than in its bipolar antecedent. It is not uncommon in GaAs ICs to mix signals using the theoretical quadratic I-V characteristic of a single FET. However, owing to short channel effects and other physical phenomena, the non-ideal characteristics of physical FETs will produce additional, often strong, cross-modulation terms, which will limit mixer linearity. A CMOS mixer with this topology, if operational at all at 1 GHz, will be even more nonlinear than the GaAs FET circuits.

An altogether different tack was taken in this design by letting the mixer produce a sampled data, discrete-time analog output signal. This may then be encoded by an A/D converter for use in a digital receiver. The design stems from the realization that the track and hold (T/H) circuit block is itself a mixer: it mixes the input signal with the sampling clock. A T/H clock frequency equal to the baseband sample rate thus directly samples the RF to produce a discrete-time analog downconverted signal at the output (Fig.1). Extracting narrowband information about the modulation by subsampling a high frequency carrier is finding use in instrumentation [6], and even in communication circuits [7].

Closed-loop T/H circuits using op amps are commonly used in CMOS to attain high linearity and cancel charge feedthrough from MOS switches (Fig.2). A properly designed T/H presents a simple *open-loop RC* circuit to the input in *track* mode, and applies feedback to the sampling capacitor only in hold mode. The track mode bandwidth is limited by the ON resistance of FET switches, FET capacitance, and the sampling capacitor. This will be well above 1 GHz in a 1-μm CMOS technology. After faithfully tracking an RF signal, the sampling capacitor acquires its instantaneous value when the sampling switch turns OFF. The acquisition aperture of the FET switch determines the effective sampling bandwidth. In the limiting case when an infinitely fast clock edge turns it OFF, the aperture is the time it takes to empty an inverted FET channel of carriers. This may be as fast as a few tens of picoseconds in a FET of 1-μm channel length. When the transition from track to hold mode is made with the proper clock phases, the charge feedthrough from the switches on to the sampling capacitor is either entirely removed, or appears as an input common mode jump, without causing any distortion on the held sample in either case [8].

There are several advantages in using a subsampling mixer over a conventional mixer in a wireless receiver. As the LO is at the baseband sampling frequency, and not close in frequency to the RF, there is almost no spurious radiation caused by LO leakage through the antenna. Furthermore, this subsampling mixer requires a large LO drive on the switches for linear operation, but unlike a conventional commutating mixer where a high frequency amplifier must boost the LO level, here a rail-to-rail voltage swing is readily obtained at the baseband LO frequency.

Circuit Design

The mixer is a closed-loop track and hold which operates in two frequency ranges widely spaced apart between these two modes. The op amp must settle fully within 10 ns to support a 50 MHz sample rate, and for low distortion, its DC gain must be about 60 dB. A super-cascode topology (Fig.3) best satisfies these requirements [9]. NMOS common source auxiliary amplifiers boost the DC gain of the cascode FETs. While meant to operate nominally at 5 V, the amplifier is fully functional at 3 V. FET sizing is optimized for the shortest settling time when driving the specified capacitances at the input, output, and in feedback. SC common-mode (CM) feedback applied to the tail current source forces the bias value of the output nodes to a reference CM voltage.

In track mode, the differential input and output terminals of the op amp are shorted by switches, thus disconnecting it entirely from the signal path (Fig.2). The desired input CM reference level is applied to the input terminals. Thus in track mode, the RF signal encounters a circuit consisting only of three switches and two 500 fF capacitors. The switch FETs are sized to obtain a low ON resistance (75 Ω) when driven by full logic levels at their gates, enabling the capacitor to track the input voltage. At the transition from track to hold, NMOSFET S1 and PMOSFETs S2 turn OFF first, followed by S3. Charge feedthrough from S1 causes a signal independent jump at the op amp CM input, which may drive the input terminals outside the CM input range, leading to a long recovery time. This problem is alleviated by sizing switches S2 so that their signal independent negative charge feedthrough approximately cancels the charge from S1. Uncancelled charge will however not cause any distortion. The *signal dependent* feedthrough on to the sampling capacitors from S3 when it turns OFF is removed by the negative feedback once the op amp output has settled. Thus, an accurate sample of the RF waveform appears at the op amp output.

The switches are driven by clock buffers consisting of a cascade of two CMOS inverters. Simulated falltime of the clock signals is 500 ps nominal, 800 ps worst-case over process and temperature, which is adequate for the sampling aperture sought. The circuit requires only two clock phases, one of which is delayed to produce the third clock. The clocks applied to sampling switches S1 and S2 must be well balanced.

Although the core of the mixer consumes a very small amount of power, its output is incapable of driving the 50Ω impedance of the measuring instruments. Therefore, an integral part of the design is an on-chip buffer to drive this impedance which is at least as linear as the mixer. A balanced differential pair buffer driving off-chip at the drains, degenerated by a polysilicon resistor at the sources, and enclosed within the op amp feedback was used (Fig.4). This unity gain buffer also converts the balanced mixer output to a single-ended signal to drive the instrumentation. While the mixer dissipates only 12 mW, the buffer dissipates 100 mW and it may swing as large as 1 V differential signal into 50Ω. On-chip 50Ω polysilicon resistors are also used to terminate the differential input pads.

Experimental Results and Discussion

The mixer occupies an active area of 0.4 × 0.65 mm on the die, and is fabricated in a double metal, single poly CMOS process with 1-μm gate feature size (Fig.5). The sampling capacitance is implemented between the two layers of metal. All measurements reported below are made with a 5 V power supply.

Using a mixer clock frequency (LO) which is a non-integer divisor of the input RF, a beat frequency is produced at the mixer output. Measurements are made at LO frequencies up to 120 MHz, and a variable output beat frequency as large as 50 MHz. Distortion in the discrete-time resetting buffered output is measured on a spectrum analyzer. All the key harmonics lie in the buffer passband. The measured IP3 at a 50 MHz sample rate is about +27 dBm with a 900 MHz RF input (Fig.6), and fundamental component at the output is compressed by 1 dB at an input power greater than +12dBm. These are some of the best results to our knowledge for an IC mixer in any technology operating in this RF band [1-3, 10], a comparable performance only available in discrete passive mixers using a transformer coupled diode bridge. The measured mixer conversion loss is 6 dB. As the mixer samples wideband noise across a 1 GHz input bandwidth, but then aliases that noise into a relatively narrowband output (determined by half the sample rate), the noise figure is 18 dB, very close to prediction. In the intended application, the mixer will be preceded by an RF amplifier with a 14 dB gain and 6 dB noise figure.

We have described a new use of a CMOS SC track and hold circuit as a downconversion mixer in a radio receiver. As the mixer plays a key role in determining the interference rejection of spread spectrum receivers, we foresee new possibilities arising from these striking results to fabricate receivers operating in certain frequency ranges, such as in the unlicensed 902-928 MHz band, entirely as single-chip CMOS ICs.

Acknowledgement: This research was supported by ARPA, Rockwell International, and the State of California MICRO Program.

1. Wilson, J.F., et al., "A Single-Chip VHF and UHF Receiver for Radio Paging", *IEEE J. of Solid State Circuits*, v. 26, pp. 1944-1950, December 1991.
2. Tokumitsu, M. and M. Muraguchi, "A Very Low Power Dissipation Front-End MMIC for L-Band Receivers", in *22nd European Microwave Conference*, Espoo, Finland, 1992.
3. Imai, Y., M. Tokumitsu, and A. Minakawa, "Design and Performance of Low-Current GaAs MMICs for L-Band Front-End Applications", *IEEE Trans. on Microwave Theory and Techniques*, v. 39, pp. 209-215, February 1991.
4. Chang, J. Y-C., A. A. Abidi, and M. Gaitan, "Large Suspended Inductors on Silicon and their use in a 2-μm CMOS RF Amplifier", *IEEE Electron Device Letters*, v.14, pp. 246-248, May 1993.
5. Chadwick, P.E., "High Performance Integrated Circuit Mixers", in *IEE Conf. on Mobile Radio Systems and Techniques*, pp. 1-9, London, 1984.
6. Ballo, D.J. and J.A. Wendler, The Microwave Transition Analyzer: A New Instrument Architecture for Component and Signal Analysis, in *Hewlett-Packard Journal*, October 1992, pp. 48-62.
7. Weisskopf, P.A., Subharmonic Sampling of Microwave Signal Processing Requirements, in *Microwave Journal*, May 1992, pp. 239-247.
8. Lin, Y.-M., B. Kim, and P.R. Gray, "A 13b, 2.5 MHz Self-Calibrated Pipelined A/D Converter in 3-μm CMOS", *IEEE J. of Solid State Circuits*, v. 26, pp. 628-636, April 1991.
9. Bult, K. and G. Geelen, "A Fast-Settling CMOS Op Amp for SC Circuits with 90 dB DC Gain", *IEEE J. of Solid State Circuits*, v. SC-25, pp. 1379-1394, December 1990.
10. Takenaka, T., S. Hara, and T. Tokumitsu, "A Miniaturized Broadband MMIC Mixer", in *GaAs IC Symposium*, pp. 193-196, 1989.

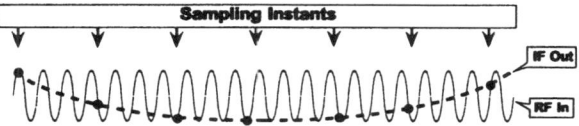

Fig.1. Samples of the RF input taken at an offset subharmonic frequency produce a discrete-time IF output.

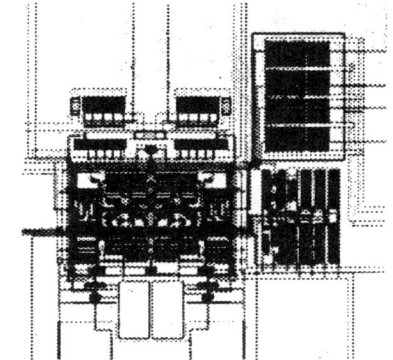

Fig.5. Microphotograph of active area of mixer. 500 fF double-metal sampling capacitors evident on bottom.

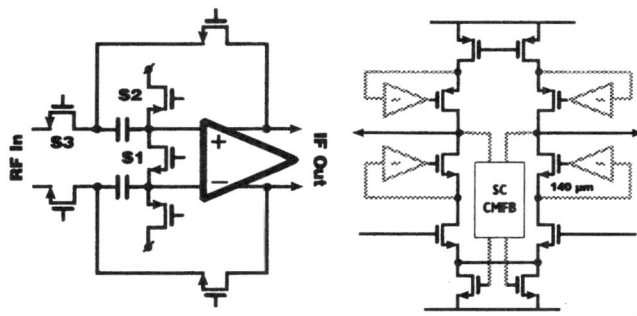

Fig.2. The downconversion mixer is a SC track and hold circuit with a very large tracking bandwidth.

Fig.3. Super cascode op amp circuit.

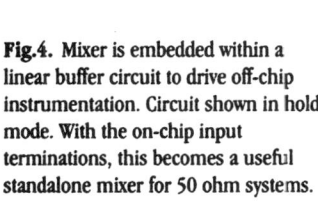

Fig.4. Mixer is embedded within a linear buffer circuit to drive off-chip instrumentation. Circuit shown in hold mode. With the on-chip input terminations, this becomes a useful standalone mixer for 50 ohm systems.

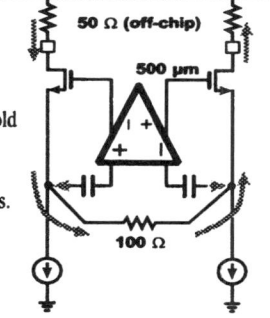

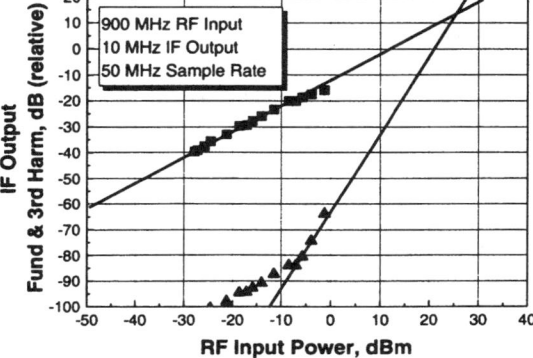

Fig.6. Measured mixer nonlinearity as third-order intercept (IP3). Discrete-time analog signal was directly applied to spectrum analyzer. Small inputs limited by noise, higher inputs by distortion in signal source.

A 1.5 GHz Highly Linear CMOS Downconversion Mixer

Jan Crols, *Student Member, IEEE* and Michel S. J. Steyaert, *Senior Member, IEEE*

Abstract—A CMOS mixer topology for use in highly integrated downconversion receivers is presented. The mixing is based on the modulation of nMOS transistors in the triode region which renders an excellent linearity independent of mismatch. With two extra capacitors added to the classical cross-coupled MOSFET-C lowpass filter structure, GHz signals can be processed while only a low-frequency opamp is required as output amplifier. The downconversion mixer has an input bandwidth of 1.5 GHz. The measured third-order intercept point (IP3) of 45.2 dBm demonstrates the high linearity. The mixer has been implemented in a 1.2 μm CMOS process. It takes up 1 mm^2 of total chip area and its power consumption is 1.3 mW from a single 5 V power supply.

I. INTRODUCTION

THE NEED FOR highly integrated receivers in the 1–2 GHz range has grown considerably with the introduction of new wireless telecommunication services like digital cellular telephone. Key issues in the design of receivers for these applications are the quality of the signal reception, the level of integration, the power consumption and the cost prize. Several realizations of the RF part for these receivers have been presented in the past [1]–[4]. They are all realized in either a GaAs or Si bipolar process. The use of the bipolar technology is often prefered over GaAs for its lower cost prize and equally good performance in the 1–2 GHz range. A further integration, i.e., combining the analog RF, the analog LF and the digital part on one chip, will be the next goal. The analog LF and the digital part are preferably realized in a CMOS technology. Combining these with the RF part would require the use of a BiCMOS technology and although these technologies are ever more used in telecommunication applications, they are not suited for the realization of receivers for frequencies above 1 GHz. The quality of the bipolar devices is in todays BiCMOS processes not sufficient. The value of their f_t and more important of their r_b is not so good as what is available in bipolar only processes. At this time, using CMOS is the only way to obtain a further and full integration. With the ongoing further decreasing of CMOS gatelengths this becomes ever more feasible [5]–[8].

In this paper a full CMOS downconversion mixer is presented. It is a key building block for the realization of a fully integrated CMOS 1 GHz zero-IF receiver. The presented downconversion mixer is not only important for the fact that it is implemented in CMOS, its linearity is also much better than any bipolar or GaAs realization that has been previously presented. A CMOS 1 GHz downconversion mixer has been proposed before [5], but it was based on the use of subsampling. Such a circuit can only cope with smallband input signals, resulting in the need for the use of a high-quality HF filter which can not be realized in an integrated way. The mixer which is presented in this paper is a true double balanced multiplier, capable of downconverting broadband input signals by multiplication with a sinusoidal local oscillator.

The second section of this paper describes the topology of the presented downconversion mixer. The topology is based on the use of four cross-coupled CMOS transistors operated in the triode region and connected to a virtual ground point [6]. Two important capacitors of 25 pF have been added to this topology on the virtual ground points. With these capacitors it becomes possible to realize a very high input bandwidth (as high as 1.5 GHz) with the use of a low-frequency opamp (10 MHz). The third and fourth part of this text describe the design aspects. The fifth and sixth part are respectively on the noise capabilities of the presented structure and the relationships between linearity, noise and power consumption. The last two parts describe the practical realization and the measurement results.

II. MIXER TOPOLOGY

The linearity of an RF mixer is in most cases rather limited. The Gilbert topology is the most common used in a bipolar technology [9]. Its operation is based on a translinear configuration, i.e., the use of the exponential voltage to current conversion of the bipolar transistor. Techniques like predistortion and emitter degeneration are necessary to obtain a reasonable linearity. Imperfections in this structure combined with a limited matching will render a third-order intercept point (IP3) which can only be slightly larger than 0 dBm [10]. In CMOS a double balanced structure which cancels out the quadratic term of the MOS transistor can be used [11], [12]. These mixer have not only a limited linearity which highly depends on matching, even more important is their limited frequency range. The input transistors of these mixers can only be biased with a relatively small $V_{GS} - V_T$ in order to keep them in saturation at all times. The result is large input transistors which limits the maximal achievable input bandwidth to about 100 MHz.

A better solution is to use the transistors in the linear region, preferably with a large $V_{GS} - V_t$. In this way a small R_{on} can be realized with a small transistor, allowing a high input bandwidth. This property has been used to realize high frequency GaAs commutating mixers [13]. The

---

Manuscript received December 20, 1994; revised April 11, 1995.
The authors are with the Katholieke Universiteit Leuven, ESAT-MICAS, Kardinaal Mercierlaan 94, B-3001 Heverlee, Belgium.
IEEE Log Number 9412406.

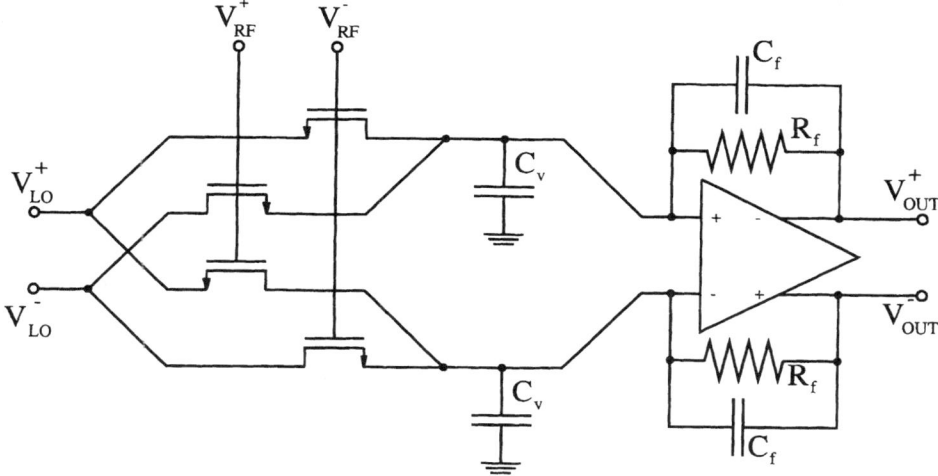

Fig. 1. The mixer topology with the extra capacitor on the virtual ground nodes.

transistors are being used as pass-transistors and the linearity of these mixers does therefore not depend on the voltage to current conversion characteristic of the transistors. The linearity is mainly determined by the speed limitation of the pass-transistor and by the generation of spurious signals during switching. This technique can also be used in CMOS. The output signal after commutation is however still a high-frequency signal and it can therefore not be further processed in CMOS. However, receivers only require a downconversion mixer and this means that the output bandwidth may be limited. A solution based on subsampling with a switched-capacitor structure has been proposed in [5]. The MOS-transistors are used as pass-transistors and an IP3 of 27 dBm has been reported.

A CMOS mixer with very high linearity can be realized by using the linear voltage to current characteristic of the MOS transistor in its triode region [6]. The use of a double balanced structure cancels out the common-mode dc biasing signals and the nonlinear dependence of g_{DS} on V_{DS}. The remaining problem is still the further processing of the high-frequency output current. This limits in [6] the bandwidth to 200 MHz at the cost of a high power consumption and a reduced linearity of the output stage. The mixer presented in this paper is based on this topology (see Fig. 1). There are however two very important capacitors added on the virtual ground nodes of this topology (the capacitors C_v in Fig. 1). Indeed, the output stage, the opamp and the feedback resistors, which convert the output current of the mixing transistors back into a voltage, must, in a downconversion mixer, only be able to produce low-frequency output signals. However, in order to let the input structure operate correctly for all frequencies, there may not be a signal on the virtual ground nodes of the mixer at any time. This is normally done with the feedback structure over the opamp which creates the virtual ground at its inputs. The transistors in the input structure would operate as pass-transistor for high-frequency signals when the frequency capability of the opamp would not be high enough. Therefore the capacitors C_v have been added. They make sure that all high-frequency currents injected to the virtual ground nodes are filtered out and not converted into voltages. The opamp still generates the virtual ground for low-frequencies. By using this structure, it becomes possible to split the design of the input structure and the opamp. The input structure can now be optimized for operation at very high frequencies (more than 1 GHz), while the opamp can be designed for low-frequency operation (a few MHz).

With a perfect virtual ground the currents through the modulated transistors are

$$
\begin{aligned}
I_{DS,1} &= \beta_1 \cdot \left(V_{RF}^+ - V_{LO,DC} - V_{Tn1} - \frac{V_{LO}^+ - V_{LO,DC}}{2} \right) \\
&\quad \cdot (V_{LO}^+ - V_{LO,DC}) \\
I_{DS,2} &= \beta_2 \cdot \left(V_{RF}^- - V_{LO,DC} - V_{Tn,2} - \frac{V_{LO}^- - V_{LO,DC}}{2} \right) \\
&\quad \cdot (V_{LO}^- - V_{LO,DC}) \\
I_{DS,3} &= \beta_3 \cdot \left(V_{RF}^+ - V_{LO,DC} - V_{Tn,3} - \frac{V_{LO}^- - V_{LO,DC}}{2} \right) \\
&\quad \cdot (V_{LO}^- - V_{LO,DC}) \\
I_{DS,4} &= \beta_4 \cdot \left(V_{RF}^- - V_{LO,DC} - V_{Tn,4} - \frac{V_{LO}^+ - V_{LO,DC}}{2} \right) \\
&\quad \cdot (V_{LO}^+ - V_{LO,DC}).
\end{aligned} \quad (1)
$$

The bulk effect gives in first order a linear change around the bias point which is cancelled with the double balanced structure [6]. Assuming perfect matching and no bulk-effect, the output signal is then

$$
\begin{aligned}
V_{out}^+ - V_{out}^- &= R_f \cdot ((I_1 - I_4) - (I_3 - I_2)) \\
&= \beta \cdot R_f \cdot (V_{RF}^+ - V_{RF}^-) \cdot (V_{LO}^+ - V_{LO}^-). \quad (2)
\end{aligned}
$$

Mismatch between the input transistors has two effects. β and V_T mismatches both result in the appearance of residual dc-offset voltages. These offset voltages either appear directly on the output of the mixer or they result in direct feedthrough of the RF and LO signal to the output caused by multiplication

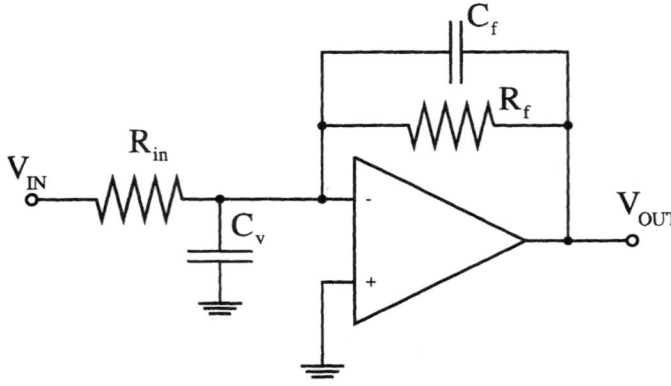

Fig. 2. The negative feedback configuration.

of these signals with the offset voltage. The RF and LO signal are however high-frequency signals. They will therefore be strongly suppressed by the added capacitors C_v. The second effect, caused by β mismatch, is the appearance of a quadratic V_{LO} component in the output signal.

$$V_{out}^+ - V_{out}^- = \beta \cdot R_f \cdot (V_{RF}^+ - V_{RF}^-)(V_{LO}^+ - V_{LO}^-) + \Delta\beta \cdot R_f \cdot (V_{LO}^+ - V_{LO}^-)^2. \quad (3)$$

This explains why the RF signal is best applied to the gates of the modulating transistors. A quadratic V_{RF} component would be highly unwanted. The squared RF signal has frequency components at twice its center frequency and at the baseband. The high-frequency components are filtered out, but the baseband components will degrade the wanted baseband signal. The bandwidth of this parasitic baseband signal is equal to the bandwidth of the RF signal (e.g., 100 MHz), but most of its power will be situated at the lower frequencies, in a band equal to the correlation bandwidth of the RF signal, which is about equal to the bandwidth of a transmission channel (e.g., 200 kHz). The squared LO signal results only in an extra dc component at the output of the opamp. This dc signal can also be a problem. It can be as large as the wanted signal and in that case it will saturate the succeeding filters. There are however techniques available to suppres the dc component without generating to much distortion. One of them is the use of a DSP which measures the dc component and then suppresses it, using a dynamic and nonlinear algoritm [14].

III. MIXER DESIGN

The dc biasing levels of the RF and LO signal must be chosen carefully. There will be a lot of distortion when the modulating transistors are not kept in the triode region at all times. The smallest possible level that can appear at the gates (i.e., $V_{RF,DC} - V_{RF,AC}$) must be at least a V_T higher than the largest possible source level (i.e., $V_{LO,DC}$). Otherwise the transistors will be turned off during a mixing period. Saturation of the modulated transistors will appear when the largest drain-source voltage V_{DS} (i.e., $V_{LO,AC}$) becomes higher than the smallest $V_{GS} - V_T$ (i.e., $V_{RF,DC} - V_{RF,AC} - V_{LO,DC} - V_T$). However, saturation does not directly result in distortion. The cross-coupled double balanced structure makes sure that all quadratic components in the voltage to current conversion characteristic of the modulated transistors are cancelled out [9], [10]. The LO dc level is taken to be 1.15 V, making the maximal LO signal that can be applied 4.6 V_{ptp} differential (i.e., 17.2 dBm). An RF dc level of 3.85 V allows an input signal of 4 times $V_{RF,DC} - V_{LO,DC} - V_T$, which gives 8 V_{ptp} differential or 22.0 dBm for the 1.2 μm CMOS process which has been used for the mixer which is presented here. These are very high values for an RF mixer, resulting in a topology which renders an excellent linearity. Saturation will occur when $V_{RF} + V_{LO} > 8$ V_{ptp}. The applied LO signal is for this reason limited to 2.5 V_{ptp}, so that an RF signal of 5.5 V_{ptp} still can be applied without driving any of the modulated transistors into its saturation region.

IV. FREQUENCY DOMAIN BEHAVIOR

The well known negative feedback configuration of Fig. 2 suppresses any signal at the virtual ground node when a perfect opamp is used. An opamp has however always a limited gain-bandwidth GBW and dc-gain A_0. With a limited GBW becomes the transfer function from the input to the virtual ground (A_0, C_f, and C_v are not taken into account)

$$\frac{V_v}{V_{in}} = \frac{\dfrac{R_f}{R_{in}} \cdot j\dfrac{\omega}{2\pi \cdot \text{GBW}}}{1 + \dfrac{R_f + R_{in}}{R_{in}} \cdot j\dfrac{\omega}{2\pi \cdot \text{GBW}}} \approx \frac{A \cdot j\dfrac{\omega}{2\pi \cdot \text{GBW}}}{1 + A \cdot j\dfrac{\omega}{2\pi \cdot \text{GBW}}}. \quad (4)$$

The transfer function from the input to the output is

$$\frac{V_{out}}{V_{in}} = \frac{-\dfrac{R_f}{R_{in}}}{1 + \dfrac{R_f + R_{in}}{R_{in}} \cdot j\dfrac{\omega}{2\pi \cdot \text{GBW}}} \approx \frac{-A}{1 + A \cdot j\dfrac{\omega}{2\pi \cdot \text{GBW}}}. \quad (5)$$

Normally, the conclusion to be drawn from these equations is that the GBW must be a certain factor higher than the highest frequency component that has to be processed. This is also true if the circuit would be used as lowpass filter with an extra capacitor C_f in the feedback loop. However, here in the mixer topology is the situation different. There is a very large spectrum of input currents to the virtual ground node, basically starting from dc up to twice the LO frequency (more than 2 GHz), but the wanted signal takes up only a few hundred kHz situated at the baseband. The output bandwith BW, given in (5) as GBW/A, only has to be 500 kHz or more. By designing it to be 1 MHz this specification is fulfilled independent of absolute parameter variations. Equation (5) shows that this limited bandwidth can be realized by using an opamp with a small GBW. In combination with a feedback capacitor C_f can the bandwidth be lowered further and can it be positioned more accurately. The use of an opamp with a small GBW poses however a problem. As (4) reflects, beyond the BW is the input signal directly transferred to the virtual ground node. It is not suppressed anymore by the opamp via the feedback construction. The solution for this problem can

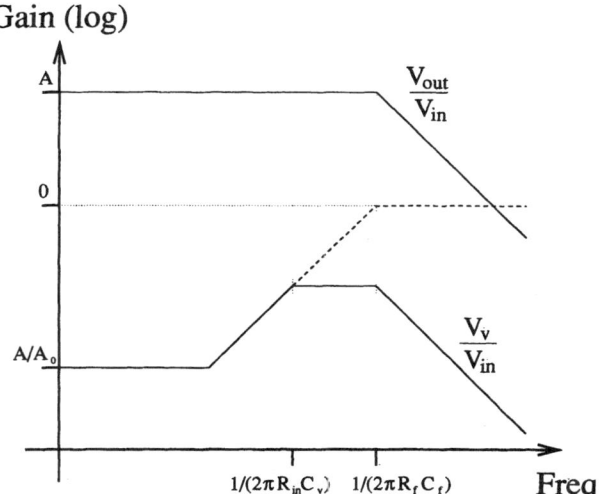

Fig. 3. Transfer function of the mixer used as amplifier.

be found in adding extra capacitance to the virtual ground nodes. This introduces an extra pole in both (4) and (5) situated at $1/(2\pi \cdot R_{in}C_v)$. In order to suppres the signals appearing on the virtual ground without changing the output bandwidth, its value must be positioned between $1/(2\pi \cdot R_f C_f)$ and the BW. Fig. 3 displays the transfer function from the input to the virtual ground and the output.

V. NOISE, POWER CONSUMPTION, AND OPTIMIZATION

The noise figure (NF) of the HF mixer is a very important parameter in receiver design. In a receiver is the mixer positioned directly after the LNA and a mixer with a high NF can only be used in combination with a high-quality LNA. In this case the LNA must have a high gain which, in turn, can only be allowed when the antenna signal is first filtered with a high Q HF filter. The problem with mixers is that its NF can not be made arbitrarely small. Their NF can not even be in the proximity of the NF's that can be achieved with LNA's. This is due to the intrinsic nature of the mixing process.

The two main noise sources in the presented mixer are the modulated input transistors in the triode region and the input stage of the LF opamp. The thermal output noise density generated by these devices is (with the factor 2 introduced by the differential operation)

$$dv_{out}^2 = 2 \cdot 4kT \cdot \left(\frac{2}{3} \cdot \frac{1}{g_{m,in,eq}} \cdot (R_f \cdot 2g_{DS})^2 + 2g_{DS} \cdot R_f^2 \right) \cdot df. \quad (6)$$

Equation (6) gives the output noise density for the mixer biased in its static operation point. The modulated transistors have however, under transient conditions, always the same impedance ($2g_{DS}$) to the virtual ground nodes. For this reason is (6), for the presented topology, also a good measure for the output noise density under transient operation conditions. The equivalent input noise can be found by dividing this expression by the conversion gain G. The conversion gain is defined as the ratio of the output signal over the input signal ($V_{out}^+ - V_{out}^-/V_{RF}^+ - V_{RF}^-$) and follows from (2):

$$G = \frac{(V_{out}^+ - V_{out}^-)_{ptp}}{(V_{RF}^+ - V_{RF}^-)_{ptp}}$$
$$= \frac{1}{\sqrt{2}} \cdot \frac{1}{\sqrt{2}} \cdot \beta \cdot R_f \cdot (V_{LO}^+ - V_{LO}^-)_{ptp}. \quad (7)$$

The first $\sqrt{2}$ appears because the multiplication is performed with a sine and not a square wave. This implies that the rms value of the LO signal has to be used. The second $\sqrt{2}$ is necessary because only the low-frequency mixing product is regarded as wanted. Deviding (6) by (7) results in

$$dv_{in}^2 = \frac{1}{G^2} \cdot dv_{out}^2 = \left(\frac{2}{R_f \cdot \beta \cdot (V_{LO}^+ - V_{LO}^-)_{ptp}} \right)^2 \cdot dv_{out}^2$$
$$= 8kT \cdot \left(\frac{4g_{DS}}{\beta \cdot (V_{LO}^+ - V_{LO}^-)_{ptp}} \right)^2$$
$$\cdot \left(\frac{2}{3} \cdot \frac{1}{g_{m,in,eq}} + \frac{1}{2g_{DS}} \right) \cdot df$$
$$\equiv 8kT \cdot \left(2 \cdot \frac{2g_{DS}}{\Delta g_{DS}} \right)^2 \cdot \left(\frac{2}{3} \cdot \frac{1}{g_{m,in,eq}} + \frac{1}{2g_{DS}} \right) \cdot df. \quad (8)$$

The factor $4g_{DS}/\Delta g_{DS}$ is the extra term intrinsic to any mixer or double balanced structure. It is equal to the ratio between the gain of the mixer used as single balanced amplifier and the conversion gain A/G. For low noise, this term should be as low as possible. The noise can also be lowered by using a larger $g_{m,in,eq}$ and g_{DS}, but this is at the expense of a higher power consumption. The minimal value for this extra term is found when the swing of the LO signal is maximal.

$$\frac{A}{G} = 2 \cdot \frac{2g_{DS}}{\Delta g_{DS}} = \frac{4 \cdot \beta \cdot (V_{GS} - V_T)}{\beta \cdot (V_{LO}^+ - V_{LO}^-)_{ptp}}$$
$$= 4 \cdot \frac{V_{RF,DC} - V_{LO,DC} - V_T}{(V_{LO}^+ - V_{LO}^-)_{ptp}} \quad (9)$$
$$\left(\frac{A}{G} \right)_{min,RF@gates} = \frac{4 \cdot (V_{RF,DC} - V_{LO,DC} - V_T)}{2 \cdot V_{LO,DC}} \quad (10)$$
$$\left(\frac{A}{G} \right)_{min,LO@gates} = \frac{4 \cdot (V_{RF,DC} - V_{LO,DC} - V_T)}{2 \cdot (V_{RF,DC} - V_{LO,DC} - V_T)} = 2. \quad (11)$$

So, according to (11) for the case in which the LO signal is applied to the gates of the modulated transistors, this means that the mixing function adds at least 6 dB to the NF. For the rest is the NF directly determined by the conductivity of a single modulated transistor (g_{DS}) and by the transconductance value of the opamp input stage ($g_{m,in,eq}$), just like in any other amplifier with negative feedback. The noise of the opamp input stage can be made sufficiently small without requiring too much power by using large input transistors (and a small $V_{GS} - V_T$). This is possible because the opamp only has to be LF and there is already standing 25 pF (C_v) at the input nodes. The value of g_{DS} can not be made arbitrarely large because this conductor has to be driven by the LO.

From (10) it might seem that the NF can be made arbitrarely small when the RF signal is applied to the gates by taking $V_{\text{RF,DC}} - V_T < 2 \cdot V_{\text{LO,DC}}$. This implies however a reduction of the input swing and in this way the dynamic range (DR) at the input is not improved. It is not only necessary to minimize the NF of a mixer. It is the input signal capability (i.e., the DR) which must be compared with the power consumption. A good measure for the performance of a mixer is therefore the DR per Watt. Here this gives

$$\left(\frac{\text{DR}}{P}\right)_{\text{RF@gates}} = \left(\frac{S}{N}\right)^2 \cdot \frac{1}{P}$$
$$= \frac{(V_{\text{RF,DC}} - V_{\text{LO,DC}} - V_T)^2}{8kT \cdot \left(\frac{A}{G}\right)^2 \cdot \frac{1}{2g_{\text{DS}}} \cdot \text{BW}}$$
$$\cdot \frac{1}{(V_{\text{LO}}^+ - V_{\text{LO}}^-)^2 \cdot g_{\text{DS}}}$$
$$= \frac{1}{16 \cdot 4kT \cdot \text{BW}}. \quad (12)$$

Equation (12) takes does not take the power consumption and the noise of the opamp into account. Only a low-frequency opamp is needed and its power consumption and noise can therefore be made sufficiently small. $4kT \cdot \text{BW}$ is the intrinsic noise power of a signal with bandwidth BW and the noise of any electronic buiding block with input bandwidth BW can thus never be lower than this value. The factor 1/16 can herefore be defined as the power efficiency for this mixer. It is independant of the choosen input swing, noise level, input bandwidth, or power consumption. In fact, this power efficiency depends, for the presented topology and in first order, only on the topology itself and it can be defined for many other building blocks likes LNA's, amplifiers, filters, A/D converters and any other type of mixer.

$$\eta = \frac{S_{\text{input}}^2}{P_{\text{total}}} \cdot \frac{P_{\text{intrinsic noise}}}{N_{\text{input}}^2}$$
$$\left[\frac{V^2}{\text{Watt}} \cdot \frac{4kT}{V^2/\text{Hz}} = \text{dimensionless}\right]. \quad (13)$$

The power efficiency of a high-quality low-frequency amplifier can be almost 50%. The power efficiency of the presented mixer topology, not taking the power efficiency of the LO signal source into account, is almost 6% and this is, for high-frequency mixers, a very high value.

Equation (12) is slightly different when the LO signal instead of the RF signal is applied to the gates of the modulated transistors

$$\left(\frac{\text{DR}}{P}\right)_{\text{LO@gates}} = \frac{S}{N} \cdot \frac{1}{P} \frac{V_{\text{LO,DC}}^2}{8kT \cdot \left(\frac{A}{G}\right)^2 \cdot \frac{1}{2g_{\text{DS}}} \cdot \text{BW}}$$
$$\cdot \frac{1}{4 \cdot V_{\text{LO,DC}}^2 \cdot g_{\text{DS}}}. \quad (14)$$

And only when A/G is minimal this reduces to

$$\left(\frac{\text{DR}}{P}\right)_{\text{LO@gates}} = \frac{1}{16 \cdot 4kT \cdot \text{BW}}. \quad (15)$$

The main difference is that in this case the efficiency of 6% is only obtained with the maximal LO signal. The efficiency of 6% is in the situation of (12) always obtained, independant of the amplitude of the LO signal. This is caused by the fact that, when the LO signal is applied to the sources of the modulated transistors, a lower conversion gain is compensated by a lower power consumption for the LO signal source. This is another reason why the topology with the RF signal applied to the gates is preferred over the topology in which the LO signal is applied to the gates of the modulated transistors.

VI. REALIZATION

The mixer has been designed and realized in a 1.2 μm CMOS process to illustrate the high linearity and high input frequency capabilities of the proposed topology. The RF and LO signal are externally made differential by means of baluns. They are, as stated in Section III, chosen to be biased at, respectively, 3.85 V (for the RF signal, applied at the gates) and 1.15 V (for the LO signal, applied at the sources and drains) because this renders a $V_{\text{GS}} - V_T$ for the modulated transistors of 2.5 V. Such a large value for the $V_{\text{GS}} - V_T$ results in an excellent linearity and a high input bandwidth. The applicable voltage swing for the RF input is peak-to-peak four times this $V_{\text{GS}} - V_T$. The large $V_{\text{GS}} - V_T$ makes it also possible to realize a large g_{DS} with a small transistor, which gives small parasitic capacitances (less than 20 fF) and thus a high input bandwidth. The W/L of the modulated transistors is 6, their g_{DS} is 1 mS. These four modulated transistors are the only high frequency part on the chip and they take up very little area. Because of their small parasitic capacitances is the on-chip RF to LO crosstalk, which is mainly determined by the absolute mismatch between C_{GS} values, also very small. Hence, RF to LO crosstalk is mainly caused by off-chip crosstalk and crosstalk between the bonding wires. The input bandwidth is also completely determined by the bonding pads capacitances and as a result extremely high frequency performances can be achieved.

The opamp topology is shown in Fig. 4. It is a load compensated folded cascode structure succeeded by a source follower which performs buffering and level shifting. The opamp runs on a single 5 V power supply. The output dc level is 1.15 V. It is kept on this level with a standard type common-mode feedback. The GBW of the opamp is 100 MHz, but because of the partial feedback the second nondominant pole is situated at only 30 MHz. The feedback resistors over the opamp are 60 kΩ, which makes the conversion gain very high. The conversion gain is almost 20 dB for a 12 dBm LO signal. An extra lowpass filter of 1 MHz is implemented with the feedback capacitors C_f (2.5 pF). This small output bandwidth and the high conversion gain are required by the zero-IF structure. The extra filtering and amplification that is performed in the output stage of the mixer relaxes the

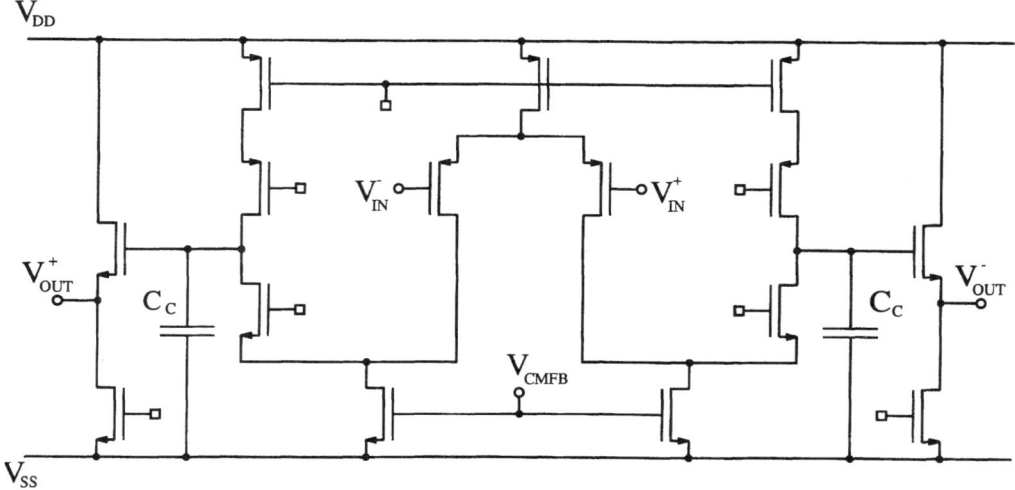

Fig. 4. The opamp topology.

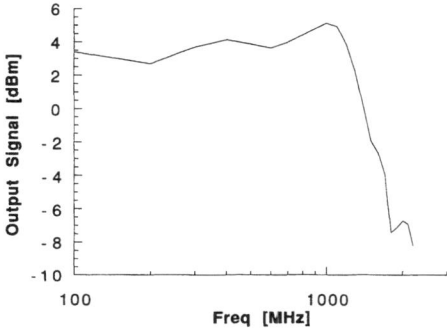

Fig. 5. The input bandwidth measurement.

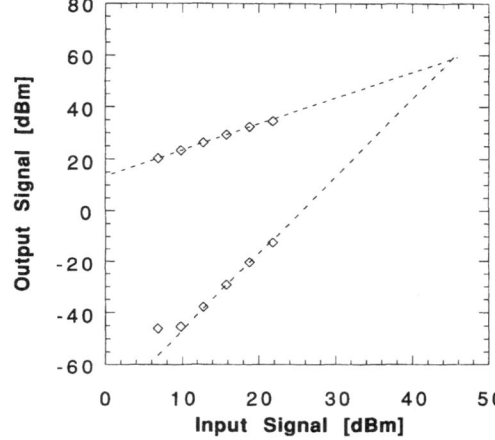

Fig. 6. The IP3 measurement.

specifications for the output stage of the opamp and for the succeeding stages of the reciever.

VII. RESULTS

The realized mixer is measured via external high frequency baluns for both the RF and LO input. These baluns are transformers which convert a 50 Ω signal source into two symmetrical 100 Ω signals sources superimposed on a common-mode dc bias voltage. Fig. 5 shows the measured input bandwidth. It is from dc to 1.5 GHz. The input bandwidth is measured at 6 dB attenuation, because both input signals are falling off at the same time. The measured 1.5 GHz maximum input bandwidth is equal to the specified bandwidth of the baluns which have been used to generate the differential input signals. The output bandwidth is determined by the untuned value of $1/R_f C_f$. For the application in mind it must be at least 500 kHz and it is designed to be 1 MHz so that it can cope with the large absolute processing value spreads. The measured output bandwidth is 780 kHz. The measured conversion gain is 18 dB for a 12 dBm differential LO signal.

The IP3 measurement, shown in Fig. 6, proves that the linearity is very high. The IP3 is 45.2 dBm. This value does not depend on the magnitude of the LO signal. An input signal of 22 dBm, the theoretic value for the maximal applicable input swing, gives an IM3 of -46.4 dB. The IM3 measurement was performed with the downconverted carriers lying out of the output band (at 10 MHz) and the IM3 product lying in the passband (at 200 kHz). This made it possible to measure the distortion of the input stage and not the output stage for signals which would normally produce an output signal of up to 40 dBm. The measured noise figure (NF, compared to a 50 Ω noise source) is 32 dB. This is mainly caused by the input stage of the opamp which has not been optimized for low noise. By using large input transistors for the low-frequency opamp, which has no effect on the high frequency performance of the mixer, the noise figure can easily be improved down to 24 dB. The intermodulation free dynamic range (IMFDR3) of this mixer is 59.6 dB. The mixer is designed in a 1.2 μm CMOS process. The measured power consumption is 1.3 mW which is all taken up in the low-frequency opamp. The total chip area is 1 mm^2. The modulated transistors only take up 300 μm^2 of this area. A microphotograph of the realized chip is shown in Fig. 7.

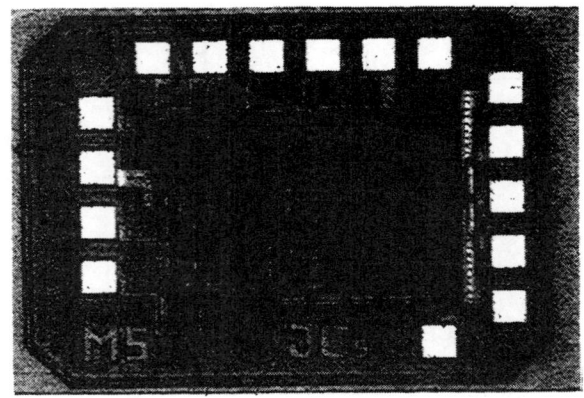

Fig. 7. Chip microphotograph.

VIII. Conclusions

In this paper it is shown how the low resistance of small MOS transistors in the triode area can be used to design a high frequency downconversion mixer. A continuous-time multiplier has been presented for use in zero-IF receivers. The large $V_{GS} - V_T$ of the triode transistors makes a low noise mixer with a high input bandwidth possible. The measured input bandwidth is more than 1.5 GHz. Its structure is based on the cross-coupled triode-MOSFET-C filter structure. The high frequency performances have been achieved due to the introduction of two extra capacitors placed at the virtual ground nodes. These capacitors reduce the opamp specifications to low frequency CMOS performances. In this paper it has been shown how the cross-coupled triode MOS-transistors render a very high linearity which is independent of mismatch. The measured IP3 is 45.2 dBm differential. This very high value for GHz downconversion mixers has been achieved due to the use of triode transistors in combination with the introduced virtual ground capacitors.

References

[1] J. Sevenhans, A. Vanwelsenaers, J. Wenin, and J. Baro, "An integrated Si bipolar transceiver for a zero IF 900 MHz GSM digital mobile radio front-end of a hand portable phone," in *Proc. CICC,* May 1991, pp. 7.7.1–7.7.4.
[2] M. D. McDonald, "A DECT transceiver chip set," in *Proc. CICC,* May 1992, pp. 10.6.1–10.6.4.
[3] D. Sallaerts, D. Rabaey, A. Vanwetsenaers, and M. Rahier, "A 270 kbit/s 35 mW modulator IC for GSM cellular radio hand-held terminals," in *Proc. ISSCC,* San Francisco, Feb. 1990, pp. 34–35.
[4] J. Sevenhans *et al.*, "An analog radio front-end chip set for a 1.9 GHz mobile radio telephone application," in *Proc. ISSCC,* San Francisco, Feb. 1994, pp. 44–45.
[5] P. Y. Chan, A. Rofougaran, K. A. Ahmed, and A. A. Abidi, "A highly linear 1-GHz CMOS downconversion mixer," in *Proc. ESSCIRC,* Sevilla, Sept. 1993, pp. 210–213.
[6] B.-S. Song, "CMOS RF circuits for data communications applications," *IEEE J. Solid-State Circuits,* vol. SC-21, no. 2, pp. 310–317, Apr. 1986.
[7] R. G. Meyer, "Intermodulation in high-frequency bipolar transistor integrated-circuit mixers," *IEEE J. Solid-State Circuits,* vol. SC-21, no. 4, pp. 534–537, Aug. 1986.
[8] J. N. Babanezhad and G. C. Temes, "A 20-V four quadrant CMOS analog multiplier," *IEEE J. Solid-State Circuits,* vol. SC-20, no. 6, pp. 1158–1168, Dec. 1985.
[9] H. Song and C. Kim, "A MOS four-quadrant analog multiplier using simple two-input squaring circuits with source followers," *IEEE J. Solid-State Circuits,* vol. 25, pp. 841–848, June 1990.
[10] K. Kanazawa *et al.*, "A GaAs double-balanced dual-gate FET mixer IC for UHF receiver front-end applications," in *IEEE MTT Int. Microwave Symp. Dig.,* 1985, pp. 60–62.
[11] D. Rabaey and J. Sevenhans, "The challenges for analog circuit design in mobile radio VLSI chips," in *Proc. AACD Workshop,* Leuven, Mar. 1993, vol. 2, pp. 225–236.

TA 8.1: A Fully Integrated 900MHz CMOS Double Quadrature Downconverter

Jan Crols, Michiel Steyaert

Katholieke Universiteit Leuven, Heverlee, Belgium

Zero-IF downconversion topology is being used more and more in receivers for mobile wireless telecommunications applications. The main advantage over the conventional IF downconverters is the much higher degree of integration. However, the zero-IF topology has some major drawbacks. First, baseband configuration makes zero-IF topology highly sensitive to parasitic dc-signals and products of self-mixing of the RF and LO signal. Second, performance is reduced by the limited matching between the two parallel downconversion paths. Effects of mismatch, i.e. phase and amplitude errors, degrade the signal quality because they result in a reduced mirror signal suppression.

The low-IF receiver is, like the zero-IF receiver, also a multipath topology that can be implemented in a highly-integrated way. It uses an IF of a few hundred kHz and is therefore not sensitive to parasitic baseband signals like DC offset voltages and self-mixing products. In these topologies the phase error of the HF quadrature generator is the main cause of mirror-signal crosstalk. It is far more important than the amplitude and phase errors in the mixers and the LF part. This fully-integrated CMOS quadrature downconverter is for use in high quality low-IF receivers. It has <0.3° phase error without any tuning or trimming.

The quadrature downconverter differs in many ways from conventional realizations. Conventional realizations use an RC-CR network for quadrature generation of LO signal and two Gilbert multipliers for downconversion [1,2]. Mirror signal suppression in these quadrature-demodulator structures is mainly determined by the phase error of the quadrature generator and is equal to $\tan(\Delta\varphi)$ (with $\Delta\varphi$ being the phase error). Generation of the quadrature signal is based on 90° phase-difference between a pole and zero with the same cut-off frequency. The RC-CR network must be closely matched and tuned. A phase-error of less than 1° can be achieved only with extra tuning and trimming [2].

The downconversion is performed with a double quadrature structure, shown in Figure 1. In this structure both RF and LO signal are put in quadrature and mixed down with four CMOS HF multipliers of which the output signals are combined two by two. The cross-coupling and combination of the four mixing products suppress amplitude and phase errors. The mirror signal crosstalk due to the phase errors of the quadrature generators of this structure is $\tan(\Delta\varphi_{RF}) \cdot \tan(\Delta\varphi_{LO})$, the square of the value of conventional structures. This is the same for amplitude mismatch and hence, the double quadrature downconverter is highly insensitive to mismatches in the quadrature generators. The phase mismatch between the downconversion mixers is approximately given by $\tan(\Delta\varphi) \approx f_{LO} \cdot \Delta f_{BW}/f^2_{BW}$. f_{BW} is the input bandwidth of the mixers. The phase error can be made low if the mixers have a large input bandwidth. The amplitude error between mixers is determined by the mismatch in dc conversion gain. It gives a mirror signal crosstalk equal to $\Delta G/2G$ and therefore the main remaining source of mirror signal crosstalk. However, this effect can be eliminated by a closely-matched digital AGC operation on the LF I- and Q-signals in the DSP that further processes and demodulates these signals.

The double-quadrature structure requires two quadrature generators with a relatively good amplitude and phase matching in a broad passband. Broadband quadrature generators eliminate frequency tuning to compensate for absolute parameter variations. The conventional RC-CR structure does not have good broadband amplitude matching and cannot be used. Instead, two-stage sequence asymmetric polyphase filters, shown in Figure 2, are used [3]. One stage of a sequence asymmetric polyphase filter passes positive sequences and suppresses negative sequences at $1/2\pi RC$. The response to positive and negative sequences for the two-stage filter are given in Figure 3. A quadrature signal can be generated with these filters by applying only an I-signal. This is the same as applying both a positive and a negative sequence at the same time. The remaining positive sequence is the wanted quadrature signal. Two two-stage sequence asymmetric polyphase filters are implemented on the chip. Their passband is designed to range from about 600MHz to 1.2GHz. The signals at 900MHz are always in quadrature, independent of process variations. Amplifier stages are added after the polyphase filters to compensate for their 6dB signal loss.

On the chip is first a single-ended to differential conversion performed on the RF and LO signal. The single-ended-to-differential converter and resistor-loaded amplifiers are shown in Figure 4. The dummy transistor M2 balances the positive and negative output node limiting the phase error to about 3° at 1GHz. The mixers use a highly-linear CMOS topology for downconverters that requires only low-frequency opamps [4].

A micrograph of the 6mm² chip is shown in Figure 5. It is realized in a standard 0.7µm CMOS process. It operates on a single 5V power supply voltage. The measured input bandwidth is 900MHz. The measured output bandwidth is 3.09MHz. The IP3 is 27.9dBm and the measured noise figure for a 12dBm LO signal is 24.0dB. The measurements of phase and amplitude errors as a function of frequency are given in Figures 6 and 7. The phase-error is < 0.3° in the 500 to 900MHz band. In this band the amplitude error is situated at 0.5dB. These phase and amplitude errors result, after a digital AGC operation, in a mirror signal suppression of 46dB.

References

[1] Steyaert, M., R. Roovers, "A 1-GHz Single-Chip Quadrature Modulator," IEEE J. of Solid-State Circuits, vol. 27, no. 8, pp. 1194-1196, Aug., 1992.

[2] Sevenhans, J., et al., "An Analog Radio front-end Chip Set for a 1.9 GHz Mobile Radio Telephone Application," ISSCC Digest of Technical Papers, pp. 44-45, Feb., 1994.

[3] Gingell, M.J., "Single Sideband Modulation Using Sequence Asymmetric Polyphase Networks," Electrical Communication, vol. 48, pp. 21-25, 1973.

[4] Crols, J., M. Steyaert, "A Full CMOS 1.5 GHz Highly Linear Broadband Downconversion Mixer," Proc. ESSCIRC, pp. 248-251, Sept., 1994.

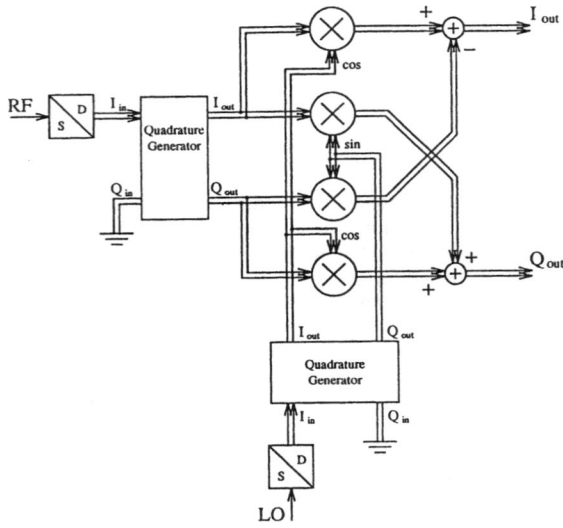

Figure 1: Double quadrature downconverter block diagram.

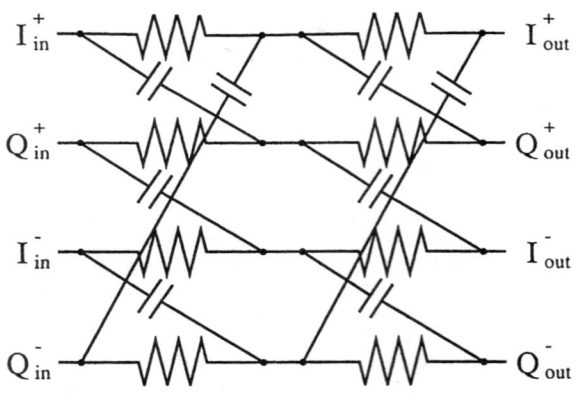

Figure 2: The two-stage sequence asymmetric polyphase filter used as quadrature generator.

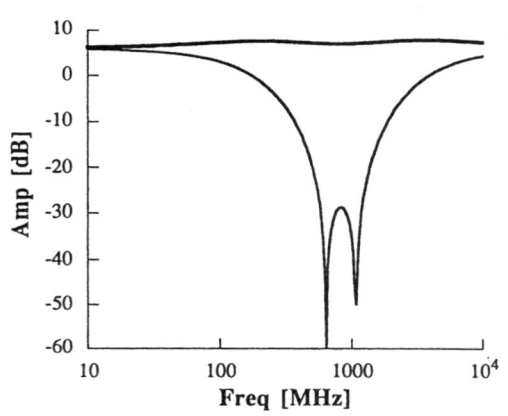

Figure 3: Frequency response of the two-stage polyphase filter for a positive and negative sequence.

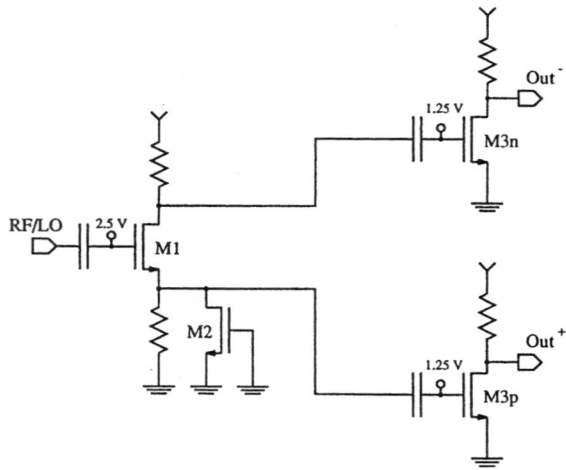

Figure 4: The single-ended to differential converter.

Figure 5: See page 352.

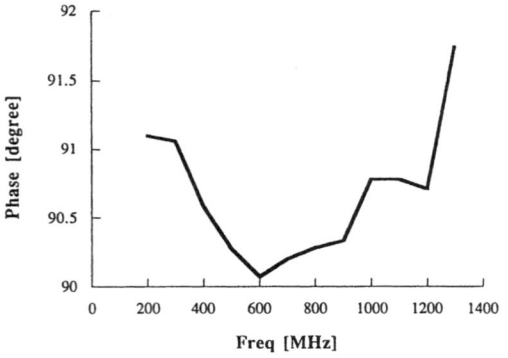

Figure 6: Phase error versus frequency.

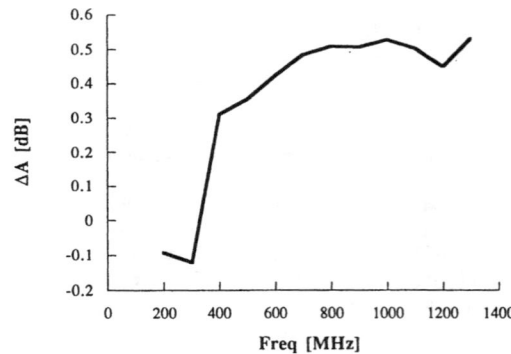

Figure 7: Amplitude error versus frequency.

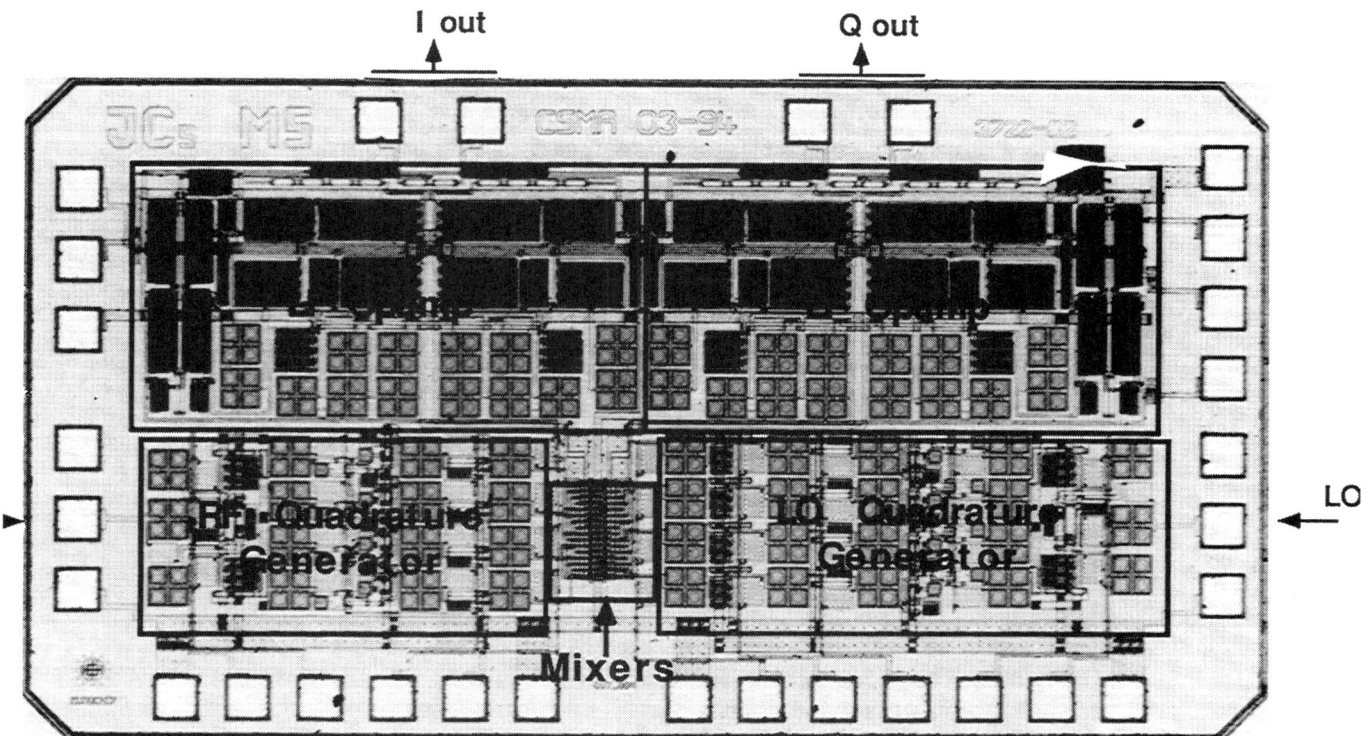

Figure 5: Chip micrograph.

Design Techniques for 1GHz Downconversion ICs Fabricated in a 1μm 13GHz BiCMOS Process

William D. Mack and Robert G. Meyer
Philips Semiconductors - Communications Products Group
Sunnyvale, CA 94086 USA

ABSTRACT

The integration of a low-noise amplifier (LNA) with a wide dynamic-range mixer has been accomplished by employing an advanced BiCMOS process. The LNA features a 2dB noise figure with 16dB of gain at 900MHz and is internally matched to 50 ohms. A through mode is implemented with a FET switch to handle large signals and save power. The amplifier is stabilized versus temperature and supply voltage. The mixer is a single-balanced common-base design to achieve wide dynamic range. The RF and LO ports are matched to 50 ohms and the IF port is high impedance to simplify matching to SAW IF filters. The input third-order intercept is +6dBm, which allows the downconverter to handle large interfering signals without affecting the receiver performance. The overall design is ESD protected by a new active clamping network, and the $1.31 \times 1.16 mm^2$ die fits in a 14 pin surface-mount package. A special shielding arrangement is used on chip in combination with coplanar circuit board layout to achieve 46dB of antenna isolation from the local oscillator. The overall supply current is 13mA from +5V.

1. Introduction

In this paper new design practices, layout techniques, and model limitations will be shown using a 1GHz downconverter IC as an example. The primary application is replacement of discrete component front-ends in 800-1200MHz cellular and cordless phones, including spread-spectrum designs. The advances in BiCMOS process technology and the explosion in growth in personal communication networks of all types has made the integration of "front-end" amplification and mixing functions feasible at frequencies of 1-2GHz, but there are numerous pitfalls to the be avoided if a successful design is to be produced in time for the end product's market window.

In this paper a discussion of a 1μm BiCMOS process will be followed by the general specification of the 1GHz downconverter IC. Discussion of the LNA, PCB, and mixer designs follows. Details of the networks that protect the IC from electrostatic damage will conclude the design. Measurement data and discrepancies between simulated and measured data will be reviewed.

2. QUBiC BiCMOS Process

In the figure below a cross-section of Philips QUBiC process is shown [1].

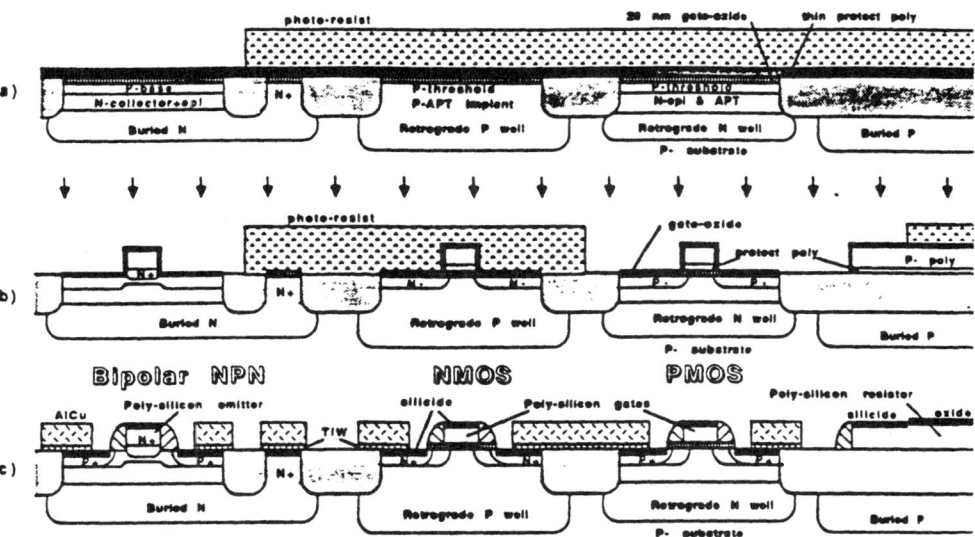

Figure 1: Process Cross-Section

This technology is largely bipolar based and includes N and P type buried layers, an N type epitaxial grown silicon layer, and recessed, planarized field oxide. The buried layers function as retrograde wells, reduce collector, well and substrate resistance, suppress latch-up, and help increase the field threshold and punch through voltages. The field oxide is planarized to facilitate subsequent processing and reaches well into the buried layers to reduce the gain of the parasitic substrate transistor. A collector plug implant and drive provides the low resistance contact to the N type buried layer.

After stripping a sacrificial oxide, 200Å gate oxide is grown, immediately followed by the deposition of a thin polysilicon layer to protect this oxide against the potentially damaging effects of the subsequent processing steps. The N and P channel devices receive separate low energy threshold adjust and high energy anti-punch-through implants to set the threshold voltage and control sub-threshold currents.

A combination of a high-energy phosphorous and low-energy boron implants are used to create a narrow and steep base profile. The phosphorus implant increases locally the concentration of the epitaxial layer and counter-dopes the tail of the base implant. This reduces the width of the base and retards the base push-out at high currents which enhances the speed of the bipolar transistor with only a minor sacrifice in breakdown and Early voltage. The thin polysilicon and gate-oxide stack is removed locally with the base implant mask still in place, as shown in Figure 1a. Next, polysilicon is deposited and N and P type regions are formed by arsenic and boron implantation, respectively. The polysilicon is etched to leave emitters, gates and resistors. This etch also removes some of the exposed monosilicon in the bipolar base areas. A short oxidation forms a protective thermal oxide and drives the arsenic into the monosilicon. This is the final high temperature treatment except for rapid thermal annealing.

The NMOS transistor receives a phosphorus lightly-doped-drain region to reduce the susceptibility to hot-electron induced degradation. The PMOS and bipolar transistor receive a P type implant, which serves as a lightly-doped-drain region and secures a sufficient base link-up, as illustrated in Figure 1b. Next, oxide side-wall spacers are created around the polysilicon structures. The subsequent source/drain and base contact implants are activated with a rapid thermal anneal to keep the junctions shallow. Platinum silicide is formed on all exposed silicon for low series and contact resistance and Schottky diodes.

Al/Cu straps with TiW barrier metal are used to contact all active base, collector, source, and drain regions, as illustrated in the schematic cross-section in Figure 1c. These straps also serve as local interconnect. After glass deposition, contacts are etched to the Al/Cu straps and the polysilicon. Narrow emitters and gates are contacted remotely, while wider emitters can be contacted on top of the active device. Two layers of planarized global interconnect and a lateral TiW fuse complete the BiCMOS processing.

The final process achieves high yields and good reliability and features 0.85μm L_{eff} MOSFETs and 1μm 13GHz F_T NPNs. The minimum NPN (emitter =1x2μm^2) has Cje=Cjc=6fF zero bias, and r_b=1250Ω at a peak F_T current of 400μA. High density capacitors can be fabricated with the thin gate-oxide, and low capacitance polysilicon resistors are available.

3. Downconverter Specification

The NE/SA600 front-end was targeted to meet an overall power gain of 13dB, overall noise figure (NF) of 3dB, and overall input third-order intercept point (iIP$_3$) of -13dBm at a 13mA supply current from 5V. All ports were to be internally matched to 50Ω except for the high impedance intermediate frequency (IF) output. Maximum local oscillator (LO) to antenna input isolation was desired. A power savings mode with high-overload capability was also desired.

4. Low-Noise Amplification

The requirement for -13dBm iIP_3 at the LNA input dictated the use of a high level mixer with wide dynamic range. Such mixers generally have low gain, so a two stage LNA was chosen to achieve sufficient system gain and to overcome noise in the mixer. A simplified schematic of the LNA is shown in Figure 2.

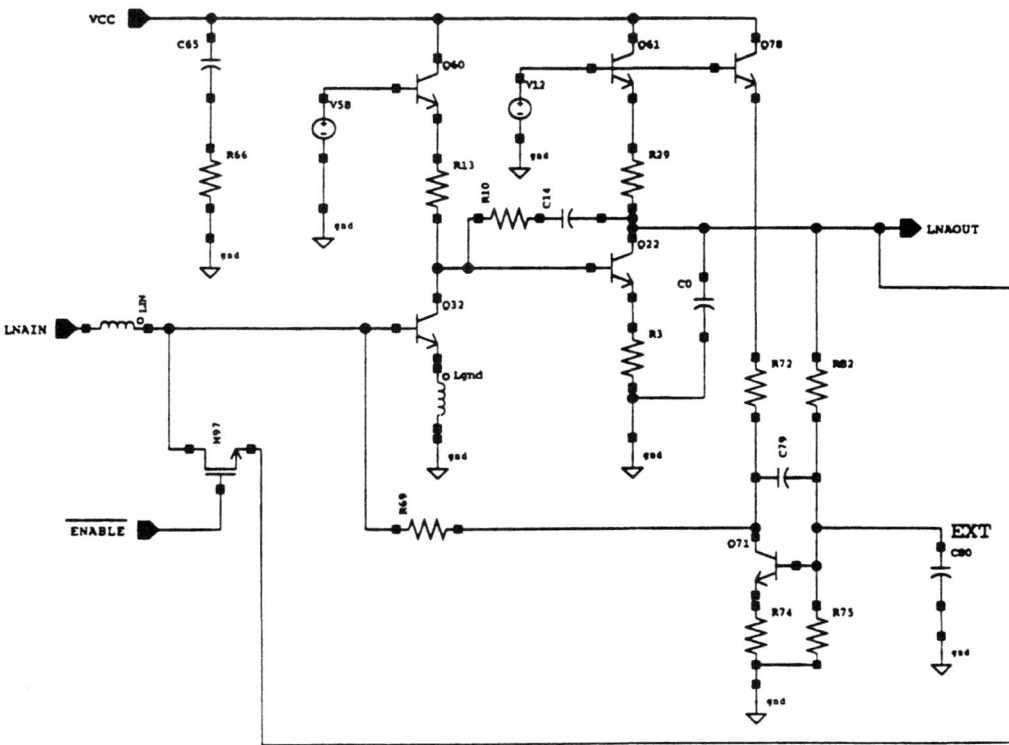

Figure 2: Simplified LNA Schematic

The LNA is a two-stage design incorporating feedback to stabilize the amplifier. The first common-emitter NPN stage Q32 uses noiseless local feedback via the emitter bondwire to match the input to 50 ohms. The second stage NPN Q22 uses shunt and series resistive feedback to stabilize the gain and develop a 50Ω match at the output. The NPNs Q60/61/78 power down the amplifier when the ENABLE pin is low, and also turn on the through mode NMOS FET M97. The amplifier gain in an SO14 package is 16dB at 900MHz when enabled, and -7dB when disabled. This dual-gain-state approach can be used in bang-bang control systems to achieve a low-gain, high-overload front-end as well as the more usual high-gain, low-overload setting. The bias for the two-stage amplifier is stabilized by a shunt feedback means via NPN Q71, with an external bypass capacitor of (typically) 0.01µF compensating the loop (C80). The DC drive voltages to the NPNs Q60/61/78 are designed to minimize the variation of amplifier gain with temperature and V_{CC}.

The NF of the LNA is primarily determined by the NPN Q32, and it is therefore sized quite large, to 36X the minimum size to minimize r_b. Also, the bias current in Q32 is set at 2.5mA to minimize shot noise in Q32. The overall LNA draws 9mA including the additional overhead current for temperature and supply leveling of gain and the through mode functionality. The lack of degeneration of Q32 also sets the distortion performance, with approximately 32mV zero-peak causing 1dB gain compression. This corresponds to P_{-1dB}=-20dBm. IP_3 at the input is about 10dB higher than P_{-1dB} at the input, approximately -10dBm. The impedance matches were simulated to be approximately -20dB, but the input match depends heavily on the exact value of the emitter lead inductance (bond wire + lead frame inductance). Further discussion of the input match will be addressed in the measurements section.

5. Through Mode Switching

When the ENABLE pin is low the through mode NMOS FET is turned on and the LNA is powered down. Powering down saves 9mA and allows very large signals to be handled by the LNA without overload. With the 100/1µm NMOS device gate at 5V and drain/source near ground, the on resistance is approximately 60Ω. This causes a mismatch loss, but increases IP_3 at the input to over +20dBm. When the ENABLE pin is high, the FET is off and the off-isolation can be a concern. It is known that the accuracy of most MOS models at 1GHz is poor, and in particular drain to source isolation is not accurately predicted. Fortunately, the required level of reverse transmission is limited by the S12 of the LNA, which is -42dB, and not by the isolation of the NMOS FET.

6. Wide-Dynamic-Range Mixing

The mixer is a single-balanced topology designed to draw very low current, typically I_{CC}=3.2mA, and provide a very high input third-order intermodulation intercept point, typically iIP_3=+6dBm. Figure 3 shows a simplified schematic diagram.

The single-balanced topology is used to save DC current at the expense of LO feedthrough. Since IF filtering can often reject LO signals this is a useful tradeoff and saves 3.2mA compared to a double-balanced mixer.

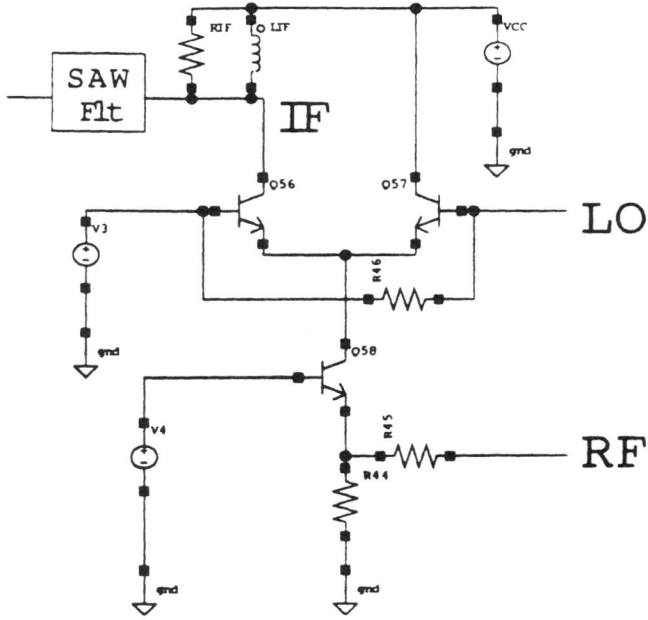

Figure 3: Simplified Mixer Schematic

The common-base stage at the RF input is padded by a 30Ω resistor R45 to achieve the high overload characteristic. P_{-1dB} at the input is -4dBm (zero to 200mV peak) and iIP_3 is +6dBm. The high overload comes at the expense of gain, since the common-base stage has unity gain. The power gain of the overall mixer is 0dB when the load impedance at IF is 500Ω and the input padding resistor is included. However, it is common to quote the power gain as -3dB since the 500Ω net IF impedance is made up of a 1KΩ filter matching resistor and a 1KΩ filter, with the power divided between the resistor and the filter. RF and LO port impedances are nearly 50Ω resistive, due to the resistors R45 and R46. The IF output is an open collector. The open-collector output allows direct interfacing with high impedance IF filters, such as surface acoustic wave (SAW) filters without the need for external step-up transformers

The basic mixer is functional from DC to well over 2.5GHz, but RF and LO return losses degrade below 100MHz due to the fact that internal AC bypassing at the bases of Q56 and Q58 is not effective at these frequencies. The IF output can be used from DC to 500MHz or more, although typically the intermediate frequency is in the range 45-120MHz in many 900MHz receivers. The NF of the mixer is determined by the input padding loss, common-base transistor NF, and base resistances of the switching pair. The NPNs in the switch were chosen as large as possible to minimize noise, consistent with the need to fully switch at 1GHz. The NF is sensitive to the LO drive level, with drive levels of 0dBm to +10dBm providing the best performance. A clean LO source without noise at the IF frequency is needed to minimize mixer NF.

The purpose of the inductor from IFout to V_{CC} is to set the midpoint of the IF swing to be V_{CC}. Without this inductor the part is sensitive to overload at the output under low V_{CC} (V_{CC}=4.5V) and hot temperature conditions, although the inductor can be deleted in many applications.

7. Physical Layout Considerations

The layout of the IC is shown in Figure 4. The LNA first stage and second stages are separated by over 1mm and use grounds on different sides of the SO14 package. This is to eliminate feedback due to inductive coupling that can make the amplifier oscillate. The interstage aluminum interconnects are routed on top of a special silicon shield. The LO input is well away from the LNA input and also uses the special shield. The line/shield is made up of second-layer metal, 2.2µm of silicon dioxide, and a low resistance N+ conductor. The net effect of the shield is to rout LO signal current that would be injected into the substrate due to parasitic capacitance of the aluminum harmlessly off to the LO ground pin. The shield is contacted at least every 50µm to maintain a low impedance on the shield. Additionally, all high-frequency ports use the shield arrangement to further minimize signal coupling. Extensive use of 220Å oxide capacitors provides lowpass filtering and bypassing of supply and internal bias voltages. A simple 1.25V bandgap reference and comparator cell are used to provide both CMOS and TTL compatibility on the digital ENABLE input.

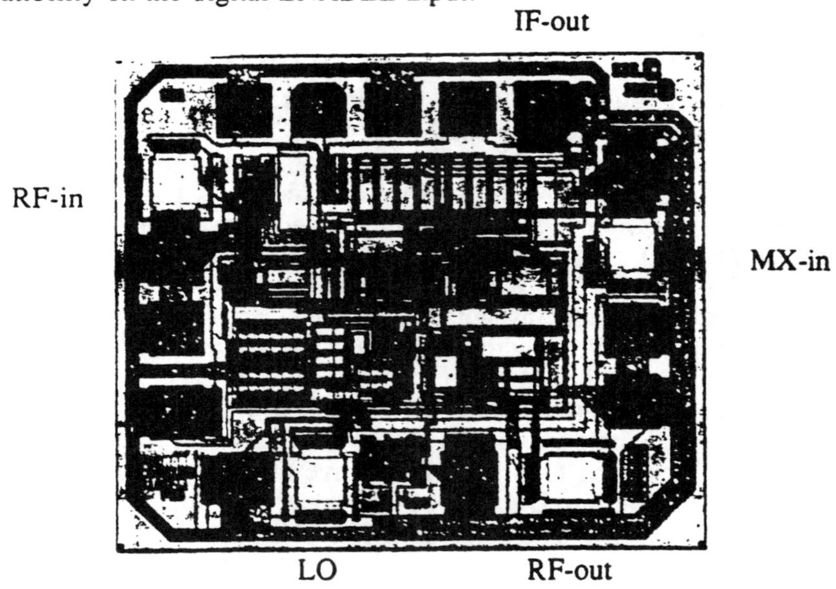

Figure 4: Micro-Photograph

8. Printed Circuit Board Interfacing

A standard fiberglass printed circuit board (FR4) was chosen with 62mil thickness. Fifty ohm lines were implemented with 60 mil wide coplanar transmission line with 20 mil separation to the ground plane. Coplanar transmission lines are narrower than microstrip lines for a given PCB media, and thus mate better to the narrow pitch (50mil) of the SO14 package. Ground plane was used extensively on top of the board as well as on the backside with extensive use of through holes. Ground plane under the SO14 package helped reduce the package bond wire inductance. It was found that boards without through holes had several resonances in the 1-2GHz range and degraded bandwidth due to extra ground inductance. Transmission lines were two-sided tapered at the SO14 pins, and launches were 90 degree SMA type. A full three dimensional simulation of the board layout was carried using the Philips tool FACET [2]. The input to FACET is the board geometry and the output is a SPICE-like simulation file. Using the FACET output the effect of various board layout techniques were studied to help optimize the 900MHz performance.

Alternate pins were designated as ground pins to minimize coupling effects, and the LO and LNA inputs were physically separated by four pins with their transmission lines routed perpendicular to each other.

9. Measurements

An HP8753C network analyzer was used to measure all the amplifier s-parameters, while Racal-Dana 9087 signal generators and an HP8568B spectrum analyzer were used to measure the distortion and overload characteristics. NF was measured with the HP8970A noise figure meter. Measurements showed 16dB gain with a 10dB input match, 15dB output match, -40dB isolation, with a 2.2dB NF, all at 900MHz. Supply current and gain variation with temperature (-0.8dB/100°C) and gain variation with frequency (-1.4dB /100MHz) matched design simulations and show the effectiveness of the leveling circuitry.

The degradation of the matches and slight degradation of the NF compared to the simulated results were traced to a coupling effect of the IC substrate and the SO14 package bond-wires and leadframe. The re-simulated input impedance is plotted on the Smith chart in Figure 5 with the measured data shown as small circles. Figure 6 shows a simplified diagram of the mutual inductance parasitic and ground capacitance coupling through the substrate that causes this shift in impedance. The S11 measurement is done in a microwave test fixture that is calibrated at the SO14 package leads. To predict this partial resonance at 700MHz during the design phase requires accurate IC-substrate and IC package models.

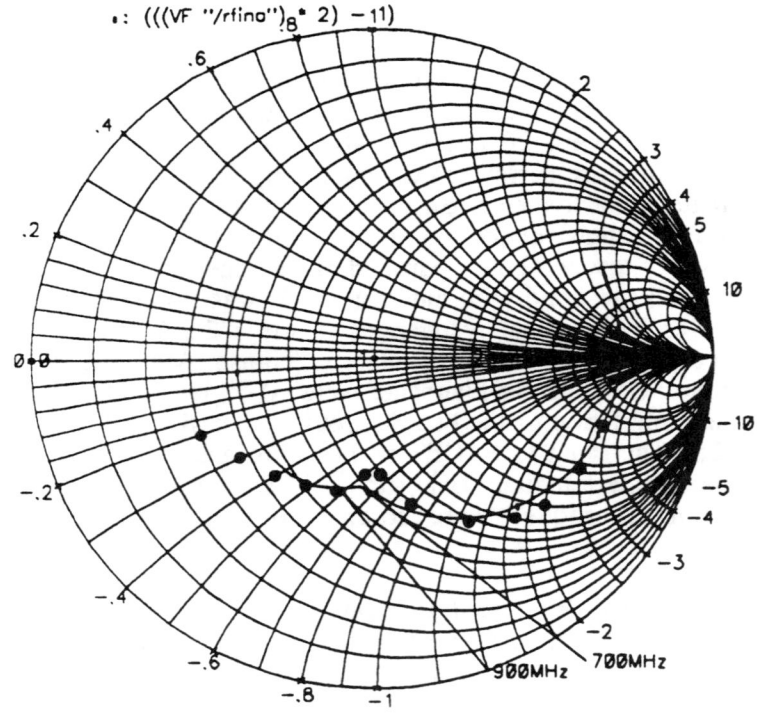

Figure 5: S_{11} on the Smith Chart

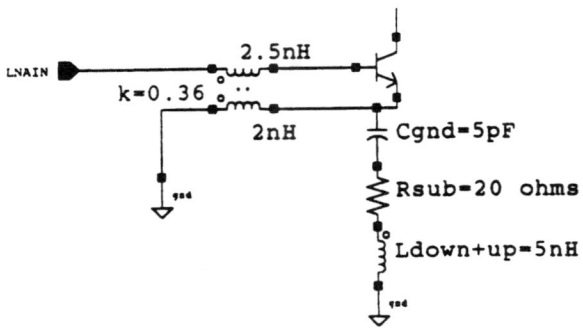

Figure 6: LNA Base-to-Emitter Parasitic Coupling

Measured P_{-1dB} was -20dBm and iIP_3 was -10dBm with a 1MHz input frequency (channel) spacing. The iIP_3 intercept point was found to be a function of channel spacing due to second-order intermodulation effects that occur in the LNA. This is due to the high gain in the two-stage LNA at low frequencies, and this effect can be minimized by using small value coupling capacitors on the LNA input and output (*eg*. 100pF).

The Racal-Dana 9087 signal generators and the HP8568B spectrum analyzer were used to measure the gain, distortion and overload characteristics of the mixer. Generally, the IF output collector from the mixer was ac coupled into the spectrum analyzer 50Ω load. This setup allows simple measurements of the mixer performance but does not fully emulate the actual application since the IF output generally feeds a high impedance IF filter. Since the target application for this IC has an IF frequency at 100MHz, the discrepancies between the 50Ω measurement and the actual application are negligible.

Measurements showed a gain of -3.4dB, IP3 at the input of +6dBm, and a NF of 15-16dB, all with a 0dBm LO drive level. The inability to simulate mixer NF with a SPICE based simulator meant that hand calculations were required for the mixer NF. The predicted NF was 11dB, well below the measured value. New techniques for calculating mixer NF have since been derived [3] which predict a 16dB NF.

10. Electrostatic Discharge (ESD) Network Design

Designing an ESD network on a thin gate-oxide BiCMOS process without degrading the IC performance at 1GHz is a very challenging task. A new network was designed to cope with this problem, but it is useful to first review ESD basics before proceeding to the ESD network design.

There are several test methods used for ESD evaluation, but the most common are the Human Body Model (HBM) and Machine Model (MM). Generally all products manufactured must pass 2000V HBM and 200V MM without damage, although targets of 4000V HBM and 400V MM are typical. Increased ESD performance comes with a cost in terms of increased die area, decreased yield, and more electrical overstress (EOS) susceptibility.

The HBM simulates the discharge that can occur when an individual touches a device. The HBM is a capacitor of 100pF charged to a specified voltage, which is then discharged into the device through a 1500Ω resistor. This model originates from measurements of capacitance/resistance of a human being (approximately 17pF per foot of height, and 100-5000 ohms depending on skin moistness) and dates originally to studies of explosive sensitivity of various gas mixtures found in underground mines. The MM simulates the discharge generated when a metal object (*eg*. a screwdriver) touches an IC pin. The MM is a 200pF capacitor charged to a specified voltage, which is then discharged into the device without a series limiting resistor. The fast rise time of this test makes it sensitive to package/board/tester lead inductance, and therefore not very repeatable. Figure 7 details the HBM and MM test circuits.

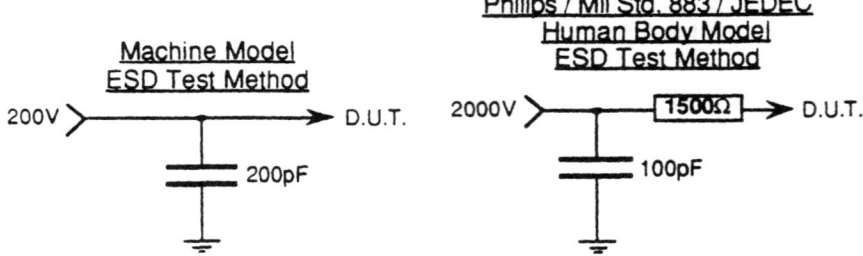

Figure 7: Models of Common ESD Test Methods

On a BiCMOS process the bipolar device is sensitive to emitter-base junction Zenering and collector-emitter LV_{ceo} breakdown. The MOS gates are easily damaged if V_{gd} or V_{gs} exceed the gate-oxide rupture voltage. Gate-oxide capacitors are analogously sensitive. NMOS devices are particularly sensitive to drain-source voltages that exceed the parasitic (NPN) latchback voltage.

For the QUBiC process, the bipolar NPN $BV_{ebo}=8V$ and $LV_{ceo}=8V$. Generally, diodes are added across NPN base-emitter junctions to avoid the Zener breakdown mechanism, and often the base impedance is low on NPN devices and their collector-emitter breakdown is limited by BV_{ces}, and not by LV_{ceo}. Since $BV_{ces}=18V$ and this breakdown mechanism is not inherently damaging, the NPN devices are generally the strong part of a BiCMOS process with respect to ESD damage.

Both PMOS and NMOS devices are sensitive to gate-oxide rupture ($BV_{ox}=22V$), however there is a beneficial mechanism due to the time it takes to generate enough hole current to damage the oxide. For very shorts pulses, such as those of the HBM or MM, the gate-oxide rupture voltage is increased beyond the static gate-oxide breakdown voltage [4]. Even with this aiding effect, the gate-oxide is still very fragile on CMOS inputs or any other circuit that has both terminals of a gate oxide connected to external pins of the IC. Perhaps the weakest link in a BiCMOS process is the NMOS transistor on CMOS outputs. This transistor has $LV_{dsg}=12V$ (drain-to-source with the gate grounded). In this case, or any other case where the drain and source come to external pins, the LV_{dsg} breakdown can be permanently damaging, usually leaving a drain-to-source filament akin to a "Zener-zap" filament.

Figure 8 shows the complicated I-V curve for a QUBiC NMOS transistor breakdown. I is the drain current and V is the drain-source voltage (with the gate and substrate shorted to the source). Above approximately 15 volts avalanche current is generated in the device, causing leakage. At a drain-source voltage Vt1, the device "snaps-back". This voltage is called the first trigger voltage and is due to parasitic bipolar transistor action. The lateral bipolar NPN parasitic has a basewidth equal to the NMOS gate length.

Between Vh1 and Vt2 the device passes current uniformly over its width, but at Vt2 thermal effects come into play and the device enters second breakdown. Generally, operation in second breakdown causes permanent damage, with destruction occurring for slightly higher currents.

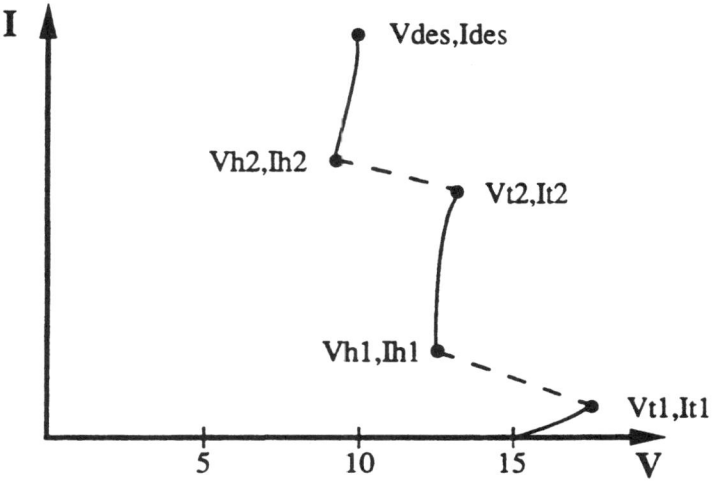

Figure 8: NMOS I-V Breakdown Characteristic

The general means to protect an IC on the QUBiC process involves connecting a diode from each input or output to V_{CC}, a diode from each input or output to GND, and a low impedance clamp from V_{CC} to GND. Figure 9 details the basic network. The design of the clamp is detailed in the next section.

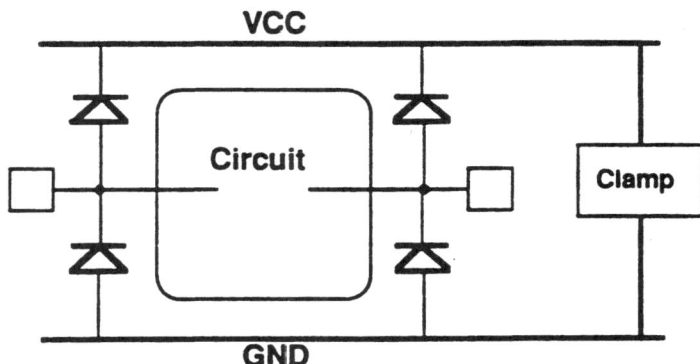

Figure 9: ESD Protection Network

For the QUBiC BiCMOS process, the availability of twin buried layers yields an excellent input-to-ground diode [5]. This diode has its p-n junction below the silicon surface, and is thus very robust. Unfortunately, the doping level in the p+ buried layer was not high enough to use this structure for the diode to V_{CC}, and therefore the collector-to-base junction of an NPN transistor was used. The collector-base junction is a surface junction and more sensitive to damage if allowed to breakdown. One of the functions of the V_{CC}-GND clamp is to prevent avalanching of this diode.

The proper geometry for shunt diodes was found to be a long stripe with wide metal traces as opposed to a multi-finger geometry. This is because of debiasing effects along the diode at current densities $>10^6$ A/cm^2, as well as blowout of metal traces. Trace blowout occurs at approximately 2×10^8 A/cm^2 of metal cross sectional area during HBM testing of narrow leads. Early multi-finger diode designs were found to blowout at 3.5KV, where single-stripe wide-metal easily survives 10KV HBM pulses.

Active Clamp

The idea of the active clamp is to emulate a discrete Zener diode with an active circuit [6]. The clamp includes a bandgap reference set to approximately 9V that is used to sense the V_{CC}-GND potential. The clamp turns on a large ballasted NPN to sink any ESD current that occurs. To clamp the fast edges found on ESD pulses the power-to-ground clamp is actuated by charge-dumping capacitors. The schematic diagram is shown in Figure 10.

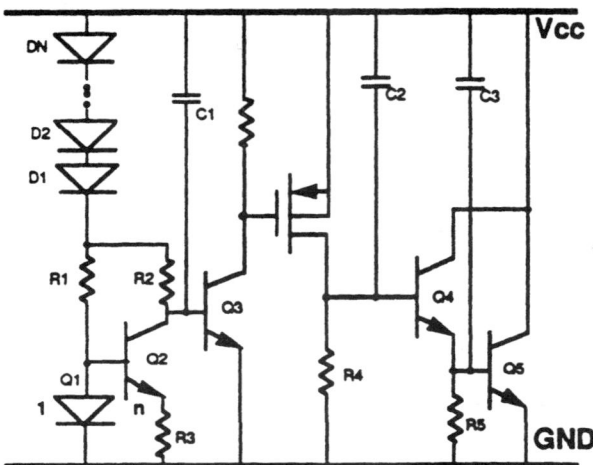

Figure 10: Active Clamp Schematic

During an ESD event the V_{CC}-GND voltage will experience a rapid increase which is coupled via C3 to the NPN Q5. Q5 turns on within 100ps and sinks the ESD current non-destructively to ground. If the ESD pulse lasts a little longer, then C2 will turn on Q4 which forms a high current Darlington with Q5. The clamp voltage at this time depends on the dynamic impedance of the Darlington, which can be designed to be well under 1Ω. The transistors Q4 and Q5 are ballasted to avoid thermal runaway. If the pulse lasts more than 200ns the bandgap reference will turn on and clamp the V_{CC}-GND potential to 9V. This voltage is set by Q1-Q3, diodes D1-DN, and the resistors R1-R3. Note that the ESD current is routed through large vertical NPN transistors from collector to emitter, which is inherently non-damaging for the NPN's.

Proper design and compensation of the bandgap reference maintains the maximum V_{CC}-GND voltage at the bandgap reference voltage or less, while the low dynamic resistance of the shunt diodes minimizes voltage drop. This power-to-ground circuit is called the "crowbar" because of its clamping nature. After the ESD pulse, the crowbar and diode network return unharmed to their normally OFF state.

To simulate an ESD event, a capacitor with initial conditions and a nonlinear controlled source can be used to emulate HBM discharge with a SPICE based simulator. Simulation of a discharge of 5KV into a test amplifier circuit at time t=0+ is shown in Figure 11. The 5μA current source charges up the 1GΩ resistor to 5KV and this is imposed on the 100pF capacitor. At time=0+ Vsw closes and G1 does as well, dumping the charge into the nodes selected.

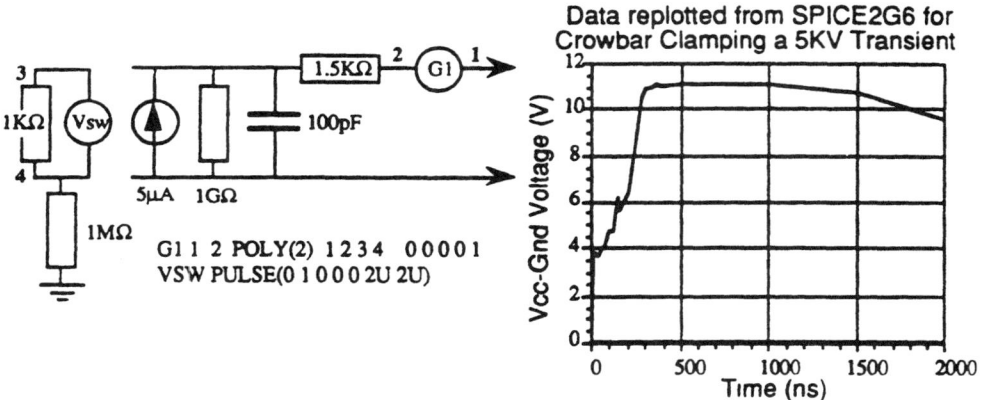

Figure 11: HBM Emulation and V_{CC} Transient from SPICE

Notice that even with a 5KV HBM input the crowbar clamp maintains on-chip V_{CC}-GND voltages under 11V (the bandgap reference voltage for this particular circuit), which is well under the 22V static gate-oxide breakdown voltage in the QUBiC process. Other pin-pin combinations will rise above the 11V level due to drops across the forward-biased diodes and metal interconnect, thus ohmic drops in the V_{CC} and GND metalization should be minimized for best ESD performance.

Application of the ESD Method to the Downconversion IC

Shunt diodes of $2 \times 100 \mu m^2$ size were connected from each pin to V_{CC} and bias ground. A 9.7V crowbar clamp was used from Vcc to bias ground. The other four ground pins were connected with back-to-back diodes to the bias ground and the mixer Vcc pin was connected with back-to-back diodes to V_{CC}.

Performance

The measured HBM ESD failure thresholds were typical in the 4-10KV region for most pin pairs and polarities. The MM ESD failure thresholds were generally in excess of 350V. A weak combination showing less than 1KV on the HBM test was initially detected and traced to a very small NMOS device used in the power-down circuitry that had its drain and source coming out to external pins. The correction for this problem was to eliminate this device since it turned out to be redundant. Had it been necessary, then a series drain ballasting resistor would have been used to limit the current into the device during an ESD transient.

11. Summary

The capabilities of an advanced, production worthy BiCMOS process have allowed the integration of a low-noise amplifier and single-balanced mixer to meet the low-cost demands of commercial cordless and cellular phones. Combinations of on- and off-chip shielding have yielded more than 45dB of antenna isolation from the local oscillator. Measurements have shown some weak areas of SPICE based simulation, particularly the inability to simulate mixer noise figure, and intermodulation intercept points for closely spaced input frequencies. The need for an accurate IC substrate model and a IC package model including mutual inductance was highlighted. To assemble a production worthy commercial design also takes extensive knowledge of ESD design techniques. Overall, the ability to integrate the "front-end" functions of a receiver on the QUBiC process will lead to higher integration levels since most active functions in a phone can now be integrated.

12. References

[1] J.L. DeJong, et.al. , "Single polysilicon advanced super high-speed BICMOS technology," IEEE Proceedings of the Bipolar Circuits and Technology Meeting, Sept. 1989.

[2] R.Milsom, et.al., "FACET - a CAE system for RF analogue simulation including layout", 1987 IEEE Design Automation Conference.

[3] C. Hull, "Analysis and optimization of monolithic RF downconversion receivers", Phd. Dissertation, University of California, Berkeley, 1992.

[4] Y. Fong and C. Hu, "The Effects of high electric field transients on thin gate oxide MOSFETs," 1987 EOS/ESD Symposium.

[5] W. Mack and R. Lane, "Protection device utilizing one or more subsurface diodes and associated method of manufacture, " U.S. patent 4,736,271, Apr. 1988.

[6] W. Mack and R. Meyer, "New ESD protection schemes for BiCMOS processes with application to cellular radio designs," IEEE Circuits and Systems Conference, May 1992.

A 1 GHz CMOS RF Front-End IC for a Direct-Conversion Wireless Receiver

Ahmadreza Rofougaran, James Y.-C. Chang, Maryam Rofougaran, and Asad A. Abidi, *Fellow, IEEE*

Abstract—An integrated low-noise amplifier and downconversion mixer operating at 1 GHz has been fabricated for the first time in 1 μm CMOS. The overall conversion gain is almost 20 dB, the double-sideband noise figure is 3.2 dB, the IIP3 is +8 dBm, and the circuit takes 9 mA from a 3 V supply. Circuit design methods which exploit the features of CMOS well suited to these functions are in large part responsible for this performance. The front-end is also characterized in several other ways relevant to direct-conversion receivers.

I. INTRODUCTION

MOTIVATED by the growing needs for low-power and low-cost wireless transceivers, mainstream IC technologies are competing to integrate more RF functions onto a single chip. Bipolar circuits dominate integration at 1 GHz today, followed by GaAs IC's. As recent results demonstrate, CMOS too is a viable contender in this frequency range [1]–[6]. If CMOS is shown to perform in certain important respects as well as circuits in other established technologies, and it successfully merges analog and digital blocks, its use at RF may become as compelling as it is in baseband circuits.

To date, most research on CMOS RF circuits shows the feasibility of elementary RF building blocks, such as stand-alone tuned amplifiers [1], mixer IC's [2], [5], and oscillators [7]. The next development step calls for the integration of these building blocks into subcells, comparable in function to currently available small-scale RF IC's in bipolar or GaAs technology. The most common example of such an IC is an RF low-noise amplifier (LNA) combined with a downconversion mixer, often labeled the *front-end* for an RF receiver. Integrated front-ends are widely used because they combine all the RF signal processing on one chip, often requiring only a small overhead of off-chip components, and they produce an amplified signal translated down to a conveniently low intermediate frequency (IF) at the output. Thereafter, it is relatively simple to implement IF and baseband circuits for the rest of the receiver. The work reported here is the first implementation of a 1 GHz front-end in CMOS.

The front-end of a wireless receiver must meet several exacting specifications. First is *sensitivity*. The input noise of the front-end must be sufficiently low to enable it to detect

Manuscript received January 23, 1996; revised March 5, 1996. This work was supported by the U.S. Advanced Research Projects Agency, Rockwell International, Texas Instruments, Harris Semiconductor, Advanced Micro Devices, Hewlett-Packard Company, and the State of California MICRO Program.

The authors are with the Integrated Circuits & Systems Laboratory, Electrical Engineering Department, University of California, Los Angeles, CA 90095-1594 USA.

Publisher Item Identifier S 0018-9200(96)04472-1.

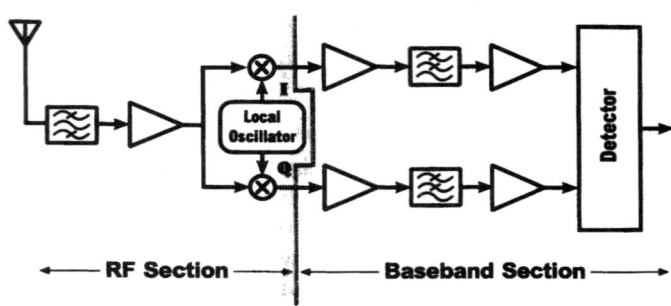

Fig. 1. A direct-conversion receiver suitable for FSK modulation.

weak input signals. The front-end gain must be high enough to overcome the noise contributions of later circuits, which may otherwise degrade the receiver sensitivity. Second, a front-end with a wide *input dynamic range* can tolerate large undesired signals nearby in frequency to a weak desired signal, which may otherwise, through intermodulation distortion, create energy at frequencies overlapping the desired channel. Third, the *RF input impedance* of the front-end must be a good match to the antenna characteristic impedance over the frequency band of interest.

The LNA and mixer together determine the performance of the front-end. For instance, although a large LNA gain is desirable as mentioned above, too large a gain may overload the mixer and compromise dynamic range. On the other hand, the gain must be large enough to overcome the fundamentally higher mixer noise. It is also desirable to connect the LNA in some simple way to the mixer input, without a power-hungry RF buffer. The front-end design is influenced by its intended use, as discussed below. Therefore, good performance is only obtained through careful *co-design* of the front-end building blocks.

This front-end is intended for a direct-conversion, or zero-IF, frequency-shift keying (FSK) receiver [8], which simplifies how the blocks are connected together (Fig. 1). In a superheterodyne receiver, a passive filter—usually a second preselect filter of the type connected to the antenna—follows the LNA [9] to suppress the amplifier noise in the image of the RF input band, where there is no signal. Without this filter, the downconverted signal must contend with downconverted noise from both the signal-bearing and the idle sidebands. In a direct-conversion receiver, on the other hand, the local oscillator (LO) is centered in the desired channel, so useful signal energy, and noise, occupy *both* upper and lower sidebands. As there is now no idle sideband to be filtered, the LNA is directly connected to the downconversion mixer. Therefore, when a

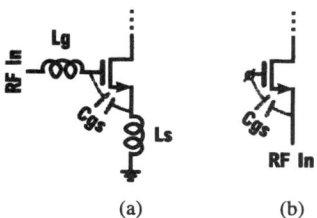

Fig. 2. Candidate FET LNA input stages. (a) Common-source stage, with lossless matching network. (b) Common-gate stage.

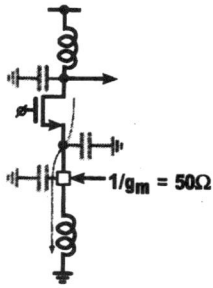

Fig. 3. Low-noise amplifier with tuned load and an off-chip tuning inductor at the input port.

modulation such as FSK permits use of direct-conversion [10], the receiver may be integrated with a need for very few off-chip components. There are several advantages to this, mainly small physical volume, less wasted power in buffering high-frequency signals off-chip, and lower assembly costs. However, the receiver requires a vector baseband signal path, consisting of two branches downconverted in quadrature to prevent the signal-bearing image sidebands from aliasing on one another.

II. CIRCUIT DESIGN

A. Low-Noise Amplifier

The LNA must simultaneously attain high RF gain, low input noise, and a good input match to 50 Ω. These requirements are often interdependent in a simple circuit, and may require iterative design for all to be fulfilled. The following discussion on a CMOS LNA design covers input impedance matching first, then input noise, and finally voltage gain.

The gate of a FET fabricated in 1-μm technology is capacitive to frequencies beyond 1 GHz.[1] However, a lossless *matching network*, consisting only of inductors and capacitors can transform the FET input into a pure resistance over some frequency band of interest. The most common matching network for FET's consists of a series feedback inductor, L_s, in the FET source, and another inductor, L_g, in series with the gate to tune out the capacitance C_{gs}, resulting in an input resistance $g_m L_s / C_{gs}$ where g_m is the FET transconductance [Fig. 2(a)] [11]. This method is preferred over resistor feedback because the matching network introduces no noise of its own. However, loss in practical inductors L_g and L_s will tend to degrade noise figure. With sufficiently good inductors, though, a noise figure well below 3 dB may be obtained with this technique, ultimately limited by such transistor imperfections as nonzero gate resistance.

When a noise figure of 3 dB is acceptable, as it is in our receiver, it is simpler to regulate the input impedance with a common-gate input stage [12] [Fig. 2(b)]. For the sake of discussion, first suppose that the load resistance at the drain is much less than the FET r_{ds}, and that $g_m r_{ds} \gtrsim 10$. Then the input resistance at the FET source is $1/g_m$. At 1 GHz, the FET C_{gs} and parasitic input capacitance, C_P, due to the bonding pads and external strays significantly shunt this resistance. Therefore, to achieve a good impedance match, the size and bias of the FET are selected for $1/g_m = 50$ Ω, and an inductor tunes out the shunt capacitance by parallel resonance in a frequency band around 1 GHz. As the capacitance at the LNA input is to be tuned, it makes good sense to do so with a grounded off-chip low-loss inductor, which also carries the LNA bias current.

Fortuitously, a FET with a small-signal channel resistance of 50 Ω produces a lower thermal noise current than a linear resistor of the same value [13]. The noise current spectral density in the FET is $4kT\gamma g_m$ A^2/Hz, where $\gamma \simeq 0.67$ owing to the distributed inversion layer. Thus, ruling out any other noise sources in a matched LNA, the noise figure due to the FET alone is $10 \log 1.67 = 2.2$ dB. In a short-channel FET biased at unfavorable conditions, hot-electron effects may augment this [14] to raise the noise figure. Flicker noise in the FET is unimportant at RF.

A tuned load peaks the frequency response of the LNA in the band of interest (Fig. 3), in effect transforming the inherent lowpass characteristic of the amplifier to a bandpass. The load also helps to reject out-of-band signals and noise. However, the LNA passband is seldom sufficiently flat and narrow for RF *preselection*, that is, to suppress image channels and out-of-band interferers. Rather, a sharply tuned discrete filter, such as SAW or dielectric resonator, is inserted before the LNA for this purpose. Discrete filters usually operate at a characteristic impedance of 50 Ω or 75 Ω. As explained above, no preselect filter need follow this LNA, nor the RF buffer required to drive the filter.

The tuned load, therefore, comprises an inductor resonating with the FET drain capacitance, C_D, and the sum, C_P, of the input capacitance of the subsequent downconversion mixer and any other parasitics. Another advantage of the common-gate stage is that the somewhat large C_{GD} of the FET returns its current to a fixed bias, rather than to the input node as it would in a common-source amplifier. This current undergoes rapid phase-shifts with frequency in an RF tuned amplifier, and makes it difficult to design an input matching circuit.

The inductance, L, to tune this total capacitive load to the resonant frequency ω_0, in our case 2π Grad/s, is

$$L = 1/\omega_0^2 (C_D + C_P). \qquad (1)$$

In most modern FET's, the drain junction capacitance $C_D \simeq C_{GS}$. Furthermore, the unity-current gain frequency, $\omega_T =$

---

[1]This assumes the FET is laid out sensibly. The gate resistance of a FET with a certain channel width is lowered by an interdigitated gate, whereas without such a layout, this resistance may dominate the FET input impedance at high frequencies.

g_m/C_{GS}. The inductance may then be expressed as

$$L = \frac{1}{\omega_0^2(g_m/\omega_T + C_P)}. \quad (2)$$

If inductor loss, as modeled by a series resistance R_S, limits the impedance of the tuned load at resonance, then using (2), the voltage gain of the common-gate stage is

$$\begin{aligned}\text{Gain} &= g_m \frac{(\omega_0 L)^2}{R_S} \\ &= \frac{\omega_T}{1 + \omega_T C_P/g_m} \frac{L}{R_S}. \end{aligned} \quad (3)$$

This form makes it clear which parameters are within the circuit designer's reach to determine RF gain. ω_T mainly depends on FET channel length, but is also controlled by gate bias. However, even with infinite ω_T, parasitic-related quantities will limit the maximum achievable gain to

$$\text{Max Gain} = \min\left(\frac{g_m}{C_P}\frac{L}{R_S}, g_m r_{DS}\right). \quad (4)$$

In the common-gate amplifier, the desired input impedance sets g_m. Thus, a large parasitic capacitance at the drain means a smaller achievable gain, unless the loss in the load inductor is somehow lowered to boost the gain. The relative quality of the inductor, L/R_S, depends on how it is physically realized, and there are limits to how large this may be in practice. For instance, at 1 GHz the L/R_S is 4 nH/Ω for a discrete 10 nH chip inductor meant for RF applications [15]. This argues for an on-chip inductor load, because it is unlikely that a discrete off-chip inductor can overcome, simply because of a higher quality, the RF gain loss due to the large parasitic capacitance of the bond pads, bondwires, package leads, and board traces.

The LNA fulfills its specifications at the price of power dissipation. At low values of gate drive voltage, long-channel FET laws fairly well describe the dependence of bias current, I_D, on the transconductance

$$I_D = \frac{1}{2}g_m(V_{GS} - V_t). \quad (5)$$

Similarly, at low gate drives, the ω_T of a FET theoretically follows the dependence

$$\omega_T = \frac{g_m}{C_{GS}} = \frac{\mu}{1 + \theta(V_{GS} - V_t)} \frac{(V_{GS} - V_t)}{L^2} \quad (6)$$

where θ captures how the inversion-layer mobility, μ, degrades with gate electric field [16]. While (5) is readily verified by measurement, there is little data in the literature on CMOS ω_T to validate (6). Therefore, the s-parameters of a single, large 1-μm NFET were characterized on a Cascade™ probe station, from which ω_T was deduced. The measured data (Fig. 4) shows that $\omega_T = 2\pi f_T$ conforms closely to (6) in the $V_{GS} - V_t$ range of interest. This curve serves as an important design aid.

Using (2), the load inductance may be calculated which tunes the LNA to a certain frequency in the absence of any significant parasitics. For instance, if the FET is biased at an f_T of 5 GHz, then the LNA requires a 40 nH inductor load to achieve a peak at 1 GHz in its frequency response. The inductor may be implemented on-chip as a square spiral in

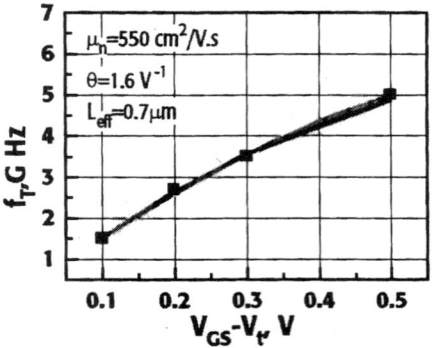

Fig. 4. Measured f_T versus $V_{GS} - V_t$ for MOSIS 1-μm NMOSFET (points) fitted to simple expression (line).

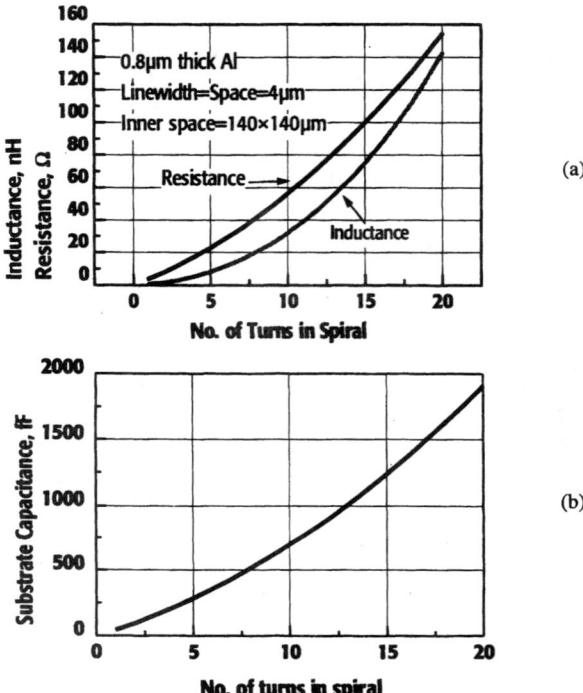

Fig. 5. Design curves for square spiral on-chip inductor. (a) Inductance and resistance of spiral versus number of turns and (b) capacitance of inductor (assuming it is an equipotential) to substrate through 1 μm thick field oxide.

Metal-2, with the return conductor in Metal-1. Greenhouse's formula accurately specifies how the inductance grows with the number of turns [17], while the series resistance accumulates with the number of squares of metal comprising the spiral [Fig. 5(a)]. The spiral grows outwards from a 140 μm square hole in the middle. Any turns within the hole would enclose relatively little magnetic flux, but would contribute a nonnegligible unwanted resistance. These curves show that in the inductance range of 40 to 50 nH, the relative quality, L/R_S, of the spiral is about 0.7 nH/Ω. It is seen from (3), (5), and Fig. 4 that the LNA may achieve a gain as large as 20 dB at 1 GHz in the absence of parasitic capacitance, while drawing 1.5 mA of current per FET.

The main impediment to a practical implementation of the tuned amplifier arises from the parasitic capacitance of a 50 nH spiral inductor to the semiconductor substrate. This is so large through the typical 1 μm-thick field oxide [Fig. 5(b)] that the spiral self-resonates at 700 MHz, and at 1 GHz appears as a capacitive, rather than inductive, load on the LNA. This is why it is generally believed that medium- to large-value inductors may only be integrated on semi-insulating substrates, while on a standard silicon substrate inductors no larger in value than 5 to 10 nH are usable at 1 GHz [18].

In earlier work, we have described a method to eliminate the parasitic capacitance under the spiral inductor by selectively removing the underlying silicon substrate [1]. This leaves the spiral encased in a layer of oxide suspended above an air-gap. Inductors as large as 100 nH may be fabricated in CMOS, whose self-resonance frequency, now limited by the small fringing capacitance through the air gap to the distant ground plane, lies beyond 2 GHz. This maskless, post-fabrication etching of the substrate does not require any foundry modifications in the 1-μm CMOS process available through MOSIS. Via holes which will expose the surface of the silicon after fabrication and passivation surround the spiral. A selective etchant removes the exposed silicon at a much higher rate than it does oxide and metal. After sufficient exposure to the etchant, a deep cavity forms under the inductor, while the remaining active area on the chip is left intact. A passivating coat protects the exposed silicon on the sides and back of the chip from inadvertent etching.

The substrate was removed in the earlier work by a liquid-phase, anisotropic etchant [1], resulting in a spiral inductor surrounded by large trapezoidal openings, and attached by four oxide bridges to the rest of the CMOS substrate. This has since been replaced by a gas-phase, isotropic etchant. Through small circular openings surrounding the spiral, the etchant now excavates hemispherical pits whose radius increases with exposure time. Etching is stopped when the multiple pits coalesce into one large pit with a depth of roughly half of one side of the spiral (Fig. 6). This suspended inductor enables wholly-integrated RF components in CMOS. In addition to the LNA, it is also used in the local oscillator, power amplifier, and even as a low quality bandpass filter. As a survey in a recent publication shows [19], no simpler method has yet been found to realize large-value integrated inductors.

It was assumed in the earlier analysis that the impedance, Z, of the LNA tuned load is much less than the FET r_{ds}. When this is not so, the expressions for the gain and input impedance must be modified to

$$\text{Gain} = \frac{g_m Z}{\left(1 + \frac{Z}{r_{ds}}\right)}; \qquad r_{\text{in}} = \frac{1}{g_m}\left(1 + \frac{Z}{r_{ds}}\right). \qquad (7)$$

The complete LNA is a balanced circuit (Fig. 7). A power-conserving, passive balun converts a single-ended antenna signal into a balanced input drive to the LNA. A printed-circuit balun may even be integrated into the transceiver case. The bias, VG, at the common-gate FET's regulates the LNA input impedance. An on-chip scaled-down replica circuit stabilizes this impedance against variations in process and temperature as follows. An op-amp drives the VG bias voltage of the replica to servo the dc value of $1/g_m$ to an off-chip reference

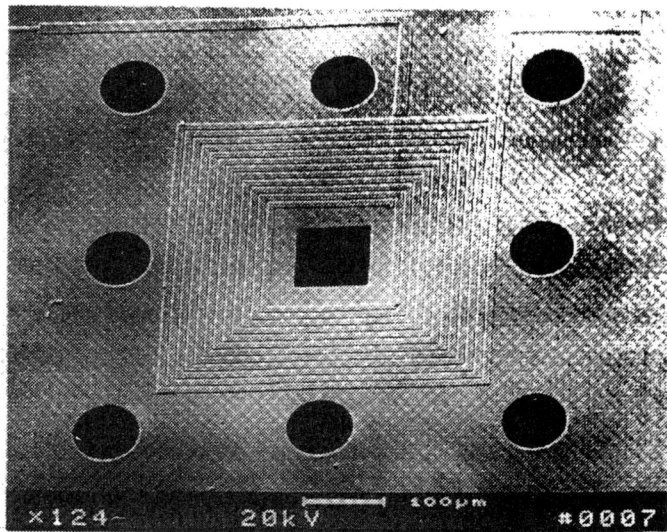

Fig. 6. Suspended 50 nH spiral inductor over a hemispherical cavity etched underneath through surrounding via holes.

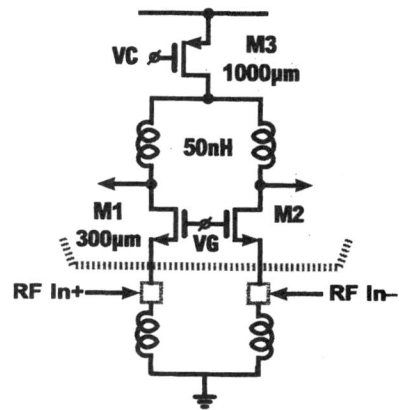

Fig. 7. Low-noise amplifier circuit diagram.

resistance, and the same voltage is then applied to the main LNA, thereby regulating its input impedance to within the FET matching in the replica and the main circuits. As the feedback loop bandwidth is well below 1 MHz, the op-amp does not contribute any RF noise. The inductor loads on each half of the circuit share a common top-node, which connects to the 3 V power supply through a triode-region PFET. By adjusting the gate voltage, VC, the dc voltage drop across this PFET resistor may be changed, and this sets the dc level at the LNA output.

The LNA FET's are biased at $V_{GS} - V_t = 0.35$ V, and drain 2.2 mA each. Taking into account the capacitive load of the mixer, the LNA requires 50 nH load inductors to obtain a peak at 1 GHz in its frequency response. From (4) and Fig. 4, this means that in the absence of parasitics ($C_P = 0$), the peak RF gain is at most 26 dB.

B. Downconversion Mixer

The amplified signal at the LNA output is converted down in frequency for further amplification, channel-select filtering, and detection. The frequency mixer is an integral part of the RF front-end.

There are two fundamentally different ways to implement a mixer in CMOS. The first, somewhat unconventional method, uses a MOSFET as a wideband analog switch. A passive track-and-hold circuit, consisting of a 1-μm FET switch and a grounded capacitor, is designed for a track-mode bandwidth of greater than 1 GHz. This follows the waveform of a modulated 1 GHz carrier and samples it at a much lower rate, which must be at least twice the *modulation* bandwidth. An op-amp feedback circuit clocked at this low sample rate buffers the held output, and removes signal-dependent charge injected by the switch. When the sample rate is an integer submultiple of the carrier frequency, the interpolated samples directly downconvert the RF signal to dc. A prototype evaluated at a 900 MHz RF [2] shows very good linearity, but fundamentally suffers from a large noise figure, because while tracking the narrowband signal, it also tracks and aliases wideband noise. This, and the difficulty of buffering such a switched mixer to the inductive load of an integrated LNA, make it inappropriate in a sensitive receiver.

The second, more conventional, mixer resembles the Gilbert analog multiplier. It consists of a linear RF voltage-to-current (V-I) converter, or RF transconductor, whose output current is commutated by the local oscillator (LO). As commutation conserves the total current, it downconverts a fraction of the RF current to the IF, and the remaining RF current upconverts around one or more harmonics of the LO. The voltage conversion-gain of the active mixer is independently set by choice of the transconductance and load resistance. The internal current conversion-loss penalizes mixer noise, as is analyzed in a later section.

A good mixer is highly linear, and its input-referred noise does not overwhelm the amplified noise of the preceding LNA. The mixer handles larger signals than the LNA, and therefore its nonlinearity must be lower by at least a factor of the LNA gain if it is not to become the bottleneck to receiver dynamic range. This is why the following discussion concentrates on mixer nonlinearity, as the LNA, with the choice of bias voltages, is not the bottleneck to front-end linearity.

Third-order intermodulation distortion in a double-balanced mixer may cause two large adjacent-channel signals to create energy at spurious frequencies coincident with a weak desired channel. The linearity of a front-end is specified by the input-referred third-order intercept point (IIP3) [20]. Often this is set by the static and dynamic nonlinearity in the RF V-I converter of the mixer, or by the static nonlinearity in the mixer load.

Linear MOS transconductance circuits have been studied extensively in the context of continuous-time active filters [21] operating at frequencies up to tens of MHz. These circuits exploit the property that the dominant second-order nonlinearity in a MOSFET circuit cancels in balanced differential inputs and outputs. For instance, if two identical common-source FET's conforming to the classic long-channel I-V characteristics are biased at some V_{GS} [Fig. 8(a)] and excited differentially by a large signal V_{in} whose amplitude is less

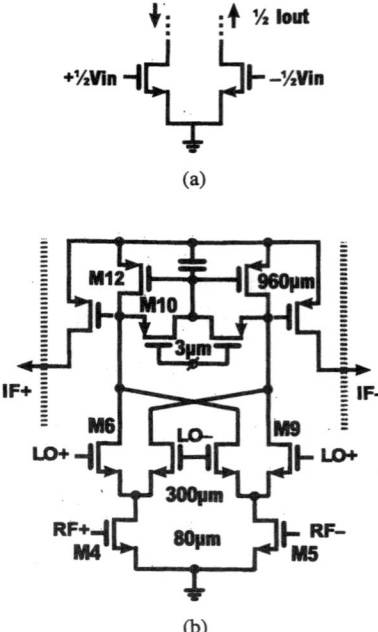

Fig. 8. (a) Linear MOS transconductor. (b) Downconversion mixer circuit diagram.

than $2(V_{GS} - V_t)$, then the differential output current

$$I_{out} = \mu C'_{ox}(W/L)(V_{GS} - V_t)V_{in} \qquad (8)$$

depends *linearly* on V_{in}, and the bias $V_{GS} - V_t$ sets the transconductance [22].

Residual third-order nonlinearity produces a small distortion in the transconductor. Its large signal handling is limited by clipping when an input swing of $V_{GS} - V_t$ turns off one of the FET's in the circuit. These sources of static nonlinearity are expected to govern the mixer up to and beyond 1 GHz, as no significant nonquasi-static effects are likely to set in given the short carrier transit time in the 1-μm channel. Dynamic nonlinear currents which grow with frequency will flow in any voltage-dependent FET capacitance and might even become the significant form of distortion at 1 GHz. The MOSFET, however, is benign in this respect, as its main capacitance, C_{GS}, is relatively independent of bias for $V_{GS} > V_t$ and behaves like a linear capacitor in the saturation region of operation.

The MOS downconversion mixer is a balanced circuit [Fig. 8(b)] comprising a linear common-source FET transconductor (as opposed to a differential pair in the bipolar Gilbert multiplier), four commutating FET switches, and a high-swing load consisting of a center-tapped FET resistor across pull-up current sources. Common-mode feedback from the center tap biases the current sources at a well-defined voltage. The LNA output is directly connected into the differential mixer input.

The mixer attains its peak conversion gain when a sinewave of at least +5 dBm (1 V ptp) is applied to the commutating switches $M6$–$M9$. This also lowers the total front-end noise figure. While it is obvious that incomplete commutation leads to conversion loss, what may not be evident is how it also degrades noise figure. The transconductor FET's and the loads clearly contribute noise in the mixer. In addition, the commutating switch FET's also contribute noise at discrete time intervals over the switching cycle. The time-varying aspects of the circuit must be understood to estimate the magnitude of this contribution. Over most of the duty cycle, one FET in each switched pair appears as a resistor in series with transconductor, while its companion FET is OFF. The series FET contributes very little noise, determined only by the finite output impedance of the transconductor FET's at 1 GHz. However, during a zero-crossing of the LO, the two companion switch FET's carry comparable drain currents, and act, for the purposes of noise analysis, like a differential pair amplifier. They then contribute a short burst of balanced noise current to the mixer load with a spectral density proportional to the switch FET transconductance, elevating the average mixer output noise [23]. A larger LO drive to the mixer forces zero crossings with a greater slope, and as the switches now dwell for a shorter fraction of the period in the high noise condition, the overall mixer noise is lowered.

It is a good assumption that under large LO drive, the mixer commutates the RF transconductance current with a square wave. Simple expressions may now be derived for the mixer *conversion gain* and its equivalent *input noise*. Referring to Fig. 8(b), suppose a unit sinusoidal input voltage of frequency ω_{RF} is linearly converted to a current, and commutated by the switches at ω_{LO}, which amounts to multiplying the sinusoidal current by a square wave, $sq(\omega_{LO}t)$, alternating between $+1$ and -1. The current flowing into the loads, I_{IF}, is then

$$I_{IF} = g_{m4} \sin \omega_{RF} t \times sq\omega_{LO}t$$
$$= g_{m4} \sin \omega_{RF} t \times (4/\pi)\left(\sin \omega_{LO}t + \frac{1}{3}\sin 3\omega_{LO}t + \cdots\right)$$
$$= (2/\pi)g_{m4}\cos(\omega_{RF} - \omega_{LO})t \quad (9)$$

where the square wave is expanded as a Fourier series, and the term containing the downconverted frequency at $\omega_{RF} - \omega_{LO}$ is retained. Equation (9) shows a current conversion loss of at least $2/\pi (=0.64)$ through this mixer. The overall mixer voltage gain is

$$\text{Mixer Gain} = (2/\pi)g_{m4}R_L \quad (10)$$

which may be adjusted to any reasonable value by the load resistance, R_L, attached to the low-frequency output node. Noise due to the mixer loads is referred to the input in the following way: the noise current is divided by the *conversion loss*; the noise spectrum is *translated* from ω_{IF} to ω_{RF}; and the noise is distributed equally between *two image sidebands* around ω_{LO} which will downconvert to the same ω_{IF}.[2] Suppose that the transconductor and load FET's ($M12, M13$) [Fig. 8(b)] produce only thermal (white) noise, that the commutating switch FET's contribute no noise, and that the noise due to the spectral density FET resistor loads ($M10, M11$) is negligible. Then the noise referred to the mixer input is

$$\hat{v}_n^2 = 4kT\gamma\left(\frac{1}{g_{m4}} + \frac{1}{2}\left(\frac{\pi}{2}\right)^2 \frac{g_{m12}}{g_{m4}^2}\right). \quad (11)$$

[2] This assumes that the conversion gain is equal from both sidebands, which is a very good approximation for low ω_{IF}.

As $M4$ and $M12$ share the same bias current, it follows that with the FET sizes used in this circuit

$$\frac{g_{m12}}{g_{m4}} = \sqrt{\frac{\mu_P W(M12)}{\mu_N W(M4)}} \approx 2. \quad (12)$$

From (11) and (12), this means that noise voltage referred at the mixer input is $1.8\times$ larger than if the input FET's ($M4$) were the sole source of noise. In the case of direct-conversion where $\omega_{IF} = 0$, flicker noise in the load will further elevate, indeed dominate, the mixer input noise.

The NFET's at the mixer input are biased at $V_{GS} - V_t = 0.75$ V, and drain 2.4 mA each from the 3 V supply. From (11), the referred input noise voltage level of this mixer is about 3.5 nV/$\sqrt{\text{Hz}}$, which is overwhelmed by LNA output noise of roughly 12 nV/$\sqrt{\text{Hz}}$ due to its own FET's. We conclude that the mixer does not appreciably degrade the noise figure of the front-end.

The $V_{GS} - V_t$ of the pull-up PFET current sources is also 0.75 V, and as the signal level at the mixer input is $1.4\times$ at the output, clipping will commence at the transconductor input. It may be shown with straightforward calculus that the rms value of a sinewave falls by 1 dB when it is clipped to 81% of its undistorted amplitude. This predicts that at the above bias points, the conversion gain of the front-end will compress by 1 dB due to clipping at the mixer input when a balanced sinewave of about -4 dBm is applied to the LNA.

For the purposes of standalone testing, open-drain PFET's connect to the mixer output to drive off-chip loads. They require an additional bias current, which is high enough to enable them to drive a large power level into the 50 Ω impedance of measuring instruments. However, in the intended on-chip use, they will be radically scaled down in size to drive a small capacitive load. The signal from the mixer output onwards is at a low or even a zero IF, so considerations of noise, not bandwidth, set the power dissipation in the subsequent stages.

III. IMPLEMENTATION AND MEASUREMENTS

A. Layout Issues

The RF front-end, comprising the LNA and mixer, is laid out for fabrication in a standard two-level metal, 1-μm CMOS process offered by MOSIS. The two on-chip spiral inductors in the LNA dominate the 1.3×2 mm active area of the die (Fig. 9). There are two features of note on this layout. First, circular via-holes surround the inductors to expose the silicon substrate after fabrication and passivation. Second, the LNA input pads are unusual. The standard MOSIS pad consists of 100-μm squares of Metal 2 and Metal 1 shorted together. This pad is, however, unsuitable for RF applications, because it is capacitively coupled through the oxide to the nonzero spreading resistance of the grounded silicon substrate. At 1 GHz, the pad impedance is mainly resistive and about 50 Ω. This parasitic resistance upsets the input impedance matching, but more seriously, it is a significant source of thermal noise. In fact, due to the pad alone, the LNA noise figure would

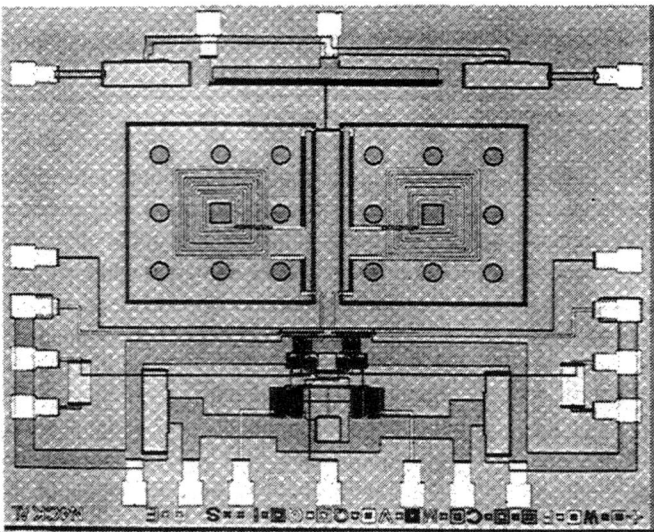

Fig. 9. Die photograph of front-end IC, consisting of LNA and downconversion mixer.

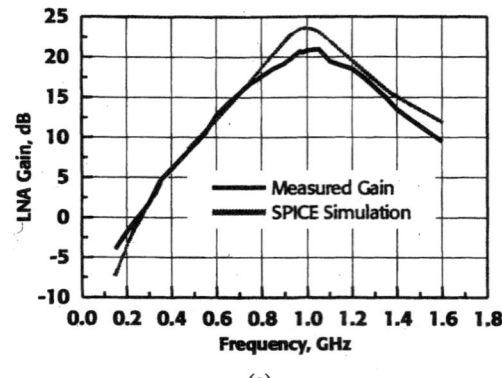

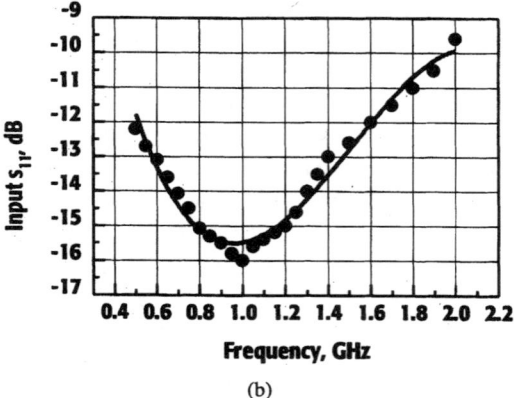

Fig. 11. Measured characteristics of low-noise amplifier. (a) Gain versus frequency, compared with SPICE simulation. (b) Input reflection coefficient versus frequency.

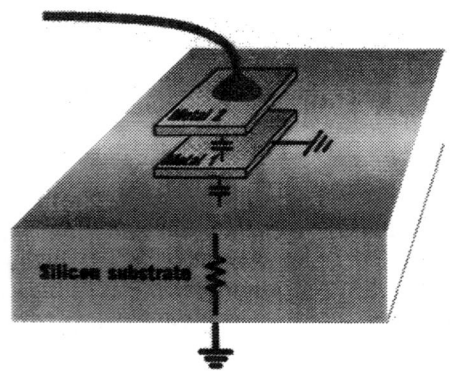

Fig. 10. Structure of RF input pad to LNA. Metal 1 shields the input signal from spreading resistance of substrate.

be 3 dB before any FET noise is taken into account. This problem always appears on silicon substrates at RF, although its magnitude varies with substrate doping. It has been noted before in discrete RF bipolar transistors [24].

The pad is modified to shield the RF input from this parasitic resistance. Instead of shorting the two layers of metal, Metal 1 is connected to a high-frequency ground, and the RF input pad is composed of Metal 2 alone, which also routes the signal from the chip periphery to the input FET's (Fig. 10). The pad impedance is now purely the capacitance between Metal 2 and Metal 1, and is no longer a source of noise. The off-chip inductor tunes out the pad capacitance with the various other capacitances that appear at the input node. This is a satisfactory though fragile solution, in that the pad is more susceptible to damage through the weak intermetal dielectric during wirebonding, and it cannot tolerate the loading of standard electrostatic protection devices. An alternative is to short the two layers of metal, but then place a heavily diffused layer under them which is connected either to ground, or stays at virtual ground when a balanced signal is applied to adjacent pads.

B. Experimental Results

All the following measurements are made with the chip mounted in a ceramic microwave package, to which a low-loss, power-conserving balun applies a balanced RF stimulus from a single-ended signal source. A low frequency power-combiner converts the balanced output into a single-ended signal for measurement. In most cases, the LO is offset from the RF input to produce a 10 MHz IF, which makes it easy to use ac-coupled instruments. However, the circuit is also evaluated in selected ways as a direct-conversion front-end. The LNA frequency response [Fig. 11(a)] is deduced from measurements on the overall conversion gain, by accounting for the mixer gain from transient simulations. SPICE, with standard static FET models and a simple three-element inductor model [1], predicts the LNA response very well. The scattering parameter s_{11} measures the input reflection coefficient, and thus the quality of the LNA input impedance match [Fig. 11(b)]. An s_{11} of -16 dB at 1 GHz is satisfactory for many applications, and implies a net input resistance of about 70 Ω, consistent with the FET g_m.

Another relevant attribute of the front-end is its ability to handle large signals without harmful distortion. Two equal-strength, closely-spaced tones around 1 GHz are applied to the LNA, and the strength of the corresponding downconverted tones f_1, f_2 is measured, as well as the strength of the spurious tones at $2f_1 - f_2$ and at $2f_2 - f_1$ produced by third-order

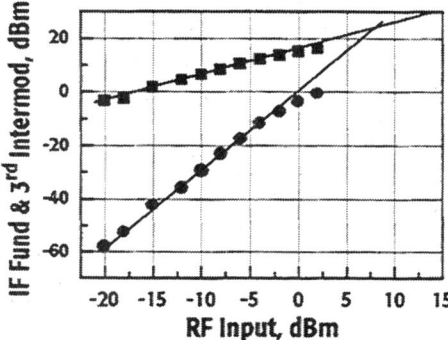

Fig. 12. Linearity of combined LNA and mixer, measured with two-tone input in LNA passband.

intermodulation. Whereas the main tones grow proportionally with the RF input amplitude, the distortion terms grow as the third power, and when extrapolated on logarithmic scales, the spurious tones grow as large as the main downconverted tones at an input of +8 dBm (Fig. 12). This is the input-referred third-order intercept point (IIP3). In practice, the LNA gain will collapse at a lower input, and although the IIP3 is the standard for specifying linearity, the 1-dB compression point better expresses the limits to linear operation. This is the input level, about −3 dBm for this circuit, at which conversion gain falls by 1 dB because of the onset of clipping. This is very close to the −4 dBm predicted above.

It is something of a challenge to measure the noise-figure of a direct-conversion receiver. A noise figure meter such as the HP 8970B can only measure noise at 10 MHz IF or above, yet its accuracy is unmatched by any other technique, because it simultaneously measures output noise and conversion gain. This meter is therefore first used to calibrate the noise-figure of the front-end at a 10 MHz IF. It measures the output IF noise levels in response to two reference levels of wideband noise spanning 10 MHz to 1.6 GHz. The correct double-sideband (DSB) noise figure is obtained when there is no RF preselect filter [20], because just as in the intended use for direct-conversion, *both* image sidebands around the LO contribute noise, as well as providing the input stimulus for gain measurement. Then, after grounding the input of the front-end through a 50 Ω resistor and applying a 1 GHz LO, the output noise spectral density is measured on a spectrum analyzer in the frequency interval from 0 to 10 MHz. If necessary, a wideband amplifier with a known gain and a bandwidth of more than 10 MHz is used to boost the front-end noise above the measurement floor of the spectrum analyzer. Any inaccuracy in the figure for double-sideband conversion-gain of the system is corrected by matching the input-referred noise spectral density at 10 MHz with the reading from the noise-figure meter. The noise-figure is then extrapolated to an arbitrary low frequency from the readings on the spectrum analyzer.

A DSB noise figure of 3.2 dB is measured on the noise-figure meter at a 10 MHz IF, which compares very favorably with the theoretical 2.9 dB noise figure of a single FET with a common-gate input impedance of 70 Ω [Fig. 13(a)]. In typical fashion of narrowband amplifiers, the noise figure rises as the

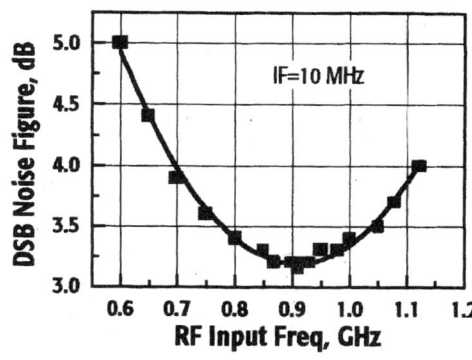

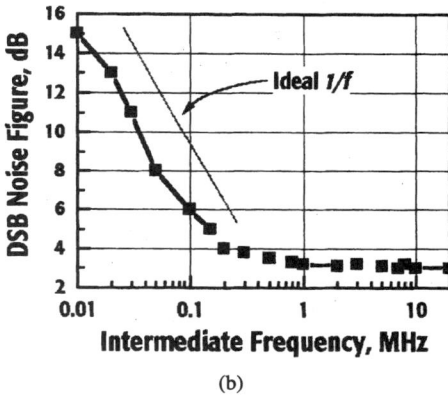

Fig. 13. Noise characteristics of front-end. (a) Direct noise figure measurement at 10 MHz IF. (b) Noise figure deduced from output noise spectrum at frequency offsets from zero IF.

matching deteriorates away from the tuned frequency of the input matching circuit—in this case, 0.9 GHz—and this is exacerbated by the declining LNA gain on either side of 1 GHz [Fig. 11(a)], when the mixer contributes more noise. Noise figure at low IF is deduced from spectrum analyzer measurements. Flicker noise now appears [Fig. 13(b)], contributed by the mixer pull-up current sources ($M12, M13$), and the PFET output buffers ($M14, M15$). The load FET's ($M10, M11$) do not carry a bias current, and therefore contribute no flicker noise. After direct downconversion in the intended receiver [8], the peaks in the spectrum of a 160 kb/s FSK signal lie at ±160 kHz, where flicker noise raises the front-end DSB noise figure to about 4.5 dB.

Second-order nonlinearity in the baseband section of a front-end also impairs a direct-conversion receiver [10], [25]. It detects the envelope of any unwanted AM signal in the LNA passband, and creates spurious energy at dc which overlaps, and may even overwhelm, the downconverted desired signal at dc. A fully-balanced circuit mitigates this to a large extent. The residual distortion is characterized by downconverting a 918 MHz carrier, which is amplitude-modulated at a 100% index by a 150 kHz tone, to an 18 MHz IF, and observing the spurious energy appearing at 150 kHz. The spurious tone grows with the second power of the carrier amplitude, and the magnitude of this nonlinearity is characterized by a second-order intercept, which is about +25 dBm referred to the input

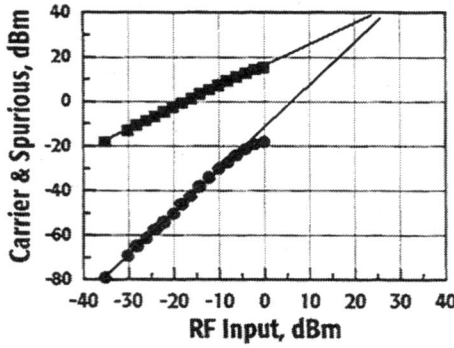

Fig. 14. Second-order intercept plots growth of detected envelope at dc versus AM signal strength. This is of importance in a direct-conversion receiver.

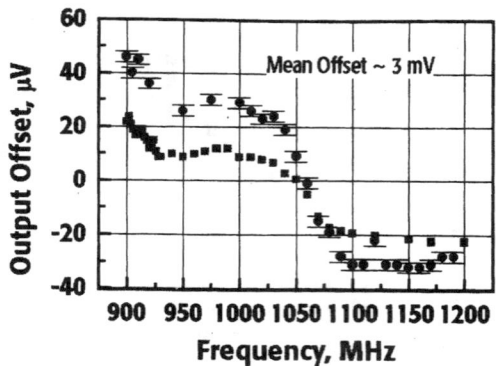

Fig. 15. Change in output dc offset as LO frequency is varied, as measured on two samples of the front-end IC.

TABLE I
RF FRONT-END IC SPECIFICATIONS

RF Input	1 GHz nominal (intended for 902–928 MHz)
LNA Gain	22 dB at 1 GHz
Mixer Gain	−3 dB
DSB Noise Figure	3.2 dB at 0.9 GHz
	4.5 dB at 160 kHz
IIP3	+8 dBm
Current Drain	9 mA from 3 V
Technology	1-μm CMOS
Active area	1.3×2 mm

of this circuit (Fig. 14). This is much higher than the IIP3, and therefore is not expected to be a major limitation to receiver dynamic range.

Finally, the dc offset at the output of a zero-IF receiver appears in the middle of the downconverted signal spectrum. This offset will overwhelm the received signal, unless removed with capacitive coupling or nulled by some form of digital offset estimation [10]. These remedies are, however, only partly effective in the case of a *dynamic offset*. For instance, the offset may consist of a component due to self-downconversion of LO leakage [10], and in a frequency-hopped receiver such as ours [8], this will inevitably change with the instantaneous LO frequency. The output offset has been measured on two IC's over a broad sweep of the LO (Fig. 15). Interestingly, it remains constant to within 5 μV over the sweep from 950 to 1025 MHz, but then changes rapidly by up to 60 μV in the vicinity of 1050 MHz. Over the 902–928 MHz ISM band, the output offset changes by about 10 μV. The methods to remove offset described above cannot easily track the variations produced by an LO hopped by a pseudonoise code, but it is expected that the spectrum of the varying offset will be noiselike, and given the small size of the dynamic offset, it will only slightly raise the receiver noise floor. It is not clear what causes the offset to depend so sharply on frequency, but parasitic resonances on-chip and in the package are suspected.

IV. CONCLUSION

A 1 GHz RF front-end IC, comprising a low-noise amplifier and downconversion mixer, has been designed and fabricated in 1-μm CMOS. This will be integrated with the baseband portions of a direct-conversion receiver, all sharing a common CMOS substrate. We have applied a design style which uniquely exploits CMOS capability to implement key RF functions. Combining this with a new on-chip inductor technology, and taking into account the receiver architecture, we have demonstrated a fully integrated 1 GHz front-end in a modest 1-μm CMOS process (Table I) which, in some respects, exceeds the performance of similar circuits fabricated in other well-established RF technologies [26]. We are hopeful that this work opens new vistas for today's predominant IC technology, CMOS, in an application believed to have ubiquitous importance in the future.

REFERENCES

[1] J. Y.-C. Chang, A. A. Abidi, and M. Gaitan, "Large suspended inductors on silicon and their use in a 2-μm CMOS RF amplifier," *IEEE Electron Device Lett.*, vol. 14, no. 5, pp. 246–248, 1993.
[2] P. Y. Chan, A. Rofougaran, K. A. Ahmed, and A. A. Abidi, "A highly linear 1-GHz CMOS downconversion mixer," in *European Solid-State Circuits Conf.*, Sevilla, Spain, 1993, pp. 210–213.
[3] M. Rofougaran, A. Rofougaran, C. Olgaard, and A. A. Abidi, "A 900 MHz CMOS RF power amplifier with programmable output," in *Symp. on VLSI Circuits*, Honolulu, 1994, pp. 133–134.
[4] A. Rofougaran, J. Y.-C. Chang, M. Rofougaran, S. Khorram, and A. A. Abidi, "A 1 GHz CMOS RF front-end IC with wide dynamic range," in *European Solid-State Circuits Conf.*, Lille, France, 1995, pp. 250–253.
[5] J. Crols and M. Steyaert, "A 1.5 GHz highly linear CMOS downconversion mixer," *IEEE J. Solid-State Circuits*, vol. 30, no. 7, pp. 736–742, 1995.
[6] ———, "A fully integrated 900 MHz CMOS double quadrature downconverter," in *Int. Solid-State Circuits Conf.*, San Francisco, pp. 136–137, 1995.
[7] J. Craninckx and M. Steyaert, "A CMOS 1.8 GHz low-phase-noise voltage-controlled oscillator with prescaler," in *Int. Solid-State Circuits Conf.*, San Francisco, 1995, pp. 206–207.
[8] J. Min, A. Rofougaran, H. Samueli, and A. A. Abidi, "An all-CMOS architecture for a low-power frequency-hopped 900 MHz spread-spectrum transceiver," in *Custom IC Conf.*, San Diego, CA, 1994, pp. 379–382.
[9] A. A. Abidi, "Low-power radio-frequency IC's for portable communications," *Proc. IEEE*, vol. 83, no. 4, pp. 544–569, 1995.

[10] ——, "Direct-conversion radio transceivers for digital communications," *IEEE J. Solid-State Circuits*, vol. 30, no. 12, pp. 1399–1410, 1985.
[11] R. E. Lehmann and D. D. Heston, "X-band monolithic series feedback LNA," *IEEE Trans. Microwave Theory Tech.*, vol. MTT-33, no. 12, pp. 1560–1566, 1985.
[12] T. Okanobu, H. Tomiyama, and H. Arimoto, "Advanced low-voltage single chip radio IC," *IEEE Trans. Consumer Electron.*, vol. 38, no. 3, pp. 465–475, 1992.
[13] P. R. Gray and R. G. Meyer, *Analysis and Design of Analog Integrated Circuits*, 3rd ed. New York: Wiley, 1993.
[14] A. A. Abidi, "High frequency noise measurements on FET's with small dimensions," *IEEE Trans. Electron Devices*, vol. ED-33, pp. 1801–1805, 1986.
[15] M. Sakakura and S. Skiest, "Ultra-miniature chip inductors serve at high frequency," in *J. Electron. Eng.*, pp. 48–51, Dec. 1993.
[16] P. K. Ko, "Approaches to scaling," in *Advanced MOS Device Physics, VLSI Electronics: Microstructure Science*, N. G. Einspruch and G. S. Gildenblat Eds. San Diego, CA: Academic, vol. 18, p. 12, 1989.
[17] H. M. Greenhouse, "Design of planar rectangular microelectronic inductors," *IEEE Trans. Parts, Hybrids, Packag.*, vol. PHP-10, no. 2, pp. 101–109, 1974.
[18] N. M. Nguyen and R. G. Meyer, "A silicon bipolar monolithic RF bandpass amplifier," *IEEE J. Solid-State Circuits*, vol. 27, no. 1, pp. 123–127, 1992.
[19] J. N. Burghartz, M. Soyuer, K. A. Jenkins, and M. D. Hulvey, "High-Q inductors in standard silicon interconnect technology and its application to an integrated RF power amplifier," in *Int. Electron Devices Mtg.*, Washington, DC, 1995, pp. 29.8.1–29.8.3.
[20] S. A. Maas, *Microwave Mixers*, 2nd ed. Boston: Artech House, 1993.
[21] Y. P. Tsividis, "Integrated continuous-time filter design—An overview," *IEEE J. Solid-State Circuits*, vol. 29, no. 3, pp. 166–176, 1994.
[22] K. Bult and H. Wallinga, "A class of analog CMOS circuits based on the square-law characteristic of an MOS transistor in saturation," *IEEE J. Solid-State Circuits*, vol. SC-22, no. 3, pp. 357–365, 1987.
[23] R. G. Meyer and W. D. Mack, "A 1-GHz BiCMOS RF front-end IC," *IEEE J. Solid-State Circuits*, vol. 29, no. 3, pp. 350–355, 1994.
[24] N. Camilleri, J. Kirschgessner, J. Costa, D. Ngo, and D. Lovelace, "Bonding pad models for silicon VLSI technologies and their effects on the noise figure of RF NPN's," in *Microwave & Millimeter-Wave Monolithic Circuits Symp.*, San Diego, CA, 1994, pp. 225–228.
[25] C. Takahashi, R. Fujimoto, S. Arai, T. Itakura, T. Ueno, H. Tsurumi, H. Tanimoto, S. Watanabe, and K. Hirakawa, "A 1.9 GHz Si direct conversion receiver IC for QPSK modulation systems," in *Int. Solid-State Circuits Conf.*, San Francisco, 1995, pp. 138–139.
[26] TriQuint Semiconductor, "TQ9203J downconverter application note," p. 7, 1994.

A 100 MHz IF Amplifier/Quadrature Demodulator for GSM Cellular Radio Mobile Terminals

I. A. Koullias, S. L. Forgues, and P. C. Davis

AT&T Bell Laboratories
2525 North 12th Street
Reading, Pennsylvania 19612-3566

Abstract -A circuit including a 100 MHz IF amplifier with a digitally controlled gain of 0-to-45 dB, a quadrature phase-shifter, balanced mixers, and active/sleep mode capability is described. The measured gain and quadrature phase variation was less than 1 dB and 1° respectively. The 92 mW circuit was fabricated using a 4.0 GHz complementary bipolar process.

INTRODUCTION

GSM (Groupe Special Mobile) is a new European digital cellular radio standard which uses frequency division multiplexing similar to standard radio. However, each "station" contains not one, but eight, digitally-encoded voice-band channels that are time-division multiplexed onto a 270 kbits/sec baseband [1]. In the receive path of a GSM mobile terminal, as shown in Figure 1, the RF amplifier and mixer section down-converts a GMSK (Gaussian Minimum Shift Keying) signal to IF. The IF GMSK signal is then amplified and demodulated to baseband by an IF amplifier/quadrature demodulator IC. The output of this circuit drives the baseband signal processor.

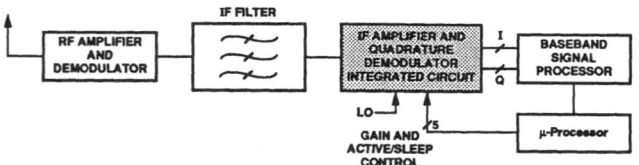

FIGURE 1 - GSM RECEIVE PATH BLOCK DIAGRAM

A "ping-pong" receive/transmit approach for each multichannel station is used, whereby the receiver and transmitter are alternately switched on for short bursts. The burst-mode operation presupposes a new approach to AGC which requires digital gain control. In a conventional radio system, the desired analog control voltage is stored on a capacitor. In GSM, it is stored in memory by a microprocessor so the receiver may start each burst of a series at the correct gain level. The gain is changed between bursts to avoid the "clicks" of digital AGC. As required by the GSM system, digital control provides the ability to measure the absolute level of the received RF signal. The accuracy is limited by the tolerance of the gain, which is allocated to be ±3 dB for the IF amplifier/demodulator portion of the receiver path.

In this paper, an integrated circuit that provides the ability to perform digital AGC and GMSK demodulation is presented. The IC includes a 100 MHz digitally-controlled gain IF amplifier that provides 0-to-45 dB of gain, selectable in 3 dB steps, and a quadrature demodulator. In addition, an active/sleep mode control is provided.

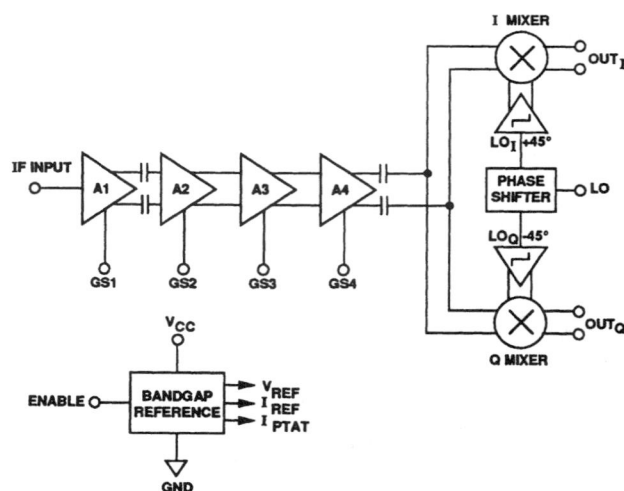

FIGURE 2 - IF AMPLIFIER/QUADRATURE DEMODULATOR BLOCK DIAGRAM

CIRCUIT ARCHITECTURE

The circuit block diagram is shown in Figure 2. The digitally-controlled-gain IF amplifier consists of four dual-gain-state differential amplifiers with gains of 24/0 dB, 12/0 dB, 6/0 dB, and 3/0 dB. The series combination, A1, A2, A3, and A4, provides a total gain of 0 dB to 45 dB selectable in 3 dB steps. The amplifier string has a bandwidth of 100 MHz. On-chip capacitive coupling eliminates offset amplification effects.

The IF signal is down-converted to baseband with a quadrature demodulator. A phase-shifter provides ±45° local oscillator (LO) signals to drive the in-phase (I) and quadrature (Q) mixers. The limiting amplifier preceding each mixer provides a constant amplitude signal to minimize gain sensitivity to amplitude variations of the externally provided LO signal. The I and Q mixers are designed to have a fixed conversion gain of 6 dB.

The active/sleep mode is controlled by a digital enable input on the bandgap reference which determines the biasing for the whole circuit when in the active mode. In the sleep mode, all the bias currents, including those required by the reference and enable circuitry, are turned off. Therefore, the circuit does not dissipate any power in the sleep mode.

IEEE 1990 Bipolar Circuits and Technology Meeting, pp. 248-251, September 1990.

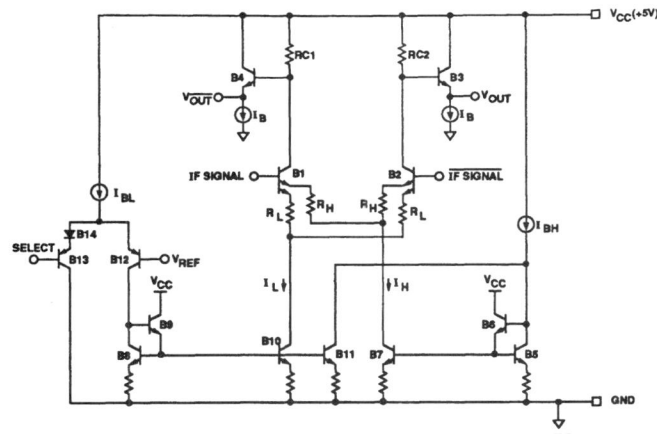

FIGURE 3 - DUAL-GAIN-STATE DIFFERENTIAL AMPLIFIER

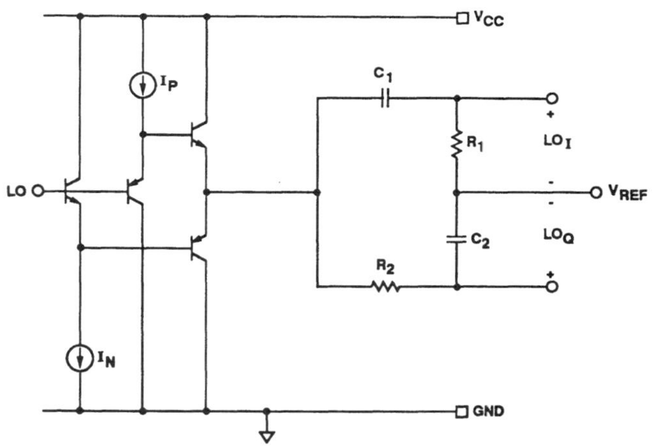

FIGURE 5 - QUADRATURE PHASE SHIFT CIRCUIT

Circuit Design

A. Digitally-controlled-gain Differential Amplifier

The dual-gain-state differential amplifier is shown in Figure 3. Since the gain of this circuit is a function of one of the resistor pairs R_L or R_H, then the two gain states are determined by the selection of the tail current sources I_L or I_H. When SELECT is low, I_{BH} is mirrored to B7 and the gain is set high. If SELECT is high, I_{BL} is mirrored to B10 and B11. When B11 is on, I_{BH} is sunk to ground. Therefore, B7 is off, B10 is on, and the low gain state is selected.

The use of high gain vertical PNPs in the current steering circuit provides better control of the currents that affect the gain in the differential amplifier. The tail currents are designed to be proportional to absolute temperature. Therefore, the gain is temperature independent because the temperature dependency of r_e is cancelled.

It can be shown that the undistorted input signal range is equal to the undistorted output signal range divided by the gain. Therefore, using this topology, the lowest gain state allows a large undistorted input signal, while the highest gain state has a lower input noise resulting in a very large dynamic range. The measured 6 dB and 51 dB normalized gain vs. IF input level is shown in Figure 4 to indicate the wide dynamic range of the digitally-controlled-gain differential amplifier string.

B. Quadrature Demodulator

The quadrature phase-shifter schematic is shown in Figure 5. In this circuit, as long as R_1 matches R_2 and C_1 matches C_2, the theoretical phase difference between the LO_I output and the LO_Q output is 90° over a large variation of temperature, processing, and frequency. In the fabricated circuit, a 93° average phase difference was achieved over four processing lots with a one-sigma variation of 0.5°. The average phase difference may be adjusted to 90° with a metal mask change. A push-pull buffer, using high-gain and high-f_T PNP and NPN transistors, drives the phase-shifter with a low output impedance. A similar buffer drives V_{REF} providing low-impedance biasing for the limiting amplifiers.

The I and Q mixers consist of Gilbert-cell balanced mixers [2, 3]. Figure 6 shows a limiting amplifier and the Q mixer (the I circuit is identical to the Q circuit). The 6 dB mixer conversion gain remains constant if the signal applied to the bases of transistors B1, B2, B3, and B4 is large enough to switch the transistors completely on or off, but not so large as to saturate the transistors. Therefore, the limiting amplifier (B7, B8, and B15 through B18) decreases the sensitivity of the mixer gain to variations of the phase-shifter output due to changes in LO amplitude, LO frequency, and process parameters.

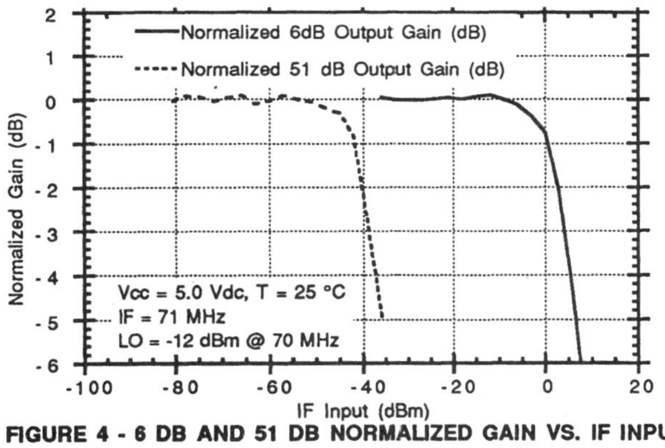

FIGURE 4 - 6 DB AND 51 DB NORMALIZED GAIN VS. IF INPUT

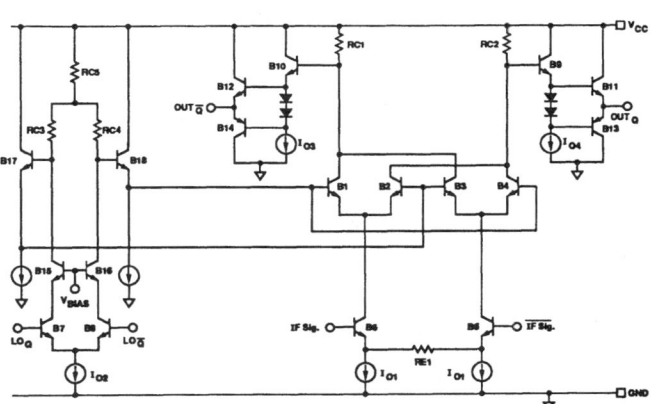

FIGURE 6 - Q MIXER AND LIMITING AMPLIFIER SCHEMATIC DIAGRAM

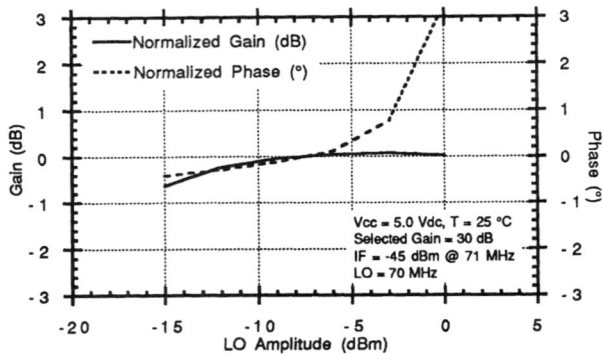

FIGURE 7 - NORMALIZED GAIN AND PHASE VS. LO AMPLITUDE

Figure 7 shows the experimental gain and phase sensitivity of the entire circuit to changes in LO amplitude. The graph shows that gain variation is less than 1 dB over a range of LO amplitudes from -15 dBm to 0 dBm while the change in phase is less than 3° over the same range of LO amplitudes.

Each mixer also has two push-pull output buffers to provide high output drive capability.

C. Digitally-controlled Bandgap Reference

The reference circuit, shown in Figure 8, consists of a bandgap reference from which the temperature independent voltage V_{REF} and current I_{REF} are derived. In addition, the current I_{PTAT} (Proportional to Absolute Temperature), used to bias the IF amplifiers, is also generated from this reference. With a high enable input (active mode), the start-up transistor B1 is activated to turn on the reference. B1 then turns off after the reference is fully on. In the sleep mode (low enable input), the start-up circuit is disabled. Next, with the emitter of B2 pulled low, B2 turns on and the bandgap reference turns off. This in turn, causes B2 itself to turn off. Thus, all devices are off during the sleep mode.

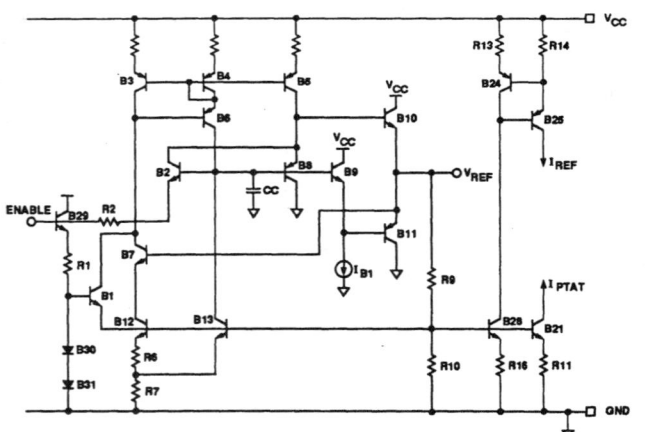

FIGURE 8 - REFERENCE CIRCUIT WITH ACTIVE/SLEEP MODE ENABLE INPUT

D. Technology

The circuit was fabricated using AT&T Microelectronics' ALA202 CBIC-U Complementary Bipolar Linear Array. This device provides 136 NPNs and 86 PNPs with f_T's of 4.0 GHz and 2.75 GHz, respectively [4,5]. Total component utilization was 74%. A 16-pin SOG package was used.

EXPERIMENTAL RESULTS

The experimental results in Figure 9 show that the measured gain-error vs. selected gain is well within the ±3 dB gain error limit over processing variations. The gain error was determined from measurements of packaged devices obtained over four separate processing lots where the one sigma variation for any gain setting is less than ±0.25 dB and the absolute error is less than ±1 dB at an ambient temperature of 25 °C

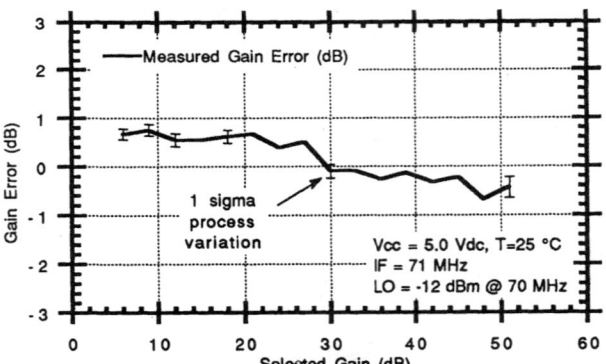

FIGURE 9 - MEASURED GAIN ERROR VS. SELECTED GAIN

Figure 10 shows how the I and Q gains are affected by changes in operating frequency. The gain in this figure is normalized to the measured gain at the normal operating frequency of 70 MHz. The low frequency rolloff is due to the finite size of the coupling capacitors while the high frequency rolloff is due to the bandwidth of the IF amplifiers.

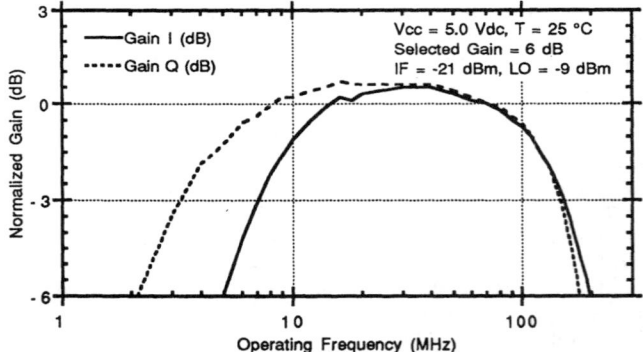

FIGURE 10 - NORMALIZED GAIN VS. OPERATING FREQUENCY

The power dissipation of the circuit was measured to be 92 mW with a single +5 V power supply. A photomicrograph of the chip is shown in Figure 11.

CONCLUSIONS

A complementary bipolar integrated circuit with a digitally-controlled-gain 100 MHz IF amplifier and GMSK quadrature demodulator has been described. The digital selection of IF amplifier gains from 0 to 45 dB in 3 dB steps with a gain error of less than ±1 dB has been achieved. In addition, a phase error variation of less than ±1° in the quadrature demodulator has been demonstrated. The circuit has applications for use in digital cellular telephony with specifications suited for mobile cellular terminals that comply with the GSM digital cellular standard.

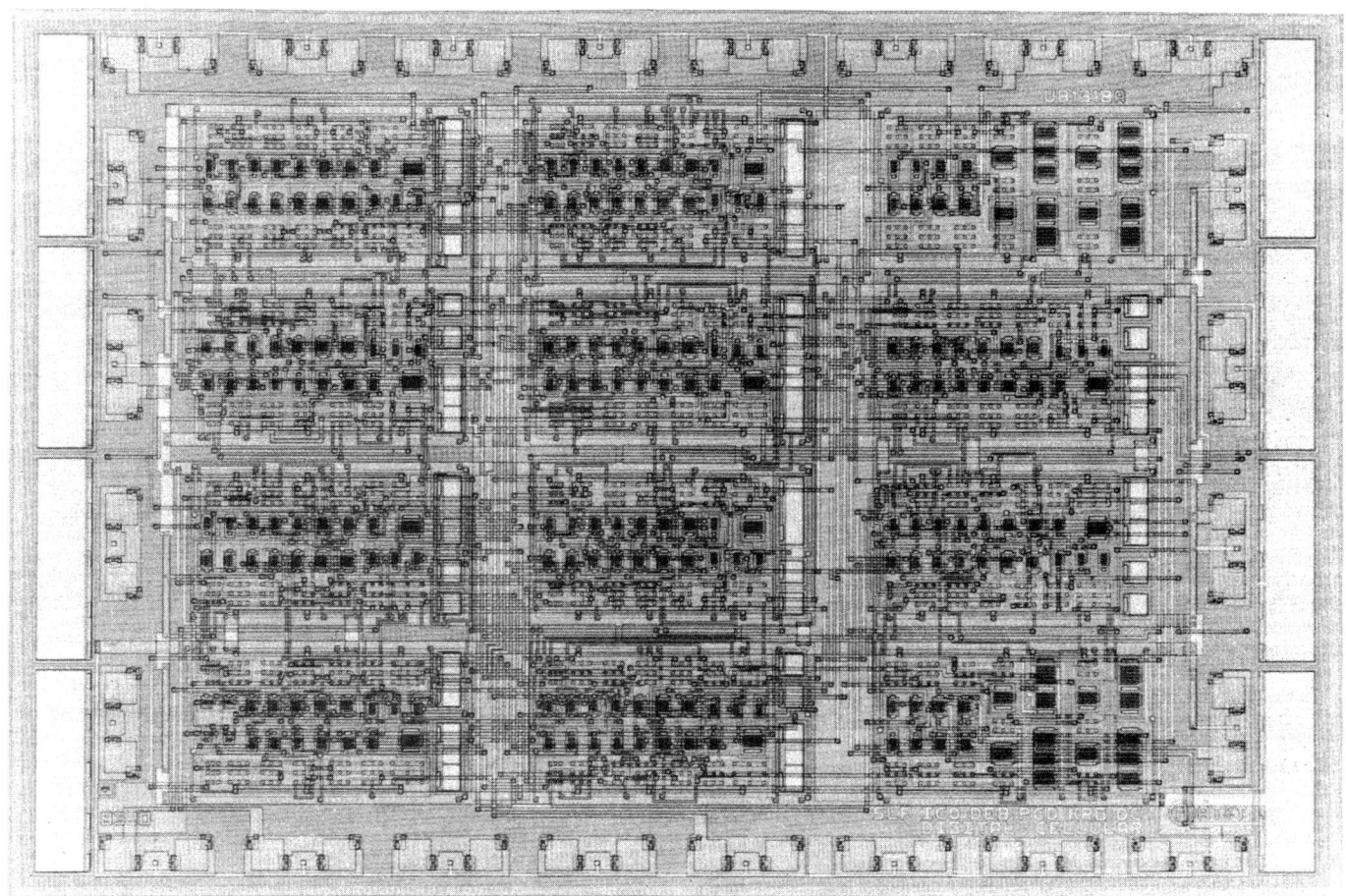

FIGURE 11 - CHIP PHOTOMICROGRAPH

ACKNOWLEDGEMENTS

The authors wish to thank David Bien for his contributions to the definition of this circuit, Guy Shovlin for his testing support, and Dionne Cohen for the chip layout. The support of Karl Gardner and Steven Parks is also gratefully acknowledged.

REFERENCES

[1] GSM Recommendations 5.05, Transmission and Reception, Version 3.5.0.

[2] Gilbert, B., "A Precise Four-Quadrant Multiplier with Subnanosecond Response", IEEE Journal of Solid-State Circuits, Vol. SC-3, pp. 365-373, December 1968.

[3] Gray, P.R. and Meyer, R.G., "Analysis and Design of Analog Integrated Circuits", Second Edition, John Wiley & Sons, Inc., 1984

[4] AT&T Microelectronics ALA201/202 UHF Linear Array Preliminary Data Sheet, AT&T Microelectronics, March 1988.

[5] B. W. McNeill, "A High-Frequency Complementary-Bipolar Array for Fast, Analog Circuits", CICC 87 Digest of Technical Papers, pp. 635-638, March, 1987

WP 2.4: A Cellular Analog Front End with a 98dB IF Receiver

Lorenzo Longo, Raouf Halim, Bor-Rong Horng, Ken Hsu, Danny Shamlou

Rockwell International Corporation, Newport Beach, CA

To date, mixed-signal CMOS devices designed for cellular telephony have adopted conventional receiver architectures such as shown in Figure 1 [1, 2]. They consist of dual baseband A/D converters with performance limited to approximately 60 to 75dB dynamic range. Thus, these architectures require the following components in the preceding IF strip. Since the overall system dynamic range is on the order of 140dB, a large amount of programmable gain (80-100dB) is required in the preceding RF/IF strip to preserve the system noise figure [3]. External filters at both the first and second IF frequencies are required to provide full channel selectivity before the A/D conversion is performed. Also, dual analog quadrature mixers are required to convert the receive signal from the second IF to baseband. The mixed-signal CMOS device presented in this paper substantially reduces the programmable gain, filtering, and external components required in the IF strip. This is accomplished by performing an IF A/D conversion and providing 98dB dynamic range.

The dual-mode cellular analog front end supports the North American IS-54 standard. A block diagram of the device is shown in Figure 2. The receiver includes a passive anti-aliasing filter, a bandpass $\Sigma\Delta$ modulator with programmable gain, digital quadrature mix to baseband and dual decimation filters. Each channel of the dual baseband transmitter consists of a 10b DAC, a sixth-order switched-capacitor filter, and a second-order continuous smoothing filter. The auxiliary section includes two 8b DACs and an 8b monitor ADC.

A block diagram of the receiver architecture used with the mixed-signal device is shown in Figure 3. It differs from conventional architecture starting at the second IF, where a simple bandpass filter is required as opposed to a narrow-band ceramic filter. This simple bandpass filter provides anti-aliasing for the bandpass $\Sigma\Delta$ modulator, described in Reference 4. It has an IF of 1.8MHz, a sampling frequency of 7.2MHz and the channel bandwidth is 30kHz. 12dB of programmable gain is added to the front end of the modulator to increase the dynamic range of the receiver to 98dB for a negligible increase in power and die area. This high dynamic range reduces the amount of programmable gain required preceding the device, thus reducing system power dissipation and cost.

Once the IF A/D conversion is performed, the 1b modulator output is mixed to baseband, producing the in-phase and quadrature-phase components. This function is illustrated in Figure 4. Since the sampling frequency (7.2MHz) is four times the IF (1.8MHz), the phase of the local oscillator (LO) can be chosen so the LO has three convenient values: 0, +1 or -1, thus simplifying the digital quadrature mixer to a few gates and preserving a 1b input for the dual decimation filters. Also, the output of the mixers can be decimated to 3.6MHz since every second sample is zero. This reduces power dissipation of the decimators by approximately 50% and also reduces their complexity by decreasing the required word lengths. The Sinc^3 filter is required to attenuate the high-frequency quantization noise produced by the bandpass $\Sigma\Delta$ modulator so that decimation can be done without a significant loss in SNR. This filter is also used to attenuate interfering signals and therefore eliminates a narrow band filter at the second IF.

The transmitter, shown in Figure 3, uses a conventional baseband architecture with dual paths for the I and Q components. The input data is applied to a fully-differential 10b D/A converter that uses a mixed capacitor and resistor ratio architecture. The DAC uses 10x oversampling in analog mode and 8x oversampling in digital mode to allow for image rejection. Its output is filtered by a fully-differential sixth-order switched-capacitor Bessel filter. Images from the switched-capacitor filter are filtered by a second-order Rauch filter, that also drives off-chip.

The auxiliary section of the device includes two 8b D/A converters and an 8b monitor A/D converter. The auxiliary DACs are for automatic frequency control and power amplifier control. These DACs use the same capacitor/resistor ratio architecture used in the 10b transmit DACs. The monitor ADC has a four-to-one input multiplexer and is for functions such as transmitter calibration and battery voltage monitoring. This ADC utilizes a successive-approximation architecture.

The device is implemented using a 1μm double-poly CMOS technology. A micrograph of the device is shown in Figure 5. The chip uses a single 5V supply and includes an on-chip differential bandgap reference and a 2.5V reference generator. Power dissipation is 100mW. The device is controlled through a four-pin serial interface that includes the following signals: clock, frame pulse, data in and data out.

The receiver performance is evaluated by applying a 1.803MHz sinewave, windowing the 16b digital output of the I or Q channel with a Dolph-Chebyshev window and performing a 4096 point FFT. In analog mode, receiver output rate is 80kHz. The noise and distortion is integrated over 40kHz bandwidth. At baseband, the bandwidth of interest is approximately 15kHz, so noise above 15kHz is attenuated by the decimation filter. Measured performance curves for three gain settings (0, 6 and 12dB) are shown in Figure 6. These curves indicate a dynamic range of 98dB. The differential bandgap generates ±0.6V references that are doubled using capacitor ratios. Thus, the modulator 0dB input level is 2.4V peak differential.

Figure 7 shows the output spectrum of the receiver for a 1.803MHz sinusoidal input signal at -20dB for the 12dB gain setting. This output spectrum, shown from zero to 40kHz, illustrates the digital baseband mixing function with the output at 3kHz. Also note that the dc offset is as low as the noise floor. This is due to the fact that the dc offset of the bandpass $\Sigma\Delta$ modulator is mixed to 1.8MHz and filtered by the decimator.

References

[1] Lakshmikumar, K. R., et al., "A Baseband Codec for Digital Cellular Telephony," IEEE J. of Solid-State Circuits, vol. 26, no. 12, pp. 1951-1958, Dec. 1991.

[2] Lakshmikumar, K. R., et al., "A Modem/Codec for Cellular Telephony," ISSCC Digest of Technical Papers, pp.148-149, Feb., 1993.

[3] Koullias, I. A., et al., "A 900MHz Transceiver Chip Set for Dual-Mode Cellular Radio Mobile Terminals," ISSCC Digest of Technical Papers, pp.140-141, Feb. 1993.

[4] Longo, L., Horng, B., "A 15b 30kHz Bandpass Delta-Sigma Modulator," ISSCC Digest of Technical Papers, pp. 226-227, Feb., 1993.

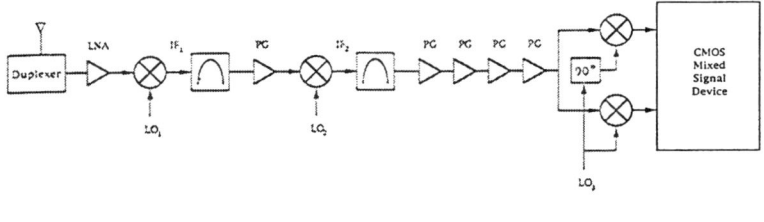

Figure 1: Conventional receiver architecture.

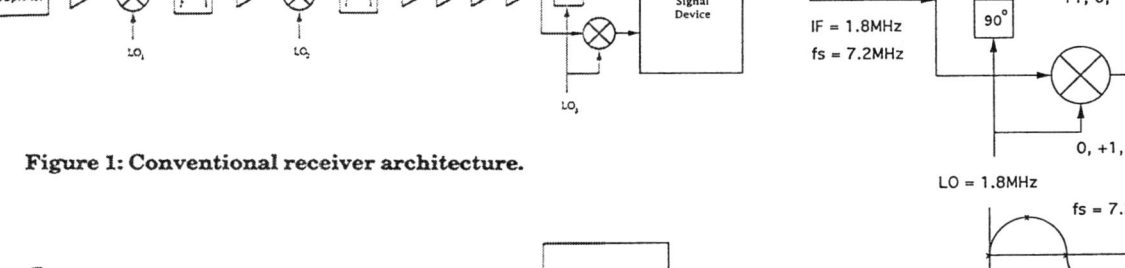

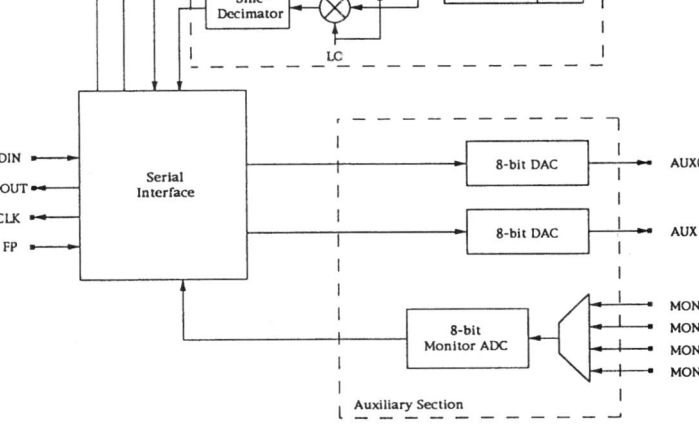

Figure 2: Receiver architecture for this device.

Figure 4: Digital quadrature mix to baseband.
Figure 5: See page 305.

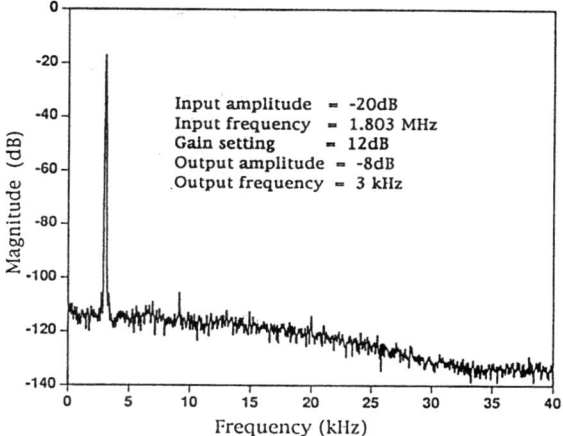

Figure 6: Measured S/(N+D) curve.

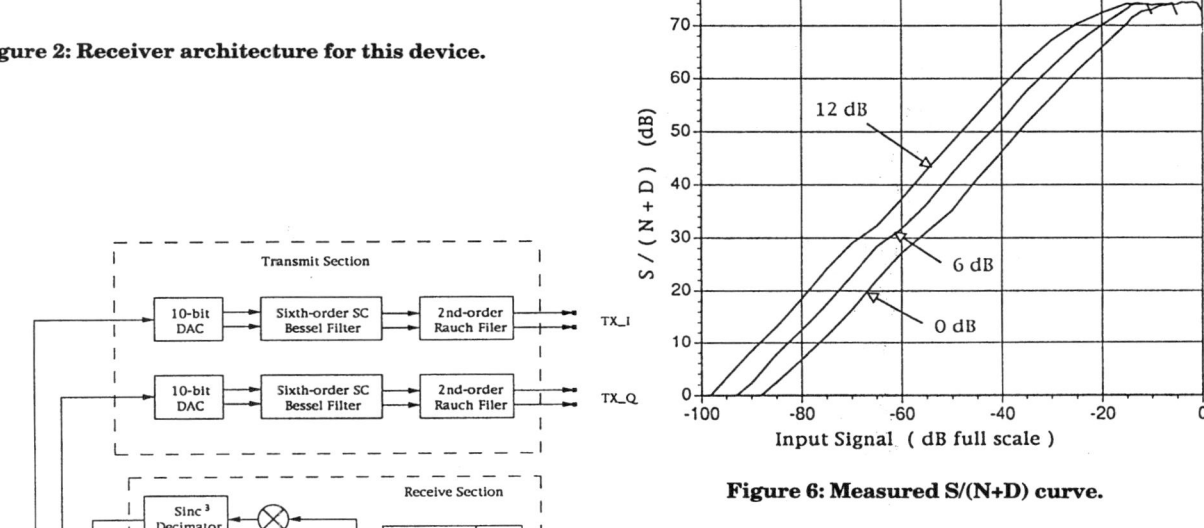

Figure 3: Chip block diagram.

Figure 7: Measured receiver output spectrum.

Figure 5: Chip micrograph.

A 270-kb/s 35-mW Modulator IC for GSM Cellular Radio Hand-Held Terminals

JOHAN J. J. HASPESLAGH, MEMBER, IEEE, DANNY SALLAERTS, PETER P. REUSENS, ARNOUL VANWELSENAERS, R. GRANEK, AND DIRK RABAEY, MEMBER, IEEE

Abstract —The modulator IC is a new mixed analog/digital transceiver component in a chip set, which is designed for the hand-held terminals of the pan-European 900-MHz Groupe Special Mobile (GSM) digital cellular radio network. The device generates the Gaussian minimum shift keying (GMSK) modulation and converts the received signal to 8-b words after filtering. The modulator IC uses digital waveform generation and a quadrature signal representation. This device is implemented in a 1.5-μm CMOS technology. The power consumption is less than 35 mW from a 5-V supply.

I. INTRODUCTION

IN 1991, a new pan-European digital cellular radiotelephone system will be introduced in Europe. This new system operates in the 900-MHz band and provides end-to-end digital transmission channels. It was designed according to the recommendations of the Groupe Special Mobile (GSM) from CEPT. This digital system requires a very aggressive design in order to compete with the next-generation hand-held portables developed for the analog cellular radio networks. Furthermore, the portable system requires complex digital processing to maintain performance in rapidly varying multipath propagation effects and in a high amount of interfering signals, which are characteristic of the transmission environment in mobile communications. For optimum performance, moreover, the communication system uses frequency hopping modulation, complicated coding methods, and data processing.

This paper will describe the most important aspects of a modulator circuit used in the new GSM system. First, the concept of the radio-frequency environment in which the circuit is used will be explained, focusing on the differences in today's existing systems. Next, the architecture and different functions of the modulator circuit are pointed out. In Section IV details of the digital and analog processing in the transmission mode are discussed. Next, the receiving mode, which is mostly based on analog processing, will be highlighted. Measurement results and technology features will be presented in Section VI.

II. CONCEPT OF THE MOBILE COMMUNICATION SYSTEM

Fig. 1 presents the simplified block diagram of the hand portable terminal. Digital from end to end, the communication channel starts along the transmit path with a GSM codec which directly interfaces to a microphone. Digitized voice samples are coded and compressed to a 13-kb/s data stream by the GSM speech transcoder implemented in the baseband section. Then a channel encoder builds a robust protection around the compressed data stream. This consists of cyclic redundancy check bits, convolutional coding, and data packet interleaving to optimize the transmission reliability. These encoded data are fed to the modulator IC in bursts with a transmission rate of 270 kb/s. They are translated into Gaussian minimum shift keying (GMSK) modulation signals by digital waveform processing [1]. GMSK leads to a narrower bandwidth. Up-conversion to the 900-MHz band is done in the transmit radio section. In the modulator the bit sequences are converted to phase trajectories with Gaussian impulse characteristics in a balanced quadrature representation (I and Q channels).

The receiver stage performs a down conversion from the 900-MHz signal. The RF block contains the mixers, automatic gain control, and prefiltering. The analog signal is further accurately filtered and converted to the digital domain by the modulator IC. In receive mode the modulator IC functions as a front end ahead of the digital demodulator. This demodulator in the baseband section removes distortion due to the Doppler shift and multipath fading and restores the data packets of 270 kb/s, which are reassembled to a 13-kb/s data stream in the channel decoder. Finally, the speech information is regenerated by the GSM speech transcoder and converted to analog by a codec in the baseband section. The analog signal is then sent to an earpiece or loudspeaker. The total system is supervised by a control section which takes into account all the operation modes of the terminal in order to optimize the total power consumption.

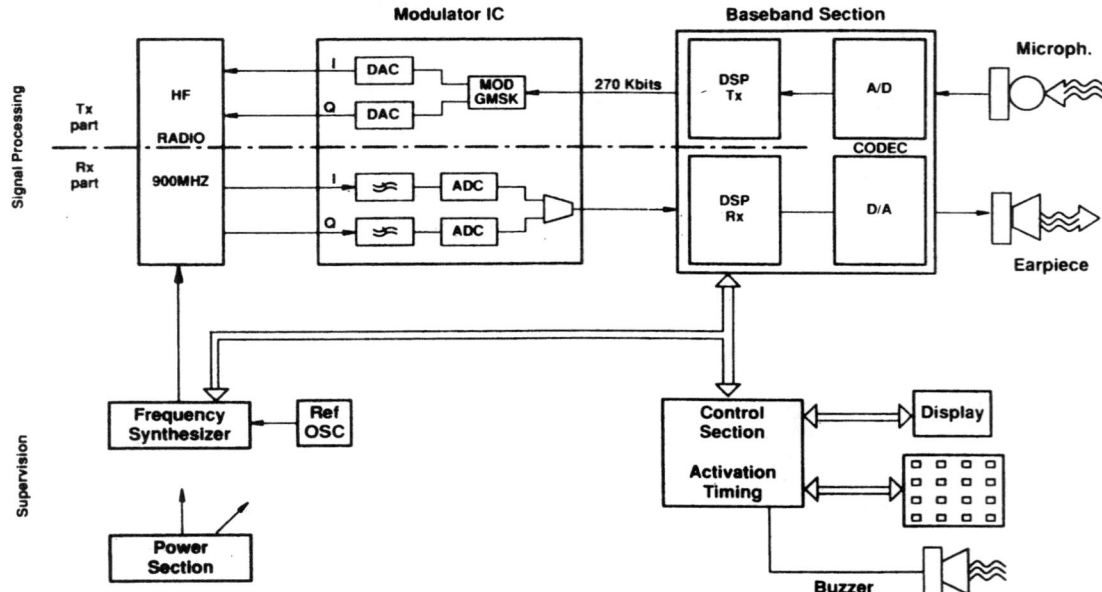

Fig. 1. Hand-head terminal block diagram.

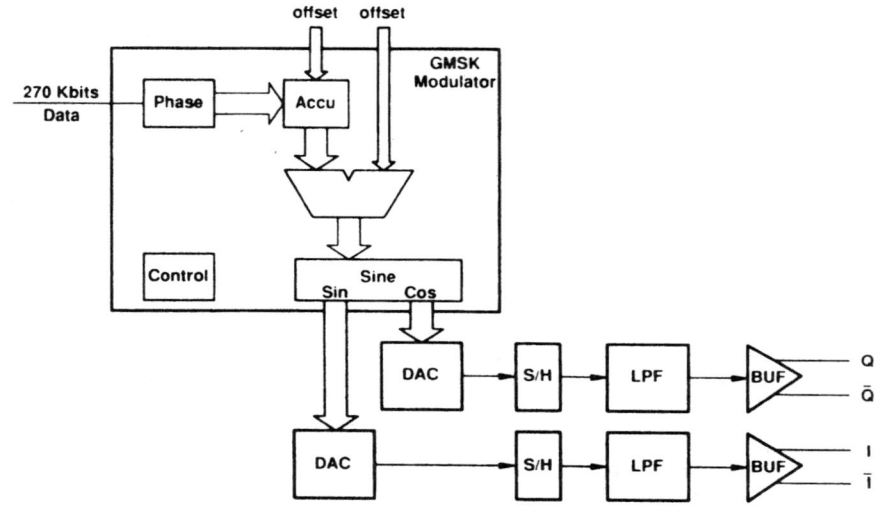

Fig. 2. Modulator IC: transmit block diagram.

III. Architecture of the Modulator Circuit

The modulator chip is designed as a central component for the radio front end for mobile terminals: the prime goal was to reduce the number of components drastically in order to increase the battery life by lowering the power consumption of the terminals and to decrease their volume. As opposed to the currently installed analog systems, the GSM system supports maximal integration of the radio components by multiplexing transmit and receive operation. In this way, interference between strong transmit levels and weak receive levels is excluded. All the necessary trimming functions have been included in the modulator circuit to minimize the number of discrete components. The modulator performs different functions for the transmit and receive paths of the terminal. For the transmit path we have the following functions (Fig. 2):

1) the circuit performs the GMSK modulation by calculating the in-phase (I) and quadrature (Q) signals with 8-b resolution from the 270-kb/s data;
2) the IC corrects phase and frequency inaccuracies occurring elsewhere in the transmit path by feeding in offsets during phase and frequency calculations;
3) both digital I and Q signals are converted to the analog domain at a sampling frequency of 2166 kHz;
4) the signal is finally smoothed and converted to balanced signals before being sent to the mixer of the HF section to generate the antenna signal.

In addition to its transmit functions, the modulator also performs a number of receive functions (Fig. 3):

1) in order to limit adjacent channel interference, which may not be more than a 200-kHz distance from the carrier frequency, the demodulated I and Q compo-

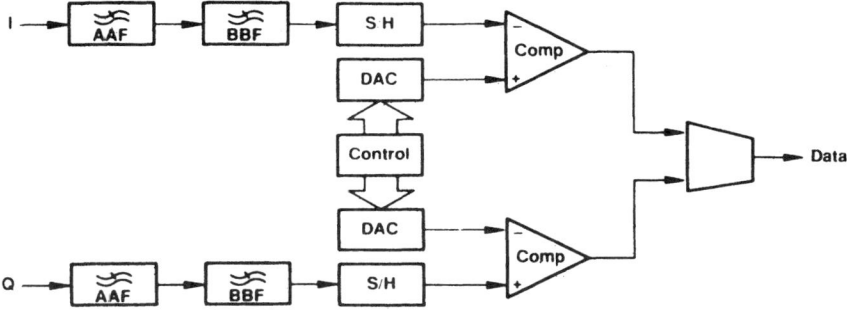

Fig. 3. Modulator IC: receive block diagram.

nents are filtered by a selective fourth-order filter built as a switched capacitor (SC) filter (BBF);
2) the receive I and Q signals are converted to an 8-b digital format at a rate of 270 kHz. In subsequent circuits, this information is further equalized and demodulated by digital signal processing techniques to reobtain the original bit stream.

Furthermore, to improve die size, parts of the transmission path have been reused in the receiving path. This approach is an immediate result of the fact that receive and transmit operations are multiplexed in time and never occur simultaneously. The interface between the digital processor circuits and the radio circuits is limited to a single component. This improves electromagnetic compatibility. Fig. 4 represents the combined TX and RX sections in the modulator IC.

Special attention has been paid to the master-clock interface. The clock signal at the input of the IC has only a small amplitude (1 V_{p-p}) in order not to generate interference with the radio-frequency environment on the printed board circuit. Therefore, a sensitive clock input buffer is needed to regenerate the 26-MHz master clock for digital levels.

The modulator IC circuit operation consists of six different modes. In power-down, no actions are performed and the circuit only consumes a few microamperes. In transmit mode, the transmit path is active while the receive path is in power-down. The reverse situation occurs in receive mode operation. In order to facilitate component screening and factory testing of the printed circuit boards, three additional loop modes have been implemented: they essentially loop back digital data from the GMSK modulator to the digital output at the receiver side and analog input data to the analog outputs. These features make it possible to check all the chip connections and performances when the PCB's are assembled, and provide a means to implement self-check of the mobile terminal.

Timing signals for the receive and transmit circuits are provided by a control unit which also controls the power-up and power-down switching, the digital interface timings, and the test modes. The control unit itself is operated by the master control unit of the terminal system.

IV. TRANSMISSION MODE

A. Digital Processing in TX Mode

The main functions in the GMSK waveform generation are performed in a digital arithmetic unit. The digital data are first differentially encoded. For the requested performance in the GSM recommendations, only 3 b have to be taken into account to select the appropriate frequency path (Fig. 5). The modulated data can be expressed as follows (see Appendix):

$$\chi(t) = \frac{2 \cdot E_c}{T} \cdot (\cos(f_c \cdot t) \cdot \cos(\Phi(t)) - \sin(f_c \cdot t) \cdot \sin(\Phi(t))).$$

In the modulator IC the terms $\cos(\Phi(t))$ and $\sin(\Phi(t))$ are generated where $\Phi(t)$ is a function of the incoming data. The modulation with the carrier f_c is generated in the HF section. In the actual implementation (Fig. 6), the digital data which are transmitted are first converted to a frequency trajectory. This approach is advantageous because it allows the tuning of the local oscillator frequency without the use of analog trimming components. Adding a constant value to the calculated frequency will shift the spectrum of the transmit signal with a determined value. The output of the frequency adder is integrated by the phase accumulator; its output value reflects the instantaneous phase of the carrier signal. The quadrature signal components I and Q are obtained from a look-up table for the sine and cosine implemented in PLA's. The sine function is generated using a two-stage approximation technique [2]. An additional phase correction can be done by adding a phase offset to the Q signal. This allows for the correction of imperfections in the phase-splitting circuit and quadrature modulator, which are further ahead in the transmit path. A numerical phase accuracy of 16 b is used to guarantee a close approximation of the ideal GMSK waveform. Using the possibilities of phase correction, the phase relationship between the I and Q signals can be controlled with an accuracy of 0.175°. The frequency error of the local oscillator (expected < 4500 Hz) can be compensated with steps as small as 33 Hz.

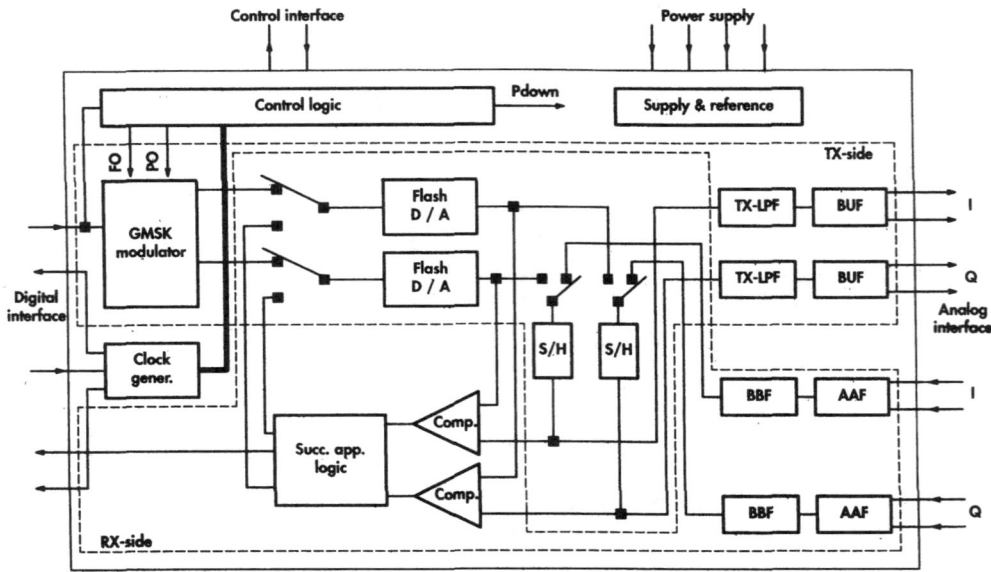

Fig. 4. Block diagram of the modulator IC.

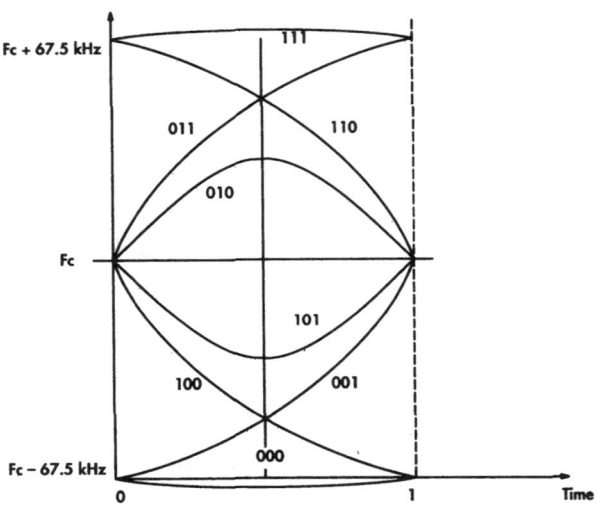

Fig. 5. Frequency paths.

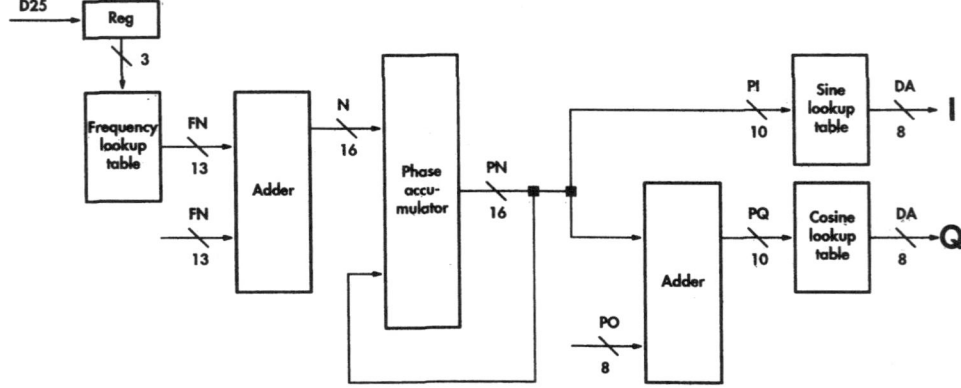

Fig. 6. GMSK waveform generator.

B. Analog Processing in TX Mode

The digital data which represent the phase of the modulated signal in the form of a I and Q signal are presented to the D/A converter with an 8-b accuracy (Fig. 2). The data rate is 2.166 Msamples/s. The D/A is also used in receive mode. The output of the converter is fed to a sample and hold (S/H). This allows filtering of the analog signal with a second-order low-pass filter with a flat delay characteristic. The filter is optimized to minimize possible AM distortion of the received signal. The output buffer circuits realize a high noise rejection at higher frequencies, in order to prevent spurious modulation components in the transmit signal.

The output driver performs three functions. In the first place it delivers a 2-V_{p-p} signal to the bipolar double-balanced mixer of the HF section. Secondly, it provides a low impedance and good power supply rejection on the connections to the mixer in order to prevent noise from reaching the mixer. This would otherwise give rise to unacceptable modulations of the carrier signal. Thirdly, the driver must convert the single-ended signal from the S/H to a complementary signal. For our purpose it is very important and necessary that the two complementary signals have the same amplitude and phase behavior. Otherwise, this would lead to spectral distortions in the GMSK modulation.

Fig. 7 shows the principle of the single-to-double-ended conversion. As can be seen, only one output is fed back and therefore follows the input signal. The symmetrical output is generated by the common-mode loop. The common-mode loop has in this case a double function; it generates the symmetrical output signal but also controls the output common-mode level. To realize the conversion with the required performances in amplitude and phase, the differential mode as well as the common-mode loop must at least have equal amplitude and phase characteristics.

Fig. 8 shows the total circuit of the output driver. The common-mode signal is fed back at nodes 3 and 4. Then by appropriate scaling of the transistors $CM1$ to $CM4$ with respect to $IT1$ and $IT2$, an equal input transconductance can be realized for the differential as well as for the common-mode path. As both loops have the same capacitive load, we obtain an equal GBW and gain. Therefore, only slight differences will exist in amplitude and phase.

The output stage is a modified Castello stage able to track the biasing currents in transistors $OS1$ and $OS4$ with the common-mode regulating signal on nodes $n6$ and $n7$. Its quiescent current is 300 μA. Its drive capacity is 1 kΩ and 20 pF.

V. THE RECEIVING MODE

The receive I and Q signals from the HF section are filtered by a fourth-order switched capacitor (SC) filter, preceded by a continuous-time second-order filter. System simulations showed that a legendre filter characteris-

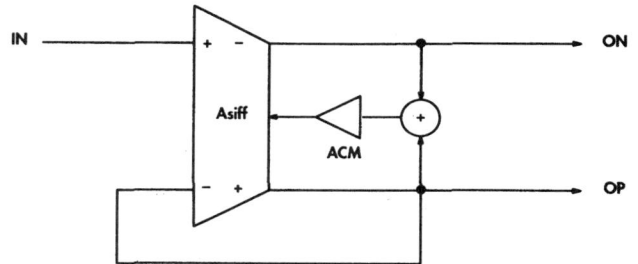

Fig. 7. Principle of the single-to-double-ended conversion used in the output buffer converter.

tic [4] is an adequate solution because of its good roll-off behavior near the 3-dB frequency. In receive mode, the D/A converters and S/H circuits, used in the TX mode, are rearranged to form a successive approximation A/D converter.

A. The Switched Capacitor (SC) Filter

The legendre filter is integrated as a SC ladder filter. It uses a 2.166-MHz sampling clock. The passband -3-dB frequency is 90 kHz. To obtain good results at this rather high clock frequency, folded-cascode operational amplifiers with a gain bandwidth of 20 MHz are used. The output of the filter is fed into an S/H circuit which operates in the RX mode at 270 kHz. The S/H output goes directly to the comparator of the A/D converter (Fig. 3).

B. The Analog-to-Digital Converter

Switched-capacitor techniques have also been used to implement the D/A converters. The reasons to choose this type of converter are twofold. First, a good matching and accuracy of the I and Q channels on the order of 0.1 dB must be obtained to prevent phase errors in the I, Q vector plane. Secondly, the circuit may not consume in power-down mode during nonactive periods. In order to save area the converter is not totally built as a binary ratioed capacitor array. The four least significant bits have been realized by scaling through a coupling capacitor (CU) (Fig. 9) [3]. Furthermore we opted for a sign and magnitude converter. The total area consumption is therefore equivalent to only a 22-unit capacitor in comparison to a two's complement converter which would need 30 units. The D/A converter operates at 2.166-MHz rate for the conversion cycle. In receive mode it performs the 8-b successive approximation A/D conversion at 270 ksamples/s.

VI. MEASUREMENT RESULTS AND IC CHARACTERISTICS

Fig. 10 shows the vector diagram generated by the modulator in the case when a repetitive digital pattern is presented at the inputs. Each data sample generates a 90° phase shift with a Gaussian impulse response. This test

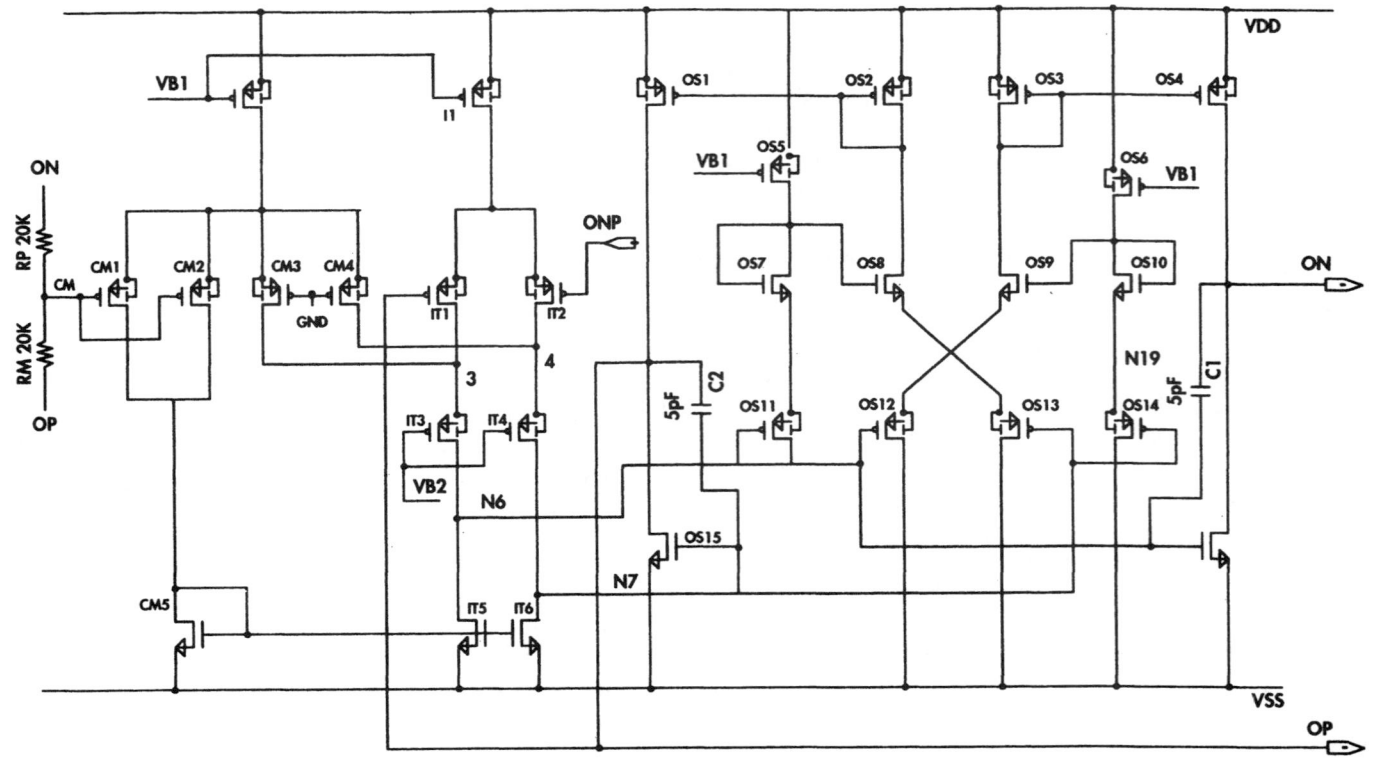

Fig. 8. Output buffer converter.

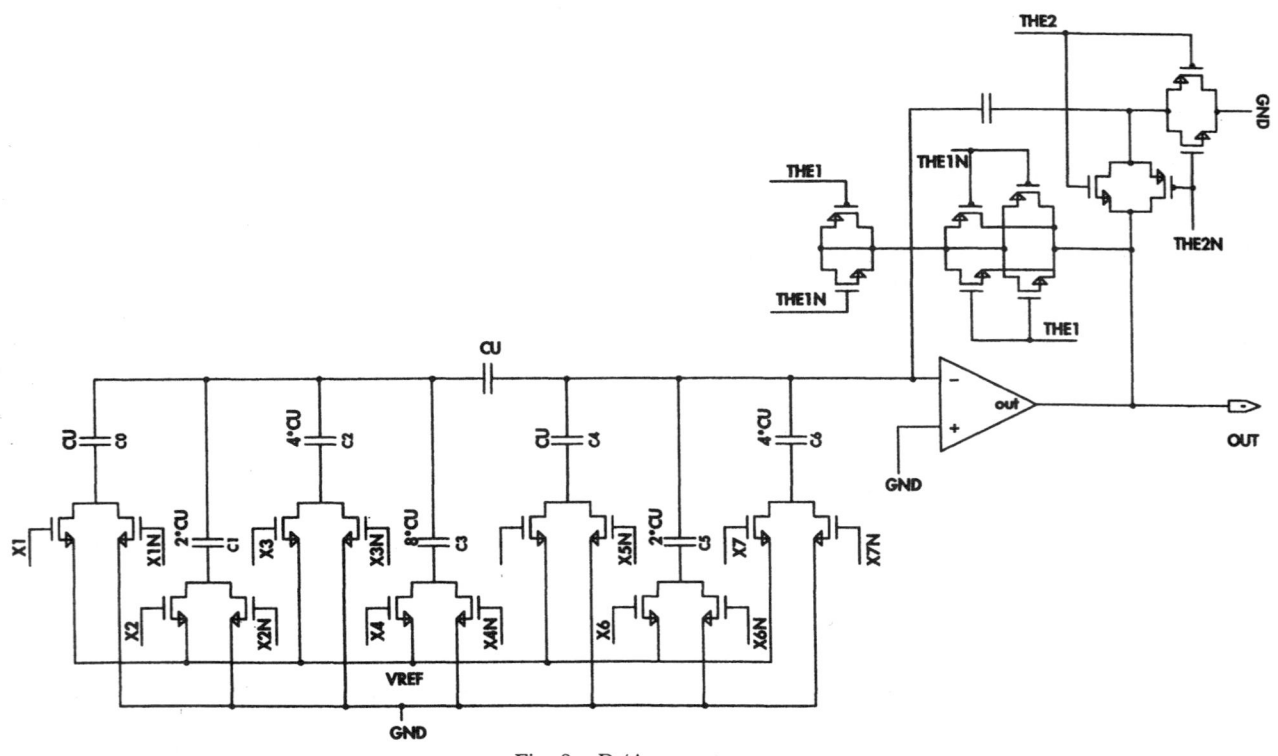

Fig. 9. D/A converter.

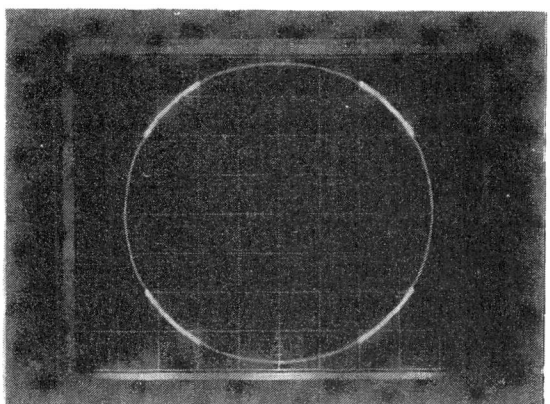

Fig. 10. Vector diagram of the GMSK modulation.

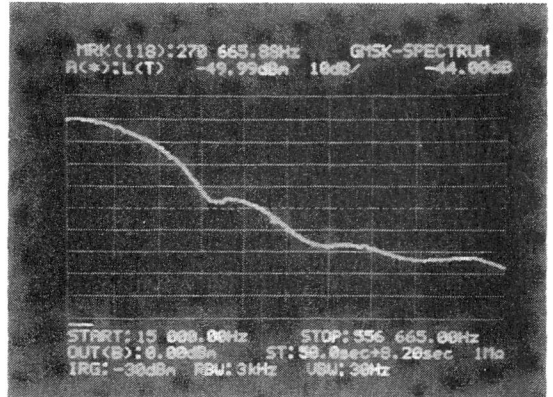

Fig. 11. Measured GMSK spectrum at the modulator output.

was used to verify the correctness of the frequency offset circuitry when modulation is applied. The GMSK spectrum measured with a random modulation pattern is presented in Fig. 11. We remark the first lobe ending at a frequency of 200 kHz according to the GSM specifications. The measurements were done on one of the I and Q signals in the baseband.

The chip has been designed using different approaches according to specific areas. The analog circuits are realized using a full-custom methodology with a handcrafted layout. For the digital parts, a semicustom approach was used. PLA's have been used to implement the frequency tables and the trigonometric conversion tables. The modulator IC has been designed in a 1.5-μm, double-metal, double-poly CMOS technology. The die size measures 4.4×5 mm^2 (Fig. 12). The chip operates from a single 5-V supply with a peak dissipation of 140 mW. Taking into account that the chip is only active for 25% in normal burst mode, the average power consumption is 35 mW. The chip is packaged in a 24-pin PLCC package well suited for the printed circuit boards of the hand-held set.

VII. CONCLUSION

All baseband functions of a mobile terminal for GSM radio, such as the generation of the quadrature modulation components, D/A conversion and A/D conversion,

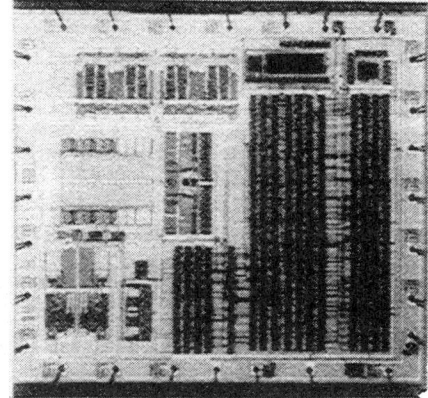

Fig. 12. Die photograph of the modulator IC.

and additional filtering, have been integrated in one component. The circuit features are a fully trimmable quadrature modulator together with a very low power consumption and little board space use. Therefore it is suitable in mobile and hand portable terminals for GSM radio. In addition, extended testability has been implemented to facilitate testing in all production phases.

APPENDIX

The GMSK waveform generation can be described as follows. The digital data are first differentially encoded and 3 b are taken into account to select the appropriate phase or frequency path (Fig. 5). The modulating data excite a linear filter with an impulse response defined by

$$g(t) = h(t) * \text{rect}\left(\frac{t}{T}\right)$$

where the function rect( ) is defined as

$$\text{rect}\left(\frac{t}{T}\right) = \frac{1}{T}, \quad \text{if } |t| < \frac{T}{2}$$
$$= 0, \quad \text{otherwise}$$

with T = duration of one input data bit, which is $= 1/270$ kHz or 3.692 μs. The rect( ) function is necessary to limit the otherwise infinite impulse response of the Gaussian filter to one symbol period. The impulse response $h(t)$ is defined as follows:

$$h(t) = \frac{1}{\sqrt{2\pi} \cdot \sigma \cdot T} \cdot e^{(-t^2/2\sigma^2 T^2)}$$

where

$$\sigma = \frac{\sqrt{\ln(2)}}{2\pi \cdot B \cdot T}$$

and $B \cdot T$ is the relative bandwidth, which in our case is 0.3. The instant phase of the accumulated signal is then

$$\Phi(t) = \sum_i \alpha_i \cdot \frac{\pi}{2} \int_{-\infty}^{t - iT} g(\tau) \, d\tau.$$

The modulated RF carrier may be expressed as

$$\chi(t) = \frac{2 \cdot E_c}{T} \cdot \cos(f_c \cdot t + \Phi(t) + \Phi_0)$$

where E_c is the energy per modulating bit, f_c is the center frequency, and Φ_0 is the starting phase at the beginning of the burst. The emitted signal can then be written as

$$\cos(f_c \cdot t + \Phi(t)) = \cos(f_c \cdot t) \cdot \cos(\Phi(t)) \\ - \sin(f_c \cdot t) \cdot \sin(\Phi(t)).$$

REFERENCES

[1] GSM Recommendation 05.04, *Modulation*, Vers. 3.02; GSM Recommendation 05.05. *Transmission and Reception*, Vers. 3.5.0.
[2] D. Sunderland *et al.*, "CMOS/SOS frequency synthesizer LSI circuit for spread spectrum communication," *IEEE J. Solid-State Circuits*, vol. SC-19, no. 4, pp. 497–505, Aug. 1984.
[3] Y. S. Yee, L. M. Terman, and L. G. Heller, "A two stage weighted capacitor network for D/A–A/D conversion," *IEEE J. Solid-State Circuits*, vol. SC-14, no. 4, pp. 778–781, Aug. 1979.
[4] B. D. Rakovich and M. V. Popovich, "Explicit expression for the characteristic function of generalized legendre filters," *Circuit Theory and Applications*, vol. 6, pp. 363–373, 1978.

A CMOS Limiting Amplifier and Signal-Strength Indicator

S. Khorram, A. Rofougaran, and A. A. Abidi

Integrated Circuits & Systems Laboratory
Electrical Engineering Department
University of California
Los Angeles, CA 90024-1594

Introduction

Mobile radio handsets require an on-chip circuit to measure the received signal-strength. This information is sent to the base station to regulate the transmitted power level, and to determine cell-handoff. As strong attenuation and fading may cause the received signal power to span 6 to 8 decades, the measurement circuit must have a semi-logarithmic input-output characteristic to encompass this dynamic range.

The design of logarithmic amplifiers is now a well-established, although specialized, art. Unlike older implementations, these amplifiers no longer derive a logarithmic characteristic from the PN junction I-V characteristic; instead, they use the *successive-detection* architecture [1], wherein the sum of equally-weighted taps along a cascade of identical clipping amplifiers approximates the logarithm as a piecewise-linear function (Fig.1). A DC measure of the amplitude of a bipolar signal such as a sinewave is obtained by rectifying each tap, and lowpass filtering the sum of the rectified taps. These measurements will be accurate to the degree the actual amplifier input-output characteristic conforms to the ideal logarithm, and how stable it remains over the operating temperature range. Both these qualities are captured by the deviation of the actual characteristics from the ideal logarithm, specified as a gain ripple in dB.

Although all commercially available monolithic log amps today are bipolar ICs [2-5], CMOS is equally well-suited to implement the successive-detection architecture. We report here on the design and performance of such a logarithmic amplifier, which is part of a monolithic all-CMOS spread-spectrum 900 MHz wireless transceiver [6]. In the intended use, a received 160 kb/s binary-FSK signal is amplified at RF, directly downconverted to DC, and applied to the logarithmic amplifier after channel-select filtering. The amplifier provides two useful outputs. First, the limited output from the cascade of clipping amplifiers contains the data encoded as signal *phase* in the zero-crossings. This is directly applied to a correlating 1-bit digital detector for binary-FSK [6], without need for AGC. Second, the circuit produces a logarithmic signal-strength measurement to an accuracy of 1 dB over a 80 dB dynamic range, which a slow 7-bit A/D converter digitizes for uplink to the base-station.

The salient features of this logarithmic amplifier are: operation to 2.7-V supply, 0.75 mA current drain with a 5 MHz bandwidth, 80 dB dynamic range, 84 dB limiting gain, and implementation in 1-µm CMOS.

Circuit Design

The useful dynamic range of a successive-detection amplifier is limited on the upper end when the input causes the first stage in the cascade to clip, and on the lower end when all stages are in the linear region. In practice, if the amplifier noise causes the last stage to limit, it will determine the lower end. The amplifier characteristics will accurately conform to the logarithm over this dynamic range when a large number of clipping amplifier blocks, each with a relatively small gain, is cascaded. For instance, with a gain of 10 dB per stage and an ideal rectifier, the ripple in the DC logarithmic characteristic is 0.3 dB [4]. We use a cascade of 7 direct-coupled, balanced clipping amplifiers, each with 12 dB gain, and sum the output of 8 rectifiers along the cascade (Fig.1). Each stage consists of an NMOS differential pair with NMOS loads, and its gain, therefore, depends mainly on device-ratio, and to a large extent is process- and temperature-independent (Fig.2). The input and output common-mode levels are the same. A four-quadrant MOS multiplier rectifies the balanced signal at each tap in the cascade (Fig.3). Single-ended output currents taken from all 8 rectifiers are summed into an on-chip 10kΩ polysilicon resistor, and filtered by an off-chip 1-µF capacitor. All circuits are designed for operation at 2.7 V supply.

A well-designed bias source is important in attaining stable logarithmic characteristics. An on-chip replica circuit generates bias voltages for the current sources in the amplifier and rectifier (Fig.3).

Experimental Results

An off-chip *RC* circuit with a cutoff frequency of 3 Hz is connected in negative feedback between the limiting output and amplifier input, to stabilize the bias and suppress DC offsets. All measurements are made with a variable-amplitude 455 kHz sinewave capacitively-coupled into the amplifier. This is the standard 2nd IF in FM cellular telephones, and a sufficiently high frequency to cover the baseband spectrum of 160 kb/s binary-FSK.

The amplifier output is observed to limit on its own noise, as is customary in limiting amplifiers of high sensitivity. The signal-strength output conforms well to the logarithm over more than 80 dB, with a minimum detectable signal of –80 dBm (32 µV amplitude), which is also close to the equivalent input noise voltage (Fig.4). A ripple of ±0.5 dB is observed at all tempera-

---

Research supported by ARPA, Rockwell International, Texas Instruments, Harris Semiconductor, and the State of California MICRO Program.

tures, but owing to a small temperature-dependence of gain, the error curve acquires a slope which exacerbates the total error to ±1 dB at 85°C relative to the room-temperature characteristic (Fig.4(b)). This is acceptable for the intended application, and a considerable improvement over the ±3 dB ripple in a previously published CMOS logarithmic amplifier [7].

A ±1V ptp differential square-wave with well-defined zero-crossings is observed at the limiting output across the entire dynamic range. The group-delay through the cascade must also stay relatively constant over this range [5], otherwise the detected SNR, which is sensitive to phase of the received signal, will suffer. We measure this to vary by less than 100 ns across all amplitudes (Fig.5), which the resultant 16° peak phase-error will cause only a 0.2 dB loss in SNR.

The 1-μm CMOS N-well IC occupies an active area of 1 sq mm, and drains a total current of 0.75 mA from 2.7 V.

References

1. R.S. Hughes, *Logarithmic Amplification with Application to Radar and EW*. Dedham, MA: Artech House, 1986.

2. D. Anderson and R.J. Zavrel, RF ICs for Portable Communications Equipment, in *Electronic Components and Applications*. v. 7, no.1, January 1985, pp. 37-44.

3. Motorola, MC3362 Low-Power Dual Conversion FM Receiver, in *Communications Device Data Book*. 1992, pp. 2/60-2/63.

4. Analog Devices, AD640 DC-Coupled Demodulating 120 MHz Logarithmic Amplifier, in *Special Linear Reference Manual*. 1992, pp. 3/31-3/46.

5. Analog Devices, IF ICs Offer 3V Operation for Digital and Analog Cellular Standards, in *Microwave J*. v. 37, no.10, October 1994, pp. 144-148.

6. J. Min, A. Rofougaran, H. Samueli, and A.A. Abidi, "An All-CMOS Architecture for a Low-Power Frequency-Hopped 900 MHz Spread-Spectrum Transceiver", in *Custom IC Conf.*, pp. 379-382, San Diego, CA, 1994.

7. K. Kimura, "A CMOS Logarithmic Amplifier with Unbalanced Source-Coupled Pairs", *IEEE J. of Solid State Circuits*, v. 28, no.1, pp. 78-83, 1993.

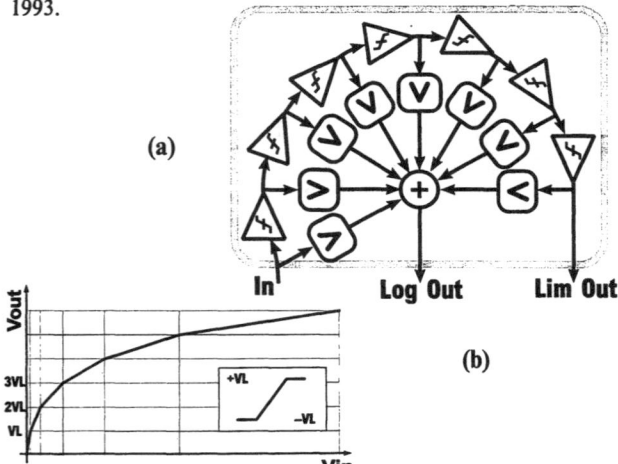

Fig.1. (a) Principle of successive-detection logarithmic amplifier. Yields both logarithmic and limited outputs. (b) Piecewise-linear approximation to log function. VL is clipping-level of each amplifier.

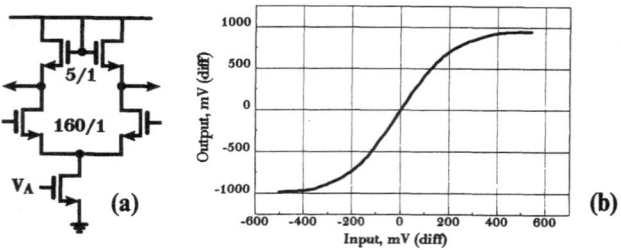

Fig.2. (a) Single clipping amplifier, and (b) its measured input-output characteristic.

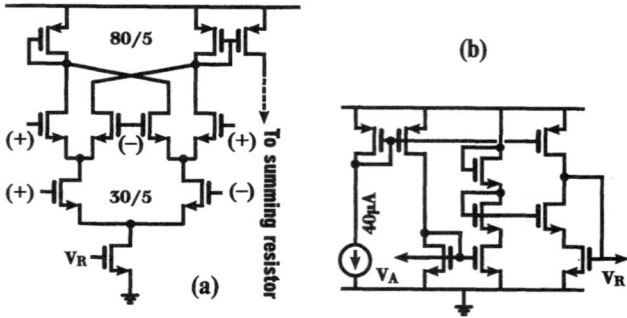

Fig.3. (a) Single full-wave rectifier. (b) Replica-bias circuit.

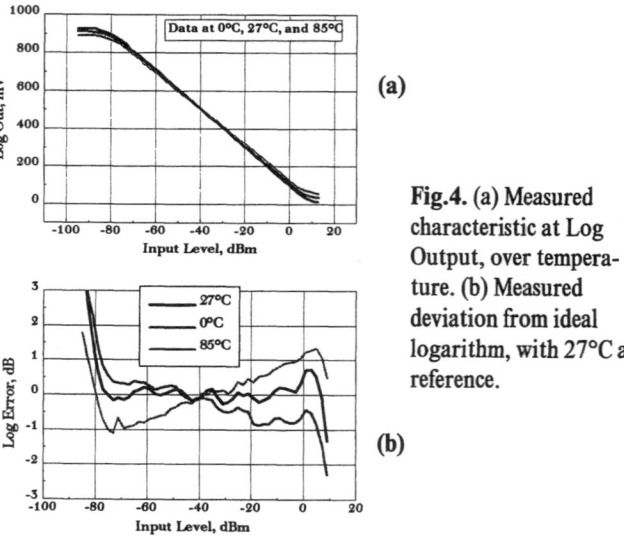

Fig.4. (a) Measured characteristic at Log Output, over temperature. (b) Measured deviation from ideal logarithm, with 27°C as reference.

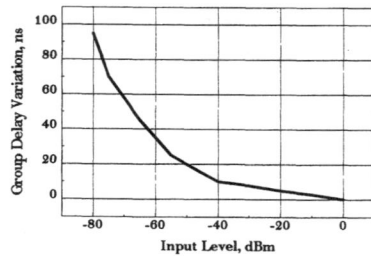

Fig.5. Variation in group-delay through Limiting Amplifier, over the complete dynamic range.

Characterization of a microwave silicon single-chip direct conversion RF transceiver

Werner Simburger * **, Herbert Knapp * **, Peter Weger **

* Institut fur Nachrichtentechnik und Hochfrequenztechnik, Technische Universitat Wien Gußhausstr. 25, E389, A-1040 Vienna, Austria Tel.: +43 1 58801-3530, Fax.: +43 1 5870583, E-Mail: sekinthf@email.tuwien.ac.at.

** Siemens AG, Corporate Research and Development, Microelectronics, Munich Otto-Hahn-Ring 6, D-81739 Munich, Germany Tel.: +49 89 636-0, Fax.: +49 89 636 47069, E-Mail: peter.weger@zfe.siemens.de.

Abstract

This paper presents the detailed microwave characterization of a highly-integrated silicon bipolar single-chip RF transceiver. The chip offers a direct conversion architecture with a standard I/Q baseband interface for both receive and transmit paths. A limiter/discriminator is integrated for comparative heterodyne demodulation application. The on-chip voltage-controlled oscillator, phase shifter (divider by two) and the dual-modulus prescaler form the RF-section of the frequency synthesizer.

Introduction

In most cases the architectures of today's commercially available RF chip sets for personal communications are based on the heterodyne or dual-conversion principle (dual-conversion with I/Q baseband interface). The incoherent direct conversion concept is currently not widely used, though the saving of the image rejection filter and the IF channel-selectivity filter is a major production advantage.

The reasons for this are isolation problems of direct modulation in the transmit direction as well as the mass production and practical performance issues related to direct conversion in the receive direction. These difficulties are partly caused by distribution of the RF-functions over several chips in current chip set architectures. Previously, the DC offset shift problem of the baseband I/Q signals, caused by any change of the gain setting of the RF preamplifier, has been identified.

To diminish this problem the stray cross-coupling effects between LO-signal and preamplifier/mixer have been minimized by a VCO-frequency at twice the operation frequency of the transceiver. Choosing a by two divider as a wide-band phase shifter two benefits are attained: a) the LO-signal does not mix with the RF input to produce unwanted DC-offset; b) a wide range of input frequencies can be conveniently handled without adaptation of the 90° phase shifter. A drawback of this method is the requirement of a high VCO-frequency.

By means of a concrete first-silicon realization of a wide-band RF transceiver the concept of direct conversion was investigated [1], [2].

This paper focuses on an improved second run. The chip was characterized by measurement of the most important parameters in mobile radio such as frequency response, intermodulation distortion, noise figure, dc-offset in baseband, LO-leakage at the RF input and power consumption.

To abate nonlinear distortion of the transmitted and received signals, linearized cross-quads were investigated and implemented in the circuit design. The measurement results will be discussed by means of details in circuit design.

A critical direct conversion example, investigates the LO-leakage radiation at the RF-input versus the LO frequency. At every setting of the variable-gain preamplifier (VGA) the LO leakage is below—74dBm, indeed a low value.

Another critical parameter is the DC offset versus gain setting of the RF amplifier and also versus LO-frequency, at the quadrature baseband output. Depending on the gain setting, the DC-offset can be made small compared to the maximum output signal level.

RF Transceiver Architecture

Fig. 1 shows the RF transceiver architecture. The receiver section consists of a low noise preamplifier (LNA) with a balanced input, followed by a variable gain amplifier (VGA), and the quadrature down-converter (QDC) including open collector baseband signal buffers. All outputs are balanced.

In the case of direct conversion operation mode the RF input pin can be connected to a pre-filter with a low quality factor, because no image rejection filtering is necessary. The channel selection is achieved by low pass filters at the quadrature down-converter outputs. The QDC is highly flexible with respect to modulation because of it's complex-valued baseband signals. Both, linear and constant envelope modulation formats can be handled by this architecture. The filtered baseband signal can be demodulated by any analog or digital baseband processor.

An additional limiter/IF amplifier and FM discriminator has been included for purpose of comparison.

The transmitter circuit contains a quadrature up-converter (QUC), current summing, and a medium power amplifier (MPA).

The quadrature up- and down-converters in the RX/TX circuits share the voltage controlled oscillator (VCO) and the phase-shift circuitry via a multiplexer. The phase shifter is realized by a static 2:1 frequency divider, halving both frequency and phase of clock signals with 180° phase shifts at twice the LO frequency.

Receiver Section

RF Frontend

A simplified schematic diagram of the LNA is shown in Fig. 2. The amplifier consists of an emitter-coupled pair with emitter degeneration. Input matching is performed with negative feedback. The gain of the LNA is about 10dB, with a noise figure of 3dB.

Variable Gain Amplifier and DC-Block

The variable gain amplifier offers two gain settings and a differential dc-block function to avoid propagation of amplified dc offsets [3]. The VGA consists of two differential amplifiers AMP1 and AMP2, connected in parallel.

Fig. 3 shows the schematic circuit of the VGA. Each partial differential amplifier consists of two linearized cross-quads, with a different transconductance [4], [5]. The BIAS1 and BIAS2 signals are used to select the desired gain.

A drawback of this method is the deterioration of the overall noise figure.

Quadrature Down Converter

The quadrature down-converter consists of two Gilbert multipliers, buffered by an open collector linearized cross-quad circuit.

Fig. 4 shows the schematic diagram of a Gilbert mixer stage, including the open collector output buffer. The two additional transistors of the linearized buffer perform cancellation of the nonlinear dynamic resistance of the output transistors by adding a dynamic resistance with the same absolute, but negative value, caused by the cross connections.

Limiter/FM-Discriminator

A block diagram of the limiter amplifier and FM detector is shown in Fig. 5. The limiter consists of five stages which are directly coupled. Each stage consists of an emitter coupled pair and an emitter follower output buffer. The level detection is accomplished by including a full wave rectifier in each limiter stage and summing the output currents.

The FM-detector consists of a multiplier using a Gilbert cell. The filtered output is applied to a level converter which provides a TTL/CMOS compatible signal.

Frequency Control Section

Voltage Controlled Oscillator

The basic principle of the VCO is the negative resistance concept (Fig. 6). An external ceramic resonator is connected to the negative impedance converter (NIC). Two transistors, connected as reversed diodes, form a varactor. The voltage across the resonator is buffered and fed to an automatic gain control (AGC) circuitry. The output of the AGC circuit is fed to an emitter-coupled pair, which controls the bias operating point of the NIC.

The closed loop control guarantees a short transient response. Further, the steady state amplitude is stabilized against changes in supply voltage, temperature and resonator quality factor.

Local Oscillator Phase Splitting using a Frequency Divider by 2

The phase splitter consists of two D-type latches and forms a master-slave D-flipflop. The master flipflop is negative edge triggered, and the slave is positive edge triggered caused by it's negated clock signal. The negated feedback from the slave output to the master D-input results in an output frequency half the clock frequency. Hence, the phase difference between the master output and the slave output is 90°, but only if the duty factor of the CLK-signal is 1:1. Further, the rise time of the clock signal should be small as possible, otherwise during the zero crossings the latches are in an undefined state.

Two major benefits follow from this: The transceiver can operate at any frequency, set by an external resonator to the VCO, from near DC to the upper frequency limit of the VCO; and coupling from the VCO to the R.X path is minimized, which diminishes the DC offset. The integrated dual modulus prescaler and an external low frequency PLL manage VCO fine tuning. A drawback of this method is the requirement of a high VCO frequency.

Another promising solution for the phase shifter could be an integrated RC all pass, where the phase coherence is primarily dependent only on the ratio of resistor and capacitor values. [6], [7], [8] and [9] are several proposals to this subject.

Dual Modulus Prescaler

The dual modulus prescaler permits selection of a 64/65 or 128/129 divide ratio. The Modulus Control input and prescaler output are compatible to standard CMOS PLL circuits.

The prescaler input is connected to the VCO and therefore it's frequency range has to reach up to twice the operation frequency of the transceiver. An additional external input is included to simplify testing.

Transmitter Section

Quadrature Up-Converter

The quadrature up-converter consists of two linearized Gilbert multipliers, including a buffer amplifier for the baseband signal. Fig. 7 shows the schematic diagram of the inphase channel of the QUC. It consists of a linearized buffer amplifier and a Gilbert multiplier, including an emitter-follower as output buffer. If the current gain of the transistors is greater than 50...100, then the linearized cross-quad offers a higher signal-to-intermodulation ratio and a lower dc current consumption, compared to an emitter-coupled pair with emitter degeneration and the same gain [5]. The emitter-coupled pair of the Gilbert multiplier is linearized, using the same scheme.

The quadrature channel is identical, except for the collector load resistors and the output emitter-follower of the Gilbert cell. The current sum is performed at the collector load resistors of the inphase-channel mixer.

Medium Power Amplifier

The MPA consists of two driver stages and an emitter-coupled pair with emitter degeneration, as output stage. The driver stages form a cross-quad with an output emitter-follower.

To simplify measurement and testing, 50Ω on-chip load resistors are connected to the collectors of the output stage. An improvement of the power efficiency, requires an external matching network (BALUN), instead of the on-chip 50Ω load.

Experimental Results

The RF transceiver was implemented on one 5k subarray of a 50k transistor array built from different types of cells for realization of basic RF functions, and fabricated in a 0.8μm/25GHz silicon bipolar technology.

To perform initial tests, the chip was mounted on a ceramic RF-package (Fig. 8), and inserted in an engineering test fixture. An evaluation printed circuit board, where the chip is bonded on the substrate directly, was used to evaluate the VCO performance.

Electrical Characteristics

Table 1 shows a short performance summary. At the RF amplifier, two gain settings are available. Fig. 9 shows the conversion gain of the RX-circuit versus frequency.

The frequency response and noise figure do not show the full potential of the technology, due to restrictions imposed by the use of a transistor array. Compared to a full-custom design, line lengths are much longer, and only a limited number of transistor types are available.

The baseband dc offset, due to stray cross-coupling from the LO section to the RF-train of the quadrature down converter is shown in Fig. 10. Depending on gain setting, the dc offset changes only about several mV. The -1dB-compression output voltage is about ±500mV. The static dc offset of the I- and Q-channel differ around 7mV.

Fig. 11 shows the intermodulation characteristics of the RX-circuit at 200MHz. The signal-to-intermodulation ratio at the compression region is 32/25dB, depending on gain setting.

Fig. 12 shows the LO leakage radiation at the RF-In and RF-In input. Over the full frequency range of operation, the LO leakage does not exceed—74dBm.

The frequency response of the TX train, including QUC and MPA, is shown in Fig. 13. An input voltage of 600mVpp results in an output power of two times—3dBm at 1.8GHz.

The intermodulation performance of the quadrature up-converter/medium power amplifier is shown in Fig. 14. The signal-to-intermodulation ratio at the compression region (at 1.8GHz) is about 20dB.

Fig. 15 shows the VCO output spectrum at a frequency of 2GHz. The phase noise is about -85dBc/Hz at Δf = 100kHz. The tuning range, using the on-chip varactors, is about 2%. The upper frequency limit is determined by packaging, because the transformation of the negative resistance of the NIC to the external resonator circuit fails at high frequencies.

The input sensitivity of the prescaler, using the external input, is shown in Fig. 16. The maximum frequency of operation is about 4.3GHz.

Fig. 17 shows the demodulated data at the output of the on-chip limiter/discriminator (bottom). The RF-signal at 1.8GHz is modulated with pseudo random data at a data rate of 1.152Mbit/s (top), using Gaussian Minimum Shift Keying. This modulation format is used by the DECT standard (Digital European Cordless Telephone Standard).

Conclusions

We have demonstrated that the architecture of the phase shifter and VCO RF-part determines the success of the direct conversion concept. By judicious choice of circuit design and technology direct conversion can be achieved successfully. We have demonstrated an important step in that direction.

Acknowledgement

Support of the Forschungsforderungsfonds fur die gewerbliche Wirtschaft, FFF, Wien, Austria, is gratefully acknowledged.

We wish to thank Dr. E. Bonek and Dr. L. Treitinger for their helpful encouragements.

References

[1] Weger, P., Simburger, W., Knapp, H., Leslie, T. C., Rohringer, N., Popp, J., Schultes, G., Scholtz, A. L., and Treitinger, L., "Completely Integrated 1.5 GHz Direct Conversion Transceiver", in Proceedings of the 1994 Symposium on VLSI Circuits Digest of Technical Papers, pp. 135-136, Hawaii, USA, June 09-11 1994. IEEE.

[2] Bonek, E., Schultes, G., Kreuzgruber, P., Simburger, W., Weger, P., Leslie, T.C., Popp, J., Knapp, H., and Rohringer, N., "Personal Communications Transceiver Architectures for Monolithic Integration", in Proceedings of the IEEE International Symposium on Personal, Indoor and Mobile Radio Communications, pp. 363 368, The Hague, The Netherlandes, September 19-23 1994. IEEE, invited paper.

[3] John F. Wilson, "A Single-Chip VHF and UHF Receiver for Radio Paging", IEEE Journal of Solid-State Circuits, vol. 26, pp. 1944-1959, December 1991.

[4] Dennis L. Feucht, Handboo1c of Analog Circuit Design, Academic Press, Inc., San Diego, California, 1990.

[5] Simburger, W., Knapp, H., Schultes, G., and Scholtz, A. L., "Comparison of Linearization Techniques for Differential Amplifiers in Integrated Circuit Design", in Proceedings of the 7th Mediteranean Electrotechnical Conference 1994, MELECON'94, pp. 1222-1225, Antalya, Turkey, April 12-14 1994. IEEE, Vol. III.

[6] Mahfoudi, M. and Alonso, Jose I., "A MMIC Broadband Active Quadrature Phase Shifter for Mobile Communications", in Proceedings of the 7th Mediteranean Electrotechnical Conference 1994, MELECON'94, p. 601 ff., Antalya, Turkey, April 12-14 1994. IEEE.

[7] Sevenhans, J., Haspeslagh, D., Delarbre, A, Kiss, L., Chang, Z., and Kukielka, J.F., "An integrated Si bipolar RF transceiver for a zero IF 900 MHz GSM digital mobile radio single chip RF up and RF down

converter of a hand portable phone", in IEEE International Solid-State Circuits Conference, pp. 44 ff., WP2.8. IEEE, 1994.

[8] Steyaert, M. and Roovers, R., "A 1GHz Single-Chip Quadrature Modulator", IEEE Journal of Solid-State Circuits, vol. 27, pp. 1194 ff., August 1992.

[9] Tsukahara, Tsuneo, Ishikawa, Masayuki, and Muraguchi, Masahiro, "A 2V 2GHz Si-Bipolar Direct-Conversion Quadrature Modulator", in IEEE International Solid-State Circuits Conference, pp. 40 ff., WP2.6. IEEE, 1994.

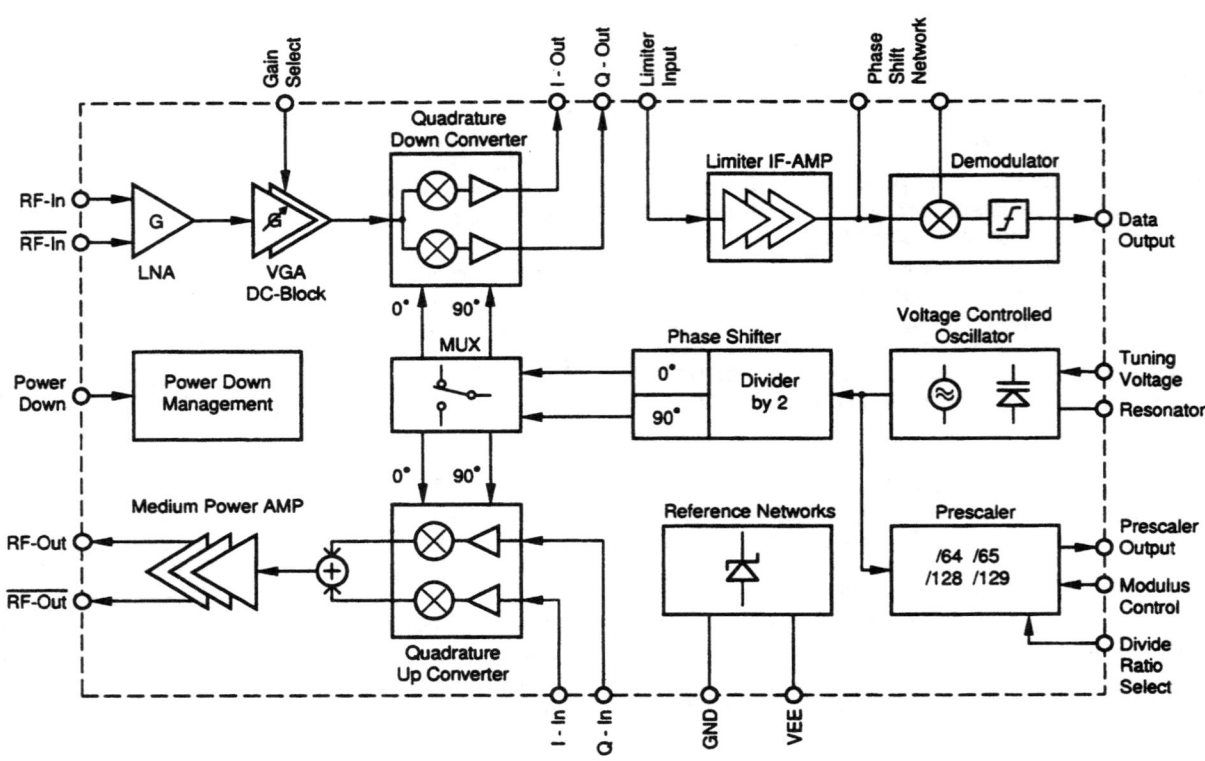

Figure 1: RF transceiver block diagram.

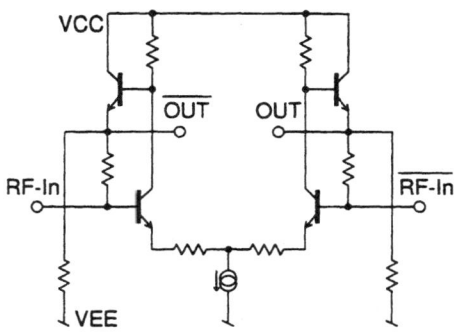

Figure 2: Simplified schematic diagram of the LNA

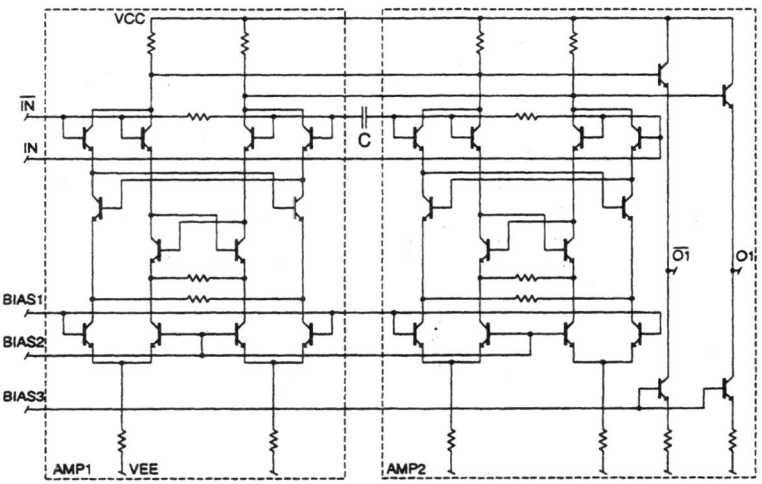

Figure 3: Schematic circuit of the VGA

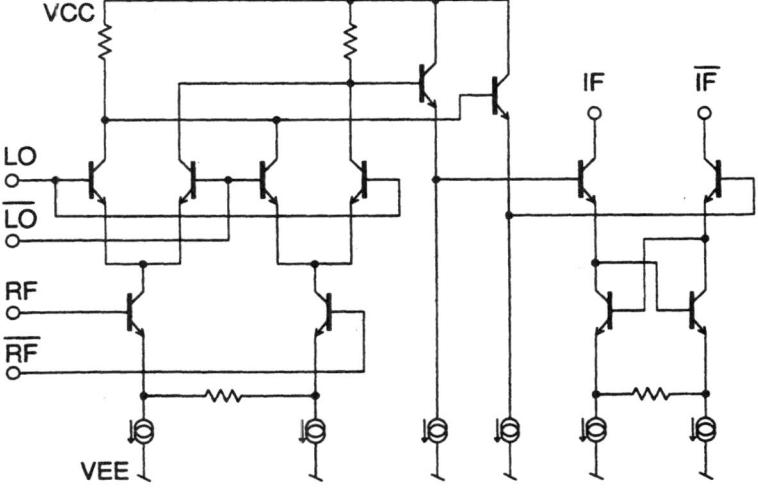

Figure 4: Schematic diagram of half the QDC

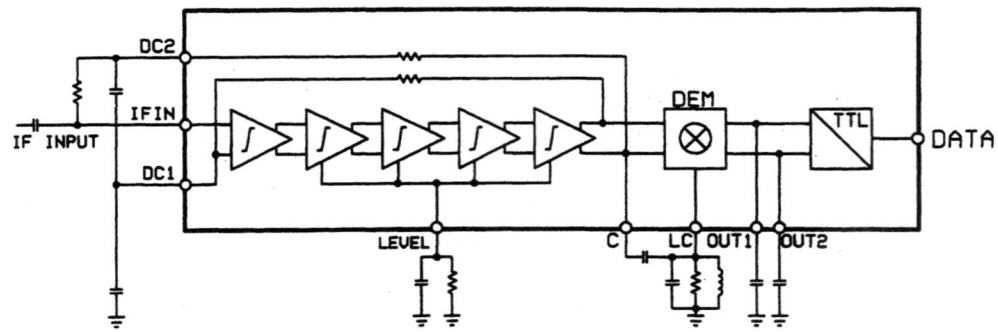

Figure 5: Block diagram of the limiter/discriminator

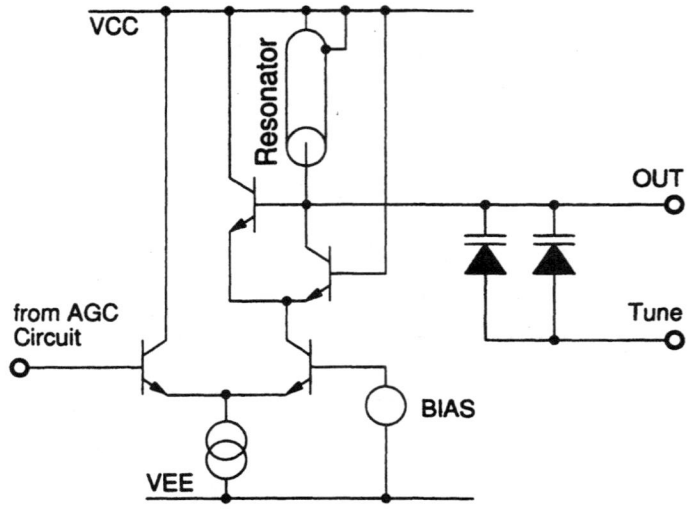

Figure 6: Simplified schematic diagram of the VCO

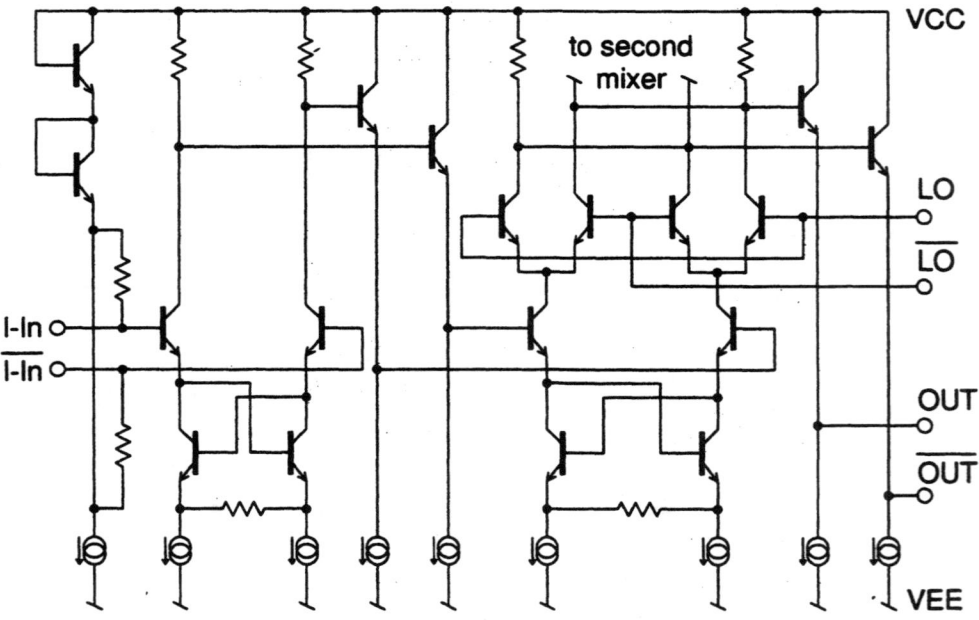

Figure 7: Schematic diagram of half the QUC

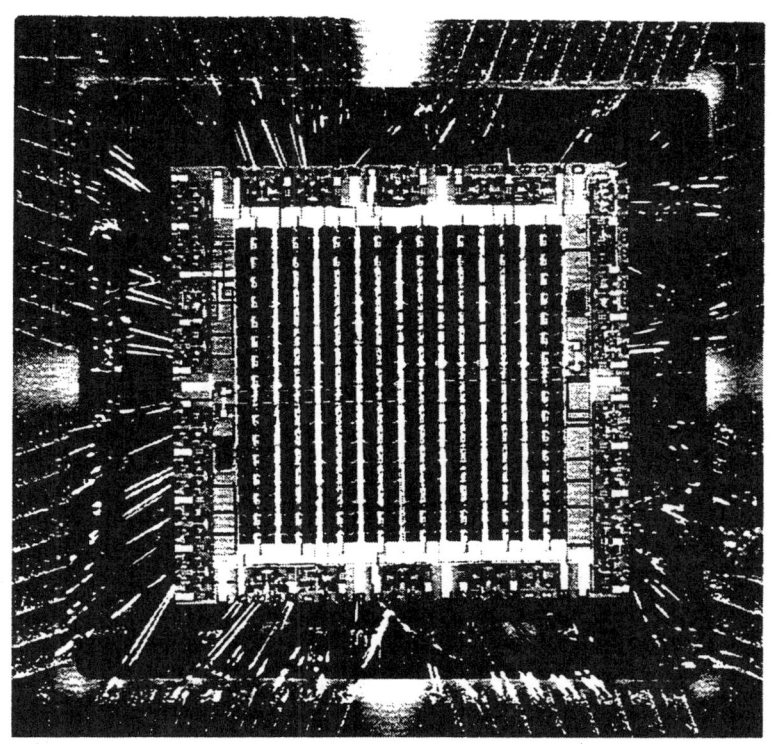

Figure 8: Transceiver chip bonded to an RF-package

Parameter	Value
RX Section (at 1.8GHz)	
Gain	$+6/+23$ dB
-1dB Input Gain Compr.	$-14/-30$ dBm
3rd-order Input Intercept	$\approx -6/\approx -22$ dBm
Noise Figure	27/12 dB
TX Section (at 1.8GHz)	
Conversion Gain	600mV$_{pp}$ at baseband input results in 2×-3dBm RF power
-1dB Output Gain Compr.	-4 dBm
3rd-order Output Intercept	$+4$ dBm
Current Consumption (at 4...6V)	
LNA/QDC	9mA
QUC/MPA	54mA
VCO	12mA
Prescaler	20mA

Table 1: RF transceiver performance summary

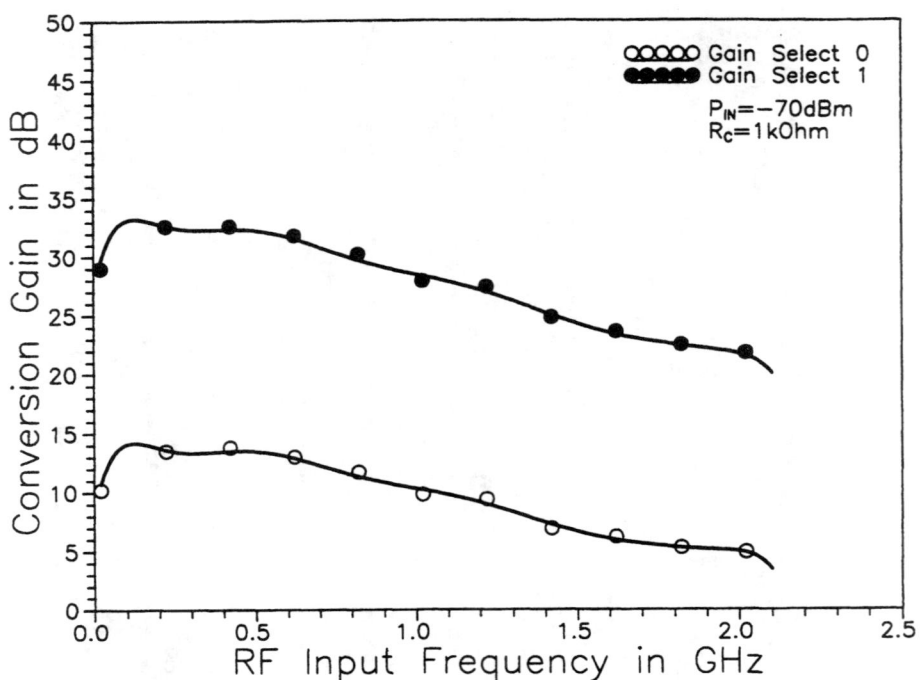

Figure 9: Frequency response of the RX train

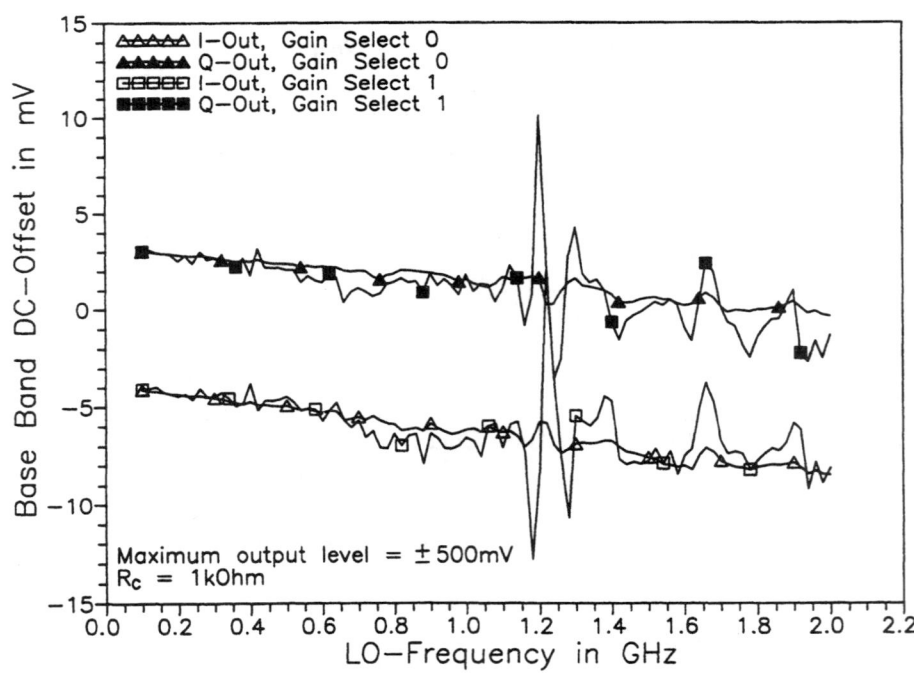

Figure 10: Baseband dc offset versus LO frequency

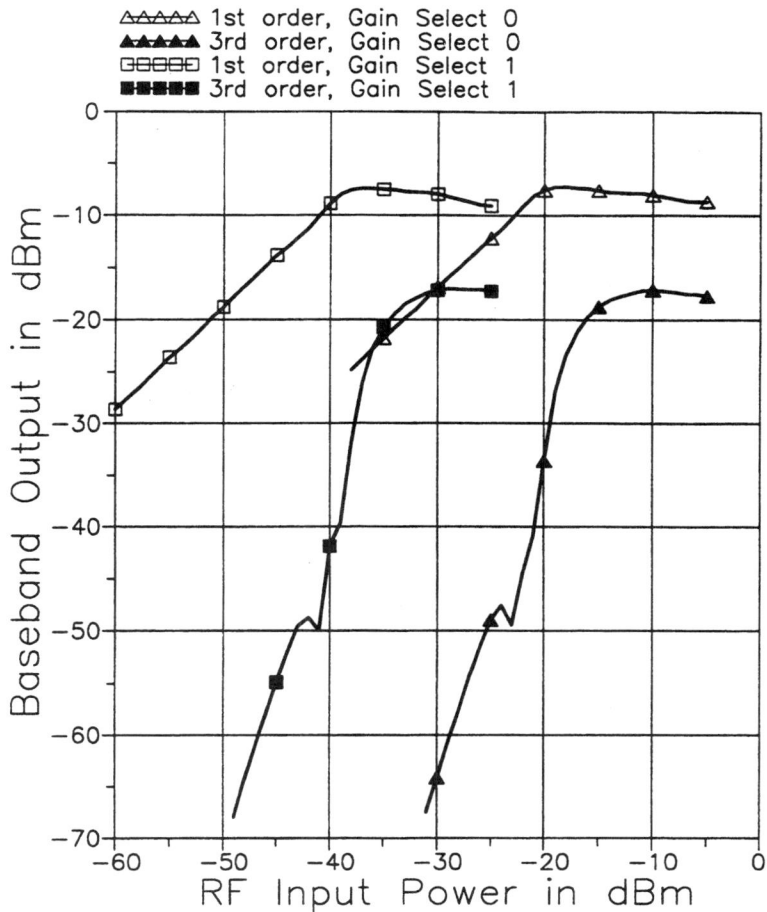

Figure 11: RX intermodulation characteristics at 200MHz

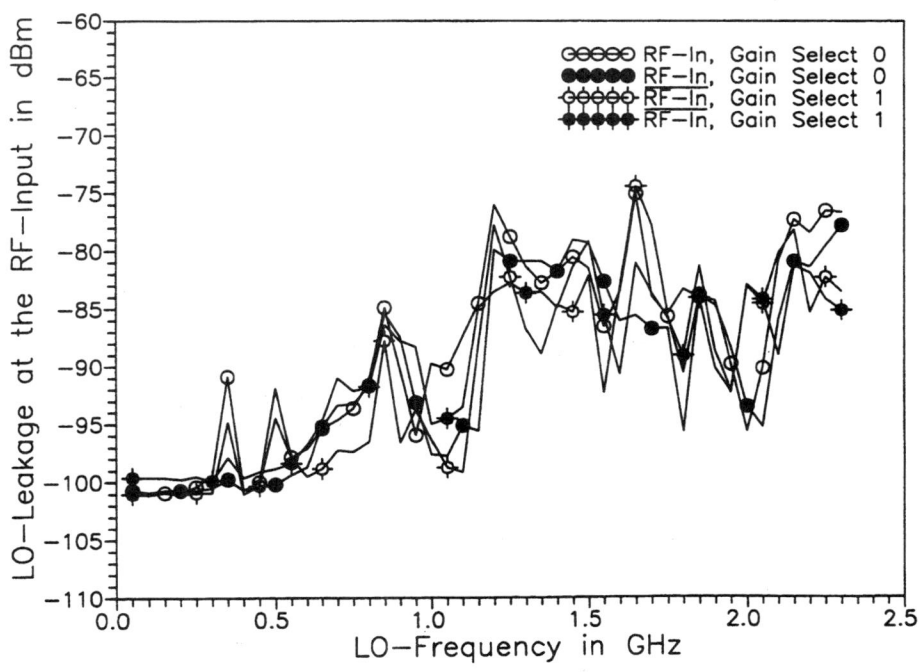

Figure 12: LO leakage radiation at the RF-input

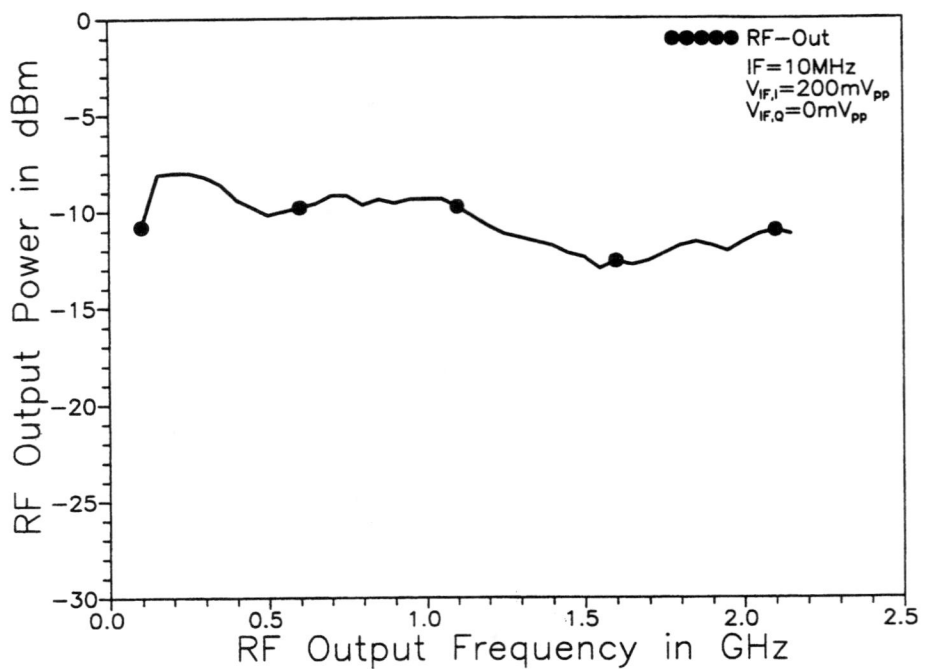

Figure 13: TX frequency response

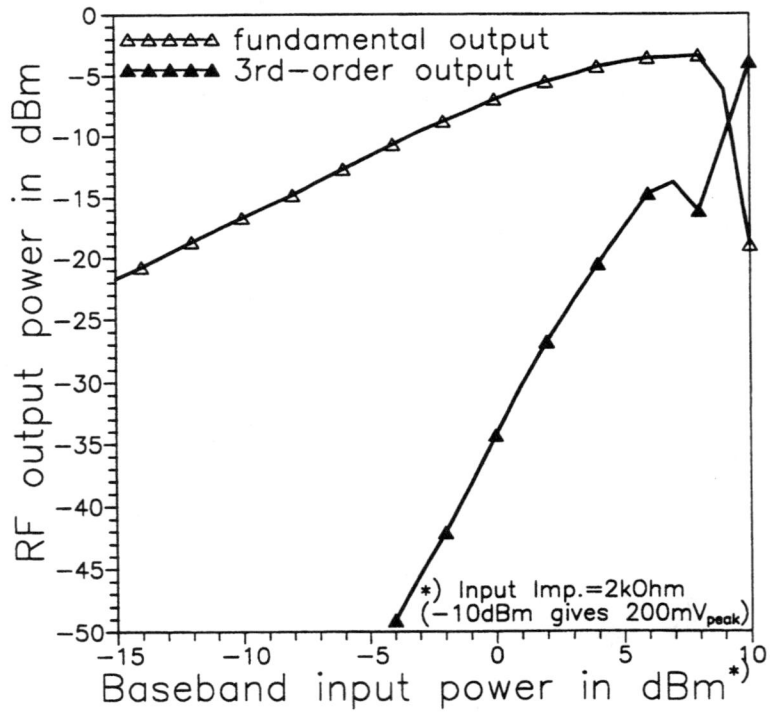

Figure 14: TX intermodulation characteristic at 1.8GHz

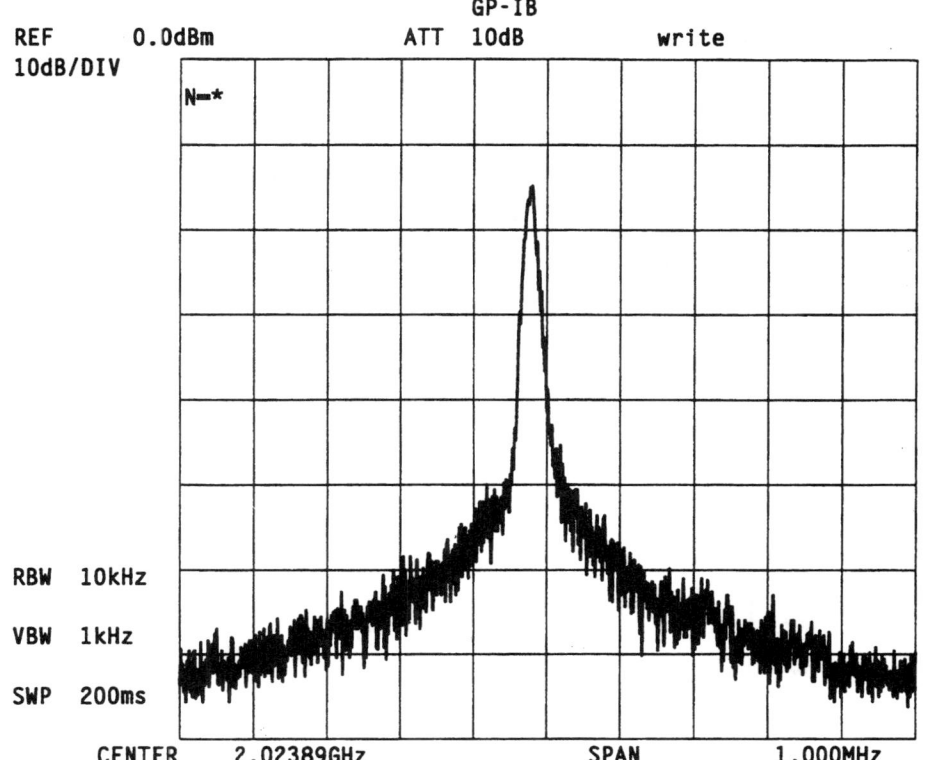

Figure 15: VCO spectrum

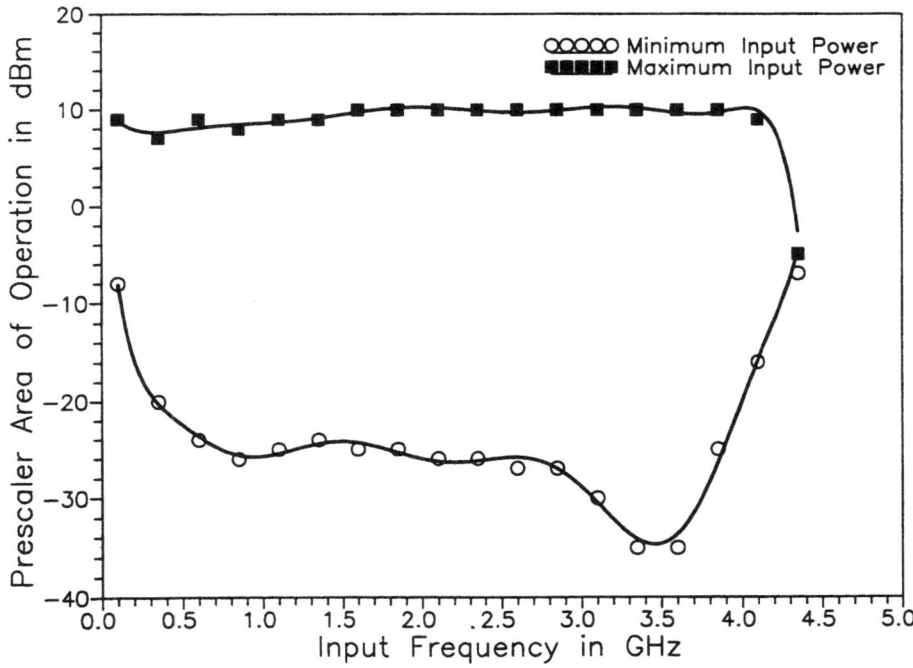

Figure 16: Prescaler input sensitivity

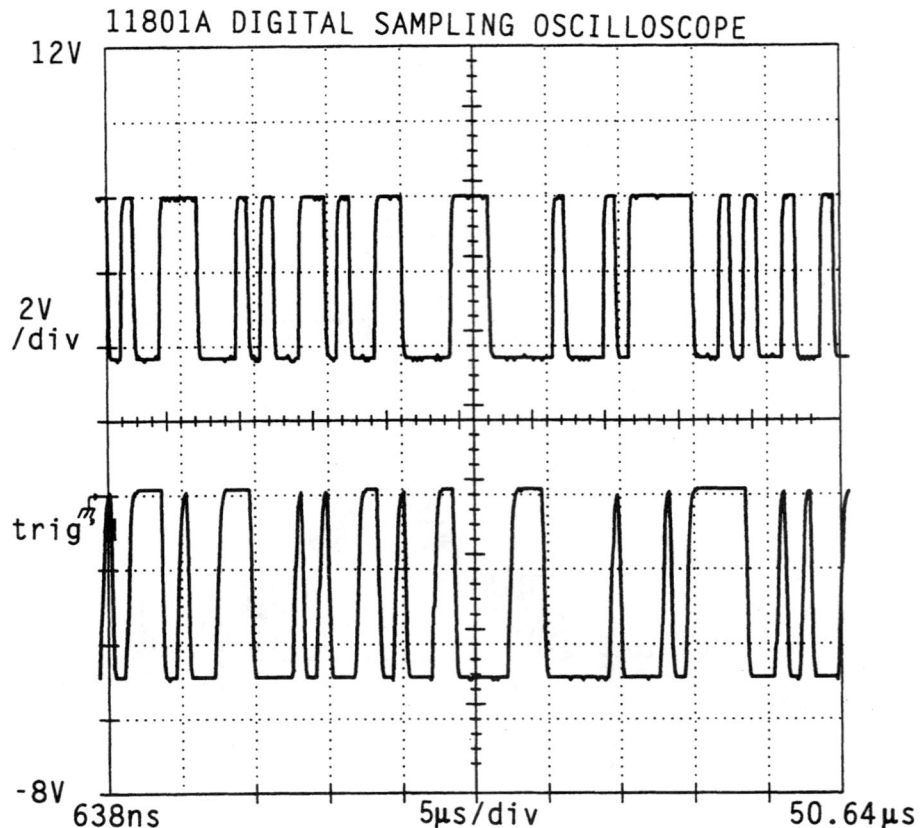

Figure 17: Demodulated DECT data (1.152Mbit/s)

A Single-Chip GaAs RF Transceiver for 1.9-GHz Digital Mobile Communication Systems

Kazuya Yamamoto, *Member, IEEE*, Kosei Maemura, Nobuyuki Kasai, Yutaka Yoshii, Yukio Miyazaki, Masatoshi Nakayama, Noriko Ogata, Tadashi Takagi, *Member, IEEE*, and Mutsuyuki Otsubo

Abstract— A 1.9-GHz single-chip GaAs RF transceiver has been successfully developed using a planar self-aligned gate FET suitable for low-cost and high-volume production. This IC includes a negative voltage generator for 3-V single voltage operation and a control logic circuit to control transmit and receive functions, together with RF front-end analog circuits—a power amplifier, an SPDT switch, two attenuators for transmit and receive modes, and a low-noise amplifier.

The IC can deliver 22-dBm output power at 30% efficiency with 3-V single power supply. The new negative voltage generator operates with charge time of less than 200 ns, producing a low level of spurious outputs below −70 dBc through the power amplifier. The generator also suppresses gate-bias voltage deviations to within 0.05 V even when gate current of −144 μA flows. The IC incorporates a new interface circuit between the logic circuit and the switch which enables it to handle power outputs over 24 dBm with only an operating voltage of 3 V. This transceiver will be expected to enable size reductions in telephones for 1.9-GHz digital mobile communication systems.

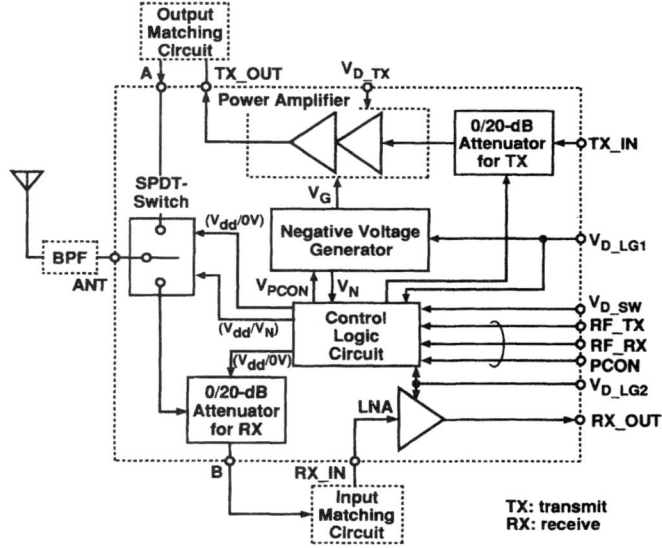

Fig. 1. Block diagram for single-chip RF transceiver.

I. INTRODUCTION

THE growing market for digital cellular and digital cordless telephones in Europe, North America, and the Far East has produced a strong demand from manufacturers of advanced digital mobile communication systems such as the Japanese personal handy phone system (PHS), for fewer RF/IF devices, devices with lower operating voltage, smaller chip size, and lower cost [1].

Regarding IF devices, some of which include RF circuits, there are many reports of 3-V operation transmitter, receiver, and transceiver chips by using Si-bipolar, BiCMOS, or CMOS circuit technologies [2]–[11]. However, these Si-chips have no power amplifier. For RF devices integrating the front-end components of portable phones, there are a few GaAs RF transceivers available in the 2.4-GHz industrial, scientific, and medical (ISM) frequency band [12]–[15]. However, these chips need both positive and negative power supplies as high as ±5 V, because a GaAs MESFET power amplifier usually requires a negative power supply for its gate biasing. Considering the requirements for the PHS, there are no reports of RF transceivers which can provide an output exceeding 19 dBm in 1.9-GHz band, although several RF monolithic microwave integrated circuits (MMIC's) operating on low supply voltage of 3 V have been reported [16]–[22]. Thus, to the best of the author's knowledge, there are no reports of an RF transceiver suitable for the PHS with 3-V single power supply.

This paper reports a newly developed single-chip GaAs RF transceiver for 1.9-GHz digital mobile communication systems such as the PHS. This transceiver integrates analog front-end circuits including a power amplifier, a low-noise amplifier, and so on, and is fabricated by using a planar self-aligned gate FET for low-cost and high-volume production. Also integrated are an on-chip negative voltage generator with a new circuit configuration suitable for biasing a power amplifier and a control logic circuit with a new interface circuit useful for improving switch power performance. These FET and circuit techniques produce a chip capable of transmitting an output power of 22 dBm with only 3-V single-voltage power supply. This paper describes the FET structure, the circuit design, and the characteristics of the IC.

II. TRANSCEIVER ARCHITECTURE

Fig. 1 shows a block diagram of the single-chip GaAs RF transceiver. This IC contains both analog and digital circuits. The analog circuits consist of a power amplifier (PA), an SPDT

Manuscript received April 10, 1996; revised August 5, 1996.
K. Yamamoto, N. Kasai, T. Takagi, and M. Otsubo are with the Optoelectronic and Microwave Devices Laboratory, Mitsubishi Electric Corporation, Itami, Hyogo 664, Japan.
K. Maemura, Y. Yoshii, and Y. Miyazaki are with the High Frequency and Optical Semiconductor Division, Hyogo, Japan.
M. Nakayama and N. Ogata are with the Information Technology R&D Center, Kanagawa, Japan.
Publisher Item Identifier S 0018-9200(96)08082-1.

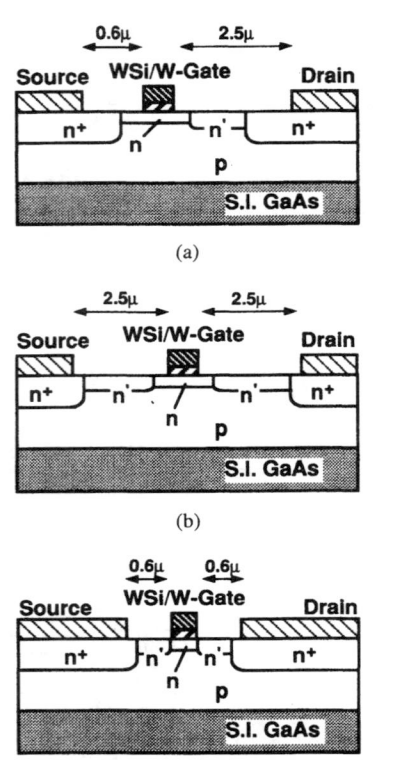

Fig. 2. Schematic cross-sections of SAG FET's for RF transceiver: (a) power FET, (b) switch and attenuator FET, and (c) low-noise amplifier and digital circuit FET.

TABLE I
TARGET SPECIFICATION FOR SINGLE-CHIP RF TRANSCEIVER

Operating frequency		1895.15~1917.95MHz
PA	Output power	> 19dBm
		(ACP< -55dBc at ±600kHz offset)
TX-ATT	Attenuation level	0/20dB two step
LNA	Noise figure	< 2dB
SW & RX-ATT	Insertion loss	< 1.5dB or 20dB (0/20dB step)
	Power handled	> 19dBm
NVG	Small ripple and small deviation in gate voltage	
	Charge time	< 1µs (negative voltage)
	Current	< 5mA
Logic	Current	< 1mA
Total current	Transmit mode	< 200mA
	Receive mode	< 10mA
Power supply		3V single-voltage
Chip size		≈4mm²

switch (SW), two attenuators for transmit and receive modes (TX-ATT and RX-ATT), and a low-noise amplifier (LNA). The digital circuits consist of a negative voltage generator (NVG) to enable single voltage operation and a logic circuit to control these analog circuits and the NVG.

The main features of the IC are: 1) 3-V single voltage operation by integrating an on-chip NVG suitable for biasing a PA; 2) incorporation of a control logic circuit capable of toggling each circuit in the transceiver between transmit and receive modes, each mode having an attenuation state controlled by three signals (RF_TX, RF_RX, and PCON); and 3) low current consumption in receive mode by operating the NVG only when transmitting.

These features are achieved through three new techniques: 1) a 3-V operation power FET fabricated using a planar self-aligned gate (SAG) FET process [23]; 2) a newly developed NVG with low output voltage deviation and ripple, and high-speed operation to enable the NVG to shut down when not transmitting; and 3) a control logic circuit with a new interface circuit that enhances switch power performance.

III. FET STRUCTURE

For GaAs power FET's, recessed gate FET's are widely used to obtain high breakdown voltage [24], [25]. However, the recess etching process makes it difficult to control threshold voltages of these FET's, and their DC/RF performances are not very uniform. To achieve better uniformity, we have previously developed a planar SAG power FET operating on a low voltage of 3.3 V shown in Fig. 2(a) [23]. This power FET utilizes an asymmetric structure with a WSi/W-double layer gate to achieve breakdown voltage of over 8 V. It can deliver a saturated power density of more than 240 mW/mm, which is useful for reducing gate-width of the final stage FET in the power amplifier. The FET also has the advantages of large-scale integration, lower cost, and enables higher volume production than ordinary recessed gate FET's.

As described in Section II, the chip incorporates many functional blocks as well as power amplification. Considering the breakdown voltage, transconductance, gain, and uniformity of threshold voltage required for the circuits other than the power amplifier, we have employed two symmetric FET's. The two FET's have different gate lengths and different gate-to-source and gate-to-drain spaces. The FET for the switch and attenuators, as shown in Fig. 2(b), has long gate-to-source and gate-to-drain spaces to achieve a high breakdown voltage of more than 10 V. To obtain high transconductance and gain, the FET for the low-noise amplifier and the digital circuits has short spaces shown in Fig. 2(c). For the digital circuits, enhancement and depletion mode FET's (E-FET and D-FET) with typical threshold voltages of -0.4 V and 0.15 V are additionally used to enable low current operation.

IV. CIRCUIT DESIGN AND MEASUREMENT RESULTS

We designed a target specification so that the transceiver would be appropriate for PHS units [1]. The main target values of the IC are listed in Table I. Circuit design focused on chip size reduction to keep costs down. The design which satisfied these specifications is described in detail below, particularly with regard to the circuit techniques, 2) and 3), introduced in Section II. Measurements were taken for a shrink small outline 30-lead package (SSOP 30).

A. Power Amplifier with TX-Attenuator

Fig. 3 shows the circuit configuration of the PA with the TX-ATT. The PA is a two-stage amplifier with a final stage FET gate-width as small as 2.4 mm. As shown in Fig. 3, the gate-bias voltages are provided by the NVG. Interstage

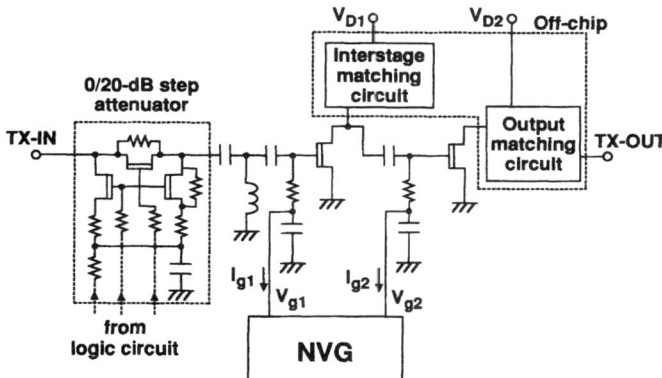

Fig. 3. Circuit configuration of power amplifier with TX-attenuator.

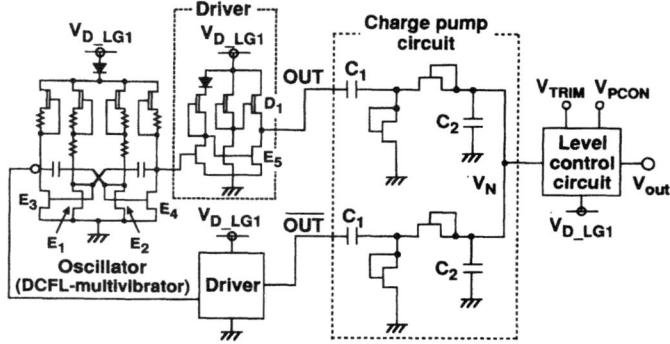

Fig. 5. Circuit configuration of negative voltage generator.

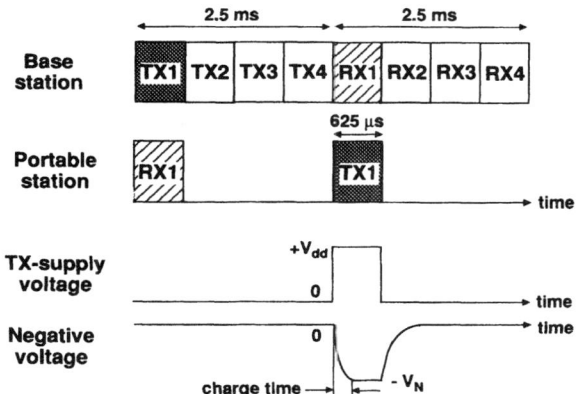

Fig. 4. Timing chart for charged negative voltage.

and output matching circuits are laid out off-chip to reduce both chip size and matching loss. A spiral inductor and several metal-insulator-metal (MIM) capacitors are formed on the chip for ac-coupling and the input matching circuit of the PA. The source and load impedances of the final stage FET were optimized for power, efficiency, and good distortion characteristics [26]. The TX-ATT employs a π-type 0/20-dB step attenuator to toggle between attenuation and no-attenuation states in transmit mode [27]. This attenuator is capable of keeping an input VSWR low for both attenuation and no-attenuation states. The gate control and pull-up bias voltages for the TX-ATT are controlled by the control logic circuit.

B. Negative Voltage Generator

A GaAs MESFET PA often needs an NVG for its negative gate biasing. Most NVG's comprise a CMOS IC operating with low-frequency of kHz-order and off-chip capacitors with large capacitances of μF-order, so they have large current capacity of mA-order. However, as shown in Fig. 4, because such a conventional NVG takes a long time to charge a sufficient negative voltage, it is unsuitable for high-speed operation that would allow the generator to be operated only during transmission as specified in the PHS standards [1]. For biasing the PA, an NVG would also require lower deviation and ripple in the output voltage.

As shown in Table I, therefore, an on-chip full NVG integrated in the IC requires the following characteristics: 1) high-speed operation to permit activating only during transmission; 2) small gate-bias voltage ripple to prevent the output of spurious signals from the PA; and 3) suppression of the gate-bias voltage deviation caused by the gate current flowing when the PA's gain is compressed, to maintain its gate-bias condition. These characteristics should be achieved by an NVG with small current capacity, because the on-chip capacitors fabricated on GaAs substrate have capacitances as small as pF-order.

We have developed the new NVG to meet these requirements. Fig. 5 shows the circuit configuration of the NVG. This NVG comprises an oscillator, two drivers, two charge pump circuits, and a level control circuit. Since these two charge pump circuits are complementarily operated, the ripple in the generated negative voltage, V_N, is greatly suppressed without requiring large capacitances such as off-chip capacitors or a high-frequency oscillator. The level control circuit converts the voltage, V_N, into the desired gate-bias voltage, V_{out}.

As shown in Fig. 5, we have employed a direct-coupled FET logic (DCFL)-multivibrator for the oscillator because a complementary signal pair is easy to generate, and the multivibrator can be connected directly to the DCFL-driver. In the multivibrator, two E-FET's for inverters, E_3 and E_4, are biased by E-FET's E_1 and E_2. With this configuration, any variations in gate-to-source voltages due to change in supply voltage affect these FET's equally, so the multivibrator can oscillate with a wide range of supply voltages. For the PHS bandwidth of about 23 MHz as shown in Table I, the circuit parameters of the multivibrator were determined to give an oscillation frequency of about 70 MHz. For the output buffer of the driver, the D-FET, D_1, and E-FET, E_5, which operate complementarily, were utilized to provide as large an output voltage swing as possible.

Sufficient reduction in the spurious output is important even if the spurious signals, which are generated by the NVG operating at about 70 MHz and are mixed with a 1.9-GHz-band desired signal, would be mostly removed through the BPF shown in Fig. 1. This is because they would affect unexpectedly some peripheral circuits such as a frequency synthesizer and up-/down-mixers in RF and IF blocks. Therefore, we gave careful consideration to this problem.

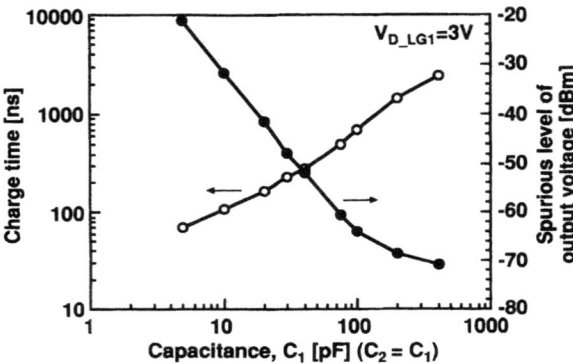

Fig. 6. Simulated results of charge time and spurious level versus capacitance of charge pump circuit.

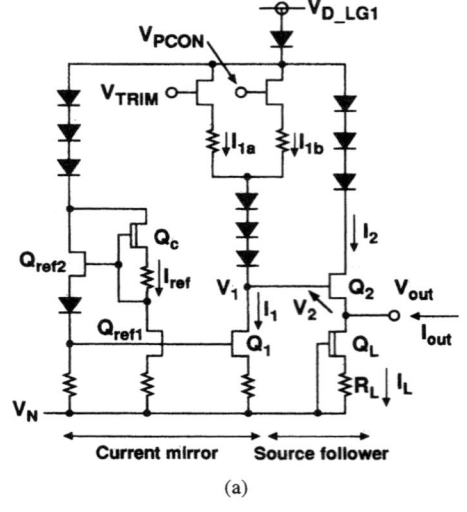

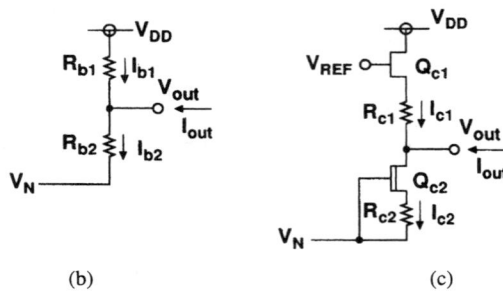

Fig. 7. Basic configurations of (a) proposed level control circuit, and (b), (c) conventional level control circuits.

SPICE simulations of the spurious level due to the output voltage, V_{out}, and of the time required to charge a sufficient negative voltage were used to determine the capacitances for C_1 and C_2. The results are shown in Fig. 6, indicating that C_1 and C_2 both need to be between 20 and 100 pF to achieve charge time below 1 μs and spurious levels of less than -40 dBm (corresponding to less than -60 dBc in consideration of the PA's output of over 20 dBm). The capacitances for both C_1 and C_2 were set to 20 pF after considering chip size and allowing a sufficient margin for charge time. Simulation was used to compare ripple of our configuration with that of a conventional circuit [28] with the same frequency and capacitances, indicating ripple reduction of at least 90%.

To reduce the deviation in the output voltage V_{out} due to the output current I_{out}, we propose the basic circuit configuration shown in Fig. 7(a) for the level control circuit. Fig. 7(b) and (c) shows the conventional configurations of the level shift circuits [28]. In Fig. 7(a), V_N is the negative voltage generated from the charge pump circuits, and V_{TRIM} and V_{PCON} are the reference voltages for trimming the gate-bias voltage of the PA and reducing the drain current dissipated there. This circuit configuration features the combination of the current mirror circuit and the source follower. The circuitry can keep the voltage V_{out} almost constant independently of the current I_{out} while I_{out} is not as large as I_L [29]. In addition, since the branches for the currents I_{1a} and I_{1b} are combined with the branch for the constant current I_1, a lower output voltage V_{out} can be obtained by changing V_{PCON} to a lower voltage in an attenuation state. This function is useful for reduction in drain current consumption of the PA when transmitting in an attenuation state.

Fig. 8 shows the simulated result of the output characteristics of the NVG for the proposed circuit of Fig. 7(a) and conventional circuits of Fig. 7(b) and (c). In this figure, the conventional circuit curves, 1 and 2, represent the level shift circuits shown in Fig. 7(b) and (c). The simulation was performed with currents I_L, I_{b2}, and I_{c2} equal to each other at 210 μA. Over the range of output current below 210 A, the proposed circuit suppresses voltage deviations to within 0.1 V even when the output current increases. The simulation indicated that our proposed NVG would satisfy all the specified requirements.

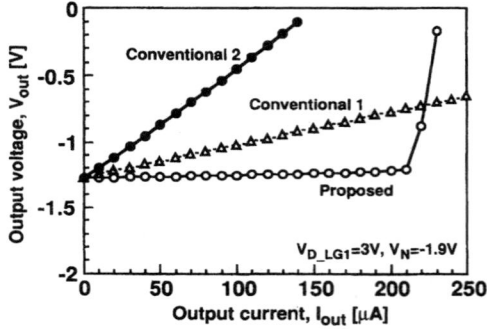

Fig. 8. Simulated result of level control circuits.

C. Measurement Results for Transmitter

Fig. 9(a) shows the input and output characteristics of the IC measured with a drain-bias voltage of 3.0 V and an input frequency of 1.9 GHz. In Fig. 9(b), the gate-bias and charged negative voltages are plotted against the gate-current of the final-stage FET for the same measurement. The PA was biased with class AB operation (about $0.2I_{dss}$). As shown in Fig. 9(a), the PA, associated gain of which is 25 dB, delivers 22-dBm output power and 30% power-added efficiency with the ± 600-kHz-offset adjacent channel power leakage, ACP, of -55 dBc or less. This efficiency is 5% higher than that of the previously reported MMIC two-stage power amplifier which operated on 3-V single power supply and was tested for the application to the PHS [17]. Fig. 9(b) shows that the NVG

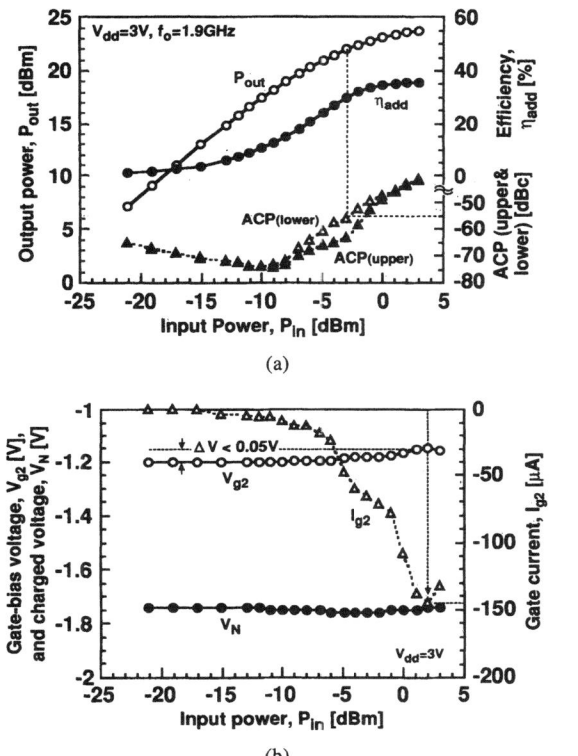

Fig. 9. (a) Output characteristics and (b) gate-bias characteristics versus input power of power amplifier with TX-attenuator biased by the negative voltage generator.

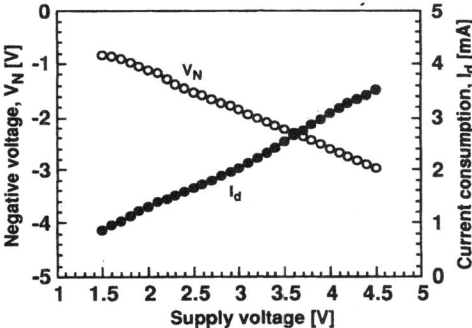

Fig. 10. Supply voltage dependence of negative voltage generator.

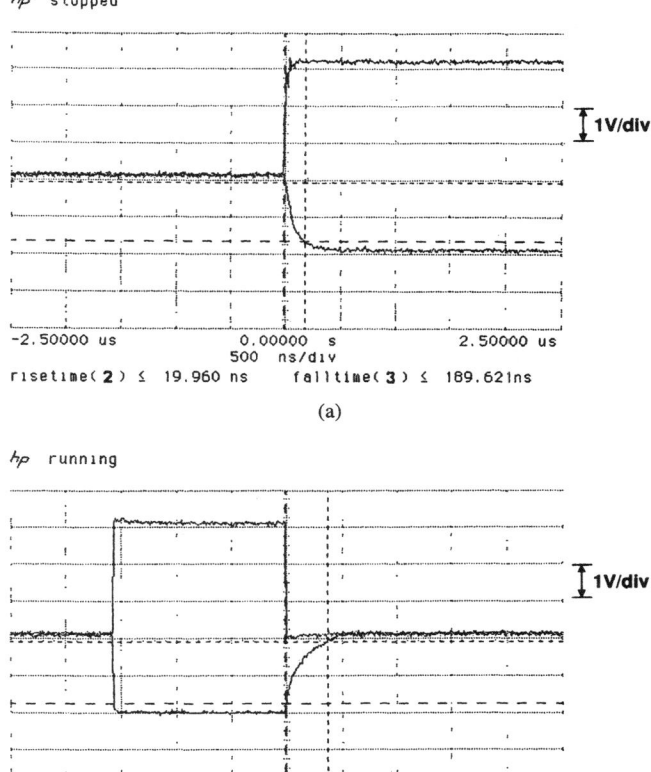

Fig. 11. Transient response of negative voltage generator: (a) charged- and (b) discharged-negative voltages.

generates the charged negative voltage V_N of about -1.75 V, and that the deviation in the final-stage FET gate-bias voltage V_{g2} is suppressed to within 0.05 V even when the gate current I_{g2} of -144 μA flows. These findings suggest that our proposed circuit configuration is useful for suppressing output voltage deviation.

Fig. 10 shows the supply voltage dependence of the NVG. Due to the multivibrator with its improved self-bias circuit, the NVG can operate over a wide range of supply voltages from 1.5 to 4.5 V. Its 2.04-mA current consumption with 3-V power supply is only about one-fourth that of the previously reported IC [28].

Transient response of the NVG is shown in Fig. 11(a) and (b). Pulse duration of power supply is 625 μs, as specified in the PHS standards (Fig. 4). Fall and rise times for charging and discharging the negative voltage are 190 ns and 152 μs, respectively. The fall time agrees well with the simulation results shown in Fig. 6. This response is fast enough to permit activation only during transmit mode, thereby achieving a significant reduction in current consumption.

Fig. 12(a) and (b) shows the output spectra of the PA measured for 1.9-GHz $\pi/4$-DQPSK modulated signal and a 1.9-GHz continuous signal. From Fig. 12(b), it can be seen that the NVG produces an extremely low level of spurious signals below -70 dBc, indicating that the spurious levels are adequately suppressed.

Fig. 13 shows measured output power, power-added efficiency, associated gain, and ACP versus supply voltage. With ACP below -55 dBc, output power of more than 21 dBm, associated gain of more than 24 dB, and power-added efficiency of more than 25% are obtained for supply voltages from 2.7 to 3.3 V. In both attenuation and no-attenuation states, output power and drain current characteristics of the PA are shown in Fig. 14. With attenuation, output power drops to 1.7 dBm at the 0/20-dB step TX-ATT, corresponding to gain degradation of 20.3 dB. The drain current of the PA decreases from 174.6 to 38.9 mA due to the level control circuit, as predicted. This decrease in the attenuation state is appropriate for battery-saving in practical applications.

D. Control Logic Circuit

Fig. 15 shows the block diagram of the control logic circuit. This circuit controls the analog circuits and the NVG in accor-

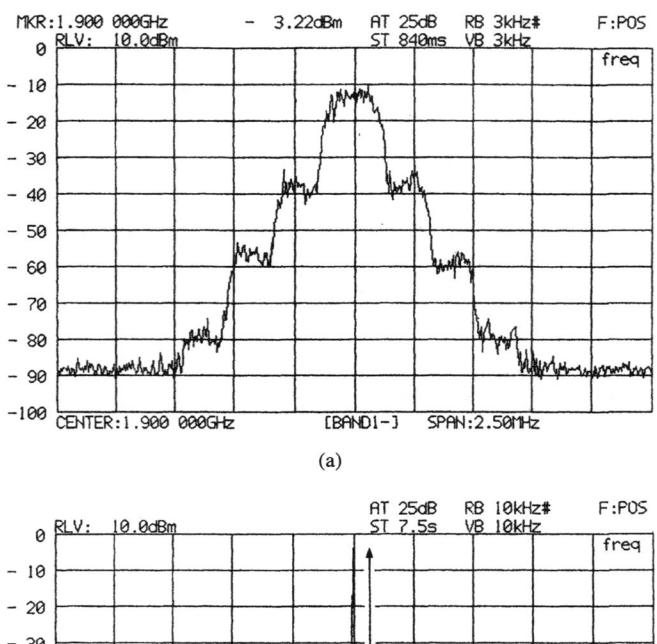

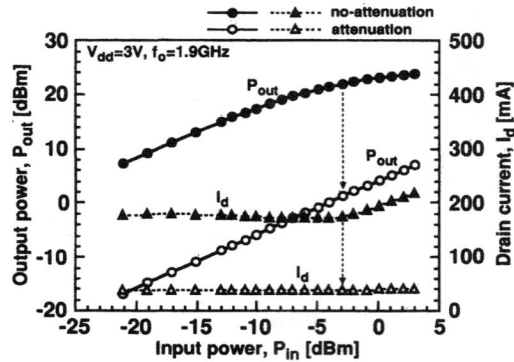

Fig. 14. Output power and drain current with and without attenuation.

Fig. 12. Output spectra for (a) $\pi/4$-DQPSK modulated signal and (b) continuous signal.

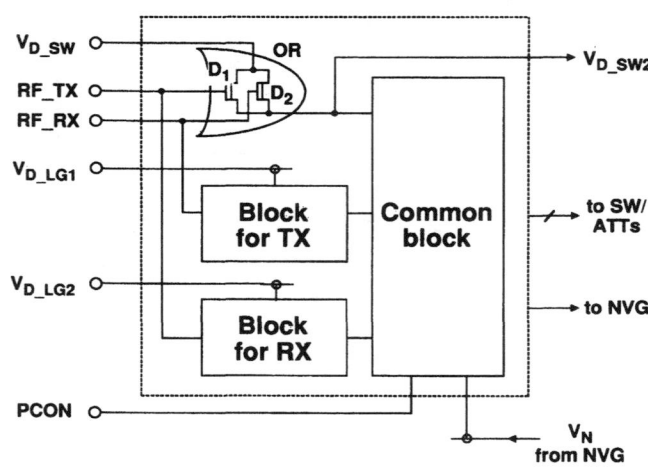

Fig. 15. Block diagram for control logic circuit.

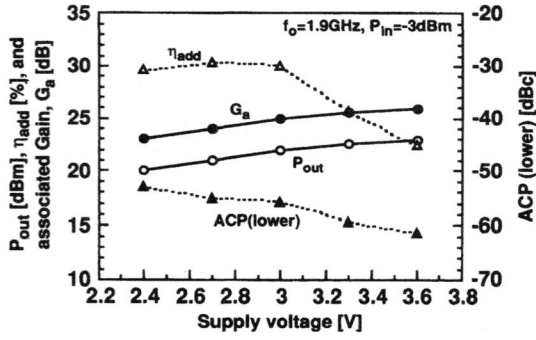

Fig. 13. Output power, efficiency, associated gain, and ACP versus supply voltage.

dance with transmit and receive modes, each of which has an attenuation state controlled by a radio signal strength indicator (RSSI) in a baseband IC. Mode switching is controlled by three signals (RF_TX, RF_RX, and PCON). A DCFL circuit with a resistor is employed as the basic gate to ensure low power dissipation.

As shown in this figure, power supply connections (V_{D_LG1}, V_{D_LG2}, and V_{D_SW}) are separated so that only the part of the circuit necessary for operation of a particular mode can be activated. The common block, necessary for both transmit and receive modes, is supplied through an OR gate whose inputs are transmit and receive signals (RF_TX and RF_RX). The D-FET's D_1 and D_2 employ a 100-μm wide gate-width to avoid voltage drop. The output of the OR gate is also available to the pull-up voltage of the SW and ATT's, V_{D_SW2}.

As a result, current consumptions in the transmit and receive modes are as low as 0.67 mA and 0.80 mA. During sleep mode, a current of less than 0.1 mA is dissipated in the logic circuit. That is low enough to enable a CMOS IC to be used for driving the logic circuit. A new interface circuit is also incorporated in the logic circuit to enhance the switch power performance. This is described in more detail later.

E. Switch and RX-Attenuator

Fig. 16 shows the circuit configuration of the SW and RX-ATT, together with the interface circuit between the output buffers in the logic circuit and the SW. To reduce chip size, we have adopted the conventional switch which consists of shunt/series FET's F_1 and F_2 for transmit and a single FET F_3 for receive [27]. A π-type 0/20-dB step attenuator, whose circuitry is shown in Fig. 3, is employed as the RX-ATT. The drain and source voltages of FET's in the SW and RX-ATT

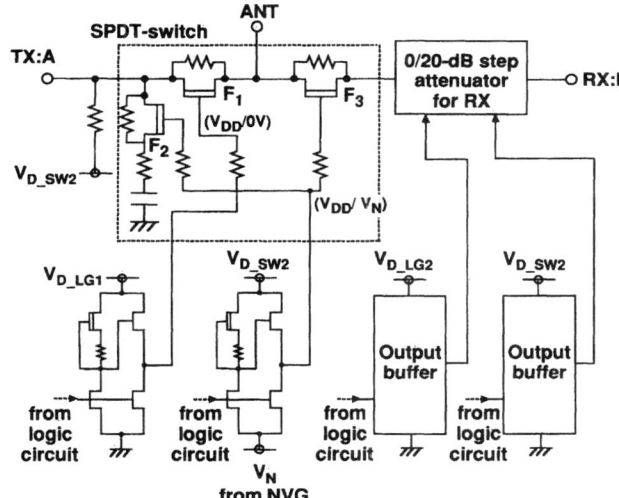

Fig. 16. New interface configuration between the output buffers in the logic circuit and the switch.

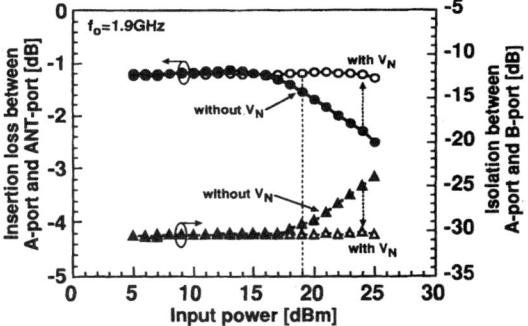

Fig. 17. Transfer characteristics measured in transmit mode.

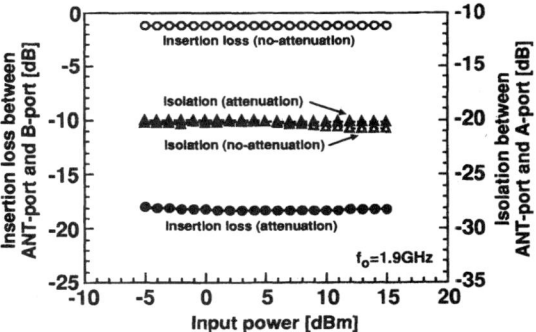

Fig. 18. Transfer characteristics measured in receive mode.

are pulled up to a supply voltage of about 3 V through the voltage V_{D_SW2}.

However, when the SW and RX-ATT operates with gate voltages of 3 V or 0 V in transmit mode, the isolation between TX port A and RX port B is insufficient to deliver the output power of more than 19 dBm required in the PHS [1]. This is because the gate-source voltages applied to the FET's F_2 and F_3 which are about -3 V, are not high enough. For this reason, conventional switch circuity has problems with handling high power with low supply voltages.

To enable the switch to handle high power levels, we have incorporated a new interface circuit onto the chip. As shown in Fig. 16, both positive and negative voltages, V_{dd} and V_N (for example, 3 V and -1.7 V), are supplied to the output buffer for driving F_2 and F_3. When the NVG operates in transmit mode, the gate-source voltages of F_2 and F_3 become as high as $V_N - V_{dd} (\cong -4.7$ V). Therefore, the SW remains off against higher power levels. In receive mode when the SW handles very low power, all output buffers operate with V_{dd} and 0 V, because the NVG goes to sleep and V_N outputs about 0 V. The interface circuit, therefore, readily improves the ability of the switch to handle high power.

F. Measurement Results for Switch and RX-Attenuator

Figs. 17 and 18 show the transmit and receive transfer characteristics measured for the circuit shown in Fig. 16. As shown in Fig. 17, the insertion loss between TX(A) and ANT ports is less than 1.2 dB, and the isolation between TX(A) and RX(B) ports is greater than 30 dB. Neither insertion loss nor isolation with negative voltage V_N are degraded over the input power range from 2 to 24 dBm in comparison to those without V_N. These results indicate that performance is satisfactory for the PHS, proving the effectiveness of the new interface circuit. When measurements were taken, the VSWR's at ports TX(A) and ANT were as low as 1.5 and 1.4.

In receive mode, as shown in Fig. 18, the insertion losses between ANT and RX(B) ports are less than 18.2 dB and 1.2 dB in attenuation and no-attenuation states, respectively. The isolation between ANT and TX(A) ports is greater than 20 dB regardless of attenuation states. The VSWR's at the ANT port were suppressed to be less than 1.5 and 1.3 in attenuation and no-attenuation states, because the π-type attenuator maintains good return losses for both states.

Thus, these characteristics are sufficiently good for realizing an RF unit of a digital cordless telephone, taking account of the output power of the PA and a loss of the external BPF shown in Fig. 1. The insertion losses are, however, a little greater than those of the previously reported IC's which have only a switching function and operate with control voltages of 3 V and 0 V, though the switch power performance is nearly comparable to that of their IC's [21], [22]. We consider that the losses would be mainly caused by the attenuator integrated together with the switch and the inductances of leads and bonding-wires in the package.

G. Low Noise Amplifier

The circuit configuration of the LNA is shown in Fig. 19. It is a self-biased single-stage amplifier utilized to permit the NVG to go to sleep when not transmitting (including receive mode). The gate-width of the FET is 300 μm. As shown in Fig. 19, the bypass capacitor for self-biasing is formed on the chip to avoid the influence of lead-inductance. In order to permit easily the connection with a next-stage circuit such as a down-mixer, the output matching circuit is fabricated on-chip by using a spiral inductor, MIM capacitors, and ion-implanted resistors. In contrast, the input matching circuit and the self-bias resistor are fabricated off-chip to reduce chip size and matching loss. This topology produces flexibility for noise matching conditions and current consumption, as well as chip-size reduction.

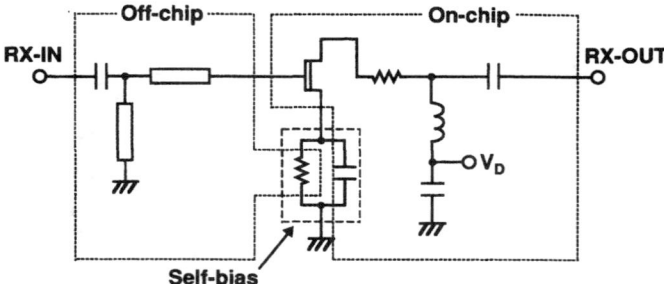

Fig. 19. Circuit configuration of low-noise amplifier.

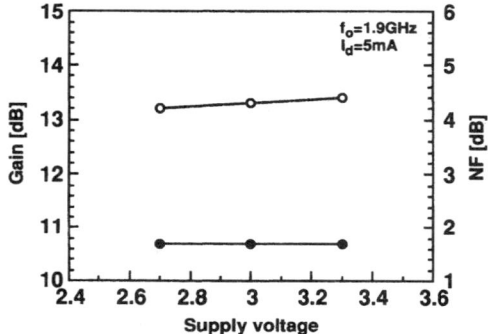

Fig. 20. Gain and NF versus supply voltage.

Fig. 20 shows gain and noise figure measurements versus supply voltage. With noise matching conditions of $I_d = 5$ mA biased with an off-chip self-bias resistor, gain of over 13.2 dB and noise figure of 1.7 dB are obtained for the range of supply voltages from $V_d = 2.7$ to 3.3 V. Then, an input third-order intercept point, IIP_3, of -2 dBm was obtained. The noise figure of 1.7 dB is better by 0.5 dB, compared with that of the previously reported self-biased single-stage LNA, though the gain is a little lower than that of the LNA due to no use of a dual-gate FET [16].

Finally, a chip micrograph of the single-chip RF transceiver and a photograph of its package are shown in Fig. 21(a) and (b). The chip size is 1.95 mm × 2.1 mm, and the package is an SSOP 30 as mentioned earlier. Table II summarizes the electrical properties of the transceiver. Total current consumptions are 177.3 mA, 5.8 mA less than 0.1 mA in transmit, receive, and sleep modes. In transmit mode with attenuation, current consumption is 41.6 mA, only a fourth of that with no attenuation. These current consumption figures are sufficiently small for digital cordless telephone applications.

V. CONCLUSION

We have successfully developed a 1.9-GHz single-chip GaAs RF transceiver using a planar self-aligned gate FET suitable for low-cost and high-volume production. By integration of a negative voltage generator and a control logic circuit together with RF front-end analog circuits, this IC is capable of 3-V single voltage operation and can deliver 22-dBm output power at 30% efficiency. The new negative voltage generator operates with charge time of less than 200 ns, achieving low spurious output below -70 dBc and small gate voltage deviation within 0.05 V even when the gate current of -144 μA flows. Despite a low supply voltage

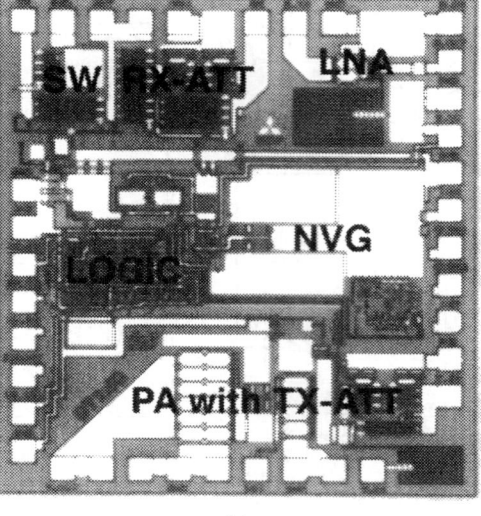

(a)

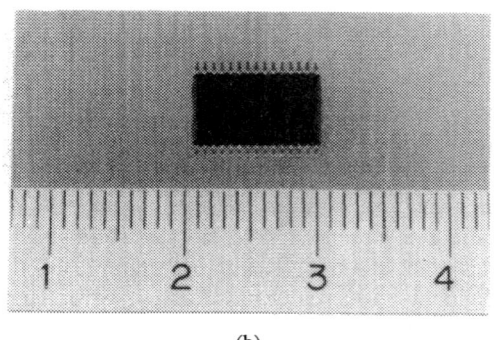

(b)

Fig. 21. (a) Chip micrograph and (b) photograph of package.

TABLE II
ELECTRICAL PROPERTIES OF SINGLE-CHIP RF TRANSCEIVER

Transmit mode		
PA & TX-ATT	Output (High)	22dBm (η_{add}=30%, ACP < -55dBc)
	(Low)	1.7dBm (-20.3dB down)
SW & RX-ATT	Insertion loss	<1.2dB
	Isolation	>30dB
	Power handled	>24dBm
NVG	Spurious level	<-70dBc
	Deviation in Vg	< 0.05V (Ig < 140μA)
	Charge time of V_N	t_f(charge)=190 ns, t_r(discharge)=152μs
Receive mode		
SW & RX-ATT	Insertion loss	<1.2dB (no-attenuation state)
		18.2dB (attenuation state)
	Isolation	>20.0dB
		(both attenuation and no-attenuation)
LNA	Linear gain	13.3dB (I_d=5mA)
	Noise figure	1.7dB
Current	Transmit (High)	177.3mA (174.6mA + 2.04mA + 0.67mA)
	(Low)	41.6mA (38.9mA + 2.04mA + 0.67mA)
	Receive	5.8mA (5mA + 0.80mA)
	Sleep	<0.1mA
Power supply		3.0V single-voltage

of 3 V, high power of over 24 dBm is handled due to the introduction of a new interface circuit between the logic circuit and the switch. Moreover, sufficiently low transmit and receive

current consumptions of 177.3 mA and 5.8 mA are suitable for practical portable telephone applications. We expect that this transceiver will enable size reductions in telephones for 1.9-GHz digital mobile communication systems.

REFERENCES

[1] F. McGrath, K. Jackson, E. Heaney, A. Douglas, W. Fahey, R. G. Pratt, and T. Begnoche, "A 1.9-GHz GaAs chip set for personal handyphone system," *IEEE Trans. Microwave Theory Tech.*, vol. 43, pp. 1733–1744, July 1995.
[2] V. Thomas, J. Fenk, and S. Beyer, "A one-chip 2GHz single superhet receiver for 2Mb/s FSK radio communication," in *IEEE ISSCC Dig. Tech. Papers*, Feb. 1994, pp. 42–43.
[3] J. Sevenhans, D. Haspeslagh, A. Delarbre, L. Kiss, Z. Chang, and J. F. Kukielka, "An analog radio front-end chip set for 1.9 GHz mobile radio telephone application," in *IEEE ISSCC Dig. Tech. Papers*, Feb. 1994, pp. 44–45.
[4] C. Marshall et al., "A 2.7 V GSM transceiver IC's with on-chip filtering," in *IEEE ISSCC Dig. Tech. Papers*, Feb. 1995, pp. 148–149.
[5] T. Stetzler, I. Post, J. Havens, and M. Koyama, "A 2.7 V to 4.5 V single-chip GSM transceiver RF integrated circuit," in *IEEE ISSCC Dig. Tech. Papers*, Feb. 1995, pp. 150–151.
[6] S. Heinen, S. Beyer, and J. Fenk, "A 3.0 V 2 GHz transmitter IC for digital radio communication with integrated VCO's," in *IEEE ISSCC Dig. Tech. Papers*, Feb. 1995, pp. 146–147.
[7] C. Takahashi, R. Fujimoto, S. Arai, K. Itakura, T. Ueno, H. Tsurumi, H. Tanimoto, S. Watanabe, and K. Hirakawa, "A 1.9 GHz Si direct conversion receiver IC for QPSK modulation systems," in *IEEE ISSCC Dig. Tech. Papers*, Feb. 1995, pp. 138–139.
[8] H. Sato, K. Kashiwagi, K. Niwano, T. Iga, T. Ikeda, and K. Mashiko, "A 1.9 GHz single-chip IF transceiver for digital cordless phones," in *IEEE ISSCC Dig. Tech. Papers*, Feb. 1996, pp. 342–343.
[9] D. H. Shen, C.-M. Hwang, B. Lusignan, and B. A. Wooley, "A 900 MHz integrated discrete-time filtering RF front-end," in *IEEE ISSCC Dig. Tech. Papers*, Feb. 1996, pp. 54–55.
[10] A. Hairapetian, "An 81 MHz IF receiver in CMOS," in *IEEE ISSCC Dig. Tech. Papers*, Feb. 1996, pp. 56–57.
[11] C. D. Hull, R. R. Chu, and J. L. Tham, "A direct-conversion receiver for 900 MHz (ISM band) spread-spectrum digital cordless telephone," in *IEEE ISSCC Dig. Tech. Papers*, Feb. 1996, pp. 344–345.
[12] L. M. Devlin, B. J. Buck, J. C. Clifton, A. W. Dearn, and A. P. Long, "A 2.4 GHz single chip transceiver," in *Dig. IEEE Microwave and Millimeter-Wave Monolithic Circuits Symp.*, 1993, pp. 23–26.
[13] T. Apel, E. Creviston, S. Ludvik, L. Quist, and B. Tuch, "A GaAs MMIC transceiver for 2.45 GHz wireless commercial products," in *Dig. IEEE Microwave and Millimeter-Wave Monolithic Circuits Symp.*, 1994, pp. 15–18.
[14] M. S. Wang, M. Carriere, P. O'Sullivan, and B. Maoz, "A single-chip MMIC transceiver for 2.4 GHz spread spectrum communication," in *Dig. IEEE Microwave and Millimeter-Wave Monolithic Circuits Symp.*, 1994, pp. 19–22.
[15] B. Khabbaz, A. Douglas, J. DeAngelis, L. Hongsmatip, V. Pelliccia, W. Fahey, and G. Dawe, "A high performance 2.4 GHz transceiver chip-set for high volume commercial applications," in *Dig. IEEE Microwave and Millimeter-Wave Monolithic Circuits Symp.*, 1994, pp. 11–14.
[16] S. Tanaka, E. Hase, A. Nakajima, K. Sugano, T. Fujioka, Y. Imakado, K. Fujiwara, T. Okamoto, Y. Shigeno, K. Sato, I. Arai, M. Yamane, C. Kusano, K. Sakamoto, J. Nakagawa, and M. Koya, "A 3 V MMIC chip set for 1.9 GHz mobile communication systems," in *IEEE ISSCC Dig. Tech. Papers*, Feb. 1995, pp. 144–145.
[17] M. Nagaoka, T. Inoue, K. Kawakyu, S. Obayashi, H. Kayano, E. Takagi, Y. Tanabe, M. Yoshimura, K. Ishida, Y. Kitaura, and N. Uchitomi, "High-efficiency monolithic GaAs power MESFET amplifier operating with a single low voltage supply for 1.9-GHz digital mobile communication applications," in *IEEE Microwave Theory Technology Symp. Dig.*, 1994, pp. 577–580.
[18] S. Makioka, K. Tateoka, M. Yuri, N. Yoshikawa, and K. Kanazawa, "A high efficiency GaAs MCM power amplifier for 1.9 GHz digital cordless telephones," in *Dig. IEEE Microwave and Millimeter-Wave Monolithic Circuits Symp.*, 1994, pp. 51–54.
[19] T. Kunihisa, T. Yokoyama, H. Fujimoto, K. Ishida, H. Takehara, and O. Ishikawa, "High efficiency, low adjacent channel leakage GaAs power MMIC for digital cordless telephone," in *Dig. IEEE Microwave and Millimeter-Wave Monolithic Circuits Symp.*, 1994, pp. 55–58.
[20] M. Muraguchi, M. Nakatsugawa, H. Hayashi, and M. Aikawa "A 1.9 GHz-band ultra low power consumption amplifier chip set for personal communications," in *Dig. IEEE Microwave and Millimeter-Wave Monolithic Circuits Symp.*, 1995, pp. 145–148.
[21] H. Uda, T. Yamada, T. Sawai, K. Nogawa, and Y. Harada, "High-performance GaAs switch IC's fabricated using MESFET's with two kinds of pinch-off voltages and a symmetrical pattern configuration," *IEEE J. Solid-State Circuits*, vol. 29, pp. 1262–1269, Oct., 1994.
[22] S. Kusunoki, T. Ohgihara, M. Wada, and Y. Murakami, "SPDT switch MMIC using E/D-mode GaAs JFET's for personal communications," in *Dig. IEEE GaAs IC Symp.*, 1992, pp. 135–138.
[23] N. Kasai, M. Noda, K. Ito, K. Yamamoto, K. Maemura, Y. Ohta, T. Ishikawa, Y. Yoshii, M. Nakayama, H. Takano, and O. Ishihara, "A high power and high efficiency GaAs BPLDD SAGFET with WSi/W double-layer gate for mobile communication systems," in *Dig. IEEE GaAs IC Symp.*, 1995, pp. 59–62.
[24] J. L. Lee et al., "3.3 V operation GaAs power MESFET's with power added efficiency for hand-held telephones," *Electron. Lett. 28*, vol. 30, no. 9, pp. 739–740, 1994.
[25] Y. Ota et al., "Application of heterojunction FET to power amplifier for cellular telephone," *Electron. Lett. 28*, vol. 30, no. 9, pp. 906–907, 1994.
[26] M. Nakayama, T. Umemoto, Y. Itoh, and T. Takagi, "1.9 GHz high-efficiency linear MMIC amplifier," in *1994 Asia-Pacific Microwave Conf. Proc.*, 1994, pp. 347–350.
[27] M. Nakayama et al., "A 1.9 GHz single-chip RF front-end GaAs MMIC for personal communications," in *Dig. IEEE Microwave and Millimeter-Wave Monolithic Circuits Symp.*, 1996, pp. 69–72.
[28] M. Nishida, T. Sawai, S. Murai, T. Higashino, K. Honda, and Y. Harada, "A fully monolithic negative voltage generator IC for gain controlled MMIC amplifier," in *1994 Asia-Pacific Microwave Conf. Proc.*, 1994, pp. 343–346.
[29] K. Yamamoto, K. Maemura, Y. Ohta, N. Kasai, M. Noda, H. Yuura, Y. Yoshii, M. Nakayama, N. Ogata, T. Takagi, and M. Otsubo, "A GaAs RF transceiver IC for 1.9 GHz digital mobile communication systems," in *IEEE ISSCC Dig. Tech. Papers*, Feb. 1996, pp. 340–341.

A 1.9-GHz Single Chip IF Transceiver for Digital Cordless Phones

Hisayasu Sato, Kenichi Kashiwagi, Kazuhito Niwano, Tetsuya Iga, Tatsuhiko Ikeda, Koichiro Mashiko, *Member, IEEE,* Tadashi Sumi, and Koji Tsuchihashi

Abstract—A 1.9 GHz IF transceiver for the Japanese standard personal handy-phone system (PHS) is fabricated in a 0.8-μm BiCMOS process with 20 GHz npn. A down-mixer, up-mixer, variable attenuator, quadrature modulator, first and second PLL, and second VCO are included in the 3.4 × 3.0 mm^2. The chip draws 24 mA in receive mode and 44 mA in transmit mode, operating from 3.0 V. A total vector error of 4% for the $\pi/4$ QPSK PN9 pattern includes the up-mixer and the dual PLL's.

I. INTRODUCTION

PORTABLE communication equipment, such as cellular phones and digital cordless phones, are trying to replace as many discrete devices as possible with high-density IC's to be competitive in size, weight, power dissipation, and price. Although single-chip transceivers for UHF or 900 MHz have been already reported, it is still difficult to achieve that integration level for L-band cordless phones [1]–[3]. This is because high-frequency operation increases the interblock signal interference and prevents high-level integration.

The Japanese personal handy-phone system (PHS) is an L-band digital cordless phone. Its handset can be used as a home cordless phone and a microcell mobile phone. The PHS commercial service started in July 1995. The PHS features a long battery life and a small handset because the microcell system needs a lower power output (20 mW) than cellular phones. However, as the transceiver part in the first version PHS handset consists of silicon/GaAs discrete devices and

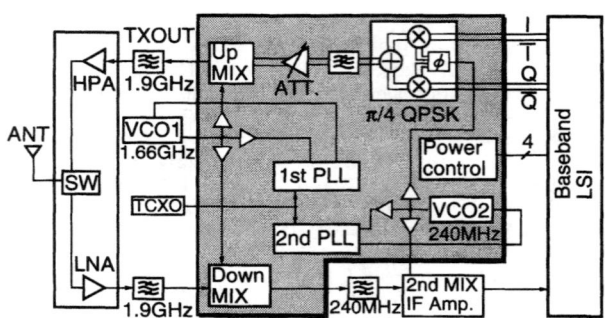

Fig. 1. IF transceiver block diagram.

several IC's, its advantage was not achieved. Our motivation is to implement the IF part on a silicon chip and to reduce current draw from 3 V batteries.

This paper describes a 1.9 GHz single chip IF transceiver IC for digital cordless phones. The chip has been implemented in a 0.8-μm BiCMOS process. In the following sections, the chip architecture and block level description are presented. The subcircuit details and layouts are then discussed. Finally, experimental results are given.

II. CHIP ARCHITECTURE

Fig. 1 shows a block diagram of the IF transceiver we have developed. The chip integrates two programmable synthesizers, an up-mixer, an attenuator, a $\pi/4$ QPSK modulator, a down-mixer, and a second VCO. The first VCO, the second mixer, and IF amplifier are the only off-chip devices in the IF part. The RF front end is a GaAs MMIC consisting of a high-power amplifier, a low-noise amplifier, and a switch.

It is very important for portable equipment to lower the cost of system component. The best solution will be to integrate all transceiver parts on a silicon chip, if silicon devices could satisfy the requirements. Unfortunately, with present process technology, the power added efficiency of a silicon power amplifier at 1.9 GHz is lower than that of GaAs amplifier, and a silicon LNA is not good enough for noise figure [3]. Therefore, front-end amplifiers are off-chip. The IF transceiver was designed to have good connectivity with the in-house GaAs front-end IC [4]. Recently, L-band silicon LNA's with noise figures of less than 3 dB have been reported [5], [6]. In the near future, LNA will be successfully integrated on a silicon transceiver.

Although the double conversion architecture requires an additional second mixer and IF amplifier, low-power, less-

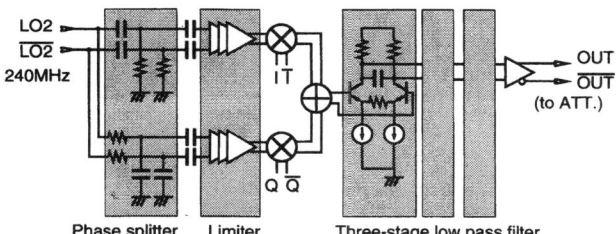

Fig. 2. Quadrature modulator.

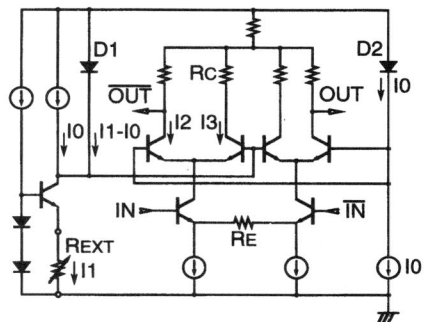

Fig. 3. Attenuator circuit.

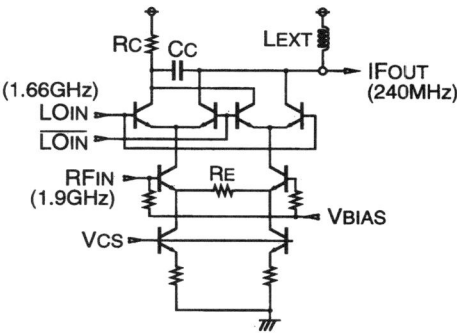

Fig. 4. Down-mixer.

expensive IF chips are commercially available. Thus these parts are off-chip. Also, the double conversion is easily applied to other systems such as digital European cordless telephone (DECT) and global systems for mobile communications (GSM).

An indirect up-modulator is employed for the transmit path. The indirect scheme requires a low-pass filter to suppress the unwanted second-local signal harmonics. A simple RC filter is placed after the modulator. This on-chip low-pass filter eliminates the needs for all off-chip devices on the transmit path from I/Q input to TXOUT.

To adjust the output power of TXOUT and to compensate temperature dependence of the high-power amplifier on the RF chip, a variable attenuator circuit is inserted on the transmit path.

The chip integrates two programmable synthesizers. The first and second local signals are distributed through buffer circuits to reduce interference from the PLL's. The first local signal (LO1) is provided to the up- and down-mixers. The second local signal (LO2) goes to the quadrature modulator and the second mixer. Four-bit control signals are for power management. These signals control the transmit path, the down-mixer, the first local buffer, and the second VCO.

III. CIRCUIT DESIGN

A. Quadrature Modulator

Fig. 2 shows a schematic of the quadrature modulator. The second local signal of 240 MHz is input to the phase splitter through the second buffer. The phase splitter is of the differential RC/CR type followed by limiter circuits [7]. The RC phase splitter consumes less power than the frequency divider type. The ac coupling reduces the phase error caused by the base current of the limiter. The RC/CR splitter produces a constant 90° phase shift regardless of frequency and process variation. The amplitude imbalance between I channel and Q channel due to process variation is reduced by the limiter. The double balanced mixers mix and sum the phase split local signals and differential I/Q signals.

The PHS system requires attenuation of the second-order ($2f_{LO2}$) and third-order ($3f_{LO2}$) harmonics by at least 30 dB. The raw (no-filtered) output signal of the $\pi/4$ QPSK modulator has second- and third-order harmonics of -25 and -20 dBc. Therefore, additional attenuation of 5 and 10 dB, respectively, is required. The three-stage RC filter is chosen to produce the suppression against process variation.

B. Attenuator Circuit

The attenuator circuit in Fig. 3 adjusts the output level of the up-mixer and temperature dependence of the gain. An external resistor R_{EXT} is used for the gain adjustment. The gain is adjusted in the range of about 10 dB. The current ratio of two diodes, D1 and D2, decides the gain. Therefore, the gain of the attenuator is given by

$$\text{Gain} = \frac{2Rc}{Re} \cdot \frac{I2}{I2 + I3}$$
$$= \frac{2Rc}{Re} \cdot \frac{I0}{I1}$$

where $I2 : I3 = I0 : (I1 - I0)$, and $I1 = V_{BE}/R_{EXT}$. Temperature dependence of the gain can be tuned by the circuit parameters. This attenuator circuit works to cancel the negative temperature dependence of GaAs HPA.

C. Up-Mixer and Down-Mixer

The up-mixer and down-mixer employ double-balanced mixers to suppress local oscillator leakage. The up-mixer converts the 240 MHz IF signal to a 1.9 GHz RF signal. The up-mixer has a differential output buffer and an emitter follower. The emitter follower is designed to be almost 50 Ω. Thus, no external matching circuit is needed.

The down-mixer illustrated in Fig. 4 converts 1.9 GHz RF signal to 240 MHz IF signal. Since an emitter follower circuit consumes more power, an open collector type is adopted. An off-chip inductive load is connected to the output. The capacitor Cc and the resistor Rc make a low-pass filter. This suppresses the undesired spurious signals and thus improves linearity.

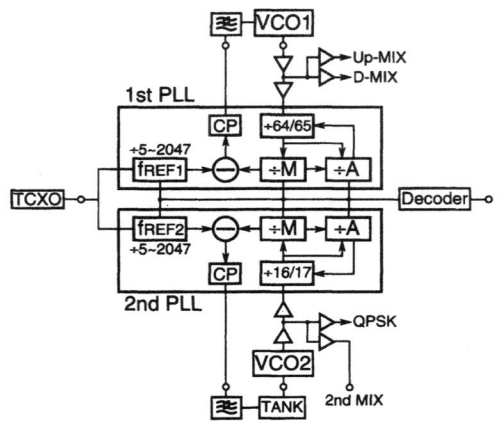

Fig. 5. Dual PLL synthesizers.

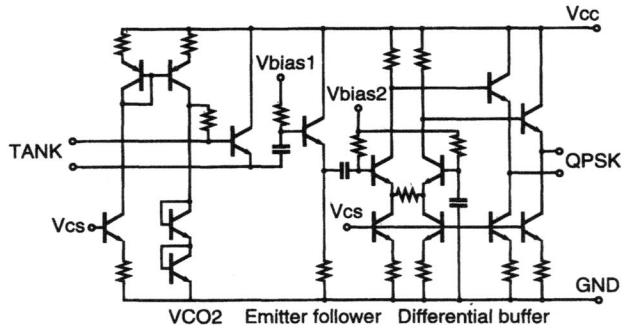

Fig. 6. Local buffer circuit.

E. Synthesizers

The chip integrates two programmable synthesizers. Fig. 5 shows a block diagram of the synthesizers. The first and second local signals are distributed through buffer circuits in Fig. 6. The buffer circuits consist of two-stage buffers: an emitter follower and a differential amplifier. The second VCO is a Colpitts oscillator with an external tank circuit. The frequency ranges from 1–2 GHz for the first local phase-locked loop (first PLL), and from 100–400 MHz for the second local PLL (second PLL). The PLL's require 22-b serial data for programming. The first 2 b and the following 3 b are used for power management and shift register selection, respectively. Four shift registers are provided to set the first and second local frequency and their reference frequencies. The remaining 17 b are used for setting the dividing ratio. In the PHS specification, the reference frequency for the first PLL is 300 kHz (PHS channel spacing).

Since the PLL operates intermittently in the time division multiple access (TDMA) system, short lockup time is required when the power control signal wakes up the PLL. In a conventional PLL shown in Fig. 7(a), reference frequency, f_{REF}, and divided VCO frequency, f_0 are asynchronous signals because power control signal, S_0, goes up independently. Thus a phase comparator may regard the frequency of f_0 as half of f_{REF} in the worst case, even though these signals have the same frequency. Initial phase error causes long and unfixed lockup time.

Fig. 7(b) shows a self-synchronized PLL which is newly adopted to shorten the lockup time. An additional timing generator produces a control signal, S_1, from S_0 and f_{REF}. S_1 is distributed to the phase comparator, a programmable divider, and a prescaler. Since S_1 is synchronized with f_{REF}, initial phase error between f_{REF} and f_0 is minimized. (When $f_{REF} = f_0$, initial phase error becomes zero.) The timing generator shown in Fig. 8 consists of an N-bit counter. The N-bit counter makes appropriate delay time for waking up of the prescaler. After N count, S_1 goes up at falling edge of f_{REF}. Therefore, the phase comparator, the programmable divider, and the prescaler operate synchronously with f_{REF}. The N count delay of around 10 μs is negligibly compared with lockup time.

F. Reference Circuit

Each analog block has a reference-voltage generator with a power control pin to manage the power dissipation during each operational mode of transmit, receive, standby, and sleep. The reference circuit shown in Fig. 9 provides the voltage Vcs for constant-current sources of analog blocks. The circuit is a self-bias type and includes a startup circuit, a pnp current mirror, a bandgap reference, and an emitter follower. The voltage Vcs is counterbalanced against the variations of the supply voltage. The supply voltage dependence of Vcs is 1.6 mV/V.

IV. PROCESS TECHNOLOGY AND LAYOUT

The chip is fabricated with 0.8-μm BiCMOS process based on our CBiCMOS process [8]. The process consists of 20 GHz npn and CMOS transistors, lateral pnp transistors, and MIM capacitors. Fig. 10 shows a cross sectional view of the transistors. A minimum feature size is 0.8 μm. The effective emitter width of the npn transistor is 0.5 μm. The npn transistor structure is double polysilicon self-aligned type. Since CMOS transistors are compatible with our standard 0.8-μm CMOS process, the well-established cell libraries are available. A polysilicon resistor and an emitter electrode are formed with n^+ polysilicon at the same process step. Two-layer polysilicon structure makes metal-insulator-metal (MIM) capacitors. The npn base electrode is used for a lower electrode of the capacitors. An upper electrode is fabricated with an additional p^+ polysilicon. The device parameters are summarized in Table I.

A photomicrograph of the IF transceiver chip is shown in Fig. 11. The chip size is 3.4 $\times$ 3.0 mm^2. The right half of the chip is dual PLL's, and the left half is analog blocks. The first local and second local buffers are placed between PLL's and analog blocks. Signal isolation is the most important issue in high frequency circuits. In particular, the interference of PLL's degrades the performance of the analog blocks. The most effective way to signal isolation is to separate power lines, as well-known, and to keep a signal apart from the others. A quasi-separation method, such as two separated ground lines connected near a ground pad, will not be always effective for microwave region, because the ground pad is not a good ground at high frequency. Each of the analog blocks is

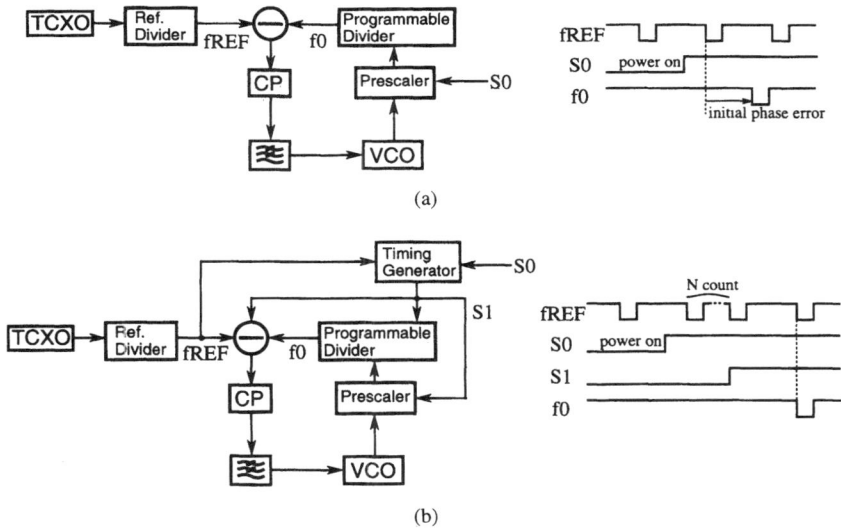

(a)

(b)

Fig. 7. Self-synchronized PLL.

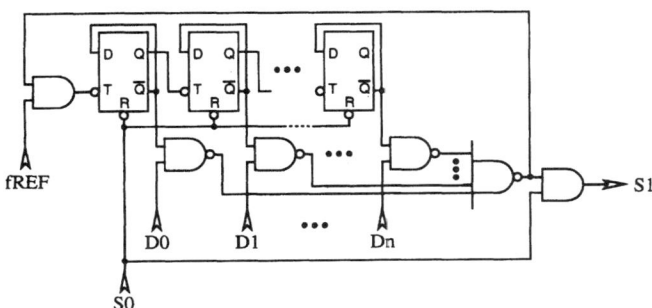

Fig. 8. Timing generator.

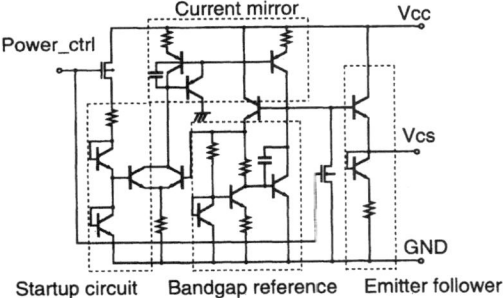

Fig. 9. Reference circuit.

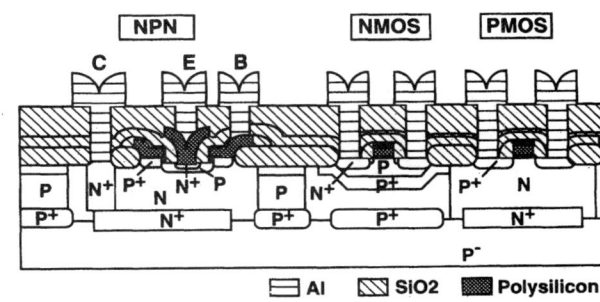

Fig. 10. Cross sectional view of transistors.

TABLE I
DEVICE PARAMETERS

NPN	Se = 0.5 x 5.7 µm²	
	fT = 20 GHz	
	fmax = 24 GHz	
	CTC = 9 fF	Rb = 155 Ω
NMOS	Ln = 0.8 µm	
	Tox = 18nm	
	Vth = 0.76V	
PMOS	Vth = -0.78V	
Resistor	n-polysilicon	150Ω/□
Capacitor	2poly-MIM	2.7fF/µm²

V. EXPERIMENTAL RESULTS

Fig. 13 shows the output spectrum of the up-mixer at 1.9 GHz and 3.0 V power supply. The center frequency is 1.9 GHz. The desired RF output has 24 kHz offset. The attenuator circuit adjusts the RF output to −12 dBm. The I/Q level is 380 mV$_{pp}$ with $\pi/4$ QPSK modulation (all-0 pattern). The first and second local signals are 1.66 GHz and 240 MHz, respectively. The carrier suppression of −35 dBc and the image suppression of −41 dBc are obtained. The third-order intermodulation (IM3) suppression is −70 dBc. The total vector error for PN9 pattern is 4.4% including the PLL's. The amplitude error is 0.1 dB and the phase error is 2.3°. The vector error in a burst mode becomes 4.5%. Fig. 14 shows the

surrounded by its own power supply, ground lines, and the P$^+$ shield line connected to the subground, as shown in Fig. 12. The shield line is directly connected to the external ground through its own pad. The P$^+$ contact on the ground line is for measures against latch-up.

Signal isolation to the pin at a distance of several hundred microns is about 40 dB at 1.9 GHz with ground line shield. The P$^+$ shield line improves additional 5 dB. When the pins are separated from each other by over 1 mm, for example, down-mix output and VCO2 output, signal isolation is 55 dB at 1.9 GHz and 82 dB at 240 MHz, respectively.

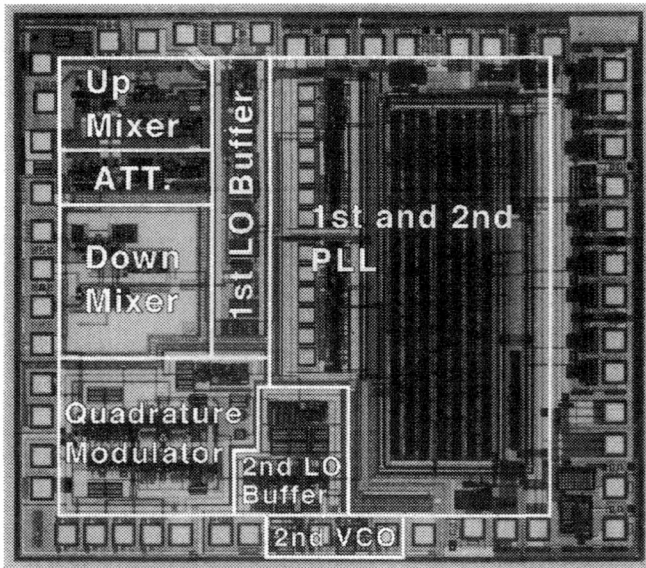

Fig. 11. IF transceiver photomicrograph (3.4×3.0 mm^2).

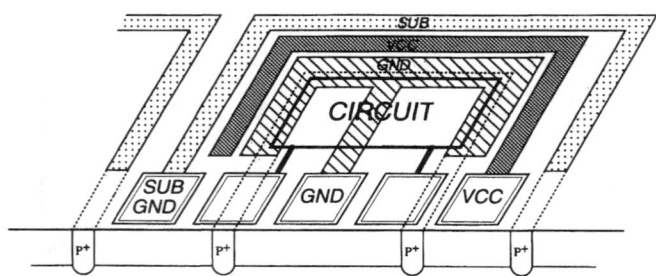

Fig. 12. Layout of power line and p$^+$ shield line.

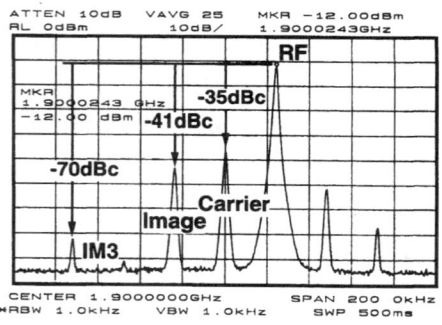

Fig. 13. TXOUT output spectrum (all-0 pattern) at RF = 1.9 GHz, $I/Q = 380$ mV$_{pp}$ at 24 kHz.

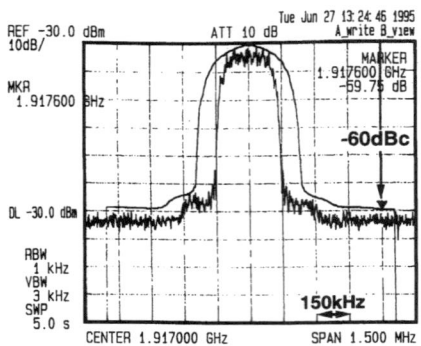

Fig. 14. TXOUT output spectrum (PN9 pattern) at RF = 1.9 GHz, $I/Q = 380$ mV$_{pp}$.

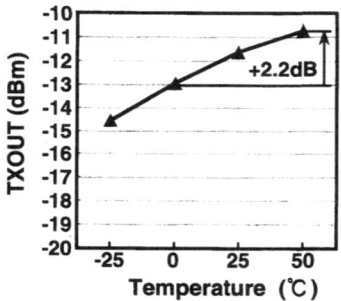

Fig. 15. TXOUT versus temperature.

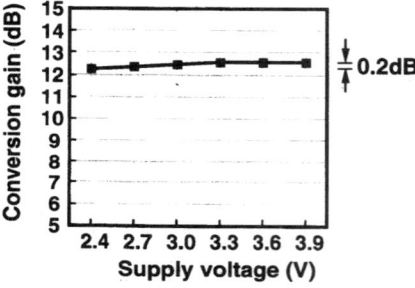

Fig. 16. Conversion gain versus supply voltage at $RF = 1.9$ GHz with -40 dBm, $LO = 1.66$ GHz.

output spectrum of the up-mixer with $\pi/4$ QPSK modulation (PN9 pattern). The adjacent channel power at 600 kHz spacing is -60 dBc. The up-mixer is usable up to -9.5 dBm output power. The GaAs power amplifier requires -12 dBm output for the PHS specification.

Fig. 15 shows the relation between the temperature and the output power of TXOUT under the condition of $V_{cc} = 3$ V, the external gain control resistor $= 5$kΩ, and the I/Q level $= 380$ mV$_{pp}$ at 24 kHz. The output power increases by 2.2 dB as temperature increases from 0–50°C. Thus the negative temperature dependence of GaAs HPA is compensated.

Fig. 16 shows the relation between the supply voltage and conversion gain of the down mixer. The RF input is 1.9 GHz with -40 dBm. The gain variation is only 0.2 dB as the supply voltage changes from 2.7 to 3.6 V.

The first PLL was evaluated with an external VCO module and a loop filter. Fig. 17 shows the output spectrum of the first local signal. The center frequency is 1.66 GHz. A reference frequency is 300 kHz. The peak shows a leak of the reference frequency. The spurious level of -92 dBc is obtained at 600 kHz spacing. The phase noise is -77 dBc/Hz at 2 kHz offset. Channel selection lockup time (1684.8–1651.2 MHz) is 0.91 ms. Power-on lockup time (sleep mode to operational mode) of 1.1 ms is constantly obtained because of the self-synchronized PLL. In the conventional PLL, lockup time varies from 1.1 to 2.4 ms.

Total current consumption of the chip is 24 mA in receive mode and 44 mA in transmit mode at 3.0 V. These values are almost one quarter of present systems which consists of

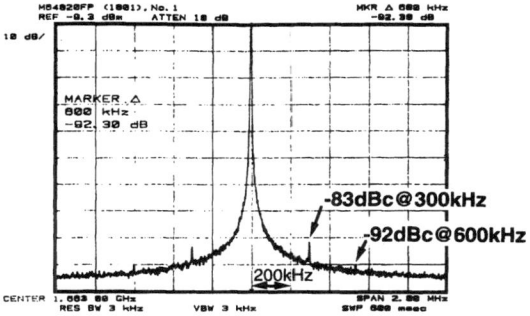

Fig. 17. First local output spectrum at 1.6638 GHz.

TABLE II
MEASUREMENT RESULTS

Transmit	Vector error	4.4%
	Amplitude error	0.1dB
	Phase error	2.3°
	Image suppression	-41dBc
	Adjacent channel power	-60dBc
	Occupied band width	247kHz
Up-Mixer Down-Mixer	1dB compression	-7dBm
	Conversion gain	12dB
	NF (DSB)	12dB
	IP3 (input)	-7dBm
	Transmit mode current	44mA
	Receive mode current	24mA
	Sleep mode current	<1µA

GaAs chips and discrete devices. Sleep mode current with all circuits cut off is less than 1 μA.

Table II summarizes the characteristics of the IF transceiver chip. All measurement data were done with the device in a 48 pin QFP package.

VI. CONCLUSION

A single chip IF transceiver has been developed using a 0.8-μm BiCMOS process. The chip successfully operates at 1.9 GHz. The measurement data proves the chip satisfies the requirements of the PHS specifications. The negative temperature dependence of the GaAs high power amplifier is compensated. Total vector error of 4% for the $\pi/4$ QPSK PN9 pattern includes the up-mixer and the dual PLL's. The chip draws 24 mA in receive mode and 44 mA in transmit mode, operating from 3.0 V. The low current consumption extends the talk and standby time of existing digital cordless phones.

ACKNOWLEDGMENT

The authors would like to thank M. Yamawaki, T. Nakashima, and S. Kubo for wafer processing and N. Kato for his encouragement.

REFERENCES

[1] T. Stetzler et al., "A 2.7–4.5 V single-chip GSM transceiver RF integrated circuit," *IEEE J. Solid-State Circuits*, vol. 30, no. 12, pp. 1421–1429, Dec. 1995.
[2] T. Tsukahara et al., "A 2-V 2-GHz Si-bipolar direct-conversion quadrature modulator," *IEEE J. Solid-State Circuits*, vol. 31, no. 2, pp. 263–267, Feb. 1996.
[3] C. Takahashi et al., "A 1.9 GHz Si direct conversion receiver IC for QPSK modulation systems," in *ISSCC Dig. Tech. Papers*, Feb. 1995, pp. 138–139.
[4] K. Yamamoto et al., "A GaAs RF transceiver IC for 1.9 GHz digital mobile communication systems," in *ISSCC Dig. Tech. Papers*, Feb. 1996, pp. 340–341.
[5] J. R. Long et al., "A 1.9 GHz low-voltage silicon bipolar receiver front-end for wireless personal communications systems," *IEEE J. Solid-State Circuits*, vol. 30, no. 12, pp. 1438–1448, Dec. 1995.
[6] N. Suematsu et al., "L-band internally matched Si-MMIC low noise amplifier," in *IEEE MTT-S Int. Microwave Symp. Digest*, June 1996, pp. 1225–1228.
[7] K. Yamamoto et al., "A 1.9-GHz-band GaAs direct-quadrature modulator IC with a phase shifter," *IEEE J. Solid-State Circuits*, Oct. 1993, vol. 28, no. 10, pp. 999–1000.
[8] T. Ikeda et al., "A high performance CBiCMOS with novel self-aligned vertical PNP transistors," in *Proc. 1994 BCTM*, Oct. 1994, pp. 238–241.

A Direct-Conversion Receiver for 900 MHz (ISM Band) Spread-Spectrum Digital Cordless Telephone

Christopher Dennis Hull, Joo Leong Tham, *Member, IEEE,* and Robert Ray Chu

Abstract—A single chip direct-conversion receiver for use in the 900 MHz ISM (industrial-scientific-medical) band has been fabricated in a 25 GHz f_T silicon-bipolar process. The receiver has a maximum gain of 105 dB with more than 80 dB of gain-control range. Integrated quadrature mixers and phase-shift networks are included. The typical phase error of the I/Q outputs is less than 1 degree. A class-AB servo amplifier is used to reduce dc offset while maintaining a turn-on time of 200 μs. An integrated filter is included to improve channel selectivity. The integrated circuit operates on a supply voltage range of 2.7 to 5.0 V and draws 50 mA of supply current.

I. INTRODUCTION

DIRECT conversion (or zero intermediate-frequency) is an attractive architecture in radio design because it eliminates the need for expensive filters at radio frequencies (RF) for image rejection and at intermediate frequencies (IF) for channel selection [1]. Instead, an inexpensive baseband low-pass filter is used for channel selection, and no image rejection filter is required. An RF preselect filter is still required prior to the low-noise amplifier (LNA) to attenuate out-of-band signals, since the front-end has little selectivity.

In order to facilitate implementation of a direct conversion architecture, a direct-sequence spread-spectrum digital code that has negligible energy below 30 KHz is used for modulation so that a combination of ac coupling and dc servo feedback may be used to block the propagation of dc offsets in the baseband chain. Adaptive differential pulse-code modulation (ADPCM) at a rate of 32 KHz is used for voice coding. This code is then spread over a 600 KHz bandwidth (per sideband) in order to reduce the transmitted power spectral density (which is limited by the Federal Communications Commission in the 902–928 MHz ISM band) and improve immunity to multipath distortion and fading. Spread-spectrum systems have the additional advantage of enhanced security. The channel spacing is 1.2 MHz, and there are 21 nonoverlapping channels in the 902–928 MHz ISM band. The system is operated using time-division duplex (TDD) at a frame rate of 250 Hz. The frame rate was chosen to avoid excessive echo, which may be audible, while allowing sufficient time for the receiver and transmitter sections to be switched on and off.

Manuscript received May 28, 1996; revised July 24, 1996.
C. D. Hull and J. L. Tham are with Rockwell International, Newport Beach, CA 92658 USA.
R. R. Chu was with Rockwell International, Newport Beach, CA 92658 USA. He is now with Maxim Integrated Products, Beaverton, OR 97005 USA.
Publisher Item Identifier S 0018-9200(96)08083-3.

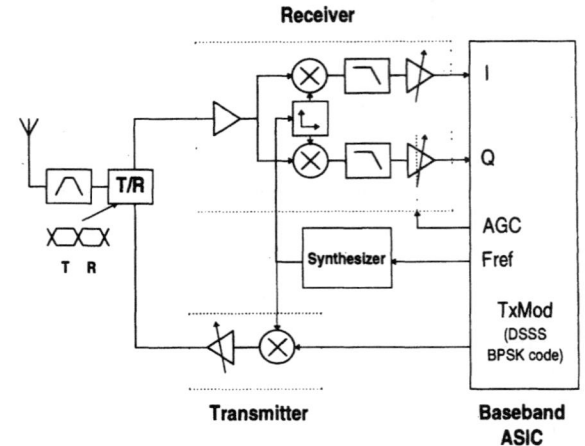

Fig. 1. Digital cordless telephone: transceiver block diagram.

II. SYSTEM ARCHITECTURE

A block diagram of the full transceiver is shown in Fig. 1. The transceiver consists of five integrated circuits: the receiver, transmitter, and synthesizer in silicon bipolar technology, and a baseband application-specific integrated circuit (ASIC) and a coder/decoder integrated circuit (CODEC IC) in CMOS technology.

The receiver includes a preselect filter, low-noise amplifier (LNA), dual mixers driven by quadrature local-oscillators signals, built-in quadrature phase-shift networks, external low-pass filters, and a variable-gain baseband amplifier with digital automatic gain control (AGC). The baseband ASIC controls the gain of the receiver and attempts to keep a constant signal level at the receiver output. The digital word representing the gain setting gives the ASIC an indication of the received signal strength (RSSI). The interface between the receiver chip and the ASIC is differential, which improves immunity to noise pickup from digital switching in the ASIC. Immunity to second-order distortion products is also improved since no differential to single-ended converter is required [2].

The complete synthesizer, including a fully integrated voltage-controlled oscillator (VCO), was optimized for operation in the ISM band. It has 600 KHz frequency steps (1/2 channel) and has a phase noise level of −123 dBc/Hz at 1.2 MHz offset. Spurious responses due to the 600 KHz comparison frequency for the synthesizer are less than −70 dBc (typically −110 dBc).

A third RF integrated circuit is used for the transmitter. It includes an upconversion mixer and power amplifier with RF output power levels of 0, +10, and +20 dBm.

IEEE Journal of Solid-State Circuits, vol. 31, no. 12, pp. 1955-1963, December 1996.

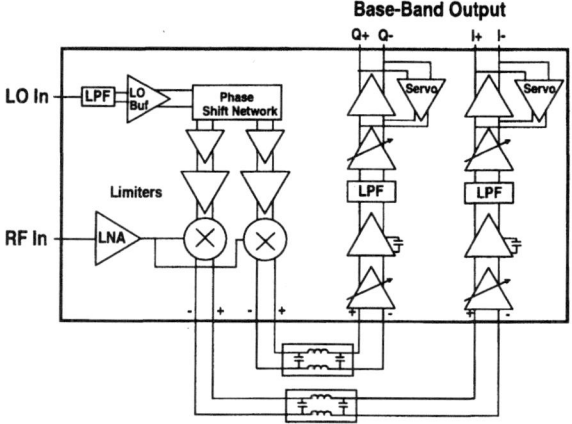

Fig. 2. Receiver: integrated circuit architecture.

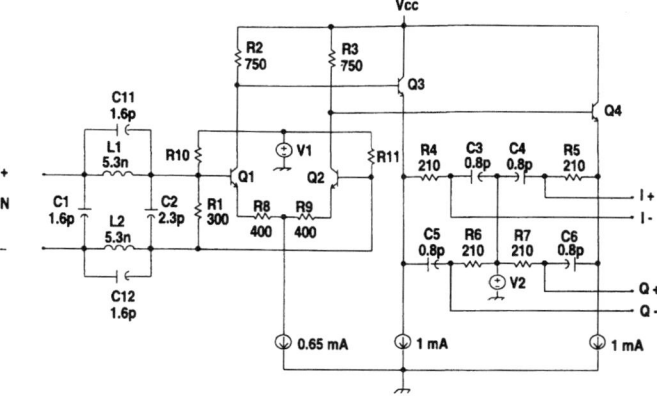

Fig. 3. Low-pass filter and phase-shift network.

The CODEC implements all of the voice coding and decoding, and the baseband ASIC controls the transmit/receive timing, modulation and demodulation of the spread-spectrum signal, and all other digital functions in the radio.

The receiver architecture is illustrated in Fig. 2. The LNA drives the RF input ports of two identical mixers. The local oscillator (LO) input ports of these mixers are driven in quadrature phase. An external LO drives an integrated low-pass filter and is then buffered. The LO buffer drives the 90 degree phase-shift network, which is followed by limiters to stabilize the amplitude and set the appropriate common-mode voltage for the mixers. The output of the mixer is filtered by an external low-pass, three-pole differential Butterworth filter with 700 KHz bandwidth for channel selectivity.

The baseband stages consist of a cascade of a variable-gain stage, a fixed-gain stage, an integrated Sallen and Key low-pass filter, another variable-gain stage, and finally a fixed-gain stage. The fixed-gain stage has a wide common-mode input range and a fixed common-mode output voltage, while the variable gain amplifier stage has a narrow common-mode input range and a widely varying common-mode output voltage (depending on the gain setting). The alternation of variable and fixed stages improves the tradeoff between noise and distortion.

The second gain stage is capacitively degenerated (with an external capacitor) to block propagation of dc offsets. The integrated filter adds an additional 35 dB of selectivity at 3.6 MHz (three channels away). The fourth gain stage has a servo-amplifier around it to minimize the dc offset at the output. The maximum total gain of the baseband stages is 88 dB with more than 60 dB of gain-control range. A gain-control buffer is used to give the desired gain-control sensitivity of 0.08 dB/mV.

III. LO PHASE-SHIFT CIRCUIT

The LO phase-shift circuit consists of an integrated low-pass filter, buffer, differential R-C phase-shift network, and limiters.

The low-pass filter, buffer, and phase-shift network are depicted in Fig. 3. The low-pass filter is realized with integrated spiral inductors $L1$ and $L2$ and capacitors $C1$, $C2$, $C11$, and $C12$. It is a differential three-pole, two-zero design. The complex zeros, determined by $C11$ and $C12$, are placed near 1.8 GHz to attenuate the second harmonic of the LO signal. Harmonic energy in the local-oscillator signal needs to be minimized, as it will distort the zero crossings at the output of the phase-shift network, and thus generate additional phase error at the output of the limiters. Transistors $Q1$–$Q4$ form a linearized buffer which drives the phase-shift network with a nominal amplitude of 200 mV. Degeneration resistors $R8$ and $R9$ reduce the harmonic distortion in the buffer, which is necessary to maintain phase accuracy. Transistors $Q3$ and $Q4$ are biased with sufficient current to drive the phase-shift network linearly.

The phase-shift network consisting of $R4$–$R7$ and $C3$–$C6$ forms a differential version of the conventional RC-CR network [3]. A differential network desensitizes the performance of the circuit to parasitics such as ground-lead inductance. The phase response of this network is depicted in Fig. 4. Note that the phase error increases with frequency, but is within one degree of quadrature from 300 MHz to 2 GHz. The frequency response of the phase-shift network is depicted in Fig. 5. Note that the amplitudes are matched at only one frequency (nominally 915 MHz). Because of process variations, an amplitude mismatch of up to 30% may occur.

To stabilize the output amplitude, a limiter is used (Fig. 6). The limiter consists of two stages. The first stage (consisting of $Q1$ and $Q2$) is an emitter-coupled pair with emitter degeneration. The gain of this stage is maximized by using large values for $R3$ and $R4$ (limited by bandwidth considerations). Bypass capacitor $C1$ and resistor $R5$ form a supply filter and prevent common-mode oscillations. The nominal output amplitude of the first stage is 300 mV.

The second stage (consisting of $Q3$–$Q6$) is made up of emitter-followers and an undegenerated emitter-coupled pair. The two stages are ac coupled to minimize propagation of low-frequency noise. In a direct-conversion mixer, low-frequency noise at the LO input port will feedthrough to the baseband output if the mixer is not perfectly balanced. This could substantially degrade the noise performance of the mixer. Emitter followers ($Q3, Q4$) reduce the loading on the first stage to maximize the bandwidth. Resistor $R10$ and capacitor $C4$ form an additional supply filter to increase the margin

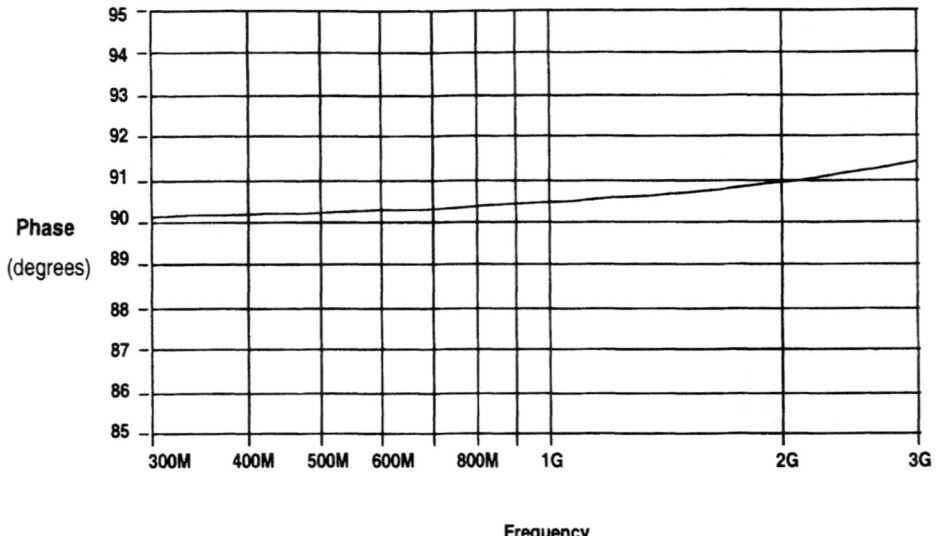

Fig. 4. Quadrature phase-shift network: simulated phase delta versus frequency.

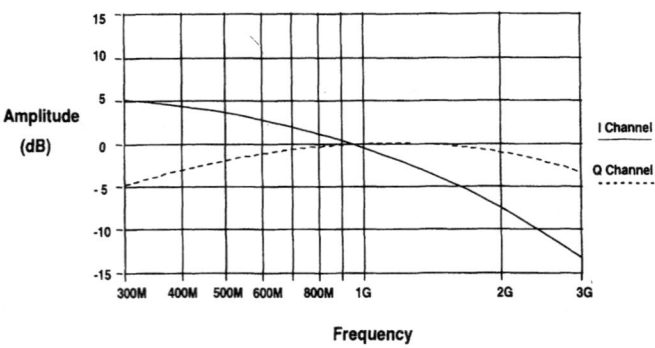

Fig. 5. Phase-shift network: simulated amplitude response versus frequency.

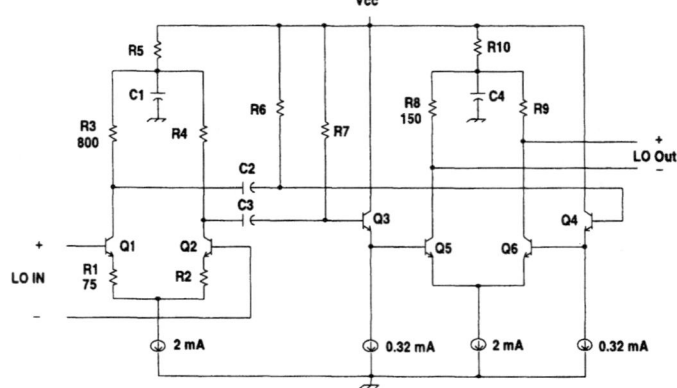

Fig. 6. Limiter.

of common-mode stability. The output amplitude versus input amplitude for the second stage is depicted in Fig. 7. Note that for input amplitudes in excess of 250 mV, the output amplitude is stabilized near 260 mV.

When a nonlinear circuit (such as the second stage of the limiter) is driven by a large signal, the phase delay becomes a function of input amplitude as plotted in Fig. 8. The reduction in phase delay with increasing amplitude can be intuitively understood as an increase in the speed of an inverter with increasing drive level. The nonlinear relationship between the phase delay and input amplitude creates a type of distortion known as amplitude-modulation to phase-modulation conversion (AM-PM conversion). For this particular design, with a nominal voltage of 260 mV, a 10% input amplitude mismatch into the second stage of the limiter will produce an output phase delay mismatch of one degree. If better phase accuracy is required, a cascade of phase-shift networks with split tuning can be used to achieve better amplitude balance into the limiters (as presented by Crols [4]).

IV. LNA AND MIXER

The LNA is depicted in Fig. 9. The input device is formed by $Q2$ and $Q3$. Transistor $Q3$ is biased at a small fraction of the current through transistor $Q2$. Transistors $Q5$ and $Q6$ are common-base stages which effectively cascode the input transistors. Transistor $Q4$ and $Q5$ form a current steering switch. When the voltage at the base of $Q4$ is low, the full bias (and signal) current of $Q2$ flows through $Q5$ to the output. When the voltage at the base of $Q4$ is high, the current in $Q2$ flows to the supply and only the smaller current flowing through $Q3$ and $Q6$ reaches the output. In this way, an accurate gain step is created, and it is set by the ratio of currents in $Q2$ and $Q3$. Integrated spiral inductor $L1$ along with $C5$ and $R5$ form a parallel L-C resonator with peak impedance of 250 Ω (nominal). Resistor $R5$ reduces the Q of the resonating network and also reduces the gain variation of the device over process, frequency, and temperature. Capacitor $C2$ is the output blocking capacitor. Transistors $Q1$ and $Q9$ set the bias levels for transistors $Q2$ and $Q3$. Resistor $R4$ presents a high impedance to RF signals so that the signal current flows to the signal transistors rather than the bias circuitry. A large value of $R4$ is optimum for noise and RF de-coupling; however, an excessive value will make the bias current very sensitive to dc beta matching between transistors $Q2$ and $Q1$. Transistors

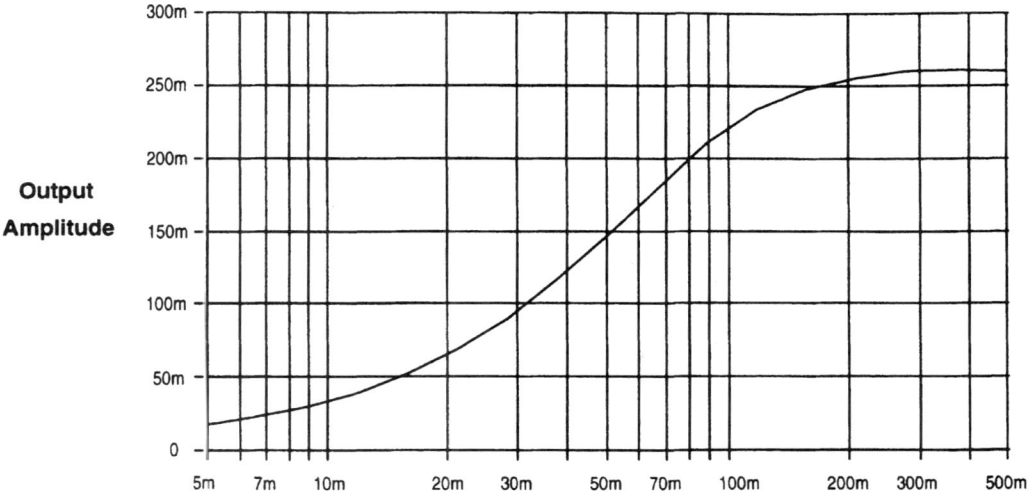

Fig. 7. Second stage of limiter: simulated output amplitude versus input amplitude.

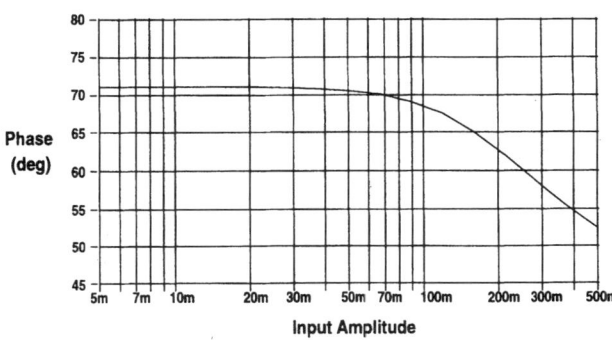

Fig. 8. Limiter: simulated phase delay versus input amplitude.

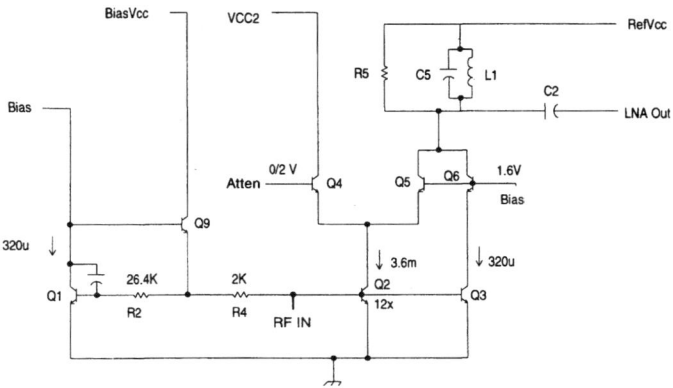

Fig. 9. Low noise amplifier.

$Q5$ and $Q6$ are bypassed with a 5 pF capacitor which must be connected to a different ground than the emitters of transistors $Q2$ and $Q3$. Failure to separate the grounds may generate spurious oscillations in the multi-GHz range due to crosstalk caused by the common ground inductance of the wire-bond and package.

Due to the low Q of the resonator at the LNA output, the front-end has a nominal bandwidth of approximately 300 MHz, and has little selectivity. It is difficult to achieve selectivity at RF frequencies in an integrated fashion because of component tolerances and achievable quality factors (especially for the integrated inductors).

The mixer, shown in Fig. 10, is a conventional double-balanced cross-coupled "quad" mixer (sometimes referred to as a "Gilbert type mixer"). The mixer is resistively degenerated ($R1$–$R4$) with 160 mV for linearity. The LO input is biased 300 mV below V_{cc} with a 260 mV peak differential swing. The 260 mV swing is sufficient to fully switch the top four transistors without substantially reducing the voltage swing available at the output.

The bases of the input transistors ($Q1, Q2, Q7, Q8$) are biased at 900 mV below V_{cc}, which provides a nominal bias

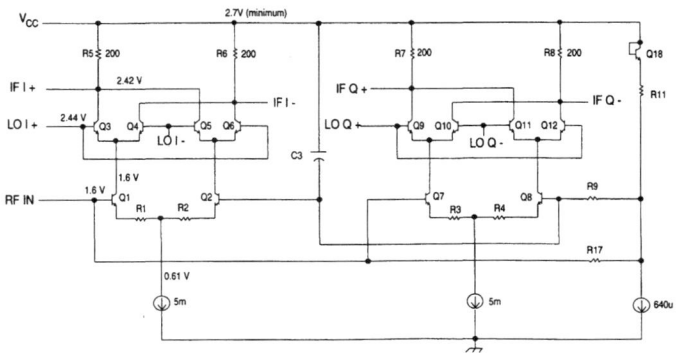

Fig. 10. Quadrature downconversion mixer.

voltage of zero for the collector-base junction. Hence, an input amplitude up to 350 mV can be accommodated without saturating the input transistors. At a 2.7 V supply voltage, there is 610 mV of headroom across the current sources, which is sufficient to maintain a high output impedance.

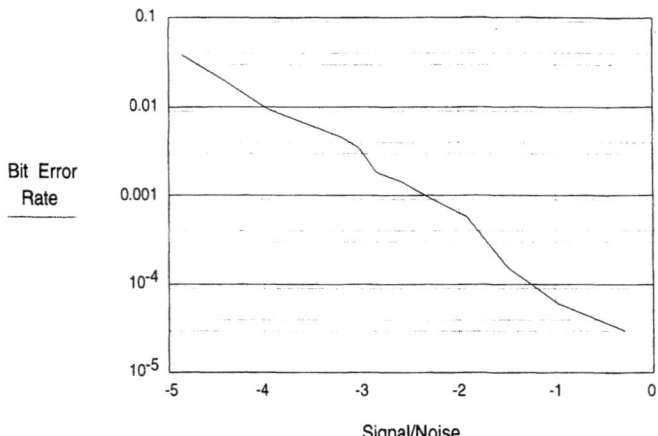

Fig. 11. Baseband ASIC: measured bit-error rate versus signal/noise ratio.

TABLE I
MEASURED PERFORMANCE OF RECEIVER FRONT-END

Parameter	Performance
Voltage Gain @915 MHz: High Gain Mode	20.9 dB
Voltage Gain @915 MHz: Low Gain Mode	−1.6 dB
Input IP3 (high gain mode)	−12.6 dBm
Input IP3 (low gain mode)	−8.0 dBm
Input P1dB: high gain mode	−21 dBm
Input P1dB: low gain mode	−14 dBm
LO to RF Isolation	58 dB
Double-Sideband Noise Figure	2.9 dB
I/Q Amplitude Imbalance	0.7 dB
I/Q Phase Imbalance	0.7 deg

Capacitor $C3$ bypasses the unused input of the differential pair to ground. Since the LNA output is directly dependent on the voltage of the V_{cc} line, referencing the mixer input to it will reduce sensitivity to noise on the supply. Resistors $R5$–$R8$ set the differential output impedance to 400 Ω. Transistors $Q1, Q2, Q7$, and $Q8$ are made large in order to reduce the noise contribution from the base ohmic-resistance. The specific device size was chosen as a compromise between noise and linearity considerations. Larger devices give lower parasitic resistance and higher parasitic capacitance. Parasitic resistance adds additional noise to the circuit, while parasitic capacitance tends to increase the distortion [5]. A similar set of tradeoffs exist for the switching transistors ($Q3$–$Q6, Q9$–$Q12$).

V. PERFORMANCE OF RF FRONT-END

The performance of the RF front-end is summarized in Table I. Note that a double-sideband noise figure of less than 3 dB has been achieved. The noise figure of the LNA is estimated to be 1.8 dB (it is not directly measurable), and the remainder of the noise is attributed to the mixer (the contribution of the baseband amplifier is negligible). Because of the spreading gain of the modulation format, a carrier/noise ratio of 0 dB is adequate to achieve a demodulated bit-error rate less than 0.01% (See Fig. 11). Assuming a bandwidth of 750 KHz, a 3 dB noise figure implies a sensitivity at the LNA input of −112 dBm. However, losses from the transmit/receive switch and preselect filter reduce the sensitivity to approximately −109 dBm when referenced to the antenna.

When the LNA is in the high-gain state, the input intercept and compression points are determined by the mixer. Subjective field tests indicate that the 1 dB compression point of −21 dBm allows for sufficient immunity from unwanted interfering signals. In the low-gain mode, the third-order intercept increases to −14 dBm and is determined by the LNA. The low-gain mode is used when a large amplitude signal is incident on the receiver input.

The mixer is designed for 0 dB of nominal conversion gain (of the voltage) and thus the voltage-gain of the front-end is due entirely to the LNA. Note that the power gain is 9 dB less than the voltage gain due to the 400 Ω output impedance.

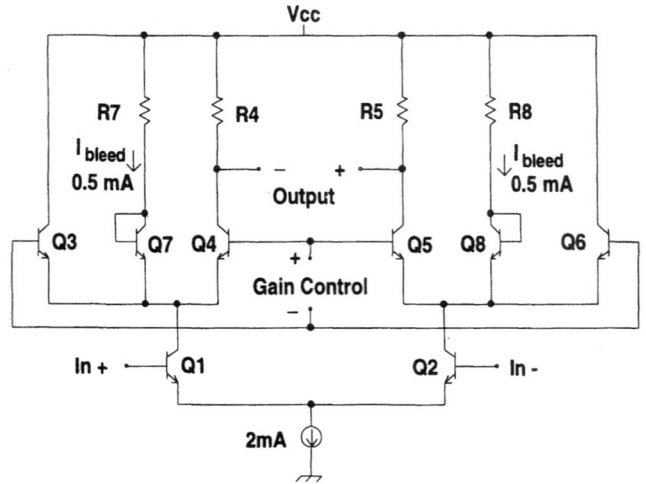

Fig. 12. Variable gain amplifier: input stage.

In a direct-conversion application, the LO to LNA input isolation is critical, since any LO signal incident on the input will mix against itself to produce a dc offset at the output of the mixers. An excessively large dc offset at the output of the mixers would cause distortion in the first stage of the variable-gain baseband amplifier. The LO to RF isolation of 58 dB is achieved by careful layout, pinout, and package selection. A thin-quad flat package with 48 pins was selected for this application. It has the benefit of leads on four orthogonal sides of the package. To minimize crosstalk, the LO input is orthogonal to the LNA input. In addition, the return path for the LO input current is an adjacent pin and bond-wire, thus minimizing the radiated magnetic fields.

VI. BASEBAND AMPLIFIER: VARIABLE-GAIN STAGES

The first and third stages of the four-stage baseband amplifier use the classical current-steering approach, shown in Fig. 12. The gain-control voltage is set by an external buffer (consisting of a differential amplifier). For gain-control voltages in excess of 100 mV, transistors $Q3$ and $Q6$ are off, and the circuit reduces to a differential cascode amplifier. As the gain-control voltage is decreased, transistors $Q3$ and $Q6$ begin to divert current away from $Q4$ and $Q5$, reducing the gain. Transistors $Q7$ and $Q8$ together with resistors $R7$ and $R8$ reduce the current flowing through the top quad ($Q3$–$Q6$) relative to the bottom pair ($Q1$ and $Q2$). This results in two

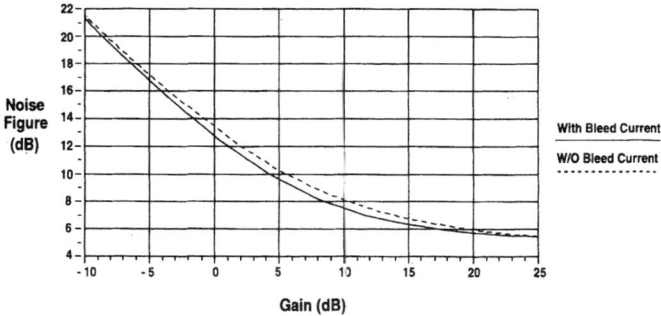

Fig. 13. VGA stage: simulated noise.

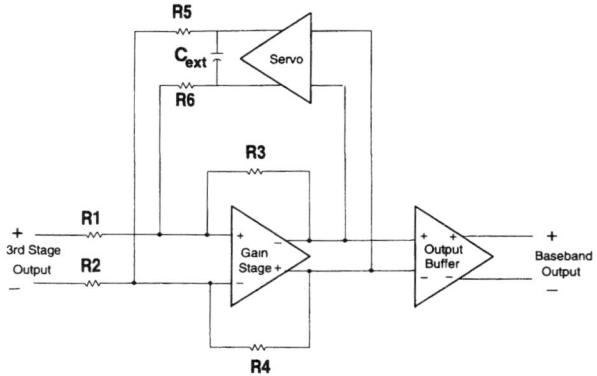

Fig. 14. Baseband amplifier: block diagram of output stage with servo.

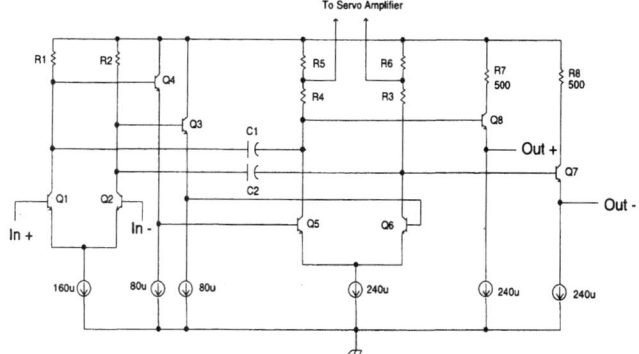

Fig. 15. Baseband amplifier: output stage.

advantages. First, reducing the current in the quad transistors reduces their output noise contribution. Second, reduced bias current in the load resistors ($R4$, $R5$) allows for increased gain without requiring additional headroom [6]. The use of a diode, $Q7$, in series with a resistor, $R7$, allows for proper temperature and process tracking of the currents.

An undesirable characteristic of this variable gain design is an increase in the common-mode voltage at the output when the gain-control voltage is decreased. The second and fourth stages of the baseband amplifiers are designed as fixed-gain stages with wide common-mode input ranges to accommodate the varying common-mode output voltage of the first and third stages.

The noise figures of the VGA with and without the bleed current through $R7$ and $R8$ are depicted in Fig. 13. Note that the reduction in current in the top quad transistors has reduced the noise figure by 1 dB for low-gain settings. The improvement is much more substantial if emitter degeneration is used for $Q1$ and $Q2$.

VII. BASEBAND AMPLIFIER: FIXED-GAIN STAGES

The second and fourth stages of the baseband amplifier are designed as fixed-gain stages. The second stage is made up of emitter-follower buffers and an emitter-coupled pair.

A block diagram of the fourth stage with resistive shunt feedback and active servo feedback is shown in Fig. 14. Resistors $R1$ and $R2$ represent the output impedance of the third gain stage. Resistors $R3$ and $R4$ provide resistive shunt feedback, which broadens the bandwidth and stabilizes the gain. In conjunction with the servo-feedback amplifier and external capacitor, resistors $R5$ and $R6$ provide feedback at low frequencies. This limits the closed-loop dc gain and reduces the dc offset at the output. A moderate value of loop-gain was chosen as a compromise between offset and speed considerations. A large value for the dc loop-gain would necessitate a large value of C_{ext} (for constant high-pass corner frequency) and would slow the large-signal settling performance of the baseband chain. A small value of dc loop gain would not adequately reduce the output dc offset.

The schematic of the fourth gain stage is shown in Fig. 15. The stage is designed as a differential operational amplifier. Cascaded emitter-coupled pairs ($Q1, Q2, Q5, Q6$) are buffered by emitter followers ($Q3, Q4$). The load resistors $R3$–$R6$ are split in order to allow an output signal with higher common-mode voltage to drive the servo feedback amplifier. Emitter followers ($Q7, Q8$) buffer the output and are designed to drive up to 50 pF of external capacitance. Resistors $R7$ and $R8$ limit the current in $Q7$ and $Q8$ under short-circuit (fault) conditions. Capacitors $C1$ and $C2$ generate Miller compensation for the amplifier. The voltage for the servo-feedback amplifier is taken from the bottom of resistors $R5$ and $R6$.

VIII. DC SERVO AMPLIFIER

The servo amplifier, shown in Fig. 16, consists of two stages. The first stage formed by $Q1$–$Q8$ is a Class AB amplifier first proposed by Hearn [7]. Unlike the conventional emitter-coupled pair, the differential output current continues to increase nearly linearly for input voltages in excess of 50 mV. In the absence of emitter degeneration ($R1$–$R4$), the transconductance characteristic of this input stage is described by a hyperbolic sine. The presence of $R1$–$R4$ along with the high-current reduction in beta of the lateral PNP's ($Q5$–$Q8$) limits the output current to prevent excessive current flow.

The second stage of the servo amplifier is a set of current mirrors ($Q10, Q13, Q14$, and $Q9, Q11, Q12$) that charge or discharge an external capacitor at both of its terminals. When the pull-down current in $Q13$ increases, the current in $Q20$ decreases. At the same time, the current in $Q12$ decreases, and the current in $Q21$ increases. This allows completely symmetrical charging and discharging of the capacitor. The voltage across this capacitor is fed back via a large shunt

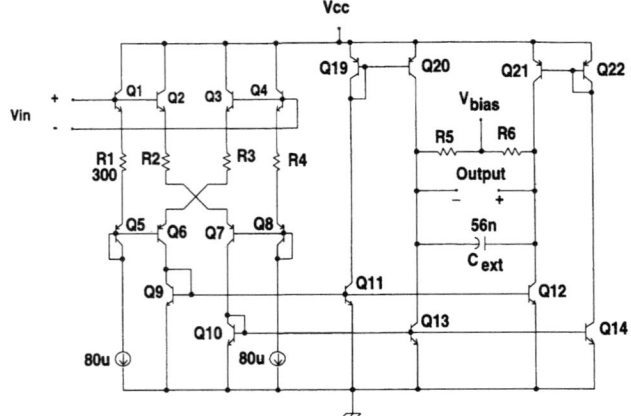

Fig. 16. Class AB servo amplifier.

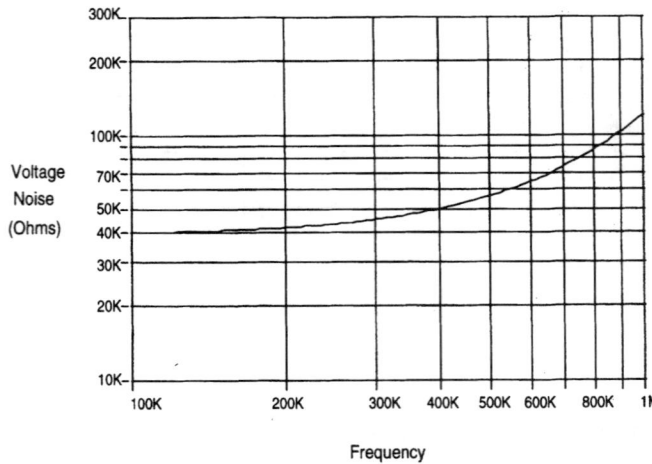

Fig. 18. Sallen and Key filter: simulated equivalent input noise resistance.

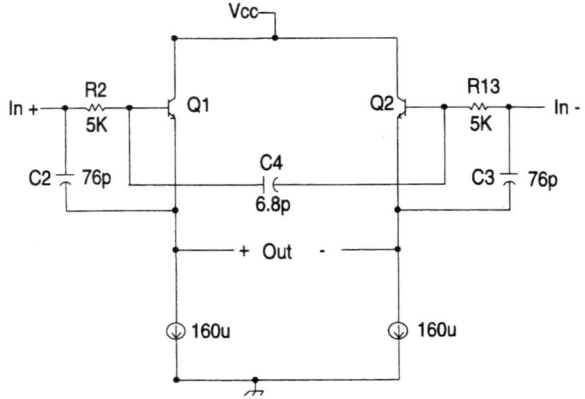

Fig. 17. Sallen and Key Filter.

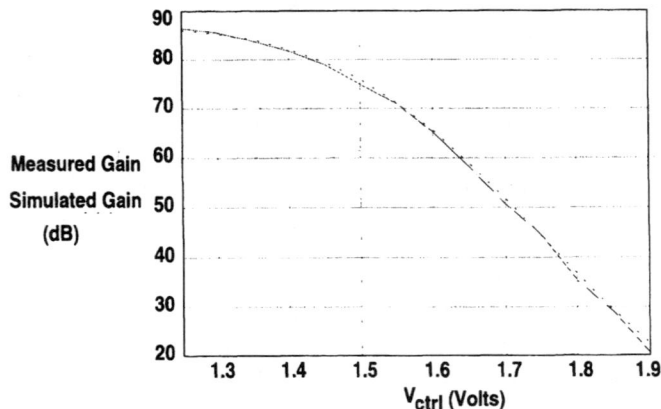

Fig. 19. Variable-gain amplifier chain: gain versus control voltage.

resistor to the input of the fourth gain stage of the baseband amplifier.

The class-AB action allows for reduced settling time in the baseband amplifier for use in time-division duplex systems. During initial turn-on, when the gain of the baseband amplifier is at its maximum setting (88 dB), the output stage is easily saturated by small offset voltages at the input of the first stage. When the output of the baseband amplifier is saturated, the servo-loop is broken because the small signal gain approaches zero. The settling time is determined by the slew rate on the external capacitor. Since the output current of the class AB stage is much larger than that of a conventional stage (for the same bias current), the slew rate is enhanced; hence the settling time is reduced.

IX. Integrated Low-Pass Filter

The integrated low-pass filter is depicted in Fig. 17. The filter topology is a differential two-pole Sallen and Key filter which, when cascaded with a third passive pole at the output of the third stage, forms a three-pole Butterworth filter at 900 KHz. Emitter-followers $Q1$ and $Q2$ are used for the purpose of buffering as well as level shifting. Capacitors $C2, C3$, and $C4$ along with resistors $R2, R13$, and the output impedance of the second baseband gain stage set the pole frequencies of the low-pass filter. The large value of resistance of $R12$ and $R13$ substantially degrades the noise floor of the filter. However, decreasing the resistance value would require an increases in size of the already large 76 pF capacitors.

The equivalent input noise resistance of the filter is depicted in Fig. 18. Note that a typical in-band noise resistance value on the order of 50 K Ω implies that the total voltage gain in front of the filter must be substantially more than 30 dB or else the filter will contribute significantly to the receiver noise floor. The large noise resistance precludes using this type of filter structure directly after the mixer.

X. Performance Summary of Baseband Amplifier

The baseband amplifier gain versus control voltage (both simulated and measured) is shown in Fig. 19. The measured and simulated results agree within 2 dB over the entire gain-control range.

The frequency response of the baseband amplifier is shown in Fig. 20. The high-pass corner is determined by the servo feedback capacitor and the emitter coupling capacitor in the second stage of the baseband amplifier. The low-pass corner is determined by the integrated Sallen and Key filter.

The measured settling time performance is shown in Fig. 21. Note that the dc settling time of the amplifier is about 200 μs.

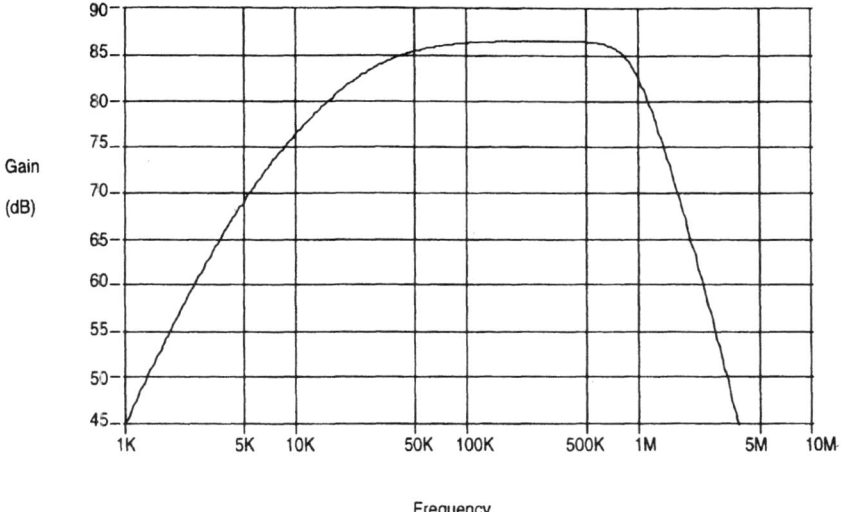

Fig. 20. Variable-gain amplifier: simulated frequency response.

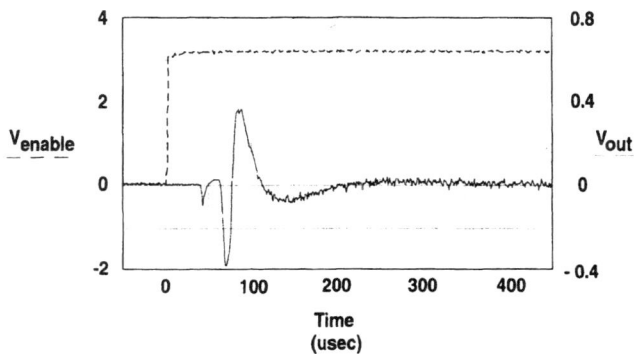

Fig. 21. Variable-gain amplifier chain: measured settling characteristics.

TABLE II
PERFORMANCE OF VARIABLE-GAIN BASEBAND AMPLIFIER

Parameter	Performance
Maximum Voltage Gain	88.2 dB
Minimum Voltage Gain	13.2 dB
3 dB Bandwidth	800 KHz
Relative Attenuation @ 3.5 MHz Output	45 dB
Input IP3 (Minimum Gain State)	−27 dBV
Output IP3 (Maximum Gain Sate)	+7 dBV

The servo loop generates a closed-loop high-pass corner frequency of 22 KHz.

The performance of the complete baseband amplifier is summarized in Table II. The IP3 for the minimum gain state is referred to the input since this is insensitive to the actual gain of the device. For similar reasons, the IP3 for the maximum-gain state is referred to the output. The gain-control range achieved is greater than 65 dB. By combining this with the 23 dB gain-step in the front-end, a total gain-control range of 88 dB has been achieved. This is close to the amount previously reported by Takahashi [2].

XI. CONCLUSION

The die photograph is shown in Fig. 22. The LNA is in the bottom-left corner. Note the presence of the spiral inductor for

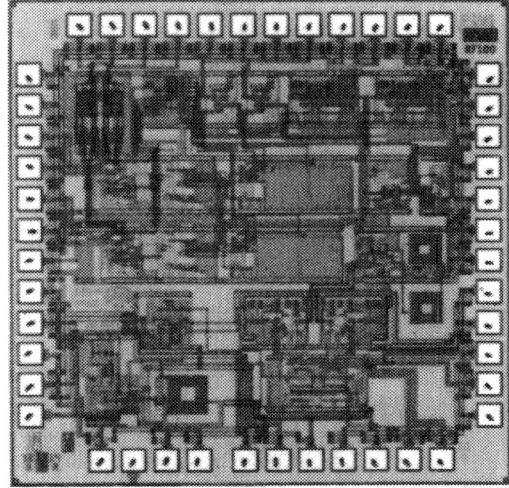

Fig. 22. Die photograph of receiver.

the output tuning. The bottom-right corner contains mixers, with symmetrically placed LO buffers just above them. The layout symmetry is required to maintain the accurate 90 degree quadrature phase relationship. On the middle of the right-hand side of the device is the LO filter and phase shift network. Note the presence of the two spiral inductors. The top portion of the die includes the first and second stage of the baseband amplifier, as well as the servo amplifier. In the middle section of the chip there are four large capacitors which form the major part of the Sallen and Key filter. The third and fourth stages of baseband amplifier are on the left side above the LNA.

A monolithic implementation of a direct-conversion receiver has been discussed. The fully integrated die needs only four external components for filtering. Performance sufficient for cordless telephony is achieved with a 2.7 V minimum supply voltage and 50 mA of current. The advantages of reduced filter cost of direct-conversion have been demonstrated. Techniques for minimizing settling time and LO to RF isolation have been discussed. Direct conversion appears to be the optimum architecture for a low cost receiver when the transmitted code has negligible low-frequency energy.

REFERENCES

[1] A. A. Abidi, "Direct-conversion radio transceivers for digital communications," *IEEE J. Solid-State Circuits*, vol. 30 no. 12, Dec. 1995.
[2] C. Takahasi and R. Fujimoto, "A 1.9 GHz Si direct conversion receiver IC for QPSK modulation systems," in *ISSCC Dig.*, 1994.
[3] M. McDonald "A 2.5 GHz BiCMOS image-reject front-end," in *ISSCC Dig.*, 1993, pp. 144–145.
[4] J. Crols and M. Steyaert, "A fully integrated 900 MHz CMOS double quadrature downconverter," in *ISSCC Dig.*, 1995, pp. 136–137.
[5] C. D. Hull, "Analysis and optimization of monolithic RF downconversion receivers," Ph.D. dissertation, Univ. California Berkeley, 1992.
[6] R. G. Meyer and W. D. Mack, "A DC to 1-GHz differential monolithic variable-gain amplifier," *IEEE J. Solid-State Circuits*, vol. 26, pp. 1673–1680, Nov. 1991.
[7] W. E. Hearn, "Fast slewing monolithic operational amplifier," *IEEE J. Solid-State Circuits*, vol. SC-6, pp. 20–24, Feb. 1971.

A 2.7-V 900-MHz CMOS LNA and Mixer

Andrew N. Karanicolas, *Member, IEEE*

Abstract— A CMOS low-noise amplifier (LNA) and a mixer for RF front-end applications are described. A current reuse technique will be described that increases amplifier transconductance for the LNA and mixer without increasing power dissipation, compared to standard topologies. At 900 MHz, the LNA minimum noise figure (NF) is 1.9 dB, input third-order intercept point (IIP3) is −3.2 dBm and forward gain is 15.6 dB. With a 1-GHz local oscillator (LO) and a 900-MHz RF input, the mixer minimum double sideband noise figure (DSB NF) is 5.8 dB, IIP3 is −4.1 dBm, and power conversion gain is 8.8 dB. The LNA and mixer, respectively, consume 20 mW and 7 mW from a 2.7 V power supply. The active areas of the LNA and mixer are 0.7 mm × 0.4 mm and 0.7 mm × 0.2 mm, respectively. The prototypes were fabricated in a 0.5-μm CMOS process.

I. INTRODUCTION

THE demand for portable wireless communications systems is driven by the expansion of personal and commercial wireless services. As a result, the design of portable handsets follows trends that include lower cost, longer battery life, smaller size, and lower weight. These trends increase the focus on RF IC implementations that traditionally rely on bipolar or GaAs technologies for 1–2 GHz RF realms. At the same time, fine-line CMOS technologies continue to evolve as the demand for higher speed and more complex digital systems grows. The high-speed attribute of fine-line CMOS technology offers opportunity for RF IC implementation. Monolithic RF IC system integration is an important factor in meeting the design needs of portable communications applications [1]. Integration of RF building blocks into a CMOS monolithic system has the advantage of leveraging efforts in CMOS frequency synthesizer, IF, and baseband processing functions. This paper presents the design of a 2.7-V 900-MHz CMOS low-noise amplifier (LNA) and mixer. A current reuse technique will be described that achieves the same amplifier transconductance for the LNA and mixer but at a decreased power dissipation compared to standard topologies.

II. DESIGN OBJECTIVES

The overall specifications for the LNA and mixer were derived from the front-end requirements of a North American Digital Cellular (NADC) handset. The primary design goal for the LNA and mixer is a total current consumption of 10 mA at a 2.7 V power supply. Fig. 1 shows the intended application of the LNA and mixer in a superheterodyne receiver front-end. As an external image-reject filter is employed, the LNA output must be capable of driving a 50 Ω load. The mixer

Manuscript received June 25, 1996; revised August 9, 1996.
The author was with the AT&T Bell Laboratories, Holmdel, NJ 07733 USA. He is now an independent consultant in Morganville, NJ 07751 USA.
Publisher Item Identifier S 0018-9200(96)08108-5.

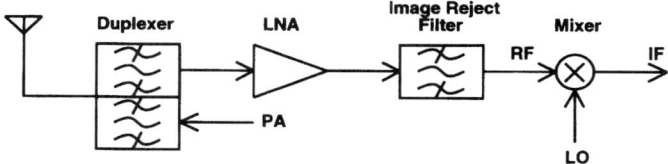

Fig. 1. Superheterodyne receiver front-end application of LNA and mixer.

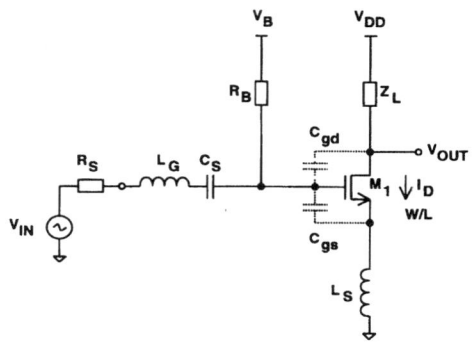

Fig. 2. Basic NMOS LNA design.

IF output is an open-drain design to provide the option of driving either an impedance matching network for a high impedance IF filter or a buffer that in turn drives an IF filter. The primary CMOS process attributes are: drawn channel length of 0.5 μm, 90 Å gate-oxide thickness, three-level metal, and two-level nonsilicided polysilicon. Attention to layout is needed to minimize parasitic interconnect resistance and overlap capacitance that can degrade circuit performance.

III. LNA DESIGN

Fig. 2 shows a schematic of a basic NMOS LNA. For simplicity, the bias network is composed of V_B and R_B to set the desired gate bias voltage for M_1. The input is coupled to the gate of M_1 with coupling capacitor C_S. The input is matched to $R_S = 50$ Ω by using inductors L_G and L_S. The matching condition can be shown to occur when

$$\omega^2 C_{gs}(L_G + L_S) \approx 1 \quad (1)$$
$$L_S \approx R_S C_{gs}/g_{m1} \quad (2)$$

where C_{gs} is the gate-source capacitance and g_{m1} is the transconductance of device M_1. With these conditions, the LNA noise factor F can be shown to be

$$F \approx 1 + (8\omega^2 C_{gs}^2 R_S)/(3g_{m1}). \quad (3)$$

The effect of the gate-drain capacitance C_{gd} is neglected in these first-order approximations. Since g_{m1} appears in the

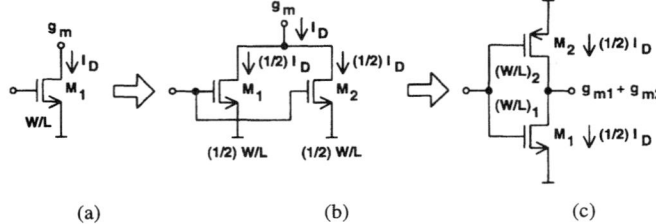

Fig. 3. Illustration of current reuse method.

denominator of (3), a large g_{m1} is typically needed to reduce the noise figure NF, where $NF = 10\log_{10}(F)$. The LNA in Fig. 2 is capable of achieving the desired NF specification, on the order of 2 dB. However, a large drain current I_D may be necessary. To relax this design tradeoff, a current reuse technique will be described.

Fig. 3 illustrates the current reuse method. The goal is to achieve the g_m and $\omega_T = g_m/C_{gs}$ of a single device with less current. Fig. 3(a) shows a single NMOS device that has aspect ratio W/L with drain current I_D. Fig. 3(b) shows two NMOS devices in parallel. Each of these devices has aspect ratio $(1/2)W/L$ and drain current $(1/2)I_D$. Thus, the transconductance of the compound device in Fig. 3(b) is the same as the transconductance of the device in Fig. 3(a). In Fig. 3(c), a PMOS device is substituted for device M_2 in Fig. 3(b). The total transconductance in Fig. 3(c) is $g_{mt} = g_{m1} + g_{m2}$. With $(W/L)_1 = (W/L)_2 = (1/2)W/L$, the input capacitance of Fig. 3(c) is $C_{gs1} + C_{gs2}$ which is nearly equal to C_{gs} in Fig. 3(a). Since the mobility of the PMOS device is lower, g_{mt} will be lower than g_m. For example, if $\mu_p \approx 0.5\mu_n$, then $g_{mt} \approx 0.85 g_m$. This reduction in g_{mt} for this design results in 0.2 dB increase in NF, which is tolerable considering that the corresponding drain current is reduced by a factor of two.

The desired LNA design approach employs a cascade connection of two transconductance amplifier stages. A two-stage LNA design is motivated by the desire to relax the design requirements imposed by a one-stage design. A primary design consideration is driving a 50 Ω load while achieving high forward gain and low reverse gain. Low reverse gain is desired to provide sufficient isolation and to simplify input and output port matching.

Fig. 4 shows a schematic of the two-stage LNA design. In this design, stage 1 is used to achieve the overall LNA forward gain while stage 2 is used as a unity-gain buffer. The output of stage 1 is directly coupled to the input of stage 2. Bias amplifiers are used for each stage to establish the operating points for the respective stages. External networks N_S and N_L match the LNA input and output ports to 50 Ω, respectively. The RF input is applied at V_{RF}, thus driving the gates of M_1 and M_2 in the first stage. Since an external image reject filter is intended for use between the LNA output and the mixer RF input, the LNA output is capable of driving a load resistance R_L of 50 Ω. The operation of a single stage is described, as the stage topologies are identical. Focusing on the first stage in Fig. 4, devices M_1 and M_2 are configured such that the transconductance of the stage is $g_{mt} = g_{m1} + g_{m2}$, where g_{m1} and g_{m2} are the transconductances of devices M_1 and M_2, respectively. Capacitor C_B bypasses the source of M_1 to ground at high frequencies.

In the first stage shown in Fig. 4, a bias feedback amplifier is used to set the dc output voltage V_{OUT1} of the stage to the bias reference V_{B1} [2]. Devices M_{3-7} are used to steer bias current into devices M_1 and M_2. The bias reference I_{REF} and the current mirror composed of devices M_8 and M_2 are used to establish the desired bias current in devices M_1 and M_2. The bias feedback loop is completed with a low-pass filter, composed of R_X and C_X, that provides the dc output voltage V_{X1} from V_{OUT1}. The low-frequency pole contributed by the filter dominates the bias feedback amplifier loop transmission to achieve a high phase margin for the loop. Direct coupling is utilized between the output of the first stage and the input of the second stage. The bias reference V_{B1} sets the dc output voltage V_{OUT1} for the first stage and thus sets the dc input voltage of the second stage. The second stage bias current is thus determined. The second stage bias feedback amplifier is used to set the dc output voltage V_{OUT2} to bias reference V_{B2}. In this design, $V_{B1} = V_{B2} = V_A$, where V_A is the dc input voltage of the first stage determined by I_{REF}.

IV. MIXER DESIGN

Fig. 5 shows a basic NMOS mixer design. A single device M_1 is used as a transconductance amplifier and a mixing cell composed of devices M_{2-3} is used to perform the chopping function. The mixer in Fig. 5 is capable of achieving the desired double sideband noise figure (DSB NF) specification, on the order of 8 dB, however, a large drain current I_D may be necessary. To relax this design tradeoff, the current reuse technique is employed.

Fig. 6 shows the desired mixer transconductance amplifier design. The design is similar to a single stage of an LNA. For simplicity, the bias network is composed of V_{B1-2} and R_{B1-2} to set the desired gate bias voltage for M_{1-2}. The input is coupled to the gates of M_{1-2} with coupling capacitor C_{S1-2}. The input is matched to $R_S = 50$ Ω by using inductors L_G and L_{S1-2}. A matching condition similar to that of the LNA in Fig. 2 can be derived for the mixer. Fig. 6 shows a mixer cell interposed between the drains of devices M_1 and M_2. The drain current is thus reused among M_2, the mixer cell and M_1. The mixer cell design is shown in Fig. 7. Devices M_{1-4} compose the main mixer cell driven by the differential local oscillator (LO) inputs V_{LO1} and V_{LO2}. The drain currents I_{D1-2} are steered through devices M_1 and M_3 as shown in Fig. 7(a), or through devices M_2 and M_4 as shown in Fig. 7(b), as a function of the LO phase.

Fig. 8 shows a schematic of the mixer. An external network N_S matches the mixer RF port to 50 Ω. The RF input is applied at V_{RF}, driving V_{RF1} and V_{RF2}, and thus the gates of M_5 and M_6, in phase. Devices M_5 and M_6 are configured as a transconductance amplifier with $g_{mt} = g_{m5} + g_{m6}$, where g_{m5} and g_{m6} are the transconductances of devices M_5 and M_6, respectively. With an input V_{RF} applied, the drain currents of M_5 and M_6 differ by $g_{mt}V_{RF}$. This difference current is then chopped by the mixer cell resulting in the desired intermediate frequency (IF) current at the mixer output ports, V_{OUT1} and

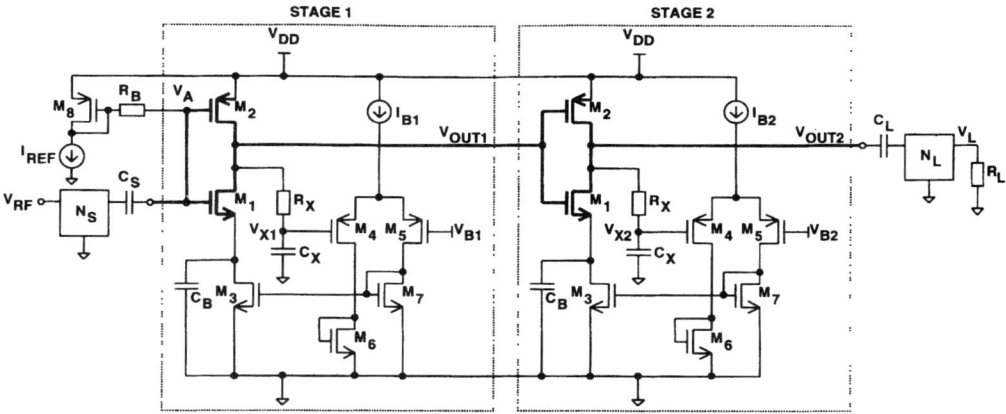

Fig. 4. Schematic diagram of the CMOS LNA.

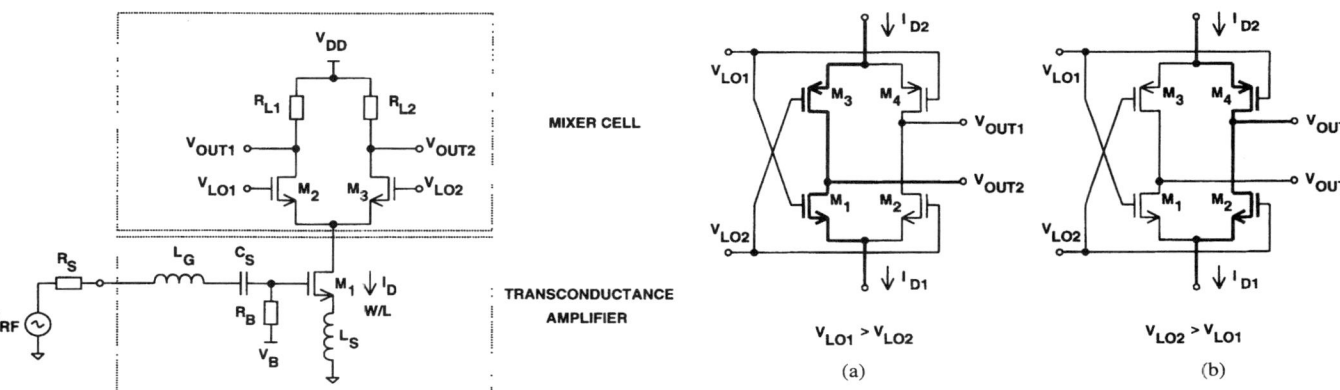

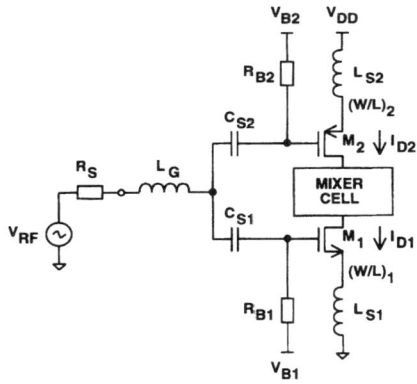

Fig. 5. Basic NMOS mixer design.

Fig. 6. CMOS mixer transconductance amplifier design.

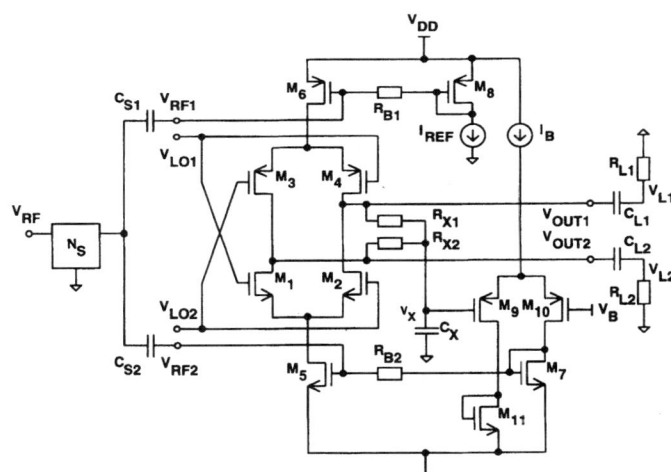

Fig. 7. CMOS mixer cell design.

Fig. 8. Schematic diagram of the CMOS mixer.

V_{OUT2}. The high impedance mixer outputs are intended to drive an external high impedance load.

The mixer biasing approach is similar to the technique used for the LNA stages. A common mode feedback amplifier is used to set the dc common mode output level of the mixer, V_X, to the bias reference, V_B. A differential pair and current mirror composed of devices M_5, M_7, M_{9-11} are used to steer bias current into the mixer cell. Bias reference I_{REF} and the current mirror composed of devices M_8 and M_6 are used to establish the desired bias current in the mixer cell. The feedback loop is completed with a low-pass filter, composed of R_{X1}, R_{X2}, and C_X, that provides the dc common mode level V_X from the outputs V_{OUT1} and V_{OUT2}.

V. EXPERIMENTAL RESULTS

The LNA and mixer IC's are individually packaged and tested. The measured performance of the LNA is summarized in Table I. The measured performance of the mixer is summarized in Table II. The LNA and mixer NF measurements are

TABLE I
SUMMARY OF LNA MEASUREMENTS

LNA	Measured Parameters
Supply voltage	2.7V
Power dissipation	20mW
Frequency	900MHz
NF (50 Ω)	2.2dB
Min. NF	1.9dB
Forward gain $|S_{21}|$	15.6dB
Reverse gain $|S_{12}|$	-32.4dB
Input IP3	-3.2dBm
Input 1dB compression level	-15.2dBm

TABLE II
SUMMARY OF MIXER MEASUREMENTS

Mixer	Measured Parameters
Supply voltage	2.7V
Power dissipation	7mW
RF frequency	900MHz
LO frequency (0 dBm)	1GHz
DSB NF (50 Ω)	6.7dB
Min. DSB NF	5.8dB
Power conversion gain	8.8dB
Input IP3	-4.1dBm
Input 1dB compression level	-16.1dBm
LO-RF feedthrough	-59.6dB
LO-IF feedthrough	-34.4dB

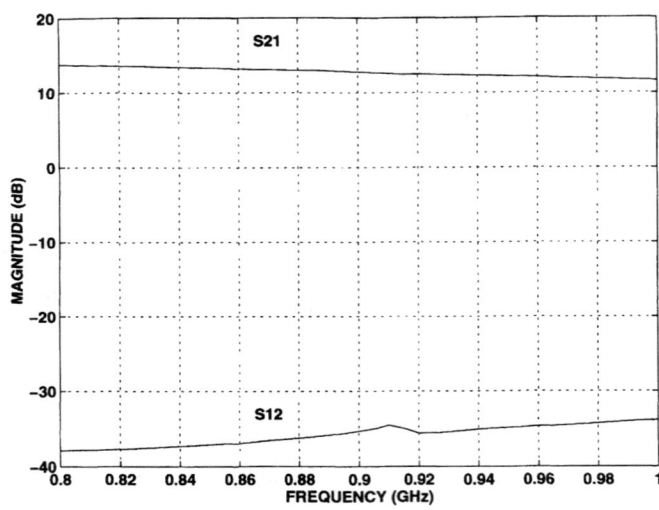

Fig. 10. Measured LNA gain magnitudes $|S_{21}|$ and $|S_{12}|$, swept from 800 MHz to 1 GHz, without external matching networks.

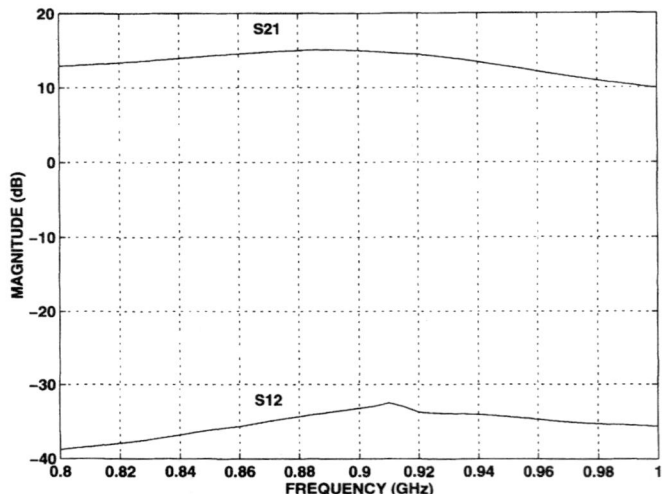

Fig. 11. Measured LNA gain magnitudes $|S_{21}|$ and $|S_{12}|$, swept from 800 MHz to 1 GHz, with external matching networks.

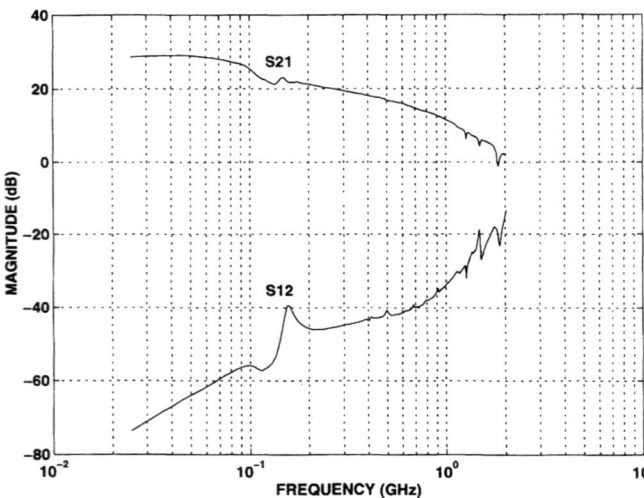

Fig. 9. Measured LNA gain magnitudes $|S_{21}|$ and $|S_{12}|$, swept from 25 MHz to 2 GHz, without external matching networks.

referred to 50 Ω at a frequency of 900 MHz. The DSB NF measurements are reported for the mixer. The single sideband (SSB) NF of the mixer can be estimated to be 3 dB higher than the DSB NF if the mixer conversion gain at frequencies LO$-$IF and LO+IF are similar [3], as in the mixer presented in this work.

The wideband frequency response measurements in Figs. 9 and 10 are performed without matching networks. Fig. 9 shows the measured LNA forward and reverse gain magnitudes, $|S_{21}|$ and $|S_{12}|$, respectively, swept from 25 MHz to 2 GHz. Fig. 10 shows a zoomed-in measurement of Fig. 9 from 800 MHz to 1 GHz.

Figs. 11 and 12 show measurements performed with matching networks for the LNA input and output ports. Fig. 11 shows the measured LNA $|S_{21}|$ and $|S_{12}|$, swept from 800 MHz to 1 GHz. Fig. 12 shows the measured LNA output spectrum when a two-tone RF input at 899.5 MHz and 900.5 MHz is applied. The RF input power level is -31 dBm for each tone.

Fig. 13 shows measurements performed with a matching network for the mixer RF port. Fig. 13 shows the measured mixer IF output spectrum when a two-tone RF input at 899.5 MHz and 900.5 MHz is mixed with a LO frequency at 1 GHz. The RF input power level is -29 dBm for each tone and the LO power level is 0 dBm.

The external matching networks are implemented with manual slide-screw tuners that feature locking verniers and

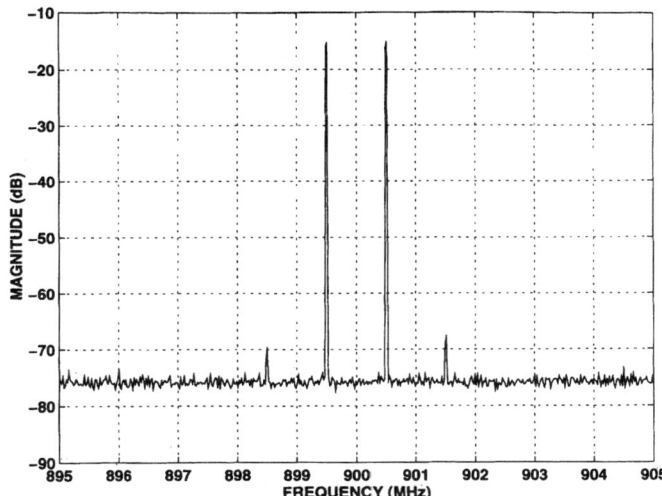

Fig. 12. Measured LNA output spectrum, RF input power level −31 dBm at 899.5 MHz and 900.5 MHz, with external matching networks.

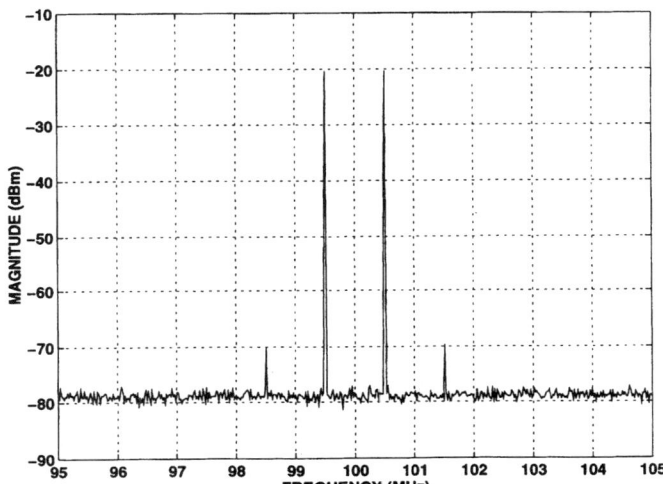

Fig. 13. Measured mixer IF output spectrum, RF input power level −29 dBm at 899.5 MHz and 900.5 MHz, LO input power level 0 dBm at 1 GHz, with external matching network at the RF port.

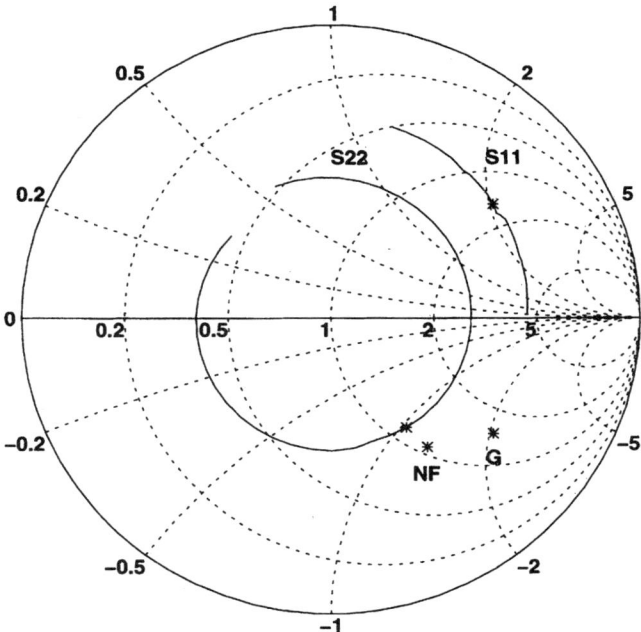

Fig. 14. Measured LNA Smith chart for S_{11} and S_{22}, swept from 800 MHz to 1 GHz. Source reflection coefficient measured at 900 MHz for maximum gain, $\Gamma_S = \Gamma_G$, and for minimum NF, $\Gamma_S = \Gamma_{NF}$.

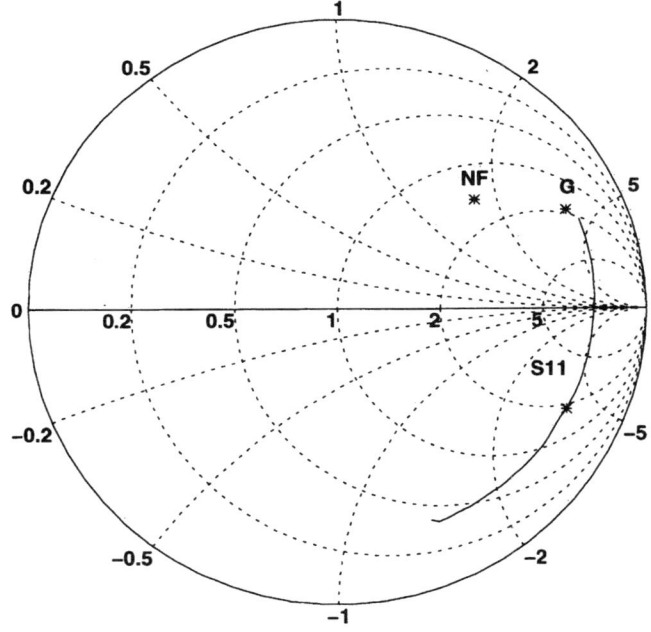

Fig. 15. Measured mixer Smith chart for the RF port S_{11}, swept from 800 MHz to 1 GHz. Source reflection coefficient measured at 900 MHz for maximum gain, $\Gamma_S = \Gamma_G$, and for minimum NF, $\Gamma_S = \Gamma_{NF}$.

micrometers for repeatability. The bondwires that are used for the V_{DD} and V_{SS} power supply connections provide real-part contribution to the LNA input impedance and the mixer RF port input impedance [2]. The measurements for Figs. 12 and 13 are repeated for different input power levels, below the respective input referred 1 dB compression levels, in order to predict the input IP3 extrapolation.

Fig. 14 shows a Smith chart for the measured LNA S_{11} and S_{22}, swept from 800 MHz to 1 GHz. Fig. 15 shows a Smith chart for the measured mixer RF port S_{11}, swept from 800 MHz to 1 GHz. Referring to Figs. 14 and 15, when the input reflection coefficient is set to point "G," $\Gamma_S = \Gamma_G$, maximum gain occurs. When the input reflection coefficient is set to point "NF," $\Gamma_S = \Gamma_{NF}$, minimum noise figure occurs [4]. When $\Gamma_S = \Gamma_{NF}$, the input return loss, or reflection coefficient magnitude, for the LNA is $RL = -11$ dB and for the mixer RF port is $RL = -7$ dB.

The LNA and mixer designs utilize external coupling capacitors at the input and output ports. The prototypes are measured in thin quad flat pack (TQFP) packages. Fig. 16 shows a micrograph of the LNA and Fig. 17 shows a micrograph of the mixer. The active areas of the LNA and mixer are 0.7 mm × 0.4 mm and 0.7 mm × 0.2 mm, respectively.

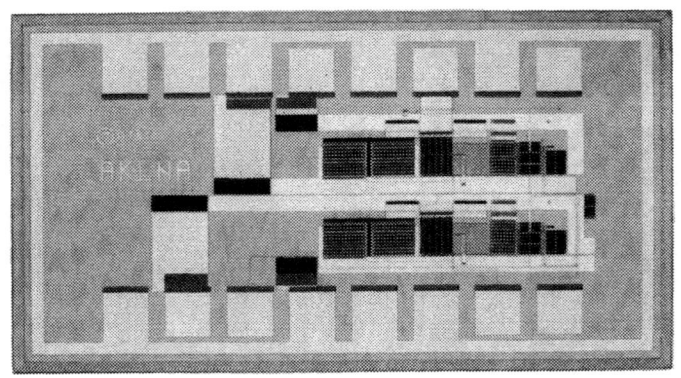

Fig. 16. Micrograph of the LNA.

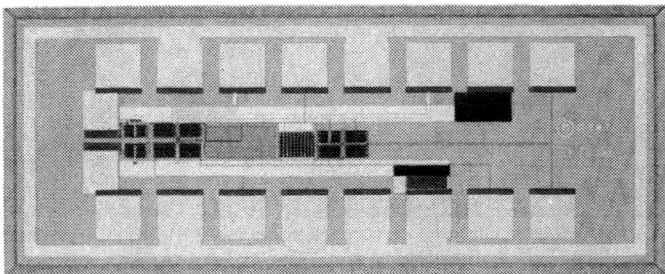

Fig. 17. Micrograph of the mixer.

VI. Conclusions

A two-stage LNA design with a 50 Ω output drive capability has been described. The LNA stages utilize a current reuse method in order to reduce the current consumption required to attain a performance level compatible with NADC. A mixer utilizing a current reuse method is described. The mixer transconductance amplifier is based on an LNA stage design and the mixer chopping cell is compatible with the current reuse method. Experimental results indicate that CMOS technology can be used to build an LNA and mixer while meeting output drive, noise, and linearity requirements for portable cellular systems.

Acknowledgment

The author thanks W. Garner for technical discussions, R. Pummer for layout, V. Archer for CAD support, Y. Ota for packaging and K.-H. Lee for processing support.

References

[1] T. D. Stetzler et al., "A 2.7–4.5 V single chip GSM transceiver RF integrated circuit," *IEEE J. Solid-State Circuits,* vol. 30, pp. 1421–1429, Dec. 1995.
[2] R. G. Meyer and W. D. Mack, "A 1-GHz BiCMOS RF front-end IC," *IEEE J. Solid-State Circuits,* vol. 29, pp. 350–355, Mar. 1994.
[3] S. A. Maas, *Microwave Mixers.* Dedham, MA: Artech, 1986.
[4] T. T. Ha, *Solid-State Microwave Amplifier Design.* Malabar, FL: Krieger, 1991.

SP 22.3: A 12mW Wide Dynamic Range CMOS Front-End for a Portable GPS Receiver

Arvin R. Shahani, Derek K. Shaeffer, Thomas H. Lee

Stanford University, Stanford, CA

At submicron channel lengths, CMOS is an attractive alternative to silicon bipolar and GaAs MESFET technologies for use in wireless receivers. A 12mW global positioning system (GPS) receiver front-end, comprising a low noise amplifier (LNA) and mixer implemented in a standard 0.35µm digital CMOS process, demonstrates the aptitude of CMOS for portable wireless applications.

A block diagram of the receiver front-end test is shown in Figure 1. The system consists of a LNA, mixer, buffer (for testing purposes only), and associated bias circuitry. Note that the LNA output and mixer input are taken off-chip to facilitate testing of each block individually.

To achieve a wide dynamic range, the noise figure of the LNA and linearity of the mixer are of primary concern. Figure 2 shows the circuit schematic of the LNA. A differential architecture is chosen in anticipation of eventual integration of a complete GPS receiver. Its common mode rejection eases the task of rejecting interference from other on-chip elements, such as a DSP core. This choice results in increased amplifier noise for a given power dissipation compared to a single-ended implementation. This increase is mitigated by the high ω_T of the process, that permits acceptable noise performance despite a differential implementation.

The input stage, consisting of $M_{1,2}$ and $L_{1,2}$, uses inductive source degeneration to generate a real term in the differential input impedance equal to $\omega_T(L_1+L_2)$ that can be matched to the off-chip transmission lines from the image-reject filter. By resonating the input capacitance of the amplifier, the minimum possible noise figure can also be achieved [1]. A minimum exists due to the presence of induced gate noise in MOS devices, as described in Reference 2. The source degeneration inductors, $L_{1,2}$, are implemented as on-chip spiral inductors.

The output signal from $M_{1,2}$ flows through cascoding devices $M_{3,4}$ and on to a parallel-tuned output stage formed by $M_{7,8}$. The tuning inductors, $L_{3,4}$, are 3nH on-chip spirals. They serve the additional purpose of sharing the bias current between the first and second stages, leading to reduced power consumption. $M_{5,6}$ are ac coupling capacitors, while R_8 and R_9 provide simple pull-up biasing for the gates of $M_{7,8}$. This arrangement allows signal voltages at the gates of the output stage to swing above the supply, which is advantageous for low-voltage operation. Note that resistors $R_{7,8}$ are off-chip 50Ω loads.

To bias devices M_1-M_4, an active common-mode feedback maximizes the use of limited supply headroom. Through the action of a simple operational amplifier, formed by M_{12}-M_{16}, the V_{gs} of the input devices is used as a reference for biasing the cascode stack. This permits the amplifier to operate reliably on a 1.5V supply, independent of process, supply, and temperature variations.

The forward gain (S21) and noise figure of the LNA are plotted in Figure 3. The LNA exhibits an S21 of 17.7dB and a noise figure of 3.8dB at 1.575GHz. This is achieved with 12mW dissipation from a 1.5V supply. Note that a single-ended implementation of the amplifier would have a 2.3dB noise figure. Although suffering increased noise, the differential architecture permits a reverse isolation of -52dB, unattainable in a single-ended version.

A simplified mixer schematic is shown in Figure 4. Note that some of the biasing details have been omitted. The local oscillator and its inverse, $\overline{LO}$, control which pair of transistors M_1-M_4 is on, and thus the polarity with which the load is presented to the RF port. It is this switching of polarity that establishes mixing. A unique feature of this mixer is the reactive termination of both the RF and IF ports. A capacitive IF termination offers the advantage that it contributes no noise to the mixing process. At the RF port, a reactive tank is used to filter broadband noise from the source resistance and parasitic tank resistance, as well as the switches. Bond wires are used to achieve a good Q in addition to an innovative capacitor layout described in the next paragraph. A further improvement is the addition of inductors in series with the input source resistance to form an L-match together with part of the tank capacitance. This boosts the signal voltage and enhances filtering by increasing the effective source resistance.

Area-efficient linear capacitors are generally unavailable in standard digital process technologies (such as the one used here). However, deep submicron processes do allow small spacing between metal lines on any given interconnect layer, leading to high coupling capacitance. While this property is normally considered undesirable, it is exploited in this design to augment the ordinary parallel plate capacitance.

By operating the transistors in the linear region, good mixer linearity is achieved. The high impedance at the mixer IF port afforded by the on-chip environment improves linearity by reducing the signal currents in the MOS switches. Additionally, a large linear tank capacitor dominates the transistor non-linear parasitic capacitances.

The results of a two-tone IP3 measurement on the mixer are shown in Figure 5. Note that the impedance level at the mixer output is not 50Ω. Indeed, the impedance is complex, so the mixer output quantities are expressed in dBV rather than dBm. The mixer has a -3.6dB voltage conversion gain, with an input-referred IP3 of +10dBm and an input 1dB compression point of -5dBm. Measured mixer SSB noise figure is 10dB. For these measurements, the LO drive amplitude is equivalent to the voltage swing associated with -3.5dBm into 100Ω. However, the LO port presents a nearly capacitive reactance, and hence actually dissipates little power.

The results of experimental measurements are summarized in Table 1. The LNA-mixer combination exhibits the best performance achieved to date of any CMOS implementation in this frequency range at this power level.

Acknowledgments:

The authors thank D. Dobberpuhl and Digital Equipment Corporation for supporting this work, and Vitesse Semiconductor for providing high-frequency packaging.

References:

[1] Shaeffer, D. K., T. H. Lee, "A 1.5V, 1.5GHz CMOS low noise amplifier," Symposium on VLSI Circuits Digest of Technical Papers, 1996, pp. 32-33.

[2] van der Ziel, A., *Noise in Solid State Devices and Circuits*, John Wiley & Sons, New York, 1986.

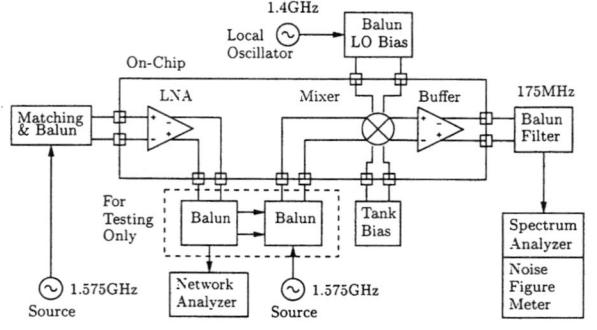

Figure 1: Block diagram of the GPS front-end test setup.

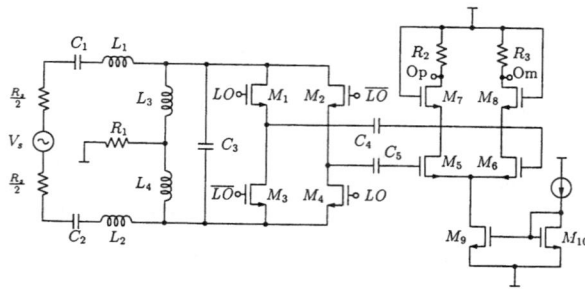

Figure 4: Schematic of mixer circuit with test buffer.

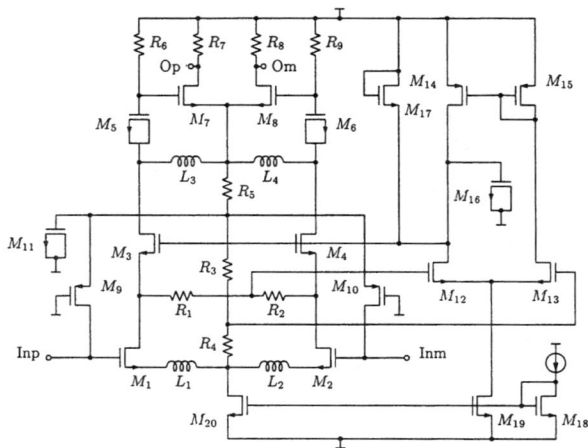

Figure 2: LNA circuit schematic.

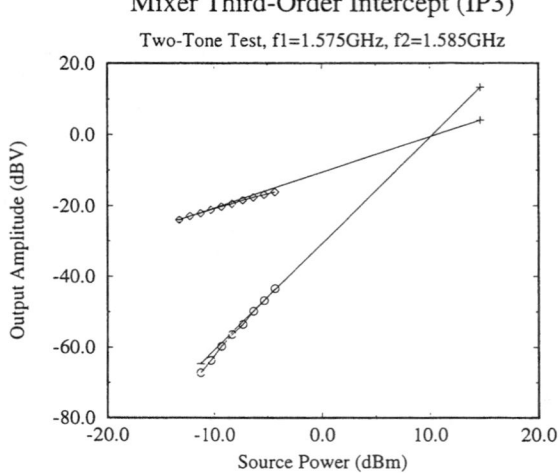

Figure 5: Mixer two-tone IP3 measurement.
Figure 6: See page 487.

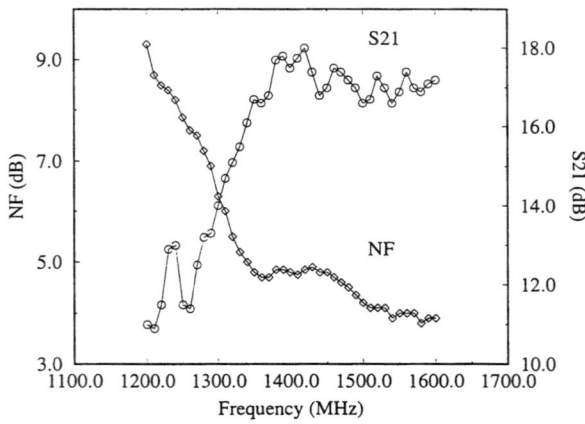

Figure 3: LNA forward gain (S21) and noise figure.

Low-noise amplifier	
Frequency	1.575GHz
Noise figure	3.8dB
S21	17.7dB
S12	$\leq$ -52dB
IP3 (input)	-6dBm
1dB compression (input)	-20dBm
Power dissipation	12mW

Mixer	
LO frequency	1.4GHz
LO amplitude	300mV ($\approx$3.5dBm in 100Ω)
Voltage conversion gain	-3.6dB
IP3 (input)	10dBm
1dB compression (input)	-5dBm
Noise figure (SSB)	10dB

Supply voltage	1.5V
Technology	0.35µm CMOS
Die area	0.84mm²

Table 1: GPS front-end performance summary.

Figure 6: Die micrograph.

4

TRANSMITTERS AND TRANSCEIVERS

THE quest for ever-lower production costs for mobile transmitters and receivers has led to considerable efforts to combine as many as possible of the electronic functions of these systems into single-chip transceivers. The transmit and receive path, can share passive external filters and various on-chip filter functions. Additional savings in hardware and power dissipation can be realized by the sharing of synthesizers, local-oscillator buffers, and bias stabilization circuits. Design challenges to these realizations include modeling and minimization of interaction between functional blocks and the switching or selection of various signal-processing functions.

An integrated Si bipolar RF transceiver for a zero IF 900 MHz GSM digital mobile radio frontend of a hand portable phone.

Jan Sevenhans (1), Arnoul Vanwelsenaers (2), J. Wenin (1), J. Baro (2)

(1) Alcatel Bell Telephone, Antwerp, Belgium
(2) Alcatel Radio Telephone, Colombes, France

Abstract

A single chip solution for a GSM 900 MHz RF TX quadrature modulator and RX quadrature demodulator was designed and fabricated in a 9 gHz Silicon bipolar technology.
The transmit modulator is a direct upconverter and the receive demodulator operates in a zero IF architecture.
This device is one of the key-components of a pocket size handportable phoneset for application in the GSM pan European digital mobile radio system. The consumption from the battery is 25 mA in RX mode and 45 mA in TX mode.
The circuit is powered from a 5 V regulated supply voltage.

1. INTRODUCTION

The success of GSM handportable phone sets depends meanly on the level of integration of the system functional building blocks in the minimum number of VLSI circuits.
Today several research labs are publishing work on GSM phone sets consisting of a number of VLSI's. Usually one has a CMOS die for the voice transceiver and the speech codec, two or 3 CMOS devices for the GSM system control, the decoding of the digital voice data from the quadrature I and Q vector phase diagram and another CMOS die to do the A/D and D/A and the channel filtering of the GMSK shaped in phase I and the quadrature Q phase vectors. On top of that, today for the RF-front-end the available components use separate silicon for the RF receiver and transmitter.

The GSM system development is growing towards the feasibility of a pocket size handportable phone by exploiting the state of the art sub micron CMOS technologies for the digital GSM system circuitry and the baseband I and Q analog filters and A-D/D-A conversion.

The further integration of the CMOS baseband and digital circuitry is only limited by the minimum dimensions of CMOS technology.
On the other hand there is the RF-front end operating at 900 MHz. In a pocket size handportable phone we cannot afford to have another twin chip set for the RF transmitter and the RF receiver.

In this paper we describe the single chip solution for our zero IF approach to the GSM RF-transceiver.
The device consumes 25 mA in receive maximum gain mode and 45 mA in transmit to obtain a receive maximum conversion gain of 25 dB, an IP3 of 0 dBm and a receive mixer noise figure of 18 dB.
In transmit mode the circuit delivers 0 dbm on a balanced output to drive the power amplifier of the 2 W pocket size handportable phone set.

2. SYSTEM ARCHITECTURE

* To obtain the overall 6 dB system noise figure we use an 18 dB gain low noise amplifier with a 2,7 dB noise figure.
  The LNA has 2 gain modes to provide a 10 dB gain switch for high input levels.
* The receive mixers perform a direct down conversion of the phase modulated GMSK 935/960 MHz input signal and the quadrature demodulation of the in phase and the quadrature base-band phase vectors. The band width of the I and Q vectors is limited to 360 kHz using a first order preblocking filter.
  The conversion gain of the I and Q RX mixers is 6 dB.
* Following the mixers we have a gainstage on each of them. These post mixer amplifiers (PMA) provide an extra 19 dB gain in the bipolar RF-frontend before we hit the CMOS base band receive filtering. In the PMA we have another 2 MHz preblocking filter to suppress interfering of the wanted base band signals with high blocking signals.
* The I and Q RX-mixers are driven by quadrature Local oscillator (LO) signals. The quadrature Lo signals are derived on chip from a balanced Lo input.
* The transmit section uses two active mixers to compose the phase modulated GSM 890/915 MHz RF signal. To obtain the 0 dBm drive signal for the power

amplifier the Tx-mixer balanced output signal is fed into an on-chip pre-power amplifier (PPA). The PPA is operating in the linear mode to avoid intermodulation in the transmit signal.

3. RX-MIXERS

The design of the RX mixers is dictated by the requirements on noise and distortion limitations.

The GSM spec is demanding the RF front end to cope with blocking signals up to -23 dBm. Since we have 17 dB gain in the low noise amplifier (LNA) of the RF front end, the RX mixers are dealing with -6 dBm blocking levels at 3 MHz distance from the carrier. To bring the 1 dB compression of the differential input pair of the gilbertcel running on a 2 x 1,5 mA tailcurrent up to this level, we need $\pm$ 20 dB emitter degeneration. Then these 200 ohm emitter resistors are the limitation for the achievable noise figure : 18 dB.

The input transistors of the RX-mixers have an R_{BB} of 25 ohm in the 9 GHz fT poly silicon emitter technology used for this application.
This results in a noise figure of 14,06 dB. On top of this we have an extra 3 dB on the noise performance because of the image noise in the single side band phase modulated RF-signal.

4. THE POST MIXER AMPLIFIERS

Before the I and Q vectors can attack the CMOS channel filtering circuitry we have to provide them an extra gain of 19 dB on top of the 6 dB of the mixers.
But before we do this we must consider the GSM blocking levels.
For the handportable receiver, they are specified at 3 MHz away from the carrier on a -23 dbm maximum level. It should be clear that we need to filter these blocking levels down by $\pm$ 15 dB before we deliver them to the PMA gain operating on the 5 V rail to rail power supply.
This 15 dB down filtering of the 3 MHz blocking levels is demanding a first order pole at $\pm$ 300 kHz.
This pole is realised by the capacitors on QB, NQB and IB, NIB, each of them respectively cutting the spectrum of the I and Q phase vectors above 300 kHz.

These first order blocking filters are configurred using external capacitors to allow for moderate collector resistor values in the mixer.

5. THE QUADRATURE PHASE GENERATOR

The main restrictions for the design of the phase shifter are noise, phase accuracy, gain matching and drive capability. The main limitation is power consumption. In the literature one finds mainly two options :

the digital option were you first do a frequency doubling and afterwards a division by 2 in a Johnson Counter to obtain the 2 quadrature Lo signals;
the RC/CR option using a RC-integrator and, a CR-differentiator as a +45°/-45° phase shifter and a limiting amplifier to follow restoring the 0° and 90° Lo signal level to obtain equal amplitude Lo drive level on both RX and TX I and Q mixers.
The digital option has the inherent advantage of the correct phase relation of the quadrature Lo-signals resulting from the devide by 2. But to obtain low phase noise (-148 dBc) this approach is very power hungry.
For this reason we have used the RC/CR approach and to reach the phase accuracy requirement of the GSM specification, we provide a phase adjust correction input. The phase correction voltage applied here is derived from a phase error observation as the phase trajectories are converted into RX-binary data.

6. THE TX-MIXERS

In the Tx-mixers the design was driven by power economising considerations mainly. We have to provide a 0 dBm TX signal level required by the external 2 W power amplifier. Run the mixers at a high

Summary of the measured performance parameters

RX	Supply current in max gain mode	25 mA
	Noise Figure of mixers + PMAs (max gain mode)	18 dB
	I/Q gain matching	0,14 dB
	I/Q conversion gain voltage	16 x
	I/Q quadrature phase error for OV phase adjust input	1 °
	I/Q offset voltage for 150 mVp differential Lo drive and RX input level - ∞ to -10 dBm	20 mV $\pm$ 5 mV
TX	Supply current	45 mA
	Balanced output power level	0 dBm
	Image rejection	40 dBs
	Carrier rejection	36 dBs

current level to obtain the 0 dBm output level on the balanced mixer open collector would have been a 50 % waste of power.
Instead of running 2 mixers at the required 0 dBm power level it appeared to be more battery-friendly to run the mixers at a moderate tail current (dictated by noise and distortion requirements) and spend the nominal power in one on chip pre-power amplifier, driving a balanced open collector RF TX signal over a balun transformer into the power amplifier.

7. CONCLUSION

A single chip solution for a GSM 900 MHz RF TX quadrature modulator and RX quadrature demodulator was designed and fabricated in a 9 gHz Silicon bipolar technology.
The transmit modulator is a direct upconverter and the receive demodulator operates in a zero IF architecture.
This device is one of the key-components of a pocket size handportable phoneset for application in the GSM pan European digital mobile radio system. The consumption from the battery is 25 mA in RX mode and 45 mA in TX mode.
The circuit is powered from a 5 V regulated supply voltage.

8. ACKONWLEDGEMENTS

The authors express many thanks to H. Rokos, P. Gaussen, P. Denny, R. Baldey and Rick Van Vliet for their contribution to the realised component.

9. REFERENCES

- Monolitic Silicon RF circuity. Current achievements and future potential.
  P. H. Saul
  ESSC/RC'90 p. 25-28.

- Quadrature mixer imbalances in digital TDMA Mobile radio receivers.
  A. Baier, 1990 International Zurich Seminar on digital communications, p 181-194.

- A 3 gbit/s Bipolar Phase shifter and AGC Amplifier.
  H.M. Rein et al., 1989 IEEE International Solid state circuit conference.

- DECT front end high frequency amplifier.
  Risto Pitkänen, VLSI Design for mobile/cellular Telephony, Finland 18-23 June 1990.

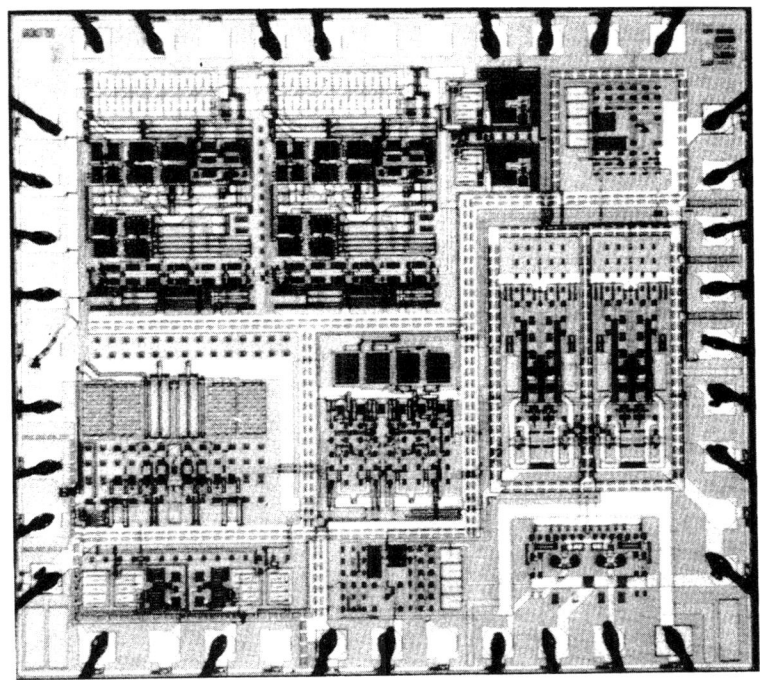

Photomicrograph of the single chip RF GSM 900 MHz quadrature modulator-demodulator (Chip area is 2,5 mm x 2,5 mm).

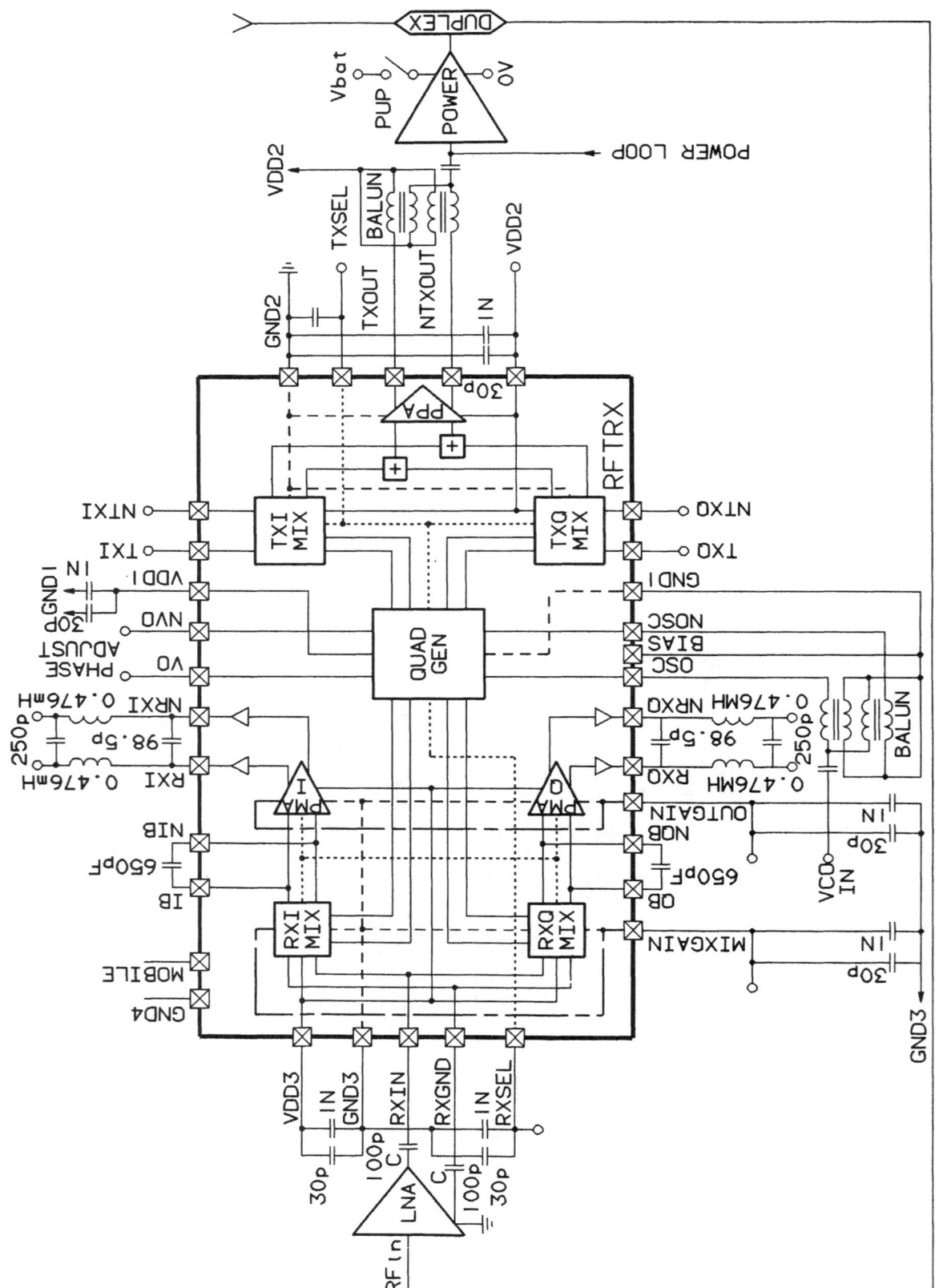

Schematic Diagram of the single chip integration of a zero IF 935/960 MHz quadrature demodulator and a zero IF 890/915 MHz quadrature modulator.

Didier Haspeslagh (1), Joan Ceuterick (1), Lajos Kiss (1), Jacques Wenin (1),
Arnoul Vanwelsenaers (2), Clothilde Enel-Rehel (2)

(1) Alcatel Bell, Antwerp, Belgium
(2) Alcatel Radiotelephone, Colombes, France

Abstract

The BBTRX (base band transmission and reception) is a multi functional, CMOS, mixed analog/digital VLSI circuit conceived to support the different ASIC's on the radio sub-assembly in the GSM handportable terminal. It interfaces the zero IF 900 MHz RF quadrature modulator/demodulator chip and the base band GMSK modulator/demodulator circuits as well as the synthesizer and control unit.
The BBTRX also controls the TX RF power amplifier output level and the burst shaping.
The base band transceiver is a mixed analog - digital design, developed in a 1.2 um analog cmos technology with a 38 mm2 die area.

INTRODUCTION

To reduce the cost and power consumption (battery life) of a GSM hand portable terminal, one should use CMOS technologies for base band analog and Si bipolar for the RF front end. To achieve that goal a single bipolar RF quadrature zero IF modulator - demodulator (900 MHz RF frequency) [1] is developed together with a low power CMOS base band transceiver (BBTRX).
The main function of the BBTRX is to support the link between the RF modulator/demodulator and the GMSK modulator/demodulator [2,3] supervised by the system control function. Additional hardware is included to control the operation of the various radio sub-assembly ASIC's.

The main functions of the BBTRX are 1) channel filtering and attenuation of the adjacent and blocking channels , 2) providing means for predictive automatic gain control, 3) controlling mismatches between the I and Q channels, 4) offset cancelling, 5) controlling the power amplifier bursts, 6) generating the activation signals.

SYSTEM ARCHITECTURE

A block diagram of the base band circuit is shown in figure 1.
The main blocks are :

1) Receive part :

Two balanced I and Q channels, coming from the RF chip, are channel filtered and amplified.
A 5th order analog filter filters the adjacent channels, while a distributed input RC filter provides a continuous roll off for the higher blocking channel frequencies.
The gain of the receive path (I and Q channel) can be set from -21 dB up to 41 dB with a step of .2 dB. Gain mismatches between the I and Q channel can be adjusted separately.
Different offset cancelling means are provided to overcome DC offsets generated by the zero IF demodulation and can be adjusted to a value less than 20 mV on the RX output, regardless the gain setting.
The noise performance achieved on the RX path, all filters and gain stages included, is 50 nV/sqrt(Hz), at a maximum gain of 41 dB.
An overall picture is shown in figure 2.

2) Transmit part :

The transmit part is divided into two parts.
The first circuit will filter (4 th order active filter) and amplify the transmit signals on the 2 balanced I and Q paths after GMSK modulation.
Offset cancelling on both paths is included to avoid DC modulation in the

quadrature modulator.

The second circuit is part of the gain loop controlling the power amplifier. Its provides the control signal needed to shape the transmitted power spectrum. The slopes of the bursts are fully programmable due to the use of a digital to analog convertor in the feedback loop. The basic principle is shown in figure 3.

3) Activation generation :

Out of two control signals, the BBTRX generates all activation signals for all radio ASIC's like Low Noise Amplifier, Power amplifier, RF quadrature modulator/demodulator. Activation sequence can be programmed by the control part to provide optimum activation timing for the different ASIC's and mastering power consumption on the radio board.

4) Data processing :

Out of the control part of the radio card, data and control signals are sent towards the BBTRX. This data is received, checked on its validity by a CRC check and then stored in the appropriate registers.

The control data path serves also the synthesizer with the correct information sent by the control unit or will send the emergency frequency control data when receiving data fails.

5) Alarm generation :

The BBTRX will evaluate and create the alarm signals when reflected power is too high, or if an error occurred during transmitting the control data.

The alarms are processed and sent to the control part, while the synthesizer will receive emergency control data.

TEST RESULTS

The device is processed in a 1.2 uM Cmos technology. Although the circuit is a mixed analog / digital design, excellent performance results were obtained. A summary of the most important analog parameters are presented in table 1.

The most critical design parameters for a system described above are the noise figures (systems S/N ratio), the mismatch between the I and Q paths (GMSK modulation) and the power consumption (battery autonomy).

CONCLUSION

A high performance, mixed analog / digital device, used for GSM base band application is designed in a 1.2 um CMOS technology. The total chip area is 38 mm2. A chip microphotograph is shown in figure 4.

The component is powered with a single 5 V supply, with a load of the battery of only 15 mA (peak value).

Excellent analog performances were measured in this complex mixed analog/digital design.

The BBTRX is used in the Alcatel handportable terminal as a multi functional device connecting the high frequency quadrature modulator as well as the power amplifier and low noise amplifier to the GMSK modulator/demodulator, the synthesizer and the control unit on the radio card board.

ACKNOWLEDGEMENTS

The authors express many thanks to M. van Paemel, B. Verstraeten, P. Ampe (Alcatel Bell) and L. vander Voorde (Alcatel Mietec) for their contribution to the realisation of the component.

REFERENCES

[1] " An integrated Si bipolar RF transceiver for a zero IF 900 MHz GSM digital mobile radio front end of a hand portable phone. "
J. Sevenhans et al., IEEE 1991 Custom Integrated Circuits Conference, 7.7.1

[2] " A 270 Kb/s 35 mW modulator IC for GSM cellular radio hand-held terminals"
J. Haspeslagh et al., IEEE-JSSC, december 1990 pp 1450-1457

[3] " A DSP based ASIC for GSM digital signal processing "
A. Chateau et al. , Mobile Radio Conference, Nice, november 1991

RX	power consumption	14 mA
	noise performance (41 dB gain)	50 nV/sqrt(Hz)
	noise performance (-21 dB gain)	800nV/sqrt(Hz)
	I/Q gain mismatch	< .05 dB
	I/Q phase mismatch	< 1 DG
	input offset cancellation range gain 41 dB	+- 150 mV
	inter modulation suppression third order intercept point	< -50 dB
TX	power consumption	12 mA
	noise performance	100nV/sqrt(Hz)
	I/Q gain mismatch	< .05 dB
	I/Q phase mismatch	< 1 DG

Table 1. : analog performance parameters

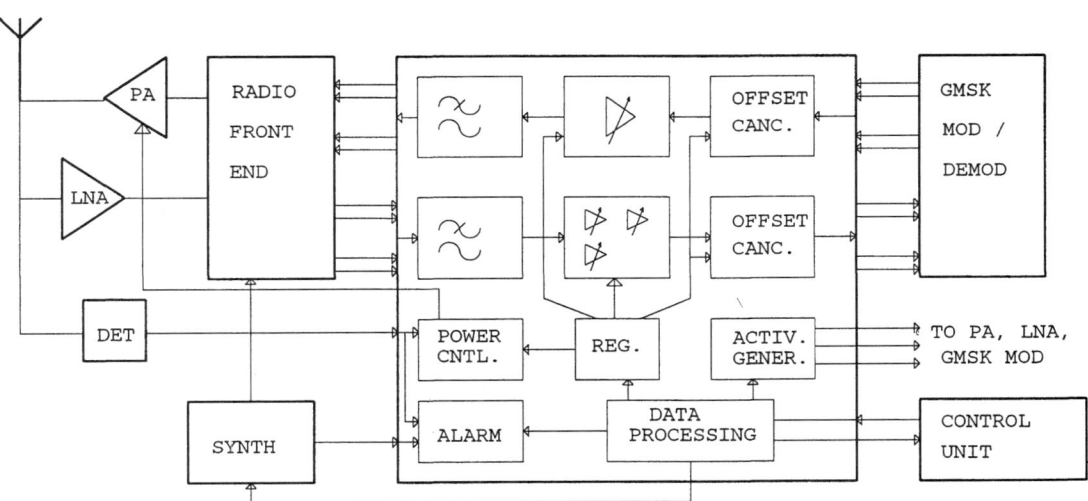

FIGURE 1 : BBTRX BLOCK DIAGRAM

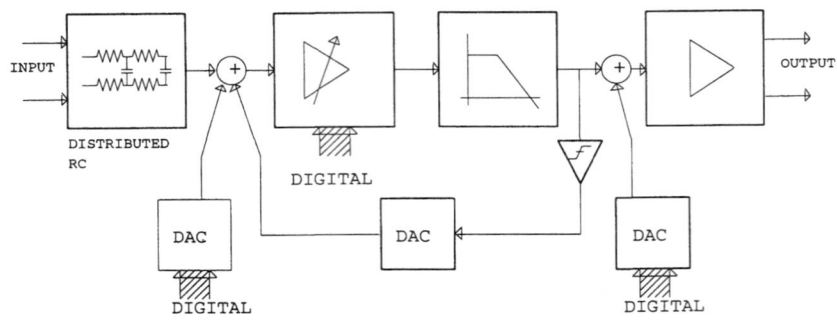

FIGURE 2 : RX CIRCUITS APPLIED IN BOTH I AND Q CHANNEL

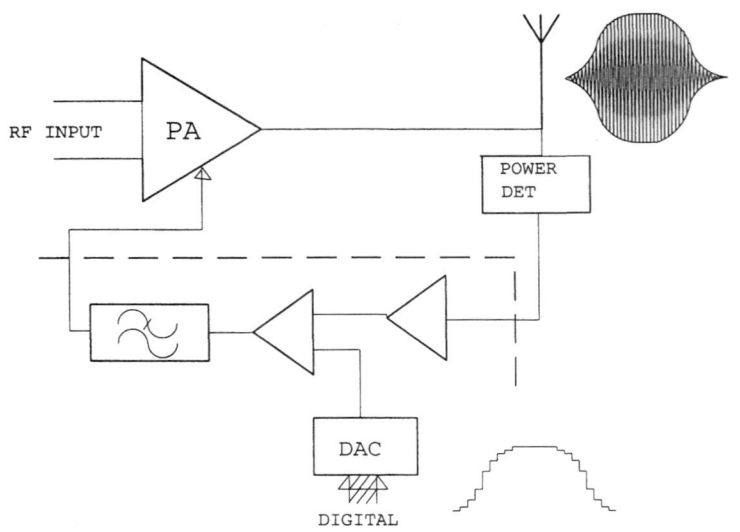

FIGURE 3 : BBTRX TX POWER AMPLIFIER CONTROL LOOP

figure 4

BBTRX : chip microphotograph

A 2.7V 800MHz-2.1GHz Transceiver Chipset for Mobile Radio Applications in 25GHz f_t Si-Bipolar

W. Veit, J. Fenk, S. Ganser, K. Hadjizada, S. Heinen, H. Herrmann, P. Sehrig
Siemens AG, Semiconductor Group, HL BR HF 1
Balanstr. 73
München, D-81541

Abstract

A 800MHz-2.1GHz, 2.7V transceiver chipset for mobile radio which includes transmitter with direkt conversion quadratur modulator, IF and RF frequency synthesizing, receiver with LNA, mixer, programmable 80dB-IF-amplifier and quadratur demodulator is reported. The ~4mm², 25GHz f_t bipolar chips consume < 250mW.

Introduction

In mobile communication there is an increasing demand for light weighted, small sized and low power dissipating equipment. Therefore semiconductor devices with very high integration level and reduced power consumption also for the RF part are needed. In addition these systems on silicon have to be flexible enough to fit for different mobile radio systems.
This paper describes a 2.7V transceiver chip set for digital mobile phone systems up to 2.1 GHz. Figure 1 shows the block diagram of a mobile radio with baseband and RF part.

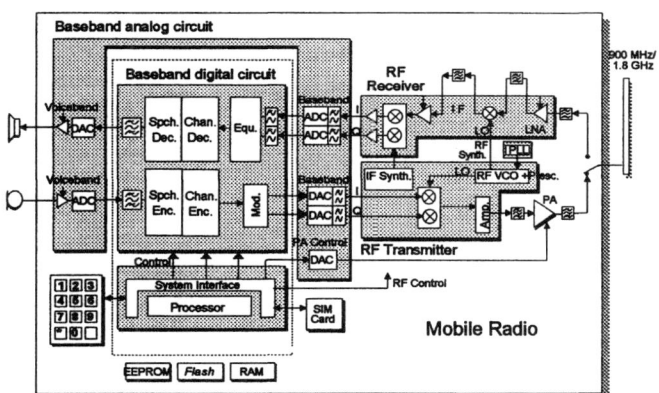

Fig.1 Block diagram of a mobile radio

The RF part consists of a transmitter circuit wich performs direkt modulation, a heterodyne receiver and a PLL circuit for the RF VCO synthesizer.
The baseband digital and PLL circuit are realized in CMOS technology. The baseband analog circuit uses BICMOS technology. The RF transmitter and receiver are designed in the bipolar technology B6HF with a transit frequency of 25 GHz and a minimum emitter size of 0.8 um.

The RF chip set has the following features:
- Various modulation schemes, such as PM, PSK, QAM, QPSK, GMSK etc. with application in vektor modulated digital mobile cellular systems as GSM, PCN, DAMPS, PDC, WLAN etc.
- RF transmitter with direkt modulation up to 2.1 GHz, complete IF synthesizer, RF VCO with prescaler and transmit mixer.
- RF receiver with switchable low noise preamplifier and first mixer with input frequency range up to 2.5 GHz, programmable 80 dB gain controlled IF amplifier with I/Q demodulator or IF sampling
- Supply voltage range from 2.7V to 4.5V and temperature range from -30°C to 85°C
- Low power consumption < 250 mW for RF transmitter and receiver at 2.7V supply
- Very high grade of integration using a 48 pin package with 0.5mm pitch and 9mm x 9mm foot print.

RF Transmitter Circuit

The transmitter circuit (fig. 2) covers a frequency range from 800MHz up to 2.1GHz selectable via alu mask at 2.7V power supply. The circuit is divided into a digital part with fix PLL for the IF oscillator, prescaler for the RF oscillator and an analog part with transmit mixer, RF and IF oscillator and direct-conversion quadratur modulator.

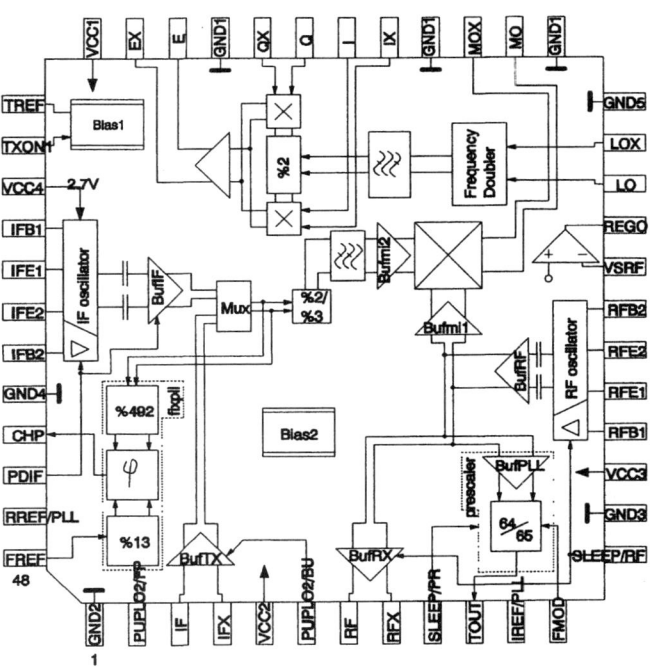

Fig. 2 RF transmitter block diagram

The chip has four different voltage supplies with separate grounds to improve decoupling of the various functional blocks and to give higher application flexibility. In fig. 3 the layout of the transmitter is shown.

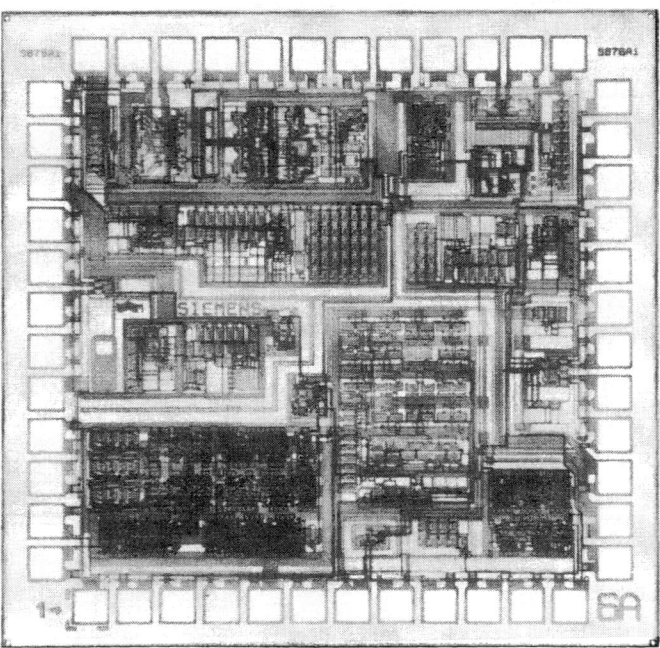

Fig. 3 Die photograph of the transmitter

The digital circuits are placed at the bottom part of the chip. The analog part including the modulator is located at the top. Between the four seperate supplied blocks an additional ground (GND5) can be seen which works as an extra shielding.

The fix PLL contains a 492 divider for the IF oscillator, a 13 divider for system clock and a phase detector with charge pump. It is optional to use the IF oscillator on the transmit or on the receive chip. With a lock in time of 250 usec a C/N ratio of 85 dBc/Hz is reached in 2kHz offset from the carrier.
The RF synthesizer consists of on chip RF oscillator with 64/65 prescaler and off chip CMOS PLL.
The RF oscillator has a tuning range of 1126MHz to 1206MHz in GSM and 1546MHz to 1634MHz in PCN application. The RF synthesizer has a C/N ratio of -88dBc/Hz within the loop filter bandwith.

The RF and IF synthesizer signals are mixed in the transmit Gilbert cell. In front of the transmit mixer a divider with metalmask selectable ratio of 2 or 3 is used to perform different frequency plans for 900MHz and 1.8GHz systems. After the selectable divider the signal is low pass filtered to reduce high order products. The output signal of the mixer has to be filtered with external circuitry to drive the quadratur modulator input. The mixer output power is about -10 dBm at 1.8GHz with matching to the filter and modulator input.
The LO signal at the input of the quadratur modulator is first high pass filtered then doubled and divided by 2 to get 90° phase shifted LO signals. This internal phase shifting is independend of temperatur variation and spread of technology and needs no internal and external tuning. This concept provides very accurate generation of I and Q LO signals with a phase error less than 1 degree. The two orthogonal LO signals are fed to two Gilbert multipliers for quadratur modulation. The outputs of the Gilbert cells are combined and fed to a linear amplifier to get -3 dBm output power with 200 Ohm load at 1.8 GHz.
For 2.7V supply it is necessary to use the Gilbert cell with resistors to ground instead of current sources to ensure linear work. The bias voltage for the modulation inputs (I, IX, Q, QX) is internally generated and available at the TREF pin. This voltage is used by the baseband analog circuit as DC reference voltage for the modulation signals. That DC voltage determines the current in the Gilbert multiplier and is temperature compensated so that high linearity is achieved.

The RF transmitter can be powered down from 60mA to 10uA. Six power down pins are used to power down the main blocks seperately:
- modulator, mixer part
- IF oscillator
- prescaler
- fix pll
- RF oscillator
- IF input buffer BufTX

RF Receiver Circuit

The receiver circuit (fig. 4) is a single-chip single-conversion heterodyne receiver up to 2.5 GHz with 2.7V power supply.

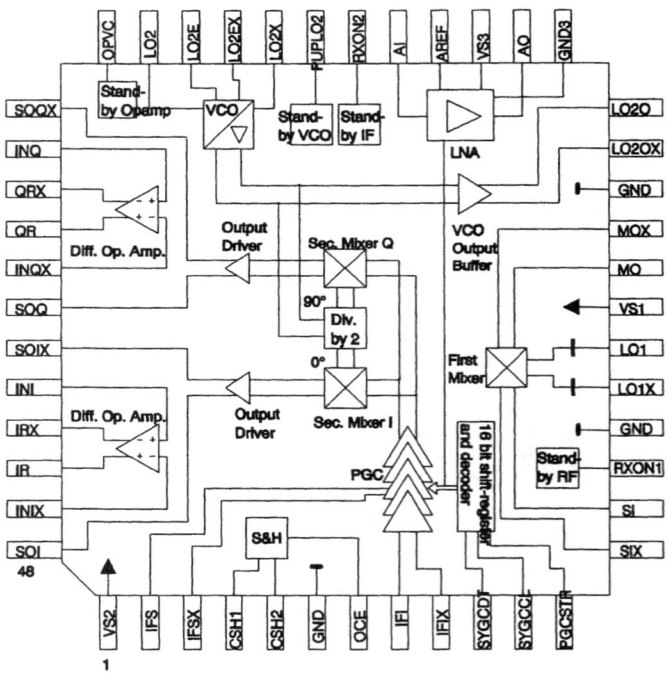

Fig. 4 RF receiver block diagram

The RF part includes a gain switchable low noise preamplifier (LNA) and double balanced first mixer. The IF part consists of a programmable gain controlled IF amplifier (PGC), an I/Q demodulator with phase shifting circuitry and two differential operational amplifiers for baseband signal processing. Instead of I/Q baseband output, IF sampling output is also available by metalmask option. The receiver also contains an

IF oscillator (VCO) with output buffer in order to use the fix PLL on the transmitter chip.

A photomicrograph of the chip is shown in Fig. 5.

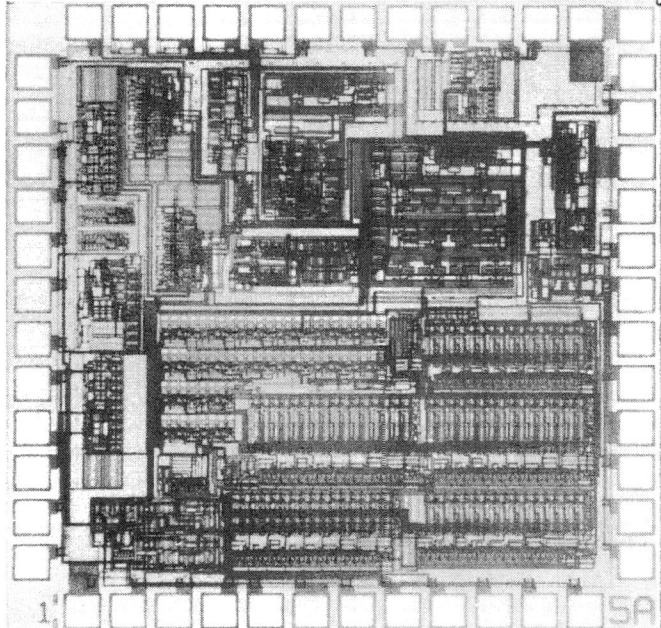

Fig. 5 Die photograph of the receiver

The circuit has three separated supply voltages for the low noise amplifier, the first mixer and the IF part to increase decoupling. The low noise amplifier has an extra ground disconnected from the other one. The RF part is placed at the right top of the layout. The IF circuits are located in the bottom region.

The chip can be powered down from 33 mA to 10 uA with four different power down pins to switch off the RF part, the IF part, the IF oscillator with buffer and the baseband OpAmps separately.

The signal from the antenna is first filtered and then amplified in the gain controlled low noise preamplifier.
The gain steps are -5dB and +15dB for -10dBm and -102dBm received signal. A noise figure of 2.5dB at 1.8GHz is achieved. After amplification the signal is additional filtered with an external image filter and then fed to the first mixer and mixed down to the IF frequency. The RF oscillator signal is fed to the first mixer from the transmitter chip. The open collector output of the mixer generates a differential current wich is filtered by an external IF LC tank circuit. A SAW filter following this LC circuit is used for channel filtering. The first mixer achieves 4dB gain with an IF frequency of 246MHz at 1.8GHz input frequency. The single sideband noise figur is 13dB and the intercept point is -3dBm with 1.8GHz input frequency.
The filtered IF signal is then fed to a five stage programmable gain controlled IF amplifier (PGC) with 80dB range (fig. 6). This amplifier is programmable via a three wire bus which is also used for the CMOS PLL. The gain steps are 2dB from -10dB to +70dB. The input stage is a common base configuration with 50 Ohm impedance and a selectable gain of 0dB or 24dB. This is for optimizing S/N. Each of the following four stages consists of eight differential amplifiers in parallel and a

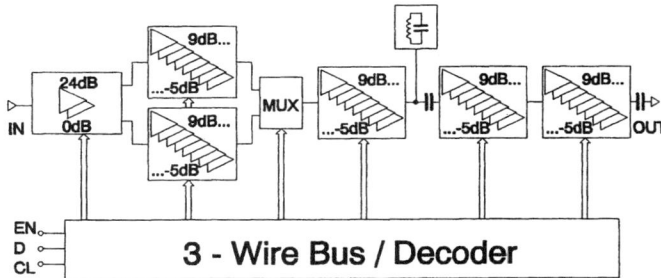

Fig. 6 Programmable gain controlled amplifier (PGC)

common load. Each differential pair performs a gain step to the next of 2dB. A 16 bit word is used for programming. 1 bit is reserved for switching the low noise preamplifier.
After the gain controlled amplifier the IF signal is fed to the I/Q demodulator. The IF oscillator signal is 90° phase shifted with a divider by 2. Then the orthogonal carriers are fed to the I/Q demodulator which consists of two Gilbert multipliers. After demodulation the I/Q baseband signals can be filtered with an active baseband filter formed by the differential operational amplifiers on chip and external resistors and capacitors.
The receiver also contains a sample and hold circuity to compensate DC offset after the active baseband filter.
Alternatively a metal mask option allows the IF signal after the gain controlled amplifier to be mixed down to a second IF in the range of 5MHz to 15MHz. This second IF drives an external second channel filter with 300 Ohm input impedance. After that filter IF sampling and digital demodulation can be done in baseband circuitry.

Measured Results

Samples of the RF transmitter and receiver have been analysed including temperature studies.

The output spectrum with single sideband modulation of the RF transmitter at 2.1GHz and 2.7V supply is shown in fig. 7. The I/Q modulation signal was a 67 KHz sine/cosine with 1Vpp. The output level with 200 Ohm load was -3.5dBm.

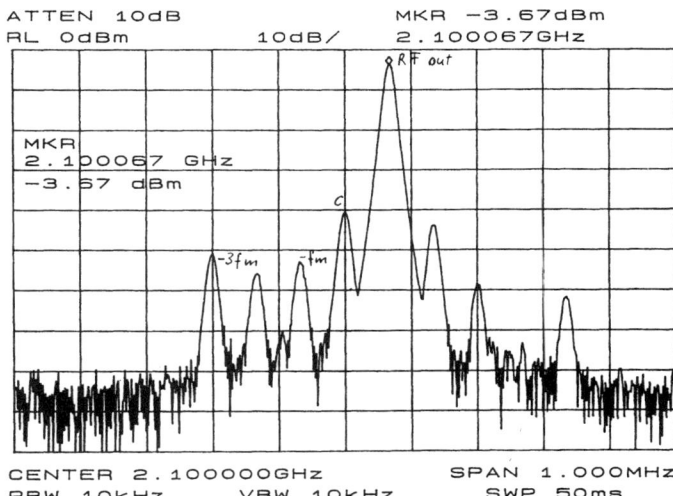

Fig. 7 Transmitter output spectrum with single sideband modulation

The carrier (c) was 37dB, the unwanted sideband (-fm) was 48dB and the 3. order produkt (-3fm) was 46dB suppressed. The RF transmitter works down to 2.4V supply at -35° celsius. The output noise is -141dBm/Hz at 25MHz offset frequency.

In fig. 8 the result of a system test at 2.7V supply with the baseband circuitry is shown. The center frequency is 1.75GHz and the I/Q signal drive level was 900mVpp. Due to the high linearity of the I/Q modulator the output spectrum passes the transmit mask for pcn with sufficient margin.

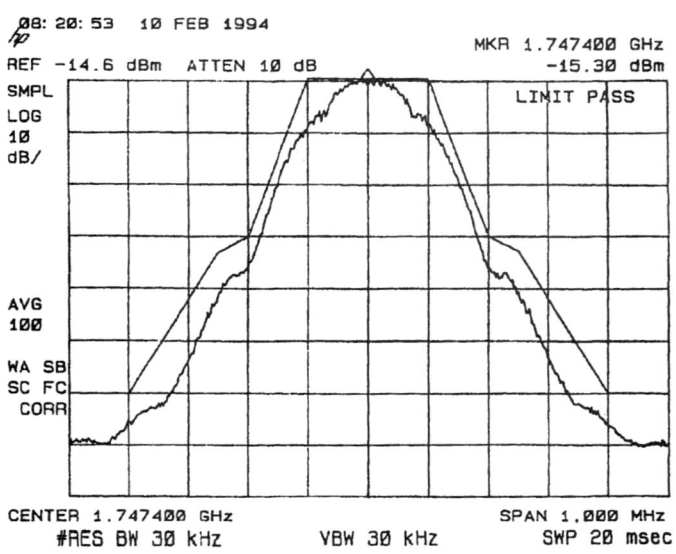

Fig. 8 Transmitter output spectrum with GMSK modulation

The gain and 1dB compression point of the programmable IF amplifier of the RF receiver is shown in fig. 9.
The gain can be programmed from -10dB to +70dB in 2dB steps. The maximum input signal for 1dB compression is -10dBm due to limitation of the first stage.

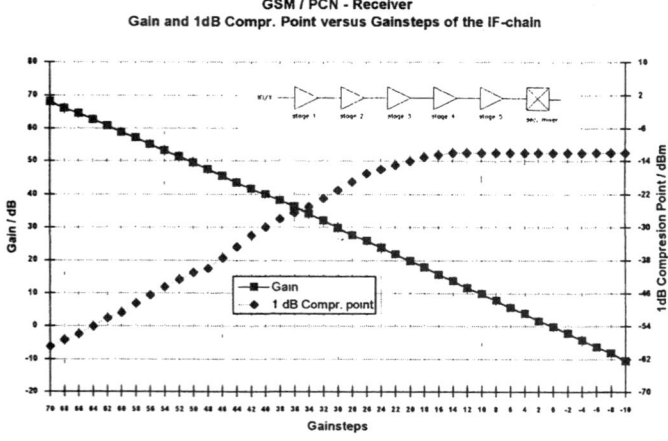

Fig. 9 Gain and 1dB compression point of IF amplifier

The low noise amplifier (LNA) of the receiver can be switched between high and low gain to increase dynamic range (fig. 10). This is done with 10 transistors in parallel of which 9 can

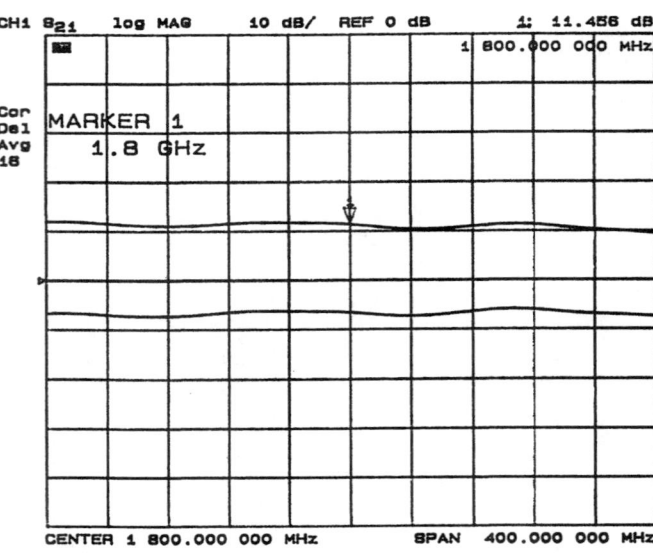

Fig. 10 High and low gain of LNA

be switched off. This results in a 20dB gain step.

Summary

The measurement results of the RF transmitter and receiver show that this circuits with one basic architecture are able to cover any digital mobil radio system up to 2.1GHz with supply voltages down to 2.7V, reduced external components and costs.

Acknowledgements

The authors wish to thank Mr. Moliere and his team from SIEMENS Mobile Radio Division for very helpful discussions and the layout team for their excellent work.

References

[1] Fenk et al., An RF Front-End for Digital Mobile Radio IEEE Bipolar Circuits and Technology Meeting 1990, p. 244-247

WP 2.8: An Analog Radio Front-end Chip Set for a 1.9GHz Mobile Radio Telephone Application

J. Sevenhans, D. Haspeslagh, A. Delarbre, L. Kiss, Z. Chang, J. F. Kukielka[1]

Alcatel Bell Telephone, Antwerp Belgium
[1]Alcatel Standard Electrica, Madrid Spain

A single-chip BiCMOS analog radio for mobile telephony is a strategic utopia today because few foundries provide a BiCMOS technology with sufficient bipolar f_T for the low-noise mixers in the receiver and transmitter phase modulators and demodulators at 1.9GHz. The solution presented in this paper as shown in Figure 1 is to use a radio silicon bipolar chip for the high frequency circuits of the transceiver and synthesizer and standard low cost CMOS for low-frequency transceiver analog signal processing and synthesizer functions.

Sensitivity requirements for today's mobile radio terminals are in the range of -80 to -105dBm. Considering the bandwidth of the receive signals these sensitivity levels require a low-noise amplifier, low noise mixers and low noise CMOS baseband signal processing, putting equally ambitious noise constraints on CMOS and bipolar in proportion to the physical limitations of each.

Phase noise requirements on the transmitter are equally stringent mainly because mobile radio concept wants to serve more than one user in the same room or office. So transmitter noise of one user should be a fraction of the minimum sensitivity receive power of other users in the room. For the same reason, synthesizer performance must provide a 1.9GHz local oscillator signal with low phase noise in the wanted signal band and even better in the wide-band spectrum.

The design of a frequency synthesizer, as shown in Figure 2, is a trade-off between settling time and spurious harmonics. For 500µs settling time, a current output digital-to-analog convertor is included. A 5b code provides a reasonable direct settling position for the VCO and the charge pump only has to complete further adjustment of the local oscillator frequency. The R_2I product is copied to the C_3, C_4, C_5 ground node by a follower. The 1.9GHz VCO frequency is divided by a 64/65 prescaler in the bipolar component and further reduced to the 1,728MHz reference in the CMOS counters. The synthesizer further consists of a reference crystal with 10ppm frequency accuracy, a divider circuit to obtain the reference frequency for the CMOS phase/frequency detector, a charge pump and a discrete component loop filter with the traditional pole to integrate the charge pump current impulses and a zero for a stable proportional-integrating control system (Figure 2).

An additional 3rd-order RC loop filter is included to suppress spurious outputs generated by the antibacklash pulse. This pulse is needed to avoid a phase detector dead zone, so that no spurious harmonics are generated in the band. However, the effect of the antibacklash pulse on the spurious output is reduced if amplitude and shape of the UP current pulse closely matches that of the DOWN current pulse. When the charge pump switches are in series with the current sources as shown in Figure 2, control pulse feedthrough is too large for a good match. Therefore, the charge pump circuit of Figure 3 is used. By proper design of the current mirror factors, it is possible to compensate almost completely the antibacklash pulse, an I_{DOWN} pulse, by an I_{UP} pulse, so that only a small I_{OUT} pulse generates spurious output components.

The receiver is a trade-off between noise performance, linearity requirements and power consumption for the 1.9GHz bandwidth. Phase accuracy is most important. For this reason the quadrature generator provides a phase-adjust input, to tune the 90° local oscillator phase relation. The bipolar receive mixers are based on degenerated Gilbert cells driven in switching mode in the local oscillator to obtain low noise performance. The linearity of these cells is dominated by blocking signals more than 60dB above the desired receive signals in present mobile radio system specifications. Power consumption of the receiver active mixers is dictated by this linear range and the degeneration resistors that are the main contributors to mixer thermal noise figure.

The quadrature generator uses passive RC and CR networks for the rough 90° phase shift. For $\Delta\Phi<0.5°$ accuracy, an active phase corrector uses 2 multipliers, α and β in Figure 4, and two limiting amplifiers, A and B, to make the tunable vector sums $(\alpha + B)$ and $(\beta + A)$. These tunable vector sums are drive signals for the receivers and transmitter Gilbert cell mixers. In the receive mixers a 6dB gain setting is controlled by digital input. An additional 6dB gain setting is used in the post-mixer base-band amplifiers (PMA). These bipolar low-noise base-band amplifiers boost the receive signal before it enters the CMOS receive circuitry with 40nV/√Hz equivalent input noise.

The overall noise figure of the bipolar Gilbert cell mixers is 15dB referred to a 50Ω input source. This explains the need for 20dB external LNA gain to obtain an overall 8dB noise figure. In the transmitter, a vector phase modulator with an I and Q baseband quadrature input and the two local oscillator drive signals are used in +45° and -45° phase relation. To optimize power consumption, I and Q modulators run at low bias current with a design compromise in the f_T versus I_C. The real output power necessary at the input of the external 50Ω antenna driver is provided by the on chip prepower amplifier (PPA) shown in Figure 5. The transmit mixers produce less than -148dBc/√Hz phase noise.

The 8 mm² bipolar chip consists of a quadrature generator, receive and transmit mixers and amplifiers, and 64/65 prescaler. The 40mm² CMOS chip consists of the low frequency part of the frequency synthesizer and the baseband signal processing. Zero IF architecture is used for both the Rx and Tx channels.

Acknowledgment

The authors express thanks to F. Bonjean, E.D. Kowalski, J.L. Aguinaco, W. Huysmans, M. Van Paemel, P. Vanleene, P. Ward, R. Van Vliet, G. Konstadinidis for contributions to the design, measurement and state-of-the-art technology.

References

[1] Crabbe, E. F., et al., "73 GHz Self Aligned Si-Ge-Base Bipolar Transistors with Phosphorus-doped Polysilicon Emitters," IEEE Electron device letters, Vol. 13, No. 5, May, 1992.

[2] Nagata, S., et al., "A GaAs MMIC Chip-Set for Mobile Communications Using On-Chip Ferroelectric Capacitors," ISSCC Digest of Technical Papers, pp. 172-173, Feb., 1993.

[3] McDonald, M., "A 2.5 GHz BICMOS Image-Reject Front-End," ISSCC Digest of Technical Papers, pp. 144-145, Feb.,1993.

[4] Brodersen, R. W., et al., "Design Techniques for Portable Systems," ISSCC Digest of Technical Papers, pp. 167-169, Feb.,1993.

[5] Villanueva, A., et al., "Wireless Functionality for PBX," Proceedings of Teleco 91, pp 37-45, May, 1991.

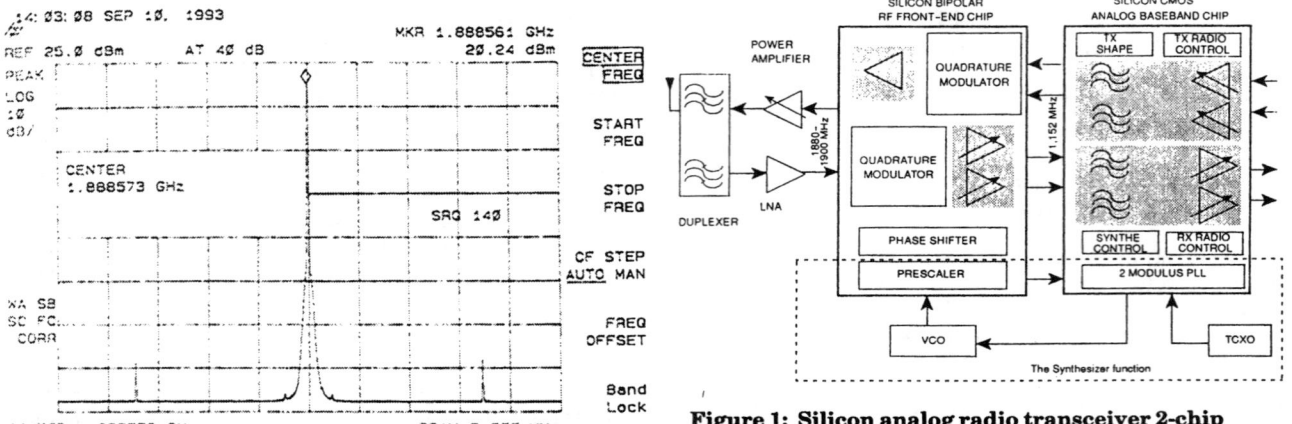

Plot 1: Measured synthesizer output spectrum.

Figure 1: Silicon analog radio transceiver 2-chip solution.

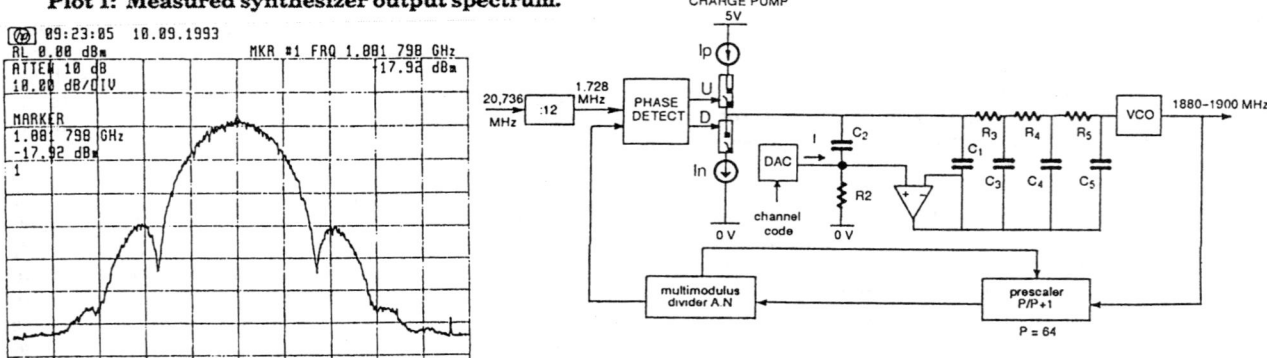

Plot 2: Spectral measurement of radio transmitter spectrum of DECT handportable terminal.

Figure 2: Synthesizer block diagram.

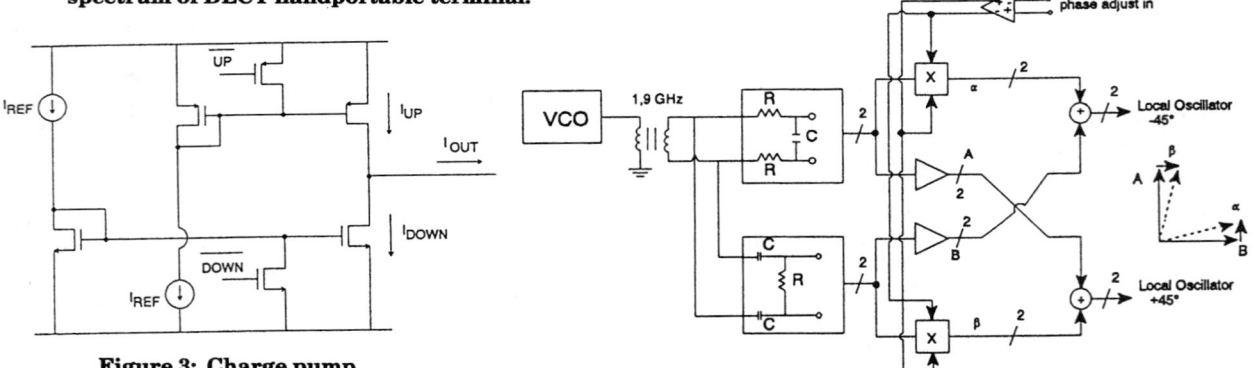

Figure 3: Charge pump.

Figure 4 (above): Local oscillator quadrature generator.

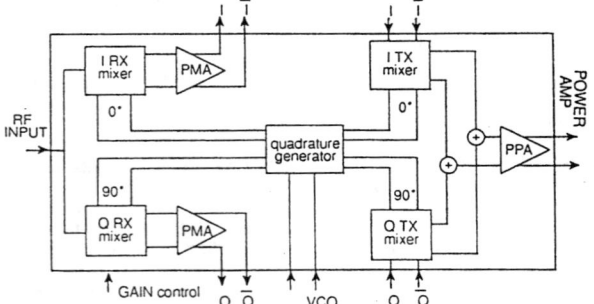

Figure 5 (left): RF front-end transceiver block diagram. Quadrature generator shown in Figure 4 provides +45° and -45° local oscillator drive to transmit and receive phase modulator and demodulators.

Figures 6 and 7: (Chip micrographs) See page 307.

336

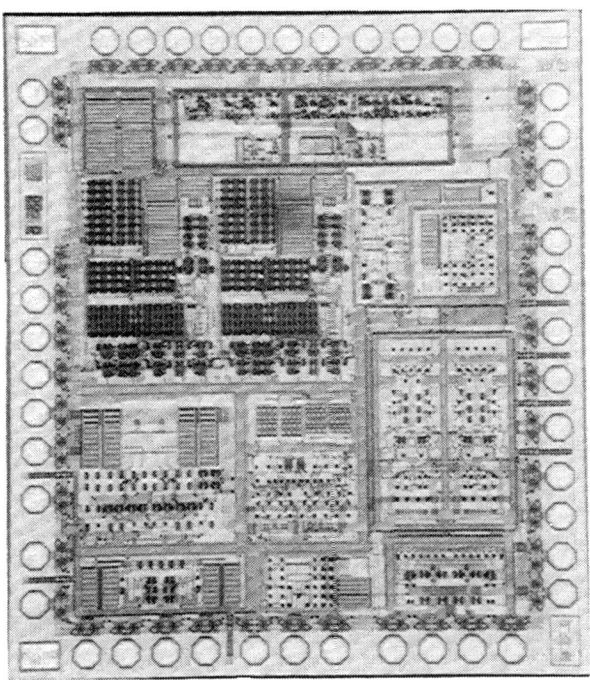

Figure 6: Radio bipolar front-end chip.

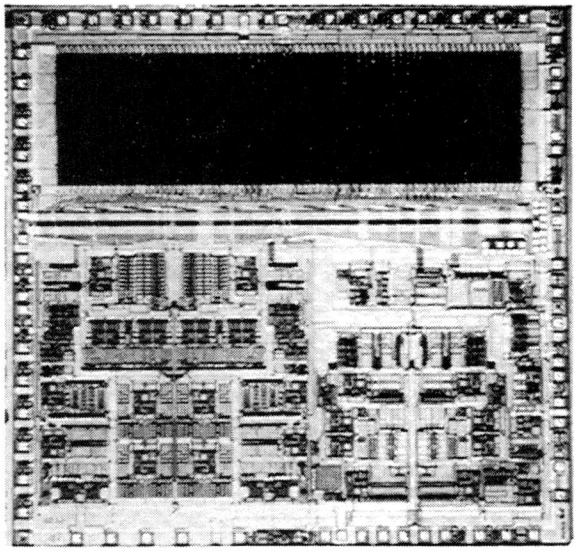

Figure 7: CMOS 30MHz synthesizer counters control logic and baseband processing; 40mm^2.

TP 9.2: A 900MHz Transceiver Chip Set for Dual-Mode Cellular Radio Mobile Terminals

Iconomos A. Koullias, Joseph H. Havens, Irving G. Post[1], Peter E. Bronner[1]

AT&T Bell Labs/[1]AT&T Microelectronics, Reading, PA

A chip set for compact transceivers meeting IS54 dual-mode cellular telephone standards consists of a direct up-conversion modulator and a double conversion receiver in a 12GHz super-self-aligned bipolar process. Operated from a 5V power supply, the total power dissipation is 310mW. Both circuits can be put into a power-down mode with a maximum off-current of 10μA. The modulator chip is 4.1mm^2, and the receiver, 4.9mm^2.

The modulator block diagram is shown in Figure 1. The advantage of direct up-conversion over indirect up-conversion is that there is only one local oscillator (LO) and no intermediate frequency (IF) filtering is required, but care must be taken to avoid producing in-band spurious signals. In this implementation the inputs are a single-ended local oscillator (LO) and balanced I (in-phase) and Q (quadrature) baseband signals. Nominal maximum output power is 0dBm. The local LO power requirement is -10dBm typical. The baseband inputs are high impedance and require 2.5V_{pp} differential for maximum RF output.

The LO input is buffered and sent to an R-C phase splitter to produce two signals approximately 90° out of phase. The two signals are passed through the Havens amplitude and phase corrector circuit to insure that the inputs to the mixer are equal in amplitude and 90° out of phase, with small deviation [1]. The LO signals are then mixed with the I and Q baseband signals to produce the single-sideband suppressed-carrier RF output signal. The output stage consists of a low-distortion voltage amplifier followed by a power gain stage with 50Ω output impedance. The corrector circuit is shown in Figure 1. Operation is based on the fact that if two equal-amplitude signals are not exactly 90° out of phase, their sum and difference will be. The pair of gain stages following the phase splitter, overdriven emitter-coupled pairs, insure that the signals into the first pair of summers are of equal amplitude. Since the correction of phase introduces amplitude imbalance, the second pair of gain stages recorrects the amplitude. The summers are linearized emitter-coupled pairs with shared collector loads. The measured phase error is 0.1±0.084° and the mixer amplitude imbalance is 0.10±0.05dB.

Part of the modulator design is concerned with avoiding the creation of spurious signals. Figure 2 shows the output spectrum generated with a single-tone baseband input (at 80kHz) as a function of baseband drive level. The desired signal is at the local oscillator frequency plus 80kHz (USB). The largest undesired signals are at the LO frequency, 2 basebands above LO (LO+2BB), and at 1 and 3 basebands below LO (LSB, LO-3BB). The behavior of LO feedthrough shows relatively flat response reflective of passive coupling. The component at LO-1BB is due to mixer imbalance and the LO phase error. The component at LO+2BB is due to feedback of the modulator output to the LO input and increases at a rate of 2dB per dB increase of output level. The output-signal component at the LO input is remixed with the baseband signal. This component decreases with increasing LO level. Finally, the signal at LO-3BB is due to intermodulation distortion in the output amplifier. Since the mixers multiply the baseband signals with a square-wave LO signal, Fourier analysis shows that the next largest component out of the summer is at frequency 3LO-1BB. Second-order intermodulation distortion mixes this signal with the main component at LO+1BB:

$$(3LO-BB)-2(LO+1BB) = (LO-3BB)$$

The signal increases at 3dB per dB increase in output power. Translation of the distortion components based on the IS54 per-channel frequency spectrum shows -26dBc adjacent channel and -45dBc alternate channel noise requirements are met.

The receiver, shown in Figure 3, uses a two-IF architecture with the second IF made of a dual path; one for the Analog (FM) and one for the digital (π/4-shifted DQPSK) demodulation. This has the advantages of RF image rejection and selectivity.

The RF mixer amplifies and demodulates the RF-band, centered around 881MHz, to the first IF (typically near 80MHz) where the first channel-selective filtering is performed. The RF mixer power gain (single-ended output into a 50Ω load) is measured at 10.1dB and can be attenuated by 26dB using a digital control input. This attenuation feature prevents strong wanted channels from overdriving later stages. The first-IF mixer amplifies the signal by 32dB and demodulates it to the second-IF (usually 455kHz). After filtering, the signal can follow two paths; the analog and/or the digital. The analog section is made of a 46dB amplifier, a 72dB limiter, an FM demodulator, and a received signal strength indicator (RSSI). The RSSI output is a dc voltage whose magnitude is a linear function of the received wanted signal power. The digital section includes a 60dB AGC amplifier and a quadrature demodulator. The gain of the AGC amplifier, expressed in dB, is a linear function of the differential AGC input voltage. Also, the AGC amplifier has highest sensitivity (lowest noise) when the gain is set to its maximum value (i.e. for weak RF signals) while remaining linear at low gain (i.e. for strong RF signals). The 0/90° phase splitter accepts a LO signal whose frequency is four times the second-IF frequency. The advantage of this is that the design is simple, accurate (0.1°, 0.05dB), and broadband and eliminates the adverse effects of LO leakage. To save power, two digital (ENBA, ENBD) inputs can be used to shut down analog, digital, or both paths.

The RF mixer, shown in Figure 4, is single-balanced. The impedance of the common-emitter input stage matches a 50Ω source by the low and slightly inductive input impedance of the feedback circuit (A1, R4). The two-tone third-order output intercept point (IP3) is +14.1dBm, and return loss (S11) -14dB. The analog IF amplifier/limiter and RSSI functions are made of a string of six blocks, each including an amplifier and a full-wave rectifier (FWR) (Figure 5). The amplifier is made of two stages in parallel so the overall transfer function is piece-wise linear. This improves the linearity of the RSSI with fewer gain/FWR stages. The FWR is made of a differential stage with one input connected to the input common-mode point and the other to the common-emitter node of the 20dB amplifier. The AGC amplifier, shown in Figure 6, is made of a 4x4 array of gain stages. The gains of each amplifier in the same row are equal and that of each succeeding row are a fraction of the previous. Each row is enabled by steering the bias currents to it using an analog decoder under the control of the AGC input. The AGC amplifier has a 60dB dynamic gain range with 2dB integral non-linearity.

Reference

[1] Havens, J. H., Japanese Patent Application No. 910752, Sept. 1992.

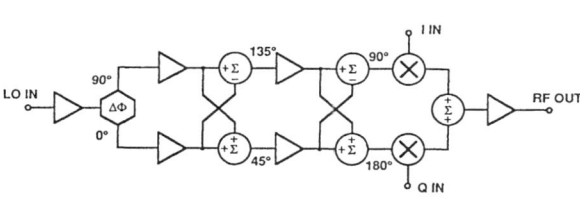

Figure 1: Single-sideband suppressed-carrier modulator.

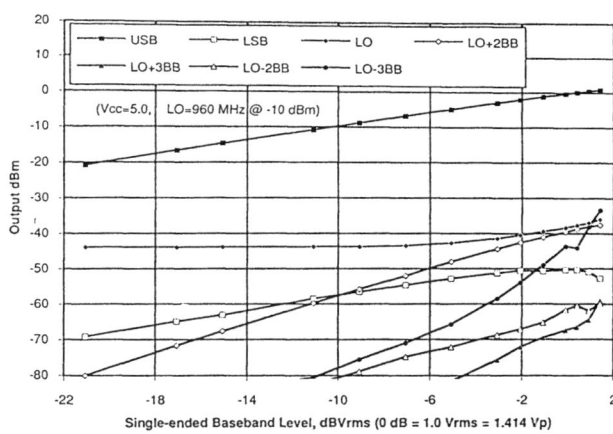

Figure 2: RF output products dependence on baseband levels.

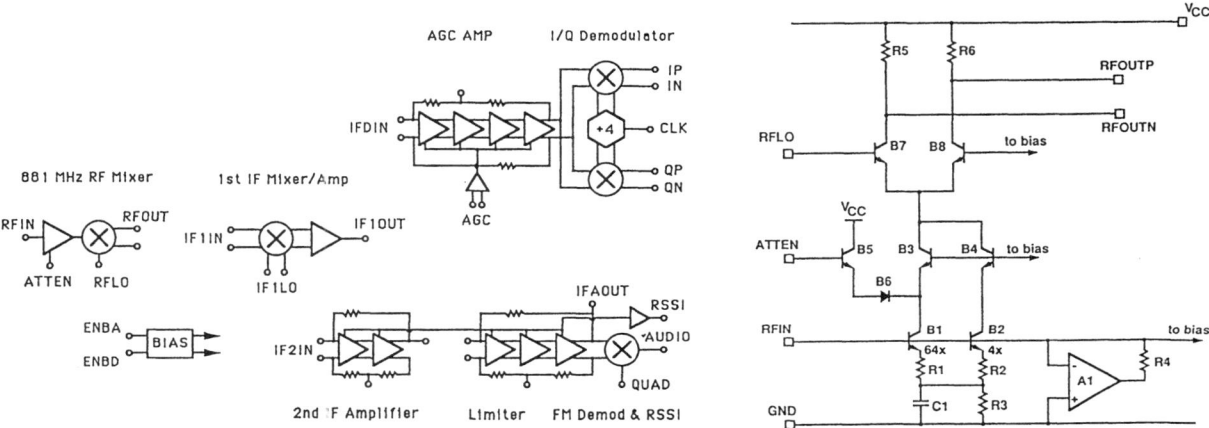

Figure 3: Receiver block diagram.

Figure 4: RF mixer with attenuation control.

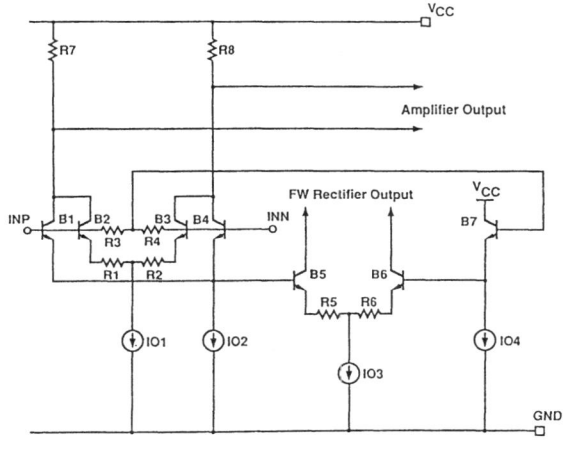

Figure 5: Piece-wise linear amplifier and full-wave rectifier.

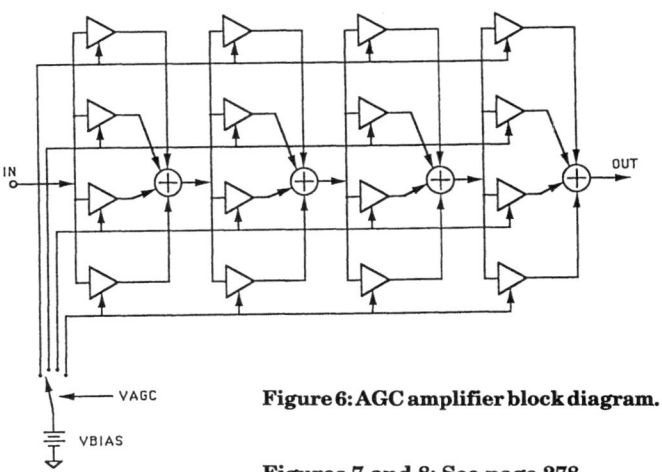

Figure 6: AGC amplifier block diagram.

Figures 7 and 8: See page 278.

Figure 7: RF modulator chip micrograph.

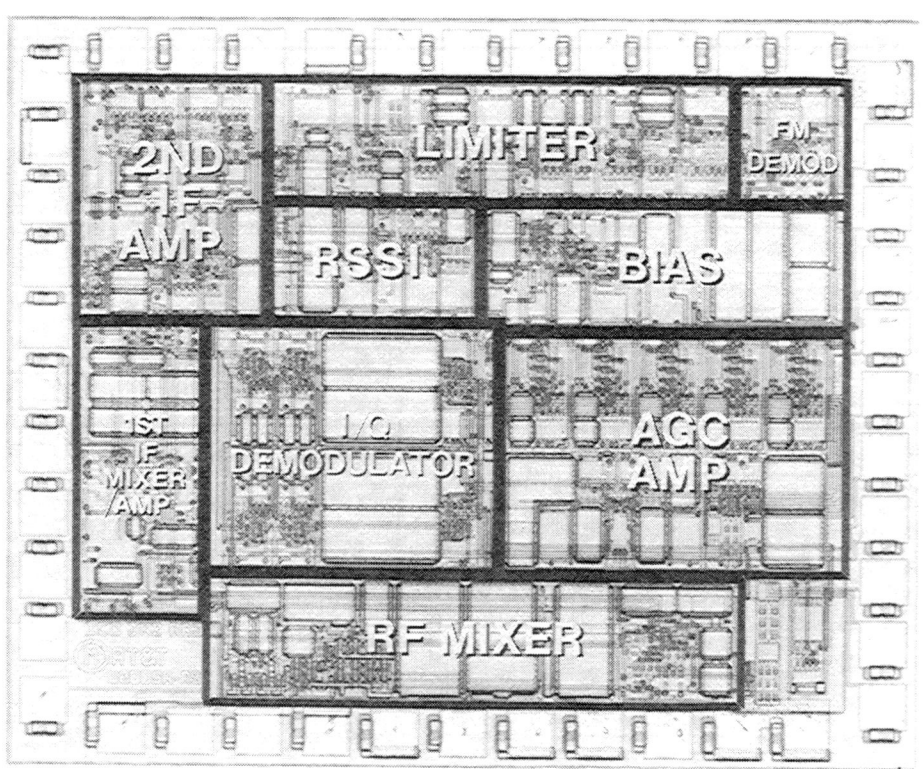

Figure 8: RF receiver chip micrograph.

TA 8.8: A 2.7V to 4.5V Single-Chip GSM Transceiver RF Integrated Circuit

Trudy Stetzler, Irving Post[1], Joseph Havens[2], Mikio Koyama[2]

AT&T Microelectronics, Reading, PA
[1] GP Microdesign, Laureldale, PA
[2] AT&T Bell Laboratories, Tokyo, Japan

This paper describes a 2.7V to 4.5V single chip GSM transceiver IC. The addition of a power amplifier, LNA and filters are all that is required to implement a GSM terminals radio section (Figure 1). This chip includes two fixed-frequency synthesizers (for transmit and receive IF frequencies), a programmable frequency agile UHF synthesizer (for channel selection in both receive and transmit modes), RF mixer, single IF down conversion with quadrature demodulation and variable gain for the receive path, and a direct-up modulator with offset oscillator to avoid spurious signals and oscillator pulling (Figure 2). The circuit also contains control circuitry for separate receive, transmit, and UHF synthesizer modes to minimize power consumption. The IC is implemented in 12GHz bipolar technology with a 1.5µm minimum feature [1].

The signal from the antenna is amplified and filtered using an external LNA and bandpass filters. The signal is then input to the on-chip RF mixer and converted down to the 71MHz IF frequency. The RF local oscillator is provided by the on-chip UHF synthesizer. This RF mixer has 8dB of gain, an 11dB noise figure, and an input third-order intercept of 0dBm. The output of this mixer drives an external SAW filter that provides channel filtering prior to the IF section.

The IF signal is then input to a 0-60dB programmable gain amplifier, programmable in 4dB steps. This signal is fed to the quadrature demodulator to be converted to baseband. The 71MHz local oscillator is generated from the 284MHz fixed frequency PLL using a divide-by-4 Johnson counter. This provides isolation and a 90° phase shift that is independent of duty cycle. These LO signals are fed to two Gilbert cell mixers that provide an additional 5dB gain. The intercept point for the entire IF section is +13dBm (when referenced to a 50Ω load) at the baseband output. The noise figure at the IF input is 7dB in the high gain setting. There is also an overload indication provided to prevent system AGC lock-up due to fading conditions.

One of the hardest requirements of the transmitter is reducing the spurious frequency content at the output. Since the VCO is on the same chip as the modulator, an offset-mixer with a direct-up modulator eliminates VCO pulling and reduces spurious frequencies. The transmitter frequency is derived from a 117MHz TX oscillator signal with the UHF oscillator to produce the final TX local oscillator (LO) signals. The unwanted sideband from the mixer is filtered with an inexpensive external LC filter to attenuate the upper sideband by at least 15dB.

The phase-splitter is a simple RC type splitter followed by a single stage phase-corrector circuit. The phase-splitter with correction circuit produces a phase error of less than 1° and less than 0.1dB gain error [2].

The double-balanced mixers use feedback in the I/Q input buffers to improve linearity and increase gain of the mixers. This nearly doubles the signal available at the output of the mixers, improves the signal-to-noise ratio, reduces the carrier feed-through, and means less gain is needed in the RF output buffer.

The RF output buffer is a simple emitter-coupled pair followed by an emitter-follower output. Headroom is improved by using integrated coupled inductors in the load of the emitter-coupled pair. An external RC termination provides control over the output SWR and further reduces variation of output level. The buffer delivers 0dBm into a 50Ω load (see Figure 3).

The synthesizer portion of the chip consists of two fixed-frequency synthesizers at 117MHz and 284MHz, and one programmable synthesizer that covers the 996MHz to 1032MHz range. This UHF synthesizer is mixed with the 117MHz for transmit mode and the 71MHz IF frequency for receive mode to cover the EGSM standard. The UHF oscillator is programmed with 8b via a serial 3-wire bus. Its synthesizer divider chain consists of a 64/65 prescaler, a 77 to 80 divider, and a 0 to 63 divider for the prescaler modulus control (Figure 2). The UHF synthesizer requires 1ms from power up to settle to within 100Hz, and 0.5ms to settle when switching between channels or modes.

The phase comparators for all three synthesizers use the same architecture. The UHF reference frequency is the same as the GSM channel spacing of 200kHz and the reference frequency for the fixed-frequency synthesizers is 1MHz. The 13MHz master clock is divided by 13 and then divided by 5 to produce references for the phase comparators. The charge pumps for the synthesizers are of the switched-current type and drive a single-ended external loop filter (Figure 4). The oscillator resonators consist of external parallel LC combinations with single varactor tuning. Each oscillator consists of a simple emitter-coupled pair, fed by a current source, with positive feedback.

To conserve power, the circuit has several enable modes, including a UHF-synthesizer-only mode. Total supply currents in transmit and receive modes are 60mA and 55mA, respectively, at 2.7V. The supply current for the UHF synthesizer is 25mA.

The output spectrum with single-sideband modulation of the RF transmitter at 915MHz and 2.7V supply is shown in Figure 5. The I/Q modulation signal is a 65kHz sine/cosine with 1Vpp level. The output is 0dBm into a 50Ω load. The carrier is -32dBc, the unwanted sideband is -40dBc, and the output IM3 is < -40dB. Figure 6 shows the transmitter output at 2.7V with a GMSK input signal applied. The center frequency is 915MHz, and the I/Q drive level is 1Vpp. The output spectrum passes the GSM transmit mask due to the high linearity of the I/Q modulator and sufficient phase noise and accuracy of the synthesizers. The receive baseband phase error is 1.2° and the magnitude error is 1.6%rms. Maximum input signal for 1dB compression of the RF mixer is -8dBm, and for the IF section is -3dBm. The die micrograph is shown in Figure 7.

References

[1] BEST-1 Product Development Guide, AT&T Microelectronics.

[2] Koullias, et al., "A 900MHz Transceiver Chip Set for Dual-Mode Cellular Radio Mobile Terminals," ISSCC Digest of Technical Papers, pp. 140-141, Feb., 1993.

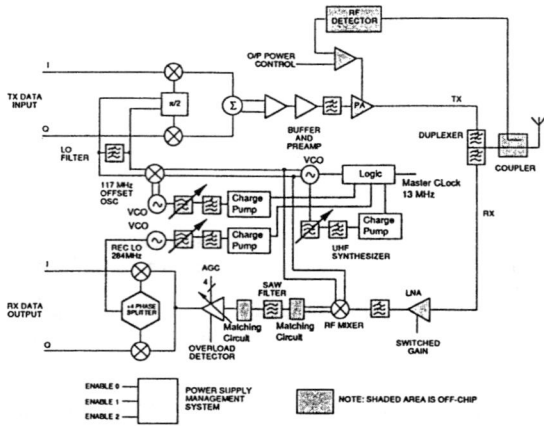

Figure 1: System block diagram.

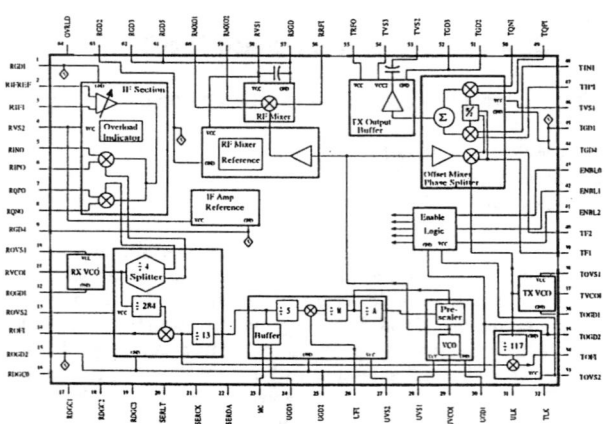

Figure 2: Block diagram and pinout.

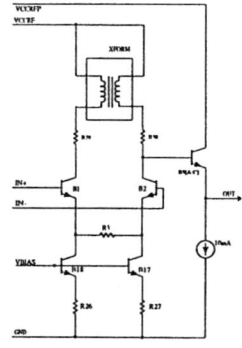

Figure 3: Transmit-output buffer.

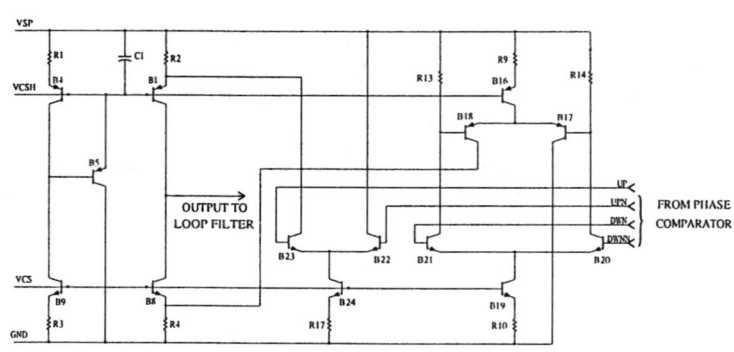

Figure 4: Charge pump schematic.

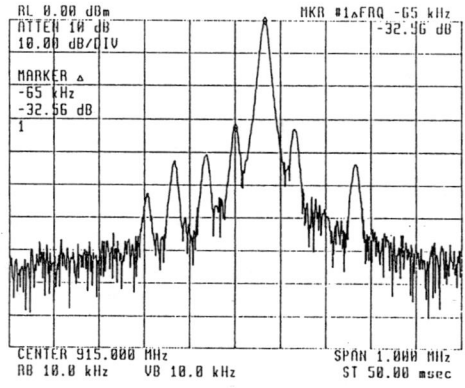

Figure 5: Transmit-output spectrum with single-sideband modulation.

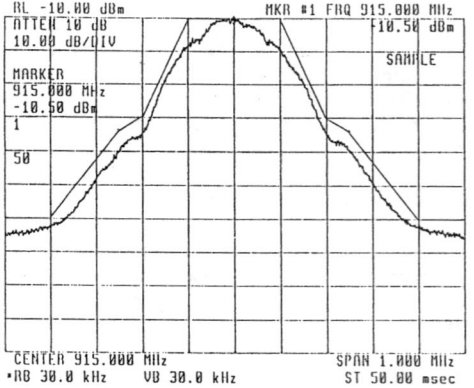

Figure 6: Transmit-output spectrum with GMSK modulation.

Figure 7: See page 354.

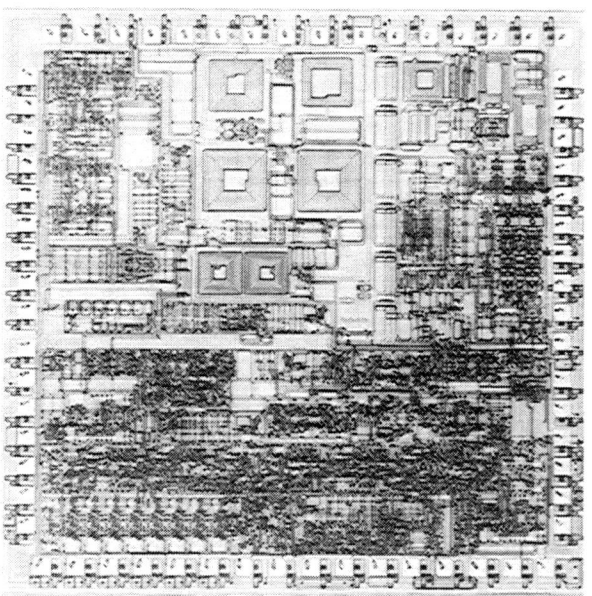

Figure 7: GSM transceiver chip micrograph.

TA 8.7: A 2.7V GSM Transceiver ICs with On-Chip Filtering

Chris Marshall, Farbod Behbahani[2], Winfrid Birth, Ali Fotowat[2], Thilo Fuchs, Rainer Gaethke[2], Emil Heimerl, Sheng Lee[2], Paul Moore[1], Saeed Navid[2], Erich Saur

Philips Semiconductors, Technology Centre for Mobile Communications, Nürnberg, Germany
[1]Philips Research Laboratories, Redhill, UK
[2]Philips Semiconductors, Sunnyvale, CA

A pair of low-power (2.7-5.5V) ICs implements small GSM transceivers with a minimum of external components. Key aspects are the integration of the RF amplifiers, image rejection mixing, baseband channel selectivity filters and peripheral functions on-chip. 1.5nF/mm² capacitors and both bipolar and CMOS devices are used in the 13GHz f_T process to mix analog and digital functions.

A single superhet architecture is used for both transmitter and receiver, one IC containing the RF and the other the IF circuitry (Figure 1) [1]. Both circuits and the common synthesizer and VCOs are time-multiplexed between transmit and receive. (GSM is a TDMA system.) An IF such as 400MHz is envisaged for both transmit and receive to ease the filtering of spurious mixing products and allow applications in the 2GHz band.

The RF IC includes two LNAs so cheap small lossy filters can be used. Both draw 3.6mA, and have input and output return loss of 10dB and 15dB. As shown in Table 1 the LNAs and double-balanced mixer provide good sensitivity and dynamic range. Furthermore, by using CMOS switches, both LNAs can be switched off (saving power) and reconfigured as attenuators to provide 4 precise AGC gain/loss steps over a 59dB range. The LNA stage isolation is high despite the limited 13dB transistor base-collector isolation by re-applying Vcc to the collector during power-down. Package parasitics and the substrate losses are included in the design simulations to enhance the modelling accuracy, and the LNA circuit is temperature and voltage compensated, resulting in just ±0.2dB gain variation between -40 and +85°C and between 3.0 and 5.0V at 900MHz. 31dB of RF image rejection filtering (for a 400MHz IF) is provided on-chip by the gain roll-off of the two LNAs and the mixer input stage.

A relaxed specification IF SAW filter protects the second mixer and baseband circuitry from strong signals, and so keeps current consumption down. Most selectivity is provided by the integrated continuous-time analog baseband filtering, that has the response shape shown in Figure 2.

The baseband signals first pass through integrated fixed first-order passive pre-filters with a nominal cut-off of 110kHz to attenuate strong stop-band signals.

This is followed by a fully-balanced bipolar transconductor-based gyrator equivalent of a current-driven doubly-terminated fifth-order low-pass LC filter. Inductors, source and load resistors are replaced by transconductors, with bias currents set by an off-chip resistor. The filter circuitry is evolved from that described previously [2]. Higher-order linearized transconductors using three long-tailed pairs increase large-signal capability and are used in a fully-balanced design, mainly to minimize second-order intermodulation products but also to improve the ultimate (feed-through-limited) stop-band. Current drive into the transconductor filter increases robustness against strong out-of-band signals by more than 10dB. The highly-linear input circuis consists of a degenerated long-tailed pair formed from two folded Darlington circuits and is shown in simplified form in Figure 3.

The filter has a transitional, Gaussian to 6dB shape, combining good unwanted signal rejection with little group delay distortion. The shape is essentially independent of temperature, process, supply voltage, and local on-chip component variations. The cut-off frequency of the filter is determined by the external current-setting resistor and the on-chip capacitance values, with a temperature variation of 50ppm/°C and process spread of ±10%. GSM receiver requirements are met with untrimmed filters with a nominal cut-off set to 80kHz. The area of the pair of transconductor filters and their bias circuitry is about 1mm², and the current consumption, 375µA.

Particular attention is paid to baseband DC offsets, as they lie within the GSM signal spectrum. Symmetrical circuitry is used throughout, and the IF oscillator is run at twice the IF signal frequency, with the divider on-chip, to reduce DC produced by self-mixing effects. The divider also then provides an accurate pair of 90° local oscillator signals, and forms the first stage of the synthesizer divider chain. Residual DC offset is corrected by digitally-controlled current sources that, together with the synthesizer, test modes etc., are under software control over a standard 3-wire serial bus.

In transmit, the quadrature I, Q signals are modulated onto the IF using the same quadrature oscillator circuit as the receiver, followed by a highly-linear IF output stage. The transmit IF signal is then regulated in the RF circuit to 50mV by a level-control amplifier to drive the RF mixers at their maximum linear level. This allows generous tolerances in the IF (harmonic removing LC) filter and also suppresses parasitic AM. The simplified circuit is shown in Figure 4. With an input level of 50mV Q1-2 operates linearly, the gain in Q3-4 drives Q7-8 into limiting operation, and by suitable choice of R9-10 the filtered outputs of the two full-wave rectifiers Q3-4 and Q9-10 balance, independent of frequency (70-500MHz), temperature and process spread.

The image reject up-mixer by virtue of 20dB sideband suppression combines efficiency and low levels of spurii, and its output is then amplified linearly to drive the final switching mode output stage. The resulting accurate +8dBm ±0.5dB output drives a lossy filter and the Power Amplifier directly; other output powers can be selected by suitable choice of the external current reference resistor. Figure 5 confirms that the spurii are well below the -37dBc design objective.

The four low-drop supply-voltage regulators provide independent 100mA and 30mA supplies for internal or external circuitry, with thermal overload detection and cascaded band-gap references to keep the tolerance of the regulator output voltage within 5%. All IC circuits are powered up or down in 10µs or less to provide maximum energy efficiency. The total current consumption of the circuits including regulation and IF synthesizer is 51mA in receive mode with maximum gain, and 105mA when in transmit delivering +8dBm. The RF and IF ICs have areas of 8.3mm² and 11mm², and are both in TQFP48 packages. Figures 6 and 7 are die micrographs of the two ICs.

References

[1] Part numbers SA1620 and SA1638 respectively.

[2] Calder, D.W.H., "Audio Frequency Gyrator Filters for an Integrated Radio Paging Receiver", Int'l Conf on Mobile Radio Systems and Techniques, York, UK, pp. 21-6, Sept. 10-13, 1984. IEE Conf Publication no. 238, ISBN 0-85296297-5.

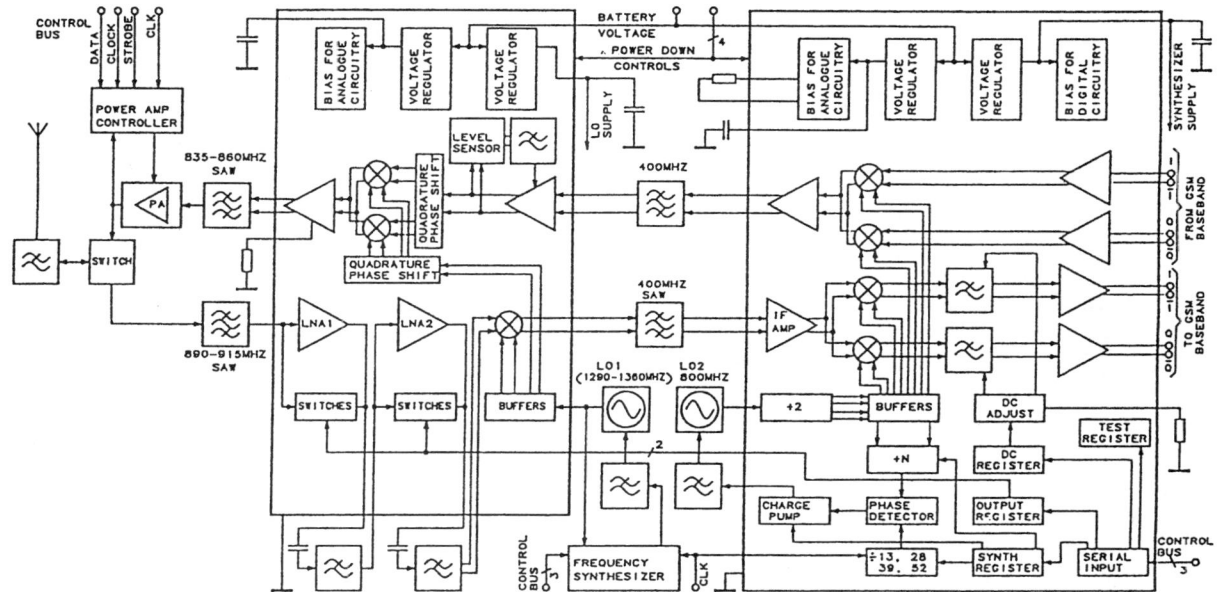

Figure 1: Full block diagram of integrated GSM transceiver solution.

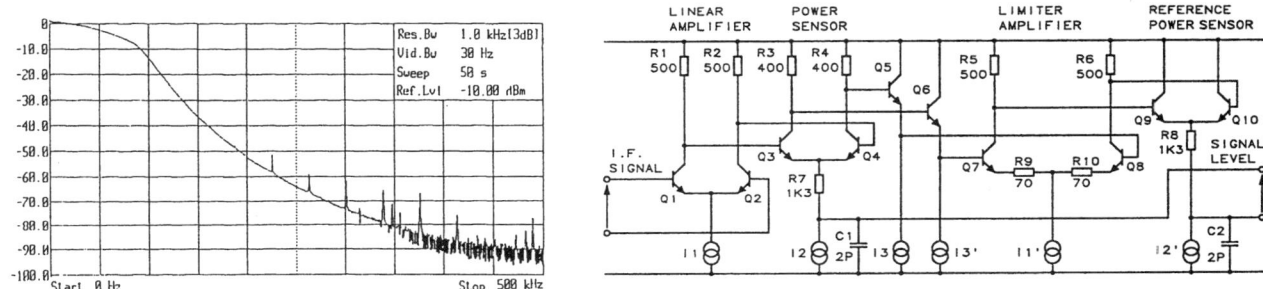

Figure 2: Measured transfer function of baseband filtering.

Figure 4: Level control loop sensor circuitry.

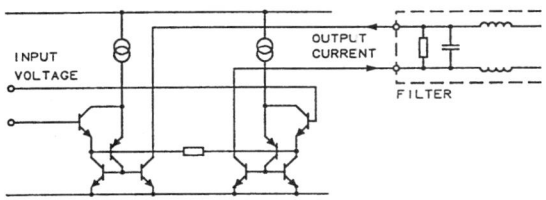

Figure 3: Circuit of gyrator filter current input.

	LNA1	LNA2	Mixer
Noise Figure	1.7dB	2.2dB	9.5dB
Input IP3	-3dBm	-5dBm	0dBm
Gain, AGC=11	+10.5dB	+10dB	8.5dB
Gain, AGC=10	+10.5dB	-8dB	8.5dB
Gain, AGC=01	+10.5dB	-20dB	8.5dB
Gain, AGC=00	-11.5dB	-27dB	8.5dB

Table 1: Measured RF component performance.

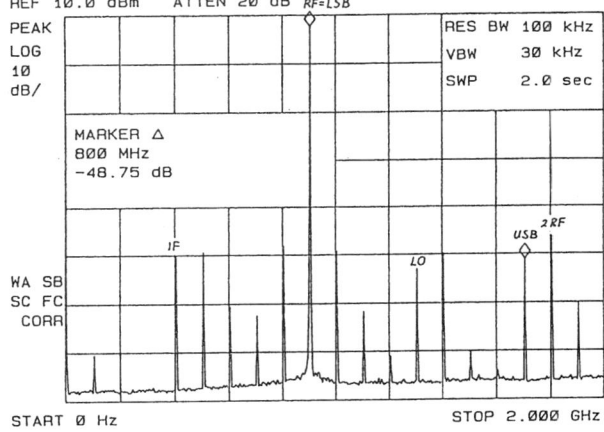

Figure 5: Measured spectrum out of transmit part.

Figures 6 and 7: See page 354.

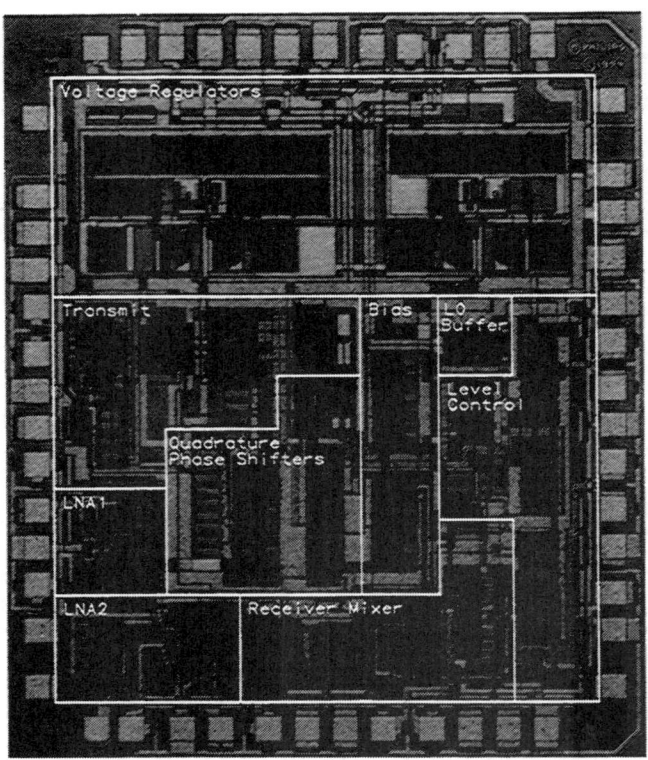

Figure 6: RF circuit chip micrograph.

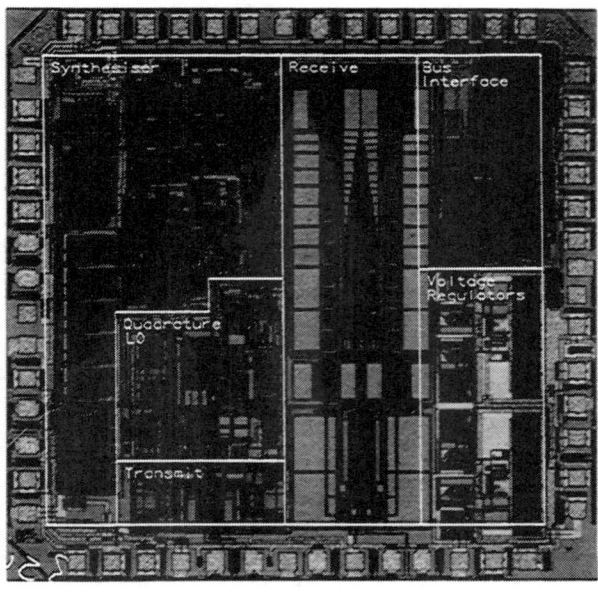

Figure 7: IF circuit chip micrograph.

A 2.4GHz SINGLE CHIP TRANSCEIVER

L M Devlin, B J Buck, J C Clifton, A W Dearn, A P Long,

GEC-Marconi Materials Technology Limited
Caswell, Towcester, Northants
NN12 8EQ

Abstract

A single chip, GaAs transceiver is described. The circuit includes all transmit and receive switching, amplification, frequency conversion and level shifting. An on chip oscillator is also provided. Low receive current of 30mA from a +5V supply and a standby current of less than 0.5mA, make this an ideal component for battery powered operation.

Introduction

The Industrial, Scientific and Medical (ISM) frequency band includes the frequency range 2.4 - 2.483GHz. In the USA, unlicensed operation using spread spectrum modulation at transmitter powers of up to 1W is permitted over this band. This paper describes a transmit/receive front end for a 2.4GHz wireless communications transceiver, the entire circuit of which has been integrated onto a single GaAs MMIC. A photograph of the 3.3mm x 5.2mm chip, which is available in an SSOP-28 style plastic package, is shown in Figure 1.

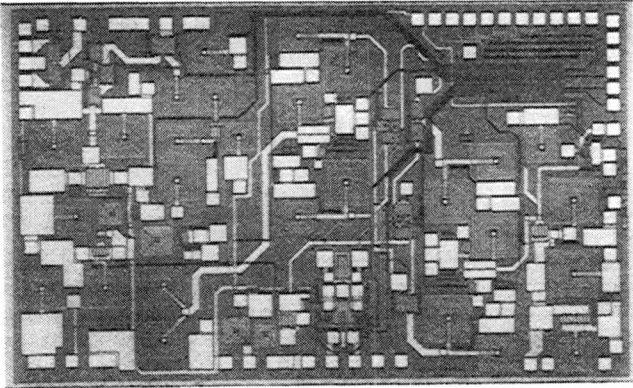

Figure 1   Photograph Of The Transceiver MMIC

Transceiver Architecture

A block diagram of the complete transceiver is shown in Figure 2. The circuit can be switched between receive, transmit and standby states. In receive mode, RF input signals are downconverted to differential IF signals. Although designed specifically for the 2.4-2.5GHz band, RF signals between 1.9GHz and 2.6GHz can be accommodated. The off chip filters can be selected to suit the band of interest.

In transmit mode, the IF input signal can be between 100MHz and 600MHz. The balanced input is upconverted to a single ended signal at the RF frequency. The circuit has been designed to provide a constant output power for a wide range of IF signal levels. A switched attenuator has been included to allow a 10dB step in the output power level.

The frequency of the VCO, and hence the IF frequency, is selected by appropriate choice of an off-chip resonator. Local oscillator frequencies of between 1.4GHz and 2.7GHz are available. A diversity switch has also been included to allow antenna selection. DC supply to the chip is +5V and -5V, with

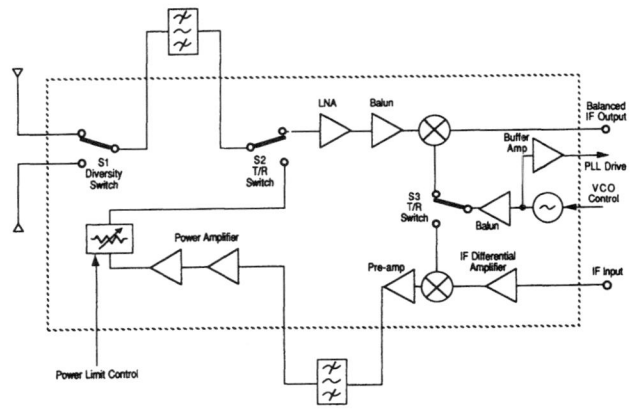

Figure 2   Block Diagram Of The Transceiver

complementary 0V/-5V switching. The -5V supply takes less than 1mA of current, regardless of transceiver operating mode. Typical current requirements from the +5V supply are 30mA in receive mode and 220mA in transmit mode. A standby state is also available and requires a current of less than 0.5mA.

In addition to the complete transceiver chip, all of the sub-circuits have been manufactured as individually measurable components. The design and measured performance of these subcircuits is described below. The circuits were realised on the standard GMMT F20 depletion mode GaAs MMIC process.

Sub-Circuit Design and Measurements

LNA: The Low Noise Amplifier (LNA) is a two stage design with series inductive feedback to allow good noise figure performance with a well matched input[1]. A stacked bias arrangement is used to help reduce current consumption. Instead of biasing the drain of each FET at +5V and the source at 0V, the arrangement shown in Figure 3 is used. This allows the +5V to be shared between the FETs and the current to be re-used. The RF On Wafer (RFOW) measured s-parameters of a

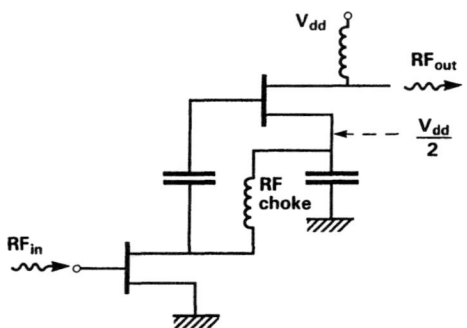

Figure 3   Stacked Bias Arrangement

typical LNA are shown in Figure 4. Gain is 17.5dB ±0.5dB from 2 - 3GHz. The input match is better than 15dB and the output match is better than 13dB. Measured noise figure is 2.5dB at 2.4GHz. Total current consumption is 6mA from a +5V supply.

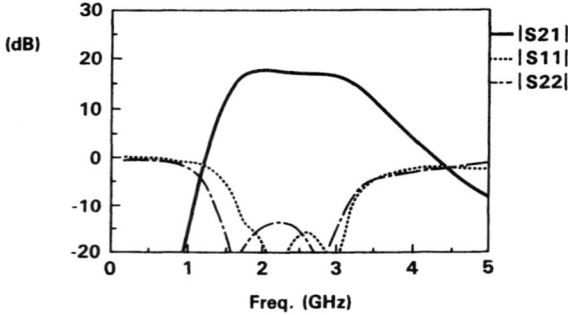

Figure 4   RFOW Measured s-parameters of LNA

Switches: T/R and diversity switches all use simple series mounted FETs[2]. The drain and source of each FET is biased at zero volts DC and the control signal is applied to the gates. One common design of Single Pole Double Throw (SPDT) switch is used throughout. The measured "on" case insertion loss at 2.4GHz is typically 0.7dB with an "off" case isolation of 20dB.

Mixers: A quad ring of zero biased FETs is used to realise a balanced conductance mixer[3]. When driven with differential inputs, excellent balance is achieved with a conversion loss of 6dB.

VCO: A Clapp type Voltage Controlled Oscillator (VCO) is used[4]. The oscillation frequency can be set between 1.4GHz and 2.7GHz by an external resonator. An off chip varactor allows an instantaneous voltage controllable tuning range of 150MHz. For RFOW testing of the subcircuit an inductor/capacitor combination was included on-chip, for use as an alternative to the off-chip resonator.

Active Baluns: Active baluns are used for both the RF and LO signals. Each balun uses a common gate stage and a common source stage of amplification to provide an equal amplitude split with 180° phase difference[5]. Figure 5 shows the RFOW measured s-parameters of a typical LO balun and Figure 6 shows the insertion phase between the input and each of the two outputs. A gain of 1dB at 2.1GHz and terminal matches of better than 20dB is achieved. The amplitude difference is only

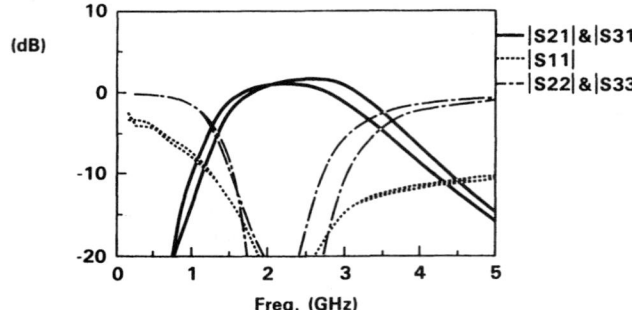

Figure 5   RFOW Measured s-parameters of LO Balun

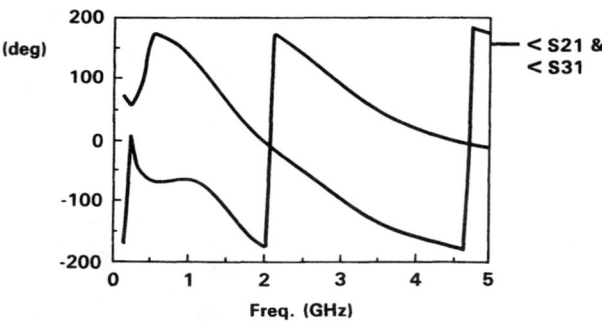

Figure 6   RFOW Measured s-parameters of LO Balun

0.15dB and the phase difference is 186°. Stacked bias has been used to allow operation with only 6mA of current from a +5V supply. The RF balun is of the same design but with a centre frequency of 2.4GHz.

Buffer Amplifier: The buffer amplifier is used to provide a low level output for phase locking of the LO. It must present minimal loading to the VCO, provide high isolation and be able to operate into any load from 50Ω to an open circuit. This has been achieved by using a small, single finger FET, biased through a 50Ω resistor in the drain. The reverse isolation is more than 40dB at 2.4GHz and the output match is better than 14dB. An insertion loss through the buffer of 12.5dB ensures the required low level of output power is delivered. Current consumption of this component is only 1.5mA.

Differential Amplifier: With the chip in transmit mode, the IF input is into a two stage differential amplifier[6,7]. Active biasing is used throughout in order to minimise chip area. The differential input impedance to the circuit is 800Ω. When driven with differential signals from a source of the same impedance, the gain of the amplifier is 20dB over a frequency range of 100MHz to 500MHz.

Pre-amplifier: The output of the transmit mixer is amplified by the pre-amplifier prior to passing off chip, through the transmit filter and into the power amplifier. A low level of gain is required to balance the gain budget through the transmit chain. The input to the pre-amp is resistivity matched with reactive matching at the output. The amplifier exhibits a flat 5dB of gain from 2 - 3GHz. Input return loss is greater than 12dB and output return loss is greater than 17dB.

Power Amplifier: A two stage power amplifier is used to increase the level of the transmit signal[4]. Figure 7 shows the small signal s-parameters of the amplifier. The gain is 23dB at 2.4GHz with input and output return losses of greater than 13dB.

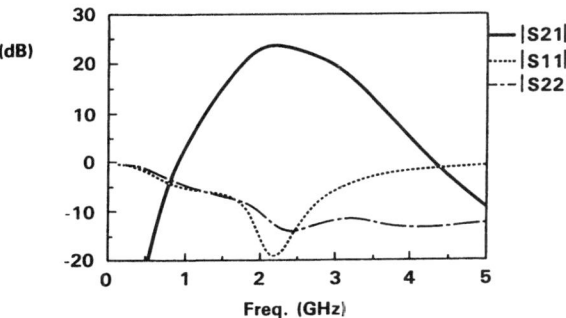

Figure 7 RFOW Measured s-parameters of Power Amplifier

Switched Attenuator: A 10dB switched resistive attenuator[8] is positioned before the common T/R port in the transmit path. This allows the output power level to be switched by 10dB. The small signal gain through the power amplifier, attenuator and T/R switch on the complete transceiver chip has been measured with the attenuator in both states. Figure 8 is a plot of this and the 10dB gain difference shows the accuracy of the switched attenuator. Because the attenuator is positioned after the power amplifier it gives an accurate 10dB step in output power level, regardless of amplifier compression.

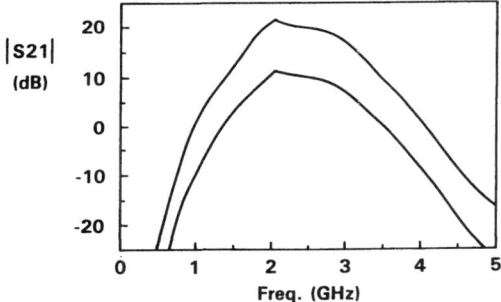

Figure 8 Measured Gain Of Power Amp, Attenuator and T/R Switch Chain, Two States

Transceiver Measurements

Measurements have been made on the complete transceiver chip. These were made on an unpackaged device in a purpose built jig. A spectral plot of the buffer amplifier output is shown in Figure 9. The output power level is -12dBm with a phase noise

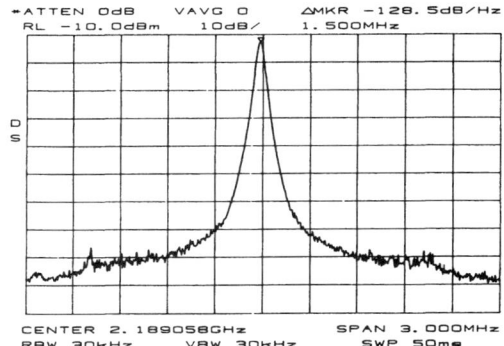

Figure 9 LO signal at Buffer Amplifier Output

of -122dBc/Hz at 1MHz off carrier. This signal is used to drive the phased lock loop of the transmit/receive circuit.

Figure 10 shows the measured receiver conversion gain and double sideband noise Figure versus IF frequency. This is for a fixed LO frequency of 2.035GHz with the IF varying from 50MHz to 500MHz. The slight roll off with increasing IF frequency is a result of losses in the chip and jig IF paths and in the balun used to combine the differential IF signals.

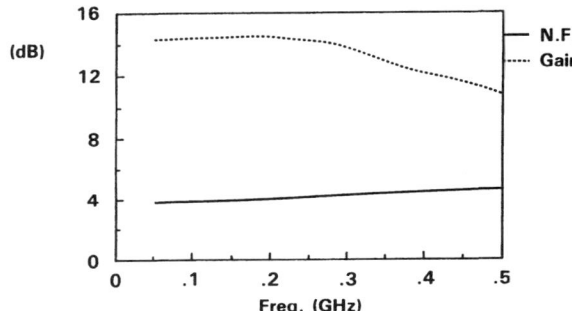

Figure 10 Measured Receiver Gain and DSB Noise Figure

The power transfer characteristic through the power amplifier, attenuator and T/R switch chain has been measured and is shown in Figure 11. A saturated output power capability of +21dBm is demonstrated. In practice the chip is designed to operate in saturation with a constant output power level. This reduces chip to chip variation, improves efficiency and allows a

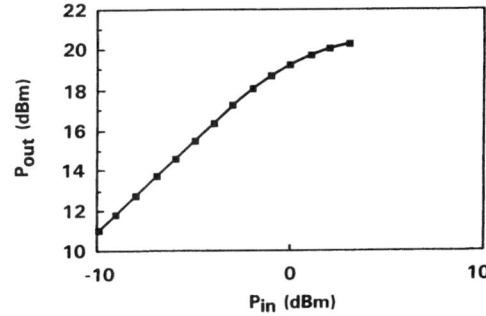

Figure 11 Power Transfer Characteristic Of Power Amp Attenuator and T/R Switch Chain

large tolerance to the range of IF input signals. Figure 12 shows the small signal gain versus IF frequency through the entire transmit chain from differential IF input to T/R common port output. This was measured with the LO frequency fixed at 2.035GHz and shows a gain of 38dB ±1dB for IF frequencies between 100MHz and 500MHz.

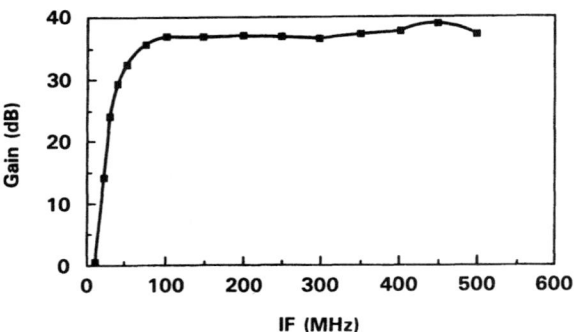

Figure 12   Small Signal Gain Of Transmit Chain Versus IF Frequency

Future Enhancements

Several enhancements to the circuit architecture have now been implemented and the modified design is currently being manufactured. In particular the input referred third order intercept point (IP3) has been increased from the present -10dBm to -3dBm. This was achieved by replacing the active baluns with passive networks, increasing the gain and linearity of the LNA, including on-chip diplexer filters to terminate properly the mixer IF ports and introducing a receive chain IF amplifier to balance the gain budget and match the output to a higher impedance. A further consequence of using passive baluns is that a single mixer is now used for both transmit and receive modes of operation. A block diagram of the modified transceiver architecture is shown in Figure 13.

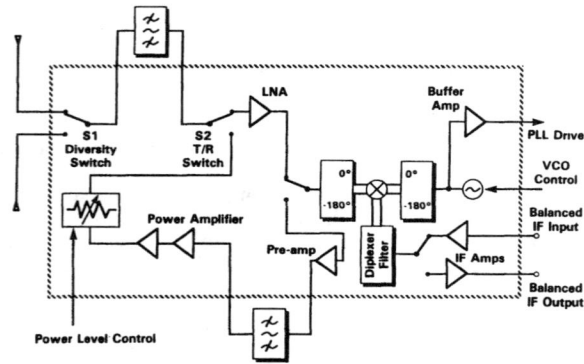

Figure 13   Enhanced Transceiver Architecture

Conclusions

A single chip GaAs transceiver to cover the 2.4 - 2.483GHz ISM band has been described. Receive gain is 13dB with differential IF outputs and a double sideband noise figure of 4dB. Current consumption in receive mode is just 30mA from a +5V supply. A standby mode is available with a current consumption of less than 0.5mA. Transmit mode offers a constant output power level, switchable by 10dB, for a large range of IF input levels. These features combine to give a component which is ideally suited to spread spectrum Wireless LAN applications. The chip is available in a low cost SSOP-28 style plastic package (Figure 14).

Figure 14   Transceiver MMIC in SSOP-28 Plastic Package

References

1. M Murphy, M.T., "Applying the Series Feedback Technique to LNA Design", Microwave Journal, November 1989, pp 143 - 152.

2. Ayasli, Y., "Microwave Switching With GaAs FETs", Microwave Journal, November 1982, pp 61 - 74.

3. Boulouard, A., "Simple Analysis Method Simulates Switching Mixers", Microwaves & RF, June 1989, pp 103 - 107.

4. Vendelin, G.D., Pavio, A.M. and Rohde, U.L., "Microwave Circuit Design", John Wiley & Sons, Inc 1990.

5. Hiraoka, T. et al, "A Miniaturized Broad-Band MMIC Frequency Doubler", IEEE Transactions On Microwave Theory and Techniques, Vol 38 No 12, December 1990, pp 1932 - 1936.

6. Gray, P.R and Meyer, R.G. "Analysis and Design of Analog Integral Circuits", Second Edition, John Wiley & Sons, Inc 1984.

7. Robertson, I.D and Aghvami, A.H "Ultrawideband Biasing of MMIC Disturbed Amplifiers Using Improved Active Load", Electronics Letters Vol 27, No 21, 10th October 1991, pp 1907 - 1909

8. Clifton J.C and Arnold J., "Digitally Controlled GaAs MMIC Attenuator For Active Phase Arrays", Proceedings of the 22nd European Microwave Conference 1992, Vol 1, pp 231 - 235.

FA 7.2: The Future of CMOS Wireless Transceivers

Asad Abidi, Ahmadreza Rofougaran, Glenn Chang, Jacob Rael, James Chang, Maryam Rofougaran, Paul Chang

Integrated Circuits & Systems Laboratory, UC Los Angeles, CA

CMOS has made steady inroads into ICs for wireless applications. Building blocks implementing the RF and baseband circuits in a 900MHz wireless transceiver have been described [1]. New uses have been discovered in the RF context for some of the well-known properties of CMOS analog circuits, such as the cancellation of quadratic nonlinearity in balanced circuits, the capability of switched-capacitor circuits to handle large signals without distortion, the use of FET switches to commutate signals at RF, and the large signal swings possible in a transistor with an insulating gate [1, 2]. 1µm CMOS circuits have been shown to operate at 900MHz, while 0.6µm channel lengths suffice at 2.4GHz. As feature size scale down to 0.35µm, 5GHz CMOS front-ends will be within reach.

Building blocks alone do not account for the current interest in CMOS for RF applications. The more compelling reason is the opportunities CMOS affords for *large-scale integration*. Modern wireless transceivers will increasingly blend digital blocks into conventional analog front-ends for frequency synthesis, adaptivity, multi-mode operation, and sophisticated detection. This raises questions such as how well digital CMOS circuits can co-exist on the same substrate as the radio front-end, or whether there is sufficient on-chip isolation in a low-cost package to guarantee stable operation of a receiver with more than 100dB of baseband gain, or how the power amplifier modulates the on-chip local oscillator. The future of CMOS transceivers may well depend on satisfactory answers to these questions. This paper presents design techniques to mitigate these problems in a *single-chip 900MHz spread-spectrum transceiver* implemented in 1µm CMOS, and measurements of the transceiver to validate their effectiveness (Figure 1).

The transceiver operates in time-division duplex, that is, at any instant of time it either transmits a data packet spread by frequency hopping across the ISM band, 902-928MHz, or it receives such a packet. The digital agile frequency synthesizer is active in both modes. The transmitter must deliver the modulated RF signal to the antenna relatively free of spurious tones arising from distortion, of the unwanted sideband, and of leakage from the local oscillator (LO). All these undesirable spectral components may fall within the ISM band, and therefore cannot be filtered. This requires linear upconversion mixers with high accuracy and matching, and very small parasitic on-chip coupling between the LO and the power amplifier (PA). The single-chip radio achieves excellent performance in all respects (Figure 2). A fixed-frequency 915MHz LO with inherently accurate quadrature outputs upconverts a digitally-synthesized baseband spread-spectrum, whose quadrature accuracy is susceptible only to the small gain mismatch in the two DACs. Four balanced passive FET switch mixers upconvert each baseband component by both quadrature LO phases, and an R-C polyphase filter suppresses the opposite sideband appearing at the 3rd LO harmonic. To maintain equal loading on I and Q signal lines, the balanced filter outputs in quadrature are each applied to a first PA buffer, and then are combined in the second PA buffer to select one sideband (Figure 3). The quasi-differential PA drives the antenna through a balun. The on-chip LC resonator and balanced oscillator circuit are responsible for the low level of LO leakage into the output spectrum. The spurious tones in the spectrum arise from the small residual nonlinearity in the DAC buffer and switch mixers.

Another measure of the high on-chip isolation is that the PA, when switched across its full output range, pulls the unlocked LO frequency by only 220PPM.

First, the various blocks in the receiver chain must be scaled and buffered to balance the overall noise figure against the total dynamic range (Figure 4) [1]. Second, the receiver front-end must be immune to parasitic feedback from the back end, where the signal is amplified by 122dB. This is achieved here with a fully balanced signal path from the RF input to the limiting amplifier output, that lowers back-end signal injection into the supplies and substrate, and provides common-mode rejection at the front-end. Third, the RF front-end must be isolated from the large LO signal, that lies within its passband and may block the weak received signal. The LO is placed far away from the receiver front-end. The LO signal in the PA buffers is electrically isolated from the receiver by turning off the first buffer during receive mode, and physically isolated by trenches between the receiver front-end and PA buffers etched with the same process that suspends the large spiral inductors [1]. The resulting dc offset due to LO self-downconversion is well below the receiver compression point, and is rejected in the limiter.

Additional problems unique to this CMOS zero-IF direct-conversion receiver are caused by dc offset due to FET mismatch in the baseband sections, flicker noise, and pattern noise created by parasitic self-downconversion during frequency hopping. The baseband receiver circuits are direct-coupled, and a negative feedback loop around the limiting amplifier with a large off-chip capacitor removes an accumulated dc offset as large as ±1V, while providing the full 84dB of limiting gain above 300Hz [1]. In the absence of RF input, the circuit limits on the amplified noise from the preceding stages. The overall receiver characteristics are probed by spectral analysis of the limiter output, taking into account the FM capture effect. The limiter output spectrum is further bandpass filtered by the digital correlating FSK detector. Figure 5a shows the limiter output spectrum with a -105dBm (1.2µV rms) RF single-tone receiver input, the minimum detectable signal for 10^{-6} BER. The received spectrum is free of spurious tones, and comprises only thermal noise shaped by the channel filter. This is proof of the on-chip isolation between receiver blocks. Figure 5b shows the limiter output spectrum with zero input signal, but with the hopping frequency synthesizer disabled and enabled, respectively. In the first case, flicker noise contributed by the baseband CMOS circuits appears below 40kHz, but is rejected in the correlating detector. In the second case, parasitic self-downconversion of the hopping LO adds a component to the offset that changes, due to various resonances, with the instantaneous frequency. This recurring pattern raises the noise spectrum, but with 33 hops at 20khop/s, the enhanced noise still lies below the detector passband.

This work shows that CMOS is well-suited, beyond building blocks, to entire mixed analog-digital radios on a single chip.

References

[1] Abidi, A. A., "CMOS-Only RF and Baseband Circuits for a Single-Chip 900 MHz Wireless Transceiver," *Bipolar Circuits & Tech. Mtg.*, Minneapolis, pp. 35-42, 1996. This and related publications downloadable at http://www.icsl.ucla.edu/aagroup.

[2] Steyaert, M., et al., "RF Integrated Circuits in Standard CMOS Technologies," *European Solid-State Circuits Conf.*, Neuchâtel, Switzerland, pp. 11-18, 1996.

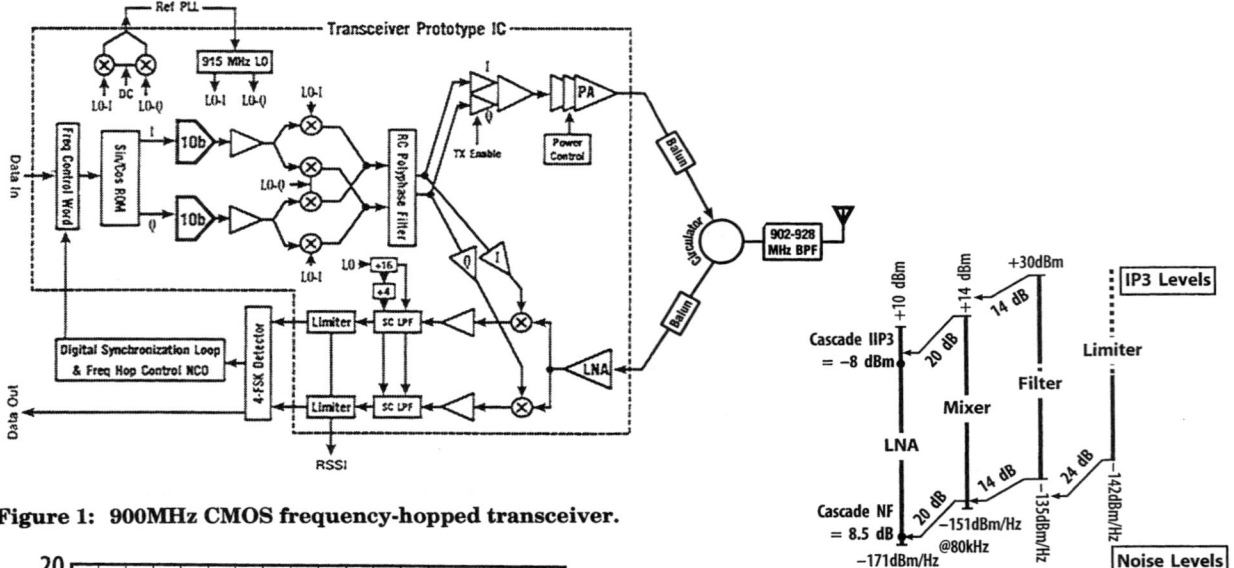

Figure 1: 900MHz CMOS frequency-hopped transceiver.

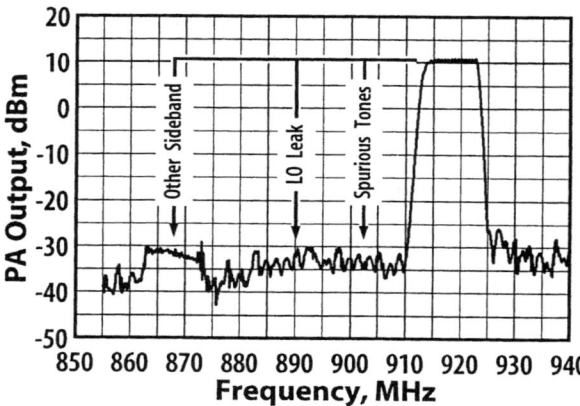

Figure 2: Spread-spectrum TX output (unfiltered).

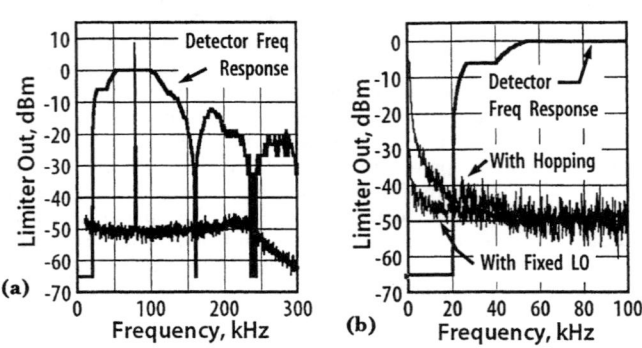

Figure 4: Dynamic range in receiver chain.

Figure 5: Limited output spectrum (a) with, and (b) without RF input.
Figure 6: See page 440.

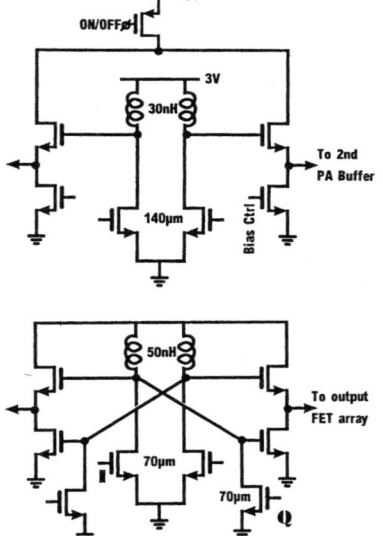

Figure 3: Two buffer stages prior to power amplifier.

Receiver RF sensitivity	-105dBm at 10dB SNR
Total receiver IIP3	-8.5dBm
Total receiver IIP2	+22dBm
Channel filter	220kHz passband, 57MHz stopband
Active RX, TX current	120mA, 100mA
PA output range	-17 to +14dBm
PA output noise floor	-138dBm/Hz
IC technology	1.0µm CMOS, p-epi on p+ substrate
LO phase noise	-101dBc/Hz at 100kHz
LO pulling by PA	<200kHz
LO-to-RX-in coupling	-92dBm with PA off
RX s_{11}, TX s_{22}	-11dB, -15dB
Frequency hop rate	150khop/s (max), 20khop/s (typ)
Supply	3V / 3.5V
Operating band	902-928MHz
Data rate	160kb/s at 4-FSK
Active chip area	10.5x7.3mm²

Table 1: Transceiver performance summary.

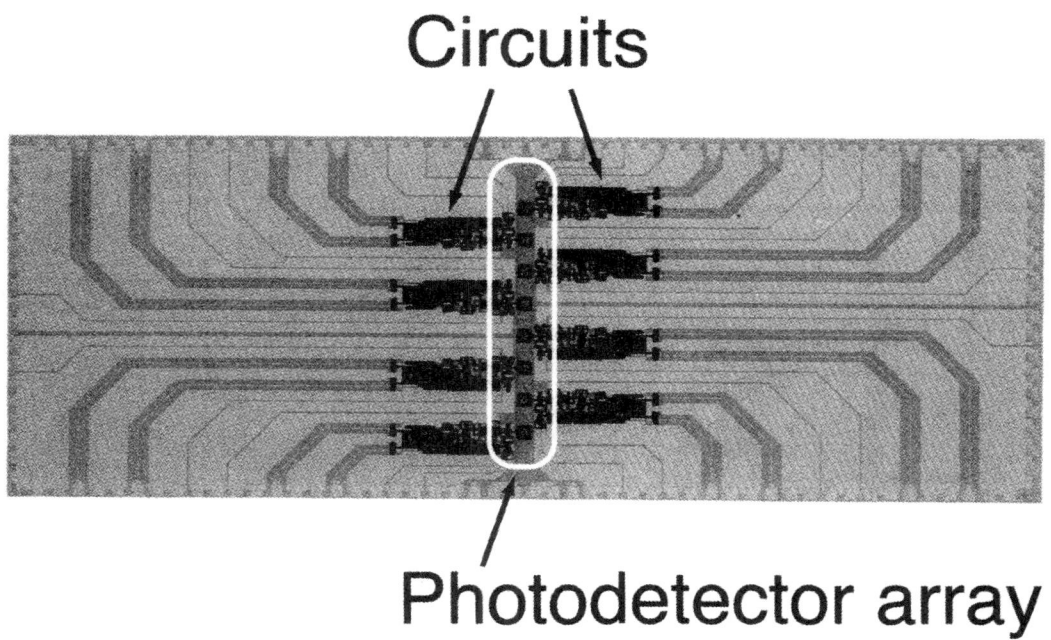

Figure 6: SiGe OEIC chip view. Chip is 3x8mm².

SA 18.3: A 1.9GHz Wide-Band IF Double Conversion CMOS Integrated Receiver for Cordless Telephone Applications

Jacques C. Rudell, Jia-Jiunn Ou, Thomas B. Cho[1], George Chien, Francesco Brianti[2], Jeffrey A. Weldon, Paul R. Gray

Dept. of Electrical Engineering and Computer Sciences, UC Berkeley, CA/[1]San Francisco Telecom / Level One Communications/[2]SGS Thomson, San Jose, CA

A number of recent efforts have concntrated on highly-integrated radio receivers using a low-cost silicon process such as CMOS [1, 2]This prototype monolithic CMOS receiver combines RF and baseband functionality by taking the carrier signal at the LNA input and producing a 10b digital baseband waveform. A wide-band intermediate frequency double conversion (WBIFDC) architecture eliminates the need for external narrow-band IF filters.

The experimental chip includes an LNA, an image-rejection mixer, and two baseband signal-processing paths, each of which includes a 2nd-order Sallen and Key anti-alias filter, an 8th-order switched-capacitor filter and a 10b pipelined ADC. The receiver meets the specifications of the digitally enhanced cordless telephone (DECT) standard. The device achieves -90dBm receiver reference sensitivity, -7dBm input IP3, with 198mW overall power dissipation from a 3.3V supply. The design is in 0.6μm double-poly, triple-metal TSMC CMOS process.

The wide-band IF double conversion architecture modulates all of the RF spectrum passing through the RF filter directly to baseband in its entirety, where channel selection is done by low-pass signal processing as in the case of direct conversion (Figure 1). No bandpass filtering is performed at the IF frequency. In contrast to direct conversion, translation takes place in two steps, using two oscillators and two sets of mixers. This provides two principal advantages over direct conversion systems: no oscillator operates at the RF input frequency, and the tuning of the receiver uses the second low-frequency LO. Because the first LO is fixed, easier trade-offs may be obtained with regard to LO phase noise. As in the case of direct conversion, channel selection can be performed at baseband, where digitally-programmable filter implementations can potentially enable more multi-standard capable receiver features.

To reduce the effects of high-frequency coupling, the entire analog RF and baseband signal paths are fully differential. The ADC output drivers are realized with a differential source coupled buffers that reduce digital substrate coupling effects into the analog section of the chip.

The low f_T and gm/Id ratio of CMOS devices limit performance of traditional broadband LNA designs. A tuned narrow-band technique passively enhances voltage gain of the LNA and performs impedance matching [3]. This approach relaxes the conventional trade-off between noise figure and power dissipation and allows linearity to be traded for noise figure. The LNA is implemented as a single-stage differential common source amplifier with on-chip spiral inductors (Figure 2). The LNA input network utilizes bondwire inductor L1 and spiral inductor L2 to enhance the Q of the input network and perform impedance matching. Spiral inductor L3 tunes the output node while improving both the gain and the image rejection of the receiver.

Similar to conventional heterodyne systems, the wide-band IF double-conversion architecture (WBIFDC) requires image rejection. In the DECT standard, approximately 80dB rejection is required in the image-band. The image rejection is accomplished through combination of a front-end external RFfilter, a tuned LNA, and a new image-rejection technique achieved by the double-conversion configuration. This scheme utilizes both the in-phase and quadrature-phase of the local oscillators to realize a broadband image reject function. A complex modulation from RF to IF is performed (Figure 1). The IF signals are mixed to baseband where in-phase and quadrature channels are generated from both of the I&Q IF channels. By properly combining these four baseband signal paths, the correct phase can be obtained for constructive interference of signals above the first local oscillator frequency, while to first-order any RF signals below LO1 are cancelled. Passive components used in conventional image-rejection mixers are eliminated from the signal path. The magnitude of image rejection is limited by gain matching of the mixer paths and by LO deviation from quadrature. To mitigate effects of low-frequency noise at baseband and dc components resulting from self mixing from IF to baseband, two offset cancellation current DACs are used at the image-rejection mixer output.

The individual mixer units are realized with a doubly-balanced CMOS active mixer (Figure 3). The cascode devices M3 and M4 improve isolation between the mixer LO and IF/RF input terminals. Current sources M11 and M12, along with a common-mode feedback loop comprised of M13, M14, and M15, set the output common-mode voltage. Two triode region pMOS devices, M9 and M10, determine both the loading and gain of the mixer cell. The conversion gain is adjustable from 0dB to 10dB by modulating the current through diode-connected device M16.

In the receiver, 55dB overall image rejection is obtained using externally supplied LOs with an on-chip phase shifting network adjusted for quadrature. Loss of the balun is estimated to be 2dB in the image-band while 8dB of rejection is contributed by the LNA, leaving approximately 45dB of image rejection coming from the mixer portion of the receiver.

Aggressive dynamic range requirements are placed on the baseband circuits because the IF filter is removed in the WBIFDC radio architecture. After a mix from IF to baseband, a 2nd-order Sallen and Key filter performs anti-aliasing before subsequent sampling into an 8th-order switched-capacitor filter meeting the bandwidth requirements for a single DECT channel (700kHz BW). In addition to providing channel filtering, the gain is variable from 6dB to 48dB in 6dB increments. The baseband channel filter output is then sampled by a 10b pipelined 10MHz ADC. Capacitor scaling techniques optimize the filter and ADC for noise and power [4].

The prototype is packaged using a chip-on-board technology The backside of the chip is attached directly to the board using a conductive epoxy with bondwires running from chip pads directly to landing zones on the board. Single-ended-to-differential conversion is achieved using an external balun placed in front of the LNA and LO input ports. Experimental results are shown in Tables 1 and 2. A -90dBm receiver reference sensitivity is obtained by measuring the ADC output signal-to-noise ratio. 10dB of SNR is assumed to meet the DECT BER requirement of 10^{-3}. The 3rd-order intermodulation intercept is -7dBm referred to the LNA input (Figure 4).

Acknowledgments:

Support provided by ARPA, California MICRO program, NSF, Rockwell International, GEC Plessey, Xerox, Harris, Philips, and National Semiconductor. Fabrication donated by TSMC and Level One Communications. The authors thank T. Weigandt, R. Meyer, R. Brodersen, and B. Boser for advice and support.

References: See page 476.

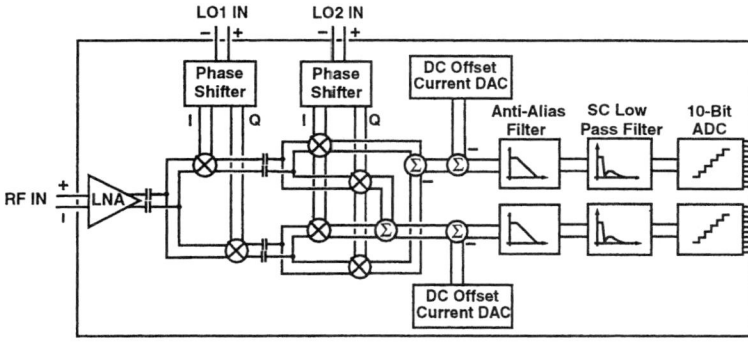

Figure 1: Wide-band IF double conversion receiver architecture.

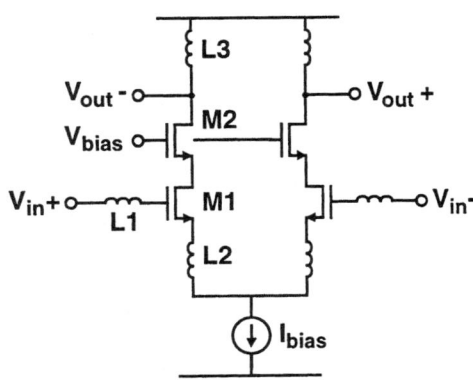

Figure 2: Narrow-band inductively tuned LNA.

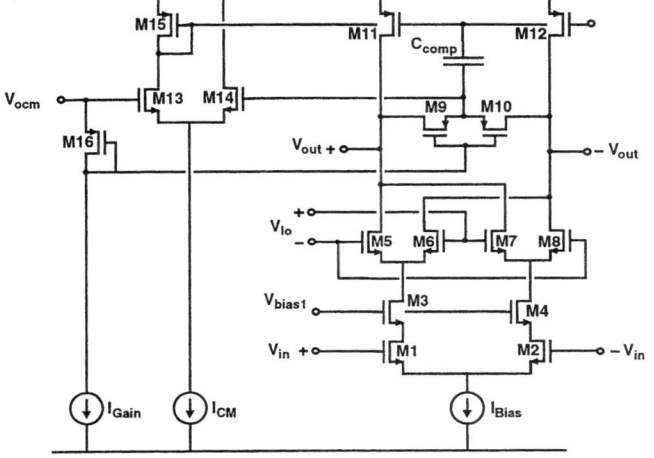

Figure 3: Variable gain active mixer with common-mode feedback.

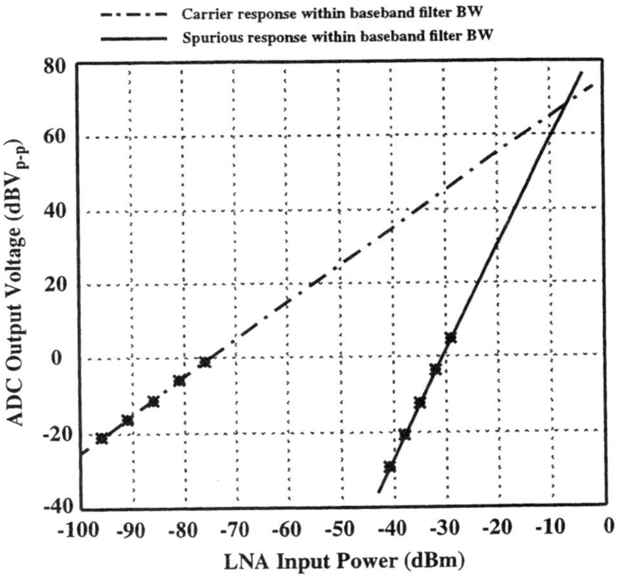

Sensitivity	−90dBm*
Input IP3	−7dBm**
Receiver image rejection	55dB
Pob3dB (max. gain setting)	−33dBm***
P-1dB (min. gain setting)	−24dBm
Max. receiver gain	78dB
Min. receiver gain	26dB
Supply voltage	3.3V
LO1	1.7GHz
LO2	182 - 197MHz
Carrier frequency	1.882 - 1.897GHz
Active chip area (including bias)	15mm²

* −83dBm sensitivity required for DECT.
** −27dBm input referred IP3 required for DECT.
*** Blocker 1 DECT channel from carrier.

Table 1: Receiver performance features.

Figure 4: Receiver two-tone (spaced 2 & 4 DECT channels from carrier) test for 3rd-order intermodulation.

LNA	41 mW
RF to IF mixers	17 mW
IF to baseband mixers	34 mW
Baseband filters	66 mW
ADCs	40 mW
Total chip	198 mW

Table 2: Receiver power dissipation.

References:

[1] Rofougaran, A., et al., "A 1GHz CMOS RF Front-End IC for a Direct-Conversion Wireless Receiver," IEEE J. of Solid-State Circuits, pp. 880-889, July 1996.

[2] Crols, J., M. Steyaert, "A Single-Chip 900 MHz CMOS Receiver Front-End with a High Performance Low-IF Topology," IEEE J. of Solid-State Circuits, pp.1483-1492, Dec.,1995.

[3] Shaeffer, D., T. Lee, "A 1.5V, 1.5GHz CMOS Low Noise Amplifier," Symp. on VLSI Circuits Digest of Technical Papers, pp. 32-33, June, 1996.

[4] Cho, T., et al., "A Power-Optimized CMOS Baseband Channel Filter and ADC for Cordless Applications," Symp. on VLSI Circuits Digest of Technical Papers, pp.64-65, June, 1996.

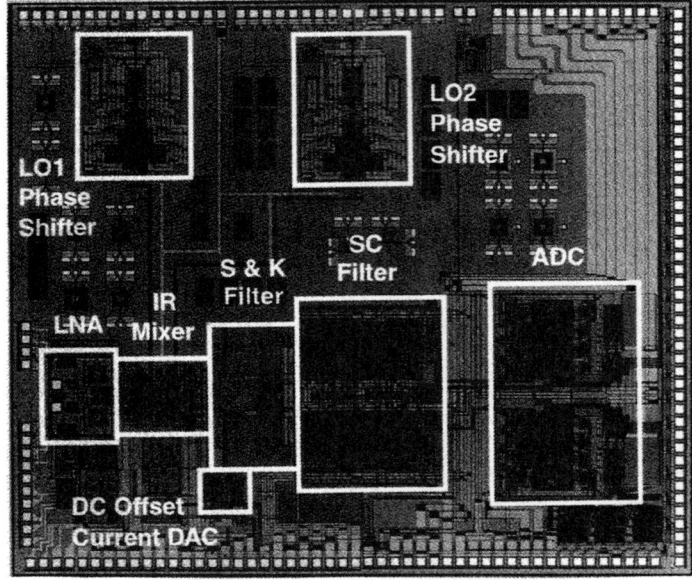

Figure 5: Receiver chip micrograph.

SA 18.6: A 2.5GHz BiCMOS Transceiver for Wireless LAN

Robert G. Meyer[1], William D. Mack, Johannes J.E.M. Hageraats

Philips Semiconductors, Sunnyvale, CA
[1]EECS Dept., University of California, Berkeley, CA

A low-cost Si IC contains most of the elements required to realize the RF transmit and receive paths of a 2.5GHz WLAN transceiver meeting 802.11 requirements. External filters and a power amplifier complete the RF front-end. Chip architecture is shown in Figure 1. The 2.73mm² die is mounted in a 24-pin plastic package (TSSOP24). A low-noise amplifier (LNA) is switchable between high (14dB) and low-gain (-13dB) modes. High-gain mode with 2.2dB NF, P-1dB=-15dBm input and IP3=-3dBm input is utilized in weak signal conditions. Low-gain mode with P-1dB=+5dBm input and IP3=+17dBm input can be selected in moderate-to-strong signal conditions. The signal goes off-chip from the LNA to an external image-rejection filter and is then input to the receive mixer via pin 19, that is also shared by the upconversion mixer output. This allows use of the same external filter in both transmit and receive modes. Both mixers are driven by an on-chip local oscillator (LO) buffer that has full-frequency and half-frequency input options, selectable via the LO switch at pin 11. An on-chip frequency doubler in the LO buffer with on-chip LC filters deliver a full-frequency LO signal to the selected mixer with half-frequency LO inputs.

The receive mixer differential output is converted to single-ended operation with an external balun. The transmit mixer accepts single-ended or differential inputs and uses on-chip LC filters together with a medium power buffer to deliver +3dBm output (P-1dB) at 2.45GHz to a separate power amplifier.

Pin 17 selects transmit or receive mode. In receive mode LNA, receive mixer and selected LO buffer option are powered up while transmit mixer and output buffer are powered down. If transmit mode is selected, LNA and receive mixer are powered down while the transmit path is activated with the selected LO buffer option. IC current from a single 3V supply is 34mA typical in transmit and 21mA typical in receive. The chip is powered down by the chip-enable at pin 13 with 1μA typical sleep-mode drain. Chip enable/disable, transmit/receive and receive/transmit switching times are 1μs typical using 50pF external coupling capacitors.

A schematic of the LNA is shown in Figure 2. In high-gain mode the amplifying device is Q_{43} (72x) biased to 4mA. To temperature compensate the gain, Q_{43} is biased to a PTAT current derived on-chip from I_1. The low-gain parallel path via emitter follower Q_{46} is connected to the same output pin through matching elements, that emulate the output impedance of the high-gain path when the low-gain path is selected. When the high-gain path is disabled by turning on M_{128} and other shunt MOSFETS (not shown), M_{49} connects Q_{46} to the input and M_{62} is turned on to provide input matching. When the high-gain path is active, current-source I_2 is off and R_{117} pulls the emitter of Q_{46} high, turning it off. Thus the switching at the LNA output is simply realized by hardwiring the two paths in parallel and alternately activating either one.

A simplified schematic of the LO buffer is shown in Figure 3. When the full frequency input is selected, current source I_3 is on and activates differential pair Q_{52}-Q_{53}. Current sources I_4 and I_5 are held off. The full frequency LO input at 2.1GHz is amplified by Q_{52}-Q_{53} and the on-chip LC loads, that, including parasitics, are tuned to 2.1GHz. Balanced operation is advantageous in that V_{cc} becomes a common-mode ac ground, and supply-line parasitic inductance (not well-controlled) does not affect the LC tuned circuit. The resulting LO voltage at Vx is buffered by followers and amplified by additional LC tuned differential amplifier stages that drive transmit and receive mixers with optimum LO voltage swing relatively independent of external LO drive. For half-frequency LO inputs, current source I_3 is switched off and I_4 and I_5 are activated. The four devices Q_{44}, Q_{45}, Q_{47}, and Q_{48} form a frequency-doubling circuit that feeds full frequency LO currents to the tuned signal path described previously [1]. The on-chip inductors provide substantial gain boost, improved output voltage headroom and attenuation of out-of-band signals and noise.

The Gilbert-quad-based down-conversion mixer has inductive degeneration on the input differential pair provided by on-chip 1nH spiral inductors (Figure 4). The degeneration improves overload performance and input matching with only small degradation in mixer noise figure. To operate with 2.7V Vcc, a 100Ω resistor is used for biasing in lieu of an active current source. This gives lower noise and better impedance characteristics at 2.4GHz for rejection of common-mode signals. The second stage of LC-boosted buffering in the LO path is provided by Q96 and Q97. The 300Ω shunt resistors drive the quad with an optimized 600mVpp swing over external LO input from -13dBm to +5dBm. The mixer exhibits 8dB conversion gain, 11dB SSB NF, and +3dBm input IP3.

The receive mixer input node and transmit mixer buffer output node share the same package pin. Traditional microwave design techniques would require physical realization of a single-pole double-throw switch to select either one of these two paths. Such switches have insertion losses that adversely impact performance and often must be realized as expensive stand-alone elements. Here the capabilities available on-chip in monolithic form are used to realize this function simply with minimal performance degradation. The up-conversion mixer is similar to the schematic of Figure 4 except for the use of resistive emitter degeneration in the input pair (noise figure is less important in the transmit path and the low IF signal frequency means that the inductors required could not be realized on-chip). Since the desired output signal of the transmit mixer is at 2.45GHz, on-chip spiral inductors can form LC tuned loads at the quad output to select the 2.45GHz component and attenuate all other outputs. This signal is passed to single-ended medium-power buffer shown in Figure 5. This results in 8.0dB image rejection and 20dB LO suppression. Output device Q_2 is biased to 10mA with 25Ω emitter degeneration. This stage delivers +3dBm output (P_{-1dB}) at 2.45GHz to a 50Ω load at pin 19. Receive input and transmit output are connected to the same pin and the desired function is selected by powering down via M_{27}, M_8 and other pull-down MOSFETS in the bias circuits (Figures 4, 5). A single 3nH external inductor acts as a pull-up in transmit mode and as a matching element in either transmit or receive mode. Switching is with minimum complexity and negligible gain, power, and noise degradation.

The BiCMOS process used for this design contains an extensive set of devices applicable to mixed-signal designs (Figure 6). The 20GHz f_T npns are used in the critical high-frequency signal paths, while the 0.7μm L_{eff} CMOS transistors are used extensively for switching and power-down modes. The process contains 250 and 2000Ω/sq. poly resistors, 2fF/μm² oxide capacitors, Schottky diodes, and lateral pnps. Minimum npn emitter is 0.7x1.0μm² with peak f_T at 210μA. The process supports both medium and high-frequency (GHz range) designs powered from 1.0V to 5.5V in high-volume production. Figure 7 is a die micrograph.

Acknowledgments:

The authors thank C. Conkling for inspiring this project, Philips

BiCMOS Technology Development team for making the QUBiC2 process production worthy, and K. McAdams for measurements.

Reference:

[1] Kimura, K., H. Asazawa, "Frequency Mixer with a Frequency Doubler for Integrated Circuits," IEEE Journal of Solid State Circuits, vol. 29, no. 9, pp. 1133-1137, Sept., 1994.

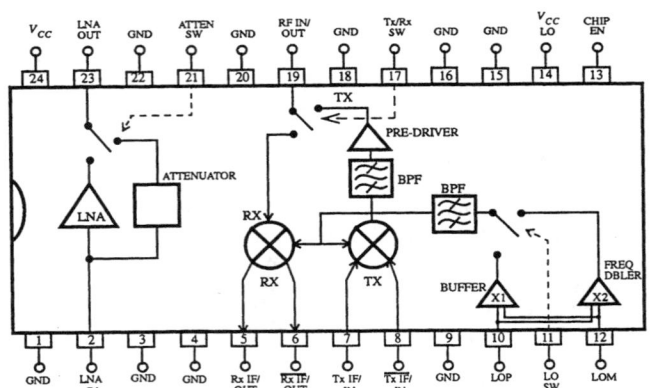

Figure 1: Transceiver architecture and pinout.

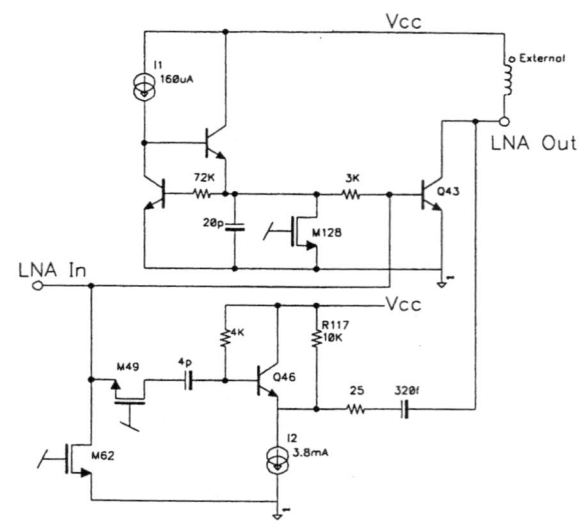

Figure 2: Simplified LNA schematic.

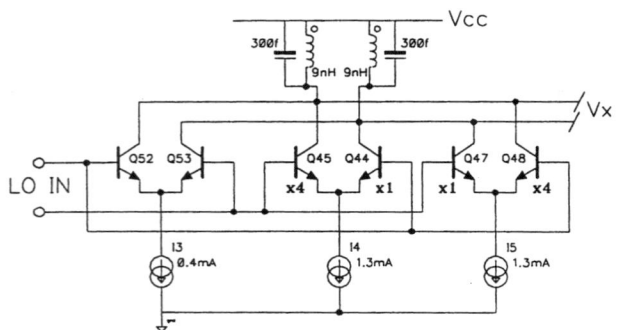

Figure 3: Simplified LO buffer schematic.

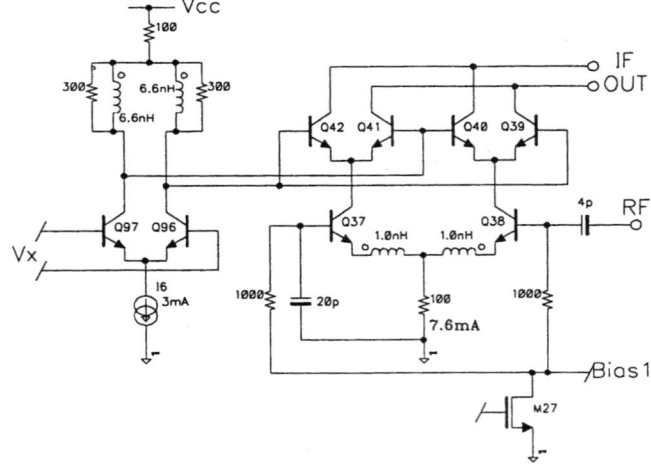

Figure 4: Down-conversion mixer schematic.

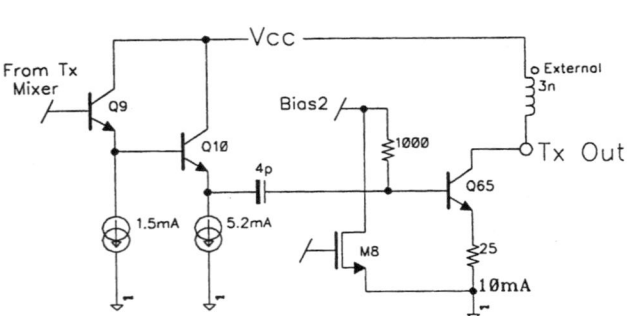

Figure 5: Transmit buffer schematic.

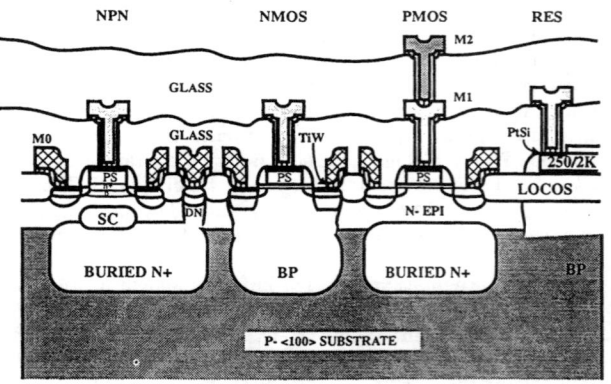

Figure 6: BiCMOS process cross-section.
Figure 7: See page 477.

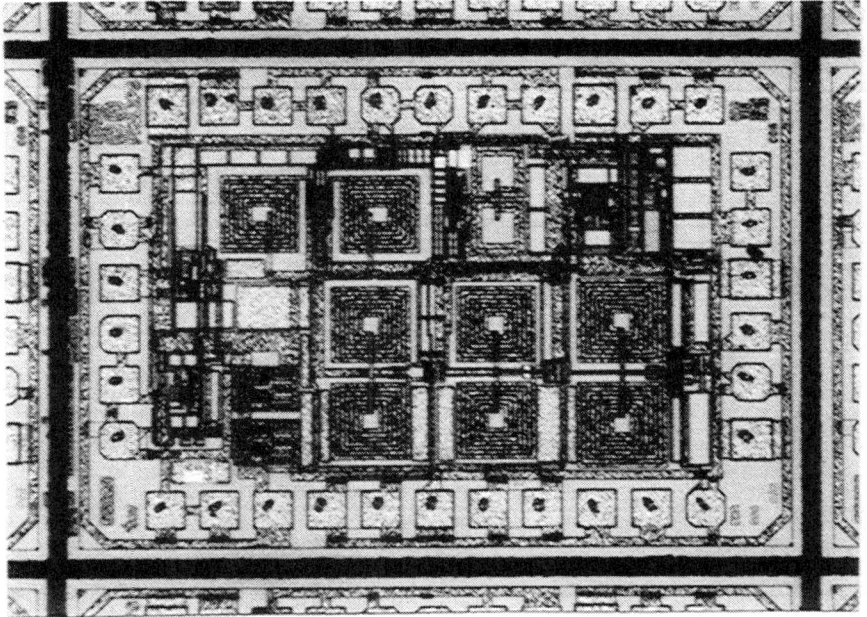
Figure 7: Die micrograph.

SA 18.5: A 2.7V DECT RF Transceiver with Integrated VCO

Geoffrey C. Dawe, Jean-Marc Mourant, A. Paul Brokaw

Analog Devices Inc., Wilmington, MA

Digitally-enhanced cordless telephone (DECT) is a standard for residential cordless telephony and wireless local loop. Components for this market must be low-cost and highly-integrated. Figure 1 shows the block diagram of the DECT radio frequency integrated circuit (RFIC). This IC, combined with an IFIC, baseband converter, low-noise amplifier, and power amplifier form the basis for a DECT telephone. The RFIC is fabricated on a high-speed (25GHz f_T) bipolar process [1].

The radio frequency (RF) architecture includes an upconverter transmit chain, a single downconverting receiver, a voltage controlled oscillator (VCO) and local oscillator (LO) distribution network, a low drop-out voltage regulator, and a mode controller. The four operating modes are: receive, transmit, synth (where the VCO is active only) and standby. Within the IC, all RF/IF blocks work with a supply voltage as low as 2.7V at -45°C. Supply biasing for all functions (except the VCO) is provided by the regulator that can be connected directly to the battery voltage. To avoid frequency pulling during mode switching, the VCO supply comes from the regulator on the IFIC.

In receive mode, the incoming signal flows from the antenna through an external LNA onto the IC where it is mixed with the low side local oscillator (externally phase locked) down to 110MHz. The receive mixer is a class AB micromixer shown in a simplified schematic form in Figure 2 [2]. Within the micromixer, the incoming signal is first split in phase and amplified by the common base, cascode parallel combination, then chopped by the switching core. The input transistors are large (10x min) to minimize noise contributions due to base resistance. The core transistors are driven by a 150mVp signal from the VCO and LO distribution network. The mixer is biased using a translinear loop and Vcc is supplied through resistors. A differential IF amplifier loads the mixer output and provides some low-pass filtering. The balanced output signal is converted to single ended by an external 1:1 transformer with a 150Ω load. In receive mode, the IC draws 37mA at 2.7V and 27°C. The receiver RF performance is summarized in Table 1.

In transmit mode, the 130MHz IF, from the IFIC, is upconverted by an degenerated Gilbert cell mixer. The differential RF output is buffered by a cascoded differential amplifier with a tuned load. The tuned load has the advantage of bandpass filtering. The IF leakage, and to some degree the LO leakage will not be amplified by the buffer. Since the cascode differential amplifier requires 3 V_{be}s and runs at a high current level, there is little or no headroom for standard resistive loads. The tuned load consists of on-chip 2nH spiral inductors resonated by 3.5pF capacitors. Resonating the inductors allows for the use of smaller value (and thus physical size) inductors while still realizing a high impedance. This high impedance is typically limited by the Q factor of the components. In this case the spiral inductors exhibit a Q of about 5 and thus limit the resonant impedance to about 100Ω.

The differential output of the cascode buffer is transformed to single ended by an on-chip passive balun. This approach consumes no dc, is less sensitive to grounding parasitics than active approaches and is reasonable in size because the coupled coils are resonated. The balun transforms the load from 50Ω to 150Ω, increasing the gain and efficiency of the cascode amplifier. The balun was designed using closed form equations and confirmed with an electromagnetic simulator and measurements. The balun loss, especially in a silicon process, is not negligible and is modeled carefully.

The output of the balun drives an external ceramic filter and the signal is then routed back onto the IC, where it is amplified by a high-dynamic-range driver. The LO leakage or the image can produce, with the RF, a third-order term that is in-band and unfilterable. A standard way to measure this non-linearity is through the third-order intermodulation intercept point. To linearize the amplifier transfer function over a wide input range, output currents and voltages can be increased, or for a given power, feedback techniques can be used. Resistive feedback is undesirable due to parasitic capacitance reducing the resistance and increasing power consumption. A lossless feedback techniques using the package parasitics is used here. Shown in Figure 3, the input and the output leads of the amplifier are placed close together. Capacitive coupling and mutual inductance provide the lossless feedback. A single stage with carefully modeled parasitics, ensures the correct level of feedback and stability. Measured output third-order intercept is in excess of +24dBm at 1.9GHz with 3V/10mA biasing. The full transmit chain draws 32mA at 2.7V and 27°C. Its overall performance is summarized in Table 2.

The VCO is a differential pair with positive feedback. Large transistors (10x min) running at peak f_T in the amplifier minimize noise contributed by base resistance and provide sufficient loop gain. Low-frequency feedback via emitter degeneration resistors improves phase noise at low offsets. The off-chip tank circuit consists of a printed inductor and an abrupt junction varactor diode. The VCO provides a minimum of 150mVp of drive to either mixer core and -17 dBm to the outside. The tuning bandwidth is 1.7-1.8GHz over 0.9 to 1.2V control voltage. The measured single sideband phase noise is shown in Figure 4. A level of -85dBc/Hz is seen at 10kHz offset from the carrier. The critical specification for DECT at an offset of 4.7MHz, -136dBc/Hz, is met by this design. The VCO and LO buffer stages draw 10mA at 2.7V and 27°C.

Acknowledgments:

The authors thank W. Foley, R. Gibson, T. Murphy and B. Murphy for help in testing and characterizing the part and thank B. Gilbert for the micromixer idea and A. O'Rourke, P. Katzin, P. Foote, and M Gilbert for contributions.

References:

[1] Garone, K. O. P., et al., "A Low Cost and Low Power Silicon npn Bipolar Process with nMOS Transistors (ADRF) for RF and Microwave Applications," IEEE Transactions on Electron Devices, vol, 42, no.10, pp. 1831-1840, Oct.,1995.

[2] Gilbert, B., "BG-1868 - GSM IF Chip Update," Analog Device Inc., Internal Memorandum 23, Aug., 1995.

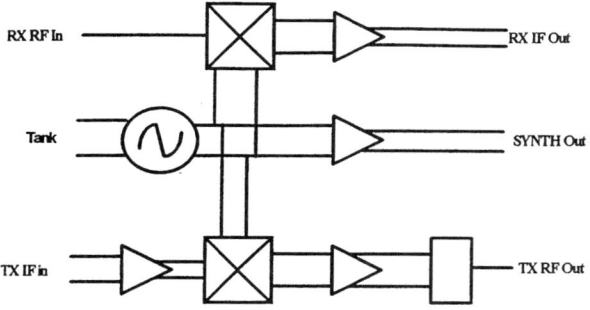

Figure 1: IC block diagram.

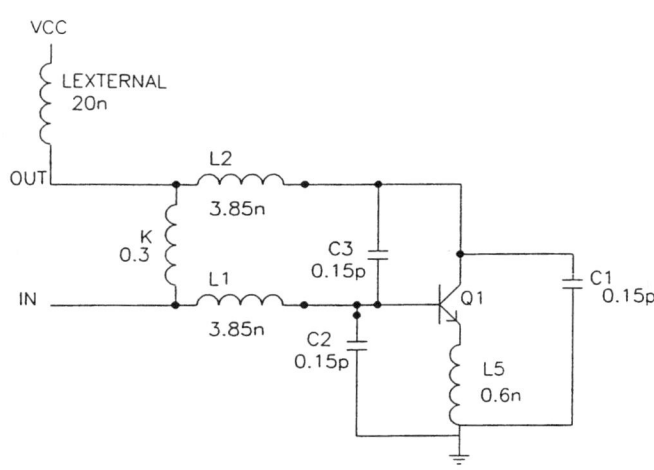

Figure 3: Transmit driver amplifier simplified schematic.

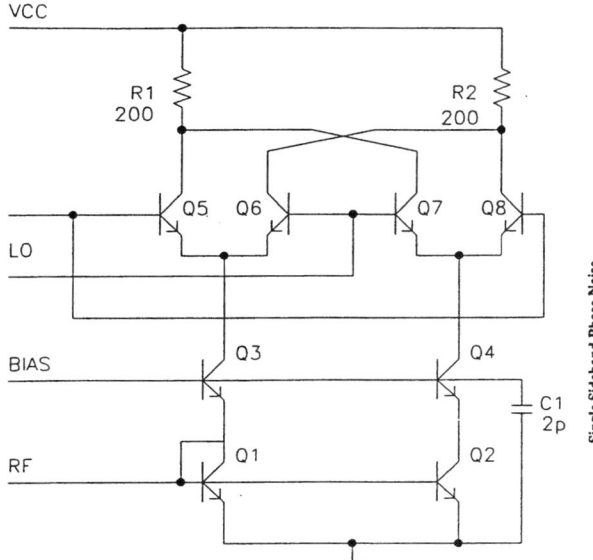

Figure 2: Receive mixer simplified schematic.

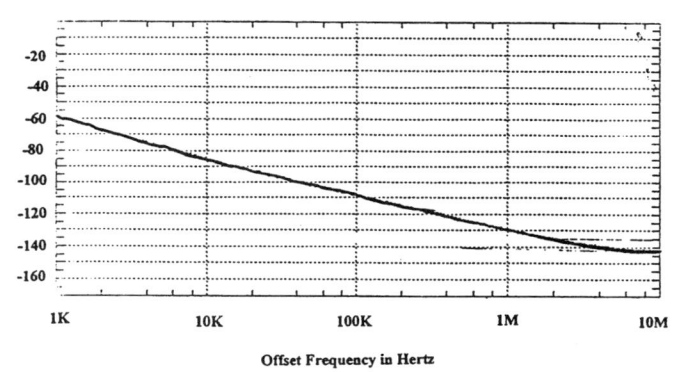

Figure 4: VCO phase noise spectrum (1.7GHz carrier).
Figure 5: See page 476.

Power conversion gain	17dB
Single sideband noise figure	15dB
Input P1dB	-14dB
Input third-order intercept point	-4dB
RF port in-band return loss	-15dB
IF port in-band return loss	-25dB
LO leakage at RF port	-31dB
LO leakage at IF port	-33dB
RF-IF isolation	41dB
RF x 2LO spurious	-44dB
2RF x 2LO spurious	-72dB
3RF x 3LO spurious	-44dB
RF x 3LO spurious	-68dB

Table 1: Receiver performance.

Upconverter	
RF output power	-15dBm
LO-RF leakage	-15dBc
2LO -12 RF leakage	-69dBc
LO + 2IF spurious	-40dBc
LO + 3IF spurious	-33dBc
Driver Amplifier	
Power gain	12dB
Output P1dB	7dBm
Output third-order intercept point	24dBm
Input in-band return loss	-25dB
Output in-band return loss	-12dB

Table 2: Transmitter performance.

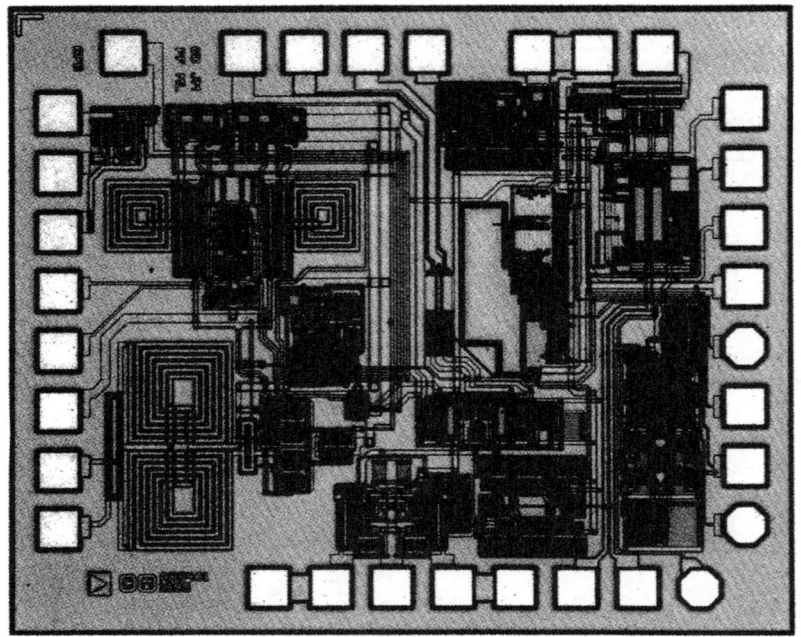

Figure 5: Chip micrograph.

SA 18.4: A 2.7V 2.5GHz Bipolar Chipset for Digital Wireless Communication

Stefan Heinen, Karim Hadjizada, Udo Matter, Werner Geppert, Volker Thomas[1], Stephan Weber[1], Stefan Beyer[1], Josef Fenk[1], Ernst Matschke[1]

Siemens AG, Microelectronics Design Ctr. Semiconductors, Düsseldorf, Germany
[1]Siemens AG, Semiconductor Group, RF IC Center, München, Germany

The chipset targets primarily the digital cordless european telecommunication (DECT) standard, that applies to business as well as to the consumer market. The transceiver chipset shown in Figure 1 for digital data transmission is realized on the 25GHz bipolar Siemens B6HF process. The transmitter IC integrates the complete synthesizer including PLL, VCOs, frequency doubler, supply voltage regulation and supporting switching functions. Due to the open-loop architecture, the input impedance variation of the switched power amplifier results in frequency deviation of the VCO caused by load pulling. The required isolation can not be achieved within a single IC package. Previous solutions use an external discrete buffer stage [2]. To increase the integration level, the buffer amplifier has been realized on the RX-IC Figure 2. The advantage of this configuration is increased isolation due to the fully-differential setup. The chipset architecture is suitable for FSK digital mobile radio portable and base stations, especially cordless phone system from 1.6GHz up to 2.5GHz including DECT, wireless LAN, wireless local loop, 2.4GHz ISM band frequency hopping system or any other system using FSK modulation. The TX design is optimized with respect to power supply, low current consumption, frequency stability and phase noise, whereas the RX-design is optimized for sensitivity and input intercept point simultaneously.

The DECT system is a TDMA system using GFSK modulation with 12 RX and 12 TX slots per frame. There are 10-channels available in the frequency band from 1880MHz up to 1900MHz. The transmitter architecture using the devices is given in Figure 1. A blind slot approach reduces the settling time requirements of the synthesizer. During the active slot, the synthesizer loop is opened, allowing direct modulation of the TX-VCO. In parallel the PA is switched on resulting in 200mVpp supply drop. The complete RF path of the chipset has been optimized to cover this operating condition. The VCOs are supplied with VCC4=2.4V generated by the on-chip low-drop regulator chain using external pnp BJTs. 95dB ripple rejection typically is achieved at 1kHz. In order to minimize load variation, the first two limiting buffer stages after the VCOs are also supplied by the regulated supply VCC4. The VCOs are Colpitts oscillators using external stripline or ceramic resonators. Two configurations are supported. The first using one VCO for TX and the other one for RX operation, to avoid a tuning range of about 150MHz. The other option is to use the built in bandswitch support for switching the electrical length of the external stripline resonator.

The analog chain from the VCOs via the frequency doubler to the output stage is based on the design already presented in Reference 2. Special care is taken to reduce noise injected by the bias circuitry. The 1/f noise and shot noise of the bias current will be converted into side bands of the carrier. The phase noise performance of the integrated VCOs within the first 2 adjacent channels does not cause any problem, because the DECT system specification are easily fulfilled [2]. DECT requires -131dBc/Hz maximum phase noise within the 3rd adjacent channel. Using a ceramic resonator a phase noise of at least -140dBc/Hz is achieved in the 3rd adjacent channel, 9dB better than the DECT requirement. The B6HF technology allows realization of a 32/33 prescaler with an operating frequency range up to 2.5GHz with 2.5mA current. This low value allows connection of the prescaler input to the doubler output. Compared to previous DECT solutions this results in 1728MHz phase detector reference frequency. From a system point of view the settling time requirement is easier to fulfill. A better signal to noise ratio is achieved. A crucial point of the open loop modulation scheme is to avoid frequency offset generated by switching off the charge pump. To avoid any effect on the loop filter voltage, the charge pump must be inactive at switching time (Figure 3). The charge pump is triggered by the rising edge of the phase detector input signal. In the DECT configuration the reference divider ratio is always an even number. Therefore the reference divider can be implemented to assure a 50% duty cycle at the input of the phase detector. In this configuration it is guaranteed that the charge pump is inactive at the falling edge of the reference divider output signal. If the open-loop mode is programmed via the 3-wire bus interface, the switching time is delayed until the next falling edge of the reference divider output, as indicated by the block diagram Figure 4. This technique results in low frequency offset with no transient due to open-loop switching as shown in Figure 5.

The output stage is an open collector. Using an external resistor the output power is adjustable up to 0dBm. The output current achieves a constant power with respect to supply voltage and temperature. The RX IC includes the double-balanced mixer, the TX-buffer stage and the complete IF circuitry. First measurements of the RX chain using an external discrete B6HF LNA BFP405 show -95dBm sensitivity at -17dBm input intercept point referred to the antenna. This is possible due to the 3dBm B6HF mixer intercept point and 11dB SSB noise figure at 12dB gain. The TX-buffer delivers an adjustable output power typically up to 5.5dBm, that is sufficient to drive the PA directly. The limiter and coincidence demodulator are optimized for low variation with respect to supply voltage and temperature. The demodulator characteristic given in Figure 6 shows no dependence on supply voltage in the operating range at 110MHz. The characteristic is clipped by the available supply voltage only at the upper limit. A fully integrated 4-pole baseband filter follows the demodulator in order to optimize the bit error rate. The sample and hold circuit compensates for dc offsets. This means generation of an optimum comparator threshold voltage. The chip set is optimized for maximum integration level with respect to DECT radio requirements. Therefore, the complete radio might be realized with about 120 components, if an integrated power amplifier is used.

References:

[1] Thomas, V., J. Fenk, S. Beyer, "A one-chip 2GHz Single Superhet Receiver for 2Mb/s FSK Radio Communication," ISSCC Digest of Technical Papers, pp. 42-43, Feb., 1994.

[2] Heinen, S., S. Beyer, J. Fenk, "A.3.0V 2GHz Transmitter for Digital Radio Communication with Integrated VCOs," ISSCC Digest of Technical Papers, pp. 146-147, Feb., 1995.

[3] ETSI, "Radio equipment and systems. Digital European Cordless Telecommunications Common Interface. Part 2: Physical Layer," DE/RES 300I-2, 1992.

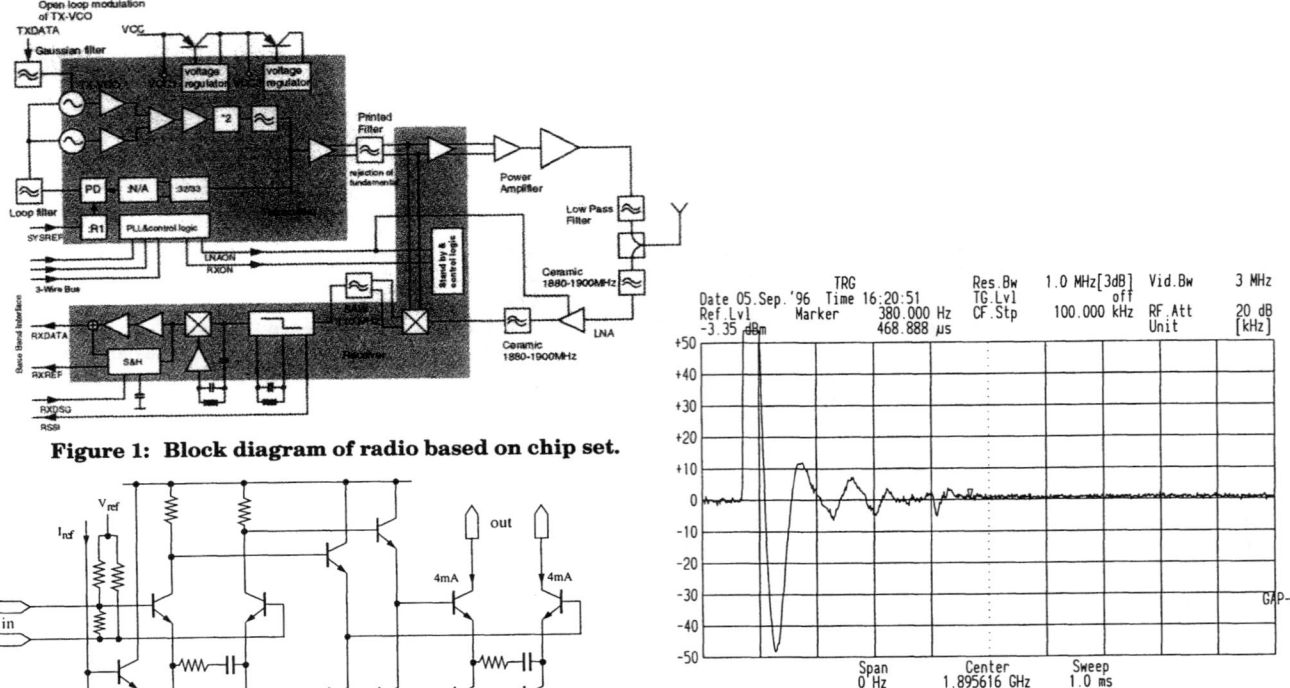

Figure 1: Block diagram of radio based on chip set.

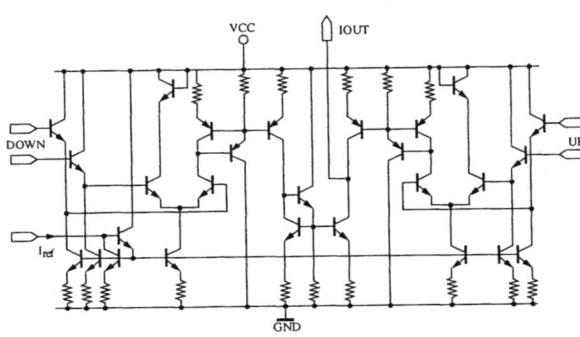

Figure 2: Transmitt buffer amplifier on RX-IC.

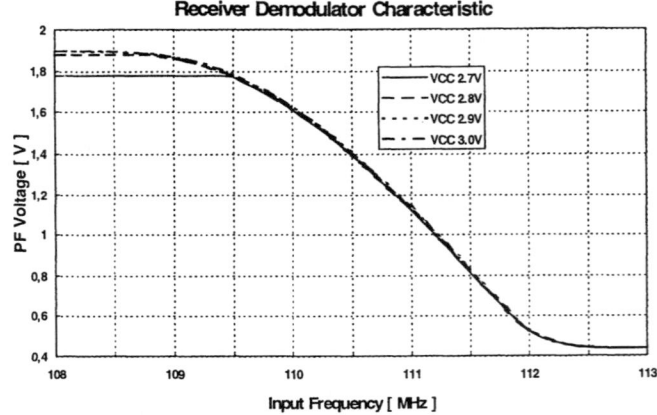

Figure 5: Frequency variation from open loop switching.

Figure 3: Charge pump.

Figure 6: Demodulator characteristic vs. supply voltage.

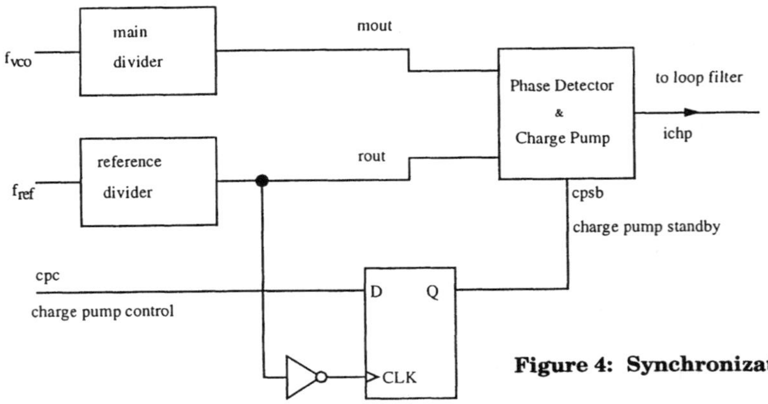

Figure 4: Synchronization of switching to open loop.

SA 18.2: A 2.7V GSM RF Transceiver IC

Kiyoshi Irie[1], Hiroaki Matsui[1], Takefumi Endo[2], Kazuo Watanabe[2],
Taizo Yamawaki[3], Masaru Kokubo[3], Julian Hildersley[4]

[1]Semiconductor Technology Development Center, SICD/[2]Analog Digital Mixed LSI Engineering Dept., Multi Purpose Semiconductor Products Operation, SICD/ [3]Central Research Lab., Hitachi, Ltd., Yokohama, Japan
[4]The Technology Partnership plc, Cambridge, UK

This 2.7V radio-frequency transceiver IC is intended for small, low-cost GSM handsets [1]. All that is required to implement a GSM terminal radio section is this IC, a power amplifier module, dual-synthesizer IC, SAW filters, and other peripheral discreet components. Figure 1 shows the system block diagram. This chip includes quadrature modulator phase-locked loop frequency translator with offset-mixer in the transmitter path, and a double-superhet receiver that consists of LNA with active-bias circuits, two Gilbert cell mixers, programmable gain linear amplifier, and quadrature demodulator. The circuit also contains frequency dividers with on-chip VHF VCO to simplify 2nd LO design. Power-control functions are provided for independent transmit and receive operation. The IC is implemented in pure bipolar technology with f_T=15GHz, r_{bb}'=150Ω, and 0.6μm features (0.6μm BiCMOS process). Table 1 summarizes the IC characteristics.

In transmit, this IC replaces the conventional up-conversion mixer with an offset phase-locked loop (OPLL) to eliminate RF filters and achieve low-cost GSM handsets. An OPLL is a PLL with a down-conversion mixer in the feedback path. The important difference between a synthesizer and the OPLL is that frequency modulation of the reference input is reproduced at the output without scaling.

A GMSK modulated signal is generated at a fixed 270MHz IF by a quadrature modulator. This signal is used as the reference input to the phase comparator in the OPLL that converts it to the final frequency in the 880-915MHz band. The modulator consists of a pair of Gilbert cell mixers, fed with quadrature phase (0°, 90°, 180°, and 270°) carriers derived from an integrated 540MHz oscillator with a frequency divider. To maintain good performance, the modulator uses transistors with 12 times basic emitter size to improve Vbe matching. The current output phase comparator allows the loop filter to be built from simple passive components and achieves the stable OPLL settling characteristics. Limiters in the inputs of the phase comparator keep its input level constant and a band-gap reference (BGR) circuit maintains its output level constant. This achieves less than 10% variation of loop bandwidth.

When RF filters are eliminated, one of the most significant problems is the noise transmitted in the GSM receive band because of the stringent GSM specification. To overcome the Tx noise problem, this IC uses the OPLL formation of a tracking band pass filter around the modulated signal. This is because the loop cannot respond to phase variations at the reference that are outside its closed loop bandwidth. The Tx noise is -170.1dBm/Hz when the loop bandwidth is 1.6MHz. This meets the GSM specification.

The spectrum from the OPLL output, with single-sideband modulation at 902MHz is shown in Figure 2. The I,Q modulation signal is a 67.7kHz continuous sinusoidal wave with 1Vpp level. The output from the Tx VCO, is about 0dBm into a 50Ω load. The carrier is -65.8dBc, the upper sideband (USB) is -42.4dBc, and the output IM3 is < -50dBc. When the I,Q inputs are GMSK modulation signals, 511b pseudo-random sequence (pn9), the phase trajectory errors are 1.8° (rms) and 4.1° (peak) at the Tx VCO output.

In receive, the IC includes an LNA in a 15GHz bipolar process. The gain and noise figure of the LNA is a significant factor in achieving the sensitivity level required for the GSM mobile phone system. Figure 3 shows schematic diagram of LNA. The matching network (Ns, NL) of LNA consists of external components to optimize operation for other high-frequency applications, for instance DCS 1800. To improve intermodulation characteristic, the LNA consists of one RF transistor to amplify (Q1: reduced $r_{bb'}$ type cell, 0.6μm * 10). In this IC, the emitter of Q1 is directly connected to GND for maximum performance by reducing parasitic capacitance, and the LNA has an active-bias circuit for Q1. The active-bias has a comparator that compares Vref to Vo (= Vcc - R1 * Icc) and outputs bias voltage Vb for the base of Q1. The evaluation results are shown in Figure 4, for a 3.0V supply. The LNA achieves 2.4dB noise figure and 13.2dB power gain at 50Ω load. Input 1dB compression point and input IP_3 are -9.0dBm and +4.1dBm, respectively. The LNA permits use of an external high-performance transistor to improve NF and gain.

The GSM system requires an adaptive gain amplifier (AGC) for suitable output for baseband. The IC has -30 to +50dB programmable linear-gain amplifier. Figure 5 shows measured gain versus control voltage (Cont. AGC). The linearity error is less than ±0.8dB at any 20dB window. To control the power gain over 80dB, the AGC has a linearizer that transforms control voltage into exponential bias current with thermal dependency provided by band-gap reference and amplified temperature circuits. By using this technique, the gain variation with temperature is kept within ±3dB for -20 to +85°C.

Signal line width and PAD location are laid out to isolate blocks, and to improve high-frequency characteristics. Particular attention is paid to reducing parasitic capacitance to GND. The design simulation includes package, PAD, and line parasitics to enhance modeling accuracy. A micrograph of IC is shown in Figure 6. The total supply current in transmit, receive, and idle (power saving) mode are 32mA, 48mA, and 1μA, respectively, at 3.0V. The 13mm^2 IC is in LQFP48 package.

Acknowledgments:

The authors appreciate marketing support of T. Hosoi, M. Guthrie, and technical cooperation of G. Aspin, and R. Annett and the contributions of all participating in this project.

Reference:

[1] Stetzler, T., et al., "A 2.7V to 4.5V Single-Chip GSM Transceiver RF Integrated Circuit," ISSCC Digest of Technical Papers, pp.150-151, Feb., 1995.

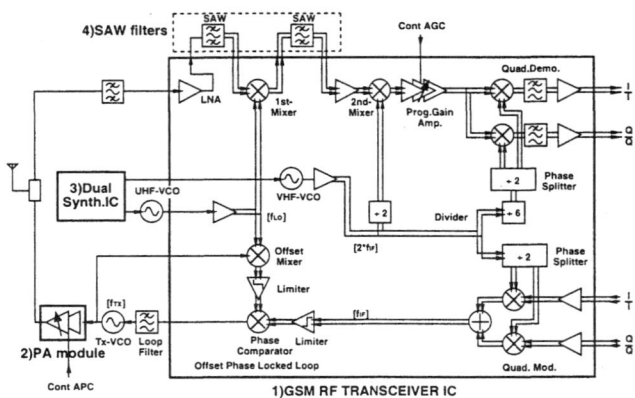

Figure 1: System block diagram.

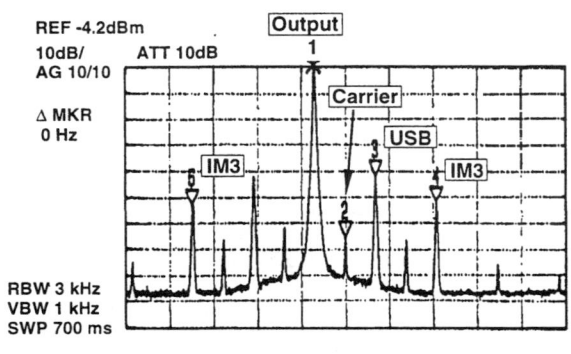

Figure 2: Transmit-output spectrum.

Multi Marker List

No.2:	67.7kHz	-65.84 dBc
No.3:	135.4 kHz	-42.38 dBc
No.4:	270.8 kHz	-52.47 dBc

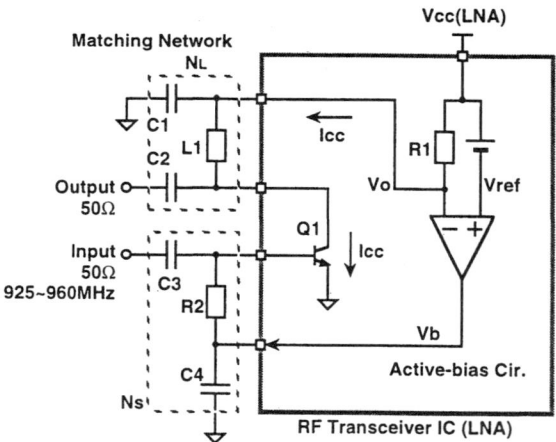

Figure 3: LNA schematic diagram.

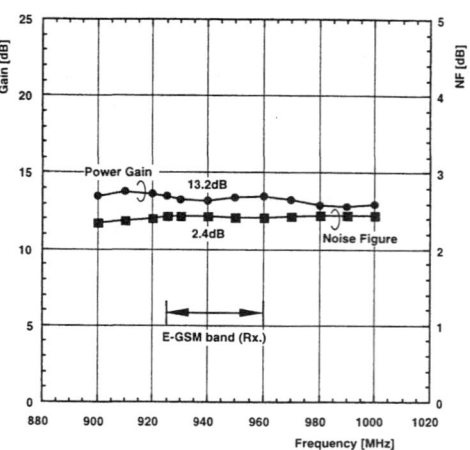

Figure 4: Measured LNA gain and NF.

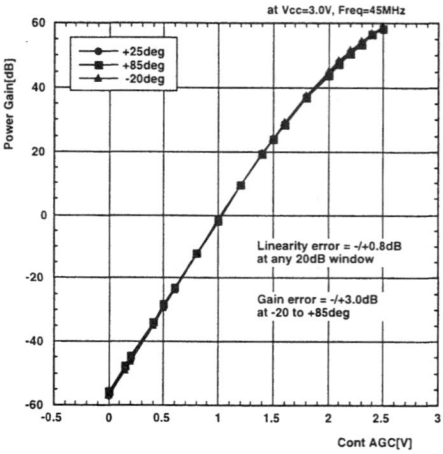

Figure 5: Measured linearity of programmable gain amp.
Figure 6: See page 475.

Process technology	15GHz, 0.6μm double-poly-Si Bipolar
Package	LQFP-48 (Low profile QFP)
Operating frequency range	Rx.: 925-960MHz
	1st-IF: 90~300MHz, 2nd-IF: 26~60MHz
	Tx.: 880~915MHz
	IF: 45~350MHz, IF-VCO: 90~700MHz
Operating power supply voltage	Phase comp. and Tx. VCO: 2.7~5.5V
	Other blocks: 2.7~3.6V
Operating temperature range	-20~+75°C
Current	Rx. mode: 48mA typ.
(including power saving control)	Tx. mode: 32mA typ.
	Idle mode: 1μA max.
TX. Offset PLL	Tx. noise=-170.1dBm/Hz (Δ20MHz)
(offset-mixer, limiter,	Carrier=-65.8dBc
phase comparator)	USB=-42.4dBc
with Quad. modulator	IM3<50dBc
	Lock up time=9μs
	Phase error=1.8°(rms)
Rx. LNA	G=13.2dB
	NF=2.4dB
1st mixer	CG=6.7dB at(400Ω+400Ω),
	differential load
	SSB NF=9.6dB
	Input IP3=-0.2dBm
2nd mixer	CG=16.0dB
Prog. gain amp.	Gain control range=-30dB to +50dB
	Linearity=±0.8dB (any 20dB window)
Quad. demodulator	on-chip LPF
	dc offset=30mV
Other frequency divider	Operating frequency=~600MHz
/Phase splitter	Function= ÷2, ÷6,×2*,×2*(*:90° shift)
VHF VCO	Operating frequency=~700MHz
Power control function	3b

Table 1: IC characteristics.

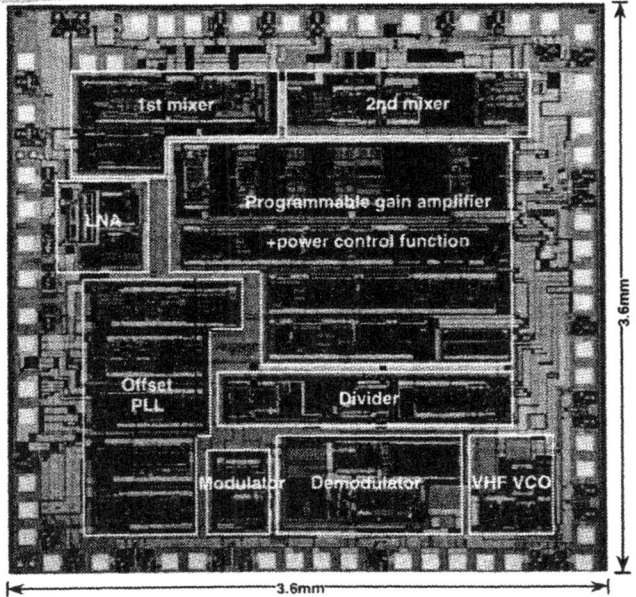

Figure 6: IC micrograph.

5

POWER AMPLIFIERS AND RF SWITCHES

THE power amplifier in a typical transceiver is a major factor in determining battery power drain and thus, the crucial operating time between battery recharges. A class B power amplifier has a maximum theoretical power-added efficiency of p/4 or 78% [1]. Practical power amplifiers can achieve efficiencies of the order of 65%, meaning that almost two-thirds of the battery power consumed by the stage is turned into useful RF power to the antenna and only one-third is wasted as lost heat in the circuit. For transmitter output powers in the 1-W range, every percentage point improvement in efficiency obviously is crucial and of great concern to the circuit and system designers. Major challenges of power amplifier design, apart from achieving high efficiency, include protecting the circuit against overvoltage damage if the load is removed and matching the extremely low output impedance of the power device (often less than 10 W at 1 GHz) to the external 50-W load presented by the antenna. The achievement of high efficiency in a power amplifier requires that the output voltage on the power device swing down to a low value when the device current pulses to a maximum; this downswing in turn requires devices whose frequency response remains high under these conditions. Power amplifier performance, like LNA performance, thus depends heavily on the parameters of the process used to fabricate the circuit. These and other issues related to RF power amplifier design are treated in this selection of papers.

REFERENCE

[1] P. R. Gray and R. G. Meyer, *Analysis and Design of Analog Integrated Circuits,* 3rd ed., New York: John Wiley & Sons, 1993. Ch. 5.

A Theoretical Analysis and Experimental Confirmation of the Optimally Loaded and Overdriven RF Power Amplifier

DAVID M. SNIDER

Abstract—Although the conventional Class B approach to RF amplifier design yields high output power and reasonable collector efficiency (78.5 percent at maximum output power), neither the power nor the efficiency are optimum, and both are dependent on RF drive level. This paper presents an analysis of appropriately selected collector voltage and current waveforms which determine the load impedance at the fundamental and harmonically related frequencies; these conditions define the Class B "optimum efficiency" case with 100 percent collector efficiency and 1.27 times the conventional Class B value of output power. If the RF drive level is increased, and the collector voltage and current waveforms are appropriately selected so that the amplifier is overdriven, a different load impedance is determined; these conditions define the "optimum power" case with 1.46 times the conventional Class B value of output power and 88 percent collector efficiency. The "optimum power" case has the added advantage that the output power and collector efficiency are essentially constant over a predetermined range of drive level.

Finally, the theory is verified by the construction and testing of a UHF power amplifier having a power output of 46 watts and an overall dc to RF conversion efficiency of 65 percent with a 1 dB for 10.5-dB insensitivity of output power to RF drive.

I. Introduction

AN IMPROVEMENT in both collector efficiency and output power over that described by conventional Class B[1] considerations can be realized if the load impedance at the fundamental and harmonically related frequencies presented to the output terminals of a tuned power amplifier stage are appropriately selected. Theoretical collector efficiencies of 100 percent at 1.27 times the conventional value of output power are possible. Furthermore, if the amplifier is overdriven, a different load impedance can be derived so that 1.46 times the conventional value of output power can be achieved with 88 percent collector efficiency. This technique has the added advantage that the output power and collector efficiency are essentially constant over a range of drive level.

To gain an intuitive understanding of the procedure, consider the textbook Class B waveforms shown in Fig. 1, where $I_c(\theta)$ and $V_c(\theta)$ have the arbitrary peak values I_s and V_{cc}, respectively, and θ is in radians. Here, $2V_{cc}$ does not necessarily equal the device breakdown voltage (V_{br}) and I_s does not necessarily equal the device saturation current (I_{sat}). That is to say, the constraints on output power are not those imposed physically by the device (V_{br}, I_{sat}), but by the yet-to-be-designed external circuit. Let the collector current waveform $I_c(\theta)$ remain unchanged, but select some new collector voltage waveform $V_c(\theta)$ so that the fundamental component of the voltage waveform is greater than the classical Class B value. If $V_c(\theta)$ is symmetrical about V_{cc}, then the dc input power is the same as the Class B case, but the fundamental output power is increased. In particular, if $V_c(\theta)$ is allowed to approach a square wave symmetrical about V_{cc} of peak value V_{cc}, it will be shown that in the limit the collector efficiency approaches 100 percent. Since the collector voltage and current waveforms have been specified, the load impedance is determined. This is called the "optimum efficiency" case.

Next, the RF drive will be increased by some amount forcing $V_c(\theta)$ and $I_c(\theta)$ to be as shown in Fig. 2, determining a different load impedance. Note that the dc

---

Manuscript received January 18, 1967; revised June 6, 1967.
The author is with the Massachusetts Institute of Technology, Lincoln Laboratory, Lexington, Mass. (Operated with support from the U. S. Air Force.)
[1] T. S. Gray, *Applied Electronics*. New York: Wiley, 1957, pp. 403–404.

input power has increased, since I_{dc} of this new $I_c(\theta)$ is greater than the textbook case. However, the collector efficiency increases faster than the dc input power. It will be shown that if the RF drive is increased by 5.2 dB, an output power of 1.47 times the conventional Class B value is developed at 88 percent collector efficiency.

For the following analysis, an ideal device is assumed so that $V_{ce\,sat}$ equals 0 volts, h_{fe} = constant, and so forth.

II. THE OPTIMUM EFFICIENCY CLASS B TUNED POWER AMPLIFIER

The collector voltage and current waveforms for optimum efficiency Class B operation are shown in Fig. 3. Here the dc input power is held constant and only $V_c(\theta)$ is allowed to change. A Fourier analysis[2] must be made of $I_c(\theta)$ and $V_c(\theta)$ to find the magnitudes and signs of the sine and cosinusoidal terms. The coefficients of the Fourier expansion of $I_c(\theta)$ are

$$I_{A0} = \frac{I_s}{\pi} \quad (1)$$

where I_{A0} is the dc component of the collector current waveform and

$$I_{A1} = 0 \quad (2)$$

$$I_{AN} = \begin{cases} \frac{I_s}{\pi}\left[\frac{1}{1+n} + \frac{1}{1-n}\right], & n \text{ even} \quad (3) \\ 0, & n \text{ odd} \quad (4) \end{cases}$$

where I_{AN} is the peak value of the nth harmonic cosinusoidal term of the Fourier expansion of $I_c(\theta)$. Also

$$I_{B1} = \frac{I_s}{2} \quad (5)$$

where I_{B1} is the peak value of the fundamental component and

$$I_{BN} = 0 \quad (6)$$

where I_{BN} is the peak value of the nth harmonic sinusoidal term of the Fourier expansion of $I_c(\theta)$. The coefficients of the expansion of $V_c(\theta)$ are

$$V_{A0} = V_{cc} \quad (7)$$

where V_{A0} is the dc component of the collector voltage waveform and

$$V_{AN} = 0, \quad \text{ALL } n \neq 0 \quad (8)$$

$$V_{B1} = V_{cc}\left[\frac{2K\theta_1}{\pi} - \frac{K\sin 2\theta_1}{\pi} + \frac{4\cos\theta_1}{\pi}\right] \quad (9)$$

where V_{B1} is the peak value of the fundamental component. Also

[2] F. B. Hildebrand, *Advanced Calculus for Applications*. New Jersey: Prentice-Hall, 1965, pp. 221–226.

$$V_{BN}(\text{odd } n) = V_{cc}\left[\frac{2K\sin(\theta_1 - n\theta_1)}{\pi} - \frac{2K\sin(\theta_1 + n\theta_1)}{\pi} - \frac{4\cos n\theta_1}{\pi}\right] \quad (10)$$

$$V_{BN}(\text{even } n) = 0 \quad (11)$$

where V_{BN} is the peak value of the nth harmonic sinusoidal term of the Fourier expansion of $V_c(\theta)$. Noticing that

$$K = 1/\sin\theta_1, \quad (12)$$

as θ_1 approaches zero $V_c(\theta)$ approaches a square wave and the values of the fundamental components of the collector current and voltage are

$$I_{B1} = I_s/2 \quad (13)$$

$$V_{B1} = \frac{4V_{cc}}{\pi} \quad (14)$$

giving an output power at the fundamental frequency of

$$P_{\text{out}}(\text{RF}) = \frac{V_{cc}I_s}{\pi}. \quad (15)$$

However, the dc input power is

$$P_{\text{in}}(\text{dc}) = \frac{V_{cc}I_s}{\pi}. \quad (16)$$

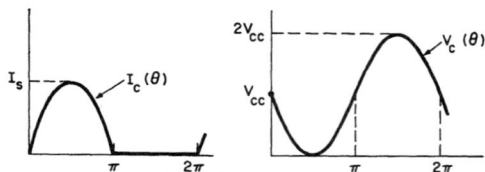

Fig. 1. Class B collector current and voltage waveforms.

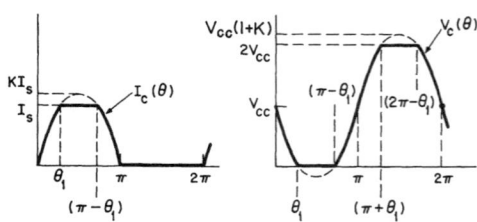

Fig. 2. Overdriven Class B collector current and voltage waveforms

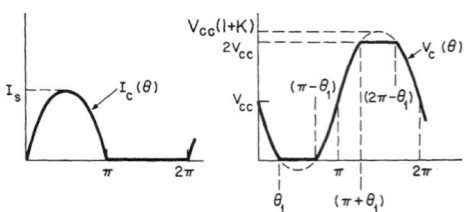

Fig. 3. Optimum efficiency Class B collector current and voltage waveforms.

Therefore, the theoretical collector efficiency approaches 100 percent as $V_c(\theta)$ approaches a square wave. The impedance conditions at the collector to common terminals (output) are necessarily

$$Z_1 = \frac{8}{\pi} \frac{V_{cc}}{I_s} \quad (17)$$

so that the fundamental load impedance is all real and

$$Z_n = \frac{0}{I_{AN}} = 0, \quad n \text{ even} \quad (18)$$

$$Z_n = \frac{V_{BN}}{0} = \infty, \quad n \text{ odd} \quad (19)$$

so that for 100 percent idealized collector efficiency, the following conditions must be fulfilled:

1) The load impedance must be a perfect short circuit at even harmonics and a perfect open circuit at odd harmonics.
2) The idealized model unit must be valid.
3) $Z_1 = R_1 = \dfrac{8}{\pi} \dfrac{V_{cc}}{I_s}$

$Z_n = 0\,\Omega, \quad \text{even } n$

$Z_n = \infty\,\Omega, \quad \text{odd } n$.

Any deviation from these conditions will detract from this ideal.

Finally, the definition of R_L for the classical Class B case is

$$R_L = \frac{2V_{cc}}{I_s} \quad (20)$$

therefore

$$Z_1 = R_1 = \frac{4}{\pi} R_L. \quad (21)$$

One can compare these results with conventional Class C operation. The design approach of this paper gives 100 percent collector efficiency with an output power which is higher than conventional Class B. In conventional Class C operation, 100 percent collector efficiency occurs at zero output power.

III. The Optimum Power Class B Tuned Power Amplifier

For the optimum power, or overdriven, case both the dc input power and the collector waveforms are allowed to change. However, for a fair comparison of Class B, optimum efficiency Class B, and the overdriven case, the peak values of $V_c(\theta)$ and $I_c(\theta)$ must not exceed V_{cc} and I_s, respectively. The waveforms for this mode of operation have been shown in Fig. 2. The coefficients of the Fourier expansion of $V_c(\theta)$ have already been derived in (8) through (11), and are obviously a function of θ_1 or K (overdrive). The coefficients of the expansion of $I_c(\theta)$ that apply to this mode of operation have been derived in a similar fashion and are

$$I_{DC} = I_s \left[\frac{(\pi - 2\theta_1)}{2\pi} + \frac{K}{\pi} - \frac{K \cos \theta_1}{\pi} \right] \quad (22)$$

$$I_{AN} = \begin{cases} 0, & n \text{ odd} \\ f(K, \theta_1), & n \text{ even} \end{cases} \quad (23)$$

$$I_{B1} = \frac{I_s}{2} \left[\frac{2K\theta_1}{\pi} - \frac{K \sin 2\theta_1}{\pi} - \frac{4 \cos \theta_1}{\pi} \right] \quad (24)$$

$$I_{BN}(\text{even } n) = 0 \quad (25)$$

$$I_{BN}(\text{odd } n) = \frac{I_s}{2} \left[\frac{2K \sin(\theta_1 - n\theta_1)}{\pi(1-n)} \right.$$
$$\left. - \frac{2K \sin(\theta_1 + n\theta_1)}{\pi(1+n)} + \frac{4 \cos n\theta_1}{n\pi} \right]. \quad (26)$$

The function $f(K, \theta_1)$ can be evaluated but is not needed for this analysis. The expression for RF output power at the fundamental frequency as a function of θ_1 is therefore

$$P_{\text{out}}(\text{RF}) = \frac{V_{cc} I_s}{4\pi^2} [2K\theta_1 - K \sin 2\theta_1 + 4 \cos \theta_1]^2. \quad (27)$$

The dc input power is also a function of θ_1 so that

$$P_{\text{in}}(\text{dc}) = \frac{V_{cc} I_s}{\pi} \left[\frac{(\pi - 2\theta_1)}{2} + K - K \cos \theta_1 \right] \quad (28)$$

and the out-of-band impedances are simply

$$Z_n = \frac{2V_{cc}}{I_s} = R_L, \quad \text{odd } n \quad (29)$$

$$Z_n = 0, \quad \text{even } n. \quad (30)$$

Since a resistive load R_L must be presented to the device at the odd harmonic frequencies, power will be dissipated at these frequencies so that

$$P_{\text{out}}(n)_{\text{odd } n} = \frac{V_{cc} I_s}{\pi^2} \left[\frac{K \sin(\theta_1 - n\theta_1)}{(1-n)} - \frac{K \sin(\theta_1 + n\theta_1)}{(1+n)} \right.$$
$$\left. + \frac{2 \cos n\theta_1}{n\pi} \right]^2. \quad (31)$$

Although the output power obviously continues to increase with increasing K, the collector efficiency does not. Therefore, the "optimum output power" is defined as that power which corresponds to maximum collector efficiency. As mentioned before, that power is 1.47 times the classical Class B value and occurs when $K = 5.2$ dB, corresponding to a collector efficiency of 88 percent.

The reader should notice that the optimum efficiency design approach is not a special case of the optimum power approach, although the derivations are similar, since the application arises in different situations. If RF drive power, V_{cc}, and other parameters are all reasonably constant, and the only design criterion is overall efficiency, then the optimum efficiency approach is implied.

If some protection against changes in these system parameters is desired, as in remote or satellite applications, then the optimum power approach is implied.

If the power gain, G_0, of any device is defined as that gain which occurs when operated according to classical Class B considerations with an output power P_0, then the new effective gain, G_{eff}, in the overdriven case is

$$G_{eff} = \frac{P_{out}}{K^2 \frac{P_0}{G_0}}$$

$$= \frac{G_0}{K^2}\left[\frac{2K\theta_1}{\pi} - \frac{K \sin 2\theta_1}{\pi} + \frac{4\cos\theta_1}{\pi}\right]^2. \quad (32)$$

The amplifier is overdriven according to Fig. 2 due to the choice of Z_n to achieve an increase in power and efficiency over that indicated in the conventional Class B design. The insensitivity due to overdrive is not that of a "hard limiter," but is characterized by (32). Hence, a theoretical insensitivity of RF output power of 0.5 dB for a change in input drive power of 5.2 dB per stage is possible.

A computer program was written to solve equations (27), (28), (31), and (32) for values of K ranging from 0 to 20 dB (power) in 0.1-dB steps. The results are discussed in Section V.

IV. Modified Class A Operation

If the same technique is applied to a tuned amplifier initially biased Class A, with the stipulation that the quiescent operating point must not change from the values determined by classical Class A[1] definitions, then an improvement in output power is possible. Since the dc input power is fixed, maximum efficiency occurs with maximum output power. According to the waveform analysis already completed, if the drive power to a Class A amplifier is fixed, and only the load impedance presented to the output terminals (collector to common) is appropriately selected so that $V_c(\theta)$ is allowed to approach a square wave symmetrical about V_c [Fig. 4(a) and (b)] then it is easily shown that the improvement in output power from the textbook value of $V_c I_c / 2$ is

$$P_{out}(\text{RF}) = V_c I_c \left(\frac{2}{\pi}\right) \quad (33)$$

if

$$Z(1) = \frac{V_c}{I_c}\left(\frac{4}{\pi}\right) \quad (34)$$

and

$$Z_n = \infty \quad n \text{ odd}. \quad (35)$$

For the more interesting case (Fig. 5), where the RF drive level is also allowed to vary (the overdriven case), we have already performed the analysis necessary to write the design equations.

$$P_{out}(\text{RF}) = \frac{V_c I_c}{2}\left[\frac{2K\theta_1}{\pi} - \frac{K\sin 2\theta_1}{\pi} + \frac{4\cos\theta_1}{\pi}\right]^2 \quad (36)$$

$$Z_1 = \frac{V_{B1}}{I_{B1}} = R_L = \frac{V_c}{I_c} \quad (37)$$

$$Z_n = R_z, \quad n \text{ odd}. \quad (38)$$

The reader can be misled into believing that the improved Class A efficiency comes from using some nonlinear characteristic of the amplifier. This is not so; all of the impedances in this device model are linear, although frequency dependent. "Class A" is used here in the sense that the operating point in the $I_c - E_c$ characteristic is stable. This definition differs from those definitions which assume that the "Class A" amplifier generates no harmonics.

It should be noticed that, unlike the overdriven Class B case, here both the output power and collector efficiency continue to increase with overdrive (K) so that in the limit as

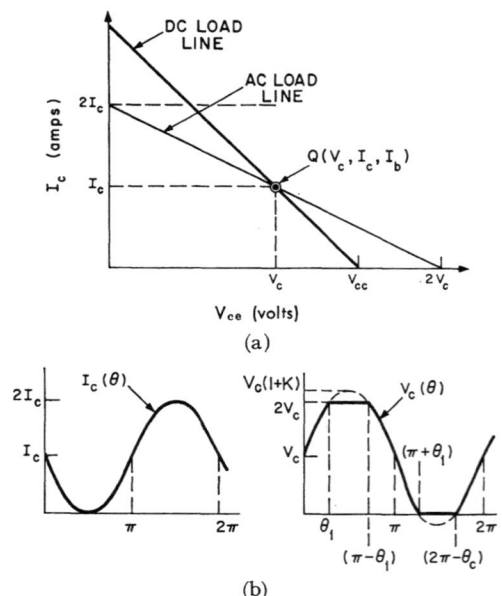

Fig. 4. (a) Class A dc and ac load line and quiescent operating point. (b) Class A collector current and voltage waveforms for optimal loading only.

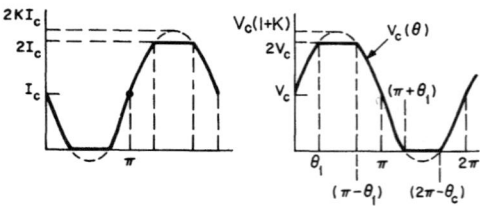

Fig. 5. Optimum efficiency and optimum power Class A collector current and voltage waveforms.

$$K \to \infty, \qquad \theta_1 \to 0$$

$$P_{\text{out}}(\text{RF}) = \frac{V_c I_c}{2} \left[\frac{2}{\pi} - \frac{2}{\pi} + \frac{4}{\pi} \right]^2 = V_c I_c \left(\frac{8}{\pi^2} \right). \quad (39)$$

What is of interest, then, is where the overall efficiency of any stage is maximum for a given G_0. Here, G_0 is defined as that value of power gain which occurs when the stage is operated according to conventional Class A considerations. A computer program was written to solve for $P_{\text{out}}(\text{RF})$, $P_{\text{out}}(n)$, G_{eff}, collector efficiency (n_c), and overall stage efficiency (n_t) for values of K ranging from 0 to 20 dB in 0.1-dB steps. These results are discussed in Section V.

V. Computed Data and Design Procedure

To best explain the design procedure and computed curves of various amplifier stage characteristics for different values of K (Figs. 6 to 12), a typical transmitter design goal is specified.

RF output power = $P_{\text{out}}(\text{RF})$ = 10 W

RF input power = $P_{\text{in}}(\text{RF})$ = -10 dBm

Output frequency = 250 MHz

Many devices are available which are capable of delivering 10 watts at 250 MHz, and a typical value for G_0 at this level is 8 dB. K can now be determined for the overdriven Class B case from Fig. 6.

$$G_0 = 8 \text{ dB}$$

$$n_t = 73.6 \text{ percent}$$
$$K = 1.8 \text{ dB}$$

From Fig. 7

$$P_{\text{out}} = n_c (P_{\text{in}} \text{ dc})$$
$$P_{\text{in}} \text{ dc} = 10W/0.85 = 11.65 \text{ W}$$

From Fig. 8

$$G_{\text{eff}} = 7.14 \text{ dB}, \quad K = 1.8 \text{ dB}$$

If we use this procedure down to a low RF drive level (say, 100 mW) where there is insufficient drive power for Class B operation, the overall dc to RF conversion efficiency so far will be approximately 70 percent. The RF input/output and efficiency characteristics for the power amplifier section of the transmitter can be calculated by using Figs. 6 to 9. The power dissipated at the third, fifth, and seventh harmonic frequency can be read directly from Fig. 13.

The curves shown in Figs. 8, 10, 11, and 12 can be used in a similar way to design an overdriven Class A preamplifier section. A typical value for G_0 at these levels is 11 dB.

VI. Experimental Verification of Theory

A power amplifier capable of delivering 45 watts at approximately 250 MHz into a 50-ohm load was designed, constructed, and tested for the space environ-

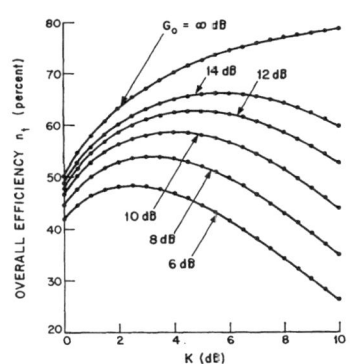

Fig. 6. Overdriven Class B operation. Overall efficiency versus K for different values of G_0.

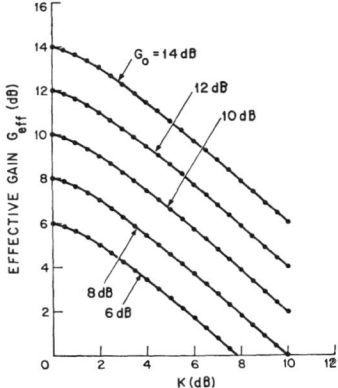

Fig. 8. Overdriven Class A or Class B operation. Effective gain versus K for different values of G_0.

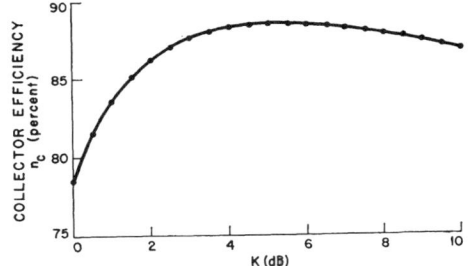

Fig. 7. Overdriven Class B operation. Collector efficiency versus K.

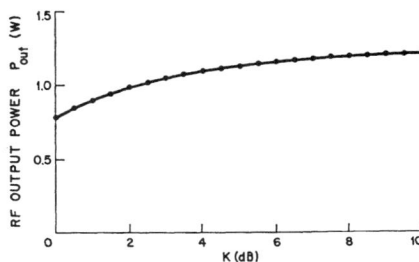

Fig. 9. Overdriven Class B operation. Output power versus K.

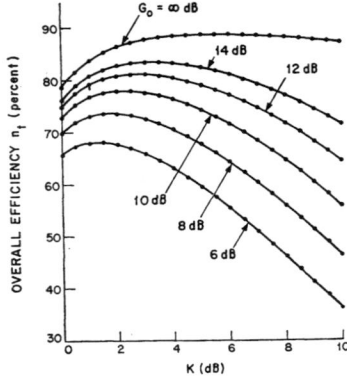

Fig. 10. Overdriven Class A operation. Overall efficiency versus K for different values of G_0.

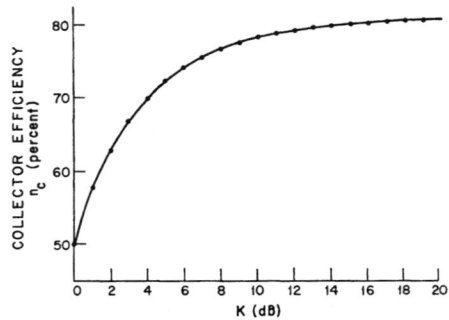

Fig. 11. Overdriven Class A operation. Collector efficiency versus K.

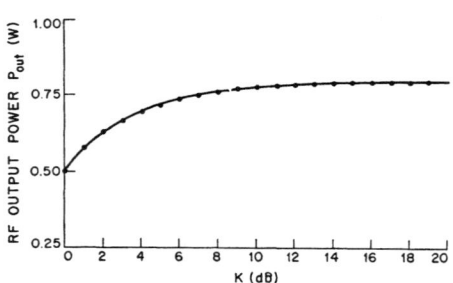

Fig. 12. Overdriven Class A operation. Output power versus K.

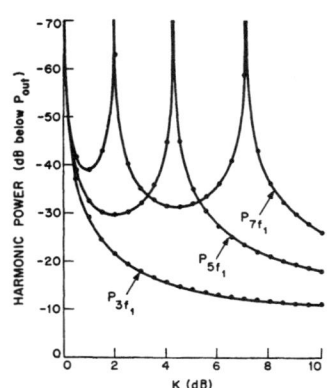

Fig. 13. Overdriven Class A or Class B. Harmonic output power versus K.

ment, according to the theory developed. Four 11.50-watt power amplifiers were paralleled in the output stage, using 3-dB hybrids to develop this power reliably with commercially available devices. At center frequency, the air-line hybrids have an insertion loss beyond 3 dB of approximately 0.07 dB.

An output stage amplifier schematic diagram is shown in Fig. 14. Since the input impedance is very low (typically $0.7 + j3$) compared to the effective value of the collector to base capacity C_{ob}, C_{ob} is effectively from collector to ground. Therefore, at the operating frequency, C_{ob} cannot be neglected and is by design a component of the output matching network. The effective value of C_{ob} is approximately twice the minimum value. At the second harmonic, the design equations call for a short circuit from collector to ground. However, at this frequency, $X_{C_{ob}}$ is essentially an RF short circuit, so that an external short circuit need not be added. For this design, the higher order impedance terms were neglected. A comparison of the theoretical and measured performance of the amplifier is given in Fig. 15(a).

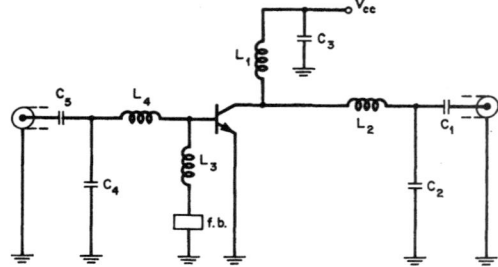

Fig. 14. Output stage schematic diagram.

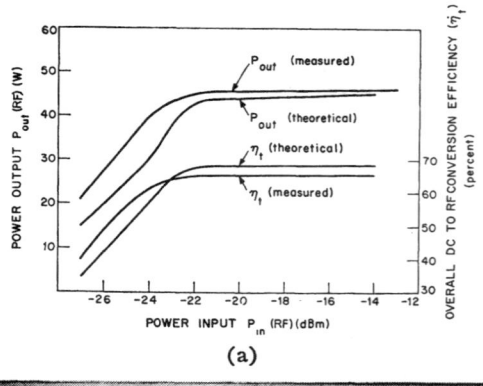

(a)

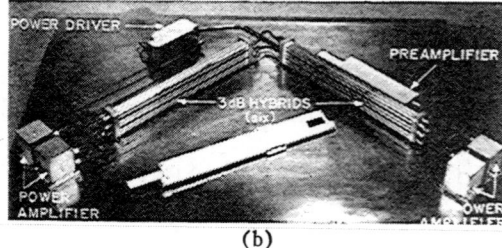

(b)

Fig. 15. (a) Fundamental output power and overall efficiency versus RF drive (theoretical and measured curves). (b) 45 W, 250 MHz amplifier and preamplifier.

VII. Summary and Conclusion

An improvement in both efficiency and output power beyond the values described by classical definitions can be realized by simply controlling the in- and out-of-band load impedances presented to a device originally biased for Class A or Class B operation. The effect on output power of variations of RF drive level with time, temperature, aging, and radiation damage can be significantly reduced by controlling the harmonically related load impedances and overdriving. Finally, the theory has been verified by the construction and testing of a 46-watt, 250-MHz amplifier, which exhibited an insensitivity of output power to RF drive level within 1.2 dB of the calculated theoretical value, and an overall dc to RF conversion efficiency within 3 percent of the theoretical value.

A complete tabulation of the calculated design data and computer program is available.[3]

Acknowledgment

The author gratefully acknowledges the many useful discussions with Dr. A. I. Grayzel and L. Hoffman of the M.I.T. Lincoln Laboratory. Finally, he wishes to thank R. E. Dolbec of the Lincoln Laboratory for his assistance in the fabrication and testing of the experimental amplifier.

[3] D. M. Snider, "A theoretical analysis and experimental confirmation of the optimally loaded and overdriven rf power amplifier," Tech. Rept. 428, M.I.T. Lincoln Laboratory DDC 647799, November 7, 1966.

Correction

The abstract below replaces the one printed with the paper, "The Physical Mechanism of the Chargistor in Terms of Minority Carrier Exclusion and Injection," by T. F. Shao and L. P. Hunter, which appeared on pp. 306–313 of the June, 1967, issue of this TRANSACTIONS.

Abstract—The chargistor, invented by Yu,[1] is here found to consist of a somewhat different structure. Its mechanism of operation can be given in terms of both minority carrier exclusion and high-level injection rather than in terms of space-limited currents as suggested by Yu. Three types of electrical behavior are distinguished. Characteristic curves, potential plots, and transconductance data are given.

INVITED PAPER Special Issue on Mobile Communications

High Efficiency Transmitting Power Amplifiers for Portable Radio Units

Toshio NOJIMA[†], Sadayuki NISHIKI[†] and Kohji CHIBA[†], *Members*

SUMMARY High efficiency amplifier construction techniques are investigated focusing on UHF band transmitting power amplifiers intended for cellular portable telephones and the state of the art amplifiers are presented. First, it is shown that high efficiency amplifiers are indispensable to attain pocket sized portable units through a theoretical analysis using a simple model. When about 1 W of transmitting power is required, it is desirable for the amplifier to operate with an efficiency of over 40%. Secondly, the switching mode scheme is described as the most effective technical means to achieve high amplifier efficiency. State of the art switching mode amplifiers, the Harmonic Reaction Amplifier (HRA) and the Linearized Saturation Amplifier with Bidirectional Control (LSA-BC), are presented as examples of nonlinear and linear amplifiers respectively. Basic operation mechanisms are shown. Experimental HRA and LSA-BC are constructed to determine their practically attainable efficiencies. Power-added efficiencies of 75% and 40% are recorded from a 1.7 GHz band 3 W HRA for CW and a 1.5 GHz band 1 W LSA-BC for $\pi/4$ QPSK respectively. These values indicate that these types of amplifier can be applied to pocket sized portable radio units.

1. Introduction

Decreasing the transmitting power and power consumption of each circuit is the most effective way of achieving small sized and light weight portable radio units for mobile telephone systems (referred to as portable units hereafter). This is because the battery is the largest device in any portable unit and decreases in the total power consumption yield corresponding battery volume and weight reductions. With respect to the transmitting power, about 1 W is the maximum transmitting power in the current Japanese NTT mobile telephone system[1]. Likewise, it seems that not much more than 1 W will be required for the digital mobile system under development[2], although smaller powers are preferable.

A typical circuit configuration of the transmitter section used in the current system is shown in Fig. 1. A transmitting power amplifier comprised of two or three transistors yields about 30 dB of amplification gain. Since there is a typical transmission loss of about 2.5 dB at the isolator and duplexer[1], a maximum output power of about 1.8 W is required from the power amplifier to obtain 1 W transmitting power. On the other hand, total maximum power consumption of the portable unit excluding the power amplifier should be less than about 1 W. This means that most of the power is used by the power amplifier during transmissions and, consequently, high efficiency power amplifiers are indispensable to decrease total power consumption. In other words, amplifier construction techniques for highly efficient operation are needed to miniaturize portable units.

This paper investigates the circuit construction techniques that can realize high efficiency amplifiers for portable units. First, relations between efficiencies and portable unit sizes are roughly analyzed. Second, it is shown that the switching mode scheme is the key to achieve high efficiency and its operation mechanism is shown. The HRA and the LSA-BC are presented as the best practical amplifiers utilizing the switching mode scheme. The former is a saturation amplifier while the latter is a linear one. Their basic operation principles are shown. Finally, experiments are performed to investigate their maximum practical performance. As a result, it is confirmed that these amplifier construction techniques yield the high efficiencies required by pocket sized portable units.

2. Power Amplifier Efficiency Required for Portable Units

The power conversion efficiency η of an amplifier from DC supply to RF output is generally defined as

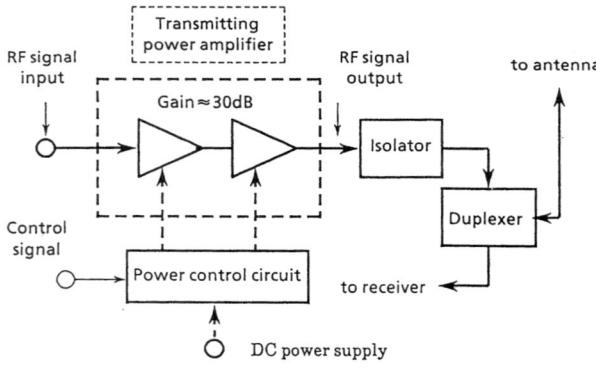

Fig. 1 Typical circuit configuration of RF transmitter section used in cellular portable telephones.

Manuscript received February 5, 1991.
† The authors are with NTT Radio Communication Systems Laboratories, Yokosuka-shi, 238-03 Japan.

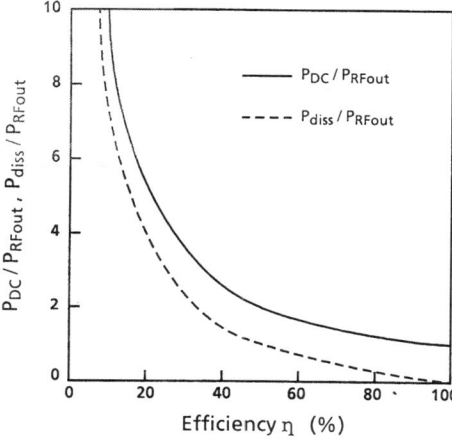

Fig. 2 P_{DC}/P_{RFout}, P_{diss}/P_{RFout} vs. efficiency η.

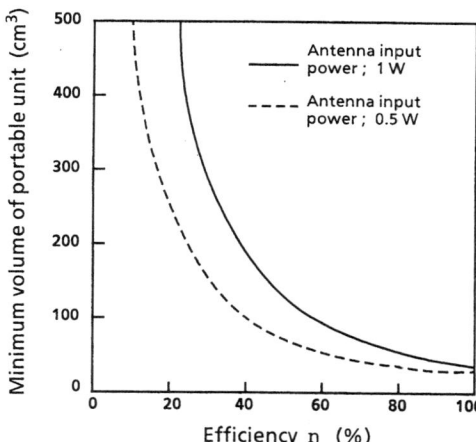

Fig. 3 Portable unit volume vs. efficiency η (ambient temperature; 25°C, temperature increase; below 15°C, case shape; cubic).

$$\eta = (P_{RFout}/P_{DC}) \times 100 \, \% \quad (1)$$

where P_{RFout} and P_{DC} are the RF output power and DC power supply of the amplifier respectively. If the power gain of the amplifier is large enough to neglect the RF input power, $P_{DC} - P_{RFout}$ is the internal power dissipation. If the normalized internal power dissipation is referred to as P_{diss}/P_{RFout}, its relation to efficiency η is given as

$$P_{diss}/P_{RFout} = (P_{DC} - P_{RFout})/P_{RFout}$$
$$= 100/\eta - 1. \quad (2)$$

Figure 2 shows the relation between the normalized DC power supply P_{DC}/P_{RFout} and amplifier efficiency η, together with the relation between the normalized power dissipation P_{diss}/P_{RFout} and η. For example, the DC power supply can be reduced by 66% if a 20% efficiency amplifier is replaced with a 60% efficiency amplifier at the same RF output power. Therefore, if the required RF output power is 1 W and the output circuit loss is 2.5 dB as mentioned in Sect.1, the DC power supply required for a 60% efficiency amplifier is 3 W against 9 W for a 20% efficiency device. If the total power consumption of the circuits other than the transmitting amplifier is 1 W, a 2.5 times longer transmitting time is available from a portable unit that uses a 60% efficiency amplifier. It is assumed that the same batteries are used.

Another factor is the heat generated by the internal power dissipation which can increase case temperatures of portable unit to uncomfortable levels. A desirable upper limit to the internal power dissipation can be determined using parameters such as portable unit volume, acceptable temperature increase, ambient temperature, heat radiation area of the portable unit, and so on[1]. Figure 3 shows an example of theoretically estimated relations between required minimum portable unit volume and the amplifier efficiency η to keep the temperature increase below 15°C during transmissions. The output circuit loss and total power consumption excluding transmitting amplifier are again assumed to be 2.5 dB and 1 W respectively. In addition, the black body radiation model is employed in the calculations. These results indicate that more than 40% efficiency is required to attain portable units having a volume of 200 cm³. Thus, it is indispensable for portable units to use high efficiency transmitting amplifiers.

3. High Efficiency Saturation Amplifiers

Roughly speaking, in order to achieve efficiencies over 40% in the frequency bands above UHF, it is necessary to operate the amplifiers at the so called saturation level. Accordingly, the input-output amplitude nonlinearity becomes significant. An amplifier operating at its saturation level usually generates a lot of undesirable output distortions, when the input signal contains AM components. Therefore, it has been almost impossible for high efficiency transmitting amplifiers to amplify AM signals. This is the reason why constant envelope modulation schemes such as FM and GMSK have been widely utilized in mobile communications. One of the most practical technical methods to achieve high efficiency at saturation level is to drive the amplifier using the switching mode scheme[3]. This technique minimizes the time integration of the product of output voltage and current functions by driving the amplifier like a high speed switch. Under ideal operation conditions, 100% efficiency is available with this technique.

There are basically two amplifier operation techniques that provide high efficiency. That is, (1) the technique that minimizes the current conduction angle, such as class-C amplifier[4], and (2) the technique that

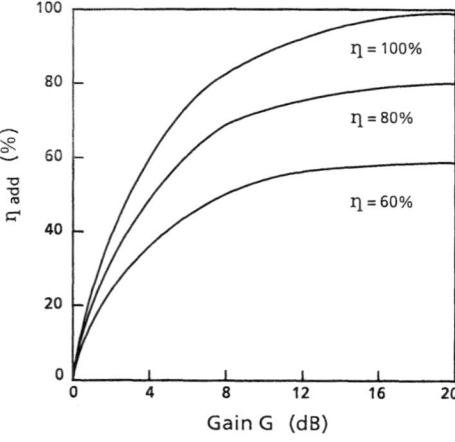

Fig. 4 η_{add} vs. gain G.

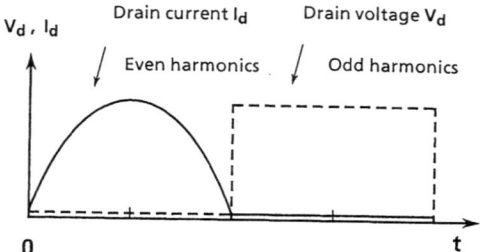

Fig. 5 Ideal switching mode.

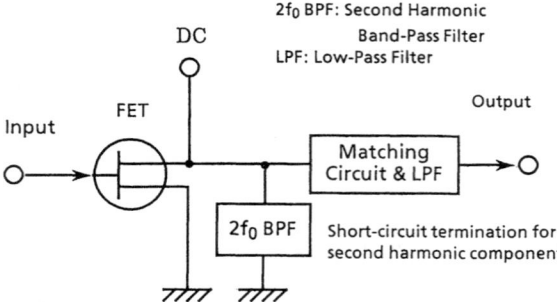

Fig. 6 Fundamental circuit configuration of class-F amplifier.

utilizes the switching mode. The latter is realized by controlling voltage and current wave forms at the output terminal of an active device, such as the class-F amplifier[5] and the HRA[6]. Because of its circuit simplicity, technique (1) has been widely utilized. To date, however, only 40% efficiency has been achieved with class-C amplifiers in UHF band portable units[4].

Here, the power added efficiency η_{add} is used to estimate the net efficiency that takes account of the input RF power. It is defined as

$$\eta_{add} = [(P_{RFout} - P_{RFin})/P_{DC}] \times 100\%$$
$$= \eta \times (1 - 10^{-G/10}). \quad (3)$$

Figure 4 shows the relation between η_{add} and power gain G for three collector or drain efficiencies η. From this figure, it is almost impossible for amplifiers with small power gains such as class-C amplifiers to obtain high η_{add} values, no matter how high η is.

3.1 Class-F Amplifier

On the other hand, technique (2) has not been used until recently, although the original idea was developed some time ago[3]. Using this technique, power added efficiencies over 60% have been attained with class-F amplifiers[5] including commercial units[7]. Figure 5 shows the ideal output voltage and current wave forms realized with this technique when an FET is used. The drain terminal voltage and the current wave forms are rectangular-half- and sin-half-waves respectively. Therefore, their product function is constantly zero, which means that zero power dissipation is realized. The rectangular-half-wave is constructed from fundamental and only odd number harmonic frequency components while the sin-half-wave is composed of fundamental and only even number harmonic components. Therefore, in order to obtain this operation condition, the class-F amplifier must be constructed as shown in Fig. 6 and driven at a class-B biased saturation condition. The output circuit connected to the FET drain terminal has frequency characteristics of matched impedance at fundamental, open-circuited at odd-harmonics and short-circuited at even-harmonics. Since it is required that this termination condition is established accurately at the FET-chip output point, construction and adjustment of the output circuit is quite difficult in microwave amplifiers. The HRA presented here overcomes these problems for conventional class-F amplifiers.

Further, it should be noted here that theoretically there is another method of obtaining switching mode operation. The method controls the output voltage and current wave forms so that they become sin-half- and rectangular-half-waves respectively. However, since it is difficult for the microwave transistors to generate large third order current component, this method is not practical for obtaining high efficiency microwave amplifiers.

3.2 HRA

The fundamental circuit configuration of the HRA is shown in Fig. 7. The HRA is basically composed of two FETs (FET-1 and FET-2), an input power divider, output power combiner, input matching circuits, output matching circuits, second harmonic interconnection circuit and second harmonic reflection filters. The output matching circuits are designed to have complex conjugate relations with the output impedances of the FET chips at both fundamental and second harmonic frequencies. This design enables the FETs to produce effective power output for the second

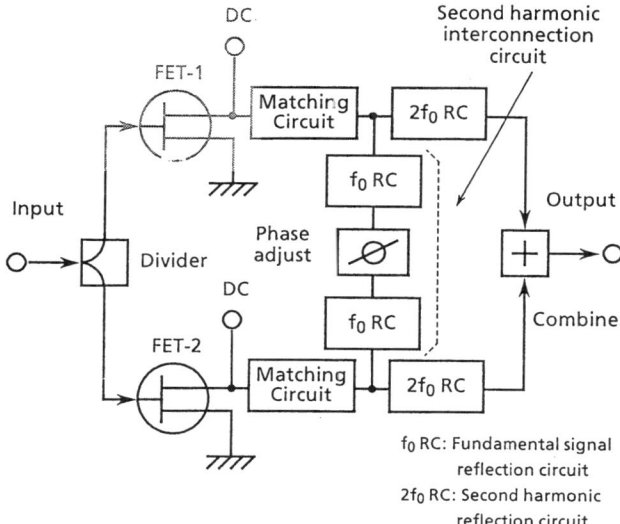

Fig. 7 Fundamental circuit configuration of HRA.

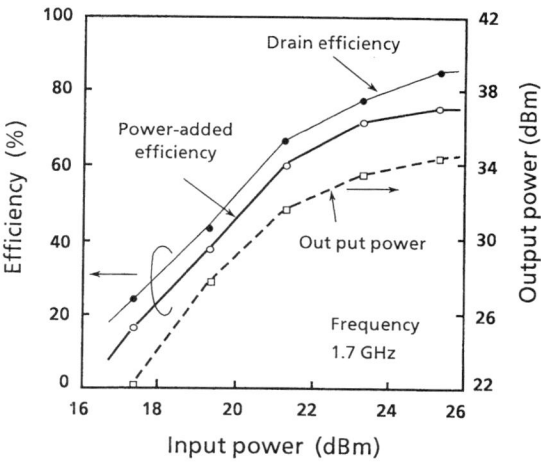

Fig. 8 Efficiency and output power vs. input power of a 1.7 GHz HRA.

harmonic as well as fundamental component. The second harmonic interconnection circuit is composed of fundamental frequency reflection filters and transmission line. Since this circuit is inserted between the output terminals of the two FETs' output matching circuits, there is a second harmonic standing wave generated between two FET outputs. However, as the length of the transmission line can be easily controlled through pattern design process, it is possible to induce electric walls precisely at the FET drain output terminals for just the second harmonic component. The electric wall termination is equivalent to the short-circuit termination. As a result, the FET output termination conditions needed to attain the switching mode are realized. Here, it should be noted that the harmonic component powers of orders higher than second are usually negligibly small in microwave FETs, and therefore it is practical to construct the amplifier considering only the second harmonic component.

In comparison with the class-F amplifier, the HRA has following features: (1) The output termination condition needed to obtain maximum efficiency can easily and precisely be created by adjusting the second harmonic interconnection circuit electric length. This is because the second harmonic termination condition can be controlled independently of the matching condition for fundamental frequency output. On the other hand, in conventional class-F amplifies, the output matching circuit impedance can not be determined independently of the second harmonic termination circuit making precise independent adjustment extremely difficult. (2) More stable operation is available for high gain amplifiers. No standing waves can be excited on the second-harmonic interconnection circuit if there is no input signal. Accordingly, each FET output is terminated with a matched impedance for the no signal input condition. The class-F amplifier, however, is always short-circuited and this makes its operation unstable when the amplifier gain is large. (Experiment)

Figure 8 shows typical input-output power and efficiency characteristics of an experimental 1.7 GHz band HRA[6]. Simple single-stage LC resonance circuits were used as filters. GaAs FETs with f_T of about 9 GHz were employed as active devices. A 75% power added efficiency (85% drain efficiency) is achieved at the 3 W output power point. In addition to this example, several HIC HRAs were experimentally constructed and yielded power-added efficiencies of 65% at 1 GHz 20 W output[8], and 70% at 2 GHz 5 W[9] were attained.

4. Linear Amplifier

Recently, enhanced digital mobile communications systems have been strongly demanded to provide large subscriber capacities and a wide variety of communication services. In Japan, a narrow band digital mobile system employing roll-off $\pi/4$ QPSK is now under development[2]. Since this modulation scheme requires a linear transmission system, a high efficiency linear amplifier is needed instead of the usual saturation amplifiers.

Figure 9 shows theoretical efficiency characteristics versus output power back-off for HRA, class-B and class-A ideal amplifiers. On-resistance and mutual conductance of all transistors are assumed to be zero and linear respectively. It is also assumed that the HRA is driven to ideal class-B biased operation and its input-output characteristics are linear in the signal output levels below saturation. If no distortion generation is required, the highest available efficiencies should occur at the value of the specific back-off point. This back-off value depends upon the input signal and is usually equal to the signal peak factor. The signal

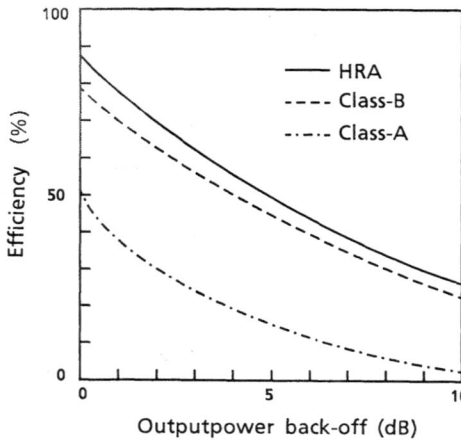

Fig. 9 Ideal efficiency vs. output power back-off of linear amplifiers (saturation level=0 dB back-off).

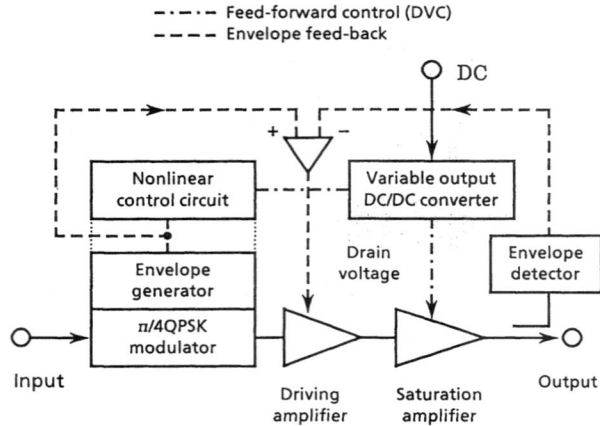

Fig. 10 Fundamental circuit construction of LSA-BC.

peak factor is defined as the ratio of average to peak power. Therefore, if the peak factor is 3 dB for roll-off $\pi/4$ QPSK for example, efficiencies of 60% and 20% are available from the ideal HRA and the ideal class-A amplifier respectively. In practical cases, however, attainable efficiencies would be much smaller, because a large back-off is required to avoid the influence of nonlinearities in the vicinity of the saturation level.

To develop highly efficient and linear amplifiers, at least three methods have been proposed so far: (1) Control the drain or collector supply power of a saturation amplifier in proportion to the signal envelope using high efficiency DC amplifiers. This action changes the saturation level and enables the amplifier to operate as an amplitude linear amplifier while retaining its high efficiency. The DVC[10], the LSA-BC[11] and the class-CS[12] can be categorized into this method. (2) Utilization of quasi linear amplifiers such as a class-F amplifier driven under a class-AB or class-A biased operation. (3) Applications of nonlinear compensation circuits such as feed-back controlled predistortion[13],[14]. It seems difficult for method (2) to attain high efficiency and linearity at the same time. Method (3) is not applicable due to its circuit construction complexity. Consequently, method (1) is the most promising technique at present.

4.1 LSA-BC

Figure 10 shows the fundamental circuit configuration of the LSA-BC[11] to be attached to a $\pi/4$ QPSK modulator. This circuit was designed for a digital portable radio transmitter. The LSA-BC is constructed with a nonlinear control circuit having an envelop generating function, a variable output DC/DC converter, and an envelope negative feed-back loop in addition to the usual transmitter circuits. The output voltage of the DC/DC converter is controlled dynamically by the nonlinear control circuit. This nonlinear

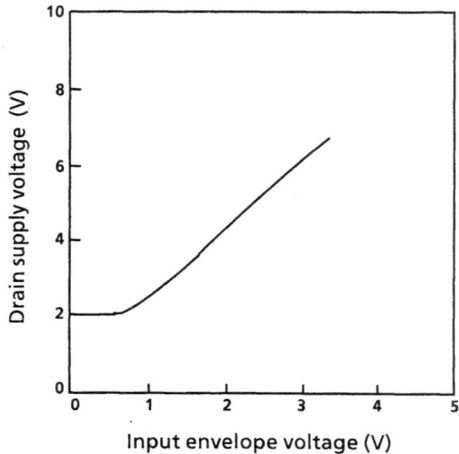

Fig. 11 Voltage transfer function of nonlinear control circuit (including DC/DC converter).

feed-forward control, which is DVC, follows a nonlinear scalar transfer function from the envelope voltage to the drain supply voltage as shown in Fig. 11. The transfer function is preliminarily determined by the measured characteristics of a typical practical amplifier. Roughly speaking, if only amplitude nonlinearity is considered, the errors in specified drain voltage shown in Fig. 11 must be limited to less than 0.2% to attain -60 dB distortion reduction for example. This was calculated under the assumption that distortion components are generated through the same mechanism as amplitude modulation which gives rise to double sideband components. Although this precise nonlinear control may be accomplished with an over 10 bit D/A converter, employment of the so called negative envelope feed-back as an additional amplitude error correction method is quite effective to improve residual distortion reduction and to stabilize the feed-forward control. Since, the principal linearization is achieved with the feed-forward control and only about 10 dB of feed-back gain is required, stable

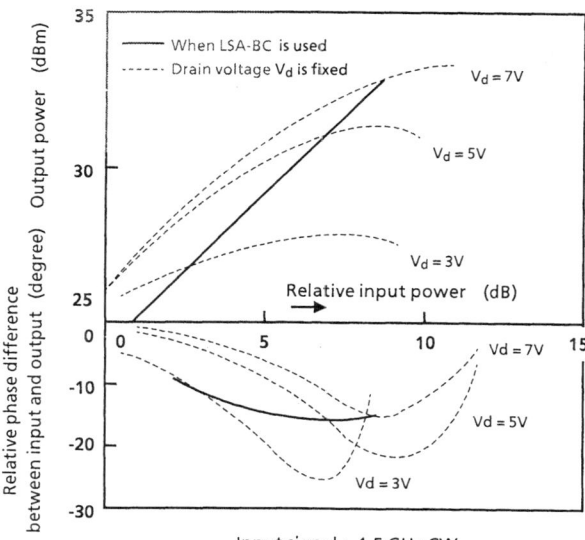

Fig. 12 Input output power and phase difference characteristics of LSA-BC (when LSA-BC is active and inactive).

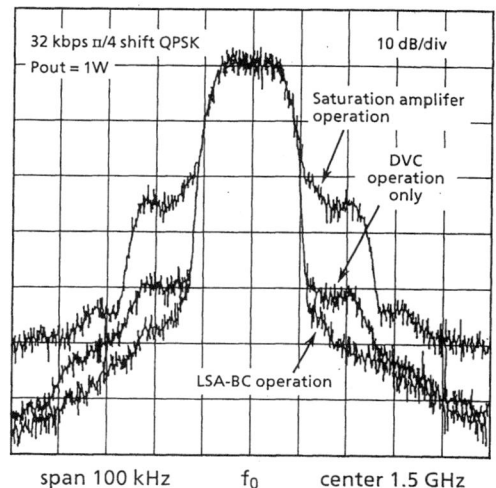

Fig. 13 Output signal spectra of LSA-BC.

operations can be obtained with the feed-back circuit. Thus, the use of both feed-forward and negative envelope feed-back, named Bidirectional Control (BC), is the technical key to achieve high efficiency and low distortion at the same time.

(Experiment)

LSA-BC amplification linearities and efficiency characteristics were examined using an experimentally constructed model. The circuit configuration is basically the same as shown in Fig. 10. A single stage 1.5 GHz band GaAs FET class-F amplifier was used as the high efficiency saturation amplifier. The DC/DC converter employs a switching frequency of 400 kHz to achieve dynamic output voltage control of the input signal envelope at frequencies from DC to 30 kHz. The measured voltage gain between control input signal and drain supply is over 10 dB. The average power-added efficiency is about 65% as the drain supply voltage ranges from 1.4 V to 7 V[15]. The nonlinear control circuit was temporarily constructed with commercially available ICs. In the modulator section, the data transmission rate is 32 kbps and the cosine roll-off factor is 0.5 dB. The negative envelope feed-back loop gain is about 10 dB.

Figure 12 shows examples of input-output power and phase difference characteristics measured with CW input operation. The solid lines are for the LSA-BC. The dotted lines show the corresponding curves when the drain supply voltages were kept constant. With LSA-BC, the input-output amplitude linearity is much improved at input power ranges of more than 10 dB. In the same way, AM/PM conversion with LSA-BC is improved by more than 50% from the constant drain voltage characteristics. Figure 13 shows output signal spectra for $\pi/4$ QPSK amplification. Three sets of

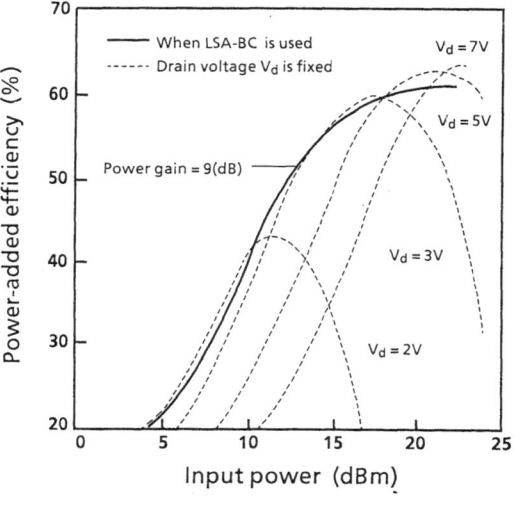

Fig. 14 Power-added efficiency vs. input power of LSA-BC RF section (when LSA-BC is active and inactive).

experiments were carried out: with BC, only feed-forward control, and without either BC or feed-forward control. Since, the radio channel occupation bandwidth is assumed to be 12.5 kHz, from Fig. 13, the adjacent channel interference due to the nonlinear intermodulation products is about -50 dB, -40 dB and -25 dB, for the three control schemes respectively. These results reveal that more than 25 dB of improvement is obtained even if only feed-forward control is used. The interference levels at the next adjacent channel are improved by over 10 dB with either feed-forward control or BC, compared with the case when neither is used.

Figure 14 shows the measured characteristics of power-added efficiency versus input power at the RF amplifier section under CW input operation. The solid line shows the measured efficiency of LSA-BC, where a constant power gain of 9 dB was maintained. Since $\pi/4$ QPSK has a peak-factor of about 3 dB, it is estimated that efficiencies over 60 % are possible. This is because an input point of about 18 dBm can be chosen as the average input power. Thus, providing that the power consumption of the nonlinear control circuit is less than 100 mW, the total efficiency of LSA-BC is approximately given by the product of RF amplifier and DC/DC converter efficiencies. Here, the efficiency of the DC/DC converter is 65%, and accordingly it is estimated that a total efficiency of abou 40% can be achieved. This level of efficiency was confirmed with the experimental model.

5. Conclusions

It was clarified that the use of high efficiency transmitting amplifiers is imperative for small size portable units which can be carried in most pockets. Amplifier construction techniques to achieve high efficiency were investigated. State of the art, HRA and LSA-BC amplifiers were presented and their high efficiency performance was experimentally confirmed.

The conclusions derived from the investigations conducted in this paper and the expectations for the evolution of related technologies in the near future can be summarized as follows :

(1) With respect to the saturation amplifiers used in the UHF bands, including up to the 2 GHz band, over 70% efficiency is available with HRA, or class-F amplifiers that have precisely constructed second harmonic termination. Moreover, the realization of high f_T and extremely low on-resistance FET devices will make it possible to achieve even higher efficiencies.

(2) The LSA-BC is one of the most promising linear amplifiers for efficiencies over 40% with an output power of about 1 W. In fact, it seems possible for LSA-BC to attain over 50% efficiency by improving the efficiency of each circuit device including DC/DC converter. Moreover, by using recent IC fabrication technologies, it will be possible to miniaturize the DC/DC converter and the nonlinear control circuit. Therefore, it is expected that the first digital portable unit will be constructed using a LSA-BC amplifier.

Acknowledgements

S. Tomisato, T. Takami and Y. Yamao assisted in the development of the experimental circuits. It is through their distinguished research works that the authors have been able to develop novel techniques on high efficiency amplifiers. Here, the authors publicly note their significant contributions.

References

(1) Sasaki A., Urabe S., Nishiki S. and Taga T.: "Basic considerations of portable radio unit for land mobile telephone system", Trans. IEICE, **J69-B**, 12, pp. 1787-1794 (Dec. 1986).
(2) Nakajima N., Kuramoto M., Kinoshita K and Utano T.: "A system design for TDMA mobile radios", IEEE 40th VTC Conf. Record, pp. 295-298 (May 1990).
(3) Tyler V. J.: "A new high-efficiency high-power amplifier", Maruconi Review, **21**, 130, pp. 96-109 (3rd Quarter 1958).
(4) Kawakami Y., Ohsawa O., Akiyama M. and Uenishi S.: "An 800-MHz band high gain and high efficiency power amplifier", 1981 Natl. Conv. Rec., IEICE, 2148.
(5) Chiba K. and Kanmuri N.: "GaAs FET power amplifier module with high efficiency", Electro. Lett., **19**, 24, p. 1025 (Nov. 1981).
(6) Nishiki S. and Nojima T.: "Harmonic Reaction Amplifier —a novel high-efficiency and high-power microwave amplifier", IEEE, MTT-S Digest, DD-5, pp. 963-966 (June 1987).
(7) Kaneko Y. and Ohkubo N.: "A high efficient power amplifier module", 1986 Spring Natl. Conv. Rec., IEICE, 2345.
(8) Nojima T. and Nishiki S.: "High efficiency GaAs FET Harmonic Reaction Amplifier (HRA)", 1989 Spring Natl. Conv. Rec., IEICE, SC-9-5.
(9) Nojima T. and Nishiki S.: "High efficiency microwave Harmonic Reaction Amplifier", IEEE, MTT-S Digest, LL-3, pp. 1007-1010 (May 1988).
(10) Nojima T. and Nishiki S.: "High efficiency quasi-microwave linear amplifier with Drain Voltage Control (DVC) scheme", 1987 Spring Natl. Conv. Rec., IEICE, 2223.
(11) Chiba K., Nojima T. and Tomisato S.: "Linearized Saturation Amplifier with Bidirectional Control (LSA-BC) for digital mobile radio", IEEE, GCOM'90, pp. 1958-1962 (Dec. 1990).
(12) Koch M. J. and Fisher R. E.: "A high efficiency 835 MHz linear amplifier for digital cellular telephony", IEEE 39th VTC Conf. Record, pp. 295-298 (May 1990).
(13) Akaiwa Y. and Nagata Y.: "Highly efficient digital mobile communication with a linear modulation method", IEEE Trans. J. Sel. Area Commun., **SAC-5**, pp. 890-895 (June 1990).
(14) Saleh A. A. M. and Salz J.: "Adaptive linearization of power amplifier in digital radio systems", Bell Syst. Tech. J., **62**, 4, pp. 1019-1033 (April 1983).
(15) Endo H. Yamashita T. and Sugiura T.: "A wide-band amplifier with two-loop control", IEICE Technical Report, **PE90**-26 (1990).

A UHF BAND 1.3W MONOLITHIC AMPLIFIER WITH EFFICIENCY OF 63%

T.Takagi, Y.Ikeda, K.Seino, G.Toyoshima,
A.Inoue, N.Kasai, and M.Takada

Electro-Optics and Microwave Systems Laboratory
Mitsubishi Electric Corporation
5-1-1 Ofuna, Kamakura, Kanagawa 247, JAPAN

ABSTRACT

A UHF band 1.3W 4-stage monolithic amplifier with an efficiency of 63% has been developed for transmitters of mobile communications. To obtain a high efficiency, a novel miniaturized second harmonic terminating circuit was devised, having a parallel resonant circuit comprised of lumped elements. The size of the amplifier is 8.6 X 5.8mm.

INTRODUCTION

Optimally loaded and overdriven RF power amplifiers[1,2,3,4] are very useful for transmitters of mobile communications because of high efficiency characteristics. In the amplifiers, to obtain high efficiency, even and odd harmonic waves have been terminated in a short and an open circuits, respectively[1]. In most of the conventional power amplifiers with high efficiency, a 1/4-wavelength short stub[2] or a 1/8-wavelength open stub[3] at the fundamental frequency is utilized to make a short circuit for even harmonic waves. However, wavelength becomes so long in the UHF band that their sizes become large. From the point of view that the second harmonic wave has a predominant effect on the efficiency of high power amplifiers, we devised a novel miniaturized second harmonic terminating circuit with a parallel resonant circuit comprised of lumped elements. The parallel resonant circuit becomes an open circuit at the second harmonic wave. With the use of the novel second harmonic terminating circuit, a 4-stage monolithic power amplifier has been developed. A maximum drain efficiency of 63% has been obtained with a saturated output power higher than 31dBm in the UHF band. The size of the amplifier is 8.6 X 5.8mm.

CIRCUIT DESIGN

For the design of a high efficiency power amplifier, we measured the load conditions providing a maximum output power and the load conditions providing a maximum drain efficiency. The load conditions for the fundamental and the second harmonic waves were measured with the use of a conventional load-pull and an active second harmonic load-pull[5], respectively.

Fig.1 shows a measurement set-up of the active second harmonic load-pull. The second harmonic wave generated by a frequency doubler is tuned by a phase shifter and a variable attenuator, and is injected to a FET under test through a broadband circulator and an output matching circuit. This method has an advantage in that the load condition for the second harmonic wave can be varied without affecting the load condition for the fundamental wave. Fig.2(a) shows output power versus $\angle \Gamma x$ for three values of $|\Gamma x|^2$, where Γx is the load reflection coefficient for the second harmonic wave. The FET under test has a gate width of 3.18mm and is matched to 50 Ω at the fundamental frequency. It is clear from Fig.2(a) that the output power changes by injecting second harmonic wave and that the maximum output power is 27.6dBm for the condition of $\angle \Gamma x = -70°$ and $|\Gamma x|^2 = 0$dB, which is realizable by using passive circuit elements. Fig.2(b) shows power-added efficiency versus $\angle \Gamma x$. From Fig.2(b), the maximum power-added efficiency is 74.2% for the same condition.

Fig.3 shows a measurement set-up of the load-pull for the fundamental wave. A second harmonic reflection circuit is located between a FET under test and a tuner. So, this method has an advantage in that the load conditions for the fundamental wave can be varied without affecting the load conditions for the second harmonic wave. Fig.4 shows output power and drain efficiency for various values of the load impedance for the fundamental wave under the condition that the load impedance for the second harmonic wave is adjusted to the condition that the drain efficiency becomes maximum. The FET under test has a gate width of 7.28mm and the measurement was done at 920MHz. From Fig.4, a maximum output power of 32.2dBm was obtained, when the drain efficiency was 59.8%.

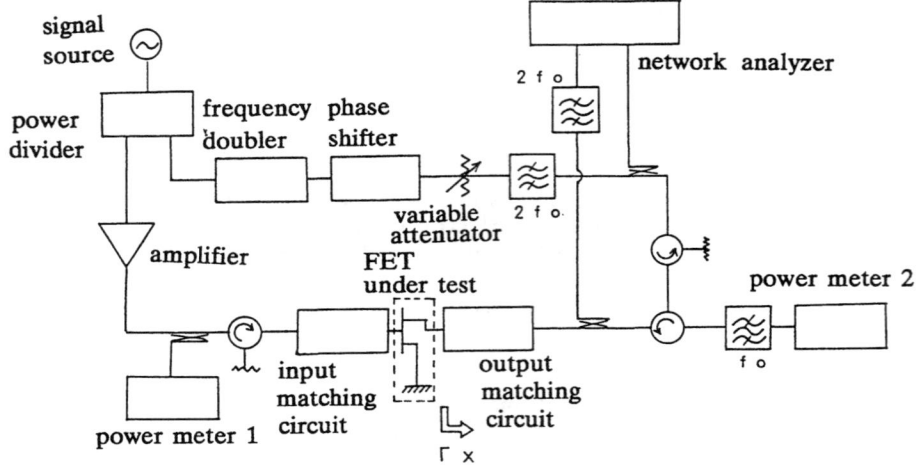

Fig.1 Measurement set-up of the active second harmonic load-pull.

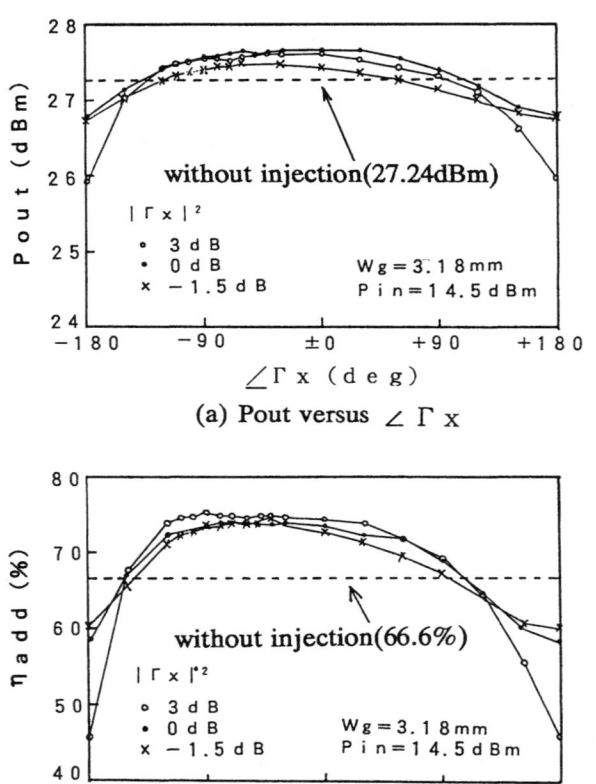

(a) Pout versus $\angle \Gamma x$

(b) η add versus $\angle \Gamma x$

Fig.2 Output power (Pout), power added efficiency (η add) versus Γx characteristics measured by the active second harmonic load-pull.

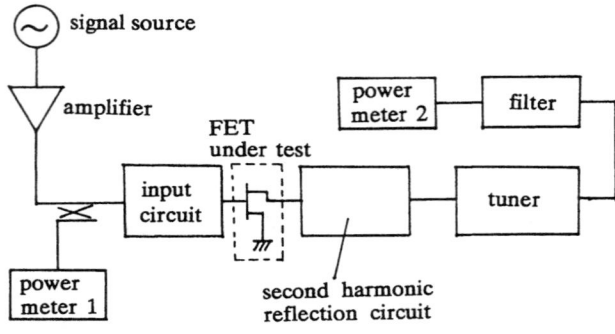

Fig.3 Measurement set-up of the load-pull for the fundamental wave.

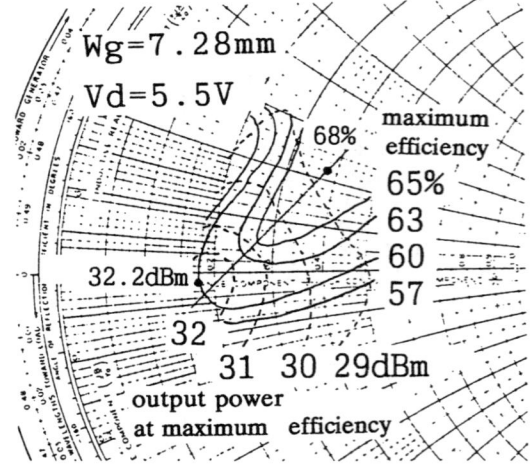

Fig.4 Experimental results of the load-pull for the fundamental wave.

On the other hand, a maximum drain efficiency of 68.2% was obtained, when the output power was 28.4dBm.

Based on the measured data for the fundamental and the second harmonic waves, a 4-stage monolithic power amplifier with a novel second harmonic terminating circuit has been designed.

Fig.5 shows a schematic diagram of the amplifier. The number of amplifier stages was determined to be four to get a large signal gain greater than 40dB and the gate width of the final stage FET was determined to be 7.28 mm to get a saturated output power higher than 31dBm from the results of Fig.4. The final stage FET is optimally loaded for the second harmonic wave by using a devised second harmonic terminating circuit which is made of a phase-adjusting transmission line and a parallel resonant circuit comprised of lumped elements. The parallel resonant circuit becomes an open circuit, and the phase-adjusting transmission line makes an optimum load reflection coefficient for the second harmonic wave. Moreover, this amplifier employs a quarter-wave open stub for the third harmonic wave in the output matching circuit to prevent the third harmonic dissipation.

From Fig.4, the load condition for the fundamental wave was designed to be Z=13+j2.5 Ω at 920MHz, where the second harmonic terminating circuit was designed to be a reflection coefficient of $\angle \Gamma_x = -120°$ and $|\Gamma_x|^2 = 0$dB.

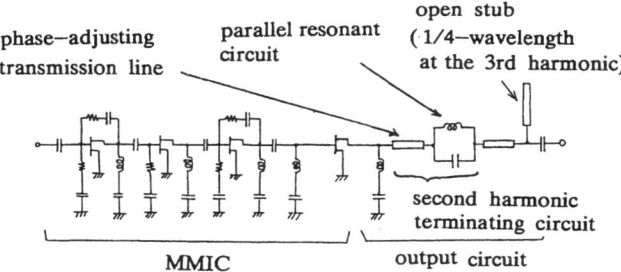

Fig.5 Schematic diagram of 4-stage monolithic power amplifier with a novel second harmonic terminating circuit.

EXPERIMENTAL RESULTS

Fig.6 shows a photograph of the 4-stage monolithic power amplifier. The output matching circuit of the final stage amplifier was fabricated on an alumina substrate to reduce a circuit loss and to achieve a higher output power and efficiency. The other circuit was fabricated on GaAs substrates. The overall circuit size is 8.6 X 5.8mm.

Fig.7 shows a measured output power of the amplifier. The saturated output power is higher than 31dBm for the frequency range from 840MHz to 1GHz at a bias condition of Vd=5.5V and Idso=0.1 Idss (deep "class AB" operation). Fig.8 shows a measured drain efficiency of the amplifier at 920MHz. The measured maximum drain efficiency is 63% at 31.2dBm output power. Fig.9 shows measured harmonic levels of the amplifier relative to the fundamental level. The second and third harmonics are kept below -50dBc and -40dBc, respectively. Fig.10 shows predicted and measured small signal gains of the amplifier. They are in good agreement. The measured small signal gain is higher than 50dB for the frequency range from 840MHz to 1GHz.

Fig.6 The potograph of 4-stage monolithic power amplifier.

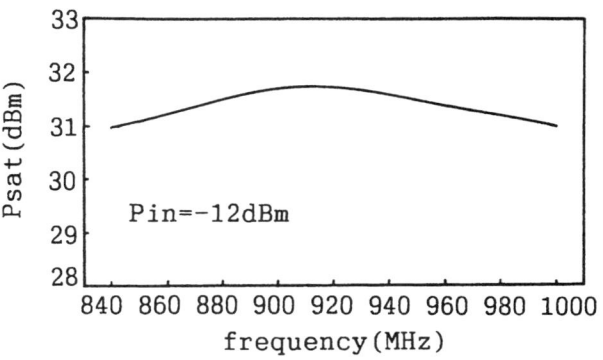

Fig.7 Saturated output power (Psat).

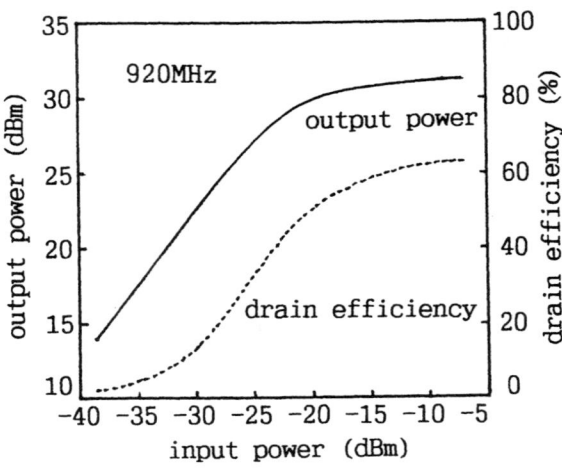

Fig.8 Output power, drain efficiency.

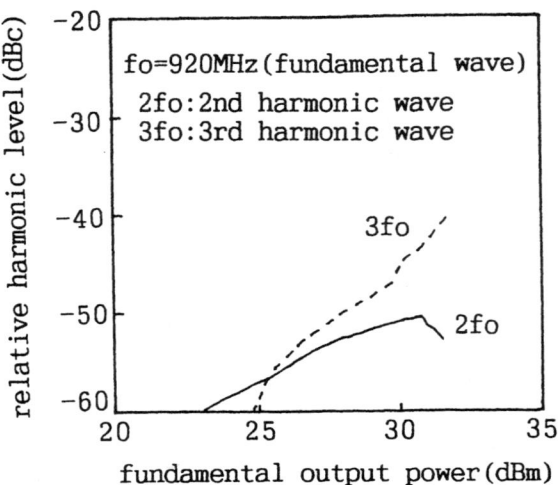

Fig.9 Harmonic levels.

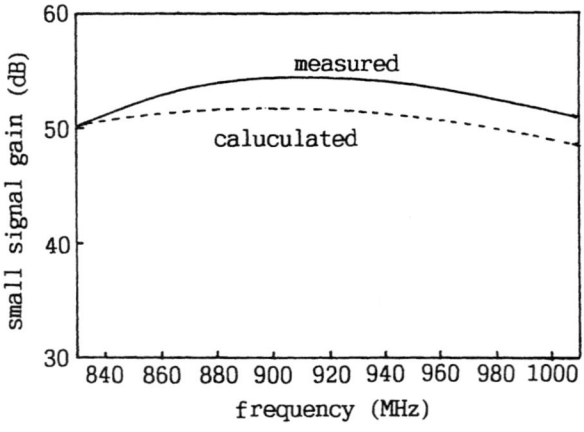

Fig.10 Small signal gain.

CONCLUSION

A UHF band high efficiency, 4-stage monolithic power amplifier with a novel miniaturized second harmonic terminating circuit has been developed. With the use of a parallel resonant circuit comprised of lumped elements for terminating the second harmonic, it achieves a maximum drain efficiency of 63%, a saturated output power of higher than 31dBm in the UHF band, and occupies an area of 8.6 X 5.8mm.

REFERENCES

[1] D.M.Sinder, " A Theoretical Analysis and Experimental Confirmation of the Optimally Loaded and Overdriven RF Power Amplifier," IEEE Trans., vol.ED14, No.12, Dec. 1967.
[2] B.D.Geller et al.,"Quasi-Monolithic 4-GHz Power Amplifier with 65-Percent Power-Added Efficiency," IEEE MTT-S Digest, pp. 835-838, 1988.
[3] K.Chiba and N.Kanmuri, " GaAs FET Power Amplifier Module With High Efficiency ," Electronic Letters, vol.19, No.24, pp.1025 -1026, November 1983.
[4] T.Nojima and S.Nishiki, " High Efficiency Microwave Harmonic Reaction Amplifier ," IEEE MTT-S Digest, pp.1007-1010, 1988.
[5] Y.Ikeda et al.,"High Efficiency Operation of FET Using Second Harmonic Injection Method ," The 3rd Asia-Pacific Microwave Conference Proceedings, pp685-688, 1990.

HIGH PERFORMANCE INTEGRATED PA, T/R SWITCH FOR 1.9 GHZ PERSONAL COMMUNICATIONS HANDSETS

P. O'Sullivan, G. St. Onge, E. Heaney, F. McGrath, C. Kermarrec*

M/A-COM IC Design Center
100 Chelmsford St., Lowell MA 01851
(508)453-3100
* Analog Devices

ABSTRACT

This paper will present a state of the art power amplifier and T/R switch GaAs MMIC product for the 1.9 GHz Japanese PHP system. The design consists of a single die (1.8mm x 1.2mm) in a 28 pin QSOP surface mount plastic package. The two stage power amplifier exhibits 28dB gain with greater than 44% PAE at a P1dB of 23dBm. The adjacent channel interference level is better than 60dBc at +/-600kHz when tested at this rated output power. The T/R switch exhibits 1dB insertion loss, 30dB isolation and OIP3 of greater than 44dBm when tested with two 0.25 watt tones.

INTRODUCTION

This product was developed for the portable handsets used in the Japanese Personal Handy Phone (PHP) system operating at 1.9 GHz.[1] The block diagram and layout are shown in Figure 1a and 1b. This system is scheduled to be on line in 1994 with projected volumes of 13 million units per year by the year 2000. The consumer nature of the market dictates the use of low-cost surface mount plastic packaging. For this reason, a 150 mil body 28 lead QSOP was chosen for this product because it is inexpensive, takes minimal board real estate, has acceptably low package parasitics and is suitable for high volume surface mount manufacturing. The π/4 DQPSK modulation scheme used in the PHP system requires tradeoffs between the power added efficiency and linearity performance in the power amplifier design. The major challenge in the switch design is to maintain low insertion loss and high isolation while achieving sufficient linearity so that the good adjacent channel interference (ACI) characteristics of the power amplifier are not degraded.

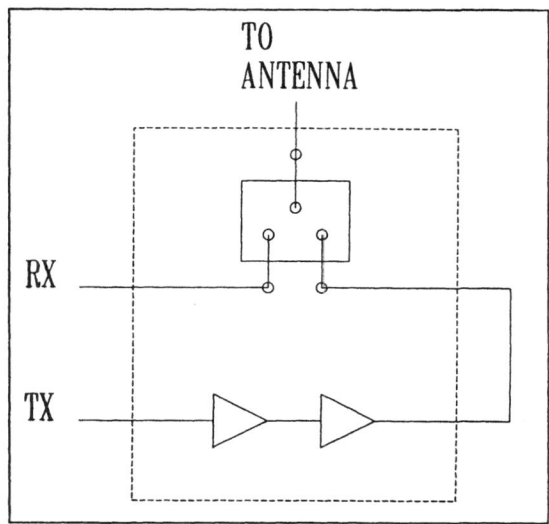

Figure 1a. Block Diagram

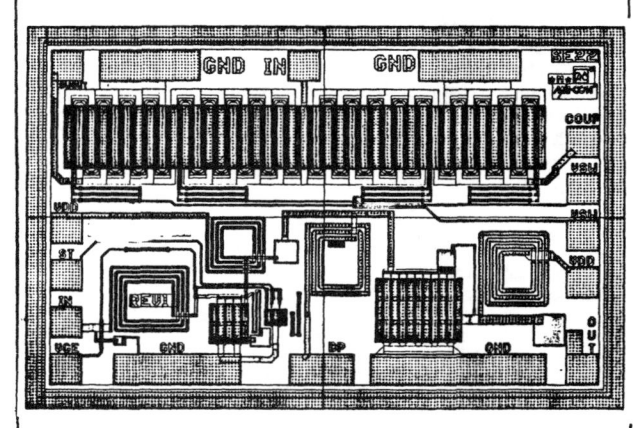

Figure 1b. MMIC Layout

MESFET PROCESS

The amplifier, T/R switch MMIC was fabricated on a 0.5um ion implanted MESFET process that employs a buried-P layer and e-beam gate lithography. The process has a pinchoff of 1.2 volts, an Idss of 180mA/mm and a breakdown voltage in excess of 10 volts. The use of a buried-P layer allows high values of power gain to be achieved at low bias levels. At 15% Idss bias the transconductance is typically 150mS/mm. The buried-P layer coupled with a low pinch-off voltage and a low knee voltage allows high values of power added efficiencies at low drain voltages for output powers of up to about 0.5 watts. The process also uses high Q inductors (typically 30 at 1.9GHz) thus allowing the low-loss matching networks to be incorporated directly on the MMIC.

POWER AMPLIFIER DESIGN

The amplifier consists of a two stage reactively matched design that incorporates the package parasitics into the matching structures (Fig.2).

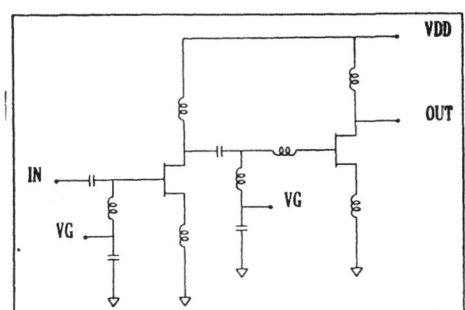

Figure 2. Power Amplifier Schematic

GaAsIc Symposium, pp. 33-35, October 1993.

The supply voltage is +4.8 volts and quiescent current is only 25mA/mm. The design was based on extensive FET characterization using load-pull /source-pull measurement techniques together with time domain non-linear simulation. The load-pull /source-pull measurements were performed under both single tone and π/4 DQPSK modulation conditions. Because π/4 DQPSK modulation is not constant envelope, the maximum output power is substantially higher than the average. The optimal input and output matching impedances are therefore different than for a single tone. For the output power and ACI requirements it was found that the optimum output conductance of the required output periphery was very close to 20mS. Therefore, a single inductor was all that was required to tune the output FET capacitance. The interstage circuit transformation ratio was kept deliberately low by properly sizing the driver FET (500um) and using a series feedback technique for the output FET. As the PHP system is time division multiple access (TDMA) based, provision is also made on-chip for a standby mode. This was implemented using a series switch in the bias line which reduces current draw to less than 50uA in standby-mode.

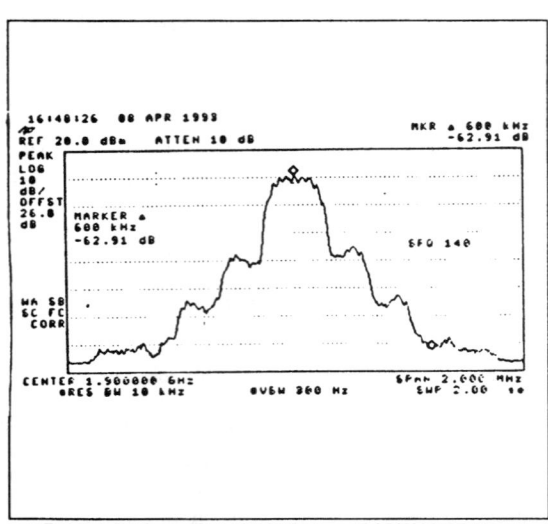

Figure 4. PA Output Spectrum Under π/4 DQPSK Modulation

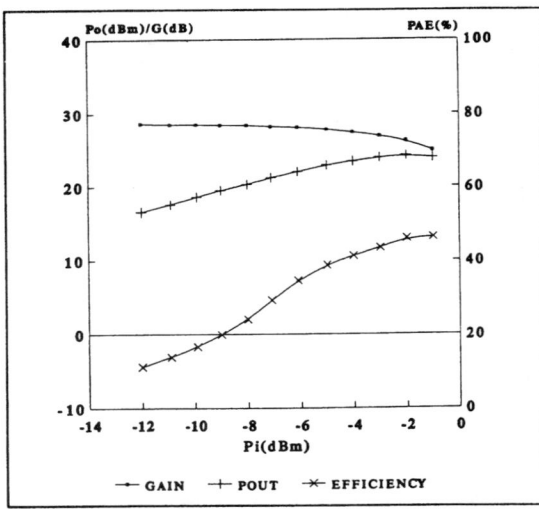

Figure 3. Power Amplifier Measured Data

Figure 3 shows measured results at 1.9GHz from the 2-stage power amp in a 28 lead QSOP plastic package mounted on an FR4 PC Board. Bypassing chip capacitors are the only external components required. The output power at 1dB compression is 23 dBm, power added efficiency is 44%, small signal gain is 28 dB with a quiescent current of less than 65mA from a 4.8v supply. These results are in good agreement with the simulated data, due mainly to the effort placed in the FET and package modeling. Figure 4 shows the output envelope of the power amplifier when tested at rated power using a nine bit pseudo-random code driving the π/4 DQPSK modulated source.

T/R SWITCH DESIGN

The FET process used to optimize the power amplifier efficiency and linearity is also very well suited for use in low insertion loss switches (low Ron). The intermodulation requirements of the system are beyond the performance normally achievable with GaAs FET switches controlled from very low voltage.

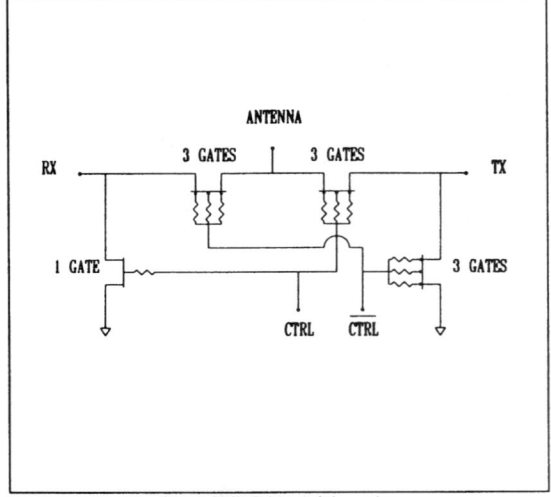

Figure 5. T/R Switch Schematic

To overcome this problem, a unique multigate FET structure was employed that effectively reduces the ac voltage that appears between the gate and drain or source terminals and therefore extends the power handling capability of the switch. Figure 5 shows the T/R switch schematic. The on state resistance of the multigate FET is about 2.5 times that for a single gate FET of the same periphery. To compensate for the higher unit resistance, larger FETs must be used. Because the source-drain spacing for the multigate FET is larger than that for a single gate FET, the unit off state capacitance is lower than for a single gate FET. The decrease in isolation that results from the larger FETs needed to decrease the insertion loss is therefore minimal. The power amplifier is connected to one arm of the switch through an external component. The common port is connected to the antenna through a filter. Only three of the four FETs in the switch are subjected to high power (voltage) levels. The shunt FET on the receive arm can therefore be a standard single gate design without degrading the overall linearity of the switch. The asymmetric design allowed the insertion loss and isolation performance to be different in the transmit and receive paths to better meet the system requirements. The parasitics of the QSOP 28 package were incorporated into the switch design to ensure good match. The periphery of the switch FETs were optimized so that their off state capacitance was resonated with the series inductance of the package leads. Finally, a non-linear simulation was performed on the switch derived from the linear simulation to ensure that design would meet the distortion specifications. The periphery of the series FETs was then increased to reduce the current density and the linear and non-linear simulations repeated. This process was repeated until an acceptable compromise between linearity and isolation/match was reached. An exceptional OIP3 of +48dBm has been measured from a 4v control voltage. Table 1 summarizes the measured switch performance which again agrees well with simulated performance.

Insertion Loss (dB)	<1dB
Isolation	30dB
Control Voltages	0/-4V
OIP3 (Pin=23dBm)	48 dBm

Table 1. Switch Performance

CONCLUSION

A single chip implementation of a high efficiency power amplifier and a high linearity switch has achieved. The chip operates in an QSOP 28 lead surface mount plastic package. A standby mode has been included for use in TDMA systems. The circuit simultaneously meets both the output power and the adjacent channel interference requirements for the PHP system.

ACKNOWLEDGMENT

The authors wish to thank Russ Pratt, Ted Begnoche, Scott Mitchell and Bill Foley for their patience in testing and characterizing this product and Evelyn Miller for the preparation of this paper.

REFERENCES

(1) C. Kermarrec, etal, "High Performance, Low Cost GaAs MMICs for Personal Handy Phone Applications at 1.90 GHz", 19th International Conference on GaAs and Related Compounds, Kurazawa Japan, Sept. 1992.

(2) F. McGrath, etal, "Multi-Gate FET Power Switches", Applied Microwave, Summer 1991, pp. 77-86.

A 3.5 V, 1.3 W GaAs Power Multi-Chip IC for Cellular Phones

Masahiro Maeda, Masaaki Nishijima, Hiroyasu Takehara, Chinatsu Adachi, Hiromasa Fujimoto, and Osamu Ishikawa

Abstract—A GaAs power multi-chip IC (MCIC) operable at a voltage of 3.5 V designed for cellular phones has been developed. The MCIC is able to deliver an output power of 1.3 W with a power-added efficiency of 60% in a frequency range from 890 to 950 MHz. This consists of two GaAs MESFET's, three GaAs passive matching chips, and a printed circuit board on which biasing networks are disposed. These are mounted on an aluminum nitride (AlN) package, occupying a half volume of conventional power hybrid IC's, i.e., only 0.4 cc. In order to improve the low voltage operation characteristics, a GaAs power MESFET operable at a low voltage of 3.5 V with an output power of 32 dBm and a power-added efficiency of 65% is developed, and microstrip lines having high impedance characteristics are incorporated also in order to minimize the conductor loss of matching network. The MCIC would be highly useful to develop compact cellular phones with advanced characteristics.

I. INTRODUCTION

THE incorporation of miniaturized amplifier operable at the lowest voltage is essential to miniaturize cellular phones, because reduction of the number of battery cells is very effective in order to reduce the total size and weight of the cellular phone. Motorola had developed a 1.9 GHz power amplifier (PA) operable at a low voltage of 3 V, realizing an output power of 28 dBm and a power-added efficiency of 50% [1], and others reported a higher voltage of 4.7 V or 5.8 V is necessary to obtain an output power of 31 dBm at 930 MHz [2]–[5].

Although GaAs power hybrid IC's (HIC's) and monolithic GaAs IC's (MMIC's) are presently available, miniaturization of HIC is found difficult and low cost MMIC operable at UHF band is hard to realize. On the other hand, the incorporation of external passive output-matching network in GaAs power MMIC had been found effective to improve the output power and power-added efficiency by reducing the output network loss, and to reduce the chip cost by improving the yield per-wafer (i.e., by reducing the size of MMIC chip) [6]–[8].

However, since the MMIC of this construction is provided with on-chip input and interstage matching circuits, a relatively large device size has to be provided, particularly in the UHF band. Therefore, the MMIC cannot be manufactured at an adequately cost-effective rate.

The design priorities shown below and summarized in Table I are considered to solve these problems. As a result of this,

Manuscript received January 15, 1994; revised May 11, 1994.
The authors are with Semiconductor Research Center, Matsushita Electric Industrial Co., Ltd., Osaka 570, Japan.
IEEE Log Number 9404683.

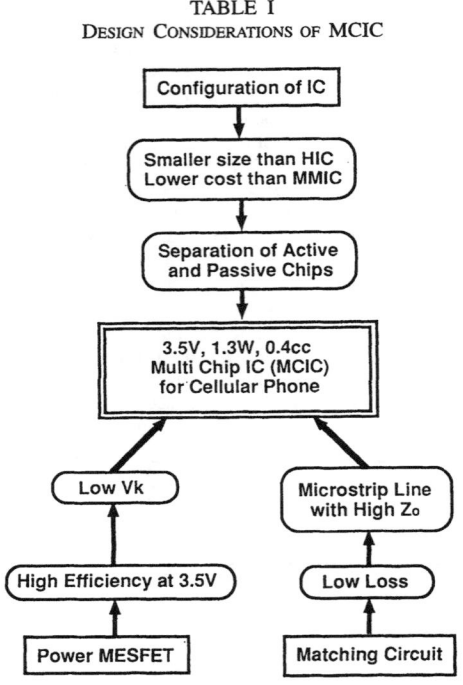

TABLE I
DESIGN CONSIDERATIONS OF MCIC

a GaAs power multi-chip IC (MCIC) operable at a voltage of 3.5 V has been developed.

A. IC Configuration

In developing an MCIC having a size smaller than an HIC and a cost lower than an MMIC, the active chips (FET) and the passive chips (matching networks) are separately mounted in MCIC. Considering the required fabrication processes and yield, the cost of an active chip is far higher than that of a passive chip. Because of a higher per-wafer yield of the active chip, the MCIC employing active chips without matching circuits can be obtained at a relatively lower cost. The cost in this case is still lower than that of the MMIC provided with on-chip input and interstage matching circuits.

B. Power MESFET

The drain voltage swing can be enhanced by attaining a low knee voltage (Vk) of I–V characteristics and a high breakdown voltage. By attaining these, a GaAs power MESFET operable at a low voltage and delivering a higher output power and a higher efficiency can be obtained. By optimizing the device

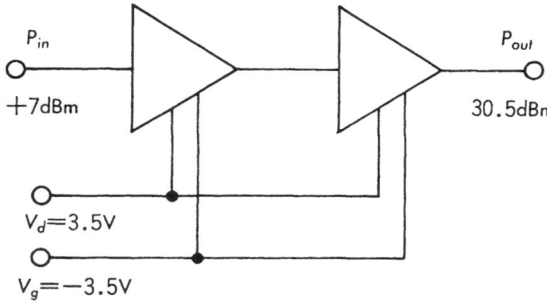

Fig. 1. A block diagram of the MCIC.

dimension and the fabrication processes, a MESFET satisfying these requirements and operable at a low voltage 3.5 V and capable of delivering a higher output power of 32 dBm with a power-added efficiency of 65% has been developed. The voltage of 3.5 V is lower than the previously reported 4.7 V by 1.2 V.

C. Matching Circuit

The characteristics of GaAs power amplifier can be improved by minimizing the conductor loss of GaAs matching circuit, and this is possible by developing a microstrip line having high impedance characteristics without increasing the width of the conductor line.

II. MCIC DESIGN

A two-stage power MCIC employing a 200 mW-class FET and a 1.5 W-class FET has been developed. Fig. 1 shows a block diagram of the MCIC, and Fig. 2 shows a photograph of the MCIC mounted within a package of 10.7×11.6 mm^2, having a total volume of 0.4 cc which is only a half of that of a conventional HIC.

A. RF Block

As shown in Fig. 2, the MCIC consists of two GaAs MESFET's, three GaAs passive matching chips, and a printed circuit board on which biasing networks are fabricated. These are mounted on an AlN package, and are interconnected by bonding wire. In this configuration, modification of the operating frequency can be easily accomplished by changing the passive matching chips.

The sizes of input, interstage, and output passive matching chips are 1.1×2.0 mm^2, 1.4×2.0 mm^2 and 1.9×2.0 mm^2, respectively. The ratio between the passive chip area and the active chip area is 5 : 1, and this means a substantially larger chip area occupied by the former. FET chips having a thickness of 150 μm are employed here in order to minimize the thermal resistance to the package.

Fig. 3 shows an equivalent circuit of MCIC. The circuit enclosed within a dotted area is partitioned into independent circuits. Within the output matching network, the load impedance for the fundamental wave is set at $Z = 6.5 + j1.0$ Ω at 930 MHz. In order to eliminate the second harmonic, a reflection circuit for the second harmonic utilizing a series

Fig. 2. A photograph of the MCIC. (Package size: 10.7×11.6 mm^2).

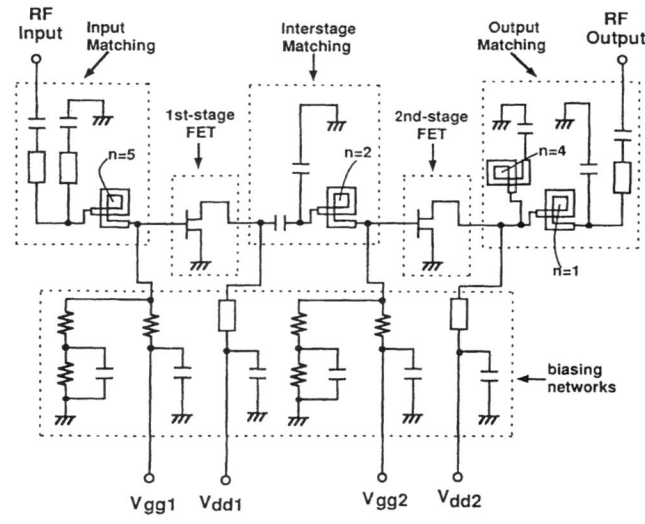

Fig. 3. An equivalent circuit of the MCIC.

resonance circuit is provided in the output matching network. The drain of the second-stage FET is shorted for the second harmonic by this. The circuit consists of a spiral inductor of 5.2 nH and a MIM capacitor of 1.5 pF.

B. DC Block

DC biasing networks are fabricated on both surfaces of a semi-flexible printed circuit board in order to reduce the size of MCIC. Fig. 4 shows a photograph of the circuit board made of PPO (poly phenylen oxide) having a dielectric constant of 10.5 and a size of 8.6×3.2 mm^2. On the upper surface of the board (Fig. 4(a)), the gate biasing networks consist of chip resistors/capacitors with a size of 1.0 mm $\times$ 0.5 mm (1005-type). The gate bias is supplied through bleeder resistors of which values are determined to provide a threshold voltage of the FET in MCIC. On the rear surface (Fig. 4(b)), $\lambda/4$ transmission lines for the drain bias are fabricated, and these

Fig. 4. Appearance of printed circuit board for dc bias. (a) Gate bias circuit on the upper surface. (b) Drain bias circuit on the rear surface.

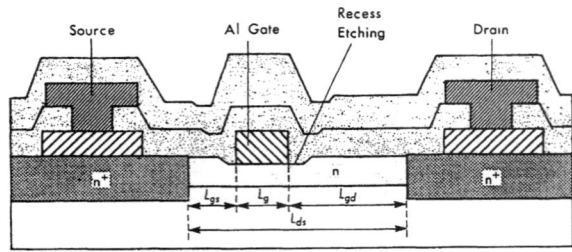

Fig. 5. A cross-sectional view of a GaAs MESFET.

are covered by a film resist for insulation. The width and the thickness of the transmission line are 150 μm and 30 μm, respectively. The resistance of the transmission line is 0.1 Ω. The voltage drop in the line is only 0.05 V for an operating drain current of 500 mA. This voltage drop is only 1/10 of that of a line with a thickness of 3 μm formed on GaAs. Since the gate and drain networks are provided with 1000 pF grounding capacitors, no external components are needed.

C. Package

The package is made of AlN instead of CuW in order to reduce the fabrication cost. AlN is a ceramic of a high thermal conductivity of 150 W/(mk) which is ten times of that of conventional alumina (Al$_2$O$_3$). Therefore, the thermal diffusion can be enhanced and the channel temperature of power FET can be kept at low by using a AlN package, attaining a high reliability and stable operation.

III. POWER MESFET

A. Fabrication Process

Fig. 5 shows a cross-sectional view of developed power FET fabricated by using an ion-implanted channel recessed-gate MESFET technology. An active channel is formed by implantation of Si ions at 80 KeV at a dose of 3.9×10^{12} cm^{-2}. An n^+ layer is formed by implantation of Si ions through insulation films. An annealing cap (SiO$_2$/WSiN) consisting of a SiO$_2$ base layer formed by thermal-CVD and a top layer of refractory metal of WSiN, is deposited. As shown in Fig. 6, the SiO$_2$ film of the annealing cap is not removed after the annealing process, and is utilized as a passivation film in the new process. However, in the conventional process, the SiO$_2$ film on GaAs is deposited again after removing the SiO$_2$ film used in the annealing process. The gate electrode with a gate-length (Lg) of 0.6 μm is formed by a conventional photo-lithography method.

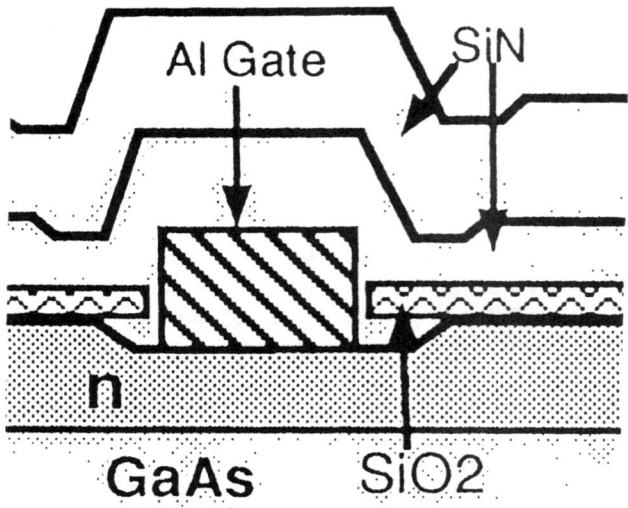

Fig. 6. A cross-sectional view around the gate electrode.

B. DC Characteristics

The gate breakdown voltage must be at least twice of the gate bias voltage. Considering bias conditions of $Vds = 3.5$ V and $Vgs = -2.5$ V, the gate-source breakdown voltage ($BVgs$) and the gate-drain breakdown voltage ($BVgd$) should be more than 5 V and 12 V, respectively. Therefore, $BVgs$ and $BVgd$ are set at 7 V and 14 V, considering a design margin of 2 V.

In order to optimize the device dimension, both the conventional FET's and the new FET's of different dimensions are fabricated by using the process explained in Section 3-1, and are evaluated. The relationships between $BVgs$ and the gate-source spacing (Lgs), and between $BVgd$ and the gate-drain spacing (Lgd) at a leakage current of 400 μA are shown in Fig. 7. The gate-length (Lg) of conventional FET's and new FET's are set at 1.0 μm and 0.6 μm, respectively, while the total gate-width (Wg) of these FET's are set at 2 mm.

By employing the new fabrication process, substantial increases of both $BVgs$ and $BVgd$ are obtained, and at the same time, shorter Lgs and Lgd are accomplished without sacrificing the breakdown voltage. The value of Lgd is reduced from 2.0 to 1.0 μm while holding $BVgd$ at a voltage of more

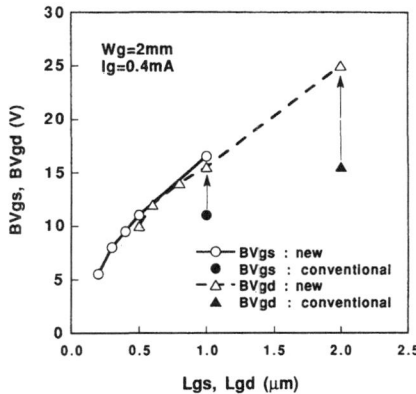

Fig. 7. Relationships between Lds and $BVgs$, and between Lgd and $BVgd$.

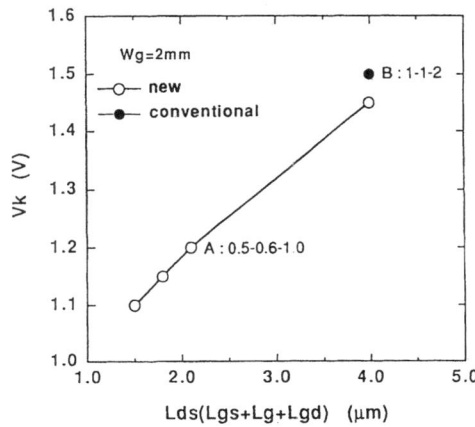

Fig. 8. A relationship between Lds ($Lgs + Lg + Lgd$) and Vk.

TABLE II
DIMENSIONS AND FABRICATION PROCESSES OF TWO FET'S

	Lgs-Lg-Lgd (μm)	process
Type-A	0.5-0.6-1.0	New
Type-B	1-1-2	Conventional

than 14 V. The required $BVgs$ of 7 V is obtained by setting Lgs at 0.3 μm. However, Lgs in this case is set at 0.5 μm allowing a margin of alignment.

The values $BVgs$ and $BVgd$ of a Type-A FET having $Lgs - Lg - Lgd$ of 0.5-0.6-1.0 (μm) are 10 V and 15 V respectively, attaining the desired breakdown voltages. The breakdown voltage of a Type-B FET having $Lgs - Lg - Lgd$ of 1-1-2.0 (μm) is found to be the same as that of a Type-A FET. Table II shows the dimensions and the fabrication processes of both Type-A and -B FET's.

A relationship between the knee voltage (Vk) of I–V characteristics and the source-drain spacing ($Lds = Lgs + Lg + Lgd$) of the FET is shown in Fig. 8. In here, the shorter the value of Lds, the lower the value of Vk. While the Type-A FET ($Lds = 2.1$ μm) has a low Vk of 1.2 V, the Vk

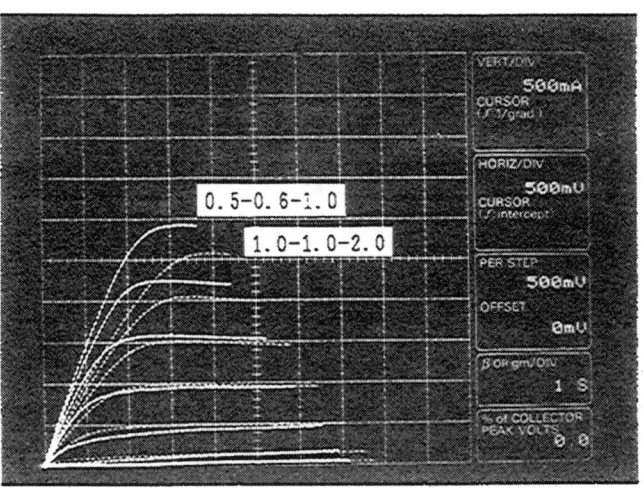

Fig. 9. I–V characteristics of two FET's.

of the Type-B FET ($Lds = 4.0$ μm) is 1.5 V. Thus, Vk is independent of the process, but depends on the device dimensions.

In order to confirm these facts, Type-A and -B FET's having the same gate-width (Wg) of 12 mm are fabricated. Fig. 9 shows I–V characteristics of these two FET's. The value of Vk of the Type-A FET is 1.2 V which is lower by 0.3 V than that of the Type-B FET. The improvement of Vk by 0.3 V is due to the shorter Lds of the Type-A FET. By this, the Type-A FET is able to realize a higher efficiency even at a low operating voltage of 3.5 V. The increases of $Idss$ and the transconductance (Gm) are due to the shorter Lg. The Type-A FET is able to deliver a higher output power due to these increases. No increase of drain-conductance due to a short channel effect is found in the Type-A FET.

C. RF Characteristics

Fig. 10 shows characteristics of the output power ($Pout$) and the power-added efficiencies (ηadd) of Type-A and -B FET's with Wg of 12 mm. The measurements are carried out at a condition of $Vd = 3.5$ V, $f = 950$ MHz, $Pin = 20$ dBm, and Iidle $= 50$ mA. The output power delivered by Type-A and -B FET's are 32 dBm and 31.2 dBm with power-added efficiencies of 65% and 60%, respectively. The developed Type-A FET shows an output power higher by 0.8 dB and an efficiency higher by 5%, over those of the Type-B FET.

The characteristics of 200 mW-class and 1.5 W-class FET's using the new process and dimensions are comparatively summarized in Table III.

IV. REDUCTION OF CIRCUIT-LOSS

Reduction of conductor loss of the matching network is possible by increasing both the width (Wm) and the thickness (Tw), and by decreasing the length (Lm) of microstrip line. However, since the utmost value of Wm is 100 μm from the aspect of size reduction of MCIC, a new method to reduce the conductor loss without increasing the value of Wm has to be developed.

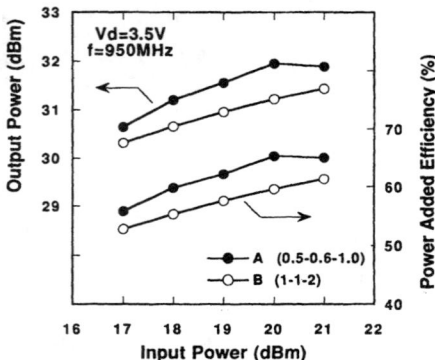

Fig. 10. Pin − Pout characteristics of two FET's.

TABLE III
CHARACTERISTICS OF TWO FET'S

Device	1st-stage FET (200mW-class)	2nd-stage FET (1.5W-class)
Gate Length Lg	0.6 μm	0.6 μm
Finger Length Fg	100 μm	150 μm
Number of Finger	20	80
Total Gate Width Wg	2 mm	12 mm
Vds	3.5V	3.5V
Frequency	950 MHz	950 MHz
Operation	Class AB	Class B
Input Power	7.0 dBm	20.0 dBm
Output Power	22.0 dBm	32.0 dBm
Power Added Efficiency	55 %	65 %

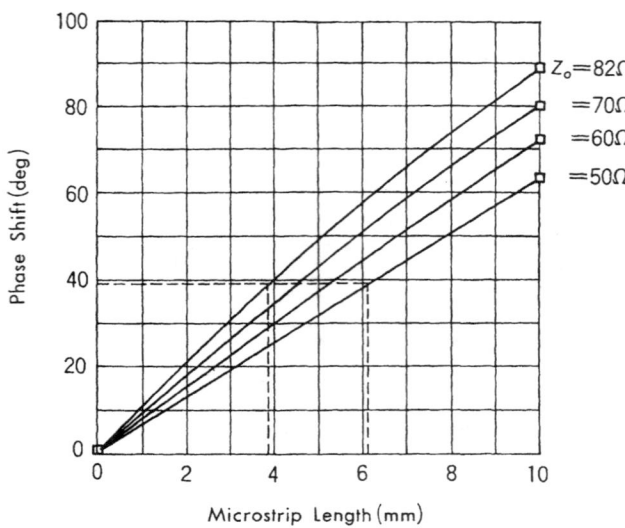

Fig. 11. Simulated relationship between the length and the phase shift as a function of Z_o of the microstrip line.

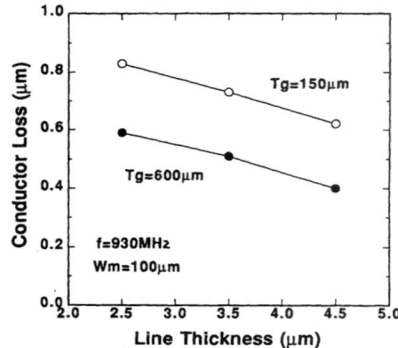

Fig. 12. Measured conductor losses of output matching networks.

Fig. 11 shows a simulated relationship between the length (Lm) of the microstrip line and the phase-shift as a function of the characteristic impedance (Zo) of the microstrip line, determined at a frequency of 1 GHz and a line thickness of 3.5 μm. This proves that the characteristic impedance has to be higher when the length of the microstrip line is reduced to obtain the same phase shift. A phase-shift of 40° has to be provided to match the optimum load impedance ($6.5 + j1.0$ Ω) of the second-stage FET to 50 Ω. When Zo is increased from 50 Ω to 82 Ω, the value of Lm is decreased from 6 mm to 4 mm.

In determining the conductor loss, actual output matching networks made of two different thicknesses of GaAs substrate ($Tg = 150$ μm or 600 μm) on which the output matching networks are fabricated, and three different thicknesses of the microstrip line ($Tw = 2.5$ μm, 3.5 μm, or 4.5 μm) are considered. These networks are designed to have a load impedance of $6.5 + j1.0$ Ω ($\pm 0.4 \pm j0.3$ Ω) at 930 MHz. Two-port S-parameters are measured by a vector-network analyzer, and the conductor losses are derived from the equation shown below.

$$\text{Loss} = |S_{21}|^2 / (1 - |S_{11}|^2).$$

Fig. 12 shows measured conductor losses of the output matching networks fabricated on GaAs. An increase of the value of Zo from 50 Ω to 82 Ω is brought by an increase of the thickness of the GaAs substrate (Tg) from 150 μm to 600 μm when the line width (Wm) is fixed at 100 μm. By increasing the value of Tg from 150 μm to 600 μm (Zo from 50 Ω to 82 Ω), the conductor loss can be reduced by $0.2 \sim 0.3$ dB. Thus, the network loss can be reduced by employing a microstrip line of high-impedance, by which the length (Lm) of microstrip line is considerably reduced and the desired phase-shift is accomplished.

In this MCIC, the output matching chip having the value of Tg of 600 μm ($Zo = 82$ Ω) is used in order to reduce the conductor loss. The thicknesses of the input and internal GaAs chips are 150 μm because the losses of these networks are negligible to those of the output network. As a result of this, an improvement of total efficiency of MCIC from 57% to 60% has been obtained.

Fig. 12 shows also a dependency of Tw on the conductor loss. By increasing the value of Tw from 2.5 μm to 4.5 μm, the conductor loss can be reduced. The thickness of the microstrip

TABLE IV
CHARACTERISTICS OF MCIC

Package Size :	10.7 mm x 11.6 mm
Volume :	0.4 cc
Frequency :	890~950 MHz
Vdd1, Vdd2 :	3.5 V
Vgg :	-3.5 V
Input Power :	7.0 dBm
Output Power :	> 31.1dBm (1.3 W)
Total Efficiency :	60 %

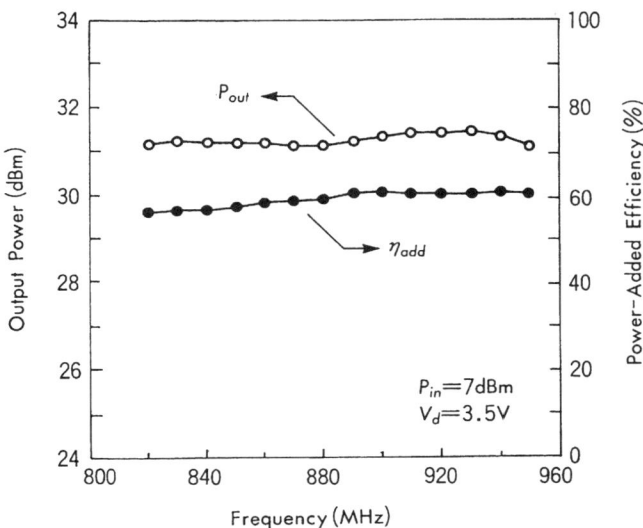

Fig. 14. Output power and power-added efficiency of the MCIC.

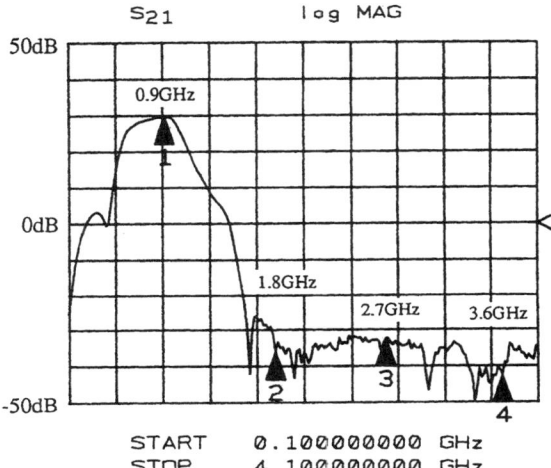

Fig. 13. Small signal gain (S_{21}) of the MCIC.

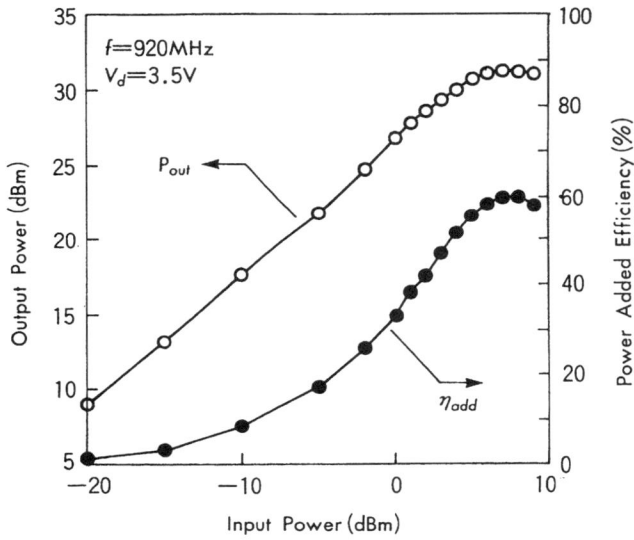

Fig. 15. $Pin - Pout$ characteristics of the MCIC.

line (Tw) is set at 4.5 μm since the mass production of lines having a thickness of more than 4.5 μm is difficult.

V. CHARACTERISTICS OF MCIC

Fig. 13 shows small-signal gain characteristics of MCIC. The reflection of the second harmonic is found at about 1.8 GHz. In evaluating the large-signal characteristics, the second harmonic level is found less than −40 dBc. This is due to the reflection circuit disposed in the output matching network. Fig. 14 shows characteristics of the output power and the power added efficiency of an MCIC operated at a condition of $Vd = 3.5$ V and $Pin = 7$ dBm. An output power of about 31.1 dBm (1.3 W) with a power added efficiency of 60% is achieved in a frequency band covering from 890 to 950 MHz. Fig. 15 shows characteristics of the output power and the power-added efficiency of MCIC operated at a condition of $Vd = 3.5$ V and $f = 920$ MHz. The maximum output power of 31.3 dBm is obtained when $Pin = 7$ dBm. The characteristics of MCIC are summarized in Table IV.

VI. CONCLUSION

A GaAs power multi-chip IC (MCIC) operable at a voltage of 3.5 V, yielding an output power of 1.3 W with a power added efficiency of 60% in a frequency band covering from 890 to 950 MHz, has been developed. The MCIC occupies a volume of only 0.4 cc which is one-half of that of a conventional power HIC. By improving the fabrication process, a substantial increase of breakdown voltage of GaAs MESFET is accomplished, and by this, both shorter Lgs and Lgd are realized without sacrificing the breakdown voltage.

Based on these developments, a MESFET operable at a voltage of 3.5 V with a high efficiency, has been developed. A considerable reduction of the conductor loss without increasing the line width has been accomplished by incorporating high-impedance microstrip lines into the output matching network.

The developed MCIC would be very effective to design the miniaturized and lightweight cellular phones operable at the 900 MHz band.

ACKNOWLEDGMENT

The authors wish to express their sincere appreciation to Dr. T. Takemoto, T. Onuma, Dr. M. Inada and A. Tamura for their encouragement. The authors are indebted greatly to Y. Ikeda, M. Miura and K. Kanaya for their earnest technical assistance.

REFERENCES

[1] D. Ngo et al., "Low voltage GaAs power amplifiers for personal communication at 1.9 GHz," *MTT-S Int. Microwave Symp. Dig.*, pp. 1461–1464, 1993.
[2] O. Ishikawa et al., "Advanced technologies of low-power GaAs ICs and power modules for cellular telephones," *GaAs IC Symp. Dig.*, pp. 131–134, 1992.
[3] Yutaka Hirano and Jun Fukaya, "Microwave high power GaAs FET," *1991 Microwave Workshop and Exhibition Dig.*, Japan, pp. 319–324, Sept. 1991.
[4] Yorito Ota et al., "High efficient very compact GaAs power modules for cellular telephones," *MTT-S Int. Microwave Symp. Dig.*, pp. 1517–1520, 1992.
[5] O. Ishikawa et al., "Cellular tele-communication GaAs power modules," *Appl. Microwave Mag.*, vol. 4, no. 3, pp. 83–88, Fall 1992.
[6] T. Takagi et al., "A UHF band 1.3 W monolithic amplifier with efficiency of 63%," in *Proc. Microwave and Millimeter-Wave Monolithic Circuit Symp.*, 1992, pp. 35–38.
[7] J. J. Komiak, "Design and performance of an octave band 11 Watt power amplifier MMIC," *IEEE Trans. Microwave Theory Technol.*, vol. 38, no. 12, Dec. 1990, pp. 2001–1006.
[8] M. Gat et al., "A 3 Watt high efficiency C-band power MMIC," *GaAs IC Symp. Dig.*, pp. 331–334, 1991.

A 900 MHz CMOS RF Power Amplifier with Programmable Output

M. Rofougaran, A. Rofougaran, C. Ølgaard, and A. A. Abidi

Integrated Circuits & Systems Laboratory
Electrical Engineering Department
University of California
Los Angeles, CA 90024-1594

Introduction

The power amplifier module constitutes the largest current drain on a wireless transceiver during transmit mode. In future cellular networks for digital wireless transceivers, the base station will adaptively regulate the transmitted *power levels* of each transceiver to enable the largest possible number of users to share a wireless channel. This requires a high-efficiency power amplifier with a digitally selectable output level spanning a wide range. The power amplifier reported here is intended for use in an all-CMOS frequency-hopped spread-spectrum transceiver operating in the 902-928 MHz band. It delivers a controllable power between 20-µW and 20-mW from a 3-V supply to the antenna.

Circuit Design

To deliver 20-mW into a typical 50 Ω antenna load, the power amplifier must apply a single-ended voltage swing of 2.8 V peak-to-peak across the load. With a 3-V power supply, this requires the use either of an impedance transformer, or of a differential drive to the antenna. The latter is used here. The quasi-differential power amplifier consists of a preamplifier stage, followed by a binary-weighted array of driver FETs selected by switches. The balanced output of an upconversion mixer drives the two identical halves of the power amplifier.

For high power-conversion efficiency, the driver FETs are biased in Class-C mode close to cutoff at a $V_{GS}-V_t \approx 200$ mV. This bias point was chosen after extensive simulations, to balance the specifications on maximum delivered power and the capacitance of the driver FETs. The gate-bias of these FETs is adjusted by an adaptive common-mode bias control voltage, which determines the level shift in a source follower between the preamplifier and driver FETs (Fig.1). An inductor load biases the preamplifier output to the supply voltage of 3-V, and allows the signal swing to exceed the supply. This on-chip inductor serves three vital functions. First, by tuning the capacitances of the 1-µm MOSFETs, it peaks the gain about 900 MHz. Second, it makes it possible to deliver a large signal to the driver stage. Third, as the signal swing exceeds the supply, it compensates for the source follower V_{GS} voltage drop which is exacerbated by the body effect on NMOSFETs. The source follower width is scaled with the driver FET width. A PMOSFET switch in series with the drain enables each source follower, and the corresponding driver FET. This method of switching least loads the high-frequency signal path. A five-element

Research supported by ARPA, Rockwell International and the State of California MICRO Program

array of binary weighted source followers and driver FETs, whose inputs and outputs in parallel, are connected to one preamplifier (Fig.2). The preceding upconversion mixer, to be integrated on-chip, will drive the preamplifier with a constant envelope sinewave. A 5-bit word applied to the PMOSFET select switches sets the power delivered to the antenna load across a 30 dB range.

The off-chip lossless matching network (Fig.2) between this IC and the antenna was an integral part of the power amplifier design. A high-frequency chip inductor tunes the output capacitance of the driver FET array and the pad capacitance of the IC for maximum power delivery into a 50 Ω load. This will also form a lowpass filter above the 902-928 MHz band of operation to attenuate harmonics created by the Class-C power amplifier. A three pole-pair dielectric resonator bandpass filter with a 26 MHz passband centered at 915 MHz is inserted between the amplifier output and antenna to suppress out-of-band noise and spurious components, while precisely defining the band of transmission. In a frequency-hopped system, only one tone is applied to the power amplifier at a time, so intermodulation products are negligibly small.

Experimental Results and Discussion

The power amplifier was fabricated in a 1-µm CMOS process, and including two on-chip spiral inductors it occupies about 1 sq mm of active area (Fig.3). The IC was mounted in a microwave package, in very close proximity to the two chip inductors used in the output matching network. For purposes of standalone testing, the inputs of the circuit were terminated on-chip with 50 Ω resistors, although in the complete system a voltage will drive the input. The two differential outputs were combined in a balun for single-ended measurements. In effect, this corresponds to the power amplifier driving a 100 Ω balanced antenna, such as a loop.

Measurements of the frequency response show that the on-chip inductor, which does not embody the sharp tuning characteristics of the suspended inductors we have reported previously [1], provides a wide gain characteristic nominally centered around 900 MHz. The off-chip matching network followed by the dielectric resonator filter precisely defines the transmit channel (Fig.4). With a constant input voltage, the output power may be swept over the expected 30 dB range under control of the digital word (Fig.5); power-level switching is essentially instantaneous. The power conversion efficiency of the amplifier approaches the designed 30% at maximum power, but drops off at low power levels where it is in any case unimportant. This efficiency is comparable to what has been reported in other power amplifiers fabricated in more complex technologies operating at 900 MHz or 2 GHz [2-4]. At 1-dB compression, this prototype delivers 32 mW (15 dBm) (Fig.6).

The first 900 MHz CMOS power amplifier operating at 3 V is reported. It exploits unique features of standard CMOS to provide digitally selectable power control over three orders of magnitude, to attain Class-C operation, and it exploits on-chip inductors to obtain high-frequency performance. This is intended to be integrated with an upconversion mixer and its preceding components as part of an all-CMOS wireless transceiver.

1. Chang, J.Y.-C., A.A. Abidi, and M. Gaitan, "Large Suspended Inductors on Silicon and their use in a 2-micron CMOS RF Amplifier", *IEEE Electron Device Letters*, v. 14, pp. 246-248, May 1993.
2. Yoshida, I., et al., "Highly Efficient 1.5 GHz Si Power MOSFET for Digital Cellular Front End", in *Int'l Symp. on Power Semiconductor Devices*, pp. 156-157, Tokyo, 1992.
3. Kermarrec, C., et al., "High Performance, Low Cost GaAs MMICs for Personal Phone Applications", in *Int'l Symp. on GaAs and Related Compounds*, Karuizawa, Japan, 1992.
4. Murai, S., et al., "A High Power-Added Efficiency GaAs Power MESFET Operating at a Very Low Drain Bias for use in L-Band Medium-Power Amplifiers", in *GaAs IC Symposium*, pp. 139-142, 1992.

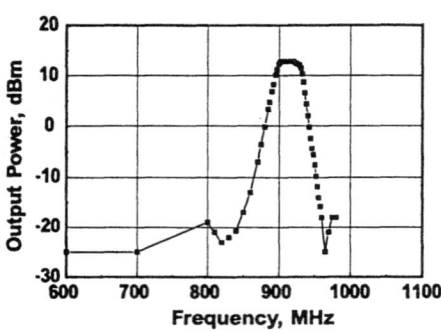

Fig.4. Measured frequency response of output power. Matching network and off-chip dielectric resonator filter define channel selectivity.

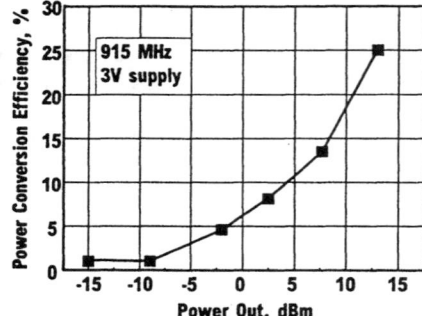

Fig.5. Measured power output vs. digital selection, at constant input voltage, shows sought 30 dB range and efficiency approaching 30%. Efficiency is the ratio of power delivered to load to the power drawn from the supply.

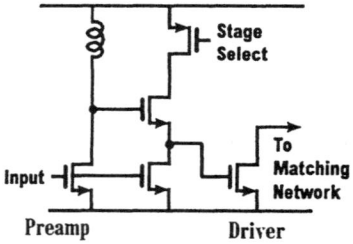

Fig.1. Single stage (half of quasi-differential) of power amplifier, showing preamplifier, level-shift, and output driver. Level shift is controlled by input common-mode.

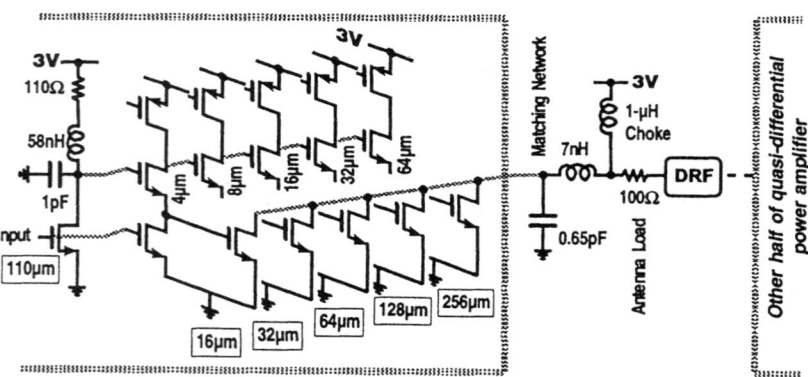

Fig.2. One half of complete power amplifier. Inputs and outputs of binary-weighted array of FETs are in parallel; FETs are selected by PMOSFET switches. Antenna is driven differentially by two such circuits. DRF is dielectric resonator filter. Bond pad capacitance and wirebond inductance are part of the matching network.

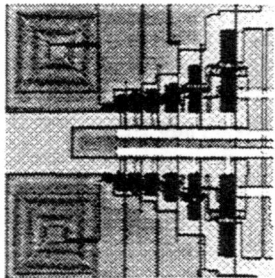

Fig.3. Detail of chip microphotograph. On-chip 60 nH inductors are 0.25 mm on a side. Outputs of binary-weighted array of driver FETs in quasi-differential amplifier are merged.

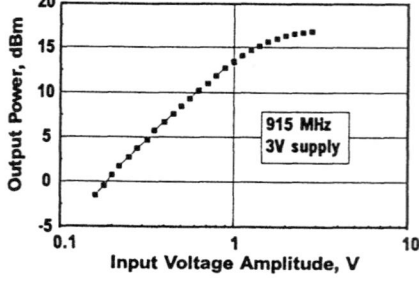

Fig.6. Measured compression characteristic of power output.

Highly Efficient UHF-Band Si Power MOSFET for RF Power Amplifiers

Isao Yoshida and Mineo Katsueda, *Members*

Central Research Laboratory, Hitachi, Ltd., Kokubunji, Japan 185

Shigeo Ohtaka and Yasuo Maruyama, *Nonmembers*, and Takeaki Okabe, *Member*

Semiconductor and Integrated Circuits Division, Hitachi, Ltd., Takasaki, Japan 370-11

SUMMARY

A low-voltage/high-efficiency Si power MOSFET has been developed for power amplifiers in mobile wireless telephone. Using submicrometer metal gate and self-aligned drain contact structures, the on-voltage and the output capacitance were reduced and the cutoff frequency was increased. As a result, compared with the conventional device, the power-added efficiency was increased by 10 percent. The power-added efficiency was 60 percent at 1.5 GHz when the output power was 1 W.

With 50-kHz offset in digital modulation, leakage power between adjacent channels of 50 dBc, and output power of 0.8 W, the power-added efficiency was 53 percent.

It was confirmed that the Si MOSFET had an excellent linear power amplification characteristic in the ultrahigh-frequency (UHF) band.

Key words: Si power MOSFET; submicrometer metal gate; self-aligned drain contact; power-added efficiency; digital modulation.

1. Introduction

Si power-MOSFETs have been used as low-loss and high-speed devices in switching power supplies, automobile OA equipment, and mobile telephones. Si power-MOS modules have been used in mobile telephones in automobiles. However, in portable mobile telephones, GaAs FET modules have been used because of their high efficiency [1]. To utilize Si power-MOSFETs in much wider areas, efficiency in the high-frequency region must be improved.

Morita et al. reported for the first time a UHF-band aluminum electrode power MOS [2]. We reported a power MOSFET for audio applications [3] and succeeded in drastically increasing the frequency region of this device by replacing the polycrystalline silicon gate electrode with a molybdenum gate electrode [4]. In addition, using the microfabrication process developed for the UHF-band MOSFET, the operating frequency of the power MOSFET was increased to over 1 GHz [5].

In 1983, a high-efficiency power MOSFET was developed for cellular telephones in the 860-MHz band [6]. Recently, because of the expansion of the automobile

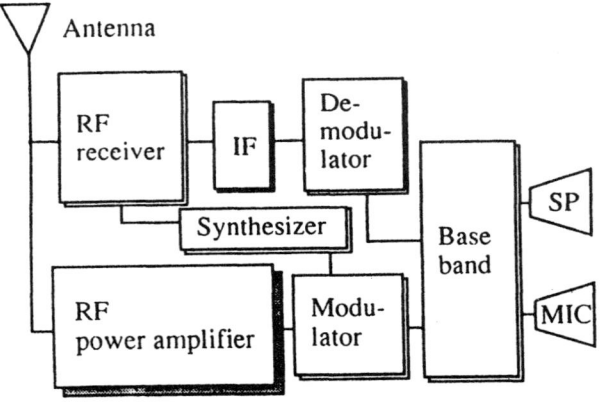

Fig. 1. Block diagram of a mobile telephone.

telephone market, not only power MOSFET devices but also power MOS modules with multiple MOS devices have been developed.

On the other hand, in 1986 Ishikawa et al. reported power MOS devices which could be operated at 2.45 GHz [7]. One hundred-watt class power MOS devices for satellite applications (860-MHz band) have also been developed [8, 9]. However, the power-added efficiency of Si power MOS devices was less than 50 percent when it was operated at a frequency over 1 GHz. This was much lower than that of GaAs FETs. In particular, no low-voltage power Si MOSFET which could be operated at voltages lower than 10 V was available.

Since 1990, we have been working on the development of low-voltage/high-efficiency power MOS devices for the UHF band [10, 11]. This paper describes the design technology of low-voltage/high-efficiency power MOS devices for the UHF band which can be used for the power amplifier in mobile telephones.

2. Technology for High Frequency/High Efficiency

2.1. Configuration of mobile telephone

Figure 1 shows a block diagram of a mobile telephone. In the mobile telephone, low power consumption is essential to prolong the conversation time. Therefore, high efficiency of the power amplifier at the front end is very important. About 90 percent of the power is consumed in the power amplifier during the transmission of the signal. Therefore, an increase of the power-added efficiency is a key factor in prolonging the battery life, which also prolongs the conversation time.

2.2. Analytical model for high-frequency/high-efficiency operation

The improvement of the power-added efficiency in a high-frequency power MOS is one of the most important aspects.

The analytical equations are given as follows [9]:

Power-added efficiency
$$\eta_{add} = \eta_d (1 - 1/G_p) \quad (1)$$

Drain efficiency
$$\eta_d = k\pi (1 - V_{on}/V_{dd}) \delta \quad (2)$$

Power gain $\quad G_p = k\pi (f_T/f) V_{dd}$
$$(1 - V_{on}/V_{dd}) \delta \quad (3)$$

Transfer efficiency $\delta = 1 - (\omega C_{oss})^2 R_{sub} R_L$
$$/(1 + (\omega C_{oss} R_{sub})^2) \quad (4)$$

where V_{on} is the on-voltage, V_{dd} is the power supply voltage, C_{oss} is the output capacitance, R_{sub} is the substrate resistance, R_L is the load resistance, f_T is the cutoff frequency, f is the operating frequency, k is the bias constant, and $\omega = 2\pi f$.

The power-added efficiency is given by the ratio of high-frequency output power to the dc input power, that is, the high-frequency power efficiency. To increase the power-added efficiency, it is necessary to increase both the drain efficiency and power gain.

The drain efficiency is the ratio of the high-frequency output power to the dc input power and can be increased by reducing the on-voltage. In particular, since it is determined by the ratio of the on-voltage to the power supply voltage, the requirement of the on-voltage becomes stricter as the supply voltage is reduced. The power gain can be increased by reducing the output capacitance which influences the cutoff frequency and transfer efficiency. Therefore, the improvement of the power-added efficiency of a power MOS device depends on the reduction of the on-voltage, increase of cutoff frequency, and the reduction of output capacitance.

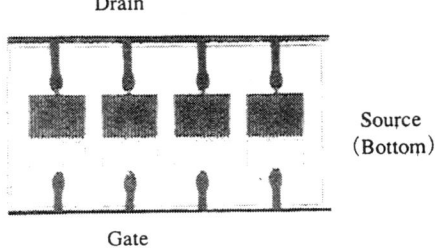

Fig. 2. Microphotograph of a power MOSFET.

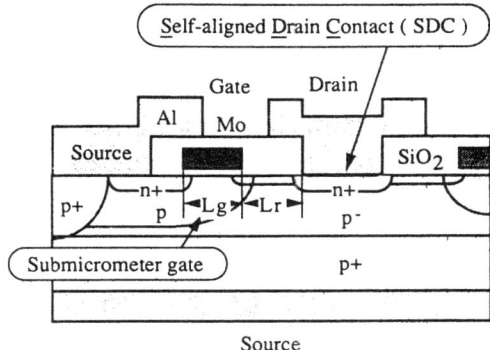

Fig. 3. Cross sectional view of a power MOSFET.

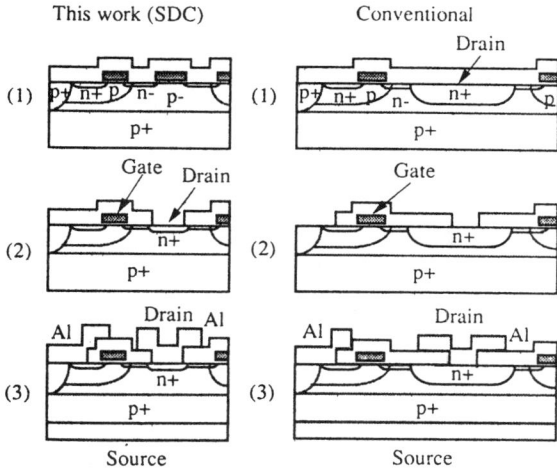

Fig. 4. Fabrication process.

3. Device Design

3.1. Design of device structure

As described in the foregoing, to increase the power-added efficiency of a power MOSFET, the reduction of on-voltage, increase of cutoff frequency, and reduction of output capacitance must be accomplished.

Figure 2 shows a photograph of a power MOSFET, and Fig. 3 shows the cross section of the major section of a cell. In this device, a molybdenum (Mo) gate electrode and unique source-substrate structure were adopted. In addition, to reduce the on-voltage and to increase the cutoff frequency, the submicron gate structure was used. To reduce the output capacitance, the self-aligned drain contact (SDC) was adopted. SDC was achieved by determining the positions of the drain region and contact and the length of the offset region using the self-aligned technique. This method is advantageous in forming the submicron gate structure with small scattering of breakdown voltage. The metal gate is essential to reduce the gate resistance.

The length of the fingers in the stripe pattern must be designed so that the product of the gate resistance and input capacitance does not influence the device performance. In this device, it was 250 μm. The source electrode on the chip surface is connected electrically to the source region and highly doped base region, and the base region is connected to the highly doped substrate. This structure is advantageous in reducing the source inductance and is one of the unique features of this Si power MOSFET. The p/p^+ substrate structure secures the safe operating region in the actual operation [11].

3.2. Fabrication process

Figure 4 shows a comparison between the SDC fabrication process and conventional fabrication process. In the conventional method, the drain junction is formed before forming the gate electrode. The gate and drain contact regions are not positioned by self-alignment, resulting in the large junction area. On the other hand, in the SDC method, the drain junction is formed after forming the gate electrode and drain contact region and their positions are self-aligned using a virtual gate electrode. As a result, the junction area is small.

Figure 5 shows an SEM photograph of the cross section of the device cell. The 0.8-μm gate electrode and offset region are observed clearly. The two-layered aluminum electrode and the protection layer also are

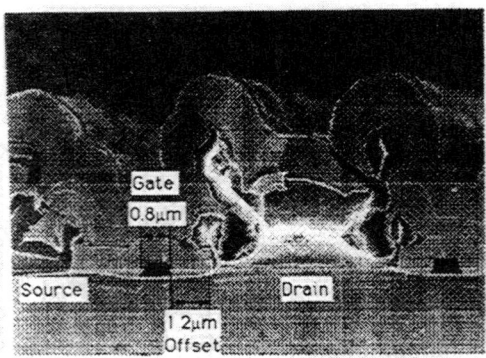

Fig. 5. SEM photograph of a cell.

shows one of the results. To suppress the punchthrough phenomena between the source and drain, the base region was boron-implanted. The cutoff frequency-gate length relationship was obtained using an ion dose where the cutoff frequency was calculated using the ratio between the mutual conductance and gate input capacitance. By observed. Hence, as described already, the drain junction area is reduced by the SDC method. Consequently, the output capacitance is reduced drastically.

3.3. Characteristic simulation

To optimize the structure of a power MOSFET with a submicron gate structure, the device characteristic was simulated by three-dimensional analysis. Figure 6

Table 1. Device parameters and dc characteristics

	This work	Conventional
L_g	0.8μm	1.4μm
L_r	1.2μm	2.5μm
BV_{ds}	20V	35V
I_{dmax}	6A	5A
R_{on}	0.3Ω	0.5Ω
g_m	1.3S	0.7S
C_{iss}	42pF	55pF
C_{oss}	15pF	22pF
C_{rss}	2pF	2pF

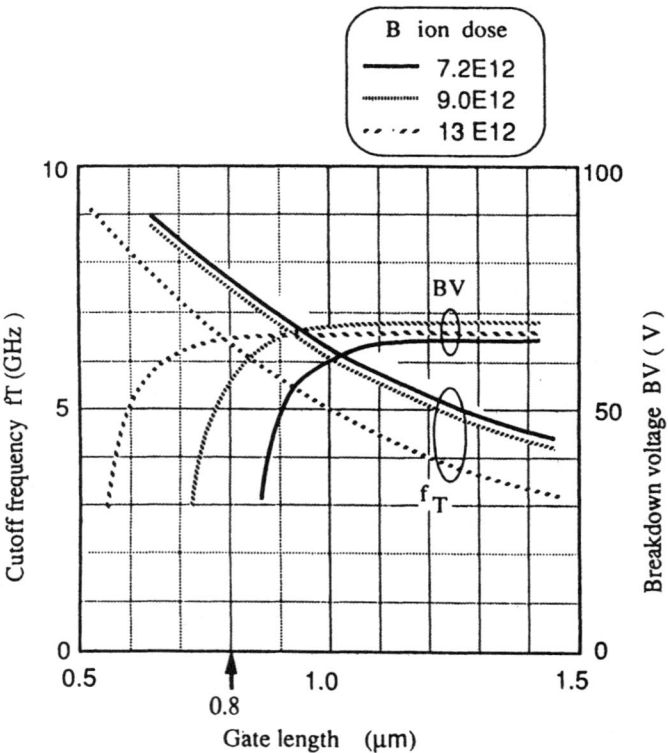

Fig. 6. Simulated cutoff frequency and breakdown voltage.

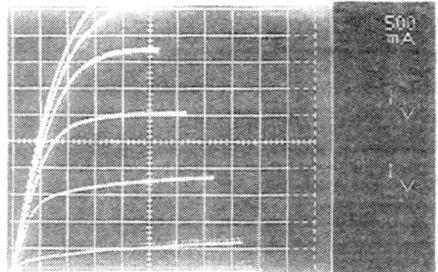

Fig. 7. Current-voltage characteristics of a fabricated device.

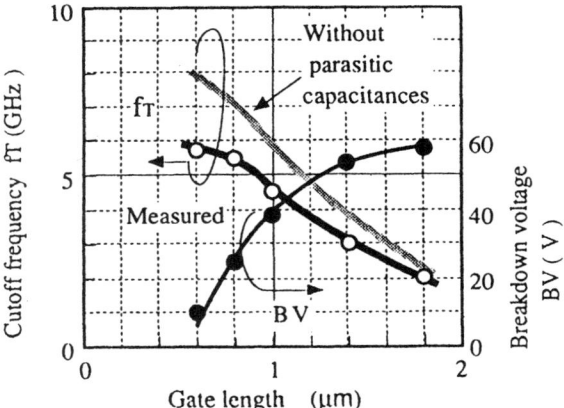

Fig. 9. Cutoff frequency and breakdown voltages of fabricated devices.

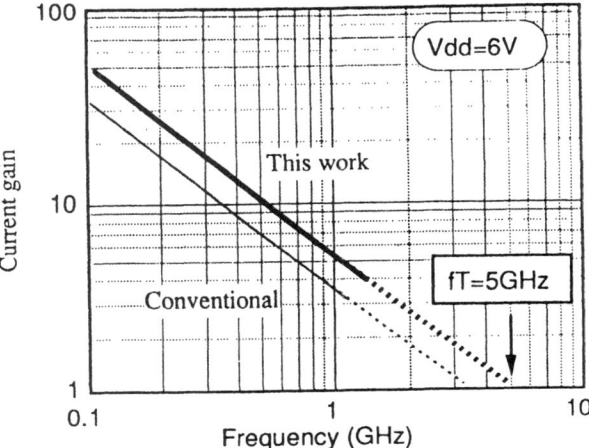

Fig. 8. Cutoff frequency of fabricated devices.

optimizing the boron dose, the breakdown voltage and frequency degradations could be avoided even if the gate length was less than 1 μm. The simulation indicated that the optimum dose of boron was $1 \times 10^{13}/cm^2$ when the gate length was 0.8 μm.

4. Experimental Result and Discussion

4.1. High-frequency/high-efficiency characteristics

Table 1 shows the dc characteristics of the test device and conventional device. The device constants for the test device were as follows. The gate length (L_g) was 0.8 μm, offset gate length (L_r) was 1.2 μm, and gate width was 3.2 cm. The maximum current was 6 A, the on-resistance was 0.3 Ω, and the mutual conductance was 1.3 S. Compared with the properties of a conventional device, those of the test device were 20 to 70 percent better. In addition, the output capacitance of the test device was 15 pF, which was about 40 percent lower than the output capacitance of the conventional device.

Figure 7 shows the dc current-voltage characteristics where the maximum gate voltage is 3.5 V in steps of 0.5 V. The on-voltage was 0.6 V when the current was 2 A.

Figure 8 shows the cutoff frequencies (f_T) of the test device and conventional device; f_T was obtained by using the current gain vs. frequency characteristic; and f_T for the test device was 5 GHz while that of the conventional device was 3.5 GHz, indicating that f_T of the test device was about 50 percent higher. These results agreed with those calculated by using the mutual conductance and input capacitance.

Figure 9 shows the experimental results of the f_T and L_g relationship; f_T increases with a decrease of L_g. When L_g was 0.8 μm, f_T exceeded 5 GHz. When L_g was 0.6 μm, f_T saturated because of the degradation of mutual conductance due to the punchthrough phenomenon. On the other hand, when L_g was less than 0.6 μm, the breakdown voltage was lower than 10 V. The low breakdown voltage occurred because the sum of the junction widths of the source and drain approaches the channel length, and the effective channel length became almost 0.1 μm. When L_g is shorter than 1 μm, f_T is influenced easily by parasitic capacitance such as pad capacitance. If these obstacles are removed, f_T of an Si power MOSFET can be higher than 10 GHz. It has been reported

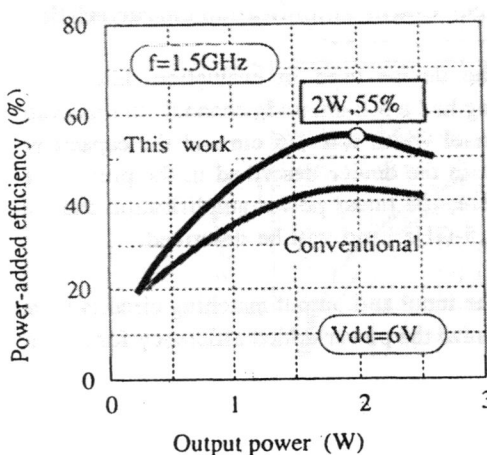

Fig. 10. Power-added efficiency vs. output power.

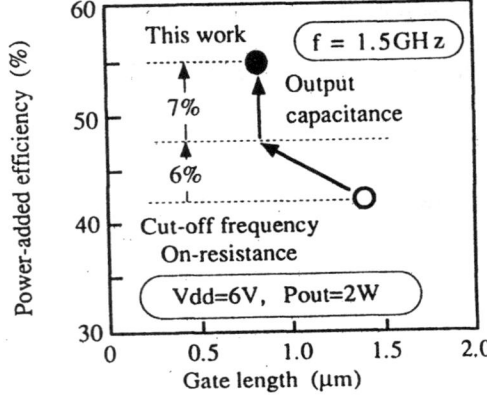

Fig. 11. Improvement of power-added efficiency.

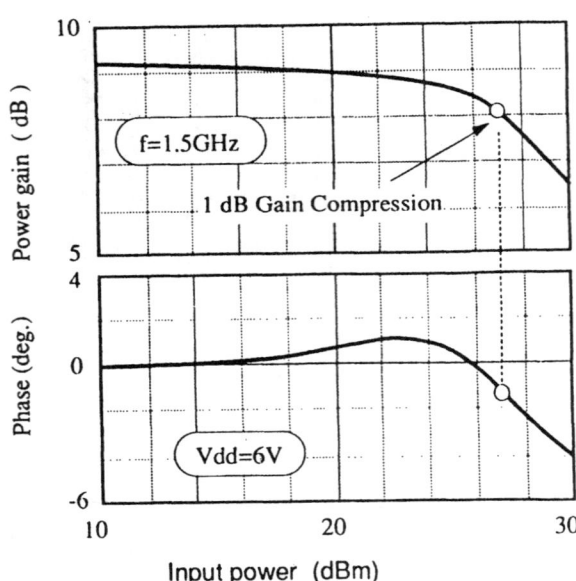

Fig. 12. AM-PM transformation characteristics.

that f_T of a MOSFET with a 0.1-μm gate was as high as 89 GHz [12].

Figure 10 shows the output power dependence of power-added efficiency of the test device where the property of a conventional device fabricated in the same lot also is shown. The operating frequency was 1.5 GHz and the supply voltage was 6 V. When the channel width was 3.2 cm, the output power and the power-added efficiency were 2 W and 55 percent, respectively. The power-added efficiency of 55 percent was 13 percent higher than that of a conventional device which was 42 percent.

Figure 11 shows the gate length dependence of power-added efficiency for an output power of 2 W.

Compared with the performance of a conventional device with L_g of 1.4 μm, the power-added efficiency of this test device was improved due to the increase of cutoff frequency and the decrease of on-voltage (by about 6 percent) and the reduction of the output capacitance (by about 7 percent).

The high-frequency power gain and phase transformation characteristic (AM-PM transformation characteristic) are important when a power MOS is used in a power amplifier. The input power dependencies of these characteristics are shown in Fig. 12. The input power dependence of power gain shows the linearity (linear amplification characteristic).

The 1-dB gain compression point is of interest, that is, the phase transformation characteristic represents the phase change caused by the saturation of the amplification of a device in a large amplitude amplification. The phase change should be as small as possible. This power MOS amplifier was optimized for a supply voltage of 6 V. The phase change was about 1° at 1-dB gain compression. These results indicate that this Si power MOSFET has a good linear amplification characteristic.

4.2. Low-voltage characteristic

The results described in the foregoing were obtained under operation at 1.5 GHz using a supply voltage

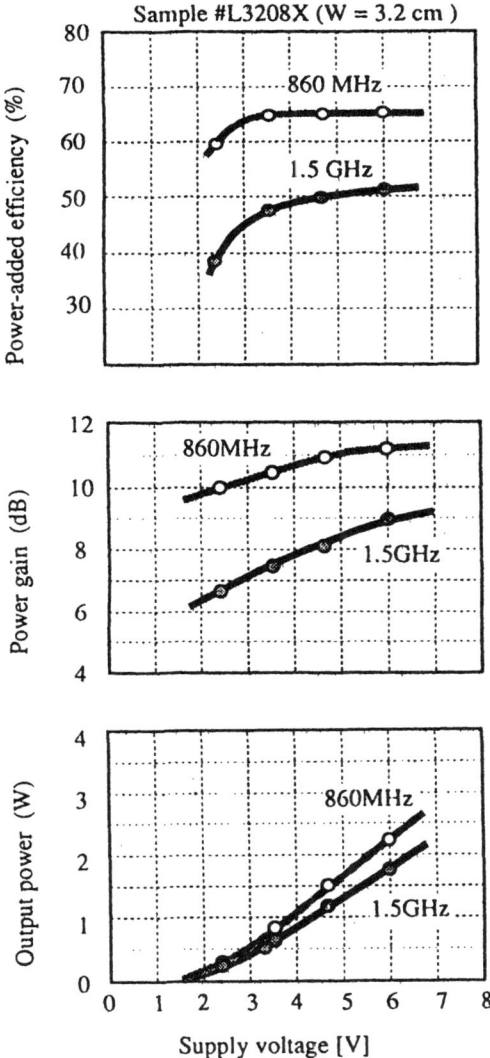

Fig. 13. Supply voltage dependences of power-added efficiency, power gain, and output power.

of 6 V. Figure 13 shows the operation for 860 MHz using a supply voltage of 2.4 V. The device was tuned for the supply voltage. Since the device was tuned to maximize the power-added efficiency, it was proportional to the square of the power supply voltage. As a result, the degradation of the power-added efficiency was small even when the supply voltage was decreased. In the operation at 1.5 GHz, it was 50 percent at a supply voltage of 6 V while it was 48 percent at a supply voltage of 3.6 V. In addition, when the supply voltage was decreased, the on-voltage must also be decreased. In the operation at 860 MHz using a supply voltage of 3.6 V, the power-added efficiency was 65 percent and the power gain was over 10 dB.

4.3. Linear amplification characteristic

The device used in evaluation described in the foregoing had a 3.2-cm wide channel. In this evaluation, the channel width was 1.6 cm and the capacitance was half that in the device described in the previous evaluation. Here, the linear power amplification characteristic in the 1.5-GHz band will be described.

The input and output matching circuits were tuned to maximize the power-added efficiency for the nonmodulation situation (that is, digital modulation is carried out). Then the measurement was carried out under digital modulation by $\pi/4$ shift quadrature phase-shift keying QPSK where the idling current under nonmodulation was 70 mA, and it was 150 mA under modulation to enhance linearity.

Figure 14 shows the results obtained with and without modulation where the operation frequency was 1.5 GHz and the supply voltage was 6 V. Figures 14(a), (b), (c) and (d) show the behavior of power-added efficiency, power gain, output power, and leakage power between neighboring channels, respectively. In Fig. 14(a), the output power increases proportionally with an increase of input power but it saturates when input power exceeds about 1 W. The increase of output power under modulation is larger than that under nonmodulation because of the difference in the power gain, as shown in Fig. 14(b), caused by the difference of the idling current. That is, when the idling current is increased, the gate bias voltage increases and the device is operated in the region of higher mutual conductance. Under digital modulation, the power gain is over 11 dB when the output power is 0.8 W.

In Fig. 14(c), both power-added efficiencies are over 50 percent. Under nonmodulation, it is as high as 60 percent when the output power is 1 W. On the other hand, the leakage power between the neighboring channels at the offset frequencies of 50 and 100 kHz tend to decrease with an increase of output power. In the figure, the point at which leakage power is 50 dBc at an offset frequency of 50 kHz is shown.

Under this condition, the output power of the device is about 0.8 W. The corresponding power gain (Fig. 14(b)) and power-added efficiency (Fig. 14(c)) are 11 dB and 53 percent, respectively. These values indicate that the linear amplification characteristic of this Si power MOS is excellent and is comparable with that of a GaAs FET at 900 MHz [14]. Although a detailed study must be carried out in the future, the excellent performance of

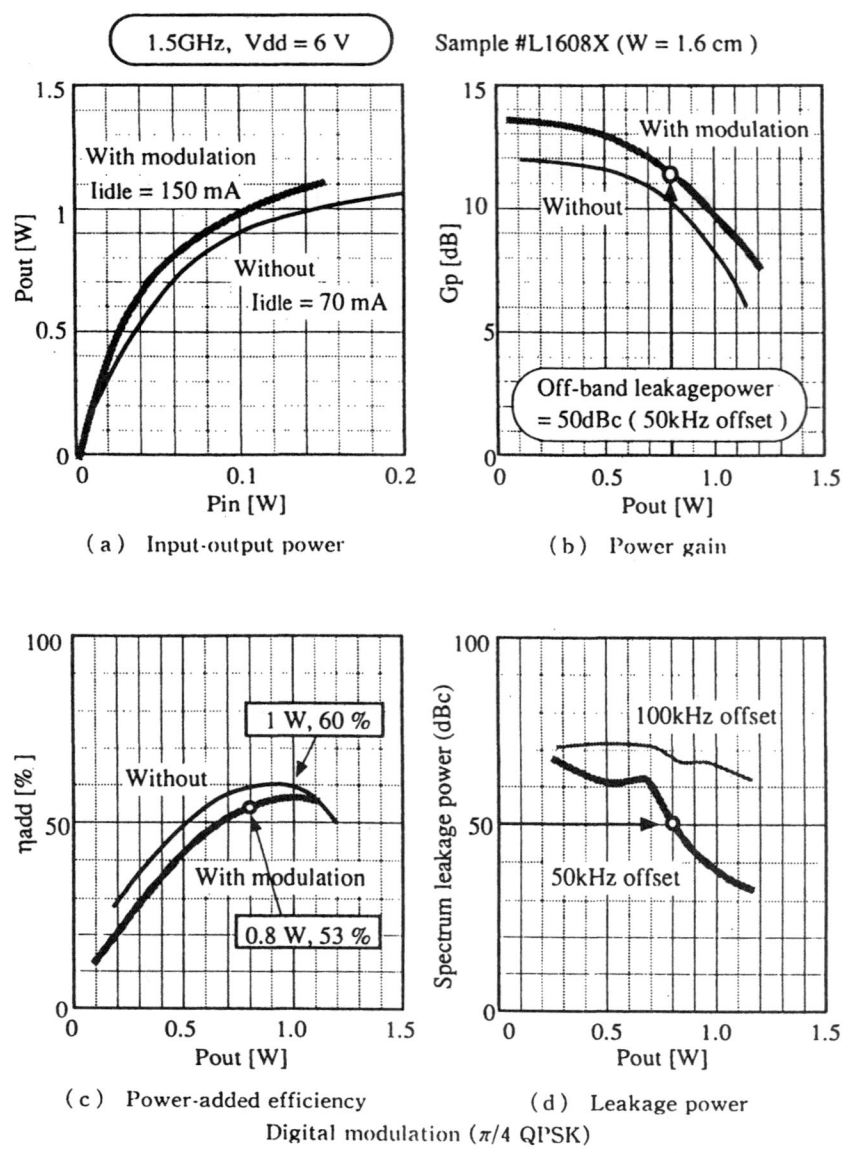

Fig. 14. Input/output characteristics.

this power MOSFET is believed to be due to the linearity of the current-voltage characteristic. As described in the foregoing, the Si power MOSFET has an excellent linear power amplification characteristic at 1.5 GHz.

5. Conclusions

A technology to realize a low-voltage and high-efficiency Si power MOSFET for mobile telephones has been established.

(1) This technology relies on the combination of a submicron metal gate and self-aligned drain contact structure. Because of this device structure, the on-voltage was decreased, the operation frequency was increased, and the output capacitance was reduced. As a result, the power-added efficiency was more than 10 percent higher than that of the conventional device.

(2) The typical high-frequency parameters at 1.5 GHz were the output power of 1 W, power-added efficiency of 60 percent, and power gain of 11.5 dB when the device was operated at the supply voltage of 6 V.

(3) When the offset frequency was 50 kHz in the 1.5-GHz digital modulation, the leakage power between the adjacent channels and power-added efficiency were 50 dBc, 0.8 W, and 50 percent, respectively.

These results indicate that an Si power MOSFET can provide an excellent linear power amplification characteristic in the UHF band.

In the near future, a detailed analysis of the linear amplification characteristic will be carried out. The development of a power MOS module for mobile telephones also is an important subject.

Acknowledgement. The authors thank Dr. Nagata, Dr. Seki, Mr. Sekine, Mr. Nakagoshi, Mr. Shimizu, Mr. Ohnishi, all of Central Research Laboratory, Hitachi, Ltd., and Mr. Masuda, Mr. Shimizu, and Mr. Fujita, all of the Semiconductor and Integrated Circuit Division, Hitachi, Ltd., for their valuable suggestions and discussions. They also appreciate the cooperation of Mr. Nagura and Mr. Haruyama, Hitachi, Eastern Section Semiconductor Corp., and Mr. Takei, Mr. Yanokura, and Mr. Seki, Hitachi Aerospace R & D Division in the device fabrication.

REFERENCES

1. Nojima, Yamao, and Kobayashi. Present status and future of high-frequency circuits for mobile telephones. I.E.I.C.E., **72**, 9, pp. 1007-1013 (1989).
2. Y. Morita, H. Takahashi, H. Matayoshi, and M. Fukuta. Si UHF MOS high-power FET. IEEE Trans. Electron Devices, **ED-21**, 11, p. 733 (Nov. 1974).
3. I. Yoshida, M. Kubo, and S. Ochi. A high-power MOSFET with a vertical drain electrode and meshed gate structure. IEEE J. Solid-State Circuit, SC-11, 4, p. 472 (Aug. 1976).
4. H. Ikeda, K. Ashikawa, and Urita. Power MOSFETs for medium-wave and short-wave transmitter. IEEE Trans. Electron Devices, **ED-27**, p. 330 (Feb. 1980).
5. T. Okabe, H. Itoh, and M. Nagata. A microwave silicon MOSFET. IEDM, Late News, p. 825 (1980).
6. H. Itoh, T. Okabe, and M. Nagata. Extremely high-efficient UHF power MOSFET for handy transmitter. IEDM, **4.7**, p. 95 (1983).
7. O. Ishikawa, H. Hamada, and H. Esaki. A 2.45-GHz power LD-MOSFET with reduced source inductance by V-grooved connection. IEDM, **6.7**, p. 166 (1985).
8. O. Ishikawa and H. Esaki. A high-power, high-gain VD-MOSFET operating at 900 MHz. IEEE Trans. Electron Devices, **ED-34**, 5, pp. 1157-1162 (May 1987).
9. Katsueda, Takei, Fujita, and Okabe. Power MOSFET for UHF Band. I.E.I.C.E. (C-II), **J72-C-II**, 12, pp. 1074-1081 (1989).
10. Katsueda and Okabe. 1.5-GHz Band Power MOSFET. 1990 Fall National Meeting of I.E.I.C.E., C-481.
11. I. Yoshida, M. Katsueda, S. Ohtaka, Y. Maruyama, and T. Okabe. Highly efficient 1.5 GHz Si power MOSFET for digital cellular front end. Proc. of 1992 ISPSD, Late News, Tokyo, **8.1**, pp. 156-157 (May 1992).
12. I. Yoshida, T. Okabe, M. Katsueda, S. Ochi, and M. Nagata. Thermal stability and secondary breakdown in planar power MOSFETs. IEEE Trans. Electron Devices, **ED-27**, 2, pp. 395-398 (Feb. 1980).
13. Y. Ran-Hong et al. 89-GHz f_T room-temperature silicon MOSFETs. IEEE Electron Device Lett., **13**, 5, p. 256 (May 1992).
14. Nojima and Nishiki. 900-MHz-band High-Efficiency Linear Amplifier. 1992 All National Meeting of I.E.I.C.E., C-253.

Analysis of Phase Characteristics of a GaAs FET Power Amplifier for Digital Cellular Portable Telephones

Toshio Ishizaki, Hikaru Ikeda, Yoshishige Yoshikawa, and Tomoki Uwano, *Members*

Materials and Components Research Laboratory, Matsushita Electric Industrial Co., Ltd., Kadoma, Japan 571

SUMMARY

The GaAs FET power amplifier in a Class AB operation used for digital cellular portable telephones must be studied for reduction of the transfer phase variation to attain high performance. However, the load impedance dependence of the phase characteristics, which is of significance, has not been analyzed in detail.

In this paper, for the analysis of phase characteristics, a new π-type FET equivalent circuit is proposed in which a negative conductance is introduced in place of the current source. By means of this equivalent circuit, an analysis and a simulation are carried out for the load impedance dependence of the phase charac- teristics. An increase of the drain-gate conductance and a decrease in the drain conductance accounts for the inverted V-shape phase-shift performance. The dependence on the load impedance also is illustrated. Further, the experimental results confirmed the correctness of the analysis since the phase variation is found to be a function of the drain-gate current.

Key words: GaAs FET; power amplifier; phase characteristics; equivalent circuit.

1. Introduction

Recently, the demand for an analog cellular telephone has skyrocketed. However, it is considered that the digital cellular telephones using a digital modulation system are more promising in the future because of higher channel capacity and better confidentiality. Both in Japan and in North America, TDMA systems using π4 shift QPSK signals have been introduced in practice. To transmit $\pi/4$ shift QPSK signals without error and without interfering with an adjacent channel, it is necessary to make the distortion of the power amplifier small. Especially, due to phase modulation, the distortion of the phase characteristics is a problem of which very little is known [1, 2].

Usually, in a battery-operated portable telephone, low-voltage operation and low-power consumption are required. Since a GaAs FET power amplifier can realize a high-power added efficiency at a low-voltage operation, it is suitable for portable telephones. Although a Class A operation of an FET is desired which has a low distortion, a Class AB operation is used from the viewpoint of efficiency. Hence, reduction of the distortion of a Class AB amplifier has become a subject of study to improve the performance of a power amplifier for the digital cellular portable telephone.

To date, numerous attempts have been reported for the analysis of a large signal operation of a GaAs FET amplifier. Willing et al. [3] has derived experimentally the bias voltage-dependent device model parameters of a GaAs FET and obtained the output power levels of the fundamental wave and harmonics from the time domain analysis. From the dc characteristics of a GaAs FET, Tajima et al. [4, 5] derived a large signal equivalent circuit model that can be treated in the frequency domain and carried out the analysis of the saturation output level. Materka and Kacprzak [6] reported extraction of the device model parameters under a large signal operation and a method for computing the power gain based on the harmonic balance method.

All these reports are concerned with the power gain and none of them discussed the phase characteristics of the amplifier. Also, Para reported the phase characteristics of a power limiter and indicated the variation of the input capacitance as a reason when the gate is swung into the positive direction by the input signal [7]. However, as demonstrated by the experimental results we obtained, the phase characteristics of the amplifier depend on the load impedance [1, 8]. To date, no detailed analysis has been carried out on the phase characteristics of a power amplifier and no clear explanation has been given with respect to the dependence on the load impedance.

The objective of this paper is to clarify by an analysis the dependency on the load impedance of the phase characteristics of the GaAs FET power amplifier for the digital cellular portable telephone.

In this paper, first, a new FET equivalent circuit is proposed which does not contain a current source by means of the introduction of a negative conductance. Based on this equivalent circuit, a theoretical analysis of the phase characteristics is carried out. Next, by means of a simulation, the input-output characteristics are computed so that the load impedance dependence of the phase variation is analyzed. Finally, measurement is carried out for an actual FET to verify the results of the analysis.

2. Analysis Model

2.1. Large signal operation FET equivalent circuit

The newly proposed FET equivalent circuit model is based on the model by Tajima et al. [4, 5] modified here to be more suitable for explanation of the load characteristics. The equivalent circuit is shown in Fig. 1.

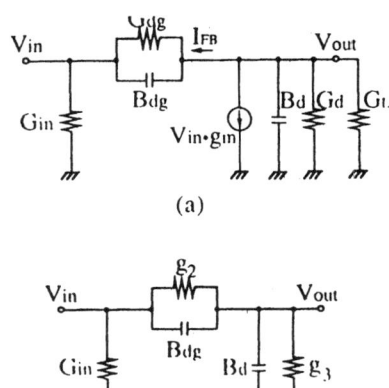

(a) Including current source
(b) Eliminating current source

Fig. 1. New equivalent circuit model for a GaAS FET.

In this equivalent circuit, only the analysis of the load impedance dependence is of interest. By limiting the input power level to a small value not to generate a nonlinearity on the gate side, the Schottky junction on the gate side is represented by a simple linear conductance. From the previous analysis, it is known that the effect of the nonlinearity of the gate side can be expressed by diode characteristics. Hence, this is not considered in this paper. Also, the parasitic impedances caused by the bonding and packaging are neglected.

The uniqueness of Tajima's equivalent circuit is in the fact that the nonlinearity of the circuit is represented in the frequency domain by means of the Fourier transform. The equivalent circuit in this paper is based essentially on the same concept. In this method, the calculations are extremely simple and a solution is obtained within a short time. Also, an analytical investigation is possible.

Figure 1(a) indicates the equivalent circuit model including a current source. From this, the equivalent circuit without a current source shown in Fig. 1(b) is derived. The equivalent circuit in Fig. 1(b) has a form identical to that reported in [8]. However, to explain the phase characteristics qualitatively, the equivalent circuit in [8] is specialized to a conceptual circuit representation. On the other hand, the equivalent circuit in the present paper is expanded in such a way that the circuit parameters are derived based on the rigorous mathematical modifications of the expressions. Both the amplitude and the phase can be evaluated quantitatively. To this end, a new concept not found in [8] is introduced. This new concept is to treat g_2 as a negative conductance.

2.2. Derivation of the circuit equation

Figure 1(a) is obtained by extracting a portion of Tajima's equivalent circuit after simplification. The method for extracting each parameter value is identical to Tajima's method. For this circuit, the Fourier transform is applied to the nonlinear element so that the latter is expressed as a linear element for the fundamental frequency component. Hence, the value of the nonlinear element is a function of the input power level. The input voltage V_{in} and the output voltage V_{out} are expressed in terms of the feedback current I_{FB} due to the fundamental component of the drain-gate current and are

$$V_{in} = V_{out} - \frac{I_{FB}}{G_{dg} + jB_{dg}} \quad (1)$$

$$V_{out} = -\frac{V_{in} \cdot g_m + I_{FB}}{G_d + G_L + jB_d} \quad (2)$$

where

g_m: transconductance,

G_{dg}: drain-gate conductance,

G_d: drain conductance,

G_L: load conductance,

B_{dg}: drain-gate susceptance, and

B_d: drain susceptance containing the load susceptance.

Here, $g_m > 0$, $G_{dg} \geq 0$, $G_d > 0$, and $G_L > 0$. Also, $B_{dg} > 0$ if the drain-gate element is only capacitive, whereas B_d can be either positive or negative depending on the load. In addition, $G_{dg} \approx 0$ under a small signal operation such that no breakdown occurs between the drain and the gate. Of these parameters, only g_m, G_{dg}, and G_d are inherently nonlinear and their values depend on the input power. The voltage transfer function is given by

$$\frac{V_{out}}{V_{in}} = \frac{-g_m + G_{dg} + jB_{dg}}{G_{dg} + G_d + G_L + j(B_{dg} + B_d)} \quad (3)$$

Also, the gain G and the phase ϕ can be expressed in terms of the input conductance G_{in} as

$$G \equiv \frac{G_L |V_{out}|^2}{G_{in} |V_{in}|^2} = \frac{G_L}{G_{in}} \cdot \frac{(-g_m + G_{dg})^2 + B_{dg}^2}{(G_{dg} + G_d + G_L)^2 + (B_{dg} + B_d)^2} \quad (4)$$

$$\phi = \tan^{-1}\left[\frac{B_{dg}(G_{dg} + G_d + G_L) - (B_{dg} + B_d)(-g_m + G_{dg})}{(-g_m + G_{dg})(G_{dg} + G_d + G_L) + B_{dg}(B_{dg} + B_d)}\right] \quad (5)$$

where, if one lets

$$g_2 \equiv -g_m + G_{dg} \quad (6)$$

$$g_3 \equiv g_m + G_d + G_L \quad (7)$$

then the voltage transfer function in Eq. (3) becomes

$$\frac{V_{out}}{V_{in}} = \frac{g_2 + jB_{dg}}{g_2 + g_3 + j(B_{dg} + B_d)} \quad (8)$$

Here, g_m is a sufficiently large number so that a gain of the FET is obtained. Since the drain-gate conductance G_{dg} is almost zero under a small signal operation, $G_{dg} \ll g_m$. Also, under a large signal operation, G_{dg} increases and a gain compression as shown in Eq. (4) occurs. However, the gain compensation is less than 10 dB in the cases cited in this paper. Hence, by assuming that $G_{dg} < g_m$, it is sufficient to use the defined regions of g_2 and g_3 as $-g_m \leq g_2 < 0$ and $0 < g_m \leq g_3$. Here, the meaning of the negative conductance g_2 is physical equivalence of a current source generating energy. Also, the gain G and the phase ϕ can be expressed as follows:

$$G = \frac{G_L}{G_{in}} \cdot \frac{g_2^2 + B_{dg}^2}{(g_2 + g_3)^2 + (B_{dg} + B_d)^2} \quad (9)$$

$$\phi = \tan^{-1}\left[\frac{-B_d g_2 + B_{dg} g_3}{g_2(g_2 + g_3) + B_{dg}(B_{dg} + B_d)}\right] \quad (10)$$

On the other hand, if the voltage transfer function of the circuit in Fig. 1(b) is computed, the result is identical to Eq. (8) for the voltage transfer function expressed with g_2 and g_3 in Fig. 1(a). Hence, it is confirmed that the equivalent circuit of the FET containing a current source can be transferred to a circuit without containing a current source by introduction of the negative conductance g_2. As a result, the characteristics of the FET, an active device, can be expressed by the π-type equivalent circuit made of resistors and capacitors only.

3. Analysis of Phase Characteristics

3.1. Interpretation of the circuit equation

In general, the phase difference between the input and output signals of a power amplifier is different in the small signal operation and in the large signal operation. As the input power is increased, the phase may advance from that for the small signal condition or it may retreat and a phase lag may occur. In this section, by means of the circuit equation, the phase variation versus that of the drain-gate conductance G_{dg} and the drain conductance G_d is investigated qualitatively.

First, the operand of the inverse tangent function $\tan^{-1}$ in Eq. (10) to represent the phase ϕ is defined by F:

$$F \equiv \frac{-B_d g_2 + B_{dg} g_3}{g_2(g_2 + g_3) + B_{dg}(B_{dg} + B_d)} \quad (11)$$

First, the phase change due to a variation of the drain-gate conductance G_{dg} is studied. The derivative of F with respect to G_{dg} is

$$\frac{\partial F}{\partial G_{dg}} = \frac{dg_2}{dG_{dg}} \cdot \frac{\partial F}{\partial g_2}$$

$$= \frac{\partial F}{\partial g_2}$$

$$= \frac{B_d g_2^2 - 2B_{dg} g_2 g_3 - B_{dg} g_3^2 - B_{dg} B_d (B_{dg} + B_d)}{\{g_2(g_2 + g_3) + B_{dg}(B_{dg} + B_d)\}^2} \quad (12)$$

For simplicity, let us consider the case in which g_m is sufficiently large so that $g_m \gg G_{dg}$, G_d, G_L, $|B_{dg}|$ and $|B_d|$. Then

$$\frac{\partial F}{\partial G_{dg}} = \frac{(B_{dg} + B_d) g_m^2}{\{-g_m(G_{dg} + G_d + G_L) + B_{dg}(B_{dg} + B_d)\}^2} \quad (13)$$

Hence, in the case where G_{dg} increases as the input power is increased, the phase advances as the input power is increased if $(B_{dg} + B_d) > 0$ whereas the phase retreats as the input power is increased if $(B_{dg} + B_d) < 0$.

Next, let us study the phase variation due to a change in the drain conductance G_d. The derivative of F with respect to G_d is

$$\frac{\partial F}{\partial G_d} = \frac{dg_3}{dG_d} \cdot \frac{\partial F}{\partial g_3}$$

$$= \frac{\partial F}{\partial g_3}$$

$$= \frac{(B_{dg} + B_d)(g_2^2 + B_{dg}^2)}{\{g_2(g_2 + g_3) + B_{dg}(B_{dg} + B_d)\}^2} \quad (14)$$

Hence, in the case where G_d increases with the incident power, the phase advances with an increase of the incident power if $(B_{dg} + B_d) > 0$ whereas the phase retreats with an increase of the incident power if $(B_{dg} + B_d) < 0$. On the other hand, in the case where G_d decreases as the incident power is increased, the phase retreats with an increase of the incident power if $(B_{dg} + B_d) > 0$ whereas the phase advances with an increase of the incident power if $(B_{dg} + B_d) < 0$.

If the load impedance dependence is considered, the value of the drain-gate conductance G_{dg} or the drain conductance G_d varies due to the effect of the voltage amplitude of the output signal. Hence, the degree of conductance variation due to an increase of the input power depends not only on the drain susceptance B_d containing the load susceptance but also on the load conductance G_L. From the foregoing results, the load impedance dependence of the phase characteristics is explained.

3.2. Setting of equivalent circuit parameters

Based on the equivalent circuits in Fig. 1, the input-output characteristics of the amplifier are simulated. First, the static characteristic of the FET is assumed as a model for the dc characteristic of the FET. The representation of the dc characteristic is the following derived by Tajima [4]:

$$\left.\begin{aligned} I_{ds}(V_{ds}, V_{gs}) &= I_{d1} \cdot I_{d2} \\ I_{d1} &= \frac{1}{k}\left[1 + \frac{V'_{gs}}{V_p} - \frac{1}{m} + \frac{1}{m}\exp\left\{-m\left(1 + \frac{V'_{gs}}{V_p}\right)\right\}\right] \\ I_{d2} &= I_{dsp}\left[1 - \exp\left\{-\frac{V_{ds}}{V_{dss}} - a\left(\frac{V_{ds}}{V_{dss}}\right)^2 - b\left(\frac{V_{ds}}{V_{dss}}\right)^3\right\}\right] \\ k &= 1 - \frac{1}{m}\{1 - \exp(-m)\} \\ V_p &= V_{po} + pV_{ds} + V_\phi \\ V'_{gs} &= V_{gs} - V_\phi \end{aligned}\right\} \quad (15)$$

The FET assumed in the simulation is a GaAs power FET with its gate width of 12 mm, gate length of 1 μm,

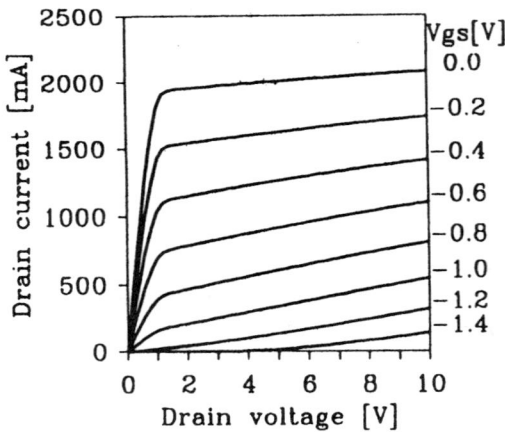

Fig. 2. Assumed dc characteristics of an FET.

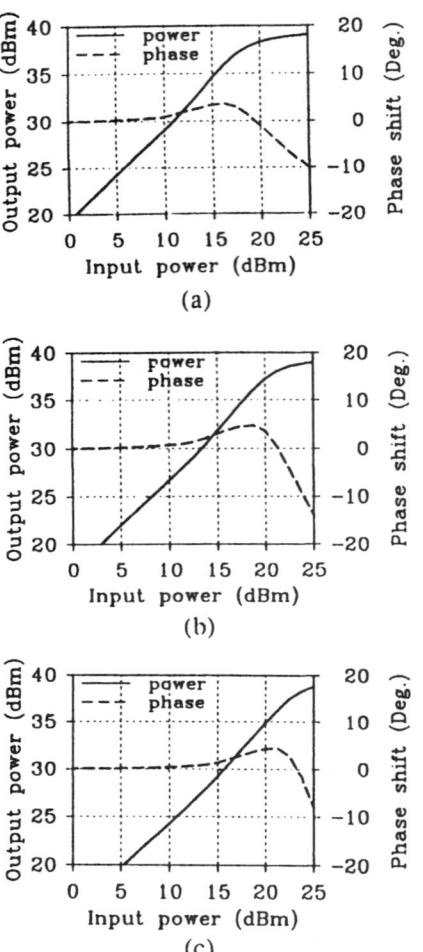

Fig. 3. Simulation results of input-output characteristics.

and drain-source saturation current I_{dss} of 2 A. The corresponding parameter values are V_ϕ = 0.3 V, V_{po} = 1.25 V, V_{dss} = 0.7 V, I_{dss} = 2600 mA, p = 0.05, m = 3.0, a = -0.2, and b = 0.6. The obtained dc characteristic is shown in Fig. 2.

Next, the breakdown between the drain and the gate is assumed as

$$i_{dg} = i_r \{\exp(\alpha_r \cdot V_{dg}) - 1\} \quad (16)$$

$$V_{dg} = V_{dgo} + v_{dg}\cos(2\pi ft) \quad (17)$$

based on the diode model by Materka [6]. Here, i_{dg} is the drain-gate current, i_r and α_r are model parameters, V_{dgo} is the dc bias voltage between the drain and the gate, and v_{dg} is the amplitude of the fundamental wave between the drain and the gate. The model parameter values are i_r = 0.5 μA and α_r = 0.55 [1/V].

Here, the drain-gate conductance G_{dg} is

$$G_{dg} = \frac{2f}{v_{dg}} \int_0^{1/f} i_{dg}\cos(2\pi ft)dt$$

$$= \frac{2i_r\exp(\alpha_r \cdot V_{dgo}) \cdot I_1(\alpha_r \cdot v_{dg})}{v_{dg}} \quad (18)$$

where I_1 is the first-order modified Bessel function of the first kind. From Eq. (18), it is found that the drain-gate conductance G_{dg} increases monotonically with the input power in correspondence with the breakdown between the drain and the gate.

As other equivalent circuit parameters, the drain-gate capacitance C_{dg} = 1.0 pF and the drain-source capacitance C_{ds} is included in the load. Also, the input conductance is G_{in} = 50 mS.

3.3. Simulation of the phase characteristics

The simulation was carried out at a frequency of 950 MHz. Also, at the bias point, the drain-source voltage V_{ds} = 6 V and the gate-source voltage V_{gs} = -1.2 V. Then the drain current was 170 mA.

First, the initial values of each nonlinear element are set. They are converged by a repetitive computation to the values noncontradicting the circuit voltage and current values [4]. The results of this simulation with changing load admittance are shown in Figs. 3(a) to (c). The corresponding admittances are: (a) (20 - j30) mS;

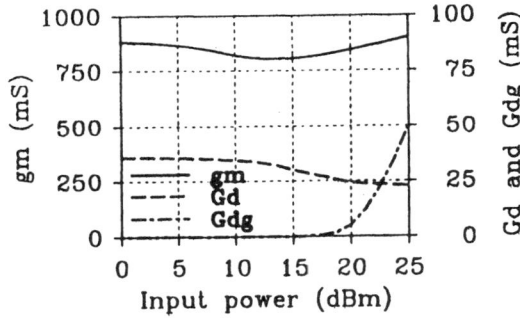

Fig. 4. Variations of nonlinear parameters.

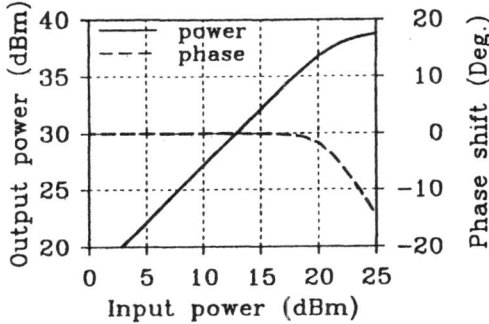

Fig. 5. Dependence of input-output characteristics on drain-gate conductance (G_{dg}); (G_d, g_m: fixed).

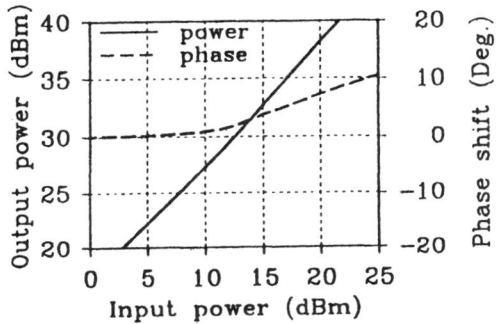

Fig. 6. Dependence of input-output characteristics on drain conductance (G_d); (G_{dg}, g_m: fixed).

(b) $(20 - j60)$ mS; and (c) $(20 - j90)$ mS. The load susceptance includes the capacitance between the drain and the source.

In Fig. 3, it is seen that the gain compression appears in the amplitude characteristic and the inverted V curve in the phase characteristic appears due to the retreat of the phase as the input power is increased. The shape of the variation curve is different depending on the load admittance.

In Fig. 4, the changes of the nonlinear parameters g_m, G_d, and G_{dg} with the input power are shown in the case of Fig. 3(b). The values of these elements under a small signal condition are g_m = 880 mS, G_d = 36 mS, and G_{dg} = 0 mS. The drain conductance G_d starts decreasing as the input power exceeds 10 dBm and becomes almost constant once it reaches 25 mS at 20 dBm. The drain-gate conductance G_{dg} increases suddenly as the input power exceeds 18 dBm and becomes 50 mS at 25 dBm.

For comparison with the analysis results in section 3.1, $(B_{dg} + B_d)$ is computed. The drain susceptance B_d in the case of Fig. 3(b) is -60 mS and the drain-gate susceptance B_{dg} is computed to be 6 mS from the drain-gate capacitance C_{dg} = 1.0 pF. Hence, $(B_{dg} + B_d)$ = -54 mS so that $(B_{dg} + B_d)$ is negative. The analysis results in section 3.1 indicate that the phase advances due to reduction of the drain conductance G_d as the input power is increased and retreats by the increase of the drain-gate conductance G_{dg}. This result coincides with the simulation result.

Further, to confirm the foregoing, the input-output simulation was carried out by fixing all but one nonlinear parameter. The results are shown in Figs. 5 and 6.

Figure 5 shows the input-output characteristic due to the nonlinearity of the drain-gate conductance G_{dg} when the drain conductance G_d and the transconductance g_m are forced to be fixed to the small signal values. In Fig. 5, saturation in the amplitude characteristic is seen as well as the phase retreat due to increase of the input power at the value larger than 18 dBm. Hence, it is confirmed that the saturation of the amplitude characteristic and the retreat of the phase are caused by the increase of the drain-gate conductance G_{dg}.

Also, Fig. 6 shows the input-output characteristic caused by the nonlinearity of the drain conductance G_d when the drain-gate conductance G_{dg} and the transconductance g_m are forced to be set at the small signal values. Although no amplitude saturation is seen in Fig. 6, phase advance can be seen as the input power is increased beyond 10 dBm. Hence, the phase advance is confirmed to be caused by the reduction of the drain conductance G_d.

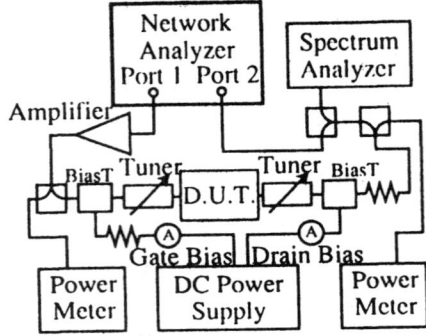

Fig. 7. Measurement setup for input-output characteristics.

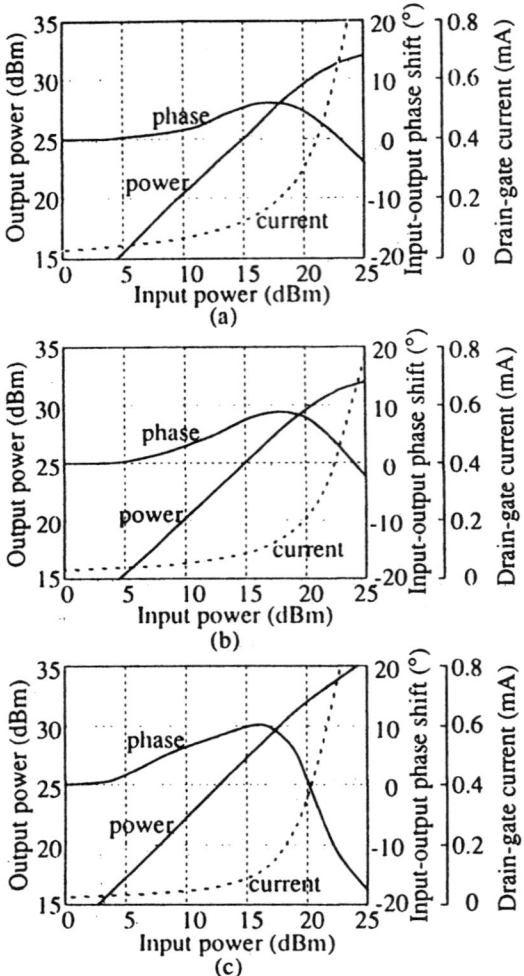

Fig. 8. Measurement results of input-output characteristics.

4. Experimental Results

Figure 7 shows the measurement setup of the FET input-output characteristics. A vector network analyzer is used in the measurement. The sinusoidal signal from the signal source in Port 1 is amplified to a needed driving power level by an ultralow distortion high output power amplifier and is provided to the FET to be measured. The output is taken out by a directional coupler and is sent to Port 2. The input and output matching of the FET was carried out by means of impedance tuners made of trimmer capacitors inserted in parallel at several locations in an air-type triplate transmission line. The bias voltages were applied to the FET via bias T's. The matching on the input side was adjusted to a maximum gain and is fixed. Each measurement instrument was controlled automatically by a computer while the measured data are taken into the computer for corrections of the loss and phase shift in the measurement setup.

In the measurement setup in Fig. 7, the current meter in a gate circuit measures the sum of the drain-gate current flowing out from the gate terminal and the gate-source current flowing into the gate terminal. Here, the gate-source starts flowing only when the voltage amplitude of the input signal swings into the positive value whereas this paper assumes an input power level small enough not to cause nonlinearity on the gate side. In practice, it was confirmed experimentally that the gate-source current starts flowing suddenly at a much larger input power level.

The FET used in the measurement is the one fabricated by an ion-implementation process and has a gate width of 12 mm, a gate length of 1 μm, and a drain-source saturation current of I_{dss} = 2 A. Also, the drain-gate capacitance is C_{dg} = 0.84 pF, the drain-source capacitance C_{ds} = 2.24 pF, the transconductance g_m = 1000 mS, and the drain conductance G_d = 63.5 mS. The measurement was carried out at a frequency of 950 MHz. The drain-source bias voltage was V_{ds} = 6 V while the gate-source voltage V_{gs} was adjusted such that the drain current under no signal is 100 mA.

Figures 8(a) to (c) show the measured input-output characteristics. The corresponding load admittances are: (a) (63 + j15) mS; (b) (87 + j34) mS; and (c) (160 - j81) mS. These all are shown in Fig. 9. Further, these load admittances are the values seen toward the load from the tip of the terminal lead of the package and do not contain the parasitic capacitance and inductance due to the package and the lead conductors. It is estimated

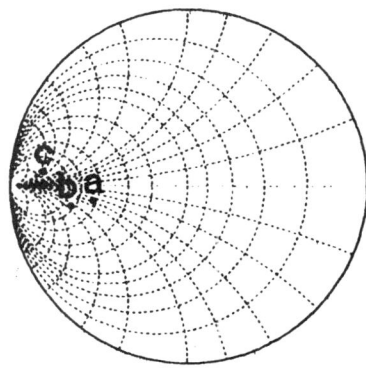

Fig. 9. Load admittances for measurement.

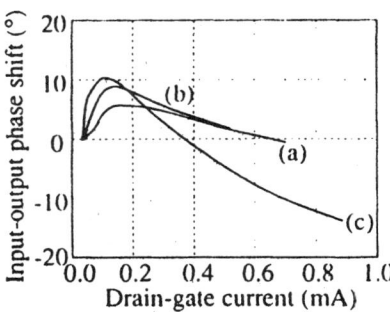

Fig. 10. Dependence of phase shift variation on drain-gate current.

that the actual load admittance seen toward the load from the drain electrode of the FET chip is located on the upper half-circle on the admittance Smith chart in Fig. 9 due to the effect of the parasitic capacitance, particularly the inductance.

Each graph in Fig. 8 shows the input-output characteristic of the identical FET. Due to the difference in the load admittance, the slope of the phase variation and the location and the magnitude of the maximum point are significantly different. The current between the drain and the gate also increases as the input power is increased and the rate of increase depends on the load admittance.

The relationship between the drain-gate current and the phase variation was studied. Figure 10 shows the phase characteristics in Fig. 8 rewritten here by taking the drain-gate current along the horizontal axis. From Fig. 10, it is confirmed that the phase variation has correlation with the drain-gate current and is a function of the drain-gate current with a load admittance as a parameter.

5. Conclusions

The phase transfer characteristic of a Class AB GaAs FET power amplifier was studied in terms of the change of the load impedance. First, by introducing a negative conductance, a new FET equivalent circuit was proposed in which no current source is included. By means of this equivalent circuit, a theoretical analysis of the phase characteristic becomes possible. From the results of the analysis and the simulation, it is found possible to explain that the phase variation becomes an inverted V-shape and its shape depends on the load impedance if the drain-gate conductance and the drain conductance are taken into consideration as the nonlinear elements. Based on the actual measurement results of an FET, it was confirmed that the phase variation depends on the drain-gate current; and the hence on the breakdown between the drain and the gate.

A future topic of investigation is accurate quantitative evaluation by using actual FET device parameters at high frequencies.

Acknowledgement. The authors thank Dr. M. Nagasawa, Director of Components and Devices Research Center, and Dr. T. Ishida, Director of Materials and Components Research Laboratory, Matsushita Electric Industrial Co., Ltd.

REFERENCES

1. H. Ikeda, Y. Yoshikawa, H. Kosugi, and T. Ishizaki. Phase characteristics of a power amplifier for digital mobile communications. 1991 I.E.I.C.E. Spring Convention, C-55.
2. Y. Yoshikawa, H. Ikeda, Y. Tsujimoto, and T. Ishizaki. Analysis of symmetric spread of $\pi/4$-shift QPSK signal spectrum caused by nonlinear amplification. I.E.I.C.E. '92 Fall Convention, C-43.
3. H. A. Willing, C. Rauscher, and P. Santis. Technique for predicting large-signal performance. IEEE Trans. Microwave Theory and Tech., MTT-**26**, 8, pp. 1017-1023 (Dec. 1978).
4. Y. Tajima, B. Wrona, and K. Mishima. GaAs FET large-signal model and its application to

circuit designs. IEEE Trans. Electron Devices, **ED-28**, 2, pp. 171-175 (Feb. 1981).

5. Y. Tajima and P. D. Miller. Design of broadband power GaAs FET amplifiers. IEEE Trans. Microwave Theory and Tech., **MTT-32**, 3, pp. 261-267 (March 1984).

6. A. Materka and T. Kacprzak. Computer calculation of large-signal GaAs FET amplifier characteristics. IEEE Trans. Microwave Theory and Tech., **MTT-32**, 2, pp. 129-135 (Feb. 1985).

7. T. Parra, M. Gayral, O. Llopis, M. Pouysegur, J. F. Sautereau, and J. Graffeuil. Design of a low-phase distortion GaAs FET power limiter. IEEE Trans. Microwave Theory and Tech., **MTT-39**, 6, pp. 1059-1062 (June 1991).

8. H. Ikeda, T. Ishizaki, Y. Yoshikawa, T. Uwano, and K. Kanazawa. Phase distortion mechanism of a GaAs FET power amplifier for digital cellular application. 1992 IEEE MTT-S Digest, M-4.

LOW VOLTAGE, HIGH POWER T/R SWITCH MMIC USING LC RESONATORS

Tsuneo Tokumitsu, Ichihiko Toyoda, and Masayoshi Aikawa

NTT Radio Communication Systems Laboratories
1-2356 Take, Yokosuka-shi, Kanagawa 238-03, Japan

ABSTRACT

A novel T/R switch for high-power/ low-distortion operation at low control voltage is proposed. LC-resonant switches composed of inductors, capacitors, and switching FET's are incorporated in the TX and RX arms to provide a reverse control scheme which removes the RF voltage limitation in the transmit mode. An LC-resonant T/R switch with total periphery of 2.88 mm exhibits 3-rd IMR less than -40 dB for input power up to 28 dBm when controlled at 0V/-2V.

INTRODUCTION

The maximum input power of conventional series/shunt FET T/R switches are limited by the "off" state shunt FET in the TX arm and series FET in the RX arm due to RF voltage limitation, as well as by the "on" state series FET in the TX arm due to RF current limiting. The linearity when the TX arm is "on" (Transmit mode) depends strongly on the RF voltage swing across drain/source and gate of the "off" state FET. Therefore, higher breakdown voltage FET's, higher control voltage, and RF voltage distribution by stacking FET's are necessary for conventional T/R switches to offer high power/ low distortion operations[1-5].

This paper proposes a novel T/R switch that uses an FET-switchable LC resonator composed of spiral inductors, MIM capacitors, and switching FET's, that is incorporated in the TX and RX arms. The most significant advantage of the switchable LC resonator is to provide a reverse control scheme: the LC-resonator is "off" when the switching FET's are "on" and vice versa, effectively removing the RF voltage swing in the "off" state FET's in conventional T/R switches. A 1.9 GHz LC-resonant

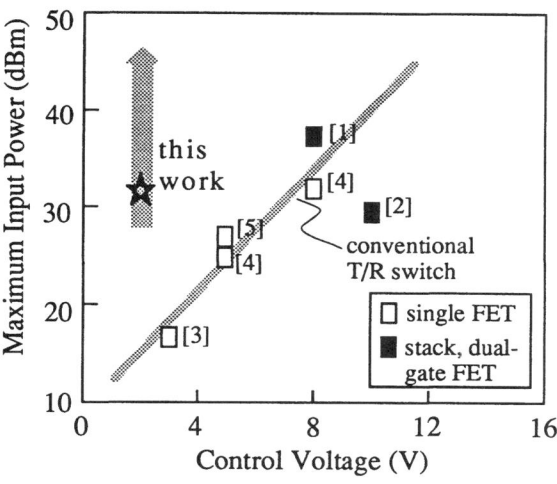

Fig. 1 Maximum input power comparison between the proposed LC-resonant T/R switch and conventional series/shunt FET T/R switches. Total FET peripheries of the proposed and referred switches are: 2.88 mm; 5.2 mm [1]; 3.5 mm [2]; 0.65 mm [3]; 4.7 mm [4]; 2.4 mm [5].

T/R switch MMIC with total GaAs FET periphery of 2.88 mm exhibits third IMR less than -40 dB for transmit power up to 28 dBm at 0V control voltage. Another control voltage of -2V is used for the receive mode.

MAXIMUM INPUT POWER COMPARISON

Figure 1 shows the maximum input power comparison at 1 dB gain compression between the proposed and conventional FET T/R switches. The maximum input power is a factor relating to the linearity of T/R switches. An LC-resonant T/R switch permits an RF input power greater than 30 dBm in the transmit mode when controlled at 0V/-2V. The achieved maximum input power is 5 dB higher than conventional T/R switches with the same FET periphery, and 15 dB higher than conventional switches operating at similar control voltages (-3V/0V). A wider FET periphery in the proposed T/R switch offers higher RF input power: whereas, conventional T/R switches do not increase the capability because of RF voltage limiting even when using wider FET peripheries.

T/R SWITCH DESIGN

Figure 2 shows a typical T/R switch scheme composed of switches SW-a, SW-b, SW-c, and SW-d. Since switching FET's in SW-a and SW-c are "off" in the transmit mode for conventional series/shunt FET T/R switches, the RF voltage swing between drain/source and gate has to be less than $|V_b - V_p|/2$ for linear operation of the T/R switch. Therefore, a higher breakdown voltage FET and higher control voltage are required as transmit power increases.

The problem is solved by incorporating an LC resonant switch which provides SW-a and SW-c with on-state switching FET's in the transmit mode. A proposed FET-switchable LC-resonant circuit is shown in Fig. 3, where a pair of FET switches (FET SW1, SW2) are combined with an inductor and capacitors. The LC-resonant circuit is open ("off") between ports ① and ② when the FET switches are in the "on" state because of the parallel resonance of inductor L and capacitor C_1. On the other hand, the circuit allows the signal to pass ("on") between ports ① and ② when the FET switches are in the "off" state because of the series resonance of inductor L and capacitors C_2 and C_s. When the FET switches are ideal and include no parasitics, the values of C_1 and C_2 are designed equal in order to provide the same resonant frequency in both FET states of the switching operation. However, since the stray capacitance, C_s, between FET drain and source is not negligible, capacitors C_1 and C_2 are designed to satisfy the following equation:

$$C_2 + C_s + C_1 C_s / (C_1 + C_s) = C_1.$$

There are three other LC-resonant circuit schemes considered. The circuit in Fig. 3 is used because of the smallest insertion loss.

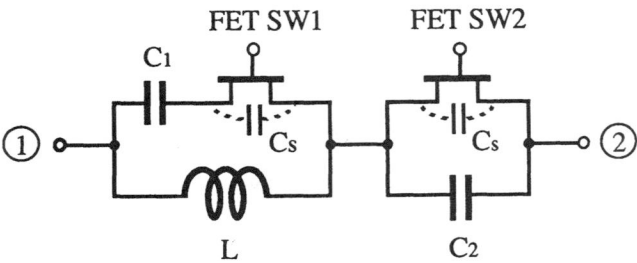

Fig. 3   FET-switchable LC-resonant circuit scheme

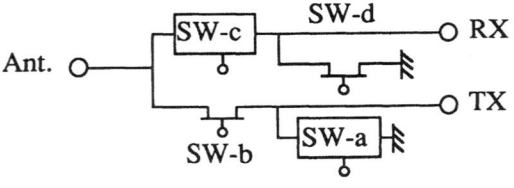

SW-a &c	State of FET's in SW-				Required FET perf.
	a	b	c	d	
Proposed LC resonator	ON	ON	ON	ON	Current proof
Conventional FET switch	OFF	ON	OFF	ON	Current & Voltage proof

Fig. 2   A typical T/R switch scheme and state of the switching FET's in SW-'s for proposed and conventional T/R switches in Transmit mode

Figure 4 shows a T/R switch incorporating the FET-switchable LC resonator. The LC resonators are used as a shunt switch in the TX arm and as a series switch in the RX arm. The series switch in the TX arm and the shunt switch in the RX arm are single FET switches. The LC resonators are positioned in places where a large RF voltage swing is applied in the transmit mode. In as much as all FET's in the transmit mode are in the "on" state (Vcont=0V), the RF voltage swing across the FET drain and source is negligibly small, that is, the T/R switch operates free from the RF voltage swing. A control voltage as small as twice the FET pinch-off voltage (Vp) is enough to switch from the transmit mode to the receive mode. Therefore, the maximum transmit power can be increased only by increasing the switching FET periphery and without any device improvement in breakdown voltage.

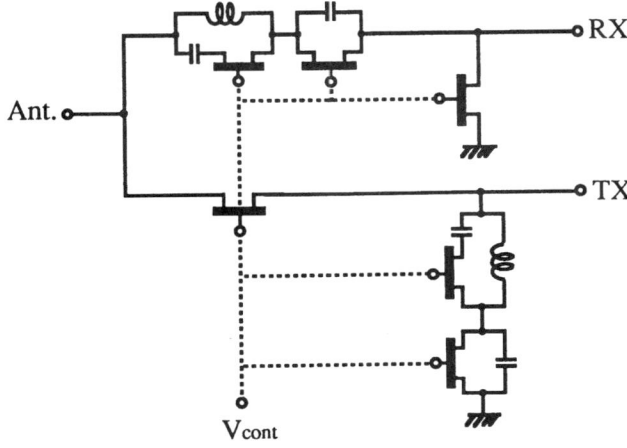

Fig. 4 Circuit scheme of a T/R switch incorporating FET-switchable LC-resonant circuits

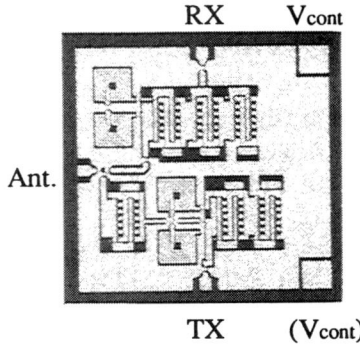

Fig. 5 Photograph of a fabricated LC-resonant T/R switch MMIC. Six switching FET's each of which has an FET periphery of 0.48 mm, a pair of 7 nH spiral inductors and MIM capacitors are integrated on a 2 mm x 2 mm x 0.6 mm GaAs chip.

The T/R switch MMIC is fabricated using uniplanar MMIC technology as shown in Fig. 5. The Q factor of the spiral inductor at 1.9 GHz is about 17. The RF current through the switching FET's of the LC resonator does not exceed the RF current through the switching FET's in a conventional T/R switch when the inductor value is less than 8.5 nH, because the loaded Q of the resonator is less than 1 at the inductor values.

MEASURED RESULTS

The linearity of the LC-resonant T/R switch in a two-tone measurement is shown in Fig. 6 compared with that of a conventional FET T/R switch, in which a pair of stacked FET's are used instead of the LC resonators in Fig. 4. These T/R switches are controlled at 0V or -2V/0V and the total FET periphery of the T/R switches is 2.88 mm. The LC-resonant T/R switch (solid lines) is linear enough to exhibit 3rd-IMR less than -40 dB for an input power up to 28 dBm. Whereas, with the conventional switch (dashed lines) the linearity degrades above 16 dBm input power. Even when controlled at -5V/0V, the linearity is degraded by the RF voltage swing above 24 dBm.

Figure 7 shows the measured frequency response of the proposed T/R switch in the transmit mode. An insertion loss of less than 1.5 dB, an isolation greater than 35 dB, and a return loss of better than 15 dB were measured between 1.8 GHz and 2.0 GHz. The frequency range is tuned to 1.9 GHz applications. Figure 8 shows the frequency response in the receive mode measured at control voltages of -2V (black lines) and -5V (gray lines). Very little difference between the two control voltages is observed.

The power handling capability of the proposed T/R switch in the transmit mode is shown in Fig. 9. The output power curve is linear for an input power up to 32 dBm, and the isolation does not show any degradation at the highest input power level.

CONCLUSION

The ability of the LC-resonant T/R switch to control high levels of power with low distortion even at low levels of DC voltage makes it ideally suited for transmit/receive switch applications in hand-held, battery powered communications equipment. Furthermore, this ability is believed to offer high power switchings at millimeter-wave frequencies where breakdown voltage of active devices is low.

ACKNOWLEDGMENTS

The authors would like to thank Dr. Kenji Kohiyama and Dr. Kozo Morita for their helpful discussions and continuous encouragement.

REFERENCES

[1] M. J. Schindler and T. E. Kazior, " High power 2-18 GHz MMIC T/R switch," Applied Microwave, pp. 90-94, Summer 1991.

[2] P. Benkoph, M. Schindler, and A. Bertrand, " A high power K/Ka-band monolithic T/R switch," IEEE Microwave and Millimeter-Wave Monolithic Circuit Symposium Dig. 1991, pp. 15-18.

[3] M. J. Schindler and A. M. Moris, "DC-40 GHz and 20-40 GHz MMIC SPDT switches," IEEE Microwave and Millimeter-Wave Circuit Symposium Dig. 1987, pp. 85-88.

[4] DC-6 GHz GaAs SPDT switch (SW-200): product of Adams Russel.

[5] DC to 6-GHz SPDT FET switch (TGS8704): product of Texas Instruments.

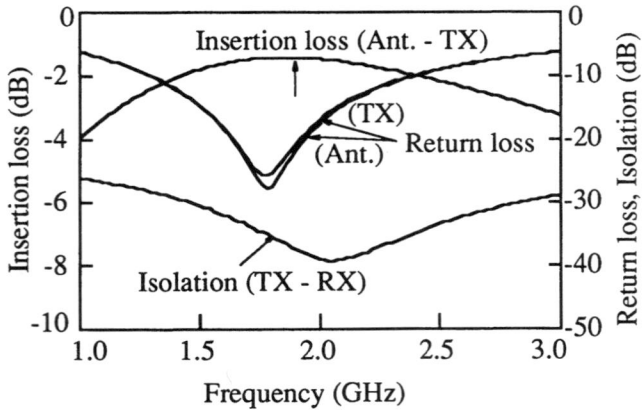

Fig. 7 Frequency response of the LC-resonant T/R switch in Transmit mode. V_{cont} is 0V.

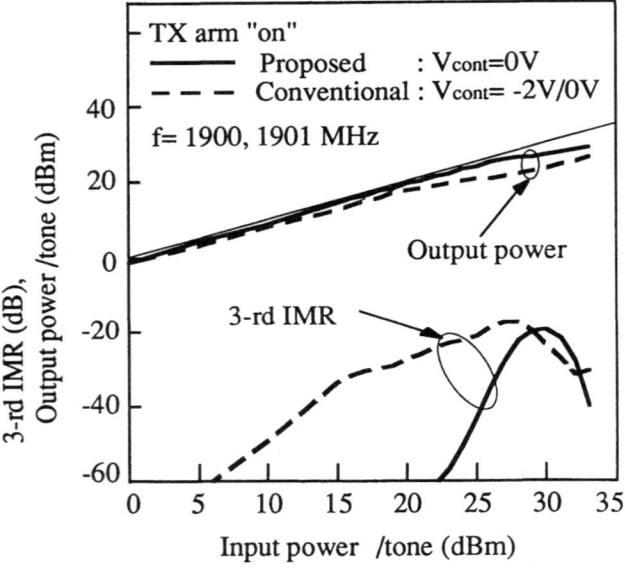

Fig. 6 Comparison of linearity between the proposed LC-resonant T/R switch and a conventional T/R switch with stacked FET's in Transmit mode. The FET periphery of both switches is 2.88 mm.

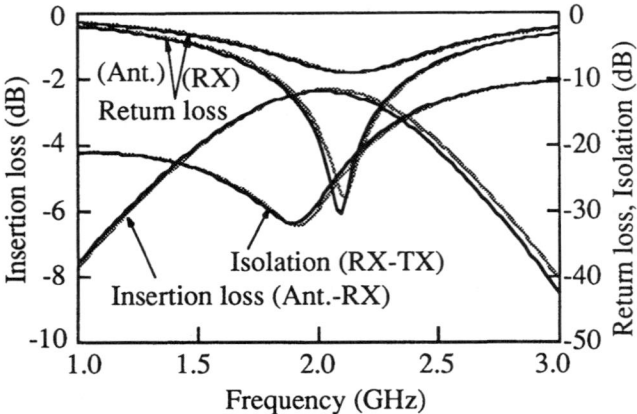

Fig. 8 Frequency response of the LC-resonant T/R switch in Receive mode. V_{cont} is -2V (black line) and -5V (gray line).

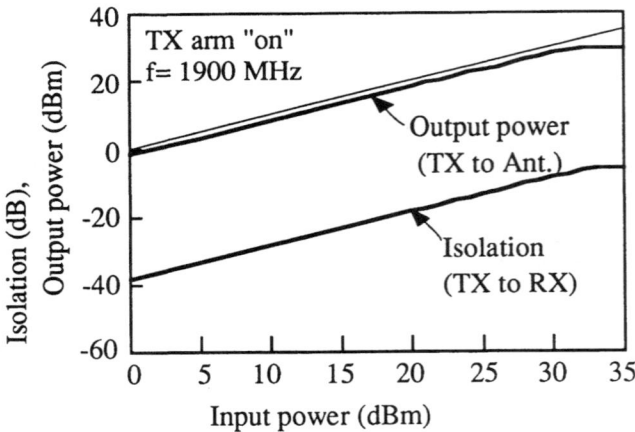

Fig. 9 Power handling capability of the proposed LC-resonant T/R switch in Transmit mode

WP 2.3: A GaAs High-Power RF Single-Pole Double-Throw Switch IC for Digital Mobile Communication System

Kazuo Miyatsuji, Shunsuke Nagata, Noriyuki Yoshikawa, Kazutune Miyanaga, Yoshiro Ohishi, Daisuke Ueda

Electronics Research Laboratory, Matsushita Electronics Corp. Osaka, Japan

Much effort has been devoted to integrate the receiving/transmitting switch since time divided multiple access (TDMA) became widely used for digital mobile communication systems. Conventional Si pin-diode switches require increased current bias as the transmit power is increased to maintain the ON state. A recently-developed GaAs monolithic switch IC can be operated with nearly zero power dissipation. However, there is distortion of the waveform as the transmit power is increased. This limits the power handling capability. Another disadvantage of the GaAs switch IC is that negative voltages are needed to control the ON and OFF states.

The high-power GaAs monolithic RF switch IC reported here handles over 5W (P1dB: 37dBm) with insertion loss less than 1.0dB using a circuit to feed forward the signal to the control gate. Positive voltage switching is achieved by integrating large coupling capacitors using high dielectric material, barium strontium titanate (BST), at inputs and outputs.

The RF single-pole double-throw (SPDT) switch IC is mounted in a SSO10, where better than 25dB isolation is achieved at a frequency of 1GHz.

The basic function block for TDMA system is shown in Figure 1, where receiving and transmitting occurs alternately. In general, low insertion loss is essential in both transmission (Tx) and reception (Rx) during the ON-state, and high isolation characteristics during the OFF-state. GaAs material is suited for this purpose since it has high electron mobility and low parasitic capacitance. However, the GaAs RF switch ICs developed previously have limited power-handling capability. This limit is not determined by FET maximum drain current or breakdown voltage, but by their threshold voltage. The power limitation of the switch originates from the negatively-biased control voltage being exceeded by the RF voltage swing on the 50Ω line at the input.

To overcome this drawback, synchronized voltage superposition to the control gate is used. Figure 2 shows the basic circuit of a single switching unit with high power-handling capability. The basic operation of the switch is as follows.

ON: FET1 is open-circuited, FET2 is short-circuited,
        RF signal is transferred to output.
        RF signal is also provided to control gate of FET1.

OFF: FET1 is short-circuited, FET2 is open-circuited,
        RF signal is isolated from the output terminal.

The notable feature of this circuit is the feed-forward capacitor, Cf, and diode connected in series between the gate of FET1 and RFin. This provides the superposed synchronous voltage to the gate of FET1. Figure 3 shows the simulated RF waveform with and without feed-forward capacitor. The sinusoidal waveform is undistorted when the feed-forward capacitor is inserted. The diode keeps a negative peak voltage to the gate of FET1 limiting loss of transmitted power.

Figure 4 shows the SPDT switch circuit. The input and output terminals are capacitively coupled to the basic switch units to operate the IC with positive control voltages. The large capacitors needed for low-loss signal coupling are provided by ferroelectric BST capacitor technology [1].

Figure 5 shows a cross section of this IC. The BST capacitors are formed using a sol-gel technique. The BST capacitor has a dielectric constant more than 40 times larger than that of a conventional SiN capacitor. A roll-off frequency over 2GHz is attained when the mole fraction of barium is 0.7. Capacitor variation on the same wafer tracks better than 5%. MESFETs have a Ti/Al recessed gate with LDD structure. The resultant gate length is 0.8µm. Resistors and diodes use the implanted channel region and Ti/Al Schottky junction.

A chip micrograph is shown in Figure 6a. The chip is 1.0x1.05mm^2. Owing to the BST capacitor technology, four large bypass capacitors of each 100pF are integrated in relatively small area. The package is 10-pin SSO type with 0.5mm lead spacing (Figure 6b). The package is the smallest reported.

Figure 7 shows 1dB compression point (P1dB) of the IC as a function of control voltage. P1dB over 37dBm is achieved for 5V control voltage. Loss and isolation characteristics of the IC are measured as shown in Figure 8. Insertion loss less than 1.0dB and isolation over 25dB are achieved at 1.0GHz. Performance of the SPDT switch is summarized in Table 1.

Acknowledgment

The authors thank G. Kano, M. Kazumura, and K. Itoh for encouragement, and C. Paz de Araujo, University of Colorado, Colorado Springs for research on and synthesis of BST material.

Reference

[1] Nagata, S., et al., "GaAs MMIC Chip-set for Mobile Communication System," ISSCC Digest of Technical Papers, pp. 172-173, Feb., 1993

Insertion loss	0.9dB
Bandwidth (loss<1dB)	0.5-3GHz
Isolation	27dB
P1dB	37dBm
Chip size	1.05x0.9mm^2

Table 1: SPDT switch performance

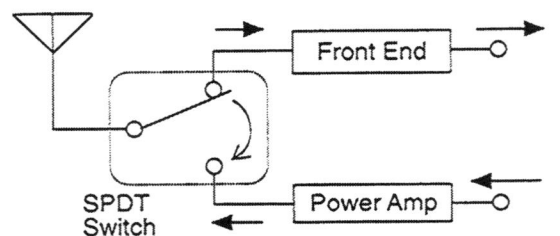

Figure 1: Basic function block used for TDMA system.

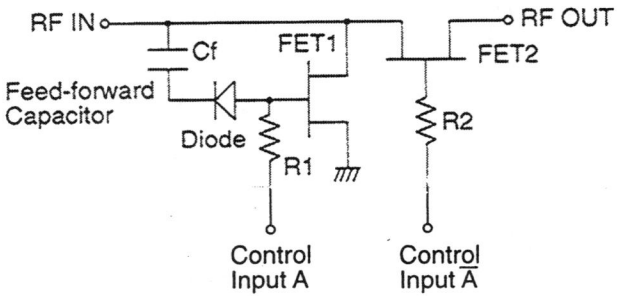

Figure 2: Basic circuit configurations of a single switch.

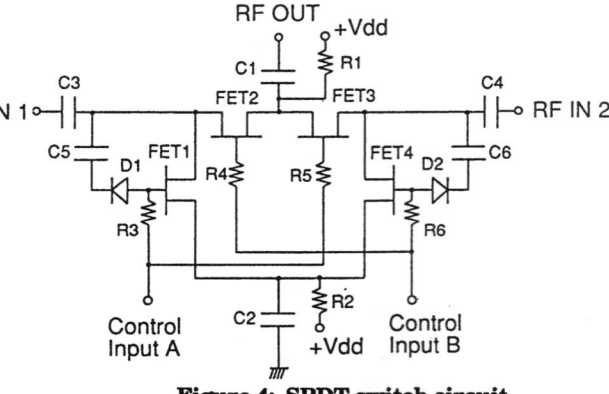

Figure 4: SPDT switch circuit.

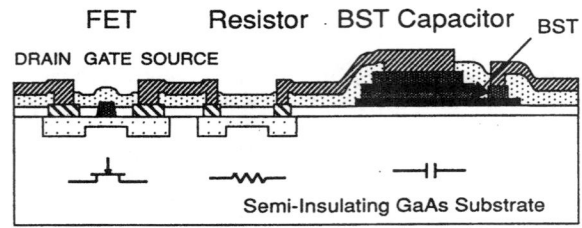

Figure 5: GaAs MMIC cross section with BST capacitor.
Figure 6: See page 305.

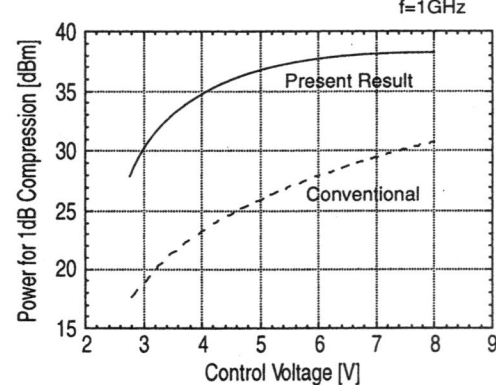

Figure 7: 1dB compression point (P1dB) vs control voltage.

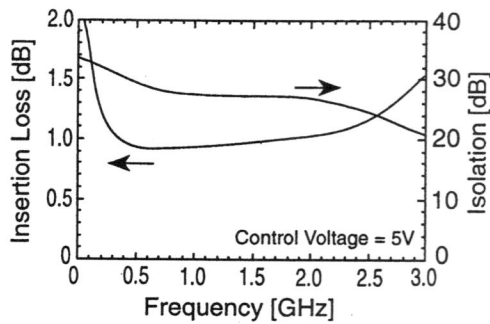

Figure 8: IC insertion loss and isolation characteristics.

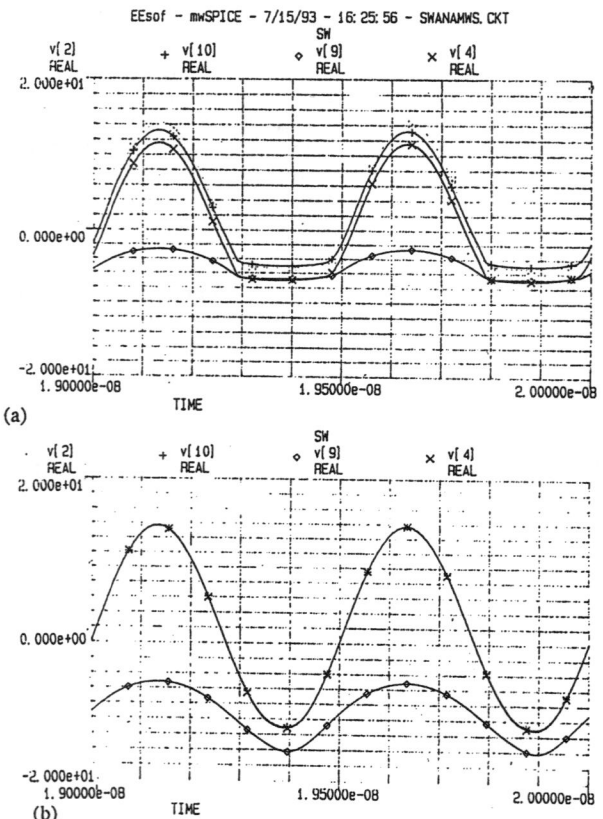

Figure 3: Simulated RF waveforms (a) without, (b) with feed-forward capacitor.

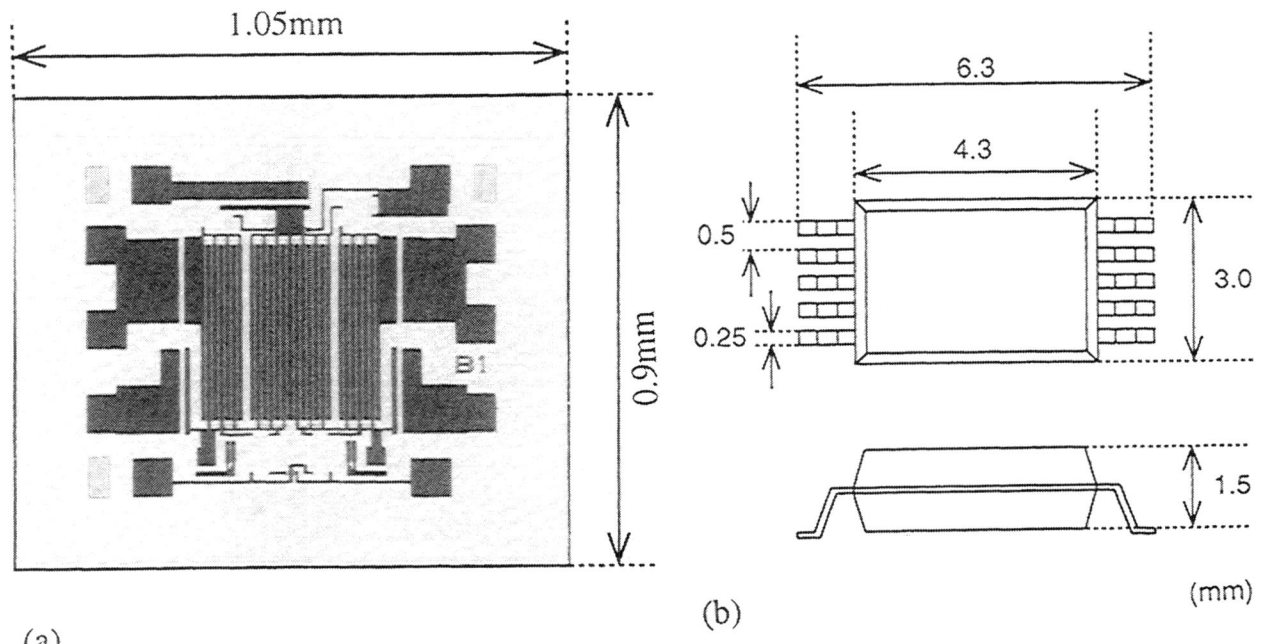

Figure 6: (a) IC micrograph, (b) SSO package outline.

6

Oscillators

THE noise performance of a receiver is determined by the parameters of several components. The noise figures (NF) of the front-end LNA and mixer are the major contributors under small-signal conditions. However, close-in interfering signals reaching the mixer can convert phase-noise sidebands from the local oscillator (LO) into the desired signal passband. Thus, the phase noise performance of the LO is an important specification, as are its frequency stability, tuning range, dc power requirements and RF output power. Most oscillator parameters improve significantly as the Q of the oscillator tank circuit is raised, and thus the choice of the resonator used to tune the oscillator is very important. Off-chip resonators in general have the highest Q, but interfacing from on-chip active devices to off-chip tank circuits at frequencies in the gigahertz range is quite difficult. Package parasitics and interface issues make this a challenging problem. In addition, the external elements consume valuable board space. There is thus great incentive to realize completely monolithic oscillators with all elements on chip, and this is an area of very active current research. Methods of modeling and calculating oscillator phase noise together with discussion of oscillator design approaches are described in this section.

A Simple Model of Feedback Oscillator Noise Spectrum

Introduction

This letter contains brief thoughts on the following points.
1) The relationships among four commonly used spectral descriptions of oscillator short-term stability or noise behavior.
2) A heuristic derivation, presented without formal proof, of the expected spectrum of a feedback oscillator in terms of known oscillator parameters.
3) Some experimental results which illustrate the validity of the simple model.
4) Comments on the effect of nonlinearity, specific spectral requirements for several applications, choice of resonator frequency and active element, and expected spectrum characteristics of several oscillator types.

Spectral Models of Phase Variations

Consider a stable oscillator whose measurable output can be expressed as

$$v(t) = A \cos [\omega_0 t + \phi(t)].$$

It is common to treat $\phi(t)$ as a zero-mean stationary random process describing deviations of the phase from the ideal. The frequency domain information about phase or frequency variations is contained in the "power" spectral density $S_\phi(\omega_m)$ of the phase $\phi(t)$ or, alternatively, in the "power" spectral density $S_{\dot\phi}(\omega_m)$ of the frequency $\dot\phi$. By analogy to modulation theory, we use ω_m to mean the modulation, video, baseband, or offset frequency associated with the noise-like variations in $\phi(t)$. The units of $S_\phi(\omega_m)$ are radians²/cps bandwidth or dB relative to 1 radians²/cps BW; $S_{\dot\phi}(\omega_m)$ is expressed in (radians/sec)² per c/s BW [1] [2]. The two are related by $S_{\dot\phi}(\omega_m) = \omega_m^2 S_\phi(\omega_m)$.

$S_{\dot\phi}(\omega_m)$ can also be expressed in terms of the equivalent rms frequency deviation Δf_{rms} in a given video bandwidth. Further, subject to the limitations that $\overline{\phi^2} \ll 1$ (small total modulation index) and that AM $\ll$ FM components, the normalized RF power spectrum $G(\omega - \omega_0)$ is identical to the two-sided spectrum of the phase $S_\phi(\omega_m)$; i.e., RF sidebands relative to the carrier are down by $S_\phi(\omega_m)$ expressed in decibels relative to 1 radian²/BW.

Relation to Oscillator Internal Noise

A basic requirement on an oscillator noise model is that it show clearly the relationship of the spectrum of the phase $S_\phi(\omega_m)$ to the known or expected noise and signal levels and resonator characteristics of the oscillator. A simple picture can be constructed using a model of a linear feedback oscillator. Minor corrections to the results are necessary to account for nonlinear effects which must be present in a physical oscillator. Assume a single resonator feedback network of fractional bandwidth $2B/\omega_0 = 1/Q$, where Q is the operating, or loaded, quality factor. For small phase deviations at video rates which fall within the feedback half-bandwidth $\omega_0/2Q$, a phase error at the oscillator input due to noise or parameter variations results in a frequency error determined by the phase-frequency relationship of the feedback network, $\Delta\theta = 2Q\dot\phi/\omega_0$. Thus, for modulation rates *less* than the half-bandwidth of the feedback loop, the spectrum of the *frequency* $S_{\dot\phi}(\omega)$ is identical (with a scale factor) to the spectrum of the uncertainty of the oscillator input phase due to noise and parameter variations. This uncertainty will be denoted $\Delta\theta(t)$, and its two-sided power spectral density $S_{\Delta\theta}(\omega_m)$.

For modulation rates *large* compared to the feedback bandwidth, a series feedback loop is out of the circuit. At these modulation rates, the power spectral density of the output *phase* $S_\phi(\omega)$ is identical to the spectrum of the oscillator input phase uncertainty $S_{\Delta\theta}(\omega_m)$.

For a physical oscillator the spectrum $S_{\Delta\theta}(\omega_m)$ of the input phase uncertainty $\Delta\theta(t)$ is expected to have two principal components. One component is due to phase uncertainties resulting from additive white noise at frequencies around the oscillator frequency, as well as noise at other frequencies mixed into the pass band of interest by nonlinearities. The second component is due to parameter variations at video frequencies which affect the phase (such as variations in the phase shift of a transistor due to carrier density fluctuations in the base resistance). The additive noise component of $S_{\Delta\theta}(\omega_m)$ is identical to the spectral density of the noise voltage squared relative to the mean square signal voltage. For white additive noise, this component is flat with frequency. For a feedback oscillator with an effective noise figure F, the two-sided $S_{\Delta\theta}(\omega) = 2FKT/P_S$; P_S is the signal level at oscillator active element input.

The video spectrum of parameter variations is found typically to have a power spectral density varying inversely with frequency (a $1/\omega_m$ or $1/f$ spectrum). The total power spectral density of oscillator input phase errors is of the form $S_{\Delta\theta}(\omega_m) = \alpha/\omega_m + \beta$, where α is a constant determined by the level of $1/f$ variations and β is $2FKT/P_S$ for two-sided spectra.

To find $S_\phi(\omega_m)$ or $S_{\dot\phi}(\omega_m)$, we use the fact that

$$\text{for} \quad \omega_m < \frac{\omega_0}{2Q} \quad S_\phi(\omega) = \left[\frac{\omega_0}{2Q}\right]^2 S_{\Delta\theta}(\omega_m)$$

$$\omega_m > \frac{\omega_0}{2Q} \quad S_{\dot\phi}(\omega) = S_{\Delta\theta}(\omega_m).$$

A suitable composite expression is

$$S_\phi(\omega_m) = S_{\Delta\theta}\left[1 + \left(\frac{\omega_0}{2Q\omega_m}\right)^2\right].$$

This yields an asymptotic model for $S_\phi(\omega)$ shown on log-log scales in Fig. 1.

The model can be summarized as follows.

$S_\phi(\omega_m)$ decreases with ω_m
  at 9 dB/octave up to the point where $1/f$ effects no longer predominate.
  at 6 dB/octave from that point up to the feedback loop half-bandwidth.
  at 0 dB/octave above that frequency up to a limit imposed by subsequent filtering.

$S_{\dot\phi}(\omega)$ decreases at 3 dB/octave up to the first breakpoint, is flat with frequency up to the feedback baseband bandwidth, and increases at 6 dB octave above that point.

The case where $1/f$ effects predominate only for frequencies small compared with the feedback loop bandwidth is shown here as an

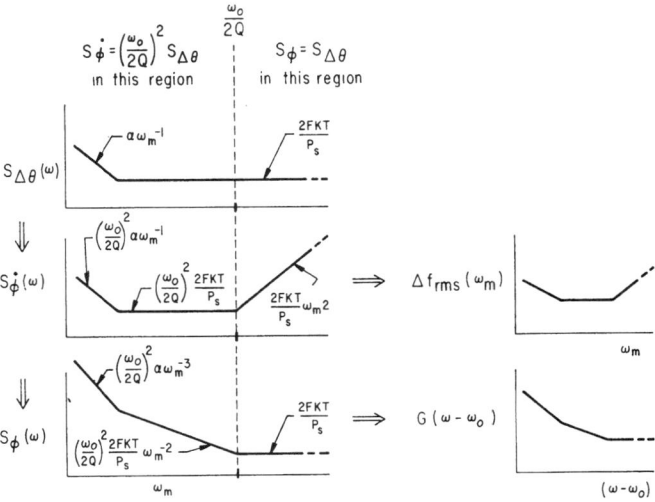

Fig. 1. Derivation of Oscillator Spectra. The logical sequence leading from oscillator parameters to spectrum characteristics is presented here. The power spectra of output phase or frequency are derived from the spectrum of input phase uncertainties and from the oscillator feedback bandwidth. The calculable constants of the oscillator are FKT, P_s, and $\omega_0/2Q$; the $1/f$ constant α is not accurately predictable but can be inferred from data. The amplitude spectrum of frequency deviation and the RF spectrum can be derived as shown, subject to limitations discussed in the text.

Manuscript received December 10, 1965; revised December 29, 1965

example. For a high-Q oscillator, $1/f$ effects in $S_{\Delta\theta}$ can predominate out to a modulation rate exceeding $\omega_0/2Q$; in this case there is no 6 dB per octave region in $S_\phi(\omega)$. A similar spectrum results where large additive noise in following amplifier stages or measuring equipment obscures the oscillator internal noise, except at very low modulation frequencies.

Note that there is a portion of the curve $S_\phi(\omega_m)$ which is proportional to $1/\omega_m^2$, leading to a $1/\omega_m$ or $1/f$ variation for rms phase deviation. This is often confused with the true $1/f$ effects associated with parameter variations leading to the $1/f$ portion of the curve for $S_{\Delta\theta}(\omega_m)$ and $S_\phi(\omega_m)$. These two are not the same thing; "$1/f$" refers to a power spectral density rather than an amplitude spectrum.

In practice, the measurable $S_\phi(\omega_m)$ is always modified by subsequent bandlimiting filtering and by additive noise contributed by following amplifiers. It is conceivable that, for a two-terminal oscillator, the filtering action of the resonator eliminates the additive phase noise component for $\omega_m > (\omega_0/2Q)$.

Experimental Verification

Measurements were taken on a stable microwave signal source[1] designed to have a spectral purity limited only by the oscillator, which was a 100 M/s crystal oscillator. This unit employs two large-jump step recovery diode multipliers with amplification between them. The data are presented in Fig. 2 in comparison with a model derived from the following constants:

Feedback bandwidth = 16 kc/s
P_S = −4 dBm
F = 9 dB
KT = −174 dBm in 1 c/s BW
Multiplication ratio = 100 = 40 dB
$N^2 2FKT/P_S$ = +40+3+9−174+4 = −118 dB.

This leads to an asymptotic value for $S_\phi(\omega_m)$ of −118 dB relative to 1 radian^2/BW in 1 c/s bandwidth, i.e., a carrier-to-sideband ratio of 118 dB. The "$1/f$" region (9 dB/octave) constant α is estimated for best data fit.

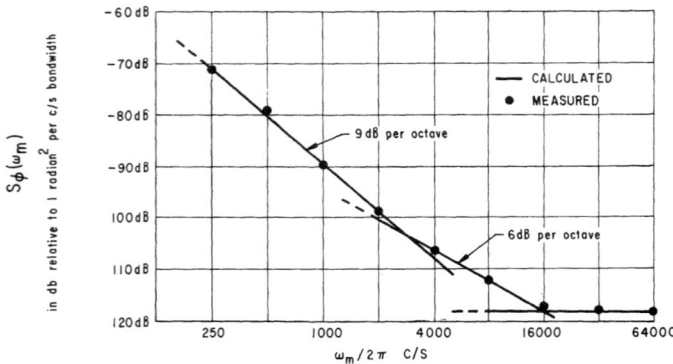

Fig. 2. $S_\phi(\omega_m)$ for Stable Microwave Signal Source. The data presented here is the average of two independent measurements which were in excellent agreement. These measurements were made at X Band on the multiplied output of a 100 Mc/s voltage controlled crystal oscillator having a 16 kc/s feedback half-bandwidth. Since this bandwidth can be reduced by a considerable factor without exceeding the present state of the art, the data is not intended to represent ultimate attainable levels, but rather serves as an illustrative example. The $1/f$ constant is chosen for best data fit. Slopes and other calculated parameters are derived from known oscillator characteristics.

Nonlinear Effects

The data was based on an estimated transistor noise figure of 9 dB. This was taken high to account for nonlinear mixing of noise at third harmonic and higher frequencies which is mixed into the pass band of interest by second harmonic periodic parameter variations

---
[1] 9.5 Gc/s Solid State Local Oscillator PN 31-007191, manufactured by Applied Technology, Inc., Palo Alto, Calif. Measurements are average of values measured by the author and D. J. Healey, III, Westinghouse Corp., Baltimore, Md., using Spectra Electronics SE-200 and Westinghouse proprietary noise test sets.

caused by the nonlinearity. The excellent fit of the data implies that this degradation of effective noise figure may well be an adequate description of the effect of nonlinearity.

Video Frequency Range of Interest

A number of applications which have been dealt with in this issue of the Proceedings may be summarized in terms of the video frequency range of interest. Space systems and Doppler radar applications are of particular interest to the author. For these two, interest lies in the range of a few c/s up to 100 kc/s. Space applications typically concentrate on the range where, for a crystal oscillator, $S_\phi(\omega_m)$ is proportional to $S_{\Delta\theta}(\omega_m)$ [3], while Doppler radar applications place additional emphasis on the region above the oscillator feedback loop bandwidth [4]. Both applications typically require microwave systems which employ multiplication from the oscillator frequency.

Choice of Oscillator Frequency for Crystal Oscillator-Multiplier

It is of interest to inspect the effect of oscillator frequency upon the output spectrum of an oscillator-multiplier system having a fixed output frequency. Two assumptions which aid the calculation are a) constant oscillator input signal-to-noise ratio, and b) resonator Q varying inversely with the oscillator frequency ω_0. Under these assumptions a comparison of two oscillator frequencies yields the following results.

1) For $\omega_m < (\omega_0/2Q)$, of the lower frequency oscillator, the multiplied output $S_\phi(\omega_m)$ is identical for either choice.
2) For $\omega_m \gg (\omega_0/2Q)$, the output $S_\phi(\omega_m)$ varies as the square of the multiplication ratio (i.e., inversely as the square of the oscillator frequency).

This can be verified by a simple graphical construction.

Choice of Active Element in a Transistor Oscillator

It is apparent that $1/f$ variations and nonlinearity can have significant deleterious effects on the attainable low levels of $S_\phi(\omega_m)$. In the light of suggestions by O. Mueller that microthermal effects [6] contribute to $1/f$ noise in transistors, it is suggested that AGC oscillators using large area transistors having high power capabilities may provide simultaneous improvements in $1/f$ level and in nonlinear effects.

Spectrum Characteristics of Microwave Solid State Sources

The spectrum model given here allows simple prediction of spectrum shape and level for microwave sources of the types discussed by Johnson et al [5]. Comparison with their data shows good agreement—their measurements for crystal oscillator units extend to $\omega_m \gg (\omega_0/2Q)$, while microwave oscillators are characterized by Q factors such that, for the measurements cited, $\omega_m < (\omega_0/2Q)$.

Acknowledgment

The author is pleased to acknowledge helpful discussions with members of IEEE Subcommittee 14.7, of which this letter may be considered a brief summary. Prepublication access to all of the papers contained in this issue is also freely acknowledged. The influence of L. S. Cutler, J. A. Mullen, and W. L. O. Smith has been of special value in the preparation of this correspondence.

D. B. Leeson
Applied Technology, Inc.
Palo Alto, Calif.

References

[1] L. S. Cutler and C. L. Searle, "Some aspects of the theory and measurement of frequency fluctuations in frequency standards," this issue, page 136.
[2] E. J. Baghdady, R. N. Lincoln, and B. D. Nelin, "Short-term frequency stability: characterization, theory, and measurement," Proc. IEEE, vol. 53, pp. 704–722, July 1965.
[3] R. L. Sydnor, J. J. Caldwell, and B. E. Rose, "Frequency stability requirements for space communications and tracking systems," this issue, page 231.
[4] D. B. Leeson and G. F. Johnson, "Short-term stability for a Doppler radar: requirements, measurements, and techniques," this issue, page 244.
[5] S. L. Johnson, B. H. Smith, and D. A. Calder, "Noise spectrum characteristics of low-noise microwave tubes and solid-state devices," this issue, page 258.
[6] O. Mueller, "Thermal feedback and $1/f$-flicker noise in semiconductor devices," 1965 Internat'l Solid State Circuits Conference, Digest, p. 68 ff.

Low current oscillator design for 900MHz GSM applications

Mobile radio applications demand low enough current consumption of all devices to enable battery operation and integration into a compact unit. Stefan Beyer and Georg Lipperer from Siemens, Munich, Germany describe one example, a phase locked VCO for GSM.

An oscillator circuit operating at 890-950MHz with low current consumption (Vb=5V; I=4 to 5mA typically) is described. The VCO, consisting of a PNP transistor used in common base mode and a frequency-linear varactor diode, works together with a prescaler circuit, the PMB2312, and a phase locked loop integrated circuit, the TBB206. These devices represent a complete phase locked loop. To modulate the RF-signal, the output is connected to a modulator circuit, a PMB2200.

The special characteristics required of the circuit include low current consumption and a tuning range of 120MHz, the large tuning range relaxing critical specifications in production.

Design targets

The main target of this design was to build an application circuit for the PLL-circuits TBB206 and PMB2312 but it was decided that the oscillator should fulfil the GSM-specifications for the pan-European cellular radio system, both in terms of the frequency spectrum and in current consumption.

The result is a PLL-circuit, which combines a suitable low noise VCO with the low current buffer stage, the circuit diagram of which is shown in figure 1. The overall current consumption of the VCO and buffer is indeed typically only 4.5mA at a supply voltage of 5V. For this discussion the PLL-circuit has been divided into the three functional blocks: VCO, buffer stage and the PLL-IC with prescaler.

The PNP-Transistor BFT92, used in

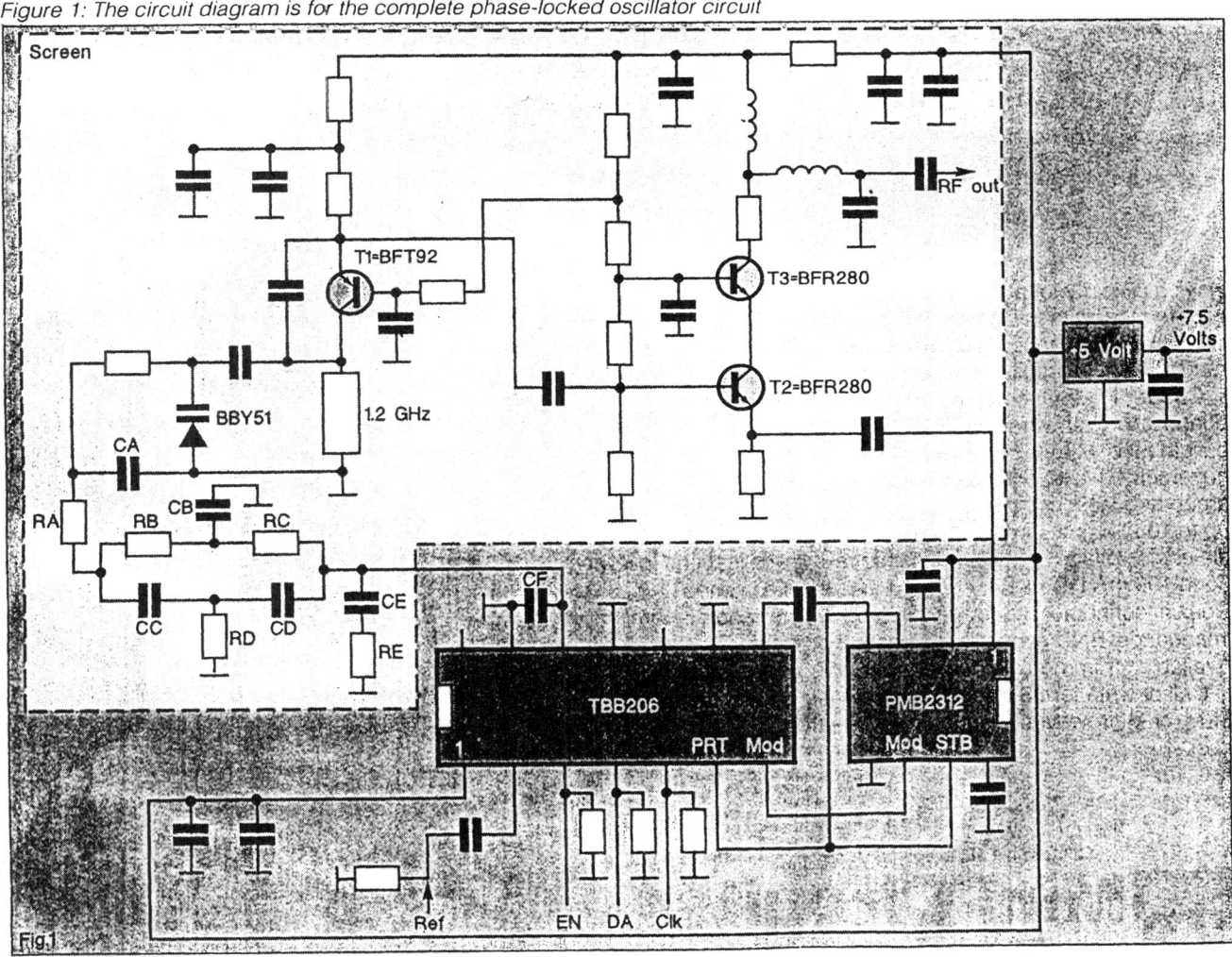

Figure 1: The circuit diagram is for the complete phase-locked oscillator circuit

Reprinted with permission from *Microwave Engineering Europe*, S. Beyer and G. Lipperer, "Low Current Oscillator Design for 900MHz GSM Applications," pp. 35-41, October 1991. © Miller Freeman Technical Ltd.

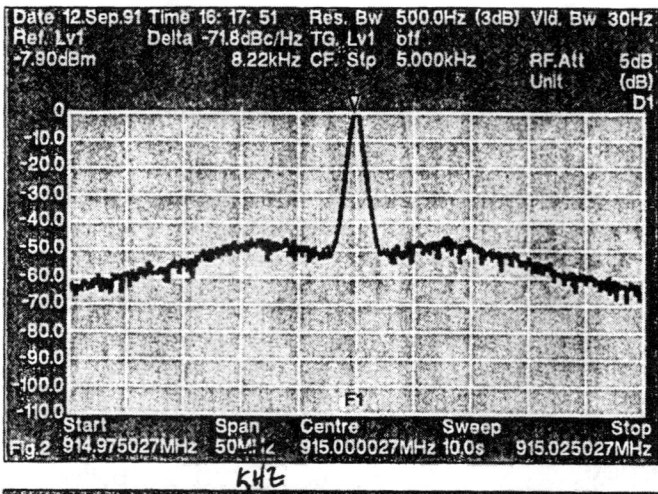

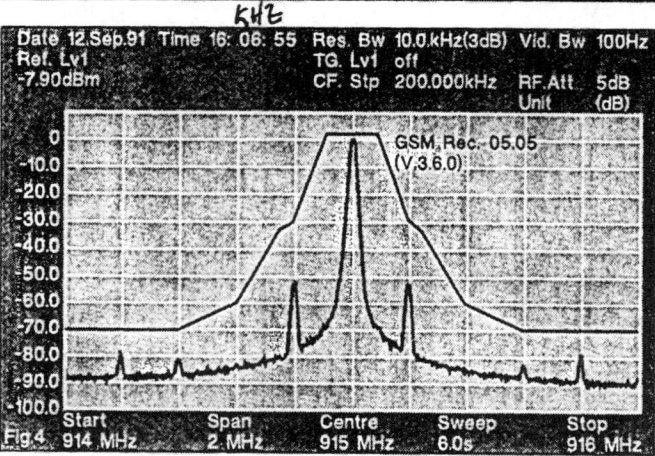

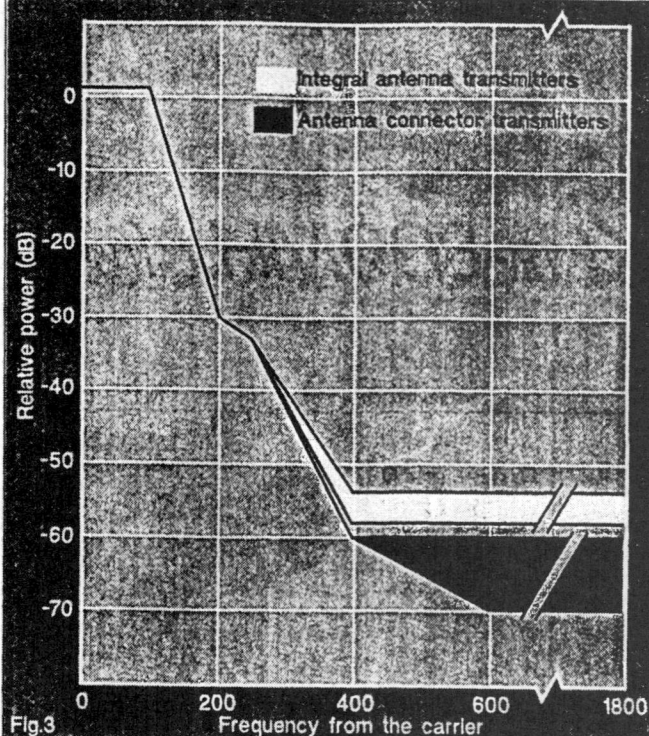

Figure 2(top left): Phase noise characteristic close to the carrier

Figure 3(above): Spectrum in the GSM specification

Figure 4(left): Comparison of the measured spectrum with the GSM envelope

common base, was chosen because the collector can be grounded via the resonator. Inductors in the collector path which are usually used with npn-transistors are not required. Such inductors in the collector path induce a feedback from the modulator to the oscillator, as the mutual inductance causes pick up of the modulated signal and hence causes remodulation.

Collector current

The BFT92 requires a collector current of 1.8mA. The frequency determining element of the oscillator is a 1.3GHz coaxial ceramic resonator with its surface completely soldered to ground.

The particularly frequency linear varactor diode BBY51 is coupled via a 10pF capacitor to the collector of the BFT92 and, in addition, to the "hot-end" of the dielectric shorted $\lambda/4$ ceramic resonator. Of course this represents a load for the oscillator circuit, which is however compensated with additional feedback from smaller 1.5pF capacitor. The internal collector-emitter capacity would suffice for oscillator operation, however the second small capacitor improves the amplitude-frequency characteristic and is used to optimise phase noise behaviour.

The loop filter (RA-RE, CA-CF) is directly connected to the varactor diode BBY51. The loop filter has little influence on the oscillator characteristic, but significantly defines the control behaviour of the PLL. By saving the cross-current for the oscillator transistor BFT92, the use of one biasing circuit for oscillator and buffer stage also reduces current consumption.

A further advantage of this oscillator circuit is the large tuning range of 120MHz, which renders circuit alignment in production unnecessary. The RF is decoupled via 2.2pF to the buffer stage.

The buffer stage

The main target for a buffer stage is high gain at low current consumption. This was realized by cascading two BFR280s. The BFR280 was chosen because of its high F_t at low current. The buffer stage needs 2mA. A good isolation between the two ports (43dB typically) can also be achieved because the emitter of T3 has low impedance.

Feedback from the modulator to the synthesizer is avoided. Such feedback would result in an undesired expansion of the GSM spectrum.

The PLL-IC with prescaler

The output for the prescaler PMB2312 is at -15dBm and ensures uncritical prescaler operation. The modulator is driven with a level of -10dBm. This is necessary because the modulator IC has its noise optimum at an input level of -10dBm.

The series inductor and shunt capacitor in the RF output path perform an impedance transformation (with low pass filter characteristic) to the required 50Ω.

The TBB206 and PMB2312 represent the heart of the PLL. The TBB206 is a complex PLL-circuit in silicon CMOS-technology for processor controlled frequency synthesis.

The function (single or dual-modulo operation) and the selection

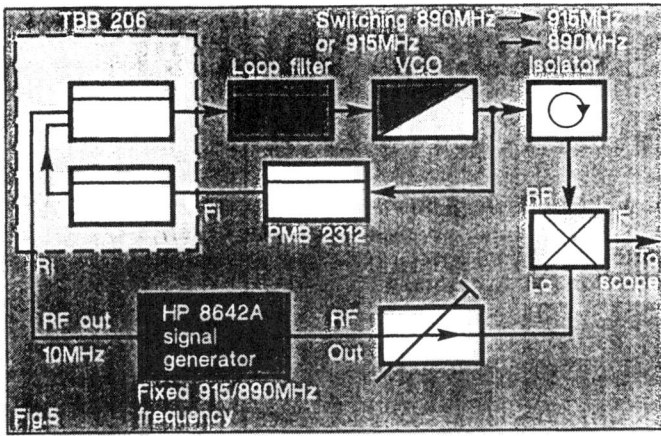

Figure 5: The block diagram shows the method for switching time measurement

of the divider ratio are controlled via a 3-wire bus consisting of clock data and enable (CL, DA, EN).

For initialisation it is necessary to send all set-up data. The bus-interface is organized for fast frequency change: One change of channel needs only 22 bits in this application.

The TBB206 supports current reduction with a standby mode. In this mode it is possible to switch off the divider or the preamplifier without losing any settings. The frequency reference is coupled into pin RI. Reference frequencies up to 30MHz are possible. The pin FI is the VCO frequency input. This input is also usable up to 30MHz.

The phase detector signal is present at the output pin PD. It features short anti-backlash pulses to avoid a dead zone. The PD output is a real current source. It is possible to build up a loop filter by use of passive elements (RA-RE,CA-CF) only. The advantage is that expensive components are not required and additional noise sources are avoided in this sensitive part of the circuit. Polarity and current range of the PD output are programable via the 3-wire bus.

The open-drain output LD supplies the Lock Detect signal. The pulse width is directly proportional to the phase error of the closed loop plus the duration of the anti-backlash pulse. In order to avoid extreme phase errors during a frequency step, the new data for the N+A and the R counter are taken over synchronously (programable counter has reached zero). This guarantees that the control process starts with the phase error "zero".

The loop was calculated to fulfil the GSM Rec. 5.04. The PLL is designed to meet a lock-in time of 500µs between any channels in a 25MHz range. The additional filter ensures the required spectral purity of the LO-signal. The 1.1GHz prescaler PMB2312 supports the division ratio 64/65 and 128/129. Its current consumption is typically 5.5mA at 25°C with a supply voltage of 5V.

Measurements

The output spectrum of the oscillator have been measured with a Rohde & Schwarz FSBS spectrum analyzer (100Hz-5GHz). Figure 2 shows the phase noise characteristic close to the carrier. In GSM Rec. 05.05(vers. 3.6.0) Annex 1 the desired spectrum characteristics are published and these are shown in figure 3. Copying the demanded values of the GSM specification in the measured output spectrum it can be seen that the GSM requirement is fulfilled (figure 4).

Figure 5 shows the method used for the switching time measurement. The following measurement devices were used: digital oscilloscope Le Croy 9450; signal generator HP8642A; passive mixer Anzac MDC 149; constant impedance adjustable phase line isolator Microtek I90L10A1

The delay line is adjusted to the dc-zero of the mixer's IF-port. Two switching times have been measured. Switching from 890MHz to 915MHz (figure 6) and from 915MHz to 890MHz (figure 7). In figures 6 and 7 the upper trace shows the data transmitted by the 3-wire bus to the TBB206. The data is valid when transmission is complete. The lower trace displays the difference phase between the HP8642A and the synthesizer due to the systematic programming shown in trace 1. The delta-t is the time between which data is valid and the lock in condition of 5 degrees peak.

Conclusions

Although originally designed for GSM applications, the circuit can also be used in many other applications. The VCO (and buffer stage) is able to work up to 2GHz with only a few modifications (resonator, feedback-capacity, operating point). Thus it has potential as a universal, low current oscillator concept.

Figure 6: Switching time for the maximum frequency excursion, an increase from 890MHz to 915MHz

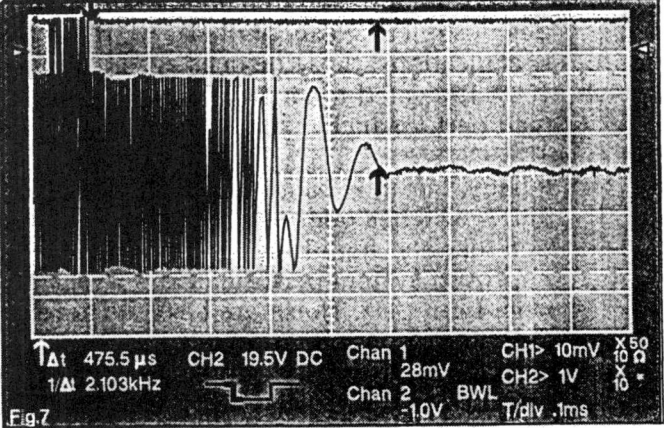

Figure 7: Switching time for reduction in the frequency from 915MHz to 890MHz

Design of a Low-Phase Noise VCO for an Analog Cellular Portable Radio Application

Tomoki Uwano and Toshio Ishizaki, *Members*

Materials and Components Research Laboratory, Matsushita Electric Industrial Co., Ltd.,
Kodama, Japan 571

Yoshihiro Nakagawa, *Member*, and Toshiaki Nakamura, *Nonmember*

Matsushita Nittoh Electric Co., Ltd., Kyoto, Japan 610-03

SUMMARY

A voltage-controlled oscillator (VCO) with a small power consumption and an excellent signal-to-noise (C/N) ratio was developed for a portable telephone. In the domestic analog portable cellular portable telephone, C/N is required to be more than 107 dBc/Hz at an off-carrier frequency of 12.5 kHz. To realize this value, the noise behavior was studied for a 900-MHz oscillator using a small-signal, high-frequency silicon transistor.

First, the relationship between the transistor noise and the C/N, the operating voltage and current dependence of the C/N and the relationship between the collector breakdown voltage and the oscillator output were studied experimentally. A transistor was selected which is most suitable for a low-noise VCO at a low operating current of 3 to 7 mA.

Next, by using this transistor, the relationship between the unloaded Q of the resonant circuit and the C/N was studied. It was found that Q of 60 is needed to obtain a C/N with a 3-dB margin. On the other hand, with a view to miniaturizing the resonator, a dielectric coaxial resonator with a material having a relative permittivity of 90 was studied experimentally. A Q of 200 was realized with a resonator having a cross-sectional size of 1.8×1.8 mm^2. By combining it with a varactor diode, a Q of more than 60 was attained. Based on these results, a low-noise VCO was realized which has dimensions of $10 \times 10 \times 3$ mm and is operated with a power supply voltage of 3 V and a current consumption of 10 mA.

Key words: VCO; phase noise; cellular portable telephone; oscillator; dielectric resonator.

1. Introduction

The cellular telephone using the 800-MHz band has many frequency channels. To selectively receive these channels and to generate a transmission signal by mixing with modulated intermediate frequency signals, a voltage-controlled oscillator (VCO) is used. The VCO in the cellular portable telephone in the initial phase was made of a large planar transmission line circuit for attaining a low-noise operation and is placed in a large package with a high degree of shielding to suppress spurious radiation [1–3]. This VCO was developed as a modularized RF component and became one of the key RF components for the cellular telephone. Recently, since the cellular portable telephone is used widely, miniaturization and low power consumption have became a design goal.

By its nature, the VCO for cellular portable telephone has a variable frequency range of more than 25 MHz at 8 to 900 MHz. In general, the resonators used in the variable resonant circuit in this frequency band are made mainly of planar transmission lines such as strip lines [4-6]. On the other hand, the fundamental performance requirement for the VCO in a cellular portable telephone is a high signal-to-noise (C/N) ratio, a shock and vibration resistance, and a low EMI. Especially, the C/N is the most fundamental character to determine the magnitude of the noise interference to adjacent channels.

With regard to studies on the C/N of an oscillator, the analysis by Leeson is well known [7]. This is a fundamental theory for understanding the mechanism of the oscillation noise. Also, as experimental investigations, there are reports on the broadband low-phase noise oscillator [8], on the analysis of the noise near the oscillation signal in a GaAs FET oscillator [9], and on the reduction of the $1/f$ noise and improvement of the C/N by means of the process improvement of the transistor [10]. In these reports, the analysis by Leeson is confirmed experimentally.

In the VCO for hand-held telephones, it is required to obtain a desired C/N at the lowest possible current with a small resonator. To realize such a VCO, selection and performance improvement of transistors and improvement of the resonator are indispensable. To this end, it is necessary to investigate the relationship of the C/N to the transistor performance and the unloaded Q (Q_u) of the resonator circuit under the practical circuit condition. For such an analysis, CAD programs deriving the characteristics of the transistor oscillator from the nonlinear analysis have been developed [11]. However, they are not sufficient for practical design due to lack of accuracy.

In this paper, the relationship of the C/N to the noise characteristics of the transistor and the miniaturization of the resonator is studied experimentally and the results are applied for the design of VCOs. First, the relationship of the phase noise of the oscillator with the noise, dc gain, operating current and voltage of the transistor is investigated experimentally. The relationship between the oscillation output and the collector breakdown voltage is studied. In place of a planar transmission line resonator, a coaxial resonator is used for experimental study of the limitation of miniaturization.

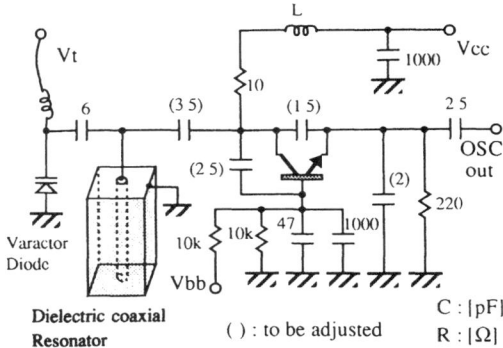

Fig. 1. A low-noise VCO test circuit.

Based on these results, a low-noise VCO of small size, low power consumption, and resistance to vibration and EMI was realized by means of this coaxial dielectric resonator.

2. VCO Experimental Circuit

Figure 1 shows the circuit of the low-noise VCO used in the experiment. The oscillation circuit is a common base Colpitts circuit. There are capacitive feedback circuits between the collector, emitter, and base; and a resonant circuit made of a dielectric coaxial resonator is inserted between the collector and the base. The resonant circuit consists of a resonator with a varactor diode connected in parallel so that the resonant frequency changes by an applied voltage. The coaxial resonator is made of a ceramic with its relative permittivity ε_r of about 90 [12]. The cross section is either square or circle and its one side or diameter is 1.8 to 10 mm. Depending on the size, Q_u of the resonator varies.

This oscillator has a conventional circuit configuration in which the coaxial resonator constitutes a Clap-type resonant circuit. The bias resistors and capacitors are determined experimentally such that the oscillation is stable for temperature variation at the operating voltage of $V_{ce} = 3$ V with a low level of harmonics and without any extraordinary oscillation. In this circuit, the C/N characteristics and the output of the oscillation are measured. In the measurement, the feedback circuit and the coupling capacitance are adjusted so that the C/N is maximized. Hence, an experimental investigation can be carried out with attention paid to the C/N. Measurement of the C/N due to Q_u of the resonant circuit also is carried out.

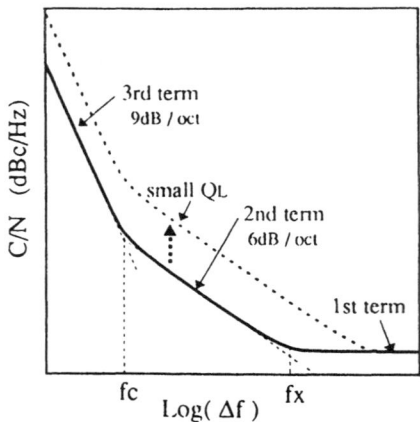

Fig. 2. Curves of Eq. (1).

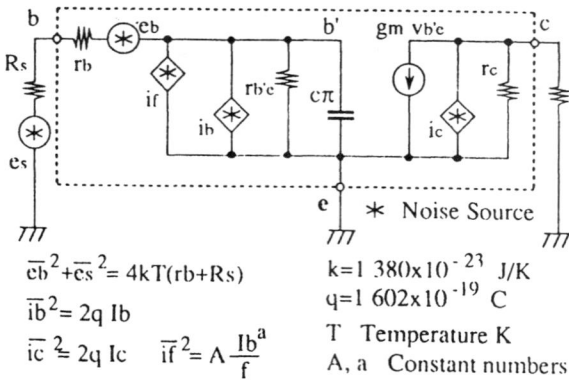

Fig. 3. A hybrid π-equivalent circuit of a transistor including noise sources.

3. Transistor Noise and VCO Phase Noise

As an oscillation transistor, a small-signal RF silicon transistor typically is used. The phase noise of the oscillator is caused by the white noise and $1/f$ flicker noise generated in the transistor. First, the relationship of the phase noise to the noise figure (NF), low-frequency noise, and the dc gain is studied experimentally [13].

The C/N of an oscillator can be expressed from the feedback model by Leeson as follows [7, 14]:

$$C/N(\Delta f) \approx -10 \log\left[\frac{P_n}{2P_s}\left\{1 + \frac{1}{\Delta f^2}\left(\frac{f_0}{Q_L}\right)^2 + \frac{f_c}{\Delta f^3}\left(\frac{f_0}{Q_l}\right)^2\right\}\right] \quad (1)$$

where f_0 is the oscillation frequency; Δf is the off-carrier frequency from the oscillation signal; f_c is the frequency at which the spectrum of the $1/f$ flicker noise of the transistor becomes equal in magnitude to the white noise at a low frequency (baseband); Q_L is the loaded Q of the resonant circuit; P_s is the oscillation output; and P_n is the white noise density. The first and second terms in Eq. (1) are related to the white noise at low frequencies, while the third term is related to the flicker noise. The curve given by Eq. (1) is shown in Fig. 2.

Figure 3 shows an equivalent circuit of the transistor including the noise generated by the transistor. When the base terminal is loaded with R_s, the input equivalent noise voltage en of the transistor at the terminal is given by [15]

$$en^2 = 4kT(rb + R_s) + 2qI_b(rb + R_s)^2 + \frac{2qI_c(rb + R_s + rb'e)^2}{\beta^2} + A\frac{I_b^a}{f}(rb + R_s)^2 \quad (2)$$

where I_b and I_c are the base and collector currents and β indicates the dc gain

What is investigated for the VCO for the cellular telephone is the C/N value at the frequency separated from the oscillation signal by the channel spacing (which is, in general, more than 10 kHz, e.g., 12.5 kHz). On the other hand, it was found that f_c of the silicon transistors used in the experiment is typically in the 3 to 7 kHz range in the oscillation mode. Hence, from Fig. 2, it is found that the phase noise at the frequency separated by the channel spacing is caused by the white noise at low frequencies. Note that the white noise at high frequencies is related directly to the NF of the transistor. The oscillation phase noise caused by the white noise at high frequencies are constant independently of the off-carrier frequency. If it is assumed that this is added equally in Fig. 2, then such is related only to the noise floor at frequencies above f_x. However, if the frequency dependence of the white noise is small in Eq. (2), the NF characteristics can be used as a reference for selecting the transistor of a low-phase noise oscillator.

The relationship between the NF and the C/N was studied experimentally. Figure 4 shows the measured results. The experimental circuit is as shown in Fig. 1, and the conditions are listed in Fig. 4. From Fig. 4, it is found that the C/N increases in general as the NF is decreased. In Fig. 4, A and B for which the C/N values

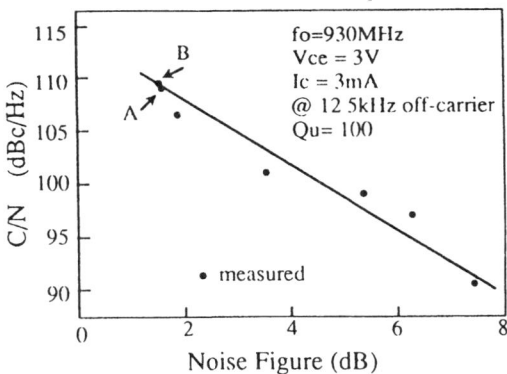

Fig. 4. C/N's vs. transistor noise figures.

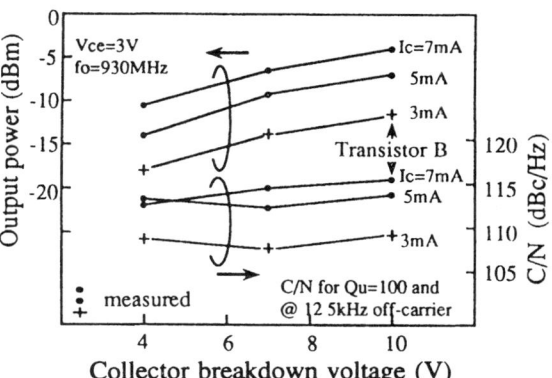

Fig. 6. Output powers and C/N's vs. collector breakdown voltage.

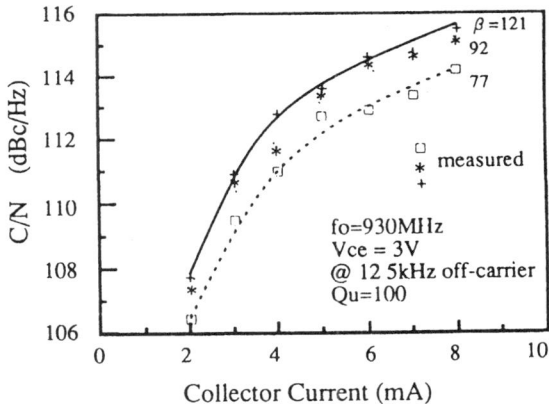

Fig. 5. Noise performances vs. collector currents regarding β's.

Table 1. Main electrical characteristics of Transistor B

Item	Characteristic
Collector cutoff current	1.0 μA (1)
Emitter cutoff current	1.0 μA (1)(2)
DC current gain	50 ~300
Gain bandwidth product	7 GHz
Collector capacitance	0.55 ~ 1.0 pF (1)(3)
Forward transfer gain S_{21}	11.5 dB (4)
Noise figure	1.1 ~ 2.0 dB (4)(5)

Conditions: V_{cb} or V_{ce} = 10 V, I_c or I_e = 20 mA, (1) I_e or I_c = 0, (2) V_{eb} = 1, (3) f = 1 MHz, (4) f = 1 GHz, (5) I_c = 7 mA.

are almost equal indicate transistors made with different processes. The oscillation output of B is larger by about 2 dB than that of A, and hence the absolute noise power is larger by the same amount. When the low-frequency noise e_n of the transistors A and B at 12.5 kHz is measured with R_s = 1 to 5 kΩ, the value for B is larger by 2 to 4 dB. The difference was 2 dB with R_s = 5 kΩ. Notice that Eq. (2) is a function of β and the collector current I_c. Hence, with the transistor B, the C/N versus I_c was measured with different values of β. The measured results are shown in Fig. 5, where it is found that the C/N increases as β becomes larger. This is because for identical values of I_c (and hence identical oscillation powers), I_b becomes smaller as β becomes larger and hence the second and subsequent terms in Eq. (2) become smaller and the low-frequency noise decreases. This fact was confirmed experimentally.

On the other hand, it is more desirable that the oscillator output be as large as possible with a low current. To obtain a large output, it is necessary that the oscillation power efficiency of the transistor be high. This power efficiency is related to the collector breakdown voltage and hence this was investigated experimentally. Figure 6 shows the measured results of the oscillator output and the C/N for three different transistors with different breakdown voltages including Transistor B. Other electrical characteristics such as the gain bandwidth product and chip sizes of these transistors are

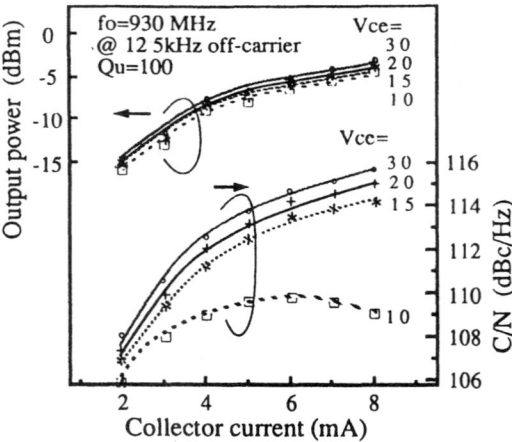

Fig. 7. Output power and C/N dependences on collector voltages and currents.

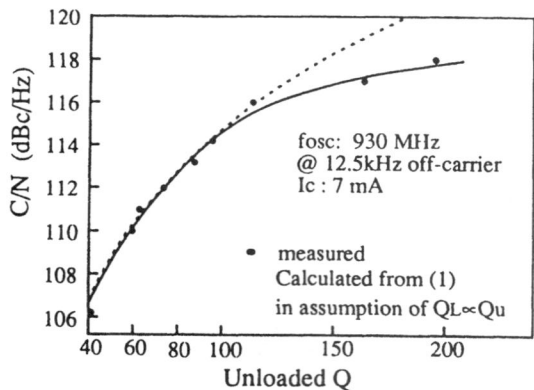

Fig. 8. Measured results of C/N's regarding resonance circuit Q_u's.

almost identical. In the figure, the oscillation power increases but the C/N has little change as the breakdown voltage is increased.

From the foregoing experimental investigations, it was found that Transistor B provides the best C/N and the largest oscillation output among the RF transistors in the range of the operating current of 3 to 7 mA. The main electrical characteristics of this transistor are shown in Table 1.

Figure 7 shows the experimental results of the output power and the C/N as a function of the collector current of the oscillator using Transistor B measured with different values of V_{ce}. The measurement was carried out by adjusting the circuit parameters to maximize the C/N at V_{ce} = 3 V and I_c = 7 mA. From the results of Fig. 7, it is possible to select the operating point of the transistor suitable for the objective of the VCO.

4. Resonant Circuit Q_u and C/N

Equation (1) indicates the relationship between the C/N near the oscillation frequency and the Q_L of the resonant circuit. Figure 2 shows the curve of Eq. (1) where the curves of 9 dB/oct and 6 dB/oct indicate the third and second terms, respectively. The noise floor is determined by the first term. In the experiment mentioned in the previous section, it was confirmed that the C/N characteristics follow this curve. Here, $1/Q_L$ is given by the sum of $1/Q_u$ and $1/Q_e$ (external Q). Generally, in the oscillation mode with Q_u less than 100 Q_u is smaller than Q_e so that Q_L is almost equal to Q_u. Then the characteristic of 12.5 kHz is on the curve of the second term. As Q_u is reduced (and so is Q_L), the C/N curve is shifted upward as indicated by the dotted line. The frequency f_x at which the curves of the first and second terms intersect is the one away from the oscillation signal by $f_0/2Q_L$. This value is usually several megahertz.

Figure 8 shows the measured results of the relationship between Q_u of the resonant circuit and C/N. In the measurement, Transistor B was used. The curve for Eq. (1) with the coefficients obtained by curve-fitting to the measured values is shown by assuming that Q_u is equal to Q_L if it is less than 100.

The C/N value needed for the domestic analog cellular system is 107 dBc/Hz at 12.5 kHz off-carrier computed from the specification. If a characteristic margin of 3 dB due to environmental change is taken into account, the VCO requires more than 110 dBc/Hz. Hence, the necessary resonator Q_u is more than 60 from Fig. 8. With a standard strip line, it is possible to obtain Q_u of more than 100. However, such a value cannot easily be attained if the line width and the configuration are modified for miniaturization and a varactor diode is attached.

With a view to miniaturizing the resonator, the characteristics of the dielectric coaxial resonator were studied. In this configuration, a ceramic of its ε_r of about 90 is used and thick silver films are used for electrodes. The thick silver films are obtained by firing silver paste with a thickness of 20 to 30 μm at 800°C. Since adhesive is contained in the paste, the conductivity is less

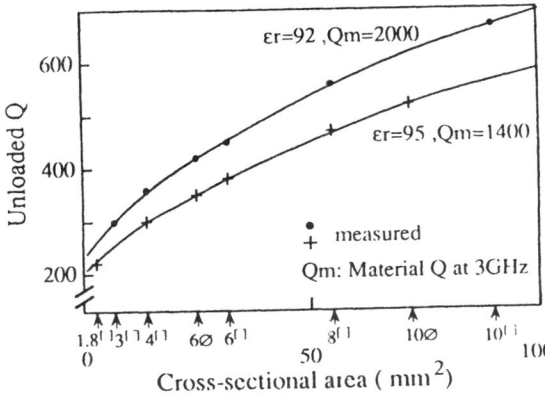

Fig. 9. Unloaded Q's of coaxial resonators regarding their sizes.

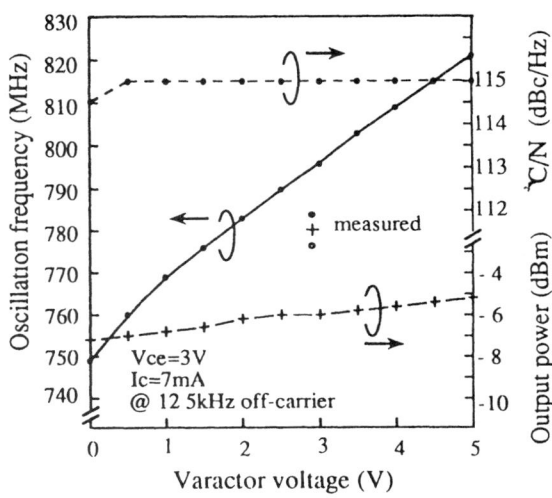

Fig. 10. Total performances of a low-noise VCO in terms of varactor voltage V_t.

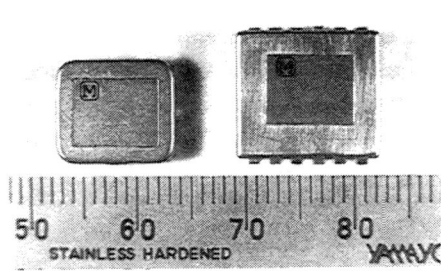

Fig. 11. Picture of a small VCO (left) and a frequency synthesizer.

than one-half that of the solid metal. While the cross section of the resonator is either square or circular, the cross section of the inner conductor is always circular. Figure 9 shows the measured results of the relationship between Q_u and the cross section of the coaxial resonator with two different Q-value dielectrical materials. The diameter of the inner conductor of the coaxial resonator is 0.3 times the length of one side of the square or the outer diameter of the circle. From the figure, it is found that the value of Q_u is determined by the cross-section size regardless of the shape of the coaxial resonator. Based on this finding, a square configuration is employed for the coaxial resonator due to packaging efficiency. From Fig. 9, Q_u of 200 was obtained for a resonator with a side length of 1.8 mm. A resonant circuit with Q_u of more than 60 can be constructed if this resonator is combined with a varactor diode with a Q of more than 100. Hence, it is possible to realize a small VCO with excellent performance with such a small dielectric resonator.

5. Experimental Results of VCO

From the foregoing experimental investigations, Transistor B used in the measurement for Figs. 7 and 8 was selected for the design of a low-noise VCO. The circuit for the VCO is shown in Fig. 1. The capacitance in the feedback circuit and the coupling capacitance to the dielectric resonator were adjusted properly. The Q of the varactor diode used is about 200. Figure 10 shows the measured results of the oscillation frequency, output power and C/N as a function of the applied voltage V_t of the varactor diode. The coaxial resonator used in the circuit has a square cross section with a side length of 1.8 mm (for which Q_u is about 200). The battery voltage of the cellular telephone is, in general, 4.8 to 6 V. As an operation with this power supply voltage, Fig. 10 indicates that the frequency change is 25 MHz for V_t changing in the range of 1 to 3 V. This result is sufficient for the cellular telephone. Also, it was confirmed that the off-carrier frequency versus C/N and follows the curve in Fig. 2 with f_c of 6.0 kHz.

As a circuit for the radio equipment, VCO is combined with a PLL-IC to form a frequency synthesizer. The output of the VCO is connected to the PLL-IC. The input impedance of the IC varies due to its switching operation. To prevent this variation from affecting the oscillation frequency of the oscillation circuit, a buffer amplifier is needed. Usually, the VCO is modularized including such an amplifier. Although the buffer amplifier is indispensable for configuring a synthesizer, it is not desirable to increase the power consumption for a VCO for a hand-held telephone. To this end, as a design with a moderate output power while a necessary C/N is obtained without increasing the current consumption, a cascode connection of the buffer amplifier transistor with the oscillator transistor has been considered.

The photograph in Fig. 11 shows the external view of the commercialized VCO. To this VCO, a buffer amplifier transistor is cascode connected. An output of -3 dBm and a C/N value of more than 110 dBc/Hz are realized with a power supply voltage of 3 V and a current consumption of 10 mA. The component on the right is a synthesizer module containing this VCO and a PLL-IC contained in one package. The current consumption is 18 mA.

6. Conclusions

The effects of the noise generated in the transistor and Q_u of the resonant circuit on the phase noise and C/N of the oscillation signal were studied experimentally. From this result, a transistor with a low operating current of 3 to 7 mA which is most suitable for the low-noise VCO was selected. To obtain the required C/N of more than 110 dBc/Hz with this transistor, the necessary value of Q_u of the resonant circuit is more than 60. This value was realized by a combination of a miniature dielectric coaxial resonator (Q_u of 200) 1.8 mm^2 and a varactor diode with a Q of more than 100. In comparison with a strip line resonator, the dielectric coaxial resonator has considerably fewer resonant frequency variations influenced by mechanical distortion and vibration. In addition, the electromagnetic energy is confined inside the dielectric and hence EMI due to unwanted radiation is small. As a result, a VCO for hand-held cellular telephone was realized. It is small in size with dimensions of $10 \times 10 \times 3$ mm, consumes a small power of 3 V and 10 mA, and has little spurious radiation.

Acknowledgement. The authors thank Mr. T. Ishida, Director of Materials and Components Research Laboratory for his guidance.

REFERENCES

1. E. Sasaki and K. Yanagisawa. Oscillator components for mobile subscriber station. NTT Research and Development Report, **26**, 7, pp. 2123-2136 (July 1977).
2. K. Nakabe, M. Watanabe, M. Makimoto, and S. Yuki. Configuration and characteristics of an 800-MHz frequency synthesizer. Papers of Technical Group on Microwave I.E.I.C.E., Japan, **MW78**-72 (Sept. 1978).
3. S. Nishiki and S. Yuki. An 800-MHz frequency synthesizer for land mobile radio equipment. I.E.I.C.E. Trans. (B), **J65-B**, 6, pp. 737-744 (June 1982).
4. K. Kawamoto, K. Hirota, N. Niizaki, Y. Fujiwara, and K. Ueki. Small-Size VCO Module for 900-MHz Using Coupled Microstrip Coplanar Lines. In: 1985 IEEE MTT-S Int. Microwave Symp. Dig., pp. 689-692 (June 1985).
5. K. Chang, S. Martin, F. Wang, and J. L. Klein. On the study of microstrip ring and varactor-tuned ring circuits. IEEE Trans. Microwave Theory & Tech., **MTT-35**, 12, pp. 1288-1295 (Dec. 1985).
6. M. Makimoto and M. Sagawa. Varactor Tuned Bandpass Filters Using Microstrip-Line Ring Resonators. In: 1986 IEEE MTT-S Int. Microwave Symp. Dig., pp. 411-414 (June 1986).
7. D. B. Leeson. A simple model of feedback oscillator noise spectrum. Proc. IEEE, **54**, 2, pp. 329-330 (Feb. 1966).
8. K. Maruhashi and M. Madihian. High-Power Low-Noise Ku-Band HJFET Oscillator Design. I.E.I.C.E. '93 Spring Convention, C-47.
9. J. Sone and Y. Takayama. FM noise of GaAs FET oscillators. Papers of Technical Group on Microwaves, **MW78**-47 (July 1978).
10. H. Nagayama and S. Takahashi. Development of a small low-noise voltage-controlled oscillator (VCO). Oki Electric Research and Development, **53**, 2, pp. 67-72 (April 1986).
11. MDS/RFDS Advanced Modules. Hewlett Packard (1989); Libra in EEsof's Series IV. EEsof (1988), etc.
12. J. Kato. Characteristics and applications of microwave high permittivity ceramics. Electronic Materials, **30**, pp. 112-117 (Sept. 1991).
13. Y. Nakagawa, K. Hashimoto, and T. Ishizaki. Analysis of the Relationships Between Transistor Noise and VCO Noise for the Application to Portable Telephones. I.E.I.C.E. '92 Fall Convention, C-33.
14. D. Scherer. Learn about low-noise design. Microwaves, **18**, 2, pp. 116-122 (April 1979).
15. C. D. Motcherbacher and F. C. Fitchen. Low-Noise Electronic Design, Chap. 4. Wiley-Interscience (1973).

CMOS VCOs for PLL Frequency Synthesis in GHz Digital Mobile Radio Communications

Manop Thamsirianunt   Tadeusz A. Kwasniewski*

PMC-Sierra, Inc., Burnaby, British Columbia, Canada
*Department of Electronics, Carleton University, Ottawa, Ontario, Canada

Abstract

We report on a CMOS inductorless VCO design with an emphasis on low-noise, low-power, gigahertz-range circuits suitable for portable wireless equipment. The paper considers three structures– one simple ring oscillator and two differential circuits. The design methodology followed optimization for high-speed and low-power consumption. The measurement results of three VCOs implemented in 1.2 µm CMOS technology verify the simulation predictions. The simplest VCO architecture exhibits 926-MHz operation with -83 dBc/Hz phase noise (100 kHz carrier offset) and 5 mW (5 volts) power consumption.

I. Introduction

A hi-speed, low-noise frequency synthesizer design remains a very challenging task driven by the digital mobile radio and portable communication market. The most critical parts in all RF PLL synthesizers as shown in Fig. 1 are the VCO and the frequency divider because they operate at the highest frequency compared to the rest of the system.

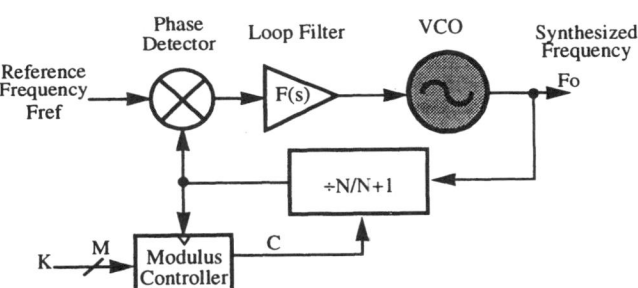

Fig. 1. RF PLL frequency synthesis building blocks.

Although, many CMOS frequency dividers have been reported which exhibit operating frequencies to a few gigahertz [1], [2], there has been no report to date on the use of on-chip GHz-range CMOS VCOs in digital mobile radio applications. Therefore most of today's RF PLL use a mixture of high-speed semiconductor technologies. The availability of the submicron MOS process has resulted in speeds of CMOS PLLs that have significantly increased compared to PLLs of the past. Recent papers indicate the potential for CMOS oscillators that operate to GHz frequencies [3], [5]. The challenge then is to develop CMOS analog-type circuits for high frequency applications.

It is the objective of this work to demonstrate that a gigahertz frequency range, milliwatt range power consumption monolithic CMOS VCO with good phase noise performance can be achieved. Three novel high-frequency CMOS VCO architectures are presented which have good potential to meet the demand for low-cost, high silicon integration for RF circuitry. When used as components in PLL frequency synthesizers, these CMOS VCOs offer phase noise performance comparable to known bipolar on-chip VCO designs [6]-[8].

II. Design Principles

General requirements for a high-quality VCO include high spectral purity, linear voltage-to-frequency transfer characteristic and good frequency stability to power supply and temperature variation. However, for mobile radio applications, the VCO must also comply with the basic requirement of low power consumption and low fabrication cost as well. While high linearity and wide tuning range may be necessary in some applications, in wireless communications where bandwidths are limited to a few tens of megahertz, tuning range and linearity do not present a problem. Also, because the feedback action of the PLL suppresses the frequency error with a magnitude that is a direct function of the open loop gain, long term frequency stability as a function of temperature is then not a crucial problem as long as the start-up procedure includes provisions to guarantee operation in a lock condition. The significant VCO properties that must be considered are spectral purity, high speed while maintaining low power consumption and suitability for circuit integration.

Most low phase-noise oscillators employ LC tank or resonant elements off chip due to the difficulty in controlling process variations which can lead to long term frequency drift. Although there is current research in the form of silicon bipolar on-chip LC VCO architectures, the achievable value of Q remains low (about 5) [6]. In addition, the inductors built on silicon occupy a substantially large area compared to other on-chip VCO designs.

IEEE 1995 Custom Integrated Circuits Conference, pp. 331-34, May 1995.

Several papers including the ones by Abidi & Meyer [7], Sneep & Verhoeven [8], and Martin [9] have shown analyses of phase noise on relaxation oscillator based on Emitter Coupled Oscillator (ECO) structures and suggested ways to reduce the causes of phase noise. However, these analyses pointed out one common inherent drawback of ECOs; the phase noise is inversely proportional to the level of current that is used to charge and discharge the timing capacitor. The product of I^2R of the collector's current indicates the power consumption of the oscillator, Consequently, the lower phase noise ECOs are only achieved with higher current.

On the contrary, a relaxation ring oscillator based on the construction of all MOS devices as illustrated in Fig. 2.a [10] does not present such limitation. Because we postulate that at the frequencies of interest, each inverter cell is a transconductance amplifier with its own circuit parasitics as shown in Fig. 2.c and d, therefore by analyzing the phase noise in terms of fluctuations in time (dt), as a function of threshold switching voltage variation, it can be shown that the jitter is directly related to MOS device sizing in a parasitic only MOS oscillator [4]. Smaller device sizes will result in higher oscillating frequencies and lower intrinsic jitter. This indicates an advantage of MOS relaxation oscillators over classical bipolar multivibrators, especially for a MOS relaxation oscillator structure that employs circuit parasitics. Parasitic MOS oscillator structures allow greater circuit integration and lower power consumption (through smaller feature size). Furthermore, since the higher operation frequency is obtained by reducing device sizes, the low phase-noise design will exhibit high operating frequencies.

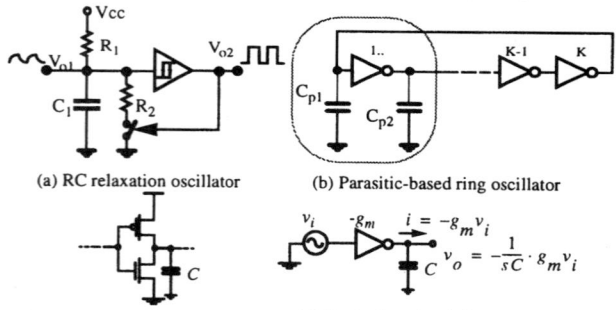

(a) RC relaxation oscillator (b) Parasitic-based ring oscillator

(c) CMOS inverter with its parasitic (d) Equivalent circuit for an integrator

Fig. 2. Generalized CMOS ring oscillator.

III. VCO Architectures

One of the primary design goals in the CMOS VCO design was to obtain a high Q-factor without the use of tank circuits. This can be achieved by using the simplest structures, such as an odd-number inverter ring oscillator based upon an RC relaxation oscillator, as shown in Fig. 2.b. However, the maximum oscillation frequency is related to device speed, and therefore MOS oscillators must be optimized for speed through architecture and layout techniques. Three VCO architectures are now considered next.

A. Modified Ring Oscillator (VCO1)[3]

The oscillator of Fig. 3 consists of a delay cell where frequency control is achieved by directly controlling the current through a series transistor of one inverter stage. The two remaining inverters are connected to form a closed loop ring. The circuit contains the least number of components required for a functional relaxation oscillator.

The unique frequency control mechanism of the oscillator exploits the short-channel effects (gate-length less than 3μm) of the controlling transistor Mn_1 to control the delay [4].

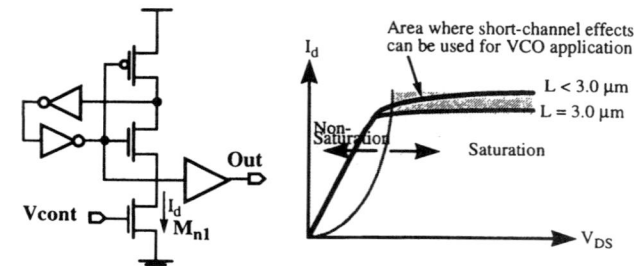

Fig. 3. Simplified three-stage modified ring oscillator.

B. Modified D-Latch Ring Oscillator (VCO2)

Using the self oscillating behavior of the D-latch circuit [1] (see Fig. 4.a), a high frequency oscillator can be realized. The arrangement needed to construct this type of oscillator is shown in Fig. 4.b, where a cross-coupled D-latch is combined with two loop inverter rings. Mp_1 and Mp_2 are used in this configuration to replace two complete inverters resulting in a pseudo-three-stage ring oscillator structure.

Since the proposed D-latch is a sequential circuit, when the Clk input (shown in the schematic of Fig. 4.b as V_{cont}) is held high (logic 1) a state transition occurs enabling the rings to start oscillation. The dc level of Clk should be held higher than the threshold level of 2.5 volts in order to start the oscillation. A further increase in the dc level of Clk results in a proportional decrease in the resistance of Mn_1 and Mn_2, and consequently, the transition time in the latch is reduced. As a result, the frequency of oscillation can be changed by varying the voltage V_{cont}. The modified D-latch oscillator allows the oscillator to operate at high speeds while maintaining good noise performance, since the configuration is a cross-coupled ring structure (differential-like structure).

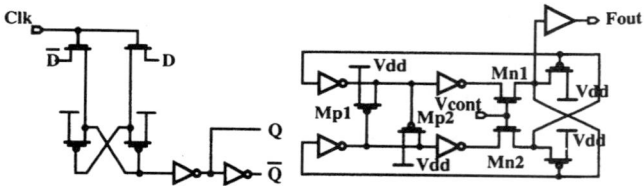

(a) Proposed new D-latch for VCO2 (b) VCO2- Pseudo three-stage-ring oscillator

Fig. 4. VCO2 modified D-latch ring oscillator.

C Fully-differential ring oscillator (VCO3)

The oscillator was constructed using a constant-current-charge-discharge scheme where the latching operation, which alternates the state of two current controlled inverters, was achieved with the use of double flip-flops [5]. Since NOR flip-flops are based on the operation of inverter logic, the architecture can be further broken down to a simple inverter-level analysis by viewing the structure as a two-loop differential ring oscillator, as shown in Fig. 5.b. The rings are formed by common three-stage ring oscillators.

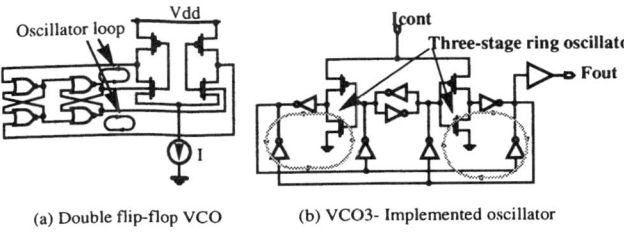

Fig. 5. VCO3 Fully balanced configuration.

Although this VCO design is based on a proposed architecture found in the literature [5], significant optimization effort was required to achieve the desired low noise and maximum oscillating frequency. A simplified schematic of the complete VCO is shown in Fig. 7.b. The VCO architecture in fact has two inherent differential structures and regenerative elements formed using cross-coupled 5-stage ring oscillators via regenerative inverter latches. These regenerative circuits also help to reduce the switching inconsistency of the waveform across the parasitic capacitance of the parasitic-effect-based relaxation oscillator.

IV. VCO Implementation and Test Results

This section describes the implementation issues and experimental evaluation of the proposed VCOs presented in the previous section. Simulation of the preliminary design of all three VCOs was carried out at the schematic level. Time domain analysis was performed in order to find the maximum possible oscillating frequency. Simulations show that optimized circuits and device sizing had to be made through an iterative process due to the dynamic interaction between the loop ring components. The circuits were laid out using a 1.2μm double-poly double-metal N-well CMOS technology. The VCOs were shielded with buried ground rings to reduce noise coupling from adjacent circuits. On-chip capacitors of 4pF were placed between Vdd and ground lines, on unused space, to decouple high frequency noise on the power supply. The layout of VCO1, VCO2, and VCO3 circuits including buffer circuits have active areas of 84 μm X 50 μm, 142 μm X 58 μm, and 127 μm X 76 μm respectively. Figure 6 shows the die photograph of the three VCOs

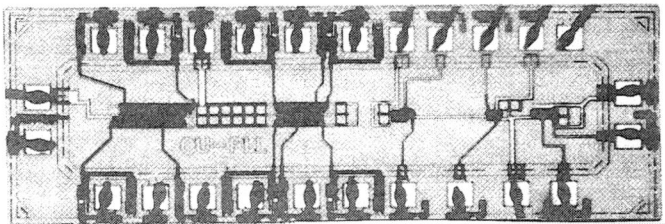

VCO2, VCO1 and VCO3 from right side

Fig. 6. Die photograph of the three VCOs.

The VCO tuning range, power consumption, phase noise response, and the effect of power supply variation were measured and compared with simulated results. The plots of the voltage-frequency transfer characteristics (or VCO tuning range) of the three VCOs are shown in Fig. 7.a. For comparison, the simulation results for each VCO are also included (dotted lines). The measured minimum-maximum oscillating frequency for VCO1, VCO2 and VCO3 occurred at 320-926 MHz, 225-606 MHz and 100-502 MHz respectively. The VCO spectral outputs are shown in Fig. 7.b.

Fig. 8.a shows the power consumed by each VCO at different oscillating frequencies. The top and middle curves, representing the power consumed by VCO3 and VCO2, show that the dissipated power is a linear function of frequency. VCO2 and VCO3 dissipate a maximum power of 14mW and 20.2mW respectively. Test results of VCO1 show that the fabricated chip operated 16 percent faster than the results obtained from simulations due to inaccurate hi-speed MOS models. VCO1's maximum power dissipation was measured at 9.4 mW. However, if the power dissipated by pad capacitance and buffer (3.92 mW, at 820 MHz) are excluded, then the predicted core circuit would consume only 5mW of total power. The VCOs were also measured for their oscillation frequencies as a function of supply voltage (Vdd) as shown in Fig. 8.b.

Measurement of phase noise was done by placing the oscillator in a phase-locked loop phase noise measurement system HP3048A. All measurements were performed through bonded IC's, packaged with a standard ceramic 68-pin PGA type. In order to characterize the performance at worst case, the frequency of oscillation was chosen to be close to the maximum possible frequency for all VCOs, that is VCO1 at 900 MHz, VCO2 at 550 MHz and VCO3 at 475 MHz. VCO1, VCO2 and VCO3 phase noise was measured and plotted on the top, middle and bottom charts of Fig. 9 respectively. The phase noise at 100 kHz offset from carrier frequencies shows that VCO1 had the highest noise levels at -83 dBc/Hz, followed by VCO2 at -87 dBc/Hz. VCO3 delivered the lowest phase noise response among the three at -90 dBc/Hz.

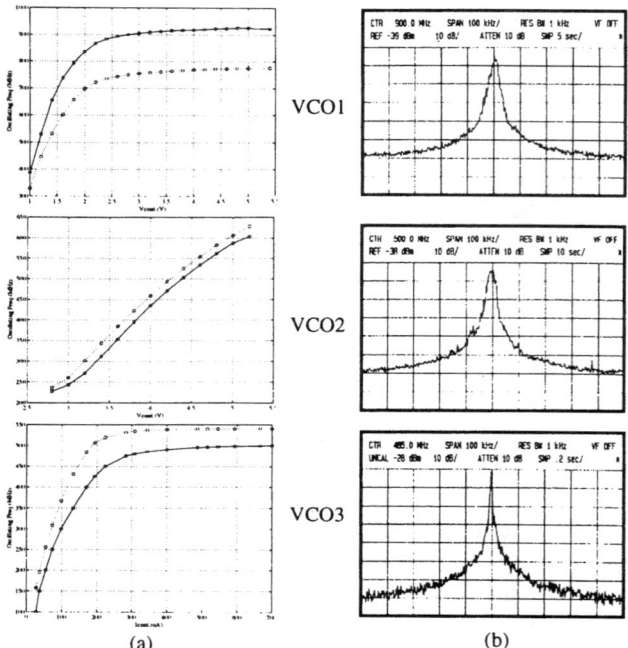

Fig. 7. (a) VCO transfer characteristic; (b) Spectra.

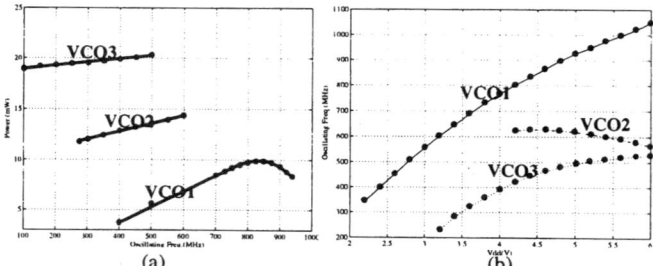

Fig. 8. (a). Power dissipation as a function of oscillating frequency, (b). Frequency of oscillation as a function of supply

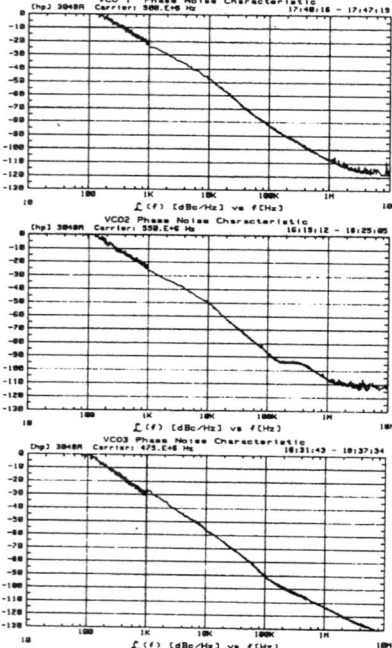

Fig. 9. Phase Noise Plots.

Table 1 Summarized results for PLL components, *Including the on-chip output buffer circuits, pad and package capacitance ($\approx$ 2.77 pF).

VCO Parameter	VCO1	VCO2	VCO3
Max. Operating Frequency (MHz)	926	606	502
Tuning Range (MHz)	606	381	402
Phase Noise @ 100KHz offset (dBc/Hz)	-83	-87	-90
Power Consumption (mW) @ Fmax	7.4*	14*	20.2*

V. Conclusion

Three high-speed CMOS VCOs were presented which simultaneously achieve high operating frequency and good phase noise characteristics through the use of several innovative architectural and circuit design techniques. These circuits, through simulation and measurement, demonstrated performance and speed in a gigahertz range which to date has not been realized in a conventional CMOS process. The proposed VCOs require no external components yet exhibit low phase noise comparable to LC on-chip oscillator. Among VCOs presented, the fully-differential-structure VCO showed the best phase noise performance of -90dBc/Hz at 100 KHz frequency offset. The simplest VCO achieved good phase noise while consuming only 7.4 mW of power. The work demonstrates the feasibility of a high-quality all-CMOS RF frequency synthesizer for mobile radio applications.

VI. Acknowledgements

The authors would like to thank the assistance of Canadian Microelectronics Corporation (CMC) for fabrication support. Also, the support of the Canadian International Development Agency (CIDA), the Telecommunication Research Institute of Ontario (TRIO) and Micronet are gratefully acknowledged.

References

[1] N. Foroudi, "CMOS High-Speed Dual-Modulus Frequency Divider for RF Frequency Synthesizers," M. Eng. Thesis, Carleton University, Ottawa, Canada, 1991.

[2] R. Rogemoser et al., "1.57 GHz Asynchronous and 1.4 GHz Dual-Modulus 1.2 μm CMOS Prescaler," IEEE Custom ICs Conf., pp. 16.3.1-16.3.4, 1994.

[3] M. Thamsirianunt, "A 1.2um CMOS Implementation of a Low-Power 900-MHz Mobile Radio Frequency Synthesizers," IEEE Custom ICs Conf., pp. 16.2.1-16.2.4, 1994.

[4] M. Thamsirianunt., "The Design and Implementation of CMOS Components for a Gigahertz Frequency Synthesizer." M. Eng. Thesis, Carleton University, Ottawa, Canada, 1993.

[5] M. Banu, "MOS Oscillators with multi-decade tuning range and gigahertz maximum speed," IEEE Jour. Solid-State Ccts., Vol. 23, No. 6, pp. 1386-1393, Dec.1988.

[6] N.M. Nguyen and R.G. Meyer, "A 1.8GHz Monolithic LC Voltage-Controlled Oscillator," IEEE Int'l Solid-State Ccts. Conf., pp. 158-159, 1992.

[7] A.A. Abidi and R.G. Meyer, "Noise In Relaxation Oscillators," IEEE Jour. Solid-State Ccts., Vol.SC-18, No.6, pp. 794-802, April, 1983.

[8] C.J.M. Verhoeven, "A High-Frequency Electronically Tunable Quadrature Oscillator," IEEE Jour. Solid-State Ccts., Vol.27, No.7, pp. 1097-1100, July, 1992.

[9] F. L. Martin, "A BiCMOS 50-MHz Voltage-Controlled Oscillator with Quadrature Output," IEEE Custom ICs Conf., pp. 27.4.1-27.4.4, 1993.

[10] A. B. Grebene, "Bipolar and MOS Analog Integrated Circuit Design," New York:Wiley,1984, ch.11.

FA 15.5: A CMOS 1.8GHz Low-Phase-Noise Voltage-Controlled Oscillator with Prescaler

Jan Craninckx, Michiel Steyaert

Katholieke Universiteit Leuven, ESAT-MICAS, Heverlee, Belgium

The two HF components for a full CMOS 1.8GHz frequency synthesizing PLL, the VCO and prescaler, are realized. The low-phase-noise oscillator employs bondwires for the high-quality on-chip inductor. A special LC-tank design enables an even further reduction of the phase noise. The prescaler uses an enhanced ECL-like CMOS D-flipflop and has a fixed division ratio of 128. The VCO and prescaler are integrated in a standard 0.7µm CMOS process.

A general circuit diagram of an LC-tuned oscillator is shown in Figure 1. Its single-sided output spectral noise density (S) at an offset $\Delta\omega$ from the carrier is given by [1]:

$$S = kTR_{eff}[1 + A + 2Q](\omega_0/\Delta\omega)^2 df$$

The effective resistance R_{eff} is first-order equal to the total series resistance in the inductor and the capacitor. There are three contributions to the phase noise, reflected in the factor [1+A+2Q]. The terms 1 and A represent the noise of the parasitic resistances and of the amplifier, respectively. A is usually equal to or larger than 1. The term 2Q is present only if an active implementation of the inductor is used. Since this term is dominant, to achieve low power consumption, it is necessary to make a high-quality passive inductor on chip.

Passive inductors on chip are usually realized as spiral inductors. However, coupling with the substrate limits the maximum operating frequency and the series resistance usually is several ohms [2, 3]. In this design the parasitic inductance of a bondwire in an IC package is employed to realize a high-quality LC-tank. Gold bondwires have low resistance (<1Ω), and inductance value approximately 1nH per mm length [4,5]. Capacitance to the substrate is, apart from the bondpads, almost non-existent.

Since phase noise is proportional to the series resistance in the LC-loop, there is a technological limit on the minimum achievable noise. To go beyond that limit, an enhanced LC-tank can be employed. The basic concept behind this improvement is the creation of an effective signal in the LC-tank loop that is larger than the signal at the output ports of the LC-tank, which is limited by the 3V power supply. By making the signal in the LC-tank loop larger, the signal-to-noise ratio becomes larger as well. This is achieved by using multiple inductors and capacitors.

The LC-tank used in this design is shown in Figure 2. Four bondwires are used to make two inductors. The capacitors are split into two parts. The first one (C1a and C2a) is a Metal/Metal capacitor and is used to shift the DC-voltage at the internal node to ground. A metal/metal capacitor is used instead of a Poly/N+ one to limit the series resistance. The other capacitors (C1b and C2b) are junction capacitors that are used to tune the center frequency with the control voltage V_C.

The circuit diagram of the complete VCO is shown in Figure 3. Implementation in CMOS is possible since the amplifier circuit is f_T-independent. Indeed, the gate-source capacitance of M1 and M2 is placed in parallel with the LC-tank, so it can be neglected compared to the LC capacitance of a few pF. The only frequency-limiting factors are the poles formed by the drain resistance and capacitance. The oscillation frequency of 1.8GHz is chosen from application considerations. Simulations indicate, however, that operation at frequencies as high as 5GHz is possible.

The prescaler operates asynchronously and has a fixed division ratio of 128. It uses a cascade of 7 divide-by-2 circuits. The high frequency D-flipflop is shown in Figure 4. To achieve the 1.8GHz operation, the current sources normally present underneath transistors M1,M2 and M11,M12 are omitted. The drawback of this is the need for a correct input DC biasing. However, this is easily accomplished by the DC biasing of the VCO itself. The result is a 20% increase in operating speed. However, this D-flipflop is still the speed-limiting factor in this circuit.

A micrograph of the IC is shown in Figure 5. The die measures 1.4x 3.7mm². The on-chip bondwires can clearly be seen. The VCO output spectrum is shown in Figure 6. The center frequency is 1.76GHz, and the resolution bandwidth is 1kHz. The chip-to-chip variation in center frequency is less than 1%. This is because the bondwire inductance is mainly determined by the horizontal length of the bondwires, which is very well controlled if chip-to-chip bonds are performed. The single-sided spectral noise density at 10kHz from the carrier is -85dBc/Hz. The VCO tuning characteristic is shown in Figure 7. The tuning range is about 4.5% for tuning voltages up to 3V and almost 7% for tuning voltages up to 10V. The VCO operates on a single 3V power supply and consumes 8mA. The prescaler takes 7mA from a 4V power supply.

References

[1] Wang, Y.-T., A. A. Abidi, "CMOS Active Filter Design at Very High Frequencies," IEEE J. of Solid-State Circuits, vol. 25, no. 6, pp. 1562-1574, Dec., 1990.

[2] Nguyen, N. M., R. G. Meyer, "A 1.8-GHz Monolithic LC Voltage-Controlled Oscillator," ISSCC Digest of Technical Papers, pp. 158-159, Feb., 1992.

[3] Basedau, P., Q. Huang, "A 1-GHz, 1.5-V Monolithic LC Oscillator in 1-µm CMOS," Proc. of the 1994 European Solid-State Circuits Conference, pp. 172-175, Sept., 1994.

[4] Greenhouse, H. M., "Design of Planar Rectangular Microelectronic Inductors," IEEE Trans. on Parts, Hybrids and Packaging, vol. PHP-10, no. 2, pp. 101-109, June, 1974.

[5] Steyaert, M., J. Craninckx, "1.1-GHz Oscillator Using Bondwire Inductance," IEE Electronic Letters, vol. 30, no. 3, pp. 244-245, Feb. 3, 1994.

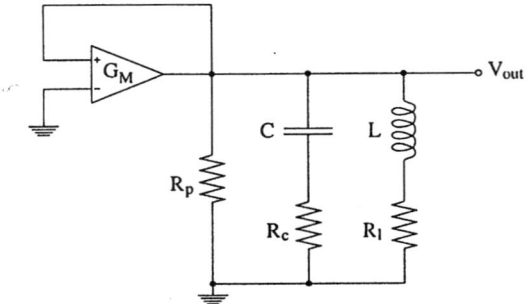

Figure 1: Basic LC-tuned oscillator circuit diagram.

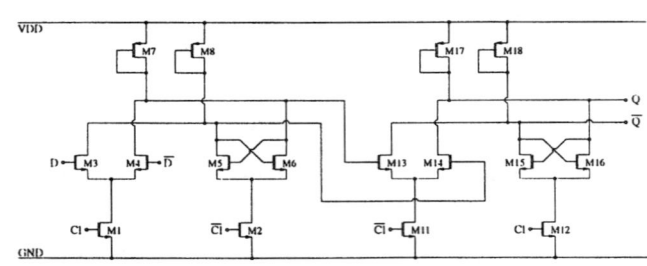

Figure 4: High-frequency D-flipflop.

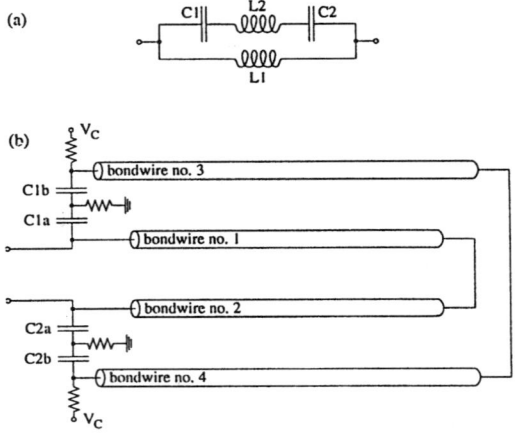

Figure 2: Enhanced LC-tank:
(a) general structure;
(b) realization with bondwires.

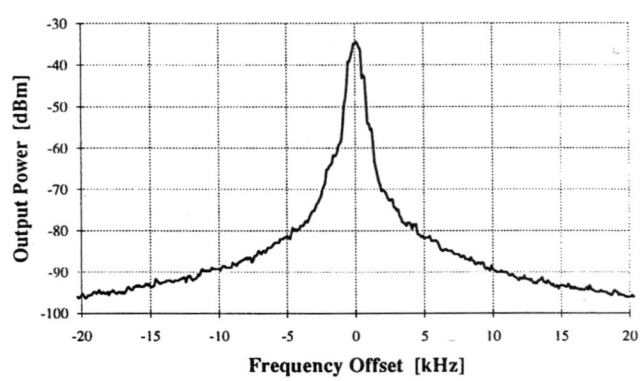

Figure 6: Output spectrum of the VCO.

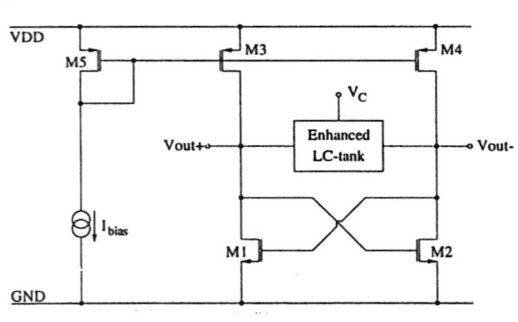

Figure 3: Circuit diagram of the VCO.

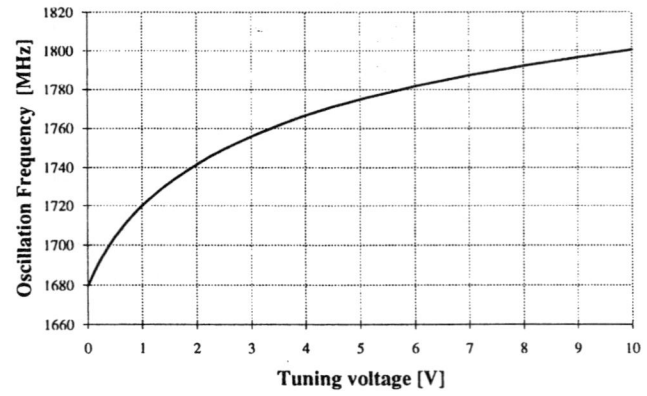

Figure 7: VCO tuning characteristic.

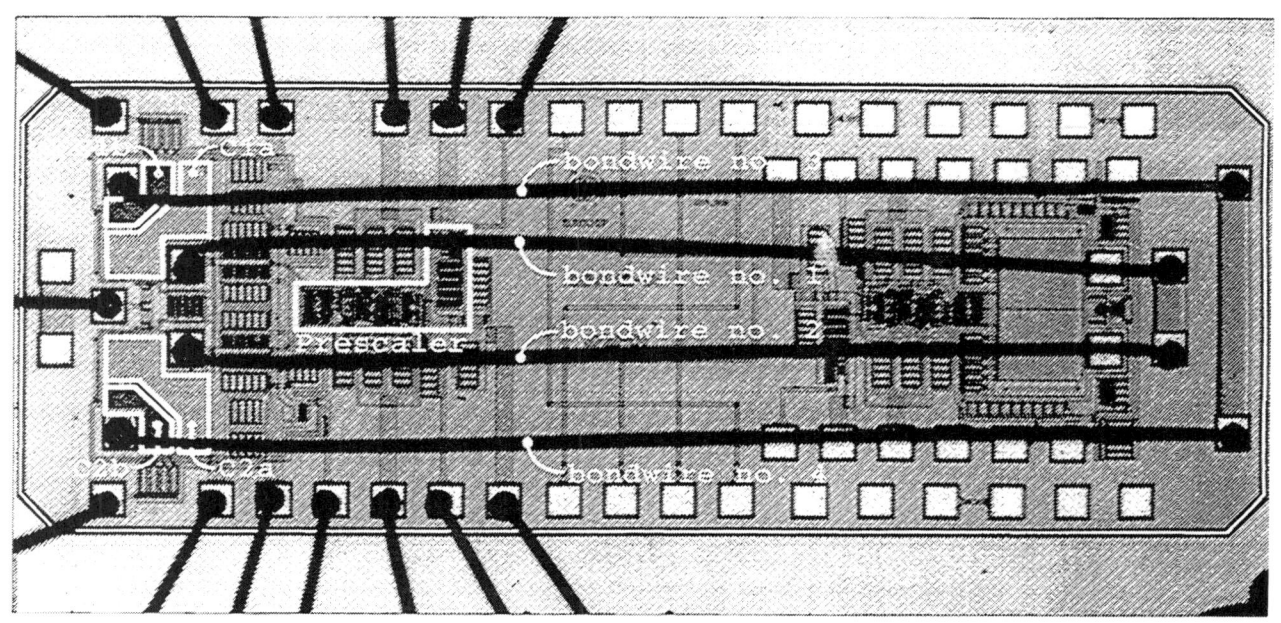

Figure 5: Chip micrograph.

WP 2.2: A Low-Power CMOS Digitally Synthesized 0-13MHz Agile Sinewave Generator*

Glenn Chang, Ahmadreza Rofougaran, Mong-Kai Ku, Asad A. Abidi, Henry Samueli

Electrical Engineering Department, Universuty of California, Los Angeles, CA

A low-power monolithic 1μm CMOS IC generates a 26MHz-wide single-sideband, frequency-hopped spread-spectrum waveform for wireless transmission in the 902-928MHz unlicensed ISM band. A direct digital frequency synthesizer (DDFS) on this IC produces 10b samples of sine and cosine waveforms whose frequency is selected with an 11b input word, followed by an on-chip 10b D/A converter (DAC) to convert the DDFS output into a sampled-data analog signal. Simplifications in architecture and circuits reduce dissipation of this 2.9x4.9mm² IC to 40mW at 40MHz with 3V supply.

A signed I-Q architecture for the frequency synthesizer halves the maximum clock frequency required to meet the specifications. Two DDFS/DAC channels produce discrete-time sine and cosine waveforms whose frequency varies from 0 to 13MHz (Figure 1). After anti-alias filtering, these sinewaves are respectively upconverted by quadrature outputs from a 915MHz local oscillator. If the two upconverted outputs are added, the output frequency ranges from 915 to 928 MHz. If they are subtracted, the frequency will cover 902 to 915MHz. Sign reversal may be implemented in the digital domain on the DDFS output. The DDFS clock must be at least three times the highest sinewave frequency to reduce anti-alias filter complexity. This DDFS/DAC thus must clock at least at 39 MHz.

The DDFS cell uses an efficient architecture to synthesize sinewaves with low spectral impurity [1]. The smallest internal and external word lengths that guarantee spurious tones at least -72.6dBc relative to the fundamental frequency are used. In response to the input control word, the output instantly steps in 19.5kHz increments over 0 to 13MHz. The DDFS ROM is 32 times smaller than in a straightforward design. Half- and quarter-wave symmetry of the sine function are exploited to generate sine and cosine waveforms from one ROM by phase shift of the argument. Internal word length is shortened by 2b because the ROM stores the difference between sine amplitude and phase. One large lookup table is replaced by smaller coarse and fine-interpolation tables.

The DAC successively bisects a reference charge between two equal capacitors through a switch controlled by the digital input, and therefore dissipates no static power. Its dynamic CV^2f dissipation gets smaller with capacitance C, whose minimum value is bounded by mismatch errors and thermal noise. At the end of conversion, a sample-and-hold buffer translates charge to voltage. If the buffer is not slew-rate limited, resampling will guarantee that no code-dependent glitches appear in the output. A conversion rate of 50MHz is obtained by pipelined operation of ten charge-redistribution stages. Increasingly significant bits of the data word are applied through delay registers to ten one-bit charge-redistribution cells in cascade, each cell consisting of one capacitor and two switches (Figure 2) [2]. Each data bit initially precharges its cell capacitor to a high or low reference voltage. When the capacitor is shorted to the previous cell, the resulting average voltage corresponds to a 2b D/A conversion. A 10b conversion is obtained after the charge traverses the ten cells. As in a CCD, the pipelined charge-transfer requires a three-phase non-overlapping clock.

---

*Research supported by ARPA, Rockwell International, and the State of California MICRO Program.

The linearity in this binary-weighted DAC is limited by capacitor mismatch, and signal-dependent stray charge injected by the MOSFET switches. The conversion rate is limited by the RC settling time in each cell. If FETs used to reduce the RC time constant are too wide, their voltage-dependent junction capacitance in parallel with the cell capacitor will add nonlinearity. Using published data on capacitor mismatch and known FET models, the cell components are optimized within these constraints, leading to a 500fF cell capacitor and a complementary switch composed of 5μm-wide nMOS and pMOS FETs. With 3V drive at FET gates, an acceptably short RC time constant is obtained when swings on the capacitor are limited to ±0.25V about a 1V common-mode level (Figure 3).

Three cells at a time are precharged by the respective data bits, followed by redistribution of charge among the cells under control of the three-phase clock. Accordingly, the input word is divided into three 3b nibbles and one MSB, and after appropriate delay in the pipeline registers, each nibble drives a 3b DAC stage. The equal-sized FETs in the complementary switch to the first-order cancel each other's charge injection. Stray charge injected by the switch at the conclusion of precharge causes a DAC gain error. Injection after charge-sharing is signal-dependent and contributes nonlinearity. This source of error is acceptably small for 10b operation.

The DAC output charge is converted by a buffer into a voltage. To attain a linearity commensurate with the DAC, the buffer is a balanced switched-capacitor unity-gain stage using a super-cascode op amp with about 60dB open-loop gain (Figure 2) [3]. Two DAC pipelines driven by complementary data apply a balanced input to the buffer, whose clocking scheme cancels the signal-dependent charge from its own sampling switches. An open-drain degenerated differential pair in the op amp feedback loop drives off-chip instruments at 50Ω. The last stage in the charge redistribution pipeline, shown driven by a data 1, boosts the common-mode voltage on the capacitors from 1V in the pipeline to 1.5V at the buffer input. The op amp sampling switches and capacitor are configured so that the buffer input itself resembles a DAC cell. The three clock phases in the DAC and the two additional phases for the buffer are derived on-chip from a single external clock (Figure 4).

At 40 MHz with 3V supply, the DDFS logic dissipates 35mW, the DAC clock generator, 4mW, and the DAC charge redistribution pipeline, less than 1mW. In this prototype a large output buffer drives off-chip loads. When scaled to drive typical on-chip capacitive loads, it is expected to dissipate 3 to 5mW. The worst-case spurious products in a DDFS system operating at a sample rate fs appear in sinewaves synthesized at frequencies fs/4 and fs/3. Spurious frequencies are below -72dBc. At low output frequencies, 3rd harmonic is -56dBc while in an output tone at a slight offset from fs/3 the largest aliased harmonic is at -50dBc (Figures 5a and 5b). These harmonics are higher than expected, and arise from a small inter-wire fringing capacitance linking DAC cells (Figure 6). Results from an IC with a revised layout with this parasitic reduced are expected to give close to 10b linearity.

References

[1] Nicholas, H.T., H. Samueli, "A 150MHz Direct Digital Frequency Synthesizer in 1.25μm CMOS with -90dBc Spurious Response," IEEE J. of Solid State Circuits, v. 26, pp. 1959-1969, Dec., 1991.

[2] Wang, F-J., et al., "A Quasi-Passive CMOS Pipeline D/A Converter," IEEE J. of Solid State Circuits, v.24, pp. 1752-1756, Dec., 1989.

[3] Bult, K., G. Geelen, "A Fast-Settling CMOS Op Amp for SC Circuits with 90dB dc Gain," IEEE J. of Solid State Circuits, v.25, pp. 628-636, Dec., 1990.

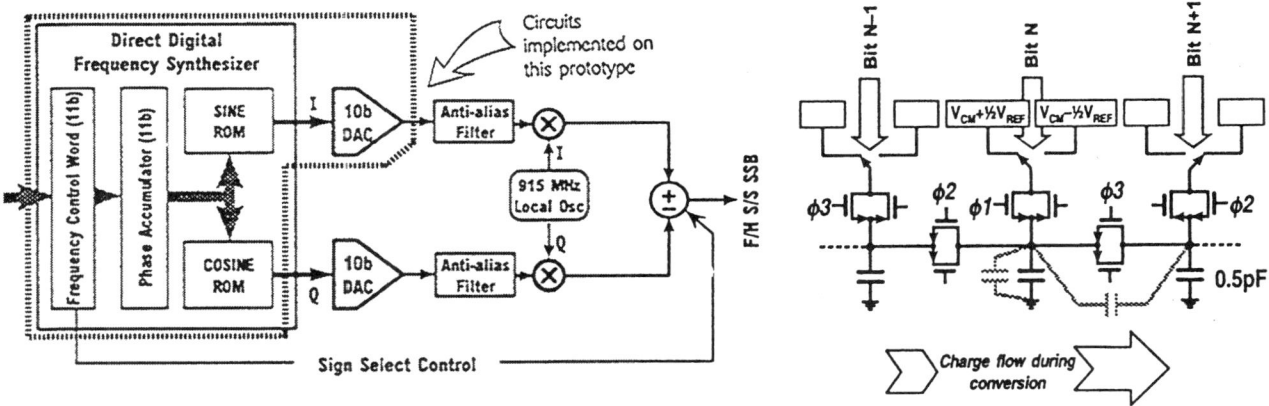

Figure 1: Block diagram of frequency synthesizer.

Figure 3: Charge-redistribution cell. Parasitic capacitance between cell capacitors (gray) causes integral nonlinearity.

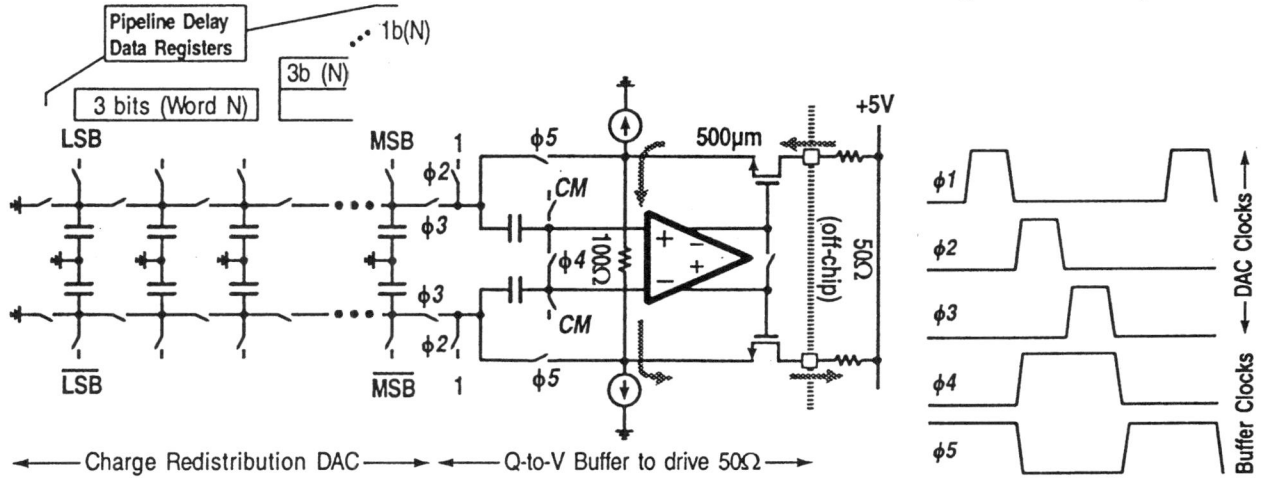

Figure 2: D/A converter uses pipelined charge redistribution in balanced configuration.

Figure 4: Clock phase required in D/A converter, buffer.

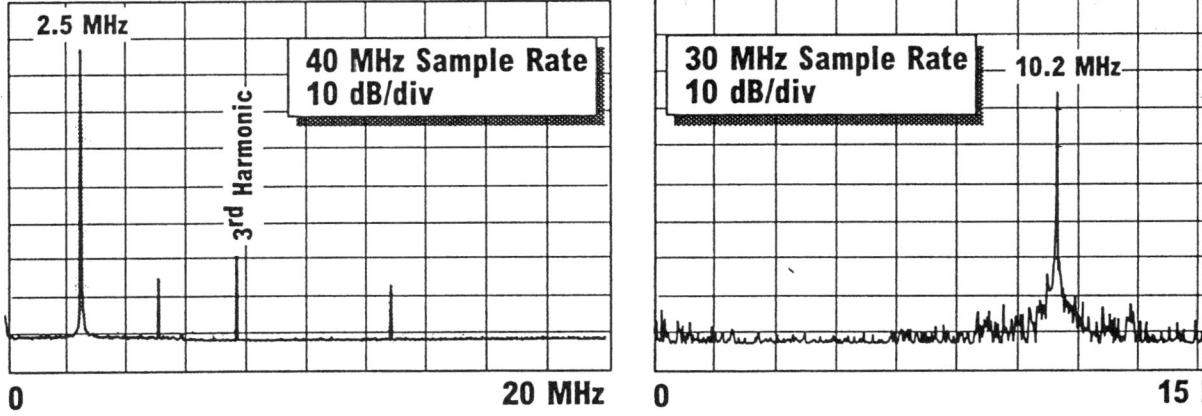

Figure 5a: 2.5MHz (low freq.) output sine spectrum.

Figure 5b: 10.2MHz sinewave spectrum ($f_s/3$ offset).
Figure 6: See page 304.

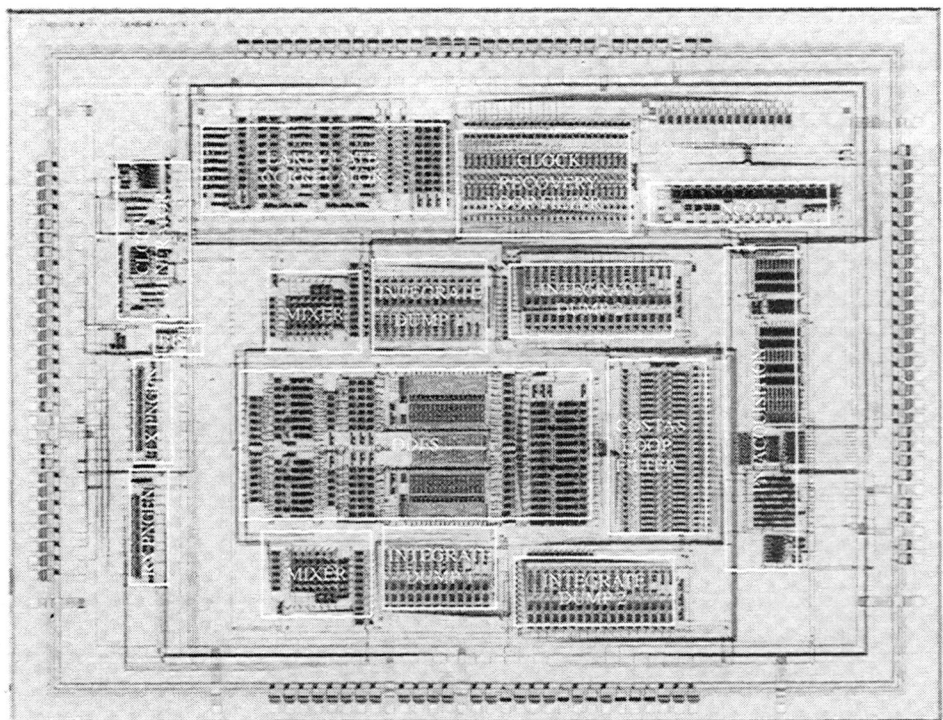

Figure 6: Transceiver chip micrograph.

SP 24.6: A 900MHz CMOS LC-Oscillator with Quadrature Outputs

Ahmadreza Rofougaran, Jacob Rael, Maryam Rofougaran, Asad Abidi

Electrical Engineering Department, University of California, Los Angeles, CA

The local oscillator (LO) in a wireless transceiver satisfies many exacting requirements. A variable frequency enables a phase-locked loop (PLL) to servo the LO to a stable lower frequency reference, or to correct frequency errors from measurements on the received signal. A low phase noise ensures little interference with nearby channels. A large LO voltage-swing means that it can drive a mixer with greater linearity. Finally, in single-sideband applications, the LO must supply precise quadrature phases. Low phase noise mandates use of a high-Q resonator to tune the LO, although most RF resonators are usually not integrable on ICs. Quadrature outputs are usually derived from RC phase-shift of a single-phase LO output, but this is susceptible to component inaccuracy and loss in LO amplitude [1,2].

This 900MHz oscillator circuit implemented in 1μm CMOS that affords modestly low-phase noise, has variable frequency has large output swing, and provides quadrature-phase outputs consists of two identical coupled oscillators, connected in such a way that they exert a mutual squelch when their relative phase is not in quadrature. The coupled oscillators synchronize to exactly the same frequency, in spite of mismatches in their resonant circuits. An early experimenter in oscillator synchronization has noted that this topology fortuitously produces a two-phase output [3].

The basic circuit building block, shown in Figure 1a, is the MOSFET LC oscillator consisting of a cross-coupled pair of FETs (M1, M2) with an inductor load. The inductor resonates with the FET gate and drain junction capacitance to determine the oscillation frequency. The negative resistance of $-1/g_m$ that the FET pair presents across the two drain terminals overcomes inductor loss. Onset of saturation in the large-signal I-V characteristic of the FET pair limits the oscillation amplitude. Two such identical oscillators, labeled A and B in Figure 1b, are coupled by FETs (M3, M4) of the same size as the main FETs (M1, M2), such that there is direct-coupling in one direction, and cross-coupling in the other.

Suppose now that the two oscillations synchronize in-phase. Then the cross-coupled path from oscillator B to A absorbs the negative-resistance current produced by M1A, M2A, and oscillator A ceases. The inductors in oscillator A pull up both drain nodes to V_{DD}, and through the cross-coupled FETs this shuts off oscillator B. The same process applies in reverse if the two oscillations are anti-phase. Therefore, the oscillations only co-exist when they synchronize in quadrature. They then acquire the unique combination of 0° at M2A, 180° at M1A, 90° at M2B, and 270° at M1B.

Almost all integrated 1GHz LC oscillators reported to date use discrete inductors, including the case when the inductors are bondwires. In this work, large spiral inductors are fabricated on the same CMOS silicon substrate as the oscillator [4]. A gaseous etchant selectively removes the substrate under the spiral in a maskless post-processing step, eliminating much of the parasitic capacitance, and extending the inductor self-resonance to several GHz. The suspended inductor after etching is shown in Figure 2. The resistance of the aluminum windings limits the Q of a large-value inductor to about 5 at 1GHz.

Two channels of four-FET switch mixers, shown in Figure 3, are also integrated on the chip to select one sideband in an upconverted quadrature baseband signal and reject the other sideband. The accuracy of LO quadrature is measured by the relative rejection of the unwanted sideband, assuming that the low-frequency baseband signal is in exact quadrature. The four output terminals of the polyphase oscillator are directly connected to the gates of the FET mixer, without an RF buffer. The oscillator absorbs the complex impedance at 900MHz seen at the gates of the mixer FETs.

The voltage at the top-rail, Figure 1c, tunes the oscillator by changing the drain bias, and thus the junction capacitance. Oscillators A and B share a common top-rail connected to the power supply through a single PFET in triode-region. The four current phasors sum to dc in this FET, whose channel resistance then sets the top-rail voltage.

The oscillator is functional at top-rail voltages as low as 1V because common-source FETs, and not a differential pair, implement the negative resistance. Although the oscillation amplitude increases with top-rail voltage, so does the negative conductance contributed by the FETs which lowers the loaded-Q of the resonator. As the phase noise depends oppositely on loaded-Q and oscillation amplitude, the two effects almost cancel one another.

The oscillator occupies an active area of 5mm², dominated by the four large on-chip spiral inductors. Frequency depends fairly linearly on PFET V_{GS} over a range of 120MHz centered at 850MHz, as plotted in Figure 4. With this sensitivity, a PLL can compensate for ±15% variation in the junction capacitance, a representative spread over process variations. Too high a sensitivity penalizes phase noise. At 1.5V nominal PFET V_{GS}, the quadrature oscillator draws 10mA from a 3V supply. The current drain of a properly-scaled version is expected to be at least three times less.

The oscillation waveform is measured at the output of the four-FET mixers with a balanced dc input applied. Owing to the inductor loads, the internal oscillation may grow as large as 7V differential peak-to-peak, limited by FET clipping. The inherent phase noise, plotted in Figure 5, is measured without embedding the oscillator in a PLL. The noise spectral density is -85dBc/Hz at a frequency offset of 100kHz from the oscillation, and remains relatively constant over the tuning range.

When this LO-mixer combination upconverts a 10MHz baseband quadrature input, the unwanted sideband lies 46.5dB lower than the wanted sideband, as shown in Figure 6. The gain in the two channels of switch-based mixers is expected to be exactly the same, implying that the LO outputs are less than 1° from perfect quadrature.

Acknowledgment:

This research was sponsored by ARPA and a consortium of semiconductor companies under the State of California MICRO Program. M. Christianson's support in test is acknowledged.

References:

[1] Sevenhans, J., et al., "An Integrated Si Bipolar RF Transceiver for a Zero IF 900 MHz GSM Digital Radio Frontend of a Hand Portable Phone," Custom IC Conf., pp. 7.7/1-4, 1991.

[2] Crols, J., M. Steyaert, "A Fully Integrated 900MHz CMOS Double Quadrature Downconverter," ISSCC Digest of Technical Papers, pp. 136-137, Feb., 1995.

[3] Millar, D. P. M., "A Two-Phase Audio-Frequency Oscillator," J. of IEE, Vol. 74, pp. 365-371, 1934.

[4] Chang, J. Y.-C., A. A. Abidi, M. Gaitan, "Large Suspended Inductors on Silicon and their use in a 2μm CMOS RF Amplifier," IEEE Electron Device Letters, Vol. 14, No. 5, pp. 246-248, 1993.

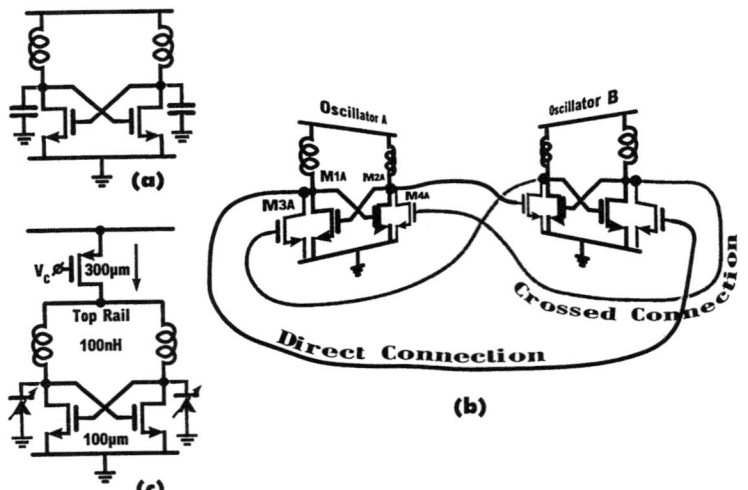

Figure 1: (a) MOS LC oscillator. (b) Identical oscillators coupled for quadrature outputs. (c) Top-rail control of oscillation frequency.

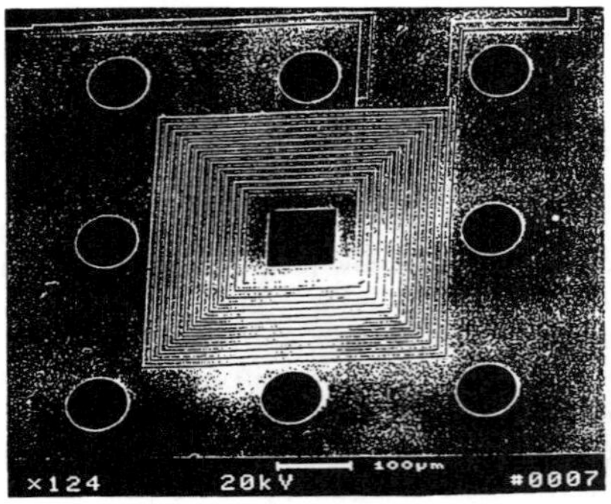

Figure 2: Spiral inductor suspended in oxide film above a cavity on a silicon substrate.

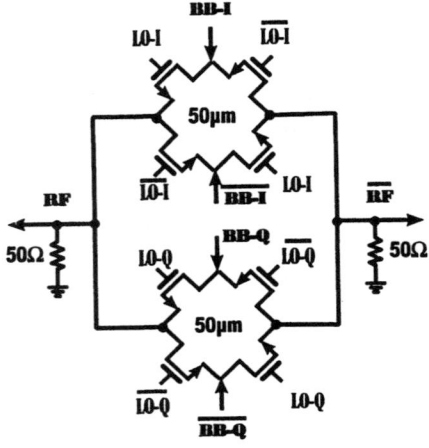

Figure 3: Four-FET single-sideband upconversion mixer integrated with oscillator for test.

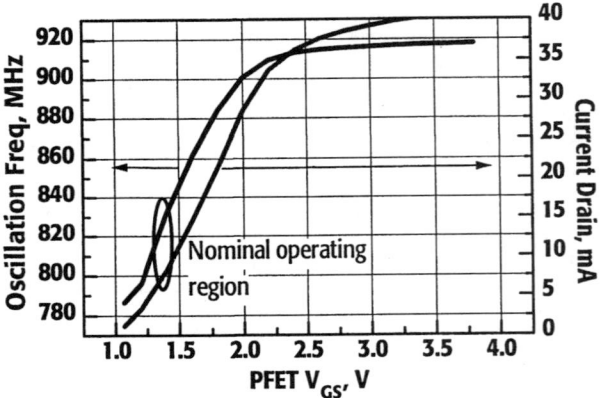

Figure 4: Measured oscillation frequency and supply current on PFET gate-bias.

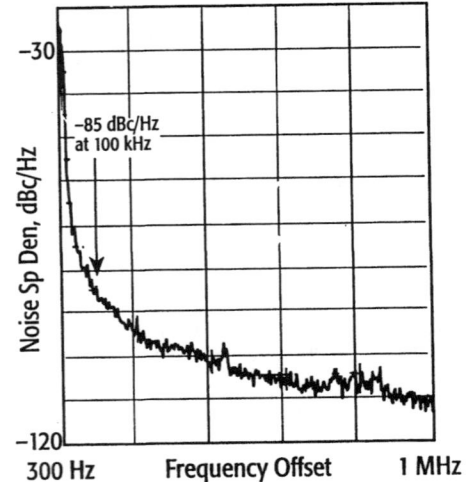

Figure 5: Measured single-sideband phase noise spectral density.

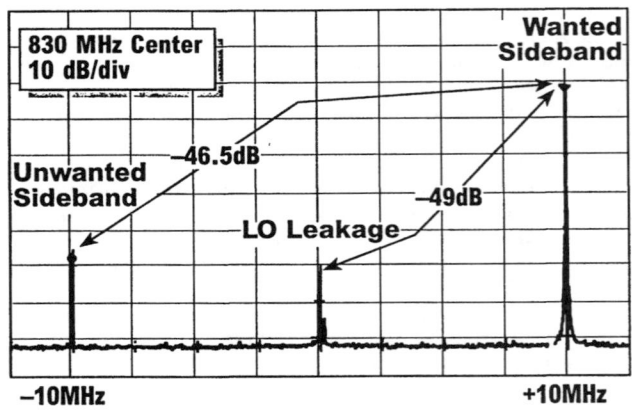

Figure 6: Measured output spectrum after upconversion of 10MHz tone.

SP 23.7: A Balanced 1.5GHz Voltage Controlled Oscillator with an Integrated LC Resonator

Leonard Dauphinee[1], Miles Copeland[1], Peter Schvan[2]

[1]Dept. of Electronics, Carleton University, Ottawa, Ontario, Canada
[2]Semiconductor Technology Access and Applications, Nortel Ltd., Ottawa, Ontario, Canada

A monolithic voltage-controlled oscillator (VCO) is presented for low-power digital radio handsets. This circuit is fabricated in a production 0.8μm 11GHz f_T BiCMOS process [1].

The VCO is based on the common-collector Colpitts circuit in a balanced configuration to provide differential output for direct compatibility with prescaler and double-balanced mixer inputs (Figure 1). The circuit utilizes an integrated LC resonator comprised of coplanar inductors, polysilicon capacitors, and on-chip varactors [2]. The fully-integrated inductors have an unloaded Q of approximately 5 at 2GHz.

The oscillator output in Figure 1 is taken from the collector as opposed to the emitter as is usual in a true common collector Colpitts circuit. This modified configuration has the advantage of isolating the output load from the LC resonant tank. Without isolation the LC tank is loaded by the next stage, and the Q of the oscillator is reduced.

The half circuit of the VCO is shown in Figure 2 for analysis. This circuit can be viewed as a negative resistance oscillator where Z_{in} is given by the equation in Figure 2 [3]. R_b comprises the inductor series resistance as well as the intrinsic base resistance, and C_π includes both intrinsic transistor capacitance and any external capacitance. The circuit oscillates when the real term of Z_{in} is negative, which occurs when the inequality given in Figure 2 is satisfied.

The ratio of C_π/C_e determines the tank voltage at which the nonlinear self-limiting effect is produced. A high C_π/C_e ratio reduces the tank loading caused by the active device and produces a larger tank voltage, thus reducing the phase noise [4]. A proper choice of this ratio is instrumental in achieving improved phase noise performance.

The measured phase noise of the circuit at 1.5GHz is -105dBc/Hz at an offset frequency of 100kHz with a supply voltage of 3.6V (Figure 3). This is the lowest phase noise reported to date for a fully integrated VCO in the 1 to 2GHz range with the exception of that reported in Reference 5 (Table 1). However, the work reported there incorporates bond wire inductors and is controversial in satisfying a fully integrated solution. The performance of the work described here surpasses the requirements of a DECT VCO (-99dBc/Hz at 1.728MHz) and meets the specifications for a CT-2 system, that requires -105dBc/Hz at an offset of 100kHz. The VCO operates with a supply voltage as low as 2.5V although there is a degradation in the phase noise performance. An enhancement to the VCO is achieved by replacing the collector resistor with an inductor or an LC parallel tank, allowing the circuit to operate at a reduced supply voltage. Such a version is now in fabrication.

The output power of the VCO from this work is -6.6dBm delivered into a 100Ω differential load. Other designs reported produce considerably less output power, and require output buffers to provide increased signal levels at the expense of increased power consumption and additional noise (Table 1) [5, 6, 7, 9, 10].

The tuning range of the VCO is approximately 150MHz as shown in Figure 4. Figure 5, obtained from spectrum analyzer measurements, demonstrates that the output power varies less than 1dB over the 10% tuning range. The power consumption is 40mW with a 3.6V supply. This power consumption includes biasing circuitry that consumes 12mW. The active die area of the circuit is approximately 0.5mm² (Figure 6).

References:

[1] Hadaway, R., et al., "A Sub-Micron BiCMOS Technology for Telecommunications," J. of Microelectronic Engineering, vol. 15, pp. 513-516, 1991.

[2] Long, J. L., M. A. Copeland, "Modeling, Characterization and Design of Monolithic Inductors for Silicon RFICs," CICC Digest of Technical Papers, pp. 185-188, 1996.

[3] Vendelin, G. D., et al., *Microwave Circuit Design Using Linear and Nonlinear Techniques*, J. Wiley, New York, chpt. 6, 1990.

[4] Leeson, D. B., "A Simple Model of Feedback Oscillator Noise Spectrum," Proceedings of the IEEE, pp. 329-330, Feb., 1966.

[5] Craninckx, J., M. Steyaert, "A CMOS 1.8GHz Low Phase Noise Voltage Controlled Oscillator with Prescaler," ISSCC Digest of Technical Papers, pp. 266-267, Feb., 1995.

[6] Soyuer, M., et al., "A 2.4 GHz Silicon Bipolar Oscillator with Integrated Resonator," IEEE J. Solid-State Circuits, vol. 31, no. 22, pp. 268-270, Feb., 1996.

[7] Aytur, T., B. Razavi, "A 2GHz, 6mW BiCMOS Frequency Synthesizer," ISSCC Digest of Technical Papers, pp. 264-265, Feb., 1995.

[8] Ali, A., J. L. Tham, "A 900MHz Frequency Synthesizer with Integrated LC Voltage Controlled Oscillator," ISSCC Digest of Technical Papers, pp. 390-391, Feb., 1996.

[9] Rofougaran, A., et al., "A 900MHz CMOS LC Oscillator with Quadrature Outputs," ISSCC Digest of Technical Papers, pp. 392-393, Feb., 1996.

[10] Nguyen, N. M., R. G. Meyer, "A 1.8 GHz Monolithic LC Voltage-Controlled Oscillator," IEEE J. Solid-State Circuits, vol. 27, no. 3, pp. 444-450, Mar., 1992.

This work was supported by grants from the Natural Sciences and Engineering Research Council of Canada (NSERC), Micronet, and the Telecommunications Research Institute of Ontario (TRIO). Fabrication and further support were provided by Nortel Ltd.

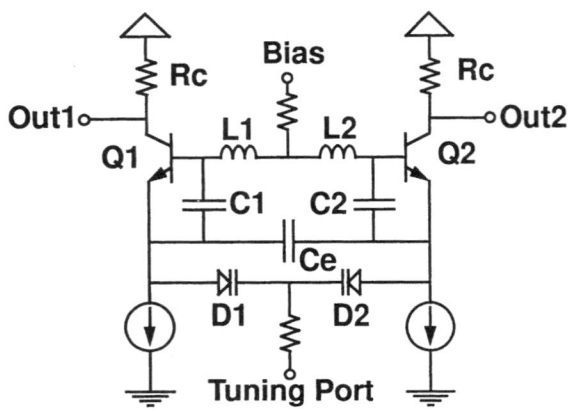

Figure 1: VCO schematic diagram.

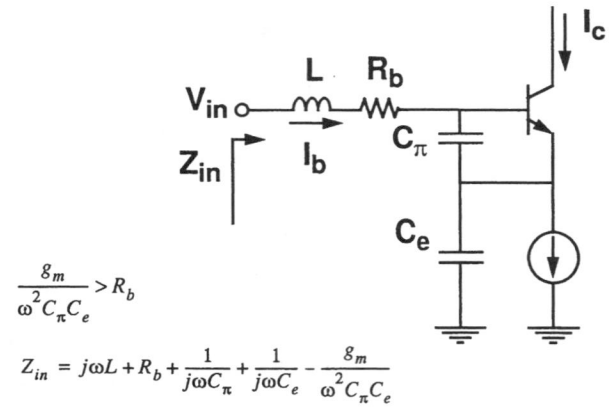

$$\frac{g_m}{\omega^2 C_\pi C_e} > R_b$$

$$Z_{in} = j\omega L + R_b + \frac{1}{j\omega C_\pi} + \frac{1}{j\omega C_e} - \frac{g_m}{\omega^2 C_\pi C_e}$$

Figure 2: Oscillator analysis.

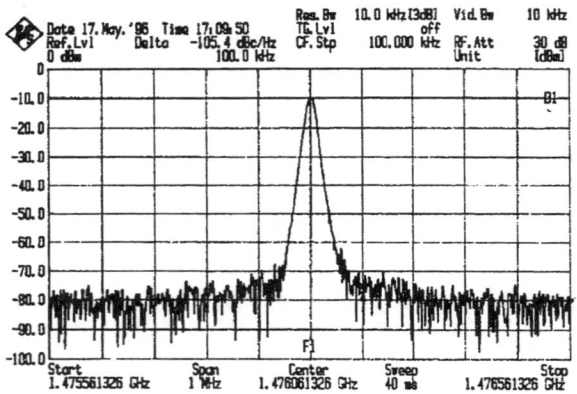

Figure 3: Phase noise performance.

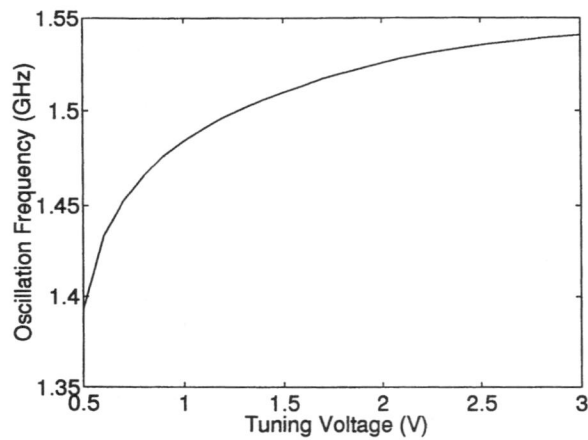

Figure 4: VCO tuning characteristic.

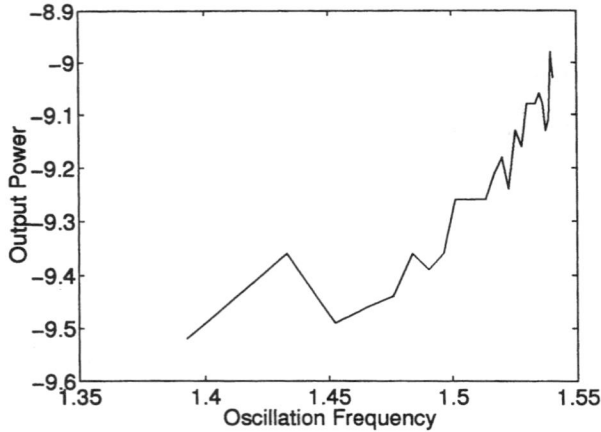

Figure 5: Output power.

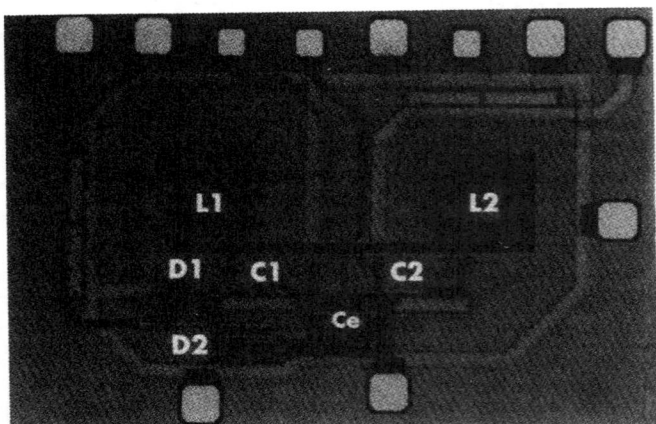

Figure 6: Chip micrograph.

f_0 (GHz)	1.476	1.760	2.360	2.048	0.913	0.85	1.75
Power (mW)	40	24	54	3	10	30	70
Pout (dBm)	-6.6(diff)	-35	-13.5	-25	-3 (diff.)	-25	-25
V_{cc} (V)	3.6	3	3.6	3	3	3	5
$L(f_m)$ dBc/Hz	-105 @ 100 kHz	-107 @ 100 kHz	-92 @ 1 MHz	-74 @ 100 kHz	-101 @ 100 kHz	-85 @ 100 kHz	-88 @ 100 kHz
Tuning Range (MHz)	150 (10%)	79.2 (4.5%)	N/A	N/A	N/A	120 (14%)	200 (11%)
Area (mm^2)	0.5	5.18	0.5	N/A			0.2
Min. V_{cc} (V)	2.5	N/A	2.6	2.7	2.7	N/A	N/A
Diff./Single	Diff.	Diff.	Single	Diff.	Diff.	Diff.	Single
Process	11 GHz BiCMOS	0.7µm CMOS	12 GHz BiCMOS	20 GHz BiCMOS	25 GHz Bipolar	1µm CMOS	10 GHz BiCMOS
Reference	This	[5]*	[6]	[7]	[8]	[9]	[10]

*Note: This VCO uses bond wire inductors and is not truly monolithic.

Table 1: Monolithic VCO performance.

SP 23.6: A 1.8GHz CMOS Voltage-Controlled Oscillator

Behzad Razavi

University of California, Los Angeles, CA
Formerly with Hewlett-Packard Laboratories, Palo Alto, CA

This paper describes the factors that limit the tuning range of monolithic LC voltage-controlled oscillators (VCOs), especially at low supply voltages, and introduces circuit techniques that alleviate this problem. Incorporating such techniques, a 1.8GHz CMOS oscillator achieves a tuning range of 120MHz with a relatively constant gain while exhibiting a phase noise of -100dBc/Hz at 500kHz offset. The actual implementation is a quadrature generator consisting of two coupled oscillators [1]. Since the two oscillators are identical, only one is considered here.

A basic negative-G_m LC oscillator, with M_1 and M_2 canceling the loss in the two tanks is shown in Figure 1a. For a given frequency of oscillation, f_{osc}, and supply current, it is desirable to maximize the value of L_1 and L_2 to increase the voltage swings and hence decrease the *relative* phase noise. This is because the equivalent parallel resistance of an inductor can be expressed as $R_p \approx (L\omega)^2/R_s$, where R_s is the series resistance. Although L and R_s scale proportionally, R_p still increases linearly with L. The upper bound on the value of L_1 and L_2 is of course determined by their self-resonance frequency, f_{SR}.

Even if the inductors are designed so that $f_{SR} > f_{osc}$, their parasitic capacitance still constitutes a significant portion of the tank capacitance. For the 10nH inductors employed here, the contribution of C_1 and C_2 to nodes X and Y is approximately equal to 200fF. Furthermore, M_1 and M_2 must be sufficiently wide to ensure reliable oscillation (W/L=50µm/0.6µm), contributing another 160fF of gate capacitance to X and Y. Additionally, in the quadrature generator, each oscillator is loaded by both the input stage of the other oscillator and an output buffer - another 150fF of gate capacitance. Since with a 10nH inductor, the total capacitance must be in the vicinity of 780fF to allow oscillation at 1.8GHz, the variable part of the tank capacitance can only be about 270fF.

The tuning range is further limited if the supply voltage is reduced. Because the capacitance-voltage characteristic of pn-junction varactors is fundamentally unscalable, a narrower range of tuning voltage translates to a smaller relative change in the capacitance. More importantly, as the voltage swings become comparable with the reverse bias across the diode, the equivalent capacitance variation decreases, especially if the diodes must remain reverse-biased.

The tuning methods described in References 1 and 2 either vary the bias current by a large factor or suffer from substantial variation in the gain of the VCO. In the oscillator of Figure 1(a), on the other hand, *IS1 provides* a constant bias current, and the common-mode voltage and hence the frequency are varied by adjusting the on-resistance of M_3. As V_{cont} increases, M_3 enters saturation, creating a sharp change in V_P and significant nonlinearity in the gain of the VCO. As depicted in Figure 1(b), another transistor, M_4, driven by a down-shifted version of V_{cont} can be added such that as M_3 saturates, M_4 is still in the triode region, but with a higher on-resistance. In other words, M_3-M_5 provide a resistance from node P to V_{DD} that varies smoothly with V_{cont}. The voltage at P is ultimately clamped by M_6 to about 1.2V.

To further increase the tuning range, the anode voltage of D_1 and D_2 can be varied in the opposite direction with respect to their cathode voltage. Illustrated in Figure 1(c), the idea is to raise V_Q as V_P drops. Since $V_Q = V_{cont} - V_{GS5} - V_{GS7}$, M_5 and M_7 are sized and biased so that for V_{cont} =3.3V, $V_Q \approx$ 0.8V. Under this condition, D_1 and D_2 experience a forward bias of a few hundred millivolts but with no significant impact on the phase noise. Also, as I_{S1} enters the triode region, transistor M_8 maintains a relatively constant tail current for M_1 and M_2. Simulations indicate that the combination of these techniques doubles the tuning range of the oscillator.

Figure 1(d) shows the complete oscillator, including the differential pair M_9-M_{10} to sense the output of the other oscillator in the quadrature circuit. In this implementation, I_{S1}=2mA and I_{S2}=0.1mA.

The quadrature oscillator is fabricated in a digital 0.6µm CMOS technology. The varactor diodes, D_1 and D_2, are implemented as n-well-p+ junctions, with the n-well strapped by n+ rings to lower the series resistance (Figure 2(a)). Shown in Figure 2(b), each inductor consists of two stacked spiral structures, achieving roughly four times the inductance of one spiral of the same area. The hole in the middle is approximately 35µm wide. No special wafer processing steps are used.

Figure 3 shows the output spectrum at 1.8GHz with a resolution bandwidth of 1kHz. The phase noise at 500kHz offset is -100dBc/Hz while each oscillator draws 2.3mA from a 3.3V supply.

Figure 4 shows the measured tuning characteristic of the VCO, indicating a frequency range of 120MHz and a gain of 114MHz/V. Table 1 summarizes VCO performance.

References:

[1] Rofougaran, A., et al, "A 900MHz CMOS LC Oscillator with Quadrature Outputs," ISSCC Digest of Technical Papers, pp. 392-393, Feb., 1996.

[2] Craninckx, J., M. Steyaert, "A 1.8 GHz Low-Phase Noise Spiral-LC CMOS VCO," VLSI Symp. Dig. Papers, pp. 30-31, June, 1996.

VCO frequency	1.8GHz
Tuning range	120MHz
Phase noise	-100dBc/Hz at 500kHz offset
Supply current	2.3mA/osc.
Supply voltage	3.3V
Technology	0.6µm CMOS

Table 1: Performance summary.

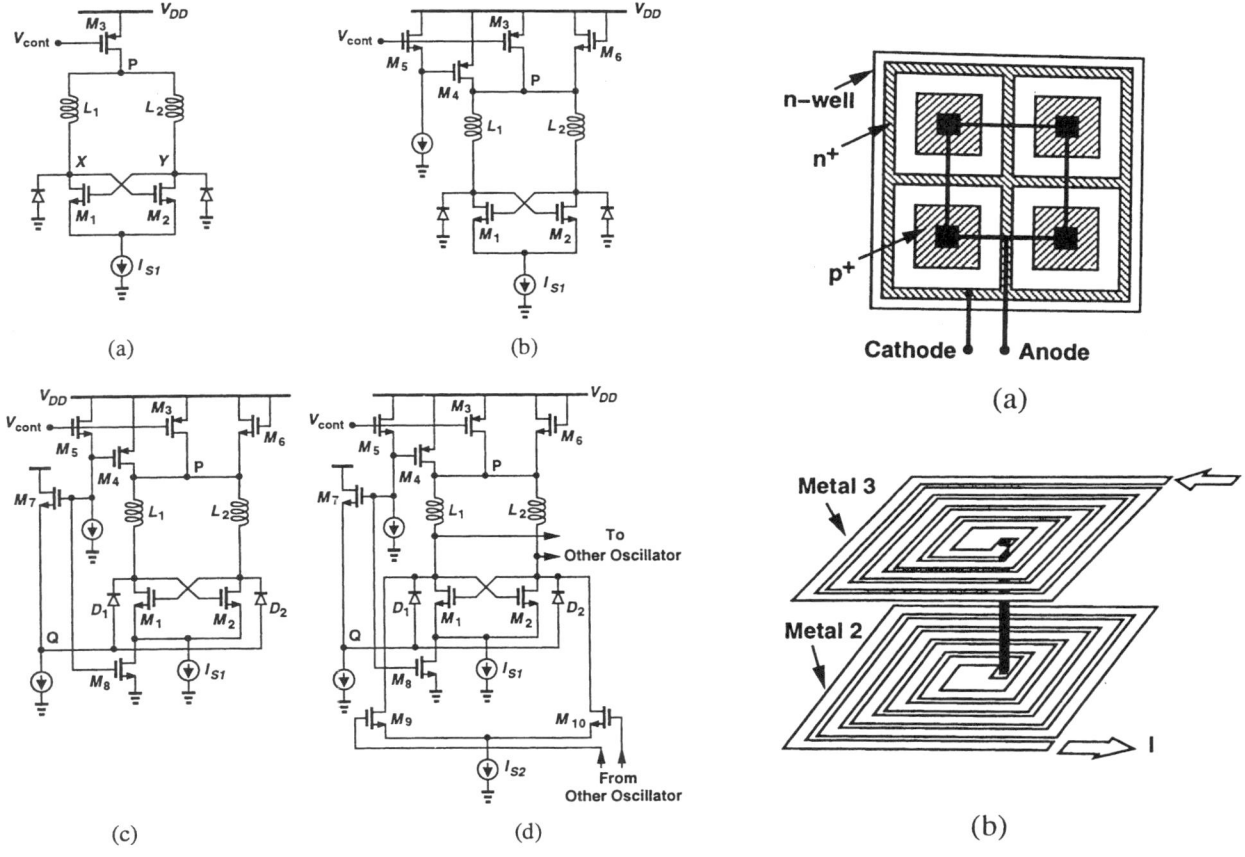

Figure 1: Evolution of VCO.

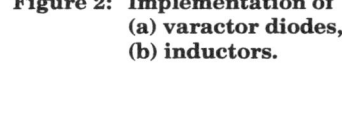

Figure 2: Implementation of (a) varactor diodes, (b) inductors.

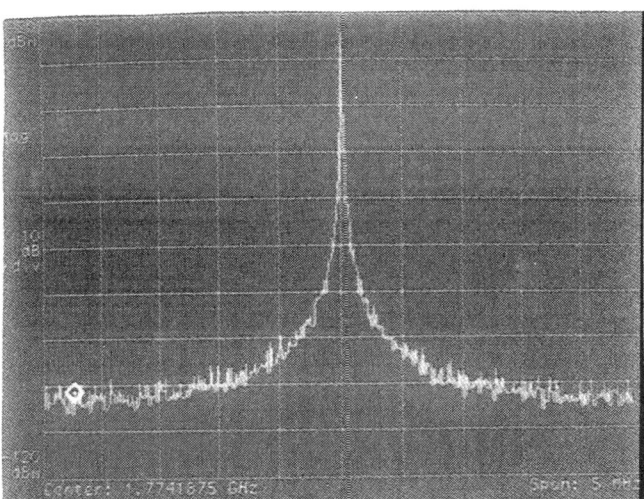

Figure 3: Measured output spectrum (horiz. 500kHz/div., vert. 10dB/div.).

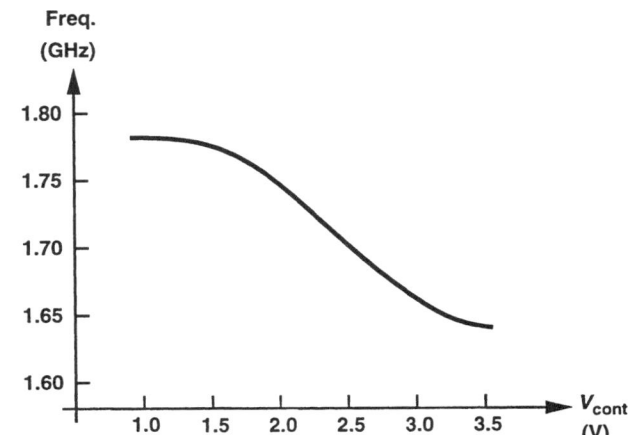

Figure 4: Measured tuning characteristic.

457

7
Components, Technology, and Modeling

FABRICATION and modeling of both active and passive elements are essential to the successful design and production of high-frequency transceivers. The modeling of bipolar transistors and GaAs FETS is a mature discipline, with numerous books available on the subject, and is not covered in this volume. The use of MOS transistors in RF applications in the gigahertz range is, however, a relatively new development. We thus have included one of the few publications available on the subject of MOS modeling for RF analog applications.

The use of on-chip inductive components is a relatively new development in Si RF ICs, and we also include several papers on that subject in this section. Such work should also be of interest to GaAs IC designers, who have been using such components for a longer period of time. The important topics of package modeling and passive discrete filter characterization are addressed here from a number of perspectives. Finally, some important general aspects of nonlinear device behavior are described in the context of intermodulation in mixers and the effect of large interfering signals (blockers) on LNA sensitivity.

Design of Planar Rectangular Microelectronic Inductors

H. M. GREENHOUSE, SENIOR MEMBER, IEEE

Abstract—Negative mutual inductance results from coupling between two conductors having current vectors in opposite directions. As a quantity in electronic circuits, negative mutual inductance is usually so much smaller in magnitude than overall inductance that it can be neglected with little effect. In the microelectronic world, however, its neglect can result in inductance values as much as 30 percent too high. This paper derives inductance equations for planar thin- or thick-film coils, comparing equations that include negative mutual inductance with those that do not. It describes a computer program developed for calculating inductances for both square and rectangular geometries, the variables considered being track width, space between tracks, and number of turns. Graphic results are presented for up to 16 turns over an inductance range of 3 nanohenries to 10 microhenries. Although details of fabrication are not included, the effects of film thickness and frequency on the mutual-inductance parameter are discussed.

INTRODUCTION

Technological progress in the areas of hybrid microelectronics and microwave integrated circuits during the past decade has seen thin-film microelectronic inductors used to an ever-increasing extent. Inductor design throughout this period, reflected in the technical literature [1]-[3] has been based largely on older theories and derivations, some dating back 100 years [4]. Now, as these inductors become smaller, the assumptions that have governed their design in the past become less valid. Nevertheless, inductor design, artwork preparation, photoreduction, and fabrication are time-consuming processes, and redesign and reprocessing must be kept to a minimum. Graphic representations of computer-made complex inductance calculations are an invaluable means toward this end.

BASIC MATHEMATICAL CONCEPTS

Self-Inductance Calculations for Straight Conductors

All theoretical equations for calculations involving planar rectangular inductors having one or more turns employ in their derivation the self-inductance of a straight conductor. The exact self-inductance for a straight conductor is[1]

$$L = 0.002\ell [\ln(2\ell/\text{GMD}) - 1.25 + \text{AMD}/\ell + (\mu/4)T] \quad (1)$$

Manuscript received May 24, 1973; revised February 1, 1974. This paper was previously published in the *Bendix Technical Journal*, pp. 7-16, Winter 1972/73.

The author is with the Bendix Communication Division, Towson, Md. 21204.

where L is the inductance in microhenries, ℓ is the conductor length in centimeters, GMD and AMD represent the geometric and arithmetic mean distances, respectively, of the conductor cross section, μ is the conductor permeability, and T is a frequency-correction parameter.

The geometric mean distance (GMD) between two conductors is the distance between two infinitely thin imaginary filaments whose mutual inductance is equal to the mutual inductance between the two original conductors. The GMD of a conductor cross section is the distance between two imaginary filaments normal to the cross section, whose mutual inductance is equal to the self-inductance of the conductor.[2] By definition, the self-inductance of a conductor is the sum of the mutual inductances of all the pairs of filaments of which it is composed. The GMD is equal to 0.7788 times the radius in the case of a circular cross section, 0.44705 times a side in the case of a square cross section, and 0.22313 times the length in the case of a straight-line cross section. Computation of the GMD for a rectangular cross section is lengthy, but its value—which is a function of the ratio between sides a and b—is easily seen to lie within a narrow range: for the limiting case where $b \to 0$, the value is 0.22313 $(a+b)$; for the limiting case where $a=b$, the value is 0.22352$(a+b)$.

The arithmetic mean distance is the average of all the distances between the points of one conductor and the points of another. For a single conductor, the arithmetic mean distance is the average of all possible distances within the cross section. In the case of a circular cross section, the AMD equals the radius; in the case of a straight-line cross section, the AMD equals one-third the length. Thin-film conductors approach the straight-line condition: as the film thickness approaches 0, the AMD of a thin-film track approaches one-third the width.

If GMD and AMD values for a circular cross section are substituted into (1), we obtain

$$\begin{aligned} L &= 0.002\ell [\ln(2\ell/0.7788r) - 1.25 + r/\ell + (\mu/4)T] \\ &= 0.002\ell [\ln(2\ell/r) - \ln 0.7788 - 1.25 + r/\ell + (\mu/4)T] \\ &= 0.002\ell [\ln(2\ell/r) - 1 + r/\ell + (\mu/4)T] \end{aligned} \quad (2)$$

which is the exact equation for a circular cross section, r being the radius. For the near-direct-current condition, T equals 1 and the equation becomes

$$L = 0.002\ell [\ln(2\ell/r) - 1 + r/\ell + \mu/4] . \quad (3)$$

[1] Though not directly stated in the literature, this equation is easily derived by combining equations (6), (8), and (211) of Grover (see reference 5). The value of T, which varies from 1 at direct current to 0 at infinite frequency, can be found for a conductor of circular cross section from Table 52, page 266 of Grover.

[2] The concept of cross-section geometric mean distance goes back to Maxwell's examples in article 692, Volume II of reference 4.

If the conductor has a magnetic permeability of 1, (3) reduces to

$$L = 0.002\ell[\ln(2\ell/r) - 0.75 + r/\ell] \qquad (4)$$

and if the length is many orders of magnitude greater than the radius, it becomes

$$L = 0.002\ell[\ln(2\ell/r) - 0.75]. \qquad (5)$$

Equations (3), (4), and (5) are supported by most authoritative sources [5]-[7].

For thin-film inductors with rectangular cross sections, (1) takes the form

$$\begin{aligned} L &= 0.002\ell \{\ln[2\ell/0.2232(a+b)] - 1.25 + [(a+b)/3\ell] \\ &\quad + (\mu/4)T\} \\ &= 0.002\ell \{\ln[2\ell/(a+b)] - \ln 0.2232 - 1.25 + [(a+b)/3\ell] \\ &\quad + (\mu/4)T\} \\ &= 0.002\ell \{\ln[2\ell/(a+b)] + 0.25049 + [(a+b)/3\ell] \\ &\quad + (\mu/4)T\} \qquad (6)\end{aligned}$$

where a and b are the rectangular dimensions of the cross section. For the near-direct-current case in which magnetic permeability is 1, (6) reduces to[3]

$$L = 0.002\ell \{\ln[2\ell/(a+b)] + 0.50049 + [(a+b)/3\ell]\}. \qquad (7)$$

As Table I indicates, the skin-depth phenomenon has little effect on thin films, and T in (6) should be considered to have a value of 1 for microwave frequencies. For thicker films and lower frequencies, corrections may be required and must be considered [5], [7], [9].

TABLE I

Variations in Frequency-Correction Parameter T for Thin Films and Microwave Frequencies

Value of T	Film Thickness	Frequency
0.9974	10,000 angstroms	10 gigahertz
0.9986	0.0025 millimeter (0.1 mil)	1 gigahertz
0.9095	0.0075 millimeter (0.3 mil)	1 gigahertz

Mutual-Inductance Calculations for Planar Coils

In the case of an L-shaped thin-film inductor, total inductance is equal to the sum of the self-inductances of the two straight segments and is less than the inductance of a single straight track of equal total length. In the case of a rectangular or square planar coil, straight conductor segments parallel other straight conductor segments and the mutual inductance between these parallel tracks contributes to the total inductance of the coil.

Fig. 1 illustrates the mutual inductance $M_{1,2}$ that results from a singularly generated current i_1. Here,

$$M_{1,2} = d\phi_{1,2}/di_1$$

[3]Equation (7) is in substantial agreement with equations derived by others (cf. [5] through [9]), the difference being that they have assumed a square cross section whereas we have assumed a rectangular cross section in which one side is many times greater than the other.

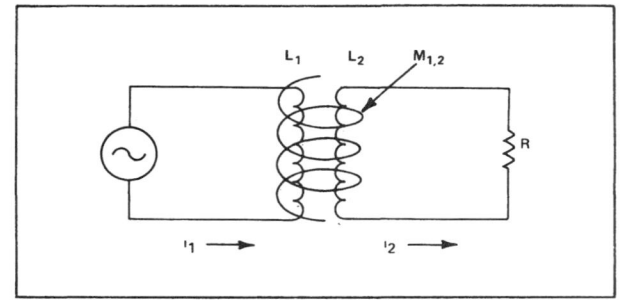

Fig. 1. Mutual inductance resulting from singularly generated current.

where $\phi_{1,2}$ is the flux common to self-inductances L_1 and L_2 that is caused by the generated current, i_2 being the induced current. Fig. 2 illustrates the mutual inductance that results from two generated currents, i_1 and i_2. In this case,

$$M_{1,2} = d\phi_{1,2}/di_1$$

and

$$M_{2,1} = d\phi_{2,1}/di_2$$

where $\phi_{1,2}$ is the flux common to self-inductances L_1 and L_2 that is caused by current i_1, and where $\phi_{2,1}$ is the flux common to self-inductances L_1 and L_2 that is caused by current i_2. When the frequencies of the two current generators are the same, the total mutual inductance M_T is equal to the vector sum of $M_{1,2}$ and $M_{2,1}$; when these frequencies differ, the instantaneous sum must be used.

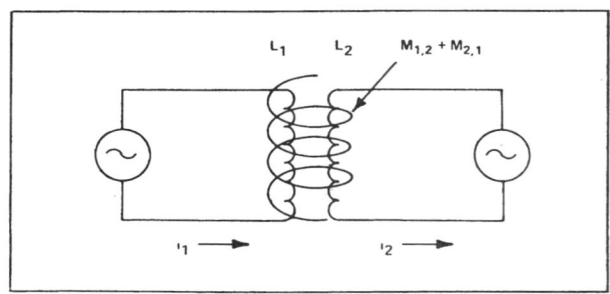

Fig. 2. Mutual inductance resulting from two generated currents.

Consider the case of the two-turn planar rectangular coil represented schematically in Fig. 3. The total inductance of this coil is equal to the sum of the self-inductances of each of the straight segments ($L_1 + L_2 + L_3 + L_4 + L_5 + L_6 + L_7 + L_8$) plus all the mutual inductances between the segments. The mutual inductance between segments 1 and 5 has a component $M_{1,5}$ cuased by the current flowing in segment 1, and a component $M_{5,1}$ caused by the current flowing in segment 5. Since the frequency and phase in both segments are identical, the total mutual inductance linking them equals $M_{1,5}+M_{5,1}$. An analogous relationship exists between segment pairs 2-6, 3-7, and 4-8; in each of these pairs, current flow is in the same direction in both segments and all mutual inductances are positive. The mutual inductance between segments 1 and 7, on the other hand, has a component $M_{1,7}$ caused by the current in segment 1, and a component $M_{7,1}$ caused by the current in segment 7. The total mutual inductance linking these two segments equals

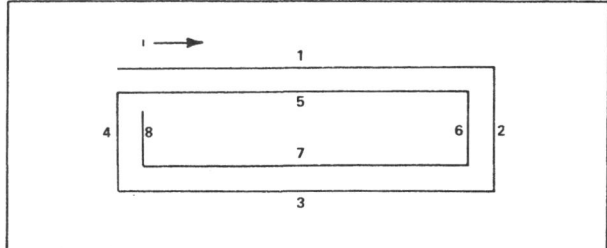

Fig. 3. Two-turn rectangular planar coil.

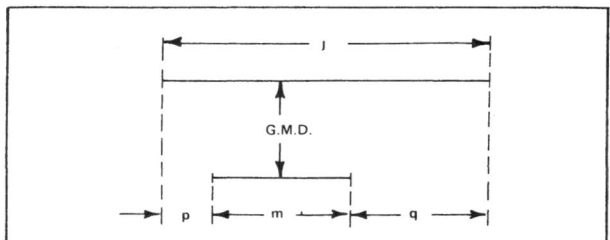

Fig. 4. Two-parallel-filament geometry.

$M_{1,7}+M_{7,1}$ but is negative because current flow in segment 1 is opposite in direction to current flow in segment 7. An analogous relationship exists between segment pairs 1-3, 5-7, 5-3, 2-8, 2-4, 6-8, and 6-4. Current magnitude is identical in all segments, with the result that $M_{a,b}=M_{b,a}$. The total inductance L_T for this two-turn coil therefore becomes

$$L_T = L_1 + L_2 + L_3 + L_4 + L_5 + L_6 + L_7 + L_8 + 2(M_{1,5}+M_{2,6} + M_{3,7} + M_{4,8}) - 2(M_{1,7} + M_{1,3} + M_{5,7} + M_{5,3} + M_{2,8} + M_{2,4} + M_{6,8} + M_{6,4}).$$

The general equation for a coil or a part of a coil of any shape is

$$L_T = L_0 + \Sigma M \qquad (8)$$

where L_T is the total inductance, L_0 is the sum of the self-inductances of all the straight segments, and ΣM is the sum of all the mutual inductances, both positive and negative. Since mutual inductance is positive when current flow in two parallel conductors is in the same direction and negative when current flow is in opposite directions, (8) can be rewritten to read

$$L_T = L_0 + M_+ - M_- \qquad (9)$$

where M_+ is the sum of the positive mutual inductances and M_- is the sum of the negative mutual inductances.

The mutual inductance between two parallel conductors is a function of the length of the conductors and of the geometric mean distance between them. In general,

$$M = 2\ell Q \qquad (10)$$

where M is the mutual inductance in nanohenries, ℓ is the conductor length in centimeters, and Q is the mutual-inductance parameter, calculated from the equation

$$Q = \ln\left\{(\ell/GMD)+[1+(\ell^2/GMD^2)]^{1/2}\right\} \\ - [1+(GMD^2/\ell^2)]^{1/2} + (GMD/\ell). \qquad (11)$$

In this equation, ℓ is the length corresponding to the subscript of Q, and GMD is the geometric mean distance between the two conductors, which is approximately equal to the distance d between the track centers. The exact value of the GMD may be calculated from the equation

$$\ln GMD = \ln d - \left\{[1/12(d/w)^2] + [1/60(d/w)^4] \\ + [1/168(d/w)^6] + [1/360(d/w)^8] + [1/660(d/w)^{10}] + \cdots\right\} \qquad (12)$$

where w is the track width.

Now consider the two-conductor geometry represented schematically in Fig. 4. Two filaments of lengths j and m, respectively, are separated by a geometric mean distance GMD. In this case,

$$2M_{j,m} = +(M_{m+p} + M_{m+q}) - (M_p + M_q) \qquad (13)$$

and the individual M terms are calculated using equation (10) and the lengths corresponding to the subscripts; that is,

$$M_{m+p} = 2\ell_{m+p}Q_{m+p} = 2(m+p)Q_{m+p}$$

where Q_{m+p} is the mutual-inductance parameter Q for GMD/$(m+p)$. Though other more general expressions are available,[4] we will limit ourselves for purposes of this paper to the use of (13) and two additional relationships:

for $p = q$,

$$M_{j,m} = M_{m+p} - M_p \qquad (14)$$

for $p = 0$,

$$2M_{j,m} = (M_j + M_m) - M_q. \qquad (15)$$

SOME COMPARATIVE CALCULATIONS

In the sections that follow, we shall calculate by several methods the inductance of a single-turn square planar coil of the type shown in Fig. 5. All segments will be assumed to be shortened at each connecting end by half the track width w, so that

$$\ell_1 = \ell_2 = \ell_3 = 0.10 - w = 0.10 - 0.01 = 0.09 \text{ centimeter}$$

and

$$\ell_4 = \ell_2 - w - s = 0.09 - 0.01 - 0.01 = 0.070 \text{ centimeter}$$

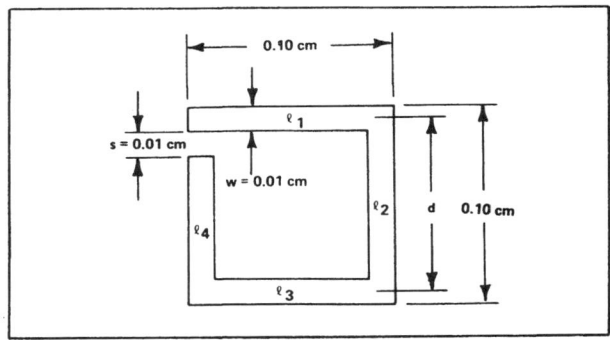

Fig. 5. Single-turn square planar coil.

[4]The derivation of (13) and these more general expressions are presented by Grover in [5].

We shall also assume that the magnetic permeability of the conductor material is 1 and that film thickness t is 0.0005 centimeter.

Expanded Grover Method

The derivations in the preceding theoretical discussion are based largely on the work of Grover [5] and produce the following calculated results. Repeating (9),

$$L_T = L_0 + M_+ - M_-$$

where $L_0 = L_1 + L_2 + L_3 + L_4$. From (7), we obtain

$$L_x = 2\ell_x \left\{ \ln[2\ell_x/(w+t)] + 0.50049 + [(w+t)/3\ell_x] \right\} \quad (16)$$

where L_x is the segment inductance in nanohenries, ℓ_x is the segment length in centimeters, w is the segment width in centimeters, and t is the segment thickness in centimeters. Substituting values into (16), we obtain

$$L_1 = 2(0.09) \left\{ \ln[(2)(0.09)/(0.01+0.0005)] + 0.50049 + [(0.01+0.0005)/(3)(0.09)] \right\}$$

or

$$L_1 = L_2 = L_3 = 0.60867 \text{ nanohenry.}$$

Similarly,

$$L_4 = 2(0.070) \left\{ \ln[(2)(0.070)/0.0105] + 0.50049 + [0.0105/(3)(0.070)] \right\}$$
$$= 0.43597 \text{ nanohenry.}$$

Then,

$$L_0 = 3(0.60867) + 0.43597 = 2.26198 \text{ nanohenries.}$$

Since the currents in parallel legs flow in opposite directions, there is no positive mutual inductance in this coil; that is,

$$M_+ = 0$$

The negative mutual inductance is equal to the sum of $M_{1,3}$, $M_{3,1}$, $M_{2,4}$, and $M_{4,2}$, or, since $M_{1,3}$ equals $M_{3,1}$ and $M_{2,4}$ equals $M_{4,2}$,

$$M_- = 2(M_{1,3} + M_{2,4}). \quad (17)$$

Going back to (12) and substituting values of 0.01 and 0.09 for w and d, respectively, yields a GMD of 0.0899 centimeter. This value and that for ℓ_1, when substituted into (11), yield a mutual-inductance parameter Q_1 of 0.4672. Now, using (10) and the fact that ℓ_1 equals ℓ_3 we can write

$$M_{1,3} = 2\ell_1 Q_1 = 2(0.09)(0.4672) = 0.084096 \text{ nanohenry.}$$

However, because ℓ_2 does not equal ℓ_4, (15) must be used to solve for $M_{2,4}$. In this case,

$$2M_{2,4} = (M_2 + M_4) - M_{0.02}. \quad (18)$$

Again using (10),

$$M_2 = 2\ell_2 Q_2$$
$$M_4 = 2\ell_4 Q_4$$
$$M_{0.02} = 2(0.02) Q_{0.02}.$$

Since ℓ_2 equals ℓ_1 and the GMD remains constant, Q_2 must equal Q_1 as calculated from (11). It follows that

$$M_2 = M_{1,3} = 0.084096 \text{ nanohenry.}$$

To obtain Q_4 and $Q_{0.02}$, however, (11) must be solved for a GMD of 0.0899 and segment lengths of 0.070 and 0.02 centimeter, respectively. Thus calculated, Q_4 is found to be 0.3770, and M_4 becomes

$$M_4 = 2(0.070)(0.3770) = 0.052780 \text{ nanohenry.}$$

Similarly, $Q_{0.02}$ is found to be 0.0110, and $M_{0.02}$ becomes

$$M_{0.02} = 2(0.02)(0.011) = 0.000440 \text{ nanohenry.}$$

Substituting these values into (18), we obtain

$$2M_{2,4} = (0.084096 + 0.052780) - 0.000440$$
$$M_{2,4} = 0.068218 \text{ nanohenry.}$$

Having determined $M_{1,3}$ and $M_{2,4}$, we can now calculate the total negative mutual inductance in the coil, as expressed by (17):

$$M_- = 2(0.084096 + 0.068218) = 0.30463 \text{ nanohenry.}$$

Finally, returning to (9),

$$L_T = L_0 + M_+ - M_-$$

we obtain for total inductance

$$L_T = 2.26198 - 0.30463 = 1.9573 \text{ nanohenries.}$$

Note that, if we were to neglect the negative mutual inductance in the coil, L_T would equal L_0 and have a value of 2.26198 nanohenries. Alternatively, if the coil were treated as equivalent to a straight conductor equal in length to the sum of the segment lengths (0.335 centimeter), the inductance value arrived at would be

$$L = 2(0.335) \left\{ \ln[(2)(0.335)/0.0105] + 0.50049 + [0.0105/(3)(0.0335)] \right\}$$
$$= 3.1479 \text{ nanohenries.}$$

Bryan Method

Bryan's equation for the inductance of a flat square coil [10], which has been referenced by Dukes [8] has the form

$$L = 0.141 a n^{5/3} \log[8(a/c)]$$

with dimensions expressed in inches and inductance in microhenries. In terms of centimeter dimensions and natural logarithms, the equation becomes

$$L = 0.0241 a n^{5/3} \ln[8(a/c)]$$

where a is outside plus inside diameter divided by 4, c is outside minus inside diameter divided by 2, and n is the number of turns.

Applying this equation to the one-turn coil represented in Fig. 5, for which

$$a = (0.10 + 0.08)/4 = 0.045 \text{ centimeter}$$

and

$$c = (0.10 - 0.08)/2 = 0.01 \text{ centimeter}$$

one obtains for total inductance

$$L = (0.0241)(0.045)(1) \{ \ln[8(0.045/0.01)] \}$$
$$= 3.8871 \text{ nanohenries}.$$

Terman Method

Terman [7] has derived two inductance equations[5] that are applicable to the simple coil under consideration. One applies to a single-turn rectangle of rectangular wire and has the form

$$L = 0.02339 \{ (S_1+S_2)\log[2S_1S_2/(w+t)] - S_1\log(S_1+g) - S_2\log(S_2+g) \} + 0.01016 \{ 2g - [(S_1+S_2)/2] + 0.447(w+t) \}$$

where S_1 and S_2 are the maximum side lengths, g is the diagonal, w is the conductor width, and t is the conductor thickness, with dimensions expressed in inches and inductance in microhenries. For the case of a square, this equation becomes

$$L = (0.02339)(2S) \{ \log[2S^2/(w+t)] - \log(S+g) \} + 0.01016 \{ 2g - S + [0.447(w+t)] \}.$$

For the coil represented in Fig. 5,

S = 0.10 centimeter = 0.0394 inch
g = (1.414)(0.0394) = 0.0557 inch
$w + t$ = 0.0105 centimeter = 0.00413 inch.

Substituting these values into the equation above, one obtains

$$L = (0.02339)(2)(0.0394) \{ \log[2(0.0394)^2/0.00413] - \log(0.0394+0.0557) \} + 0.01016 \{ [2(0.0557)] - 0.0394 + [0.447(0.00413)] \}.$$

Then

$$L = [(1.844)(10^{-3})(-0.125 - 1.022)] + [(0.751)(10^{-3})] \text{ microhenries}$$
$$= 2.403 \text{ nanohenries}.$$

Terman has also derived an equation for square coils of rectangular cross section that is good for any number of turns n. This equation,

$$L = 0.0467Sn^2 \{ \log[2S^2/(t+w)] - \log 2.414S \} + 0.02032Sn^2 \{ 0.914 + [0.2235(t+w)/S] \}$$

where dimensions are expressed in inches and inductance in microhenries, is simply a modification of the first and would yield identical results.

Other Methods

Inductance equations have been derived by Wheeler [11], Gleason [1], and Olivei [3], but they are limited to spiral geometries and cannot be applied to square or rectangular coils. The formula developed by Dill [2] for flat square geometry applies only to cases in which the coil area is completely filled.

Summary of Results

The inductances calculated for a single-turn coil by the various methods described above are summarized in Table II. The differences noted are particularly alarming when one considers that none of the methods used was derived for circular spirals and none assumed either a zero cross section or a circular cross section for the conductor. Indeed, all but the Bryan method took into consideration both the width and the thickness of the conductor. Though no direct measurements have been made on coils of the exact size represented in Fig. 5, measurements on other coils have been shown to agree with results calculated by the expanded Grover method within experimental error.[6]

As we have seen, the expanded Grover method is very lengthy and cumbersome, even for a single-turn coil. For a multiturn coil, calculations requiring as long as eight hours if performed without computer aid are not uncommon. The computer program described in the section that follows has proved an effective solution to this problem.

TABLE II
Comparison of Inductance Calculations for a Square Planar Single-Turn Coil

Calculation Method	Calculated Inductance, nanohenries
Expanded Grover Formula	1.9573
Grover Formula without Mutual Inductance	2.2620
Coil Considered a Straight Conductor	3.1479
Bryan Formula	3.8871
Terman Formula	2.403

COMPUTER PROGRAM FOR INDUCTANCE CALCULATIONS

Computer calculation of total inductance is based on (9) previously cited, namely,

$$L_T = L_0 + M_+ - M_-.$$

All straight segments of the induction coil are assigned serial numbers from 1 to Z, Z being the total number of segments. Numbering proceeds from outside to inside. Since Z need not be a multiple of 4, inductance can be calculated for coils with a resolution of a quarter turn. For a coil with four turns, Z equals 16; for a coil with 2¾ turns, Z equals 11. The data required for each calculation are the number of segments Z, the length of the first segment ℓ_1, the length of the second segment ℓ_2, the width of the conductor w, the thickness of the conductor t, the edge-to-edge distance between conductors s, and the number of complete turns n.

The computer calculates the lengths of all other segments. For even-numbered segments, it uses the expression

$$\ell_{2y} = \ell_2 - (y-1)(w+s) \qquad (19)$$

and for odd-numbered segments,

$$\ell_{2y-1} = \ell_1 - (y-2)(w+s) \qquad (20)$$

with $y \geq 2$. Then

---

[5]Equation (34) in [7] applies to a single-turn rectangle of rectangular wire; equation (60) applies to square coils of rectangular cross section.

[6]Supporting data are presented in a subsequent section.

$$L_0 = \sum_{y=1}^{Z} L_y \qquad (21)$$

L_y being calculated using a form of (7), namely,

$$L_y = 0.002\ell_y \{ \ln[2\ell_y/(w+t)] + 0.50049 + [(w+t)/3\ell_y] \} \qquad (22)$$

where inductance is in microhenries.

The number of terms contributing to M_+ increases rapidly with the number of segments in the coil. For n full turns and Z total segments, the number of positive mutual-inductance terms will be

$$4[n(n-1)] + 2n(Z-4n).$$

Since these terms have the general form

$$M_{y,(y+4n)}$$

the total positive mutual inductance may be represented

$$M_+ = \Sigma M_{y,(y+4n)} = 2[M_{y,(y+4)}, M_{y,(y+8)}, M_{y,(y+12)}, \ldots] \qquad (23)$$

where y has values from 1 through $Z-4$, n has values from 1 through the number of complete turns, and $y+4n$ has a maximum value of Z. Consider, for example, a coil having 3¼ turns, such as is diagrammed in Fig. 6. This coil, for which $n=3$ and $Z=13$, will have 30 positive mutual-inductance terms. The $M_{y,(y+4)}$ terms are $M_{1,5}$, $M_{2,6}$, $M_{3,7}$, $M_{4,8}$, $M_{5,9}$, $M_{6,10}$, $M_{7,11}$, $M_{8,12}$, and $M_{9,13}$. The $M_{y,(y+8)}$ terms are $M_{1,9}$, $M_{2,10}$, $M_{3,11}$, $M_{4,12}$, and $M_{5,13}$. The $M_{y,(y+12)}$ term is $M_{1,13}$. These 15 terms fall inside the bracket, so that the expression for total positive mutual inductance becomes

$$M_+ = 2[M_{1,5}+M_{2,6}+M_{3,7}+M_{4,8}+M_{5,9}+M_{6,10}+M_{7,11}+M_{8,12} \\ +M_{9,13}+M_{1,9}+M_{2,10}+M_{3,11}+M_{4,12}+M_{5,13}+M_{1,13}].$$

Equation (14), which is used by the computer to calculate values for these individual terms, is an exact equation for all conductor pairs except those involving segment 1. As can be seen from Fig. 6, however, pairs of the latter type—in this case, 1-5, 1-9, and 1-13—are almost symmetrical, and using a symmetrical formula for them introduces only a very small error; moreover, since this error also exists in the calculation of negative mutual inductance, it tends to cancel out of the total inductance equation. Equation (14) can be rewritten in the form

$$M_{y,(y+4n)} = M_{(y+4n)} + \left\{ \begin{array}{c} [y-(y+4n)]/2 \\ -M \{[y-(y+4n)]/2\} \end{array} \right. . \qquad (24)$$

Then, combining (10) and (24), we obtain

$$M_{y,(y+4n)} = 2\ell_{(y+4n)} + \{[y-(y+4n)]/2\} Q_{(y+4n)} \\ + \{[y-(y+4n)]/2\} -2\ell\{[y-(y+4n)]/2\} \\ Q\{[y-(y+4n)]/2\} . \qquad (25)$$

The ℓ values are calculated using (19) and (20), and Q is calculated using (11).

Negative mutual inductance results from fluxes common to segments on opposite sides of the coil. The number of terms contributing to M_- is even greater than the number contributing to M_+. For a coil having n full turns and Z total segments, it equals

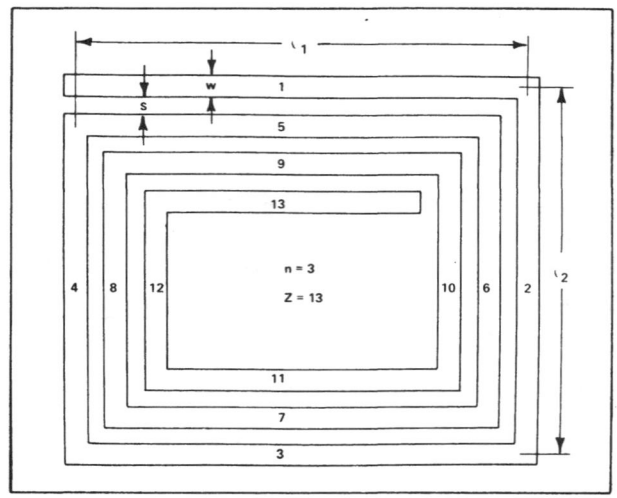

Fig. 6. Square planar 3¼-turn coil.

$$4n^2 + 2n(Z-4n) + (Z-4n-2)(Z-4n-1)[(Z-4n)/3].$$

Negative mutual-inductance terms have the general form

$$M_{y,(y+4n-2)}$$

and total negative mutual inductance may be represented

$$M_- = \Sigma M_{y,(y+4n-2)} \\ = 2[M_{y,(y+2)}, M_{y,(y+6)}, M_{y,(y+10)}, \ldots] \qquad (26)$$

where y has values from 1 through $Z-2$, n has values from 1 through the number of complete turns, and $y+4n-2$ has a maximum value of Z. The coil in Fig. 6, for which $n=3$ and $Z=13$, will have a total of 42 negative mutual-inductance terms. The $M_{y,(y+2)}$ terms are $M_{1,3}$, $M_{2,4}$, $M_{3,5}$, $M_{4,6}$, $M_{5,7}$, $M_{6,8}$, $M_{7,9}$, $M_{8,10}$, $M_{9,11}$, $M_{10,12}$, and $M_{11,13}$. The $M_{y,(Y+6)}$ terms are $M_{1,7}$, $M_{2,8}$, $M_{3,9}$, $M_{4,10}$, $M_{5,11}$, $M_{6,12}$, and $M_{7,13}$. The $M_{y,(y+10)}$ terms are $M_{1,11}$, $M_{2,12}$, and $M_{3,13}$. These 21 terms fall inside the bracket, so that the expression for total negative mutual inductance becomes

$$M_- = 2[M_{1,3}+M_{2,4}+M_{3,5}+M_{4,6}+M_{5,7}+M_{6,8}+M_{7,9} \\ +M_{8,10}+M_{9,11}+M_{10,12}+M_{11,13}+M_{1,7}+M_{2,8}+M_{3,9} \\ +M_{4,10}+M_{5,11}+M_{6,12}+M_{7,13}+M_{1,11}+M_{2,12}+M_{3,13}].$$

Values for these negative mutual-inductance terms are calculated in much the same manner as those for positive mutual inductance. Rewritten for this calculation, (14) takes the form

$$M_{y,(y+4n-2)} = M_{(y+4n-2)} + \left\{ \begin{array}{c} [y-(y+4n-2)]/2 \\ -M\{[y-(y+4n-2)]/2\} \end{array} \right. \qquad (27)$$

Then, combining (10) and (27), we obtain

$$M_{y,(y+4n-2)} = 2\ell_{(y+4n-2)} + \{[y-(y+4n-2)]/2\} \\ Q_{(y+4n-2)} + \{[y-(y+4n-2)]/2\} \\ -2\ell\{[y-(y+4n-2)]/2\} Q\{[y-(y+4n2)]/2\}. \qquad (28)$$

As in the case of the positive terms, ℓ values are calculated using (19) and (20) and Q is calculated using (11).

COMPUTER-CALCULATED INDUCTANCE DATA

Four different induction coils were fabricated by vacuum-depositing phased chromium/gold onto 99 percent alumina and then gold-plating to a thickness of 0.0005 inch (0.0127 millimeter). Values for L and Q were measured at 150 megahertz, and the corresponding L values were calculated by the Bryan, Terman, and expanded Grover methods. Results are summarized in Table III. It will be noted that Terman's method yielded values two to four times higher than the measured values. Bryan's method proved better, though the results for small coils were much too high.

Figs. 7 through 12 present plots developed from other computer-calculated inductance data for square and rectangular planar coils.[7] In Figs. 7, 9, and 11, inductance in nanohenries is plotted versus number of segments Z for various lengths ℓ_1 of the first outside segment, the number of complete turns n being equal to $Z/4$. In Figs. 8, 10, and 12, inductance in nanohenries is plotted versus segment length ℓ_1 for different numbers of complete turns n.

The irregularities in the Fig. 9 and Fig. 11 curves result from the fact that segment lengths between the outside and

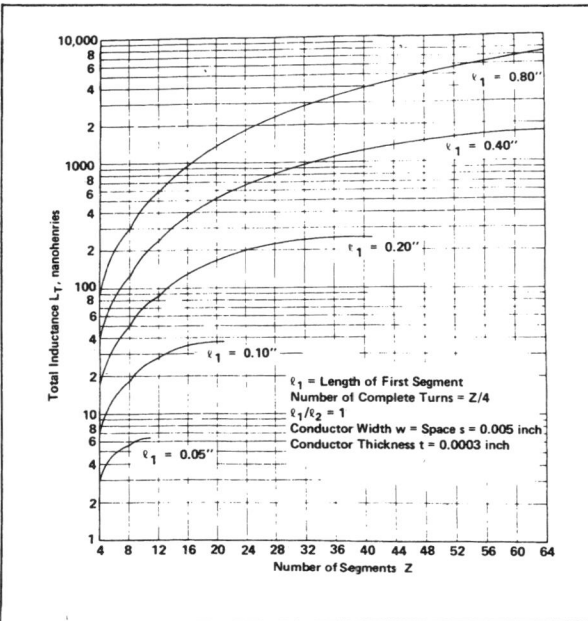

Fig. 7. Square-planar-coil inductance as a function of number of coil segments.

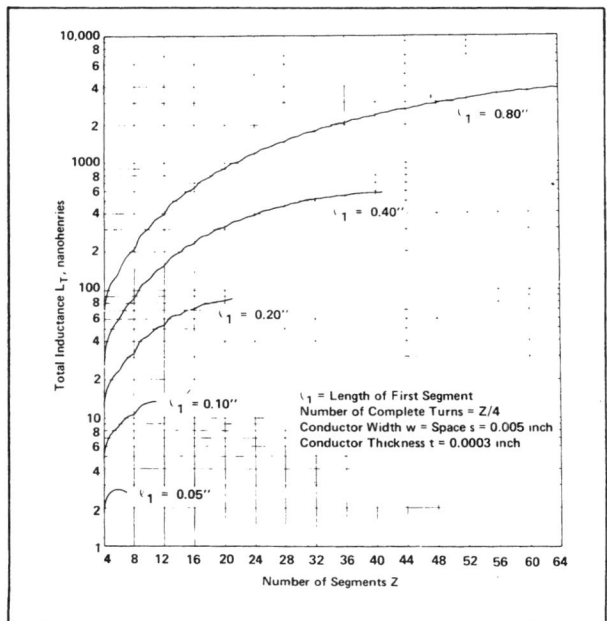

Fig. 9. Rectangular-planar-coil inductance as a function of number of coil segments ($\ell_1/\ell_2 = 2$).

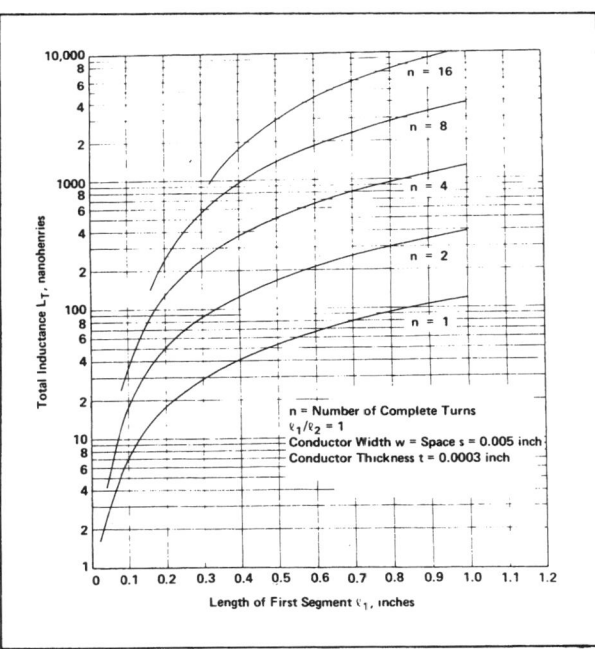

Fig. 8. Square-planar-coil inductance as a function of first-segment length.

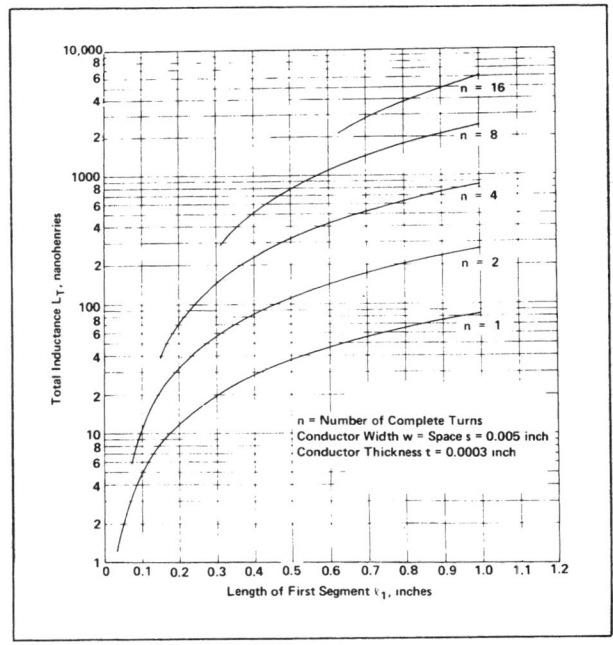

Fig. 10. Rectangular-planar-coil inductance as a function of first-segment length ($\ell_1/\ell_2 = 2$).

[7]Exploded copies of these figures with much finer resolution are available from the author upon request.

the inside of the coil do not decrease uniformly. Segments that are longer than the preceding segments contribute greater amounts of inductance and cause fluctuations in the value of L_T. Since the amount of this fluctuation constitutes a decreasing percentage of total inductance as the number of segments increases, the irregularities become less pronounced as the semilogarithmic plot progresses.

The slight irregularities in the Fig. 7 curves result from a similar phenomenon. The fact that ℓ_1 has been defined as

TABLE III

Calculated as Compared with Measured Inductance Values for Typical Square Coils

Method of Determination	Inductance, nanohenries			
	Coil A	Coil B	Coil C	Coil D
Bryan Calculation	35.6	71.4	111.4	207.3
Terman Calculation	67.13	111.7	447.6	636.2
Expanded Grover Calculation	28.33	56.84	106.68	197.88
Experimental Measurement	23.2	51.8	98.9	211.9

equal to ℓ_2 in a square coil causes an initial nonuniformity in the rate of decrease of ℓ, the pattern being

$$\ell_1 = \ell_2 = \ell_3 > \ell_4 = \ell_5 > \ell_6 = \ell_7 > \ell_8 = \ell_9 > \ell_{10} \ldots$$

Moreover, the addition of new segments does not add uniformly to the number of the positive and negative mutual-inductance terms. The addition of segments 5, 6, 9, 10, 13, 14, etc. introduces two positive and two negative terms per segment whereas with the addition of segments 7, 8, 11, 12, 15, 16, etc., the number of negative terms introduced per segment exceeds the number of positive terms by two.

Two important types of information can be readily extracted from Figs. 7 through 12. It is possible to determine by inspection not only the inductance value that corresponds to a given coil geometry, but also the various coil geometries that will yield a given inductance value. Information of the latter type is particularly valuable and is difficult to obtain in a practical manner without the aid of a computer. Even assuming that an accurate method of calculating inductance for a given geometry is available, the search for a geometry that will yield a given inductance is inevitably an iterative process that begins with a best guess. The calculation must usually be repeated a number of times before it is possible to single out a geometry the inductance value for which is close enough to permit fabrication of the coil. Using Figs. 7 through 12, however, the task becomes relatively simple.

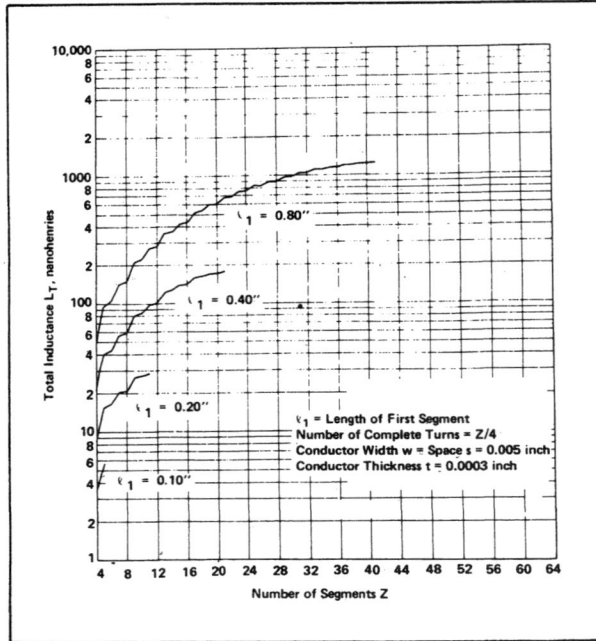

Fig. 11. Rectangular-planar-coil inductance as a function of number of coil segments ($\ell_1/\ell_2 = 4$).

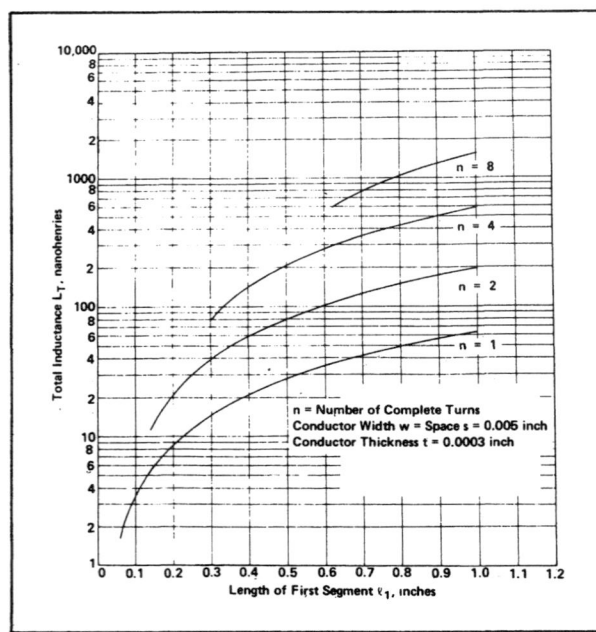

Fig. 12. Rectangular-planar-coil inductance as a function of first-segment length ($\ell_1/\ell_2 = 4$).

Assume, for example, that an inductance of 100 nanohenries is required. Fig. 7 indicates that this inductance value can be obtained using a square coil of just over four segments with an ℓ_1 of 0.80 inch (2.032 centimeters), a square coil of almost seven segments with an ℓ_1 of 0.40 inch (1.016 centimeters), and a square coil of 13 segments with an ℓ_1 of 0.20 inch (0.508 centimeter). Fig. 8 indicates that the same inductance can be obtained using a square two-turn coil with an ℓ_1 of 0.34 inch (0.864 centimeter) or a square one-turn coil with an ℓ_1 of 0.85 inch (2.160 centimeters). Examining each of the other figures in the same manner, one finds that all the geometries described in Table IV will, in fact, yield the desired inductance value. Such a tabulation provides a great deal of latitude in overall hybrid microcircuit design. Although Q values for these coils have not been thoroughly investigated, we have observed that they range typically between 20 and 45, provided that the thickness of the inductor is adequate for the frequency stipulated. A thickness of 0.0003 inch (0.0076 millimeter) is required at 3 gigahertz and a thickness of 0.0001 inch (0.025 millimeter) at 10 megahertz.

TABLE IV

Rectangular-Planar-Coil Geometries for a 100 Nanohenry Inductor

Coil Parameters			
ℓ_1/ℓ_2	ℓ_1, inches (centimeters)	Number of Segments Z	Number of Turns n
1	0.80 (2.032)	4+	
1	0.40 (1.016)	7−	
1	0.20 (0.508)	13	
1	0.17 (0.432)	16	4
1	0.34 (0.864)	8	2
1	0.85 (2.159)	4	1
2	0.80 (2.032)	4½	
2	0.40 (1.016)	8½	
2	0.20 (0.508)	16	4
2	0.46 (1.168)	8	2
4	0.80 (2.032)	5½	
4	0.40 (1.016)	11½	
4	0.33 (0.838)	16	4
4	0.59 (1.499)	8	2

CONCLUSION

The concept of negative mutual inductance has been discussed, and equations for calculating mutual inductance as well as total inductance for planar rectangular coils have been presented. A computer program designed to solve these equations has been described, and the utility of its graphic data output for coil design has been demonstrated. Sufficient detail has been presented to permit the interested reader to develop similar computer programs for other inductor types.

SYMBOLS

AMD   Arithmetic mean distance.
d Distance between track centers.
g Diagonal of coil cross section.
GMD   Geometric mean distance.
i Current.
ℓ Conductor length.
L Self-inductance.
L_0 Sum of self-inductances (total minus mutual inductance).
L_T Total inductance.
L_x Self-inductance of coil segment x.
M Mutual inductance.
$M_{a,b}$ Mutual inductance between segments a and b due to $\phi_{a,b}$.
M_T Total mutual inductance.
M_+ Sum of positive mutual inductances.
M_- Sum of negative mutual inductances.
n Number of complete turns in coil.
Q Mutual-inductance parameter.
r Radius of conductor cross section.
s Edge-to-edge distance between conductors.
S Maximum side length.
t Conductor thickness.
T Frequency-correction parameter.
w Conductor width.
Z Total number of coil segments.
μ Conductor permeability.
$\phi_{a,b}$ Magnetic flux common to segments a and b and generated in a.

ACKNOWLEDGMENT

The author is indebted to W. R. Mackey and I. G. Goldsmith for their work in the subject area. Mr. Goldsmith wrote the computer program described. Mr. Mackey was one of the first to use the resulting data in the design of inductors for microwave integrated circuits and has experimentally verified certain of the calculations.

REFERENCES

[1] F. R. Gleason, "Thin-Film Microelectronic Inductors," *Proceedings of the National Electronics Conference* **20**, 197-198 (1964).
[2] H. G. Dill, "Designing Inductors for Thin-Film Applications," *Electronic Design* (February 1964), 52-59.
[3] A. Olivei, "Optimized Miniature Thin-Film Planar Inductor Compatible with Integrated Circuits," *IEEE Transactions on Parts, Materials, and Packaging* **5** (No. 2), 71-88 (1969).
[4] J. C. Maxwell, *A Treatise on Electricity and Magnetism, Parts III and IV,* 1st ed., 1873, 3d ed., 1891; reprinted by Dover Publications, New York, N.Y. 1954.
[5] F. W. Grover, *Inductance Calculations,* Van Nostrand, Princeton, N.J., 1946; reprinted by Dover Publications, New York, N.Y., 1962.
[6] J. H. Dellinger, J. M. Miller, F. W. Grover, L. E. Whittemore, and R. S. Ould, *Radio Instruments and Measurements,* Circular C74, National Bureau of Standards, Washington, D.C., 1937, pp. 260-267.
[7] F. E. Terman, *Radio Engineering Handbook,* McGraw-Hill, New York, 1943, pp. 48-60.
[8] J. M. C. Dukes, *Printed Circuits: Their Design and Application,* MacDonald, London, 1961, pp. 120-135.
[9] E. B. Rosa and F. W. Grover, "Formulas and Tables for the Calculation of Mutual and Self Inductance," *Bulletin of the Bureau of Standards* **8** 1-238 (1912).
[10] H. E. Bryan, "Printed Inductors and Capacitors," *Tele-Tech & Electronic Industries* (December 1955), 68.
[11] H. A. Wheeler, "Simple Inductance Formulas for Radio Coils," *Proceedings of the Institute of Radio Engineers* **16**, 1398 (1928).

MINIATURE MULTILAYER SPIRAL INDUCTORS FOR GaAs MMICs

M.W. Geen, G.J. Green, R.G. Arnold*, J.A. Jenkins* and R.H. Jansen*

Plessey Three Five Group Ltd
*Plessey Research Caswell Ltd
Towcester, Northants NN12 8EQ, England

ABSTRACT

This paper presents a novel approach to reducing the area consumed by MMIC spiral inductors by exploiting a multilayer MMIC process. These miniature spiral inductors may be designed to have inductance values as high as 20nH but occupy areas of less than $0.1mm^2$. A simple analysis is presented to estimate the prime inductance and the results of geometric optimisation using a two level electromagnetic field analysis are shown. A practical equivalent circuit has been developed together with a set of design equations. A miniaturised amplifier has been produced to demonstrate the successful application of the component.

INTRODUCTION

The reduction of MMIC chips area is of prime concern for the economic production of GaAs integrated circuits since it has direct bearing on the number of chips produced from each wafer processed and therefore the cost of each chip. In recent years much effort has been directed towards the development of high packing density circuit techniques and this has been significantly aided by the development of new CAD tools to account for electro-magnetic coupling effects [1,2]. The latest developments include new packages for the analysis of two level metal structures and these have allowed some of the more complex multilayer structures which it is now possible to make on MMICs to be analysed and therefore exploited in new high packing density circuit designs. The stacked spiral inductor was first proposed by H.J. Finlay[3] in 1985. This paper reports the first comprehensive results of the microwave characterisation of these components and also the experimental verification of performance improvements predicted by the two metallisation level field-theory based analysis.

CONSTRUCTION

The miniature spiral inductor or 'stacked spiral' consists of two spirals overlayed one on top of the other, separated by a dielectric layer and joined in the centre by a via hole through the dielectric layer. Fig 1(i) shows an example where the upper and lower metal strips are the same width and directly overlayed. This is the simplest case to analyse (see below) but it is obvious that the capacitive coupling between the two coils will be a maximum. The inter coil capacitance can be reduced by offsetting the coils as shown in Fig 1(ii). Variations on both geometries have been manufactured and assessed.

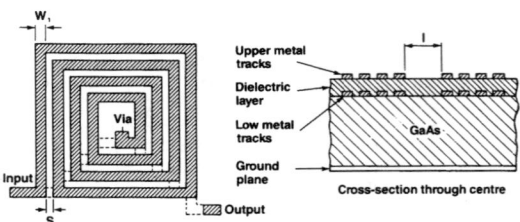

(i) Directly overlayed stacked MMIC spiral inductor

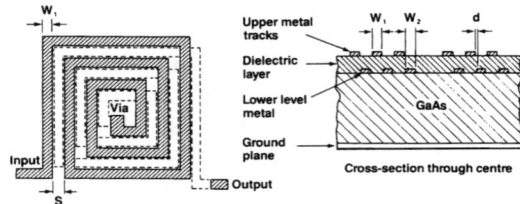

(ii) Offset stacked MMIC spiral inductor

Fig. 1. MMIC Stacked Spiral Inductor Construction

The Plessey Three Five MMIC foundry process is a multi-layer system which allows the construction of stacked spiral inductors with no additional or non-standard process steps. The lower spiral is patterned in the first level metal (M2). The upper spiral is formed in the top metal (M3) and isolated from the lower spiral by a dielectric consisting of a thin layer of Si_3N_4 and a much thicker layer of polyimide.

Inductance

The total inductance of a stacked spiral consists of the sum of the two planar spiral inductances plus their mutual inductance (Fig. 2).

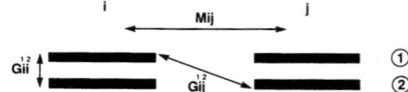

Fig. 2. Stacked Spiral Mutual Inductance Terms

Thus

$$L_{stacked} = 2 L_{spiral} + 2 \sum_{ij} G_{ij}^{12} \qquad (1)$$

where L_{spiral} is the planar spiral inductance and $2\sum_{ij} G_{ij}^{12}$ is the mutual inductance between their windings. To obtain the stacked spiral inductance therefore the interspiral mutual inductance must be found.

For the directly overlayed case, consider a spiral above a ground plane at a distance of half the stacking distance. This situation can be represented by the same stacked spiral but with the current in the mirrored spiral flowing in the opposite direction. The inductance of the planar spiral is

$$L = L_{spiral} - \sum_{ij} G_{ij}^{12}$$

Furthermore, the spiral inductance can be written as

$$L_{spiral} = \sum_i L_i + \sum_{ij \neq i} M_{ij}$$

where L_i is a winding self-inductance and M_{ij} the winding mutual inductance. Thus, the inductance of planar spiral is

$$L = \sum L_i + \sum_{ij \neq i} M_{ij} - \sum_{ij} G_{ij}^{12}$$

However since the spiral's distance from the ground plane is much smaller than the interwinding separation

$$\sum_{ij \neq i} (M_{ij} - G_{ij}^{12}) \approx 0$$

and the inductance reduces to

$$L = \sum_i (L_i - G_{ii}^{12})$$

but this is the inductance of a microstrip line, $L_{microstrip}$, with the length of the unwound planar spiral. Thus the interspiral mutual inductance is

$$\sum_{ij} G_{ij}^{12} = L_{spiral} - L_{microstrip}$$

Substituting in equation (1) this gives for the stacked spiral, the inductance

$$L_{stacked} = 4 L_{spiral} - 2 L_{microstrip} \qquad (2)$$

The planar spiral inductance can be estimated from the formulae of Grover[4]. The simple formula:

$$L_{spiral} = 0.0008 N^2 S \left[\log_e \frac{S}{b} + 0.726 + 0.1776 \frac{b}{S} + \frac{1}{8} \frac{b^2}{S^2} \right] \text{ (nH)}$$

where
- N = number of turns
- b = pN where p = W + S
- S = I + (N−0.5)p

has proved satisfactory for W = S up to 12μm. By using the simple Grover expression together with calculations of $L_{microstrip}$ based on Kuester[5], equation (2) has been shown to predict the prime inductance of a directly overlayed stacked spiral induction to within 10% of the values obtained by equivalent circuit modelling, described below.

Inter Spiral Capacitance

Due to the structure of the stacked overlay spiral, the centre point at which the strip is turned on itself at a lower level has a fixed potential. The capacitance of the structure is therefore approximately half that of the total interlevel capacitance. Using a simple parallel capacitance inductor model for the stacked spiral behaviour, an estimate of the component's first resonance can be found from the capacitance-inductance estimates. The following table shows the calculated inductance, interspiral capacitance and resonant frequency for a range of directly overlayed spiral inductors with S=W=12μm. The measured resonance frequency is shown in brackets.

Turns	Inductance nH	Capacitance pF	Resonant Freq. GHz
2	3.07	0.137	7.76 (8.5)
3	7.29	0.243	3.78 (4.15)
4	13.81	0.374	2.22 (2.4)
5	23.14	0.529	1.44 (1.55)

The discrepancy in the measured and calculated resonant frequency could be due to the effects of the negative mutual inductance of the substrate ground plane on the self inductance of the stacked spiral.

GEOMETRIC OPTIMISATION

From the preliminary analysis significant increases in inductance per unit area, between 3 and 4 times, could be expected but the high interwinding capacitance needed to be reduced. In the first instance the simple step of offsetting the upper and lower coils by a distance equal to W in the x and y directions, as shown in Fig. 1(ii), was investigated experimentally. The results obtained, (presented in the next section) showed an increase in resonant frequency of between 26% and 39% in the case of inductors with W1 = W2 = S = 12μm with less than 10% reduction in inductance for a given number of turns.

Subsequent field theory-based analysis and optimisation, using the hybrid-mode dual metal level structure look-up table generator DMMICTL (1), has shown that by introducing a gap 'd' between the offset windings a considerable reduction in parasitic capacitance can be obtained. The following table shows the maximum inter-track capacitance per unit length for a range of stacked spiral cross sections.

W_1 μm	W_2 μm	d μm	S μm	Capacitance pF/mm
12	12	0	12	0.1
12	12	2	16	0.076
12	8	2	12	0.071
6	12	3	12	0.057
4	8	6	16	0.038

Since the resonant frequency is proportional to $1/\sqrt{C}$ this type optimisation could lead to increases in resonance frequency of up to 60%.

MEASUREMENT AND EQUIVALENT CIRCUIT MODELLING

Initial S-parameter measurements were carried out in 50 ohm microstrip test jigs and de-embedding routines used to extract the inductor S-parameters. More recent measurements have been carried out using RFOW probing on full 2 inch wafers containing a wide range of inductor test structures using. Network analyser calibration was carried out using on-wafer through, open, load and short components with through-substrate vias providing grounding.

Equivalent circuit modelling of the planar spirals is generally satisfactory up to 80% of the first resonance [2]. Fitting beyond the first resonance is hampered by the asymmetry of the planar spiral due to the presence of the underpass. The stacked spiral, even though it is a more complicated structure, is in fact easier to model due to its intrinsic symmetry. The equivalent circuit is shown in Fig. 3. Each arm on the parallel network in effect covers a particular frequency range containing a resonant feature. L1 and R1 fit the true inductive behaviour up to the first resonance. The elements C3, L3 and R3 control the first dip of S11 and the first peak of S21. The elements C2, L2 and R2 control the second dip of S11 and the second peak of S21. The full model is only needed to fit the 4 and 5 turn stacked spirals: for fewer turns certain elements (e.g. C2, L2 and R2) become redundant. The capacitors to ground, C1 and C4, can be used to control the fit at high frequency. Insufficient reflection at high frequencies was overcome by addition of resistors in series with these capacitors (R4 and R5). This also allowed more asymmetry in the model (for the higher frequency end). Fig.4 shows the fit obtained for a 5 turn offset stacked spiral inductor with W = S = 12µm and d = 2µm.

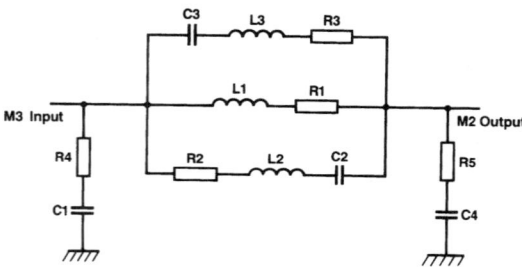

Fig. 3. Stacked Spiral Equivalent Circuit

By fitting the equivalent circuit to the measured S-parameters of a range of inductors it is possible to derive a set of design equations by polynominal curve fitting the relationship between each element in the equivalent circuit and the number of turns, N, on the inductor. The prime inductance, L1, and the parasitic elements C2, C3, R1, R2, L2 and L3 can be fitted well with a fourth order polynominal, for example for W1 = W2 = 12, d = 2µm.

$$L1 = -0.053N^4 + 0.646N^3 - 1.784N^2 + 3.28N - 1.635$$

C1 and C4 have a linear relationship to N. R3, R4 and R5 are constant. Equations such as these can be incorporated into Touchstone(TM) or Supercompact(TM) circuit files to allow circuit optimisation in terms of the geometric parameter, N.

Figure 5 shows the measured prime inductance versus inductor area for the range of geometries measured and for comparison, the inductance versus area of a planar spiral. Fig. 6 shows the advantage gained in increased resonant frequency by offsetting the windings and confirms the further increase in resonance frequency predicted by the field-theory analysis by introducing the gap, d. Fig. 5 shows that the offset spirals sacrifice some of the advantage in inductance per unit area. Fig. 7 shows the overall improvement in resonant frequency for a given inductance resulting from geometric optimisation.

APPLICATIONS

A practical example of the application of a stacked spiral inductor is shown in the miniature 1 to 6GHz feedback amplifier shown in Fig. 8. This circuit uses a 3.5 turn 10nH offset stacked spiral inductor in the drain bias line. This, together with the use of lumped element matching results in a chip size of 1mm x 1mm. The measured performance is shown in Fig. 9.

Stacked spirals are particularly well suited to use in bias circuits where the slightly higher loss and low self resonant frequency are of little importance. Other applications include lumped element matching circuits - particularly at lower microwave and radio frequencies and filter structures where one can incorporate series capacitors at the centre of the spiral to produce compact L-C-L combinations.

CONCLUSIONS

Stacked spiral inductors provide up to 4 times the inductance of equivalent area planar spiral inductors. A full field analysis package with high spatial resolution (DMMICTL) has been used to optimise the geometry and it's predictions have been verified by r.f. measurement. A practical equivalent circuit model has been developed for use in established CAD packages. Stacked spiral inductors has been successfully demonstrated in several high packing density circuits [6] one example of which has been shown.

ACKNOWLEDGEMENT

The authors wish to thank R.W.W. Charlton, D. Parker and D. Warner for the high quality MMIC processing, S. Cornelius for assistance with RFOW measurements and data processing and H.J. Finlay for his pioneering work on high packing density MMIC components.

REFERENCES

(1) Jansen, R.H et al, "Recent Developments in the CAD of High Packing Density MMICs". Proc. 1989 Microwave and Optoelectronics Conf. (MIOP), Sindelfinden, West Germany, Feb. 1989.
(2) Jansen, R.H et al, "A Comprehensive CAD Approach to the Design of MMICs up to mm-wave frequencies. IEEE MTT-36, No.2, pp208 to 219, Feb 1988.
(3) Finlay, H,J, U.K. Patent Application No.8800115.
(4) Grover, F.W, "Inductance Calculations". Van Norstrand New York 1946; reprinted by Dover Publications, New York, 1962.
(5) Kuester, E.F, "Accurate Approximations for a Function Appearing in the Analysis of Microstrip", IEEE MTT-32, No.1, Jan. 1984.
(6) Lane, A.A et al, "S and C Band GaAs Multifunction MMICs for Phased Array Radar". Proc. 1989 GaAs IC Symposium

Si IC-Compatible Inductors and *LC* Passive Filters

NHAT M. NGUYEN, STUDENT MEMBER, IEEE, AND
ROBERT G. MEYER, FELLOW, IEEE

Abstract — Passive inductors and *LC* filters fabricated in standard Si IC technology are demonstrated. Q factors from 3 to 8 and inductors up to 10 nH in the gigahertz range have been realized. Measurements on a five-pole maximally flat low-pass filter give midband insertion loss and -3-dB bandwidth close to the nominal design values of 2.25 dB and 880 MHz.

I. Introduction

Planar inductors and *LC* passive filters have been implemented in practical systems for many years using a variety of substrates. These include standard PC boards, ceramic and sapphire hybrids, and more recently GaAs IC's [1], [2]. In the early development of Si IC's, planar inductors were investigated [3] but the prevailing lithographic limitations and relatively low frequencies of operation (less than several hundred megahertz) led to their abandonment as impractical due to excessive chip-area requirement and low Q. Reflected losses from the conductive Si substrate were a major contributor to low inductor Q.

Recent advances in Si IC processing technology have prompted another look at this situation. In particular, metal width and pitch in the low micrometer range allow many more inductor turns per unit area than in the past. Also, modern oxide-isolated processes with multilayer metal options allow thick oxides to help isolate the inductor from the Si substrate. In addition, interest is growing in applications at much higher frequencies with the advent of 900-MHz communications and gigahertz-range satellite reception such as Global Positioning Satellite (GPS) and Direct Broadcast Satellite (DBS).

In this paper we describe inductors and *LC* filters fabricated in a production Si bipolar process featuring oxide isolation and active device peak f_T of 8 GHz. In the frequency range of interest (above about 1 GHz) the Q of the filter elements is quite usable (3–8) and appears to be almost totally limited by metal and contact resistance, with little effect from the Si substrate. In this regard, there is little difference between these inductors and filters and those implemented in GaAs.

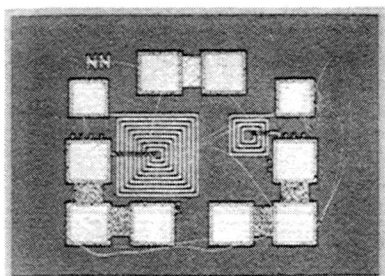

Fig. 1. Die photograph of the inductors.

II. INDUCTOR CHARACTERIZATION

Two square-spiral inductors were fabricated, measured, and characterized. A die photo of the test layout is shown in Fig. 1. Metal width was 6.5 μm with 5.5-μm spacing of 1.8-μm-thick second-metal Al. The sheet resistance of Al was 20 mΩ/$\square$ over 1.7 μm of oxide, and the parasitic capacitance from Al to the substrate was 0.016 fF/μm^2. The substrate resistivity was 14 $\Omega \cdot$cm and 500-μm-thick p-type Si. The larger inductor had nine turns with an outer dimension of about 230 μm. The smaller one had four turns with an outer dimension of about 115 μm. Measured S_{11} plots from 0.3 MHz to 3 GHz for these inductors in a 50-Ω system are shown in Fig. 2. The large inductor is self resonant at 2.47 GHz while the small one has an estimated resonant frequency of 9.7 GHz. Pad capacitance was zeroed out of the on-chip measurement. The large inductor had a measured value of 9.7 nH while the smaller was 1.9 nH. The theoretical values are, respectively, 9.3 and 1.3 nH [4]. The differences are attributed to lead inductances and possible minor imprecision in calibration of the test equipment. The series loss in the inductors deduced from RF measurements agreed very closely with measured and predicted dc series resistance, indicating that coupled loss from the Si substrate was negligible. While metal shrinkage due to photolithography and etching tolerances can affect the series-loss value, it theoretically has negligible effect on the inductance value. In fact, with a typical ± 0.2-μm metal shrinkage, the inductance value in the large inductor has been calculated to vary less than 1%.

An equivalent circuit for the square-spiral inductor is shown in Fig. 3. In this circuit, L_s models the self and mutual inductances in the second-metal segments, R_s is the accumulated sheet resistance, C_p models the parasitic capacitance from the second-metal layer to the substrate, and R_p represents the resistance of the conductive Si substrate. Coupling capacitance between metal segments due to fringing fields in both the dielectric region and the air region is neglected in this model. Such an approximation is valid because the relative dielectric constant of the oxide is small and the inductor is used at frequencies well below its self-resonant frequency. Since the structure of the square-spiral inductor is not symmetrical, the parasitic capacitance values at the inductor terminals should be different from one another. This difference, however, is only small [5] and the two capacitors are assumed the same. If the spiral inductor is treated as a lossless transmission line with a total length much smaller than the wavelength, it can be shown that C_p is approximately equal to one half the input capacitance of the open-circuited line. This gives a first-order estimate of C_p. More accurate analytical expressions for C_p can be found in [6]. The substrate resistance R_p can be derived from measured S parameters. It is interesting to note that in a GaAs inductor using microstrip lines, substrate resistance R_p is not present

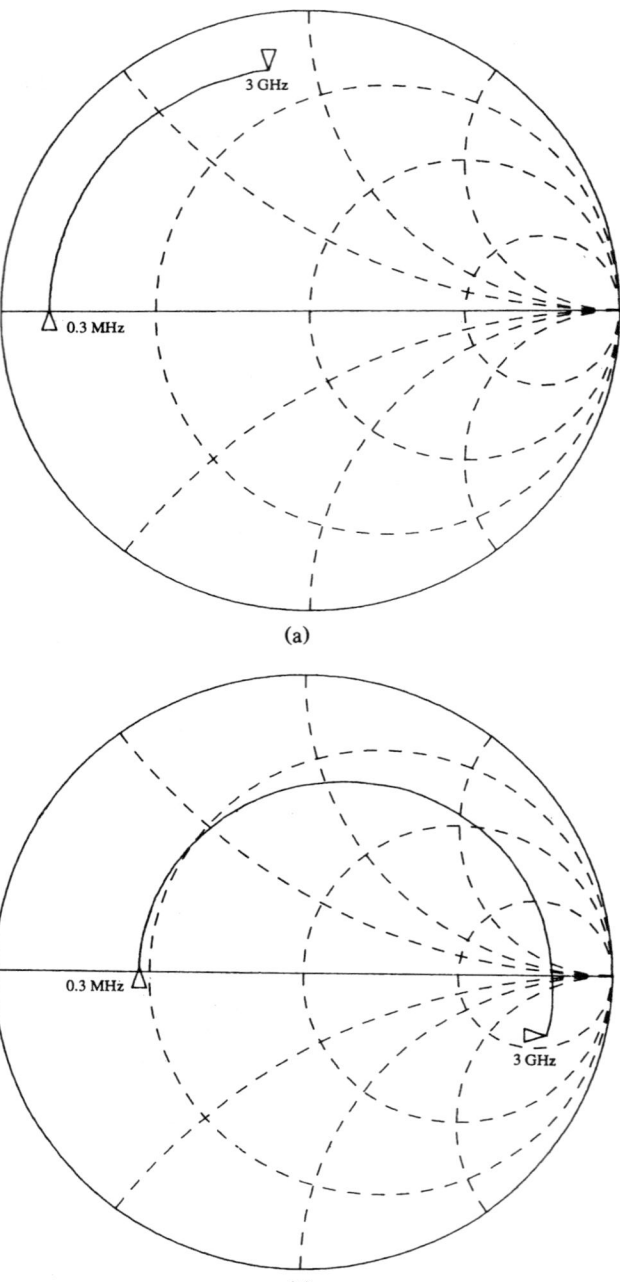

Fig. 2. Measured S_{11} plots from 0.3 MHz to 3 GHz: (a) small inductor and (b) large inductor.

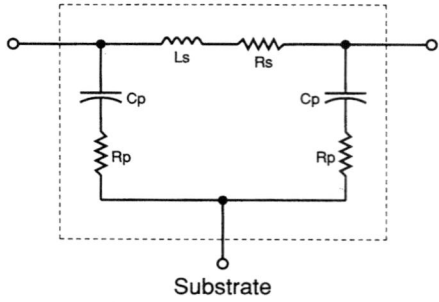

Fig. 3. Equivalent circuit of the square-spiral inductor.

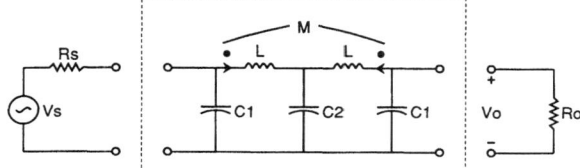

Fig. 4. Simplified circuit diagram of the five-pole maximally flat low-pass LC filter.

because the GaAs substrate acts as the dielectric layer which is in direct contact with the conductive ground plane.

If one side of the inductor is grounded, the self-resonant frequency of the spiral inductor can be derived from the equivalent circuit. It is approximately equal to

$$\omega_R = \frac{1}{\sqrt{L_s C_p}} \left[\frac{1 - R_s^2 \left(\frac{C_p}{L_s}\right)}{1 - R_p^2 \left(\frac{C_p}{L_s}\right)} \right]^{1/2}. \quad (1)$$

Beyond the resonant frequency, the inductor becomes capacitive. Frequency ω_R is limited mainly by C_p which is inversely proportional to the oxide thickness between the second-metal layer and the substrate. The frequency at which the inductor Q is maximum can also be derived. It is

$$\omega_Q = \frac{1}{\sqrt{L_s C_p}} \left\{ \frac{R_s}{2R_p} \left[\left(1 + \frac{4}{3}\frac{R_p}{R_s}\right)^{1/2} - 1 \right] \right\}^{1/2}. \quad (2)$$

The large inductor has a measured maximum Q of 3 at 0.9 GHz and the small one has an estimated maximum Q of 8 at 4.1 GHz. If the inductor is used as a floating inductor, the shunt branches in the equivalent circuit are effectively in series with one another. Equations (1) and (2) hence still hold provided that C_p and R_p are replaced by $C_p/2$ and $2R_p$, respectively. Circuit elements in the equivalent circuit for the large inductor were derived from both theory and measured S parameters. The set $\{L_s, R_s, C_p, R_p\}$ is equal to $\{9.7 \text{ nH}, 15.4 \text{ }\Omega, 590 \text{ fF}, 70 \text{ }\Omega\}$.

III. LC Filters

As a test vehicle, a five-pole maximally flat low-pass filter with nominal designed -3-dB frequency of 880 MHz and midband insertion loss of 2.25 dB was fabricated. The circuit is shown in Fig. 4 where R_s and R_o are 50-Ω off-chip resistors. Element values are $L \approx 9.7$ nH, $M \approx 0.4$ nH, $C_1 \approx 1.3$ pF, and $C_2 \approx 4.3$ pF where parasitic capacitance associated with the inductors is included in the capacitor values. Mutual inductor M exists between the inductors due to layout proximity. The capacitors were fabricated in standard form using metal over 1500 Å of oxide with an n^+ bottom plate. Since the sheet resistance of n^+ is high (20 $\Omega/\square$), the series loss in the capacitor must be minimized by reducing the ratio L/W, where L and W are, respectively, the length and width that define the capacitor area. Pads were included with the filter to allow testing but were not included in the design and were zeroed out of the on-chip measurements. This would correspond to use of such a filter in

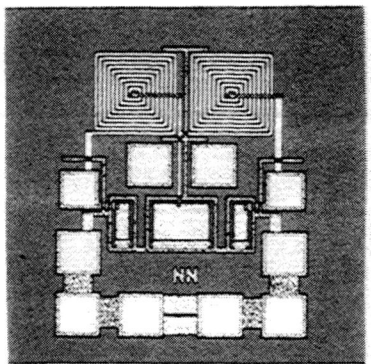

Fig. 5. Die photograph of the filter.

an on-chip environment where pads are not present. If a packaged stand-alone filter was required, pads and bond wires would have to be included in the design.

The transfer function can be derived from the simplified circuit in Fig. 4 and is given by

$$\frac{V_o(S)}{V_s(S)} = \frac{a_2 s^2 + a_0}{b_5 s^5 + b_4 s^4 + b_3 s^3 + b_2 s^2 + b_1 s + b_0} \quad (3)$$

where

$a_2 = MC_2$

$a_0 = 1$

$b_5 = (L - M)C_1(L + M)C_2 R_s C_1$

$b_4 = (L - M)C_1(L + M)C_2 \left(1 + \frac{R_s}{R_o}\right)$

$b_3 = (L + M)C_2 \frac{(L - M)}{R_o} + [2(L - M)C_1 + 2LC_2]R_s C_1$

$b_2 = 2(L - M)C_1 + LC_2 + \frac{R_s}{R_o}[2(L - M)C_1 + LC_2]$

$b_1 = 2\frac{(L - M)}{R_o} + R_s(2C_1 + C_2)$

$b_0 = \left(1 + \frac{R_s}{R_o}\right).$

As seen from (3), the mutual inductance M creates two high-frequency zeros on the $j\omega$ axis. Since M is relatively small compared to the inductance value L, its effect on the filter attenuation is significant only in the stopband. Let f_Z ($= 1/[2\pi\sqrt{MC_2}]$) denote the magnitude of the complex-conjugate zeros. It can be shown that the zeros increase the stopband loss at frequencies below f_Z but decrease the loss at frequencies above. If maximum high-frequency attenuation is desired, the value of M should be minimized by separating the two inductors far apart in the layout. If $M = 0$, (3) reduces to a simple five-pole transfer function.

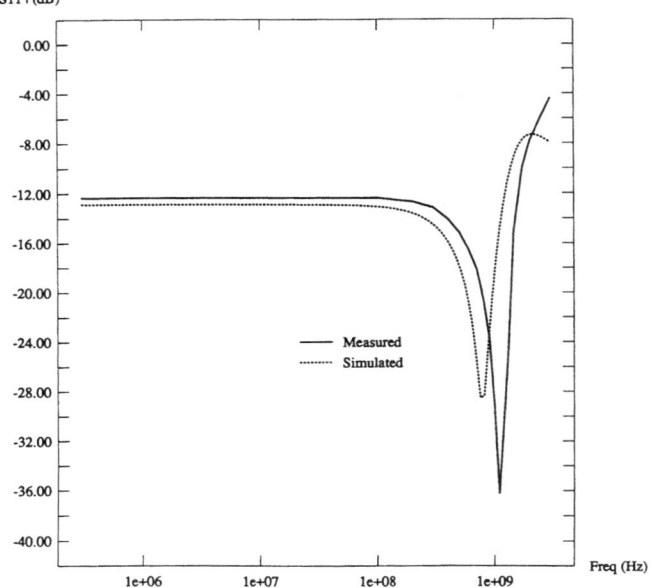

Fig. 6. Measured and simulated $|S_{11}|$ for the filter.

A die photo of the filter is shown in Fig. 5. In order to minimize the electrical coupling through the substrate, a buried p-type layer was placed around the I/O pads and around the periphery of each device. Measured $|S_{11}|$ is shown in Fig. 6 and is close to the simulated values. The filter was simulated using the 3σ limits of capacitance of $\pm 10\%$ due to process variations but assuming all capacitors tracked closely. The resulting spread of S_{21} characteristics is shown in Fig. 7 together with two measured characteristics from opposite sides of a 4-in wafer. The filter has a measured midband insertion loss of 2.4 dB and measured -3-dB frequencies of 845 and 860 MHz. Simulation with capacitor tolerances predicted the -3-dB frequencies at 830, 880, and 930 MHz with 880 MHz being the nominal design value.

Since MOS capacitors display small but finite voltage coefficients [7], the filter was checked for nonlinearity by a third-order intermodulation measurement at 500 MHz. Measurements at signal levels of $+15$ dBm indicated that the third-order intercept was better than the measurement resolution of $+42$ dBm.

IV. Conclusion

Passive inductors and LC filters useful in the gigahertz range have been demonstrated in standard Si IC processing. These elements can be used for high-frequency on-chip filtering, inductive peaking of high-frequency amplifiers, and impedance matching for low-noise amplifiers.

Acknowledgment

The authors wish to acknowledge the contributions of Bill Mack of Signetics Corporation for layout assistance and characterization, and of Signetics for HS3 fabrication.

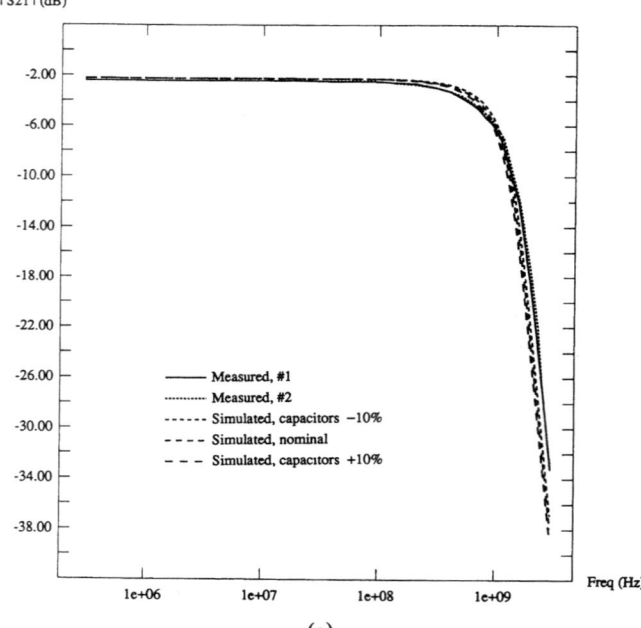

(a)

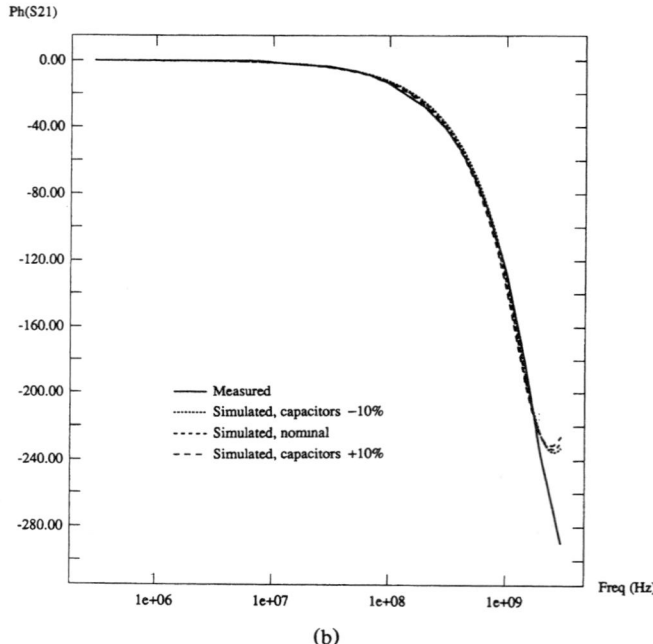

(b)

Fig. 7. Measured and simulated S_{21} characteristics of the filter: (a) magnitude and (b) phase.

References

[1] E. Pettenpaul et al., "CAD models of lumped elements on GaAs up to 18 GHz," *IEEE Trans. Microwave Theory Tech.*, vol. 36, no. 2, pp. 294–304, Feb. 1988.

[2] E. Frian, S. Meszaros, M. Cuhaci, and J. Wight, "Computer-aided design of square spiral transformers and inductors," in *1989 IEEE MTT-S Dig.*, pp. 661–664.

[3] R. M. Warner, Jr., and J. N. Fordemwalt, *Integrated Circuits*. New York: McGraw-Hill, p. 267.

[4] H. M. Greenhouse, "Design of planar rectangular microelectronic inductor," *IEEE Trans. Parts, Hybrids, Packag.*, vol. PHP-10, no. 2, pp. 101–109, June 1974.

[5] M. Parisot, Y. Archambault, D. Pavlidis, and J. Magarshack, "Highly accurate design of spiral inductors for MMIC's with small size and high cut-off frequency characteristics," in *1984 IEEE MTT-S Dig.*, pp. 106–110.

[6] R. Garg and I. J. Bahl, "Characteristics of coupled microstrip lines," *IEEE Trans. Microwave Theory Tech.*, vol. MTT-27, no. 7, pp. 700–705, July 1979.

[7] J. L. McCreary, "Matching properties, and voltage and temperature dependence of MOS capacitors," *IEEE J. Solid-State Circuits*, vol. SC-16, no. 6, pp. 608–616, Dec. 1981.

Large Suspended Inductors on Silicon and Their Use in a 2-μm CMOS RF Amplifier

J. Y.-C. Chang, *Student Member, IEEE*, Asad A. Abidi, *Member, IEEE*, and Michael Gaitan, *Member, IEEE*

Abstract—Large spiral inductors encased in oxide over silicon are shown to operate beyond the UHF band when the capacitance and loss resistance are dramatically reduced by selective removal of the underlying substrate. Using a 100-nH inductor whose self-resonance lies at 3 GHz, a balanced tuned amplifier with a gain of 14 dB centered at 770 MHz has been implemented in a standard digital 2-μm CMOS IC process. The core amplifier noise figure is 6 dB, and the power dissipation is 7 mW from a 3-V supply.

I. Introduction

THE growing needs for miniature wireless communication in the 1-GHz band have prompted interest in monolithic RF amplifiers in silicon. Modern bipolar transistors and FET's certainly have a high enough f_T to provide gain in the required narrow frequency band at these frequencies; the challenge, interestingly enough, is in the difficult fabrication of monolithic passive components. An RF amplifier employs a tuned load, to act as a secondary filter for out of band signals and noise following preselection at the antenna, but most importantly as a means to obtain gain which may be as large as that available at dc by using LC resonance to null out device and parasitic capacitances at the center frequency.

Efforts to fabricate large value spiral inductors on silicon substrates in the 1960's [1] led to the conclusion that the self-resonance caused by parasitic capacitance of these structures to the substrate would limit their use at high frequencies, and the series spreading resistance in the lossy substrate their quality factor (Q). Spiral inductors were revisited when the fabrication of tuned amplifiers became common on the semi-insulating GaAs substrate, where these limitations no longer exist. Most GaAs circuits required inductances of only a few nanohenries.

The work reported here was motivated by the need to develop a silicon *low-power* RF amplifier in the 800- to 900-MHz frequency band. A low-power amplifier must necessarily operate at a high impedance level, and therefore requires large inductors in the tuned load. Previous work [2] in a 0.3-μm GaAs MESFET technology has demonstrated a monolithic RF amplifier tuned to 1.6 GHz with a conventional on-chip inductor of about 12 nH. We set ourselves the greater challenge of obtaining a similar performance in the 700- to 900-MHz band with a 2-μm digital silicon CMOS technology. The lower frequency of operation means that the tuned load requires inductors as large as 100 nH. Lessons learned from past efforts in inductor fabrication on silicon suggested, however, that, short of a fundamental innovation, this was an impossible goal. How this challenge was met is described in the following sections.

II. A Suspended Inductor

Large-value inductors may be fabricated as aluminum spirals with many turns. As the inductance of the spiral is made larger, the capacitance to substrate increases, leading to a progressively *lower* frequency of self-resonance. Spiral inductors of 25 nH are found to self resonate at about 3 GHz on GaAs substrates [2] and on insulating sapphire substrates [3]. On the other hand, aluminum inductors only as large as 10 nH on standard silicon substrates will self resonate at 2 GHz [4], and furthermore the spreading resistance of the substrate will introduce a loss.

These characteristic problems of a silicon substrate may be overcome if the area under the inductor is made to appear locally insulating. This is most simply accomplished by selectively etching out the silicon, leaving the inductor encased in a suspended oxide layer attached at four corners to the rest of the silicon IC. There is a similarity between this technique and the practice in some GaAs technologies of suspending spiral inductors on air bridges [5], but as the typical gap under an air bridge is 3 μm, while removal of the substrate offers air gaps as large as 200 to 500 μm, inductors fabricated by our technique obtain a much lower capacitance to substrate.

A 100-nH inductor was designed using analytical formulas [6] as a 20-turn square spiral of 4-μm-wide lines of second-layer aluminum metal separated by 4-μm spaces, resulting in an outer dimension of 440 μm. Simulations on the SONNET® EM 3-D electromagnetic simulator showed that removal of the underlying substrate will cause the inductor self-resonance to move out from 800 MHz to 3 GHz (Fig. 1).

Manuscript received February 12, 1993. This work was supported by DARPA, Rockwell International, and the State of California MICRO Program. In this report, commercial computer programs and instruments are identified to specify the procedure adequately. This does not imply recommendation or endorsement by NIST, nor that the program is the best available for the purpose.

J. Y.-C. Chang and A. A. Abidi are with the Integrated Circuits and Systems Laboratory, Electrical Engineering Department, University of California, Los Angeles, CA 90024-1594.

M. Gaitan is with the Device Technology Group, Semiconductor Electronics Division, National Institute of Standards and Technology, Gaithersburg, MD 20899.

IEEE Log Number 9208809.

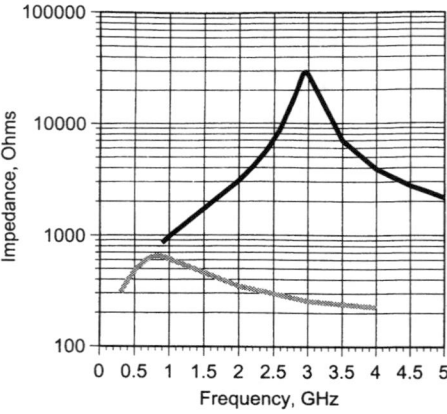

Fig. 1. Self-resonance frequency of a 100-nH inductor increases from 800 MHz to 3 GHz after removal of underlying substrate. This corresponds to a 14-fold reduction in parasitic capacitance. Impedance of inductor over silicon (gray) and over pit (black) obtained from 3-D electromagnetic simulations.

Accompanied by active circuits for an amplifier and buffers, this structure was fabricated through MOSIS as a standard n-well 2-μm CMOS IC. Using a previously described technique [7], the fabricated die were then subject to a selective EDP wet etch to remove the substrate under the inductors, while leaving the remaining circuits intact (Fig. 2). The starting areas for the etch were defined without use of an extra mask [7].

III. RF Amplifier and Experimental Results

Two identical inductors of nominal value 100 nH were used as the load of a balanced amplifier, tuned to a center frequency of 800 MHz by the junction capacitance of FET's built with 2-μm design rules (Fig. 3). The circuit attained high-frequency gain at low power at the expense of noise figure. Cascode stages were used to enhance the available gain. For ease of testing in a network analyzer, on-chip polysilicon 50-Ω resistors terminated the two inputs to ground. The outputs were buffered through common-source FET's to a 50-Ω environment off-chip. The buffer frequency response was separately measured to deembed the core amplifier.

Operating at a 3-V power supply, a peak gain of 14 dB centered at 770 MHz was obtained in the core amplifier with a 7-mW power dissipation. The measured frequency response conformed well to simulations (Fig. 4), although the center frequency was lower by 50 MHz owing to a larger FET junction capacitance than anticipated. Chips where the substrate was left intact under the inductor show a dramatically worse frequency response (Fig. 4). The measured noise figure of the core amplifier, excluding the noise contribution of the termination resistors, was 6 dB. As the substrate capacitance and loss in the suspended inductor are negligibly small, it was modeled for the purpose of circuit simulation by a simple LCR equivalent circuit, whose parameter values correspond very well to Cascade® probe measurements on a test inductor (Fig.

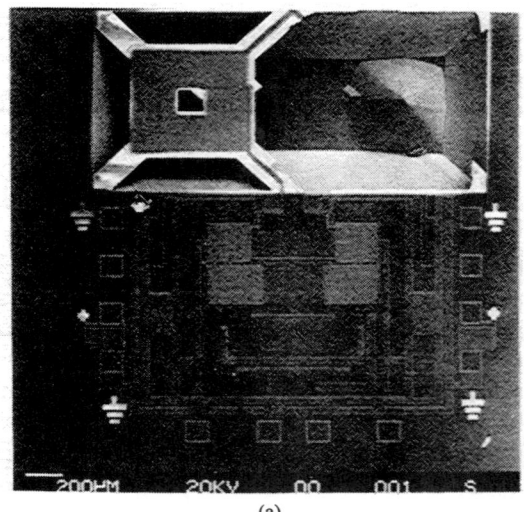

Fig. 2. (a) SEM of RF amplifier after selective substrate etch. Inductor shown is suspended on oxide layer attached to substrate at four corners. Spiral fabricated as second-level aluminum, while contact from center brought out on first level. Second inductor has been manually removed to show pit. (b) Cross section of suspended inductor and substrate after etching.

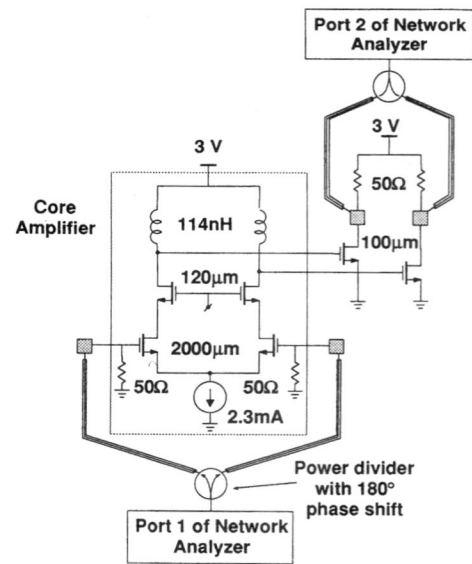

Fig. 3. Schematic of CMOS RF amplifier, including test arrangement. All FET's are 2-μm channel length. Bond pads define the chip periphery.

4). The large suspended structure is found to be mechanically very robust. No damage has yet been observed on an IC when it sustained impacts during the packaging procedure and while measurements were being taken.

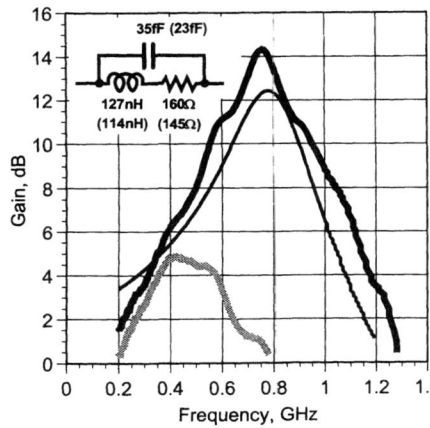

Fig. 4. Measured frequency response of core amplifier, after (black) and before (gray) substrate removal. Simulated response from SPICE (thin black) using a two-terminal LCR inductor model obtained from EM simulations. Measured parameter values in equivalent circuit (inset) compare favorably with values obtained from 3-D simulations (in parentheses).

IV. Conclusions

A use of selective etching is shown to obviate problems which were thought to plague all selective RF circuits on silicon substrates. As an added bonus, the etching requires no modifications to a standard digital CMOS IC process. The test vehicle is possibly the highest performance 2-μm CMOS amplifier reported to date, with a 14-dB gain at 770 MHz while requiring only a 3-V power supply from which it drains 7 mW.

References

[1] R. M. Warner, Ed., *Integrated Circuits: Design Principles and Fabrication.* New York: McGraw-Hill, 1965.

[2] Y. Imai, M. Tokumitsu, and A. Minakawa, "Design and performance of low current GaAs MMICs for *L*-band front-end applications," *IEEE Trans. Microwave Theory Tech.*, vol 39, pp. 209–215, Feb. 1991.

[3] M. Caulton, S. P. Knight, and D. A. Daly, "Hybrid integrated lumped element microwave amplifiers," *IEEE Trans. Electron Devices*, vol. ED-15, pp. 459–466, July 1968.

[4] N. M. Nguyen and R. G. Meyer, "Si IC-compatible inductors and LC passive filters," *IEEE J. Solid-State Circuits*, vol. 25, pp. 1028–1031, Aug. 1990.

[5] M. E. Goldfarb and V. K. Tripathi, "The effect of air bridge height on the propagation characteristics of microstrip," *IEEE Microwave Guided Wave Lett.*, vol. 1, pp. 273–274, Oct. 1991.

[6] H. M. Greenhouse, "Design of planar rectangular microelectronic inductors," *IEEE Trans. Parts, Hybrids, Packaging*, vol. PHP-10, pp. 101–109, June 1974.

[7] M. Parameswaran, H. P. Baltes, Lj. Rustic, A. C. Dhaded, and A. M. Robinson, "A new approach for the fabrication of micromachined structures," *Sensors and Actuators*, vol. 19, pp. 289–307, Sept. 1989.

High Q Inductors For Wireless Applications In a Complementary Silicon Bipolar Process

K. B. Ashby, W. C. Finley, J. J. Bastek, S. Moinian

AT&T Microelectronics, AT&T Bell Laboratories
2525 North 12th Street, Reading, PA 19612

I. A. Koullias[†]
Wireless Microsystems
2714 Tennyson Avenue, Reading, PA 19608

Abstract -- Rectangular spiral inductors with Q's as high as 12 have been built in a high-speed complementary bipolar process and characterized for use in wireless applications. An accurate broadband model has been developed for the inductors, and a test filter and mixer have been built to verify the performance of the inductors and the accuracy of the model.

INTRODUCTION

Integrated circuits for wireless applications are being driven to higher levels of integration, lower supply voltages, and minimal power dissipation by consumer desires for low cost, small size, and long battery life. It is becoming increasingly more challenging to meet these demands with existing silicon technologies which provide only transistors, resistors, and capacitors as design components. Adding inductors to the list of available components allows the designer to use matching networks, passive filtering, inductive loading, and other techniques which until recently have been unavailable on silicon integrated circuits.

Planar inductors have been used for many years in circuits with insulating or semi-insulating substrates, but because of the conductive substrate and generally low frequency of operation for silicon circuits, it was felt that integrating inductors on chip was impractical [1]. In 1990 the first of three papers was published by Nguyen and Meyer showing that it was possible to make useable inductors on silicon integrated circuits [2, 3, 4]. A 9.7nH inductor was reported with a measured maximum Q of 3 at 0.9GHz, and a 1.9nH inductor was estimated to have a maximum Q of 8 at 4.1GHz. These inductors were fabricated using aluminum metallization in a technology with a substrate resistance of 14Ω-cm. The first lumped element model for an inductor on silicon was published in that work. To our knowledge, no other measurement results for inductors on silicon have been presented to date.

In 1992 we designed and manufactured a test array of inductors using AT&T's CBIC-V2 technology. CBIC-V2 has a higher substrate resistivity (150 to 200Ω-cm vs. 15 to 20Ω-cm) than many other high-speed bipolar processes. It also provides the possibility to reduce the series resistance of the inductor (compared to an aluminum metallization process) by using a thick (5μm to 6μm) gold metallization. We felt that the combination of higher substrate resistance and lower series resistance would result in a higher Q than that reported previously. The results are the inductors that are discussed in this paper.

DESCRIPTION OF THE INDUCTORS

The key parameters in the layout of a rectangular spiral inductor are the outer dimensions of the rectangle, the width of the metal traces, the spacing between the traces, and the number of turns in the spiral. For maximum inductance per unit area, all the inductors in our test patterns are squares. The outer dimensions of the inductors are 300μm and the spacing between runners is 4μm. Three inductors of a varying number of turns were made with metal trace widths of 5, 9, 14, 19, and 24μm and one inductor was made with a trace width of 49μm.

A technique to estimate the inductance and Q for each inductor is desirable for design purposes. The technique developed by Greenhouse [5] was used to estimate the inductance of each inductor for the test array. We found this technique for estimating inductance to be accurate to within ±10%. The DC resistance of the metallization in the inductor sets an upper limit on the achievable Q of the inductor and can be estimated by multiplying the number of squares of metal by an appropriate value for sheet resistance. Although there is a frequency dependence to the effective series resistance which limits the inductor Q (to be discussed later), the DC resistance value is easy to estimate and measure. This provides an important starting place for modeling.

[†]Formerly with AT&T Bell Laboratories

INDUCTOR MODEL AND MEASUREMENT RESULTS

Fig. 1 shows the lumped element model we have developed for the inductors. L_S represents the series inductance. R_S represents the series resistance due to the metallization resistance and includes a frequency dependence due to skin effects and edge effects. C_{SUB1} and C_{SUB2} represent the capacitance from the metal layer to the substrate. R_{SUB1} and R_{SUB2} represent the resistance of the substrate. C_{SUB3} and C_{SUB4} represent the capacitance across the substrate layer.

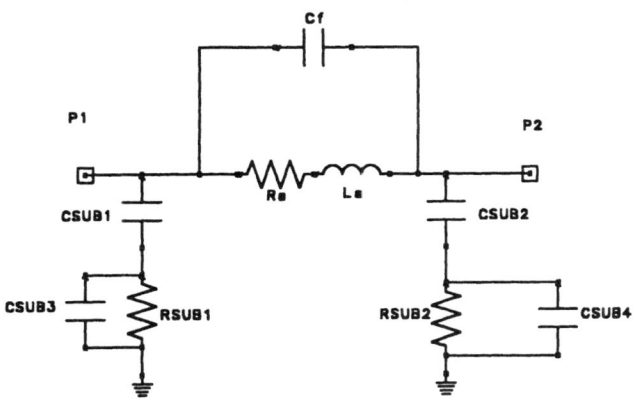

Fig. 1 Inductor Model

The topology of this model is similar to that used by Nguyen and Meyer but there are some important differences. The first is that we are modeling the frequency dependence of the series resistance. The fields in the inductor cause the current to flow along one edge of each trace. These edge effects combined with skin effects create an effective series resistance that is non-linear function of frequency. We have attempted to model this effect by fitting the data to the following equation where R_0 is the DC series resistance, f is the frequency in GHz, and K_1 and K_2 are arbitrary constants:

$$R_S = R_0(1 + K_1 f^{K_2})$$

The second difference between the models is our addition of the capacitors C_{SUB3} and C_{SUB4}. We found that these were very important to accurately fit the model to the data across a broad frequency range. As stated previously, we believe that these capacitances represent the capacitance across the substrate layer.

The inductors were measured using a network analyzer and high frequency probes. Full 2-port S-parameter measurements were taken for each inductor over a frequency range of 100MHz to 10GHz. A set of unconnected probe pads was also measured to characterize the parasitics of the pads. The pad parasitics were de-imbedded from the data by subtracting the Y-parameters of the pads from the Y-parameters of the inductor and converting back to S-parameters. The values of the model elements were fit to the data using computer optimization.

Fig. 2 and Fig. 3 show measured versus modeled S-parameter data for a 2.88nH inductor over the range of 100MHz to 10GHz. Like all the inductors in the test array, this inductor has outer dimensions of 300μm, and is made with 2 turns of 14μm metal with 4μm spacing. An inductance of 2.8nH and DC resistance of 1.3Ω (±30%) were predicted for this inductor. After optimization, the model parameters for the inductor were found to be L_S=2.88nH, R_0=1.75Ω, C_{SUB1}=1.02pF, C_{SUB2}=1.02pF, C_{SUB3}=33.5fF, C_{SUB4}=50.4fF, C_F=23.4fF, R_{SUB1}=5706Ω, R_{SUB2}=5706Ω, K_1=0.30, K_2=1.22. The self-resonant frequency of this inductor is greater than 10GHz.

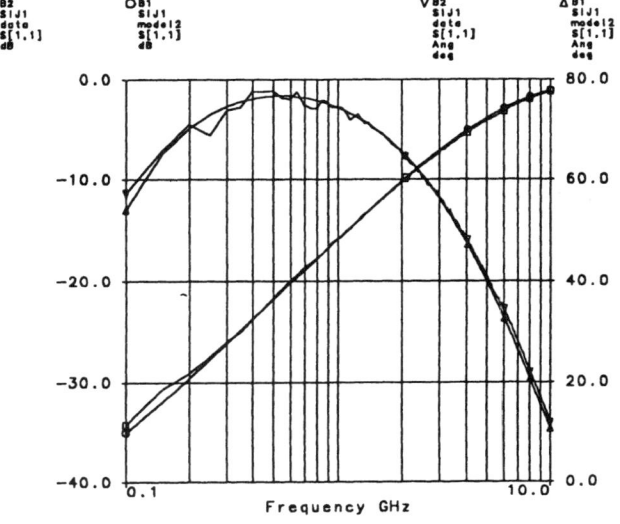

Fig. 2 Modeled and Measured S11

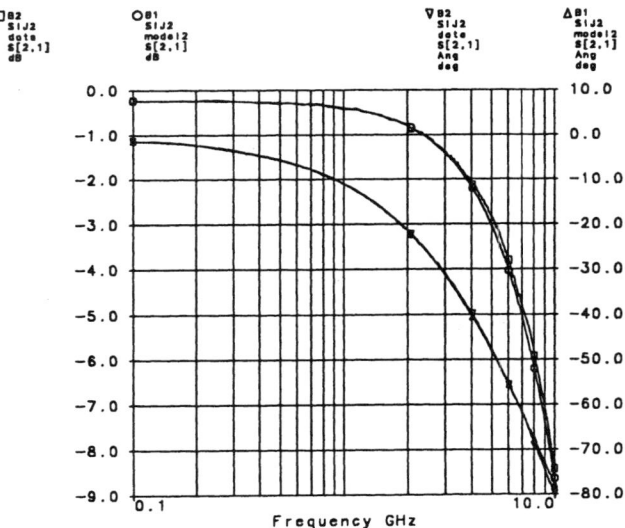

Fig. 3 Measured and Modeled S21

The Q of each inductor was found by taking the ratio of the magnitude of the imaginary part to the magnitude of the real part of the 1-port input impedance which was derived from the 2-port measurements. Fig. 4 shows measured versus modeled Q of the inductor as a function of frequency and shows the Q as seen at each port with the other port grounded. The Q reaches a peak of 12 between 3GHz and 4GHz and then begins to decrease with increasing frequency. The peaking and rolling off of the Q show that the peak Q should be measured as it cannot be accurately estimated by extrapolating from lower frequency data. We also found that the location of the peak Q varied from inductor to inductor implying it should be possible to peak the Q at a desired operating frequency by controlling the layout of the inductor. The other inductors in the test array had maximum Q's that ranged from 12 as shown for this inductor down to 6 for which occurs at about 1.5GHz, and have a self resonant frequency of 5.5GHz. Measured versus modeled results for the filter are shown in Fig. 6 The filter has a corner frequency of 400MHz, insertion loss of 2dB in the pass band, and return loss in the pass band of greater than 11dB.

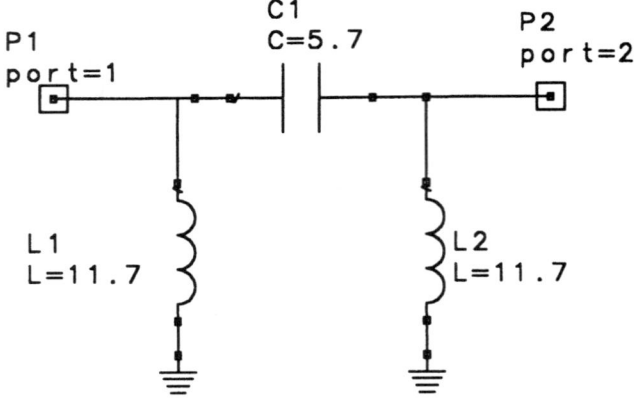

Fig. 5 Schematic of High-Pass Filter

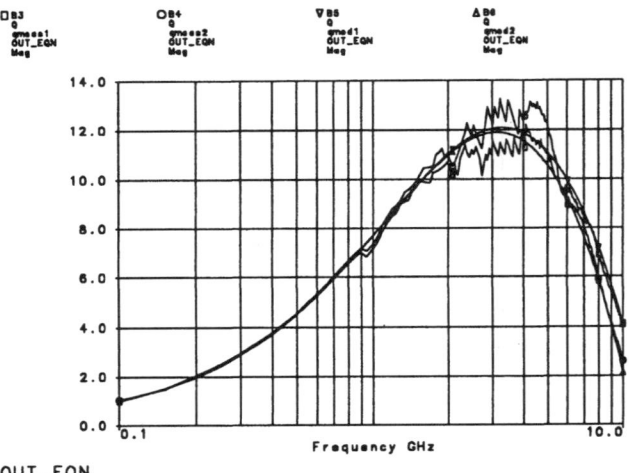

Fig. 4 Measured and Modeled Q a 34nH inductor.

The model does an adequate job of representing the specific inductors we have characterized, but in its present form it is not scalable. The ideal model would include the ability to estimate values for each of the subcircuit elements from layout parameters of the device. As indicated previously, the Greenhouse technique is adequate to estimate the series inductance to within about 10%, and the DC series resistance can be estimated fairly accurately from the number of squares of metal in the inductor. The other model parameters, however, are more difficult to estimate. We believe that a scalable lumped element model dependent upon layout parameters would be a valuable design benefit and is an area needing additional investigation.

CIRCUITS

A third-order high-pass filter with topology as shown in Fig. 5 was implemented using two 11.7nH inductors and a 5.7pF capacitor. The inductors have outer dimensions of 300μm and are made with 6 turns of 9μm metal with 4μm spacing. They have a measured maximum Q of 8.5

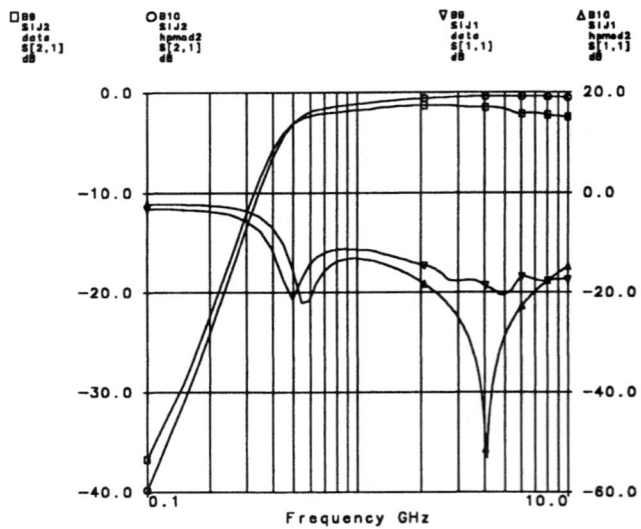

Fig. 6 Measured and Modeled High-Pass Filter Response

Fig. 7 shows the schematic for a single balanced mixer using integrated inductors. Such a mixer could be used in a receiver for wireless applications. Use of inductors provides several benefits. Inductor L1 between the bases of Q3, and Q4 acts as a high-pass filter, eliminating the need for an external filter to reduce noise at the IF frequency which is typically present on the LO input line. L1 also improves the input impedance match reducing the required LO1 drive level. Inductors L4, L2, and C1 provide on-chip matching to 50Ω, eliminating the need for external matching components. L4 also increases the IP3 of the mixer by improving the linearity of the cascode amplifier formed by Q1 and Q2. Use of inductors L2 and

L3 allow DC biasing of Q1 and form a low-pass filter with C1 to reduce RF noise on the bias line, improving the noise performance of the mixer.

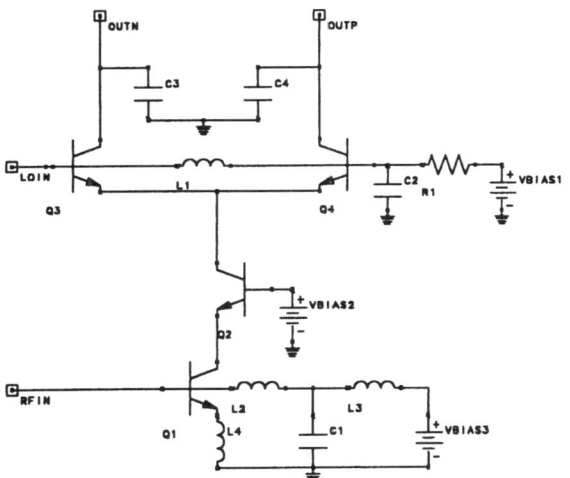

Fig 7. Single Balanced Mixer Using Inductors

The mixer uses 6.7mA from a 5.0V supply. Measurements were taken with RFIN at 890MHz and the LO level set at -10dBm for an IF output at 73MHz. All matching was internal to the chip except for a matching network on the open-collector outputs which also performed a differential to single-ended conversion. The double sideband noise figure of the mixer was measured at 10dB. The differential gain (including the matching circuit losses) was measured at 14.3dB, and the IP3 at the input was measured at 2.4dBm.

SUMMARY

We have reported measurement results for inductors on silicon with measured Q's as high as 12, the highest reported to date. We have developed an empirical lumped element model accurate over a wide frequency range and have shown test circuits confirming the performance of the inductors, verifying the accuracy of the model, and showing the design benefits of using inductors on silicon integrated circuits.

ACKNOWLEDGEMENTS

We wish to express our appreciation to Gene Markle, Terry Pearson, Bob Plummer, Terry Flemming, Jim Yoder, Russ Lamm, George Terefenko, and Joe Havens for their assistance on this project.

REFERENCES

[1] R. M. Warner, and J. N. Fordemwalt, eds., *INTEGRATED CIRCUITS, Design Principles and Fabrication*, New York: McGraw-Hill, pp. 267.

[2] N. M. Nguyen, and R. G. Meyer, "Si IC-Compatible Inductors and LC Passive Filters", *IEEE Journal of Solid-State Circuits,* vol. 25, no. 4, August 1990, pp. 1028-1031.

[3] N. M. Nguyen, and R. G. Meyer, "A Si Bipolar Monolithic RF Bandpass Amplifier", *IEEE Journal of Solid-State Circuits,* vol. 27, no. 1, Jan. 1992, pp. 123-127.

[4] N. M. Nguyen, and R. G. Meyer, "A 1.8-GHz Monolithic LC Voltage-Controlled Oscillator", *IEEE Journal of Solid-State Circuits*, vol. 27, no. 3, March. 1992, pp. 444-450.

[5] H. M. Greenhouse, "Design of Planar Rectangular Microelectronic Inductors", *IEEE Transactions on Parts, Hybrids, and Packaging*, vol. PHP-10, no. 2, June 1974, pp. 101-109.

New Development Trends for Silicon RF Device Technologies

Natalino Camilleri, David Lovelace, Julio Costa, David Ngo

Motorola, ACT, M/D M350
Mesa, AZ 85202

Abstract:

Today more than ever, low cost, high performance RF devices are in high demand due to explosive growth in the wireless communications business. As the RF performance of silicon-based technologies improve, silicon solutions are an obvious choice due to its low cost, high reliability and the ability to integrate other analog and logic functions on-chip. This paper will highlight some of the development efforts that are taking place to improve silicon device technologies for RF and microwave applications. Performance of current state-of-the-art silicon bipolar, MOSFET, TFSOI and SiGe HBTs will be discussed.

Introduction:

With the fast growth in the wireless communications business, the demand for higher performance but cost effective RF solutions is increasing. RF designers currently have several different choices of silicon and GaAs based technologies. The advantages offered by silicon technologies are low-cost - due to the volume of wafers processed, and the ability to integrate RF circuits with other analog and logic circuits. On the other hand, GaAs technologies offer high-performance devices and monolithic RF circuits. This paper we will track the progress of several silicon RF device technologies used in wireless communication components. Although device technologies based on GaAs have a lot to offer in this emerging business, this paper will only focus on the Silicon technology developments since historically the GaAs activities have been well covered by several authors within the MTT community.

A clarification of the figure of merits will be the basis of technology comparisons and will be discussed first. The technology discussion will focus around the applications and will discuss progress in discrete and integrated BJT and MOS transistors. Discussion on what technology is available today and what might be used tomorrow will follow for each application.

Device Figure of Merits:

The two most important terms in comparing device technologies are Ft and Fmax. Thanks to the development of accurate network analyzers and on wafer probing techniques these frequency figure of merits can be accurately measured. These measurements are more difficult to perform on silicon since the probe pad parasitics impair the measurement and cannot be ignored. Deembedding pad parasitics is common practice but does not yield meaningful data when used on small digital devices which have device parameters smaller than the probing pad parasitics. For more accurate measurements it is recommended that the devices under test will be sized such that the device capacitance is at least five times the pad parasitic capacitance.

Silicon technologists still debate which frequency figure of merit is best to quote, but in reality they are both important for an RF application. Ft is essentially the inverse of the transit time across the device and it shows the highest fundamental frequency component a device can generate. This term has traditionally been the most popular among the high speed digital technologists. Fmax is defined as the frequency when maximum available gain (MAG) is unity. Fmax is also referred to as the maximum frequency of oscillation. This term generates the most confusion among technologists. Fmax is best related to multiples of Ft and it gives an indication of the extrinsic device parasitics such as access resistance and capacitance. If a device had no extrinsic parasitics, Fmax could be as much as five times Ft. However such condition is not practical nor necessary. A well optimized RF device usually has an Fmax close to twice Ft. The challenge facing RF device technologists is to increase Ft while maintaining Fmax at about 2xFt. Device technologies designed for very high speed digital applications sometimes have devices with Fmax equal to or less than Ft. Such devices usually do not perform well in RF applications.

IEEE 1994 Microwave and Millimeter-wave Monolithic Circuits Symposium, pp. 5-8, 1994.

How do these frequency figures of merit relate to the actual frequency of operation? A competitive device should have an Ft that is about five to ten times higher than the operating frequency. If a device is operated at higher frequency it will have gain but other specifications such as bandwidth, noise, efficiency and linearity will suffer, thus making it impractical to use in most systems.

Base Station and Mobile Power Technologies:

These applications require some of the toughest device specifications because they require the highest possible output power and linearity. For 900MHz base stations, these devices deliver 150W of linear power using a 26V supply, and some 900MHz mobile transmitters require 45W of power using a 12.5V supply. Typical specifications of such devices is listed in Table 1.

Table 1

Device Type	Freq. (MHz)	Vcc (V)	Pout (W)	Gain (dB)	Effic. (%)	IMD (dBc)
MRF899	900	26	150	9	40	-32
MRF15090	1490	26	90	8	36	-32
MRF847	870	12.5	45	5.5	65	N/A

Power bipolar RF discrete devices have a collector contact on back side of the substrate and are packaged in ceramic modules that contain an internal matching network. The internal matching network is necessary due to the low input impedance of the power bipolar devices. The emitter-base structure of these transistors is interdigitated and gold metalization is used to counter electromigration problems that arise from the large current densities at the emitter metal fingers. The transit time of these devices (1/Ft) is dominated by the collector region which is required to withstand large voltage swings. These devices are optimized for the highest Fmax to increase gain and the lowest output capacitance to be efficient in delivering power.

Historically, improvements in the bipolar RF performance have come about by shrinking the emitter-base pitch. Transistors with emitter geometries larger than 1um suffer from emitter crowding effects due to the intrinsic base resistance, thus the power delivered is usually a function of the emitter periphery. When emitter geometries are shrunk from 2um to 1um, all the parasitic capacitances are reduced (namely input, output and feedback capacitance) which consequently increases both Ft and Fmax. Since the emitter periphery is not impacted, the power per emitter finger does not change. Shrinking lithography below 1um causes the device performance to be emitter area dependent rather that periphery dependent. Once the emitter geometry is minimized such that current crowding is eliminated, further shrinks will only reduce feedback capacitance. Since the Ft of these devices is limited by the collector transit time, future developments should focus on engineering an improved collector region for the vertical structure.

Vertical and lateral MOSFET technologies for RF applications have matured and now are slowly replacing BJTs in RF power applications. In the past, vertical MOS devices have competed with BJTs at lower frequencies but have been limited by the inherent high feedback capacitance. These vertical MOS devices have a back side drain and are assembled in ceramic packages similar to BJTs. Lateral devices like the ones described in [1,2,3] have a frequency performance very similar to BJTs, but at lower power densities. An attractive feature of lateral MOS devices is that unlike the BJT and Vertical-MOS devices, the source contact is located on the back side of the die. This eliminates the need for high conductivity ceramics used in the packages and no longer requires bond wires to connect the source to the package lead pin. Back side source contacts effectively remove the source lead inductance, a parasitic which reduces gain. Improvements in power MOS technologies typically come about by reducing the gate length while maintaining the same drain-to-source breakdown voltage.

Portable Power Technologies:

Until recently, designers favored silicon bipolar devices for most portable power applications. Currently, however, GaAs and silicon FETs have replaced most silicon BJT devices in portable cellular phone systems. This shift was due in part to the fact that less costly and more efficient power amplifiers can be built with these FET technologies. Another important factor in the selection of FETs over bipolar devices is due to PA (power amplifier) noise specifications in the receiver band of cellular systems. Since FETs have about 20dB lower noise than BJTs, the duplex filter specifications can be relaxed resulting in the use of a lower loss filter and thus a more efficient transmitter. BJT devices deliver less gain per stage than FETs, thus more gain stages are required resulting in increased complexity which leads to higher costs. The cellular industry currently uses GaAs in the smaller and higher cost systems due to its efficiency, but silicon MOS is preferred for lower cost radios. The silicon MOS solution is less costly since the die is cheaper and the system does not need to provide a negative gate bias supply. Today the best 6V GaAs PA modules are 65% efficient, whereas the best silicon MOS PA modules are 55% efficient.

As portable electronics migrate to lower operating voltages to take advantage of low voltage digital technology and reduce battery weight, power amplifier designs will need to be modified to operate as efficiently at these new voltages. Technologies such as GaAs PHEMT do not compromise any efficiency or power density at 3V. Silicon MOSFETs operate respectably at voltages as low as 3V [3] but still have room to improve. The advent of Thin Film Silicon On Insulator (TFSOI) [4] devices has led to performance improvements since these structures are constructed in such a way as to eliminate the parasitic diode junction at the output of the silicon MOS transistor.

New device tradeoffs and specifications are emerging as digital cellular systems replace existing analog cellular networks. Noise in the receiving band becomes less of an issue while new specifications such as adjacent channel coupling, linearity, and splatter become more important. In TDMA systems, where the PA is only on 1/12 of the time, the DSP chips consume most of the system's current. Therefore, the PA may not be the dominant power consumer of the over all system anymore. System designers will thus be less willing to pay for slight improvements in PA efficiency.

Silicon BJTs are still the preferred technology for linear systems and they still offer some of the best efficiency for a given -30dBc IMD specification. Unlike high-voltage bipolar devices which are collector-delay limited, BJT technologies at lower voltages have several prospects related to base and emitter engineering. With the advent of poly-silicon emitters and poly-base links in the 80's, higher doped junctions with less delay through the emitter-base were made possible. In the 90's, bipolar SiGe (silicon-germanium) technology will likely mature as key improvements in epitaxy and device technology occur. SiGe RF technology is still at its infancy; but low-voltage SiGe bipolar devices have been very encouraging. The SiGe base effectively lowers the transit time through the base of the device but does not influence the collector delay. Therefore, as the collector is widened and doped lower to increase the breakdown voltage, the SiGe base is less likely to result in an improvement over the simple Si base.

Table 2 summarizes the performance of silicon and SiGe base transistors versus the operating voltage for power amplifier applications.

Table 2

Operating Voltage (V) for PA in Class C:	A:	BVceo (V)	Silicon Ft/Fmax (GHz)	SiGe Ft/Fmax (GHz)
6	12	18	7/14	9/18
3	6	9	10/20	15/30
1.5	3	4.5	15/30	30/60

Technologies for Voltage Controlled Oscillators:

Historically, the phase noise specifications have prevented the integration of VCOs (Voltage Controlled Oscillators). The 1/f noise and the breakdown requirements of the three terminal devices are usually not compatible with VLSI processes. High-Q resonators and varactors required to meet most VCO specifications are not easily integrated into standard silicon processes. Discrete resonators can easily be attached off chip. High Q varactors are usually fabricated in a mesa-structure which makes them impractical to integrate with the epitaxial configuration of VLSI processes. With the aid of sub micron lithography, high quality lateral varactors which are compatible with VLSI technologies can be realized.

Silicon JFETs, the preferred device for VCO designs, exhibit the best 1/f noise characteristics with corner frequencies under 100Hz. Unfortunately most commercially available devices do not operate well above 500MHz, since the parasitic gate to drain feedback capacitance limits the high-frequency performance of the device. With sub-micron lithography, these devices can be made to operate at frequencies as high as 3GHz.

Currently, commercially available VCOs for 900MHz operation are fabricated with discrete BJTs. These BJTs are constructed similarly to the power devices described above. For 3V applications, a BVceo of 9V or higher is required so that the device does not operate near avalanche breakdown. The corner frequency of these BJTs is about 1KHz and it is significantly lower than the adjacent channel spacing of 25KHz where most VCO phase noise requirements are specified. For most FM systems the 1/f noise components of such devices translate into residual FM noise in the audio band from 300Hz to 3KHz. This is the reason why JFETs are preferred over BJTs for this application. Future systems will tend to decrease the adjacent channel spacing and make the VCO specifications even more difficult. Spectrum limited systems such as private mobile radios have some of the toughest specifications, since the VCO performance dictates the dynamic range of the system. Most cellular systems have less severe requirements and usually have a 10 dB worse phase noise specifications.

Technologies for Synthesizers:

The most demanding RF component in a phase-locked-loop (PLL) circuit is the prescaler, which divides the incoming reference signal from the VCO to a lower frequency reference. The rest of the PLL circuitry consists of lower frequency digital and precision analog functions. For low power consumption, the PLL ICs are usually integrated in CMOS, but CMOS prescalers have historically not been as fast or efficient as BJTs. BiCMOS is the optimum solution for this application as a performance tradeoff between MOS and BJT. Improvements in MOS technology such as TFSOI have shown that CMOS prescalers can compete well with current BJT technologies.

Technologies for Receiver Front Ends:

With the increasing number of spectrum uses, receiver front ends need to be increasingly immune to interfering signals. The first mixer in the front end typically saturates first, thus the less gain obtained from the LNA, the better the system IP3. In order to suppress image noise and provide more selectivity, most systems use a filter between the LNA and the mixer. The ideal system will use only a high gain, low noise mixer for a front end. By making use of an LO filter several cellular systems have successfully used a discrete single ended BJT mixer. The integration of RF filters between the LNA and the Mixer is not easily accomplished although research in the field of Bulk Acoustic Wave technology looks promising [5,6]. Current technology utilizes an off-chip filter for this function.

High speed bipolar technologies have been most successful in the implementation of front ends. The LNA function is fairly straight forward since the input impedance of small-geometry BJTs is easily matched without the need of complicated matching networks. High speed bipolar device technologies that were originally developed for super computers have excellent gain and noise figure characteristics. Designers using state-of-the-art bipolar technologies such as MOSAIC-5 [7] or .5um BiCMOS [8] can easily build an LNA with over 10dB gain and 1.6dB noise at 2.4GHz. Gilbert cell BJT mixers are easy to design, perform well, and are easily integrated without consuming large amounts of real estate. Current BJT technologies have Ft and Fmax greater than 15GHz and 30GHz respectively. These devices perform extremely well in front-end applications up to 3GHz. These figures of merit are easily doubled in SiGe HBTs, and such technology is expected to have excellent performance up to 6GHz.

MOS technologies can provide similar performance to that of BJTs, but due to the inherent high input impedance, MOSFET designs are more challenging and often do not offer a performance advantage. For systems which are very sensitive to higher order intercept products, MOS does offer an advantage over BJTs. One of the most promising new MOS technologies is the TFSOI process [9]. This technology has no parasitic junctions under the source and drain which make it ideal for switch applications which have historically been dominated by GaAs MESFETs.

Reference:

[1] I. Yoshida, et. al. "High Efficiency 1.5GHz Si Power MOSFET for Digital Cellular Front End", Proceedings of 1992 Int. Symp. on Power Semiconductor Devices & ICs, Tokyo, pp. 156-157.

[2] H. Itoh, T. Okabe, M. Nagata, "Extremely High Efficiency UHF Power MOSFET for Handy Transmitter", IEDM Tech. Digest, 1983, pp. 95-98.

[3] N Camilleri, et. al. "Silicon MOSFETs, the Microwave Device Technology of the 90s", IEEE MTT-S, 1993, pp. 545-548.

[4] C. Raynaud, et. at. "High-Frequency Performance of Submicrometer Channel-Length Silicon MOSFET's", IEEE Electron Device Letters, Vol 12, No. 12, Dec. 1991, pp. 667-669.

[5] K.M. Larkin, et. al. "High Q Microwave Acoustic Resonators and Filters", IEEE MTT-S, 1993, pp. 1517-1520.

[6] SV. Krishnaswamy, et. al. "Film Bulk Acoustic Wave Resonator and Filter Technology", IEEE MTT-S, 1992, pp. 153-155.

[7] V. dela Torre, et. al.. "MOSAIC V - A Very High Performance Bipolar Technology," IEEE BCTM, 1991, pp. 21-24.

[8] J. Kirchgessner, et. al.. "An Advanced 0.4um BiCMOS Technology for High Performance ASIC Applications," IEDM Technical Digest, 1991, pp. 97-100.

[9] W.M. Huang, et. al.. "TFSOI BiCMOS Technology for Low Power Applications," IEDM Technical Digest, 1993, pp. 449-452

MINIATURIZED RF-CIRCUIT MODULES FOR LAND MOBILE COMMUNICATION EQUIPMENT

Yoshikazu Mori, Hiroyuki Yabuki, Motoi Ohba,
Morikazu Sagawa, Mitsuo Makimoto, and Ichiro Shibazaki*

Matsushita Elec. Ind. Co., Ltd.
Higashimita, Tama-ku, Kawasaki
214, Japan

*Matsushita Comm. Ind. Co., Ltd.
Tsunashima-Higashi, Kohoku-ku, Yokohama
223, Japan

ABSTRACT

This paper describes newly developed compact RF circuit modules using multi-layered printed circuit boards (M-PCB's). RF circuit modules, such as a frequency synthesizer, PLL-modulator, up-converter and receiver front-end, are demonstrated to be applicable to land mobile communication equipment. RF circuit modules using M-PCB's have a structure suitable for excellent antivibration characteristics, and the size of the newly developed modules has been reduced to one quarter of that of conventional modules designed by surface mount technology.

I. INTRODUCTION

The widening popularity of mobile radio communication urges the development of compact and portable radio terminals.

Conventionally, surface mounted circuit boards, using different sized components, have been usually put to practical use in radio equipment. However, this mounting technology generates waste space which limits the compactness of radio equipment.

In order to make radio equipment more compact and portable, we have proposed the use of miniaturized RF circuit modules using M-PCB's, which have three dimensional structures embedded with RF circuit elements, such as inductors and resonators. The internal conductive layer of M-PCB's are conventionally used for DC power feeds and signal transmission in digital circuits. We focused our attention on the high density mounting properties of M-PCB's. The application of three dimensional structured M-PCB's to RF circuit modules was made possible by the introduction of newly developed split-ring resonators and the theoretical analysis of triplate spiral inductors. These modules using M-PCB's are also constructed to the same height to eliminate waste space.

This paper describes the theoretical analysis of split-ring resonators and spiral inductors, fundamental RF circuit elements embedded in M-PCB's. Furthermore, the performances of newly developed compact RF circuit modules using M-PCB's, such as a frequency synthesizer, PLL-modulator, up-converter and receiver front-end are also presented.

II. SPLIT-RING RESONATORS [1][2][3]

A. Resonator Structure

We previously introduced the split-ring resonator to reduce the size of conventional ring resonators(one wavelength in length). It is composed of a transmission line and a capacitor which connects at both ends of the line, as shown in Fig.1. The length of split-ring resonators is much less than one wavelength because of the existence of the tuning capacitance. It should be noted that this resonator structure has no RF short circuited points which often generate parasitic components.

The input admittance of the split-ring resonator from one end of the line, Y_i, is given using ABCD matrices, one expressing a transmission line and the other a capacitor.

$$[F]_T = \begin{bmatrix} \cos\theta_t & jZ_0 \sin\theta_t \\ j\sin\theta_t / Z_0 & \cos\theta_t \end{bmatrix}$$

$$[F]_c = \begin{bmatrix} 1 & 1/j\omega C_t \\ 0 & 1 \end{bmatrix}$$

$$Y_i = jY_0 \cdot \frac{Y_0 \sin\theta_t - 2\omega C_t(1-\cos\theta_t)}{Y_0 \cos\theta_t - \omega C_t \sin\theta_t}$$

where Y_0 : characteristic admittance of the line
θ_t : total electrical length of the line
C_t : tuning capacitance

An example of the calculated input impedance is shown in Fig.2. It can be seen that the split-ring resonator has both a series and a parallel resonance point and that the frequency span between them is very close. These characteristics are valuable in a resonator of RF oscillator circuits with low phase noise.

Some advantages of split-ring resonators are summarized as follows:
(1) Small size, with no Q-value degradation
(2) Close span between the series and parallel resonance points
(3) Easy adjustment of resonance frequency

B. Resonance Conditions

The resonance conditions of split-ring resonators can be calculated from the input admittance, Y_i. The results are as follows:

$$Y_o \sin\theta_t - 2\omega C_t(1 - \cos\theta_t) = 0$$

The resonance frequency can be calculated by solving the above equation, the results of which are shown in Fig.3 as a function of the resonator length.

III. TRIPLATE SPIRAL INDUCTORS [4][5]

Inductors are important RF circuit elements applicable to LPF, HPF and matching circuits. Triplate spiral inductors have many advantages and a few disadvantages, as follows:

Advantages:
(1) Reduction of stray coupling between circuits
(2) High density, three dimensional mounting

Disadvantages:
(1) Small inductance values caused by the existence of grounded planes

The embedding of spiral inductors requires precise estimations of fundamental properties, such as inductance values and self-resonance frequencies, because it is impossible to adjust the values after assembly. Fig.4 shows the structure of a triplate spiral inductor. We designed rigorous circuit models of triplate spiral inductors, which were derived from introducing three kinds of propagation mode and lumped coupling capacitance between lines.

Figs.5 and 6 show the propagation modes and an analytical model of the spiral inductor, respectively. The equations for the characteristic impedance are as follows:

$$Z_0 = \frac{30\pi}{\sqrt{\varepsilon_r}} \frac{1}{\left[\frac{w}{b-t} + \frac{A}{\pi}\right]}$$

$$Z_1 = \frac{30\pi}{\sqrt{\varepsilon_r}} \frac{1}{\left[\frac{w}{b-t} + \frac{b}{2(b-t)} \frac{CA_e}{\pi}\right]}$$

$$Z_2 = \frac{30\pi}{\sqrt{\varepsilon_r}} \frac{1}{\left[\frac{w}{b-t} + \frac{1}{\pi}\left\{\frac{b}{b-t} CA_e - A\right\}\right]}$$

$$A = \ln\left[\frac{(x+1)^{x-1}}{(x-1)^{x+1}}\right], \quad x = \frac{1}{1-(t/b)}$$

$$A_e = 1 + \frac{\ln(1+\tanh\theta)}{\ln 2}$$

$$\theta = \pi s/2b$$

$$C = 2\ln\left(\frac{2b-t}{b-t}\right) - \frac{t}{b}\ln\left[\frac{t(2b-t)}{(b-t)^2}\right]$$

Fig.7 shows the frequency responses of experimental spiral inductors. Good conformity was obtained between the calculated value and the measured data.

IV. APPLICATIONS

To realize RF modules using M-PCB's, the following design plans were taken into consideration.

(1) A triplate configuration was applied and the spacing between the grounded planes was extended as wide as possible to reduce transmission losses.
(2) Via-holes and the side-edge metalization of PCB's were effectively used for the improvement of shielding between circuits.

On the basis of the above discussion, several RF circuit modules for land mobile communication equipment were designed.

Fig.8 shows the structure of RF circuit modules using M-PCB's. M-PCB's consist of 5 metal layers. The RF components were mounted on the surfaces of the 1st and the 5th metal layers. The 2nd and the 4th metal layers were used for the grounded layers. And the 3rd (center) layer was used for embedding the RF circuit elements, such as split-ring resonators or spiral inductors. Epoxy-glass M-PCB's were employed in order to minimize copper loss despite much dielectric loss.

The trial circuit modules were as follows.

A. Frequency Synthesizer Module [6]

A frequency synthesizer module requires low phase noise and a fast locked-up time. To realize these characteristics, we introduced a voltage controlled oscillator(VCO) using embedded split-ring resonator in a frequency synthesizer module.

Fig.9 shows the circuit configuration of a trial VCO. A split-ring resonator and spiral inductors were embedded in a M-PCB. Si-bipolar transistor was used for Colpitts type oscillator with common base configuration in order to achieve low noise characteristics. A buffer amplifier used a common DC bias current with an oscillator via cascode connection, producing a total current consumption of 8.8mA under the condition of Vcc=4.3V.

A tuning range of 30MHz was obtained by the varactor voltage from 0.6 to 4.0V. Power output level was about -1dBm within the desired frequency band. Fig.10 shows the CNR and SNR characteristics of a VCO.

B. PLL Modulator Module

The PLL modulator module requires an expanded modulation frequency response to lower frequencies and a fast locked-up time. We achieved the miniaturization, even for VHF oscillators by using the embedded spiral resonator.

Figs.11 and 12 show the lock-up time characteristics and modulation frequency response, respectively, which satisfy requirements for land mobile communication equipment.

C. Up Converter Module

Fig.13 shows the configuration of an up converter module, which produces the transmitted signal by mixing the modulator signal and the local signal from the frequency synthesizer module. It consists of a mixer, BPF's and amplifiers.

The mixer is composed of a Si-bipolar transistor and a diplexer using an embedded LPF and HPF. The BPF's, which utilize split-ring resonators, are placed at the output of the mixer and the final stage of amplifiers.

This module has 4 split-ring resonators, which were used for two-stage BPF's, and 10 spiral inductors, which were used for LPF HPF and matching circuit of amplifiers, as shown in Fig.14. The power output level of this module was 10dBm under the condition of an input power of -13dBm.

D. Receiver Front-end Module

Fig.15 shows the configuration of the experimental receiver front-end module. Dual-gate GaAs FETs were used for the LNA, local amplifier and mixer. The two stages of a monolithic crystal filter was utilized as the first IF BPF.

This circuit was fabricated on an M-PCB, which used embedded spiral inductors as the input and output matching elements. The measured data show that the DC current consumption is less than 10mA, noise figure is 4dB and the gain is larger than 30dB. IM_3 and the suppression of sensitivity caused by the transmitted signal are also excellent, and suitable for land mobile communication equipment.

Fig.16 shows a photograph of the trial RF circuit modules. The size of each, including shielding cases, are given below. These modules are approximately from one quarter to one half the size of a conventional surface mount printed circuit modules.

Frequency synthesizer module : $35 \times 15 \times 7.5$ mm^3
PLL modulator module : $34 \times 15 \times 7.5$ mm^3
Up converter module : $35 \times 15 \times 7.0$ mm^3
Receiver front-end module : $33 \times 15 \times 7.5$ mm^3

V. CONCLUSIONS

We proposed compact RF circuit modules using M-PCB's. In order to realize them, a miniaturized split-ring resonator was developed and the rigorous design equations for embedded inductor was analytically derived.

Several compact RF circuit modules were designed and fabricated by using these techniques, and all satisfied the requirements for land mobile communication equipment and have been put to practical use in portable radio equipment currently in service.

ACKNOWLEDGMENT

The authors gratefully acknowledge the assistance of many colleagues at Tokyo Information and Communications Research Laboratory, Matsushita Electric Industrial Co., Ltd. and Matsushita Communication Industrial Co., Ltd..

REFERENCES

[1] I.Wolff;"Microwave Bandpass Filter Using Degenerate Mode of a Microstrip Ring Resonators," Electron.Lett., Vol.8, No.12,pp.302-303:June.1982.
[2] M.Makimoto and M.Sagawa; "Varactor Tuned Bandpass Filters using microstrip-line ring resonators," IEEE MTT-S Int. Microwave Symp., Q-4, pp.411-414:May.1986.
[3] M.Makimoto; "Microstrip-line Split-ring Resonators and their Application to Bandpass Filters," Trans. IEICE Japan, Vol.J71-C, No.7, pp.1063-1070:July.1988.
[4] H.Yabuki and M.Makimoto;"Theoretical Analysis and Fundamental Characteristics of Triplate Spiral Inductors," Trans.IEICE Japan, Vol.J72-C-1, No.7,pp.414-422:July.1989.
[5] D.Lang; "Broadband Model Predicts S-parameters of Spiral Inductors," Microwaves & RF,27, 1, pp.107-110:Jan.1988.
[6] M.Sagawa, I.Ishigaki, M.Makimoto and T.Naruse; "Dielectric Split-ring Resonators and their Application to Filters and Oscillators," IEEE MTT-S Int. Microwave Symp., W-2, pp.605-608:May.1988.

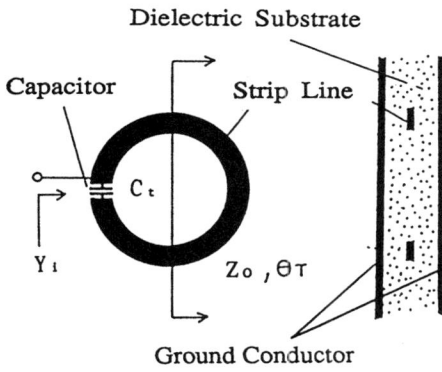

Fig.1 Basic Structure of the Split-ring Resonator.

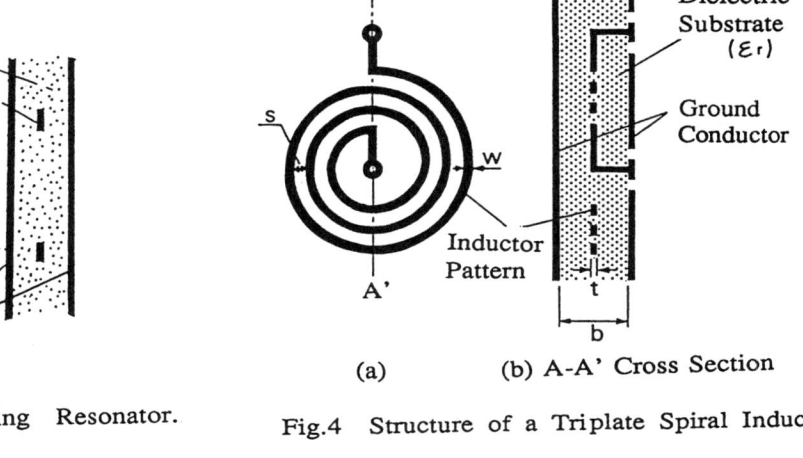

Fig.4 Structure of a Triplate Spiral Inductor.

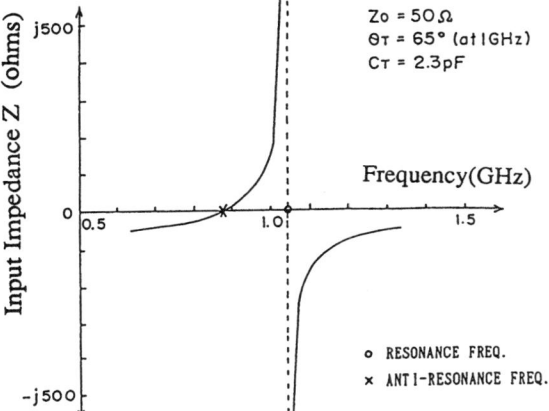

Fig.2 Calculated Results of Input Impedance of the Split-ring Resonator.

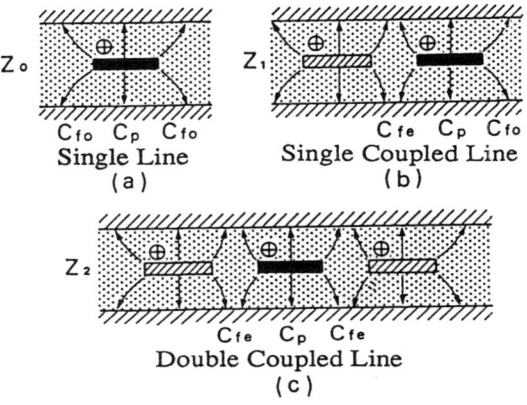

Fig.5 Propagation Modes of the Spiral Inductor.

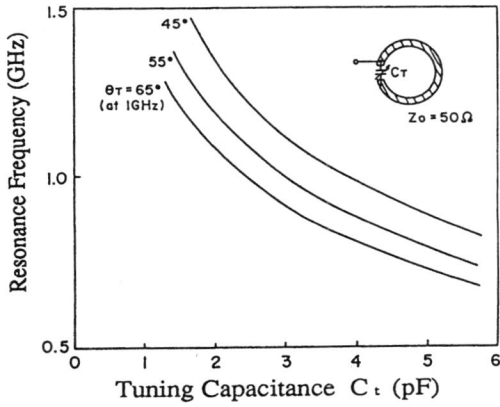

Fig.3 Calculated Results of Resonance Frequencies.

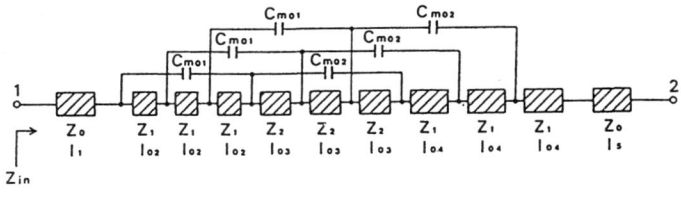

Fig.6 An analytical model of the spiral inductor.

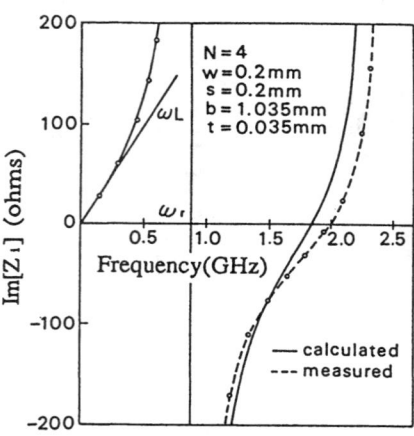

Fig.7 Frequency Responses of the Experimental Spiral Inductor.

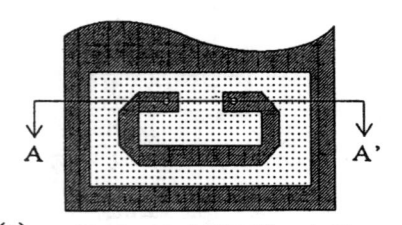

(a) Embedded RF Circuit Pattern

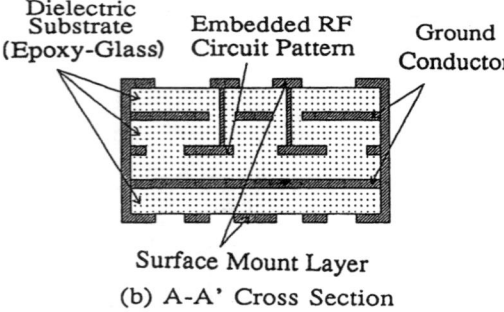

(b) A-A' Cross Section

Fig.8 Structure of RF Circuit Modules using an M-PCB.

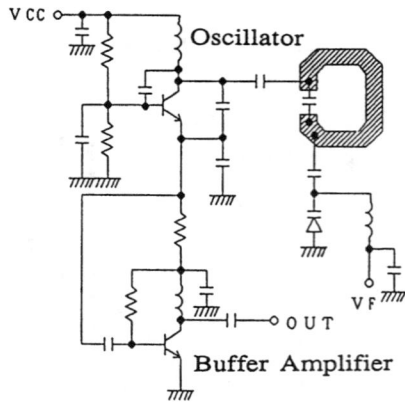

Fig.9 Circuit Configuration of a Trial VCO.

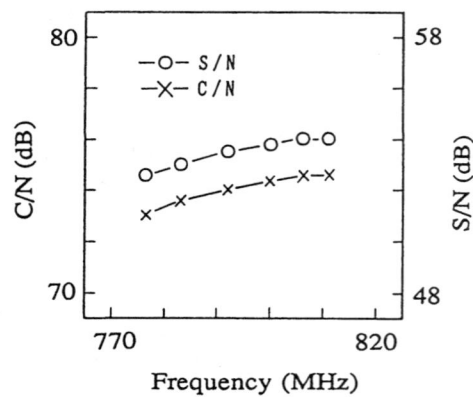

Fig.10 CNR and SNR of a Trial VCO.

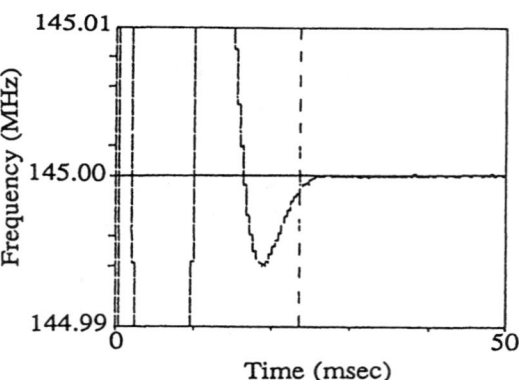

Fig.11 Lock-up time of the PLL Modulator Module.

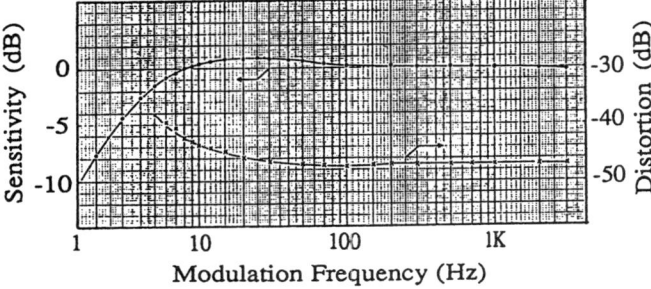

Fig.12 Frequency Responses of the PLL Modulator Module.

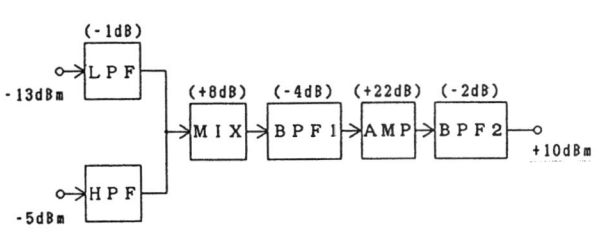

Fig.13 Configuration of an Up-Converter Module.

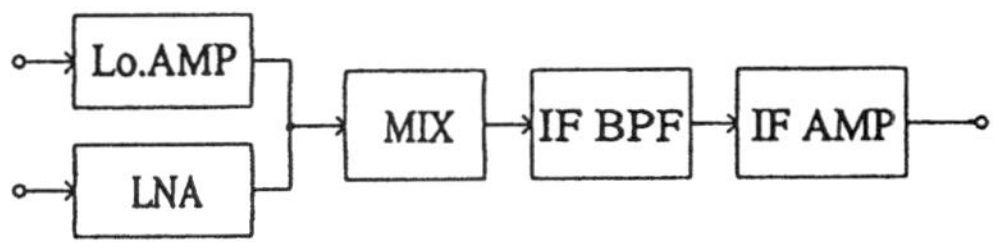

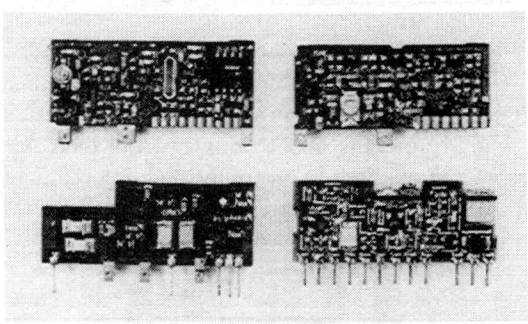

Fig.15 Configuration of the Experimental Receiver Front-end Module.

Fig.16 Photograph of Trial RF Circuit Modules.

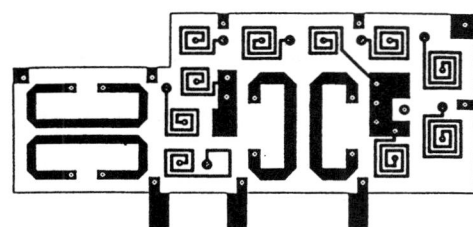

Fig.14 Structure of embedded RF circuit layer (Up-conveter Module)

Miniaturized SAW Devices for Radio Communication Transceivers

MITSUTAKA HIKITA, MEMBER, IEEE, TOYOJI TABUCHI, AND ATSUSHI SUMIOKA

Abstract—Wide applications of surface acoustic wave (SAW) technologies are investigated to reduce the volume and weight of recent and future VHF/UHF radio communication transceivers. Insertion loss and power improvements for SAW filters, one of the most successful devices within the above technologies, are examined theoretically and experimentally. By analyzing loss mechanisms of the previously developed high-performance SAW filter for mobile telephones, a new extremely low-loss SAW filter (loss of as low as 1.2–2.0 dB at 600 MHz) is achieved. The results of fundamental experiments for a SAW antenna duplexer using a new SAW filter as the transmitter filter and a semicoaxial filter as the receiver filter at 800 MHz are also presented.

I. INTRODUCTION

IN THE NEAR future, the provision of voice and data to people who are moving about or away from their wireline telephones will be required as one of the major communication means [1], [2]. The goal for such communication systems is to provide high-quality service with low-power small radio telephones and data terminals [3].

Section II below discusses several types of radio communication services, such as the local area radio network (LARN), mobile radio, radio paging, and radio data terminals. The main concern in these systems is reducing the size of radio transceiver units, that is, from fixed transceivers to hand-held (portable) or pocketable ones.

SAW technologies offer miniature devices by their very nature. The SAW wavelength is 10^{-5} times smaller than the same frequency electromagnetic wavelength [4], [5]. Section III investigates key technologies for achieving small radio transceiver units. SAW technologies also provide the most leading devices that can reduce the size of transceiver units.

SAW filters are the most successful devices, because many other types of filters, such as semicoaxial filters and *LC* filters, are most limited in their uses due to large volumes. However, SAW applications to communication equipment have been restricted because of their large insertion losses. We have previously developed a high-performance SAW filter for an 800-MHz mobile telephone transceiver [6]. This filter suppresses the spurious signals generated in a transmitter mixer. Section IV analyzes the loss mechanisms of this SAW filter. Section V investigates further loss reduction and high-power characteristics to develop a new filter configuration.

An antenna duplexer is used in two-way transceivers that simultaneously transmit and receive radio frequency signals.

Manuscript received January 4, 1988, revised June 1, 1988.
M. Hikita and T. Tabuchi are with the Central Research Laboratory, Hitachi, Ltd., Kokubunji, Tokyo, 185 Japan.
A. Sumioka is with Hitachi Denshi, Ltd., Kodaira, Tokyo, 817 Japan.
IEEE Log Number 8927113.

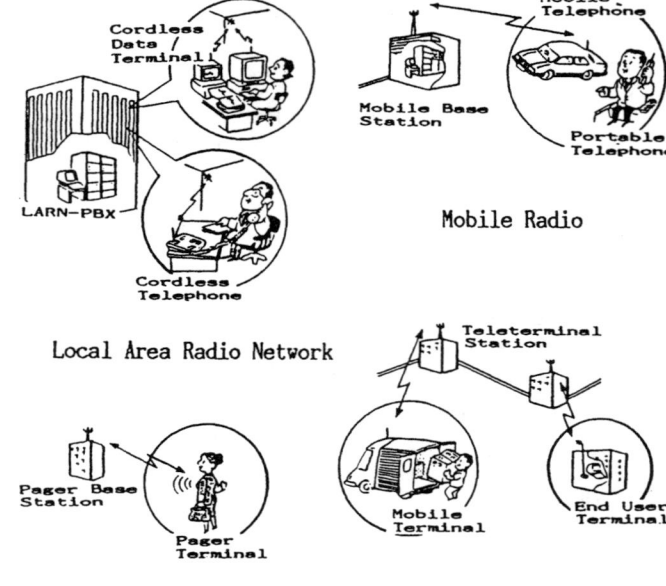

Fig. 1. VHF/UHF radio communication systems for people moving about.

New electrical characteristics for interdigital-transducer (IDT)—that is, reactive characteristics in mutual frequency bands (receiver and transmitter bands), are required in a duplexer. Section VI shows that these characteristics are possible using previously developed IDT weighting techniques. Fundamental experiments with a SAW duplexer are also presented. A new SAW filter is used as the transmitter filter and a semicoaxial filter as the receiver filter.

II. RADIO COMMUNICATIONS FOR PEOPLE MOVING ABOUT

Examples of popular VHF/UHF radio communication systems for people moving about are shown in Fig. 1. These systems require such miniature devices as SAWs, MMICs, LSIs, etc., to reduce the size of radio transceiver units. The features of these systems are as follows.

1) The local area radio network (LARN) will become an extremely useful local communication system in the near future. Entry/exit radio ports and the LARN-PBX (private branch exchange) will be installed in a large building to provide wireless voice and data transmission [1], [2]. Thus the implementation of loop radio links eliminates the need for wiring to individual telephones, data terminals, etc. The transmitter power from radio transceiver units in these systems is less than 10 dBm (10 mW).

TABLE I
TECHNOLOGIES FOR SMALL-SIZED RADIO COMMUNICATION TRANSCEIVER

Circuit elements	Miniaturization Technologies
RF (50 MHz–3 GHz)	
Antenna	Inverse F type antenna
	Microstrip antenna
	Directional antenna
Antenna duplexer filter	SAW filter- ①
Low noise amplifier	MMIC (Si, GaAs)
Mixer	
Voltage controlled oscillator (VCO)	SAW resonator- ②
	MMIC (Si, GaAs)
High-power amplifier	Harmonic control circuit (GaAs)
IF (500 KHz–500 MHz)	
Amplifier	Monolithic IC
Detector	LSI
Filter	Crystal filter
	SAW filter- ③
	DSP
Modulator	Monolithic IC
Demodulator	DSP
	SAW modulator- ④

2) Mobile radio service including cellular radio already exists in almost all advanced countries. It is expected to become worldwide in the near future. In these advanced countries the use of portable telephones is now more extensive than that of mobile telephones, which are fixed within the vehicles [3]. The transmitting power from radio transceiver units in cellular radio systems is 0.6–4 W. Battery volume and weight are severely restricted, especially in portable telephones. Thus cellular radio requires devices that are not only high-performance but also miniature.

3) Radio paging means one-way call systems which are very popular all over the world. The size of radio transceiver units has been greatly reduced. In the near future, a lightweight card-type pager will appear, in which miniature devices will play increasingly important roles.

4) Radio data terminals are one of the future means of data transmission. Teleterminal stations relay data between end user terminals and terminals of people moving about. Of course, mobile terminals and portable terminals are also possible in these systems. Therefore, the size reduction of the transceiver units is as essential here as in other systems.

III. POSSIBILITIES FOR MINIATURE SAW DEVICES

SAW technologies offer advantages in that they are small, do not need adjustments, and are highly reproducible and reliable. This section investigates possibilities for device miniaturization using SAW technologies for radio transceiver units.

Miniaturization technologies for the radio transceiver circuit elements, including SAW and others, are summarized in Table I. A schematic diagram of a well-known two-way radio transceiver unit, in which the above corresponding circuit elements are pointed out, is shown in Fig. 2.

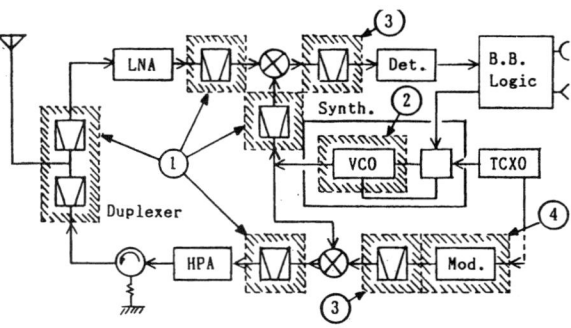

Fig. 2. Applications of SAW technologies to two-way radio transceiver.

1) For a mobile telephone transceiver unit, conventional cascade-connected semicoaxial resonators have been used in the antenna duplexer, transmitter noise suppression filter, receiver second filter, and local filter. These filters are assembled on a mother board together with other active and passive circuit elements. Thus RF filters have occupied very large circuit area. They must be replaced by SAW filters.

2) Either hybrid microwave integrated circuit (MIC) resonators with half- or quarter-wavelength, or semicoaxial resonators are used in voltage-controlled oscillators (VCOs) to generate high carrier-to-noise ratio (C/N) RF signals. A SAW-VCO with high C/N is of particular interest because of its small device size. One problem in achieving SAW-VCO's for radio communication equipment is achieving a wide-frequency variation bandwidth [7]. However, this point will be overcomed, and miniature SAW-VCOs using monolithic technologies will be available in the near future.

3) Conventional LC filters are used to suppress spurious signals in the transmitter modulation phase-locked loop (PLL) IF circuit. In general, crystal filters are used for the receiver IF circuit. However, these filters may also be replaced by SAW filters, especially when rather high intermediate frequencies are used (above 50 MHz).

4) The possibilities for SAW modulators are also pointed out. Most of the recent radio communications use FM or PM with analog modulation signals. However, digital radio communication services will appear in the near future. Code-division multiple-access communications, such as spread-spectrum communication systems, will also be available to the radio communications for people moving about. SAW-based modulators, which produce the desired pseudorandom code sequences, will become very powerful devices [8]. Of course, SAW convolvers/correlators are also very useful devices in spread-spectrum communication systems [9], [10]. However, these signal-processing SAW devices will be proposed and discussed in other papers.

Of the above SAW applications, item 1) is the main focus of this paper. We investigate low-loss high-performance high-power SAW filters that will replace not only simple semicoaxial filters but also antenna duplexers. We have already developed two types of high-performance SAW filters for a mobile telephone transceiver. One is a UHF (830 MHz) filter that can replace the semicoaxial noise suppression filter for transmitter circuits [6]. The other is a VHF (90 MHz) filter that can replace the LC filter for modulation PLL IF circuit.

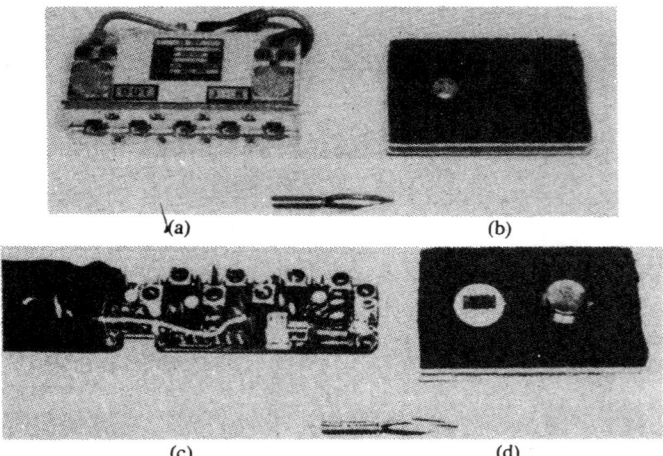

Fig. 3. Examples of VHF/UHF miniature high-performance SAW filters. (a) 830-MHz coaxial filter. (b) 830-MHz SAW filter. (c) 90-MHz coil filter. (d) 90-MHz SAW filter.

Photographs of the two filters are shown in Figs. 3(a) and (b). A UHF filter, which is mounted on a TO-5 package, has almost the same performance as that of the conventional semicoaxial filter. The size of the SAW filter is about 1/100 that of the conventional filter. A VHF filter, which is mounted on a TO-8 package, is designed to have 75 Ω pure resistance for input and output impedances. The size of the SAW filter is about 1/10 that of the conventional LC filter.

However, these developed SAW filters do not perform well enough to use in an antenna duplexer for two-way transceiver units. An antenna duplexer requires further loss reduction, high-power characteristics, and reactive IDT impedances for the filter's stopband.

The loss mechanisms for the above high-performance SAW filter are analyzed in the next section. A new filter configuration which achieves further loss reduction and high-power characteristics is then examined.

IV. Loss Analysis of High-Performance SAW Filter for Mobile Telephone

We proposed an image-impedance connection of IDTs to achieve the sharp cutoff frequency responses required for mobile telephones [6]. Three types of high-performance SAW filter configurations using the image-impedance connection of IDTs are schematically shown in Fig. 4. Fig. 4(a) shows a four-time laterally repeated structure for image-impedance connected IDTs. Fig. 4(b) shows a nonlaterally repeated structure with reflectors at both sides. Fig. 4(c) shows a configuration similar to Fig. 4(b) but using unidirectional IDTs with U-shaped multistrip couplers (MSCs) [11] instead of lateral repetition or reflectors.

We achieved a UHF filter using the configuration in Fig. 4(a) and a VHF filter using that in Fig. 4(b). Filter patterns for these filters are shown in Fig. 5. In the actual patterns, previously developed new phase weighting is introduced in image-impedance connected IDTs [6]. Frequency characteristics for these filters are shown in Figs. 6(a) and (b). An insertion loss of as low as 3.5–4.0 dB is achieved for the UHF filter, and 1.7–2.0 dB for the VHF filter.

The configuration in Fig. 4(c) also offers a high-performance broad-band filter. We have already achieved a broad-band filter with bandwidth of over ten percent (−3 dB down) and insertion loss of 3 dB using this configuration [12].

Before developing a new filter configuration for a duplexer, we analyzed the loss mechanisms of a UHF SAW filter. In general, SAW filters have three main losses. They are leakage (bidirectionality), weighting, and thin Al electrode conductivity losses. Bidirectionality loss is largest in conventional transversal SAW filters (more than 6 dB). To overcome this loss, an energy trap resonant structure with lateral repetitions is introduced in addition to SAW reflectors. To reduce weighting loss, new phase weighting, which excites SAWs with uniform wavefront distribution in the passband, is introduced to the image-impedance connected IDTs.

The loss breakdown for this filter is summarized in Table II. We estimated bidirectionality and weighting losses using a computer simulation. New phase weighting causes no increase of loss due to weighting. The Al conductivity loss was estimated by measuring temperature dependence of insertion loss. This is because other losses are much less temperature-dependent than is the loss due to thin Al (0.1 μm) film conductivity. The propagation loss was estimated separately using a propagation-loss measurement special pattern having three IDTs with different propagation lengths. Bulk wave radiation loss was also estimated using computer simulation procedures by introducing measured bulk wave radiation conductances into the equivalent circuit representation for image-impedance connected IDTs [6].

Due to the reduction of bidirectionality and weighting losses, thin Al electrode conductivity loss accounts for nearly half of all losses. Leakage (bidirectionality) loss is second.

Thus to achieve a low-loss SAW filter for a duplexer, three points must be considered: 1) further reduction of leakage loss; 2) reduction of Al electrode conductivity loss; and 3) high-power characteristics.

V. New Configuration with Low-Loss, High-Power Characteristics

A. Reduction of Leakage Loss

To reduce leakage loss as much as possible, we developed a new repetition structure for IDTs with a tapering number of finger pairs. One pair of input and output IDTs is laterally repeated. They have a tapering number of finger pairs. That is, there is a larger number of finger pairs in the center of the filter and a smaller number to both sides. This configuration is schematically shown in Fig. 7. Thus the energy of acoustic vibrations is confined within the filter.

Generally, in the repeated structure, the leakage loss decreases as the number of repetitions increases. However, this loss decrease is not very remarkable, as is shown in Fig. 8(a) [13]. For example, over nine repetitions are needed to achieve a 0.7-dB loss.

The losses of a new IDT repeated structure with only a simple taper in the number of finger pairs (half the number of finger pairs at both sides) are shown in Fig. 8(b). In this case, six repetitions are enough to achieve 0.7 dB loss.

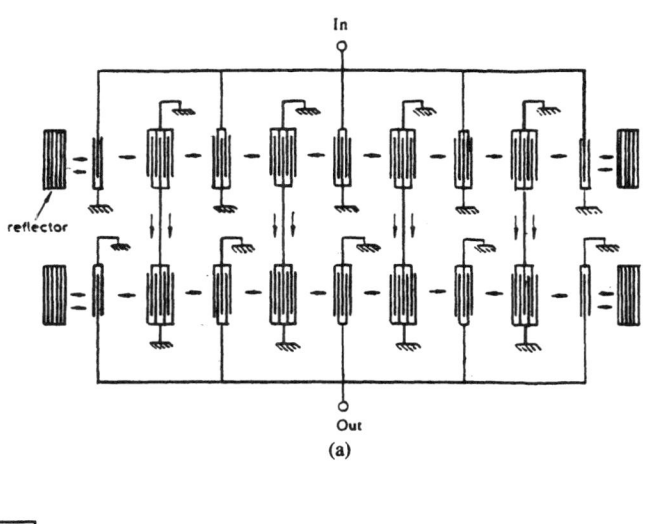
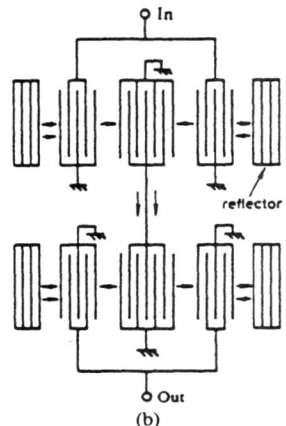
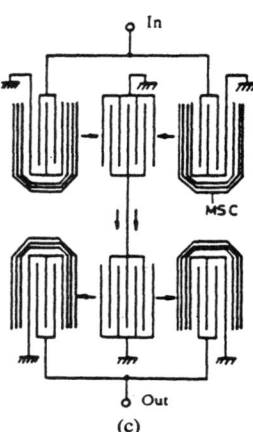

Fig. 4. Fundamental configurations for high-performance SAW filters using image-impedance connected IDTs. (a) Four-time laterally repeated structure. (b) Nonlaterally repeated structure with reflectors at both sides. (c) Nonlaterally repeated structure with unidirectional IDTs having U-shaped MSCs.

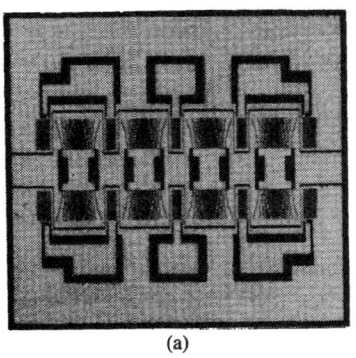
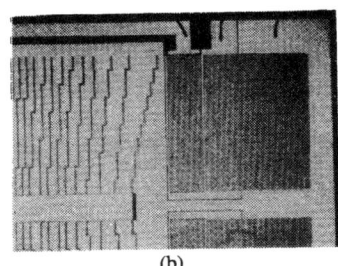

Fig. 5. Filter patterns for high-performance SAW filters. (a) UHF filter. (b) VHF filter.

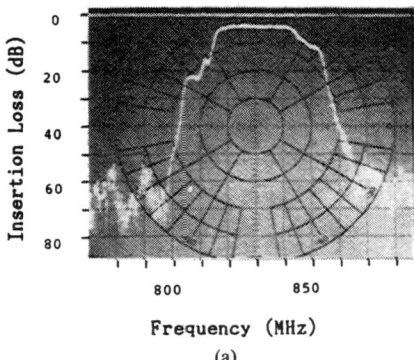
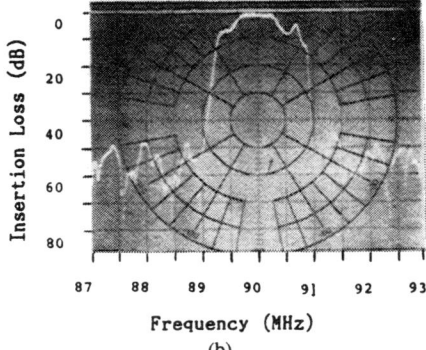

Fig. 6. Frequency characteristics for high-performance SAW filters. (a) UHF filter. (b) VHF filter.

TABLE II
BREAKDOWN OF LOSSES FOR HIGH-PERFORMANCE SAW FILTER (IN dB)

Bidirectionality loss	0.3~1.3
Weighting loss	0
Al conductivity loss	1.8~2.0
Propagation loss	0.3
Bulk wave loss	0.2~0.3
Other losses	0.1~0.4
Total loss	3.5~4.0

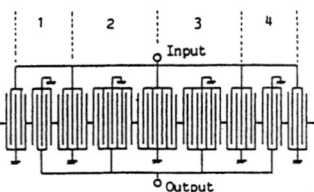

Fig. 7. New repetition structure for IDTs with tapered number of finger pairs.

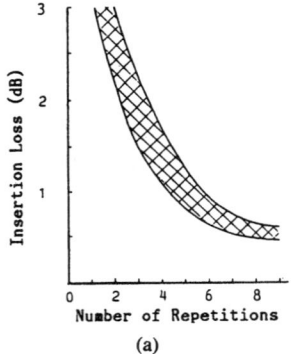

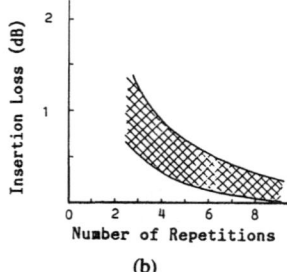

Fig. 8. Insertion loss versus number of repetitions for IDTs. (a) Nontaper in number of finger pairs. (b) Simple taper in number of finger pairs (half number of finger pairs at both sides).

B. Reduction of Conductivity Loss

Three types of conductivity loss occur in a SAW filter: 1) losses in the IDT fingers; 2) losses in lines drawn to bonding pads; and 3) losses in lead wires and matching circuits.

To reduce the first, the filter aperture must be small. From this standpoint, our repeated structure is superior to that of conventional SAW filters, because several IDTs are connected in parallel.

To reduce the second type of loss, the impedance of each repeated IDT must be increased. As for 3), it causes a 0.2-0.3-dB loss increase, by computer simulation results, which is negligible compared with 1) and 2).

From these considerations, low-conductivity losses can be

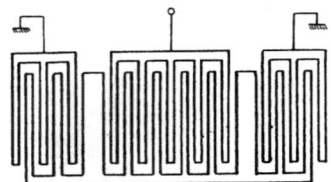

Fig. 9. High-impedanced IDT with serial connections.

realized by using a repeated structure of high-impedance IDTs having small apertures. In the repeated structure, the resistances of lines drawn from each IDT to bonding pads are several ohms. To reduce the losses caused by these resistances to less than 0.1-0.2 dB, each IDTs impedance must be two orders of magnitude higher than these resistances. We made high-impedance IDTs by dividing each IDT into several parts and introducing serial connections, as shown in Fig. 9.

C. High-Power Characteristics

The repeated structure shown in Fig. 7 provides not only low insertion losses but also high-power characteristics. This is because, within the repeated structure, acoustic energy is spread over large areas. Thus energy density in each IDT can be decreased. This configuration is very similar to that of high-power transistors or FETs.

D. Experimental Results

In the experiment, ninefold repeated structure with a simple tapering number of finger pairs (half the number of finger pairs at both sides) was used. The experimental results are shown in Fig. 10. The substrate used was 36° YX-LiTaO$_3$ (surface shear wave) [14]. The center frequency was about 600 MHz. Insertion loss of as low as 1.2-2.0 dB was obtained (Fig. 10(a)). This level is by no means inferior to that of semirigid coaxial filters. However, off-band rejection is not as good as with conventional transversal SAW filters (Fig. 10(b)). This is because the filter has no weighting.

VI. APPLICATION TO ANTENNA DUPLEXER

An antenna duplexer consists of two filters connected in parallel via appropriate transmission lines. This is schematically shown in Fig. 11. Such a duplexer provides two functions: filtering and diverging. It ensures that transmitter power does not flow into the receiver and receiver signals do not suffer attenuation by flowing into the transmitter.

Thus a duplexer requires: 1) low loss (below 2 dB), high-power characteristics for the transmitter filter; 2) low-loss (below 4 dB), high-sidelobe suppression for the receiver filter (due to the noise figure and spurious signals); and 3) reactive characteristics for input impedances of the filters in the mutual frequency bands (due to the parallel connection of the two filters).

In a conventional duplexer, each filter is constructed with semicoaxial resonators (four to seven resonators for each filter) containing high-dielectric ceramics. Thus the duplexer is the largest circuit component in a two-way transceiver, such as a portable or mobile telephone.

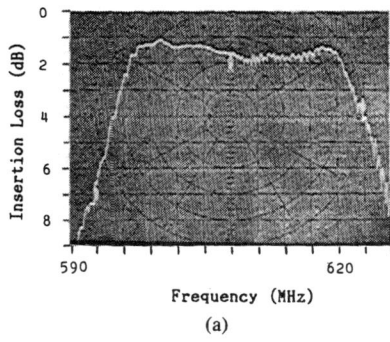 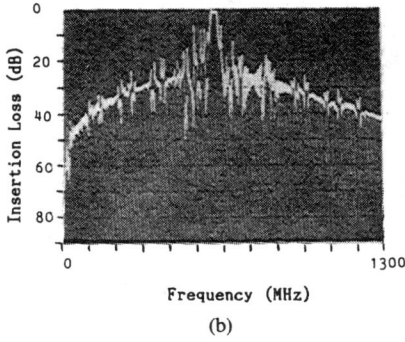

Fig. 10. Experimental results for low-loss SAW filter with new repetition structure. (a) Passband characteristics. (b) Out of band characteristics.

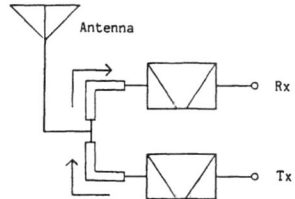

Fig. 11. Antenna duplexer.

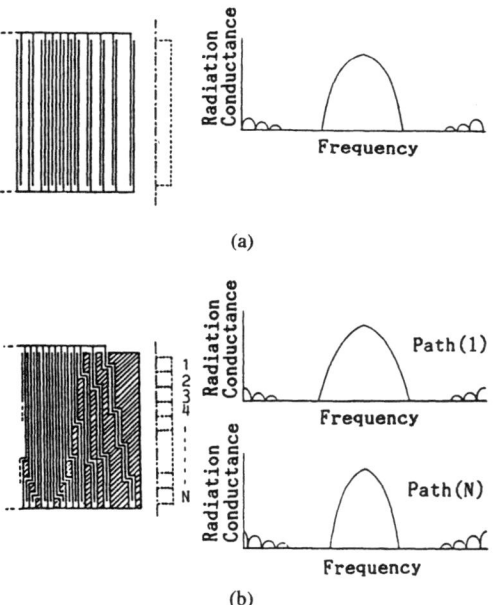

Fig. 12. Radiation conductances of IDTs versus weightings. (a) Withdrawal weighting. (b) New phase weighting.

A. Reactive Characteristics for Stopband

One of the differences between conventional SAW filters and semicoaxial filters is the impedances of the filters in the stopband. In SAW filters, the frequency characteristics in the stopband near the passband are formed by dispersion of acoustic energies into the substrate. Thus the impedance of such a filter contains rather large resistive components in the stopband. Low-loss parallel connection of two filters via appropriate transmission lines is possible for semicoaxial filters, but it is difficult for conventional SAW filters. A low-loss SAW filter having reactive characteristics in the stopband must be developed.

The main reason for nonreactive characteristics in the stopband in conventional SAW filters is apodization (amplitude weighting). This is the most popular means of weighting to synthesize the required frequency responses [5]. Our computer simulation showed that withdrawal weighting [15] can provide reactive characteristics in the stopband near the passband. This is schematically shown in Fig. 12(a). However, withdrawal weighting is not suitable for synthesizing fine frequency responses.

The new phase weighting technique [6] which we published previously can also provide reactive characteristics in the stopband. This weighting provides almost the same performance as that of amplitude weighting.

Radiation conductances of each section of the new-phase-weighted IDT are very similar to those of a withdrawal-weighted IDT, as shown in Fig. 12(b). This is because the new-phase-weighted IDT is essentially constructed with transverse superpositions of withdrawal-weighted IDTs with different weighting functions. Computer simulation shows that reactive frequency bands are closely related to weighting functions. Thus it is very important to choose the appropriate weighting functions, considering the required frequency responses and the allocations of the mutual passbands (receivers and transmitter bands).

B. Fundamental Experiments for Duplexer

Results of a fundamental experiment in parallel connection of two filters at 800 MHz are shown in Fig. 13. In this experiment, we used a SAW filter whose fundamental structure is the same as that of Fig. 10 for the transmitter filter, and a semicoaxial filter as the receiver filter. The two filters were connected to an antenna terminal via appropriate transmission lines $\simeq \lambda/4$). Insertion loss for the SAW filter was as low as 2.0 dB.

New phase weighting with a Hamming function as the weighting function was introduced in this SAW filter. The impedance of the SAW filter (measured from the antenna terminal) was about 400 Ω within the receiver band. Thus connection loss in the receiver semicoaxial filter due to energy flowing into the SAW filter was less than 0.5 dB. This loss can be improved by making the impedance of the SAW filter

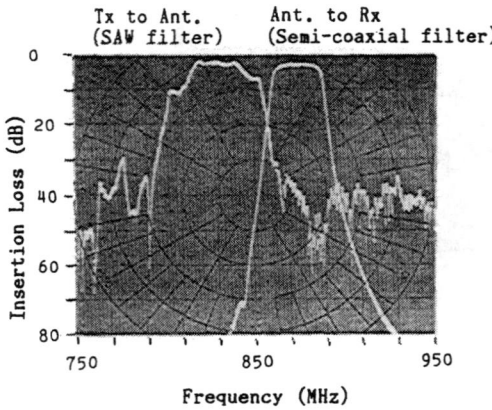

Fig. 13. Frequency characteristics for parallel connection of two filters.

higher within the receiver band. A connection loss of 0.2 dB is possible if 1 kΩ is achieved. No other effects on frequency responses due to connection are observed.

Fig. 13 implies not only the feasibility of a miniature antenna duplexer for a two-way transceiver using SAW filters, but also the possibility of new miniature functional devices combined with other circuit elements.

VII. Conclusion

Recent and future VHF/UHF radio communication systems, in which small-sized transceiver units are strictly required, are investigated. It is proved that SAW devices will be a leading technology to meet the above requirements.

Loss mechanisms of the previously published high-performance SAW filter for a mobile telephone are analyzed. Based on the results, extremely low-loss (1.0–1.2 dB at 600 MHz), high-power SAW filters with new repetition structures are achieved.

The previously contrived new phase weighting procedure provides low-loss frequency synthesization, as well as reactive impedance characteristics for IDT in the stopband near the passband of the filter. This ensures the low-loss parallel connection of two filters required in a duplexer.

Fundamental experiments for a SAW antenna duplexer using a SAW filter as the transmitter filter and a semicoaxial filter as the receiver filter show the possibilities of new miniature functional devices, which will be more important in achieving much larger integration for RF circuits in radio communication transceiver units.

Acknowledgment

The authors gratefully acknowledge the many useful discussions and considerable help of Mr. Y. Fujiwara of Hitachi Denshi Ltd., and Mr. T. Toyama of Hitachi Ltd.'s Yokohama Works. Thanks are also due to Mr. H. Kojima of Hitachi Shounan Denshi Ltd. for the Al deposition process to make devices.

References

[1] D. C. Cox, "Universal portable radio communications," *IEEE Trans. Veh. Technol.*, vol. VT-34, p. 117, 1985.
[2] D. C. Cox, H. W. Arnold, and P. T. Porter, "Universal digital portable communications: an applied research perspective," presented at the IEEE ICC 1986, Toronto, ON, Canada.
[3] S. Crump, Jr., "Cellular portable update: What's new and what's coming," *Personal Commun. Technol.*, p. 15, 1986.
[4] *Proc. IEEE (Special Issue on Surface Acoustic Wave Devices and Applications)*, vol. 64, 1976.
[5] A. J. Slobodnik, Jr., T. L. Szabo, and K. R. Laker, "Miniature surface-acoustic-wave filter," *Proc. IEEE*, vol. 67, p. 129, 1979.
[6] M. Hikita, H. Kojima, T. Tabuchi, and Y. Kinoshita, "800-MHz high-performance SAW filter using new resonant configuration," *IEEE Trans. Microwave Theory Tech.*, vol. MTT-33, p. 510, 1985.
[7] J. Ladd, C. Abdallah, and T. O'Shea, "A temperature compensated L-band hybrid SAW oscillator and resonator filter," in *IEEE Ultrasonics Symp. Proc.*, 1984, p. 191.
[8] M. A. Belkerdid and D. C. Malocha, "Amplitude-weighted quadrature phase shift keying using SAW technology," *IEEE Trans. Sonics Ultrasonics*, vol. SU-32, p. 791, 1985.
[9] L. B. Milstein, J. Gevargiz, and P. K. Das, "Rapid acquisition for direct sequence spread-spectrum communication using parallel SAW convolvers," *IEEE Trans. Commun.*, vol. COM-33, p. 593, 1985.
[10] M. E. Motamedi, M. K. Kilcoyne, and R. K. Asatourian, "Large-scale monolithic SAW convolver/correlator on Silicon," *IEEE Trans. Sonics Ultrasonics*, vol. SU-32, p. 663, 1985.
[11] F. G. Marshall, C. O. Newton, and E. G. S. Paige, "Surface acoustic wave multistrip components and their applications," *IEEE Trans. Microwave Theory Tech.*, vol. MTT-21, p. 216, 1973.
[12] M. Hikita, H. Kojima, T. Tabuchi, and Y. Kinoshita, "New low-loss, broad-band SAW filter using unidirectional IDTs with U-shaped MSCs," *Electron. Lett.*, vol. 20, p. 453, 1984.
[13] M. Lewis, "SAW filter employing interdigitated interdigital transducers, IIDT," in *IEEE Ultrasonics Symp. Proc.*, 1982, p. 12.
[14] K. Nakamura, M. Kazumi, and M. Shimizu, "SH-type and Rayleigh-type surface wave on roted Y-cut LiTaO$_3$," in *IEEE Ultrasonics Symp. Proc.*, 1977, p. 819.
[15] K. Kogan and P. Romik, "SAW bandpass filter with withdrawal weighting transducers," in *IEEE Ultrasonics Symp. Proc.*, 1980, p. 302.

RECENT DEVELOPMENT OF DIELECTRIC RESONATOR MATERIALS AND FILTERS IN JAPAN

K. WAKINO

Murata Manufacturing Co., Ltd., Nagaokakyo-shi, Kyoto 617, Japan

(Received August 29, 1988)

Since 1970s, many kinds of dielectric resonator materials with improved Q values and improved temperature coefficients have been developed. Representative of them are the systems, $(Ba,Sr,Ca)(Zr,Ti)O_3$, $MgTiO_3$-$CaTiO_3$-$La_2O_3 \cdot TiO_2$, $Ba_2Ti_9O_{20}$, $(Zr,Sn)TiO_4$, $(Sr,Ca)[(Li,Nb),Ti]O_3$, and BaO-Nd_2O_3-TiO_2. In the past 15 years, the Q values of these materials have gradually been improved.

Using these new materials, dielectric resonator filters and oscillators are being developed which are more compact and applicable to high power operation. Their main applications are cellular mobile radio systems and SHF-TV systems using direct broadcasting satellite.

This paper will present the recent development in Japan of resonator materials and filters.

INTRODUCTION

The biggest advantage of using dielectric resonator is the size reduction of microwave components that they make possible. This reduction depends on the phenomenon that the wavelength of an electromagnetic wave is shortened to $1/\sqrt{K}$ (K: relative dielectric constant) in dielectrics compared with a wavelength in free space.

The word "dielectric resonator" was described in a paper by Richtmyer in 1939, where he showed theoretically that ring shaped dielectrics could work as resonators.[1]

In the 1960s, many pioneers investigated the behavior of dielectrics at microwave frequencies and applied them to the microwave filters. For example, the microwave loss mechanism in $SrTiO_3$ was measured and discussed by Rupprecht et al.,[2] far infrared dielectric dispersion was investigated by Spitzer et al.,[3] and a measurement method for dielectric characteristics was developed by Hakki and Coleman.[4,38]

A microwave filter using TiO_2 ceramics was also designed by Cohn,[5] and the others. But this filter was not put into practical use because the temperature coefficient of the TiO_2 ceramics was large at about 450 ppm/°C.

Research and development for temperature stable dielectric resonator materials started about 1970 worldwide. And in the present decade, the Q values of these materials have been remarkably improved. Dielectric resonator materials with dielectric constants from 20 to 90 are now available.

Along with material development, new designs and techniques have been developed and applied to dielectric filters. These filters were put into practical use, and realized size reduction and cost reduction of microwave components.

The state of recent development of resonator materials and filters in Japan is described in this paper.

Dielectric Resonator Materials

Figure 1 shows the patent applications for dielectric resonator materials filed in Japan between 1970 and September 1985, which were presented by JEIDA (Japan Electronic Industry Development Association) in March 1988. 232 patents including three from foreign companies were applied for during the period. One patent from a foreign company in 1975 was added by the author, although JEIDA omitted it as it was applied for as a device patent. These patents were classified into several groups according to material systems by JEIDA. The result is shown in Figure 2.

Table I shows the characteristics of the resonator materials presented since 1970. In the table, τ_f is the temperature coefficient of resonant frequency, and is given by the following equation.

$$\tau_f = -\frac{1}{2}\tau_k - \alpha, \tag{1}$$

where, τ_k: temperature coefficient of dielectric constant frequency
   α: thermal expansion coefficient of dielectrics.
Q is the reciprocal of dielectric loss tangent, $\tan\delta$.

It is predicted from the classical dispersion theory that the dielectric constant is constant at microwave frequency and that dielectric loss increases proportionately with frequency.[20] Therefore the value (Q × freq.) is inherent in each material. Figure 3 shows the frequency dependence of K and Q for some of the material. In this figure, the value (Q × f) of (Zr,Sn)TiO$_4$ or MgTiO$_3$-CaTiO$_3$ material deteriorates at lower frequency. This is because the resonators with lower frequency have larger dimensions (for example, the size of 54 mm dia. × 24 mm thickness is needed for 1 GHz on the case of a K38 resonator, while, sizes are smaller than 10 mm dia. × 4 mm thickness for frequency higher than 6 GHz), and because the large resonators are apt to have lattice imperfections such as oxygen vacancies in

TABLE I
Dielectric resonator materials.

Material	K	Q	(ppm/°C) τ_f	(GHz) f_o	Ref.
MgTiO$_3$-CaTiO$_3$	21	8,000	0	7	6
Ba(Sn,Mg,Ta)O$_3$	25	20,000	0	10	7
Ba(Zn,Nb)O$_3$-Ba(Zn,Ta)O$_3$	30	14,000	0	12	8
Ba(Zr,Zn,Ta)O$_3$	30	10,000	0	10	9
(Ca,Sr,Ba)ZrO$_3$	30	4,000	5	11	10
BaO-TiO$_2$-WO$_3$	35	8,400	0	6	11
(Zr,Sn)TiO$_4$	38	7,000	0	7	12
Ba$_2$Ti$_9$O$_{20}$	40	8,000	2	4	13
(Sr,Li)[(Li,Nb)Ti]O$_3$	42	3,500	0	9	14
BaO-Sm$_2$O$_3$-5TiO$_2$	77	4,000	15	2	15
BaO-PbO-Nd$_2$O$_3$-TiO$_2$	90	5,000	0	1	12

Q: reciprocal of dielectric loss tangent, $\tan\delta$.
τ_f: temperature coefficient of resonant frequency.
f_o: measurement frequency.

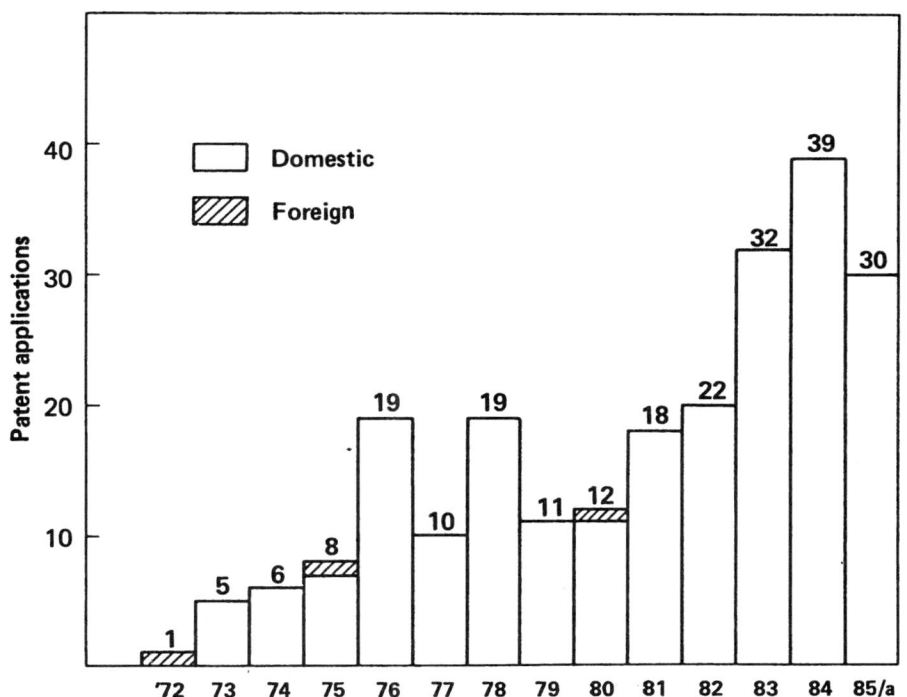

FIGURE 1 Patent applications of dielectric resonator material filed in Japan between 1970 and 1985.

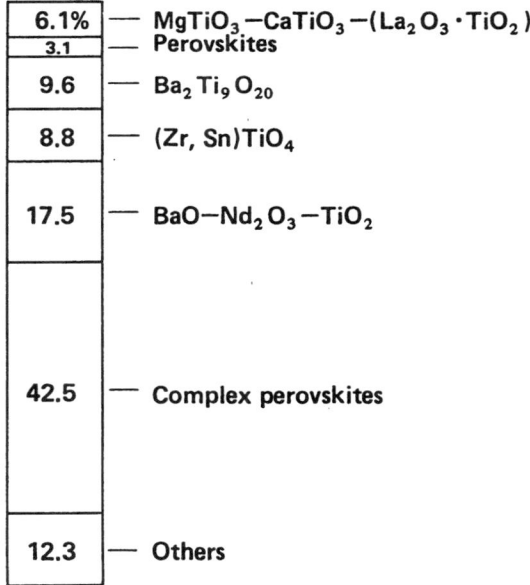

FIGURE 2 Amount ratio of patent applications classified according to material system.

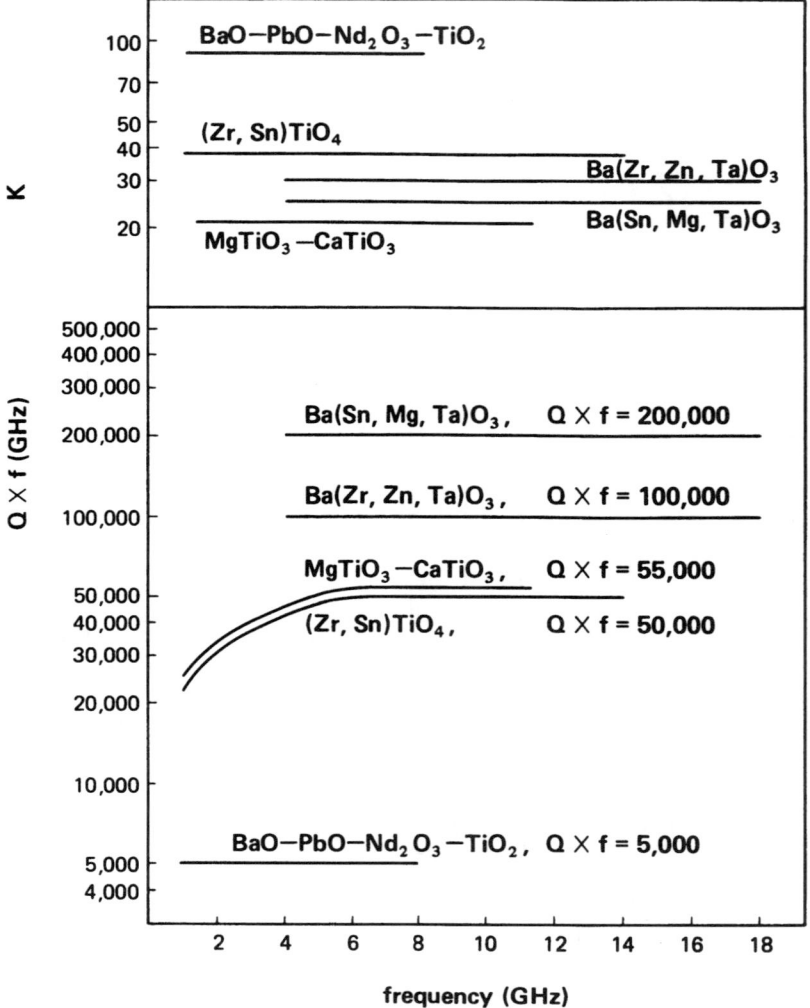

FIGURE 3 Frequency dependence of k and Q value of dielectric resonator materials.

the core of the dielectric blocks. The lattice imperfections deteriorate Q value more than expected from the equation, $Q \times f$ = constant.

Figure 4 shows the temperature dependence of the resonant frequency and dielectric loss for three resonator materials. The $(Zr,Sn)TiO_4$ material has good linearity of temperature dependence of resonant frequency compared with the other two materials. The dielectric loss for these materials increases proportionately with temperature.

The recent development in Japan for the materials shown in Figure 2 is as follows.

1) $MgTiO_3$-$CaTiO_3$ material is well known as the material for temperature com-

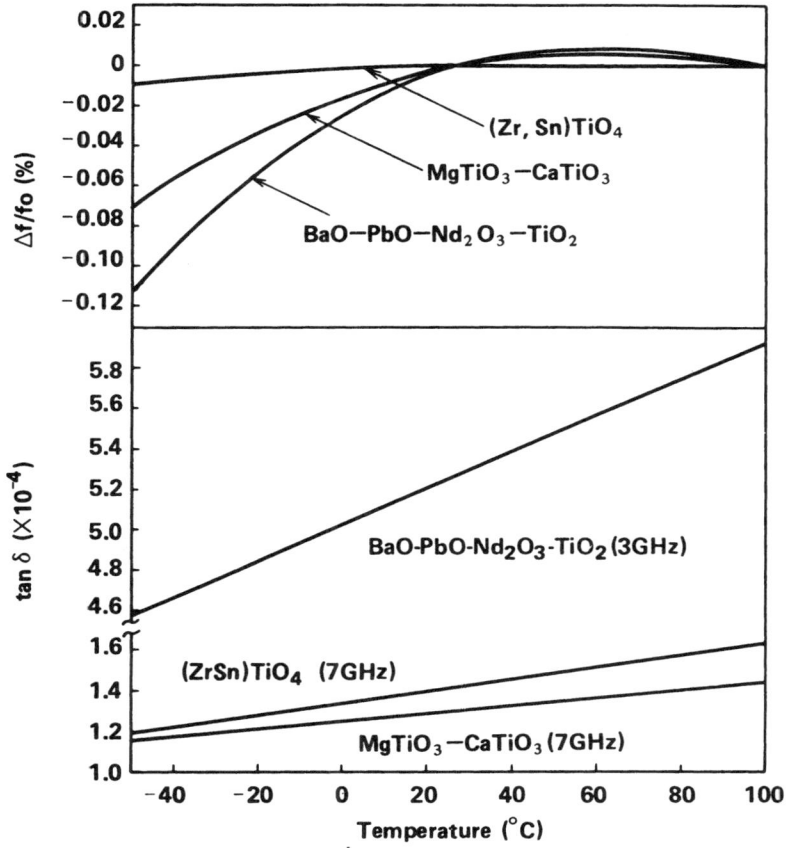

FIGURE 4 Temperature dependence of resonant frequency and dielectric loss tangent.[24]

pensating type capacitors. This material is made of a mixture of $MgTiO_3$ ($\tau_f = -50$ ppm/°C) and $CaTiO_3$ ($\tau_f = 450$ ppm/°C). Figure 5 shows the characteristics at 7 GHz of this system. Its dielectric constant can be increased by the addition of $La_2Ti_2O_7$ or $Nd_2Ti_2O_7$, although its Q value drops.

2) $(Ba,Sr)(Zr,Ti)O_3$ material is the material presented first as dielectric resonator material.[16] $(Ca,Sr,Ba)ZrO_3$ material was recently presented with improved Q value as shown in Figure 6.[10]

3) $Ba_2Ti_9O_{20}$ material was reported in 1974 as a high-K and high-Q resonator material.[13] This system was modified by the addition of WO_3 to increase Q value. The result is shown in Figure 7.

4) $(Zr_{0.8}Sn_{0.2})TiO_4$ material has a high Q value and good temperature stability. The characteristics of this system are shown in Figure 8.[6] Its phase transition was presented in 1981.[17] The Q value of this material was improved by modifications to the composition, additives and processing as shown in Figure 9. This improvement is remarkable because Q was improved at lower frequency. The reason for

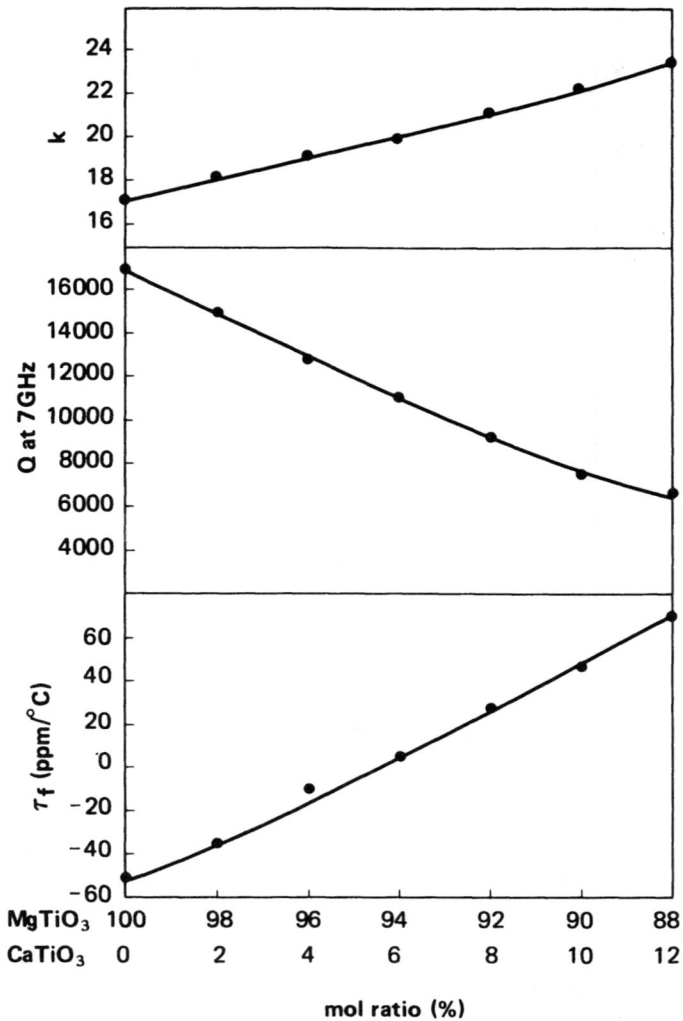

FIGURE 5 Microwave properties of the system $MgTiO_3$-$CaTiO_3$.

this seems to be the elimination of the lattice imperfections in the core of the large dielectric blocks which was mentioned previously.

5) $Ba(Zn_{1/3}Ta_{2/3})O_3$ material is now the representative material having extremely high Q value. The $Ba(Zn,Ta)O_3$-$Ba(Zn,Nb)O_3$ system was presented in 1977 (Figure 10).[18] Since then, many researchers have investigated the materials with complex perovskite structure. Among this type of materials, $Ba(Zn,Ta)O_3$ and $Ba(Mg,Ta)O_3$[19] have the highest possibility of obtaining Q values higher than 10,000 and 20,000 at 10 GHz for each. These materials are used for applications at higher frequencies, such as 10 GHz to the mm-wave region.

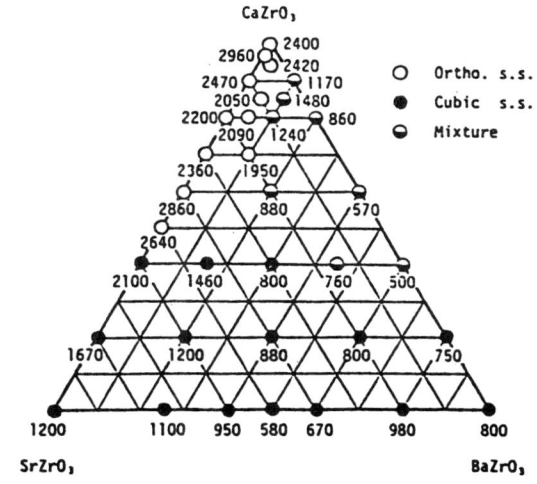

FIGURE 6 Q at 11 GHz of the system (Ca,Sr,Ba)ZrO$_3$.[10]

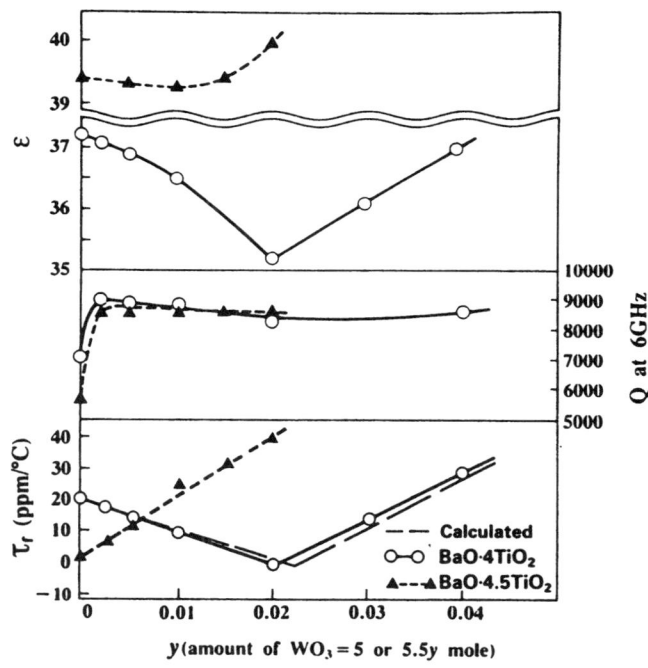

FIGURE 7 Microwave properties of the system BaO-TiO$_2$-WO$_3$.[11]

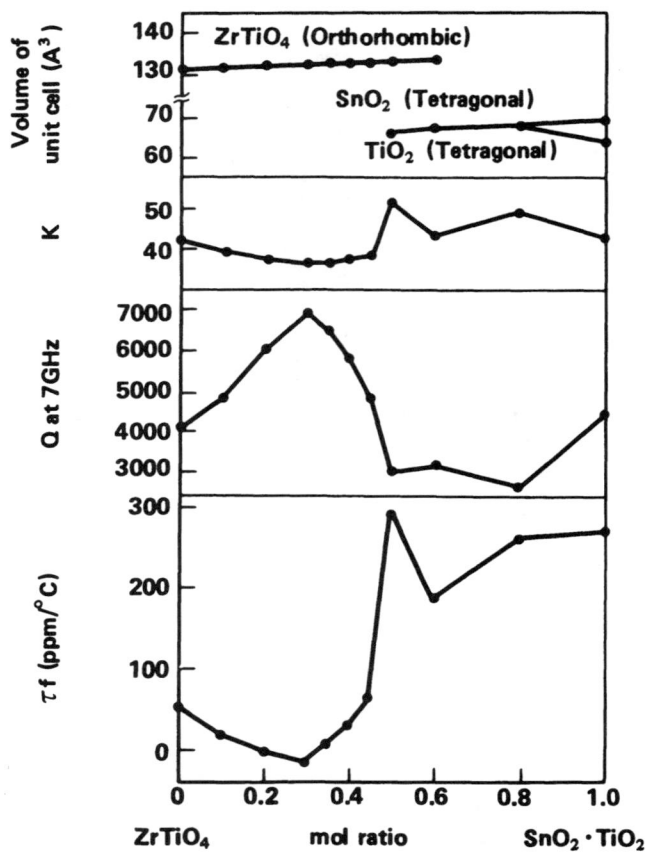

FIGURE 8 Microwave properties of the system $ZrTiO_4$-$SnO_2 \cdot TiO_2$.

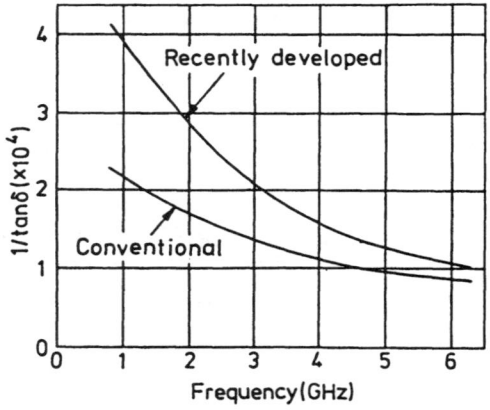

FIGURE 9 Improvement of Q value of $(Zr,Sn)TiO_4$ material.[31]

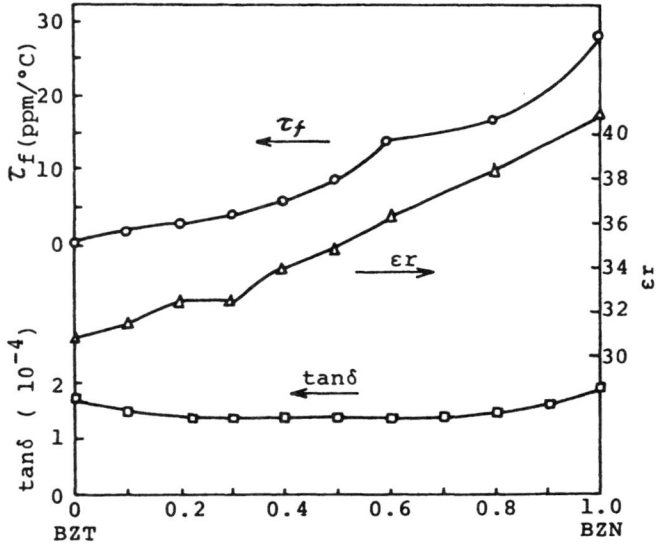

FIGURE 10 Microwave properties of the system $Ba(Zn_{1/3}Nb_{2/3})O_3$-$Ba(Zn_{1/3}Ta_{2/3})O_3$.[18]

For $Ba(Zn,Ta)O_3$ material, far infrared reflection spectra and latice vibrations were investigated and reported.[20,21]

6) BaO-PbO-Nd_2O_3-TiO_2 material has a high dielectric constant of 90 (Table II).[12] Although the material has a lower Q value compared with other materials, this material is very important for the applications at frequencies around 1 GHz. And its Q value is sufficiently high at these frequencies. The TiO_2 rich part of the system BaO-Nd_2O_3-TiO_2 was investigated and the ternary system of $BaNd_2Ti_5O_{14}$ and $BaNd_2Ti_3O_{10}$ was indicated.[22]

The BaO-Sm_2O_3-TiO_2 material is the new system which has k of 77 and Q value of 10,000 at 1 GHz which is twice as high as that of the K90 material.[15,23]

As mentioned above, dielectric resonators with dielectric constant from 20 to 90 are now available for practical applications. These resonators can be selected for various applications by utilizing their peculiarities such as high k, high Q and good temperature stability.

TABLE II
Dielectric Characteristics of BaO-PbO-Nd_2O_3-TiO_2

	composition (mol%)					
$NdO_{3/2}$	BaO	PbO	TiO_2	K	Q(at 1 GHz)	τ_f(ppm/°C)
38	7	0	55	60	1,900	0
31	9	2	58	65	3,900	0
29	9	5	57	85	4,600	0
27	9	7	57	88	5,000	0

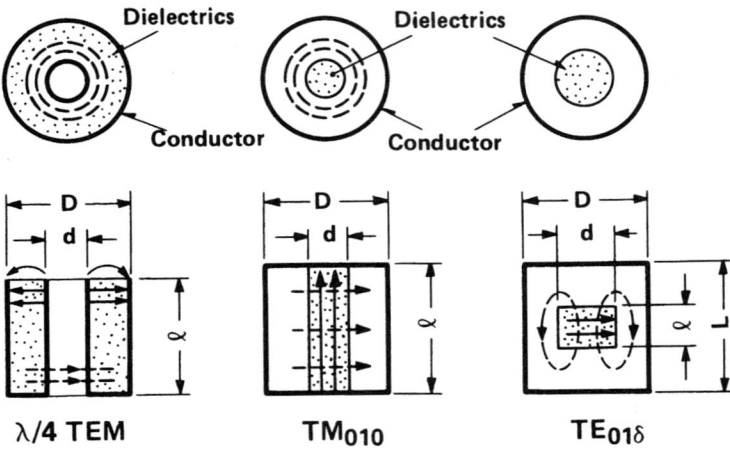

FIGURE 11 Resonant modes of dielectric resonator.[24]

Dielectric Filters

Three kinds of resonant modes which are often used for dielectric resonator filters are shown in Figure 11.[24] The TEM mode has the largest size reduction effect and the smallest unloaded Q. The $TE_{01\delta}$ mode has the smallest size reduction effect and the largest unloaded Q. Here, unloaded Q, Qo is given by the following equation.

$$1/Q_o = 1/Q + 1/Q_c, \qquad (2)$$

where Q: reciprocal of dielectric loss tangent

Qc: Q due to conductor loss.

FIGURE 12 Construction of duplexer using quarter wave TEM mode resonator.

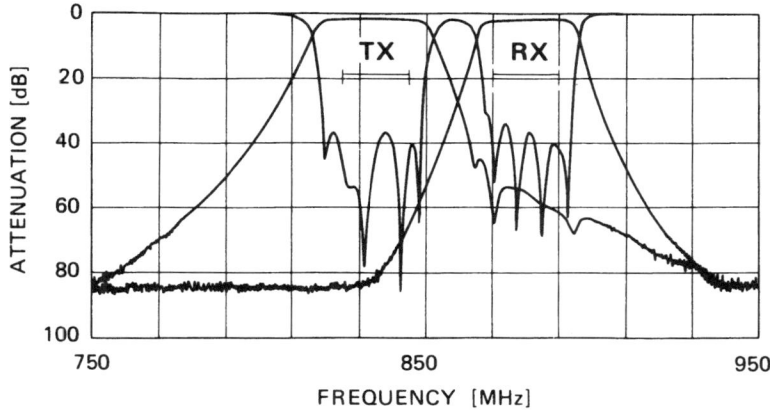

FIGURE 13 Attenuation and return loss of duplexer.[24]

$TE_{01\delta}$ mode and TEM mode filters using dielectric resonator materials have been developed since the middle of 1970s.[25,26] Recently, other modes such as the TM_{010} mode or the HE mode have begun to be used for filters.

1) Duplexer for cellular mobile radios. Cellular mobile radios used at 800 MHz band have been put into practical use. We have reported several times on bandpass filters using coaxial TEM mode resonators.[27,28]

The construction of a compact duplexer using $k = 21$ material is shown in Figure 12. The duplexer has 12 quarter wave coaxial resonators on which a copper electrode is deposited by electroless plating. This copper electrode is useful for cost reduction. The attenuation and return loss of the duplexer are shown in Figure 13. The volume of this duplexer is 55 cm³ (82 × 55 × 13 mm).[24]

2) Duplexer for cellular hand-held radio. For miniaturizing the duplexer for use in hand-held radios, the monolithic type dielectric resonator block filter was developed.

The configuration of the filter is shown in Figure 14. Inner conductors are formed

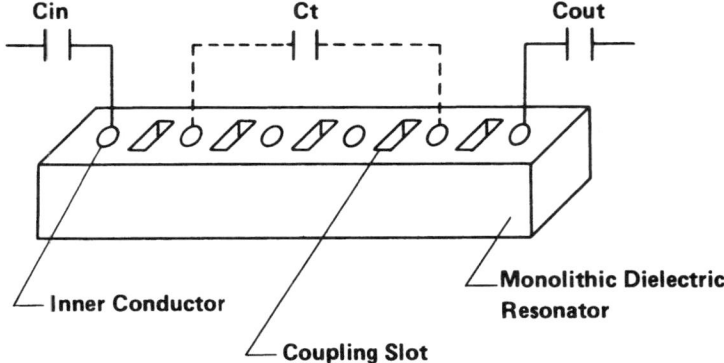

FIGURE 14 Configuration of monolithic dielectric block filter.[24]

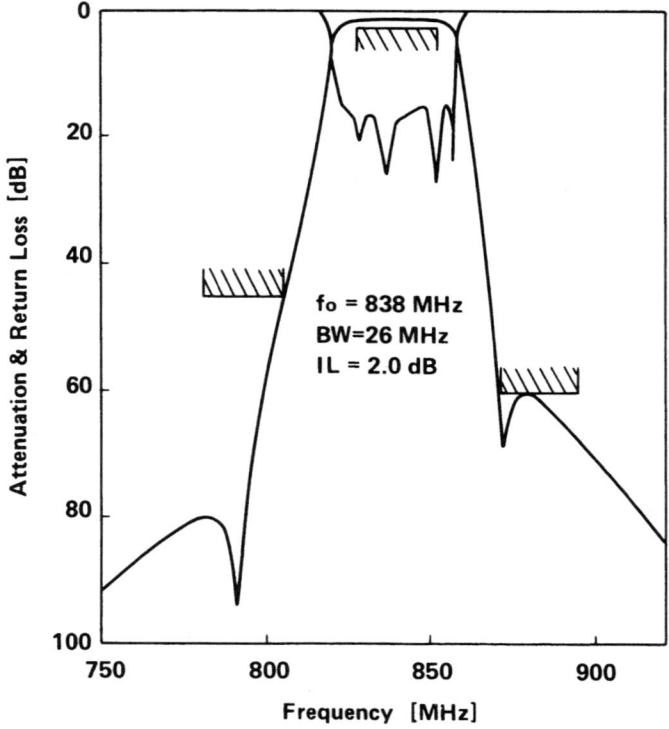

FIGURE 15 Attenuation and return loss of monolithic dielectric block filter.[24]

in circular holes and each resonator having the inner conductor is coupled through a vacant rectangular hole to the adjacent resonator. The capacitor C_t is used for making the attenuation poles and for improving the attenuation characteristics as shown in Figure 15.[24] High-k material with $k = 90$ was used for this filter which brought about the miniaturization of the filter. The volume of this filter is 13 cm^3 and that of the duplexer is about 27 cm^3.

This monolithic type duplexer was miniaturized even more (Figure 16).[29] The new duplexer has a volume smaller than 6 cm^3 which is about 1/10 of the duplexer shown in Figure 12. The smaller size monolithic blocks composed of 4 poles and 5 poles were used for Tx and Rx filters. A duplexer is made by parallel connection of these filters. Its attenuation characteristics are shown in Figure 17.[29]

3) Filters for cellular base stations. The number of the base stations for 800 MHz cellular mobile telephone systems has increased. The transmitter multiplexer was developed[30] by combining the channel dropping filter[31] and the high power antenna filters.[32,33] These filters are composed of the newly developed (Zr,Sn)TiO$_4$ resonators which has Q value higher than 40,000 at 900 MHz (Figure 9).

Figure 18 shows the construction of the channel dropping filter.[31] The shielding cavity is made of ceramics metalized with fired silver. And the thermal coefficient

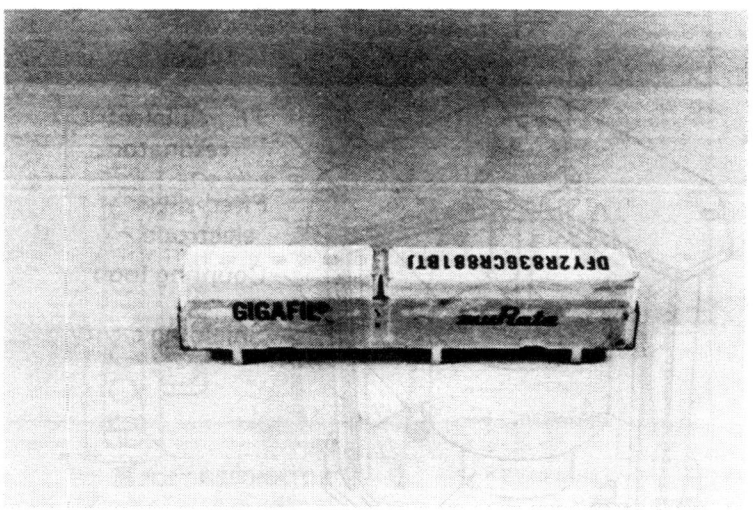

FIGURE 16 Miniaturized monolithic dielectric block filter.

of the ceramics is the same as that of the resonator ceramics. This filter has high stability versus both temperature and high power of 60 W.

Figure 19 shows the TM_{110} mode high power antenna filter for base stations.[33] The size of this filter was miniaturized to 2,500 × 140 × 60 mm, about 1/5 of the conventional comb line filter. Under the RF power of 500 W, the temperature rise of the filter case was 15°C, the increase of insertion loss was 0.03 dB. The third-order intermodulation level was less than −140 dBc which was the limit of the sensitivity of the measurement system.

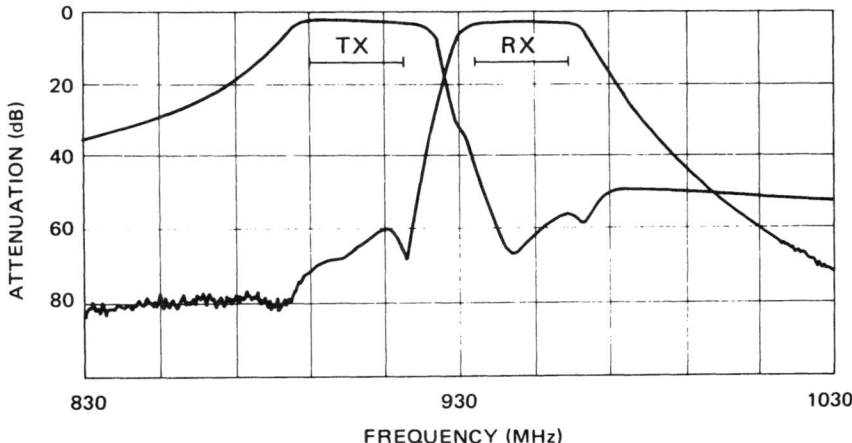

FIGURE 17 Attenuation characteristics of miniaturized monolithic dielectric block resonator.[29]

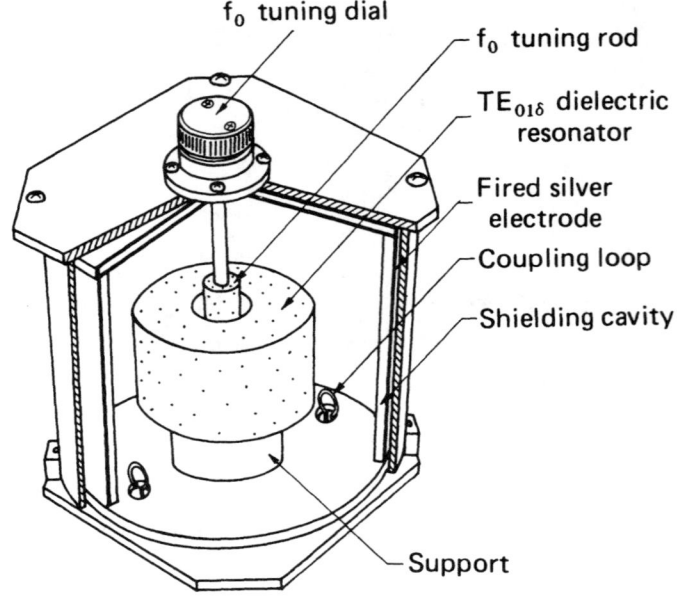

FIGURE 18 Construction of channel dropping filter for cellular base station.[31]

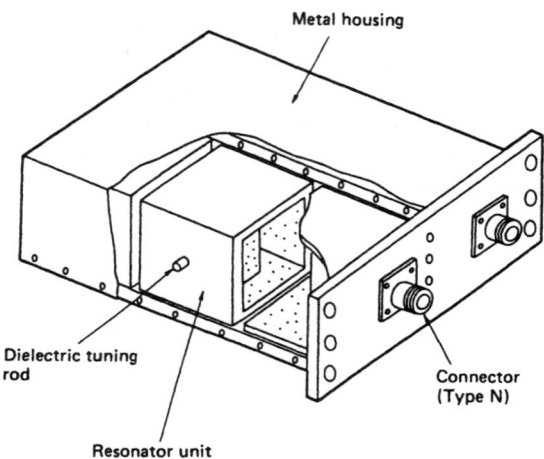

FIGURE 19 Construction of TM_{110} mode antenna filter for cellular base station.[33]

By using 16 channel dropping filters and a TM_{110} high power filter, the transmitter multiplexer was developed. The maximum insertion loss from the input port of the isolator to the output terminal of the antenna filter is 2.4 dB and the size of the multiplexer is miniaturized to W260 × D225 × H1600 mm.[30]

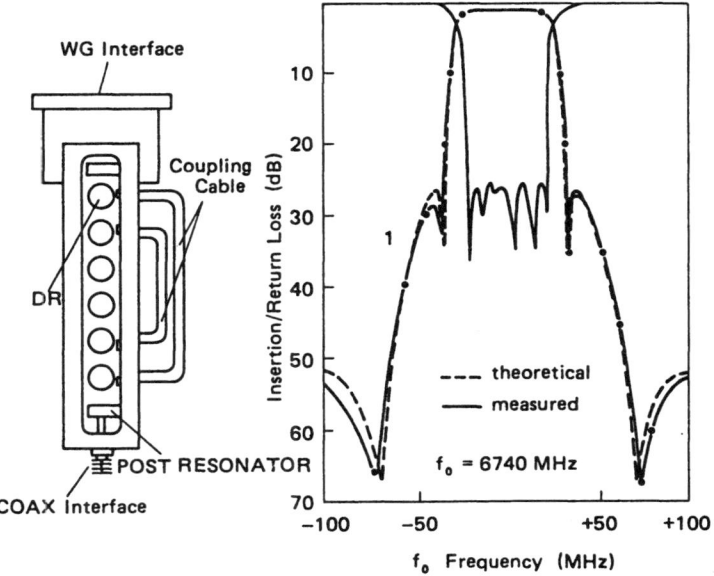

FIGURE 20 Construction and electric properties of elliptic-function type bandpass filter.[34]

Elliptic bandpass filter using $TE_{01\delta}$ mode resonators

Elliptic function type bandpass filters have improved attenuation characteristics. Figure 20 shows the construction of an elliptic type filter for SHF communications reported by Sei, *et al.*[34] The first and the sixth resonators, and the second and the fifth resonators were coupled by cables. Its attenuation characteristics with poles are shown in Figure 20. K = 30 resonators were used for this filter. The unloaded Q of the resonator is 8500, the insertion loss and temperature coefficient of the filter is 0.5 dB at 6740 MHz, and -1 ppm/°C for each.

Bandpass filters using $EH_{11\delta}$ and $TM_{01\delta}$ mode resonators

Figure 21 shows a bandpass filter using $EH_{11\delta}$ mode resonators. This is an elliptic type filter, and four- or six-pole attenuation characteristics were obtained by using 3 dielectric resonators.[35]

Figure 22 shows a bandpass filter using $TM_{01\delta}$ mode dielectric resonators. This filter has the advantages of low loss and good spurious response.[36]

Both of these filters use the K = 25 high Q resonator, and their center frequency is 12 GHz.

Bandpass filter for millimeter-wave frequency

Dielectric resonator is now used for millimeter applications. Figure 23 shows a construction of a bandpass filter for millimeter-wave frequency.[37] Three or four

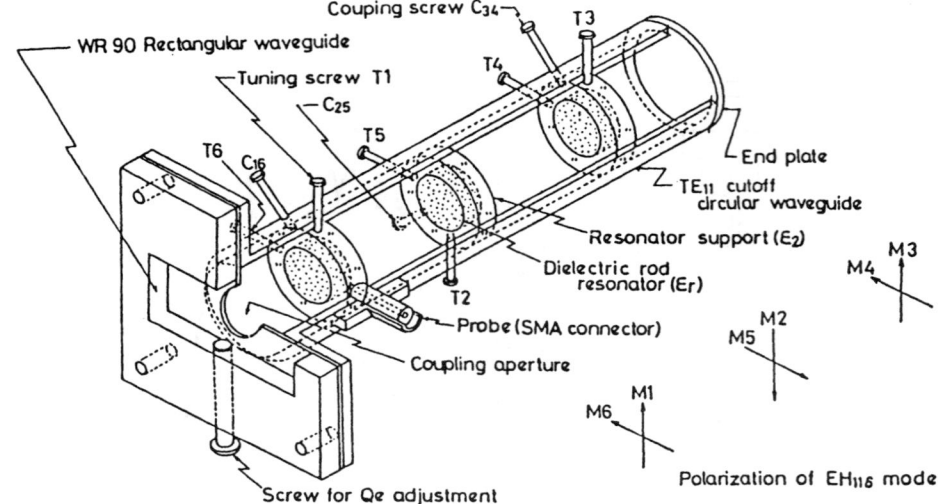

FIGURE 21 Bandpass filter using $EH_{11\delta}$ mode resonator.[35]

pole filters were designed using K = 30 dielectric resonators. The unloaded Q of the resonators were 3400, 2400 and 1500 at frequencies of 30, 50, and 90 GHz, and good filter characteristics were obtained at these frequencies.

Conclusion

Dielectric resonators have many advantages for microwave applications, such as size and cost reduction. Dielectric characteristics of resonator materials have been greatly improved for the past 15 years, and the designs of filters using these resonators are also being improved year by year.

The examples which are described in this paper are only a few of the many types. I believe that the fields of application will expand in the future and dielectric ceramics will be ready to take their part in the coming new systems.

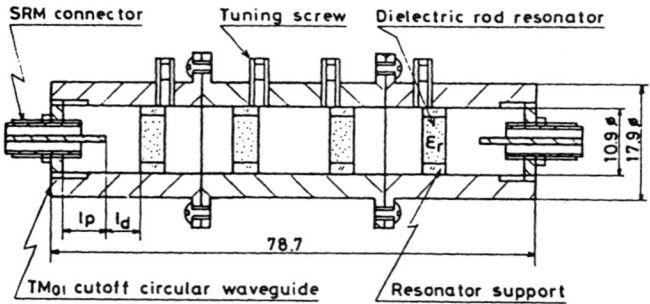

FIGURE 22 Bandpass filter using $TM_{01\delta}$ mode resonator.[36]

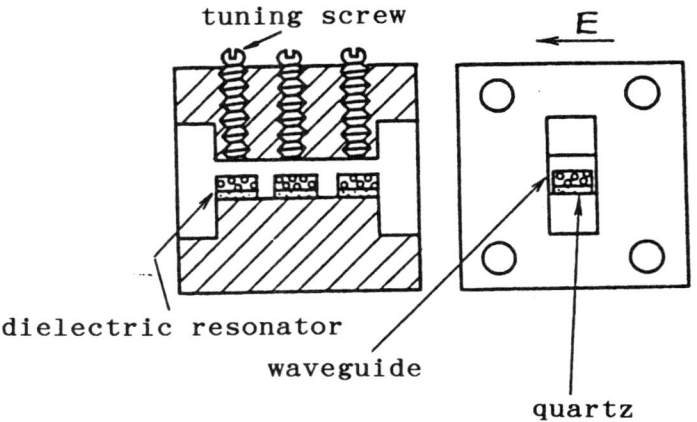

FIGURE 23 Bandpass filter for millimeter frequency.[37]

REFERENCES

1. R. D. Richtmyer, "Dielectric Resonator," *J. Appl. Phys.*, **10**, 1939, pp. 391–398.
2. G. Rupprecht et al., "Nonlinearity and Microwave Losses in Cubic Strontium-Titanate," *Phys. Rev.*, **123**, 1961, pp. 97–98.
3. W. G. Spitzer et al., "Far Infrared Dielectric Dispersion in $BaTiO_3$, $SrTiO_3$, and TiO_2," *Phys. Rev.*, **126**, 1962, pp. 1710–1721.
4. B. W. Hakki et al., "A Dielectric Resonator Method of Measuring Inductive Capacitance in the Milimeter Range," *IRE Trans. on MTT*, MTT-8, 1960, pp. 402–410.
5. S. B. Cohn, "Microwave Bandpass Filters Containing High-Q Dielectric Resonators," *IEEE Trans. on MTT*, MTT-16, 1968, pp. 218–227.
6. K. Wakino et al., "Dielectric Materials for Dielectric Resonator," 1976 Joint Convention Record of Four Institute of Electrical Engineers, Japan, No. 235.
7. H. Tamura et al., "High-Q Dielectric Resonator Material for Millimeter-Wave Frequencies," Abstracts of the 3rd U.S.: Japan Seminar on Dielectric and Piezoelectric Ceramics, Nov. 9–12, 1986, pp. 53–54.
8. S. Kawashima et al., "$Ba(Zn,Ta)O_3$ Ceramics with Low Dielectric Loss at Microwave Frequencies," *J. Am. Ceram. Soc.*, **66**, 1983, pp. 421–423.
9. H. Tamura et al., "Improved High-Q Dielectric Resonator with Complex Perovskite Structure," *J. Am. Ceram. Soc.*, **67**, 1984, C-59-C-61.
10. T. Yamaguchi et al., "(Ca,Sr,Ba) Zirconate Ceramics for Microwave Dielectric Resonator," Annual Report of Study Group on Applied Ferroelectrics in Japan, **29**, XXIX-159-1017, 1980.
11. S. Nishigaki et al., "BaO-TiO_2-WO_3 Microwave Ceramics and Crystallin $BaWO_4$," *J. Am. Ceram. Soc.*, **71**, 1988, pp. C-11-C-17.
12. K. Wakino et al., "Microwave Characteristics of $(Zr,Sn)TiO_4$ and BaO-PbO-Nd_3O_3-TiO_2 Dielectric Resonators," *J. Am. Ceram. Soc.*, **67**, 1984, pp. 278–281.
13. H. M. O'Bryan, Jr., et al., "A New BaO-TiO_2 Compound with Temperature-Stable High Permittivity and Low Microwave Loss," *J. Am. Ceram. Soc.*, **57**, 1974, pp. 450–453.
14. T. Mizutani et al., "$(Sr,Ca)[(Li,Nb)Ti]O_3$ Dielectric Material," Annual Report of Study Group on Applied Ferroelectrics in Japan, **24**, XXIV-137-896, 1975.
15. S. Kawashima et al., "Microwave Dielectric Materials and their Applications," Annual Report of Study Group on Ferroelectrics in Japan, **30**, XXX-164-1036, 1980.
16. R. C. Kell et al., "High-Permittivity Temperature-Stable Ceramic Dielectrics with Low Microwave Loss," *J. Am. Ceram. Soc.*, **56**, 1973, pp. 352–354.
17. G. Wolfram et al., "Existance Range, Structural Dielectric Properties of $Zr_xTi_ySn_zO_4$ Ceramics ($x + y + z = 2$)," *Mat. Res. Bull.*, **16**, 1981, pp. 1455–1463.
18. S. Kawashima et al., "Dielectric Properties of $Ba(Zn_{1/3}Nb_{2/3})O_3$-$Ba(Zn_{1/3}Ta_{2/3})O_3$ Ceramics," *Proc. Ferroelectr. Mater. Appl. Japan*, **1**, 1977, pp. 293–296.
19. S. Nomura et al., "$Ba(Mg_{1/3}Ta_{2/3})O_3$ Ceramics with Temperature-Stable High Dielectric Constant and Low Microwave Loss," *Jpn. J. Appl. Phys.*, **21**, 1982, pp. L624–L626.

20. K. Wakino et al., "Far Infrared Reflection Spectra of Ba(Zn,Ta)O_3-BaZrO_3 Dielectric Resonator Material," *J. Am. Ceram. Soc.*, **69,** 1986, pp. 34–37.
21. H. Tamura et al., "Lattice Vibrations of Ba(Zn$_{1/3}$Ta$_{2/3}$)O_3 Crystal with Ordered Perovskite Structure," *Jpn. J. Appl. Phys.*, **25,** 1986, pp. 787–791.
22. D. Kolar et al., "Ceramic and Dielectric Properties of Selected Compositions in the BaO-TiO_2-Nd$_2$$O_3$ system," *Bert. Dr. Keram. Ges.* **55,** 1987, pp. 346–348.
23. S. Nishigaki et al., "Microwave Dielectric Properties of (Ba, Sr)O-Sm$_2$$O_3$-Ti$O_2$ Ceramics," *J. Am. Ceram. Soc. Bull.*, **66,** 1987, pp. 1405–1411.
24. K. Wakino, "High Frequency Dielectrics and Their Applications," Proceedings of 1986 IEEE International Symposium on Application of Ferroelectrics, 1986, pp. 97–106.
25. K. Wakino et al., "Microwave Bandpass Filters Containing Dielectric Resonators with Improved Temperature Stability and Spurious Response," 1975 IEEE MTT-S, Int. Microwave Symp. Dig., pp. 63–65.
26. K. Wakino et al., "Miniaturized Bandpass Filters Using Half Wave Dielectric Resonators with Improved Spurious Response," 1978 IEEE MTT-S Int. Microwave Symp. Dig., pp. 230–232.
27. K. Wakino et al., "Quarter Wave Dielectric Transmission Line Duplexer for Land Mobile Communications," 1979 IEEE MTT-S Int. Microwave Symp. Dig., pp. 278–280.
28. K. Wakino et al., "Miniaturized Duplexer for Land Mobile Communication Using High Dielectric Ceramics," 1981 IEEE MTT-S Int. Microwave Symp. Dig., pp. 185–187.
29. K. Wakino et al., "Dielectric Resonator Materials and their Applications," Microwave Journal, June 1987, pp. 133–150.
30. T. Nishikawa et al., "16 Channel Dielectric Transmitter Multiplexer for Cellular Base Stations," 38th IEEE Vehicular Technology Conference, 1988, pp. 461–468.
31. K. Wakino et al., "800 MHz Band Miniaturized Channel Dropping Filter Using Low Loss Dielectric Resonator," Densi Tokyo Japan, No. 24, 1985, pp. 72–75.
32. T. Nishikawa et al., "Dielectric High-Power Bandpass Filter Using Quarter-Cut TE$_{018}$ Image Resonator for Cellular Base Station," IEEE Trans. on MTT, MTT-35, 1987, pp. 1150–1155.
33. T. Nishikawa et al., "800 MHz Band High-Power Bandpass Filter Using TM$_{110}$ Mode Dielectric Resonators for Cellular Base Stations," 1988 IEEE MTT-S Int. Microwave Symp. Dig., pp. 519–522.
34. H. Sei et al., "Dielectric Resonator Bandpass Filters," IECE Tech. Rep. Japan, MW86-35, 1986.
35. Y. Kobayashi et al., "Canonical Bandpass Filters Using Dual-Mode Dielectric Resonators," 1987 IEEE MTT-S Int. Microwave Symposium Dig., pp. 137–140.
36. Y. Kobayashi et al., "A Bandpass Filters Using Electrically Coupled TM$_{018}$ Dielectric Rod Resonators," 1988 IEEE MTT-S Int. Microwave Symposium Dig., pp. 507–510.
37. M. Ishizaki et al., "Millimeter-Wave Dielectric Resonator Filters," IECE Tech. Rep. Japan, MW85-66, 1986.
38. Y. Kobayashi et al., "Microwave Measurement of Dielectric Properties of Low-Loss Materials by the Dielectric Rod Resonator Method," IEEE Trans. on MTT, MTT-33, 1985, pp. 586–592.

RF Front End Circuit Components Miniaturized Using Dielectric Resonators for Cellular Portable Telephones

Toshio NISHIKAWA[†], *Member*

SUMMARY As cellular mobile communication systems' markets are expanding worldwide, miniaturization for the terminals are strongly required and accordingly the RF front end circuit components in the terminals are required to be much smaller. This markets' demand of miniaturization has provided a perfect opportunity for pushing forward with creating the components using dielectric ceramics. The miniaturization is realized by a sophisticated integration of techniques consisted of monoblock forming, multi-layer, electrode patterning and processing, MIC, electro-magnetic wave analysis oriented simulation, production and measurement based on dielectric ceramic materials. In this paper, examples of several filters and VCO developed using these techniques are picked up, the constructions and features are described, and technical prospects are discussed. They include monoblock type filters with additional coupling capacitors on the dielectric substrate, coaxial resonator type filters of rectangular coaxial dielectric resonator and its circuit theory, balanced double layer stripline type filter of substantial triplate structure using dielectrics of higher and lower permittivities, multilayer lumped constant filter and VCO by multilayer, electrode patterning and simple circuit theory. The technologies are applicable to over-all 900 MHz band mobile communication systems in the world, and these components are possibly miniaturized to less than 1 cm^3, to be small enough for smaller-than-the-present-size portable telephones application.

1. Introduction

Dielectric resonator filters have been used for RF front end circuit components to miniaturize cellular mobile telephone terminals since the early practical use. In 1978, dielectric antenna duplexers of 700 cm^3 (90×140×45 mm) have been mounted on the terminals with space diversity for service trial in the US-AMPS. In 1979, another type of an antenna duplexer of 230 cm^3 (53×195×22 mm) have been mounted on all kinds of terminals for the systems in Japan. They served as an effective trigger to put into practical use and miniaturize the systems terminals. Recently, as the markets of the cellular portable telephone systems are expanding worldwide, the terminals are required to be much smaller and the duplexer in the volume of about 1 cm^3 was realized. On the other hand, there is a voltage controlled oscillator (VCO) using a dielectric resonator for another RF circuit component. The

Manuscript received February 5, 1991.
† The author is with Murata Manufacturing Company Limited, Nagaokakyo-shi, 617 Japan.

component's volume was 4.2 cm^3 in 1986, and subsequently brought down to realize the current 0.8 cm^3.

In this paper, technical factors contributing to the miniaturization of the components are described with a few examples of filters and VCO.

2. Monoblock Type Dielectric Filters[1]

Figure 1 shows construction elements of monoblock type dielectric duplexer for the cellular portable telephone terminals. The duplexer consists of transmitting (TX) and receiving (RX) filters, and is constructed of 3 pieces of ceramic monoblock, a coupling substrate and a metal housing.

The monoblocks are metallized by fired silver on the lateral walls of the holes working as the center conductors of the resonators and the external walls of the monoblocks except one side plane working as the open ends of the resonators and the lateral walls of the holes working as the coupling between the resonators. The substrate is a group of capacitors made with photolithography treatment to pattern a silver metallized ceramic substrate and has the dielectric constant (K) of 21.

Figure 2 shows the equivalent circuit of the duplexer.

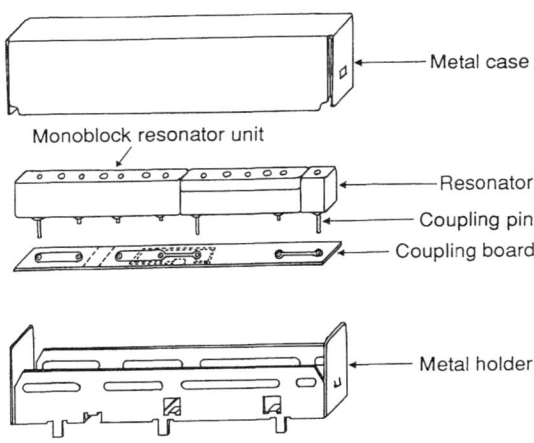

Fig. 1 Construction of monoblock duplexer.

Reprinted with permission from IEICE, T. Nishikawa, "RF Front End Circuit Components Miniaturized Using Dielectric Resonators for Cellular Portable Telephones," vol. E74, no. 6, pp. 1556-1562, 1991. © The Institute of Electronics, Information and Communication.

The transmitting (TX) filter consists of two blocks of band-pass filter (BPF) of 3 poles and band elimination filter (BEF) of mono-pole, and the receiving (RX) filter is one block of BPF of 4 poles. The material of blocks of BEF for TX filter and BPF for RX filter is dielectric ceramics having high K of 90, high stable temperature characteristics and low dielectric loss, and BPF of TX filter uses a block comprised of the ceramics of K of 21 which stacks on the same ceramics to suppress higher spurious response. The metallized blocks work as a quarter wavelength TEM mode resonators.

The coupling coefficients between the resonators are obtained by the air holes in the blocks and the capacitors on the substrate. The air holes cause the different effective dielectric constants in the even and odd modes, and produce the coupling indispensable for monoblock. The connection of the capacitors to the center conductors of the resonators in supplementary achieved by additional coupling. In the TX filter, monopole trap filter is used with BPF to get the required attenuation in the RX band, and in the RX filter, and finite attenuation poles characteristics are obtained by jumped coupling capacitances between the input and output ports to the second and third resonators using the capacitors on the substrate. A T branch of the duplexer has a coil inductor to match the impedance, and the line length to each filter is shortened by a matching design. Figure 3 shows the transmission characteristics of the duplexer. The size of the duplexer is currently about 20 cm^3, and it is possible to realize one below 2.5 cm^3 using simpler housing. Further, higher grade circuit functions of transmission characteristics are added to the duplexer by using a multilayer dielectric substrate which provides couplings.

3. Coaxial Resonator Type Filter[2]

Figure 4 shows a construction of the coaxial resonator type duplexer. The duplexer consists of TX filter with characteristics of BEF and RX filter with those of BPF. The construction elements are rectangular coaxial dielectric resonators, square plate ceramic chip capacitors, coils, dielectric substrate and metal housing. It is known that traditional lumped constant circuit technologies are available when the size of the construction elements are small enough compared with the wavelength of the specified frequency. So, the lumped elements like chip capacitors and coils can be adopted to the filter considering their advantages in sizes. The dielectric substrate has the same functions and features that mentioned in the paragraph 2.

The resonator used in the filters has following features.
(1) The unloaded Q is improved by about 10 percents comparing with conventional coaxial dielectric resonators of the same diameter.
(2) Fixing the resonator to the housing is easy and it results in improved firm electrical contact condition.
(3) Progresses in ceramic forming techniques and improvements in metallized technologies with the copper plating applied to resonators manifest in the unloaded Q being in fair compliance with the theoretical calculation results even when it comes to the small size rectangular coaxial resonator whith 3 mm square faces (less than 0.1 cm^3 in volume).
(4) Face mounting is possible with all the compris-

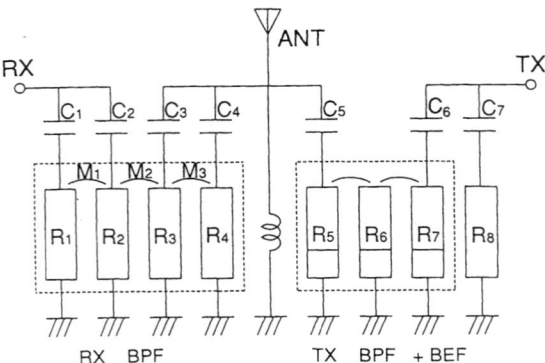

Fig. 2 Equivalent circuit of monoblock duplexer.

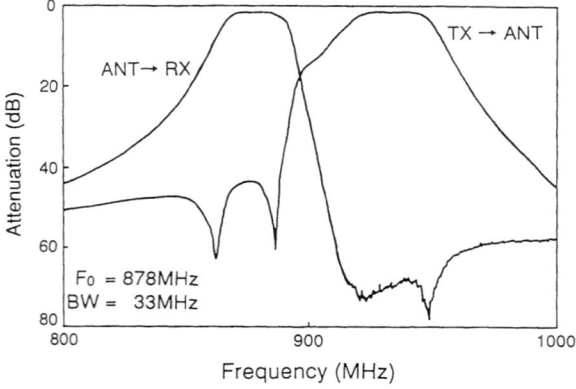

Fig. 3 Transmission characteristics of the duplexer.

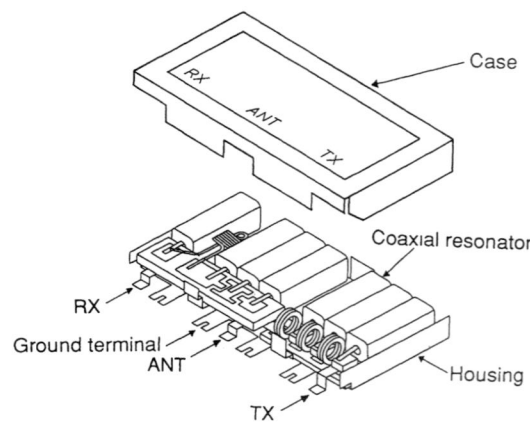

Fig. 4 Construction of coaxial resonator duplexer.

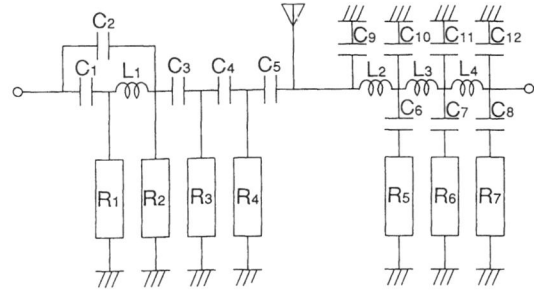

Fig. 5 Equivalent circuit of coaxial resonator duplexer.

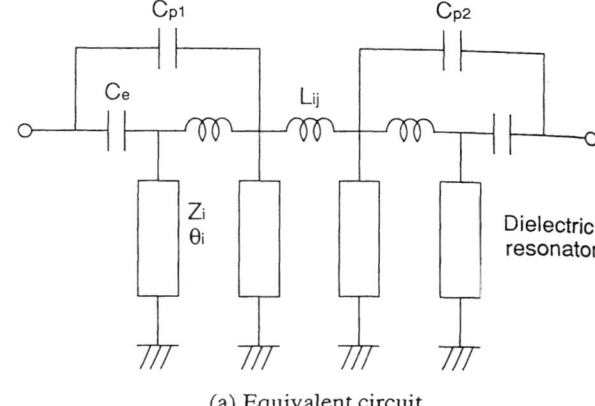

(a) Equivalent circuit

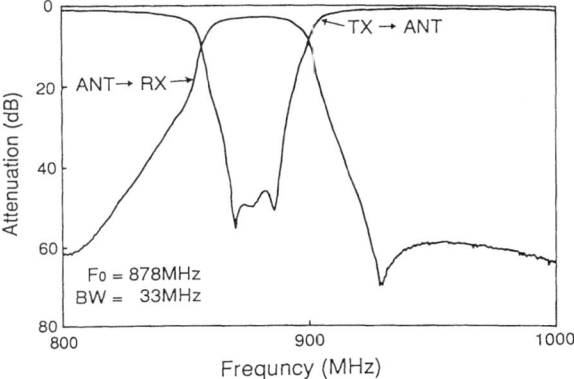

Fig. 6 Transmission characteristics of the duplexer.

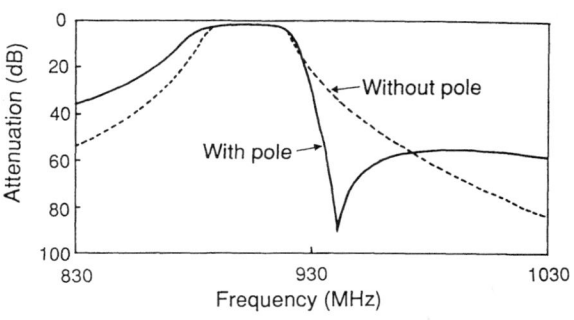

(b) Frequency characteristics

Fig. 7 Filter with a pole in higher band.

ing elements of the filter, making assembly easy and resulting in making the filter applicable to surface mount (SMT).

Figure 5 shows the equivalent circuit of the duplexer. For the purpose of making the duplexer as compact as possible, the circuit of TX filter is based on BEF constructed by capacitive-coupling the coaxial resonators. The circuit lessens the insertion loss below 1/5 that of the same-sized BPF, and suppresses the response of higher order harmonic spurious signals. The enough suppression is realized by replacing the phase shift line of BEF to lumped C-L-C π type network circuit by which the circuit elements values are optimized to work as LPF in higher order harmonics.

Transmission characteristics of a fundamental BPF don't always have symmetry in attenuation in relation to the center frequency of the pass band-width, rather, it is common that the attenuation in higher side of the out of bands is smaller than that in lower side. This RX filter improves the attenuation in higher side of the out of bands by an additional finite attenuation pole circuit. Figure 7(a) shows a basic circuit of the dielectric filter to improve an attenuation in higher out of band by a conversion from the ladder type filter circuit, and the Fig. 7(b) shows the typical effect on the characteristics. The circuit in Fig. 5 has an inductive coupling between RX port and 2nd resonator, and

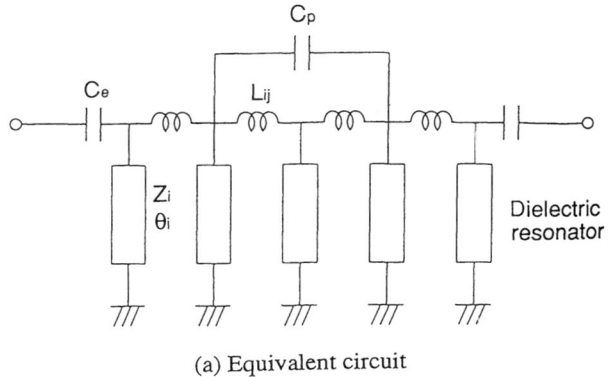

(a) Equivalent circuit

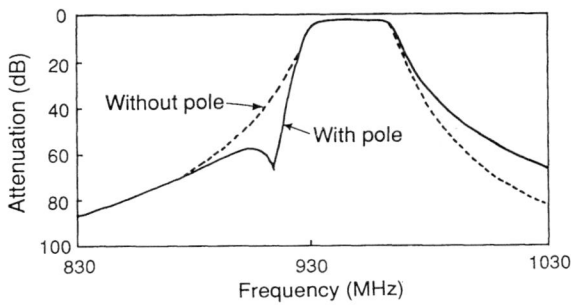

(b) Frequency characteristics

Fig. 8 Filter with a pole in lower band.

521

the attenuation curve is improved as shown in Fig. 6. On the contrary, if a greater attenuation in lower side is required strongly, it can be obtained by the circuit as shown in Fig. 8. Thus lumped element coupling circuits are constructed by using chip size elements on the dielectric substrate, and the symmetry of the transmission characteristics can be controlled flexibly with the coupling circuit.

Consequently, transmission characteristics of duplexer tell that TX filter suffers from only meager insertion loss in pass band while at the same time has sufficient attenuation in RX band with suppressing spurious responses, and RX filter has sufficient high attenuation in TX band.

Duplexer less than 2 cm³ in volume can be realized by using rectangular coaxial dielectric resonators of 3 mm square face.

4. Balanced Double Layer Stripline (BDLS) Type Filter[3]

Figure 9 shows a construction of balanced double layer stripline (BDLS) filter. Each of the sandwiching dielectric substrates has high permittivity. Both ground electrode and stripline are of silver-fired with 5 μ thickness and the filter's pattern is processed by photo-lithographic techniques. A pair of strip line are placed in a symmetrical layout with the resin plate of lower permittivity inserted between them. Short circuit pin or through hole, arranged as illustrated in the figure, extending across the entire thickness of the striplines and the resin plate combined allows the pair of sandwhching dielectric substrates to act as a block of plural resonators.

Such a structural design increases effective dielectric constant to enable miniatuarization of the filters for which small coupling coefficients are required. The equivalent circuit of the filter is shown in Fig. 10. R_j expresses a pair of stripline resonators short circuited electrically near open ends. Two short ended transmission lines (Z_{even}, Z_{odd}) express coupling part of neighboring resonators. In conventional combline type filters, the effective dielectric constant of even mode is equal to that of odd mode, but not in this case, because of the resin plate. Input and output ports are provided at suitable points of the outside resonators to get the predetermined external Q.

The mechanism of the coupling between

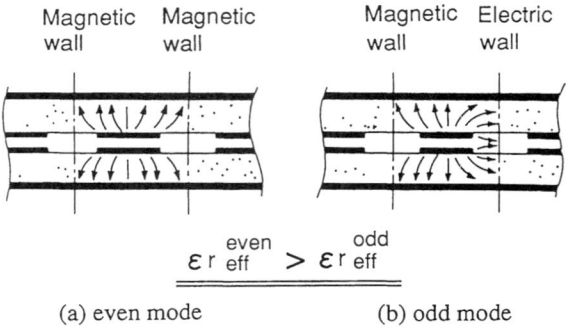

$$\varepsilon r_{eff}^{even} > \varepsilon r_{eff}^{odd}$$

(a) even mode   (b) odd mode

Fig. 11 Electric field distribution.

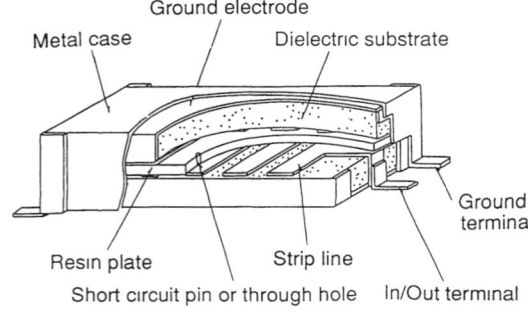

Fig. 9 Construction of BDLS filter.

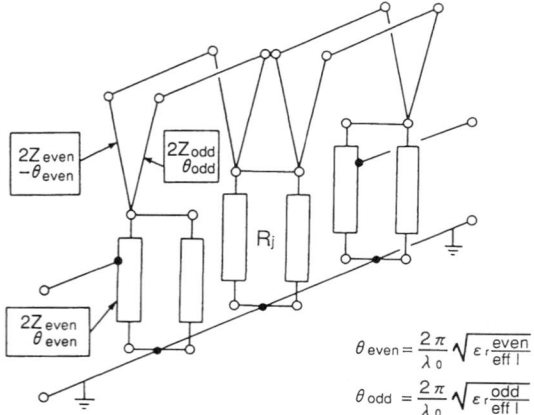

$\theta_{even} = \frac{2\pi}{\lambda_0}\sqrt{\varepsilon r_{eff\,1}^{even}}$

$\theta_{odd} = \frac{2\pi}{\lambda_0}\sqrt{\varepsilon r_{eff\,1}^{odd}}$

Fig. 10 Equivalent circuit of BDLS filter.

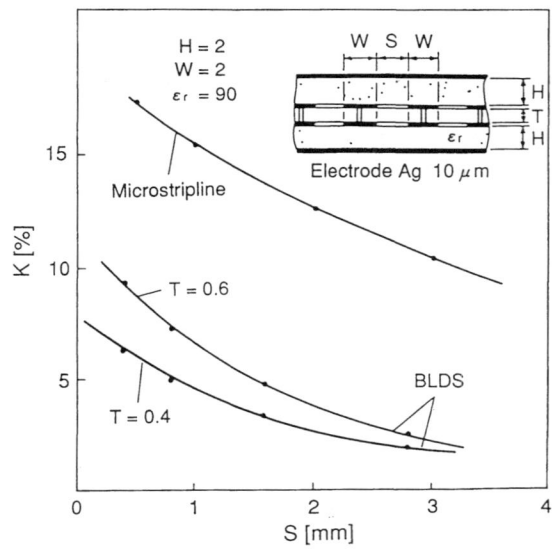

Fig. 12 Relation between coupling coefficient K and inter-line distances in microstrip type and BDLS.

resonators is explained electrical field distributions in even and odd modes as shown in Fig. 11. Under the condition of even mode, the electric energy is more concentrated in the dielectric substrates than that in odd mode.

Figure 12 shows the relations between coupling coefficient and the inter-line distance both on micro-stripline type and on BDLS type, and demonstrates the great edge of BDLS type on the micro-stripline type in term of coupling coefficient, as small as that is never achieved with micro-stripline type. This structure design of low permittivity between the pair of resonators connected by stripline is considered in principle equal to that of monoblock type. This evaluation is endorsed by the experimental findings that the unloaded Q is affected little by the resin plate thickness. In another respect, the advantage that any arbitrary pattern for this filter can be designed and circuit elements are mounted accordingly holds possibility to meet a variety of transmission characteristics requirements.

With the proto-type BDLS 4-section BPF, the center frequency and the bandwidth of which are 880 MHz, 26 MHz respectively, as small volume as 1.0 cm³ was already achieved.

5. Multi-Layer Lumped Constant (LC) Type Filter[4]

This is a chip-shaped filter to which a multi-layer forming technology and the printed coil structuring one are applied. Identical technologies are used for ceramic capacitor production. As shown in Fig. 13, this chip-shape gives the filter an advantage in SMT and the convenient aspect of this type of filter that any arbitrary internal electrode pattern can be designed holds a great possibility for realizing multipurpose filter characteristics.

Figures 14 and 15 explain simplified structure and the equivalent circuit of a typical filter developed respectively. As shown in the figures, the inductance pattern layer is placed center and is sandwiched by a set of a capacitance pattern layer, two dummy layers and a shield pattern layer on each side. This shield lessens the external electrical effects. When it becomes necessary to tune the center frequency of the filter, it is done by trimming the electrode on capacitance ajustment layer outside the shield pattern which decreases the capacitance and moves the center frequency upward.

The ceramic material used for this filter is required to have low loss, frequency temperature coefficient of 0 ppm/°C and nondeclining frequency dependency up to microwave, therefore the field proven ceramic material for multi-layer ceramic capacitor is employed and its permittivity is 10.

The insertion loss of the filter of this type is greatly influenced by characteristics of interlayer electrode. To minimize the insertion loss, Copper having high conductivity is applied to the electrode. Such a structure is realized by sintering the laminate pile under deoxidized atmosphere at lower temperature than the normal sintering temperature for microwave ceramics. With all these applications mentioned above, embedding the high Q printed coils and the capacitors of

Fig. 13 Appearance of multi-layer LC filter.

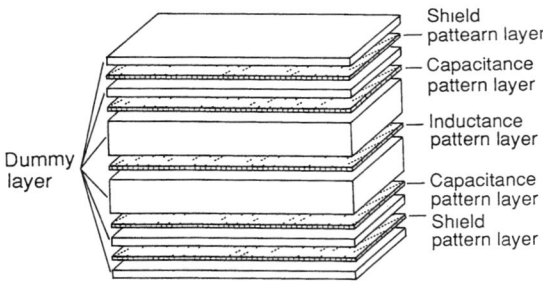

Fig. 14 Structure of multi-layer LC filter.

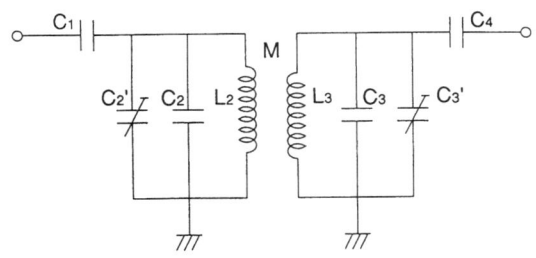

Fig. 15 Equivalent circuit of LC filter.

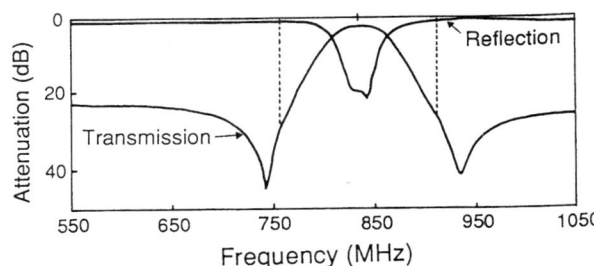

Fig. 16 Transmission characteristics of LC filter.

superior microwave characteristics as lumped constant circuit elements was materialized.

Triple-layer structure;

Cu electrode layer, Ni plating layer and Sn plating layer are applied to the external electrode to achieve rigidity, especially, on terminals.

Figure 16 shows characteristics of a 2 section filter, 0.07 cm³ in volume (5.0×5.7×2.5 mm). Noteworthy is the maximum insertion loss of 4 dB achieved under the conditions of center frequency 860 MHz, bandwidth 26 MHz. The upper-most frequency achieved so far with the filter of this type is 1 GHz and there remains great potentiality for advancing into even higher frequency range given the diversity in their structures and dimensions.

6. Multi-Layer Stripline Type Voltage Controlled Oscillator (VCO)[5]

VCO is another example of dielectric resonators application. Figure 17(a) illustrates structure of VCO in which rectangular coaxial dielectric resonator is used.

Certain phase noise (C/N ratio) level is one of the required characteristics for VCO and is affected greatly by unloaded Q of the reasonant circuit controling oscillation frequency. Resonant circuit usually consists of dielectric resonator and varactor and in this example of VCO, a quater wavelength 4 mm×4 mm rectangular coaxial resonator of dielectric constant 90, temperature coefficient 0 ppm/°C is used. Other circuit parts are surface mounted on the surface of the thick film printed alumina substrate, the overall volume of which is 1.9 cm³. Figure 17(b) illustrates structure of multi-layer stripline VCO. As shown in the figure, the hair-pin stripline embedded in the multi-layer dielectrics acts as a resonator.

The dielectrics is a microwave ceramics of permittivity 38 and superior temperature stability. By installing proper electrode pattern among layers, required circuit constant is embedded also in them.

In this structure, ground electrodes are formed on both top face and bottom, creating shielding effects from surroundings without any cover case, and moreover, SMT design concept is made simple with external electrodes on flanks just as in the case of LC filter.

Figure 18 shows typical phase noise characteristics of VCO, the specifications of which are volume 0.5 cm³, center frequency 836 MHz, variable bandwidth 26 MHz.

The phase noise level at a 50 kHz a offset satisfies the required value, because it is 110 dBc/Hz in this

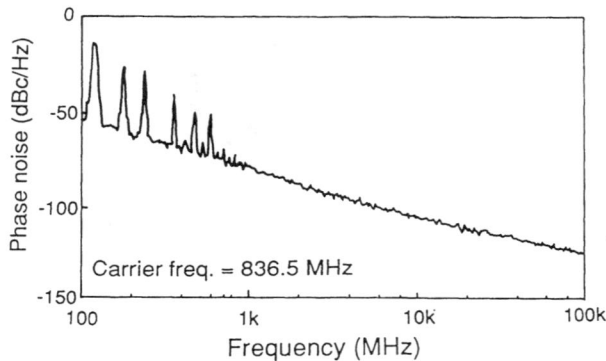

Fig. 18 Phase noise characteristics.

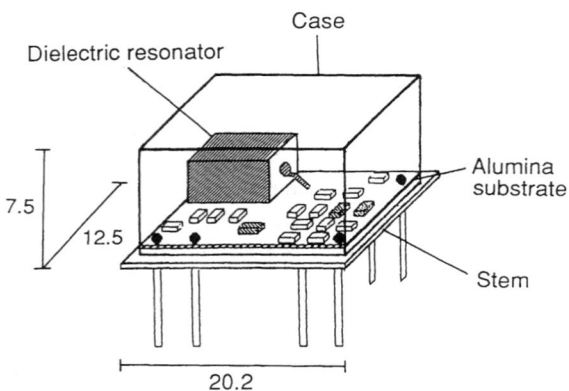

(a) With dielectric resonator

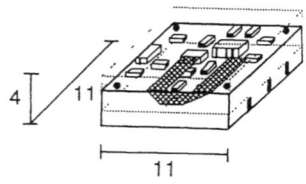

(b) With microstripline

Fig. 17 Structure of VCO.

Fig. 19 Internal view of VCO.

Fig. 20 Appearance of monoblock and coaxial duplexer filters.

specification.

Figure 19 shows the typical internal view of VCO using microstripline resonator.

7. Conclusion

Miniaturization movement of high frequency components applied to mobile telecommunication equipment are being propped up by several technologies developed with dielectric resonators which serve as effective propellant; technologies on dielectric material production, on monoblock, on multilayer, on electrode patterning and processing, on MIC and on electro-magnetic wave analysis-oriented circuit simulation, among others.

These technologies were materiallized by several components of monoblock type filters with additional coupling capacitors on the dielectric substrate, rectangular coaxial resonator type filters, balanced double layer stripline type filter, multilayer lumped constant filter and multilayer stripline type VCO with SMT design concept.

The technologies are applicable to over-all 900 MHz band mobile communication systems in the world, and these components can be miniaturized to less than 1 cm^3 to be small enough for smaller-than-the-present-size portable telephones application.

References

(1) Wakino K., Nishikawa T., Ishikawa Y. and Tsunoda K.: "800 MHz Band Elliptic Function Type Bandpass Filter Using High K. Monoblock Ceramics", IEEE Denshi Tokyo, 25, pp. 117-120(1986).
(2) Wakino K., Nishikawa T., Ishikawa Y. and Hattori J.: "800 MHz Surface Mount Type BDLS Filter", IEEE Denshi Tokyo, 26, pp. 138-142(1987).
(3) Matsumoto H., Yorita T., Ishikawa Y. and Nishikawa T.: "Miniaturized Duplexer Using Rectangular Coaxial Dielectric Resonator For Cellular Portable Telephone", The 3rd APMC Proceedings, 18-4, pp. 407-410(1990).
(4) Yamana H. and Tsuru T.: "Technical Trend Of Complex Chip Parts", The Optimum SMT Handbook, Science Forum, pp. 130-137(Nov. 1990).
(5) Inoue A.: "Development Trend Of VCO For Mobile Communication Equipment", J. I. T. Center, T-2508, pp. 185-251(Nov. 1990).

Bonding Pad Models for Silicon VLSI Technologies and Their Effects on the Noise Figure of RF NPNs.

Natalino Camilleri, Jim Kirchgessner,
Julio Costa, David Ngo, David Lovelace

Motorola, SPS-ACT, M/D M350, Mesa, AZ 85202

Abstract:

VLSI technologies such as BiCMOS and high speed ECL Bipolar are candidates for mixed mode applications which include RF receiver functions. In order for these silicon technologies to achieve low noise characteristics one needs to optimize both the active device and the signal path to the IC interface. Studies in the bonding pad parasitics indicate that these path losses can be very significant. This paper models the bonding pads and presents measured vs. modeled noise figure data for several bonding pad configurations.

Introduction:

As applications for wireless systems move up in frequency, high speed silicon ICs need to provide good interfaces up to 3GHz. Unfortunately in silicon technologies the bonding pads can load such high frequency signals thus special considerations need to be taken since the pads become a significant part of the circuit. High performance silicon VLSI processes such as BiCMOS [1] and MOSAIC5 [2] can be used for mixed mode and RF applications. The devices used in these technologies are capable of switching and amplifying signals in the 5GHz range. When designing the input and the output interface to circuits operating in the 1 to 3GHz range one needs to be careful in modeling any undesirable parasitics that become part of the circuit. This paper concentrates on modeling the bonding pad parasitics and their effects on the noise figure of NPN transistors.

First Order Bonding Pad Model:

The typical bonding pad for these technologies and its first order model are shown in Figure 1. Most pads consist of a metal layer on top of a layer of low loss dielectric (usually a mixture of oxide and nitride layers) which in turn sits on top of semi-conducting silicon. In most IC designs the silicon is mounted to the die flag of the package which is also ground for most RF applications. To first order the piece wise linear model for this structure is a capacitor for the high Q dielectric in series with a resistor for the semi-conducting silicon.

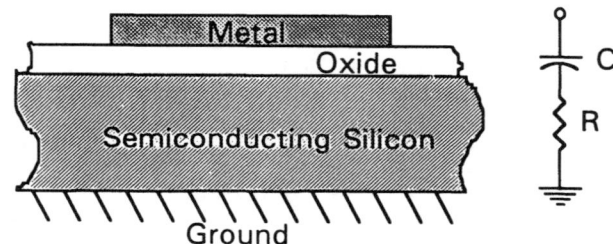

Figure 1. Typical bonding pad configuration for a silicon process. Note the equivalent circuit is a first order approximation for this structure.

The value of the capacitor ranges from .1 to .3 pF depending on the oxide thickness and the resistor varies from 5 to 200 Ohms depending on the resistivity of the substrate used. As the frequency gets higher the bonding pads acts as a resistance to ground and due to the resistive nature of the substrate power lost in the pad cannot be recovered by impedance matching.

IEEE 1994 Microwave and Millimeter-wave Monolithic Circuits Symposium, pp. 225-228, 1994.

Comprehensive Bonding Pad Model:

This simple R-C model is usually sufficient to model the pads up to 3GHz but at higher frequencies a more sophisticated model need to be used. A good model for two pads in an RF probe structure is shown in Figure 2. This model takes into account pad to pad and pad to ground capacitance from the fringing fields in the air as well as resistive coupling in the substrate. The pad to ground capacitance is also distributed to obtain a good fit. The measured vs modeled data for this structure is given in Figure 3. Although this model and its fit are impressive it is not necessary for most first order calculations below 3GHz.

Bonding Pad Design Tradeoffs:

The ideal RF pad will have a small capacitance and a low series resistance to ground. With this in mind the tradeoffs for the best bonding pad design are as follows: (1) The metal pad area needs to be as small as possible to reduce the capacitance but this is limited by the assembly design rules. (2) The oxide thickness needs to be as thick as possible thus in a multi layer metal system the final metal should be chosen and from a process integration standpoint thick field oxides should be provided. In some processes insulating trenches can be placed under the pads to lower the capacitance. (3) The silicon

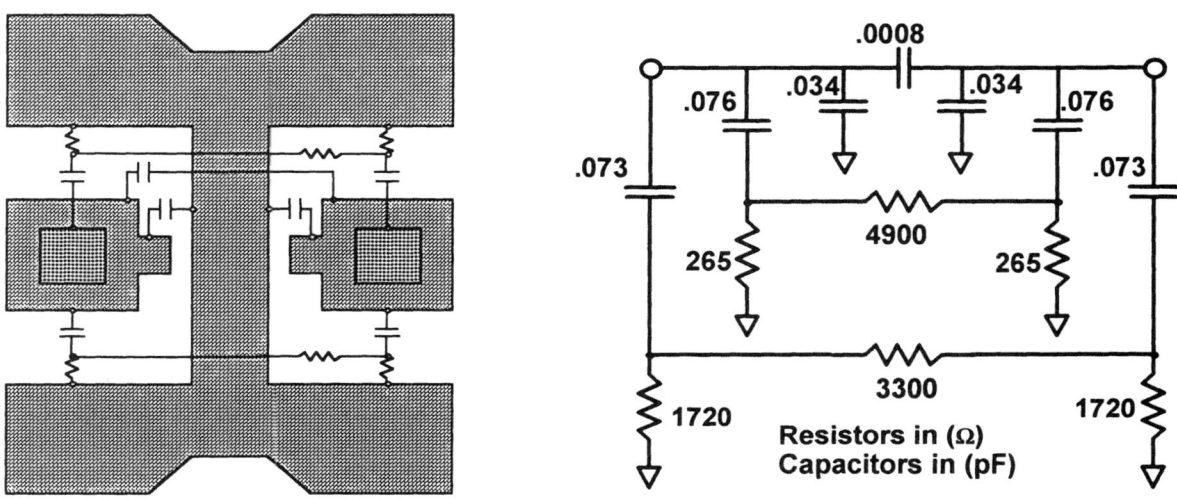

Figure 2. Comprehensive bonding pad model for a BiCMOS 100x100um pad on metal-2.

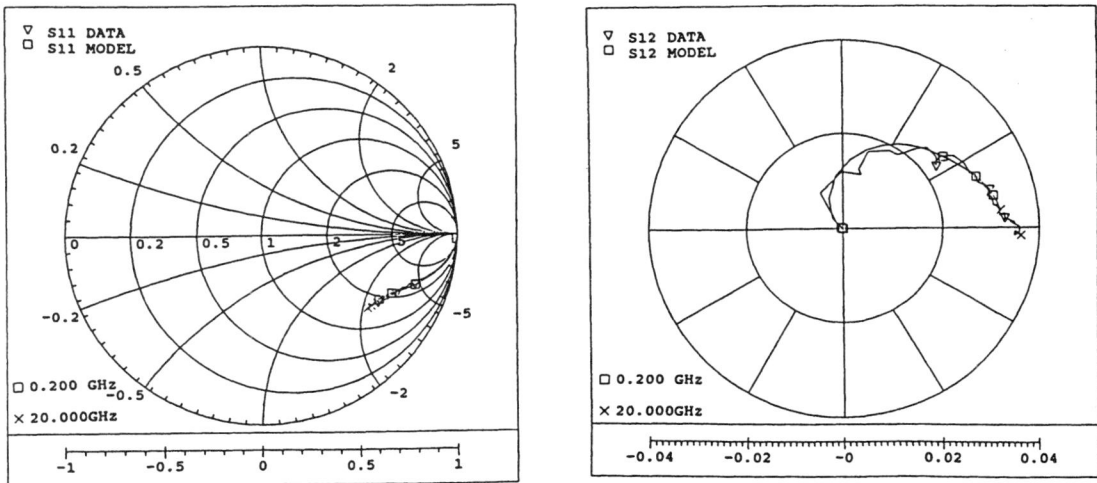

Figure 3. Measured vs modeled impedance for the pad model.

substrate should be either highly conductive to lower the resistance or semi-insulating to act as a dielectric rather than a lossy resistor. The former has been successfully implemented in processes such as MOSAIC5 and RFMOS [3] and provides low loss pads. The latter is more difficult to implement since high resistivity (2 to 10K Ohm cm) material is required. In the case of BiCMOS the silicon substrate resistivity is about 10 Ohm cm which makes the pad design more challenging. (4) Other techniques use a grounded metallic plate under the pad oxide as shown in Figure 4. This technique lowers the resistance value to create a high Q pad.

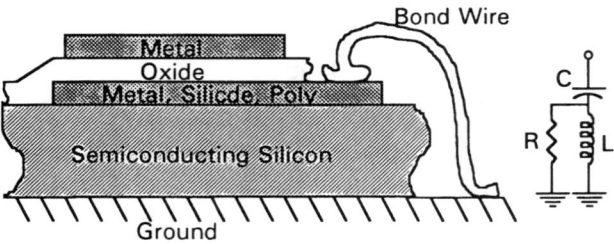

Figure 4. Making use of multi layer metal to reduce the pad resistance.

A summary of the various pads used in various Motorola technologies is shown in Table 1. This study has concentrated on the various pad options of BiCMOS and several options of this technology have been investigated. One option of unique characteristics has a layer of silicide under the oxide. This layer is connected to an adjacent bonding pad which is then grounded to the flag using a bond wires. This technique should work well given that the inductance from the surface of the IC to the die flag is designed to be minimal. An other option is to use trenches under the pad to lower the capacitance.

Bonding Pads Effects on Noise Figure:

The effects of the bonding pad parasitics on noise figure were studied on a high performance NPN used in the BiCMOS process. The devices consist of an interdigitated emitter base structure with 20x14 um or 9x20 emitter fingers. These device geometries are the typical ones used for low noise and low power NPN applications. Devices with the various bonding pads were built and then the noise figure and gain for each individual transistor was measured using a tuned circuit technique. This technique consisted of constructing a low loss circuit that was tuned for the optimum noise performance of each device type at 900MHz. The measured noise data for these tuned amplifiers using the various bonding pads is shown in Figure 5. This data shows that as the device current is reduced the bonding pad parasitic is more significant. This occurs since as the device draws less current the input impedance is higher and the pad noise resistance effect becomes larger.

An attempt to simulate these pad effects on noise figure was made. The simulation consisted of extracting an accurate small signal device model for a specific bias point. Figure 6 shows the equivalent circuit used in the simulation. This circuit was extracted from measured S-parameters using an iterative optimization technique. Then the noise was simulated using a linear simulator [4] which calculates the respective noise components. The results of the simulation for the various bonding pad configurations is shown in Figure 7. This simulation compensates for about .6dB of loss in the circuits. These results agree very well with the measured data at 900MHz. The simulation also shows the large effects the pads have on noise figure as the frequency is increased.

Table 1

Pad Description		C (pF)	R (Ohms)
MOSAIC5: 100x100um pad on Metal 2		19	32
RFMOS: 100x100um pad on Metal 1		.16	180
BiCMOS: 100x100um pad on Metal 1	(100-M1)	.32	180
BiCMOS: 75x75um pad on Metal 2	(75-M2)	.12	225
BiCMOS: 75x75um, Metal 2 with grounded silicide underlayer	(75-Gnd)	.13	42
BiCMOS: 75x75um, Metal 2 with insulating trenches under pad	(75-Tren)	.11	225

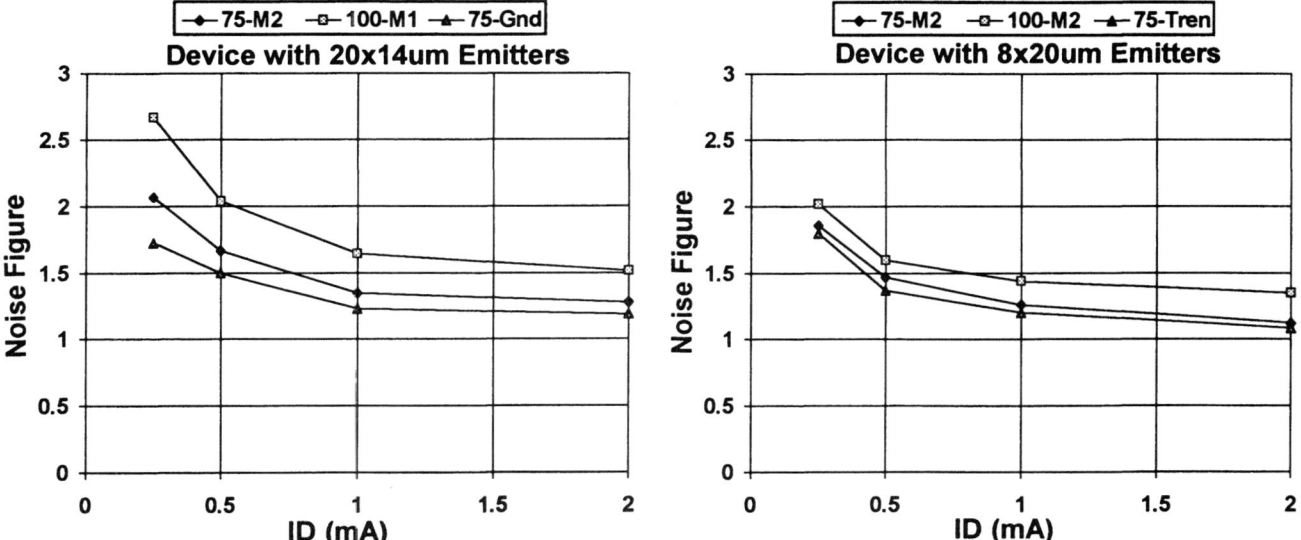

Figure 5. Measured noise data for a low noise BiCMOS NPN amplifiers with various bonding pad structures. Data taken at 900 MHz with Vcc at 1V.

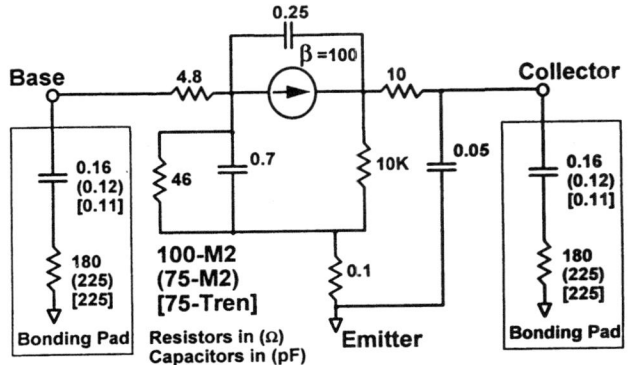

Figure 6. Equivalent circuit for a 9x20 um BiCMOS NPN at 1V, .5mA.

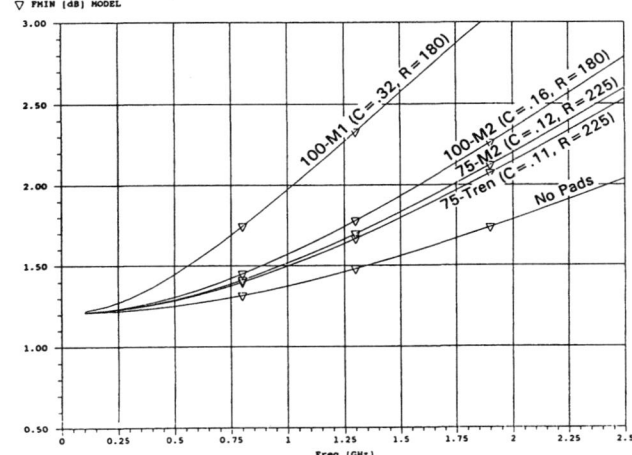

Figure 7. Modeled noise figure data for a low noise BiCMOS NPN with various bonding pad structures. Device size is 9x20um, biased at 1V, 0.5mA

Reference:

[1] J.Kirchgeccner, et. al.. "An Advanced 0.4um BiCMOS Technology for High Performance ASIC Applications," IEDM Technical Digest, 1991, pp. 97-100

[2] V. dela Torre, et. al.. "MOSAIC V - A Very High Performance Bipolar Technology," IEEE BCTM, 1991, pp. 21-24

[3] N Camilleri, et. al. "Silicon MOSFETs, the Microwave Device Technology of the 90s", IEEE MTT-S, 1993, pp. 545-548.

[4] "Microwave Harmonica", Compact Software, 483 McLean Blvd, Paterson NJ 07513

Wideband Characterization of Mutual Coupling Between High Density Bonding Wires

Hai-Young Lee, *Member, IEEE*

Abstract—Mutual coupling between grounded bonding wires for high density IC packaging has been characterized over a wide frequency range using the Method of Moments in consideration of the ohmic and radiation losses. At high frequencies, the mutual inductance greatly increases due to the radiation-enhanced mutual coupling effect. For 500-μm-long bonding wires, a minimum 200-μm separation is required to maintain a 20-dB crosstalk level at low frequencies. This wideband analysis will be useful for designing packages and interconnection layouts of high frequency IC's with increased packaging density and operating frequency.

I. INTRODUCTION

RECENT ADVANCES in integration technology and performance of integrated circuits require greater density of packaging and interconnections demanding close arrangement of bonding wires. Especially in MMIC's and OEIC's operating at high frequencies, the bonding wire becomes dominant parasitic and limits their frequency performance and packaging density [1], [2]. Theoretical characterization of a single bonding wire has been treated using static and full-wave analyses [3], [4]. The full-wave analysis shows significant radiation of the bonding wire enhanced by the slow-wave effect of ohmic loss at high frequencies. The radiation enhances mutual coupling between high density bonding wires at high frequencies. Since the mutual coupling increases crosstalk as well as the parasitic effect between bonding wires, it should be accurately characterized in consideration of the radiation in order to increase packaging density and operating frequencies. Only a simple representation of static mutual inductance is available for infinitely long wires near a ground plane [5]. The radiation effect and the fringing effect at wire edges are not considered in the static approximation. Direct measurement of the mutual coupling has difficulties of precise calibration and error modeling due to the tiny geometry and the radiation problems.

In this letter, two grounded bonding wires coupled with a separation have been characterized over a wide range of frequencies using the Method of Moments (MoM). The mutual inductance and the radiation resistance are calculated by varying the separation between bonding wires. For 500-μm-long bonding wires separated by 200 μm, the mutual inductance is about 10% of the self inductance (0.28 nH) and the corresponding crosstalk level is 20 dB. High mutual coupling enhanced by the radiation has been observed at high frequencies. This wideband characterization of the mutual coupling is expected to be helpful for design and packaging of high frequency integrated circuits with increased packaging density and operating frequency.

Manuscript received March 10, 1994.
The author is with the Department of Electronics Engineering, Ajou University, Suwon 441-749, Korea.
IEEE Log Number 9402968.

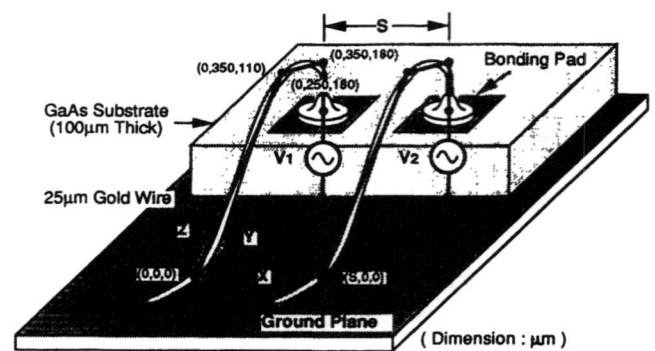

Fig. 1. Two grounded ball-bonding wires of a 25 μm diameter and a 500 μm length separated by S on a 100-μm-thick GaAs substrate.

II. ANALYSIS

Two identical ball-bonding wires shown in Fig. 1 are considered to model the mutual coupling and to see the behavior for mutual separation (S) between the bonding wires. Two voltage sources (V_1, V_2) are respectively applied to two bonding pads on a 100-μm-thick semi-insulating GaAs substrate with a ground plane. Current distributions and input impedances of the bonding wires are calculated by the MoM with incorporation of conductor loss using the Phenomenological Loss Equivalence Method (PEM) [6]. In the MoM calculation, each bonding wire of total length 500 μm is approximately linearized as shown in Fig. 1, and then the linearized bonding wire is appropriately segmented by 12 total pulses using pulse-basis and pulse-testing functions. Distributed internal impedance of the gold bonding wire is calculated over a wide frequency range using the PEM, and it is appropriately distributed on the pulse segments using the lumped impedance loading method for the MoM calculation. The perfect ground plane is substituted by the image of the bonding wires. The dielectric effect of the semi-insulating substrate is neglected for the low-impedance wires, and electric coupling between two bonding pads is not considered but can be simply calculated using the conformal mapping method.

The input impedance for in-phase excitation (Z_+) and that for out-of-phase excitation (Z_-) have been respectively calculated for an in-phase and an out-of-phase excitation of two voltage sources (i.e., $V_1, V_2 = 1$ V and $V_1 = 1$ V, $V_2 = -1$ V).

The input impedance of the bonding wires is obtained from the applied source voltage divided by the calculated source current at bonding pad. Using lumped element modeling, the coupled bonding wires can be equivalently represented by coupled inductors with series resistances associated with the ohmic and radiation losses at a given frequency. Then, the self inductance (L) and the mutual inductance (M) can be calculated, respectively, by

$$L = \text{Im}(Z_+ + Z_-)/2\omega \quad (1)$$
$$M = \text{Im}(Z_+ - Z_-)/2\omega \quad (2)$$

The series wire resistance associated with the ohmic loss and the radiation loss is obtained from the real part of the input impedance. The pure radiation resistance is obtained from the input resistance calculated by assuming an ideal conducting wire.

III. NUMERICAL RESULTS

In Fig. 2(a), shown are the self inductance and the mutual inductance of the coupled bonding wires calculated by the MoM over a wide frequency range for different bonding wire separations (S). They are compared with the static results calculated by simple static formulas [5] commonly used in CAD software. In the static calculations, the curved bonding wires are approximated by two coupled straight wires of 500 μm total length and 100 μm average height from a ground plane. Their self inductances are in very good agreement at low frequencies below 20 GHz. As shown in Fig. 3, it is observed that the current disbribution, calculated by the MoM with the incorporation of the ohmic resistance using lumped loading method, is highly uniform along the bonding wires and the static approximation is valid at low frequencies. For a very close separation of 20 μm, the static mutual inductance is underestimated compared to the MoM result due to the static approximation of straight wires and the fringing effect of the real bonding wires.

At high frequencies above 20 GHz, the self and mutual inductances greatly increase due to the radiation effect and the radiation-enhanced mutual coupling effect, respectively. In order to see the radiation effect clearly, the total resistance and the radiation resistance are shown in Fig. 2(b) for a typical 100-μm separation since the input resistance is almost independent of the wire separation. The total resistance is dominated by the ohmic resistance and the radiation resistance at low and high frequencies, respectively. The increasing radiation resistance implies the significant radiation and consequent high mutual coupling at high frequencies. The nonuniform current distribution at high frequencies shown in Fig. 3 is due to the radiation effect as well. The current disbribution increasing from the bonding pad at high frequencies is related to the radiation-enhanced magnetic coupling between the real bonding wire and its image wire. Higher magnetic coupling around the ground point effectively increases vertical current component of the real bonding wire, whereas horizontal current around the top of the bonding wire is decreased by the opposite image current. Since the mutual inductance is increasingly more rapidly to the increasing frequency than the

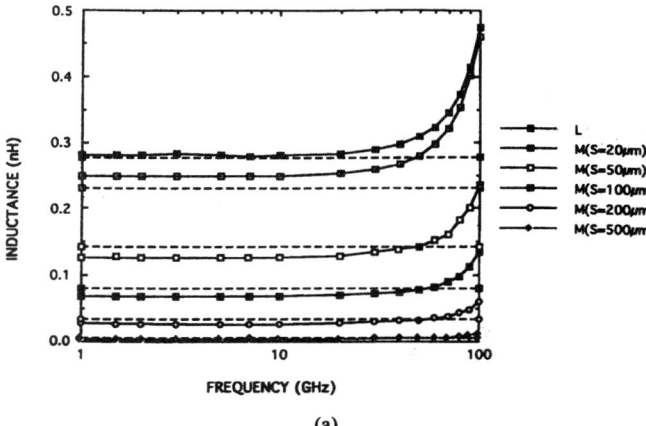

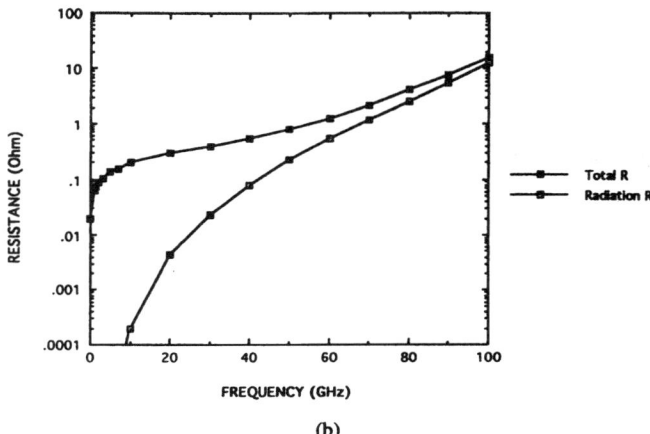

Fig. 2. (a) Self (L) and mutual (M) inductances of coupled bonding wires for different wire separations (S) calculated by the MoM (solid line) and static formulas (dashed line). (b) Total input resistance and pure radiation resistance calculated by the MoM for a 100-μm wire separation.

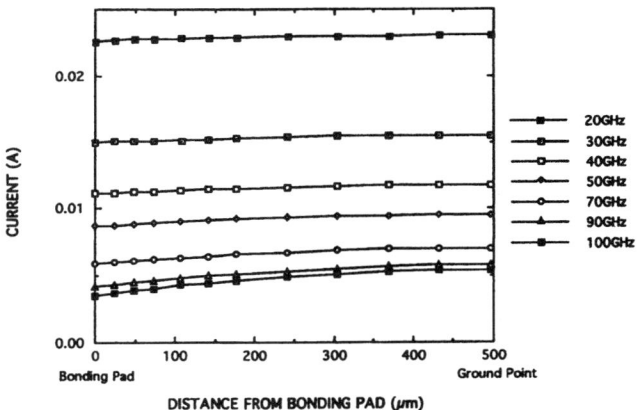

Fig. 3. Current distributions along the bonding wire axis for in-phase excitations ($V_1, V_2 = 1$ V) of different frequencies in the case of 100-μm separation.

self inductance because of the radiation-enhanced mutual coupling, high density bonding wires for high frequency integrated circuits should be accurately characterized in consideration of the radiation.

From the variations of calculated self and mutual inductances to the separation in Fig. 2(a), we can note that the bonding wires should be separated by at least 200 μm in order to keep 20-dB crosstalk factor [7] corresponding to 10% mutual inductance to the self inductance. However, the minimum separation to maintain the low crosstalk level is increasing proportionally to both operating frequency and total wire length due to the radiation-enhanced mutual coupling effect.

IV. Conclusion

The method of moments with incorporation of ohmic loss is applied to two grounded bonding wires in order to characterize the mutual coupling for a wide variation of wire separations. The calculated results show high mutual coupling, enhanced by the radiation effect, greatly increases the equivalent mutual inductance at high frequencies. For 500-μm-long bonding wires, a minimum 200-μm separation is required to maintain a 20-dB crosstalk level at low frequencies. The minimum separation is increasing proportionally to the frequency as well as the wire length due to the radiation-enhanced mutual coupling. Simple static formulas are not accurate for close wire separations due to the straight wire approximation and the fringing effect at the wire ends. These wideband results are expected to be useful for designing packages and interconnection layouts of high frequency integrated circuits with increased packaging density and operating frequency.

References

[1] *Proc. IEEE MTT-S Int. Microwave Symp., Joint Workshop on New Packaging Techniques for MMICs and Discrete Devices and Loss, Crosstalk, and Package Effects in Microwave and Millimeter-Wave Integrated Circuits*, 1991, Boston, MA.
[2] M. Nakamura, N. Suzuki, and T. Ozeki, "The superiority of optoelectronic integration for high-speed laser diode modulation," *IEEE J. Quantum Electron.*, vol. QE-22, pp. 822–826, June 1986.
[3] R. H. Caverly, "Characteristic impedance of integrated circuit bond wire," *IEEE Trans. Microwave Theory Tech.*, vol. MTT-34, pp. 982–984, Sept. 1986.
[4] H.-Y. Lee, "Wideband Characterization of a Typical Bonding Wire for Microwave and Millimeter-wave Integrated Circuits," to be published in *IEEE Trans. Microwave Theory Tech.*
[5] R. J. Mohr, "Coupling between open wires over a ground plane," *IEEE Symp. Electromagnetic Compatibility Dig.*, 1968, pp. 404–413.
[6] H.-Y. Lee and T. Itoh, "Phenomenological loss equivalence method for planar quasi-TEM transmission line with a thin normal conductor or superconductor," *IEEE Trans. Microwave Theory Tech.*, vol. MTT-37, pp. 1904–1909, Dec. 1989.
[7] C. S. Walker, *Capacitance, Inductance and Crosstalk Analysis*. Artech House, Inc., 1990.

HIGH-SPEED CHARACTERISTICS OF MULTILAYER CERAMIC PACKAGES AND TEST FIXTURES

David H. Smith and Raj M. Savara
TriQuint Semiconductor, Inc.
Group 700, P.O. Box 4935
Beaverton, OR 97076

Abstract — The high-speed electrical characteristics of a typical multilayer ceramic package are described and a model presented. The tests were conducted in a realistic environment which closely models the surface mount environment for which the packages are intended.

A dominant electrical feature of the configuration is the ground plane discontinuity at the package-to-circuit-board interface. This feature is prominant in the observed high-speed pulse response of the system.

INTRODUCTION

Multilayer ceramic technology has proven to be well suited for application to high-speed packages for GaAS ICs. These packages were initially described at this conference in 1985[1] and have proven extremely effective in digital and mixed analog/digital applications involving clock rates into the 1-2 gHz frequency range. Some typical MLC packages are illustrated in figure 1.

Figure 1: Typical high-speed MLC packages. The packages are normally mounted cavity down with leads bent in a gull wing as shown.

To allow realistic high-speed testing, test fixtures which closely emulate a high-speed system environment have also been developed. For example the fixture used in the measurements reported here is illustrated in figure 2.

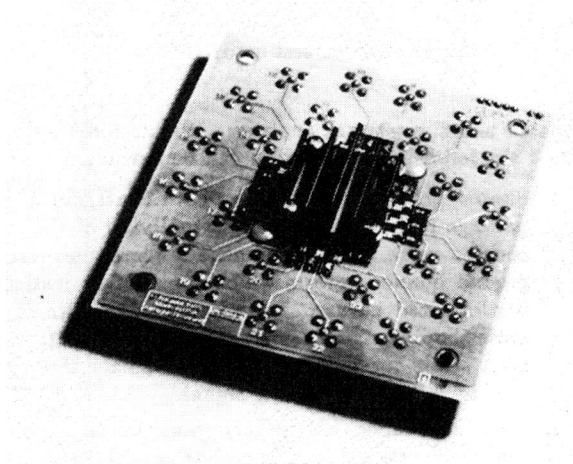

Figure 2: A test fixture for the MLC package closely emulates a systems environment.

These MLC packages incorporate many of the same refinements used at the circuit board level to improve high-speed performance, namely use of controlled impedance, 50Ω, signal lines, multiple ground connections, and on-package power and ground planes.

MEASUREMENTS

Traditional time domain measurements were performed using a Tektronix 11801 digital sampling oscilloscope. The tests were set up for observation of reflected and transmitted signals derived from the built in high speed pulse generator. A typical test setup is illustrated in figure 3.

In order to separate the effects of the various electrical interfaces measurements were performed with and without various components of the complete test setup in place. The results are illustrated in figure 4.

From this data, and from observation of the perturbations due to carefully placed small probes, one can readily deduce a relationship between the features on the TDR and physical features such as the discontinuity due to a coaxial connector. Frequency domain data was also ex-

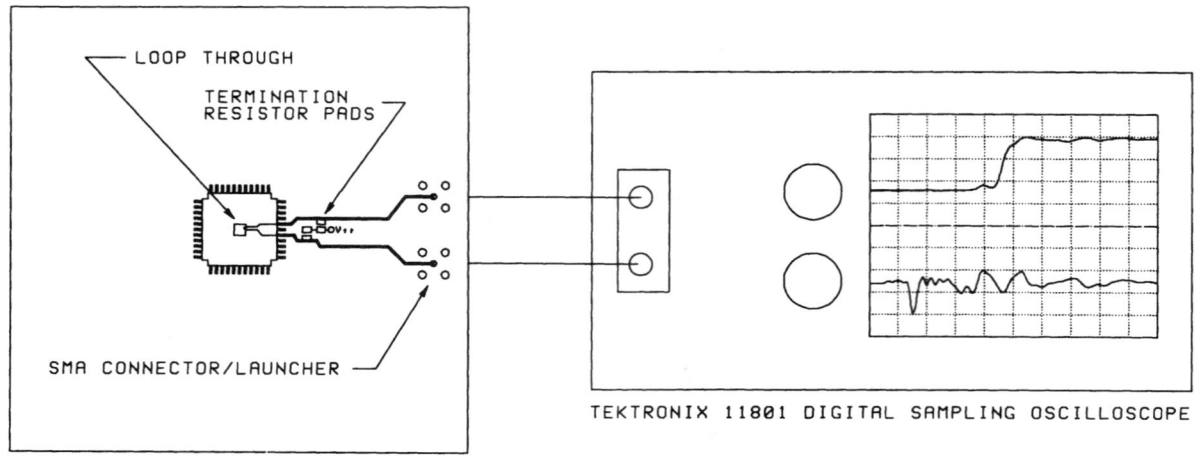

Figure 3: The test setup incorporated a Tektronix 11801 oscilloscope to perform the measurements.

tracted using a network analyzer. With reference to figure 4 the following information can be extracted:

1. Shared ground inductances of ≈1.5 nH/pin at the package pins accounted for virtually all of the cross coupled noise. This phenomena is sometimes termed ground bounce, a term which does not do justice to it's localized nature. For adjacent signal lines, the cross coupled signal remained below 10 db up to 2.0 gHz, while for widely separated signals the signal remained 20 db below the signal.

2. The inductance of package pins in signal lines, though necessary in a complete model, was determined to be somewhat less important in controlling signal distortion than the inductances of ground pins.

3. A capacitive discontinuity of approximately 0.2 pF was identified at the coaxial to microstrip transition for the circuit board/test fixture, with a capacitive discontinuity of similar size near the solder pad for the package pin.

4. Dribble-up effects of up to 5-10% due to skin effect losses in the connecting transmission lines was observed in the measurements, but in most cases signal quality was overwhelmingly dominated by the parasitic capacitance and inductance effects, including multiple reflection effects.

CIRCUIT BOARD INTERFACE MODEL

Figure 5 shows a typical geometry at the point where the package pins contact the circuit board. Note that far from the package pins, the return current is confined to the area directly under the signal trace. At the package pin transition, the return currents from multiple signal lines are forced to flow through ground pins and vias which are both more remote from the signal line and must be shared with other signals.

SPICE SIMULATION

A SPICE model for the measurement was constructed based on the discussion in previous section. The circuit

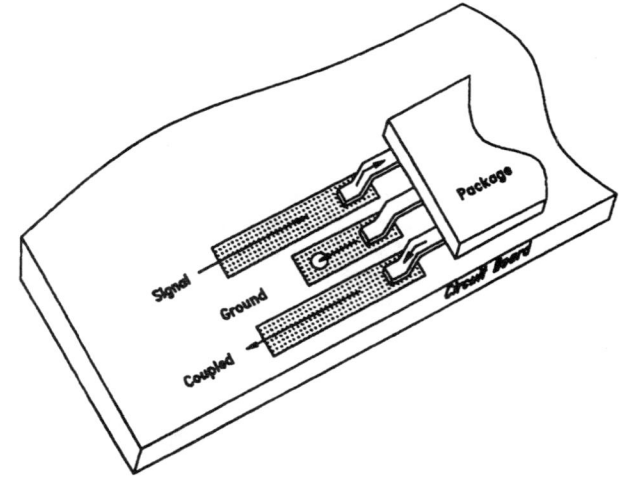

Figure 5: At a typical board-to-package all return current must flow through the common ground pins, vias, etc.

is depicted in figure 6 with the simulation results shown in figure 7. Notice the close similarity to the measured results.

CONCLUSIONS

Using pulsed reflection and transmission measurements as a diagnostic tool, we have shown how to develop a full SPICE model for a high-speed MLC package in a systems environment. A prominent feature of the model is the shared ground present at the package pins.

References

[1] D. H. Smith, T. G. Bowman, R. Lind, and T. S. Riley, "New approaches to packaging for high-speed gaas ic applications," in *IEEE GaAs IC Symposium Technical Digest*, pp. 151-154, 1985.

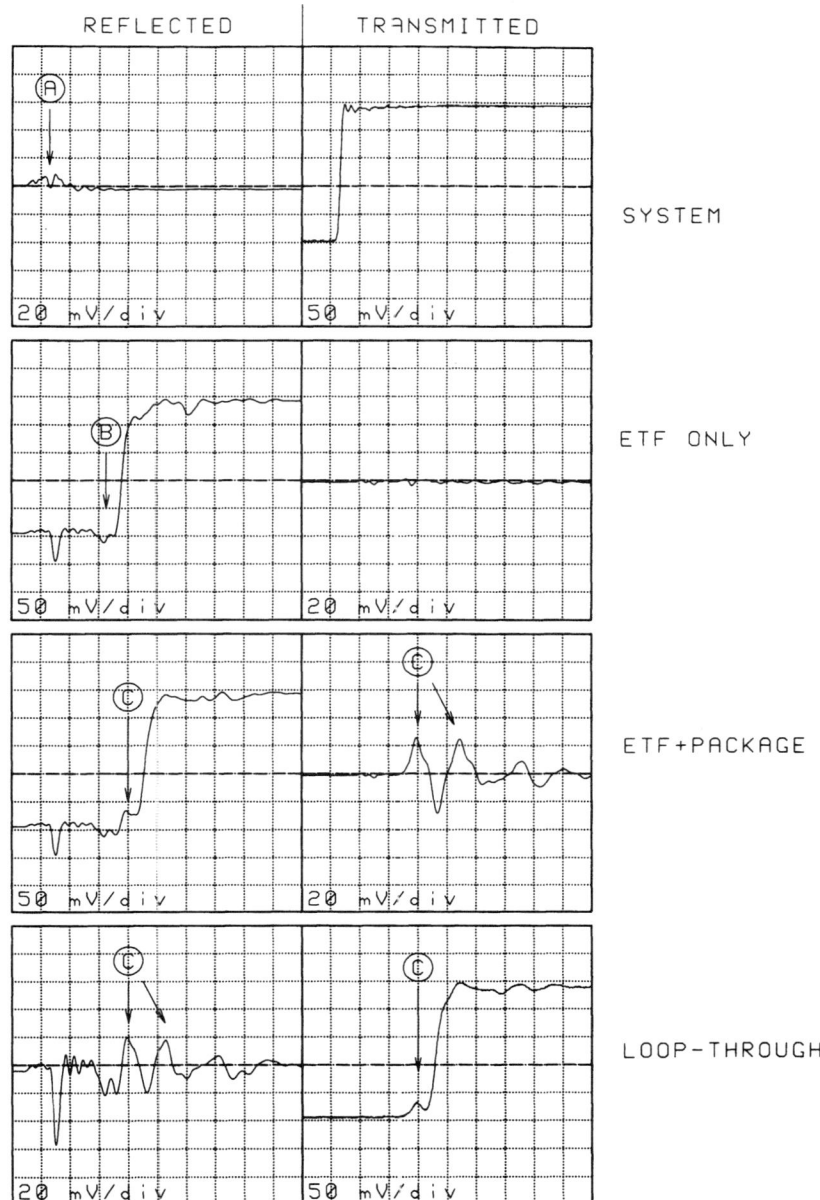

Figure 4: Time domain measurements for various test conditions reveal much about circuit model. The time scale is 200 ps/div for all traces, and the horizontal scale is noted for each trace. *Top Row*: Test board is replaces with an SMA "barrel" to show system response. *Second Row*: Test fixture only (no package) *Third Row*: Test fixture plus package (no loop through). *Bottom Row:* Test fixture plus package with loop through inside the die cavity. Flagged features result from: A) SMA connector. B) Pad for termination resistor. C) Shared grounds.

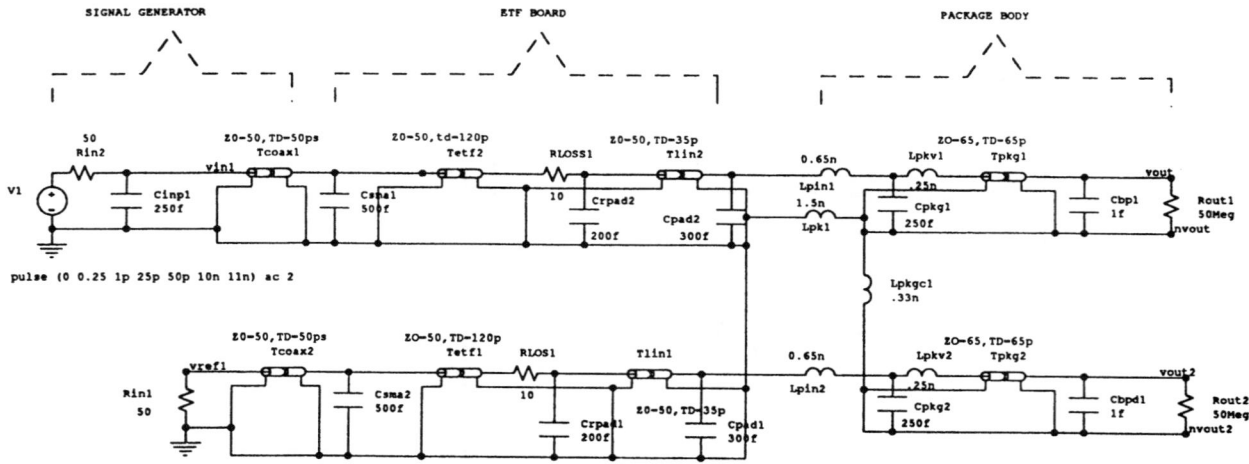

Figure 6: A fully developed SPICE model for the package-plus-test fixture can become quite complex as illustrated here. This circuit provides a close representation of the two signal measurement depicted in figure 4

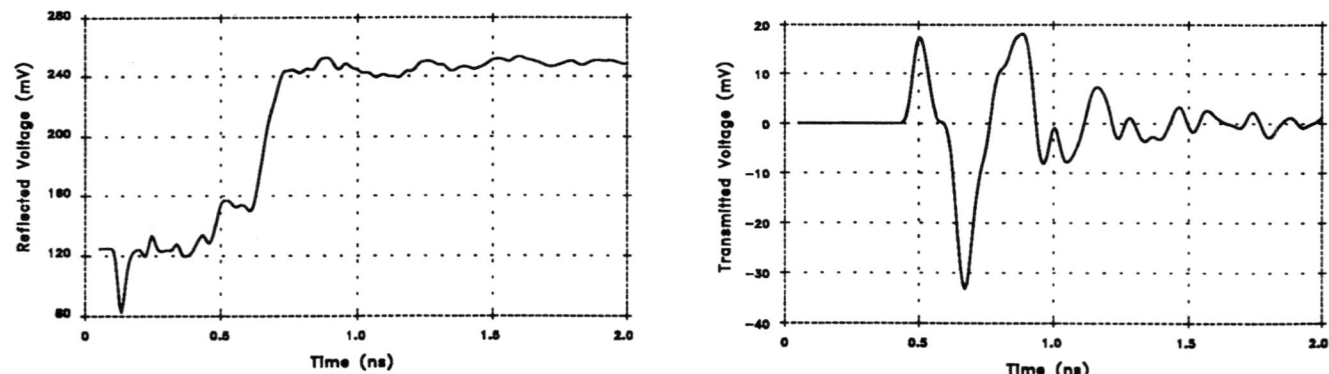

Figure 7: Spice simulation results closely approximate the results of pulsed measurements reported in figure 4.

An Equivalent Circuit for a Microwave Surface Mount Package

David W. Hughes and **David M. Jackson**
Georgia Institute of Technology
Atlanta, GA

An electrical equivalent circuit has been developed for a microwave surface mount package. The self-consistent procedure used for the creation of this model is described and comparisons between the measured and the calculated results are presented.

Introduction

An electrical equivalent circuit has been developed for a microwave surface mount package. These packages are useful for housing monolithic microwave integrated circuits (MMICs). The electrical characteristics of such MMICs are frequently perturbed by their enclosure in a microwave chip package. However, knowledge of the equivalent circuit for the package allows the intrinsic performance of the MMIC to be de-embedded from measurements on the chip-package ensemble.

Often the package is placed in some sort of a fixture in order to conduct certain microwave evaluations. In essence, the package permits electrical access to the chip while the fixture constitutes the interface between the packaged chip and external circuitry.

This article illustrates how proper modeling of the surface mount package and the associated microwave fixture allows the determination of the intrinsic performance of an MMIC embedded in such hardware.

The Microwave Fixture

The selected microwave fixture[1] is specified for operation from 0 to 12 GHz and has eight 3.5-mm input/output connectors. The entire assembly is solderless and relies on elastomeric pressure plates to keep the electrical contact regions in the proper physical alignment. An internal cavity machined into the integral circuit board is sized to accommodate the surface mount package. Figure 1 is a schematic diagram of the evaluation fixture.

A series of tests was performed in order to confirm the specifications, such as network analyzer reflection measurements, on each of the eight input/output ports on the assembly. The resulting data showed that each port performs similarly and that there appears to be a well-matched transition between the coaxial connectors and the respective microstrip lines on the internal circuit board.

Also, the repeatability of the solderless connections between the microwave fixture and the surface mount package accommodated by the hardware was investigated. Figure 2 shows transmission data on a packaged through line that has been removed and re-inserted into the fixture six times.

During operation, any of the eight input/output connectors not being used must be terminated. Figure 3 shows that leaving the unused terminals open during measurements causes reflections.

Finally, tests were performed to determine the upper frequency limit at which the microwave fixture can be used with convenience. Although it is specified for operation up to 12 GHz, the performance of the particular fixture obtained for this study seems to degrade some-

[Continued on page 136]

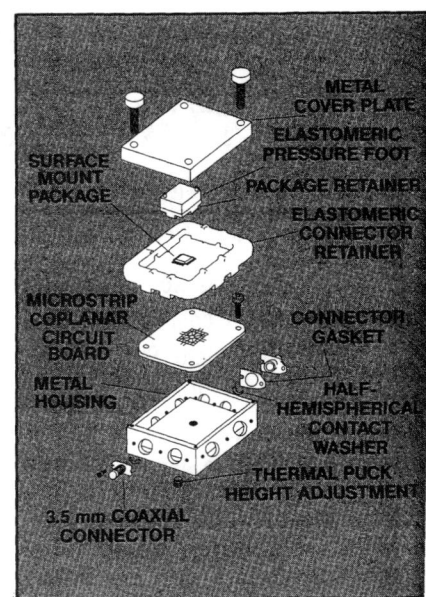

Fig. 1 The microwave evaluation fixture.

what above 10 GHz. As a result, measurements and modeling with the available hardware were limited to 9.5 GHz in order to ensure the validity of the corresponding results.

Calibration of the Microwave Fixture

When placing the surface mount package into the fixture and connecting the fixture to a network analyzer, the corresponding measurement does not yield only data on the surface mount package. It results in the measured performance of the package being perturbed by the presence of the fixture. If the fixture has well-matched transitions between the surface mount package and the outside circuitry, it is possible to eliminate its influence on the measurement.[2]

The first step toward this goal is to calibrate the test instrument itself at the ends of the cables that will be connected to the fixture. For example, a network analyzer-cable combination can be calibrated using the coaxial shorts, opens, sliding loads, fixed loads and throughs commonly available. A shorting block is used to fill the fixture cavity designed to accommodate the microwave package. The resulting network analyzer data makes a lengthy excursion around the Smith chart. This is a result of the time delay experienced by the signal in traveling from the initial calibration plane at the cable end through the fixture to the edge of the shorting block and back again. By adjusting the network analyzer to extend the reference plane, the Smith chart trace can be collapsed into a dot representing a short circuit. This indicates that the calibration plane is coincident with the edge of the shorting block. As a consequence, it is now possible to perform network analyzer measurements referenced to the edge of the package.[2] For this paper's hardware and network analyzer, a reference plane extension of 124.5 pS collapsed the trace to a dot.

By calibrating the network analyzer at the ends of the cables and extending each calibration plane by 124.5 pS, the delay added by the test fixture can be eliminated. Thus, it is possible and convenient to perform the desired measurements directly on the enclosed surface mount package.

The Surface Mount Package

The surface mount package[3,4] has two microwave signal lines arranged in a ground-signal-ground configuration along each edge of the square package. Thus, microwave engineers have easy access to eight signal leads. A schematic diagram of the surface mount package is shown in Figure 4.

Inside the surface mount package is a thin-film ceramic hybrid circuit board attached to both a Kovar leadframe and a heat-sinking button. The coplanar waveguide connects through a via hole to a microstrip region running across the top side of the circuit board. Wires can be bonded from near the ends of the microstrip sections to the corresponding pads on the adjacent MMIC.

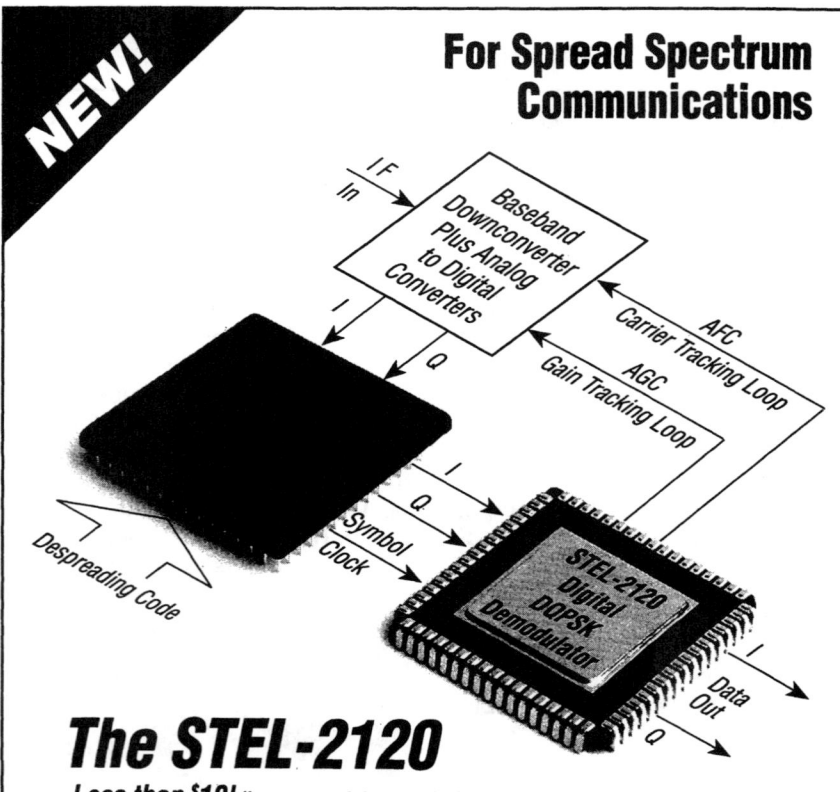

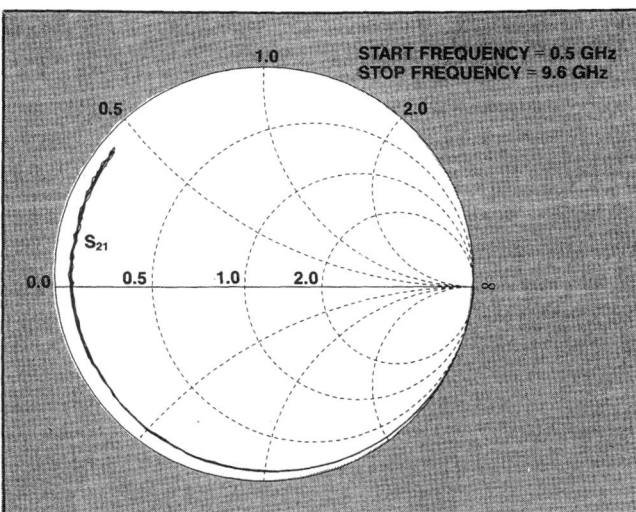

Fig. 2 Transmission data on a packaged through line that has been removed and re-inserted into the fixture six times.

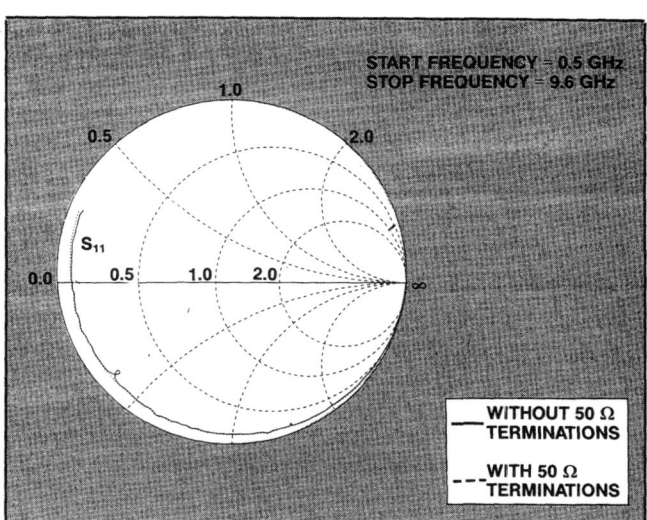

Fig. 3 A comparison of measurements on the microwave fixture with and without 50 Ω terminations on the unused connectors.

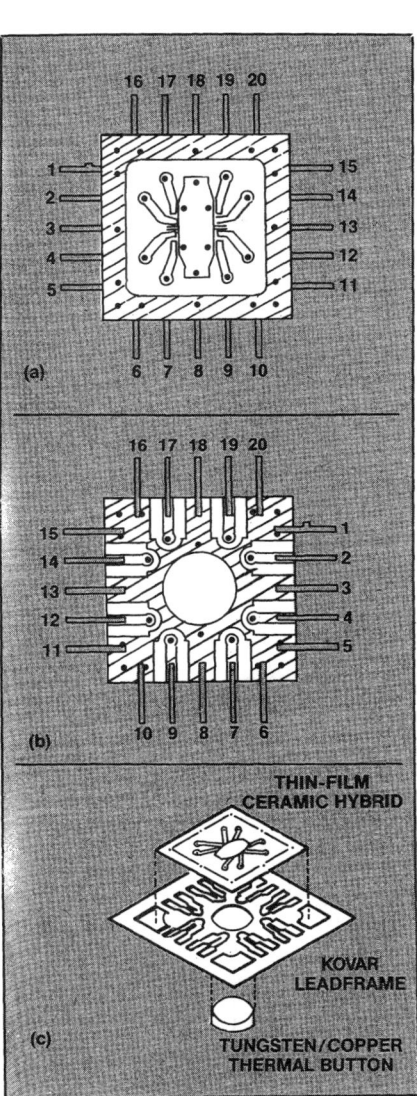

Fig. 4 A schematic diagram of the surface mount package; (a) top view, (b) bottom view and (c) exploded view.

†Touchstone is trademark of EEsof Inc.

The surface mount package is designed for hermetic assembly. A gold-tin solder preform is available to aid in the attachment of the metallic lid to the body of the package. Network analyzer measurements reveal that the presence or absence of a lid has a negligible effect on the measured electrical characteristics of the package.

An Electrical Equivalent Circuit for the Surface Mount Package

The first step in the generation of an equivalent circuit for the surface mount package is to derive the topology of an electrical schematic diagram representing the physical geometries inherent in the package.[5] Perusal of Figure 4, augmented with optical measurements of the respective line widths and spaces, suggests that the circuit schematic diagram shown in Figure 5 can be used to describe the corresponding hardware. The nomenclature in Figure 5 corresponds to that employed in the Touchstone† program.

The physical origin of each of the components included in Figure 5 is detailed in Table 1. The electrical connections to this package are assumed to be made immediately adjacent to the package body itself. Similarly, it is presumed that the surface mount package makes the transition smoothly to an appropriate fixture. Measurements on the surface mount package-microwave fixture combination show this to be the case and, thus, the usual discontinuity capacitance at this interface can be neglected. Similarly, there is negligible crosstalk between signal leads for the frequencies of interest.

Determination of the values for the components in Figure 5 involves a combination of manual calculations, microwave measurements and computer simulations. For example, the inductance of model element L_1 is calculated from the self-inductance formula.[6]

$$L_1 = 5.08 \, l \cdot \left[\ln\left(\frac{2l}{w+t}\right) + \left(\frac{1}{2}\right) + \left(\frac{2}{9}\right)\left(\frac{w+t}{l}\right) \right]$$

where

L_1 = self-inductance of the conductive stripe (nH)
l = length of the conductive stripe (inches)
w = width of the conductive stripe (inches)
t = thickness of the conductive stripe (inches)

For the utilized surface mount package, $l = 0.019''$, $w = 0.015''$, $t = 0.0001''$ and, thus, $L_1 = 0.15$ nH.

The values for components R_0 and C_1 were determined by using a network analyzer to measure the S-parameters of the empty package. Subsequently, the package equivalent circuit was entered into Touchstone. The computer routine was used to adjust the values of R_0 and C_1 until the simulated S-parameters for the equivalent circuit matched those measured with the network analyzer.

Conclusion

An electrical equivalent circuit has been developed for a surface mount package. The elements in this model correspond to the physical configuration of the hardware. Numerical values for these elements were determined by a self-consistent, iterative method using measured data augmented with computer calculations. The resulting electrical model accurately predicts the

[Continued on page 144]

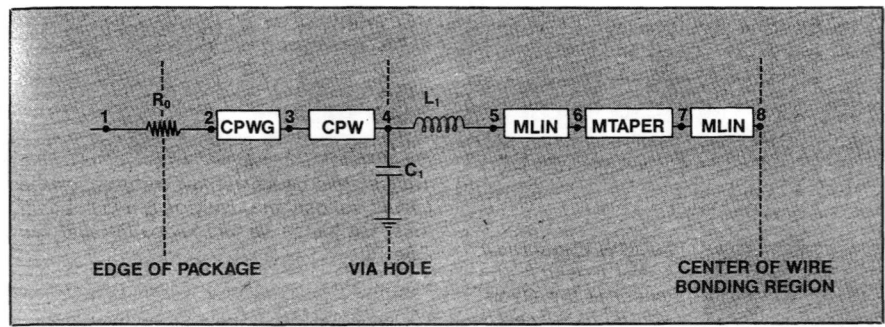

Fig. 5 An equivalent circuit for the surface mount package using Touchstone nomenclature.

TABLE I ELEMENTS IN THE SURFACE MOUNT PACKAGE MODEL		
Symbol	Typical Value	Description
—	relative permittivity = 9.5 substrate thickness = 0.015" metal thickness = 0.0001" type of metal = gold	Gold metalization on an alumina substrate
RO	1.5 Ω	Contact resistance between the microwave fixture and the surface mount package
CPWG	stripe width = 0.015" gap = 0.0215" stripe length = 0.041"	Portion of the coplanar waveguide with a lower ground plane
CPW	stripe width = 0.015" gap = 0.0215 inches stripe length = 0.030"	Portion of the coplanar waveguide without a lower ground plane
C1	0.05 pF	Capacitance of the via region transition
L1	0.15 nH	Self-inductance of the stripline without a lower ground plane
MLIN	stripe width = 0.015" stripe length = 0.020" (pins 2, 4, 12 and 14) = 0.021" (pins 7, 9, 17 and 19)	Portion of the microstrip line with a lower ground plane
MTAPER	stripe initial width = 0.015" stripe final width = 0.010" stripe length = 0.023" (pins 2, 4, 12 and 14) = 0.015" (pins 7, 9, 17 and 19)	Tapered portion of the microstrip line with a lower ground plane
MLIN	stripe width = 0.010" stripe length = 0.005"	Portion of the microstrip line with a lower ground plane and which is assumed to terminate at the center of the wire bonding region

performance of the package up to at least 9.5 GHz. Knowledge of the equivalent circuit for this package can be used to de-embed mathematically the intrinsic performance of a chip from measurements on the chip-package assembly.[7,8]

Acknowledgment

It is a pleasure to acknowledge the capable assistance of Geoffrey Herrick and John Wright at Tektronix and Mike Harris at the Georgia Tech Research Institute during the preparation of the manuscript. The described microwave surface mount package was the Tektronix TPAK 142 and the described microwave fixture was the TriQuint ETF 9000. ∎

References

1. Anon., "12 GHz Microwave Test Fixture for Surface-Mount Package Testing of MMICs," *Microwave Journal,* Vol. 31, No. 2, Feb. 1988, pp. 202-203.
2. D.W. Hughes, C.T. Rucker, R.K. Feeney and D.R. Hertling, "Calibrating RF Test Fixtures," *RF Design,* Vol. 9, No. 9, Sept. 1986, pp. 41-43.
3. G. Herrick and K.E. Jones, "Surface-Mount Pack Houses GaAs MMICs," *Microwaves & RF,* Vol. 25, No. 6, June 1986, pp. 157-159.+
4. K.E. Jones, "Surface Mountable Microwave IC Package," US patent 4,626,805, Dec. 2, 1986.
5. D.W. Hughes, H.M. Harris and C.T. Rucker, "An Equivalent Circuit for a 70 mil Microstrip Package," *Microwave Journal,* Vol. 29, No. 8, Aug. 1986, pp. 97-98.+
6. P.M. Rostek, "Avoid Wiring-Inductance Problems," *Electronic Design,* Vol. 22, No. 25, Dec. 6, 1974, pp. 62-66.
7. R.F. Bauer and P. Penfield, Jr., "De-embedding and Unterminating," IEEE Trans. on Microwave Theory and Techniques, Vol. MTT-22, No. 3, March 1974, pp. 282-288.
8. R. Lane, "De-embedding Device Scattering Parameters," *Microwave Journal,* Vol. 27, No. 8d, Aug. 1984, pp. 149-150."

FOR USE WITH GaAs MMIC AMPLIFIERS

Stephen R. Smith and Michael T. Murphy

M/A-COM Microelectronics Division, IC Design Center

ABSTRACT

A test methodology will be presented which combines the advantages of on-wafer RF probing with a TRL calibration to create a completely de-embeddable, novel "test fixture" capable of electrically characterizing most any style package or device. This scheme has been used to characterize many of the currently available microwave packages in order to identify appropriate packages for our MMIC amplifier products which cover frequencies up to 12 GHz. In addition, the technique has been employed to characterize injection molded plastic packages and to evaluate non-probeable MMICs.

INTRODUCTION

Most package vendors have very little microwave design and characterization capability. Their limited characterization efforts typically involve the use of poor fixturing, which obscures the true frequency response of the package. Companies specializing in fixturing, while investing considerable mechanical engineering effort, expend far less on electrical considerations, often producing fixtures inadequate for use at microwave frequencies. Consequently, there is very little microwave performance data available from package vendors.

Therefore, to evaluate and identify candidate packages for each of the amplifiers in our MMIC amplifier product line required developing fixturing for each package style considered. A novel fixturing approach was designed and implemented, which not only eliminates the need for expensive, package specific fixtures, but also overcomes the frequency limitations of traditional connectorized, plunger style fixtures. Additionally, a rigorous calibration method was developed which allows complete fixture de-embedding.

This test methodology is applicable to practically any style device. Table 1 lists the package styles investigated. Through this work, proper electrical characterization of commonly used packages has indicated useful frequency ranges broader than expected by even the package manufacturers. This finding has allowed us to use low-cost packages for frequency applications where our competitors typically resort to high priced custom packages.

DESIGN APPROACH

To eliminate the need for expensive, device specific, traditional fixtures and overcome their frequency limitations, an RF probeable ceramic substrate was designed as the interface to the device-under-test (DUT). Figure 1 illustrates this coplanar probe to microstrip transition. It is a 50 ohm line fabricated on 10 mil thick alumina, with an 8 mil pitch, ground-signal-ground (G-S-G) probe pattern at one end. The two ground pads are connected to the substrate backside with 8 mil diameter plated vias. The G-S-G pattern can be probed using commercially

Package Description	Manufacturer
5 lead, ceramic	Kyocera
6 lead, ceramic	Kyocera
leadless, 6 port, ceramic	StratEdge
7 lead, ceramic	Kyocera
8 lead, ceramic	Kyocera
8 lead, glass	Mini-Systems
8 lead, glass, ground straps	Mini-Systems
leadless, 8 port, ceramic	Oxley
leadless, 10 port, ceramic	Alcoa

Table 1. Summary of Packages

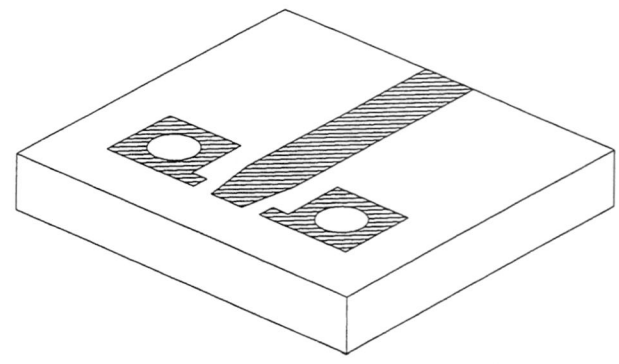

Figure 1. Probeable Ceramic Substrate

available microwave probes on a standard microwave probe station. The opposite end of the substrate can be bonded to a test port of the DUT.

To complete the "test fixture", only a thin brass block is required to serve as the mounting surface for the ceramic substrates and the DUT. If necessary, the brass block could be machined to compensate for any difference in height between the substrate and DUT test port. To fixture practically any DUT, all that is needed is a brass plate and the probeable ceramic substrates. Figure 2 shows the configuration used for characterizing our MAAM71200-H1, a packaged 7-12 GHz GaAs MMIC low noise amplifier.

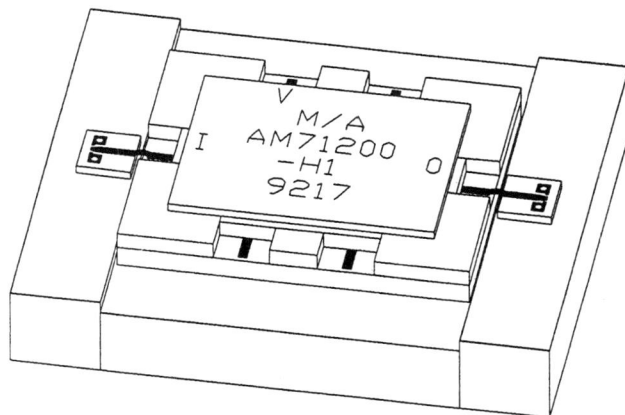

Figure 2. Fixtured MAAM71200-H1

To de-embed this "test fixture", a set of through-reflect-line (TRL) standards was employed. A "zero-length" through, a short, and two delay lines were fabricated. These standards, shown in figure 3, are used with the common TRL de-embedding algorithm. This allows any measurement made with the probeable ceramic substrates to be de-embedded to yield data for only the DUT with connecting bonds. Bond wires can also be de-embedded by first characterizing and modeling them using this same "probeable ceramic" technique. For this work, multiple bond wire and ribbon lengths were characterized to generate fully scalable bond models.

To demonstrate the package characterization method, the evaluation of a standard Kyocera 8 lead ceramic flatpack will be examined. Figure 4 shows how one feedthrough structure in the wall of this package was tested. Package leads were cut close to the package body, and the ceramic substrates were mounted flush to the package ports. Two short 3 mil wide gold ribbons bond the substrates to the package. Similarly, sealed packages with leads internally terminated with 50 ohm chip resistors were tested to determine the cross-coupling between

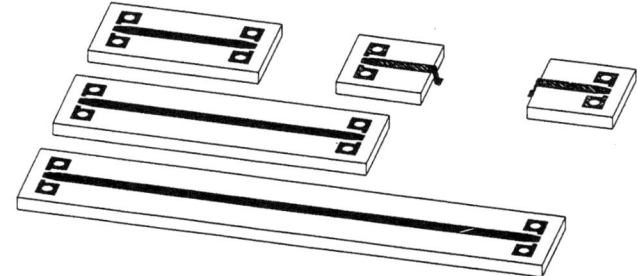

Figure 3. TRL Calibration Standards

opposite and adjacent leads. Through lines within sealed packages were also measured. With this data the true electrical performance of the package was determined and models for the feedthrough and coupling were developed. This information allows the identification of an appropriate package for existing MMIC products and provides an accurate model for incorporating package effects into future design work.

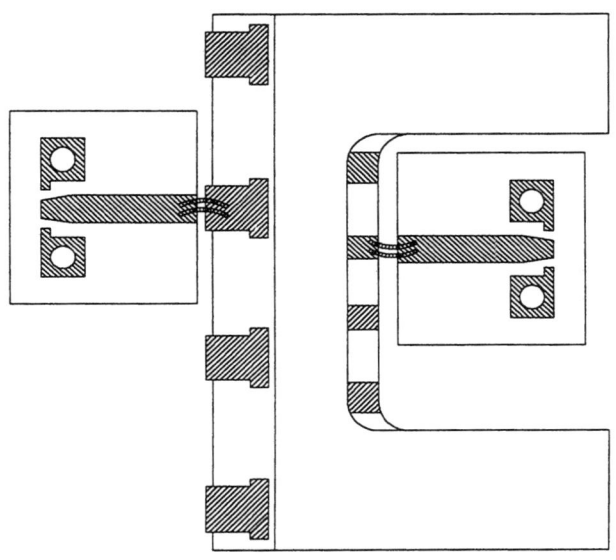

Figure 4. Fixtured Feedthrough

EXPERIMENTAL RESULTS

The feedthrough walls of each package listed in table 1 have been tested and modeled. This feedthrough data alone largely indicates the useful frequency range of each package. Figure 5 shows the frequency response for the feedthrough of the 8 lead ceramic flatpack. This package, previously thought to be useful only at lower frequencies, demonstrates excellent performance

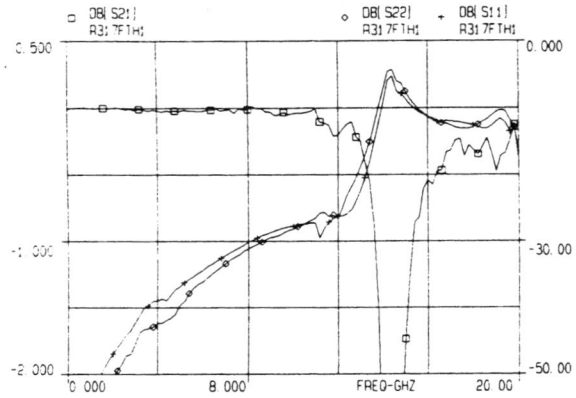

Figure 5. Feedthrough Frequency Response

well into X-band before resonating. Based on this result, we assembled our 2-8 GHz GaAs MMIC amplifier into this package. The performance of this packaged amplifier, part number MAAM28000-A1, is shown in figure 6.

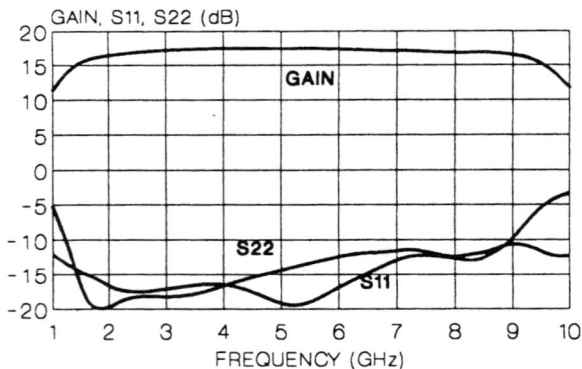

Figure 6. MAAM28000-A1 Performance

Using the de-embedded feedthrough data, a Y-parameter extraction followed by a constrained optimization was performed to derive the feedthrough model shown in figure 7. Figure 8 shows the measured versus modeled insertion loss and input return loss for this package feedthrough. The model simulates the feedthrough performance closely over the useful frequency range of the package.

Coupling effects between package ports were also measured and modeled. A Y-parameter extraction showed that the coupling could be attributed to equivalent capacitance values.

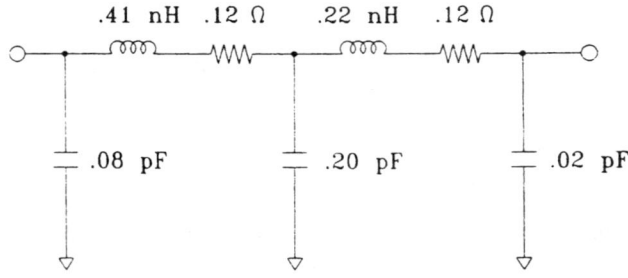

Figure 7. Feedthrough Model

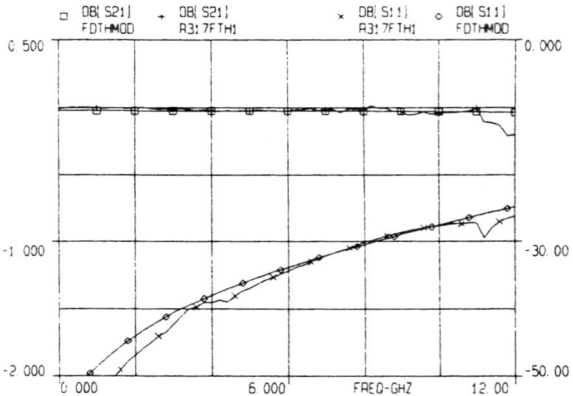

Figure 8. Measured vs. Modeled Performance

In the case of the 8-lead ceramic flatpack, coupling between adjacent ports along one side of the flatpack can be represented by a .03 pF capacitance. Between alternate ports along the same side the coupling capacitance is nominally .003 pF. Coupling between internally terminated ports on opposite sides of the flatpack was modeled with a .007 pF capacitor. This coupling model accurately predicts the measured input to output isolation, as illustrated in figure 9, over the package's useful frequency range.

Characterizing the packages in table 1 produced interesting results. The five relatively inexpensive packages (the 5, 7, and 8 lead flatpacks) are commonly used for fairly low frequency applications. However, as detailed above, the 8 lead ceramic flatpack, supplied by Kyocera, exhibits excellent performance into X-band. Mini-Systems' 8 lead glass flatpack also exhibits excellent performance into X-band, and their version with ground straps has similar performance through C-band. The Kyocera 5 and 7 lead ceramic flatpacks, often used in switching applications, have higher insertion loss and lower return loss, but demonstrate reasonably good performance into X-band and C-band, respectively. The Oxley manufactured leadless 8 port ceramic package has excellent performance through C-band.

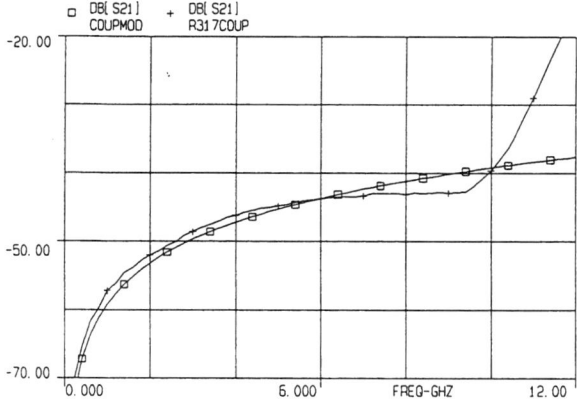

Figure 9. Package Isolation

The remaining three packages shown in table 1 are all advertised for high frequency applications. Of these, StratEdge's leadless 6 port ceramic flatpack exhibits the best performance through 20 GHz. The Alcoa 10 port ceramic package also works reasonably well up to 20 GHz. Kyocera's leaded version of the 6 port ceramic package demonstrates reasonably good performance to 16 GHz.

At least one suitable package was chosen for each of the small signal amplifiers, and one of the power amplifiers, in our GaAs MMIC amplifier product line. Table 2 lists all the packaged amplifiers now offered as standard products. This test method was also used to characterize the lead parasitics of the SOP and SSOP plastic packages. That data has been incorporated into the design of several new products specifically targeted for high volume, low cost, commercial applications. Bond wires, bond wire pairs, and ribbons have also been characterized with this test method, resulting in scalable, empirically derived models. In addition, this test methodology is widely employed in our engineering test lab to RF probe MMICs which are otherwise not RF probeable.

P/N (MAAM-)	Function	Package Style
02350-A1, D1	.2-3.5 GHz IFA	8 lead, ceramic or glass
12000-A1, D1	1-2 GHz LNA	8 lead, ceramic or glass
23000-A1, D1	2-3 GHz LNA	8 lead, ceramic or glass
37000-A1	3-7 GHz LNA	8 lead, ceramic
71200-H1	7-12 GHz LNA	leadless, 6 port ceramic
28000-A1	2-8 GHz WBA	8 lead, ceramic
28010-A1	2-8 GHz WBA	8 lead, ceramic
26100-B1	2-6 GHz PA	7 lead, ceramic

Table 2. Packaged Amplifier Products

CONCLUSION

A novel fixturing and test methodology has been designed and implemented which allows accurate microwave frequency characterization of virtually any device. This approach has been used to evaluate many of the currently available microwave packages. Appropriate packages have been identified for our GaAs MMIC amplifiers, resulting in many new standard products. Models for package feedthrough structures, plastic packages, and bond wires and ribbons have all been developed using this method.

ACKNOWLEDGMENTS

The authors thank Scott Mitchell and Ted Begnoche for testing these devices, Brenda Milinazzo for assembling them and Bill Fahey for helping to prepare this paper.

Frequency Limitation on an Assembled SO8 Package

F. NDAGIJIMANA, J. ENGDAHL, A. AHMADOUCHE, J. CHILO
LABORATOIRE D'ELECTROMAGNETISME MICROONDES ET
OPTOELECTRONIQUE
LEMO, B.P. 257, 38016 Grenoble Cedex - FRANCE
Phone (33) 76 87 69 76, Fax (33) 76 46 56 36

ABSTRACT

An electrical modelling and simulation concept for high speed assembly including SO8 package and Integrated switch is presented. Using the Method of Moments (MoM), electrical parameters are extracted for each part of the whole assembled package, and equivalent circuits are proposed. The resulting global network is analysed using a nodal simulator such as SPICE. Isolation and Insertion losses are calculated for a signal frequency up to 5 GHz. From our modelling and simulation results, the influence of the mutual coupling between strips or between the wire bondings are shown, the importance of the path connection to ground is pointed out, then required modifications of the connecting layout can be proposed. Our original concept, demonstrated on the SO8 package, is applicable on any package where the propagation effects can be neglected.

INTRODUCTION

Recent developments in silicon technology enable Integrated Circuits (I.C.) to operate up to 1 Gbit/s. At the same time the cost of using GaAs technology has been substantially reduced, allowing today fabrication of low cost GaAs I.C. It is desirable to encapsulate these latter circuits in low cost packages, yet having sufficient electrical performances to ensure proper functioning at clock frequencies up to some giga-Hertzs.

One way to accomplish this goal is to include the electrical performance constraints of the low cost package and to perform full system analyses while designing the circuit. This approach is only justified for ultra fast speed GaAs I.C. (10 Gbit/s). Another way is to caracterize the electrical performances of standard packages and to select those that fill the requirements. The latter approach is well-suited for our applications (max 4 Gbit/s) and it is demonstrated in this paper on an SO8 package.

THE ANALYSED STRUCTURE :

The analysed structure is based on the plastic SO8 package used for CMS applications (fig.1) assembled with a Single Pole-Double Throw (SPDT) switch integrated circuit.

The SPDT switch :

The switch (1 input, 2 outputs) is represented in figure 2a, and the equivalent electrical network is given in figure 2b. The electrical parameters R_{on}, C_{off} are determined experimentally, by direct on wafer measurements.

At 2 GHz, we obtain :   - Insertion losses = 1.2 dB
                        - Isolation losses = 35 dB
From these experimental data, the following values for R_{on} and C_{off} parameters are deduced :

R_{on} = 12 Ω C_{off} = 136 fF

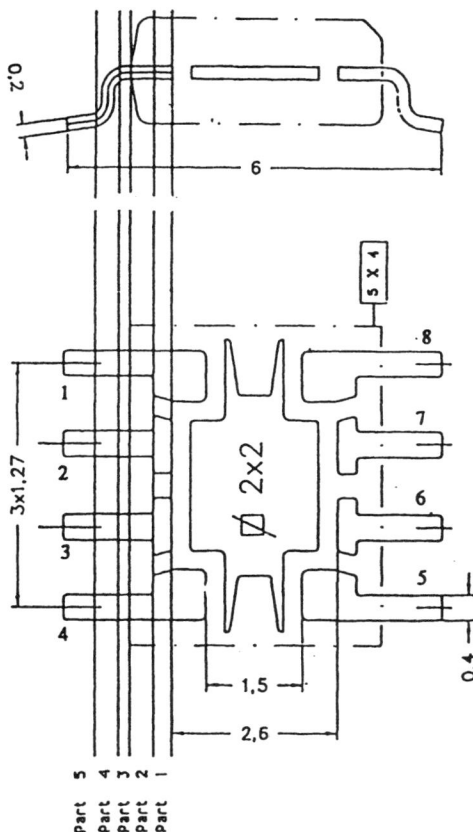

Figure 1 : The SO8 package

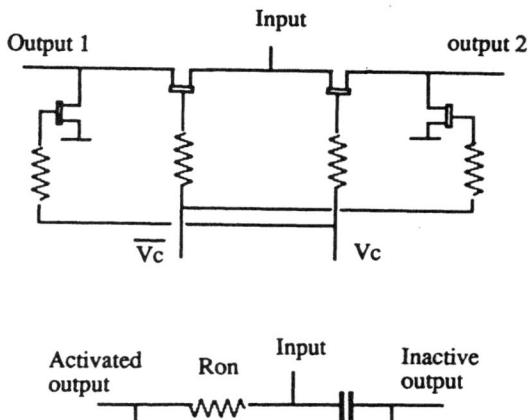

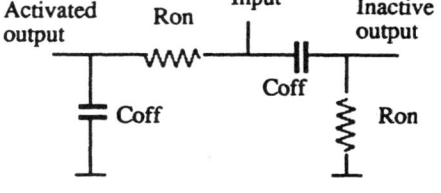

Figure 2. The SPDT switch : a - Electrical circuit ,
b - Equivalent network.

The lead frame :

A typical transversal cross-section of the structure is given in figure 3; it shows the position of the IC with respect to the main ground.

According to the maximum frequency used (4 GHz), the dimensions of the structure (some mm) are much smaller than the wavelength (7.5 cm). To perform the electrical simulations, an equivalent lumped network of the package (consisting of inductances and capacitances) is elaborated. Due to the coupling effects, these elements are described with matrices. Assuming that there is no coupling between the set of pins 1--4 and the set of pins 5--8, we obtain 4-order symmetrical matrices [L] and [C]. As shown in figure 1b, the pins of the package are divided in 5 elementary parts having approximately homogenous transversal cross-sections. For each part, the [L] and [C] matrices are calculated from the geometrical and the technological parameters.

For exemple, the transversal cross-section of part 1 (length l= 0.3 mm) is shown in figure 4. The [L] and [C] matrices are calculated using the Boundary Integral Technique [1] associated with the Moment Method[2]. These matrices are defined for the circuits so as to be compatible with the nodal electrical simulator SPICE.

From the obtained results we notice that :

a)-the inductance of the strip 2 (L_{22}=393 pH/mm) is lower than the inductance of the strip 1 (L_{11}=413 pH/mm); this is due to the proximity effect of strips 1 and 4, which reduces the inductive effect and increases the capacitive effect of strips 2 and 3.

b)-the coupling coefficient between strips 1 and 2 is important ($k=L_{12}/L_{11}$=0.44), however the influence is limited since the length of this part (l=0.3 mm) is small.

In the same way as described above, the matrices [L] and [C] of each part are calculated and the results are given in figure 5.

From these results a SPICE file can be defined, describing the electrical behaviour of the set of strips 1--8 of the SO8 package. Notice that part 5 does not need to be investigated because it is assumed to be included in the

50Ω driving microstrip line. Important is also to notice that part 4 is perpendicular to the ground plane. Consequently, the inductive effects are the predominant ones and thus we may neglect the capacitive effects of this part.

The wire bondings :

A quasi-static technique using the MoM in 3D domain enables an accurate calculation of the self and the mutual inductance of the bondings. The typical profile used for the wire bonding modelling is shown in figure 6. The self inductance depends on the length of the bonding and its position with regard to the ground plane, while the mutual inductances depend on the bondings position

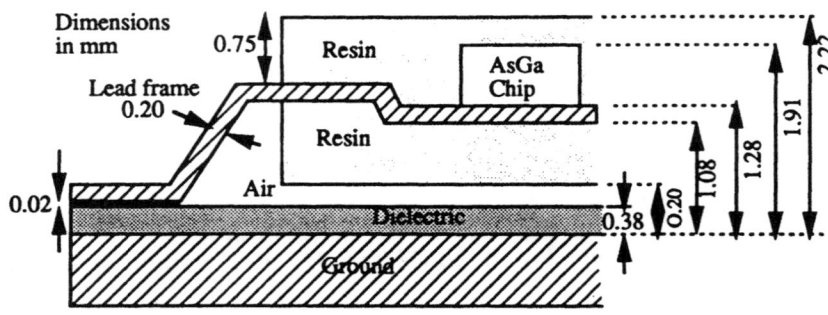

Figure 3 : Dimensions in the transversal cross-section of the S08 package

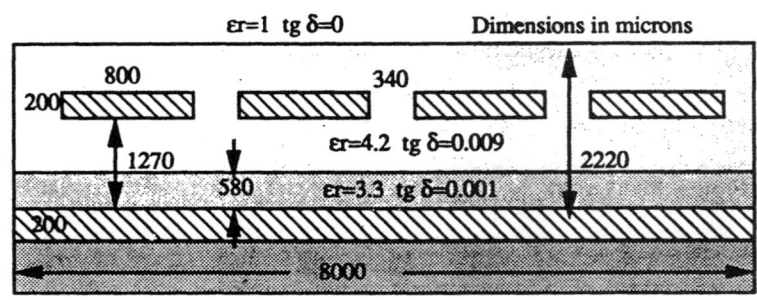

Figure 4. The transversal cross-section of the part 1.

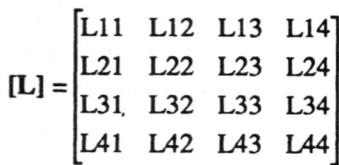

Figure 5-a : [L] and [C] matrices.

Figure 5-b : Calculated values for [L] and [C] matrices.

	[L] (pH)	[C] (pF)
Part 1 : l=0.3mm	123.8 54.3 27.4 16.2 54.3 117.9 51.5 27.4 27.4 51.5 117.3 54.3 16.2 27.4 54.3 123.8	19.11 19.74 0.75 0.03 19.74 11.79 18.75 0.75 0.75 18.75 11.79 19.74 0.03 0.75 19.74 19.11
Part 2 : l=0.4mm	213.3 63.5 26.8 15.3 63.5 209.5 60.6 26.8 26.8 60.6 209.5 63.5 15.3 26.8 63.5 213.3	22.50 11.80 0.92 0.36 11.80 16.20 10.84 0.92 0.92 10.84 16.20 11.80 0.36 0.92 11.80 22.50
Part 3 : l=0.2mm	110.7 34.7 15.9 8.6 34.7 110.0 34.5 15.9 15.9 34.5 110.0 34.7 8.6 15.9 34.7 110.7	3.32 1.26 0.18 0.08 1.26 2.54 12.20 0.18 0.18 12.20 2.54 12.20 0.08 0.18 12.20 3.32
Part 4 : l=1mm	275.8 38.1 10.1 3.8 38.1 275.8 38.1 10.1 10.1 38.1 275.8 38.1 3.8 10.1 38.1 275.8	0

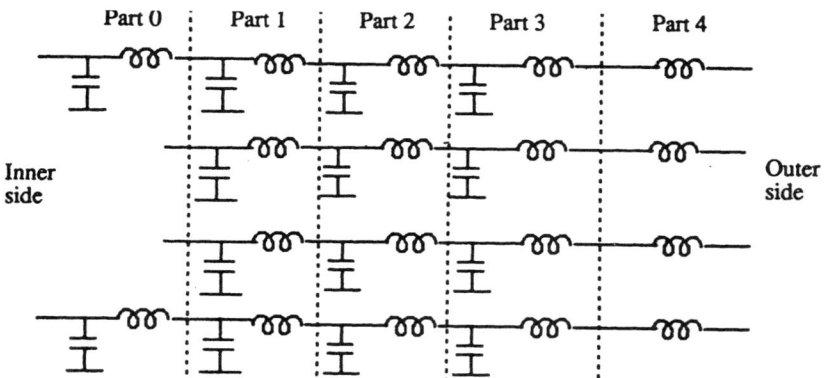

Figure 5-c :
Electrical network for 1/2 package (mutual capacitances and inductances not included)

relative each other. Consequently, the calculated [L] matrix varies somewhat from one assembly to another.

A typical calculated value of the self inductance is roughly 1100 pH, and typical value of a mutual inductance is around 10 pH. This leads to a coupling factor around 1%. In the next section, we will discuss the origin of the strong global coupling effect (30%) that is observed in experimental results.

THE STUDY OF THE WHOLE ASSEMBLY

We supposed that the signal is applied on pin 2 and pin 8 is the path "ON" and pin 5 is the path "ÔFF"(Fig.7). We also include in our simulation the inductances of the via holes (51 pH). The DC supply is supposed to be ideal and the power consumption is simulated with 10 KΩ resistances.

The investigation is done within two steps :

1°-the isolation and the insertion losses are calculated assuming that the global coupling is reprented by a mutual coefficient between the wire bondings (L_{B5} and L_{B8}).

2° the exact coupling factor between the wire bondings is determined, and the main coupling is found to be due to the inductive connexion of the die to the ground.

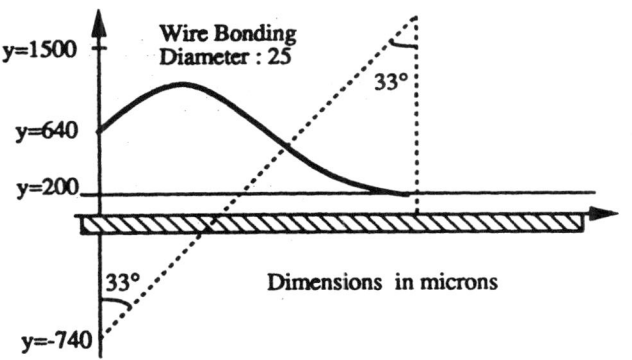

Figure 6 : The typical profile for a wire bonding

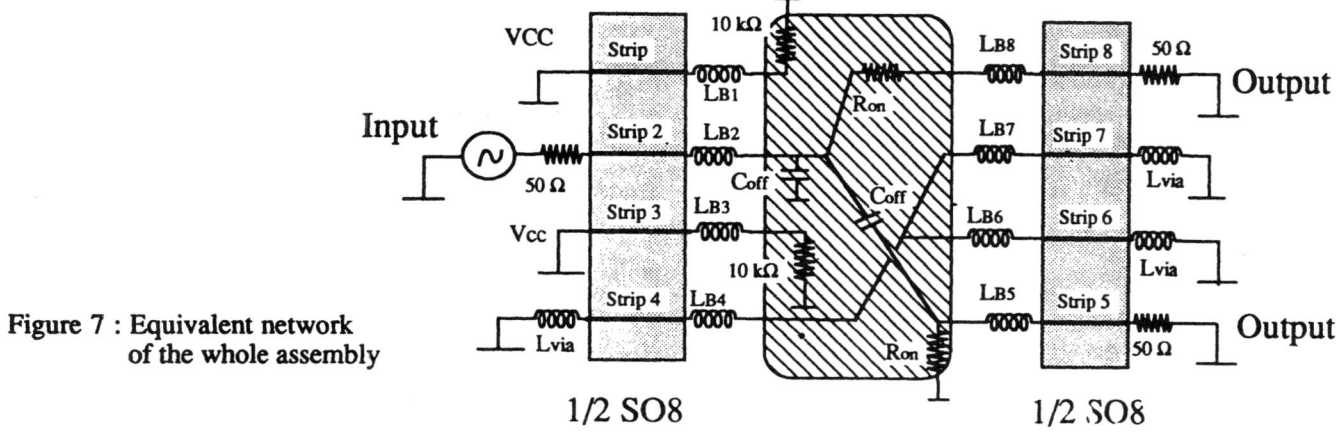

Figure 7 : Equivalent network of the whole assembly

The insertion losses :

The insertion losses (transmission along the path "ON") is extracted using the equivalent network, as a function of the frequency, and for different values of the mutual coefficient between the wire bondings (K_B).

Fom the results shown in figure 8-a, we conclude that the insertion losses are little sensitive to the coupling coefficient K_B.

At 2 GHz the global device has 1.4 dB losses for $K_B=0$, the switch has 1.2 dB losses (fig.2), and the SO8 package introduces only 0.2 dB additional losses. This performance seems acceptable for our application. Electrically, the major part of the inductance of the path "ON" is due to the bonding effects. If we neglect the capacitive effects, the following (simplified) electrical equivalent circuit can be deduced :

In this expression, the factor $1/(1+a)$ is due to R_{on} (SPDT resistance), and $1/(1+j\tau\omega)$ is due to the inductive effect. Lt represents the total inductance (package + bondings) and it is about 1800 pH. According to the value of R_{on} and R_c, we obtain : $a \simeq 0.12$. The insertion losses due to the package + bonding are written in the form :

$$I_{loss}(dB) = 10 \log(1+ \tau^2\omega^2)$$

At 2Ghz we obtain $I_{loss} \simeq 0.17$ dB; the theoretical simulation gave $I_{loss} \simeq 0.2$ dB.

Isolation :

The isolation is defined as the transmission along the path "OFF", and the results are given in figure 9-a; the isolation is as expected very dependent on the mutual coefficient K_B.

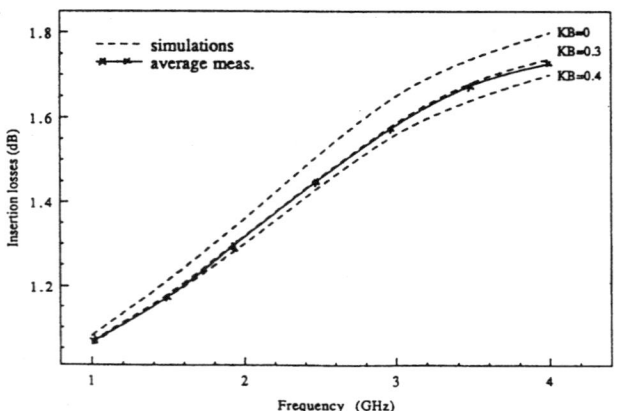

Figure 8 : a) Insertion losses of the whole assembly

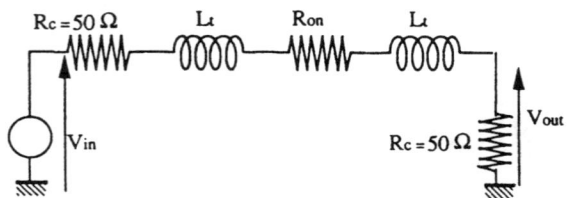

Figure 8 : b) Equivalent circuit for the insertion losses analysis

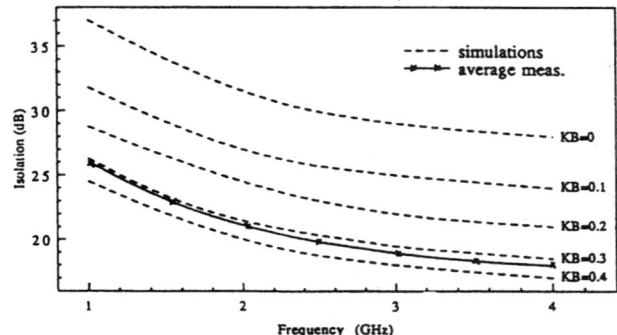

Figure 9 : a) Isolation of the whole assembly

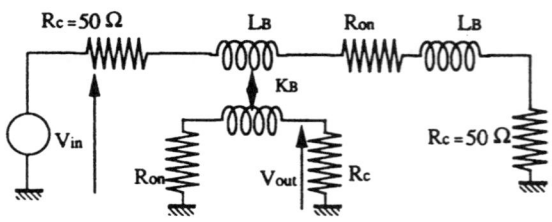

Figure 9 : b) Equivalent circuit for the isolation analysis

Assuming that $L\omega \ll R_c$ is respected, using a first order expansion, we deduce :

$$\frac{V_{in}}{V_{out}} \simeq \frac{1}{1+a} \cdot \frac{1}{1+j\tau\omega}$$

$$a = \frac{R_{on}}{2R_c}, \quad \tau = \frac{L_t}{R_c} \cdot \frac{1}{1+a}$$

We can notice that at 2 GHz the device has about 22 dB of isolation for $K_B=0.35$. The switch having 35 dB of isolation, hence, the S08 package isolation reduction is 13 dB. While the isolation of the chip is due to the capacitor C_{off}, the isolation of the assembly mainly depends on the coefficient K_B.

The electrical diagram in figure 9-b shows the simplified circuit that was used while perfoming the analytical isolation analysis.

We obtain, using a first order approximation :

$$\frac{V_{in}}{V_{out}} \cong \frac{\omega L_B}{R_c} \cdot K_B \cdot \left(\frac{R_c}{R_c + R_{on}}\right)^2$$

For $K_B = 0.35$, we obtain Vout/Vin = -14 dB; the rigourous simulation gave : -13 dB.

Influence of the inductive ground connection:

In the sections above, the die and the chip were supposed to be connected directly to the ground (figure 10-a). However, the actual mounted SPDT is connected to ground via the die (the fictitous ground plane) and through the lead frame as shown in figure 10-b.

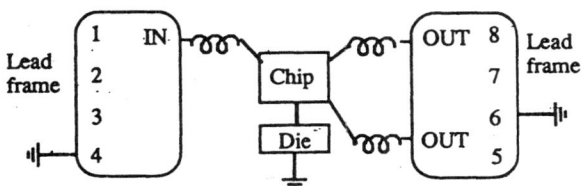

Figure 10 : a) Direct connexion of the die to the ground

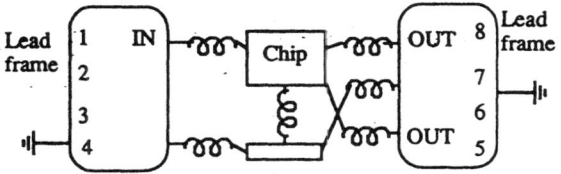

Figure 10 : b) Actual connexion of the die

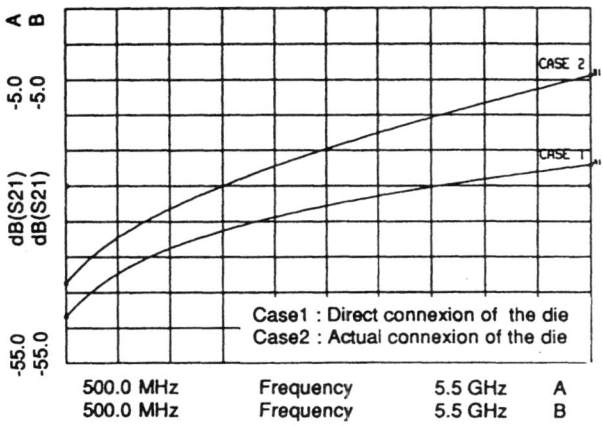

Figure 11 : Influence of die connexion on the isolation

The global resulting equivalent networks for the two cases mentioned earlier have been simulated. Figure 11 shows the isolation for the two cases. It can be noticed that the isolation is unrealistically good for case 1, i.e. when the die and the chip are assumed to be connected directly to the signal ground. The isolation of the actual mounting (case 2) reveals a curve that is in better agreement with experimental results; 20 dB at 4 GHz.

CONCLUSION

The agreement obtained between the experimental results and the simulations show how accurate our modelling concept is. On the investigated assembly, the results show that the S08 package is suitable for applications up to 2 GHz (1.4 dB of insertion losses and 22 dB of isolation losses).

The analysis of the insertion losses shows that the inductive bondings are the main sources of this limitations. It is a must to reduce these inductive bonding effects when higher frequencies are planned (reduction of the length and the distance to the ground).

The simulation reveal the impact of the inductances connecting the chip to the die, and the die to the principal ground plane through the lead frame. The best way to improve the electrical performances of the mounted IC is therefore to reduce these inductances, may be by direct connecting the die to the main (global) ground plane.

REFERENCES

[1] T. ITOH
Numerical Techniques for Microwave and Millimeter-Wave Passive structure
J Wiley and Son

[2] R.F HARRINGTON
Field Computation by Moment Method
Collier-Mc Millan New York 1968

Multifunction Silicon MMIC's for Frequency Conversion Applications

KEVIN J. NEGUS, MEMBER, IEEE, AND JAMES N. WHOLEY

Abstract —Recent advances in silicon bipolar IC technology have produced devices with 10-20 GHz f_T and f_{max} and excellent yields at MSI levels (≈ 100 devices). Thus cost-effective multifunction silicon MMIC's can now be developed for many commercial RF/microwave systems. In this work, the modeling, design, and testing of two silicon MMIC's for frequency conversion applications are illustrated in detail. The first product is a wide-band frequency doubler with conversion gain, 20 dBc rejection of harmonics, and a 2 GHz bandwidth. The second product is a wide-band vector demodulator (or image reject mixer) that utilizes an on-chip digital frequency divider to generate 0° and 90° LO phases from 0.05-1.5 GHz. Both products operate from a single 5 V supply, are load-insensitive, require no external baluns, and are packaged in tiny 180 mil hermetic packages. These frequency conversion MMIC's and others currently under development have been prototyped on the analog silicon transistor array starCHIP™-1, which is also described.

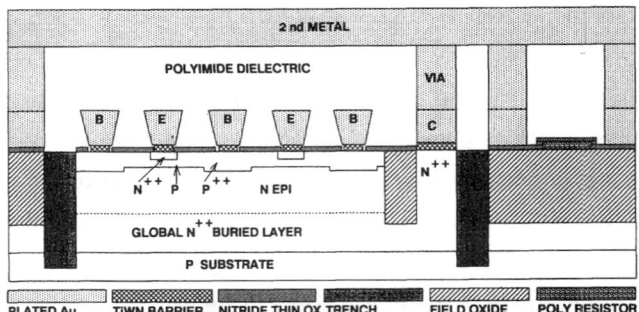

Fig. 1. Cross section of the ISOSAT-II process.

I. INTRODUCTION

SILICON BIPOLAR IC technologies have now produced many monolithic microwave products for applications well above 1 GHz [1]–[3]. In general, silicon MMIC's can offer higher reliability, reduced size, reduced power consumption, and lower cost (in sufficient volumes) when compared with traditional hybrid circuit approaches. Silicon MMIC's have also proved to be more cost-effective solutions than equivalent GaAs MMIC's for most commercial applications below 5 GHz [1].

The silicon MMIC's presently available have consisted primarily of single-function generic RF/microwave components. Multifunction MMIC's are now becoming a production reality as microwave silicon bipolar processes mature and yields increase. Designing and testing these multifunction MMIC's present many new challenges. For example, the multiple functions must ideally be dc-coupled on-chip to reduce the packaging complexity. Also, RF wafer sort testing is generally required due to the poor dc–RF correlation of many complex multifunction MMIC's.

The design of multifunction silicon MMIC's is very dependent on accurate circuit simulation of both the devices and packages since "breadboarding" with hybrid approaches has poor correlation to the final MMIC performance. Even with accurate device and package models though, the design and manufacture of fully customized silicon MMIC's require substantial nonrecurring costs and development time. Thus, the frequency conversion MMIC's described in this work have been prototyped initially on Avantek's first semicustom product, the starCHIP™-1 analog transistor array. Semicustom silicon MMIC arrays offer a cost-effective compromise for small volumes which combines many of the performance advantages of full-custom IC's with the flexibility of hybrid circuits.

II. PROCESS OVERVIEW

Fig. 1 gives an overall cross section of transistors, resistors, and interconnects available with the ISOSAT™-II technology. The bipolar devices feature 0.6 μm nitride self-aligned emitters on a 2 μm emitter–base pitch. The fully ion-implanted structure has shallow emitters and active base widths below 100 nm. These devices have a peak f_T of 10 GHz and a peak f_{max} of 20 GHz for a standard collector profile that allows typical $BV_{CEO} > 15$ V and $BV_{CBO} > 25$ V. Other available collector profiles for digital applications (or 5 V analog) feature f_T to 20 GHz with $BV_{CEO} \sim 5$ V.

Parasitics are minimized throughout the process. A global buried layer and deep collector plug keep collector resistance low. The global buried layer also provides a good RF ground plane at a depth of only a few micrometers below the metal interconnections on the die. Polyimide-filled trench isolation minimizes collector-substrate capacitance and a 2-μm-thick field oxide greatly reduces the parasitic capacitance of first metal and thin-film polysilicon resistors. The thick field oxide isolates the parasitic collector-base sidewall junction and increases the device breakdown voltages. This process features a

---

Manuscript received December 14, 1989; revised March 12, 1990.
The authors are with Advanced Bipolar Products, Avantek, Inc., 39201 Cherry St., Newark, CA 94560.
IEEE Log Number 9036831.

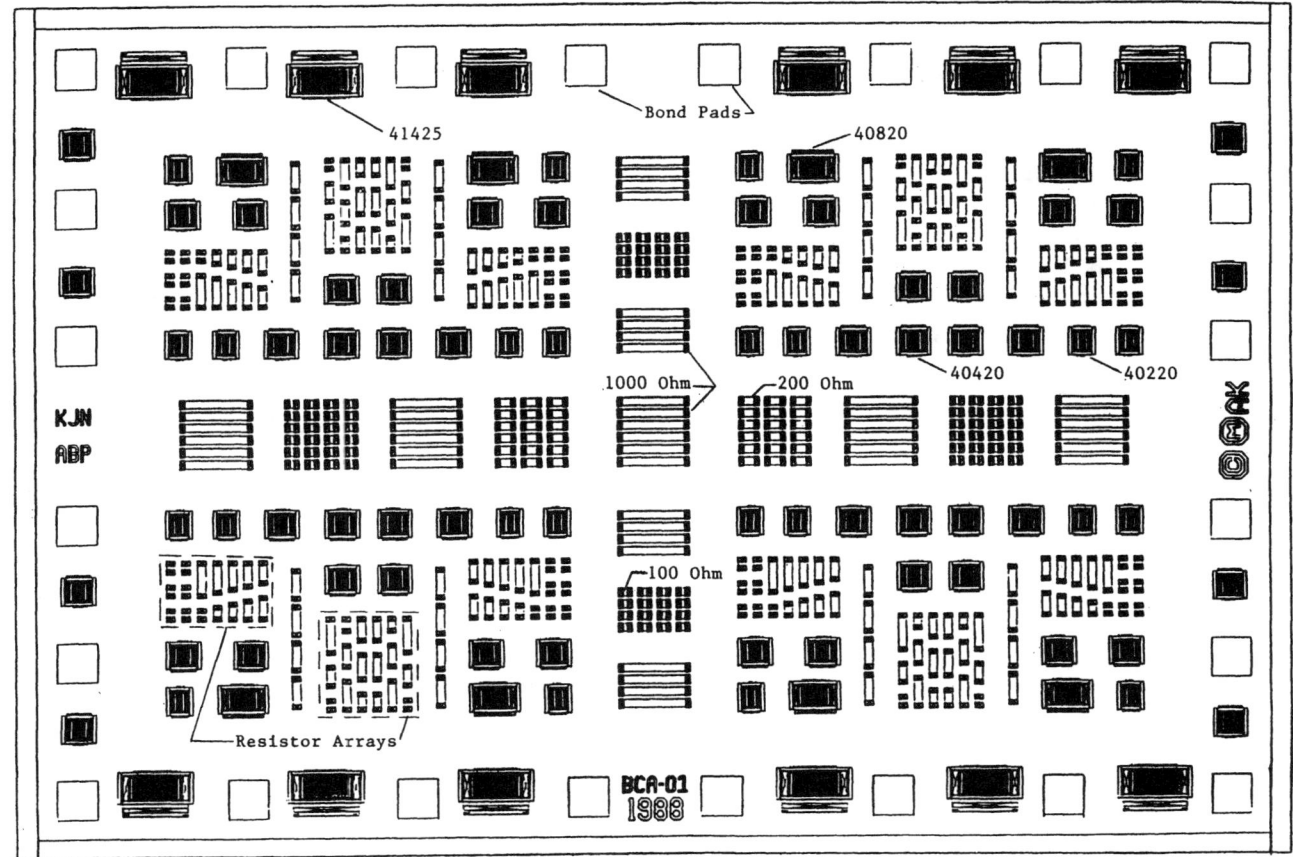

Fig. 2. Layout of the starCHIP-1 analog transistor array.

TABLE I
THIN-FILM POLYSILICON RESISTOR DATA

Value (Ω)	Number on starCHIP^{tm} 1	Tolerance ±%	Temp. coeff. ppm/°C	Matching ±%
100	144	15	-800	1
200	52	12	-800	.7
1000	46	10	-800	.5
Misc. 125–500	152	12	-800	.7

TABLE II
TYPICAL HIGH-FREQUENCY TRANSISTOR DATA

Device Type	Number on starCHIP^{tm} 1	I_C (mA) for peak f_T	C_{jc} (fF) at $V_{CB}=0$	r_b (Ω) at peak f_T
40220	24	2.5	60	45
40420	48	5	100	20
40820	8	10	190	9
41425	12	25	400	4

TABLE III
DC TRANSISTOR DATA (ALL CURRENTS REFER TO 40420 DEVICE)

Parameter	Measurement Condition	Minimum Value	Typical Value	Maximum Value
h_{FE}	$I_C = 2$ mA, $V_{CE} = 4$ V	40	100	150
V_A	$I_C = 2$ mA, $V_{CE} = 4$ V		25 V	
V_{BE}(ON)	$I_C = 5$ mA, $V_{CB} = 0$ V		820 mV	
BV_{CS}	$I_C = 10$ μA	30 V	35 V	
BV_{CBO}	$I_C = 10$ μA	20 V	25 V	
BV_{CEO}	$I_C = 10$ μA (base open)	12 V	15 V	
BV_{EBO}	$I_E = 10$ μA	1.5 V	2 V	

"true" second metal capability (as opposed to air bridges) for layout ease while maintaining low parasitic capacitance due to its thick polyimide dielectric spacer.

III. THE starCHIP-1 ARRAY TOPOLOGY

The silicon semicustom array starCHIP-1 contains 92 high-speed, low-noise npn transistors and 394 low-parasitic, thin-film resistors. Some thin-film capacitance can also be realized on-chip.

As shown in Fig. 2, the 40 × 60 mil die is organized into four symmetric tile arrays separated by tightly packed resistor/capacitor arrays and surrounded by 24 bond pads and 20 extra transistors. Each tile array contains 18 npn transistors and three variable resistor arrays. Resistors are easily combined in series and/or parallel to produce desired values. Typical resistor data are given in Table I. Practical resistor values vary from 5 Ω to 10 kΩ. Four different transistor sizes are available, as summarized in Table II. The transistors are arranged for easy layout as differential pairs, Gilbert cells, or Darlingtons. Some typical dc device characteristics are given in Table III for the most common transistor on the array (the 40420). Thin-film, on-chip capacitance is available by using metal 1 and the 1000 Ω polysilicon resistors separated by Si_3N_4 and SiO_2. Practical values of capacitance are 0.2–5 pF with a dielectric strength of at least 40 V.

Adequate component spacing and a full two-level metal capability (2 μm line/space in metal 1) permit easy routing and high component utilization. The 24 low-para-

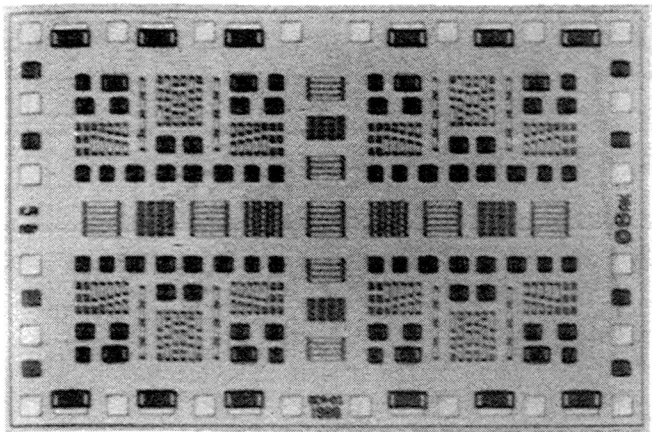

Fig. 3. Photograph of a blank starCHIP-1 die.

Fig. 4. Photograph of the 180 mil hermetic packages.

TABLE IV
TYPICAL SPICE GUMMEL–POON MODEL PARAMETERS FOR 40420 DEVICE

Parameter	Value	Parameter	Value
BF	100	TF	12 ps
IS	.16 fA	CJE	230 fF
VAF	25 V	CJC	100 fF
IKF	15 mA	CJS	130 fF
BR	10	RB	30 Ω
XTB	1.8	RC	50 Ω

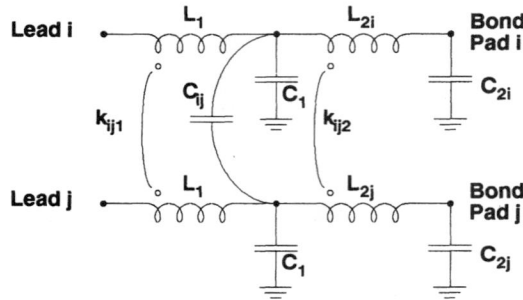

Fig. 5. Electrical model of the 180 mil hermetic package.

TABLE V
TYPICAL LUMPED-ELEMENT MODEL VALUES FOR 180 MIL PACKAGE

Parameter	Value	Condition
L_1	1.5 nH	Package base AC grounded
L_{2i}	.8 nH	Single bond wire, 1 mm long
k_{1ij}	.3	Adjacent leads
k_{1ij}	.05	Opposite leads
k_{2ij}	.2	Adjacent bonds
k_{2ij}	.07	Opposite bonds
C_1	200 fF	Package base AC grounded
C_{2i}	50 fF	2 mil bond pad
C_{ij}	20 fF	Adjacent leads
C_{ij}	3 fF	Opposite leads

sitic 2 mil bond pads allow for flexible pin-outs, multiple in-package capacitors, and easy probing of critical internal circuit nodes. An SEM photo of the blank array is given in Fig. 3.

Design freedom is naturally restricted somewhat by fixed device sizes available on a given transistor array. For the starCHIP-1 array, ample variation in resistor values permits easy optimization of the available device sizes for peak RF performance. However, the power dissipation for a fully utilized array will inevitably be 50–100 mA and peak output powers will be on the order of 0–5 dBm. Lower power dissipation parts or higher output powers can be attained by transferring the design either to other arrays or to a full-custom layout.

IV. DEVICE AND PACKAGE MODELING

In the design of silicon MMIC's, both small- and large-signal analyses are routinely required [4]. Thus accurate device and package models are needed in a format applicable to a nonlinear circuit simulator such as SPICE.

The silicon bipolar devices on the array are modeled by the extended Gummel–Poon model of SPICE [5]. Some important parameters for this model are summarized for the 40420 device in Table IV. The dc parameters are determined using extraction and optimization software on data collected by a dc source/monitor unit. The ac parameters require the addition of a vector network analyzer and capacitance meter.

Layout parasitics can be readily calculated for interconnects and thin-film resistors and back-annotated into the SPICE netlist of a given design. Typical values are 0.017 fF/μm^2 for metal 1 or polysilicon resistors and 0.006 fF/μm^2 for metal 2. Thus, a 400 Ω resistor as implemented on the array has a parasitic capacitance of about 7 fF and a 200 μm metal 1 line 4 μm in width has a parasitic capacitance of about 13 fF. Although these typical parasitics are significant, comparison with the device junction capacitances of the 40420 as shown in Table IV indicates that device parasitics dominate most designs on the array.

The products described in this work are packaged in the eight-lead 180 mil glass–metal package shown in Fig. 4. For circuit simulations, a simple electrical model of lumped elements is used as illustrated in Fig. 5. In this model the package leads and bonding wires are modeled by series inductors and shunt capacitors with coupling as shown. This model could be made more accurate with

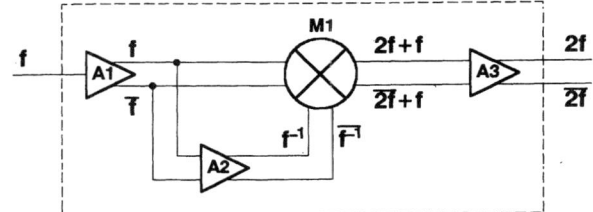

Fig. 6. Block diagram of wide-band frequency doubler.

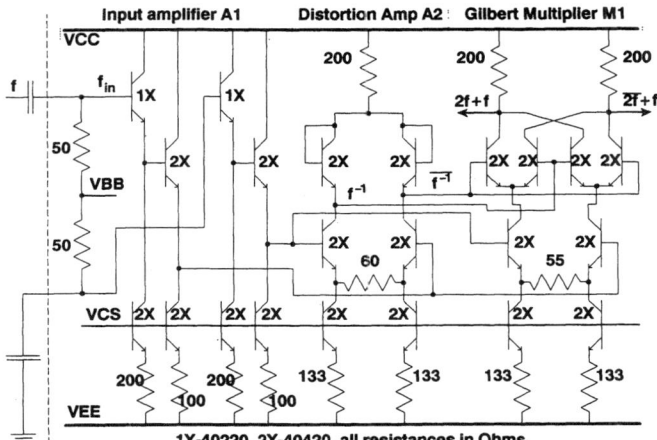

Fig. 7. Circuit schematic of frequency doubler input stage, distortion amplifier, and Gilbert cell multiplier.

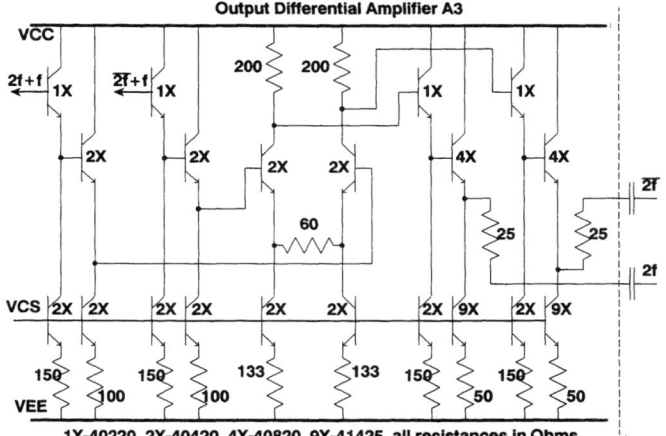

Fig. 8. Circuit schematic of frequency doubler output stage.

extra elements (especially for frequencies above 5 GHz) but at the expense of increased computer simulation time. Fortunately, this simple model adequately predicts the most important effects of impedance mismatch, signal isolation, and effective ground loop delays. Actual elemental values for the model shown in Fig. 5 can be determined through numerical simulation and/or measurements on a network analyzer. Some typical model values for the 180 mil package of Fig. 4 are given in Table V.

V. Wide-Band Frequency Doubler

The basic conceptual operation of a wide-band frequency doubler is shown in Fig. 6 [6], [7]. In this circuit an input amplifier A1 provides impedance matching, single to differential conversion, and appropriate dc level shifting for this monolithic design. The distortion amplifier A2 provides the inverse distortion of the LO quad in the Gilbert cell multiplier M1. Thus, the multiplication of f^{-1} and f by M1 produces an output $2f$ which is linearly related to f in power. A final output amplifier A3 filters out the substantial common mode signal f and provides power gain and output impedance matching. The component functions from a single power supply VCC = 5 V and draws approximately 80 mA.

The actual circuit schematic for the input stage A1, distortion amplifier A2, and Gilbert cell multiplier M1 is given in Fig. 7. Note that for implementation on the starCHIP-1 array, the device size choices are limited and most design effort is spent on choosing the appropriate resistor values for device sizes chosen, as shown in Fig. 7.

The input signal f is ac coupled by an external capacitor to a differential converter centered by the on-chip bias VBB = VCC − 1.3 V. An in-package but off-chip bypass capacitor of 50 pF is used for input termination. For low-frequency operation this point is connected to a package lead for a larger external bypass capacitor to VEE = 0 V. Current sources throughout the design are based on the bias line VCS shown in Fig. 7. Both VBB and VCS are generated on-chip using a modified band gap regulation circuit for excellent temperature and supply stability.

The distortion amplifier is similar to a standard linear differential amplifier except that active loads are used instead of resistors. The use of devices with collectors and bases shorted as shown in Fig. 7 provides a "predistortion" on the input signal which is ideally the inverse of the distortion that the nonlinear LO quad of the Gilbert cell multiplier will place on its input signal. The 200 Ω resistor above the active load devices serves only as a level shifter to keep the LO quad devices out of saturation and to minimize their collector–base capacitance. The multiplier ideally produces $2f$ components and a dc offset on the 200 Ω load resistors, but the common-mode switching of the emitter-coupled pairs introduces a substantial f component, as indicated in Figs. 6 and 7.

Several design approaches can be taken to improve the ratio of $2f$ to f output power. One approach is to increase the emitter resistors of these differential amplifiers and thus reduce the relative amount of current being switched common-mode by the LO quad devices. Unfortunately this also reduces the $2f$ output gain. The gain can be restored by increasing the load resistors but this reduces the bandwidth of the multiplier and creates saturation problems for the LO quad due to dc offsets. The approach in this design is to optimize the multiplier for $2f$ bandwidth and then use a differential output amplifier A3 (as illustrated in Fig. 8) to reject the common-mode signal f which is present in the otherwise differential outputs of the multiplier. The output amplifier of Fig. 8 also provides substantial $2f$ power gain and output impedance matching, as shown.

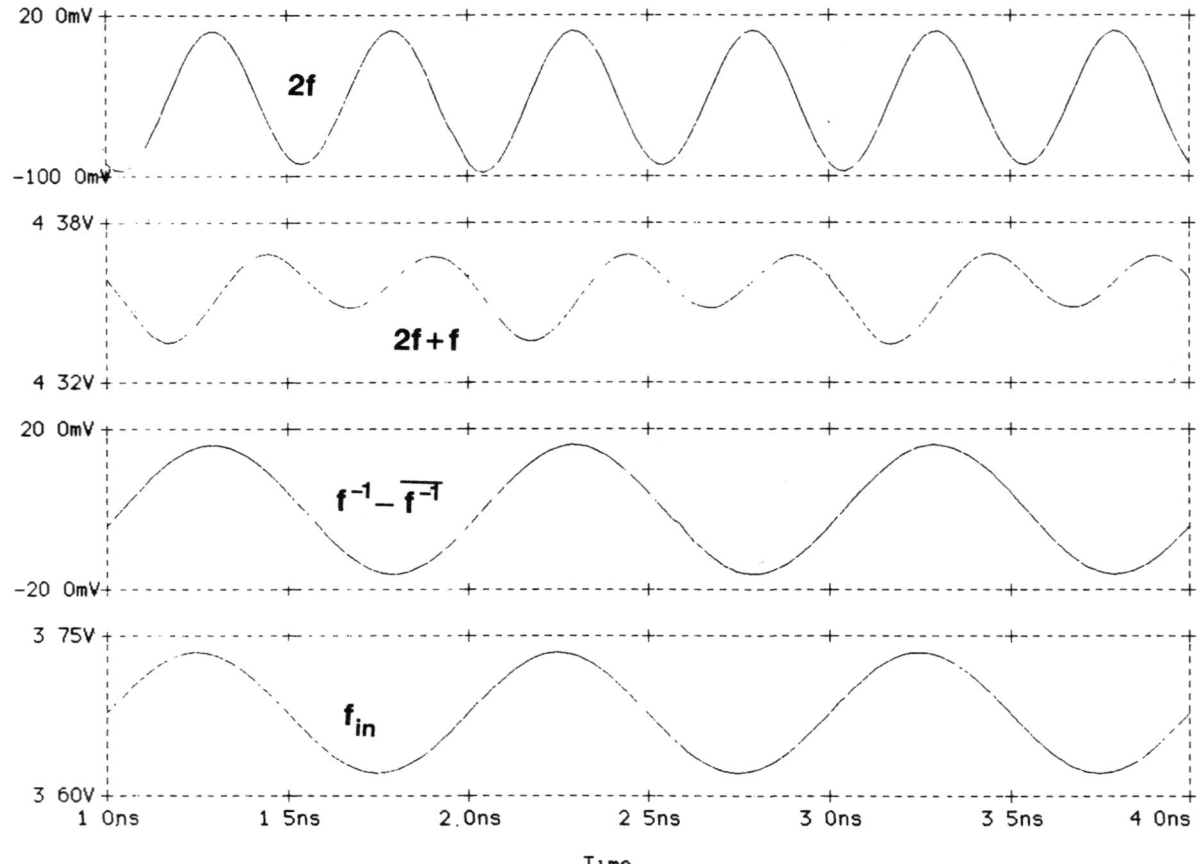

Fig. 9. SPICE-generated waveforms of frequency doubler.

The operation of the wide-band frequency doubler can be better understood by considering the SPICE-generated internal waveforms of Fig. 9. The bottom graph shows the actual input waveform seen by the doubler, f_{in}, versus time for an input signal of about -17 dBm at 1 GHz. The second graph shows the differential distortion amplifier output ($f^{-1} - \overline{f^{-1}}$ with reference to Figs. 6 and 7). Note that the inverse distortions must correlate extremely well since this signal is significant relative to thermal voltage V_T and thus highly nonlinear. The third graph is one of the two outputs of the Gilbert cell multiplier, which obviously contains a substantial f component in addition to the $2f$ component. The top graph is one of the differential outputs as delivered to a 50 Ω load (about -15 dBm). Clearly the output differential amplifier successfully removes most of the common-mode f component from the output. In summary, the simulations predict a 2 dB conversion gain, > 15 dBc rejection of the f, $3f$, and higher order components, and a 3 dB bandwidth of about 2.5 GHz output frequency.

This design has been successfully manufactured and assembled. Measured results of a typical part are given in Fig. 10, where the f, $2f$, and $3f$ output powers on a single output channel are plotted versus $2f$ output frequency for an input power of -10 dBm. The 3 dB bandwidth is seen to be approximately 2 GHz and at low frequencies the rejection of unwanted harmonics is greater than 20 dBc.

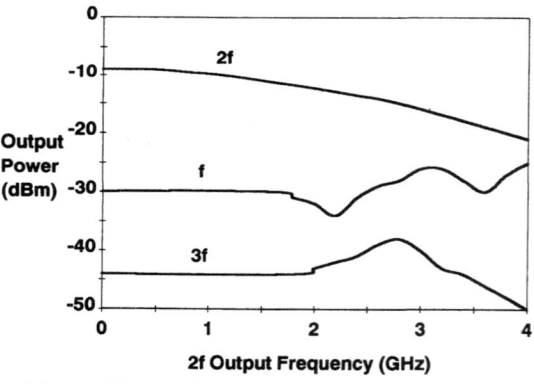

Fig. 10. Measured harmonic output powers of frequency doubler versus $2f$ frequency for constant input power of -10 dBm.

Wafer sorting this complex multifunction MMIC required RF testing since the dc to RF correlation was very poor.

VI. WIDE-BAND VECTOR DEMODULATOR

A second multifunction silicon MMIC implemented on the starCHIP-1 array is the vector demodulator or image reject mixer illustrated in Fig. 11 [8]. In this circuit, a input of frequency $2f_{\text{LO}}$ is provided through a buffer amplifier to the clock input of a master/slave D flip-flop connected as an asynchronous toggle. The flip-flop divides the frequency by 2 and generates the four quadrature phases of f_{LO}. These in turn differentially drive two

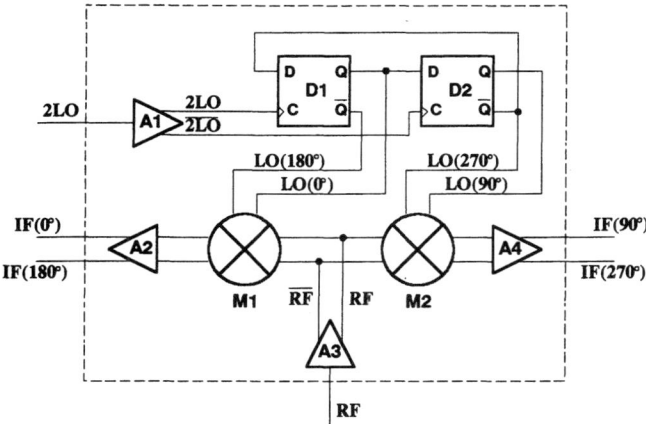

Fig. 11. Block diagram of wide-band vector demodulator.

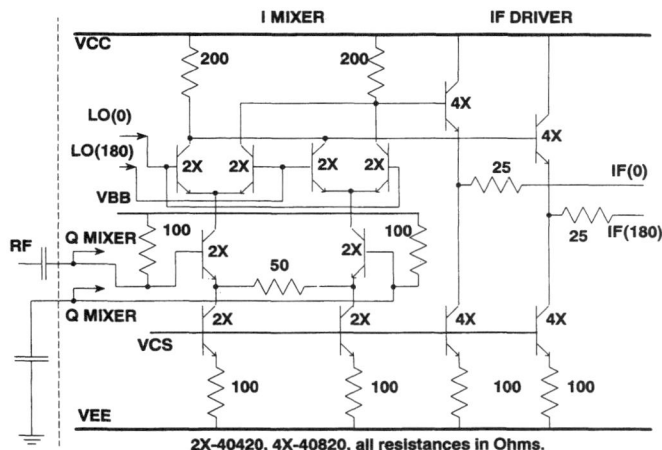

Fig. 13. Circuit schematic of Gilbert cell active mixer and IF output stage for vector demodulator.

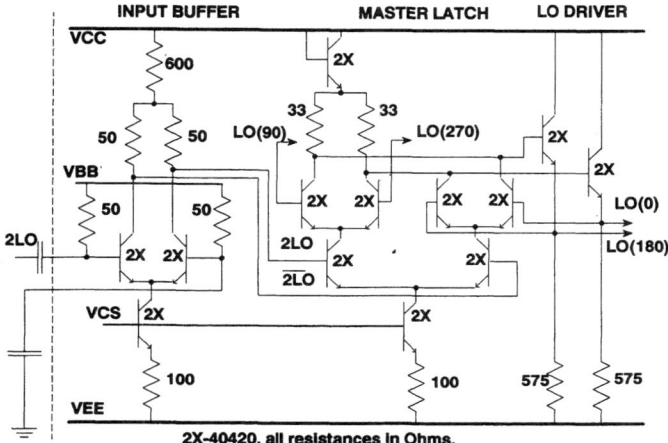

Fig. 12. Circuit schematic of $2f_{LO}$ input buffer, master D latch, and LO driver for vector demodulator.

identical active mixers that ultimately produce the four quadrature IF outputs. Critical parameters are matching the mixer conversion gains and obtaining ideal quadrature separation of the outputs.

Fig. 12 shows the schematic of the $2f_{LO}$ input buffer and the master D latch of the flip-flop (slave latch is identical). The VBB and VCS bias lines are generated on chip by voltage dividers and resistor-ratioed current mirrors. The $2f_{LO}$ buffer provides single to differential conversion, gives excellent limiting for a wide range of input powers, and achieves wide-band matching through 50 Ω resistors in shunt with the device bases. Necessary dc level shifting is accomplished by a resistive drop in the common-mode collector segment. The differential-mode collector resistors are kept small to provide sharp pulse edges for the clock input of the static frequency divider.

The master D latch shown in Fig. 12 (and the equivalent slave latch) are of standard emitter-coupled logic (ECL) configuration. Emitter follower outputs are used to provide adequate high-frequency drive to the LO inputs of the active mixers. Delays in the latches are significantly affected by the time constant $3R_L C_{jc}$, where R_L is the collector load resistance (33 Ω in Fig. 12) and C_{jc} is the collector-base capacitance of the 40420 device. The excellent matching of the load resistors on a single MMIC thus helps to minimize quadrature phase errors.

Fig. 13 shows the in-phase (I) active mixer driven by the LO(0°) and LO(180°) signals. The identical quadrature phase (Q) mixer (not shown) is driven by the LO(90°) and LO(270°) signals. These mixers are based on the Gilbert cell structure [9], where an RF signal is provided to a differential pair and multiplied by ±1 at f_{LO} to produce double balanced mixer characteristics. As with the $2f_{LO}$ input buffer, shunt resistors are used to maintain wide-band impedance matching of the RF input. Emitter followers provide power gain for 50 Ω operation with wide-band matching provided by series resistors as shown in Fig. 13. Conversion gain balance of the two mixers is closely related to the ratios of the collector load resistor to the emitter resistor and to the current source resistors. Again the single-chip implementation of these multiple functions enhances the performance due to the excellent matching of thin-film resistors on a single silicon MMIC.

This vector demodulator has been successfully realized on a single array. The layout complexity approaches the practical maximum density for this particular array. Power supply (VCC−VEE) is only 5 V and total current is typically 80 mA. RF wafer testing was used out of necessity to sort the MMIC's. Typical packaged performance parameters were < 0.5 dB mixer gain imbalance and < 5° quadrature error, which corresponds to > 25 dB image rejection. The $2f_{LO}$ input power can be as low as −20 dBm with minimal performance degradation and the $P_{\text{IN-1dB}}$ is approximately 0 dBm for the RF port.

The vector demodulator in combination with the previously described frequency doubler can operate as a wide-band image reject mixer, as illustrated in Fig. 14. The measured conversion gain of each IF channel is plotted versus RF frequency in Fig. 14 for a fixed IF frequency of 70 MHz. Both the frequency doubler and the master/slave D flip-flop limit the LO frequency to approximately 1.5 GHz (or $2f_{LO}$ = 3 GHz). The measured 3 dB IF bandwidth was typically 850 MHz. Excellent gain balance and quadrature were obtained, as shown by the time-domain

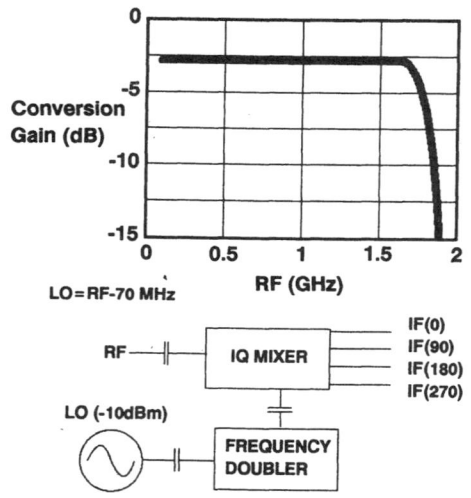

Fig. 14. Measured RF–IF conversion gain per channel for frequency doubler and vector demodulator combination as shown.

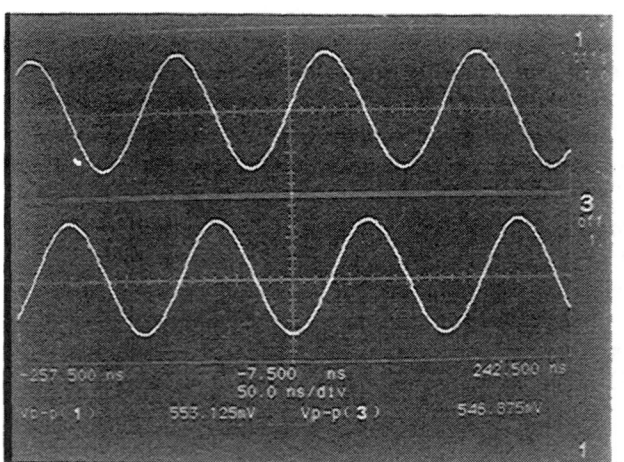

Fig. 15. Measured time-domain IF(0°) AND IF(90°) waveforms for vector demodulator.

measurements of the IF(0°) and IF(90°) channels in Fig. 15. These multifunction silicon MMIC's greatly simplify receiver architectures that presently utilize two-stage up-conversion/down-conversions to eliminate image signals.

VII. Comparison with Full-Custom MMIC's

As noted throughout this paper, both multifunction MMIC's were implemented on a semicustom array. Design and layout of each MMIC required approximately three to four weeks of one engineer's time. Fabrication time (including ordering the three customization masks) was approximately three to four weeks as well. Thus the total cycle time for prototyping MMIC's on the semicustom array was about seven weeks. In contrast, full-custom MMIC's typically require two to three months for design and layout (due to the greater degrees of freedom available) and at least three months to order all 11 masks and fabricate the initial wafers. The main advantages of semicustom arrays are thus their faster development time and much lower nonrecurring expenses.

Full-custom design and layout offer several performance and manufacturing advantages. Because device sizes are not restricted, the design can be better optimized for reduced power consumption. Overall RF performance may be marginally improved owing to lower layout parasitics. The full-custom layout can also produce a substantially smaller die size in many cases. In addition, yield can be enhanced by using devices of lower performance (and higher yield) for noncritical areas of the circuit such as biasing and current sources. The substantially increased initial costs of full-custom design can be recovered in the long term if sufficient production volumes are required. Although each MMIC design must be examined individually, the break point in most cases is on the order of 10 000 parts. In some cases, prototyping one or two design iterations on a semicustom array may still be advantageous before committing to a full-custom MMIC.

VIII. Conclusions

The application of advanced silicon bipolar IC technology to multifunction MMIC's has been demonstrated in this work. Both the wide-band frequency doubler and the wide-band vector demodulator offer compact size, a single 5 V power supply, and good input/output impedance matching. These multifunction silicon MMIC's are ideal for high-volume commercial frequency conversion applications owing to their inherent low cost on a high-yielding, mature silicon process technology. The example MMIC's in this work also illustrate the utility of the starCHIP-1 analog transistor array for systems below 5 GHz.

Acknowledgment

The authors acknowledge the excellent work performed by M. Dutta and the entire Pilot Line at Advanced Bipolar Products in fabricating the wafers used to prototype these MMIC's.

References

[1] C. P. Snapp, "Practical silicon bipolar ICs for RF, microwave and lightwave applications," in *Proc. Wescon/89*, Nov. 1989, pp. 282–287.
[2] J. Wholey, I. Kipnis, and C. P. Snapp, "Silicon bipolar double balanced active mixer MMICs for RF and microwave applications up to 6 GHz," in *IEEE MTT-S Int. Microwave Symp. Dig.*, June 1989, pp. 281–285.
[3] I. Kipnis, J. F. Kukielka, J. Wholey, and C. P. Snapp, "Silicon bipolar fixed and variable gain amplifier MMICs for microwave and lightwave applications up to 6 GHz," in *IEEE MTT-S Int. Microwave Symp. Dig.*, June 1989, pp. 109–112.
[4] I. Kipnis and A. P. S. Khanna, "Large signal computer analysis and design of silicon bipolar MMIC oscillators and self-oscillating mixers," *IEEE Trans. Microwave Theory Tech.*, vol. 37, pp. 558–564, Mar. 1989.
[5] I. Getreu, *Modeling the Bipolar Transistor* Tektronix Part No. 062-2841-00, Beaverton, OR, 1976.
[6] A. Bilotti, "Applications of a monolithic analog multiplier," *IEEE J. Solid-State Circuits*, vol. SC-3, pp. 373–380, Dec. 1968.
[7] P. R. Gray and R. G. Meyer, *Analysis and Design of Analog Integrated Circuits*. New York: Wiley, 1984, pp. 598–600.
[8] H. Kikushi, S. Konaka, and M. Umehira, "GHz-band monolithic modem ICs," *IEEE Trans. Microwave Theory Tech.*, vol. MTT-35, pp. 1277–1282, Dec. 1987.

[9] B. Gilbert, "A precise four-quadrant multiplier with subnanosecond response," *IEEE J. Solid-State Circuits*, vol. SC-3, pp. 365–373, Dec. 1968.

Intermodulation in High-Frequency Bipolar Transistor Integrated-Circuit Mixers

ROBERT G. MEYER, FELLOW, IEEE

Abstract — Intermodulation in bipolar-transistor double-balanced mixers at high frequencies is analyzed theoretically and by computer simulation. The dependence of the distortion on a relatively few normalized parameters is illustrated. Computed results are compared with measurements on a monolithic quad mixer.

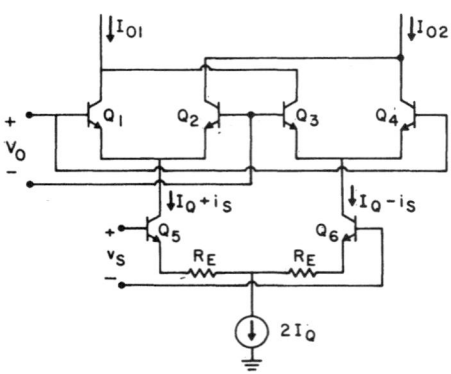

Fig. 1. Double-balanced quad mixer using bipolar transistors.

I. INTRODUCTION

THE DOUBLE-BALANCED mixer configuration shown in Fig. 1 is widely used for frequency conversion systems fabricated using bipolar-transistor integrated-circuit technology [1]–[5]. This circuit is attractive because a large applied oscillator voltage V_0 causes the transistor quad Q_1–Q_4 to behave as a set of almost perfect switches. The signal input v_s is converted to a current i_s by the differential pair Q_5, Q_6 which can have arbitrarily large linearizing resistors R_E (limited by noise figure degradation and gain loss). The current i_s is then switched back and forth by the quad switch, ideally producing frequency conversion with no distortion (such as intermodulation) of the signal input.

In practice it is not possible to achieve instantaneous switching of the quad, so that there is a time period in each cycle when all four devices Q_1–Q_4 are on and the current i_s may be subject to distortion from the nonlinear device transfer characteristics. At low frequencies, even this possibility causes few problems as the exponential base–emitter nonlinearities of Q_1–Q_4 tend to cancel [6], leaving only very small distortion caused by extrinsic base and emitter resistances during the switching interval. At high frequencies, however, charge storage in the quad devices and the influence of their base resistances causes significant distortion in the ideal mixing process. The most important manifestation of this is the creation of third-order intermodulation products in the converted output signal.

In this paper the mechanisms of intermodulation creation in the basic mixer of Fig. 1 are examined, and methods are derived for estimating distortion from device parameters and signal levels.

II. LARGE-SIGNAL ANALYSIS

The circuit of Fig. 1 may be analyzed at high frequencies by considering one pair of transistors in the quad and assuming this is driven by the oscillator voltage V_0 at the base and fed by an ideal current source at the emitter. This is shown in the large-signal equivalent circuit of Fig. 2 and is justified by the symmetry of the quad and the fact that any residual distortion due to Q_5, Q_6 can be considered independently. Note that in the equivalent circuit of Fig. 2 the effects of r_π and r_e are neglected. Simulation has shown this to be a good approximation at high frequencies. Also the effect of C_μ in the quad is small and it also is neglected. Although base resistance r_b is a function of I_C, the variation is typically not large and equal and constant values of r_b are assumed for the two devices. The large-signal input capacitance C_π represents charge storage in the device due to emitter–base depletion capacitance C_{je} and base charging capacitance so that

$$I_{B1} = \tau_1 \frac{dI_{C1}}{dt} + C_{je} \frac{dV_1}{dt} \qquad (1)$$

$$I_{B2} = \tau_1 \frac{dI_{C2}}{dt} + C_{je} \frac{dV_2}{dt}. \qquad (2)$$

Note that equal and constant transit times τ_1 are assumed in the two devices, as are equal and fixed values of C_{je}. The former assumption restricts the validity of the analysis to I_C levels below high-current f_T roll-off. The assumption of constant C_{je} is justified by simulations showing that almost the same distortion is generated if the actual C_{je} is

---

Manuscript received August 13, 1985; revised January 16, 1986. This work was supported by the U.S. Army Research Office under Grant DAAG299-84-K-0043.

The author is with the Department of Electrical Engineering and Computer Sciences and the Electronics Research Laboratory, University of California, Berkeley, CA 94720.

IEEE Log Number 8608916.

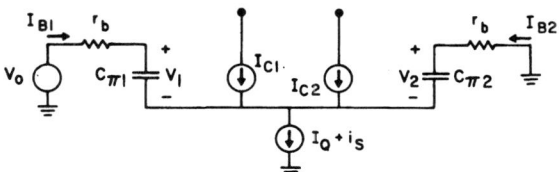

Fig. 2. Large-signal high-frequency equivalent circuit of Fig. 1.

replaced by a constant value equal to the bias value. This is plausible because of the relatively slow variation of C_{je} with bias voltage.

The analysis proceeds as follows. The transistor transfer characteristics are

$$I_{C1} = I_S e^{V_1/V_T} \quad (3)$$
$$I_{C2} = I_S e^{V_2/V_T} \quad (4)$$

where equal values of I_S are assumed and $V_T = kT/q$. From Fig. 2

$$V_0 = I_{B1} r_b + V_1 - V_2 - I_{B2} r_b \quad (5)$$
$$I_Q + i_s = I_{B1} + I_{C1} + I_{B2} + I_{C2}. \quad (6)$$

Substituting (1)–(4) in (5) we find

$$\frac{V_0}{V_T} = \tau_1 r_b \frac{I_Q}{V_T} \frac{d}{dt}\left(\frac{I_{C1}}{I_Q}\right) - \tau_1 r_b \frac{I_Q}{V_T} \frac{d}{dt}\left(\frac{I_{C2}}{I_Q}\right)$$
$$+ C_{je} r_b \frac{d}{dt}\left(\ln \frac{I_{C1}}{I_Q} \frac{I_Q}{I_{C1}}\right) + \ln \frac{I_{C1}}{I_Q} \frac{I_Q}{I_{C2}}. \quad (7)$$

Assuming a sinusoidal local oscillator waveform $V_0 = V_{OM} \cos \omega_0 t$ and normalizing time to $t' = \omega_0 t$ in (7) we find

$$\frac{V_{OM}}{V_T} \cos t' = \omega_0 \tau_1 r_b \frac{I_Q}{V_T} \frac{d}{dt'}\left(\frac{I_{C1}}{I_Q}\right)$$
$$- \omega_0 \tau_1 r_b \frac{I_Q}{V_T} \frac{d}{dt'}\left(\frac{I_{C2}}{I_Q}\right)$$
$$+ C_{je} r_b \omega_0 \frac{d}{dt'}\left(\ln \frac{I_{C1}}{I_Q} \frac{I_Q}{I_{C2}}\right) + \ln \frac{I_{C1}}{I_Q} \frac{I_Q}{I_{C2}}. \quad (8)$$

Substituting (1)–(4) in (6) we find

$$1 + \frac{i_s}{I_Q} = \omega_0 \tau_1 \frac{d}{dt'}\left(\frac{I_{C1}}{I_Q}\right) + \frac{I_{C1}}{I_Q} + \omega_0 \tau_1 \frac{d}{dt'}\left(\frac{I_{C2}}{I_Q}\right)$$
$$+ \frac{I_{C2}}{I_Q} + \omega_0 C_{je} \frac{V_T}{I_Q} \frac{d}{dt'} \ln \frac{I_{C1}}{I_Q} \frac{I_{C2}}{I_Q}. \quad (9)$$

III. Computer Simulation

Third-order intermodulation IM_3 in the mixer is defined as the ratio of the amplitude of the third-order intermodulation product in the output to the intermediate frequency (i.f.) fundamental signal in the output. Thus if ω_0 is the local oscillator frequency and ω_{s1} and ω_{s2} are frequencies of two input signals in i_s, then the i.f. fundamentals in the output have frequencies $(\omega_0 - \omega_{s1})$ and $(\omega_0 - \omega_{s2})$. Usually $\omega_{s1} \approx \omega_{s2} \approx \omega_0$. The IM_3 products in the output have frequencies $(\omega_0 - 2\omega_{s1} + \omega_{s2})$ and $(\omega_0 - 2\omega_{s2} + \omega_{s1})$.

Computer simulation of the circuit of Fig. 1 was performed using the program SPICE. Ideal current inputs $(I_Q \pm i_s)$ were applied at the emitters of the quad and the output was taken as $(I_{01} - I_{02})$. Simulation showed that the same values of IM_3 were present in I_{01} and I_{02} individually, as expected. The large-signal analysis routine in SPICE was used to generate a time-domain waveform of 15 000 points which was then Fourier analyzed. Careful checks using known waveforms as inputs verified the ability of the program to compute accurately IM_3 values of -90 dB. This requires appropriate manipulation of the program tolerance limits and care to allow transients to decay to adequately low levels.

Equations (8) and (9) together describe the large-signal high-frequency behavior of the mixer. A general closed-form solution of these equations is not available but an approximate and practically useful solution can be found using computer simulation. Equations (8) and (9) show that the normalized output currents I_{C1}/I_Q and I_{C2}/I_Q depend on a relatively few normalized parameters:

$$\frac{I_{C1}}{I_Q}, \frac{I_{C2}}{I_Q} = f_1\left(\frac{V_{0M}}{V_T}, \frac{i_s}{I_Q}, \omega_0 \tau_1, A, B, C\right) \quad (10)$$

where

$$A = \omega_0 \tau_1 r_b I_Q / V_T \quad (11)$$
$$B = \omega_0 C_{je} V_T / I_Q \quad (12)$$
$$C = \omega_0 r_b C_{je}. \quad (13)$$

Computer simulation and examination of the circuit together with (8) and (9) shows that IM_3 is only very weakly dependent on factors $\omega_0 \tau_1$ and C and further that the functional dependence on the remaining variables of IM_3 in the output current can be approximately expressed as

$$IM_3 \propto \left(\frac{I_{SM}}{I_Q}\right)^2 f_2\left(\frac{V_{0M}}{V_T}\right)[f_3(A) + f_3(B)] \quad (14)$$

where

$$i_s = I_{SM} \cos \omega_{s1} t + I_{SM} \cos \omega_{s2} t. \quad (15)$$

Note that IM_3 in the mixer varies as the square of the signal input amplitude I_{SM}, just as in amplifiers. Thus when distortion is calculated for one particular amplitude of input signal, distortion values for any other amplitude can be derived from this if all else is constant. By convention, equal amplitudes are assumed for all components in i_s.

Equations (8), (9), and (14) show two major sources of intermodulation in the mixer. First, if $C_{je} = 0$ then (9) becomes a linear equation and from (8) the parameter

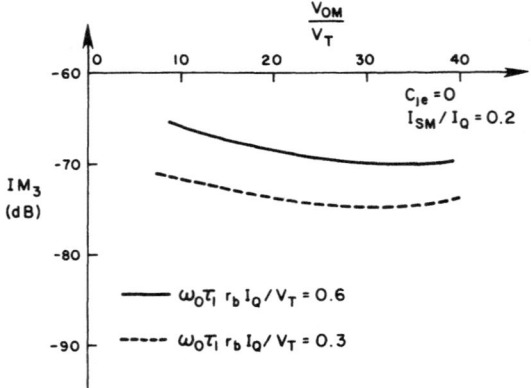

Fig. 3. Computed values of IM_3 versus normalized local oscillator voltage amplitude with $C_{je} = 0$ and $I_{SM}/I_Q = 0.2$.

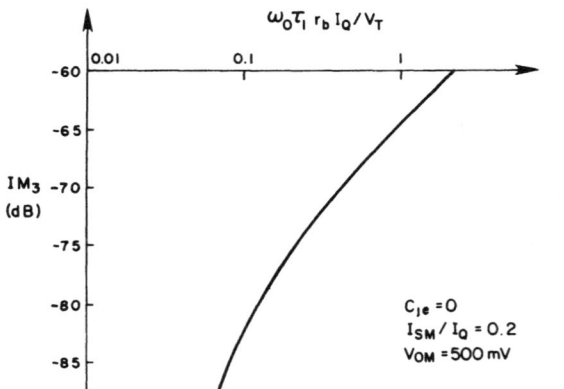

Fig. 4. Computed values of IM_3 versus $\omega_0 C_{je} V_T / I_Q$ with $C_{je} = 0$, $V_{OM} = 500$ mV, and $I_{SM}/I_Q = 0.2$.

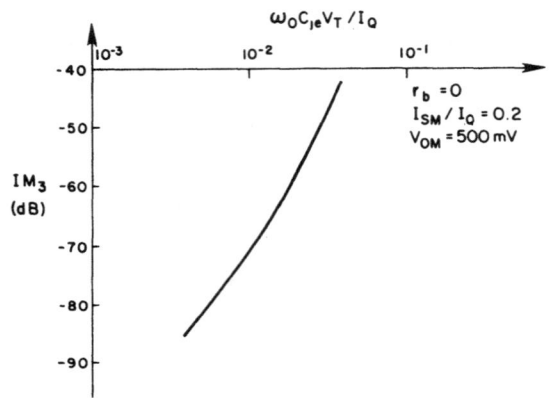

Fig. 5. Computed valued of IM_3 versus $\omega_0 C_{je} V_T/I_Q$ with $r_b = 0$, $V_{OM} = 500$ mV, and $I_{SM}/I_Q = 0.2$.

$\omega_0 \tau_1 r_b I_Q / V_T$ together with local oscillator drive V_{OM}/V_T and signal input i_s/I_Q determines the nonlinear behavior. Simulated values of IM_3 as a function of V_{OM}/V_T for two different values of $\omega_0 \tau_1 r_b I_Q / V_T$ with $C_{je} = 0$ and $I_{SM}/I_Q = 0.2$ are shown in Fig. 3. Note that for low values of V_{OM}/V_T, IM_3 increases as V_{OM}/V_T decreases because the quad transistors spend more time per cycle in the state where all four are on and thus generate higher distortion. The physical origins of the increase in IM_3 for large values of V_{OM}/V_T are not known but this effect is seen in both computed and measured data. Simulations with τ_1 and/or r_b equal to zero gave IM_3 near -100 dB with numerical limitations becoming apparent. The value of the parameter $\omega_0 \tau_1$ in (9) had almost no effect on the simulated distortion. Simulated values of IM_3 versus $\omega_0 \tau_1 r_b I_Q / V_T$ with $C_{je} = 0$, $I_{SM}/I_Q = 0.2$, and $V_{OM} = 500$ mV are plotted in Fig. 4.

The second major source of nonlinearity in the mixer is the depletion capacitance C_{je}. Simulations show that the distortion produced by C_{je} is essentially independent of both the value of τ_1 and of r_b. This is similar to the situation in a common-base amplifier at high frequencies. Note that in the mixer, when the quad devices are conducting they act as common-base stages so that this result is plausible. This result indicates that the term in (8) which depends on the factor $C_{je} r_b \omega_0$ is not an important source of distortion in the mixer. For purposes of distortion prediction, C_{cs} and C_μ of the driver transistors Q_5 and Q_6 effectively add directly to C_{je} of Q_1-Q_4. Equations (8) and (9) indicate that the factor $\omega_0 C_{je} V_T / I_Q$ is now the only factor affecting the nonlinear behavior of the circuit and this has been verified by computer simulation. Computed values of IM_3 in the mixer versus $\omega_0 C_{je} V_T / I_Q$ with $r_b = 0$, $I_{SM}/I_Q = 0.2$, and $V_{OM} = 500$ mV are shown in Fig. 5.

Simulations of typical quad mixers with complete device models (including all parasitics and allowing C_{je} to vary with V_{BE}) yield values of IM_3 that are usually close to the result obtained by simply adding the distortion products due to the two mechanisms, as represented by (14). Thus in any particular application the designer can estimate in advance whether a given circuit can meet a desired IM_3 specification. Appropriate and necessary bias levels can be estimated and, if necessary, alternative fabrication processes giving more favorable combinations of device parameters can be proposed.

IV. MEASUREMENTS

Measurements were made on a monolithic quad mixer using $R_E = 200$ Ω external to the chip. This was sufficient to make distortion due to Q_5, Q_6 negligible at the current levels used. The output was taken single ended from one common collector driving a 50-Ω load. Device parameters were $r_b = 35$ Ω, $C_{jeo} = 2.4$ pF, $C_{jco} = 0.86$ pF, and $\tau_1 = 320$ ps. (For frequencies $f_0 = 100$ MHz, $f_{s1} = 99$ MHz, and $f_{s2} = 98.9$ MHz, measured and computed values of IM_3 are shown in Figs. 6 and 7.) The computation was performed with the full device model and examination showed that both distortion mechanisms considered previously were contributing significantly. The overall agreement between computed and measured values is reasonably close and indicates that the major distortion sources and mechanisms in the circuit have been adequately modeled.

As an example of the use of Figs. 4 and 5, consider IM_3 in the test circuit for $V_{OM} = 500$ mV, $I_{SM} = 0.8$ mA, and $I_Q = 4.2$ mA. The total predicted distortion from Fig. 7 is $IM_3 = -58$ dB. The value of $\omega_0 \tau_1 r_b I_Q / V_T = 1.14$ and Fig.

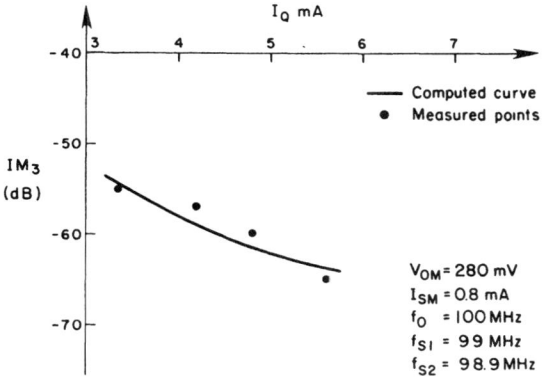

Fig. 6. Computed and measured values of IM_3 versus bias current for a monolithic quad mixer.

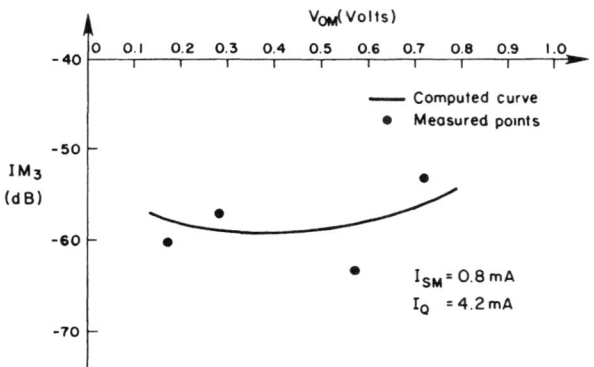

Fig. 7. Computed and measured values of IM_3 versus local oscillator voltage for a monolithic quad mixer.

4 predicts IM_3 due to r_b is -65 dB when $I_{SM}/I_Q = 0.19$. The effective value of C_{je} at the operating point is 4.1 pF, giving $\omega_0 C_{je} V_T / I_Q = 0.016$, and from Fig. 5 IM_3 due to C_{je} is -62 dB. When these values of IM_3 are directly added the result is -57.4 dB, which is very close to the value predicted from the simulation using the complete circuit model.

V. Conclusions

Intermodulation in bipolar-transistor integrated-circuit mixers has been analyzed theoretically and shown to depend on a relatively few normalized parameters. This has been verified by computer simulation. The analysis and computer simulation were able to successfully predict the magnitude and parameter dependence of intermodulation in an experimental monolithic quad mixer.

Acknowledgment

The author wishes to acknowledge reviewer's comments which significantly improved the paper.

References

[1] B. Gilbert, "A precise four-quadrant multiplier with subnanosecond response," *IEEE J. Solid-State Circuits*, vol. SC-3, no. 4, pp. 365–373, Dec. 1968.
[2] A. Bilotti, "Applications of a monolithic analog multiplier," *IEEE J. Solid-State Circuits*, vol. SC-3, no. 4, pp. 373–380, Dec. 1968.
[3] R. G. Meyer, "Integrated circuit mixers," in *IEEE NEREM Rec.*, vol. 12, 1970, pp. 62–63.
[4] C. Yamada *et al.*, "A 470 MHz 5V CATV tuner," in *IEEE ISSCC Dig.*, Feb. 1985, pp. 28–29.
[5] E. H. Nordholt, H. C. Nauta, and C. A. M. Boon, "A high-dynamic-range front end for an up-conversion car-radio receiver," *IEEE J. Solid-State Circuits*, vol. SC-20, no. 3, pp. 688–696, June 1985.
[6] W. M. C. Sansen and R. G. Meyer, "Distortion in bipolar-transistor variable-gain amplifiers," *IEEE J. Solid-State Circuits*, vol. SC-8, no. 4, pp. 275–282, Aug. 1973.

Blocking and Desensitization in RF Amplifiers

Robert G. Meyer and Alvin K. Wong

Abstract— Blocking and desensitization in RF amplifiers is analyzed and related to second and third order intermodulation performance. Methods of predicting blocking behavior are described and used to improve the performance of an existing amplifier. Measurements are compared with theoretical predictions.

I. Introduction

PREAMPLIFIERS in RF receivers are used to boost the incoming signal level prior to the frequency conversion process. This is important in order to prevent mixer noise from dominating the overall front-end noise performance. Important specifications of such RF amplifiers include noise figure, gain and third-order intermodulation intercept [1].

In typical applications, the RF receiver must tolerate large interfering signals emanating from users in adjacent channels, as well as from transmission sources which may be relatively far removed in frequency but whose large transmission power can cause significant interference problems. The influence of large interfering signals is manifested in several ways. One of these is third-order intermodulation in which two interfering signals at frequencies f_1 and f_2 (with $f_1 \simeq f_2$ and both close to the desired signal) combine in the amplifier third-order nonlinearity to produce an intermodulation product at $(2f_2 - f_1)$ which is again close to the desired signal. This intermodulation product is then processed by the receiver along with the desired signal. The resulting interference causes BER degradation in digital communication systems and audible or visible defects in analog communication systems. This is an example of desensitization in that the receiver sensitivity (ability to process very weak signals) is compromised by the presence of these intermodulation products. The relation between amplifier nonlinearity and intermodulation distortion is well-known and will not be considered further here.

Desensitization can also occur due to a single large interfering signal. This is called blocking and the interferer is called a blocker. In this paper the phenomenon of desensitization due to blocking is considered. The connection between blocking and amplifier nonlinearity is delineated and the results verified with experimental measurements.

II. Theoretical Analysis of Blocking and Desensitization in Amplifiers

The reduction in sensitivity in an amplifier caused by a large blocking signal can be attributed to two separate mechanisms. One of these is the gain compression caused by third-order nonlinearity in the circuit which occurs in the presence of large intefering signals, allowing existing noise sources in the amplifier (and in the mixer which usually follows it) to exert a larger influence, thus degrading the overall noise performance. The relationship between third-order nonlinearity and this gain compression can be analyzed assuming frequency independence of the nonlinearity [2] using a power series approach, or more generally for frequency dependent nonlinearities using Volterra series [3].

The second mechanism producing desensitization is caused by second-order nonlinearity in the circuit. In this case there is a mixing mechanism between (relatively) low-frequency noise sources in the amplifier and the interfering signal, which results in the low-frequency noise being up-converted to the desired signal frequency. This again degrades the circuit noise performance. Either one or both of these mechanisms may be significant in any given amplifier or receiver.

The phenomenon of gain compression caused by a large interfering signal acting on the amplifier third-order nonlinearity is well-known and the basic relations are summarized here for completeness. A power series representation is used for purposes of illustration and shows the same basic features as the more complex Volterra series approach.

Assume that the amplifier transfer function can be expanded in a power series as

$$V_o = a_1 V_i + a_2 V_i^2 + a_3 V_i^3 + \cdots \quad (1)$$

where V_o is the output signal (bias removed), V_i is the input signal and coefficients a_1, a_2, a_3 are frequency independent. Coefficient a_1 is the small-signal gain. Consider a small wanted signal $V_1 \cos \omega_1 t$ applied to the circuit together with a large interferer $V_2 \cos \omega_2 t$. Then

$$V_i = V_1 \cos \omega_1 t + V_2 \cos \omega_2 t. \quad (2)$$

Substituting (2) in (1) and collecting all terms at frequency ω_1 we have

$$V_o = a_1 V_1 \cos \omega_1 t + \frac{3}{2} a_3 V_1 V_2^2 \cos \omega_1 t + \cdots \quad (3)$$

where the second term comes from the third-order nonlinearity. Thus the apparent gain of the circuit is

$$a_1' = a_1 \left(1 + \frac{3}{2} \frac{a_3}{a_1} V_2^2\right). \quad (4)$$

A common method of characterizing this phenomenon is to specify the interfering signal level required to cause a -3 dB gain compression. For this case we set

$$20 \log_{10} \left(1 + \frac{3}{2} \frac{a_3}{a_1} V_2^2\right) = -3$$

and assuming a_3 and a_1 have opposite signs we find

$$V_2 = 0.442\sqrt{\frac{|a_1|}{|a_3|}}. \quad (5)$$

Thus the blocking level caused by third order effects is dependent on the ratio of coefficients a_1 and a_3. This process is related to other third-order phenomena and a similar analysis to the above shows the 1 dB gain compression signal level (wanted signal only) is $V_1 = 0.383\sqrt{\frac{|a_1|}{|a_3|}}$, or about 1 dB less than (5). Similarly the third-order intermodulation (IM_3) intercept level can be calculated as $V = 1.155\sqrt{\frac{|a_1|}{|a_3|}}$, or about 8 dB more than (5).

Desensitization can also occur from phenomena involving the second-order term in (1). Consider the small wanted signal $V_1 \cos\omega_1 t$ applied to the amplifier together with a large interferer $V_2 \cos\omega_2 t$ and a small interfering signal $V_3 \cos\omega_3 t$. Further assume that ω_2 is close to ω_1 and that $\omega_3 \ll \omega_1$. In practice, filters connected in front of the amplifier will usually ensure that practically no external signals at ω_3 (including external noise sources) are present at the amplifier active input node. However the amplifier itself has internal noise sources which act as unwanted signal inputs and are represented by $V_3 \cos\omega_3 t$. Thus

$$V_i = V_1 \cos\omega_1 t + V_2 \cos\omega_2 t + V_3 \cos\omega_3 t. \quad (6)$$

Substituting in (1), neglecting powers beyond the second and collecting the relevant terms we find

$$V_o = a_1 V_1 \cos\omega_1 t + a_2 V_2 V_3 \cos(\omega_2 \pm \omega_3)t + \cdots \quad (7)$$

Since noise is present in the circuit at all frequencies, ω_3 can always be chosen so that the term generated at $(\omega_2 \pm \omega_3)$ falls in the band occupied by the wanted signal at frequency ω_1. This interference can thus be identified as a second-order intermodulation (IM_2) effect with the amplifiers own internal noise acting as one input. The effect occurs for any values of ω_2 and ω_3 as long as $(\omega_2 \pm \omega_3) \simeq \omega_1$. However the worst case is usually found for $\omega_2 \simeq \omega_1$ and $\omega_3 \ll \omega_1$. Excessive $1/f$ noise levels in the circuit will also tend to make this phenomenon more pronounced.

From (7) the ratio of the interference to the amplitude of the wanted signal is $\frac{|a_2|}{|a_1|}\frac{V_2 V_3}{V_1}$. The second-order intermodulation intercept is $V = \frac{|a_1|}{|a_2|}$, and thus the desensitization produced by this blocking mechanism is directly related to the second-order intermodulation performance of the amplifier

III. MEASUREMENT

The theory developed above was prompted by measurements on an existing RF amplifier described elsewhere [4]. A schematic is shown in Fig. 1. This is a 2-stage amplifier (Q_1 and Q_2) with overall dc feedback via Q_6 used to stabilize the bias point. The feedback is decoupled in the kilohertz range by R_{14} and C_3. Bias currents are $I_{C1} = 2.5$ mA, $I_{B1} = 25$ μA, and $I_{C2} = 5.5$ mA with a supply voltage of 5 V. Peak process f_T is 10 GHz with base resistance for the large area input device Q_1 of 11 Ω. The input blocking signal power level

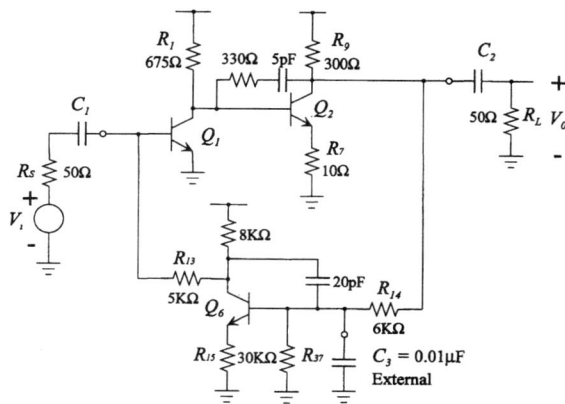

Fig. 1. RF amplifier schematic.

required to cause a -3 dB gain compression in the amplifier at 950 MHz was measured as -20 dBm at the input. The 1 dB gain compression level was also about -20 dBm and the third-order intercept was -10 dBm. These results agree closely with the theoretical predictions above for third-order blocking.

Blocking due to second-order effects was also observed in that large interfering signals at 950 MHz (offset by 3 MHz from the desired channel) representing adjacent channels caused significant increases in the background noise of the amplifier at 947 MHz. In order to predict these effects analytically, information is needed on both the noise generators of the amplifier at 3 MHz and the value of IM_2 for $f_2 = 950$ MHz and $f_3 = 3$ MHz. Equivalent input noise generators for the amplifier are conveniently generated using SPICE. Noise from all generators in the circuit is first calculated at V_o and then referred back to the input as a current generator across the base-emitter of Q_1. At 3 MHz this gave an equivalent input noise current spectral density of 25.4 pA/$\sqrt{Hz}$.

Levels of IM_2 in the circuit can be determined from simulation or from measurements. Since the measurements are quite straightforward whereas the simulation is dependent on the fine detail and accuracy of the large-signal device models, the IM_2 data was generated by measurement in the circuit shown in Fig. 2. A small signal V_{i2} at 3 MHz representing circuit noise is injected into the input via a 2 KΩ resistor. A large blocking signal V_{i1} at 950 MHz is injected at the input and the IM_2 product at V_o measured on a spectrum analyzer. For a power level of -54 dBm delivered to the circuit from V_{i2} and -23 dBm (the blocking level specification of interest) the measured IM_2 product at V_o was -60.1 dBm. The IM_2 product varied approximately 1 dB per dB variation in V_{i1} or V_{i2}, as predicted by theory. The power level of -54 dBm at 3 MHz represents 0.446 mV rms delivered to 50 Ω. Taking a Norton equivalent at the amplifier input through the 2 KΩ resistor gives an input current $I_i = 0.446$ mV/2 KΩ = 0.223 μA rms. This gave a measured output from the amplifier at 947 MHz of -60 dBm, which is 223 μV rms. Thus the transfer function of the amplifier from noise current input at 3 MHz to voltage out at 947 MHz in the presence of a -23 dBm blocker at 950 MHz is

$$\frac{V_o}{I_i} = \frac{223 \times 10^{-6}}{0.223 \times 10^{-6}} = 1 K\Omega. \quad (8)$$

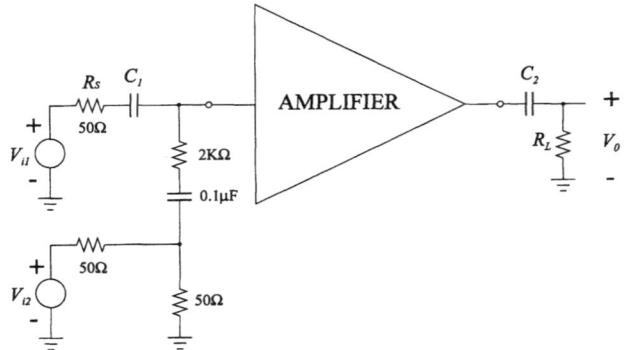

Fig. 2. Test circuit for IM_2 measurements.

Thus the SPICE simulated equivalent input noise current density of 25.4 pA/$\sqrt{\text{Hz}}$ at 3 MHz should give an output noise voltage density of 25.4 nV/$\sqrt{\text{Hz}}$.

The desensitization caused by second-order blocking can now be calculated. First, the noise floor of the amplifier with no blocker can be calculated from the measured gain (15.5 dB) and noise figure (2.4 dB) at 947 MHz. This gives an output noise 17.9 dB above the input noise in 50 Ω (which is $\sqrt{kTR} = 0.445$ nV/$\sqrt{\text{Hz}}$). Thus the output noise from the amplifier at 947 MHz is 3.5 nV/$\sqrt{\text{Hz}}$ or -156 dBm/$\sqrt{\text{Hz}}$ with no blocker present. In the presence of the 950 MHz blocker at -23 dBm the output noise is increased by the 25.4 nV/$\sqrt{\text{Hz}}$ calculated above. This is added to the amplifier output noise which is itself compressed somewhat by the third order blocking effect to 3 nV/$\sqrt{\text{Hz}}$. The rms combination of these two gives a predicted output noise of 25.6 nV/$\sqrt{\text{Hz}}$ which is -138.8 dBm/$\sqrt{\text{Hz}}$. The measured output noise under these conditions was -139.1 dBm/$\sqrt{\text{Hz}}$, a very close agreement. Note that the noise floor has risen 17 dB from the value with no blocker present.

The theoretical understanding of the blocking phenomenon described above allows investigation of methods of improving blocking performance. The equivalent input noise of the amplifier is not easily decreased but the IM_2 performance can be enhanced in several ways. One of these is to increase the bias current in Q_2 by adding a 1 kΩ pull-up resistor from the collector of Q_2 to V_{CC}. This increases the output stage bias current by several mA and improves the large signal handling capability of the amplifier. The measured IM_2 product at 947 MHz was then reduced by 10 dB to -70.0 dBm for V_{i1} of -23 dBm and V_{i2} of -54 dBm in the test described previously. A similar calculation to that described previously predicts that the noise output of the amplifier is increased by 8.0 nV/$\sqrt{\text{Hz}}$ due to second-order blocking, and that the total output noise should be -148.4 dBm/$\sqrt{\text{Hz}}$. The measured noise was -148.1 dBm/$\sqrt{\text{Hz}}$ with this modification.

The IM_2 performance can be further improved by reducing the low-frequency gain of the amplifier. This reduces the signal level in the circuit at frequency f_3 and can be accomplished by adding a bypassed resistor in series with R_7 in Fig. 1. A parallel RC network consisting of 10 Ω in parallel with 390 pF added in this way further reduced the measured IM_2 product in the circuit at 947 MHz to -78 dBm. The added noise output of the amplifier caused by second-order blocking is then predicted to be 3.2 nV/$\sqrt{\text{Hz}}$ and the total output noise -153.4 dBm/$\sqrt{\text{Hz}}$. The measured noise was -152.1 dBm/$\sqrt{\text{Hz}}$ with these modifications.

IV. CONCLUSIONS

Blocking and desensitization in RF amplifiers can be predicted from second and third order intermodulation performance and amplifier noise parameters. Good agreement is obtained between measured and predicted blocking performance in a practical amplifier, and circuit modifications to improve performance are described.

REFERENCES

[1] R. S. Carson, *Radio Communications Concepts: Analog*. New York: Wiley, 1990.
[2] K. A. Simons, "The decibel relationship between amplifier distortion products," *Proc. IEEE*, vol. 58, pp. 1071–1086, July 1970.
[3] D. A. Weiner and J. F. Spina, *Sinusoidal Analysis and Modeling of Weakly Nonlinear Circuits*. New York: Van Nostrand Reinhold, 1980.
[4] R. G. Meyer and W. D. Mack, "A 1-GHz BiCMOS RF front-end IC," *IEEE J. Solid-State Circuits*, vol. 29, pp. 350–355, Mar. 1994.

Modeling of the MOS Transistor for High Frequency Analog Design

PAUL J. V. VANDELOO, MEMBER, IEEE, AND WILLY M. C. SANSEN, SENIOR MEMBER, IEEE

Abstract—The exact knowledge of the small signal behavior of the MOS transistor is essential for the design of analog high frequency circuits. This paper presents a high frequency small signal model, capable of describing accurately the MOS transistor (in saturation) at frequencies beyond the cutoff frequency. The model is carefully compared to other models. It is shown how all the circuit elements of the model can be measured. This measurement method uses S parameter measurements, computer-controlled calibration techniques of the test setup and network analyzer, mathematical transformations of the S parameters to other linear parameters, and extraction routines to fit the data towards the HF model. The model is finally used in the design of an HF OTA to prove that the novel model is much more accurate than the classical models at higher frequencies.

I. Introduction

THE most widely used models of the MOS transistor are very simple and neglect several intrinsic elements of the MOS transistor. However, these simple equivalent circuits are found to be accurate enough for the simulation of digital circuits. Even if short channel transistors at high frequencies are used, the results are satisfactory since the neglected equivalent circuit elements are largely masked by the parasitic elements. Although the use of a simple model can result in some error in speed, it is usually only a few percent, and therefore, acceptable.

Today, the use of the MOS transistor is no longer restricted to digital applications. Many recent MOS chips combine digital and analog circuits on one piece of silicon. This trend will continue in the future since the MOS technology is an economical solution for single chip implementation of analog-digital circuits. These mixed mode circuits require a new generation of CAD tools to simulate the behavior of the analog circuit accurately at high frequencies. Indeed, many models ignore the intrinsic transcapacitances and neglect high frequency transmission line effects. However, in analog design, long channel transistors are dominated by these intrinsic capacitances and transmission line effects. Therefore, simulation with these simple models results in unrealistic prediction of gain and speed.

Manuscript received February 22, 1988; revised July 22, 1988 and November 16, 1988. The review of this paper was arranged by Associate Editor J. G. Fossum.
P. J. V. Vandeloo is with the Interuniversity Micro Electronics Center (IMEC), Katholieke Universiteit Leuven, B-3030 Leuven, Belgium.
W. M. C. Sansen is with ESAT, Katholieke Universiteit Leuven, B-3030 Leuven, Belgium.
IEEE Log Number 8826967.

In this paper, the small signal behavior of the MOS transistor is revised. Starting from the solution of a set of differential equations, which describe the channel current, a new small signal model is presented in Section III. The performance of the new model at frequencies in the neighborhood of the cutoff frequency is studied. A maximum frequency of validity of the model is indicated in Section IV. In Section V, measured S parameter data are converted to Y parameters and fitted towards the small signal model. In Section VI, the HF model is used to design a high frequency circuit. The last section clearly shows how the classical models fail to predict transfer functions accurately at high frequencies. Measurements agree with the HF model, making clear that the limit of tolerable errors of the currently used small signal models is reached and the development of novel simulation software is mandatory.

II. High Frequency Characteristics of the MOS Transistor

The dc current which flows from drain to source through the channel of the MOS transistor can (for strong inversion) be expressed in terms of the inversion layer charge Q_n, the effective mobility μ, the width W, the length L, and the channel voltage distribution V_x [1]–[3] (X is the normalized distance in the channel):

$$I_x = \frac{W}{L} \mu Q_n \frac{\partial V_s}{\partial X}. \quad (1)$$

Assuming an n-channel device (with source grounded), the inversion layer charge Q_n is expressed as

$$Q_n = C_{ox}(V_{gs} - V_T + K2(\sqrt{2\phi_F + V_{sb} + V_x} - \sqrt{2\phi_F + V_{sb}}) - V_x) \quad (2)$$

where

$$K2 = \frac{\sqrt{2\epsilon_{Si} q N_B}}{C_{ox}} \quad (3)$$

$K2$ represents the substrate doping effects, C_{ox} the gate capacitance per unit area, N_B the substrate doping, and $2\Phi_F$ the surface potential.

The small signal current i_x can be obtained by using Taylor series expansion around the bias point and retaining only the first terms. It is assumed that v_x and v_{gs} are small signals superimposed on the mean dc levels of V_x and V_{gs}. It is important to note that in most analog designs, the transistors (which determine the cutoff frequency of the circuit) are working in the saturation region. For that reason, further formulas and models are only deduced for the saturation region. The small signal current can be written as

$$i_x = \beta \frac{\partial}{\partial X}\left[(V_{cx}v_{cx} + K2(\sqrt{2\phi_F + V_{sb} + V_{dsat}} - \sqrt{2\phi_F + V_{sb}}))v_x\right] \quad (4)$$

where

$$\beta = \mu C_{ox} W/L$$
$$V_{cx} = V_{gs} - V_T - V_x$$
$$v_{cx} = v_{gs} - v_x. \quad (5)$$

Under small signal operation the following charge continuity equation must be satisfied:

$$\frac{\partial i_x}{\partial x} = -WL \frac{\partial q_n}{\partial t} \quad (6)$$

where q_n is the small signal inversion layer charge which can be obtained from (2).

$$i_g = g_m v_{gs} \frac{\frac{2}{3}\frac{j\omega}{\omega_0} + \frac{4}{45}\left(\frac{j\omega}{\omega_0}\right)^2 + \frac{2}{405}\left(\frac{j\omega}{\omega_0}\right)^3 + \cdots}{1 + \frac{4}{15}\frac{j\omega}{\omega_0} + \frac{1}{45}\left(\frac{j\omega}{\omega_0}\right)^2 + \frac{8}{8190}\left(\frac{j\omega}{\omega_0}\right)^3 + \cdots}. \quad (11)$$

The high frequency behavior of the MOS transistor is described by the set differential equations (4)–(6). The exact determination of the small signal performance of the MOS transistor is extremely difficult since the solution of the set of differential equations (4)–(6) is not straightforward. A considerable simplification results if substrate effects are negligible. In this context, this statement means that the influence of the depletion layer charge is negligible, compared to the inversion layer charge.

For the special case ($K2 = 0$), Paulos and Antoniadis [3] have solved equations (4) and (6), which can be combined as [3]

$$\frac{\partial^2 i_x}{\partial V_{cx}^2} = \frac{j\omega}{\omega_0}\frac{4V_{cx}}{(V_{gs} - V_T)^3 L} i_x \quad (7)$$

where

$$\omega_0 = \frac{g_m}{WLC_{ox}} = \frac{\mu(V_{gs} - V_T)}{L^2} \quad (8)$$

is defined as the cutoff frequency and will be used as a reference when high frequency properties of the device are discussed. A description of related quantities is given in Section IV.

Equation (7) can be solved in terms of modified Bessel functions of order $2/3$. It can be shown that the solution of the above equation is given by [4]

$$i_x(X, \omega) = g_m(1 - X)^{1/4}\sqrt{j\omega/\omega_0}$$
$$\cdot \frac{J_{1/3}(\frac{4}{3}\sqrt{j\omega/\omega_0}(1 - X)^{3/4})}{J_{2/3}(\frac{4}{3}\sqrt{j\omega/\omega_0})} \quad (9)$$

where $J_{-1/3}$ and $J_{2/3}$ are modified Bessel functions of the first kind and $2/3$, of order $-1/3$, respectively. The drain current i_d is obtained from (9) when $X = 1$. After using series expansions for the Bessel functions, the drain current can be expanded to

$$i_d = \frac{g_m v_{gs}}{1 + \frac{4}{15}\frac{j\omega}{\omega_0} + \frac{1}{45}\left(\frac{j\omega}{\omega_0}\right)^2 + \frac{8}{8190}\left(\frac{j\omega}{\omega_0}\right)^3 + \cdots}. \quad (10)$$

It is also shown in [4] that the gate current is given as

An approximate solution of (4)–(6), including substrate effects, can be obtained when the gate-to-channel and bulk-to-channel capacitive distributions are replaced by their mean values $\bar{\eta}_1$ or $\bar{\eta}_2$ (obtained by integrating the capacitance distribution function over the channel length). Using this approximation, the solution is calculated by Das [5] as

$$i_x = g_m v_g \left[\frac{\theta \cosh(\theta X)}{\sinh(\theta)}\left(\frac{\chi}{1 + \chi} - \frac{\bar{\eta}_2}{\sinh(\theta)}\right) - \frac{\theta \cosh(\theta(1 - X))}{\sinh(\theta)}\left(1 - \frac{\bar{\eta}_2}{\bar{\eta}_1 + \bar{\eta}_2}\right)\right] \quad (12)$$

with χ the body effect coefficient

$$\chi = \frac{1}{C_{ox}}\sqrt{\frac{\epsilon_{Si} q N_B}{2(2\phi_F + V_{sb})}}$$

$$\theta^2 = \frac{j\omega(\bar{\eta}_1 + \bar{\eta}_2)WLC_{ox}}{g_m}.$$

The drain current is calculated as the channel current for $X = 1$ and by using Taylor series expansions for the hyperbolic functions [5]

$$i_d = \frac{g_m v_{gs}\left[1 + \partial\left(\frac{j\omega}{\omega_0} + \frac{1}{6}\left(\frac{j\omega}{\omega_0}\right)^2 + \cdots\right)\right]}{1 + \frac{1}{3}\frac{j\omega}{\omega_0} + \frac{1}{30}\left(\frac{j\omega}{\omega_0}\right)^2 + \frac{1}{630}\left(\frac{j\omega}{\omega_0}\right)^3 + \cdots}$$

$$\partial = \frac{\overline{\eta}_2(1 + \chi)}{\overline{\eta}_1 + \overline{\eta}_2} - \chi. \quad (13)$$

Equation (13) thus shows the frequency dependence of the drain current of the MOS transistor. The influence of the bulk charge is the introduction of zeros in the right hand plane of the transfer characteristic as can be seen by comparing (13) with (10). These zeros become more important when the $K2$ of the transistor increases and the depletion layer becomes relatively more important. Furthermore, it is clear from both derivations that the transconductance decreases at a frequency of 4–6 times the cutoff frequency.

III. Small Signal Equivalent Circuits

Simulating and analyzing the small signal behavior of analog circuits requires for each circuit component an equivalent network, in a form appropriate for sinusoidal steady-state analysis. These circuits are representations of the linearized equations derived in the previous sections.

The expressions of the admittance or Y parameters can be obtained easily from (10), (11), and (13) and are of the general form of a power series numerator over a denominator series. For hand calculations and even for CAD models, the expressions are simplified by truncating them after some terms. The result is a model of corresponding order, valid up to a frequency which is usually expressed as a multiple of the cutoff frequency (8).

The model is obtained only if the first terms in the numerator and denominator of these equations are kept and all other terms are neglected, is shown in Fig. 1. A more accurate model is realized by considering one term more in the numerator and denominator. This results in the small signal circuit of Fig. 2(a) [8]. The intrinsic part of this model is a special case of the general model obtained by Bagheri and Tsividis (see [2, fig. 1 and table I]). Note, however, that our simplified HF model is accurate enough for our analog applications as will be proven later in this paper. Parasitic effects (usually referenced as the extrinsic transistor elements) are included. They are the series resistances R_g, R_d, R_b, R_s (from sheet resistances and contact resistances), depletion layer capacitances C_{bsp}, C_{bdp} (from drain and source implantations), overlap capacitances C_{gsp}, C_{gdp} (from gate-to-drain and source overlap), serial inductances L_g, L_d, L_b, L_s (from the wiring), etc. These parasitic elements are added externally to the intrinsic transistor model (Fig. 2(a)). However, the model can be simplified if one is not concerned with an exact value for parasitics and intrinsic transistor elements. In this special case, the parasitic transistor elements are

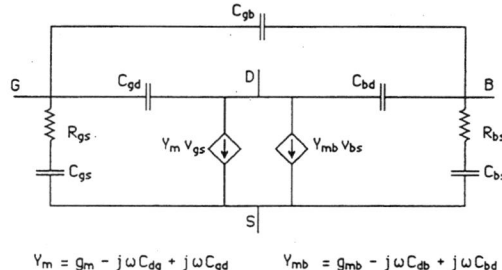

Fig. 1. First-order high frequency equivalent circuit for the intrinsic MOS transistor.

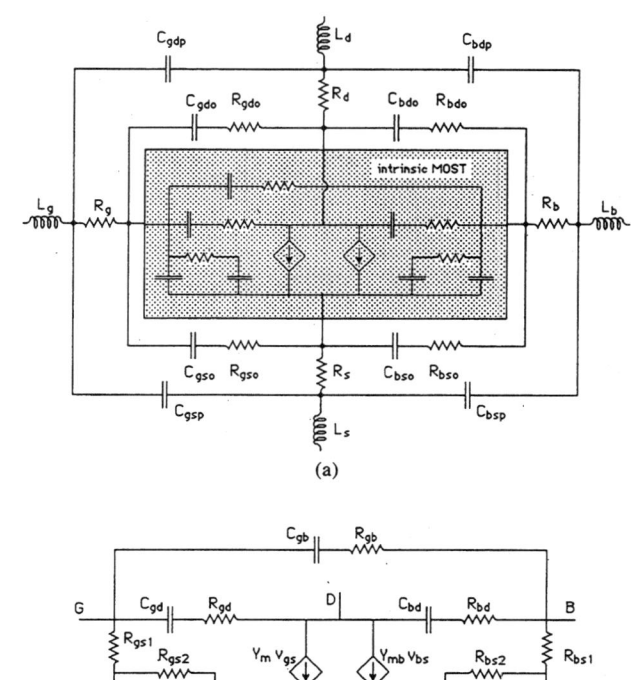

Fig. 2. The second-order HF model. (a) With parasitics is simplified and converted to (b) The layout.

combined with intrinsic transistor elements. As a result, the model of Fig. 2(b) is obtained [8].

A drawback of this model is that the different model elements do not have a close relation to the physical behavior of the transistor. For example, the capacitance C_{dg} is not only an intrinsic transistor capacitance, but part of C_{dg} is due to the extrinsic gate drain overlap capacitance. Note that the model of Fig. 2(b) is only obtained if the serial inductances and pin capacitances of the package are calibrated out in the measurement procedure and if C_{gd} and C_{bd} are both equalized to zero (which is true in the saturation region of the transistor).

When high frequency small signal parameters of the MOS transistor have to be measured, the determination of parasitics of the package and bounding pads becomes very important. For this calibration, special structures (as open and shorts) are provided on wafer. The shorts are packaged in the same package as the transistor will be packaged. A set of measured S parameters of the packaged short will be converted to impedance or Z parameters. Out of these Z parameters, a value can be obtained for the inductances. The pin capacitances can be obtained from the Y parameters of the packaged open structure. Later, the measured small signal parameters of the MOS transistor are corrected for these parasitics.

IV. EVALUATION OF THE SMALL SIGNAL MODEL AT HIGH FREQUENCIES

Since the equivalent circuits of Figs. 1 and 2(b) only take into account the first and/or second terms of the power series expansion, inaccuracies are introduced. The errors caused by these approximations can be estimated by calculating for each Y parameter at each frequency of interest

$$(Y_{ij(\text{series expr})} - Y_{ij(\text{model})})/(Y_{ij(\text{series expr})}). \quad (14)$$

The maximum of this expression over all Y parameters is used as a measure of the inaccuracy. In this formula, the $Y_{ij(\text{series expr})}$ terms are the exact Y parameters as obtained from the theoretical study (series expressions of Section II) while the $Y_{ij(\text{model})}$ terms are the Y parameters of the HF model of Fig. 2(b) as given in Table I. It is found that at a frequency of 5 times the cutoff frequency, these errors can be estimated to be ± 5 percent when two terms and ± 0.5 percent when three terms are retained.

The non-ideal representation of the truncated expressions causes a second group of errors [14]. Consider for example the drain current (truncated after two terms = first order). The pole of the numerator (in the left half plane) is represented by the transcapacitor C_{dg} which gives a zero in the right hand plane. Although this representation causes an increase of magnitude beyond the frequency of the pole (opposite from reality), the phase response beneath that frequency becomes more accurate by introducing the zero. Indeed, a zero in the right hand plane gives the same phase response as a pole in the left hand plane. Therefore, including the capacitor C_{dg} improves the frequency response at moderate frequencies but gives a bad fit at higher frequencies.

The model of Fig. 2(b) is too complex for hand calculations. Most designers simplify the model by equalizing the resistances to 0 Ω. The frequency dependency of the current sources is omitted and the capacitances are replaced by one global capacitance (e.g., $C_{gs} = C_{gs1} + C_{gs2}$). This model can be derived from theory by retaining only the first term of the numerator. All the other terms are neglected. The model obtained has been known for many years and is still widely used at high frequencies, even in recent CAD programs [11], [12]. Using charge storage theory, Ward and Dutton have recently updated this circuit towards a new quasi-static model [9], which is further worked out by Turchetti and Tsividis [1], [2] toward a non-quasi-static model. Ward and Dutton have

TABLE I
Y-PARAMETERS OF THE HF EQUIVALENT CIRCUIT OF FIG. 2

$$Y_{11} = \frac{j\omega(C_{gs1}+C_{gs2})+(j\omega)^2 C_{gs1} C_{gs2} R_{gs2}}{1+j\omega(R_{gs1}C_{gs1}+R_{gs1}C_{gs2}+R_{gs2}C_{gs2})+(j\omega)^2 C_{gs1} C_{gs2} R_{gs1} R_{gs2}}$$
$$+ \frac{j\omega C_{gd}}{1+j\omega R_{gd}C_{gd}} + \frac{j\omega C_{gb}}{1+j\omega R_{gb}C_{gb}}$$

$$Y_{21} = Y_m - \frac{j\omega C_{gd}}{1+j\omega R_{gd}C_{gd}} \quad Y_m = \frac{g_m}{1+j\omega \tau_m} + \frac{j\omega C_{gd} - j\omega C_{dg}}{1+j\omega R_{gd}C_{gd}}$$

$$Y_{31} = -\frac{j\omega C_{gb}}{1+j\omega R_{gb}C_{gb}}$$

$$Y_{41} = -Y_m - \frac{j\omega(C_{gs1}+C_{gs2})+(j\omega)^2 C_{gs1} C_{gs2} R_{gs2}}{1+j\omega(R_{gs1}C_{gs1}+R_{gs1}C_{gs2}+R_{gs2}C_{gs2})+(j\omega)^2 C_{gs1} C_{gs2} R_{gs1} R_{gs2}}$$

$$Y_{12} = -\frac{j\omega C_{gd}}{1+j\omega R_{gd}C_{gd}} \quad Y_{32} = -\frac{j\omega C_{bd}}{1+j\omega R_{bd}C_{bd}}$$

$$Y_{22} = \frac{j\omega C_{gd}}{1+j\omega R_{gd}C_{gd}} + \frac{j\omega C_{bd}}{1+j\omega R_{bd}C_{bd}} + R_{out} \quad Y_{42} = -R_{out}$$

$$Y_{13} = -\frac{j\omega C_{gb}}{1+j\omega R_{gb}C_{gb}}$$

$$Y_{23} = Y_{mb} - \frac{j\omega C_{bd}}{1+j\omega R_{bd}C_{bd}} \quad Y_{mb} = \frac{g_{mb}}{1+j\omega \tau_{mb}} + \frac{j\omega C_{bd} - j\omega C_{db}}{1+j\omega R_{bd}C_{bd}}$$

$$Y_{33} = \frac{j\omega(C_{bs1}+C_{bs2})+(j\omega)^2 C_{bs1} C_{bs2} R_{bs2}}{1+j\omega(R_{bs1}C_{bs1}+R_{bs1}C_{bs2}+R_{bs2}C_{bs2})+(j\omega)^2 C_{bs1} C_{bs2} R_{bs1} R_{bs2}}$$
$$+ \frac{j\omega C_{bd}}{1+j\omega R_{bd}C_{bd}} + \frac{j\omega C_{gb}}{1+j\omega R_{gb}C_{gb}}$$

$$Y_{43} = -Y_{mb} - \frac{j\omega(C_{bs1}+C_{bs2})+(j\omega)^2 C_{bs1} C_{bs2} R_{bs2}}{1+j\omega(R_{bs1}C_{bs1}+R_{bs1}C_{bs2}+R_{bs2}C_{bs2})+(j\omega)^2 C_{bs1} C_{bs2} R_{bs1} R_{bs2}}$$

$$Y_{14} = \frac{j\omega(C_{gs1}+C_{gs2})+(j\omega)^2 C_{gs1} C_{gs2} R_{gs2}}{1+j\omega(R_{gs1}C_{gs1}+R_{gs1}C_{gs2}+R_{gs2}C_{gs2})+(j\omega)^2 C_{gs1} C_{gs2} R_{gs1} R_{gs2}}$$

$$Y_{24} = -Y_m - Y_{mb} - R_{out}$$

$$Y_{34} = -\frac{j\omega(C_{bs1}+C_{bs2})+(j\omega)^2 C_{bs1} C_{bs2} R_{bs2}}{1+j\omega(R_{bs1}C_{bs1}+R_{bs1}C_{bs2}+R_{bs2}C_{bs2})+(j\omega)^2 C_{bs1} C_{bs2} R_{bs1} R_{bs2}}$$

$$Y_{44} = -Y_{14} - Y_{24} - Y_{34}$$

introduced transcapacitors to model the non-reciprocal charge distribution in the transistor, while Tsividis has added resistances and time constants in the transconductances g_m and g_{mb}.

The Ward model leads to a simple equivalent circuit and can be derived from our HF model by neglecting the resistances. The Ward formulation is correct for the low frequency Y parameters. However, problems arise at high frequencies since the model is quasi-static and fails to take into account the transmission line effects. Indeed, the transcapacitances will generate a current which increases infinitely with increasing frequency. This unrealistic behavior is opposite to what is actually expected. The first-order HF model (Fig. 1) does not suffer from the above limitation since the terminal currents are limited by the resistances and the frequency dependent conductances g_m and g_{mb}. So, the Ward model is an approximation to the exact solution at low frequencies, including the effect of non-reciprocity, but cannot be used at high frequencies (somewhere above ω_0). The question still arises till which frequency the models remain valid within a specified error margin.

The various equivalent circuits, mentioned in the previous paragraphs are compared with the analytical results in order to find their regions of validity [14]. The error is defined by the error function (23). The maximum fre-

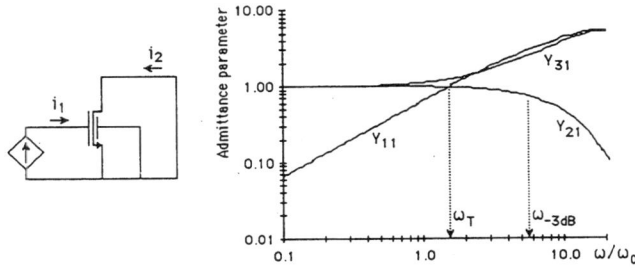

Fig. 3. Definition of ω_0, ω_T, and $\omega_{-3\,dB}$.

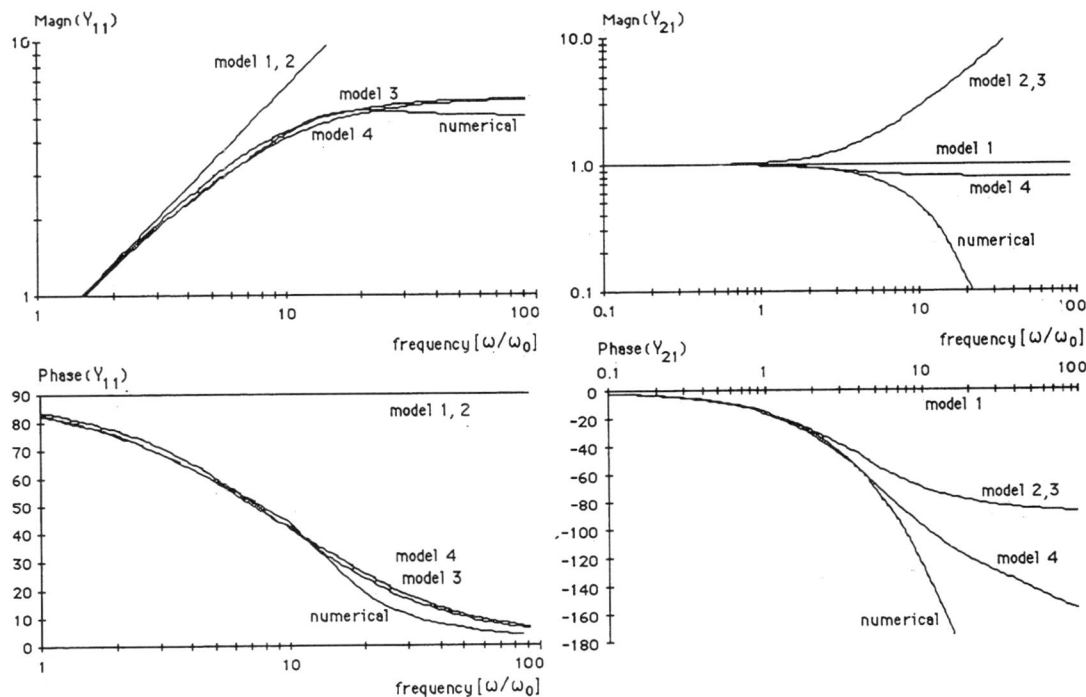

Fig. 4. Y_{11} and Y_{21} for the various models.

quency, for which the error does not exceed 0.5 dB, is retained as maximum frequency of validity. This maximum frequency of validity is usually related to the cutoff frequency of the device. When comparing data from different authors, care should be taken with the definition of the following quantities (Fig. 3).

1) The *cutoff frequency* ω_0 is defined by formula (11) and is thus rather an arbitrarily chosen quantity. However, figures for the cutoff frequency are commonly used in the literature since this quantity is closely related to the technology parameters of the device.

2) The *unity gain frequency* ω_T is defined as the frequency where the gate and drain current become equal (1). This definition implies that the logarithm of the magnitude of H_{21} equals zero at the cutoff frequency and also that the magnitude of Y_{11} and Y_{21} are identical at that frequency. Note that the definition of the bipolar transistors cutoff frequency is very similar.

3) Some authors also define a *−3-dB frequency* as the frequency where the forward transconductance (Y_m) drops over 3 dB. This frequency is only of interest when overlap capacitances at the gate are very small.

It is obvious that fixed relations between the parameters exist. By evaluating the series expansions of the drain current (10) and (13), it is found that the unity gain frequency ω_T is 1.47 times the cutoff frequency ω_0 while the −3-dB frequency equals 5.8 times the cutoff frequency ω_0. The same relationship is also found by other authors [15], following another method.

In Fig. 4 the magnitude and phase of the normalized Y_{11} and Y_{21} parameters are plotted as a function of the frequency for various models. Comparison with the analytical solution indicates a maximum radial frequency of validity (max 0.5-dB error):

1) $\omega_0/8$ for the quasi-static model;
2) $\omega_0/3$ for the quasi-static model with transcapacitors (Ward);
3) ω_0 for the first-order high frequency model (Turchetti-Tsividis);
4) $4\omega_0$ for the second-order high frequency model.

The usable frequency range of the simple quasi-static topology is smaller than that of the Ward model. Without the transcapacitor, the transconductance Y_m remains constant till infinite frequencies. This is wrong. Although the introduction of the transcapacitor increases the upper frequency limit of validity, the entire model becomes worse than the simple model at very high frequencies (several times ω_0). In general, the accuracy of all quasi-static equivalent circuits is poor at very high frequencies, irrespective of which model is used. Indeed, in small signal analysis, the quasi-static assumption implies that all admittances between terminals are modeled as a parallel circuit of a low frequency conductance (eventually 0 mho) and a capacitor ($Y_{ij} = g_{ij} + j\omega C_{ij}$). While this approximation is valid at low frequencies, this is no longer true at high frequencies since in reality the channel resistance prevents rapid changes in charge.

V. Measuring and Fitting the Small Signal Parameters of the MOS Transistor

When a linear device has to be characterized at high frequencies, S parameter measurements are preferred over the measurement of other parameters. The measurement of precise S parameters is not obvious, due to the presence of several error sources as source/load mismatch, directivity errors, reflections at the connectors, and test yig, etc. Furthermore, parasitic effects as lead inductances and pin capacitances of the package will degrade the accuracy of the S parameters of the MOS transistor. It is well known that the calibration of the instruments, test yig, and package is difficult but mandatory [19], [20]. Since most of the error sources are systematic, a calibration procedure will reduce the influence of the error terms. By understanding where the different errors come from, an appropriate model can be developed to describe their effects. Today, network analyzers are equipped with a computer and suitable software which allows the use of complex calibration routines [8], [20], [21]. The real S parameters of the device are calculated out of the measured ones, taking into account the error terms. By using such a calibration and error correction procedure, several sources of error are calibrated out.

A MOS transistor is a four terminal device. The transistor is connected to a two port S parameter test set and the transistor configuration is changed during the measurement [8]. In a first measurement, the gate is defined as port 1 and the drain as port 2. Four S parameters are measured at a given dc bias point. Next the measurement configuration is changed: the bulk is connected to port 1, while the source is connected to port 2. A second set of four S parameters is measured at the same dc bias. The eight measured S parameters are converted mathematically to Y parameters.

In the following sections, the pins of the MOS transistor are referenced with numbers in the abbreviations for the Y parameters: the gate as 1, the drain as 2, the bulk as 3, and the source as 4. For example, Y_{21} is the admittance parameter from gate-to-drain, etc. The eight Y parameters, obtained as a function of the frequency, are thus: $Y_{11}, Y_{21}, Y_{12}, Y_{22}, Y_{33}, Y_{34}, Y_{43}, Y_{44}$. Measuring the other eight missing Y parameters has such a main disadvantage that another very time consuming measurement is needed. For that reason, the missing Y parameters are calculated. It is obvious from network theorems that six linear independent relations exist between all the Y parameters. To find the eight missing parameters, two extra equations are thus needed. The first relation used is the measurement of the admittance parameter Y_{42}, which is equal to the output impedance. As second relation, a special property of the MOS transistor is used. Considering the model of Fig. 2(b) or Table I, it is obvious that Y_{13} equals Y_{31}. This relationship is used as last equation. As a consequence, the usefulness of the measurement method is limited to the special case of MOS transistors following this equality. The missing Y parameters are then calculated by using the relations: $\Sigma Y_{ij} = 0$. The final result of the measurement and calculations is thus a three-dimensional matrix of N by 4 by 4 elements (N is the number of frequency points). This three-dimensional matrix of measured data is fitted towards the theoretical model. The fitting process is built up of two parts [22], [23].

1) The data are fitted towards a general transfer function with p poles and z zeros (p and z ranges from 1 to 3).

2) These poles and zeros to the poles and zeros of the theoretical transfer functions (Table I), which results in the determination of the small signal network elements. Both steps are described next.

The curve fitting routines have to find a pole zero model of a linear system, based on the measured frequency response (linear parameters) of the device under test. Let $H(j\omega)$ be the transfer function of the device under test and let $\omega_i (i = 1 \cdots n)$ be a sequence of n frequency points at which values of $H(j\omega)$ are measured. These measured values have generally some errors. Let the true value at ω_i be denoted by $H(j\omega_i)$, while the measured value is given by $H'(j\omega_i)$. The object of the curve fitting routines is to approximate the measured transfer function $H'(j\omega)$ at ω_i by a rational function in $(j\omega)$ [22]

$$H'(j\omega_i) \approx \frac{a_0 + a_1(j\omega_i) + a_2(j\omega_i)^2}{b_0 + b_1(j\omega_i) + b_2(j\omega_i)^2}. \tag{15}$$

The a_k and b_l coefficients are real values which have to be determined so that the error function (EF) is minimized

$$EF = \sum_{i=1}^{n} w(j\omega_i) \sigma_i \overline{\sigma}_i. \tag{16}$$

The function $w(j\omega)$ is called the weight function; σ_i is the residue at each measured data point ($\overline{\sigma}_i$ is the complex conjugate of σ_i), and is defined as

$$\sigma_i = \frac{(X_i + jY_i)\big((b_0 + b_1(j\omega_i) + b_2(j\omega_i)^2\big) - \big(a_0 + a_1(j\omega_i) + a_2(j\omega_i)^2\big)\big)}{a_0 + a_1(j\omega_i) + a_2(j\omega_i)^2}.$$

The minimization of the error function uses a simple least mean square algorithm. Iterations are used in order to keep the equations which have to be solved linear [8]. In each iteration, the calculated zeros of the previous iteration are used to calculate a new value for the coefficients a_k in (15). As a result, the dominator of (16) becomes constant and different from zero. The fit routine will give a new set of coefficients a_k and b_l. If the wanted accuracy is reached, the polynomials in the denominator and numerator are factored to find the poles and zeros of the measured transfer function. As a result, the amplitude, poles, and zeros of each Y parameter of interest is known. The model parameters of the MOS transistor are determined by equating these poles, zeros, and amplitudes to those of the theoretical formulas of Table I.

It is obvious that not each Y parameter is as sensitive or as useful to extract the different transistor parameters. As example, the input admittance Y_{11} is sensitive to the resistances R_{gs1} and R_{gs2}. The parameter Y_{11} can thus be used to extract the elements R_{gs1} and R_{gs2}. However, it is much easier to fit the resistances R_{gs1} and R_{gs2} from the admittance parameter Y_{14}. Indeed, the poles of Y_{11} are a complex function of the resistances R_{gs1}, R_{gs2} but also R_{gd} and R_{gb}. For that reason, an optimum fit strategy has to be developed [22].

Define

$A(Y_{ij})$ as the amplitude of Y_{ij} if there is no zero at $\omega = 0$ or as the magnitude at $\omega = 1$ if there is a zero at $\omega = 0$,
$p_1(Y_{ij})$ as the time constant of the dominant pole of Y_{ij},
$p_2(Y_{ij})$ as the time constant of the second pole of Y_{ij},
$z_1(Y_{ij})$ as the time constant of the dominant zero of Y_{ij},
$z_2(Y_{ij})$ as the time constant of the second zero of Y_{ij}.

The different small signal network elements of the MOS transistor can be calculated easily. Consider as an example the admittance from source-to-bulk with amplitude $A(Y_{34})$, pole time constants $p_1(Y_{34})$, $p_2(Y_{34})$, and zero time constants $z_1(Y_{34})$ and $z_2(Y_{34})$. From Table I it follows that:

$$A(Y_{34}) = \omega(C_{bs1} + C_{bs2})$$
$$p_1(Y_{34})p_2(Y_{34}) = C_{bs1}C_{bs2}R_{bs1}R_{bs2}$$
$$p_1(Y_{34}) + p_2(Y_{34}) = R_{bs1}VC_{bs1} + R_{bs1}C_{bs2} + R_{bs2}C_{bs2}$$
$$z_1(Y_{34}) = \infty$$
$$z_2(Y_{34}) = \frac{C_{bs1}C_{bs2}}{C_{bs1} + C_{bs2}} R_{bs2}. \quad (17)$$

Equation (17) can be solved for the model parameters:

$$C_{bs1} = \frac{z_2(Y_{34})A(Y_{34})}{p_1(Y_{34}) + p_2(Y_{34}) - \dfrac{p_1(Y_{34})p_2(Y_{34})}{z_2(Y_{34})}}$$

$$C_{bs2} = A(Y_{34}) - C_{bs1}$$

$$R_{bs1} = \frac{p_1(Y_{34})p_2(Y_{34})}{z_2(Y_{34})A(Y_{34})}$$

$$R_{bs2} = \frac{p_1(Y_{34}) + p_2(Y_{34}) - \dfrac{p_1(Y_{34})p_2(Y_{34})}{z_2(Y_{34})}}{C_{bs2}}. \quad (18)$$

Consider the practical example of a NMOS transistor with a W/L ratio of 100 μm by 10 μm. The Y_{34} parameter, measured at 1 mA is plotted in Fig. 5. The fit results are

$A(Y_{34}) = -0.682E - 12$
$p_1(Y_{34}) = -6.08E - 9 \qquad z_1(Y_{34}) = \infty$
$p_2(Y_{34}) = -134E - 12 \qquad z_2(Y_{34}) = -1.61E - 9.$

Using these fit results in the formulas (18) leads to

$C_{bs1} = 192$ fF $\qquad C_{bs2} = 490$ fF
$R_{bs1} = 740$ Ω $\qquad R_{bs2} = 11.6$ kΩ

As second example, consider the admittance parameter Y_{21} of the same transistor (Fig. 6):

$A(Y_{21}) = 0.681E - 3$
$p_1(Y_{21}) = -258E - 12 \qquad z_1(Y_{21}) = +379E - 12.$

The low frequency intercept is the transconductance g_m of the MOS transistor ($g_m = 681$ μA/V), whereas the zero in the right hand plane is caused by the transconductance C_{dg} as can be seen in Table I. This leads to a value for C_{dg} as

$$C_{dg} = g_m * z_1(Y_{34}) = 258 \text{ fF}.$$

The determination of the unity gain frequency is somewhat more inconvenient. Remember that the unit gain frequency is defined as the frequency where the current gain equals unity. By using this definition, the fitting of f_T becomes easy. Indeed, the definition can be modified to

$$f_T = f\big|_{i_2/i_1=1} = f\big|_{(i_2/v_1/i_1/v_1)=1} = f\big|_{Y_{21}/Y_{11}=1}.$$

This means that at the frequency f_T, the admittance parameters Y_{21} and Y_{11} are equal in magnitude. The determination of f_T is thus reduced to finding the frequency where Y_{21}/Y_{11} equals 1 (Figs. 3 and 9).

Some interesting graphs can be obtained from these figures. Indeed, it is easy to obtain values for parameters as g_m, g_{mb}, C_{gs}, $\cdots$ as a function of the dc current through the MOS transistor. Other important relations of the transistor parameters versus the transistor dimension can be measured. Consider as example the unity gain frequency f_T versus the transistor length L at a constant drain current of 1 mA (Fig. 7). The theory predicts that the unity gain frequency will decrease when the length of the MOS transistor increases. For long channel transistors, the unity gain frequency is related to $1/L$ (due to velocity saturation).

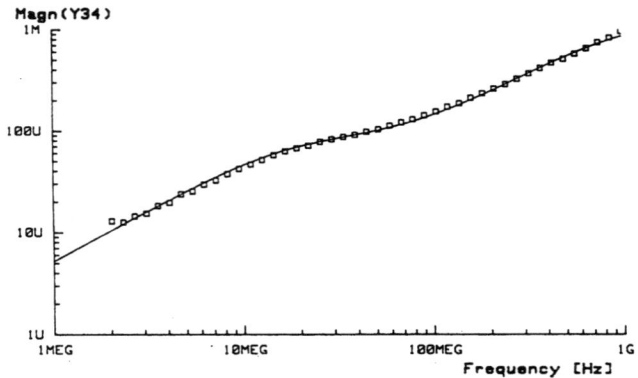

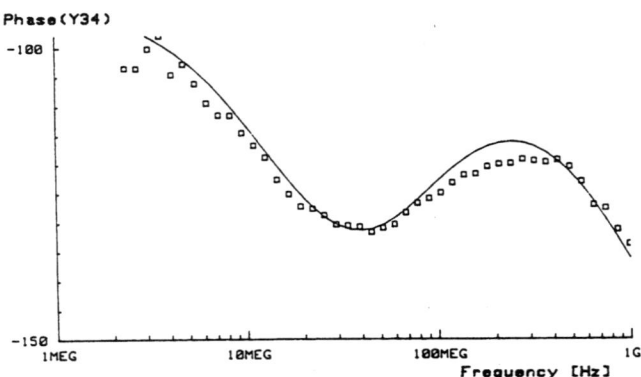

Fig. 5. Y_{34} parameter for a NMOS with W/L of 100/10 and $I_{ds} = 1$ mA.

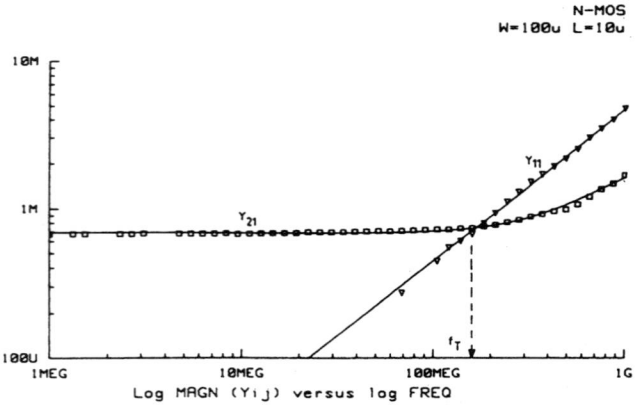

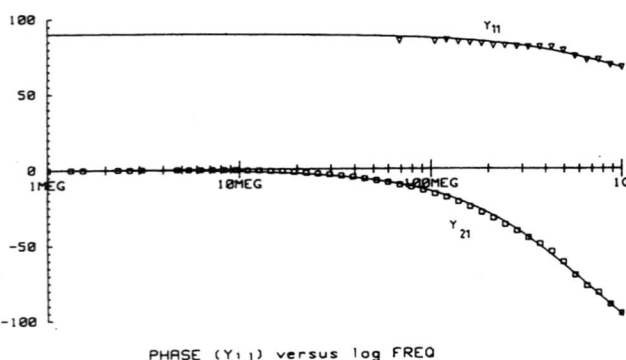

Fig. 6. Y_{21} and Y_{21} and Y_{21} parameters as a function of the frequency. The cutoff frequency f_T is the frequency where Y_{21} equals Y_{21}.

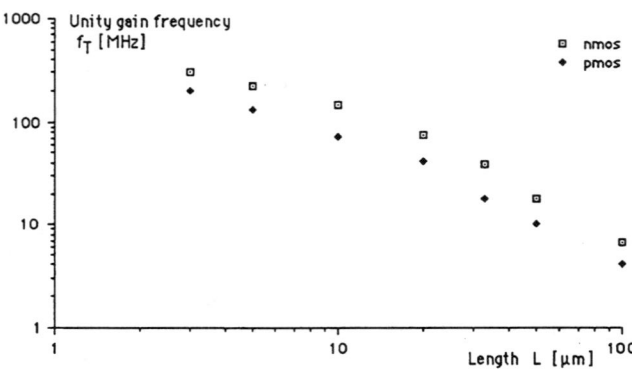

Fig. 7. The cutoff frequency against the length of the MOS transistor.

VI. Application of the Small Signal Models to the Design of an OTA

In this section, the validity of the HF small signal model is demonstrated. A fully differential folded cascode operational transconductance amplifier (OTA) is designed and realized. This circuit is well suited to discuss the validity of the high frequency MOS transistor models. Indeed, the unity gain frequency of the OTA can come relatively close to the unity gain frequency of the cascode transistor. As a consequence, the transistor models need to be accurate till at least the unity gain frequency of the cascode MOS transistor.

The complete diagram of the fully differential folded cascode is shown in Fig. 8. The operation of this folded cascode amplifier is obvious [23]. Due to a differential voltage at the input, the transistors $M1$ and $M2$ provide a small signal current which is fed into the cascode stage. The transistors $M3$ and $M4$ form the current source of the cascode stage. $M10$ and $M11$ are the cascode transistors. The load of the cascode stage is implemented by the transistors $M5$, $M6$, $M12$, and $M13$. In order to stabilize the common mode output of the OTA, two error amplifiers are added ($M17$–$M26$). These two error amplifiers will sense the output voltage, compare this voltage with the ground voltage, and produce a common mode output voltage. This common mode voltage is used to change the currents of both current sources, which act as active load. As a result, the common mode output voltage will change in the direction, opposite to the direction which has caused the activation of the common mode error correction amplifier.

The gain of the amplifier is given by

$$G = -g_{m1} R_l \frac{G_{S10}}{G_{S10} + G_{OUT1} + G_{OUT3}}$$

where

$R_L = R_{OUT13}(1 + g_{m13} R_{OUT5})$,
g_{m1} the transconductance of transistor $M1$,
G_{OUT1} ($= 1/R_{OUT1}$)   the output conductance of $M1$,
G_{OUT3} ($= 1/R_{OUT3}$)   the output conductance of $M3$,
G_{S10} the conductance seen into the source of transistor $M10$.

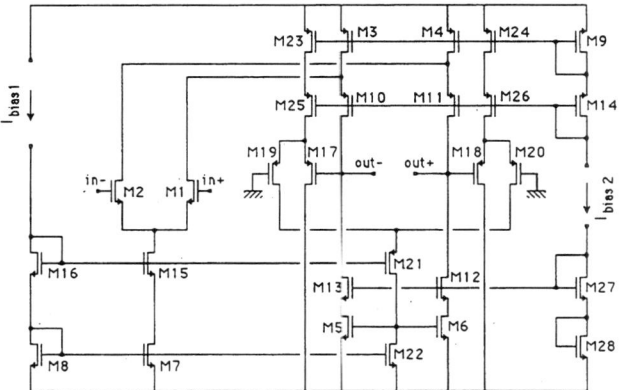

Fig. 8. Complete diagram of the differential folded cascode OTA, including the error amplifiers as developed in this paper.

G_{S10} can be calculated as

$$G_{S10} = \frac{g_{m10} R_{OUT10}}{R_{OUT10} + R_L} = \frac{1}{R_{S10}}$$

where

g_{m10} the transconductance of transistor $M10$,
R_{OUT10} the output impedance of $M10$

The frequency response of the folded cascode is calculated by using three approaches:

1) hand calculations are used to obtain an insight in the small signal behavior of the circuit;
2) the circuit is simulated with the SPICE small signal models (level 2 and level 3);
3) the circuit is analyzed with the HF small signal model.

The folded cascode has two major poles. The dominant pole p_1 is located at the output node and is given by

$$p_1 = -\frac{G_L + G_{D10}}{C_{L,TOT}}$$

where

$G_L (= 1/R_L)$ the load conductance,
G_{D10} the conductance seen into the drain of transistor $M10$.

The resistance R_{D10}, which equals $1/G_{D10}$, can be calculated as

$$R_{D10} = R_{OUT10}(1 + g_{m10}/(G_{OUT1} + G_{OUT3}))$$

where G_{OUT1} and G_{OUT3} are the output conductances of $M1$ and $M3$.

The most important nondominant pole p_2 is found at the cascode node of the OTA:

$$p_2 = -\frac{g_{m10} + G_{OUT3} + G_1}{C_p} \approx -\frac{g_{m10}}{C_p}$$

where C_p is the total parasitic capacitance at the cascode node. This capacitance is composed of the gate-to-source and bulk-to-source capacitance of transistor $M10$ and the gate-to-drain and bulk-to-drain capacitance of transistor $M3$.

The circuit of Fig. 8 was simulated by using SPICE (with MOS model level 2 and 3). On the other hand, the circuit was also simulated by using the HF model of the previous section. For this simulation, the equivalent small signal network of the OTA was described as a network of resistances, capacitances, and current sources. The value of each network element (at the dc operating current) was obtained by using the measurement method of Section V. Afterwards, SPICE was used to calculate the HF response of this network. This working method is, of course, very time consuming, but the analog designer has no alternative because of the lack of CAD tools with a build-in HF model. However, the feasibility of the model is now proven and in the near future, the HF model should be introduced in a circuit simulator as SPICE.

The measurement scheme of [24] was used to measure the OTA. The plot of Fig. 9 gives the open-loop gain in the neighborhood of the unity gain frequency of the amplifier. It is just at these frequencies that the discrepancies between the different simulations become important. It is obvious from the graph that the simulation of the amplifier by using the HF model gives the best results. This is not surprising if the transistor characteristics, as measured in Section V, are considered. Indeed, Fig. 5 clearly shows that a pole is present at a frequency of about 20 MHz. This results in a markable phase variation beyond approximately 2 MHz. The presence of this extra phase shift is observed in several Y parameters of the MOS transistor. This causes the introduction of resistances in all the branches of the small signal model of the MOS transistor. None of them is provided in the SPICE model. Moreover, the transcapacitors are not modeled either in the SPICE models. As a consequence, the simulated phase characteristic is significantly improved when the HF small signal model is used. The improvement becomes even more significant for designs where very long channel transistors are used. Indeed, the transconductance of these transistors is frequency dependent due to the time constant τ_{gm}. The extra pole, associated with this time constant, occurs beneath the unity gain frequency and gives rise to extra errors when the quasi-static models are used.

It can be concluded that the simulation of analog circuits by using simple quasistatic models (as the one in SPICE) will give accurate results for relatively low frequencies. As stated in Section IV, these simple models are valid till the frequency $f_0/8 = f_T/12$. When the transfer characteristic of the circuit has to be calculated at higher frequencies, the use of more accurate models is mandatory. In the described design of the folded cascode OTA, the f_T of the PMOS cascode transistor equals 22 MHz (at 50 μA). The simple models (SPICE level 2 and 3) are valid within 0.5 dB till 1.8 MHz, whereas the first-order high frequency model is valid till 14 MHz and the second-order high frequency model is valid till 64 MHz. From these figures, it is obvious that the estimation of the phase margin of the folded cascode (with a unity gain fre-

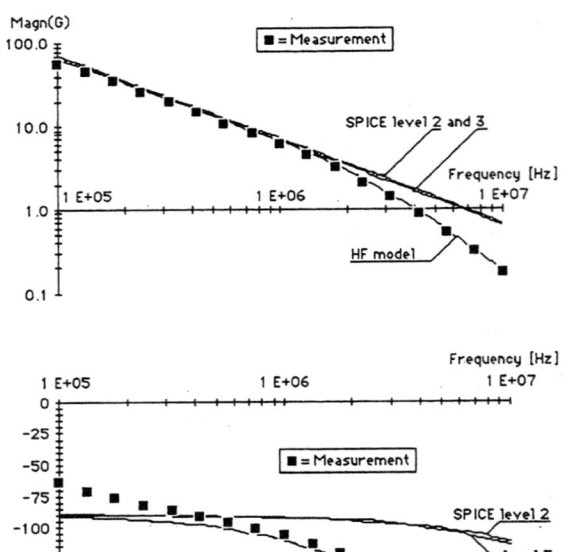

Fig. 9. Measured and simulated transfer characteristic of the folded cascode.

quency of 7 MHz) cannot be accurate if the SPICE model is used. The simulation with the HF model, however, will give precise information at the unity gain frequency of the amplifier!

VII. CONCLUSION

In this paper, a study of the high frequency characteristics of MOS transistors is presented. The result is a novel non-quasi-static high frequency model. The validity of all this model is compared with other small signal models. It is found that the quasi-static models are valid till a frequency of the unity gain frequency f_T divided by twelve. The introduction of transcapacitors will improve the validity of the model up to a frequency of $f_T/6$. Further improvement up to $2/3 f_T$ is achieved by the novel first-order HF model, whereas the novel second-order model is accurate up to $8/3 f_T$.

The main problem of the novel HF model is the measurement of all the small signal elements. S parameter measurements are chosen and special attention is paid to the calibration of the instrumentation. The true S parameters of the MOS transistor are converted to Y parameters and fitted towards the small signal models by using least square fit routines. The measurement and parameter extracting method, presented in this work, is able to fit the intrinsic MOS transistor characteristics, up to a frequency of almost three times the unity gain frequency of the transistor.

The results of these measurements were used to design a folded cascode amplifier. Although the gain bandwidth was not higher than 7 MHz for a capacitive load of 5 pF, the simulations by using the quasi-static models of SPICE give very erroneous results. The HF model, however, gives very accurate predictions for the open-loop characteristic of the OTA. This clearly demonstrates the upper limits of validity for the different small signal models.

REFERENCES

[1] C. Turchetti, G. Masetti, and Y. Tsividis, "On the small-signal behaviour of the MOS Transistor in quasistatic operation," *Solid State Electron.*, vol. 26, no. 10, pp. 941-949, 1983.
[2] M. Bagheri and Y. Tsividis, "A small signal dc-to-high-frequency nonquasistatic model for the four-terminal MOSFET valid in all regions of operations," *IEEE Trans. Electron Devices*, vol. ED-32, pp. 2383-2391, Nov. 1985.
[3] J. J. Paulos and D. A. Antoniadis, "Limitations of quasi-static capacitance models for the MOS transistor," *IEEE Electron Device Lett.*, vol. EDL-4, pp. 221-224, July 1983.
[4] J. R. Burns, "High frequency characteristics of the insulated gate field effect transistor," *RCA Rev.*, pp. 385-418, Sept. 1967.
[5] M. B. Das, "High frequency network properties of MOS transistors including the substrate resistivity effects," *IEEE Trans. Electron Devices*, vol. ED-16, pp. 1049-1069, Dec. 1969.
[6] Y. Tsividis and G. Masetti, "Problems in precision modeling of the MOS transistor for analog applications," *IEEE Trans. Computer-Aided Design*, vol. CAD-3, pp. 72-79, Jan. 1984.
[7] ——, "Relation between incremental intrinsic capacitances and transconductances in MOS transistors," *IEEE Trans. Electron Devices*, vol. ED-27, pp. 946-948, May 1980.
[8] P. Vandeloo, "Modeling of the MOS transistor for high frequency analog design," Ph.D. dissertation, Katholieke Universiteit Leuven, 1987.
[9] D. E. Ward and R. W. Dutton "A charge-oriented model for MOS transistor capacitances," *IEEE J. Solid-State Circuits*, vol. SC-13, pp. 703-707, Oct. 1978.
[10] D. E. Ward, "Charge-based modeling of capacitance in MOS transistors," Ph.D. dissertation, Integrated Circuit Laboratory, Stanford Univ., CA, June 1981.
[11] S. Liu and L. W. Nagel, "Small-signal MOSFET models for analog circuit design," *IEEE J. Solid-State Circuits*, vol. SC-17, pp. 983-998, Dec. 1982.
[12] A. Vladimirescu and S. Liu, "The simulation of MOS integrated circuits using SPICE2," Electronics Res. Lab., University of California, Berkeley, Memo. ERL M80/7, Feb. 1980.
[13] B. J. Sheu, D. L. Scharfetter, and P. K. Ko, "SPICE2 implementation of BSIM," Electronics Res. Lab., University of California, Berkeley, Memo. UCB/ERL M85/42, May 24, 1985.
[14] P. J. Vandeloo and W. Sansen, "Determination of the HF model parameters of the MOS transistor by using standard dropin test structures," in *IEEE Proc. Microelectronic Test Structures*, vol. 1, pp. 97-101, Feb. 1988.
[15] H. Khorramabadi, "High-frequency CMOS continuous time filters," Electronics Res. Lab., University of California, Berkeley, Memo. UCB/ERL M85/19, Feb. 27, 1985.
[16] K. C. K. Weng and P. Yang, "A direct measurement technique for small geometry MOS transistor capacitances," *IEEE Electron Device Lett.*, vol. EDL-6, pp. 40-42, Jan. 1985.
[17] J. J. Paulos, D. A. Antoniadis, and Y. P. Tsividis, "Measurement of intrinsic capacitances of MOS transistors," *IEEE J. Solid-State Circuits Conf.*, pp. 238-239, 1982.
[18] Hewlett Packard, "S-parameters · · · circuit analysis and design," Application Note 95.
[19] P. Vandeloo and W. Sansen, "Measuring and fitting of the small signal model of the MOS transistor for high frequency applications," *IEEE Instrum. Meas. Technology Conf.*, IEEE Catalog 88CH2569-2, pp. 232-237, Apr. 1988.
[20] Mingchen He, "A simplified error model for accurate measurement of high-frequency transistor S parameters," *IEEE Trans. Instrum. Meas.*, vol. IM-34, pp. 616-619, Dec. 1985.
[21] E. F. Da Silva and M. K. McPhun, "Calibration of microwave network analyzer for computer corrected S parameter measurements," *Electron. Lett.*, vol. 9, no. 6, pp. 126-128, Mar. 1973.
[22] W. Sansen and P. Vandeloo, "Modeling the MOS transistor at high frequencies," *Electron. Lett.*, vol. 22, no. 15, pp. 810-812, July 1986.
[23] T. C. Choi, R. T. Kaneshiro, R. W. Broderson, P. R. Gray, W. B. Jett, and W. Wilcox, "High frequency CMOS switched capacitor fil-

ters for communication applications," *IEEE J. Solid-State Circuits*, vol. SC-18, pp. 652–664, Dec. 1983.

[24] W. Sansen, M. Steyaert, and P. Vandeloo, "Measurement of operational amplifier characteristics in the frequency domain," *IEEE Trans. Instrum. Meas.*, vol. IM-34, pp. 59–64, Mar. 85.

*

8

SYSTEM APPLICATIONS

THE circuit design and device-level issues discussed in previous sections are central to the realization of the hardware for mobile communication systems. However, these various circuit components ultimately must be brought together to form a final working product. The assembly of complete chip sets and their functioning in various radio formats comprise the subject of the papers in this section.

A DOUBLE-CONVERSION BROAD BAND TV-TUNER WITH GaAs ICs

J.-E. Müller, U. Ablassmeier, J. Schelle*, W. Kellner, H. Kniepkamp

Siemens AG, Corporate Research and Development, D-8000 Munich 83, FRG
*Siemens AG, Components Group, D-8000 Munich 83, FRG

ABSTRACT

In conventional terrestrial TV tuners single down conversion is applied comprising three tuners in parallel for reception of the TV frequency band (47-860 MHz). As a consequence 12 tunable resonant circuits have to be adjusted manually for good frequency tracking.
A monolithic solution, instead, avoids tweaking of the different tuning circuits and reduces the number of components. Both measures are steps to a costeffective approach. In this paper, a double-conversion broad band tuner based on GaAs-ICs is presented. Its gain (20 dB), noise figure (7.5 dB) and in-channel cross modulation (6 mV) correspond to the state-of-the-art of conventional tuners. Based on an analysis of the measured results necessary improvements in off-channel cross-modulation are discussed.

INTRODUCTION

The TV frequency band covers four octaves. Thus, for a single-range tuner based on single down conversion a local oscillator (LO) is needed which can be tuned over four octaves, too. As this is not feasible, conventional TV-tuners consist of three tuners in parallel (47-68, 174-230, 470-860 MHz).
Fig. 1 shows the block diagram of a conventional tuner. According to the frequency the signal is fed to one of the three tuners. The signal is amplified by a selective amplifier using a three stage tunable band filter. Then the signal is mixed with an LO. The LO-frequency is tuned with a fourth resonant circuit, in order to downconvert the signal to the standard TV intermediate frequency (IF) of 36 MHz.
For good tracking altogether 12 tuning circuits must be adjusted manually. Furtheron, with this concept only a rather limited degree of integration is feasible. A monolithic solution, however, avoids tweaking of the different tuning circuits and diminishes the number of components. Fig. 2 shows a GaAs-solution based on double conversion. In contrast to the narrow band frequency tuning of conventional tuners there is no frequency selection in front of the first mixer. The entire TV frequency band is fed to the first mixer which upconverts the signal to a high IF (e.g. 2.5 GHz). The signal is then selected by a high Q fixed dielectric filter. The second mixer downconverts the signal to the standard IF of 36 MHz. With this principle, only one tuning circuit is needed compared to 12 tuning circuits in conventional tuners. Because of the high IF, the relative tuning range of the LO-frequency is reduced to about ± 15 %, such that tuning of the entire TV band in a single range is possible.

CIRCUIT FABRICATION

Active layers were fabricated by selective implantation of Si into low-Cr doped LEC material and annealing at 840° C with Si_3N_4 cap. The FET gate length was 1 µm, the source/drain spacing 5 µm. Contacts were patterned using contact lithography and a chlorobenzene assisted lift-off technique. Sputtered Si_3N_4 was used for dielectric isolation of metal layers.

EXPERIMENTAL SETUP

Fig. 3 shows the circuit diagram of the GaAs double-conversion tuner (DCT). The balanced input stage presently consists of three GaAs ICs: two broadband CGY 20-amplifiers /1/ in front of the mixer IC. The CGY 20 provide good input match over the entire TV band. The mixer upconverting the signal consists of a gain controlled amplifier followed by a double-balanced FET-mixer. The mixer is an integrated version of the double-balanced mixer described earlier (Fig. 4, /2/). The mixer output is connected to a fixed two stage dielectric bandfilter adjusted to the first IF (2.5 GHz). The second mixer downconverts the signal. It is of the same type as the first mixer.
The circuit is balanced from the first mixer's input to the second mixer's output. At the output unbalancing and matching to 50 Ohm is achieved by the 36 MHz IF-filter. Selective matching between the mixer ICs and the dielectric filter is made off chip with simple lumped and distributed elements.
For this experimental setup external local oscillators are still used. Upconverter, dielectric filter and downconverter are realized on separate PC boards, which can be connected by coaxial adapters. This procedure allows testing and optimization of the three subcircuits. Fig. 5 shows the upconverter. The two CGY-amplifiers and the mixer are mounted in a single ceramic package.

RESULTS

Both tuners show a gain of approximately 20 dB at low frequencies. In contrast to the conventional tuner the gain of the DCT decreases slowly to about 15 dB at 860 MHz. This gain decrease is due to non-optimized interstage matching, but of no importance for the tuner performance.
Fig. 6 shows the noise figure NF versus frequency for both tuners. The midband NF of 7 dB of the DCT com-

pares well with that of the conventional tuner. The slight degradation in NF towards higher frequencies can be overcome by optimization of the matching between the amplifier and the first mixer.

Because TV-signals can have big differences in amplitude level, tuners must show a high dynamic range. The large signal tuner properties are described by the undesired signal level for 1 % cross modulation in- and off-channel, respectively. The in-channel level of 6 mV for the DCT meets the state-of-the-art of conventional tuners. The measured off-channel level of 40 mV of the present setup is lower than that of conventional tuners which exceeds 100 mV. This stems from the fact that conventional tuners apply frequency selection in front of the mixer whereas the DCT does not.

DISCUSSION

The total system gain is $\approx$ 20 dB. The CGY-amplifier/mixer combination contributes 14 db gain. The dielectric filter has a loss of 2 dB and the second mixer has a gain of 8 dB. The CGY-amplifier/mixer combination has enough gain to determine the system noise figure.

The limitations corresponding the undesired signal levels in-channel and off-channel stem from different stages of the system. An undesired in-channel signal is amplified by the upconverter stage, passes through the dielectric filter with small attenuation and arrives at the downconverter stage with high amplitude. As mixers of the same type are used presently in the up- and downconverter stages the in-channel level is limited by the second mixer because of the higher amplitude level there.

On the contrary an undesired off-channel signal is strongly attenuated by the dielectric filter before arriving at the second mixer. The off-channel level is therefore limited by the first mixer.

From these arguments guidelines for optimization of system properties can be derived for the DCT. Optimum performance can be achieved with different types of mixer-ICs. For the first mixer a compromise is required between noise and cross modulation. The second mixer, however, can be optimized with respect to cross modulation alone because its noise properties have only a small influence on system noise figure. Better cross modulation can be realized with larger gate width FETs and/or feedback.

Optimization of the DCT has not been done yet. The following properties of an optimized DCT compared to conventional tuners are expected:

	DCT (broad band)	conventional tuner (selective)
noise figure	6-8 dB	6-8 dB
undesired signal level for 1% cross modulation:		
in-channel:	20 mV	10 mV
off-channel:	80 mV	>100 mV

The in-channel level of the DCT is expected to be twice that of the conventional tuner. On the other hand because of the missing preselection the DCT cannot meet the off-channel level of the conventional tuner.

CONCLUSION

A GaAs double conversion TV tuner has been presented for the first time. The performance of the realized broadband single-range TV tuner meets the state-of-the-art of conventional tuners with respect to gain (20 dB), noise figure (7.5 dB) and in-channel cross-modulation (6 mV). The off-channel cross-modulation (40 mV) can be improved.

From principle the GaAs double conversion tuner has advantages compared to the conventional tuner:

- Problems with image frequency rejection are avoided because of high IF.
- Manual adjustment of tuning circuits for good frequency tracking is not required since there is only one tuning circuit compared to 12 in conventional tuners.
- Integration is possible to a large extent. Thus the number of components is reduced to about 40 compared to about 160 in conventional tuners.
- Automatic fabrication is possible.

On one hand these results form a good basis for a cost-effective approach in tuner fabrication. On the other hand a low-cost solution for the LO and its frequency control is still mandatory.

This work has shown that new solutions for existing systems can be found by using GaAs-ICs.

ACKNOWLEDGEMENT

We wish to thank Dr. G. Wolfram for development of the dielectric filter, J. Fenk for numerous suggestions and discussions and H. Hohmann for supplying data from state-of-the-art conventional tuners.

REFERENCES

/1/ H. P. Weidlich, J. A. Archer, E. Pettenpaul, F. A. Petz, J. Huber: "A GaAs Monolithic Broadband Amplifier", IEEE ISSCC Digest of Technical Papers, pp. 182-183, Feb. 1981.

/2/ U. Ablaßmeier, W. Kellner, H. Kniepkamp: "GaAs FET Upconverter for TV-Tuner", IEEE Trans. ED-27, N° 6, pp. 1156-1159, June 1980.

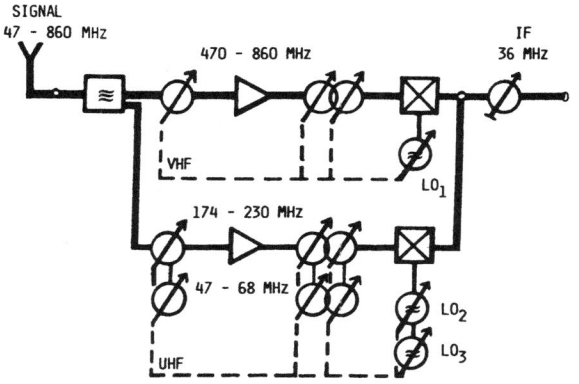
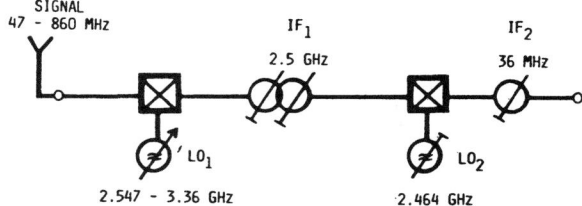

Fig. 1   Block diagram of a conventional tuner

Fig. 2   Block diagram of the GaAs double-conversion tuner (DCT)

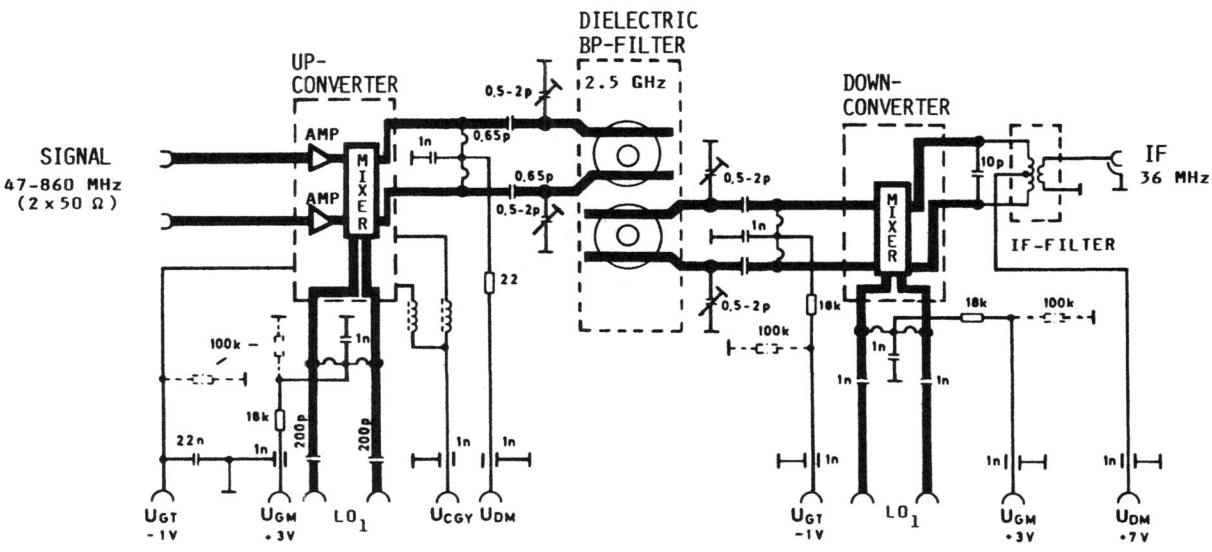

Fig. 3   Circuit diagram of the GaAs double-conversion tuner

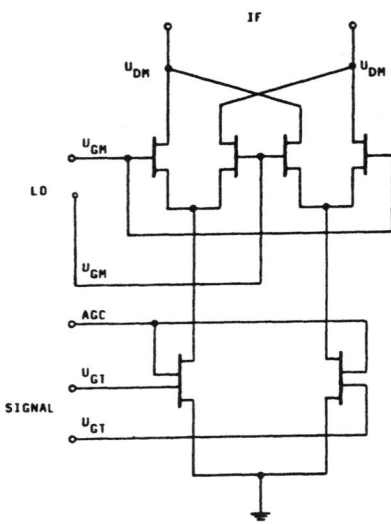

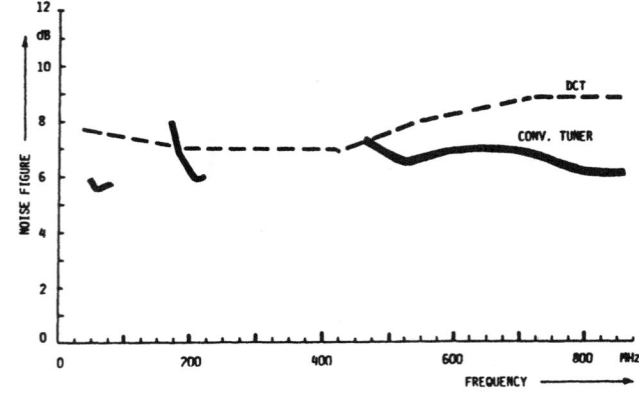

Fig. 6 Noise figure versus frequency for GaAs double-conversion tuner (DCT) and conventional tuner

Fig. 4 Double-balanced GaAs mixer IC

Fig. 5 PC board with GaAs converter stage mounted in a ceramic package

Hand-Held Portable Equipment for Cellular Mobile Telephone

By Yoshiharu TAMURA,* Tsuguo MARU,* Naoki HIRASAWA,*
Hirotoshi OKUNO,† Motoyoshi KOMODA,* Minoru HOTSUMI,*
Hirofumi MATSUMOTO* and Katsuji KIMURA‡

ABSTRACT A hand-held portable radio telephone has been developed for US and UK cellular systems. Using newly developed LSIs and other technologies, the HHP has a compact design and long operation time with a NiCd battery pack. The software implemented in the HHP is user-friendly and flexible, taking advantage of a dot matrix liquid crystal display. This paper also describes the optional equipment developed for the HHP.

KEYWORDS Cellular, Radio, Mobile, Portable, Telephone, Hand-held

1. INTRODUCTION

This paper describes the hand-held portable (HHP) radio telephone for US and UK cellular systems. Photo 1 shows the HHP, which is compact and light enough to fit in the hand. Until recently, this kind of radio telephone had the following problems to be solved before being accepted into its generation.

(1) System Adaptation Problem

Most cellular systems were originally designed to run with car mounted radios that have maximum output RF power of 3 W or more and that are allowed to use high gain antennas mounted on the roof of vehicles. However, HHPs usually have a maximum output RF power of 1 W or less, and antennas available are less effective than those of vehicle antennas. Therefore, the cell size had to be reduced for HHPs to maintain service quality even in fringe areas.

(2) Hardware Problems

Two major hardware problems had to be solved to realize an HHP that satisfied market needs. a) The size of the equipment had to be reduced by half or more. b) Standby current drain had to be greatly reduced to assure long operation time with a battery capacity of 1 Ah or less.

The next section generally describes how the cellular systems dealt with the problem and how we overcame the hardware problems. Then, the details of the approach taken to reduce the size and current drain are described. Additionally, the software implemented in the HHP, and hardware expansion with a variety of options are mentioned.

2. HHP GENERATION OF CELLULAR SYSTEMS

Today, different kinds of cellular systems are under operation in many countries as shown in Table I. The number of subscribers continue to increase, and the market growth has recently been accelerated by the introduction of transportable radio telephones. Now, cellular systems are entering the HHP generation. System adaptation to accommodate HHPs has been made and additional base stations are being installed to reduce the cell size.

The US cellular system that made the first experimental trial of HHP in Baltimore and Washington area in 1980, has already introduced HHP service. The

Photo 1 The HHP.

*Mobile Communications Division
†NEC Mobile Radio Service, Ltd.
‡System LSI Development Division

Reprinted with permission from *NEC Research and Development*, Y. Tamura, T. Maru, N. Hirosawa, H. Okuno, M. Komoda, M. Hotsumi, H. Matsumoto, and K. Kimura, "Hand-Held Portable Equipment for Cellular Mobile Telephone," no. 87, pp. 34-43, October 1987. © NEC Creative, Ltd.

UK system has been taking HHP service into consideration since the initial system design stage.

In Japan, the NTT system made technical and commercial tests for HHP in the past few years, and started HHP service in 1987. NMT 900, a cellular system running in northern Europe, also has substantial capability of HHP service.

Along with the system adaptation, new technology has enabled NEC to realize a compact HHP with reasonable operation time and with reasonable cost. The history of the cellular radio equipment is summarized in Fig. 1. Keys to the development of the HHP were,

1. Concentration of baseband circuitry and control circuitry to custom LSIs.

Table I Cellular systems.

	US cellular	TACS	NMT-900	NTT
Frequency Band Mobile → Base Base → Mobile	824 ~ 849 MHz 869 ~ 894 MHz	890 ~ 905 MHz 935 ~ 950 MHz	890 ~ 905 MHz 935 ~ 950 MHz	915 ~ 940 MHz 860 ~ 885 MHz
Number of channels	866	1000	1999	1000
Channel separation	30 kHz interleave	25 kHz interleave	12.5 kHz interleave	25 kHz
Transmitted RF power (mobile)	Class 1: 4.0 W Class 2: 1.6 W Class 3: 0.6 W	Class 1: 10.0 W Class 2: 4.0 W Class 3: 1.6 W Class 4: 0.6 W	Mobile: 6 W Portable: 1 W	Mobile: 5 W Portable: 1 W
Maximum frequency deviation	±12 kHz PM (voice) ±8 kHz FM (data)	±9.6 kHz PM (voice) ±6.4 kHz FM (data)	±4.7 kHz PM (voice) ±3.5 kHz PM (data)	±5 kHz PM (voice) ±5 kHz FM (data)
Data transmission	10 kbit/sec Manchester code	8 kbit/sec Manchester code	1200 bit/sec NRZ code	300 bit/sec Manchester code
Error correction	Majority voting, BCH	Majority voting, BCH	Hagelbarger code	Gold code, BCH
Number of subscribers as of May 1987 (approx.)	760,000	160,000	350,000 (including NMT450)	100,000

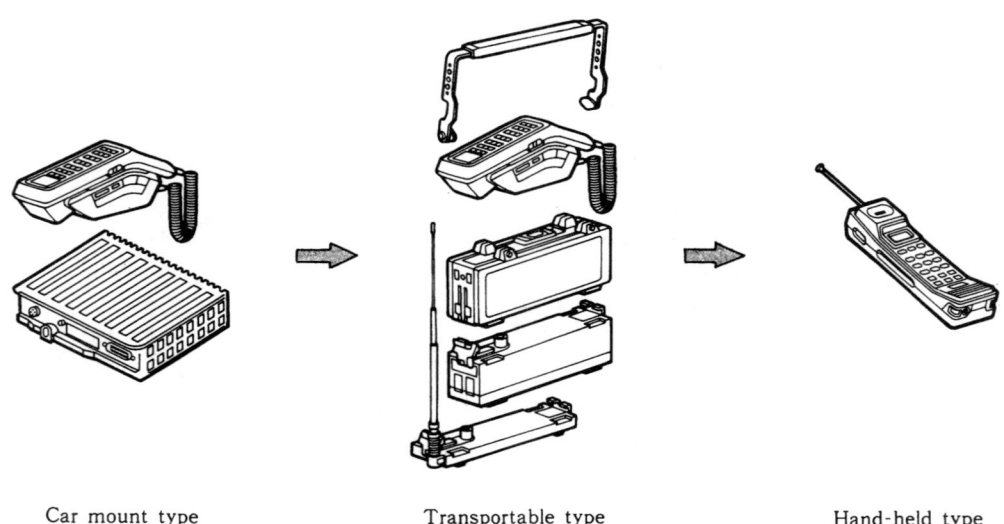

Car mount type   Transportable type   Hand-held type

Fig. 1 History of cellular radio equipment.

2. Minituarization of electronic components by microelectronics technology.
3. Complete use of surface mount technology.
4. Development of low power consumption devices and circuits.

In the following sections, details of these breakthroughs are described for each unit composing the HHP.

3. HARDWARE CONFIGURATION

Figure 2 shows the block diagram of the HHP. The HHP comprises five units as below.

(1) Baseband Unit:
The unit filters and controls all baseband signals such as voice, data, tone, and SAT (Supervisory Audio Tone).

(2) Radio Unit:
The unit includes a receiver, a transmitter, a synthesizer, and an antenna duplexer.

(3) Antenna Switch Unit:
The unit switches the antenna system and the external RF terminal.

(4) Logic Control Unit:
The unit controls all the other units to function for the cellular signalling specifications and user interface. The unit also includes keypads and an LCD (Liquid Crystal Display).

(5) Power Supply Unit:
The unit supplies all the other units with filtered DC power.

The functional partition into the five unit configuration was decided for easy maintenance and quick modification for the other cellular systems.

4. BASEBAND UNIT

The unit is assembled compactly ($25 \times 50 \times 10$: mm) with all the parts surface-mounted. As shown in Fig. 3, most of the functions of the AF unit are concentrated in the baseband LSI. The LSI, which was specially designed for the HHP, includes the following circuits.

1. Transmit audio bandpass filter.
2. Receive audio bandpass filter.
3. Pre-emphasis and de-emphasis circuit.
4. SAT detection and generation circuit.
5. Transmit audio limiter and splatter filter.
6. Low pass filter for wideband data.
7. Mixing circuit for voice, tone, data, and SAT with path switches.
8. Electronic volume control for an ear piece and a loudspeaker.
9. Electronic volume control for deviation control.

These circuits are constructed by using SCF (Switched Capacitor Filter) technique. The LSI is packed in a 44 pin flat package. Current drain of the LSI is 11 mA normally, but can be reduced to 1.8 mA when the HHP is in standby mode.

A flat-packed syllabic compander was also newly developed for 5 V operation. The comparison in Fig. 4 reveals how the baseband circuitry of a transportable radio was concentrated in a single chip.

The unit further includes a microphone pre-amplifier, an ear piece amplifier, a loudspeaker amplifier, and a voice detection circuit for discontinuous transmission (DTX). In DTX conversation mode, the HHP cuts off or reduces the transmitted power unless it detects voice to be sent. Using this technique, average power consumption is reduced.

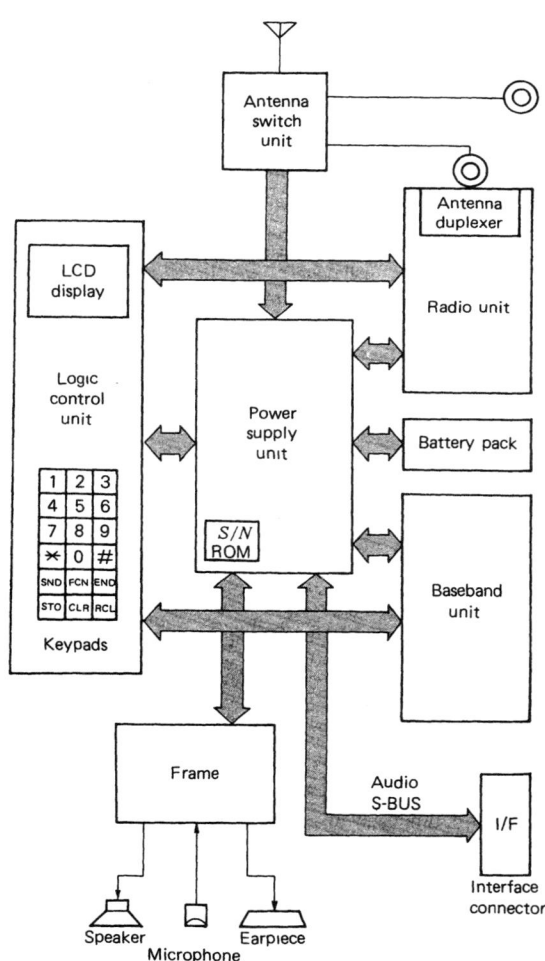

Fig. 2 Block diagram of the HHP.

Fig. 3 Block diagram of the baseband unit.

Transportable radio (Components replaced by the baseband LSI)		HHP
Resistors	121	The baseband LSI
Ceramic capacitors	24	
Tantalum capacitors	40	
Transistors, diodes	7	
ICs	18	
Hybrid ICs	4	
Others	3	

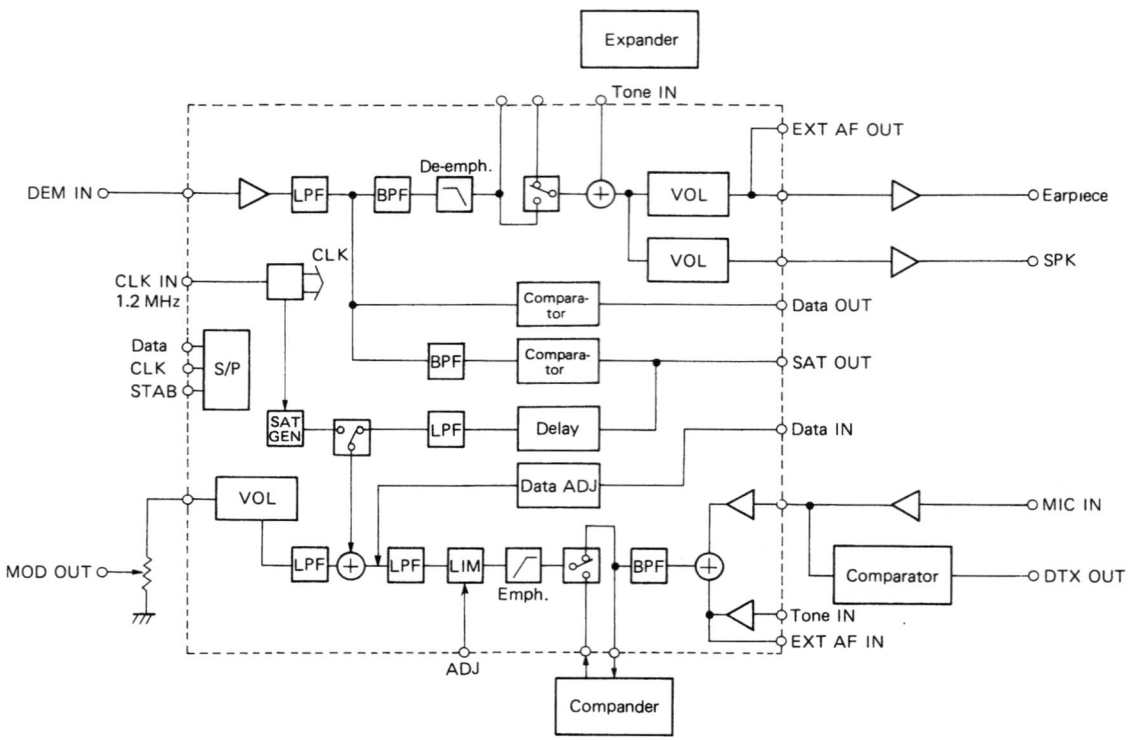

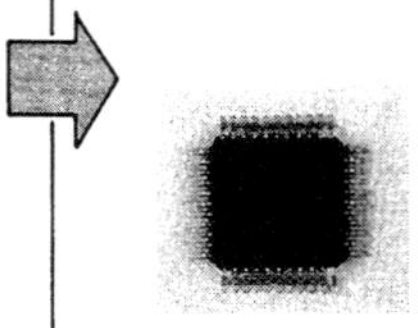

Fig. 4 Comparison of the baseband circuit.

5. RADIO UNIT

Although the unit is assembled in a single PWB (98 × 52 × 14: mm), it is composed of four sections; a transmitter, a receiver, a frequency synthesizer, and a duplexer. The overall block diagram is shown in Fig. 5.

(1) Transmitter Section

The section includes a power amplifier (PA), a power control circuit thereof, a driver amplifier, a filter, and an isolator. The PA is a hybrid module using bare transistor chips mounted on a ceramic substrate. Three-stage amplifier therein has 33 dB power gain so that only one additional stage of driver amplifier is needed between the synthesizer and the PA.

(2) Receiver Section

The receiver employs double conversion with the first IF frequency of either 88.75 MHz or 94.35 MHz for UK and US, and with the second IF frequency of 455 kHz for both. These frequencies are carefully chosen so as not to yield channel blocking or spurious response. Generally, use of hybrid modules is not cost effective, so the design concept of the receiver was to assemble it on both sides of the PWB, without using any hybrid modules.

For impedance matching, RF front end uses distributed transmission line when requested inductance is small enough, and chip inductors when relatively large inductance is needed. Thus, both the number of components and the space occupied by the components are minimized. Received RF signal amplified by the RF AMP is fed into the first mixer together with the local signal through the RF helical filter.

The RF signal is converted to the first IF frequency, and the signal, after passing through the crystal filter, is down-converted to 455 kHz by the second mixer. Then a monolithic IF amplifier with 120 dB gain is followed by the discriminator to finally demodulate the IF signal to produce the baseband signal. The IF amplifier is an 8 pin flat-packed IC that includes an amplifier with a total gain of 120 dB and a rectifier for RSSI (Received Signal Strength Indicator) output. The rectifier assures more than 100 dB dynamic range with temperature deviation of ±2.5 dB.

(3) Synthesizer Section

The receiver local frequency and transmitting frequency are generated by individual 400 MHz PLL circuits. Then both signals are multiplied by double multipliers to have 800 ~ 900 MHz band frequencies. Frequency generation by 400 MHz band PLL is employed mainly for its low power consumption. The receiver local PLL drains only 17.5 mA with carrier to noise ratio of 90 dB ($B_w = 1$ kHz, 25 kHz off the carrier). However, a 400 MHz band VCO using stripline resonator may become larger than an 800 MHz band VCO. This problem was solved by using a multilayer ceramic substrate for the 400 MHz VCO. A compact hybrid VCO was developed in a $20 \times 15 \times 10$ mm housing including a buffer amplifier and a low pass filter. Moreover, the VCO has a capability of direct modulation.

The ECL pre-scaler and the CMOS programmable divider were also specially developed in an 8 pin and a 16 pin flat pack respectively. Transmit PLL is switched off when the HHP is in standby mode in order to save power consumption. The synthesizer has 25 MHz bandwidth that covers 1000 channels for the UK system and 866 channels for the US system. Fast switching time between the lowest and the highest channels is kept within 40 msec because of the high sensitivity of the 400 MHz VCO, which is 5 MHz/volt approximately.

(4) Antenna Duplexer Section

Comparison of the duplexer used in the HHP with that used in a transportable radio is summarized in Fig. 6. Although the specifications are different, reduction of volume and weight is significant.

6. LOGIC CONTROL UNIT

As shown in Fig. 7, the unit is composed of two CPUs, memories, I/Os, and custom made peripheral LSIs. The main CPU is an 8-bit CMOS, while the slave CPU that controls the keypads and the LCD is a 4-bit CMOS microprocessor. The memories are 64K byte EPROM and 8K byte RAM. These LSIs are mounted

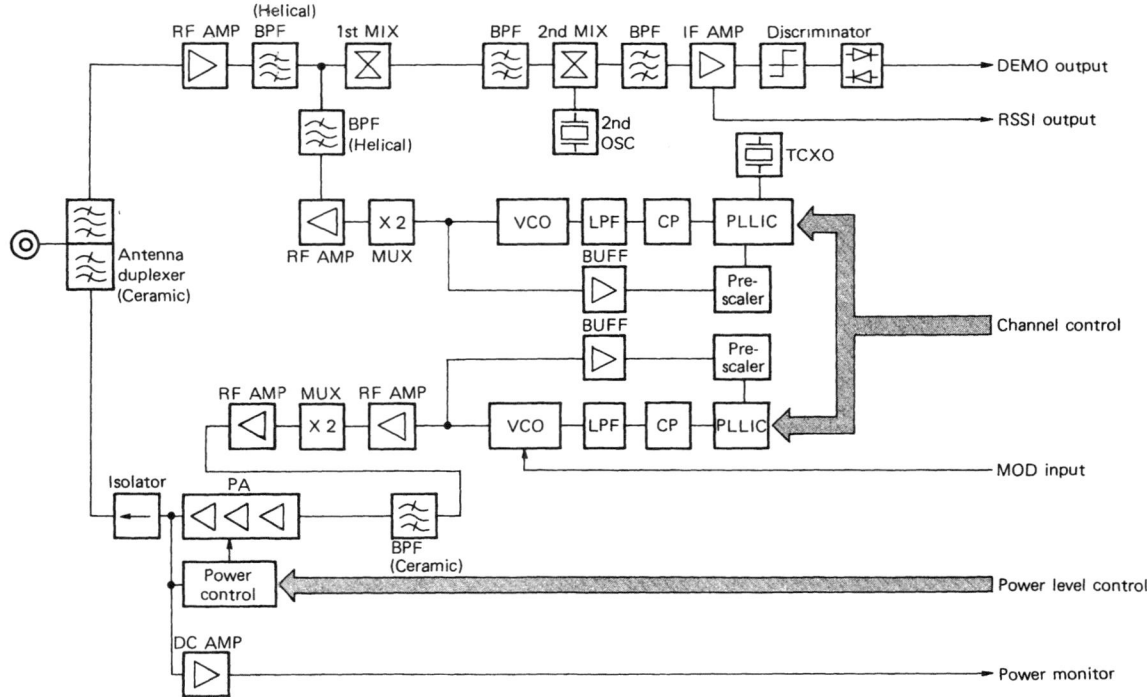

Fig. 5 Block diagram of the radio unit.

mainly on one side of the PWB, and the other side is mainly occupied by the keypad contacts and the liquid crystal display.

Transportable radios had separate PWBs for logic section and keypad/display section. The HHP as stated above, unified the two PWBs to minimize space, interconnection, and cost. The unit has a total of 160 parts, of which 93% are automatically surface-mounted which results in good productivity. The unit includes three gate arrays and one custom made LSI to manage cellular signalling protocols and user interface via the keypads and the LCD.

"STANDBY CONT" is a 4K gate array packed in a 64 pin flat pack. It controls data transmission/reception, awakes the CPU during the standby state, and controls the "S-BUS" that is a special bidirectional serial interface. The S-BUS will be described briefly later. On reception of data from a base station, the LSI establishes word synchronization, then interrupts the CPU on every 8 bit data, while otherwise the CPU stays in standby mode to save power consumption. "DIV" is a clock divider gate array (400 gate type) packed in a 44 pin flat pack. It generates clocks of various phase from two crystals of 4 MHz and 4.8 MHz. The 4 MHz clock drives the two CPUs, while the 4.8 MHz is the source of the other clocks such as 1.2 MHz for the baseband LSI. "SAT CONT" is a 2K gate array that has a SAT counter, an autonomous counter, timers, and a set of parallel I/Os. "TONE LOGIC" is a custom CMOS LSI in a 44 pin flat pack. The LSI mainly generates DTMF (Dual Tone Multi-Frequency) by synthesis of quasi sine wave. It also includes an A/D converter for RSSI signal detected in the radio unit. The keypads includes six functional keys as well as 12 numerical keys. Instead of LED displays, the HHP employs a dot matrix LCD that features 30 characters (10 digits × 3 rows). Each character is expressed by a 5 × 7 dot matrix, which gives the software good flexibility and a user-friendly interface. How the software takes advantage of the LCD will be mentioned later in the next section. In addition, the keypads and the LCD are fully backlit by low power LEDs and an EL device respectively.

The S-BUS, as shown in Fig. 8 is a bidirectional serial interface that comprises three lines; SDATA, BUSY, SCK. The S-BUS allows the master CPU to communicate with the other CPUs such as the slave CPU and those CPUs that may be connected with the

	Duplexer for transportable radio	Duplexer for the HHP
Volume	52 cc	8.6 cc
Weight	150 g	40 g
Filter configuration	TX. 5 BPF + 1 BEF RX: 6 BPF	TX. 4 BPF RX. 5 BPF + 1 BEF
Insertion loss	TX. 2 dB RX. 2.7 dB	TX. 2 dB RX. 3.5 dB
Attenuation	TX. 50 dB RX: 60 dB	TX. 42 dB RX. 58 dB
LPF for 3rd harmonic	5 LPF	5 LPF

Fig. 6 Comparison of the duplexer.

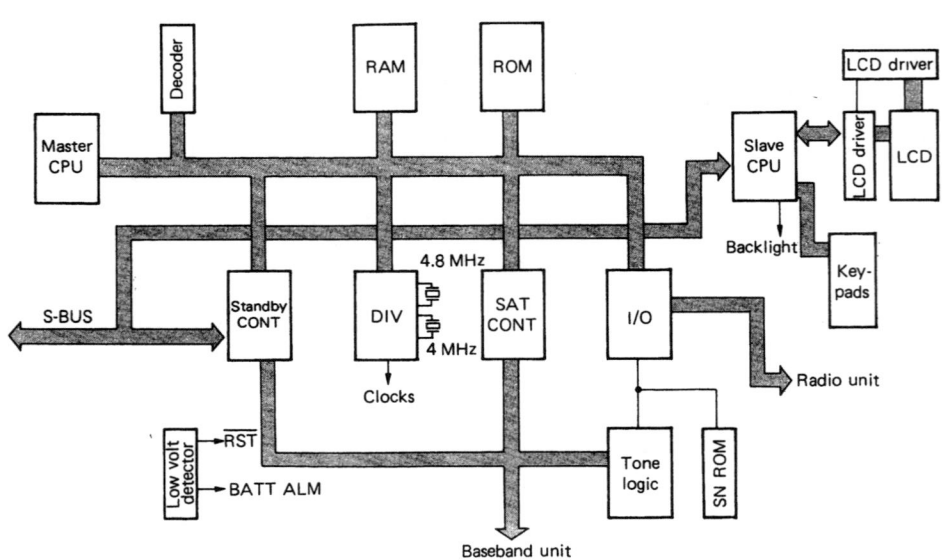

Fig. 7 Block diagram of the logic control unit.

HHP via the external interface connector.

Additionally, through the S-BUS, those CPUs outside the HHP can make communication with each other. Communications on the S-BUS are done under supervision of the master CPU, for only the master CPU can place SCK on demand of the S-BUS by pulling down the BUSY line. Collision detection and priority bus aquisition are also considered by assigning an 8-bit identification (ID) code to each CPU. The ID code should be sent and examined prior to data transmission. S-BUS protocol is well defined in physical layer and data link layer so that the bus is utilized for automatic testing and maintenance of the HHP.

7. SOFTWARE

As for software implemented mainly in the EPROM, user interface is focused here. As stated above, the HHP uses a dot matrix LCD. The 5 × 7 dot matrix can express 165 different characters including alphabet, numbers and even arbitrary characters.

Taking advantage of this, the user interface software has good flexibility. The messages that appear on the LCD during operation can be easily modified according to customer needs or system specifications. Both flexibility and user-friendliness are achieved by the dot matrix LCD. Instructions and state information can be displayed in natural sentences like "Recharge battery" or "Adjust volume of keys," etc. Another good example is the built-in instruction manual. Complicated operation to use various functions sometimes discourages the user.

Usually a written instruction manual comes with radio telephones, but still, it is inconvenient for the user to carry it around. So, the HHP implemented simplified instructions on how to use functions, which

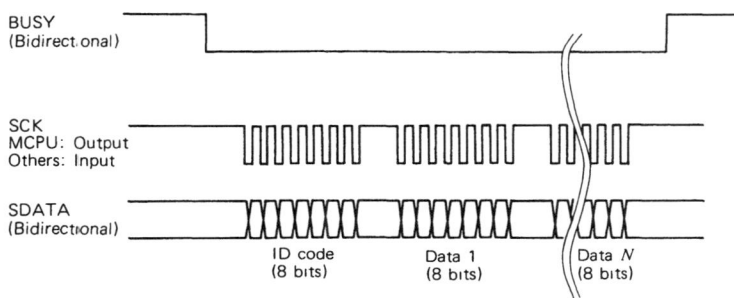

Fig. 8 S-BUS.

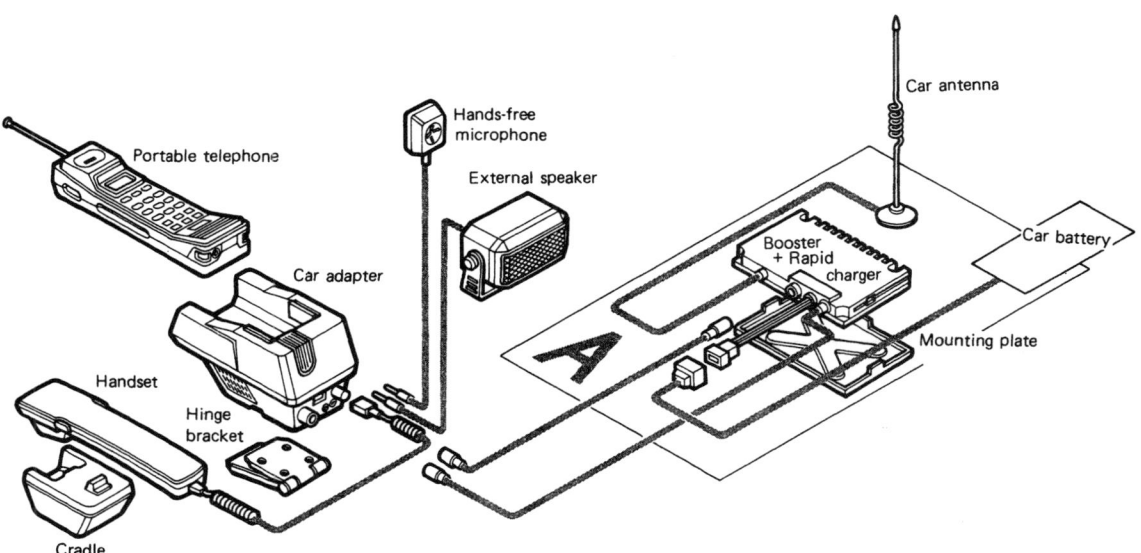

Fig. 9 Example of in-car installation.

appear on the display by pushing the FCN button for more than 0.5 sec. All the instructions can be reviewed by simply pushing the up/down button in the same manner as the scrolling of the dial memories. The up/down button also works as volume control of voice, keypad tone, and ringer tone.

In recent years, software quality has become one of the most important factors in the total quality of a product. NEC has established SWQC (Software Quality Control) approach for software production. Total software of the HHP consists of several functional blocks, each of which should meet individual specification. Each block had to pass reviews and inspections after every design phase. Even after the implementation of the software, abnormal operation tests were executed by many people who know nothing about the

Table II Specifications.*

General	EIA version	TACS version
Transmitting frequency range	825 through 845 MHz	890 through 915 MHz
Receiving frequency range	870 through 890 MHz	935 through 960 MHz
Transmitter-receiver duplex spacing	45 MHz	
Channel spacing	30 kHz	25 kHz
Number of channels	666	1000
Type of Modulation Data	Direct FSK (±8 kHz)	Direct FSK (±6.4 kHz)
Audio	Equivalent PM (with compander)	
Data speed	10 kbps Manchester code	8 kbps Manchester code
Nominal antenna impedance	50 ohms	
Power supply voltage	DC 7.2 V (nominal)	
Current drain standby	Approx. 0.05 A (7.2 V)	
Transmitting 0.6 W ERP	Approx. 0.75 A (7.2 V)	
Weight	Approx. 0.65 kg	
Dimensions	Approx. 62 × 39 × 194 mm, 470 cc	

Receiver	EIA version	TACS version
Sensitivity	−116 dBm (12 dB SINAD)	−113 dBm (20 dB SINAD)
Adjacent channel selectivity	60 dB or more (60 kHz separation)	55 dB or more
Spurious rejection	60 dB or more	
Intermodulation rejection	65 dB or more	
Hum and noise	32 dB or more	
Distorsion	26 dB or more	24 dB or more
Frequency stability	Better than ±2.5 ppm	

Transmitter	EIA version	TACS version
RF output	0.6 W ERP (controlled in six 4 dB steps)	
Spurious emission	43 dB or more	56 dB or more
Hum and noise	32 dB or more	
Distortion	26 dB or more	
Frequency stability	Better than ±2.5 ppm	

*Specifications are subject to change without notice.

HHP. Finally the software was tested in the real cellular system of UK and US individually.

8. OPTIONAL PRODUCTS

In parallel with the development of the HHP, various optional products have been developed.

Two types of AC battery chargers have been developed as below.

(1) AC Slow Charger:
   The charger recharges an HHP and a battery pack simultaneously in 8 to 10 hours.
(2) AC Rapid Charger:
   The charger recharges an HHP in 1.5 hours while a battery pack is recharged in 8 to 10 hours in parallel.

Other options are for use of the HHP in automobiles. Using the following options or combination of them, the user can convert his HHP into a car telephone.

(3) Simple Car Kit:
   Instead of a battery pack, a DC/DC converter can be attached to the HHP to take power source from a cigarette lighter outlet in a car. In this application, the HHP uses its own antenna.
(4) Basic Car Kit Using the HHP as a Handset:
   This car kit includes an in-car adapter with an RF curly cord so that a car-mount antenna can be used. The HHP is used just like a handset.
(5) Handsfree Car Kit Using the HHP as a Handset:
   This car kit is the same as (4), except that the in-car adapter has a built-in handsfree circuit and a loudspeaker.
(6) Handsfree Car Kit with an Additional Handset:
   Instead of the RF curly cord, this kit includes a slim and light handset. The kit also features a built-in handsfree.
(7) DC Rapid Charger:
   Car kits (4)-(6) can recharge the HHP in 8 to 10 hours without an additional charger. However, by using the DC rapid charger, 1.5 hour recharging is possible.
(8) RF Power Booster:
   The HHP has RF output of 0.6 W ERP. This booster can amplify the output up to 3 W. The booster includes a DC rapid charger unit identical to (7).

Figure 9 shows one example of in-car installation using option (6) with (8).

9. SPECIFICATIONS

Listed in Table II, are specifications of the HHP for UK and US cellular system.

10. CONCLUSION

A hand-held portable radio telephone has been developed. The HHPs are designed to work in UK TACS type or US AMPS type cellular systems. The compact, lightweight HHP with long operation time is realized by use of specially designed LSIs, together with low-power dissipation circuit techniques and microelectronics. The software implemented in the HHP is flexible and user-friendly, taking advantage of a dot matrix LCD. Additionally, a variety of optional equipment such as AC chargers and car kits have been developed for system expansion.

Future efforts to realize more compact HHP with longer operation time will be focussed on further introduction of LSIs and development of a high-efficiency power amplifier.

ACKNOWLEDGMENT

The authors would like to express their thanks to people in NEC Corporation, NEC America, Inc. and NEC Business Systems (Europe) Ltd. for their helpful discussions and cooperation during the equipment and LSIs development period.

REFERENCES

[1] T. Kai, et al., "Cellular Mobile Radio Equipment," *NEC Res. & Develop.*, 84, pp. 85-93, 1987.
[2] T. Kikuchi, et al., "High-Capacity Automobile Telephone System — Part 3: Equipment Outline —," *Japan Telecommun. Rev.*, 22, 3, pp. 242-253, July 1979.
[3] T. Miyamoto, et al., "Mobile Subscriber Station (MSS)," *NEC Res. & Develop.*, 57, pp. 109-117, April 1980.
[4] T. Kontani, et al., "NEC's Medium-Sized Mobile Telephone System," *NEC Res. & Develop.*, 62, pp. 1-17, July 1981.

Received June 19, 1987

* * * * * * * * * * * * * * *

Compact Size Numeric Display Pager with New Receiving System

By Koji YAMASAKI,* Shigeo YOSHIZAWA,* Yoichiro MINAMI,* Takayuki ASAI,*
Yasushi NAKANO† and Mitsuru KURODA*

ABSTRACT The authors have developed a compact size numeric display pager: its size is the smallest in the world for a top-end display pager. This pager uses many of NEC's advanced technologies, such as new receiving system called direct conversion receiving method, bipolar LSI for new receiving system, and low voltage operable CPU and LCD. This paper describes the features of this pager and details of advanced technologies such as new receiving system.

KEYWORDS Radio pager, Digital paging system, Display pager, Direct conversion receiving method, Low voltage technology

1. INTRODUCTION

The pager market is growing in all countries of the world each year. However, the decline in price of the pagers through competition, providing it with high functions and also the demand for miniaturization to improve portability are becoming increasingly stronger.

Especially, such demands are strong in the case of the numeric display pager. To deal with these exchange demands, the authors developed the smallest numeric display pager, type R3N4-12A, in the world as the top display type with a new receiving system, the direct conversion receiving method.

In the direct conversion receiving method, only one local oscillation circuit is required, and an image rejecting filter becomes unnecessary because an image frequency does not exist, compared with the double superheterodyne receiving method which NEC generally adopted in the past for overseas. So it realizes the miniaturization of equipment to make the cost lower by reduction of the number of components.

This paper first gives an outline of the R3N4-12A. Next, it describes the direct conversion receiving method in detail and the prominent points of the control and display part. Lastly, it describes the contents of the mechanical part.

---

*Mobile Media Terminals Division
†NEC Shizuoka, Ltd.

2. OUTLINE OF NUMERIC DISPLAY PAGER

2.1 Basic Description of Each Section

R3N4-12A is a small numeric pager using the direct conversion receiving method based on the Post Office Code Standardization Advisory Group (POCSAG) code on 150 MHz VHF band, and it is roughly divided into the radio section and the control and display section.

In the radio section, the FSK radio signals based on the POCSAG rule which are sent from the base station are received and demodulated, and the digital data are generated.

In the control and display section, the digital data which were generated in the radio section are decoded. And the ID number which is contained in the data is compared with its own ID number. If they are identical, an alert tone is generated.

Moreover, if the numeric message is contained in the said data, the message is displayed on the LCD.

The following is a detailed explanation of the function of each section, referring to the block diagram as shown in Fig. 1.

(1) Radio Section

This section consists of the Antenna (ANT), RF Amplifier (RF AMP), Local Oscillator (Local OSC) and Direct Conversion IC (DC IC).

ANT is a loop-type antenna. It receives radio signals on 150 MHz band sent from the base station and it sends them to RF AMP. RF AMP amplifies the radio signals and sends the signal to the Mixer circuit (MIX) in DC IC.

In Local OSC, the frequency which is the same

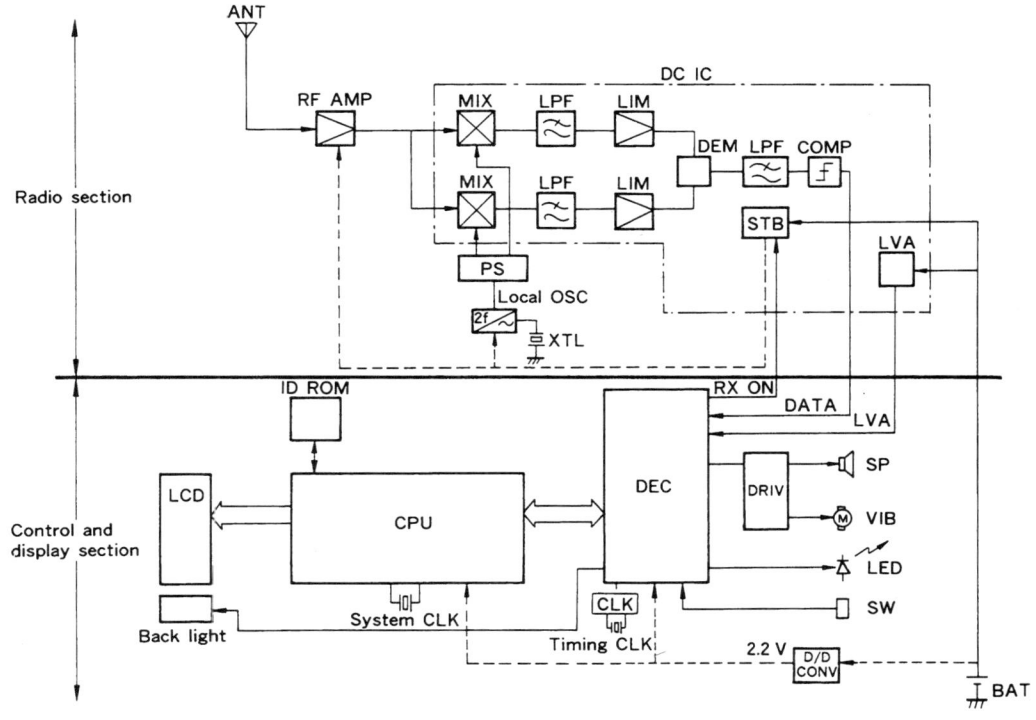

Fig. 1 Block diagram.

ANT:	Antenna	PS:	Phase shifter
RF AMP:	Radio frequency amplifier	XTL:	Crystal resonator
Local OSC:	Local oscillator	DEC:	Decoder
DC IC:	Direct conversion IC	RX ON:	Wake up signal to radio section
MIX:	Mixer circuit	DATA:	Received data
LPF:	Low pass filter	DRIV:	Driver IC
LIM:	Limiter	VIB:	Vibration
DEM:	Demodulator	D/D CONV:	DC/DC converter
STB:	Stabilization power supply circuit	SP:	Loudspeaker
COMP:	Comparator	BAT:	Battery
LVA:	Low voltage detection circuit		

as the carrier frequency is generated using the doubler to convert the frequency of the crystal oscillator. The crystal is the over tone type. The signal is input to the mixer circuit in DC IC.

DC IC contains a couple of the MIX, Low Pass Filter (LPF) and Limiter (LIM), the Demodulator (DEM), another LPF, Comparator (COMP) to produce the binary digital signal, Stabilization power supply circuit (STB) and Low Voltage detection circuit (LVA). In DC IC, first, the signal which was input from RF AMP and the signal which was input from Local OSC are mixed by MIX to get the base band signal as the difference between two signals. Then the base band signal passes LPF, LIM and DEM to be detected, and it passes LPF, COMP to be changed to a binary digital signal.

Meanwhile, the battery voltage is always monitored by LVA in DC IC. When this voltage falls below the preset voltage, it reports the low voltage condition to the control and display section. As for the details of the function of DC IC, please refer to Section 3.

(2) Control and Display Section

The control and display section contains the Decoder (DEC), CPU, ID ROM, Driver IC (DRIV) for alert, DC/DC converter (D/D CONV), LCD, back light, Loudspeaker (SP) and LED.

DEC and CPU operate on the voltage of 2.2 V which is converted by D/D CONV from the battery voltage. DEC processes the signal of binary digital data which was generated with the radio section and compares the identification number contained in the receiving signal and its own identification number stored in ID ROM, exchanging the data to/from CPU.

If they are identical, DEC puts out an alert with blinking LED generating alert tone or vibration. Also, DEC and CPU analyze the numerical message data received and displays it on LCD.

The ID ROM is an EEPROM (Electrically Erasable PROM) and stores data such as the identification number and various corresponding functions.

2.2 Features

Newly developed compact size numeric display pager R3N4-12A has the following features.

(1) Receiving Address

This pager has three addresses. ID_1, ID_2, and ID_3. ID_1 and ID_2 have four sub-addresses, respectively, and ID_3 has one sub-address. In total, nine kinds of addresses can be maintained.

Also, for alert patterns in calls, the address for each of them will be different. Therefore, through different patterns it can be recognized which address has been called.

(2) Alert Tone

The standard frequency is the single tone of 2.7 kHz but it is possible to choose from eight different kinds through ROM programming. These eight kinds are: the single tones of 1.5 kHz / 2 kHz / 2.7 kHz / 3.2 kHz, and 1.5 + 0.7 kHz / 2 + 1 kHz / 2.7 + 1.4 kHz / 3.2 + 1.6 kHz.

(3) Message Display

It is possible to display a message at once on LCD of up to 12 figures by switch operation. During message display, the time when the message was received and the message number can both be displayed at the same time.

1. Character Set

   It is possible to make a choice from two sets in which character strings are to be used in the display. This selection is completed by ROM programming.

   0, 1, ⋯ , 9, A, U, space-] [

   0, 1, ⋯ , 9, A, b, C, d, E, space

2. Time Stamp

   The pager memorizes the time when it received each message, and when the message is read it will display that reception time.

3. Display Illumination

   During a message display, it is also possible to read the message even in dark locations. This is due to the display screen being brightly lit by a LED.

(4) Message Memory

It is possible to memorize a maximum of 16 messages (total of 200 figures). Within one message, it can hold a maximum of 20 figures.

2.3 Functions

Next, the main functions of R3N4-12A are introduced.

R3N4-12A has two switches. They are the slide switch and the push switch. The slide switch has three positions in all. They are the power supply ON/OFF and an alert mode selection.

The alert mode is equipped to report calls at the times of call. The levels of sound for this alert mode are several. They are the general speaker tone, the soft tone for a speaker, and graded escalating alert tones.

Also, for giving an alert other than sound, there are many types of methods including the vibrator, the LED blinker, the message memory maker without any alert reporting and others. If two means of reporting are registered in the ID ROM beforehand, by making changes with the slide switch, it is possible to choose the alert mode.

The following is on the push switch. It is used to reset the alert and to call a message that has already been memorized.

A message can be composed of a maximum of 20 figures per message. The memory can store 16 messages in all, and a maximum of 200 figures. It is possible to display 12 figures at once with LCD. Also, this pager is equipped with a 24 hour watch. Usually, the LCD displays the present time.

However, when reading a memorized message, the so-called timestamp function can display the time at which it received the message.

2.4 Specifications

The specifications of R3N4-12A are as follows.

- Radio frequency range:
  One specified frequency within the range of 138 MHz to 174 MHz
- Paging sensitivity: 14 dBuV/m or better (tone only, front on body)

- Spurious response: 60 dB or more
- Channel spacing: 25 kHz
- Adjacent channel rejection: 60 dB or more
- Frequency stability: 0.001% (−10 to +50°C)
- Digital modulation: Frequency Shift Keying (FSK)
- Deviation: Mark carrier frequency: typ. −4.5 kHz
  Space carrier frequency: typ. +4.5 kHz
- Signal format: POCSAG code, Non Return to Zero (NRZ)
- Bit rate: 512 bps
- Alert tone frequency: 2.7 kHz (standard)
- Alert tone output: 75 to 90 dBspl at 30 cm (2.7 kHz alert tone and single alert frequency mode)
- Dimensions: Approx. 50(W) × 18.6(D) × 64(H) mm
- Power supply: One AAA 1.5 V penlight battery
- Nominal battery life: Approx. 1,000 H (alkaline battery)

3. RECEIVING UNIT

3.1 New Receiving System

(1) General

Recently, miniaturization of radio receivers has been advancing by the improvement of the integrated circuit technique.

However, taking the radio part as an example, it is the present situation that integration is impossible because the basic method of the circuit is the same and reaches the limit of the miniaturization on account of the existence of the discrete device. For example, in a superheterodyne radio receiver, it needs a large area because the high frequency, intermediate frequency filter and so on are big.

On the other hand, a direct conversion receiving method is thought of as small and lightweight. A direct conversion receiving method is that making equivalent in a carrier frequency and a local oscillator frequency. The beat of the carrier frequency and the local oscillator frequency is taking up in the mixer only a base band signal which is transmitted to a low-pass filter. This beat is the method to perform the demodulation processing after limiting the amplitude of the signal with a limiter circuit, and to get a demodulated signal.

The circuit block diagram of the new receiving system is shown in Fig. 2, and the circuit motion is explained as follows.

The reception wave frequency-modulated by a binary digital signal is amplified with a Radio Frequency Amplifier (RF AMP). Then it is divided in two and each signal is input to its Mixer circuit (MIX).

Also, the signal with local oscillator frequency is input from a local oscillator circuit to a 90-degree phase shifter. Then, it is input to each of two mixer circuits where phases are shifted to +45 degrees and −45 degrees, respectively. Taking up such circuit composition, signal with a 90-degree lagged phase is changed to the base band signal in the frequency from a mixer circuit and is output.

In the above-mentioned, the base band signal becomes a beat frequency as the carrier frequency and the local oscillator frequency agree. Next the base band signal is input to a low-pass filter where noise outside the base band is removed.

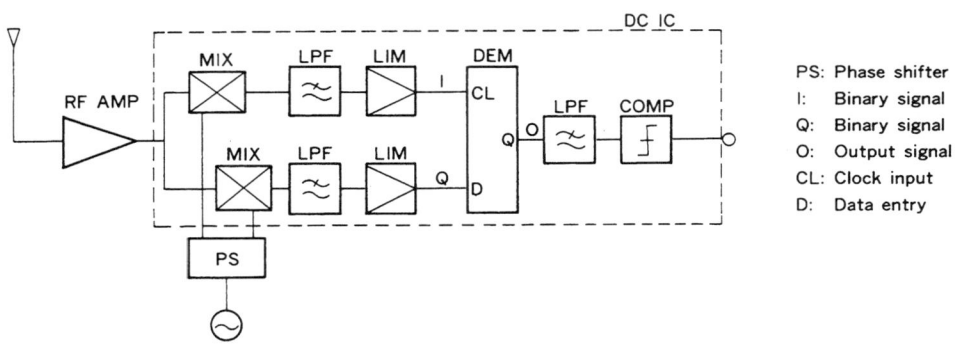

Fig. 2 Circuit block diagram.

A base band signal is input to a Limiter circuit (LIM) in each and becomes signal I, Q of the binary.

Here, simply, the Demodulation circuit (DEM) is composed of a D type flip-flop. Clock input CL of D type flip-flop and data entry D are input by signal I and Q, respectively. Figure 3 shows the input signal I, Q and output signal O.

In this way, the demodulation signal is output as a binary digital signal through the low-pass filter to remove a noise.

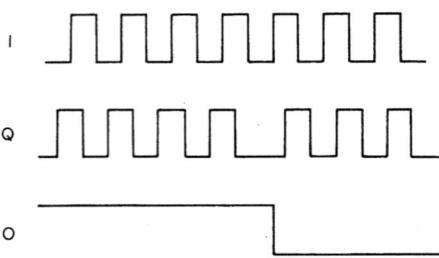

Fig. 3 Signal patterns of discriminator.

(3) Miniaturization

The receiving system embedded with mechanical filter is large in size because it is composed of many thick parts such as Crystal (XTL) filter for 2nd image frequency, ceramic filter of 455 kHz, etc. So a Printed Wiring Board (PWB), mounted mechanical filter, is unsuitable to make it into a thin type.

However, the new receiving system embedded with DC IC can be miniaturized because DC IC parts are sufficiently thin to use surface mounting. There is the advantage that a layout degree of freedom is high in a composition of two PWBs.

Photo 1 shows the difference between R3N4-12A and superheterodyne receiver.

An adjustment circuit with it becomes unnecessary because there is no mechanical filter. Also, filter adjustment becomes unnecessary. The mechanical filter embedded in superheterodyne receiver requires much adjustment work because it has many additional parts.

However, the radio receiver using DC IC realizes higher productivity, that is, reduction of production manpower, because its additional parts are less than those of a superheterodyne receiver.

Superheterodyne receiver

R3N4-12A

Superheterodyne receiver

R3N4-12A

Photo 1 External view of superheterodyne receiver and R3N4-12A.

The reduction of a component area becomes possible as mentioned above, and it becomes possible to secure antenna space and improve Q factor of antenna.

3.2 LSI Technology for New Receiving System

(1) Semiconductor Process for RX LSI

The following process technique is necessary to realize DC IC on one chip. Table I shows the basic technique elements.

First of all, in order to use the gyrator filter circuit, a capacitor used in it requires high capacitance. Use of this capacitor will enable substantial reduction of the chip area.

The NPN transistor is another important factor. Through a low current of $0.1\,\mu\text{A}$, it can operate without decreasing the current amplification factor. This contributes to low current power supply circuit, despite the high frequency transistor of about 4.5 GHz for the f_T.

The current at the power supply circuit is reduced to a lower current for the entire circuit. However, due to the increase in the resistance value, the chip area increases. Adoption of an ion implantation resist or resistor with high sheet resistivity is to counteract this occurrence (i.e., reduction of chip area).

In addition, the polysilicon resistance is used for the input circuit. This is to reduce the parasitic capacitance of the mixer circuit and also to improve its high frequency characteristics.

The discriminator circuit uses IIL (Integrated Injection Logic) because of its small size and low current operation compared to TTL (Transistor Transistor Logic).

(2) Circuit Technique for Low Voltage Operation

The authors adopted the following circuit composition to reduce the current consumption of the whole circuit.

Through a voltage stabilizer, each circuit block is supplied with a stabilizing voltage of 1.05 V output as an attempt to stabilize the circuit.

The mixer circuit is using the single balanced mixer of the resistance load type which can work even at low voltage. Figure 4 shows the mixer circuit. It continues mixing operation until 300 MHz in each mixer.

The boosting charge circuit with large capacitance stabilizes circuit motion in a short time in case of the power supply in the circuit which handles low frequency.

The operational amplifier used in gyrator filter can operate on $0.5\,\mu\text{A}$.

(3) Integrated Gyrator Filter

As the channel filter's input signal is identical with the base band signal, the frequency is very low. If a ladder-type active filter is composed, and its resistance value is set to hundreds of kΩ in order to improve the time constant, its capacitance value will be between thousands of pF and tens of thousands of pF.

This makes it impossible to integrate the capacitor on a chip. Also, the many discrete devices used in it will prevent the receiver from becoming smaller and lighter.

Similarly, the passive filter which reduces the number of circuit element results in a large inductance value L due to its low frequency. These remains the same problem related to the receiver in size and weight.

In dealing with the above, the authors use a low voltage integrated gyrator filter.

The technique lies in the gyrator filter's capacitor used for its inductance. That is, the

Table I Process technique.

Item	Specification
NPN transistor	
High frequency	4.5 GHz ($V_{ce} = 3$ V)
Low current	$0.1\,\mu\text{A}$ ($h_{FE} = 100$)
PNP transistor	7 MHz ($V_{ce} = 3$ V)
High capacitance	1.37 fF/μm^2
High sheet resistivity	1 kΩ/□

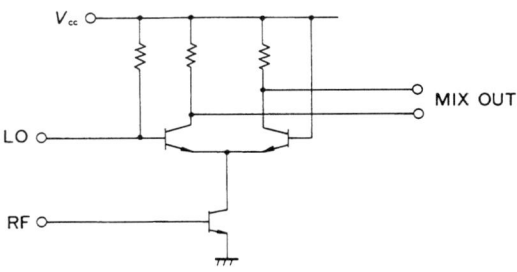

LO: Local frequency
RF: Radio frequency

Fig. 4 Mixer circuit.

transconductance characteristic of the transconductance amplifier is put to full use. The transconductance amplifier can be used to change the circuit's voltage and current.

The transconductance G can be expressed as follows:

$$G = \frac{dI_o}{dV_i}, \quad (1)$$

where

V_i: Input potential difference,
I_o: Output current.

Inductance L can be expressed similarly:

$$L = \frac{C}{G \cdot G}. \quad (2)$$

The Eqs. (1) and (2) show that a low current is needed for the transconductance amplifier in order to allow large inductance. Therefore, the process described in Section 3.2 (1) takes place under low current without suppressing the current amplification factor.

As described above, an equivalently large inductance coil can be realized on an IC chip. This is possible through a gyrator filter with a transconductance amplifier, being used as a low-pass filter.

The gyrator filter is 7th elliptic low-pass filter, and cut-off frequency is about 8 kHz. Figure 5 shows the channel filter response. It realizes 70 dB attenuation with an adjacent channel.

There have been major results in major parts of capacitor and resistor in miniaturization and lighter weight, in addition to the fact that they can now be loaded on an IC chip. This is because neither capacitor nor resistor are necessary for the RC active filter outer part.

4. CONTROL AND DISPLAY SECTION

4.1 Function of Control Section for Paging Receiver

Generally, the functions shown below are necessary for the control section of a paging receiver.

1. Synchronous control of a radio system
2. Battery saving control to the receiving section
3. Check of R IC (Address) numbers
4. Control and generation of an alert tone signal
5. Detection of switch operation
6. Display control

The hardware block diagram to realize these functions is shown in Fig. 6.

4.2 Characteristics of Control Section for Paging Receiver

The characteristics of the control section hardware and software for a paging receiver are shown below.

(1) Real-Time Processing

It realizes an interface with a radio system and a human-interface at the same time. For this reason, a nonsynchronous interruption control and real-time processing are required.

(2) Logic Noise Interference to Receiving Section

Because a radio receiving section and a control section coexist in a small area in a paging receiver, it must suppress the interference of logic noise of a control section to a receiving section.

(3) Low Power Consumption, Low Cost and Small Size

Because a paging receiver is operated by a

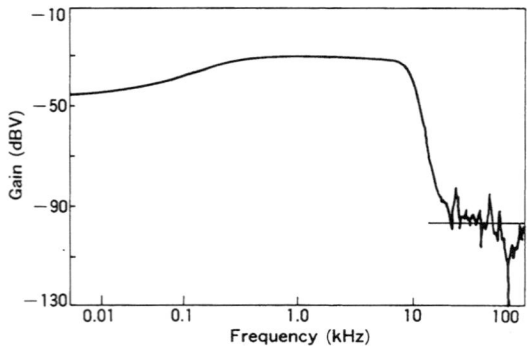

Fig. 5 Channel filter response.

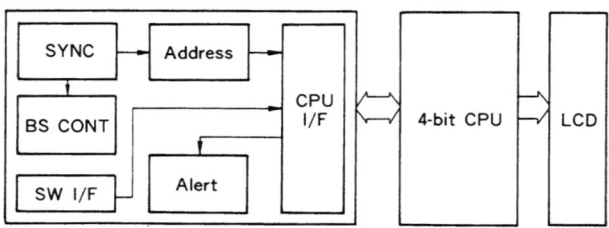

Fig. 6 Block diagram of control section.

Table II Comparison with current model.

	R3N4-12A (this model)	R3N4-5E (current model)
Dimensions ($W \times D \times H$)	$50 \times 18.6 \times 64$	$60 \times 19 \times 64$
Volume (excluding chip)	59.5 cc	73 cc
Weight (including AAA batt)	57.5 g	85 g

small-size dry battery, low power consumption characteristics are required.

And as a matter of course, it is necessary to make it small in size and low-priced.

4.3 Design of R3N4-12A

To realize the above functions and characteristics, the control section of R3N4-12A is designed as follows:

The control and display section of R3N4-12A is composed of two chips of a custom decoder LSI and a CPU. The authors examined a functional partition between the hardware block information of adequate quantity and quality, and then they adopted a method of operation of 1/3 bias and 1/4 duty.

To give it 2 V operation in this condition, the authors selected a liquid crystal material with high dielectric constant. Therefore, under a low temperature condition, the response time of this new type LCD became longer than that of the current type one. It was confirmed that there was no practical problems occurred in the paging receiver. Figure 7 shows LCD layout.

5. MECHANICAL DESIGN

5.1 Mechanical Characteristics

(1) Design Concept

Regarding the design of this equipment, the following two items were taken into account:

- The smallest size and the lightest weight in the world (as a top display pager).
  The height of the equipment parts is lower than customary by reason of a direct conversion receiving method. As a result, an assembly density realized the smallest size and lightest weight in the world.
- Cost reduction.
  This equipment achieved cost reduction through the drop assembly mechanism that makes possible automatic assembly.

Photo 2 External appearance.

Fig. 7 LCD layout.

(2) Dimensions

See Table II.

(3) External Appearance

Photo 2 shows the external appearance and its surface plane heightens the assembly density.

5.2 Other Features

(1) Large Reset Button

Location, size, shape and stroke of the reset switch have been studied to find the best solution of the human interface. Thus the selection of function can be performed accurately and comfortably.

(2) Improved Portability

Portability is the key feature of this pager. The pager housing has the mechanism to accept two types of clips, butterfly and chain, so that customers can select optional clips depending on their taste.

(3) Selection of the Housing Material

For the purpose of realizing lightweight, thin body and high strength, the housing of this equipment is made of polycarbonate resin. This has made the pager durable, rugged but lightweight even though the body thickness of the housing is from 0.7 mm to 2 mm.

6. CONCLUSION

As explained above, by use of the direct conversion receiving method, R3N4-12A can make smaller, lighter or reduce the cost compared with the conventional pager.

Moreover, by adding special functions such as the clock function, the message time stamp, and so on, authors would be further improving their pager. As authors launch their pager onto the market, they are confident in fully satisfying the needs of the overseas pager market.

ACKNOWLEDGMENTS

The authors wish to express their sincere appreciation to people in NEC for their valuable advice and cooperation during the development of this pager and advanced LSIs.

REFERENCES

[1] T. Mizukami, M. Masaki and K. Hikosaka, "NEC Digital Paging System," *NEC Res. & Develop.*, 50, pp. 1-10, July 1978.

[2] I. A. W. Vance, "Fully Integrated Radio Paging Receiver," *Proc. Inst. Elec. Eng. F*, **129**, 1, pp. 2-6, 1982.

[3] K. Nagata, M. Akahori, T. Mori and S. Umetsu, "Digital Display Radio Paging System," *NEC Res. & Develop.*, 68, pp. 16-23, Jan. 1983.

[4] K. Nagata, D. Ishii, T. Mori and T. Seo, "Slim Digital Pagers," *NEC Res. & Develop.*, 74, pp. 55-64, July 1984.

[5] D. W. H. Calder, "Audio Frequency Gyrator Filter for an Integrated Radio Paging Receiver," *IEE Conf. Mobile Radio Syst. & Techniques*, pp. 21-26, 1984.

[6] P. A. Moore, "A High-Performance Low-Power VHF Receiver Front End," *IEE Conf. Mobile Radio Syst. & Techniques*, pp. 16-20, 1984.

Received November 18, 1991

* * * * * * * * * * * * * *

Development of Advanced Mobile Telephone P3 (Personal Pocket Phone)

By Yoshiharu TAMURA,* Atushi YONEHATA,* Shinichi MIYAZAKI,*
Fumiyuki KOBAYASI,* Yuji FUKUDA† and Jiro KAMISHIRO*

ABSTRACT A handheld cellular telephone has been developed for AMPS and TACS (ETACS) cellular systems. Using newly developed LSIs and surface mount technology, the telephone (P3 series) accounted for half the volume of NEC's existing product (9 series). To ensure long talk time, automatic efficiency control technique for a GaAs FET power amplifier has been developed. Accessory expandability and transceiver controllability have been improved by upgrading the interface protocol. The phone is manufactured by a new state of the art production line, such as laser soldering and assembly robots. This paper also describes the optional products developed for the new telephone.

KEYWORDS Cellular system, Portable telephone, GaAs FET, Robot, SAW (Surface Acoustic Wave), Custom LSI, Software, CPU

1. INTRODUCTION

Today, AMPS (Advanced Mobile Phone Service) and TACS (Total Access Communications System) cellular systems are being adopted in many parts of the world. More than 3.5 million subscriber units are in service, and the number is increasing rapidly. Among the several types of the unit, the handheld type is expected to increase most significantly, because of its convenience and system adaptation to smaller cell configuration.

NEC launched its first cellular handheld telephone "9 series" in 1987. As the 9 series was one of the smallest handheld telephones (470 cc) with more than one hour of talk time, it has been one of the leading products in the market. However, today's market trend necessitates more compact and lighter telephones to satisfy various user classes.

The P3 series has been developed to meet the demand to cut down the volume by half, but with operation time and many other features the same or even better compared to 9 series. Photo 1 is a photograph of the P3, which measures 185 × 58 × 25.5 mm (270 cc), and weighs 400 g approximately. It is ergonomically designed to be as close as conventional telephone handsets for comfort and ease of use, and as thin as possible to achieve good portability. Two types of antenna, the built in flip up antenna and the external half wavelength antenna can be used. A 30 character dot matrix LCD and a loudspeaker are employed so as to keep the same user convenience as that of the 9 series.

Throughout the development many breakthroughs had to be made and they are summarized as the following four major projects.

(1) Development of small components and LSIs to reduce unit volume.

Photo 1 Photograph of the P3.

---

*Mobile Media Terminals Division
†NEC Mobile Communications, Ltd.

Reprinted with permission from *NEC Research and Development*, Y. Tamura, A. Yonehata, S. Miyazaki, F. Kobayasi, Y. Fukuda, and J. Kamishiro, "Development of Advanced Mobile Telephone P3 (Personal Pocket Phone)," no. 98, pp. 60-70, July 1990. © NEC Creative, Ltd.

(2) Development of low power circuit and low power devices to ensure sufficient operation time.
(3) Upgrading of the transceiver interface to improve controllability and accessory expandability.
(4) Adoption of new production technologies for low cost and flexible mass-production.

This paper first describes the board configuration and its blockdiagram briefly, and then each design project is focused on in the following sections. Optional products such as an in-car kit are also described.

2. BOARD CONFIGURATION AND BLOCK-DIAGRAM

Figure 1 shows the board configuration of the P3, which is composed of two major boards; LOGIC and TRX.

2.1 LOGIC

LOGIC contains the logic control section, the baseband section, the power supply control section, a keyboard matrix, and an LCD as well.

As shown in Fig. 2, LOGIC takes three CPU architecture. MCPU (IC303) with a 512K EPROM (IC301) and a 64K RAM (IC302) controls the cellular system sequence and user functions. A single chip microprocessor SCPU (IC310) is used mainly for data reception of 8 kbps or 10 kbps Manchester code.

TEL CPU (IC312) is another single chip microprocessor which controls the keyboard, the LCD, power switching, and clock functions. IC305 is a baseband processor including a syllabic compander. All the other random logic and I/Os are integrated in a standard cell LSI, IC304.

2.2 TRX

As shown in Fig. 3, TRX is further composed of TX (transmitter) and RX (receiver), both of which contain separate 900 MHz band synthesizer, TX uses a GaAs power amplifier (PA) module (IC451) in the final stage. The power gain of the PA is 24 dB at its full output with efficiency over 55%. IC452 is a PA controller including automatic efficiency control (AEC) which will be mentioned later.

RX is a double conversion receiver using quadrature FM demodulator. Dielectric resonator filters and a helical filter are used for RF bands, while a surface acoustic wave filter and ceramic filters are used for IF bands.

3. VOLUME REDUCTION

Three approaches were necessary to achieve significant reduction of the unit volume;

(1) Miniaturization of the components.
(2) Development of VLSIs unifying several LSI chips.
(3) Optimization of components and board arrangement to minimize unused space.

Table I shows examples of the components of which size has been greatly reduced. Moreover, all components without a single exception are surface mount type, which were so designed that they would withstand fall tests of 1.5 m.

Generally, the smaller components become, the more interference occurs. So, input and output impedance of the components were designed as low as possible, and packages thereof are considered to have good shielding effect. Other than these key components, use of smaller resistor chips and capacitor chips (1.6 × 0.8 mm) contributed to overall size reduction.

Especially for LOGIC board, the most effective

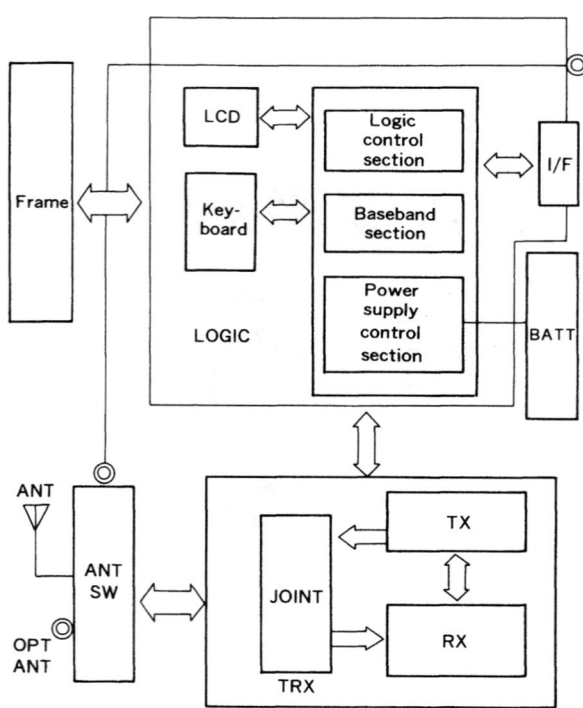

Fig. 1 Board configuration of the P3.

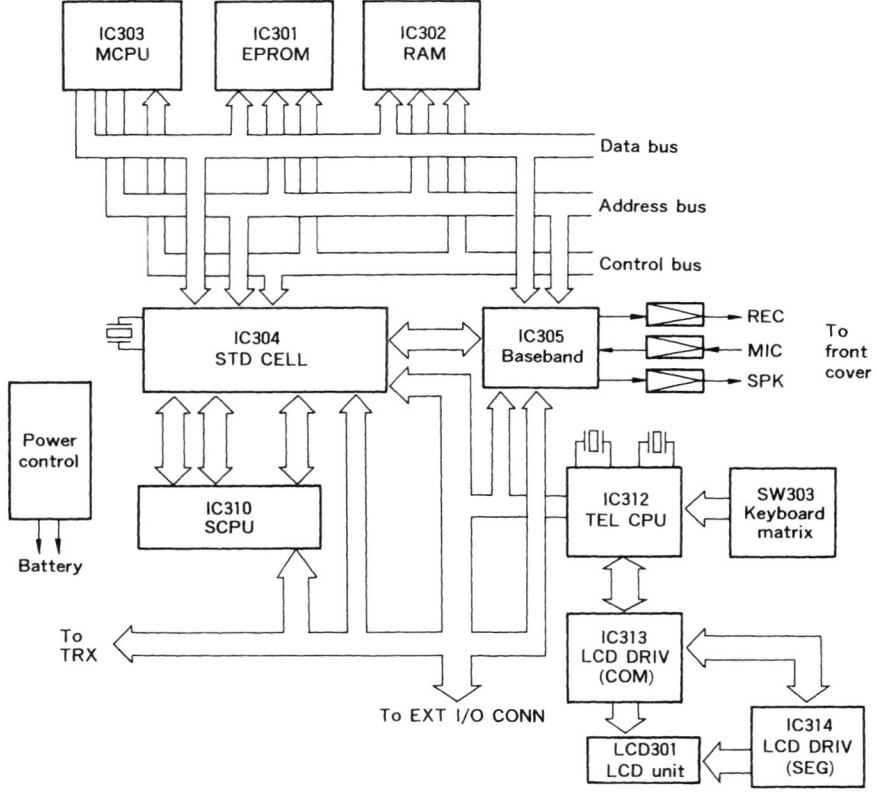

Fig. 2 Block diagram of LOGIC unit.

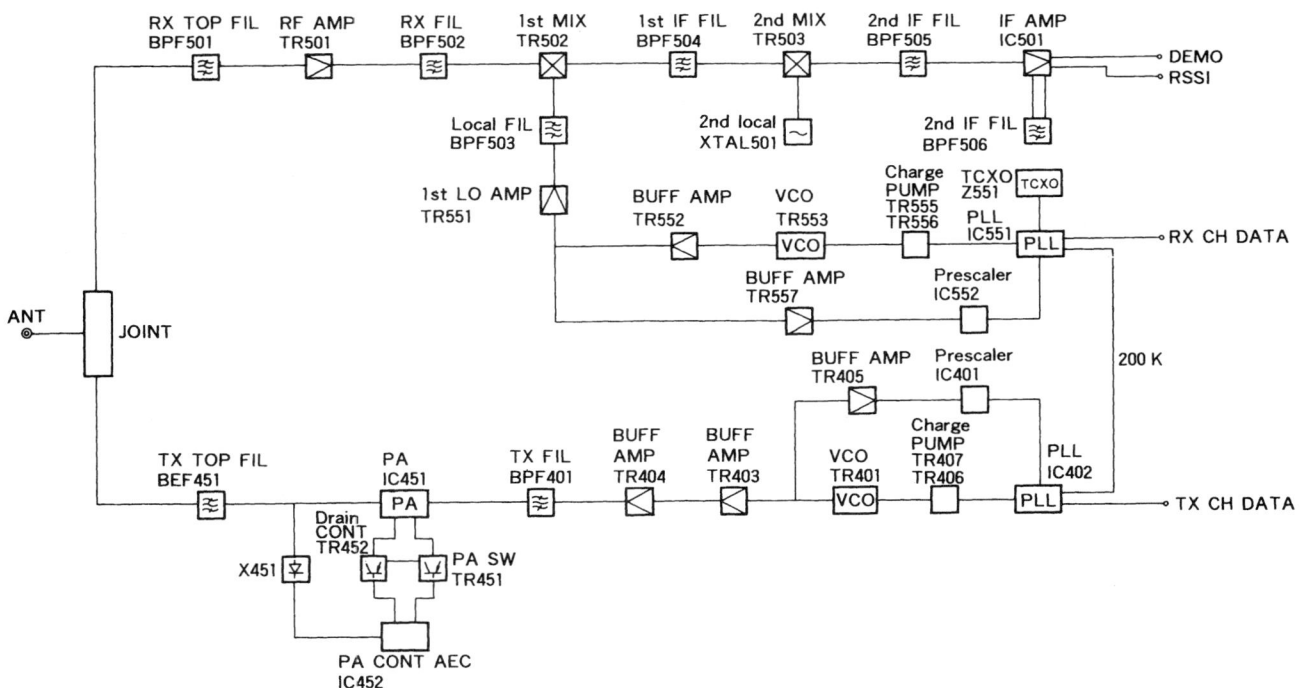

Fig. 3 Block diagram of TRX.

way to reduce volume is to reduce the number of LSIs. Using "standard cell" technique, the 100 pin VLSI IC304 reduced corresponding space to 1/3 of 9 series which used four 44 pin LSIs and a 64 pin LSI. IC305 is also a newly developed CMOS custom LSI. In the LSI, SCF (Switched Capacitor Filter) baseband filters, a SCF syllabic compander, a DTMF (Dual Tone Multiple Frequency) generator, and seven automatic level controls are concentrated. This LSI is controlled directly by the MCPU data/address bus which might cause noise problem. So, the chip layout was very carefully designed. As a result of the development of these LSIs, all the other sections except the radio section are accommodated on one surface of a single PCB (Printed Circuit Board).

Figure 4(a) is a typical components arrangement of conventional radio unit assembled in a single PCB. Relatively large components are mounted on one side and pins thereof are soldered on the other side where small components such as resistors and capacitors are "surface mounted." In this arrangement considerable space is left unused on the large component side, and this problem had to be solved for P3. Additionally, the arrangement of components and boards in P3 had to satisfy the following requirements.

(1) Maximum height of components should be as low as possible to realize the thin profile of the telephone.
(2) Board assembly should be simple enough to introduce assembly robots.
(3) Interference should not occur between TX, RX and LOGIC.
(4) PCB design should be executed independently from other boards to shorten design TAT (Turn Around Time).

Figure 4(b) is the answer to the above requirements. To minimize unused space and to satisfy (1), large components were designed as surface mount type keeping the maximum height within 4.5 mm. Requirements (3) and (4) are met by using the two same sized PCBs for TX and RX, both of which are single side surface-mounted. Finally, the overall structure was arranged to enable simple Z-axis (uni-directional) assembly to satisfy (2).

4. POWER CONSUMPTION REDUCTION

9 series had 80 minute talk time or 17 hour standby time with a 1000 mA/7.2 V NiCd battery. For P3, 700 mA/6V battery was chosen for volume reduction. Therefore, current drain had to be lowered to at least 7/10 to keep the performance. Every circuit was rechecked to cut down current drain. Major efforts were focused on the following items:

For reduction of talk time current drain;
(1) Adoption of a high efficiency GaAs power amplifier module.
(2) Development of a control circuit to derive maximum efficiency of the GaAs PA at any power level and any temperature.

For reduction of standby current drain;
(1) Development of a low power prescaler for the 900 MHz synthesizer.
(2) Furnishing all LSIs with standby mode or power down mode.
(3) Improvement of standby ratio by developing a software to efficiently utilize the three CPU architecture.

Resultant current consumption and achieved

Table I Size reduction of the components.

Item	9 series	P3 series
PA module	40 × 14 × 7 (3.9 cc)	22 × 13 × 4.5 (1.3 cc)
Duplexer (AMPS)	82 × 15 × 7 (8.6 cc)	TX: 19 × 14 × 4.5 (1.2 cc)
		RX: 28 × 14 × 5.5 (2.2 cc)
		(3.4 cc)
1st IF filter	18 × 10 × 10 (1.8 cc)	18 × 9 × 2.5 (0.4 cc)
TCXO	18 × 11.5 × 9 (1.9 cc)	18 × 11.5 × 4.5 (0.9 cc)

(Size in mm)

operation time are shown in Table II.

4.1 Automatic Efficiency Control Circuit

Recently, GaAs FETs have begun to be employed in handheld cellular telephones because of their excellent power conversion efficiency. However, with conventional control circuits, high efficiency operation is realized only in restricted conditions. Moreover, because GaAs FETs have usually the fixed bias in class A or class AB, degradation of efficiency at low power levels sometimes becomes worse than that of bipolar transistors. Then variable gate bias method was investigated for use as an AGC loop instead of the conventional variable drain bias method. But several problems were encountered such as oscillation or miss-convergence. Finally, cooperative double loop named AEC (Automatic Efficiency Control) was developed. This technique is based on the phenomenon shown in Fig. 5.

Combined current drain of the final stage FET and the drive stage FET has a minimum peak (point M) at a certain gate bias of the final FET maintaining the output power constant. The minimum peak M is always observed at any conditions of power levels, temperature, etc., but having different gate bias voltage. The double loop of AEC is so designed that the gate bias of the final stage FET converges at the point M in any

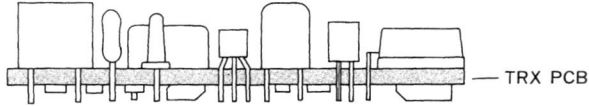

(a) Conventional arrangement of TRX (9A).

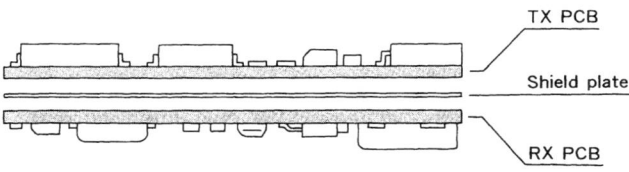

(b) New arrangement of TRX (P3).

Fig. 4 Arrangement of TRXs.

Table II Power consumption.

Series	Battery	Standby		Talk	
		Current	Time	Current	Time
9 series	7.2 V, 1000 mAH	60 mA	17 hours	750 mA	80 minutes
P3 series	6 V, 700 mAH	38.5 mA	18 hours	500 mA	80 minutes

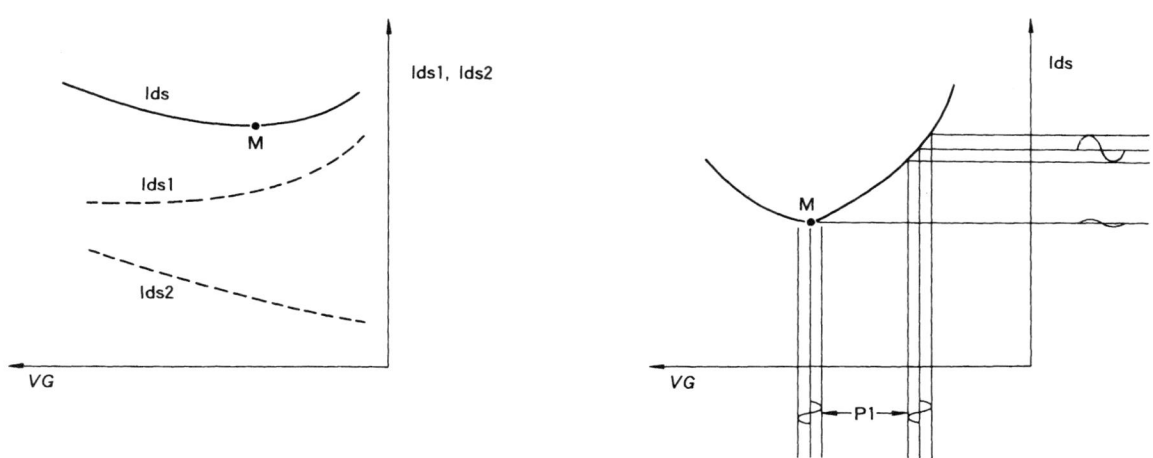

Ids1: Current drain of the drive stage FET
Ids2: Current drain of the final stage FET
Ids: Ids = Ids1 + Ids2
VG: Gate bias of the final stage FET

Fig. 5 Automatic efficiency control.

condition.

Figure 6 is the block diagram of AEC. The first loop controls the drain bias of the drive stage FET to get predetermined output levels. The loop gain is set high enough to get fast output switching. The second loop places a perturbation signal (P1) onto the variable DC gate bias of the final FET. P1 which has very low subaudio frequency results in the deviation of the current through the transistor Q1. Then it is converted to the voltage deviation P2 and fed into the full-wave rectifier to produce the error voltage E. The second loop finally works to minimize E at a slower convergence speed than the first loop. Because of the loop speed difference, the RF output level is kept constant and P2 is suppressed in the output spectrum.

Introduction of AEC has the following merits:

(1) Power efficiency is maintained at maximum under any conditions.
(2) Individual bias adjusting is eliminated.
(3) As heat generation is kept to the minimum, the PA module can be in a surface mount package without a large heat sink.

A custom CMOS LSI has been developed for AEC circuit including a DC/DC converter for negative voltage generation.

4.2 Low Power Prescaler

To reduce current drain of the receiver in standby mode, the prescaler in the synthesizer was the bottleneck. Therefore, a new bipolar process "SST" has been adopted to develop an ultralow power prescaler. Figure 7 shows the frequency characteristics of the prescaler. Current drain of 3 mA at 1 GHz is one of the lowest values except those experimentally reported.

4.3 Three CPU Architecture

Even though the LSIs in LOGIC are furnished with power down mode, sophisticated software should be developed to coordinate them.

MCPU processes the cellular sequences and complicated user functions. But in combination with the ROM and RAM, its current drain is relatively high. SCPU has been introduced to work for data reception, relieving MCPU from this task. Similarly, TEL CPU has been introduced to relieve the MCPU from handling the keyboard, the LCD, etc. The other LSIs are powered only when they need to be. As a result, MCPU standby ratio reached 95%, and the average current drain of the LOGIC is 10 mA approximately.

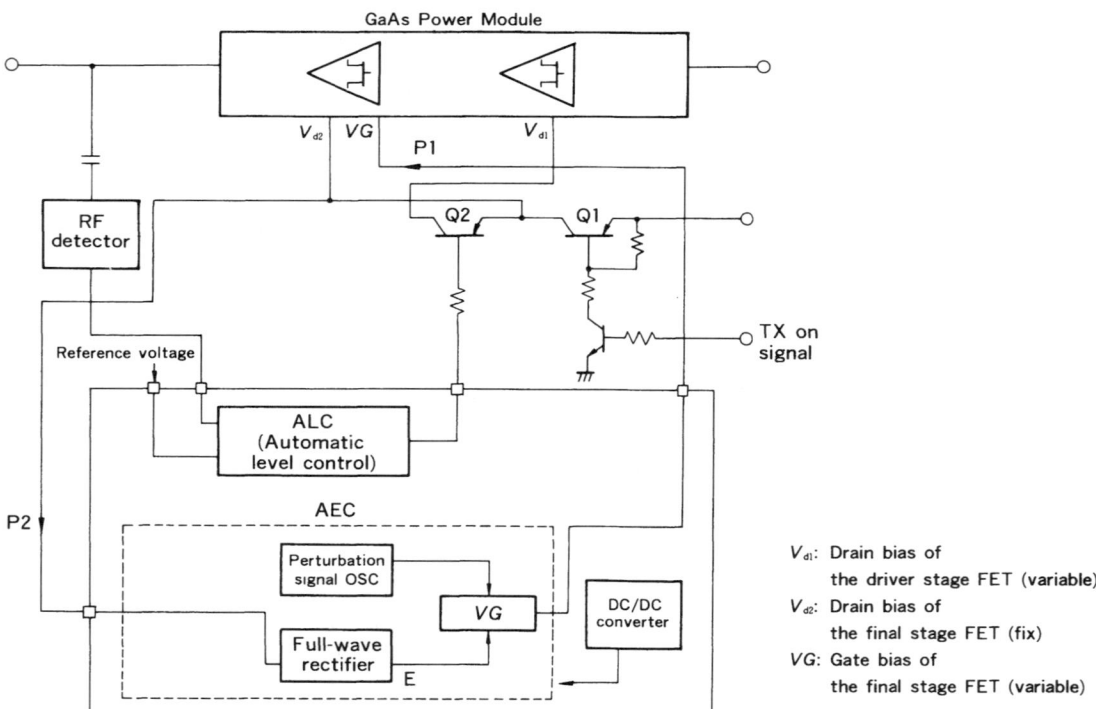

V_{d1}: Drain bias of the driver stage FET (variable)
V_{d2}: Drain bias of the final stage FET (fix)
VG: Gate bias of the final stage FET (variable)

Fig. 6 Block diagram of AEC.

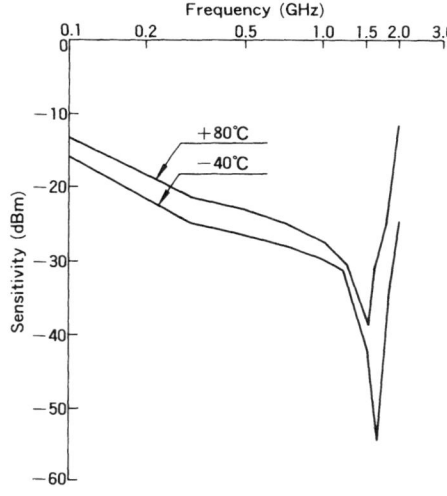

Fig. 7 Frequency characteristics of the new prescaler.

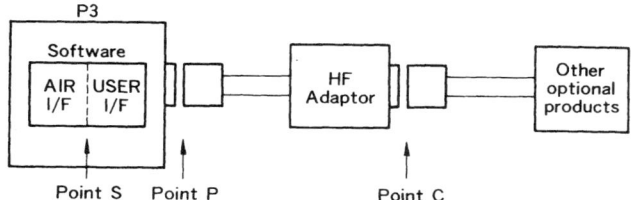

Fig. 8 NOTIS interface.

5. NEC's OPEN INTERFACE AND SOFTWARE (NOTIS)

Today, as stated earlier, many countries adopt AMPS or TACS cellular systems. Although cellular sequences (AIR I/F) are sometimes modified slightly for different countries, most of the software which executes AIR I/F does not need to be redesigned once it is developed. However, software to handle user functions (USER I/F) differs from one system to another quite frequently. Furthermore, requirements for USER I/F are becoming more complicated and multiform as the market matures.

In this situation, authors concluded that AIR I/F and USER I/F should be definitely divided by clear and optimum software interface. By doing this, TAT for modification has been shortened and maintainability of the software has been improved. The other purpose of the development of the new interface was to increase expandability of USER I/F from external equipment. Some parts of the new interface to control the telephone and USER I/F were designed to permit access by engineers, which, on a contract basis, enables them to design optional equipment. The new interface named NOTIS (NEC's Open Interface and Software) was thus developed by upgrading the old interface, S-BUS.

Hereafter, not only P3 series, but all NEC cellular telephones will be designed using NOTIS interface.

Practically, NOTIS defines three interface point; S, P and C as shown in Fig. 8.

(1) Point S

The point S is the interface between the two software; AIR I/F and USER I/F. In defining the point S, the following items were mainly considered.

a) All device drivers should be standardized as system call modules.
b) Parameters used in different modules should also be standardized and restructured.
c) Frequency and cadence of tones (alert tone, etc.) should be easily accessible and modifiable.

By these considerations, interface inside the software was very clearly defined and maintainability has been greatly improved.

(2) Point P and C

The point P and C interface includes both software and hardware. Hardware requirements are not explained here, but such conditions as the voltage and impedance as well as physical dimensions of the connectors are defined for the point P and C separately.

Logical requirements for the communication through the interface are defined in the physical layer and the data link layer, both of which are common for P and C. The logical requirements are up-compatible to those of the old S-BUS.

Data format which is commonly sent through P and C, has also been upgraded to expand controllability. Practically, additional data set has been defined to inform external equipment of P3's internal status. Some other data sets have been added for the display to accommodate optional user interface, and for the automatic adjustment system to write adjustment data.

6. PRODUCTION ENGINEERING

In parallel with the develoment of the telephone, the production engineering group conducted the project to introduce some new production technologies. Here, three examples are described which have been introduced for low cost and flexible mass-production.

(1) Complete Automatic Adjusting System

Although automatic adjustment had been used in our products before, 9 series still used ten potentiometers to set RF, AF and DC levels. By using variable gain amplifiers or A/D and D/A converters along with EEPROM, the potentiometers have been eliminated and all the adjustments are now performed automatically.

(2) Laser Soldering

To realize smaller and thinner products, various kinds of chip components and surface mount type block modules are assembled in very high density and in different forms. As these components and modules have different heat capacity and different heat proof characteristics, they can not undergo uniquely conditioned reflow soldering. Therefore, small chip components are reflow-soldered by infrared ray first, then larger components and shielding covers are soldered by local heating.

These days, several local heating methods are available; soldering iron, laser, light beam, hot ram, etc. Authors chose laser soldering for the following reasons.

a) It is suitable for local heating as the sharp beam can be focused into a small radius, which prevents undesirable heating of other components nearby.

b) Soldering time is relatively short because of the high power density.

c) Unlike other methods that need a contact with the heat source, it is a contactless heating

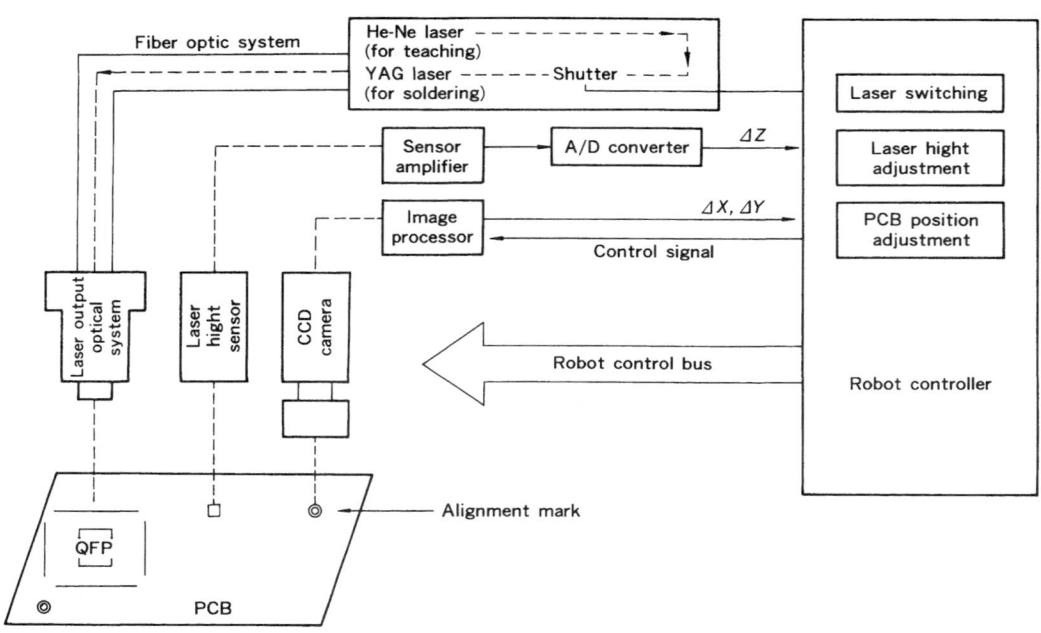

Fig. 9 System configuration of the laser soldering.

Table III Comparison of CO_2 and YAG laser.

Laser	Wavelength (μm)	Reflection by solder	Absorption		Usability of fiberoptics
			Flux	PCB	
CO_2	10.6	High	High	High	Difficult
YAG	1.06	Low	Low	Low	Easy

method which prevents discoloration of works and enables soldering at points hard to access by hand.

d) Because it is an optical system, such technology as optical fiber and beam division can be utilized. Controllability of heat power and intervals using this technology is suitable for automation.

Figure 9 shows the system configuration of the laser soldering. Here, as the laser source, YAG laser

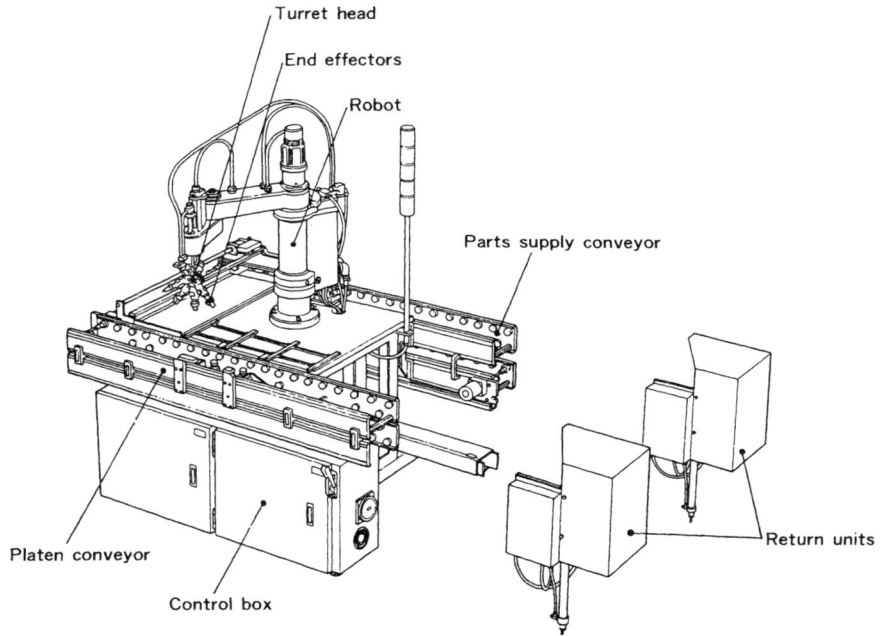

Fig. 10 Cylindrical robot.

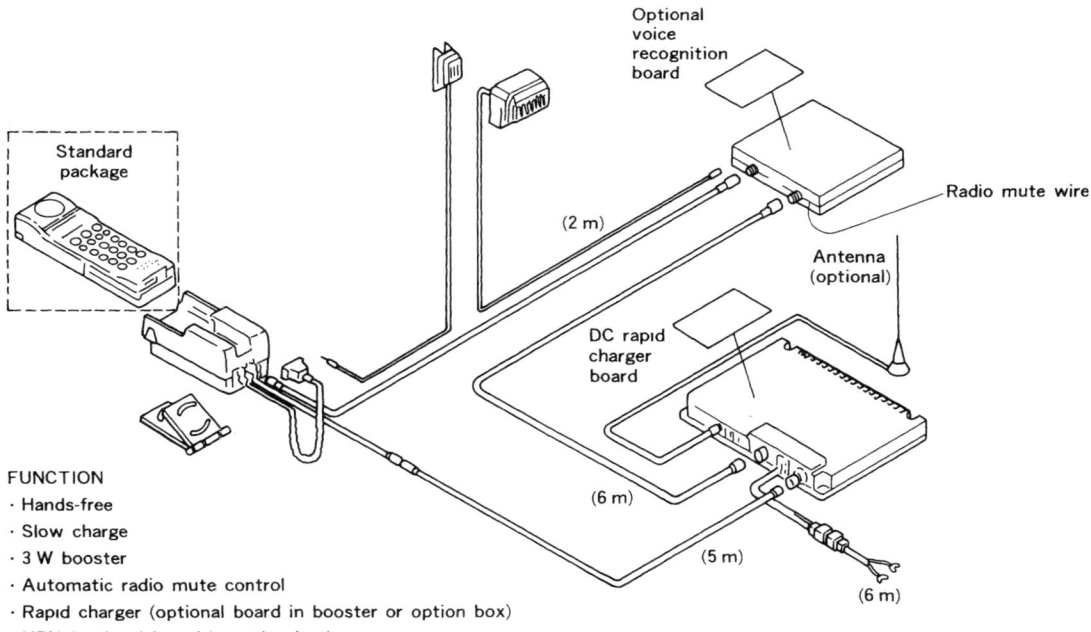

Fig. 11 Connection of hands-free car kit with booster and option box.

has been chosen by comparing the features with those of CO_2 laser (Table III).

To maintain quality of the laser soldering, the physical tolerance of components' leads is checked and characteristics of the cream solder are controlled.

(3) Assembly Robots

A new production line using assembly robots has been constructed for P3. The automated line is programmed for both board assembly and final unit assembly. Two types of robots are adopted; Cartesian robots for soldering and cylindrical robots for screwing, picking up, placing, etc. Main features of the robot line are;

a) Use of circulating conveyors (two conveyors; parts supply conveyor and platen conveyor for works).
b) The turret head can be equipped with up to six hand modules.
c) Up to six different components can be fed at one stage.
d) Each stage is standardized as 1.5 m "cell."

The line, therefore, is flexible enough to be rearranged for future products at a minimum investment without many modifications of the machines and the layout.

Figure 10 shows a cylindrical robot.

Table IV Specifications of P3.

Item	AMPS	TACS
Frequency range (TX)	824 - 849 MHz	872 - 905 MHz
(RX)	869 - 894 MHz	917 - 950 MHz
Number of channels	832 channels	1320 channels (E-TACS)
RF power output	0.6 W ERP (controlled in six 4 dB steps)	
Frequency stability	Better than ±2.5 ppm over a temperature range of −30℃ to +60℃	Better than ±2.5 ppm over a temperature range of −10℃ to +55℃
Talk time	Up to 80 minutes continuous talk time Up to 18 hours on standby time	
Power supply voltage	+6.2 Vdc nominal NiCd battery	
Dimension	Approx. 58(W) × 25.5(D) × 184(H) mm, 270 cc	
Weight	Approx. 400 g	

Table V Brief specifications of P3 for AMPS.

- Basic Specifications
  EIA standard
- Frequency Range
  TX: 824 ~ 849 MHz
  RX: 869 ~ 894 MHz
- Number of Channels
  832 channels
- RF Power Output
  0.6 W E.R.P. (controlled in six 4 dB steps)
- Frequency Stability
  Better than ±2.5 ppm over a temperature range of −30℃ to +60℃
  (−22℉ to +140℉)
- Talk time
  Up to 80 minutes continuous talk time
  Up to 18 hours on standby time
- Power Supply Voltage
  +6.2 Vdc nominal NiCd battery
- Dimensions
  Approx. 2.3(W) × 1(D) × 7.2(H) in.
  (58(W) × 25.4(D) × 184(H) mm)
  16.6 Cubic inches (270 cc)
- Weight
  Approx. 14 oz. (400 g)

7. OPTIONAL PRODUCTS

Recently, as cellular telephones are becoming more popular, optional products have become more important to satisfy various user requirements.

A variety of optional products have been developed simultaneously with the development of the P3. As for in-car use application, the basic concept of 9 series' optional products is taken over to P3, while their size has been reduced. Additionally, the "Option box" has been developed for future expansion of applications.

The Option box is designed to contain several optional boards such as a radio muting adaptor, a rapid charger and a voice recognition dialer. All these optional products are designed according to the specifications of NOTIS S, P and C as mentioned above. Figure 11 shows an example of the connection of the in-car kit including a Hands-free adaptor, a Booster, and an Option box.

8. SPECIFICATIONS

Summary is given of the specifications of the P3 for AMPS based on EIA standard, and for E-TACS (see Tables IV and V).

9. CONCLUSION

P3 series has been developed as NEC's new generation cellular handheld telephone. Several projects have been worked together and many breakthroughs have been made to accomplish the entire development. Introduction of the robot line has become possible only through the development of low profile surface-mounted components and VLSIs.

The new interface protocol NOTIS enabled the simultaneous development of the optional products and of the full automatic adjustment system. Future efforts to realize more compact telephone will be focused on further introduction of LSIs, especially in the radio section.

ACKNOWLEDGMENTS

The authors would like to express their thanks to people in NEC Corporation, NEMCOM (NEC Mobile Communications, Ltd.), NEC America, Inc. and NEC (UK), Ltd. for their helpful discussions and cooperation throughout the development.

REFERENCES

[1] T. Kai, et al., "Cellular Mobile Radio Equipment," *NEC Res. & Develop.*, 84, pp. 85-93, 1987.
[2] Y. Tamura, et al., "Hand-held Portable Equipment for Cellular Mobile Telephone," *NEC Res. & Develop.*, 87, pp. 34-43, Oct. 1987.

Received March 30, 1990

* * * * * * * * * * * * * * *

Radios for the Future: Designing for DECT

By Benny Madsen and Daniel E Fague
National Semiconductor

Nations worldwide are implementing standards that will transform today's conventional land line phone system to an untethered network of services A first step in achieving this goal is the implementation of digital telephony standards for both cellular and cordless telephones One of the first digital cordless standards to be implemented is the Digital European Cordless Telecommunications (DECT) standard, a pan-European standard designed to connect all of Europe with a common digital cordless system. This article introduces a new 2 GHz radio architecture which implements the Digital European Cordless Telecommunications (DECT) standard.

The DECT standard presents the radio designer with a number of challenges For example, a DECT phone must provide all the capabilities of a standard cellular phone at roughly twice the operating frequency, and satisfy user demand for portability: small size, lightweight, long talk and standby times These requirements make a strong case for integrated RF solutions. Table 1 shows a summary of the key specifications for DECT as well as other digital cordless standards

The DECT system is based on Time Division Duplex, Time Division Multiple Access (TDD/TDMA) This technique of duplexing uses separate timeslots on a single carrier for transmitting and receiving signals Figure 1 shows the structure of the DECT TDD/TDMA frame The complete frame is 10 ms in duration, with a 5 ms transmit and a 5 ms receive sub-frame Each sub-frame is divided into 12 timeslots of 480 bit times in duration (416 67 ms) When a communication link is made, a transmit and receive timeslot is assigned to the users Normally the same timeslot number (i e , T3 and R3) is used for both transmit and receive Multiple users are accommodated by assigning different timeslots to different users This method requires the transmitter to be turned off while receiving and the receiver to be turned off while transmitting This method of duplexing places strict requirements on the Phase Locked Loop (PLL). It must switch fast and have low spurious noise.

Applicability to the U.S.

In the United States, the FCC has proposed allocating 20 MHz within the Emerging Technologies (ET) band for "non-licensed" Personal Communications Systems (PCS), with 10 MHz allocated for voice type (isochronous) services. The U.S. can choose to modify a standard already developed, like DECT with a frequency almost identical to that proposed by the FCC, or develop its own The following explains how DECT could be modified to meet the proposed U S requirements

Modulation method — DECT employs a frequency modulation scheme, called Gaussian-shaped frequency shift keying, which could be used in the PCS ET band Frequency modulation schemes have a number of advantages, including the use of more efficient power amplifiers to lower power consumption In addition, FM techniques allow the use of limiter/frequency discriminator structures for demodulation These structures are simple to use and build, as opposed to fully linear receivers requiring sophisticated gain control.

Channel bandwidth vs. data bit rate — The DECT bit rate (1.152 Mb/s) and channel spacing (1.728 MHz) limits spectral efficiency to about 0.67 bits/sec/Hz. As a single parameter, the spectral efficiency is not sufficient for determining the capacity in systems like DECT or any other small cell system. System capacity is a function of cell size (transmitted power), frequency reuse distance, available frequency range, and spectral efficiency. The frequency reuse distance is determined by the receivers' ability to cope with interference from unwanted transmitters; interference on the same channel (co-channel) or adjacent channels. A receiver's ability to cope with same channel and adjacent channel interference is determined by the chosen modulation scheme, as well

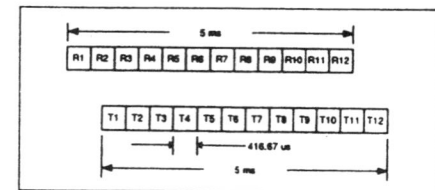

Figure 1 - TDD/TDMA frame structure for DECT.

Parameter	DECT	CT-2	USER PCS	PHP
Origin	ETSI	UK	US	Japan
Access	TDMA & FDMA/TDD	FDMA/TDD		TDMA/TDD
Modulation	GFSK	GFSK		π/4-DQPSK
Baseband Filter	$B_bT = 0.5$ (Gaussian)	$B_bT = 0.5$ (Gaussian)		$\alpha = 0.5$ (Root Nyquist)
Data Rate	1.152 Mb/s	72 kb/s	≈1.0 Mb/s	384 kb/s
FM Deviation	288 kHz	14.4-25.2 kHz		
RF Channels	0: 1897.344 MHz 9: 1881.792 MHz	1: 864.15 MHz 40: 868.05 MHz	1910-1930 MHz	1895-1911 MHz
No. of RF Channels	10	40	4	52
Channel Spacing	1.728 MHz	10 kHz	1.25 MHz	300 kHz
Synthesizer Switching Speed	30 us (BS) 450 us (HS)	1 ms (ch.-ch.) 2 ms		30 us (BS) 1.5 ms (HS)
Frequency Accuracy	50 kHz	10 kHz		3 ppm
Speech ch./RF ch. (Full/Half Rate)	12/24	1/1		4/8
Speech CODEC	32 kb/s ADPCM	32 kb/s ADPCM		32 kb/s ADPCM
Frame Length	10 ms (12 Tx + 12 Rx)	2 ms (1 Tx + 1 Rx)		5 ms (4 Tx + 4 Rx)
Peak Power	250 mW	10 mW		100 mW

Table 1 - A comparison of digital cordless telephony standards

as receiver structure and design.

The narrower 10 MHz frequency band proposed for U.S. isochronous services will require a higher spectral efficiency than DECT. This can be obtained by narrowing the (Gaussian) modulation filter or by using other modulation schemes. Narrowing the filter will limit receiver sensitivity, while other modulation schemes may sacrifice the simple transceiver structure.

Radio Design Issues

Single conversion receiver, direct modulation transmitter — Digital modulation can be done in a couple of ways. Quadrature (I and Q) modulation is highly accurate, and is the method of choice for the digital cellular standards. A quadrature down conversion provides maximum flexibility in the baseband section for any type of demodulation. However, the added circuits that accompany these types of demodulators are costly. Normally, a Digital Signal Processor (DSP) must be used in the backend, which increases current consumption, and some automatic gain control circuits must be used to ensure that the baseband A/D converter dynamic range is fully utilized. In some systems, a carrier recovery loop such as a Costas Loop must be implemented to recover the carrier's phase.

The TDD/TDMA nature of DECT allows for two simplifications that drastically reduce the radio's complexity. First, the transmitter is on for short bursts of 424 bits, or about 380 ms. This, and a moderate frequency drift tolerance, allow for an open loop, direct modulation of the voltage controlled oscillator (VCO). Second, the receiver is only on for the same short length of time. The argument can be made that in such a short burst of data, a carrier recovery loop is not required, because the phase difference of the local oscillator (LO) will not deviate much in such a short time. However, the coherent solution will suffer from the frequency drift allowed in DECT. The large frequency drift makes an unlocked LO unacceptable. But, the drift is not too large that a discriminator can't be used. In fact, the drift allowed is about 10 percent or less of the discriminator's bandwidth, making the discriminator a good choice for DECT.

Traditionally, 2 GHz radio designs were limited to at least two down conversions and one up conversion. Figure 2 shows a typical receiver. The second down conversion is necessary because limiting and demodulating at the high IF is difficult and modulating a low frequency VCO is more practical.

For DECT, National Semiconductor chose a single conversion receiver, direct modulation transmitter architecture. This is shown in Figure 3. There are several key features that make this architecture possible. The signal is received at the antenna and filtered with a low loss (typ. 1 dB) antenna filter. The filter is used on the receive side to reduce possible out-of-band interfering signals. On the transmit side the filter is used to attenuate harmonics of the transmitted signal and to reduce possible wide band noise.

The signal then passes the duplexer which can either be a fast switch or circulator. The insertion loss in the duplexer should not exceed 1 dB. From the duplexer the signal enters a low noise amplifier, image reject filter and mixer structure. National Semiconductor provides various IC solutions for receiver front ends, which combine the low noise amplifier (LNA) and mixer circuit. One of them is the LMX2216 shown in Figure 3.

Before the DECT standard was finalized, a group of component and equipment manufacturers had worked out standards for key DECT components, within the framework of a group called ECTEL (European Community Telecommunications). A standard intermediate frequency of 110.592 MHz was chosen by this group. Consequently, a number of manufacturers provide highly selective Surface Acoustic Wave filters for this frequency with low insertion loss (typ. 3 dB) and reasonable group delay characteristics. The adjacent channel selectivity of most of the filters permits FM demodulation on the output of the filter.

Two ways to do FM demodulation are: (1) the limiter/discriminator structure as shown in Figure 3; and (2) PLL tracking of the instantaneous frequency deviation. By first limiting the signal in a high gain amplifier, then mixing the signal

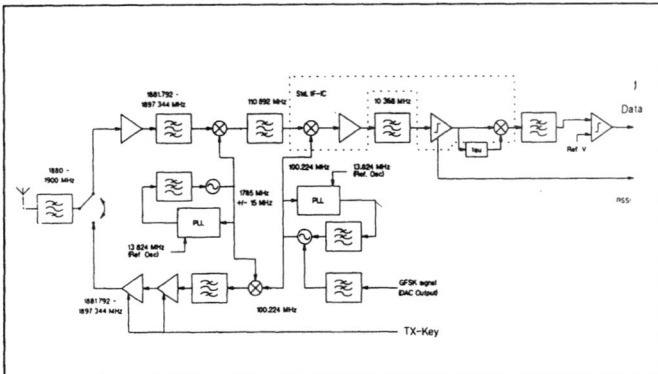

Figure 2 - A block diagram of a typical 2 GHz radio.

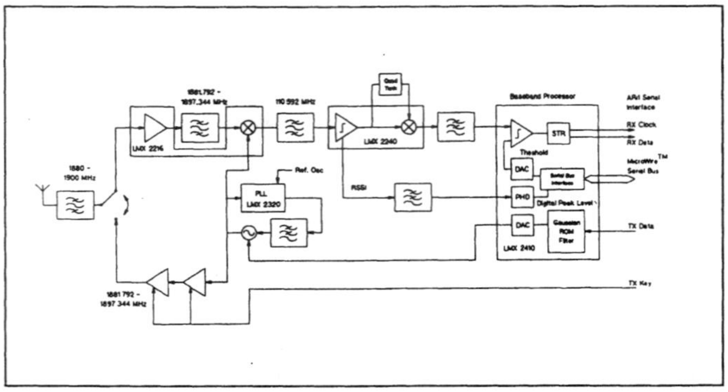

Figure 3 - Block diagram of National Semiconductor's single conversion receiver architecture.

with a 90-degree phase shifted replica of itself, the instantaneous frequency deviation is derived. The quadrature tank shifts the phase of the signal 90 degrees, but only at the center frequency. The important parameter of the phase-shifter is a large frequency dependence of the phase-shift. Normally a linear phase-shift dependency of the frequency is desirable. Thus a steep linear phase characteristic is desired. The steepness, though, is determined by the Q of the quad tank, which limits the design possibilities.

For demodulating FM signals at relatively high frequencies a phase locked loop is a common solution. The PLL tracks the instantaneous phase and thereby the instantaneous frequency deviation of the received signal. Such solutions are commonly used in satellite TV demodulators. Some of the drawbacks of PLL FM demodulator solutions are less immunity to interference (such as co-channel interference), a significantly higher level of power consumption and increased component cost.

As the demodulated signal leaves the LMX2240, a lowpass filter removes remnants of IF frequency and its harmonics, and improves the signal-to-noise ratio before the signal enters the data comparator. Depending on the signal level at the input of the comparator, the comparator can become crucial to single conversion design. National Semiconductor therefore integrated a high performance data comparator into the LMX2410 baseband processor. The output of the comparator is a binary signal, corresponding to the transmitted binary signal. To derive timing information, the signal is processed by a symbol timing recovery circuit. The symbol timing can either be based on continuously phase locking to the edges of the signal, or burst wise correlating with the known burst signal. The LMX2410 uses the digital PLL method, but simulations show that either of the methods are usable and yield about the same performance.

On the transmit side, the baseband processor includes the Gaussian filter needed to pulse shape the incoming serial binary transmit signal. The baseband processor internally regenerates the transmit clock, eliminating the need for an extra transmit clock line from the digital back-end. The Gaussian filter is implemented by ROM table lookup instead of a conventional FIR or IIR filter. This ROM table and the filter can be mask programmed to meet other filter requirements, e.g. to support U.S. standards for the PCS and ISM bands. From the filter, the signal is fed to a DAC, then directly to the frequency synthesizer VCO, so the signal directly modulates the VCO. This principle is commonly used in analog cellular and cordless telephones. In order not to introduce distortion, the filter bandwidth of the loop must be significantly lower than the lowest frequency components of the modulating signal. But a low loop frequency leads to a slow switching speed, which is undesirable in a DECT transceiver. In order to modulate the VCO directly, the phase locking needs to be disabled, corresponding to opening the loop while modulating. The principle is simple, but puts requirements on the components used. There are two basic needs. The loop must be opened quickly, without causing a sudden jump in the VCO frequency, and the VCO frequency must be stable during a data burst. The National Semiconductor PLL, LMX2320, in combination with an active loop filter, accomplishes this.

The synthesizer needs to span the DECT frequency band both transmitting and receiving. A wide VCO tuning range and a fast switching speed — in the order of 400 us (max.) for a frequency jump of 130 MHz — are needed. The modulated signal from the VCO is fed to the power amplifier. The power amplifier delivers a maximum of 250 mW peak, plus whatever loss is introduced between the PA and the antenna (typically in the order of a couple of dB). The signal is finally fed to the antenna through the duplexer and antenna filter.

Supply voltage and power consumption are also key parameters in the design of handheld and pocket equipment. Today, cordless telephones use as few as three rechargeable battery

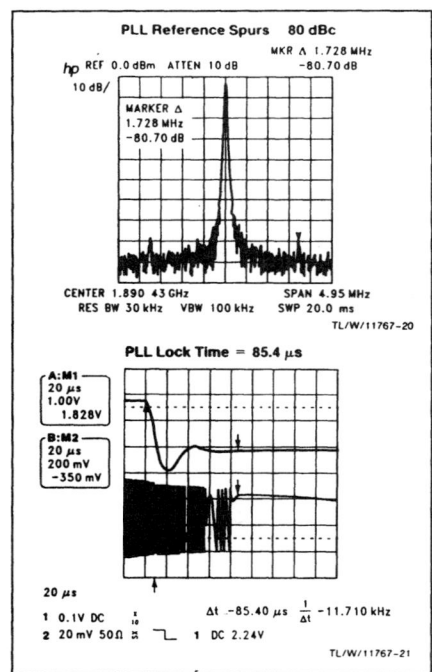

Figure 4 - Lock time and spurious noise performance of National's PLL for the DECT transceiver.

cells, either Nickel Cadmium (NiCD) or Nickel Metal Hydride (NiMH). To avoid costly voltage conversion circuits, all parts of the cordless telephones need to operate at 3 volts. The DECT chipset from National Semiconductor will operate down to 2.85 V. Several manufacturers offer 3 volt wideband VCOs with a control voltage range of 0.5 to 2.5 volts, as well as RF power amplifiers in GaAs technology, operating from a 3 volt supply. GaAs, however, requires a negative bias supply. Power amplifiers can also be built using discrete bipolar transistors. These run from a typical supply voltage of 3.6 volts, with a lower power efficiency than GaAs modules. Both solutions can be used and the choice between them depends on cost, power consumption and size requirements.

Phase Locked Loop Issues

Lock time and low spurious noise — A single conversion radio architecture requires fast synthesizer switching speed in order to transmit and receive on as many as 24 timeslots per frame. In addition, in a single conversion transceiver, the synthesizer needs to make a large jump in frequency between transmitting and receiving, typically in the order of 110 MHz.

Lock time is defined as the time it takes a PLL to switch from one frequency to another within a given tolerance. Spurious noise, also called reference spurs, is defined as the level of the reference sidetones in relation to the desired tone. Together, these two parameters form a major tradeoff in radio design. Faster lock time almost certainly means higher spurs and vice versa.

Besides designing a proper loop filter, another key to obtaining simultaneous fast switching and low spurs is a well balanced charge pump circuit on the phase detector output. This charge pump circuit must have excellent balance to reduce or eliminate reference frequency spurs, and its dead band region must be at a minimum, even at high reference frequencies, to ensure good noise performance and fast switching speed. The LMX2320 was designed to meet these requirements. For a DECT transceiver, the PLL must also have a wide tuning bandwidth at a high reference frequency. This requires a 64/65 prescaler to achieve legal divide ratios. The prescaler and PLL must be kept as low current as possible to help preserve battery life. An example of a typical DECT PLL performance is shown in Figure 4. The figure shows settling time and reference spurs for a simple passive loop filter.

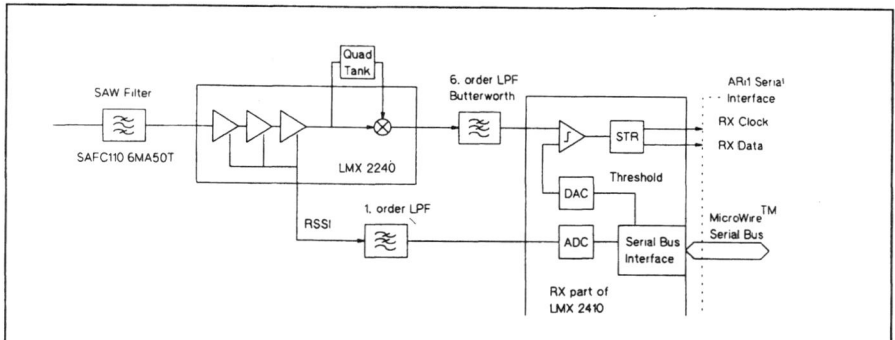

Figure 5 - Simplified block schematic for a DECT transceiver.

Open loop modulation — Another key element of this architecture is open loop modulation. In this mode, the PLL loop is opened while modulating the VCO, allowing the VCO to run free. The loop filter voltage is preserved by putting the PLL's charge pump circuit into a high impedance state. An active loop filter can be chosen to counteract possible loop discharge due to leakage current from the VCO tuning varactor diode. Alternatively, the PLL chip can be powered down, which also brings the charge pump circuit into a high impedance state. In this case, a buffer may be needed to limit an unintended jump in frequency due to possible load pulling of the VCO.

The modulating signal is added to the loop voltage at the input of the VCO, and the result is a modulated carrier. The modulating signal must have a stable midband voltage before opening the loop, in order to avoid a frequency offset during modulation. Such a frequency offset will occur if the loop is stabilized when the modulation signal is either at the negative or positive peak voltage. In these cases, as the loop is opened and the modulation commences, the resulting center frequency will be a sum of the intended center frequency set by the PLL, and an unintended frequency offset equal to plus or minus half of the peak to peak frequency deviation of the modulated signal.

The LMX2410 Baseband Processor is designed to provide an accurate midband voltage. When the loop is opened and the modulation starts, the deviation correctly swings above and below the intended center frequency. The drift in the carrier is kept to a minimum because the burst in DECT is very short. The DECT standard specifies a frequency drift of less than 13 kHz/ms. A DECT PLL based on the LMX2320 can be designed to have orders of magnitude less frequency drift than allowed by the DECT standard.

Receiver Parameters

Most of the requirements for the receiver can be derived from type approval specifications. A noise figure, gain and linearity budget can be made for the receiver using data from available components.

Receiver sensitivity — Receiver sensitivity is normally defined as the signal level needed to produce a certain signal to noise ratio, or, for digital systems, the level needed to produce a certain Bit Error Rate (BER) at the output of the receiver. In the case of DECT, the BER at the sensitivity level is 10^{-3}, at nominal temperature and supply voltage and for frequency offsets of −50 kHz, 0 kHz, +50 kHz of the received signal. The required sensitivity for all DECT equipment is −83 dBm. For equipment meant for public access use, like the U.K.'s Telepoint, the requirement is −86 dBm.

In a receiver for digital modulation, the sensitivity is composed of two figures of merit: the receiver front-end sensitivity, normally expressed by the noise figure, and the demodulator performance. The latter depends on the demodulator chosen. Circuit simulations can be used to get an initial idea of the demodulator performance. Figure 5 shows a simplified block schematic for a possible DECT receiver. The IF channel selection filter is a Murata 110.592 MHz SAW filter SAFC110.6MA50T; the limiter-discriminator circuit is a National Semiconductor LMX2240; the lowpass filter is a discrete filter; finally, the baseband processing function (i.e. bit slicing, symbol timing recovery and bit slice threshold setting circuit) is contained in a National Semiconductor LMX2410.

The BER performance for this specific

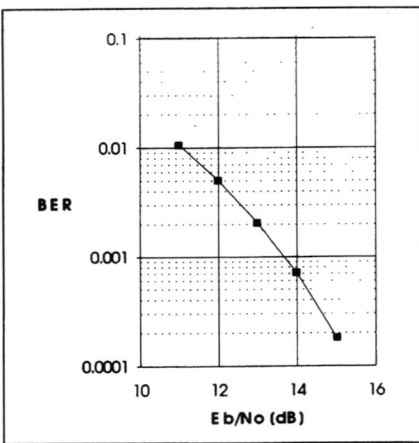

Figure 6 - Simulated BER for the DECT receiver.

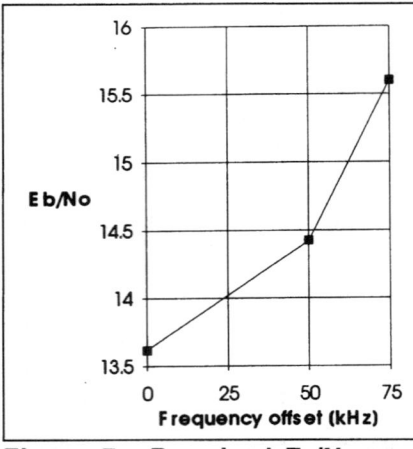

Figure 7 - Required E_b/N_o as a function of the frequency offset on the received signal.

circuit is shown in Figure 6. The implication of frequency offset is roughly indicated in Figure 7. As shown by these figures, an E_b/N_o ratio of approximately 13.6 dB is needed for a BER of 10^{-3} and needs to be increased by approximately 1 dB to compensate for a frequency offset of 50 kHz. The receiver (demodulator) could compensate for the frequency offset/error using a DAC to set the threshold level for the bit slicer (data comparator), assuming that the discriminator is sufficiently wideband. The latter will be the case in a DECT receiver, since offsets are on the order of 50 kHz, which is only a fraction of the required discriminator bandwidth.

The following equations assume ideal compensation. Not compensating for frequency offsets increases the requirements in receiver sensitivity by approximately 1 dB. The required noise figure for the receiver front-end can be derived by knowing the demodulator's E_b/N_o performance and assuming conducted measurements, as in the case of DECT. The E_b/N_o ratio equals the signal to noise ratio on the output of a filter with a noise bandwidth equal to the bit rate, r_b:

$$\frac{E_b}{N_o} = \frac{S}{r_b N_o} \quad (1)$$

Since

$$N_o = kT_oF_n \quad (2)$$

where $k = 1.38054 \times 10^{-23}$ K/J, the Bolzmann constant; $T_o = 290$K, the standard temperature; and F_n is the noise factor. Expressing the ratio E_b/N_o in decibels (dB), we get:

$$\left(\frac{E_b}{N_o}\right) dB = S - F_{dB} - 10\log_{10}(r_b kT_o) \quad (3)$$

$$F_{dB} = S_{dBm} - \left(\frac{E_b}{N_o}\right) dB + 113.4 \quad (4)$$

where S_{dBm} is the signal in dBm

Inserting the specified sensitivity for all DECT equipment, $S_{dBm} = -83$ dBm, and the needed $E_b/N_o \geq 13.6$ dB, yields

$$\begin{aligned} F_{dB} &= -83 - 13.6 + 113.4 \text{ [dB]} \\ &= 16.8 \text{ [dB]} \end{aligned} \quad (5)$$

The requirement is therefore a maximum noise figure of 16.8 dB for all DECT equipment, and similarly NF = 13.8 dB for equipment meant for Public Access.

Receiver linearity — Non-linearities in a receiver, such as gain compression, will cause signals from several transmitters to mix with each other in the receiver. The result is intermodulation products. In the worst case, these intermodulation products can end up having the same frequency as the desired RF signal. The intermodulation products can either be products from the desired signal or unwanted signals. Thus, the receiver intermodulation performance is a measure of the receiver's ability to avoid being disturbed by other DECT-like signals on other DECT channels. An ideal receiver is completely linear and would not suffer from intermodulation.

Receiver design is a trade-off between sensitivity, linearity, power consumption and component cost. Receiver linearity is normally expressed as the receiver's third-order Input Intercept Point, (IIP_3). Requirements for IIP_3 can be derived from the type approval measurement methods and specifications.

The receiver intermodulation performance is measured using three signals: a desired DECT signal and two undesired signals. The three signals are on three different DECT RF channels. The frequencies of the undesired signals are chosen such that one of their third-order intermodulation products becomes equal in frequency to the desired signal. Of the undesired signals, one is modulated with a DECT-like signal and the other is not. The intermodulation product of interest is the one that carries DECT modulation with nominal frequency deviation. The frequencies of the two undesired signals are chosen such that the intermodulation product of interest appears at the receive frequency. Hence the intermodulation product will appear as an undesired DECT interferer with the nominal frequency deviation and on the same channel as the desired signal.

The level of the desired signal is specified to be −73 dBm and the level for each of the two undesired signals is −46 dBm. The third order intercept point for a non-linear 2-port input (IIP_3) is defined as the point at which the third order intermodulation product equals the ideal linear, uncompressed output.

The intercept point can be found by :

$$P_{IIP3} = \frac{3P_{uw} - P_{im3}}{2} \quad (6)$$

with:

P_{IIP3} is the third order input intercept point (in dBm).
P_{im3} is the related intermodulation product for the two unwanted signals (in dBm).
P_{uw} is the power of each of the unwanted signals (in dBm).

P_{uw} is known from the DECT type approval documents. P_{im3} can be derived, remembering that the intermodulation product appears as a co-channel interfering signal, and DECT documents specify the co-channel interference rejection performance.

Co-channel interference rejection is defined as the ability of the receiver to cope with DECT-like interfering signals appearing at the same RF frequency as the desired signal. The co-channel rejection ratio is the ratio of the desired signal to the undesired that produces a certain BER. For a desired signal of −73 dBm, the co-channel rejection ratio is specified to be 10 dB maximum. This 10 dB ratio can be directly used, as the intermodulation performance is measured at the same level of the desired signal. Thus $P_{im3} \geq -83$ dBm. Inserting for $P_{uw} = -46$ dBm and $P_{im3} = -83$ dBm, yields a receiver input intercept point of $P_{IIP3}=-27.5$ dBm.

Table 2 shows an example of a typical RF chain budget for a DECT receiver. The design issues are to have as low a noise figure as possible, a high input intercept point and a high overall receiver gain. The latter is in order to have the limiter saturated all the way down to the sensitivity level. From the table it can be concluded that (in this particular example) the mixer sets the limit for the overall noise figure, and that the LMX2240 in combination with the IF filter sets the limit for the overall input intercept point. Both figures of merit for the receiver are met by a reasonable margin, i.e. receiver sensitivity is met by a margin of 4.2 dB and input intercept point by a margin of 4.1 dB.

# Component	Gain	NF	OIP3	#	Gain	NF	IIP3	OIP3
1 Filter	-1.0	1.0	100.0	1	-1	1.0	97.5	96.5
2 Circulator /Switch	-1.0	1.0	40.0	2	-2	2.0	42.0	40.0
3 LNA/ LMX2216B	10.0	4.7	7.0	3	8	6.7	-1.0	7.0
4 Filter	-2.0	2.0	100.0	4	6	6.8	-1.0	5.0
5 Mixer/ LMX2216B	6.0	17.0	0.0	5	12	12.3	-12.3	-0.3
6 IF SAW-Filter	-4.0	4.0	100.0	6	8	12.4	-12.3	-4.3
7 LMX2240	70.0	8.0	55.0	7	78	12.6	-23.4	54.6

System Cumulative Values

Required E_b/N_o	13.60 [dB]		Gain	78.0 dB
Sensitivity	-87.20 [dBm]		N Fig	12.6 dB
IIP3	-23.4 dBm		OIP_3	54.6 dBm

Table 2 - A typical RF chain budget for a DECT receiver.

Baseband Issues

Also facing the radio designer are baseband issues such as bit or symbol timing recovery (STR), compensation for DC drift due to carrier frequency drift through a discriminator, and the baseband filtering requirements for the transmitter. National's LMX2410 addresses all of these concerns. Symbol timing recovery is achieved through the use of an all-digital phase locked loop and a non-linear element at the input. Timing is recovered by over-sampling the input data stream (after a threshold comparator) and finding the rising or falling edge. The symbol clock phase is then adjusted to lock it to the incoming data stream's phase. The STR circuit in National's DECT solution can lock the internal clock to the received data in as few as five edges of the incoming data stream.

DC compensation of the incoming data stream is important when using a discriminator because carrier frequency drift will manifest itself as DC drift after an FM discriminator. The DECT preamble is 32 bits in length with a balance of 16 1s and 16 0s. A good way to track the DC level of the incoming signal is to monitor the duty cycle of the preamble and adjust the LMX2410's threshold DAC for the data comparator.

Baseband pre-modulation filtering is important in achieving the proper shaping of the transmit spectrum. In DECT, the pre-modulation filtering is specified as a Gaussian filter characteristic with a -3 dB bandwidth of half the bit rate ($B_bT = 0.5$). National achieves this pre-modulation filtering through the use of a ROM look-up table and an 8 bit DAC on the LMX2410. This allows an ideal Gaussian characteristic to be saved in the ROM. National's chip is also mask-programmable, which lends itself to easy adaptation for other standards.

System Issues

Three issues which face DECT phone designers are range, multipath fading (or delay spread), and voice quality. With the high bit rate, a DECT phone will have more susceptibility to multipath reflections. A normal delay spread in indoor environments is on the order of 100-200 ns. The bit time of a DECT bit is 880 ns. This means that the potential delay spread due to multipath reflections is 10-20 percent of a bit time. Typically, some method of diversity is used to combat such relatively small fractional delay spreads.

Switched antenna diversity is the simplest to implement. It involves using two antennas (usually in the base station) and measuring the received signal strengths of both antennas. The larger signal is then received for the duration of the burst. Full receiver diversity involves two complete receivers from antenna to the cyclic redundancy checker (CRC) in the baseband section. Whichever received signal passes the CRC is the one that is used for the duration of the burst.

Another, even simpler method is dual slot diversity. In this method, only one receiver is used, but two (usually successive) timeslots are used for receiving one data burst by each of the antennas. Identical information is thus transmitted in both timeslots. In the receiver, the timeslot that was received with the fewest errors is used. This method adds the performance enhancement of the dual receiver diversity principle to a single receiver architecture. However, it decreases system capacity due to the extra timeslots being used.

The range and voice quality of a DECT phone must be minimally wireline quality. The range of the phone must be long enough so that the phone is not always in hand-off mode, but short enough to allow for multiple cell (base station) sites. In the indoor office environment, a good range is about 100 meters and about 200-300 meters outdoors.

Summary

A new chip set for DECT applications is now available from National Semiconductor. Using a single-conversion receiver and direct-modulation transmitter, these components offer a low cost, low power consumption solution for DECT equipment. *For more information, circle Info/Card #260.* **RF**

About the Authors

Benny Madsen has a Ph.D in Electrical Engineering from Aalborg University in Denmark. Recently, he has worked at Dancall Radio, Denmark, developing analog and digital cordless and cellular phones. Since January 1993, he has worked at National Semiconductor Corporation as an engineering project manager in the Wireless Communications Group.

Daniel E. Fague received his M.S.E.E. from the University of California at Davis. Since 1991, he has worked at National Semiconductor Corporation in the Wireless Communications Group, specifying integrated circuits for radio modems. His interests include digital radio modems, indoor and outdoor propagation for mobiles, and digital signal processing.

They can be reached at National Semiconductor, M/S A1500, P.O. Box 58090, Santa Clara, CA 95052-8090.

Techniques for Open Loop Modulation of a Wideband VCO for DECT

Daniel E. Fague, Andy Dao, and Clayton R. Karmel
National Semiconductor Corporation
P.O. Box 58090   M/S A-1500
2900 Semiconductor Drive
Santa Clara, CA   95052-8090

Abstract

A detailed examination of open loop modulation is done. Specifically, the basic concept of open loop modulation is discussed, as well as its application to modern digital cordless telephony standards (e.g. Digital European Cordless Telecommunications). Common limitations such as VCO leakage, load pulling, and frequency pushing effects are analyzed. Practical techniques to successfully combat these imperfections are presented, along with robust implementations. Measured results from the robust hardware implementation of the techniques are presented, including documentation of the frequency jumps due to various loads and temperatures. Modulated spectra and demodulated eye diagrams are also presented to show modulation accuracy. It is shown that open loop modulation can be used as a robust modulation method with simple and low cost implementations for constant envelope systems.

Introduction

Since digital communications systems are emerging as the solution for modern telecommunications, the need to find low cost implementations of high performance circuits is ever greater. In the digital cordless telephony area, the Digital European Cordless Telecommunications (DECT) standard requires high bit rates (1.152 Mb/s), high radio frequency (2 GHz) transmissions, and fast switching between frequencies (~30 μs). These stringent requirements must be met while keeping the overall cost of the radio low enough so as to compete in the consumer residential cordless market as well as the business wireless PBX/PABX markets.

Open loop modulation is an exciting technique to meet such widely diverse requirements. In the first section, this paper will present the concept of open loop modulation and discuss some common limitations that can occur. Solutions to these imperfections are presented in the second section, along with the measured performance of open loop modulation and some analysis of the measurements. Finally, conclusions will be drawn in the final section.

Open Loop Modulation

To achieve the combination of low cost and high performance for DECT, unique circuits must be implemented. DECT is a Time Division Multiple Access/Time Division Duplex (TDMA/TDD) system. There are 24 time slots in 10 ms frames, which means each time slot is 420 μs. The modulation method for DECT is Gaussian filtered Minimum Shift Keying (GMSK) with a pre-modulation lowpass filter bandwidth of half the bit rate ($B_b T = 0.5$). Because this is a frequency modulation (FM) technique, a Voltage Controlled Oscillator (VCO) can be directly modulated by the baseband signal. There are several techniques that can be used to modulate a VCO. The most common are modulation "in the loop" and modulation "over the loop". Modulation in the loop can be used when the output signal is narrowband with respect to the loop filter, and modulation over the loop can be used when the time required to switch frequencies is relatively long.

In DECT, however, neither of these conditions is valid. The lock times of the phase locked loop (PLL) must be very fast (~30 μs) and the output signal is much wider than the loop filter bandwidth. In certain radio architectures such as single conversion receivers, there is a need for wideband VCOs to cover the IF bandwidth plus the system bandwidth. In DECT, this amounts to a 130 MHz bandwidth at 1.88-1.90 GHz. Open loop modulation is an exciting technique that allows for fast switching times and wide output signal bandwidths while delivering high performance.

In open loop modulation, the PLL is actually unlocked ("opened") for a brief time to allow the modulation to occur. Figure 1 shows a block diagram of open loop modulation. When transmission is to occur, the loop is first closed to lock the VCO to the correct carrier frequency. The modulating signal is then turned on, and the loop remains closed to re-lock to the center frequency. The modulating voltage is added to the loop filter voltage (center frequency) at either a modulation port or the tuning port. The loop is then opened, and modulation occurs. Once the modulation is finished, the loop is closed again, and the PLL can be tuned to the receive frequency.

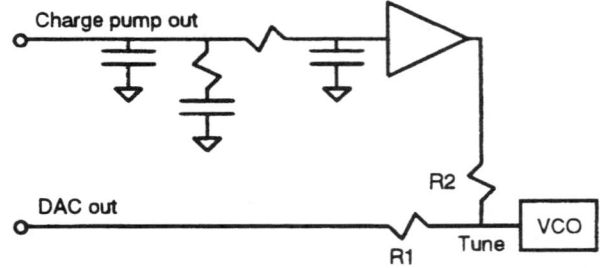

Figure 1 - Block diagram of open loop modulation

Reprinted with permission from *RF Expo West*, D. E. Fague, A. Dao, and C. R. Karmel, "Techniques for Open Loop Modulation of a Wideband VCO for DECT," pp. 274-277, 1994. © Intertec Publishing Corporation.

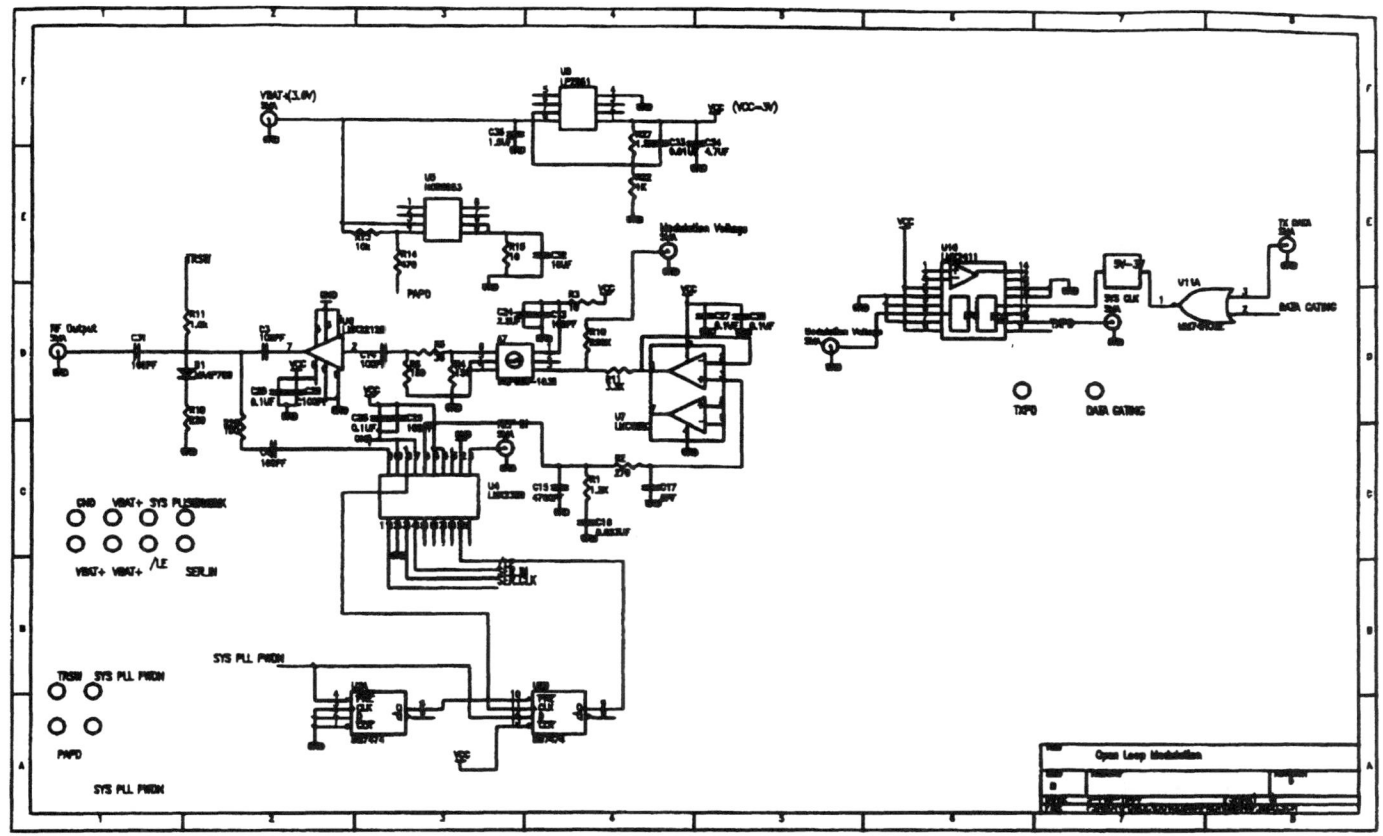

Figure 2 - Circuit diagram of the printed circuit board used to demonstrate open loop modulation performance.

There are some elements of open loop modulation that can limit performance. Some of the main ones are frequency pushing, load pulling, and frequency drift. Frequency pushing is normally described as a change in VCO output frequency caused by a change in supply voltage. Load pulling is a change in VCO output frequency that is caused by a change in the load the VCO output buffer sees. Frequency drift can be caused by RF coupling or by droop in the VCO tuning voltage. Droop in the VCO tuning voltage can be caused by leakage from the PLL charge pump, the loop filter components, or the VCO's tuning varactor.

Experimental Setup

Figure 2 shows the circuit diagram used for the open loop modulation measurements. This circuit utilizes a 2 GHz PLL with TRI-STATE™ on the charge pump output, a 130 MHz wideband VCO, a 3.0 V regulator, and a Gaussian filter ROM-DAC to shape the transmit data. The data rate used was 1.152 Mb/s (DECT bit rate). The carrier frequencies that were used to demonstrate open loop modulation performance were channels 0, 5, and 9. The circuit was implemented on a printed circuit board, and it simulated a practical DECT implementation of open loop modulation. The power amplifier load was simulated by a 10 ohm resistor and a P-channel switch (for current draw), and a PIN diode and 220 ohm resistor to ground (for load impedance change). Measurements were made by using a DECT system tester to demodulate the output of the circuit (from the PIN switch), and the frequency offsets were determined from the demodulated signal on an oscilloscope.

The circuit featured a CMOS rail-to-rail operational amplifier that was used to control leakage from the VCO's tuning varactor. The low output impedance of the op amp also allowed the summing node to be a simple resistive adder. An RF buffer with good reverse isolation was used at the output of the VCO to effectively control the load pulling caused by the power amplifier and T/R switch. A low dropout voltage regulator was used to supply the 3.0 V Vcc to the components, and this regulation of the input battery voltage plus power supply filtering limited noise and frequency pushing.

Experimental Results

Figure 3 shows the demodulated eye diagram of the open loop modulated DECT signal. The virtually perfect eye diagram of Gaussian $B_bT = 0.5$ filtered data can be seen. Figure 4 shows a time domain view of the demodulated signal after the loop was opened. A small 15 kHz jump that settles out can be seen in this time domain plot. Figure 5 shows the open loop modulated DECT signal operating in burst mode (i.e., TDMA/TDD operation). The measurement was taken with a time gated spectrum analyzer to capture a number of bursts. Note that the output spectrum of the signal is a DECT-compliant GMSK signal.

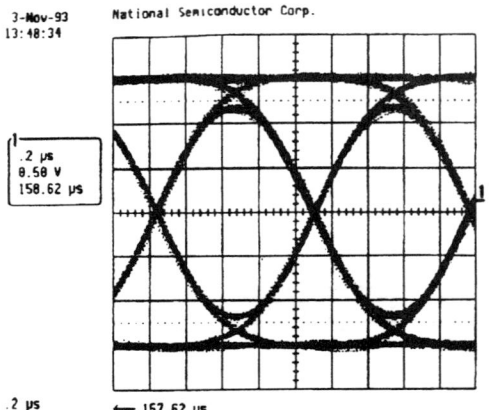

Figure 3 - Plot of the eye diagram of the demodulated output of the open loop VCO while being modulated by Gaussian filtered data.

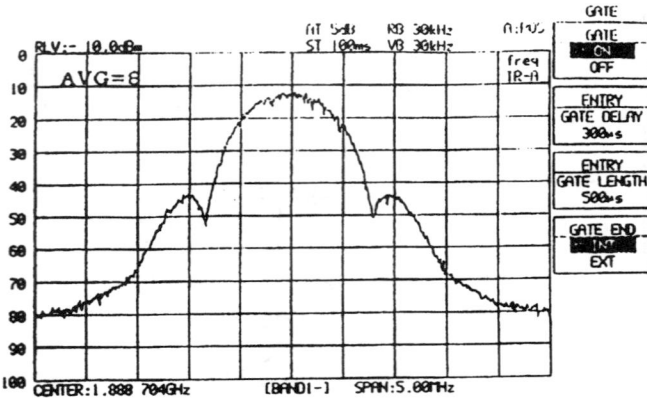

Figure 5 - Plot of the open loop modulated DECT signal in burst mode as taken by a time gated spectrum analyzer.

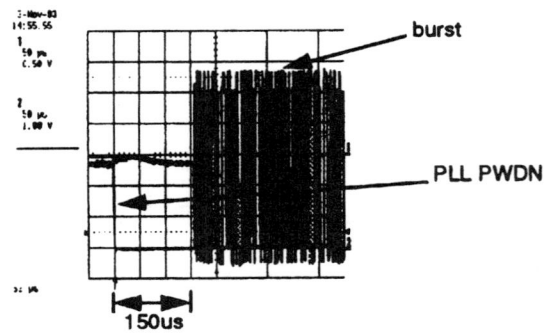

Figure 4 - Plot of the demodulated output of the open loop VCO while modulated by Gaussian filtered data.

The last three plots show the small imperfections that can occur in open loop modulation. Figure 6 shows open loop frequency pushing, pulling, and drift at 25° C on Channel 0 in the DECT band (~1.9 GHz). The lower trace is the power down signal sent to the PLL to TRI-STATE the charge pump. The upper trace shows that there is imperceptible frequency drift over a 9 ms burst. The middle trace shows a 15 kHz pushing jump while the voltage regulator recovers from the PLL powering down. This lasts about 100 µsec. The power amplifier turning on causes a 15 kHz pushing spike, and 1.25:1 load pulling causes a < 10 kHz offset. Similar curves are seen at Channel 9, the lower end of the DECT band (~1.88 GHz).

Figure 7 shows the performance of open loop modulation at -20° C. It can be seen that the only difference in performance is that the power amplifier pushing creates a slightly larger spike at the beginning of the burst, but the rest of the performance is nearly identical to that of 25° C. Figure 8 shows the performance at 60° C. In this case, the power amplifier pushes a slightly smaller spike. However, a small frequency drift is now noticed, and this is due to P-

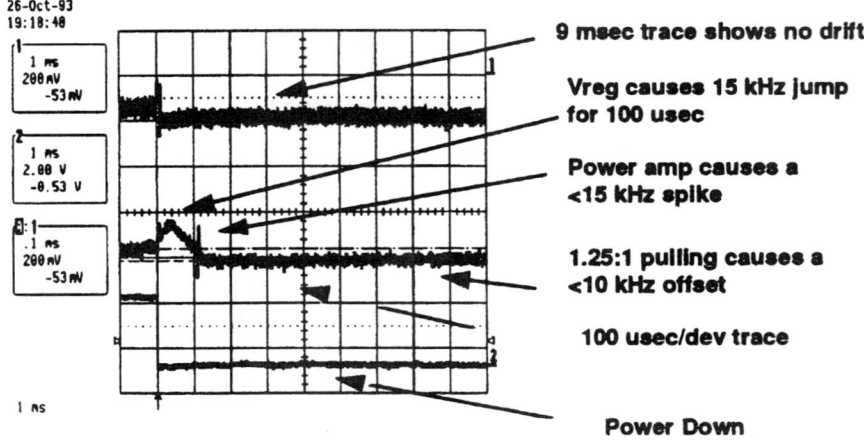

Figure 6 - Measurement of open loop frequency pushing, load pulling, and frequency drift for DECT channel 0 at 25° C.

channel leakage from the charge pump to the VCO.

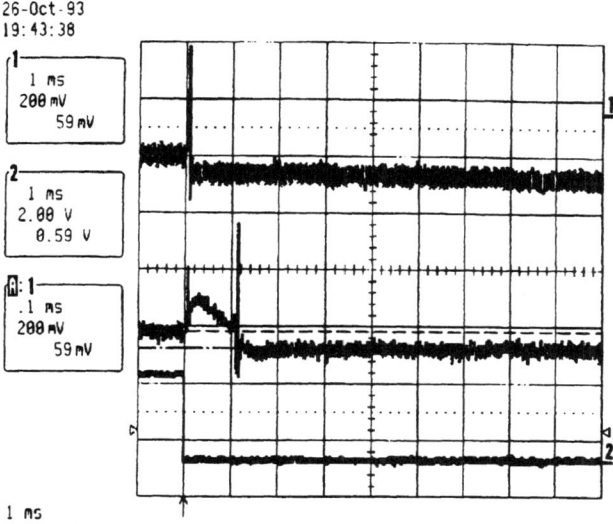

Figure 7 - Measurement of open loop frequency pushing, load pulling, and frequency drift on DECT Channel 0 at -20° C. Note that the only difference from Figure 6 is a slightly larger spike caused by the power amplifier pushing.

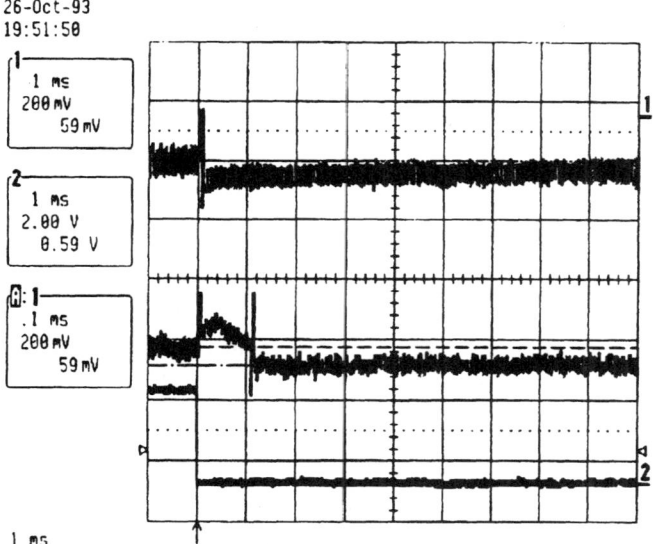

Figure 8 - Measurement of open loop frequency pushing, load pulling, and frequency drift for DECT channel 0 at 60° C. A very small drift due to leakage is seen.

Conclusions

The principle of open loop modulation for wideband VCOs has been discussed and measured experimentally. The measured results show that the methods used to reduce or eliminate frequency pushing, load pulling, and frequency drift are effective. These methods are also robust, as they can be taken over a wide range of temperature and the circuit shows virtually no degradation in performance. It can therefore be seen that open loop modulation of a wideband VCO can be used for low cost applications that still require high performance.

Acknowledgments

. The authors would like to thank Dr. Benny Madsen and Bill Keese for their help and guidance in this work.

References

[1] ETSI, Final Draft - prETS 300 175-2, Specification for DECT, Part 2: Physical Layer, May, 1992, Cedex, France.
[2] Madsen, Benny, and Fague, Daniel E., "Radios for the Future: Designing for DECT", *RF Design*, April, 1993, pp. 48-53.
[3] Fague, Daniel E., "A Fully Integrated Modulator/Demodulator for DECT", *Proceedings of the RF EXPO WEST*, March, 1993, San Jose, CA, pp. 228 - 231.
[4] K. Feher, *Advanced Digital Communications*, Prentice Hall, Inc., Englewood Cliffs, NJ, 1987.
[5] LMX2411 Baseband Processor for Radio Communications Data Sheet, National Semiconductor Corp., Lit. # 108550-001, September, 1993.
[6] LMX2320 2.0 GHz Frequency Synthesizer for RF Personal Communications Data Sheet, National Semiconductor Corp., Lit. # 108535-001, January, 1993.

Performance Evaluation of a Single Chip Radio Transceiver

Kendal McNaught-Davis Hess, *Member IEEE* and Daniel E. Fague, *Member IEEE*

National Semiconductor Corporation
2900 Semiconductor Drive   M/S D3-500
Santa Clara, CA   95052-8090

Abstract

A radio transceiver encompassing a single conversion receiver and a direct conversion transmitter has been fabricated on a single integrated circuit (IC). The architecture of the chip is optimized for use in the Digital European Cordless Telecommunications (DECT) standard. A description of the block diagram of the chip is given. In addition, the performance of the transceiver in terms of noise figure, overall sensitivity, and bit error rate (BER) is presented. A complete radio design using the single chip transceiver is done, and demonstrates that the high level of integration reduces component count and board size. The performance of the chip and the radio design is compared to the earlier generation to emphasize the improvements.

The performance of the earlier generation radio has been documented in [1]. In the previous generation radio, separate IC's were fabricated for the various radio functions. The new IC combines these four IC's onto a single chip. In addition, several functions that were discrete designs on the radio board are also integrated onto the single chip transceiver. The resulting IC provides the opportunity for highly compact radio designs which contribute to smaller, lighter personal communications equipment.

I. Introduction

Modern digital communications systems are developing at a rapid pace. New generations of mobile telecommunications equipment have increasing levels of complexity, enabling larger numbers of features and options to be offered to the end user. These new solutions also are becoming smaller in size, lighter in weight, and lower in price. One of the main enabling factors of this continuing development is the integrated circuit. While the first relied heavily on discrete components, newer generations now utilize high levels of integration in both the baseband and RF circuits. The DECT system is a good candidate for highly integrated RF front ends due to its somewhat relaxed modulation scheme (GMSK, BT=0.5), wide channel spacing (1.728 MHz), access method (time division multiple access), and duplex method (time division duplex). A single conversion receiver architecture is well proven for DECT [1], providing a solution that eliminates filters, amplifiers, and oscillators. Other architectures are possible [2,3], but the low cost and power of a single conversion receivers makes it an attractive choice.

This paper concentrates on the integration of a single conversion receiver and a direct conversion transmitter onto a single integrated circuit. The second section of this paper will describe the architecture of the single chip transceiver. Details of the various transceiver blocks will be given. Section three will present the performance of the single chip's various blocks. The fourth section will present the system performance of the single chip receiver as used in the DECT system. The performance will be compared to the earlier generation. Finally, some conclusions will be drawn.

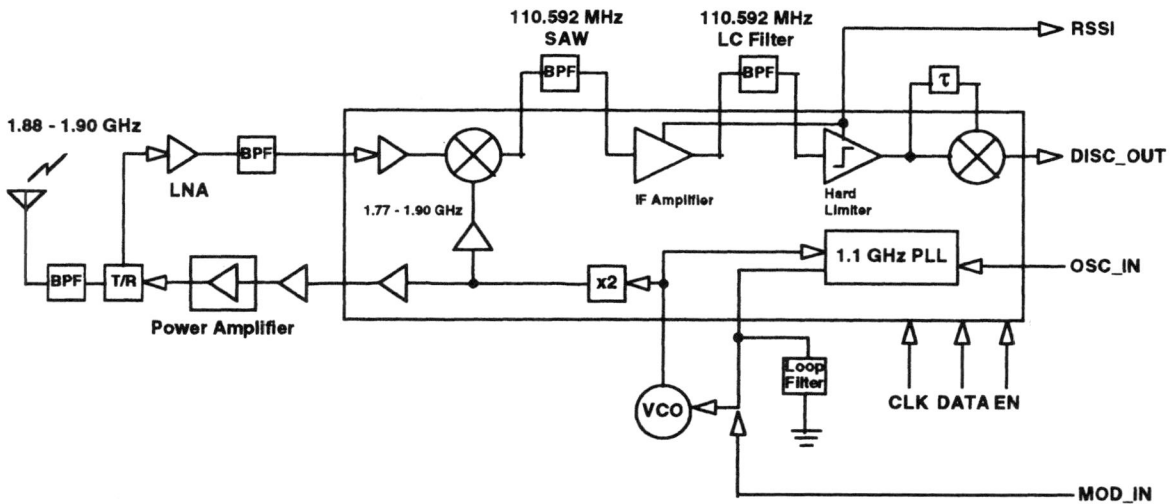

Figure 1 - Block diagram of the single chip transceiver

II. IC Architecture

The block diagram of the single chip receiver is shown in Figure 1. The chip is fabricated in a 0.5 μm BiCMOS process, allowing the monolithic integration of both analog functions (amplifiers, mixers, regulators) and digital functions (frequency synthesizer, control logic). The resulting device is very power efficient in transmit, receive and standby modes. During active mode, the chip uses 20 mA in transmit and 45 mA in receive mode. The synthesizer uses 6 mA, with an additional 3 mA coming from the transmit buffer. In power down mode, the IC draws less than 100 μA, enabling extremely long battery life when combined with an equally efficient baseband device. A summary of the current consumption is given in Table 1. For optimum drive of some GaAs power amplifiers, an external gain stage is necessary to increase the power output of the single chip transceiver. This will also allow for some matching for maximum rejection of the frequency doubler spurious products.

IC Power Status	Current Draw
Receive Mode (open loop)	45 mA
Transmit Mode (open loop)	20 mA
Synthesizer lock	9 mA
Power Down (sleep)	100 μA

Table 1 - Current consumption at 3.0 V supply voltage.

It can be seen from the transceiver architecture in Figure 1 that the chip contains the necessary functions which, combined with the appropriate off-chip active and passive components, forms a complete radio front end. For DECT, a 450mW power amplifier (250mW at the antenna after filter and switch losses) and low noise amplifier are used along with a transmit buffer and discrete VCO to provide a very high performance, low cost transceiver. The single conversion receiver limits expensive passive filters to one IF surface acoustic wave (SAW) filter and two low cost front end ceramics.

A. *The Transmitter*

The transmitter consists of a 1.1 GHz phase-locked loop, a VCO input buffer, a frequency doubler, and a transmit output buffer. The transmitter utilizes direct, open loop modulation of the 1 GHz VCO to generate a distortion-free modulated signal. The PLL is locked to half the desired DECT transmit frequency in the time slot prior to the desired transmit time slot. It is then powered down (placing the PLL charge pump in a high impedance mode) to effectively open the loop during the desired transmit slot. Low leakage from the charge pump, varactor, and loop components ensures very low drift of the open loop, free-running VCO (< 1 kHz/ms typical). This is more than adequate to meet the DECT frequency drift specification of 13 kHz/ms. Good shielding and source bypassing ensure frequency stability upon opening the loop. Performance of the open loop modulator has been documented [4].

The frequency doubler is implemented on chip as a full wave rectifier. The doubler frequency output is then amplified and filtered on chip. Additional amplification and filtering is achieved with the use of an off-chip tuned amplifier. The use of the tuned amplifier allows usage of many available DECT power amplifiers with various input drive level requirements.

B. *The Receiver*

The single chip receiver is an improved version of the single conversion receiver. The RF input consists of a low noise downconverter. The downconverter has a high intercept point (OIP3) and gain as well as low noise. For lowest noise, the downconverter can be preceded by an off-chip discrete gain stage. A comparison of the noise figures with and without the external LNA is given in Table 2. The downconverting mixer is a single-balanced RF input, and has single-ended inputs and outputs.

The downconverter output feeds into an external SAW filter centered at 110.592 MHz (112.32 MHz is also an option). The gain of the radio allows for SAWs with high or low insertion loss to be used in the receiver. The output of the SAW filter is input to the IF limiter-discriminator. The limiter is split into an IF pre-amplifier and a limiter. The output of the pre-amplifier goes off chip for additional filtering of adjacent channels. The limiter is connected internally to the discriminator mixer, and also drives off-chip to the phase-shifting LC tank circuit. The limiter has a cascaded gain of 80 dB and a limiting sensitivity of -72 dB. Table 2 summarizes the performance of the limiter.

Parameter	Measured value	Previous generation
Downconverter Noise Figure	6 dB	9.7
Downconverter + Ext. LNA NF	3 dB	4.3 dB
Downconverter Conversion Gain	16 dB	16 dB
IF Limiter Gain	85 dB	70 dB
IF Limiting Sensitivity	-72 dBm	-72 dBm

Table 2 - Comparison of single chip measured performance vs. previous generation

The output of the discriminator goes off chip for optional filtering and DC gain, then comes back on chip for DC recovery. The DC recovery circuit's performance has been documented [1]. The DC recovery circuit provides a DC voltage suitable for use as the threshold for the comparators on burst mode controllers.

III. Transceiver Measured Performance

The single chip transceiver's performance is summarized in this section. The transmitter's key performance metric is the doubler's spurious rejection since direct, open loop modulation gives near ideal performance [4]. A frequency doubler is implemented to avoid the effects of radiation on the VCO. It is also helpful in reducing the effects of load pulling on the VCO by increasing the isolation between the VCO and the power amp. The increase in current that is encountered by adding a frequency doubler is offset by the PLL drawing less current since it is running at a lower frequency. The fundamental (900 MHz) component of the doubler must be rejected by 36 dB. The second and third harmonics must be attenuated by 30 dB to meet the required output spectrum for DECT (after RF filtering). The doubler's measured spurious rejection is given in Table 3 with the calculated DECT requirement at the output of the doubler. (This assumes a filter with in-band gain of 5 dB and out-of-band gain of -30 dB is used between the doubler output and the power amp. In addition the power amp must have a gain of 26 dB for a fundamental signal of 0 dBm and harmonic gain of less than 0 dB.) It can be seen in the table below that the transmitter achieves the necessary performance.

Transmitter Output Frequency	Measured suppression	DECT Requirement
945 MHz	-25 dBc	-6 dBc
1890 MHz	-5 dB	-5 dB
2835 MHz	-26 dBc	0 dBc
3780 MHz	-32 dBc	0 dBc

Table 3 - Transmitter output spurious suppression

The receiver has several metrics, some of which have been discussed in the previous section. This section concentrates on the bit error rate (BER) performance of the transceiver. Please note that at the time of this paper's submission, the performance of the receiver had only been measured using an IF SAW filter which had very high insertion loss (20 dB) and without a first stage LNA. With this less than optimal configuration, a received signal strength of -75 dBm is required to achieve a BER of 10^{-3}. The sensitivity will be improved by using a first stage LNA with a gain of 10 dB and a noise figure of 2 dB and an IF SAW filter with less than 10 dB of insertion loss. Using a proper first stage LNA and IF SAW filter the receiver sensitivity for a BER of 10^{-3} is conservatively estimated at -93 dBm, exceeding the DECT (Generic Access Protocol) specification of -86 dBm. Figure 2 shows the expected bit error rate curve of a DECT receiver implemented with the single chip receiver. The sensitivity of the radio with the single chip transceiver is equal to or better than that of the first generation solution. Added sensitivity comes from the improved system noise figure and the increased limiter sensitivity. The single chip transceiver meets the DECT specifications. Most importantly, this integrated solution substantially reduces component count, board size, and overall system cost.

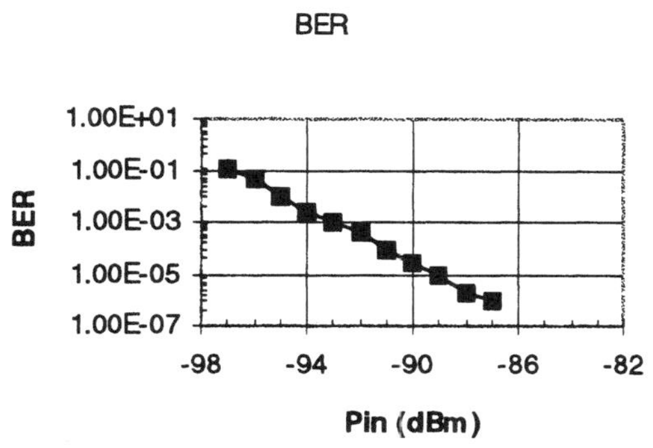

Figure 2 - BER performance of the single chip transceiver

IV. Conclusions

The architecture and performance of a single chip transceiver suitable for DECT has been presented. It has been shown that the single chip transceiver has better performance than the first generation solution. With a sensitivity of -93 dBm for a 10^{-3} BER, the receiver is a high performance solution. The single chip transceiver is therefore an excellent solution for a DECT radio transceiver.

Acknowledgments

The authors would like to thank Bill Burdette and Michael O'Hearn for their work on the device. In addition, the work of Eric Lindgren in designing the chip is acknowledged. Finally, the help of others in the Wireless Communications group is greatly appreciated.

References

[1] D. Fague, A. Dao, C. Karmel, and B. Madsen, "Performance Evaluation of a Low Cost, Solid State Radio Front End for DECT", *Proceedings of the IEEE Vehicular Technology Conference*, Stockholm, Sweden, June 8-10, 1994, pp. 512-515.

[2] B. Wuppermann, S. Atkinson, B. Fox, B. Jansen, and G. Jusuf, "A 2.7 V Two Chip Set Transceiver for DECT", *Proceedings of the Fourth International Symposium on Personal, Indoor, and Mobile Radio Communications*, Yokohama, Japan, September 8-11, 1993, pp. 407-411.

[3] B. Bjerde, J. Lipowski, J. Petranovich, and S. Gilbert, "An Intermediate Frequency Modulator Using Direct Digital Synthesis Techniques for Japanese Personal Handy Phone (PHP) and Digital European Cordless Telecommunications (DECT)", *Proceedings of the 44th IEEE Vehicular Technology Conference*, Stockholm, Sweden, June 8-10, 1994, pp. 467-471.

[4] D.E. Fague, A. Dao, and C. Karmel, "Techniques for Open Loop Modulation of a Wideband VCO for DECT", *Proceedings of the RF Expo 1994*, San Jose, CA (USA), March 22-25, 1994.

[5] ETSI, Final Draft - prETS 300 175-2, Specification for DECT, Part 2: Physical Layer, Cedex, France, May, 1992.

Chip Set Addresses North American Digital Cellular Market

By Michael M. Sera
Philips Semiconductors

Being compatible with divergent standards is not the only challenge cellular handset manufacturers must meet — flexibility, size, and price are also considerations. To help designers meet these challenges, a new chip set targeting dual-mode (IS-54) cellular telephone applications has been introduced.

Cellular phone use has dramatically changed over the past decade, from an eccentric communication device that only the privileged few could afford, to one that is now being given to family members as a security device. The cellular phone has reached the mass market, a dream come true for all cellular manufacturers and service providers. Now the problem becomes maintaining the same level of service. Users will quickly become frustrated if, from downtown San Francisco, they receive a fast busy signal. This condition is unacceptable if the cellular phone is ever to provide the same level of service as the wired system.

Note: Since this issue is only a problem in major metropolitan areas, Digital Cellular is not necessarily required everywhere today.

There are many ways to tackle this problem. Service providers could increase the cost of service during peak times to discourage use. The disadvantage is that cellular would lose the mass market appeal and make other services such as Personal Communications Systems (PCS) or Personal Communication Networks (PCN), Private Mobile Radio (PMR) and Long Range Cordless more attractive. Another alternative would be to increase the number of channels currently allocated. The disadvantage here is that other services already exist in adjacent frequencies. Instead, the cellular companies as a group, through the TIA (Telecommunications Industry Association) or CTIA (Cellular Telecommunications Industry Association), have decided to address the problem by using the existing channels more efficiently.

The current North American analog Advanced Mobile Phone Service (AMPS), specifies the use of Frequency Division Multiple Access (FDMA) with 832 channels separated by 30kHz. The cellular manufacturers have come up with several schemes to share a single channel with multiple users. The first being Narrow band AMPS (NAMPS), which takes an existing 30kHz channel and divides it into three 10kHz channels. The increase is from one user to three users per channel. However, to date, the NAMPS scheme has not been widely accepted by the cellular community.

TDMA (Time Division Multiple Access), standardized as IS-54, uses

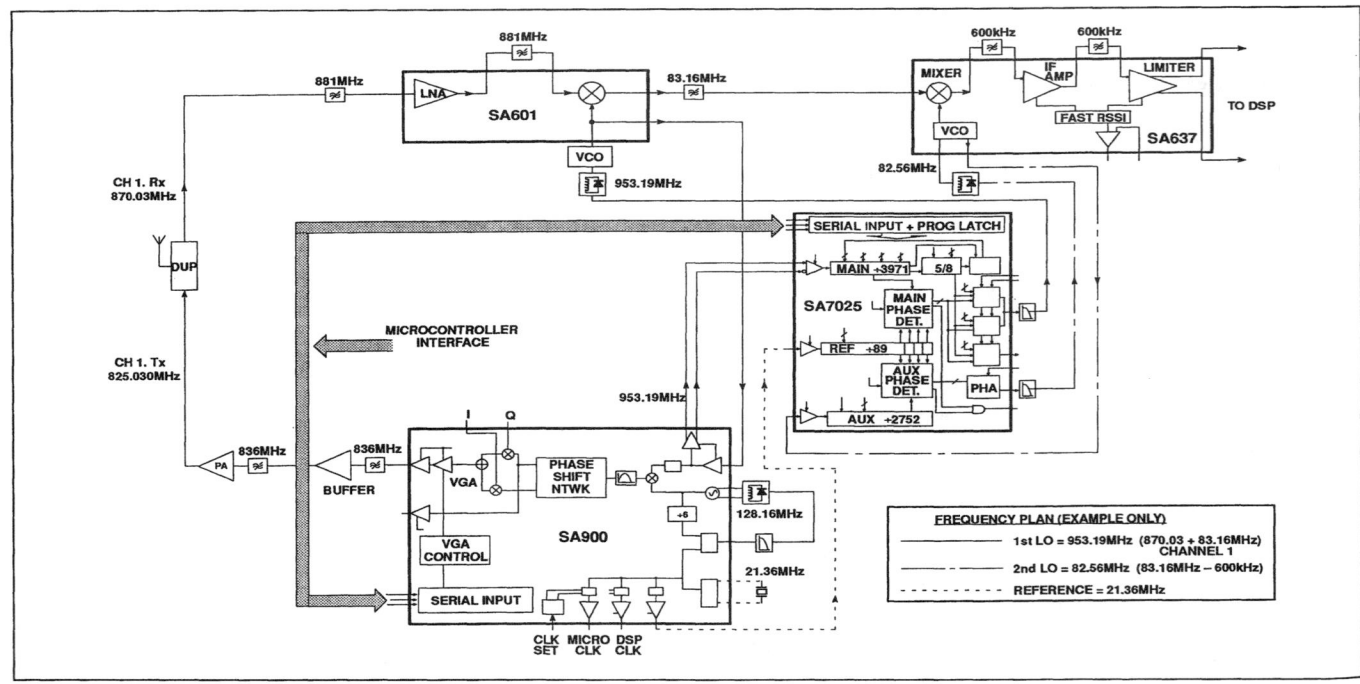

Figure. 1 Block diagram of TDMA IS-54 chip set.

Reprinted with permission from *RF Design*, M. M. Sera, "Chip Set Addresses North American Digital Cellular Market," vol. 17, no. 3, pp. 54-62, March 1994. © Intertec Publishing Corporation.

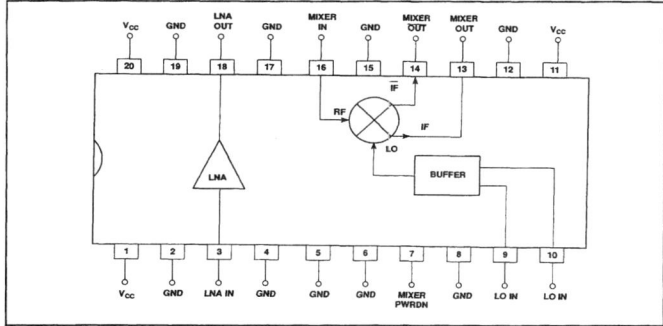

Figure 2. Block diagram of SA601.

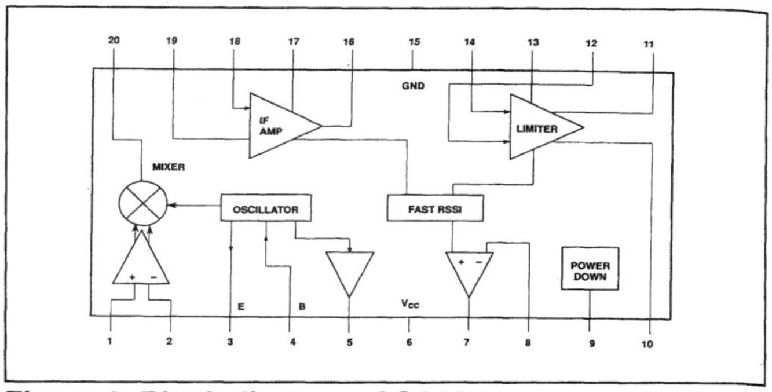

Figure 3. Block diagram of SA637.

the same 30kHz channel spacing but multiplexes users over time. Therefore at different time intervals multiple user are present on the same frequency. The increase is from one user to three users per channel. E-TDMA (Extended Time Division Multiple Access) increases this number even higher with some manufactures reporting improvements of up to ten users per channel.

The TDMA service today is still sparse. The plan is to provide service to the major metropolitan areas first. A user of a TDMA phone shouldn't worry though, the specification for TDMA, IS-54, requires that the handset and system be dual mode (i.e.; compatible with digital and analog standards). Therefore both TDMA and AMPS are supported. This allows a TDMA handset owner to have a seamless cellular network in North America regardless of the location. The major advantage for the TDMA handset in downtown San Francisco, where TDMA is supported, is that the TDMA phone is less likely to experience a fast busy signal.

The service providers in general are also offering reduced airtime costs to TDMA handset users. The digital service provider in the San Francisco Bay Area is offering approximately a 3-cent discount per minute along with a monthly discount of about $10 depending on the plan. Another advantage of the TDMA standard is that the modulation is digital and a conventional scanner will not be able to eavesdrop on calls. If the analysts are correct, TDMA will grow to over 1 million users in 3 years. This will help drive the economy of scale for this new and upcoming standard.

The North American cellular market has another digital standard called CDMA (Code Division Multiple Access) adopted by the TIA as IS-95. This scheme uses a combination of spread spectrum technology plus a coding scheme that San Diego based Qualcomm Inc. has licensed. CDMA also requires that the handsets and system support dual mode operation. Therefore, they support both CDMA and AMPS. This allows a TDMA handset to operate in a CDMA environment through the common mode of AMPS.

Chip Set Simplifies Design

Now, with a clear decision to improve capacity by using a digital scheme, we must also address the cost, performance and size issues. Users of cellular phones have become comfortable with the fact that their phones fit nicely into their shirt pocket or purse. They also enjoy the fact that their batteries are usable for the whole day. Trying to convince them to use a digital phone that is the size of a brick will not be successful. Therefore a low power, highly integrated solution is essential for any digital standard to succeed.

Philips Semiconductors has addressed this by combining the experience of their low power AMPS chip set and customer inputs. The result is a low power, highly integrated chip set for the IS-54 (TDMA) North American Digital Cellular standard (Figure 1). The chip set combines all of the necessary RF and IF functions into four integrated devices: the SA601 RF front end, the SA637 digital IF receiver, the SA7025 dual frequency synthesizer and the SA900 I/Q transmit modulator.

These devices were designed as a system and therefore have interface levels which are matched, this eliminates the need for additional buffers and interface devices. There is also a common high speed serial interface bus, making addressing the devices simpler. Additionally the frequency plan was designed to eliminate the need for an additional synthesizer and VCO loop. All of these features dramatically reduce the cost and size while improving the performance of the overall system.

These features ensure that a TDMA handset manufacturer can produce phones that will succeed in this demanding market.

How the System Works

To understand how this chip set operates let's follow a signal at Channel 1 through the receive and transmit paths in Figure 1. To begin, a received signal on Channel 1 (870.03MHz) enters at the antenna.

The signal then moves through the 881 MHz bandpass filter, to be amplified by the SA601 low noise amplifier (LNA). The low noise figure of the SA601's LNA (1.6 dB) adds very little noise to the original signal. The signal is then down converted by the SA601's mixer to the first IF (83.16 MHz for this example). The first LO is phase locked by the SA7025 main loop at the receive frequency plus the IF frequency (870.03MHz + 83.16 MHz = 953.19 MHz). Continuing down the receiver chain, the signal passes through an image reject filter centered at 83.16 MHz before entering the SA637 digital IF circuit. The SA637's mixer down converts the incoming signal to the second IF of 600 kHz. The second LO is also phase locked by the SA7025's auxiliary synthesizer. The second LO in this case is the 1st IF Frequency minus the 2nd IF frequency (83.16 MHz − 600 kHz = 82.56 MHz). The signal is then amplified and passes to the DSP/baseband processor for demodulation and decoding.

Looking at the transmit side, a carrier is first generated for Channel 1 (825.03 MHz). This is accomplished by using the same receiver first LO signal from the SA7025 at 953.19 MHz and the SA900's on-chip mixer and synthesizer loop. The SA900's synthesizer loop is fixed at the receiver's second LO frequency plus the TX/RF offset of 45 MHz (83.16 MHz + 45 MHz = 128.16 MHz). When the synthesizer loop and the first LO signals are mixed together on the SA900 the sum (1081.35 MHz) and difference (825.03 MHz) are generated. A low pass filtering function in the SA900 will pass the desired 825.03 MHz signal while rejecting the undesired sum frequency. The carrier is then phase shifted by 90° to create the quadrature for I and Q mixers. The I and Q differential data inputs are provided from the DSP. The I and Q inputs can be used to generate a variety of digital and analog modulation schemes. IS-54 requires π/4 DQPSK (digital) and FM (analog) modulation. Before leaving the SA900, the modulated signal is finally passed through a variable gain amplifier (VGA) which provides precise control of the output power level. Once off-chip the modulated signal is filtered and amplified to the desired power level before being transmitted out the antenna.

Integration and Connectivity

Although, certainly, this is not the only IS-54 solution, the Philips four-chip configuration described here is the most integrated and easy to use chip set available today. The SA900 is a good example of this; it incorporates many components that would otherwise be realized with a large number of discrete devices or single-function ICs.

Previously, a designer had to select these individual components, such as I/Q modulators, from a number of different manufacturers. Once chosen, the designer would have to make these various discrete devices and ICs work together in a system. This made both selection and design processes more complicated and time-consuming.

In contrast, the SA900 provides I/Q modulators, the phase shifter, the VGA, a filter, control logic, clock distribution and more in a single IC. Philips has also eliminated the need for two synthesizers by closely coupling the SA7025 and the SA900 so it is possible to use the main synthesizer to simultaneously generate receive and transmit signals. The integration and connectivity of the chip set promote significant cost reduction. If the functions contained in the SA900 were realized discretely, they would cost approximately $30.00. The SA900, however, sells for about half the price. In addition this integrated solution reduces the time to a final product by simplyfing the design effort. The result is a smaller, cost effective, low power phone that is ultimately more attractive to users.

The Future of NADC

Moving to a digital standard not only provides for increase in capacity, but offers the advantages of service integration. With the use of a digital modulation, other services such as data and FAX can also be handled more easily over this system.

The SA601

The SA601 (Figure 2), which translates the received RF signal to a first IF, is a RF LNA and mixer designed as a front end for high-performance low-power communication systems from 800 to 1200 MHz. The low-noise preamplifier has a 1.6 dB noise figure at 900 MHz with 11.5 dB gain and an IIP3 intercept of −3 dBm. Gain is stabilized by on-chip compensation to vary less than 0.2 dB over a −40 to +85°C range. The wide

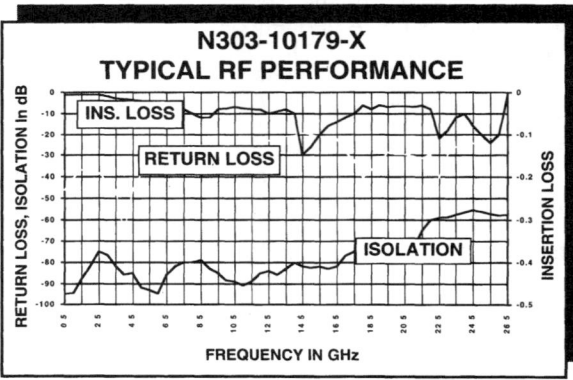

INFO/CARD 46

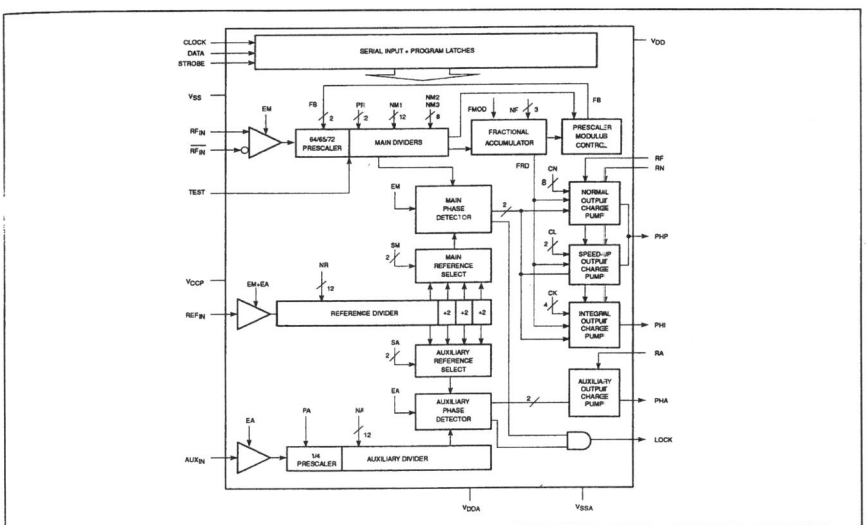

Figure 4. Block diagram of SA7025.

dynamic range mixer has 7 dB of power gain, a 10 dB noise figure, and an IIP3 of 0 dBm at 900 MHz.

The SA601's nominal current consumption from a single 3 V supply is 7.4 mA. Additionally, the mixer can be powered down to further reduce the current supply to 4.4 mA. This part addresses three main design concerns: low power requirements to extend talk time, size (it comes in an SSOP-20 configuration, the smallest 20-pin commercially available package), and performance.

The SA637

The SA637 (Figure 3) is a low-voltage high performance monolithic digital IF receiver with a high-speed received signal strength indicator (RSSI). It is composed of a mixer, oscillator with buffered output, two limiting intermediate frequency amplifiers, fast logarithmic RSSI, voltage regulator, RSSI output op amp and power down pin. It, too, comes in an SSOP-20.

This device is designed for portable digital communication applications and will function down to 2.7 V. The limiter amplifier has differential outputs with 2 MHz small signal bandwidth. The RSSI output op amp feedback pin is accessible, enabling the designer to level adjust the output or add filtering. This one chip provides everything needed to get to the limited second IF.

The SA7025

The next block in the transceiver is the SA7025 (Figure 4) low power, dual frequency fractional-N synthesizer. This IC

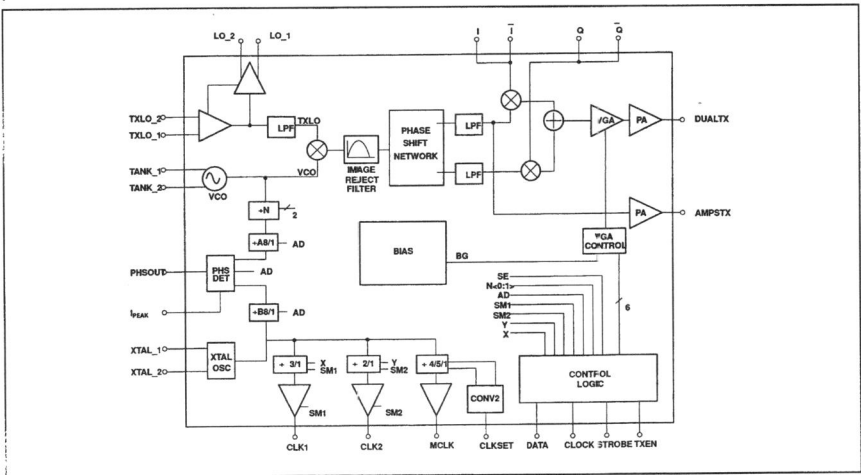

Figure 5. Block diagram of SA900.

Accessories for rf testing

PCB emissions scanner • Locates low-to-high emissions, displays color image, stores data for design corrections at development stage.

Dual-directional couplers • Seven models up to 15 kW, matched to AR amplifiers and antennas.

Ultra-broadband E-field monitor • Four-channel capability, 1 to 300 V/m, isotropic probes cover 10 kHz to 1000 MHz or 80 MHz to 40 GHz.

Broadband fiberoptic data links • Modular plug-in analog systems for acquiring and measuring interference data, stimulating EUT, displaying results in color or monochrome, 10 kHz to 1 GHz.

Fiberoptic CCTV systems • Watch performance of EUT under hostile EMI and/or EMP shielded-room conditions, in color or monochrome.

TEM cells • Half again the bandwidth of comparable-size chambers: To 750 MHz for 15-cm EUT, 375 MHz for 30-cm EUT.

Computer bus interfaces • Two models, for isolated GPIB connection or isolated TTL connection, permit remote operation of amplifiers.

RF connection kit • Things you'd search for around the lab: Cables, coax adapters, connectors, fuses, lamps, fabricated cables.

Power combiner/dividers • Combine signals from four amplifiers, or divide one signal into four outputs.

High-power rf matching transformers • Match 50-ohm input to 12.5- or 200-ohm output. Up to 2 kW cw.

Call toll-free (**1-800-933-8181**), and one of our applications engineers will answer the phone.

160 School House Road,
Souderton, PA 18964-9990 USA • Fax 215-723-5688
In Europe, call EMV: Munich, 89-612-8054;
London, 908-566-556; Paris, 1-64-61-63-29.

INFO/CARD 47

Please see us at RFEW, Booths #610, 612.

is fabricated using the Phillips QUBiC BiCMOS technology and represents another first for Philips. The SA7025 features fractional-N with selectable modulo five or eight implemented in the main synthesizer. This allows the phase detector comparison frequency to be five or eight times the channel spacing. So if the channel spacing is 30 kHz which is what is used for AMPS and IS-54, it becomes possible to use a comparison frequency of 150 kHz (modulus 5) or 240 kHz (modulus 8). The use of a higher comparison frequency moves spurs further away from the fundamental, thus allowing the use of a wider loop bandwidth filter. This results in quicker switching time as required by IS-54.

A triple modulus high frequency prescaler (divide by 64/65/72) is integrated on chip with a maximum input frequency of 1.0 GHz. Philips opted for a triple modulus prescaler because it lowers the main divider ratio providing more flexibility in synthesizing channels, which in turn eliminates the possibility of blind channels.

Programming and channel selection are realized by a high-speed three-wire serial interface. The phase detectors and charge pumps are designed to achieve 10 to 5000 kHz channel spacing. RF input sensitivity is −20 dBm which is matched to the SA900 transmit LO buffer output drive level. The SA7025 is packaged in an SSOP-20 package.

The SA900

The SA900 (Figure 5), a high performance multifunction transmit modulator, brings yet another level of integration. Fabricated in QUBiC BiCMOS technology, it features both analog (AMPS) mode and complex I/Q dual mode (IS-54) functions, a PLL synthesizer with VCO, a crystal oscillator, buffer, programmable prescalers, and Gilbert cell multiplier phase detector with programmable charge pump output. The DUALTX output can be used in DUAL mode cellular phone applications with the AMPS and NADC modulation being applied to the I/Q baseband inputs. The DUALTX output also provides 6-bit power control with 40 dB of gain range in 0.63 dB steps. In addition, the crystal oscillator buffer feeds three programmable prescaler outputs to support system clock reference needs. Programming of the devices function is realized by a high-speed, three-wire serial interface. The SA900 can be programmed into a sleep mode (low current mode providing crystal oscillator and master clock functions), a standby mode (providing crystal oscillator, master clock, system clock 1, and transmit LO buffer functions), and the AMPS mode and the DUAL mode configurations. The SA900 is packaged in a small TQFP 48-pin package.

For more information about this IC chip set, contact the author or circle Info/Card #150.   *RF*

About the Author
Michael M. Sera is Strategic Marketing Manager for RF/Wireless Communication Products at Philips Semiconductors, 811 E. Arques Avenue, P.O. Box 3409, Sunnyvale, CA 94088-3409; tel. (408) 991-4544; fax. (408 991-4800.

INFO/CARD 48

A Spread Spectrum Cordless Telephone

Kiyoshi Tanaka, *Member*

Uniden Corporation, Tokyo, Japan 104

SUMMARY

The United States industrial, scientific, and medical (ISM) band has been released for spread spectrum communication, and various applications for it are being proposed. A spread spectrum cordless telephone set has been developed for domestic use as one of its applications.

In this paper, the design outline and its characteristics are explained. The direct sequence spread spectrum technique with spreading ratio of 16 in the 902- to 928-MHz ISM band is used. To solve the near-far problem in the mobile radio communication channel, the frequency division multiple access-time division duplex (FDMA-TDD) technique is adopted where different frequencies can be assigned to each channel. Compact size and low power consumption is achieved by developing a chip in which spread spectrum processing and transceiver multiplexing can be handled on a single ASIC chip.

The voice codec uses ADPCM and provides better voice quality compared to the conventional analog cordless phones. Wider talking range compared to the conventional 46/49 MHz cordless phone is obtained by increasing the amount of transmitted power. Moreover, since the ISM band is used, interference from other sources can be avoided by changing the channel frequencies.

Key words: Spread spectrum; cordless telephone; ISM; FDMA-TDD.

1. Introduction

Spread spectrum communication has been used mainly by the military because of its antijam capability and secrecy characteristics. However, with the development of digital signal processing techniques and devices, nowadays it is being used in radio local area networks (LAN) and mobile communication such as digital cellular phone systems. Its applications to consumer communication are increasing rapidly [1-3].

In the United States, application of spread spectrum communication in the industrial, scientific, and medical (ISM) band have been permitted in compliance with Part 15 of Federal Communications Commission (FCC) regulations. Since transmitted power has a maximum limit of 1 W, which is a relatively moderate value, the domestic cordless telephone using spread spectrum can be considered as one of its applications.

Previously, the 46-MHz/49-MHz frequency band was used for the cordless phone in the United States. When the cordless phone is used for offices inside a building, then due to lower frequencies, space for wave propagation is narrow compared to the wavelength, and the talking range is very small and there is reasonable demand to increase the talking range. Moreover, since analog modulation schemes were used, voice quality was degraded easily by noise and interference, and crosstalk appeared between cordless phones which made unauthorized listening quite easy. Thus, the talking range and

Table 1. Main parameters

Frequency band	902 MHz to 928 MHz
Carrier spacing	1.024 MHz
Number of radio channels	23
Access system	FDMA
Transmission/reception duplex	TDD
Spread spectrum system	direct sequence
Spreading ratio	16
Modulation scheme	FSK
Chip rate	1.365 Mcps
Voice codec	32 Kbps ADPCM
Transmitting power	100 mW (peak power)

quality of the conventional analog cordless phones do not satisfy demand.

However, if we apply the spread spectrum technique to the cordless phone in the ISM band, then we expect the advantages of: (1) wide talking range by increasing the transmitted power; (2) improvement in secrecy; (3) better antijamming capability; and (4) better voice quality due to digital transmission.

Since various types of application are available in the ISM band, unpredicted interference from other signals should be avoided. Moreover, which type of multiple access is used for the radio channel when spread spectrum is used in cordless phones is an important topic. Again, due to limited battery power of the cordless phone, economic realization of low-power consumption so that the operating time is the maximum possible is also a problem.

In this paper, to solve the aforementioned problems, a basic design of the cordless phone using the spread spectrum technique is described. The configuration of actual equipment is described, and its main characteristics are given. Moreover, control operation of the equipment is explained.

2. Basic Design

The cordless phone developed here uses the frequency range 902 to 928 MHz of the ISM band. Its main parameters are shown in Table 1. Below, the relation of these parameters with the basic design of the cordless phone is explained.

2.1. Spread spectrum technique and ratio

Spread spectrum systems can be divided widely into the direct sequence system and frequency hopping system. Since low cost and low power consumption are important for common users having a domestic cordless phone, we have adopted a direct sequence spread spectrum system for its relatively simpler configuration. Using such a system, a baseband processing section can be realized in the form of an integrated circuit; hence, low cost and low power consumption are possible.

According to the FCC regulation, Part 15, the 6-dB bandwidth should be more than 500 kHz and processing gain should be greater than 10 dB for the direct sequence spread spectrum system in the ISM band. The cordless phone developed here satisfies this condition; and, to maintain the maximum number of radio channels related

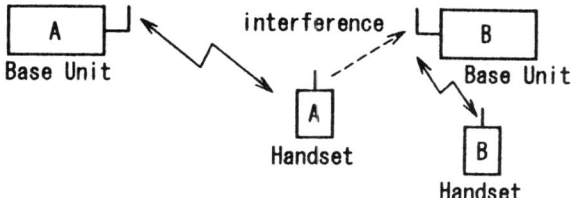

Fig. 1. Interference occurrence between two cordless phones.

to the multiple access technique explained below, we have chosen the spreading ratio of 16.

2.2. Multiple access techniques

A very common type of multiple access technique used in spread spectrum communication is code division multiple access (CDMA) where signals of a different coding scheme are multiplexed in the same frequency. It is applied in the digital cellular phone system [4] due to its efficient use of spectrum. In CDMA using the direct sequence spread spectrum technique, the so-called near-far problem is present because the signal arriving at the base unit from the far point has interference with strong signal from relatively near points. One method to solve this problem is to control the transmitted power such that the strength of the signal reaching the base unit has constant level regardless of the distance from the base unit.

In the cellular phone system, the base unit is placed in a cell and since communication between the base unit is placed in a cell and since communication between the base unit of the cell is possible with a moving body inside the cell, transmitted power control can be applied effectively. However, since the cordless phone can be used at various places such as the home, offices, etc., different handset positions are to be considered corresponding to the base unit and transmitted power control cannot be applied effectively.

Thus, as shown in Fig. 1, even if received power is controlled such that the base unit A has constant level of reception with respect to handset A, it has completely no relation with the neighboring handset B. When handset A is closer to the base unit B, then handset A is away from base unit A and, by increasing the transmitting power, interference from handset A is possible.

Moreover, since the ISM band is used, interference other than cordless phones may also occur. Since such type of interference cannot be controlled completely, its reduction is required. When strong interference is present, then, by changing the frequency, it can be reduced effectively.

In the cordless phone developed here, we have adopted the frequency division multiple access (FDMA) technique for dividing the ISM band into a number of radio channels. When a strong interference signal is present, then communication can be performed by using different radio channels and the near-far problem of the direct sequence spread spectrum system can easily be solved.

2.3. Duplexing techniques for transmission/reception

Since talking and listening is done simultaneously in both directions with cordless phones, two radio channels are required for transmission and reception. The multiplexing techniques for this transmission and reception are frequency division duplex (FDD) and time division duplex (TDD). The TDD system does not require a pair containing two frequency bands and has large freedom of frequency use. Moreover, since propagation characteristics for transmission and reception are the same and diversity can be applied only at the base unit, considerable research is being conducted in this area [5]. Since frequency of transmission and reception is the same and when a nonsynchronous mode is used, then there is the possibility of interference occurrence both at the base unit and handset; but, comparing the average interference degradation, it does not vary much from the FDD system [6].

In the cordless phone developed here, since the frequency band was limited to the ISM band, we have adopted the TDD system due to freedom of frequency setting and to avoid the requirement of the duplexer.

2.4. Modulation scheme

In the direct sequence spread spectrum system, generally the PSK modulation scheme is used where the spread spectrum coding sequence modulates the carrier. On the reception side, despreading is performed by multiplying the similar coding sequence with the received wave.

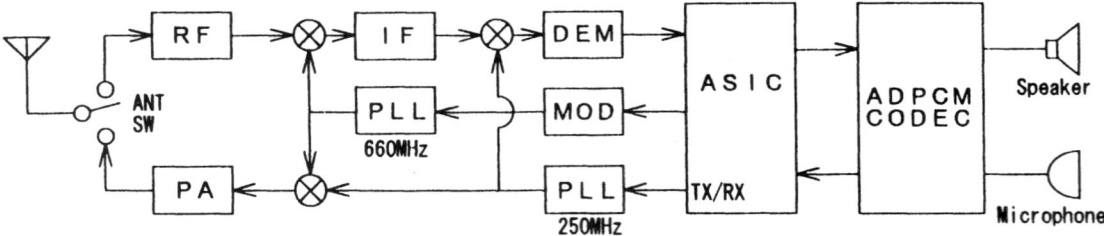

Fig. 2. Block diagram.

However, in the PSK modulation scheme, a linear transmitter power amplifier is required to suppress the spreading of the spectrum and large power consumption becomes a problem. Here, we have used FSK modulation. Using FSK, C-class transmitter power amplifier could be used and low power consumption became possible. In addition, a multiplier is not used on the receiver side and demodulation became possible by a simple discriminator. The modulation index was chosen as 0.5 so that the transmission characteristics are not degraded and the maximum number of radio channels could be obtained. Moreover, to suppress the leakage power toward the adjacent channels, baseband band filtering of $B_b T \approx 0.5$ was performed. Here, B_b is the single-sided bandwidth and T is the chip period which is the reciprocal of the chip rate of 1.365 Mcps.

2.5. Voice codec

The voice codec should have equal or better voice quality compared to the conventional analog cordless phone. Moreover, considering the efficient use of radio frequencies, the lowest possible transmission rate is desirable and it should be suitable economically for the cordless phone. As a result, we used a 32-kbit/s ADPCM codec. When interference was present in the conventional analog cordless phone, then the ratio of desired signal to the interference signal was relatively large and the beat voice was disturbing. However, by transmitting the digital voice signal, relatively low desired signal to the interference signal ratio without degrading the quality became possible.

2.6. Transmitted power

In the ISM band, a maximum transmitted power of 1 W is allowed. By increasing the transmitted power, the talking range can be increased greatly. However, since large reduction in continuous communication time compared to the conventional cordless phone is not desirable, we set it at 100 mW due to power consumption considerations.

2.7. Radio channel control

The cordless phone developed here scans a number of radio channels, selects a vacant channel, and establishes the communication. Selection of a radio channel is possible from both the base unit and the handset. The number of radio channels is 23 and due to its relation with scan time and for each group separated by ID number of cordless phone, nine channels are decided as the control channels. Scanning these nine channels and finding the vacant channel, message control is performed by a suitable radio channel and communication is established through that channel.

When communication is once established by the control channel, then any one of the 23 channels is selected by channel switching and communication is continued. Radio channel spacing is 1.024 MHz and adjacent channel attenuation is not sufficient for FSK signal of 1.365 Mcps spread coding rate. To obtain sufficient attenuation between channels, we have to use the channel that is next to the adjacent channel. Thus, the number of channels is 12 where simultaneous communication in the same area is possible.

3. Configuration

3.1. Block diagram

The block diagram of the base unit of the cordless phone developed here is shown in Fig. 2. Its main difference from the handset is the interface with telephone line; this is not a large difference.

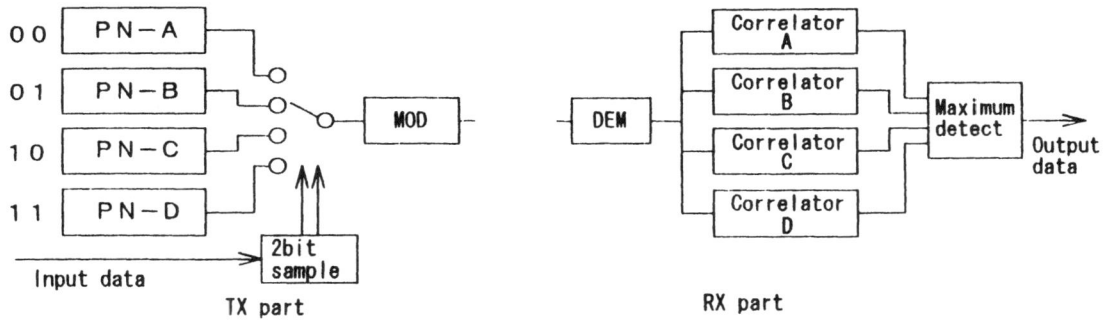

Fig. 3. Spread spectrum processing of ASIC.

The voice signal from a microphone is changed into a digital voice signal by ADPCM codec and enters into ASIC for spread spectrum processing. In ASIC, the voice and control data signals are spread according to the spread coding scheme and FSK modulated by the modulator. This FSK modulation is performed after adding the baseband band-limited signal that is obtained by using a lowpass filter at the input of VCO of the 660-MHz band PLL. The FSK modulated signal is multiplied with the 250-MHz local signal and is frequency translated to the 900-MHz signal. It is amplified by the transmitter power amplifier and, after passing through the antenna switch, is transmitted by the antenna.

On the reception side, the signal from the antenna after passing through the antenna switch is amplified by RF amplifier and is translated to the first IF signal of 250 MHz by the local signal of 660 MHz and is band-limited IF filter. Then it is translated to the second IF signal of 11 MHz by the local signal of about 240 MHz and is demodulated by the discriminator. The detected baseband signal is transformed into the signal of two logical levels and enters into the ASIC. Despreading processing is performed in the ASIC and digital voice data are obtained. The digital voice data enter into ADPCM codec and the analog voice signal is obtained at the output of the speaker.

3.2. Internal diagram of ASIC

The ASIC for spread spectrum processing is a standard cell of 0.8-μ processor where the number of gates is about 15,000. The input baseband data of 85.3 kbit/s are sampled after every 2 bits and 32-chip spread spectrum coding of four types as shown in Fig. 3 is applied to the sampled data, and we obtain the data of 1.365-Mcps at the output. Since, in the spread spectrum processing of ASIC, correlation detection of 1-bit quantized data is performed, if we make the spreading ratio constant to obtain better performance, then the chip length should be made longer. However, if the chip length is made longer, then the number of required spreading codes is increased and the length of register for correlation detection in the ASIC is increased.

Due to these reasons, 32 chips are assigned to every 2 bits. Spread coding sequence of four types adds 0 after M sequences of 31 chips and the orthogonal relation between their codes is used. These can be written from CPU at the time of initial setting.

Time division processing for TDD communication and processing of frame synchronization also are carried out in the ASIC.

At the time of reception, the demodulated data of the FM detector is changed with the transmission chip rate. However, since the clock on the receiver side is not synchronized, the demodulated data are oversampled in the ASIC and its correlation value with the spread coding sequence of four types is determined.

The 2-bit data corresponding to the coding sequence having the highest value of correlation are obtained as the baseband data. The correlation detection circuit which determines the correlation value consists of shift registers and work as a digital matched filter of two-value data. Bit synchronization is obtained from the pulse sequence of detected output.

From these baseband data having TDD frame structure, a unique word is detected and frame synchronization is performed. Control data are detected and fed to CPU whereas digital voice data are obtained and given to the ADPCM codec.

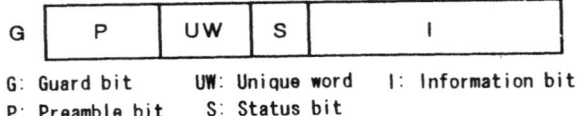

G: Guard bit UW: Unique word I: Information bit
P: Preamble bit S: Status bit

Fig. 4. Burst frame structure.

To establish the initial synchronism of TDD Frame, the ASIC also gives the burst frame signal. This burst is transmitted when synchronism is not yet established; and, when reply of the same burst from the opposite side is received, then synchronism is established and communication is started by switching the burst frame for communication.

3.3. Frame structure

The frame structure for TDD communication is shown in Fig. 4. It consists of guard bit, preamble bit, unique word, and status bit other than the digital voice data. Guard bit is the time for switching the TDD frame transmission/reception and actual data are not transmitted. Preamble bit is the bit to obtain synchronization of baseband data. Unique word is the data of fixed pattern and is used to obtain synchronization of TDD frame. Status bit is the data which perform the exchange of control data.

The period of TDD frame is 9 ms and timing signal to perform switching of transmission/reception of this period is given by ASIC to the PLL of 250 MHz.

3.4. Frequency synthesizer (PLL)

This cordless phone consists of two PLL of 250 MHz and 660 MHz.

In the TDD radio equipment, we have the problem of reception of transmitter leakage signal which has the same frequency as that of the received signal.

To avoid the occurrence of this interference, the 250-MHz PLL can perform fast frequency switching at the time of transmitter timing and receiver timing.

Switching of radio channel frequency is done by controlling the PLL of 660 MHz from CPU. This PLL also has fast frequency switching which is less than 1 ms.

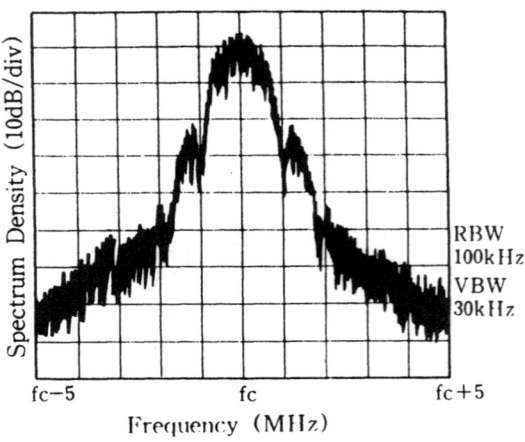

Fig. 5. Spectrum characteristics.

4. Equipment Characteristics

4.1. Spectrum characteristics

According to the FCC regulation, Part 15, 6-dB bandwidth should be greater than 500 kHz for the direct sequence spread spectrum system and it should not increase above 8 dBm for 1 s on the average for a 3-kHz bandwidth which shows that the spectrum is controlled so that it is not concentrated in a single frequency.

Figure 5 shows the spectrum characteristics of our cordless phone where no inclination is seen to any particular frequency. Moreover, due to bandlimiting of the baseband signal, sharp attenuation is seen for the adjacent channel. Since FSK modulation is used, even if the final stage of power amplifier is nonlinear, the spectrum characteristics are well maintained.

4.2. Bit error rate performance

In the cordless phone developed here, spectrum is spread by FSK modulation and, after detection by the wideband FM demodulator, correlation operation is performed. The bit error rate of this system is different from the case where spreading is performed by BPSK modulation.

For the spreading code of 32 chips with 1 bit quantized, more than 16 errors are produced after FM detection; then a bit error is considered to have been generated. If two cases exist where one out of two bits assigned to the spreading code of 32 chips is in error and one case

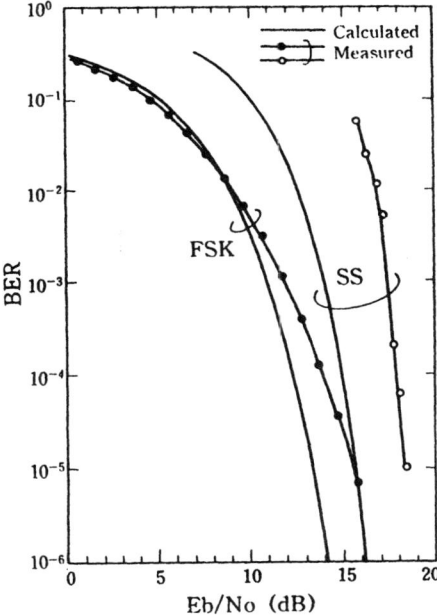

Fig. 6. Bit error rate performances.

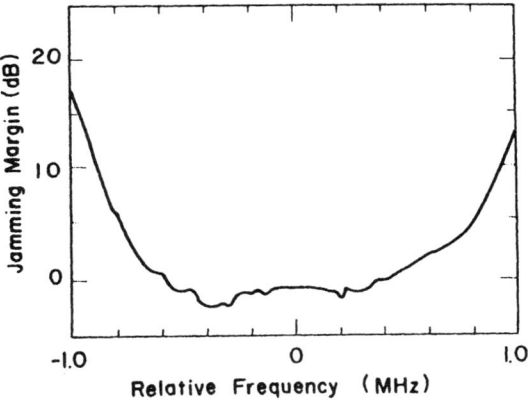

Fig. 7. Jamming margin performances.

exists where two bits are in error, then, on the average, a 4/3-bit weighting factor can be attached and error rate p_e can be expressed as follows:

$$p_e = \frac{4}{3} \sum_{k=17}^{32} {}_{32}C_k p^k (1-p)^{32-k} \quad (1)$$

Here, p is the error rate of FSK and can be expressed as follows:

$$p = \frac{1}{2} \exp(-\gamma/2) = \frac{1}{2} \exp(-8 E_b/N_0) \quad (2)$$

γ is the SNR of FSK and E_b/N_0 is the SNR for one information bit when spread spectrum is performed. Here, in the actual demodulator circuit, decision of bit is performed by comparing the outputs of four correlation detectors so three undesired correlator outputs should also be considered. Moreover, oversampling of the FM-detected chip data should also be considered. To avoid complex calculations, we have fixed the undesired output of three correlators at 16 and we use the simple model where central data of the chip are only sampled.

The bit error rate calculated from Eq. (1), and its measured values for the cordless phone are shown in Fig. 6. In the figure, the curves marked with FSK represent the bit error rate performance of FSK modulated signal (1.365 Mbit/s) after FM detection. The curves marked with SS represent the bit error rate performance of 32 kbit/s voice data when spread spectrum is applied and burst transmission is performed using TDD.

To obtain good quality of sound at the ADPCM codec, E_b/N_0 is about 17.8 dB to give 1×10^{-4} error rate and there is degradation of about 3 dB from the calculated value. The processing gain varies greatly with the decision of desired bit error rate; and it is degraded highly from 12 dB, which is the expected processing gain when spreading is performed by BPSK modulation. This degradation is due to adopting the system where correlation is performed after FM detection and also to fixed deterioration of the ASIC for spread spectrum processing.

4.3. Jamming margin characteristics

Jamming margin performance of the spread spectrum equipment is very important in finding the processing gain of the equipment. The following relationship is established between the processing gain PG and the jamming margin J/S [8]:

$$\frac{E_b}{N_J} = \frac{PG}{J/S} \quad (3)$$

Here, E_b/N_J is the ratio of bit energy to the jamming noise; E_b/N_J can be replaced by the ratio of bit energy to noise E_b/N_0. If no degradation by the equipment is considered, then the value of E_b/N_0 can be determined from the theoretical equations of the adopted

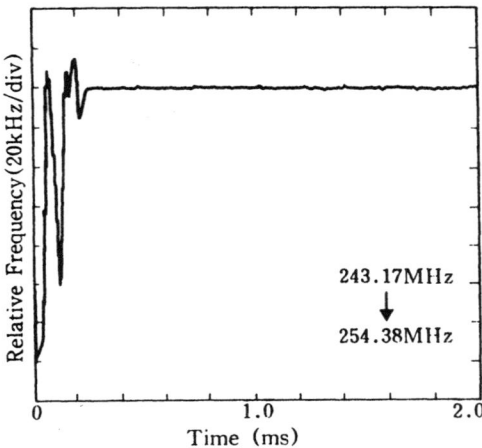

Fig. 8. Frequency switching response.

modulation system and the relation between the processing gain and jamming margin can be estimated.

Jamming margin performance of our cordless phone is shown in Fig. 7. This figure gives the jamming margin for off-tuned frequencies when the bit error rate is 1×10^{-4}.

The measurement of processing gain as required by FCC regulation, Part 15, can be done with the help of the jamming margin. Here, to determine the value of the jamming margin that is used for actual estimation and which varies greatly with off-tuned frequencies, we find that the values up to 20 percent of the worst case values have been adopted. Using this method, the values for our cordless phone became as follows:

$$J/S = -1.2 \text{ dB}$$
$$E_b/N_0 = 12.3 \text{ dB}$$
$$PG = 11.1 \text{ dB} \qquad (4)$$

4.4. Frequency switching response

Since the 250-MHz synthesizer of our cordless phone, as shown in Fig. 2, works as a carrier source at the time of transmission and works as a local oscillator to obtain the second IF of 11 MHz at the time of reception, fast frequency switching is done at the time of transmission/reception change of TDD. Frequency difference to be switched between transmission and reception is about 11.2 MHz. Figure 8 shows the switching response of this synthesizer. The guard time for switching of transmission/reception is about 350 μs and frequency switching is completed within this period.

5. Control Operation

5.1. Intermittent reception

To reduce the power consumption at waiting time, this cordless phone is capable of intermittent reception. The power supply to all the circuits is cut off except the CPU which is operating at a low-speed clock. Then the receiver section starts operation and reception is performed. Again, the power supply to all the circuits is cut off and the circuit goes into the sleep condition. Reception is performed by switching the frequencies of nine channels in a sequence and it is checked whether or not the burst signal for synchronization can be received. Time required to complete one such operation is about 100 ms which is a short time. When the burst signal for synchronization is received, then communication at that frequency is started. Current consumption at the time of reception is about 100 mA, but in the sleeping condition this current consumption is less than 1 mA. Performing intermittent reception of 1.5 to 4 s, average current consumption becomes about 3.2 mA and stand-by time of about 1 week is possible with 600-mAh battery.

5.2. Transmitted power control

To reduce the interference occurrence and to extend the communication time, the transmitted power is controlled under the required minimum limit. When the received level averaged over constant time is higher than the previously known threshold value, then the control data contained in the status bit request reduction of transmitted power on the opposite side. Conversely, when the received level is lower than threshold value, then increase in the transmitted power is requested on the other side. In this way, optimum receiver level is maintained and power consumption may be reduced.

5.3. Channel changing during communication

When interference occurs during communication, then, by changing the frequency channel, communication is continued. When degradation of voice quality due to any reason is detected, then the base unit or handset of the cordless phone sends the request for channel change on the opposite side and channel changing operation is started. Degradation of voice quality is detected by the

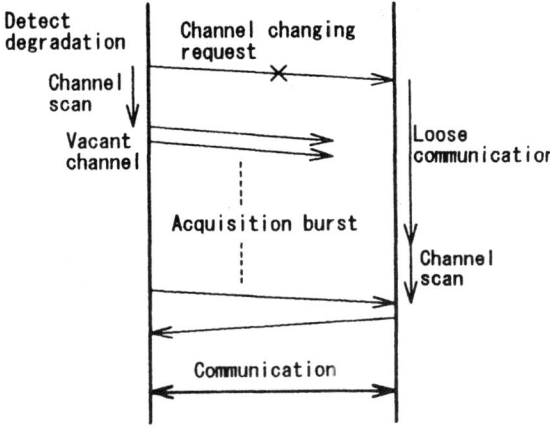

Fig. 9. Sequence flow of channel changing.

Fig. 10. Photograph of cordless phone.

average value of correlation and deviation of TDD synchronization. If interference is strong and the request for channel change cannot be communicated to the opposite side, even then the channel can be changed accurately by the following method.

Figure 9 shows the sequence of channel changing. The side requesting the channel change cuts off the communication and starts its channel change operation. The opposite side observes the time where communication is stopped; and when communication remains cut off for a fixed time, then it assumes that the channel change operation has been started. Either the handset or base unit is considered as master with the other treated as slave. The master side searches for a vacant channel during the channel change operation and transmits the burst for synchronization. The slave side continues the channel scan until it receives the synchronization burst. Communication is restarted when the slave side confirms the synchronization burst.

The ordinary channel change is completed in a few hundred microseconds, including the channel scan and reconnection, provided the control signal is transmitted without error. This value is not large for the cordless telephone system.

6. Field Test Results

A photograph of the cordless phone developed here is shown in Fig. 10. According to the field test results in the United States, communication within 1 mile (~1.6 km) is possible if communication conditions are favorable. Moreover, since transmitted power is large, a much wider covering area is obtained compared to the conventional 46/49-MHz cordless phone.

If CW interference occurs during communication, then the channel can be changed instantly and communication can be continued through a better-quality channel.

7. Conclusions

The configuration of the cordless phone using the spread spectrum technique was described. Using the spread spectrum technique in the ISM band and by increasing the amount of transmitted power, the cordless phone was developed which has great advantage in the talking range compared to the conventional analog cordless phone. Low power consumption was achieved with the help of intermittent reception and transmitted power control and stand-by time of 1 week, and communication time of 3 hours was realized.

Moreover, an economical ASIC for spread spectrum processing was developed due to simple circuits of FSK direct sequence spread spectrum system. Its cost is increased nearly 1.2 times compared to the low power output analog cordless phone operating in the same ISM band. Moreover, better quality of voice communication is achieved due to digital voice data transmission encoded by ADPCM codec. It has the capability of avoiding the interference from other sources which is the characteristic of the ISM band. Using the FDMA system and changing the channel during communication, better quality of communication is maintained.

In the future, we plan to increase the channel capacity so that it may be used for offices in large traffic environment. For this purpose, CDMA will be adopted and realization of precise power control will be taken as the problem for the future. Moreover, realization of diversity techniques to maintain the quality of communication under fading conditions is the problem for the future.

Acknowledgement. The author wishes to thank Dr. Omura of Cylink Co. for his cooperation in the development of the system. He is grateful also to all concerned staff of the company for their assistance.

REFERENCES

1. Marubashi. Recent R & D trends in spread spectrum communication. Trans. I.E.I.C.E. (B-II), **J74-B-II**, 5, pp. 176-181 (May 1991).
2. Tsubouchi. Devices and applications of spread spectrum communication. Trans. I.E.I.C.E. (B-II), **J74-B-II**, 5, pp. 189-198 (May 1991).
3. M. Nakagawa and T. Hasegawa. Spread Spectrum for Consumer Communications. I.E.I.C.E. Trans. Commun., **E74**, 5, pp. 1093-1102 (May 1991).
4. W. C. Y. Lee. Overview of Cellular CDMA. IEEE Trans. Vech. Technol., **40**, pp. 291-302 (May 1991).
5. R. Esmailzadeh and M. Nakagawa. Time division duplex method of transmission of direct sequence spread spectrum signals for power control implementation. I.E.I.C.E. Trans. Commun., **E76-B**, 8, pp. 1030-1038 (Aug. 1993).
6. K. Tanaka et al. Multiple access techniques for digital cordless telephone. 1991 Spring Nat'l Conv. I.E.I.C.E., B-342.
7. M. K. Simon, J. K. Omura, R. A. Scholtz, and B. K. Levitt. Spread Spectrum Communications, Vol. I, pp. 137-139. Computer Science Press (1985).

A 1.9-GHz GaAs Chip Set for the Personal Handyphone System

Finbarr McGrath, Karen Jackson, Eugene Heaney, Allan Douglas,
William Fahey, Russell G. Pratt, and Ted Begnoche

Abstract—The Japanese Personal Handyphone System (PHS) is representative of the latest generation of digital portable communications systems currently being deployed. Enabling technologies for these systems include high performance Radio Frequency Integrated Circuit (RFIC) chip sets. These chip sets allow all the RF transceiver functions to be included in low cost surface mount plastic packages. With the addition of filters and bypassing capacitors, the RF portion of the phone shares the same printed circuit board (PCB) as the DSP, CODEC, and Logic IC's. The availability of such highly integrated 1.9-GHz RFIC's requires the solution of many complex design, manufacturing, and test problems. This paper explains the critical issues relating to the air interface of the PHS system and how it affects the RFIC design. The chip partition, design, and performance of each subfunction is discussed relative to the requirements imposed by the air interface. The result is a highly integrated, cost effective solution that occupies the minimum board area.

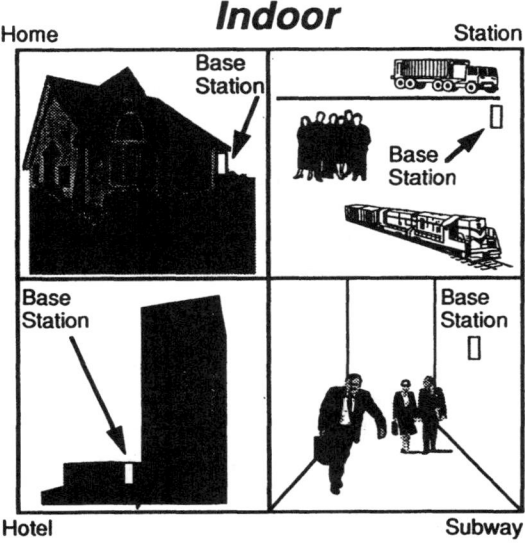

Fig. 1. PHS system concept.

I. Introduction

THE PERSONAL Handyphone System defined by Japan's Research and Development Center for Radio System (RCR) establishes the minimum requirements for the unified digital mobile telephone system. The standards defined in document #28 (RCR-28) cover the communications control methods, protocols, encryption, voice coding, and air interface [1]. The document also serves as a guideline for private systems based on the publicly established minimum requirements. The concept behind the Personal Handyphone System is to provide a low data rate, digital, two-way wireless link from any hand-held terminal at any time or place. The hand-held terminal is pocket sized and similar in weight and appearance to the smallest cellular phones. The RCR-28 standard ensures that any hand-held terminal can provide common access to home, office, and public (outdoor) locations in the coverage area. While the original concept was designed around a voice system, its digital nature makes paging, low data-rate video, fax modem, data communication, and ISDN applications possible. The standard is flexible enough to make provision for private wireless branch exchanges (WPABX), peer to peer communications or home cordless phones coexisting with the public digital cordless phone system. Thus the hand-held unit can be used in the home as a cordless phone, in the office as part of a PABX system or in the street as part of the public PHS network. Only one phone number is required. The public network provides for many of the features of cellular systems such as roaming and hand-off. The interoperability ensured by RCR-28 will allow major telephone systems operators such as Nippon Telephone and Telegraph (NTT) to coexist with manufacturers of portable consumer electronics, such as Sony Corp. Once the public infrastructure is complete the growth is expected to be explosive.

Fig. 1 shows the PHS system concept. It is essentially a microcellular system with cell sites of approximately 50-m radius. The base stations are placed on telephone poles at regular intervals in the street or private units can be purchased for the home. The average transmit power varies from +19 to +27 dBm, depending on equipment classification, and is much less than a cellular base station. Hand-off is accomplished at a slow rate as the user walks from one cell site to the next. The system does not hand-off quickly enough to be used in vehicles. The access method is TDMA/FDMA utilizing 77 RF channels from 1895.15–1917.95 MHz. The channel spacing is 300 kHz but the minimum allowed spacing for two channels that are spatially collocated is 600 kHz. In any given frequency channel four voice channels are available using a TDMA/TDD protocol. The timing protocol is shown in Fig. 2 and uses a frame length of 5 mS with four time slots for both transmit and receive of 626-μS duration. The voice transfer rate is 32 kB/s using ADPCM.

Manuscript received June 30, 1994; revised February 20, 1995.
The authors are with the M/A-COM Microelectronics Division, Lowell, MA 01851 USA.
IEEE Log Number 9412079.

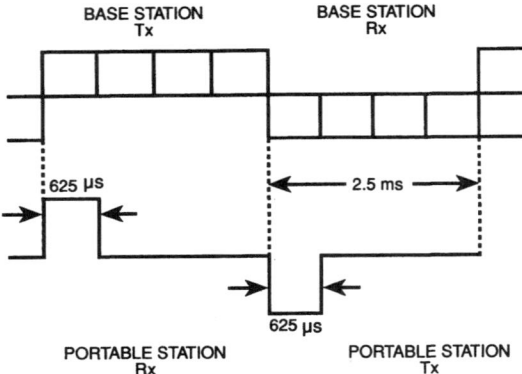

Fig. 2. Time division duplex in the PHS system.

TABLE I
PHS AIR INTERFACE REQUIREMENTS

Parameter	Performance
Access method	TDMA/FDMA
RF Frequency (MHz)	1895.15 - 1917.95
Number of Channels	77
Channel Spacing (kHz)	300
Frequency Assignment	Dynamic
Duplex Method	TDD
Frame Length Tx/Rx (ms)	5
Number of slots/frame	4
Slot Duration (microseconds)	625
Bit Time (microseconds)	2.6
Transfer Rate (kb/s)	384
Tx Power peak//avg (mW)	80/10
Power Control (dB)	20/40
Modulation	$\pi/4$ DQPSK ($\alpha = 0.5$)
ACI in 192kHz Band @ 600kHz Offset (dBc)	51
Spurious Output (dBm)	-26
Sensitivity @ BER = 1% (dBm)	-100
Intermodulation Immunity @ -97dBm and BER = 1% (dBc)	47
Circuit Modes	Tx/Rx/Sleep

As the system is intended at the outset for portable handheld applications power consumption is a premium. The main concern in developing the chip set was to implement the Air Interface standards in the most cost effective manner with minimum power consumption.

II. THE AIR INTERFACE FOR PHS

The Air Interface specifies transmitter and receiver performance allowed in the RF channel. It defines the allowable modulation technique; the spectral content of the baseband signal; the filter transfer function; and allowed carrier frequencies. During transmit the air interface determines allowable transmit power levels and limits interference into adjacent channels (ACI) and spurious output. The transient response during duplexing or burst-mode is also specified. During receive the sensitivity is specified for a given bit error rate (BER) and receiver intermodulation immunity and selectivity is defined. The key specifications of the air interface are outlined in Table I. The on air data rate including ramp bits, preamble and error correction is 384 kB/s.

During a transmit burst the average transmit power at the antenna is +19 dBm (80 mW). Once a transmit burst is complete the leakage power should not exceed 80nW, implying 60-dB on/off ratio. The transmitter should switch on or off during burst mode in 13.0 μs (5 bits) in a monotonically increasing or decreasing fashion. The transmitter should not generate any spurious levels in the RF band (1895.15–1917.95 MHz) that exceed -36 dBm. The transmitter should not generate any out of band spurious that exceed -26 dBm. In practice the unwanted lower sideband or the local oscillator leakage are the largest spurious signal levels to be filtered out.

The modulation scheme required in PHS is $\pi/4$ DQPSK. In this scheme the binary data is combined in a serial to parallel converter and differential coder to generate the impulses I_K and Q_K. The low pass filter $H(f)$ generates shaped pulses in the time domain $i(t)$ and $q(t)$, which are modulated by a single sideband modulator. This results in differentially encoded phase shifts $\Delta\Phi$ as shown in Fig. 3. While there are eight phase states possible in this scheme, only four can be chosen at any given phase transition yielding 2 bits/symbol (Fig. 4). The low pass filter $H(f)$ determines the occupied bandwidth of the system. It also determines instantaneous amplitude and phase during phase transitions. The filter transfer function (root Nyquist) is defined for

$$H(f) = 1 \quad \text{for } 0 \leq |f| < \left(\frac{1-\alpha}{2T}\right)$$

$$H(f) = \cos\left[\frac{T}{4\alpha}\left(2\pi|f| - \frac{\pi(1-\alpha)}{T}\right)\right]$$
$$\text{for}\left(\frac{1-\alpha}{2T}\right) \leq |f| < \frac{1+\alpha}{2T}$$

$$= 0 \quad \text{for } \frac{1+\alpha}{2T} \leq |f|$$

where T is the bit rate and $\alpha = 0.5$. This filter transfer function has a high spectral efficiency since the power density drops to zero at 1.5 times the ideal Nyquist rate. In practice this baseband filter is implemented digitally. The major disadvantage of this filter function is that the resulting filtered signal is not constant envelope. In fact is will vary from $+2.9$ to -11 dB about the power level at the sampling points. If the transmit chain is perfectly linear then the RF spectrum at 1.9 GHz is a perfect double sideband representation of the baseband signal. However, any transmit chain nonlinearity in amplitude or phase will generate side lobes outside the ideal Root Nyquist filter bandwidth. In this case the baseband filter no longer determines the occupied bandwidth. As the nonlinearities become more extreme, power can spread into adjacent channels at 600- and 900-kHz offset. Since this can seriously disrupt communications in other bands the system specifies the maximum allowed adjacent channel power to be -31 dBm @ 600 kHz and -36 dBm @ 900 kHz from the carrier. For +19 dBm transmit power translates to 50 and 55 dBc of adjacent channel leakage suppression. In order to achieve optimum power added efficiency (PAE) it is desirable

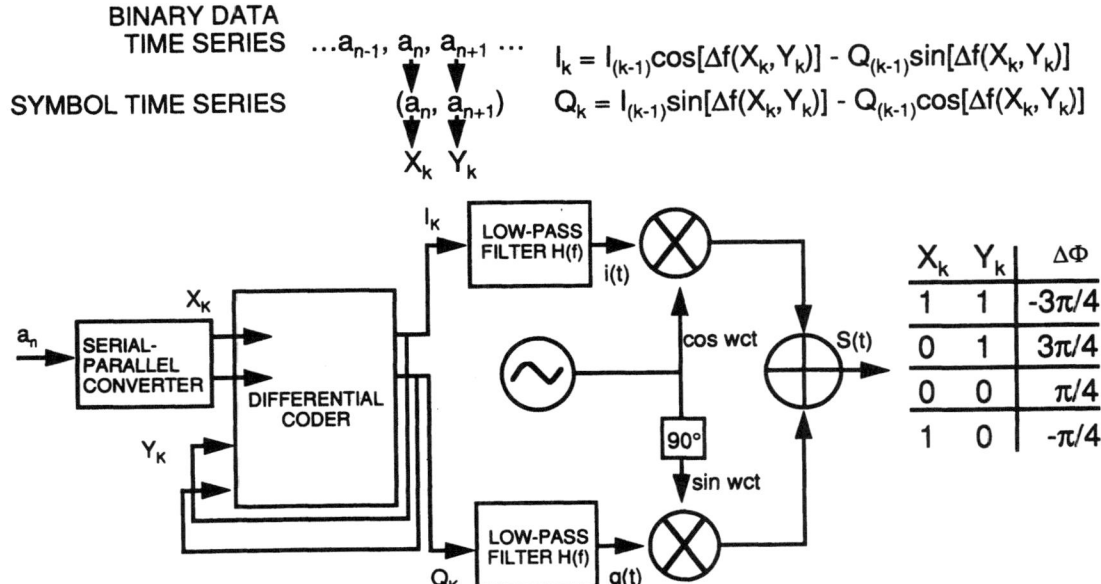

Fig. 3. $\pi/4$ DQPSK phase encoding.

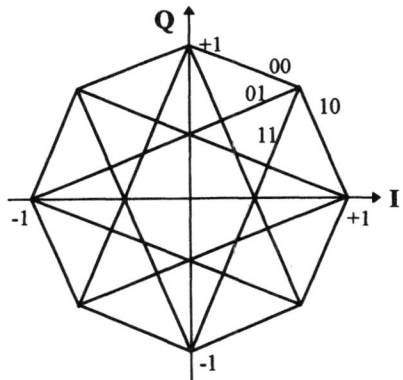

Fig. 4. $\pi/4$ DQPSK I—Q diagram.

to operate the power amplifier at the highest RF power level possible. The major design trade off in the Tx chain is between linearity and power added efficiency. As the transmit chain is compressed the maximum and minimum deviation from the ideal will vary. The system allows for +1.1 and −3 dB margin and the instantaneous power during each transmission burst is defined by Fig. 5.

The low noise amplifier, Rx Filter, and first down converting mixer determines the receiver sensitivity and intermodulation immunity of the overall receiver. In the RCR-28 document the sensitivity is defined to be less than −97 dBm when the BER = 1%. Another key receiver specification is the sensitivity in the presence of two strong interfering signals. With a reference signal set at −94 dBm and two strong interferers detuned at 600 and 1200 kHz from the reference signal the interfering signals should be at least 47 dB above the reference level when the BER drops to 1%. Fig. 6 defines these specifications graphically. If the phone is "on the hook" but not actively receiving data the receiver is switched on every 100 mS to synchronize in time and frequency with the local base station and detect any incoming calls. Thus the receiver demand on the battery is much higher than the transmitter. Receiver current consumption needs to be minimized.

III. CHIP SET DESIGN CONSIDERATIONS

Given the requirements defined by RCR-28 the chip set design considerations can be grouped into four interrelated categories:
1) Factors affecting the Transmit chain
2) Factors affecting the Receive chain
3) Electromechanical considerations
4) System timing and operating modes

A. Transmit Chain

The main goals in the Transmit chain are to meet transmit power levels and linearity, filter unwanted spurious outputs, and optimize power added efficiency. Assuming that the transmit power amplifier design can achieve 20 dB or more gain, then power added efficiency is optimized by allowing the power amplifier to operate at the highest level of nonlinearity consistent with the ACI specification. The rest of the transmit chain is designed to operate linearly. The difference between nominal Tx power and allowed out of band spurious determines the degree of filtering. In this case

$$P_{\text{out}} - P_{\text{spurious}} = 19 \text{ dBm} - (-26 \text{ dBm}) = 45 \text{ dB}.$$

At typical IF frequencies of 90 or 240 MHz there is little selectivity in the upconverter. Thus all of the filtering must come from external filters distributed in the Tx chain. The overall level of filtering needs to be as high as 55 dB for production margin. This fact precludes putting the entire transmit chain in a single package since package isolations of 55 dB are not easily achieved. A convenient break in the

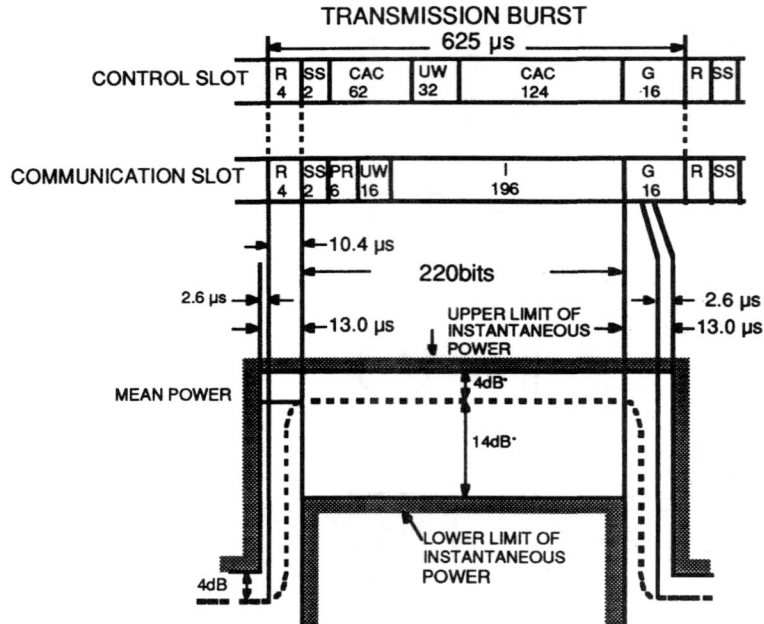

Fig. 5. PHS transmission burst.

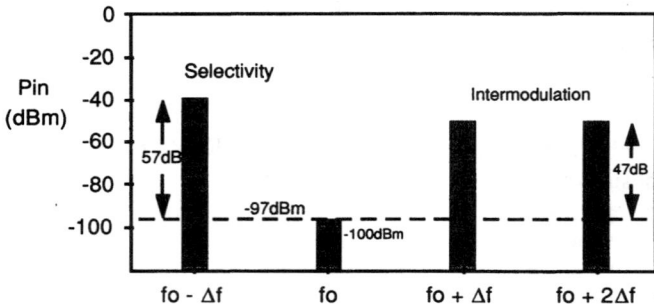

Fig. 6. Receiver specifications.

transmit chain must be found. Filtering the LO signal can be made easier by using a balanced mixer topology.

B. Receive Chain

The main challenge in the receiver design was to achieve the required sensitivity, intermodulation, and spurious rejection while operating at low supply voltage and current. The statistical nature of a digital receiver causes an increase in bit error rate (BER) as the ratio of received signal power to noise power decreases. The RCR specification requires the BER $>10^{-2}$ when -97 dBm of signal power is received. For $\pi/4$ DQPSK signal the ratio S_o/N_o should be 12 dB when the bit error rate is 10^{-2}. Thus the equivalent input noise floor is -97 dBm - 12 dBm = -109 dBm for the receiver.

As

$$N_o = kTBNF$$
$$B = 225 \text{ kHz}$$

the resulting noise figure is NF = 11.5 dB. With filter and PCB loss the noise figure required of the LNA downconverter is 9 dB. As with any multi-channel receiver, the intermodulation of signals in adjacent channels causes noise to be generated in the desired channel. This effectively reduces the carrier to noise ratio. The specification requires the system to have a BER better than 10^{-2} with -94 dBm of received power in the presence of two strong interfering signals offset at 600 kHz and 1.2 MHz. In this case there is thermal noise and intermodulation products in the received channel. We have

$$12 \text{ dB} > \frac{S^o}{N_o + I}$$

assumed that the interference is essentially noise like. Thus

$$\frac{S_o}{I} > 15 \text{ dB}$$

and the maximum intermodulation product is $(-94 - 15)$ dBm $= -109$ dBm. The strong interfering levels are defined to be no less than -47 dBm. Therefore the minimum input intercept point is

$$\text{IIP3} = P_{\text{in}} + \text{IMR}/2 = -47 + (-47 + 109)/2$$
$$= -16 \text{ dBm}$$

C. Electromechanical Considerations

In keeping with the portable nature of the application it is desirable to keep the overall PCB area to a minimum. If the chip set is to be user friendly and truly a mass production product it has to be manufactured with high volume packaging and assembly technology. In this case a fine pitch shrink small outline (SSOP) 28 lead package is chosen. The high pin count, 25-mil lead pitch, and small size (150 × 390 mils) minimize the area required. The complete assembly and packaging in high volume cost 50–75 cents. The package lead widths are 14 mils, which make for conveniently narrow 50-ohm traces

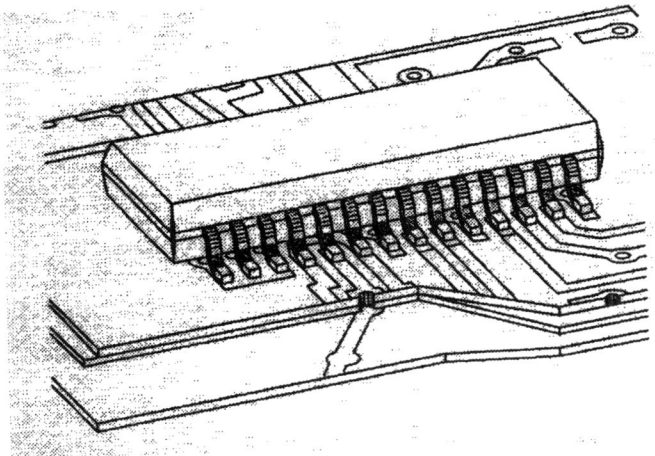

Fig. 7. SSOP 28 lead package on 4-layer FR-4 board.

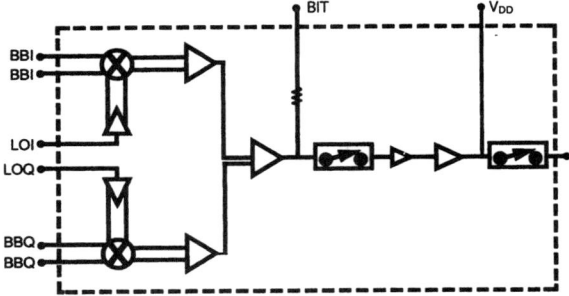

Fig. 8. QPSK modulator IC block diagram.

on multi-layer PCB's. The disadvantage of this package is its lack of a solid ground since any RF grounding must be done through the leads. This causes potential for unwanted coupling between various functions and potential oscillation. It is necessary to model the complete IC in this chip for optimum results. The package also has significant electrical parasitics at 2 GHz with lead inductances from 1–2.5 nH and coupling capacitance from 60–100 fF. All these package and PCB parasitics must be absorbed into the design. Fig. 7 shows the SSOP 28 lead package mounted on a four layer PCB. The potential for unwanted coupling can clearly be seen.

D. System Timing and Operating Modes

A chip set that consumes less power ultimately allows for longer talk time between recharging or lighter weight batteries. As digital IC's move towards 3-V logic levels the pressure to make RFIC's meet performance goals at 3 V is increasing. As supply voltage drops the natural tendency is to increase current. No measure is spared to minimize power consumption. The TDMA/TDD nature of the system means that either the Transmit or Receive chain can be powered down when the phone is not transmitting or receiving. For each handset the transmitter or receiver is active for only 625 μs out of 5 ms, so powering down results in substantial savings. Also when the receiver is not active it only needs to poll the system at a very low duty cycle. Each chip must be designed with the facility to power down under control of the system timing logic.

Since one of the heaviest components in the portable radio is the battery every effort is made to make the entire system operate from the lowest battery voltage available. In practice such systems will operate from 3.6-V nominal supply voltage or 3.0 V regulated from the 3.6-V supply. We will now discuss the IC designs in detail.

IV. QPSK MODULATOR IC

The vector I/Q modulator is the interface between the digitally shaped $i(t)$ and $q(t)$ pulses at baseband of Fig. 3 and the RF carrier $s(t)$. Theoretically it is possible to convert from baseband to RF (1900 MHz). This requires a continuous-wave carrier signal at 1900 MHz, which can leak into the modulated RF channel. In practice it is impossible to isolate this carrier since 35–40-dB isolation is required. An IF of 90–240 MHz is typically chosen. The modulator should allow the predetermined phase encoded pulses $i(t)$ and $q(t)$ to linearly modulate the carrier according to

$$s(t) = a(t)\cos(\omega_c t - \phi(t))$$
$$a(t) = \sqrt{i^2(C) + q^2(t)}$$
$$\phi(t) = \tan^{-1}\frac{q(t)}{i(t)}$$

Fig. 8 shows the block diagram of the QPSK IC. Quadrature carrier inputs are provided to the chip along with differential in phase and quadrature baseband inputs

$$i(t), \overline{i(t)}, q(t), \overline{q(t)}.$$

The carrier inputs are amplified and split to provide complimentary drive to the mixers, which then upconvert the baseband data to 90 or 240 MHz. The output of each mixer is differential and is recombined with two amplifier stages, where their single ended outputs are then combined with another amplifier stage, thereby cancelling one of the sidebands. Perfect cancellation occurs when 0° phase and 0-dB amplitude imbalance is obtained. To that end, careful attention was made to the IC layout symmetry and biasing schemes. Both LO baluns and IF combiners as well as the I/Q combiner are realized using differential amplifiers. In order to ensure the best amplitude tracking on I and Q channels one current source is used for both the IF combiners and I/Q combiner.

A FET quad topology was used for the mixer designs. Each mixer contributes to the carrier leakage and therefore careful attention must be paid to the baseband drive and biasing. Single-ended baseband drive to the mixers is possible, although differential drive results in much better carrier suppression. In the PHS system, it is necessary to dc couple the baseband inputs, and on this IC each mixer was designed for inputs equal to 2.0 V. Because dc offset equates to excess carrier leakage, it is critical to maintain less than 5 mV offset across each mixer.

The 70 dB isolation requirement is obtained using two SPST switches. Complimentary control lines of 0 V and −4 V for each switch can be tied together externally or controlled independently. The switches are designed using series and shunt depletion mode FETs whose pinch off voltage is half that of the negative rail (−2.0 V), allowing a smooth, monotonic attenuation between the series and shunt devices.

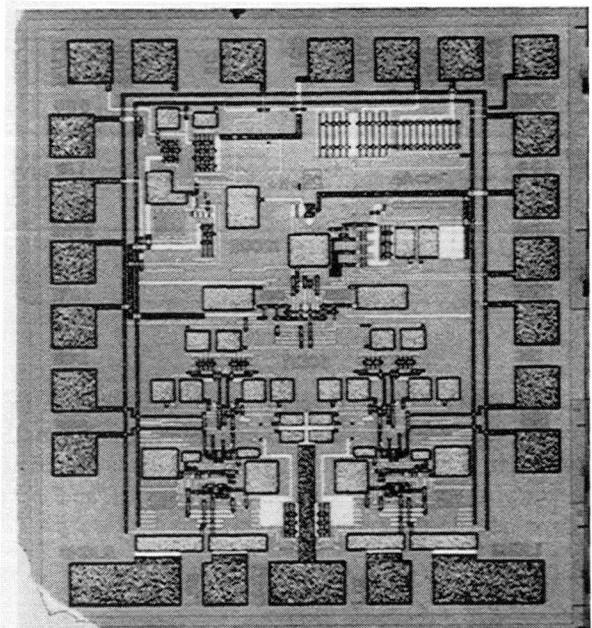

Fig. 9. QPSK modulator chip.

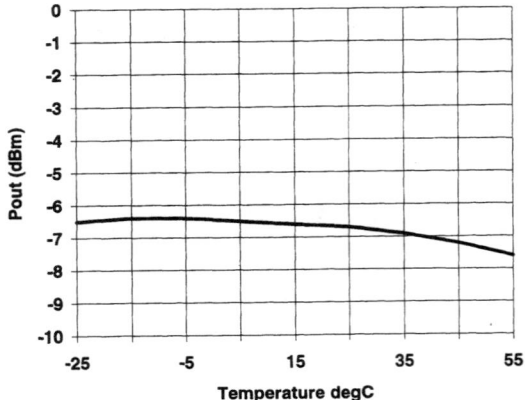

Fig. 10. Output power of modulator over temperature.

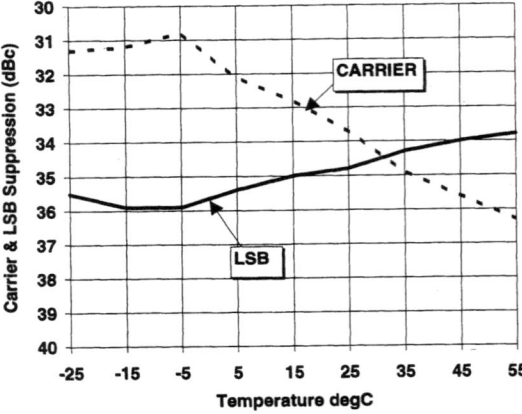

Fig. 11. Carrier suppression and LSB suppression over temperature.

TABLE II
QPSK MODULATOR IC PERFORMANCE SUMMARY

Carrier Inputs	90 to 250 MHz
LOI, LOQ	-4 dBm
Baseband Inputs	150 mV p-p
I, $\overline{IR}$, Q, $\overline{QR}$	2.000V DC
VDD	+4 V
I(VDD) Tx Mode	18 mA
I(VDD) Save Mode	0.8mA
IF Pout, SSB	-7 dBm
Carrier Suppression	>30 dBc
LSB Suppression	>32 dBc
Burst Isolation	>75 dB
ACI (adjacent channel distortion)	>60 dBc at 600 kHz >40 dBc at 300 kHz

A two-stage common source network is used for the IF amplifier design where the output impedance of the second stage is transformed to 50 Ω using an external LC network. In this way, good distortion performance can be obtained with low power consumption.

Finally, the full integration is achieved on a single die design only 2.1 mm² in size, shown in Fig. 9.

The electrical performance and required inputs are summarized in the Table II. The IC is designed to operate from a +4-V supply and consumes only 72 mW of power. In order to save the battery life, the chip can be powered down using a negative logic pin and will consume less than 4 mW of power in this state.

The typical SSB output power is −7 dBm and varies less than 2 dB over temperature as shown in Fig. 10.

Fig. 11 shows the carrier and LSB suppression over the temperature range of interest. As is indicated, both these parameters are very stable over temperature. The resulting

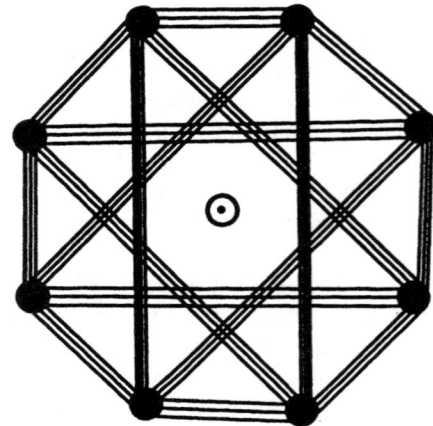

Fig. 12. RMS vector error of modulator.

RMS vector error plot is shown in Fig. 12. This plot includes the error generated by the baseband circuitry, which is roughly 2% resulting in a vector error for this design of 2.81%, significantly better than was budgeted.

Fig. 13 indicated the carrier leakage sensitivity to the dc offsets introduced at the baseband inputs. In order to maintain

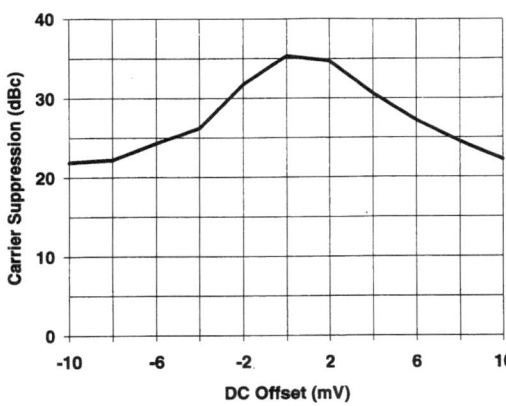

Fig. 13. Carrier leakage sensitivity to dc offsets.

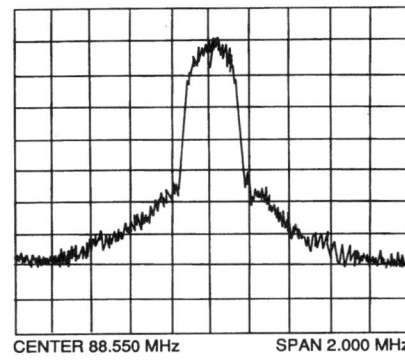

Fig. 15. Output spectrum of modulator with −7 dBm average power.

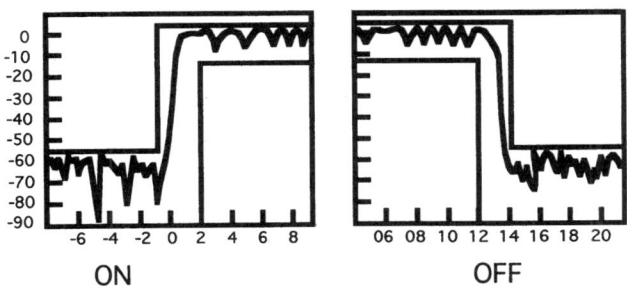

Fig. 14. Burst timing of QPSK modulator on/off.

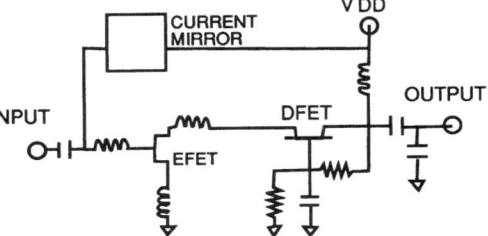

Fig. 16. LNA schematic.

an acceptable RMS vector error, carrier leakage should be better than 25 dBc resulting in a 5 mV allowable dc offset.

The burst timing sequence in Fig. 14 shows greater than 60 dBc on/off ratio achieved in a ramp up or ramp down time of less than 13.0 μS (2 symbols). There is no overlap into an adjacent time slot.

Using a PN9-drive at the baseband inputs, the output spectrum is shown in Fig. 15. For average output power of −10 dBm, the design demonstrates excellent linearity achieving 61 dBc adjacent channel power at 600-kHz offset against a 50-dBc system specification.

V. TRANSCEIVER INTEGRATED CIRCUIT

The main requirement of the downconverter is to achieve good sensitivity and dynamic range with minimum current consumption. Using a 1μm ion-implanted E/D process with low knee voltage and low maximum channel current, a complete up/downconverter was developed. The E-FET features a threshold voltage of +150 mV and I_{max} of 40 mA/mm. The D-FET features a pinch-off voltage of −0.6 V with 60-mA/mm IDSS. Using these devices a Transceiver consisting of an LNA and active downconverter mixer can be realized together with an upconverter mixer driver amplifier and step attenuator. A LO amplifier and LO switch is also realized.

The LNA sets the receiver sensitivity. It should have a good noise figure and sufficient gain to mask the mixer noise figure. The gain should not be so high that the input intercept point is degraded. To this end a cascade configuration is chosen since the gain is not excessive and the device can be series biased to conserve current. Fig. 16 shows the LNA schematic. It utilizes inductive feedback to the common source E-FET to achieve simultaneous noise and impedance matching. Capacitive feedback is used in the common-gate device to optimize linearity, gain, and stability. The input impedance for the common source stage is approximated

$$Z_{in} = Lg_m/C_{gs} + j\left(\omega L - \frac{1}{\omega C_{gs}}\right)$$

The inductance value

$$L = 50/2\pi \text{ ft}$$

where

$$\text{ft} = \frac{gm}{2\pi C_{gs}}$$

Using these equations L is selected to create a 50-ohm input real impedance. The E-FET is selected such that it provides sufficient gm to meet the required transducer gain. The key to optimizing the linearity of a low voltage cascode is to split RF voltage optimally between the FET's. By using capacitive feedback in the common gate stage the input impedance is increased by a factor of $(1 + C_{gs}/C)$. The RF voltage is dropped across the common gate input capacitance C_{gs} and the feedback capacitor C. This technique can be used to optimize the linearity. The LNA output is matched for transducer gain and linearity. The impedance transformation is provided by a single tuned LC circuit that presents a 350-Ω load impedance to the FET while matching the LNA output port to 50 ohms. The LNA achieves 13-dB gain and 2.6 dB noise figure at 3 V and 2 mA, which is equivalent to the best discrete or single function MMIC's currently available [2], [3]. The gain and noise figure also show less than 1 dB and 0.5-dB variation from −20 and 70°C. (Figs. 17 and 18).

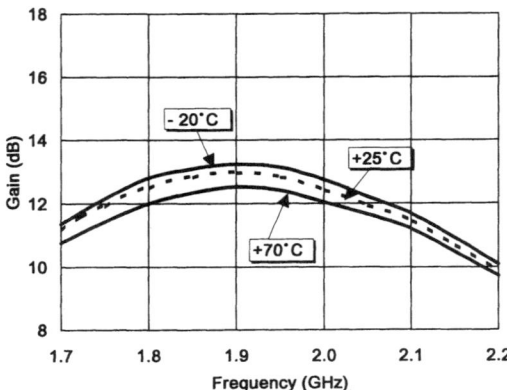

Fig. 17. LNA gain at 3 V and 2 mA.

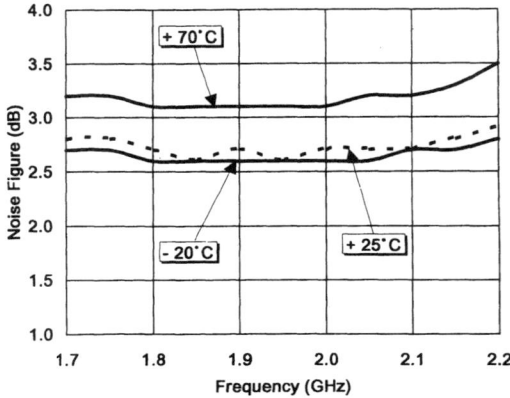

Fig. 18. LNA noise figure at 3 V and 2 mA.

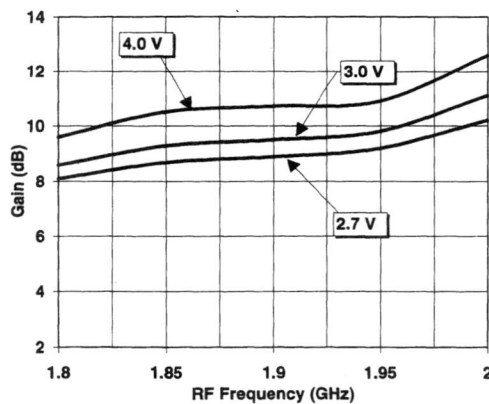

Fig. 19. Receive mixer conversion gain.

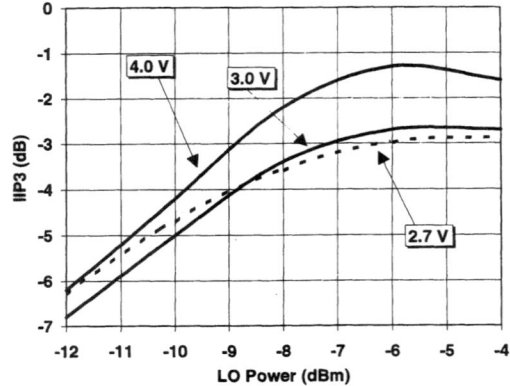

Fig. 20. Receive mixer third-order input intercept point.

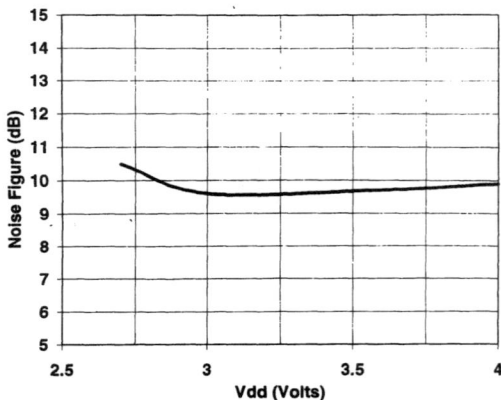

Fig. 21. Receive mixer SSB noise figure.

The receiver mixer is an active single-ended FET design. The LO and RF are summed into the gate of the mixing FET using current sources and a summing circuit. The gain of the LO and RF combiner allows low noise figure and low LO drive to be achieved simultaneously. The conversion gain at 2.7 V is 9 dB with an input intercept point of −4 dBm. The mixer noise figure is 9.5 dB. Figs. 19, 20, and 21 summarize the mixer performance.

The combined LNA and Receive Mixer assuming 2 dB loss in the receive image filter allows for an input intercept point of −14 dBm at the input to the LNA or −12 dBm at the antenna input. This allows 4-dB margin to be achieved against the system specification of −16 dBm. The receiver noise figure at the input to the LNA is 3.5 dB. Refering to the antenna input, the noise figure is 5.5 dB, which gives 6-dB production margin. This significantly increases the sensitivity of the radio over the minimum required by the system allowing for high yields in mass production. The entire receiver including LNA, receive mixer and LO buffer amplifier draws only 8 mA @ 3 V.

The main concept behind the transmit chain design is to allow reasonable levels of filtering and optimize the power added efficiency of the complete chain. Since the PA consumes most of the dc power it is optimal to drive the PA as hard as possible without degrading the adjacent channel interference (ACI) specification. Therefore all the other components in the chain must be linear in operation.

A key feature of the Transceiver is a linear transmit mixer, which is essentially a balanced Gilbert cell upconverter [4]. To maintain linearity and good LO suppression, it is necessary to operate the mixer at a level where the third order intermodulation ratio (IMR) is better than 40 dBc at the mixer output. In this case the output intercept point of the mixer is minimum of +6 dBm when operating with −14 dBm—output power. The intermodulation ratio is held at 40 dBc by this design. The LO leakage measured at the mixer output is better than

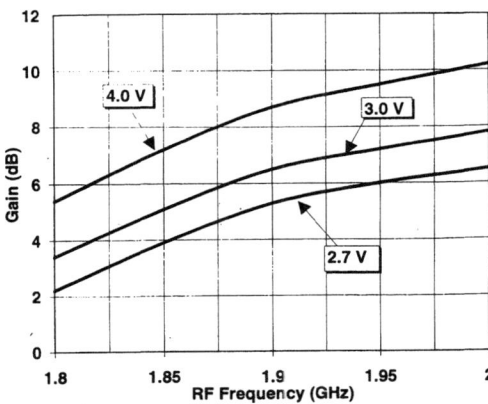

Fig. 22. Transmit mixer conversion gain.

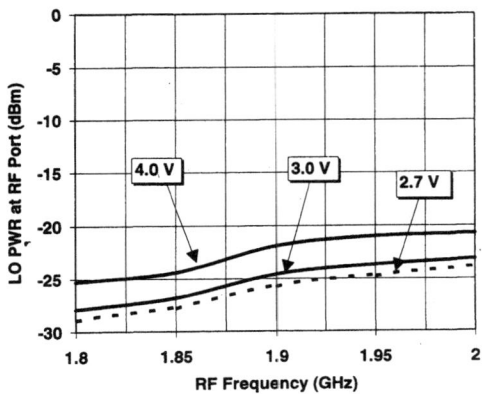

Fig. 23. Transmit mixer L_o leakage.

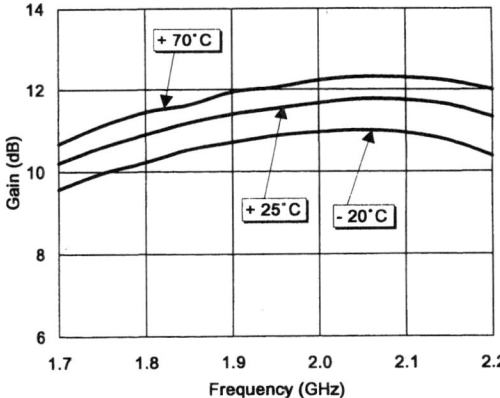

Fig. 24. Driver amplifier gain.

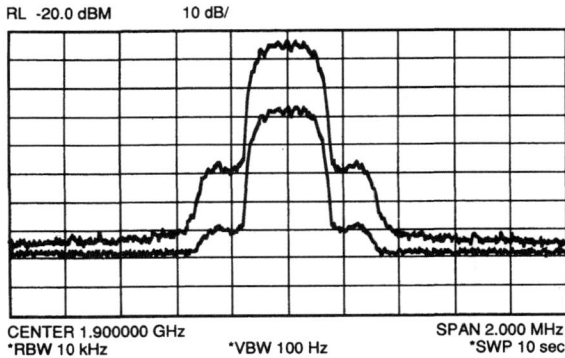

Fig. 25. Driver/attenuator output spectrum at +2 dBm and −17 dBm average power.

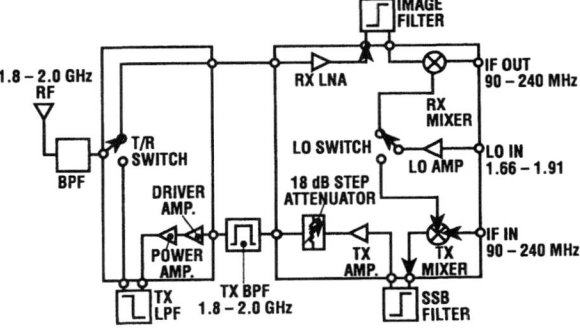

Fig. 26. RF chip set partition.

−20 dBm with −8 dBm of LO drive. Since the on chip LO amplifier supplies 8-dB gain the LO to RF isolation is better than 20 dB. Thus filtering of the LO is simplified. The mixer consumes 20 mA at +3 V. Figs. 22 and 23 show the mixer conversion gain and LO leakage.

Following the mixer in the Tx Chain is a driver amplifier with 11-dB gain drawing 10 mA of bias current. Fig. 24 shows the power gain of this amplifier. The amplifier is designed to deliver 0 dBm of drive to the power amplifier. As with the mixer and modulator the key design issue for the driver amplifier is linearity. The amplifier is driven with a $\pi/4$ DQPSK signal modulated by a 9-bit psuedonoise sequence. Even with +4-dBm output power the sidelobe regrowth under drive is minimal and the ACI level at 600-kHz offset is better than 60 dBc as seen in Fig. 25. The system specification is 50 dBc. Following the driver amplifier is a 19-dB digital attenuator required for adaptive power control. This allows received power levels to be somewhat equalized at the base station when the portable station are at different distances from it. Fig. 25 shows the output spectrum in the high and low power states.

VI. POWER AMPLIFIER/SWITCH IC

The Power Amplifier and Transmit/Receive Switch completes the chip set. Fig. 26 shows the RF partition. The transmit band pass filter reduces the image and LO levels to 45 dBc required at the antenna. The main requirements of the power amplifier are:

1) Good adjacent channel distortion;
2) Low operational voltage 3 V; and
3) High power added efficiency

This requires optimization of the active device and the circuit design. Since the $\pi/4$ DQPSK signal to be amplified is continuously varying in amplitude and phase, any MESFET nonlinearity can distort the ideal Root Nyquist spectrum, causing sidelobe regrowth. Therefore it is not just the "hard" nonlinearities at the device knee or breakdown but also the soft nonlinearities such as gm (V_{gs}, V_{ds}), $C_{gs}(V_{gs}, V_{ds})$ R_{DS}

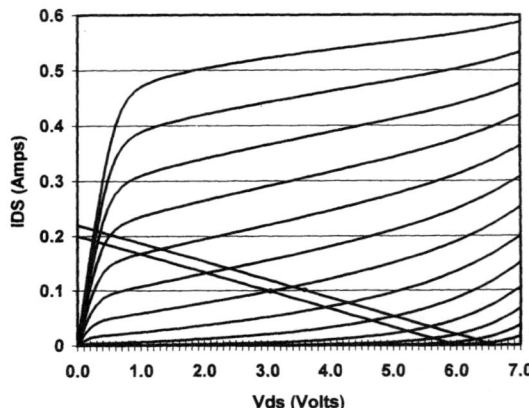

Fig. 27. I-V characteristics of 3-mm device (30 Ω load).

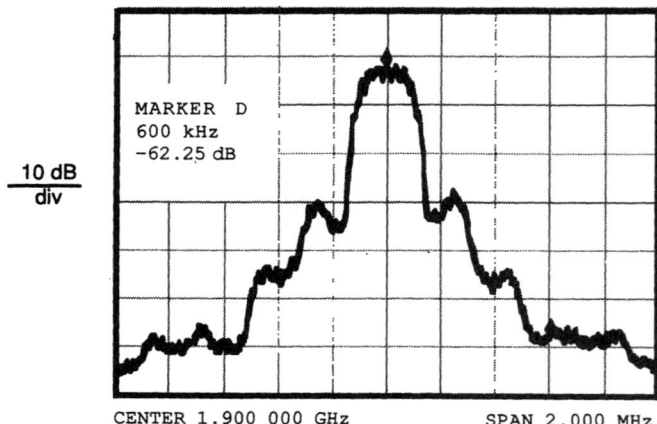

Fig. 28. Output spectrum of power amplifier +22 dBm @ 3.6 V and 130 mA.

TABLE III
POWER AMPLIFIER PERFORMANCE SUMMARY

Vdd1 / Vdd2	Pout (dBm)	ACI (dBc)	Pin (dBm)	Idd1/Idd2 (mA)	PAE (%)	Harmonics (dBc)
(4.8/5.8)	27	56	+5.8	212	39	41
4.0	23	60	1.8	138	34	48
3.6	22	61	1.1	130	34	43
3.5	22	59	1.2	130	35	42
3.4	22	57	1.7	125	37	34
3.0	21	60	1.0	110	38	42

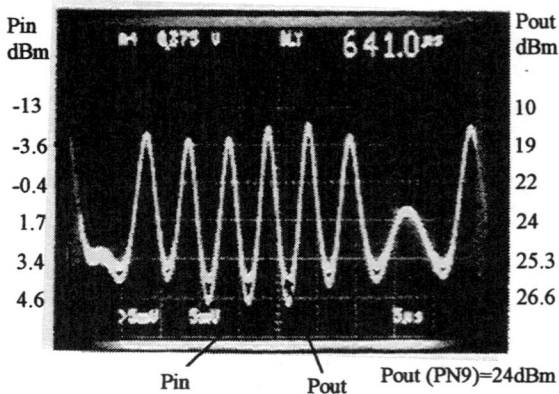

Fig. 29. Instantaneous power variation in the time domain.

(V_{gs}, V_{ds}) as the device swings over its load line. Thus linearity as a function of V_{gs}/V_{ds} is required for minimum side lobe regrowth. To optimize efficiency in the 3–3.6-V range it is desirable to have a device with the minimum "knee" voltage. A device with low pinch-off and knee voltage is thus more suitable than the traditional power device. By optimizing the load line and oversizing the device the "knee" voltage can be further reduced. Fig. 27 shows the I-V characteristics used in the final stage of the power amp design. A 3-mm device is chosen using a process with a buried Be P-layer. This helps to linearize gm at low currents. The pinch-off voltage is 1.2 V for this device the I_{DSS} is 520 mm. The quiescent bias of 100 mA (20% I_{DSS}) is chosen. Presenting a 30-Ω load line to this device allows for class AB operation. The real voltage swing is

$$V_{DD} - V_{knee} = 3.0 - 0.5 \text{ V} = 2.5 \text{ V}.$$

The peak power is estimated at 23 dBm. The intrinsic linearity of gm (V_{gs}, V_{ds}) allows for minimum distortion on the load line. It is possible to simulate the effect of other device nonlinearities such as C_{gs} (V_{gs}, V_{ds}) on the output spectrum. This approach is lengthy and requires extreme precision in device models since 5th- and 7th-order nonlinearities need to be simulated. Our approach was to load pull the device to achieve maximum efficiency band power under 60 dBc distortion levels using a $\pi/4$ DQPSK modulated source. Once the final stage FET periphery is established the two-stage design is optimized for maximum gain. The matching circuitry is realized using high Q inductors with a Q of 30. The available loss of the output matching circuits including plastic package losses is 0.5 dB. Fig. 28 shows the output spectrum of the final PA stage. It delivers +22 dBm of ($\pi/4$ DQPSK average power at 3.6 V bias with 60 dBc distortion. The overall efficiency is 34%. It is capable of delivering 21 dBm at 3 V thus allowing for 2-dB loss between the PA output and the antennae. Table III summarizes the output power and efficiency for various supply voltages. Fig. 29 shows instantaneous output power of the amplifier showing clearly compression of the peak output power.

The T/R switch uses a multigate design [5] to achieve low loss and high linearity. The insertion loss is 0.6 dB with an input intercept point of +48 dBm. Thus the switch causes no degradation in side lobe performance.

VII. MANUFACTURING THE CHIP SET

Many devices including LNA's, power amplifiers have been published recently [2], [5], [6]. This work represents the most highly integrated chip set available for PHS applications. The increased integration allows the RF portion of the phone to be realized with less board area allowing ultimately for a smaller handset. Fig. 30 shows the transceiver IC. All the functions are realized on a 3.5 mm^2 die. Fig. 31 shows the PA/switch IC which occupies 1.5 mm^2.

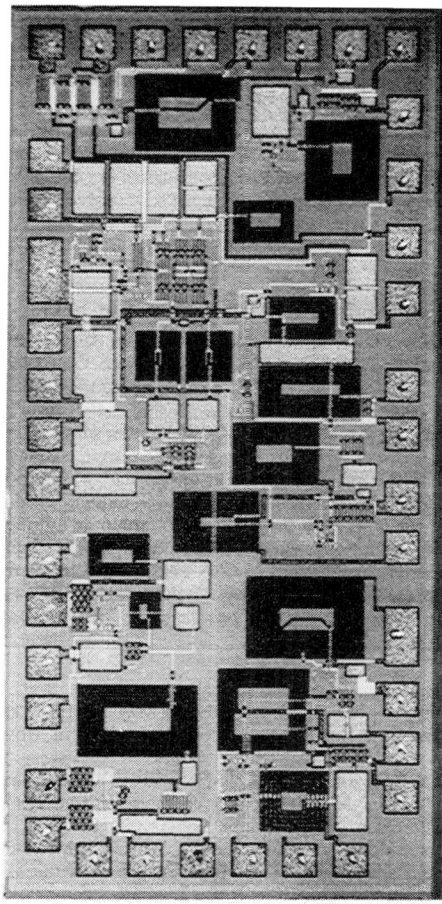

Fig. 30. Transceiver integrated circuit.

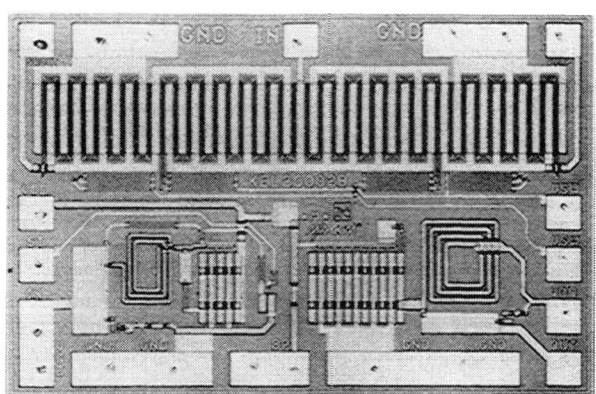

Fig. 31. Power amp/switch integrated circuit.

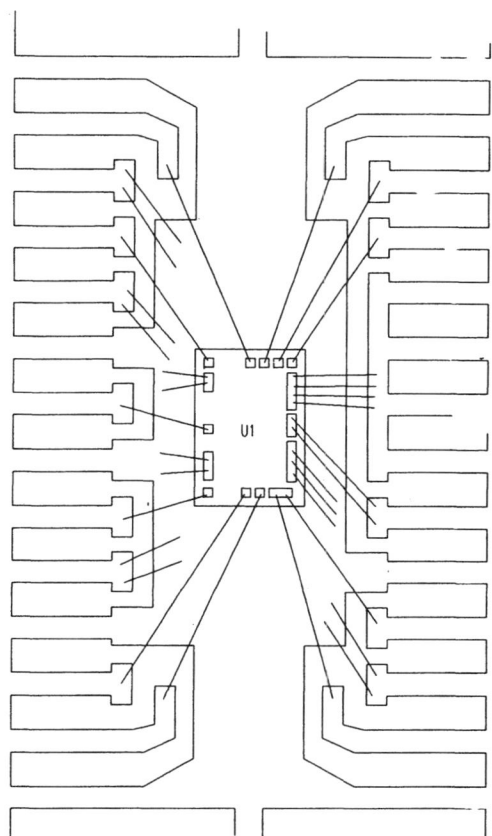

Fig. 32. Power amplifier switch assembly.

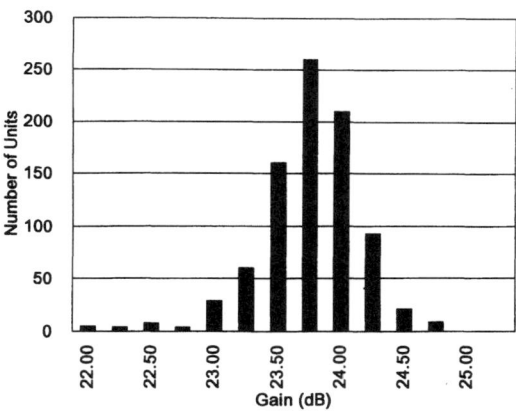

Fig. 33. Power amplifier gain statistic in production.

The chip set uses standard surface mount Shrink Small Outline (SSOP) Packages. Standard pick and place machines can be used in mass production. The packages are assembled using standard wire bonding and transfer molding techniques with a total cost of tens of cents. Fig. 32 shows the power amp and switch assembly. The use of standard packages allows for regular autohandlers to perform fully functional RF testing. Thus the complete chip set can be tested and the transceiver performance prediction before the PCB is assembled. Figs. 33 and 34 show the power amplifier gain and one dB compression point from a production lot of this power amplifier switch. Yields in excess of 90% have been achieved. As the manufacturing line matures the yield is expected to increase.

This chip set represents the standard in high volume highly integrated RFIC's currently available and is a crucial enabling technology for the PHS system. As the system is deployed in Japan usage of this kind of chip set is expected to be in the millions per year range. The PHS digital communications system is one of the major emerging global standards.

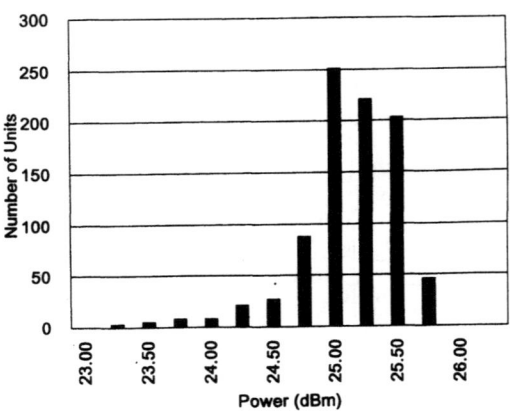

Fig. 34. Power amplifier one dB compression statistics in production.

REFERENCES

[1] Personal Handyphone System RCR Standard 28, Ver 1, Dec. 20, 1993.
[2] E. Heaney et al., "Ultra low power low noise amplifiers for wireless communications," in *GaAs IC Symp.*, 1993, pp. 49–52.
[3] M. Nakatsugawa et al., "An L-band ultra low power consumption monolithic low noise amplifier," in *GaAs IC Symp,* 1993, pp. 45–48.
[4] M. S. and J. M. Moniz, "A high performance GaAs MMIC up-conversion mixer for 2.45-GHz wireless spread spectrum communications" *IEEE MTT-S Euro. Topical Congress,* Nov. 2–5, 1994.
[5] F. McGrath, C. Kermarrec, C. Varmazis, and R. Pratt, "Multigate FET power switches," *Appl. Microwaves,* pp. 77–86, summer 1991
[6] T. Kunihisa, T. Yokoyama, H. Fujimoto, K. Ishida, H. Takehara, and O. Ishikawa, "High efficiency, low adjacent channel leakage GaAs power MMIC for digital cordless telephone," in *IEEE 1994 Microwave and Millimeter-Wave Circuits Symp.,* 1994.
[7] T. Yoshimasu, N. Tanba, and S. Hara, "An HBT linear power amplifier for 1.9-GHz personal communications," in *IEEE 1994 Microwave and Millimeter Wave Circuits Symp.,* 1994.

PAPER Special Issue on Microwave Devices for Mobile Communications

Design Study on RF Stage for Miniature PHS Terminal

Hiroshi TSURUMI[†], Tadahiko MAEDA[†], Hiroshi TANIMOTO[†], Yasuo SUZUKI[†],
Masayuki SAITO[†], Kunio YOSHIHARA[†], Kenji ISHIDA[†],
and Naotaka UCHITOMI[†], *Members*

SUMMARY A miniature transceiver, including highly integrated MMIC front-end, for 1.9 GHz band personal handy phone system (PHS) has been developed. The terminal, adopting direct conversion transmitter and receiver technology, consists of four high-density RF circuit modules and a digital signal processing LSI with 2.7 V power supply. The four functional modules are a power amplifier, a transmitter, a receiver, and a frequency synthesizer. Each functional module includes one IC chip and passive LCR components connected with solder bumps on module substrate. The experimental miniature PHS handset has been fabricated to verify the design concepts of the miniature transceiver. The total volume of the developed PHS terminal is 60 cc, including the 12 cc front-end which comprises the four RF functional circuit modules. The air interface connection with the PHS base station simulator has been confirmed.
key words: direct conversion, personal handy phone system, RF functional circuit module, bump interconnection technique

1. Introduction

Recently, various kinds of wireless communication terminals for GSM, DECT, and PHS have been developed. The principal requirements for RF stage in these terminals are as follows:
Small size – reduction of the RF stage volume will provide the terminal with sufficient space for the other general functions, such as an LCD, a vibrator, and a connector for multimedia interface.
Low cost – both components and assembly processes should be inexpensive for mass production.

The conventional superheterodyne transceiver with on-board chip-components assembly, however, has serious limitations in terms of size and cost.

This paper describes the miniaturization technologies applied to RF stage for the PHS terminal and the above requirements are discussed, focusing on the electrical design and the high-density assembly technique for the RF stage circuits.

2. Features of the RF Stage

The features of the miniature terminal developed to meet the requirements described above are as follows:
(1) Adoption of the direct converstion architecture

Manuscript received November 14, 1995.
Manuscript revised February 6, 1996.
[†]The authors are with the Research and Development Center, Toshiba Corporation, Kawasaki-shi, 210 Japan.

The direct conversion architecture [1],[2] is applied to realizing front-end ICs without off-chip components and adjustment parts. This architecture is suitable for achieving the miniature low-cost transceiver, since adherence to this design principle can eliminate bulky passive filters for both transmitter and receiver chains.
(2) Module configuration of the RF functional circuits
The RF circuit module with the bump interconnection technique has advantages in terms of assembly density and improvement of RF performance.
(3) Operation with 2.7 V power supply
GaAs MESFET power amplifier (PA) operating at high gain and high linearity with single low voltage power supply is required for commercial use, for which lithium ion rechargeable battery will be applied.

3. RF Stage Circuits and Modules

The block diagram of the transceiver RF stage is depicted in Fig. 1. The RF stage consists of four RF functional circuit modules, which are a power amplifier (PA), a transmitter (TX), a receiver (RX), and a frequency synthesizer (SYN). Each module includes one IC chip and some off-chip components.

3.1 Transmitter Chain

The transmitter chain consists of PA module and TX

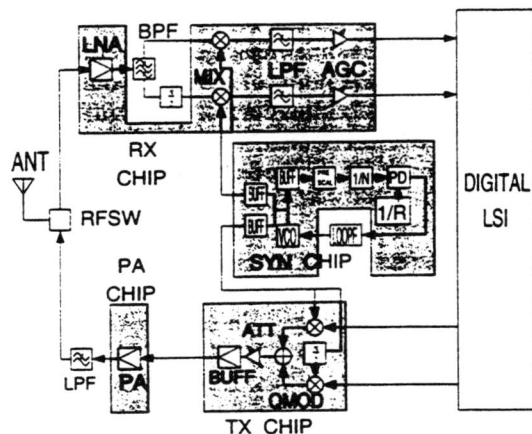

Fig. 1 Block diagram of transceiver RF stage.

Reprinted with permission from *IEICE Transactions on Electronics*, H. Tsurumi, T. Maeda, H. Tanimoto, Y. Suzuki, M. Saito, K. Yoshihara, K. Ishida, and N. Uchitomi, "Design Study on RF Stage for Miniature PHS Terminal," vol. E79-C, no. 5, pp. 629-634, 1996. © The Institute of Electronics, Information and Communication Engineers.

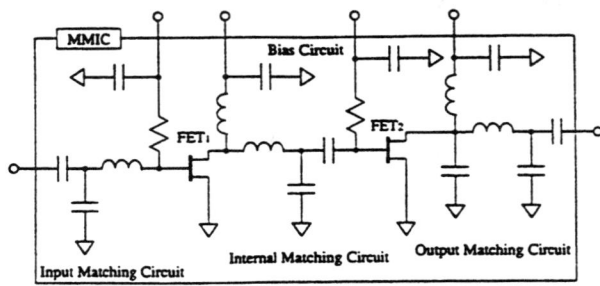

Fig. 2 Power amplifier MMIC.

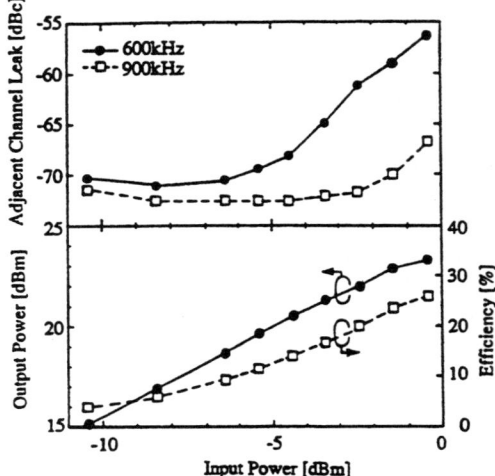

Fig. 3 Electrical performance of PA module.

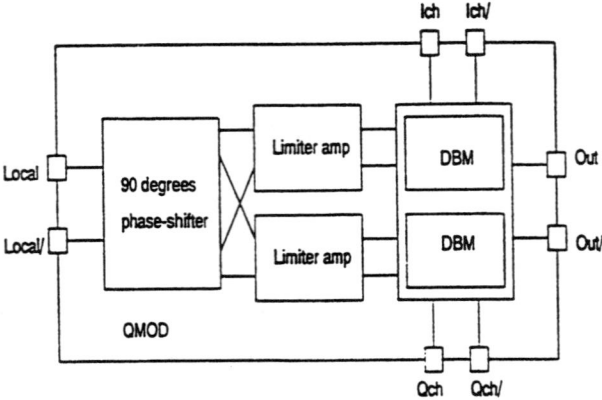

Fig. 4 Fully differential QMOD configuration.

module. The ICs of both modules are fabricated with GaAs MESFET process.

The PA IC, fabricated by the refractory WNx/W self-aligned 0.6 μm gate GaAs power MESFET [3], operates with gate bias of 0 V and drain bias of 2.7 V. The IC includes two MESFETs, MIN capacitors, resistors, and spiral inductors as shown in Fig. 2. The PA IC is mounted on alumina substrate with chip capacitors [4]. The electrical performance of the PA module is shown in Fig. 3. The required PA output power for the PHS terminal is over 22 dBm under the condition of the adjacent channel leakage of less than −55 dBc. The measured adjacent channel leakage is −61 dBc with efficiency of 20 % at the point of measured output power of 22 dBm. The dissipation current is 270 mA.

The TX IC, fabricated with 0.5 μm gate GaAs MESFET process, consists of a quadrature modulator (QMOD), a variable gain attenuator with 0 to 28 dB attenuation in 4 dB step, and a buffer amplifier with fixed gain of 16 dB [5]. The main requirement for the TX module is to achieve accurate IQ matching in the QMOD and high-accuracy gain step in the attenuator without manual adjustment.

Figure 4 shows the block diagram of the QMOD, including a π/2 phase shifter, limiter amplifiers, and a pair of double balanced mixers (DBM).

The following innovated key components are applied to TX IC.

(1) High gain π/2 phase-shifter

The current mode technique [6] reduces the amplitude error caused by the offset voltage difference in limiter amplifier pairs.

(2) High-accuracy step attenuator

The current driven MESFET switching techniques [5] reduces the threshold voltage variation in GaAs MESFET.

The QMOD, the attenuator, and buffer amplifier can be designed with the same IC process technology without degradation of the gain accuracy, which reduces the process fabrication cost.

The TX IC is mounted in the TX module with DC blocking capacitors for input/output stage. The output power is −3 dBm in the 1.9 GHz band with the dissipation current of 96 mA and the local input level of −10 dBm. The local leakage rejection ratio of over 28 dB and the vector error of less than 6 % are attained without any manual adjustment over the entire PHS bandwidth.

3.2 Receiver Chain

The direct conversion receiver (DCR) chain consists of a Si/BiCMOS chip [7] including a low noise amplifier (LNA), a pair of down converters (MIX), low-pass channel selection filters (LPF) and automatic gain control amplifiers (AGC). The DCR chip is assembled into the RX module with off-chip passive components. The outputs of IQ channels are fed to the digital LSI including A/D converter, digital root roll-off filter, and differential detector.

This paper focuses on the following three inherent problems which are of predominant interest in the development of the direct conversion receiver.

(1) Local radiation
(2) Dynamic range reduction
(3) 2nd-order nonlinear distortion

Local (LO) radiation from DCR could often interfere with nearby receivers. The LO power must be

Table 1 Electrical characteristics of RX module.

Gain Control	LNA 40 dB (2 step)
	MIX 20 dB (2 step)
	AGC 36 dB (7 step)
Power Down Control	RF Stage : LNA and MIX
	BB Stage : LPF and AGC
Output	I & Q 2ch (1 V_{p-p})
Dissipation Current	60 mA

Table 2 Bit error rate measurement conditions.

Frequency	1.9 GHz
Modulation Scheme	$\pi/4$ shifted QPSK
	(roll-off factor: 0.5)
Bit Rate	384 kbps
RF Input Level	$-100 \sim +10$ dBm
Fading Condition	static
Detector	baseband differential detection

suppressed by 65 dB to avoid the sensitivity degradation of the nearby PHS receivers within a distance of 1 m from the DCR. Both LO-RF isolation in MIX and backward gain $|S_{12}|$ reduction in LNA achieve over 65 dB LO power suppression. Furthermore, the metal cap is prepared for the RX module to increase the shield effect.

Dynamic range reduction will occur when baseband LPF or AGC is saturated by received signal, especially linear modulation signal such as $\pi/4$ shifted QPSK specified in PHS[8]. In the developed receiver section, the gain-control function is distributed in LNA, MIX, and AGC as shown in Table 1 to avoid saturation at baseband stage, resulting in an overall gain-control range of 96 dB. Each circuit-gain is changed discretely according to the received signal power detected in the baseband received signal strength indicator (RSSI) included in the digital LSI. Both the LNA and MIX operate in high-gain and low-gain mode to produce 4 different gain modes. Additionally, AGC is applied to adjust the total gain precisely with 6 dB gain step. The optimum gain mode is selected automatically to realize sufficient C/N to meet average bit error rate (BER) of less than 10^{-5} over the entire receiver input range under static condition.

2nd-order nonlinear distortion, produced mainly in MIX and LPF circuits, is not so severe in the constant envelop modulation scheme, such as GMSK specified in GSM [1] or DECT [2]. However, for the PHS direct conversion receiver, $\pi/4$ shifted QPSK will cause 2nd-order nonlinear distortion because its envelop has amplitude components. In the developed MIX circuit, firstly, the received signal is capacitor coupled to the single balanced mixer (SBM) to cut the 2nd-order distortion produced in the preceding preamplifier stage [7]. Secondly, common centroid geometry is applied in the layout to improve the mismatching of the differential pair in SBM, which causes 2nd-order nonlinear distortion Furthermore, the LPF achieves over 80 dB 2nd-order inter-modulation distortion by using the linearization technique in the gyrator circuit [7].

BER measurement was carried out under the conditions described in Table 2. The stripline circuit introduced in Sect. 5, which has two functions of RX band-pass filter and the $\pi/2$ power splitter depicted in Fig. 1, was inserted between LNSA and MIX. The output IQ signals from the RX module are fed into the digital LSI to be digitized by 8 bit A/D converter. The digitized

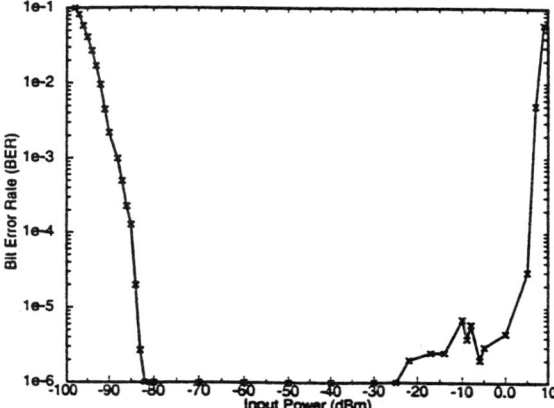

Fig. 5 Bit rate measurement result of RX module.

IQ signals are filtered by digital root roll-off filters and recovered by the following differential detector to data sequence in this LSI. The measured BER characteristic is shown in Fig. 5 where the optimum gain mode is selected according to the input signal level. Figure 5 reveals that the receiver chain achieves dynamic range of around 85 dB at BER of 10^{-5} and 95 dB at BER of 10^{-2}.

DCR will often suffer noise figure (NF) degradation compared with the superheterodyne receiver. The measured NF of the DCR chip has not reached the target necessary to preclude degradation of receiver sensitivity. Further NF improvement especially in LNA and MIX in the advanced IC version is required to meet the sensitivity specification.

3.3 Frequency Synthesizer

The frequency synthesizer IC, fabricated by Si/BiCMOS process, consists of a reference divider, a programmable divider, a phase detector, a prescaler, a voltage controlled oscillator (VCO) circuit, buffer amplifiers, and a logic circuit[9]. The IC is assembled into the SYN module with off-chip passive components for a loop filter and a VCO resonator. Figure 6 shows the block diagram of the frequency synthesizer.

The features of the developed frequency synthesizer IC are as follows:
(1) Efficient power down function is provided (the dissipation current in power down mode is less than 10μA).
(2) High frequency-stability performance against load

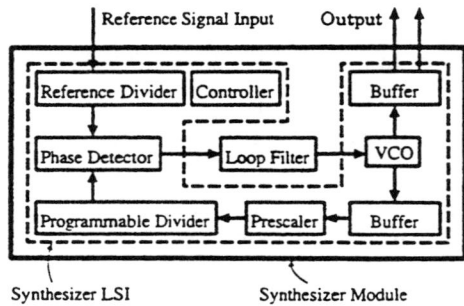

Fig. 6 Block diagram of synthesizer IC and module.

Table 3 Electrical performance of the SYN module.

Output power	-7 dBm
Switching time	3.5 msec ($\Delta F = 22.8$ MHz)
SSB phase noise	-122 dBc/Hz @ 600 kHz separation
Spurious rejection	< -80 dBc @ 600 kHz separation
Dissipation Current	28 mA

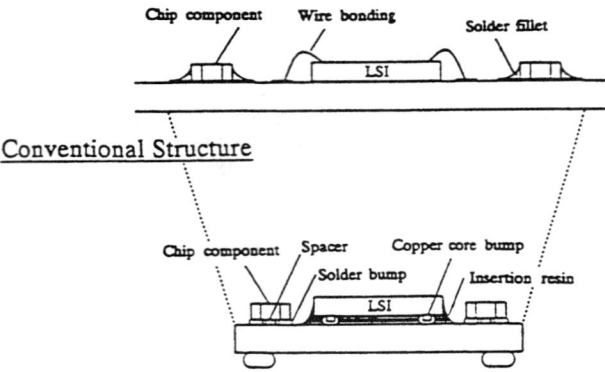

Fig. 7 Cross-sectional view of bump interconnection structure.

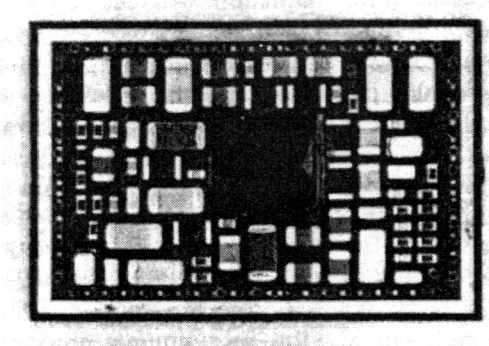

Fig. 8 Photograph of RX module.

changes is achieved by on-chip buffer amplifiers.

The electrical characteristics of the SYN module are summarized in Table 3. The required phase noise and spurious rejection in the adjacent channel for PHS application are -120 dBc/Hz and -80 dBc, respectively, and both have been achieved. The phase detector operates at 50 kHz whereas channel separation frequency in PHS is 300 kHz, since PHS carrier frequency is not a multiple of the channel separation frequency. Much switching time is expended compared with a conventional synthesizer under the above constraint. The measured switching time is acceptable for PHS application, but further improvement will be a subject for future study.

4. Module Assembly Techniques

The front-end RF functional circuits, namely PA, TX, RX, and SYN, are assembled into four RF modules, respectively. Each module consists of one IC chip, and passive LCR components connected with solder bumps on module substrates.

The bump interconnection technique for IC chip and passive LCR components are also adopted for the RX and SYN chip[10]. The cross-sectional view of the bump interconnection structure is shown in Fig. 7. The passive LCR components and the IC chip can be assembled simultaneously using conventional assembly equipment. The module size is reduced by around 40% compared with those assembled by the conventional method. Additionally, the bump interconnection technique has advantages in terms of improved RF performance due to reduction of the parasitic capacitance and inductance of the interconnection pad. A photograph of the RX module for which the bump interconnection technique is adopted is shown in Fig. 8.

A glass epoxy printed wiring board (PWB) is used for RX and SYN module as substrate so as not to increase the module cost. An alumina substrate is applied to PA and TX in view of the thermal compatibility between GaAs IC and alumina substrate.

A land grid array (LGA) structure[4] is applied for each module to minimize the assembly area on the motherboard. An additional effect of adopting the LGA structure is the reduction of electro-magnetic radiation from input and output electrodes to improve high-frequency performance on the RF stage.

In a final stage, a metal cap is prepared for each TX, RX and SYN module, and a resin cap for the PA module. For the SYN module, a further metal cap is provided on the motherboard to avoid the interference from the antenna and PA because the interference frequency is the same as SYN output.

5. Motherboard Structure

The motherboard consists of multilayers including component layer, stripline circuit layer consisting of RF filters and directional coupler, digital circuit layer, and power supply circuit layer. Each circuit layer is electrically separated by GND layers to reduce electromagnetic interference[11].

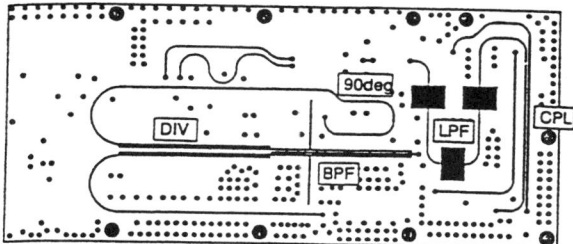

Fig. 9 Structure of stripline circuit layer.

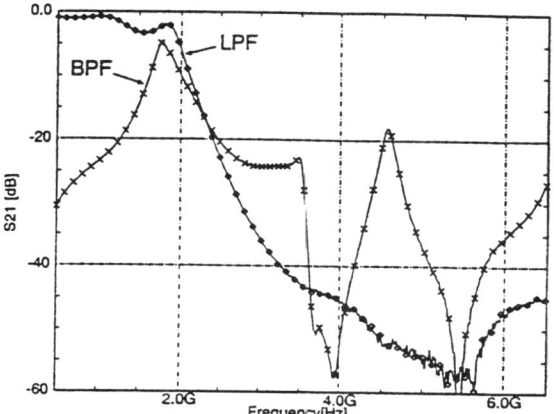

Fig. 10 Frequency response of BPF and LPF.

The RF filters can be designed in a stripline structure with rather low Q since the direct conversion architecture has inherently no image and spurious response. The stripline RX filter provides sufficient tolerance against interference signals that can cause nonlinear distortion such as intermodulation or crossmodulation in the receiver RF stage [12]. The structure of the RF filters and directional coupler is introduced below.

The filter layer contains a TX low-pass filter (LPF) and an RX bandpass filter (BPF) designed in the triplate-type stripline structure. The horizontal-sectional view of the filter layer is shown in Fig. 9. The stripline structure reduces both the terminal cost and volume compared with conventional superheterodyne transceiver with dielectric or SAW filters. Figure 10 shows frequency response of BPF and LPF. The filter layer also contains a power splitter with $\pi/2$ phase shifter for RX chain, providing gain and phase imbalance of less than 1dB and 5 degree, respectively. A directional coupler, providing 15 dB of coupling within 1 dB error, is used for the transmission power detection. The practical use of the circuits in the filter layer has been confirmed through the air interface connection test.

6. Experimental PHS Handset

The experimental miniature PHS handset has been manufactured to examine the design concepts of the miniature transceiver, using the developed four RF circuit

Fig. 11 Photograph of miniature transceiver handset.

modules, the motherboard and all other components such as digital LSIs, an antenna, interface connectors, a keyboard, a lithium ion rechargeable battery, and so on required to realize basic functions of the terminal. The total volume of the handset is 60 cc, including 12 cc RF stage. A photograph of the developed miniature transceiver handset is shown in Fig. 11. The air interface connection test with PHS base station simulator has been achieved and voice communication has been confirmed.

7. Conclusion

A miniature PHS transceiver utilizing the direct conversion technology and the RF module structure concept has been developed. Both architectures contribute to achieving miniaturization and cost reduction of the wireless terminal. The experimental miniature PHS handset has been manufactured to examine the miniature design concepts using the developed RF modules. The air interface connection with PHS base station simulator has been confirmed; however, the evaluation of the total performance based on RCR standard [8] is left as a future subject.

The front-end design concepts and techniques described here are promising for application in the development of low-cost miniature terminals for other radio communication systems such as DECT or PCS.

References

[1] J. Sevenhans, A. Vanwelsenaers, J. Wenin, and J. Baro, "An integrated Si bipolar RF transceiver for a zero IF 900 MHz GSM digital radio frontend of a hand portable phone." in Custom IC Conf., pp.7.7.1-4, 1994.

[2] J. Sevenhans, D. Haspeslagh, A. Delarbre, L. Kiss, Z. Chang, and J.F. Kukielka, "An analog radio front-end chip set for a 1.9 GHz mobile radio telephone application,"

ISSCC'94 Dig. Tech. Papers, pp.44–45, 1994.

[3] M. Nagaoka, T. Inoue, K. Kawakyu, S. Obayashi, H. Kayano, E. Takagi, Y. Tanabe, M. Yoshimura, K. Ishida, Y. Kitaura, and N. Uchitomi, "High-efficiency monolithic GaAs power MESFET amplifier operating with a single low voltage supply for 1.9 GHz digital mobile communication applications," '94 IEEE MTT-S Dig. pp.577–580, 1994.

[4] E. Takagi, N. Ono, M. Nagaoka, T. Inoue, S. Obayashi, H. Kayano, Y. Kitaura, K. Ishida, and M. Konno, "Miniaturized power amplifier module for 1.9 GHz digital mobile communication applications," Proc. Asia Pacific Microwave Conf., pp.355–358, 1994.

[5] T. Sasaki, S. Otaka, T. Maeda, T. Umeda, K. Nishihori, A. Kameyama, M. Morizuka, and N. Uchitomi, "A GaAs direct-conversion $1/4\pi$ shifted QPSK modulator IC with 0-28 dB variable attenuator for 1.9 GHz personal handy phone system," '95 GaAs IC Symposium Tech. Dig., pp.241–244, 1995.

[6] S. Otaka, T. Yamaji, R. Fujimoto, C. Takahashi, and H. Tanimoto, "A low local input power 1.9 GHz Si-Bipolar quadrature modulator without any adjustment," Proc. BCTM'94, pp.171–174, 1994.

[7] C. Takahashi, R. Fujimoto, S. Arai, T. Itakura, T. Ueno, H. Tsurumi, H. Tanimoto, S. Watanabe, and K. Hirakawa, "A 1.9 GHz Si direct conversion receiver IC for QPSK modulation systems," ISSCC'95 Dig. Tech. Papers, pp.138–139, 1995.

[8] Research and Development Center for Radio Systems: RCR STD-28, 1993.

[9] H. Yoshida, C. Takahashi, Y. Shizuki, H. Masuoka, and H. Kokatsu, "1.9 GHz synthesizer for direct conversion," Proc. IEICE Commun. Society Conf. '95 B-196 (in Japanese), 1995.

[10] T. Togasaki, T. Uchida, Y. Iseki, M. Mori, and M. Saito, "Bump interconnection for chip components and LSI chips for high density modules," Proc. ISHM'94, pp.266–271, 1994.

[11] T. Amano, H. Tsurumi, T. Maeda, and Y. Suzuki, "1.9 GHz stripline filters for direct conversion radio terminals," Proc. IEICE Commun. Society Conf. '95 B-191 (in Japanese), 1995.

[12] H. Tsurumi and T. Maeda, "Design study on a direct conversion receiver front-end for 280 MHz, 900 MHz, and 2.6 GHz band radio communication systems," Proc. IEEE VTC'91, pp.457–462, 1991.

AUTHOR INDEX

A

Abidi, A. A., 3, 118, 135, 218, 248, 273, 351, 399, 448, 451, 477
Ablassmeier, U., 581
Adachi, C., 392
Ahmadouche, A., 546
Ahmed, K. A., 218
Aikawa, M., 419
Arimato, H., 69
Arnold, R. G., 470
Asai, T., 594
Ashby, K. B., 466

B

Baro, J., 323
Bastek, J. J., 480
Bateman, A., 92, 99
Begnoche, T., 643
Behbahani, F., 344
Beyer, S., 212, 363, 431
Birth, W., 197, 344
Bonato, G. L., 173
Bonek, E., 105, 112
Bóveda, A., 173
Brianti, F., 354
Brokaw, A. P., 360
Bronner, P. E., 338
Buck, B. J., 347

C

Camilleri, N., 484, 526
Ceuterick, J., 327
Chan, P. Y., 218
Chang, G., 351, 448
Chang, J., 351
Chang, J. Y.-C., 248, 477
Chang, P., 351
Chiba, K., 370
Chien, G., 354
Chilo, J., 546
Cho, T. B., 354
Chu, 302
Cioffi, K., 167
Clifton, J. C., 340
Copeland, M., 453
Copeland, M. A., 215
Costa, J., 484, 526

Craininckf, J., 445
Crols, J., 222, 229

D

Dao, A., 620
Dauphinee, L., 453
Davis, P. C., 258
Dawe, G.-C., 360
Dearn, A. W., 347
Devlin, L. M., 347
Douglas, A., 643

E

Edwards, G. P., 122
Endo, T., 365
Enel-Rehel, C., 327
Engdahl, J., 546

F

Fague, D. E., 614, 620, 624
Fahey, W., 643
Fenk, J., 197, 212, 331, 363
Finley, W. C., 480
Forgues, S. L., 258
Fotowat, A., 344
Fuchs, T., 344
Fujimoto, H., 392
Fujimoto, K., 177
Fukuda, Y., 603

G

Gaethke, R., 344
Gaitan, M., 477
Ganser, S., 331
Geen, M. W., 470
Geppert, W., 363
Granek, R., 265
Gray, P. R., 29, 354
Green, G. J., 470
Greenhouse, H. M., 461

H

Hadaway, R. A., 215
Hadjizada, K., 331, 363
Hageraats, J. J. E. M., 357
Haines, D. M., 92, 99

Halim, R., 262
Hansen, K., 157
Haspeslagh, D., 327, 335
Haspeslagh, J. J., 265
Havens, J. H., 338, 341
Heaney, E., 183, 389, 643
Heimerl, E., 344
Heinen, S., 331, 363
Herrmann, H., 331
Hess, K. M.-D., 624
Heston, D. D., 190
Hikita, M., 494
Hildersley, J., 365
Hirasawa, N., 585
Horng, B.-R., 262
Hotsumi, M., 585
Hsu, K., 262
Hughes, D. W., 537
Hull, C. D., 302

I

Iga, T., 296
Ikeda, H., 410
Ikeda, T., 296
Ikeda, Y., 385
Imai, Y., 161
Inoue, A., 385
Irie, K., 365
Irvine, R. G., 197
Ishida, K., 655
Ishikawa, O., 392
Ishizaki, T., 410, 434
Isuchihaski, K., 285
Itoh, J., 177

J

Jackson, D. M., 537
Jackson, K., 643
Jansen, R. H., 470
Jenkins, J. A., 470

K

Kamishiro, J., 603
Karanicolas, A. N., 311
Karmel, C. R., 620
Kasai, M., 287
Kasai, N., 385
Kashiwagi, K., 296
Kasperkovitz, W. G., 55
Katoh, 177
Katsueda, M., 401
Kellner, W., 581
Kermanec, C., 389

Kermarrec, C., 183
Khorram, S., 273
Kimura, K., 585
Kirchgessner, J., 526
Kiss, L., 327, 335
Klein, J., 186
Knapp, H., 112, 275
Kniepkamp, H., 581
Kobayasi, F., 603
Kokubo, M., 365
Komoda, K., 585
Koullias, I. A., 258, 338
Koyama, M., 341
Kruezgruber, P., 112
Ku, M. K., 448
Kukielka, 335
Kuroda, M., 594
Kwasniewski, T., 441

L

Lee, H.-Y., 530
Lee, S., 344
Lee, T. H., 317
Leeson, D. B., 419
Lehmann, R. E., 190
Leslie, T. C., 112
Lipperer, G., 431
Long, A. P., 347
Long, J. R., 215
Longo, L., 262
Lovelace, D., 484, 526
Luff, G., 85

M

M. Nakayama, M., 276
Macario, R. C. V., 48
Mack, W. D., 206, 232, 357
Madsen, B., 614
Maeda, M., 392
Maeda, T., 655
Maemaura, K., 287
Magill, D. T., 122
Makimoto, M., 488
Marshall, C., 344
Maru, T., 585
Maruyama, Y., 393
Mashiko, K., 296
Matschke, E., 363
Matsui, H., 365
Matsumoto, H., 585
Matter, V., 363
McGrath, F., 183, 389, 643
Mejallu, I. D., 48
Meyer, R. G., 29, 201, 206, 232, 357, 473, 560, 564

Min, J., 118
Minakawa, A., 161
Minami, Y., 594
Mishiki, S., 370
Miyanaga, K., 423
Miyatsuji, K., 423
Miyazaki, S., 603
Miyazaki, Y., 287
Moinian, S., 480
Moore, P., 344
Mori, Y., 177, 488
Mourant, J.-M., 360
Murphy, M. T., 542

N

Nagata, S., 423
Nagueras, A., 157
Nair, V., 186
Nakagawa, Y., 424
Nakamura, T., 424
Nakayama, 287
Nakano, Y., 594
Natali, F. D., 122
Navid, S., 344
Ndagijimana, F., 546
Negus, K. J., 552
Ngo, D., 484, 526
Nguysen, N. M., 201, 473
Nishi, C., 147
Nishijima, M., 392
Nishikawa, T., 519
Niwano, K., 296
Nogueras, 157
Nojima, T., 378

O

O'Sullivan, P., 183, 389
Ogata, N., 287
Ohba, M., 488
Ohisi, 423
Ohtaka, S., 393
Okabe, T., 393
Okanobu, T., 69, 147
Okuno, H., 585
Olgaard, C., 399
Onge, G., St., 389
Otsubo, M., 287
Ou, J. J., 354

P

Pilaski, R., 85
Popp, J., 112
Post, I. G., 338, 341
Pratt, R. G., 643

R

Rabaey, D., 37, 265
Rael, J., 351, 451
Razavi, B., 456
Reusens, P. P., 265
Richards, T. H., 85
Ripollés, O., 173
Rofougaran, A., 118, 218, 248, 273, 351, 399, 448, 451
Rofougaran, M., 248, 351, 399, 451
Rohringer, N., 112
Rudell, J. C., 354

S

Sagawa, M., 177, 488
Saito, M., 655
Sakai, H., 177
Sallaerts, D., 265
Samueli, H., 118, 448
Sansen, W. M. C., 567
Sato, H., 296
Saur, E., 344
Savara, R. M., 533
Scheitlin, D., 186
Schelle, J., 581
Scholtz, A. L., 105
Schon, K. R., 197
Schultes, G., 105
Schultex, G., 112
Schvan, P., 215, 453
Sehrig, P., 197, 331
Seino, K., 385
Semburger, W., 112
Sera, M. M., 628
Sevenhans, J., 37, 323, 335
Shaeffer, D. K., 317
Shahani, A. R., 317
Shamlou, D., 262
Shibazaki, I., 488
Simburger, W., 275
Smith, D. H., 533
Smith, S. R., 542
Snider, D. M., 371
Steyaert, M., 445
Steyaert, M. S. J., 222, 229
Stetzler, T., 341
Sumi, T., 296
Sumioka, A., 494
Suzuki, Y., 655
Swanson, H., 186

T

Tabuchi, T., 494
Takada, M., 385

Takagi, T., 287, 385
Takehara, H., 392
Tamura, Y., 585, 603
Tanaka, K., 633
Tanimoto, H., 641
Tezuka, A., 177
Tham, 302
Thamsirianunt, M., 441
Thomas, V., 212, 363
Tokumitsu, M., 161
Tokumitsu, T., 419
Tomijama, H., 69
Toyada, I., 419
Toyoshima, G., 385
Tsurumi, J., 655

U

Uchitomi, N., 655
Ueda, 423
Uwano, T., 410, 434

V

Vaitkus, R., 186
Vance, I. A. W., 80
Vandeloo, P. J. Y., 567
Vanwelsenaers, A., 265, 323, 327
Veda, D., 415
Veit, W., 331
Veith, P., 105

W

Wakino, K., 501
Watanabe, K., 365
Weaver, D. K. Jr., 45
Weber, S., 363
Weger, P., 112, 275
Weldon, J. A., 354
Wenin, J., 323, 327
Wholey, J. N., 552
Wilkinson, R. J., 92
Wilson, J. F., 85
Wong, A. K., 564

Y

Yabuki, H., 488
Yamamoto, K., 287
Yamasaki, K., 594
Yamawaki, T., 365
Yamazaki, D., 147
Yonehata, A., 603
Yoshida, I., 401
Yoshihara, K., 655
Yoshii, Y., 287
Yoshikawa, N., 423
Yoshikawa, Y., 410
Yoshizawa, S., 594
Youell, R., 85

SUBJECT INDEX

A

Adaptation in receiver implementation
  power-adaptive transceivers, 31
  standard-adaptive transceivers, 32
Advanced mobile telephone
  board configuration and block diagram, 606–607
  LSIs and surface mount technology, 603–604
  NEC's open interface and software (NOTIS), 609
  power consumption reduction, 604–609
  production engineering, 609–12
  specifications, 612–13
  unit volume reduction, 604, 606
AM only radio integrated circuit (IC), 77–78
Analog and digital circuits, 120–21
Analog cellular portable radio, design of low-phase noise voltage-controlled oscillator (VCO), 434–40
Analog circuit design, 232–47
Analogue circuit design, mobile radio, 37–41

B

Balanced double layer stripline (BDLS) type filter, dielectric resonator application, 522–3
Bandpass filters, recent developments, 515–17
Base band transmission and reception (BBTRX)
  main functions, 327
  multi functional CMOS, mixed analog/digital VLSI circuit, 327–30
  processed in CMOS technology, 328
  system architecture, 327–30
Baseband unit of hand-held portable equipment, 587–88
Base station technologies, 485
BiCMOS RF front-end integrated circuit (IC), 1-GHz, 206–11
BiCMOS transceiver, 2.5 GHz, for wireless low-noise amplifier (LNA), 354–55
Binary frequency-shift keyed (FSK) modulation, wireless paging receivers, 10
Bipolar chipset, 2.7V 2.5 GHz, digital wireless communication, 364–65
Bipolar low-noise amplifier (LNA) and mixer, monolithic microstrip transformers, 215–16
Bipolar transceiver chipset for digital mobile radio, 197–200
Bipolar-transistor double-balanced mixers, intermodulation at high frequencies, 560–63
  computer simulation, 561–62
  large signal analysis, 560–61
Blocking and desensitization in RF amplifiers, 564–66
Bonding pad models for silicon VLSI technologies, 526–29
  effects on noise figure, 528–29
Bonding wires, wideband characterization of mutual coupling, 530–32
Bond wire inductors, 34
Broadcast radio receivers, integrated circuits (ICs), 7–9
Broadcast receivers, phasing method for sideband selection, 48–53

C

Carrier-frequency local oscillators with quadrature outputs in direct-conversion transceivers, 141–43
Cartesian loop transmitter with feedback, 93–94
Cartesian loop Weaver method transmitter, 99–100
Cellular analog front-end with IF receiver, 263–65
Cellular base stations, filters for, 512–514
Cellular phones
  board configuration and block diagram, 604–05
  GaAs power multi-chip IC, 3.5V 1.3W, 392–98
  hand-held portable equipment, 585–93
  LSIs and surface mount technology, 603–13
  miniaturization of RF front-end components using dielectric resonators, 519–25
  NEC's open interface and software (NOTIS), 609
  power consumption reduction, 606–609
  production engineering, 609–12
  specifications, 612, 613
  unit volume reduction, 604, 606
Cellular portable radio, analog, design of low-phase noise voltage-controlled oscillator (VCO), 434–40
Cellular telephones, digital, analysis of phase characteristics of GaAs FET power amplifier, 410–18
Cellular telephones, dual-mode, 900 MHz transceiver chip set, 338–39
Cellular telephone transceivers, integrated circuits, 11–14
Ceramic technology, multilayer, 533–36
  SPICE simulation of, 534, 536
Chip set targeting dual-mode (IS-54) cellular telephone, 628–29
  chip set operation, 629–30
  design simplification, 629
  integration and connectivity, 630
  SA601 RF front-end, 629–31
  SA637 digital IF receiver, 629, 631
  SA7025 dual frequency synthesizer, 631–32
  SA900 I/Q transmit modulator, 632, 633
Circuit approaches for integrated synthesizers, 32–34
  direct digital synthesis of LO signal, 33–34
  ring-oscillator-based VCOs and noise-optimized synthesizer architectures, 33
Circuit board interfacing, printed, 236

Circuit design, mixed A/D, sensor interface, and communication, 229–40
Circuit simulator SPICE with modified FET model and Touchstone, 161–63
CMOS 1.8 GHz low-phase-noise voltage-controlled oscillator with prescaler, 445–46
CMOS baseband chip and Si bipolar chip for mobile radio telephone, 335–37
  receiver and transmit Si bipolar mixers, 335–37
  synthesizer, 335–37
CMOS device, cellular analog front-end with 98dB IF receiver, 263–64
CMOS digitally synthesized 1-13 MHz agile sinewave generator, 448–50
CMOS double quadrature downconverter, 900 MHz fully integrated, 229–31
CMOS downconversion mixer, 1.5 GHz highly linear, 222–28
  frequency domain behavior, 224–25
  mixer design, 224
  mixer topology, 222–24
  noise, power consumption, and optimization, 225–26
  realization of high linearity and input frequency capabilities, 226–27
CMOS front-end, 12mW wide dynamic range, for portable global positioning system (GPS) receiver, 318–20
CMOS integrated circuit technology, 20
CMOS integrated receiver, 1.9 GHz wide-band IF double conversion cordless telephone, 354–55
CMOS LC-oscillator with quadrature outputs, 900 MHz, 451–52
CMOS limiting amplifier and signal-strength indicator, 274–75
CMOS low-noise amplifier (LNA) and mixer, 2.7-V 900-MHz
  front-end requirements of North American Digital Cellular (NADC) handset, 300
  mixer design, 313–14
  RF front-end applications, 312–17
CMOS radio frequency (RF) front-end IC for direct-conversion wireless receiver, 1 GHz, 249–58
  down conversion mixer design, 252–54
  implementation and measurements, 254–55
  low-noise amplifier (LNA) design, 248–52
CMOS RF power amplifier with programmable output, 900 MHz, 399–400
CMOS SC track and hold circuit as downconversion mixer, 218–21
CMOS technology
  frequency-hopped spread spectrum (FH/SS) transceiver, 118–21
  modulator IC (270-kb/s 35-mW) for GSM cellular radio hand-held terminals, 266–73
CMOS voltage-controlled oscillator (VCO), 1.8 GHz, 456–57
  PLL frequency synthesis, GHz, 441–44

Subject Index

CMOS wireless transceivers, future of, 351–53
Coaxial resonator type filter, dielectric resonator application, 520–22
Code division multiple access (CDMA) technique, 118
Common collector Colpitts circuit, VCO with integrated LC resonator, 453–54
Communication circuits, 240–48
Compatible high-Q inductors and resonators in silicon
   bond wire inductors, 34
   spiral inductors on silicon, 34–35
Control logic circuit, analog circuits and negative voltage generator, 292–93
Cordless telephone
   intermediate frequency (IF) double conversion CMOS integrated receiver, 1.9 GHz wideband, 354–55
   spread spectrum, 633–42

D

DC offset in baseband section, direct-conversion receiver problem, 140
DECT (Digital European Cordless Telecommunications)
   applicability to the United States, 614–15
   challenges for radio designers, 614
Desensitization and blocking in RF amplifiers, 564–66
Design of integrated FM stereo receiver, 62–68
Despreading and demodulating spread-spectrum signals, 131
Despreading direct-sequence spread spectrum (DS SS) signals, 132–33
Despreading frequency-hopping spread spectrum (FH SS) signals, 131–32
Deviation muting of audio signal in FM mono receiver, 58–59
Dielectric filters
   monoblock type, 519–20
   recent developments, 510–517
Dielectric resonator materials
   recent developments, 502–10
   temperature dependence, 502–5
Dielectric resonators, miniaturization of RF front-end components, 519–25
Digital cellular telephones
   analysis of phase characteristics of GaAs FET power amplifier, 410–18
   direct-conversion receivers for, 138–39
Digital cordless telephones
   direct-conversion receiver for 900 MHz ISM band, spread-spectrum, 303–11
   integrated circuits, 14–17
   single chip intermediate frequency (IF) transceiver, 1.9 GHz, 297–302
Digital European Cordless Telecommunications (DECT)
   applicability to the United States, 614–15
   challenges for radio designers, 614
Digitally-enhanced cordless telephone
   bipolar chipset, 2.7V 2.5 GHz, 363–64
   TDMA system using GFSK modulation, 363

Digitally-enhanced cordless telephone RF transceiver with integrated voltage-controlled oscillator (VCO), 2.7V, 360–61
Digital mobile communication, GaAs high-power RF single-pole double-throw switch IC, 423–24
Digital mobile communications, single-chip GaAs RF transceiver, 288–96
Digital mobile radio
   bipolar transceiver chipset for, 197–200
   CMOS VCOs for phase-locked loop (PLL) frequency synthesis in GHz, 441–44
Digital radio personal communications devices, 29
Digital-to-analog converter, 120–21
Direct-conversion concept
   intermediate frequency (IF) circuit, new radio receiver system, 147–54
   quasi-direct conversion, and low-IF receiver architectures, 30–31
Direct-conversion radio transceivers
   architecture, 136–37
   carrier-frequency local oscillators (LO) with quadrature outputs, 141–43
   components of, 141–44
   frequency-shift keying (FSK) receivers, 137–38
   mixers for direct-conversion, 143–44
   problems in receivers, 139–41
   receivers for digital cellular telephones, 138–39
   single-sideband synthesizers, 138
Direct-conversion receiver for 900 MHz (ISM band) spread-spectrum digital cordless telephone, 303–11
   baseband amplifier, variable-gain and fixed-gain stages, 307–8
   industrial-scientific-medical (ISM) band, 303
   local oscillator (LO) phase-shift circuit, 303–305
   low-noise amplifier and mixer, 305–7
   servo amplifier, 308–9
   system architecture, 303–4
Direct-conversion receivers for GSM digital cellular telephones, 18
Direct-conversion radio frequency (RF) transceiver, microwave silicon single-chip, 276–87
   electrical characteristics, 278, 282–87
   limiter/FM-discriminator, 277, 281
   low-noise amplifier (LNA), 277, 280
   quadrature down converter, 277, 280
   quadrature up-converter (QUC), 277–78, 281
   RF transceiver architecture, 276–77, 279
   variable gain amplifier (VGA) and DC-block, 277, 280
   voltage-controlled oscillator (VCO), 277, 281
Direct-conversion single-sideband synthesizers, 17–18
Direct-conversion transceiver design
   compact low-cost portable mobile radio, 99–104
   fully functional single-chip, 114–15
   receiver fading performance, 103–4
Direct-conversion transceivers and their problems
   direct-conversion receivers for GSM digital cellular telephones, 18

direct-conversion single-sideband synthesizers, 17–18
   local oscillators with quadrature outputs, 18
Direct-conversion versus heterodyne transceivers, 114
   comparison of building blocks, 113
   normalized production expenses, 113
Direct-conversion wireless receiver, 1 GHz CMOS RF front-end IC, 249–58
Direct-digital frequency synthesizer (DDFS), 21–23, 121
   sine and cosine waveform generation, 448–49
Direct digital synthesis (DDS) for generating frequency-hopping (FH) and direct-sequence (DS) signals, 130
Direct digital synthesis (DDS) for modulation and carrier generation in direct sequence (DS)-spread spectrum (SS), 131
Direct digital synthesis (DDS) for modulation and carrier generation in frequency-hopping (FH)-spread spectrum (SS), 130–31
Direct digital synthesis (DDS) of LO signal, 33–34
Direct-sequence (DS) spread spectrum (SS), 123
   comparison with frequency-hopping (FH) SS, 124
   despreading, 132–33
Direct-sequence modulation to spread spectrum of signal, 19
Display pager, numeric
   control and display section, 600–601
   description, features, functions, and specifications, 594–97
   integrated gyrator filter, 599–600
   LSI technology for new receiving system, 599–600
   mechanical design, 601–602
   receiving unit, 597–99
Distortion, direct-conversion receiver problem, 139–40
Double-balanced Gilbert-cell with resistors, 8–9
Double-conversion broad band TV-tuner with GaAs ICs, 581–84
Double-conversion system, conventional, 148
Double-superheterodyne receiver
   digital cellular and cordless telephones, 14–15
   wireless paging receivers, 10
Downconversion mixer, CMOS SC track and hold circuit, 217–21
   circuit design, 219–20
Downconversion mixer, CMOS technology, 222–28
Downconversion mixer design, 1 GHz CMOS RF front-end IC, 252–54
Downconverter, low current GaAs integrated, 186–89
Downconverter, fully integrated 900 MHz CMOS double quadrature, 229–31
Dual-gate MESFET's, 12
Dual-mode cellular telephones, 900 MHz transceiver chip set, 338–40
Duplexers for cellular mobile and hand-held radios, 511–12
Dynamic range CMOS front-end (12mW wide) for portable global positioning system (GPS) receiver, 319–20
Dynamic range problem in receiver, 4–5

666

E

Electrical characterization of packages
  design approach, 542–43
  use with GaAs MMIC amplifiers, 542–45
Electrical equivalent circuit for surface mount package, 539–41
Electrostatic discharge (ESD) network design, 245
Elliptic bandpass filter, 515
ESPICE simulator, 89

F

Feedback oscillator noise spectrum, model and experimental verification, 429–30
FET-switchable LC resonators in T/R switch, 419–22
Filtering, on-chip in 2.7V GSM transceiver ICs, 344–45
Filter method of single-sideband generation, 45
Filters
  balanced double layer stripline (BDLS) type, 522–23
  coaxial resonator type, 520–22
  dielectric monoblock type, 519–20
  dielectric, recent developments, 510–517
  integrated gyrator for numeric display pager, 599–600
  LC passive, 473–76
  multi-layer lumped constant (LC) type, 523–24
FM mono receivers
  basic features, 58–59
  block diagram, 59
  conventional, 55–57
  integrated, 59–61
  integrated mono receiver circuit, 57
  measurements on, 61–62
  tuning behavior of experimental receiver, 62
  tuning behavior of conventional mono receiver, 56
FM stereo receiver (integrated), design of
  block diagram of, 67–68
  frequency control of mono receiver, 63–65
  frequency control of stereo receiver, 65–67
  model of frequency-control system, 66
  stability of frequency-locked loop, 62–63
Four-path phasing method of sideband selection, 50–53
  cancellation method, 51–52
  choice of phasing network, 53
  equations for frequency choice, 54
  preferred component values, 52–53
Frequency control and signal muting in FM mono receiver, 60–61
Frequency control of integrated FM mono receiver, 63–65
Frequency control of integrated FM stereo receiver, 65–67
Frequency division multiple access-time division duplex (FDMA-TDD) technique, spread spectrum cordless telephone, 633–42
Frequency error between transmitter and receiver, direct-conversion receiver problem, 140–41

Frequency-hopped spread spectrum (FH/SS), 19–20
  analog and digital circuits all in CMOS technology, 120–21
  baseband architecture, 119–20
  code division multiple access (CDMA) technique, 118
  comparison with direct-sequence SS, 124
  despreading, 131–32
  RF/IF architecture, 118–19
  transceiver architecture, 118–20
  transceiver with CMOS RF power amplifier with programmable output, 399–400
Frequency limitations on assembled SO8 package, 546–51
Frequency-locked loop, stability in integrated FM stereo receiver, 62–63
Frequency-offset receiver principle, wireless pager ICs, 10
Frequency-shift keyed (FSK) demodulator, 121
Frequency-shift keyed (FSK) paging receiver, 81–84
  data-rate limitations and jitter, 82–83
  demodulator characteristics, 82
  detailed performance of system, 81–83
  predetection bandwidth, 81–82
  selectivity and spurious responses, 83
  sensitivity, 81
  spurious radiation, 83
Frequency-shift keyed (FSK) radio communication, one-chip 2 GHz single superhet receiver, 212–13
Frequency-shift keying (FSK), direct-conversion receivers, 137–38
Fully-differential ring oscillator, voltage-controlled oscillator (VCO) architecture, 443
Future public land mobile telecommunications systems (FPLMTS), 99
Future transceivers, integrated circuits, 20–23

G

GaAs, low power ICs for mobile communication, 177–82
GaAs chip set (1.9 GHz) for Personal Handyphone System (PHS), 643–44
  chip set design considerations, 645–47
  manufacturing of, 652–54
  power amplifier/switch IC, 651–52
  QPSK modulator IC, 647–49
  transceiver integrated circuit, 649–51
GaAs double-conversion tuner (DCT), 581–84
GaAs FET power amplifier for digital cellular telephones, 410–18
  analysis model, 411–12
  analysis of phase characteristics, 413–15
  input-output characteristics, 416–17
GaAs FETs, in hand-held cellular telephones, 607–608
GaAs high-power RF single-pole double-throw switch IC, 423–24
GaAs integrated downconverter, low current, 186–89
GaAs monolithic microwave integrated circuit (MMIC) amplifiers, electrical characterization of packages for, 542–45

GaAs MMIC
  circuit design, 161–63
  image rejection mixer, 175–76
  low noise amplifier (LNA), 174–75
  MMIC fabrication, 163–64, 176
  performance, 164–66
  receiver performance in RF bandwidth 1.5–2.5 GHz, 176
  small size LNA, 175
  spiral inductors for, 470–72
  technology for cellular telephones, 11–12, 23–24
GaAs power amplifier module, reduction of talk time current drain in advanced personal phone, 606
GaAs power multi-chip IC (MCIC), 3.5V 1.3W
  MCIC characteristics, 397
  MCIC design, 393–94
  power MESFET, 394–95
  reduction of circuit-loss, 395–97
GaAs RF transceiver, single-chip for 1.9 GHz digital mobile communications, 288–96
GaAs transceiver, 2.4 GHz single chip, 347–50
Gaussian minimum shift keying (GMSK) modulation, 266–69
  waveform generation, 268–69, 272–73
Gilbert-quad-based down-conversion mixer, 357–58
Global positioning system (GPS) receiver, 12mW wide dynamic range CMOS front-end, 318–20
Groupe Special Mobile (GSM), European digital cellular radio standard, 259
GSM RF transceiver IC, 2.7V, 365–66
Gyrator-capacitor elliptic filter in FSK radio paging receiver, 86
Gyrator filter, integrated, in numeric display pager, 599–600

H

Hand-held portable (HHP) equipment for cellular mobile telephone
  baseband unit, 587–88
  hardware configuration, 587
  HHP generation of cellular systems, 585–86
  HHP terminal, 266–67
  logic control unit, 589–91
  radio unit, 588–89
  software, 591–93
  specifications, 578–79
  system adaptation and hardware problems, 585
Hardware configuration of hand-held portable equipment, 586
Harmonic reaction amplifier (HRA), 380–81
Heterodyne versus direct-conversion transceivers
  comparison of building blocks, 113
  normalized production expenses, 113
High-data rate time division multiple access (TDMA) system, 114
High-efficiency transmitting power amplifiers, portable radio units, 378–84
High-integration RF transceiver technologies, 34
High-performance integrated PA, T/R switch for 1.9 GHz handsets, 389–91
High-Q inductors in Si bipolar process, 480–83

I

Image canceling circuits, 69–70
Image-reflection problem in receivers, 4–5
Image-reject downconverter, dual of Weaver's phasing method of sideband selection, 136
Image rejection receiver, GaAs MMIC for 1.5–2.5 GHz, 173–76
Image-reject mixer, 8
Incoherent detection of (G)FSK type time division multiple access (TDMA) signals, 115–16
Incoherent direct-conversion receiver, 105
  computer simulations, 109–10
  description of receiver blocks, 106–9
  principle of vectorial receiver, 105–6
Incoherent homodyne receiver, 105–11
Indoor channel, 127–28
Indoor wireless applications, 127
Inductor design, planar rectangular microelectronic, 461–69
  basic mathematical concepts, 459–63
  Bryan method, 464–65
  computer-calculated inductance data, 466–69
  computer program for inductance calculations, 465–66
  expanded Grover method, 464
Inductors, high Q, in complementary Si bipolar process, 480–83
Inductors, large suspended, on silicon for CMOS RF amplifier, 477–79
Inductors, Si IC-compatible, 473–76
Inductors, triplate spiral, miniaturized RF-circuit modules, 489, 491–92
Inductors for GaAs MMICs, miniature multilayer spiral, 470–72
Industrial, scientific, and medical (ISM) frequency band, 347
  direct-conversion receiver for 900 MHz spread-spectrum digital cordless telephone, 303
Input noise-figure (NF), 5
Input-referred intercept point (IP3), 5
Integrated circuits (ICs), low-power radio-frequency (RF)
  digital cellular and cordless telephones, 14–17
  direct-conversion transceivers and their problems, 17–18
  ICs in broadcast radio receivers, 7–9
  ICs in cellular telephone transceivers, 11–14
  ICs in wireless paging receivers, 9–11
  ICs to enable future transceivers, 20–23
  ICs with spread-spectrum wireless transceivers, 18–20
  key signal-processing issues in wireless transceivers, 4–7
  wireless design trends, 3–4
Integrated direct-conversion transceiver, 114–15
Integrated FM mono receiver, 59–61
Integrated Injection Logic (ILL), discriminator circuit of numeric display pager, 599
Integrated Si bipolar RF transceiver for zero IF 900 MHz mobile radio front-end, 323–26
Integrated synthesizers, circuit approaches, 32–34
Integration of complete FM mono receiver on single chip, 55–68

Intermediate frequency (IF) amplifier/quadrature demodulator (100 MHz)
  circuit architecture and design, 259–62
  digitally-controlled band gap reference, 261
  digitally-controlled-gain differential amplifier, 260
  Groupe Special Mobile (GSM) cellular radio mobile terminals, 259
  quadrature demodulator, 260–61
Intermediate frequency (IF) circuit with direct-conversion concept, new radio receiver system, 147–54
Intermediate frequency (IF) transceiver, 1.9 GHz single chip for digital cordless phones, 297–302
Intermodulation
  design considerations for wireless communications, 159
  high-frequency bipolar transistor IC mixers, 560–63
ISOSAT™-II process technology, 552–53

J

Japanese Personal Handyphone System (PHS), 297
  air interface for, 644–45
  design study on RF stage for miniature PHS terminal, 655–60
  GaAs chip set (1.9 GHz), 643–44
  high performance integrated PA, T/R switch for 1.9 GHz handsets, 389–91

L

Land mobile communications equipment, miniaturized RF-circuit modules, 488–93
L-band components, monolithic
  front-end applications, low-current GaAs MMIC, 161–66
  ultra-low dc power consumptions, 167–72
LC-oscillator with quadrature outputs, 900 MHz CMOS, 451–52
LC passive filters, 473–76
LC-resonant T/R switch for high-power/low-distortion operation, 419–22
LC resonator with balanced 1.5 GHz voltage-controlled oscillator (VCO), 453–54
Linear applification with nonlinear components (LINC) transmitter, 95–96
Linear CMOS downconversion mixer, 1.5 GHz, 222–28
Linearized saturation amplifier with bidirectional control (LSA-BC), 382–84
Linear modulation (LM) scheme, 99
  Cartesian loop Weaver method transmitter, 99–100
  Weaver direct conversion receiver, 101–4
Linear transceiver architectures
  adaptive predistortion of transmitter, 94–95
  linear amplification with nonlinear components (LINC) transmitter, 95–96
  receiver section, 96–97
  transmitter section, 93–96
Load-pull, conventional and active second harmonic, 385–86

Local oscillator (LO)
  phase-locked loop/voltage-controlled oscillator (PPL/VCO), 120
  phase-shift circuit in direct-conversion receiver for 900 MHz ISM band, 304–306
  with quadrature outputs, 18
Logic control unit of hand-held portable equipment, 589–91
Low-current GaAs integrated down converter circuit design, 186–87
  down converter evaluation, 188–89
  fabrication, 187
  gain, return loss, noise figure and intermodulation characteristics, 187–88
Low-current oscillator design, 900 MHz GSM applications, 431–33
Low-intermediate frequency (IF), direct-conversion, and quasi-conversion receiver architecture, 30–31
Low-noise amplification in circuit design, 234
Low-noise amplifiers (LNA)
  CMOS and mixer, 2.7-V 900-MHz, 308–13
  RF amplifier, 120
  ultra low power for wireless communications, 183–85
  X-band monolithic series, 190–96
Low-phase noise voltage-controlled oscillator (VCO), 434–40
Low-power baseband signal conversion and processing, 34
Low-power GaAs integrated circuits (ICs)
  design of, 177–79
  fabrication of dual gate MESFET's, 179–80
  performance of, 180–82
Low-power wireless communications prospects, 24
Low-voltage FM stereo/AM radio IC, 71–76
  AM band pass filter (BPF), 73
  FM low pass filter (LPF), 73
  FM MPX decoder circuit, 74
  FM RF amplifier, 72
  I, Q mixer and phase shift network (PSN), 69, 72
  microphotograph of IC, 75–76
  peripheral components, 75–76
  pulse count circuit, 74
Low-voltage silicon bipolar RF front-end for wireless receiver applications, 215–16
LSI technology for numeric display pager, 598–99

M

Mechanical design of numeric display pager, 601–2
MESFET, active device on semi-insulating GaAs substrates, 6–7
Method of Moments (MoM)
  mutual coupling between grounded bonding wires, 530–32
  wire bondings analysis of SO8 package, 548–49
Microelectronic inductor design, planar rectangular, 461–69
Microwave silicon single-chip direct conversion RF transceiver, 276–87
  electrical characteristics, 278, 282–87
  limiter/FM-discriminator, 277, 281

low-noise amplifier, 277, 280
quadrature down converter, 277, 280
quadrature up-converter (QUC), 277–78, 281
RF transceiver architecture, 276–77, 279
variable gain amplifier (VGA) and DC-block, 277, 280
voltage-controlled oscillator (VCO), 277, 281
Microwave surface mount package for housing MMICs, 537–41
electrical equivalent circuit for package, 539–41
microwave fixture, 537–38
Miniaturized RF-circuit modules for land mobile communications, 488–93
Miniaturized surface acoustic wave (SAW) devices
application of antenna duplexer, 498–500
configuration for low-loss/high-power, 496–98
loss analysis of high-performance SAW filter for mobile telephone, 496
possibilities for, 495–97
radio communications, 494–95
Mixed analog/digital circuit design, 239–48
Mixer, downconversion, CMOS SC track and hold circuit, 217–21
Mixers for direct-conversion receivers, 143–44
Mobile communication equipment, low power GaAs integrated circuits, 177–82
Mobile power technologies, 485
Mobile radio applications
transceiver (2.7V, 800 MHz-2.1 GHz) chipset, Si bipolar technology, 331–34
twin chip set, CMOS baseband chip and Si bipolar chip, 335–40
Mobile radio transceiver, overall architecture, 112–13
Mobile radio very large scale integration (VLSI) chips, 37–41
low noise radio receiver, 37–39
low noise radio transmitter, 39–40
RF-bipolar technology, 40–41
temperature-controlled reference crystal oscillator (TCXO), 40
voice interface with microphone and earpiece, 39
Mobile satellite propagation, 125–26
Mobile telephone, advanced
board configuration and block diagram, 604–605
LSIs and surface mount technology, 603–5
NEC's open interface and software (NOTIS), 609
power consumption reduction, 606–9
production engineering, 609–12
specifications, 612–13
unit volume reduction, 604, 606
Mobile transceiver architectures, 92–97
Modeling of MOS transistor for high frequency analog design, 607–8
Modified D-latch ring oscillator, voltage-controlled oscillator (VCO) architecture, 442
Modified ring oscillator, VCO architecture, 442
Modulator IC (270-kb/s 35-mW) for GSM cellular radio hand-held, 265–73
analog-to-digital converter, 270–71
Gaussian minimum shift keying (GMSK) modulation, 266–73

modulator circuit architecture, 267–70
switched capacitor (SC) filter, 270
Monoblock type dielectric filter, dielectric resonator application, 519–21
Monolithic image-rejection receiver for 1.5-2.5 GHz radio links, 173–76
Monolithic L-band components
amplifier design, 168–69
amplifier results, 169–71
device/component design optimization, 167–68
filter results, 171
subsystem module results, 171–72
voltage-controlled filter design, 169
Monolithic microwave integrated circuit (MMIC)
design optimization for minimum dc power consumption, 167–69
driven by military applications, 6–7
GaAs, 161–66
ultra-low DC power consumption, 167–72
Monolithic RF bandpass amplifier, Si bipolar, 201–5
Monolithic series feedback LNA, X-band, 190–96
MOSFET, LC oscillator with quadrature outputs, 450–51
MOSFET, low-voltage/high-efficiency Si power, 401–409
MOS transistor for high frequency analog design
application of small signal models to design of operational transconductance amplifier (OTA), 574–76
evaluation of small signal model at high frequencies, 570–72
high frequency characteristics of, 567–69
measuring and fitting small signal parameters, 572–74
modeling of, 567–77
small signal equivalent circuits, 569–70
Multi-chip IC (MCIC), 3.5V 1.3W GaAs power, 392–98
MCIC characteristics, 397
MCIC design, 393–94
Multifunctional silicon MMICs for frequency conversion applications, 552–58
Multilayer ceramic technology
high-speed characteristics, 533–36
SPICE simulation, 534, 536
Multi-layer lumped constant (LC) type filter, dielectric resonator application, 523–24
Multi-layer stripline type voltage-controlled oscillator (VCO), dielectric resonator application, 524–25
Multiple-access technique and channel crosstalk, 128–29
Mutual coupling between high density bonding wires, 530–32

N

Negative voltage generator, GaAs MESFET power amplifier, 290–91
NMOS mixer design, 313–14
Noise figure, effect of bonding pad models, 526–29
North American Digital Cellular (NADC) standard, 628–29
future of, 630

NOTIS (NEC's open interface and software), personal pocket phone (P3), 609
Numeric display pager
control and display section, 600–601
description, features, functions, and specifications, 594–97
integrated gyrator filter, 599–600
LSI technology for new receiving system, 599–600
mechanical design, 601–2
receiving unit, 597–99

O

Off-chip-filter-less system, direct conversion for, 150–53
Offset phase-locked loop (OPLL), 365–66
On-chip filtering in 2.7V GSM transceiver ICs, 344–45
One-chip 2 GHz single superhet receiver for 2Mb/s FSK radio communication, 212–13
Open loop modulation techniques for wideband VCO for DECT, 620–23
Operational transconductance amplifier (OTA), application of small signal models to design of, 574–76
Optimally loaded and overdriven RF power amplifiers
theoretical analysis and experimental confirmation, 371–77
UHF band 1.3W monolithic amplifier with 63% efficiency, 385–88
Oscillator design, low current, for 900 MHz GSM applications, 431–33
Oscillator noise spectrum, feedback, model and experimental verification, 429–30

P

Package and substrate modeling, barrier to higher integration in RF transceivers, 35
Perovskites, dielectric resonator materials, 502–10
Personal communication by satellite, 126–27
Personal Handyphone System (PHS)
air interface for, 644–45
design study on RF stage for miniature PHS terminal, 655–60
GaAs chip set (1.9 GHz), 643–44
high performance integrated PA, T/R switch for 1.9 GHz handsets, 389–91
Personal Handyphone System (PHS) terminal
experimental handset, 659
features of RF stage, 655
module assembly techniques, 658
motherboard structure, 658–59
RF stage circuits and modules, 655–58
Personal pocket phone (P3) development
board configuration and block diagram, 604–605
LSIs and surface mount technology, 603–13
NEC's open interface and software (NOTIS), 609
power consumption reduction, 606–9
production engineering, 609–12
specifications, 612–13
unit volume reduction, 604, 606

Phase characteristics analysis, GaAs FET power amplifier, 410–18
Phase-locked loop (PLL)
  frequency synthesis in GHz, CMOS voltage-controlled oscillator, 441–44
  PLL-circuit design targets for low current oscillators, 431–33
Phasing method
  architectures, 92–93
  sideband selection in broadcast receivers, 48–53
  single-sideband generation, 45
Phenomenological Loss Equivalence Method (PEM), current distributions and input impedances of bonding wires, 530
PHILPAC, Philips proprietary circuit simulator, 89
POCSAG (Post Office Code Standardization Advisory Group) code, 81–82, 594
  coded FSK transmissions, 86–88
Portable communications, 3–4
  low current GaAs integrated down converter, 186–89
  specifications, 92
Portable consumer applications, 23–24
Portable global positioning system (GPS) receiver, wide dynamic range CMOS front-end, 318–20
Portable power technologies, 485–86
Post Office Code Standardization Advisory Group (POCSAG) code, 81–82, 594
  coded FSK transmissions, 86–88
Power-adaptive transceivers, receiver implementation, 31
Power amplifier
  CMOS RF (900 MHz), with programmable output, 399–400
  conventional and second harmonic load-pull, 3785–86
  efficiency requirements for portable units, 378–79
  harmonic reaction amplifier (HRA), 380–81
  high efficiency saturation amplifiers, 379–81
  high efficiency transmitting, 378–84
  high performance, T/R switch for 1.9 GHz handsets, 389–91
  linear amplifier, 381–82
  linearized saturation amplifier with bidirectional control (LSA-BC), 382–84
  in N-well CMOS process, 120
  phase characteristics analysis of GaAs FET, 410–18
  T/R switch design, 390–91
  UHF-band 1.3W monolithic, 385–88
  UHF-band Si power MOSFET, 401–9
Power consumption (ultra-low DC), monolithic L-band components, 167–72
Prescaler with voltage-controlled oscillator, CMOS 1.8 GHz, 445–46
Printed circuit board interfacing, 244
Production engineering of personal pocket phone (P3), 609–12

Q

QPSK data modulation format, 129
Quadrature modulator phase-locked loop frequency translator, 365–66
Quadrature outputs, 900 MHz CMOS LC-oscillator, 450–51
Quasi-direct conversion, direct-conversion, and low-IF receiver architectures, 30–31
QUBiC BiCMOS process, 241

R

Radio communication transceivers, miniaturized SAW devices, 494–500
Radio design for Digital European Cordless Telecommunications (DECT), 614–19
  baseband issues, 619
  phase-locked loop issues, 617
  radio design issues, 615–17
  receiver parameters, 617–19
  system issues, 619
Radio-frequency (RF) amplifiers
  blocking and desensitization measurements, 565–66
  sub-1 V operation, 8–9
  theoretical analysis of blocking and desensitization, 564–65
Radio frequency (RF) and intermediate frequency (IF) transceiver ICs with on-chip filtering, 2.7V GSM, 344–45
Radio frequency (RF) bandpass amplifier, Si bipolar monolithic, 201–5
Radio frequency (RF)-bipolar technology, mobile radio, 40–41
Radio frequency (RF)-circuit modules, miniaturized, 488–93
Radio frequency (RF) device (silicon) technologies, development trends, 484–87
Radio frequency (RF) front-end components miniaturized
  balanced double layer stripline (BDLS) type filter, 522–23
  coaxial resonator type filter, 520–22
  monoblock type dielectric filters, 519–20
  multi-layer lumped constant (LC) type filter, 523–24
  multi-layer stripline type voltage-controlled oscillator (VCO), 524–25
Radio frequency (RF) front-end for digital mobile radio
  bipolar technology, 197
  intermediate frequency (IF)-circuit section, 199–200
  mixer and SAW-driver circuit section, 199
  receiver circuit, 198
  theoretical details, 197
  transmitter circuit, 197–98
Radio frequency (RF) front-end integrated circuit, 1-GHz BiCMOS, 206–11
  mixer schematic, 208–10
  preamplifier (LNA) design, 206–8
  system application, 206
Radio frequency (RF) integrated circuit, 2.7V to 4.5V single-chip GSM transceiver, 341–42
Radio frequency (RF) power amplifier, optimally loaded and overdriven, 371–77
  experimental verification of theory, 375–77
  modified class A operation, 374–75
  optimum efficiency class B tuned power amplifier, 372–73
  optimum power class B tuned power amplifier, 373–74
Radio frequency (RF) stage design study for miniature PHS terminal, 655–60
Radio frequency (RF) transceiver, 2.7V, with integrated voltage-controlled oscillator (VCO), digitally-enhanced cordless telephone (DECT), 360–61
Radio frequency (RF) transceiver, single-chip GaAs, for 1.9 GHz digital mobile communications, 288–96
Radio frequency (RF) transceiver IC, 2.7V GSM, 365–66
Radio frequency transceiver implementation, typical, 29–30
Radio IC, AM only, 77–78
Radio interference, design considerations for wireless communications, 158
Radio paging, single-chip VHF and UHF receiver, 85–90
  demodulator and data output, 88
  design methodology, 89–90
  intermediate frequency (IF) filtering and amplification, 86
  POCSAG coded FSK transmissions, 86–88
  RF amplifier, 86
  zero-IF or direct conversion receiver, 85–86
Radio paging receiver
  frequency-shift keyed (FSK) receiver, 81–84
  fundamentals, 80
  typical specifications, 80
  zero-IF (direct conversion) type, 80–81
Radio receivers and transmitters problems, 4–5
Radio receiver sensitivity, design considerations for wireless communications, 158–59
Radio receiver system, new
  carrier leak detector and adjustment-free system, 151–52
  characteristics of, 154
  combining direct conversion concept and intermediate frequency (IF) circuit, 147–48
  configuration, 147–50
  direct conversion for off-chip-filter-less system, 150–51
  for personal communication, 147
  selectivity of off-chip-filter-less system, 152–53
Radio transceiver, single chip, performance evaluation, 624–27
Radio unit, hand-held portable equipment, 588–89
Radio wave propagation, design considerations for wireless communications, 157–58
Receiver fading performance for direct conversion receivers, 103–4
Receiver front-end, developing technologies, 487
Receiver implementation, adaptation
  power-adaptive transceivers, 31
  standard-adaptive transceivers, 32
Receiver radio frequency (RF) design considerations for wireless communications
  intermodulation, 159
  radio interference, 158
  radio wave propagation, 157–58
  receiver block diagram, 158
  receiver sensitivity, 158–59
  spurious responses, 159–60

Resonator materials, dielectric, recent developments, 502–10
Resonators, split-ring, miniaturized RF-circuit modules, 488–89, 491
Ring oscillator, four-phase voltage-controlled, 18, 19
Ring-oscillator-based VCOs and noise-optimized synthesizer architectures, 33

S

Satellite mobile applications, 124–25
Satellite, personal communication, 126–27
Sensor interface circuits, 240–48
Servo amplifier, direct-conversion receiver for 900 MHz ISM band, 308–9
Sideband folding, 63
Sideband selection in broadcast receivers
    four-path method, 50–53
    phasing method, 48–53
    sideband separation, 48–49
    two-path method, 49–50
Signal-strength indicator and CMOS limiting amplifier, 274–75
Silicon bipolar and BiCMOS technologies, 13
Silicon bipolar chip and CMOS baseband chip for mobile radio telephone, 335–46
    receive and transmit radio Si bipolar mixers, 341–43
    synthesizer, 338–40
Silicon bipolar integrated circuit (IC) technologies, multifunctional for frequency conversion
    comparison with full-custom MMICs, 558
    device and package modeling, 554–55
    ISOSAT™-II process overview, 552–53
    starCHIP™-1 array topology, 553–54
    wide-band frequency doubler, 555–56
    wide-band vector demodulator, 556–58
Silicon bipolar monolithic RF bandpass amplifier
    circuit configuration and implementation, 203–5
    design approach, 201–3
Silicon bipolar process, high Q inductors for, 480–83
Silicon bipolar radio frequency (RF) front-end, low voltage, 215–16
Silicon bipolar radio frequency (RF) transceiver, zero-IF 900 MHz digital mobile radio front-end, 323–26
Silicon bipolar single-chip RF transceiver, highly integrated, microwave characterization, 276–77
Silicon bipolar technology, transceiver chipset (2.7V 800 MHz-2.1 GHz) for mobile radio, 331–34
Silicon CMOS and RF-bipolar technology, mobile radio, 37–41
Silicon integrated circuit (IC)-compatible inductors, 473–76
Silicon integrated circuits, future directions, 29–35
Silicon power MOSFET, highly efficient UHF band, for RF power amplifiers, 401–409
    device design, 403–405
    technology for high frequency/high efficiency, 402–403

Silicon radio frequency (RF) device technologies, development trends, 484–87
Silicon single-chip direct conversion RF transceiver, 276–77
Silicon very large scale integration (VLSI) technologies, bonding pad models, 526–29
Sinewave generator, CMOS digitally synthesized 0-13 MHz, 448–49
Single-chip GaAs RF transceiver for 1.9 GHz digital mobile communications
    circuit design and measurement results, 289–95
    control logic circuit, 292–93
    electrical properties, 295
    FET structure, 289
    negative voltage generator, 290–91
    target specifications, 289
    transceiver architecture, 288–89
Single-chip GaAs transceiver, 2.4 GHz, 347–50
Single-chip GSM transceiver RF integrated circuit, 2.7V to 4.5V, 341–42
Single-chip IF transceiver, 1.9 GHz, for digital cordless phones, 297–302
    BiCMOS process technology and layout, 299–301
    chip architecture, 297–98
    circuit design, 298–99
    GaAs MMIC RF front-end, 297–98
    Japanese Personal Handyphone System (PHS), 297
Single-chip radio transceiver, performance evaluation, 624–27
    for DECT standard, 624–27
Single-pole double-throw switch IC, GaAs high-power RF, 423–24
Single-sideband (SSB) signal generation and detection, 45–47
Single-sideband, linear modulation (LM) system, 92
Single-sideband synthesizers, direct-conversion, 138
SO8 package
    frequency limitation, 546–51
    single pole-double throw (SPDT) switch IC, 546–47
    study of assembly, 549–51
Software, hand-held portable equipment, 591–93
Specifications, hand-held portable equipment, 592–93
SPICE circuit simulator
    for design, 161–63
    of multi-layer ceramic technology, 534, 526
Spiral inductors
    for GaAs MMICs, miniature multilayer, 470–72
    on silicon, 34–35
    on silicon substrates, 21
Split-ring resonators, miniaturized RF-circuit modules, 488–89, 491
Spread spectrum cordless telephone, 633–42
    ASIC for processing, 637–38
    basic design, 634–36
    configuration, 636–38
    control operation, 640–41
    direct-conversion receiver for 900 MHz ISM band, 303–11
    equipment characteristics, 638–40

    field test results, 641
    industrial, scientific, and medical (ISM) band, 634–36
    phase-locked loop (PLL) frequency synthesizer, 638
Spread spectrum (SS) for commercial applications, 124
Spread-spectrum (SS) technology
    comparison of direct sequence (DS-) and frequency-hopping (FH-) SS, 124
    despreading and demodulating spread-spectrum signals, 131
    despreading DS-SS signals, 132–33
    despreading FH-SS signals, 131–32
    direct sequence (DS) SS, 123
    frame structure, 129
    frequency-hopping (FS) SS, 124
    generating FH and DS signals using direct digital synthesis (DDS), 130
    indoor channel, 127–28
    indoor wireless applications, 127
    mobile satellite propagation, 125–26
    modulation format, 129
    multiple-access technique and channel crosstalk, 128–29
    network control system, 129
    personal communication by satellite, 126–27
    properties and applications, 122–23
    satellite mobile applications, 124–25
    SS for commercial applications, 124
    using DDS for modulation and carrier generation in DS-SS, 131
    using DDS for modulation and carrier generation in FH-SS, 130–31
    wireless PBX system, 128
Spread-spectrum wireless transceivers, integrated circuits, 18–20
Spurious-free dynamic range (SFDR), 5
Spurious LO leakage, direct-conversion receiver problem, 139
Spurious responses of receiver, design considerations for wireless communications, 159–60
Standard-adaptive transceivers, receiver implementation, 32
starCHIP™-1 silicon semicustom array, 553–54
Sub-sampling mixer using CMOS technology, 120
Superheterodyne receiver, frequency plan of, 135
Superheterodyne receivers, wireless paging receivers, 9
Superhet receiver for frequency shift-keyed (FSK) radio communication, one-chip 2 GHz, 212–13
Surface acoustic wave (SAW) devices, miniaturized, 494–500
    application of antenna duplexer, 498–500
    configuration for low-loss/high-power, 496–98
    loss analysis of high-performance SAW filter for mobile telephone, 496
Suspended inductors on silicon for CMOS RF amplifier, 477–79
Switched-capacitor lowpass filter, 120
Synthesizer phase-locked loop (PPL) bandwidth, in receiver design, 32
Synthesizers, developing technologies, 487

T

Temperature-controlled reference crystal oscillator (TCXO), 40
Time division multiple access (TDMA) technique, 118
  standardized as IS-54, 628–29
Transceiver
  BiCMOS (2.5 GHz) for wireless LNA, 357–59
  DECT RF (2.7V) with integrated VCO, 360–61
  GaAs 2.4 GHz single chip, 347–50
Transceiver chip set
  for dual-mode cellular telephone, 900 MHz, 338–40
  for mobile radio applications (2.7V, 800 MHz-2.1 GHz), 331–34
Transceiver integrated circuit (IC), 2.7V GSM RF, 364–67
Transceiver integrated circuits (ICs) with on-chip filtering, 2.7V GSM, 344–469
Transistor noise and VCO phase noise, design of low-phase noise voltage-controlled oscillator (VCO), 436–38
Transparent tone-in-band technique (TTIB), Weaver transmitted spectrum, 97–98
Triplate spiral inductors, miniaturized RF-circuit modules, 489, 491–92
T/R switch
  high performance integrated power amplifier, 389–91
  MMIC using LC resonators, low voltage, high power, 419–22
TV frequency band, 581
TV-tuner with GaAs ICs, double-conversion broad band, 581–84
Two-path phasing method of single sideband demodulation, 49–50

U

Ultra-high frequency (UHF) band 1.3W 4-stage monolithic amplifier, 385–88
Ultra-high frequency (UHF) band silicon power MOSFET for RF power amplifiers, 401–409
  device design, 403–5
  technology for high efficiency/high frequency, 402–3

Ultra-low DC power consumption, monolithic L-band components, 167–72
Ultra-low power, low noise amplifiers (LNA), 183–85

V

Varactor controls frequency, 13
Varactor-tuned LC resonators for RF transceiver implementation, 29
Vectorial receiver, 105–9
  baseband filter, 107–8
  block diagram, 106
  decoding, 109
  digitization, 108
  direct converter, 107
  gain and frequency control, 109
  local oscillator, 107
  signal processing, 108–9
  synchronization, 109
Very large scale integration (VLSI) chips, mobile radio, 37–41
Very large scale integration (VLSI) technologies, silicon, bonding pad models, 526–29
Vestigial intermediate frequency (IF) or quasi-IF approach to receiver architecture, 31
Voltage-controlled oscillator (VCO)
  application in cellular telephone transceivers, 13
  CMOS 1.8 GHz CMOS, 455–56
  design of low-phase noise, 434–40
  developing technologies, 486
  with integrated LC resonator, 1.5 GHz, 453–54
  multi-layer stripline type, 524–25
  open loop modulation techniques for DECT, 620–23
  PLL frequency synthesis, GHz (CMOS), 441–44
  with prescaler, CMOS 1.8 GHz low-phase-noise, 445–46
VCO architectures, 442–43

W

Weaver direct conversion receiver, 101–4
Weaver method linear modulation (LM) architecture, 92–93, 96–97
  transparent tone-in-band technique (TTIB), 97–98

Wideband frequency doubler, 555–56
Wideband intermediate frequency (IF) double conversion CMOS integrated receiver, 1.9 GHz, 354–55
Wideband vector demodulator, 556–58
Wideband voltage-controlled oscillator for DECT, open loop modulation techniques, 620–23
Wireless applications
  high Q inductors for Si bipolar process, 480–83
  indoor, 127
Wireless communication
  digital, 2.7V 2.5 GHz bipolar chipset, 363–64
  receiver RF design considerations, 157–60
  ultra-low power low noise amplifiers (LNA), 183–85
Wireless low-noise amplifier (LNA), BiCMOS 2.5 GHz transceiver, 357–58
Wireless paging receivers, integrated circuits, 9–11
Wireless PBX system, 128
Wireless transceivers
  future of, 351–53
  key signal-processing issues, 4–7

X

X-band monolithic series feedback LNA
  circuit design, 191–93, 194, 195
  device characterization, 190–91
  history, 190
  radio frequency (RF) performance, 193, 195

Z

Zero-intermediate frequency (IF) 900 MHz digital mobile radio front-end, Si bipolar RF transceiver, 323–26
Zero-intermediate frequency (IF) (direct conversion) paging receiver
  detailed performance of system, 81–83
  frequency-shift keyed (FSK) demodulator, 81
  principle of operation, 81
  single-chip VHF and UHF, 85–90

ABOUT THE EDITORS

Asad A. Abidi was born in 1956. He received the B.Sc.(Hon.) degree from Imperial College, London, U.K., in 1976 and the M.S. and Ph.D. degrees in electrical engineering from the University of California, Berkeley, in 1978 and 1981.

He was at Bell Laboratories, Murray Hill, NJ, from 1981 to 1984 as a member of technical staff in the Advanced LSI Development Laboratory. Since 1985, he has been at the Electrical Engineering Department of the University of California, Los Angeles, where he is a professor. He was a visiting faculty researcher at Hewlett Packard Laboratories during 1989. His research interests are in CMOS RF design, high-speed analog integrated circuit design, data conversion, and other techniques of analog signal processing.

Dr. Abidi served as the program secretary for the International Solid-State Circuits Conference from 1984 to 1990 and as general chairman of the Symposium on VLSI Circuits in 1992. He was Secretary of the IEEE Solid-State Circuits Council from 1990 to 1991, and from 1992 to 1995 he was editor of the *IEEE Journal of Solid-State Circuits*. He has received the 1988 TRW Award for Innovative Teaching, the 1997 IEEE Donald G. Fink Award, and is co-recipient of the Best Paper Award at the 1995 European Solid-State Circuits Conference, the Jack Kilby Best Student Paper Award at the 1996 International Solid-State Circuits Conference (ISSCC), and the Jack Raper Award for Outstanding Technology Directions Paper at the 1997 ISSCC.

Paul R. Gray was born in Jonesboro, AR, on December 8, 1942. He received the B.S., M.S., and Ph.D. degrees from the University of Arizona, Tucson, in 1963, 1965, and 1969, respectively.

In 1969 he joined the Research and Development Laboratory, Fairchild Semiconductor, Palo Alto, CA, where he was involved in the application of new technologies for analog integrated circuits, including power integrated circuits and data conversion circuits. In 1971 he joined the Department of Electrical Engineering and Computer Sciences, University of California, Berkeley, as a professor. His research interests during this period have included bipolar and MOS circuit design, electro-thermal interactions in integrated circuits, device modeling, telecommunications circuits, and analog-digital interfaces in VLSI systems. He is the co-author of a widely used college textbook on analog integrated circuits. During year-long industrial leaves of absence from Berkeley, he served as project manager for telecommunications filters at Intel Corporation, Santa Clara, CA, in 1977–1978, and as director of CMOS design engineering at Microlinear Corporation, San Jose, CA, in 1984–1985. At Berkeley, he has held several administrative posts, including director of the Electronics Research Laboratory (1985–1986), Vice-Chairman of the EECS Department for Computer Resources (1988–1990), and Chairman of the Department of Electrical Engineering and Computer Sciences (1990–1993). He is currently the Dean of the College of Engineering and is the Roy W. Carlson Professor of Engineering.

Dr. Gray has been co-recipient of best-paper awards at the International Solid-State Circuits Conference, the European Solid-State Circuits Conference, and was co-recipient of the IEEE R. W. G. Baker Prize in 1980, the IEEE Morris K. Liebman Award in 1983, and the IEEE Circuits and Systems Society Achievement Award in 1987. In 1994, he received the IEEE Solid-State Circuits Award. He served as editor of the *IEEE Journal of Solid-State Circuits* from 1977 through 1979, and as program chairman of the 1982 International Solid-State Circuits Conference. He served as President of the IEEE Solid-State Circuits Council from 1988 to 1990. He is a member of the National Academy of Engineering.

Robert G. Meyer was born in Melbourne, Australia, on July 21, 1942. He received the B.E., M.Eng.Sci., and Ph.D. degrees in electrical engineering from the University of Melbourne in 1963, 1965, and 1968, respectively.

In 1968 he was employed as an assistant lecturer in electrical engineering at the University of Melbourne. Since September 1968, he has been employed in the Department of Electrical

Engineering and Computer Sciences, University of California, Berkeley, where he is now a professor. His current research interests are high-frequency analog integrated-circuit design and device fabrication. He has acted as a consultant on electronic circuit design for numerous companies in the electronics industry. He is co-author of the book *Analysis and Design of Analog Integrated Circuits* (Wiley, 1993) and editor of the book *Integrated Circuit Operational Amplifiers* (IEEE Press, 1978). Dr. Meyer was president of the IEEE Solid-State Circuits Council and was an associate editor of the *IEEE Journal of Solid-State Circuits* and of the *IEEE Transactions on Circuits and Systems*.